Membrane Gas Separation

Membrane Gas Separation

YURI YAMPOLSKII

A.V. Topchiev Institute of Petrochemical Synthesis, Moscow, Russia

BENNY FREEMAN

University of Texas at Austin, Austin, TX, USA

WILEY

A John Wiley & Sons, Ltd., Publication

This edition first published 2010

Registered office
John Wiley & Sons Ltd, The Atrium, Southern Gate, Chichester, West Sussex, PO19 8SQ, United Kingdom

For details of our global editorial offices, for customer services and for information about how to apply for permission to reuse the copyright material in this book please see our website at www.wiley.com.

Library of Congress Cataloging-in-Publication Data

Yampolskii, Yuri
Membrane gas separation / Yuri Yampolskii and Benny Freeman.
p. cm.
Includes bibliographical references and index.
ISBN 978-0-470-74621-9 (cloth)
1. Gas separation membranes. 2. Polyamide membranes. 3. Gases–Separation. I. Yampolskii, Y. P. (Yuri Pavlovich) II. Title.
TP248.25.M46F74 2010
660′.043–dc22

2010013153

A catalogue record for this book is available from the British Library.

ISBN: HB: 9780470746219

Set in 10/12 pt Times by Toppan Best-set Premedia Limited.

FSC
www.fsc.org
MIX
Paper from responsible sources
FSC® C013604

Contents

Preface

The International Congress on Membranes and Membrane Processes (ICOM) meetings are the most important events for the community of membrane science and technology. They are organized once every three years in the succession: Europe, Asia, America. The most recent ICOM2008 took place in Honolulu, Hawaii (USA). According to the opinion of the editors of this volume it was one the most interesting meetings ever attended. Probably the greatest interest was the subject of membrane gas separation. There were five sessions devoted to this sub-field of membrane science and technology, more than to any other area discussed at ICOM2008. After strong competition, 30 submitted papers were selected by the program committee as oral presentations. The material of these presentations was so interesting and relevant for the field that it occurred to us while in Honolulu that it would be worthwhile to publish a volume based on the oral presentations of the gas separation sessions. This book is conceived to give a snapshot of the current situation in membrane gas separation and related fields as described by the speakers at ICOM2008. In this regard it differs from the volume published several years ago, also by John Wiley & Sons, Ltd [1], which summarized the state of the art in this field achieved during the last decade.

The results presented in this book were obtained owing to the activity of really international group of teams from Australia, Czech Republic, France, Germany, Italy, Japan, Poland, Russia, Slovenia, United Kingdom and USA. The chapters submitted by the authors were partitioned among four Sections. Before considering the contents of these sections it should be emphasized that many chapters focus on the most relevant problems; without a solution to these problems further progress of membrane gas separation looks doubtful.

Chapters 1, 4, 6 and 8 discuss the possible ways to prevent plasticization effects that result in significant reduction in selectivity of separation of mixtures containing active (strongly sorbed) vapours. Recently it became clear that high permeability of certain polymer materials can be rationalized on the basis of the concept of inner porosity. In this regard, such polymers are similar to common porous inorganic media. In such polymers, mixed gas permeation is characterized by high selectivity, so the problem of plasticization is excluded. Transport behaviour of membranes with inner porosity is the subject of Chapters 2, 3 and 5. Another approach that permits overcoming the plasticization problem is an application of perfluorinated polymers that reveal a reduced solubility of hydrocarbon vapours. Different aspects of the use of such membrane materials are considered in Chapters 4, 6 and 7. Some of the contributions include interesting reviews of different problems of membrane gas separation (Chapters 1, 10, 14 and 15).

Section I (Novel Membrane Materials and Transport in Them) focuses on the most recent advances in development of new membrane materials and considers the transport parameters and free volume of polymeric and even inorganic membranes. Kanehashi *et al.* (Chapter 1) present a detailed review of hyperbranched polyimides, which are compared with more common cross-linked polyimides. These polymers with unusual architecture were studied in the hope that they would show weaker tendency to plasticization than conventional linear polymers. However, many representatives of this new class of polymers reveal relatively poor film forming properties due to absence of chain entanglement. Nonetheless, some promising results obtained can show directions of further studies.

The next two chapters deal with novel amorphous glassy polymers that are characterized by large free volume, high gas permeability and good combination of permeability and permselectivity. The polymer of intrinsic microporosity (PIM-1, Chapter 2) has attracted a great attention in membrane community, and at present several polymers structurally similar to PIM-1 have been prepared and characterized. This polymer has 'inner' surface area of about $700\,m^2/g$, higher than that of some sorts of carbon, and its gas and vapour solubility coefficients are the highest among all the polymers studied so far (even polytrimethylsilylpropyne). The subject of Chapter 3 is Si-containing addition-type polynorbornene, also having relatively high gas permeability. The polymers of this class have been the subject of investigation during the last decade; however, only the introduction of $Si(CH_3)_3$ groups into the monomer resulted in rather attractive properties of the polymer obtained. If somebody asked us 10 years ago which polymer structure would provide extra high permeability of the membrane materials, the answer would be: 'Polyacetylenes and, maybe, some perfluorinated polymers'. Now we see that much wider variation of polymer design can lead to large free volume and high permeability and diffusivity. This result seems to be very optimistic for further activity of synthetic polymeric chemists aimed to create new membrane materials.

The objects of the investigation by Jansen *et al.* (Chapter 4) were perfluorinated copolymers of Hyflon AD. The authors reported novel data on free volume, presented the results of computer modelling and the gas permeation parameters. It should be stressed that such comprehensive study of a polymer becomes more and more popular today if one wants to *understand* transport and sorption parameters of a membrane material. This chapter will give much 'food' for future comparisons with other perfluorinated polymers as well as conventional glassy polymers.

The aim of Chapter 5 by Thornton *et al.* was to give systematic consideration to different types of transport in porous membranes. They developed a new model that allows one to predict the separation outcome for a variety of membranes in which the pore shape, size and composition are known, and conversely to predict pore characteristics with known permeation rates.

An important event of recent years in membrane science was the discovery of a new phenomena observed when nano-particles are added into (mainly high permeability) polymer matrix: references to these pioneer works can be found in chapters of Section II: Nanocomposite (Mixed Matrix) Membranes. So it is not surprising that several presentations at ICOM2008 dealt with such systems. Golemme *et al.* (Chapter 6) investigated the system that contained perfluorinated polymers and surface-fluorinated zeolites as nano-additives. Perfluorinated polymer AF2400 with nano-additives was also the object of Chapter 7, where detailed investigation of sorption and diffusion are presented. The authors showed a good agreement between a theoretical approach based on the NELF model and the experimental data for mixed matrix membranes. While the authors of these chapters introduced nano-particles into high permeability polymer matrices, Chapter 8 by Suzuki *et al.* presents a mixed matrix system based on hyperbranched polyimides. Maybe there is more reason to introduce nano-particles in such systems, because in polyimides, it is permeability and not permselectivity that usually requires to be increased. Indeed, it was shown that permeability coefficients of the hybrid membranes increased with increasing silica content because of additional formation of free volume elements. Especially large enhancements of CO_2 permeability was combined with improved CO_2/CH_4 separation factor.

Chapter 9 takes a special place in Section II. For some time the problem of acceleration of membrane permeation of paramagnetic molecules of oxygen mixed with diamagnetic nitrogen has been discussed in membrane community, but only the team headed by Grzywna has demonstrated that it is really possible. For this purpose they introduced neodymium powder into films of ethylcellulose and polyphenyleneoxide and exposed such films to external magnetic fields. Quantitatively, the observed effects are rather modest, but the demonstration of the effect itself seems to be the main gain of this really pioneering study.

The problem of sequestration of carbon dioxide in order to tackle global warming is of utmost importance for the future of humanity. So, no wonder that several presentations at ICOM2008 tackled this subject. This is the theme of Section III (Membrane Separation of CO_2 from Gas Streams); however, to some extent the same problem is discussed in other chapters of this volume (2, 8, 14, 15). It is well known that Pebax copolymers show excellent transport parameters in separating gas mixtures containing carbon dioxide. So, Chapters 12 and 13 deal with certain modifications of this material. The polymers considered in Chapter 11 are rubbery polydimethylsiloxane and glassy polysulfone and polyimide Matrimid, but the main emphasis made in this chapter is on the effects of various minor impurities that can be presented in gaseous feedstock. Chapter 10 considers the application of ionic liquids for the separation of the mixture containing carbon dioxide. It seems to us that an attentive reader will be able to compare the data of these four chapters and make some conclusions about advances and drawbacks of various membranes for separation of CO_2.

The last section 'Applied Aspects of Membrane Gas Separation' contains three chapters. Brunetti *et al.* start their contribution with a brief review of membrane materials and membranes used in gas separation and survey the main directions of industrial applications of gas separation (hydrogen recovery, air separation, etc.). In the second part of their chapter they present a new concept for comparison of membrane and other, more traditional, methods for gas separation. Their approach includes a consideration of engineering, economical, environmental and social indicators. Something similar had been written 15 years ago [2] but this analysis is now rather outdated. White (Chapter 15) focuses on a specific but very important problem in industrial gas separation: membrane separation of natural gas. The main emphasis is on cellulose acetate based membranes that have the longest history of practical applications. This chapter also contains the results of field tests of these membranes and considers approaches how to reduce the size and cost of industrial membrane systems. The final chapter is an example of detailed engineering analysis of another membrane problem – the improvement of performance of a module for gas dehydration.

Finally, the editors wish to express their gratitude to all the contributors of this book. We also greatly appreciate the help and understanding of the publishers of this book, John Wiley and Sons, Ltd., Chichester, UK.

(1) *Materials Science of Membranes for Gas and Vapor Separation*, Yu. Yampolskii, I. Pinnau, B. D. Freeman. John Wiley & Sons, Ltd., Chichester, 2006.
(2) R. Prasad, R. L. Shaner, K. J. Doshi, Comparison of membranes and other gas separation technologies, in: Polymeric Gas Separation Membranes, Ed. by D. R. Paul, Yu. P. Yampolskii, CRC Press, Boca Raton, 1994, p. 531.

Benny Freeman
Yuri Yampolskii

Contributors

Giuseppe Barbieri, National Research Council – Institute for Membrane Technology (ITM–CNR), Via Pietro BUCCI, c/o The University of Calabria, Cubo 17C, 87030 Rende CS, Italy

Nikolai Belov, A. V. Topchiev Institute of Petrochemical Synthesis, Russian Academy of Sciences, 29 Leninsky Pr., 119991, Moscow, Russia

Paola Bernardo, National Research Council – Institute for Membrane Technology (ITM–CNR), Via Pietro BUCCI, c/o The University of Calabria, Cubo 17C, 87030 Rende CS, Italy

Adele Brunetti, National Research Council – Institute for Membrane Technology (ITM–CNR), Via Pietro BUCCI, c/o The University of Calabria, Cubo 17C, 87030 Rende CS, Italy

Andrea Bruno, Università della Calabria, Dipartimento di Ingegneria Chimica e dei Materiali, and INSTM Consortium; Via Pietro Bucci 45/A, I-87030 Rende, Italy

Peter M. Budd, Organic Materials Innovation Centre, School of Chemistry, University of Manchester, Manchester, M13 9PL, UK

Maria Giovanna Buonomenna, Università della Calabria, Dipartimento di Ingegneria Chimica e dei Materiali, and INSTM Consortium; Via Pietro Bucci 45/A, I-87030 Rende, Italy

Anja Car, Institute of Material Research, GKSS-Research Centre Geesthacht GmbH, Max-Planck-Str.1, 21502 Geesthacht, Germany

Corinne Chappey, Laboratoire 'Polymères, Biopolymères, Surfaces', FRE 3103, Université de Rouen- CNRS, 76821 Mon-Saint-Aignan Cedex, France

Jungkyu Choi, University of Minnesota, Department of Chemical Engineering and Materials Science; 421 Washington Ave. SE, Minneapolis, MN 55455, USA

Maria Grazia De Angelis, Dipartimento di Ingegneria Chimica, Mineraria e delle Tecnologie Ambientali (DICMA), Alma Mater Studiorum-Università di Bologna, via Terracini 28, 40131 Bologna, Italy

Luana De Lorenzo, Institute on Membrane Technology, ITM-CNR, c/o University of Calabria, Via P. Bucci, 17/C, Rende (CS), Italy

Enrico Drioli, Institute on Membrane Technology, ITM-CNR, c/o University of Calabria, Via P. Bucci, 17/C, Rende (CS), Italy

Maria Chiara Ferrari, Dipartimento di Ingegneria Chimica, Mineraria e delle Tecnologie Ambientali (DICMA), Alma Mater Studiorum-Università di Bologna, via Terracini 28, 40131 Bologna, Italy

Eugene Finkelshtein, A. V. Topchiev Institute of Petrochemical Synthesis, Russian Academy of Sciences, 29 Leninsky Pr., 119991, Moscow, Russia

Karel Friess, Department of Physical Chemistry, Institute of Chemical Technology, Technická 5, 166 28 Prague 6, Czech Republic

Detlev Fritsch, Institute of Polymer Research, GKSS Research Centre Geesthacht GmbH, Max-Planck-Strasse 1, D-21502 Geesthacht, Germany

Michele Galizia, Dipartimento di Ingegneria Chimica, Mineraria e delle Tecnologie Ambientali (DICMA), Alma Mater Studiorum-Università di Bologna, via Terracini 28, 40131 Bologna, Italy

Giovanni Golemme, Università della Calabria, Dipartimento di Ingegneria Chimica e dei Materiali, and INSTM Consortium; Via Pietro Bucci 45/A, I-87030 Rende, Italy

Maria Gringolts, A. V. Topchiev Institute of Petrochemical Synthesis, Russian Academy of Sciences, 29 Leninsky Pr., 119991, Moscow, Russia

Zbigniew J. Grzywna, Faculty of Chemistry, Silesian University of Technology, 44-100 Gliwice, Poland

Pingjiao Hao, Chemical and Environmental Engineering Department, University of Toledo, Toledo, OH 43606-3390, USA

Matthias Heuchel, GKSS Research Center, Institute of Chemistry, Kantstrasse 55, D-14513, Teltow, Germany

Anita J. Hill, CSIRO Materials Science and Engineering, Locked Bag 33, Clayton Sth MDC, Victoria 3169, Australia

James M. Hill, Nanomechanics Group, School of Mathematics and Applied Statistics, University of Wollongong, New South Wales 2522, Australia

Jeffery B. Ilconich, Parsons, P.O. Box 618, South Park, PA 15129, USA

Kumi Itahashi, Department of Biomolecular Engineering, Kyoto Institute of Technology, Matsugasaki, Sakyo-ku, Kyoto 606-8585, Japan

Johannes Carolus Jansen, Institute on Membrane Technology, ITM-CNR, c/o University of Calabria, Via P. Bucci, 17/C, Rende (CS), Italy

Shinji Kanehashi, Department of Applied Chemistry, Meiji University, 1-1-1 Higashi-mita, Tama-ku, Kawasaki 214-8571, Japan

Sandra E. Kentish, Cooperative Research Centre for Greenhouse Gas Technologies (CO2CRC), University of Melbourne, Parkville, Victoria 3010, Australia

Dominique Langevin, Laboratoire "Polymères, Biopolymères, Surfaces', FRE 3103, Université de Rouen-CNRS, 76821 Mon-Saint-Aignan Cedex, France

G. Glenn Lippscomb, Chemical and Environmental Engineering Department, University of Toledo, Toledo, OH 43606-3390, USA

David R. Luebke, P.O. Box 10940, Pittsburgh, PA 15236-0940, USA

Marialuigia Macchione, University of Calabria, Via P. Bucci 14/D, 87030 Rende (CS), Italy

Raffaella Manes, Università della Calabria, Dipartimento di Ingegneria Chimica e dei Materiali, and INSTM Consortium; Via Pietro Bucci 45/A, I-87030 Rende, Italy

Stéphane Marais, Laboratoire 'Polymères, Biopolymères, Surfaces', FRE 3103, Université de Rouen-CNRS, 76821 Mon-Saint-Aignan Cedex, France

Neil B. McKeown, School of Chemistry, Cardiff University, Cardiff, CF10 3AT, UK

Daniela Muoio, Università della Calabria, Dipartimento di Ingegneria Chimica e dei Materiali, and INSTM Consortium; Via Pietro Bucci 45/A, I-87030 Rende, Italy

Christina R. Myers, P.O. Box 10940, Pittsburgh, PA 15236-0940, USA

Kazukiyo Nagai, Department of Applied Chemistry, Meiji University, 1-1-1 Higashi-mita, Tama-ku, Kawasaki 214-8571, Japan

Quang Trong Nguyen, Laboratoire 'Polymères, Biopolymères, Surfaces', FRE 3103, Université de Rouen-CNRS, 76821 Mon-Saint-Aignan Cedex, France

Klaus-Viktor Peinemann, Institute of Material Research, GKSS-Research Centre Geesthacht GmbH, Max-Planck-Str.1, 21502 Geesthacht, Germany

Henry W. Pennline, P.O. Box 10940, Pittsburgh, PA 15236-0940, USA

Fabienne Poncin-Epaillard, Laboratoire Polymères, Colloïdes, Interfaces, UMR 6120, CNRS-Université du Maine, 72085 Le Mans Cedex, France

Aleksandra Rybak, Faculty of Chemistry, Silesian University of Technology, 44-100 Gliwice, Poland

Jun Sakai, Department of Biomolecular Engineering, Kyoto Institute of Technology, Matsugasaki, Sakyo-ku, Kyoto 606-8585, Japan

Giulio Cesare Sarti, Dipartimento di Ingegneria Chimica, Mineraria e delle Tecnologie Ambientali (DICMA), Alma Mater Studiorum-Università di Bologna, via Terracini 28, 40131 Bologna, Italy

Shuichi Sato, Department of Applied Chemistry, Meiji University, 1-1-1 Higashi-mita, Tama-ku, Kawasaki 214-8571, Japan

Colin A. Scholes, Cooperative Research Centre for Greenhouse Gas Technologies (CO2CRC), University of Melbourne, Parkville, Victoria 3010, Australia

Victor Shantarovich, N. N. Semenov Institute of Chemical Physics, Russian Academy of Sciences, 4 Kosygin Street, 119991 Moscow, Russia

Ludmila Starannikova, A. V. Topchiev Institute of Petrochemical Synthesis, Russian Academy of Sciences, 29 Leninsky Pr., 119991, Moscow, Russia

Geoff W. Stevens, Cooperative Research Centre for Greenhouse Gas Technologies (CO2CRC), University of Melbourne, Parkville, Victoria 3010, Australia

Chrtomir Stropnik, University of Maribor, Faculty of Chemistry and Chemical Engineering, Smetanova 17, Slovenia

Anna Strzelewicz, Faculty of Chemistry, Silesian University of Technology, 44-100 Gliwice, Poland

Julie Sublet, Institut de recherches sur la catalyse et l'environnement de Lyon, UMR 5256, CNRS-Université de Lyon 1, 69626 Villeurbanne Cedex, France

Tomoyuki Suzuki, Department of Biomolecular Engineering, Kyoto Institute of Technology, Matsugasaki, Sakyo-ku, Kyoto 606-8585, Japan

Aaron W. Thornton, CSIRO Materials Science and Engineering, Locked Bag 33, Clayton Sth MDC, Victoria 3169, Australia

Elena Tocci, Institute on Membrane Technology, ITM-CNR, c/o University of Calabria, Via P. Bucci, 17/C, Rende (CS), Italy

Michael Tsapatsis, University of Minnesota, Department of Chemical Engineering and Materials Science; 421 Washington Ave. SE, Minneapolis, MN 55455, USA

Jean-Marc Valleton, Laboratoire 'Polymères, Biopolymères, Surfaces', FRE 3103, Université de Rouen-CNRS, 76821 Mon-Saint-Aignan Cedex, France

Lloyd S. White, UOP LLC, Des Plaines, Illinois, USA

Shan Wickramanayake, Parsons, P.O. Box 618, South Park, PA 15129, USA

Yasuharu Yamada, Department of Biomolecular Engineering, Kyoto Institute of Technology, Matsugasaki, Sakyo-ku, Kyoto 606-8585, Japan

Yuri Yampolskii, A. V. Topchiev Institute of Petrochemical Synthesis, Russian Academy of Sciences, 29 Leninsky Pr., 119991, Moscow, Russia

Wilfredo Yave, Institute of Material Research, GKSS-Research Centre Geesthacht GmbH, Max-Planck-Str.1, 21502 Geesthacht, Germany

Part I

Novel Membrane Materials and Transport in Them

1

Synthesis and Gas Permeability of Hyperbranched and Cross-linked Polyimide Membranes

Shinji Kanehashi, Shuichi Sato and Kazukiyo Nagai
Department of Applied Chemistry, Meiji University, Tama-ku, Kawasaki, Japan

1.1 Introduction

Recently, the polymer science field has focused on the role of polymers as membrane materials with precise, well-ordered structures through the development of defined synthesis and analysis of polymers. Among these well-ordered polymers are the hyperbranched polymers (e.g. hyperbranched polyimides). Part of the interest in such polymers is due to the expectation that they could have different properties as compared to common linear polymers. Also, cross-linked polyimides have attracted much attention from researchers, as can be judged by a high number of publications.

In general, hyperbranched polymers have many orderly branching units whose structures are different compared to linear and randomly cross-linked polymers [1–3]. According to the Commission on Macromolecular Nomenclature of the International Union of Pure and Applied Chemistry (IUPAC), a crosslink polymer is defined as a polymer having a small region in a macromolecule from which at least four chains emanate [4]. It is formed by reactions involving sites or groups on existing macromolecules or by interactions between existing macromolecules. The word 'network' is also defined as a highly ramified macromolecule in which essentially each constitutional unit is connected to each other constitutional unit and to the macroscopic phase boundary by many permanent paths through the macromolecule, the number of such paths increasing

Membrane Gas Separation Edited by Yuri Yampolskii and Benny Freeman

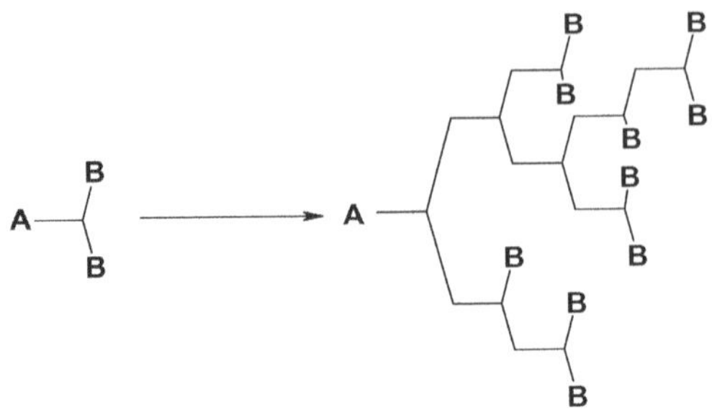

Scheme 1.1

with the average number of intervening bonds; the paths must on the average be co-extensive with the macromolecule [4]. In this chapter, we use the term crosslink polymer to describe a random cross-linked network between polymer segments.

Precisely branched polymers include hyperbranched polymers, dendrimers and dendrons. Dendrimers and dendrons are characterized by perfectly controlled structures in three dimensions such as tree branch architecture, and they have attractive features such as a well-ordered chemical structure, molecular mass, size and configuration of polymers [5]. Although the precise order of shape of hyperbranched polymers is less than that of dendrimers and dendrons, hyperbranched polymers have unique properties such as low viscosity attributed to the lack of entanglement of polymer segments, and the possibility of chemical modification in terminal functional groups such as in dendrimers [1–3].

Synthesis of hyperbranched polymers is typically performed through the self-polycondensation reaction of AB_2-type monomers (Scheme 1.1) [6,7]. The theoretical study of the random AB_x polycondensation has already been reported by Flory in 1952 [8]. He pointed out that the synthesis of hyperbranched polymers from AB_x monomers should resemble linear polymers in their elusion of infinite network (i.e. gelation) formation, which cannot occur except through the intervention of other interlinking reactions. Since then, there have only been a few experimental data made available on hyperbranched polymers; some have even been overlooked due to the fact that the use of the term hyperbranched polymers began only in the late 1980s. However, in early 1990s, hyperbranched polyphenylene was synthesized from AB_2-type monomers [9]. This marked the beginning of the reawakened hyperbranched polymer concept. A variety of hyperbranched polymers such as polyphenylene [9], polyimide [10–12], polyamide [13–15], polyester [16], polyetherketone [17] and polycarbonate [18] have been reported in recent years. It is important that hyperbranched polymers with feathers of closed dendrons can be synthesized through the self-polycondensation one-step reaction because dendrimers and dendrons are synthesized by multistep procedures (e.g. protection, coupling and deprotection cycles). Producing dendrimers and dendrons is also costly and requires complicated manufacturing processes for industrial applications.

On the one hand, linear aromatic polyimides have been generally used as electronic and aerospace materials because of their excellent mechanical strength, thermal, chemical and electronic/optic properties compared with other common amorphous polymers. Polyimides are also excellent membrane materials for gas separation due to their rigid chemical structures, allowing the production of larger functional free volume. Over the

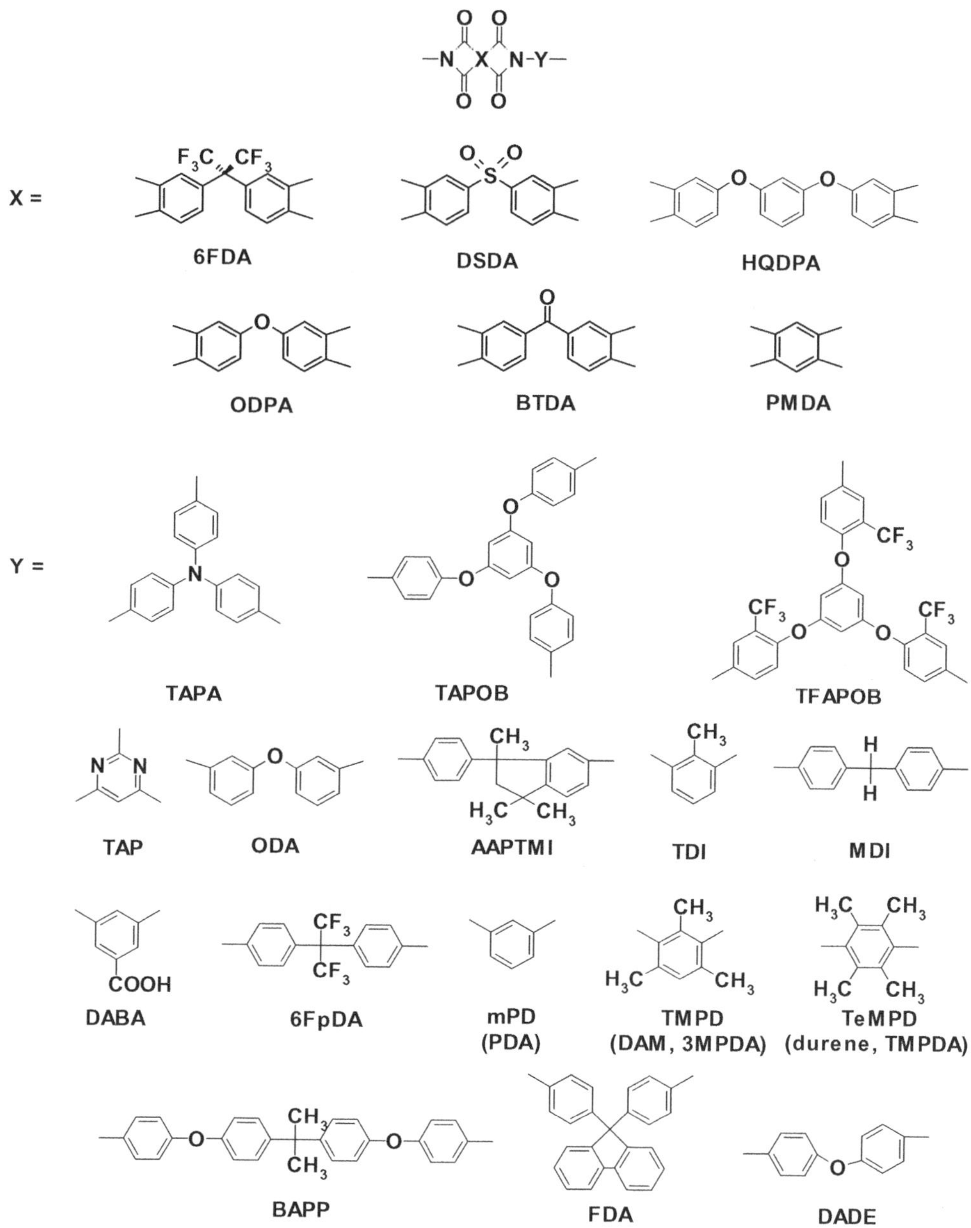

Scheme 1.2

past decades, numerous polyimides have been synthesized and their gas transport properties have been investigated. Scheme 1.2 shows the chemical structures of acid anhydrides and diamines mentioned in this chapter.

On the other hand, hyperbranched polyimides not only have the features of other hyperbranched polymers (e.g. low viscosity, good solubility) but also possess high thermal and physical stability, which is attributed to their rigid imide ring. It is commonly known that the kinds of terminal functional groups affect their physical properties, such as glass transition temperature and solubility [19]. Hyperbranched polyimides have weak polymer chain interactions (lack of entanglement) and this affects their density, dielectric constant, refractive index and other properties [19]. It is expected that they could provide an alternative to conventional polymer materials as novel functional and high-value added materials. Furthermore, hyperbranched polyimides could have a well-ordered structure compared with linear polyimides, which have a random distribution of polymer segments. Therefore, hyperbranched polyimides are expected to have favourable gas separation performance since their controlled branched structure could be advantageous in separating small molecules. Since the early 2000s, research on hyperbranched polyimides as gas separation materials has been reported, and these studies are still in progress [20–26].

Plasticization behaviour induced by condensable gases and vapours (e.g. carbon dioxide, hydrocarbons and other organic vapours) in polymer membranes is still a painful problem in polymeric membrane-based gas separation applications [27,28]. Recently, novel hyperbranched polyimides were prepared from telechelic polyimides and an attempt was made to improve its gas separation performance and physical stability by obtaining plasticization-resistant materials [29–33] (see e.g. Chapters 4, 6 and 7 of this book).

This chapter presents a review of numerous publications devoted to the concept and synthesis of hyperbranched and cross-linked polyimides. Also, gas permeation properties of these polymers are considered in detail.

1.2 Molecular Designs for Membranes

There exist different architectures of polymer macromolecules, as is shown in Figure 1.1.

Type I represents common linear polymers such as polysulfone, polycarbonate or polystyrene, for example. In glassy polymers, the movement of segments is frozen, though small-scale mobility of side groups is possible. In general they have good solubility in

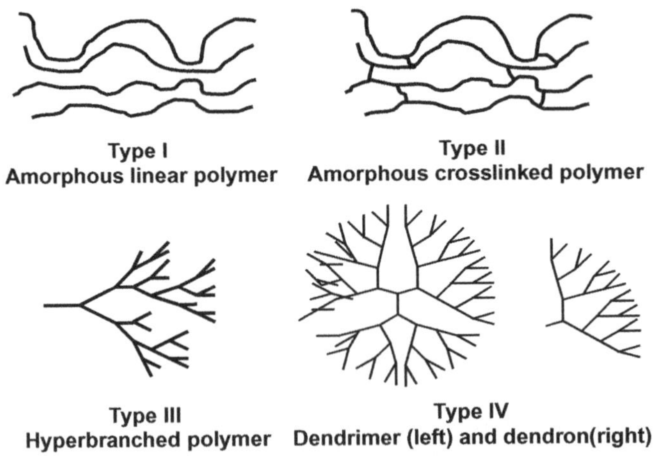

Figure 1.1 Architecture of polymers

various organic solvents; however, their gas permeation properties in the presence of organic vapours are affected by plasticization phenomena.

Type II. In randomly cross-linked polymers the solubility in organic solvents gradually decreases with the increasing degree of crosslink density. Too frequent crosslinks result in the gelation of the polymer and a decline in gas permeability while simultaneously permselectivity can increase.

Type III. Hyperbranched polymers have numerous branch units. They have low viscosity, good solubility and are capable of being chemically modified in terminal functional groups. Hyperbranched polymers have a potential to be good gas separation materials because their molecular-sized spaces between branched polymers can be controlled.

Type IV. Dendrimers and dendrons have perfectly and orderly branched tree-like structures. Their molecular mass increases with the growth of the number of generation. Dendrimers and dendrons, like common organic molecules, are perfectly controlled in terms of chemical structure, molecular mass, configuration and distribution of polymers [5]. Dendrons are well-ordered hyperbranched polymers and dendrimers are assembled from dendrons. It is expected that molecular-sized spaces between branched as well as hyperbranched polymers of dendrimers can be controlled and, therefore, could have high potential as gas separation membranes. An obvious disadvantage of dendrimers as membrane materials is their poor film-forming properties.

One of the key problems for polymeric gas separation membranes is gas and vapour-induced plasticization. The plasticization of polymers produces an enhancement of polymer chain mobility. It is a recognized fact that almost all polymeric membranes undergo swelling and plasticization under high pressure (concentration) of CO_2 and organic vapours, resulting in a significant loss in gas separation performance. One of the effective techniques against plasticization of polymers is the crosslink approach. There is a trade-off relationship between polymer crosslink density and gas permeability.

The mobility of polymer chains is larger for their polymer terminal chain ends as compared to that for the sections of macromolecules inside main chains. Therefore, plasticization may occur more easily around the polymer chain ends than in the polymer main chains. Moreover, if the number of polymer chain ends were minimized in a membrane, plasticization would be prevented. It is the hyperbranch structure that can create such behaviour in the case of rigid polymer chains.

Thus, we can state that the use of hyperbranched polyimides can enhance the resistance to plasticization of polymer membranes.

1.3 Synthesis of Hyperbranched Polyimides

Cross-linked (Type II) and hyperbranched (Type III) polyimides can be prepared for the use as gas separation membranes. There are no dendrimers and dendrons known, which would form free-standing membranes. Therefore, we focus on the synthesis of cross-linked (Type II) and hyperbranched (Type III) polyimides.

1.3.1 Amorphous Cross-linked Polyimides (Type II)

Generally, the aim of the study on crosslink polyimides is an attempt to enhance their gas selectivity and physical stability for gas-induced swelling and plasticization. Several crosslink techniques such as monoesterification and transesterification reactions of carboxylic acid, imide ring-opening reactions, grafted with epoxy reactions, UV-induced cross-linking and Diels–Alder-type cyclization reactions have been reported.

The monoesterification and transesterification reactions of carboxylic acids were performed using the following steps. The carboxylic acid-containing polyimide was monoesterificated under acid catalyst and thermal treatment, and the transesterification reaction was induced through further thermal treatment under vacuum. Many carboxylic acid-containing copolyimides have been synthesized and the crosslink reaction of the varieties of diol agents has been investigated [34–42]. The structure of cross-linked membranes could be strongly affected by structures of the diol agent and polyimide compositions and annealing temperature after membrane formation. In the case of 6FDA-TMPD/DABA (3:2) cross-linked polyimides, 1,3-propanediol can be considered as an efficient crosslink agent [42].

The decarboxylation-induced cross-linking reaction of carboxylic acid is preceded by the reaction of the phenyl radical and the elimination of the carboxylic group by high temperature annealing [43]. This decarboxylation-induced reaction is more sensitive to the reactivity of phenyl radicals rather than the effects of charge transfer complexing, oligomer and dianhydride formation. It was reported that the sites within the diamines section could be the TMPD methyl, biphenyl (between the carboxylic acid group) and at the site of cleaved CF_3 groups in 6FDA.

The imide ring-opening reaction occurred between the polyimide and primary diamine agents. Many chemical cross-linking reactions between 6FDA-based polyimides and primary diamines have been investigated [44–52]. They were carried out by immersing the polyimide membranes into the methanol solution of amine compounds. The structure of the cross-linked membranes could depend on the structures of the primary amine agents and the reaction conditions such as the reaction time and temperature. Furthermore, the gas permeation properties in 6FDA-TMPD modified by amine compounds were described [44]. The cross-linking in 6FDA-TeMPD with dendrimers such as poly-amidamine (PAMAM) and polypropyleneimide (DAB-AM) has also been reported [53–56]. There was no doubt that they took place, as the measurements of gel fraction and FTIR data showed; in addition, the degree of crosslink density increased in the order of generations G1 > G2 > G3 at the same reaction. The dielectric constant increased with the reaction time owing to the decrease in the polymer chain's mobility and free volume.

The etherification reaction of polyimides is similar to the process of the imide ring-opening reaction. It was demonstrated for the reaction of polyimides with primary diamine and epoxy agents [57,58] (for example, for 6FDA-TeMPD polyimides and tetraglycidyldiaminodiphenylmethane (TGDDM), diethyltoluenediamine (DETDA) [57], TMPDA, 1,3-phenylenediamine (PDA) and 4,4'-(9-fluorenylidene)dianiline (FDA) [58]). The density of polymers and crosslink concentration increased with the increase in epoxy content.

It is known that the crosslink reaction proceeds with participation of photo reactive benzophenone and alkyl chains under UV irradiation [59,60]. Many benzophenone-containing BTDA-based polyimides have been synthesized and their cross-linking investigated [61–69]. The same effects as discussed earlier were observed due to increases in the UV irradiation time.

Acetylene-terminated or internal acetylene imide oligomers were investigated for aerospace and electronic applications, in particular because of their good thermal and environmental stability [70–74]. In respect of membrane application the aim of these studies was an enhancement of the physical stability under high pressure CO_2, that is, resistance to plasticization. Recently, a co-polyimide was synthesized from 6FDA, TeMPD and 4,4′-diaminodiphenylacetylene (*p*-intA) having internal acetylene structure [75]. After thermal treatment at 400 °C of such a membrane (an internal acetylene membrane), the cycloaddition of a Diels–Alder-type reaction occurred, according to the results of DSC and FT-Raman spectroscopy. The cross-linked membrane was insoluble; however, no densification of the membrane was observed.

1.3.2 Hyperbranched Polyimides (Type III)

Hyperbranched polyimides can result due to the self-polycondensation reactions of AB_2-, A_2- and B_3-types. The preparation of hyperbranched polyimides involves chemical imidization of polyamic acid ester synthesized from AB_2-monomers, which are carboxylic dianhydrides containing an ether bond and a diamine [6,19,76]. Polyamic acid in combination with a condensation agent is used because it is difficult to separate the synthesized polymer from AB_2-type monomers.

For example, it is possible to prepare hyperbranched polyimides from 3,5-dimethoxyphenol and 4-nitrophthalonitrile in the presence of diphenyl(2,3-dihydro-2-thioxo-3-benzoxazolyl) phosphonate (DBOP) as a condensation agent at room temperature. Hyperbranched polyimide was obtained through thermal or chemical imidization of the precursor (polyamic acid) (Scheme 1.3) [19]. The obtained hyperbranched polyimide had a relatively great molecular mass (M_w) of about 190 000 g mol^{-1} but low intrinsic viscosity of 0.30 dL g^{-1}. Therefore, it had a compact configuration and the lack of entanglement of polymer chains. The polymer obtained via chemical imidization was soluble in aprotic polar solvents such as tetrahydrofuran (THF), while the polymer from thermal imidization was insoluble in any solvents.

Other examples of self-polycondensation of an AB_2-type monomer containing an imide-ring via etherification reactions can be found in the literature [7,10,77]. The self-polycondensation can be performed though nucleophilic etherification of silylated phenol and aryl fluoride in diphenylsulfone at 240 °C under the presence of caesium fluoride

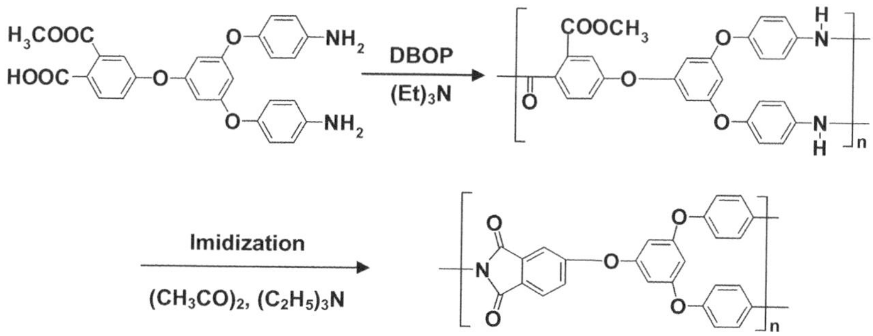

Scheme 1.3

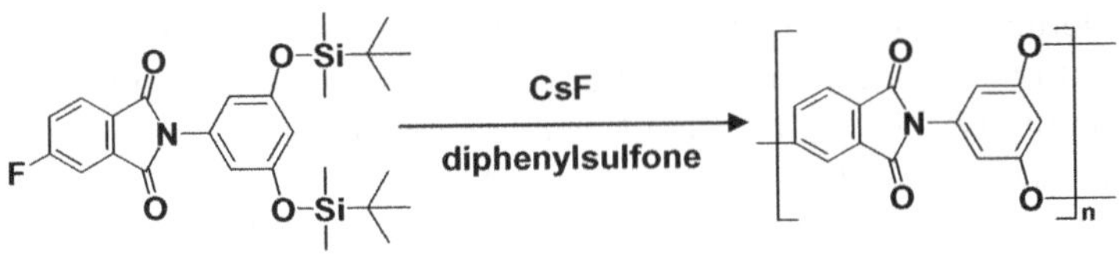

Scheme 1.4

Scheme 1.5

(Scheme 1.4) [7]. This reaction involved a nucleophilic substitution of the halogen group interacting with the electron-attracting imide-ring. It gave an increase in number average molecular mass (M_n) value from 52000 to 85000 g mol^{-1}. The hyperbranched polyetherimide was soluble in common organic solvents and showed high thermal stability.

The hyperbranched polyimides can be synthesized via using the preparation of polyamic acids from A_2- and B_3-type monomers as well. The advantage of this reaction is that it can produce the hyperbranched polyimides from commercially available aromatic dianhydride [11,20] and triamine monomers or trianhydride and diamine monomers [12,78]. Almost all reported hyperbranched polyimides for gas separation applications have been synthesized using this reaction [20–26]. However, the obtained polymers had poor mechanical properties due to high branching and the absence of chain entanglements [8]. Therefore, isolated soluble hyperbranched polyimides should be treated by poor solvents before gelation. By controlling the molar ratio, the addition sequence of each component, the monomer concentration and the imidization method it is possible to obtain polymers with different terminal functional groups (as amine-terminated and anhydride-terminated polymers). For example, the hyperbranched polyimide was prepared from 2,2-bis (3,4-dicarboxyphenyl)hexafluoropropane dianhydride (6FDA) as the anhydride and (4-aminophenyl)amine (TAPA) as the triamine (Scheme 1.5) [11]. The synthesized hyperbranched polyimide was soluble in organic solvents, had a M_w value of 37000 for amine-terminated and 150000 g mol^{-1} for anhydride-terminated product. The intrinsic viscosity was 0.76 dL g^{-1} for amine-terminated and was 1.92 dL g^{-1} for anhydride terminated polymers, respectively, higher than for the polymers synthesized via AB_2-type monomer.

Hyperbranched polymers basically have poor membrane-forming ability due to the lack of chain entanglement. Studies on the enhancement of membrane-forming ability for hyperbranched polymers suggested to use the crosslink approach at terminal functional groups and cross-linkable components [1,18]. Preparation of the crosslink polymers was reported [20] using reagents such as ethylene glycol diglycidyl ether (EGDE) and terephthaldehyde (TPA), or anhydride-terminated and 4,4-diaminodiphenyl ether (ODA) and diphenylsulfone (DDS) have also been reported.

Recently, hyperbranched polyimides were obtained by selective orderly reaction only at polymer ends using telechelic polyimides with terminal reactive groups of acetylene, vinyl and acryl [29–33].

Thus, acetylene-terminated telechelic polyimide was synthesized from 6FDA as the anhydride and 4,4′-(9-fluorenylidene) dianiline (FDA) as an amine, also using 4-(2-phenylethynyl)phthalic anhydride (PEPA) as an acetylene-terminated monomer at room temperature (Scheme 1.6) [29,30]. The obtained telechelic polyimide had $M_w = 12\,000$ g mol^{-1} and a glass transition temperature of 420 °C. The reaction using acetylene moiety is cyclotrimerization of three acetylene groups in three PEPA groups in the presence of tantalum(V) chloride catalyst with subsequent thermal treatment at 250 °C. The obtained membrane was brittle, however, making it difficult to measure the gas permeation properties. A free-standing membrane was formed from acetylene-terminated telechelic polyimide using a flexible diamine structure, 3,4-diaminodiphenyl ether (DADE) [31].

Vinyl- and acryl-terminated telechelic polyimides were also synthesized. The polymers were prepared from 6FDA, 2,3,5,6-tetramethyl-1,4-phenylene diamine (TeMPD), and *p*-aminostyrene (PAS) as a vinyl-terminated monomer, or 1,1-bis(acryloyloxymethyl) ethyl isocyanate (BEI) as a acryl-terminated monomer (Schemes 1.7 and 1.8) [32,33]. Both polymers were soluble in common organic solvents and had low M_n and viscosity.

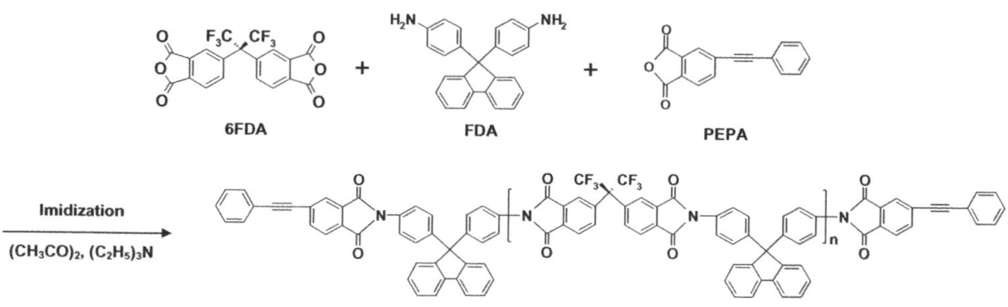

Scheme 1.6

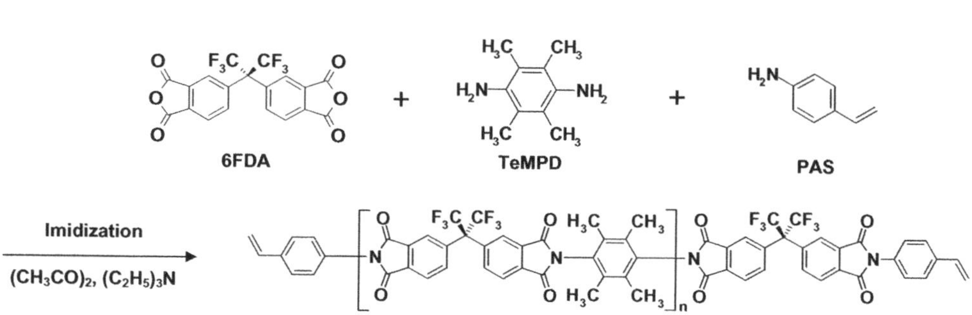

Scheme 1.7

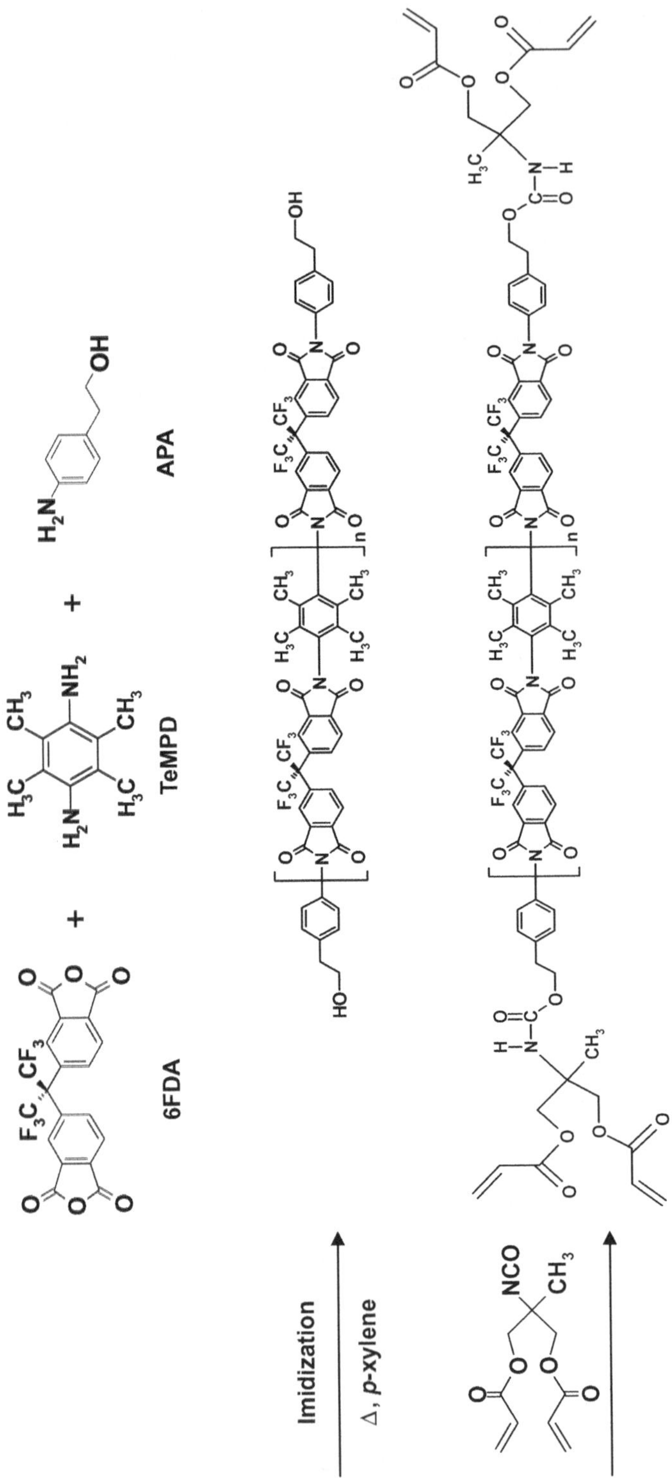

Scheme 1.8

The reaction in vinyl or acryl-terminated polyimides was preceded by UV irradiation under the presence of a photo initiator. Formation of free-standing membranes was significantly enhanced by UV irradiation.

1.4 Gas Permeation Properties

In this section, we shall consider gas permeation properties of films based on cross-linked and hyperbranched polyimides. We shall focus on CO_2 capture from the flue gases of power plants (CO_2/N_2), CO_2 separation from natural gas (CO_2/CH_4), hydrogen purification (H_2/N_2) and oxygen and/or nitrogen enrichment of air (O_2/N_2) [28]. We shall consider permeability coefficients P_i and ideal separation factors for gas pairs A and B (α(A/B)) defined as the ratio P(A)/P(B). The trade-off relationship has been recognized between gas permeability and selectivity in polymeric membranes: highly permeable polymers have low selectivity and vice versa. The general trade-off relationship between gas permeability and selectivity has been summarized by Robeson [79,80], who suggested an idea of the upper bound. Traditionally, an interest of researchers was directed to materials whose data points are located at the Robeson diagrams close to or above the upper bound lines.

1.4.1 Amorphous Cross-linked Polyimides (Type II)

The gas permeability and selectivity of cross-linked membranes prepared by monoesterification and transesterification reactions between the carboxylic acid in polyimides and diol agents are summarized in Table 1.1 [34–38, 41, 42]. These gas permeation properties are strongly affected by the structures of the diol agent, polyimide composition and the annealing temperature after membrane formation. For example, the range of variation of $P(CO_2)$ is 10–145 Barrer while the selectivity $\alpha(CO_2/CH_4)$ is in the range 29–87 at 35 °C. The 6FDA-TMPD/DABA (3:2) cross-linked polyimides with 1,3-propanediol have better gas separation performance among the other cross-linked polyimides using monoesterification and transesterification reactions: $P(CO_2)$ of cross-linked 6FDA-TMPD/DABA (3:2) is 77 Barrer and $\alpha(CO_2/CH_4)$ is 40 at 35 °C and 4.4 atm [42].

The decarboxylation-induced cross-linked 6FDA-TMPD:DABA (2:1) polyimide reveals enhanced resistance to plasticization by high pressure of CO_2 [43]. This can be a result of high annealing temperature which leads to decarboxylation of the pendant acid groups. In this process phenyl radicals are formed that are capable of attacking other portions of the polyimide macromolecules. The CO_2 permeability of the cross-linked membrane prepared by rapid quenching from above the glass transition temperature is 260 Barrer at 35 °C and 6.8 atm.

The gas permeability and selectivity of cross-linked membranes prepared by the imide ring-opening reaction between the imide ring in polyimides and primary diamine agents are summarized in Table 1.2 [45–56]. As the reaction time increases, the gas permeability gradually decreases, whereas the selectivity increases. However, excess cross-linking leads to a reduction of both permeability and selectivity. The observed transport parameters could be strongly affected by the structures of diamine agent. For example, the $P(CO_2)$ is in the range 1.9–568 Barrer and the selectivity, $\alpha(CO_2/CH_4)$ is in the range

Table 1.1 *Gas permeability coefficients and selectivity (at 35 °C) of cross-linked polyimides (Type II) using monoesterification and transesterification reaction of free carboxylic acid*

Polyimide	Cross-linkable	Feed pressure (atm)	$P(O_2)$	$P(CO_2)$	$\alpha(O_2/N_2)$	$\alpha(CO_2/N_2)$	$\alpha(CO_2/CH_4)$	Reference
6FDA-DABA	linear (Type I)	2	1.01	3.4	8.0	26.9	63.0	[34]
	Thermal cross-linked	3.74	2.69	10.40	6.7	26.0	87.0	[34]
6FDA-mPD	linear (Type I)	3.74	2.60	11.03	6.5	27.6	58.0	[34]
6FDA-mPD/ DABA (9:1)	linear (Type I)	3.74	1.71	6.53	6.9	26.1	65.3	[34]
	Ethylene Glycol	3.74	1.81	9.50	6.8	35.2	63.3	[34]
6FDA-6FpDA/ DABA (1:2)	linear (Type I)	10	–	29	–	–	45	[35]
	Ion compound ($Al(AcAc)_3$)	10	–	19	–	–	–	[36]
6FDA-6FpDA/ DABA (2:1)	linear (Type I)	10	–	29	–	–	45	[37]
	Ion compound ($Al(AcAc)_3$)	10	–	25	–	–	46	[36,37]
	Ethylene glycol	10	6.0	35	–	–	42	[37]
	Butylene glycol (140 °C)	2		42.8	6.2	44.2	47	[38]
6FDA-TMPD/ DABA (2:1)	Thermal cross-linked (130 °C)	10	–	133	–	–	29	[41]
	Thermal cross-linked (220 °C)	10	–	115	–	–	27	[35,41]
	Thermal cross-linked (295 °C)	10	–	110	–	–	30	[41]
	Ethylene Glycol (140 °C)	10	–	90	–	–	30	[35]
	1,4-cyclohexanedimethanol (140 °C)	10	–	21	–	–	30	[35]
	1,4-cyclohexanedimethanol (220 °C)	10	–	22	–	–	30	[41]
	1,4-cyclohexanedimethanol (295 °C)	10	–	79	–	–	29	[41]
	Butylene glycol (140 °C)	10	–	44	–	–	34	[35,41]
	Butylene glycol (140 °C)	2	13.7	51	4.8	17.9	37	[38]
	Butylene glycol (220 °C)	10	–	46	–	–	34	[35,41]
	Butylene glycol (295 °C)	10	–	138	–	–	30	[35,41]
6FDA-TMPD/ DABA (3:2)	1,3-propanediol (220 °C)	4.4	–	57.5	–	–	37.1	[42]
	1,3-propanediol (295 °C)	4.4	–	77.3	–	–	39.9	[42]

1 Barrer = $10^{-10} cm^3$ (STP) $cm\,cm^{-2} s^{-1} cmHg^{-1}$.

Table 1.2 *Gas permeability coefficients and selectivity (at 35 °C) of cross-linked polyimides (Type II) using imide ring-opening reaction by primary amine agents*

Polyimide	Cross-linkable	Feed pressure (atm)	$P(O_2)$	$P(CO_2)$	$\alpha(O_2/N_2)$	$\alpha(CO_2/N_2)$	$\alpha(CO_2/CH_4)$	Reference
Matrimid 5218	linear (Type I)	10	1.7	6.5	6.6	25.2	34	[45]
	p-xylenediamine (1 day)	10	1.9	7.4	6.5	25.3	36	[45]
	p-xylenediamine (7 day)	10	1.5	5.1	7.0	23.8	33	[45]
	p-xylenediamine (14 day)	10	1.4	4.7	7.0	23.5	34	[45]
	p-xylenediamine (32 day)	10	0.9	1.9	6.9	14.6	28	[45]
	linear (Type I)	2	–	5.39	–	–	36	[46]
	Poly(ethylene oxide) 2000 (1:0.2)	2	–	7.51	–	–	22	[46]
	Poly(ethylene oxide) 2000 (1:0.5)	2	–	59.16	–	–	18	[46]
	Poly(ethylene oxide) 2000 (1:1)	2	–	115.18	–	–	17	[46]
	Poly(ethylene oxide) 600 (1:1)	2	–	1.47	–	–	23	[47]
6FDA-TeMPD	linear (Type I)	10	125	456	3.7	13.5	–	[48]
	p-xylenediamine (5 min)	10	45.2	136	4.1	12.3	–	[48]
	p-xylenediamine (10 min)	10	28.9	91.8	4.4	14.0	–	[48]
	p-xylenediamine (15 min)	10	26.5	70.0	4.4	11.6	–	[48]
	p-xylenediamine (30 min)	10	13.7	30.3	4.8	10.6	–	[48]
	p-xylenediamine (60 min)	10	2.34	2.14	5.9	5.4	–	[48]
6FDA-TeMPD	linear (Type I)	10	186	612	3.36	11.0	13.6	[49]
	1,3-cyclohexanebis(methylamine) CHMA (30 min)	10	20.2	55.2	4.67	12.7	19.3	[49]
	CHMA (100 °C)	10	14.6	37.2	4.85	12.4	19.9	[49]
	CHMA (150 °C)	10	17.1	54.5	4.57	14.6	22.7	[49]
	CHMA (200 °C)	10	17.4	59.8	4.70	16.2	26.2	[49]

(continued overleaf)

Table 1.2 *(continued)*

Polyimide	Cross-linkable	Feed pressure (atm)	$P(O_2)$	$P(CO_2)$	$\alpha(O_2/N_2)$	$\alpha(CO_2/N_2)$	$\alpha(CO_2/CH_4)$	Reference
6FDA-TeMPD	linear (Type I)	10	190	640	3.5	11.8	14	[50]
	Ethylenediamine (1 min)	10	120	490	4.0	16.3	23	[50]
	Ethylenediamine (2 min)	10	30	120	4.3	17.2	32	[50]
	Ethylenediamine (5 min)	10	12	25	6.0	12.5	28	[50]
6FDA-TeMPD	linear (Type I)	3.5	186	612	3.4	11.0	13.6	[52]
	ethylenediamine (1 min 10 wt%)	3.5	108	435	3.8	15.3	21.1	[52]
	1,3-propanediamine (1 min 10 wt%)	3.5	26.4	81.1	4.9	15.2	27.8	[52]
	1,4-butanediamine (1 min 10 wt%)	3.5	55.5	218	4.2	16.5	24.7	[52]
6FDA-TeMPD	linear (Type I)	10	148.3	517	3.46	12.1	15.7	[53]
	PAMAM (20 min)	10	137.6	568	3.93	16.2	22.8	[53]
	PAMAM (60 min)	10	115.6	451	4.30	16.8	25.1	[53]
	PAMAM (24h)	10	44.1	157	5.38	19.1	35.6	[53]
	PAMAM (7days)	10	7.33	21.6	6.42	18.9	46.8	[53]
6FDA-TeMPD	linear (Type I)	3.5	186	612	3.4	11.0	13.6	[54]
	G1-DAB-AM (5 min)	3.5	130	435	4.6	15.3	23.0	[54]
	G1-DAB-AM (20 min)	3.5	74	220	5.1	15.2	24.4	[54]
	G1-DAB-AM (30 min)	3.5	70	177	5.7	14.5	24.5	[54]
	G1-DAB-AM (60 min)	3.5	50	98	6.9	13.6	21.7	[54]
6FDA-TeMPD	G0-PAMAM (1 day)	10	–	160	–	–	36	[56]
	G0-PAMAM (1 day, 120 °C)	10	–	70	–	–	37	[56]
	G0-PAMAM (1 day, 250 °C)	10	–	84	–	–	30	[56]

1 Barrer = $10^{-10} cm^3 (STP) cm cm^{-2} s^{-1} cmHg^{-1}$.

14–47 at 35 °C. The 6FDA-TeMPD cross-linked by PAMAM dendrimer has higher gas permeability among the other cross-linked polyimides using the imide ring-opening reaction [53]: $P(CO_2)$ = 568 Barrer and $\alpha(CO_2/CH_4)$ = 22.8 at 35 °C and 10 atm. PAMAM dendrimers have good gas separation performance especially for CO_2/N_2 because their amine groups have excellent affinity to CO_2 [81]. The reactions using dendrimers were studied by several authors [53–56]. The cross-linking modification by PAMAM and amine-terminated diaminobutane (DAB-AM) results in a decrease in permeabilities for most gases. In the case of the polyimide cross-linked with DAB-AM dendrimer, the maximum selectivity increases by about 400, 300 and 265% for the gas pairs of He/N_2, H_2/N_2 and H_2/CO_2, respectively [54]. The gas permeability decreases in the order of G1 > G2 > G3, which is consistent with the increasing order of the degree of gel contents.

The gas permeability and selectivity of cross-linked membranes obtained by UV irradiation of polyimides are summarized in Table 1.3 [63–69]. The gas permeability decreases with the increase in UV irradiation time and is in the range 0.2–21 Barrer for O_2 and 0.6–99 Barrer for CO_2 at 35 °C. The selectivities are 4.3–13.8 for $\alpha(O_2/N_2)$ and 21.2–111.4 for $\alpha(CO_2/CH_4)$.

The gas permeation parameters have also been reported for polymers obtained via ether reaction of epoxy and diamines [57,58], for polymer blends with acetylene-terminated oligomer [73,74] and internal acetylene polyimide [72,75]. The gas permeability of cross-linked internal acetylene-containing polymer, 6FDA-TeMPD/*p*-intA (4:1) declines from 612 to 186 Barrer, while the selectivity, $\alpha(CO_2/CH_4)$ increases from 14 to 25 at 35 °C and 10 atm [75]. Moreover, this cross-linked 6FDA-TeMPD/*p*-intA (4:1) membrane is still stable under CO_2 pressure of about 47 atm.

1.4.2 Hyperbranched Polyimides (Type III)

Almost all hyperbranched polyimides with reported gas permeation parameters were synthesized from A_2- and B_3-type monomers. The corresponding transport data are presented in Table 1.4 [20–26]. All the polymers are in a glassy state. The gas permeabilities of hyperbranched polyimides vary from 0.02 to 11 Barrer for O_2 and from 0.08 to 65 Barrer for CO_2 at 20–35 °C. The selectivities vary from 2.0 to 13 for $\alpha(O_2/N_2)$ and from 3.4 to 70.1 for $\alpha(CO_2/N_2)$.

6FDA-TAPA hyperbranched polyimides with terephthaldehyde (TPA) form a group of the most permeable polymers among those presented in Table 1.4. For example, the $P(CO_2)$ of cross-linked 6FDA-TAPA is 65 Barrer and $P(O_2)$ = 11 [20]. The gas transport properties of polymer blend with commercial polyimide, Matrimid 5218 (BTDA-AAPTMI) and P84 (BTDA-TDI/MDI), and hyperbranched polyester, Boltorn (H40) have also been reported [25]. Selectivities of Matrimide-H40 (1 wt%) containing membrane show both high and low values: thus, $\alpha(O_2/N_2)$ varies from 6.8 to 2.0. The permeability of the $P(N_2)$ of P84 with various concentrations of H40 membrane decreases with the increase in the concentrations of H40 in comparison to pure P84, while their selectivity generally are almost constant.

Figure 1.2 presents the effects of UV irradiation time on gas permeability and selectivity for acryl-terminated telechelic polyimide, 6FDA-TeMPD-BEI [33]. It is seen that the gas permeability decreases, while the selectivity increases.

Table 1.3 *Gas permeability coefficients (Barrer) and selectivity of UV cross-linked polyimides (Type II)*

Polyimide	Temperature (°C)	Feed pressure (atm)	$P(O_2)$	$P(CO_2)$	$\alpha(O_2/N_2)$	$\alpha(CO_2/N_2)$	$\alpha(CO_2/CH_4)$	Reference
BTDA-TMPD (linear (Type I))	35	10	9.84	38.3	5.5	21.4	26.4	[63]
BTDA-TMPD (1 min)	35	10	4.15	17.8	5.5	23.7	33.6	[63]
BTDA-TMPD (10 min)	35	10	0.482	1.48	13.8	42.4	–	[63]
BTDA-TMPD (30 min)	35	10	0.199	0.618	13.2	40.9	111.4	[63]
BTDA/6FDA-TMPD (1:1) (linear (Type I))	35	10	12.3	48.3	4.5	17.7	18.0	[63]
BTDA/6FDA-TMPD (1:1) (30 min)	35	10	1.74	5.58	9.6	30.8	85.2	[63]
BTDA/6FDA-TMPD (1:1) (60 min)	35	10	1.58	4.84	10.5	32.3	–	[63]
BTDA/6FDA-TMPD (1:3) (linear (Type I))	35	10	18.5	83.2	3.6	16.4	23.4	[63]
BTDA/6FDA-TMPD (1:3) (30 min)	35	10	4.57	14.9	5.9	19.3	49.3	[63]
BTDA/6FDA-TMPD (1:3) (60 min)	35	10	3.59	13.3	7.0	25.8	–	[63]
BTDA-BAPP (linear (Type I))	30	1	0.54	2.8	4.9	25.4	–	[64]
BTDA-BAPP (1 min)	30	1	0.45	2.4	7.5	40.0	–	[64]
BTDA-BAPP (5 min)	30	1	0.40	1.9	8.5	40.4	–	[64]
BTDA-BAPP (10 min)	30	1	0.26	1.9	10.4	76.0	–	[64]
ODPA-BAPP (linear (Type I))	30	1	0.49	2.7	7.0	38.6	–	[64]
ODPA-BAPP (1 min)	30	1	0.43	2.2	6.1	31.4	–	[64]
ODPA-BAPP (5 min)	30	1	0.48	2.4	7.4	36.9	–	[64]
ODPA-BAPP (10 min)	30	1	0.37	1.7	7.9	36.2	–	[64]
6FDA-TMPD (linear (Type I))	30	1	130	900	3.6	25.0	28.1	[65]
6FDA-TMPD (30 min)	30	1	14	59	6.1	25.7	42.1	[65]
6FDA-TMPD (BP 0.99 wt% 15 min)	30	1	21	99	5.4	25.4	39.6	[65]
6FDA-TMPD (BP 9.09 wt% 15 min)	30	1	5.2	21	4.3	17.5	21.2	[65]

BTDA-BAPP (linear (Type I))	30	1	0.45	2.3	7.50	38.3	–	[66]
BTDA-BAPP (10 min)	30	1	0.36	1.8	7.83	39.1	–	[66]
BTDA-BAPP (20 min)	30	1	0.36	1.8	8.37	41.9	–	[66]
BTDA-BAPP (30 min)	30	1	0.24	0.97	10.4	42.2	–	[66]
BTDA-BAPP (linear (Type I))	30	1	0.54	2.8	4.90	25.5	–	[66]
BTDA-BAPP (quench 10 min)	30	1	0.47	2.4	7.34	37.5	–	[66]
BTDA-BAPP (quench 30 min)	30	1	0.34	1.5	8.50	37.5	–	[66]
PMDA/BTDA-3MPDA (9:1) linear (Type I)	30	–	22.3	–	4.3	–	–	[69]
PMDA/BTDA-3MPDA (9:1) (2h)	30	–	4.35	–	5.7	–	–	[69]
PMDA/BTDA-3MPDA (9:1) (6h)	30	–	0.67	–	6.7	–	–	[69]
PMDA/BTDA-3MPDA (8:2) linear (Type I)	30	–	21.4	–	4.3	–	–	[69]
PMDA/BTDA-3MPDA (8:2) (2h)	30	–	9.10	–	6.3	–	–	[69]
PMDA/BTDA-3MPDA (8:2) (10h)	30	–	0.21	–	7.2	–	–	[69]
PMDA/BTDA-3MPDA (5:5) linear (Type I)	30	–	12.1	–	4.9	–	–	[69]
PMDA/BTDA-3MPDA (5:5) (2h)	30	–	7.40	–	5.6	–	–	[69]
PMDA/BTDA-3MPDA (5:5) (6h)	30	–	0.40	–	11.0	–	–	[69]
PMDA/BTDA-3MPDA (3:7) linear (Type I)	30	–	9.42	–	4.9	–	–	[69]
PMDA/BTDA-3MPDA (3:7) (2h)	30	–	1.15	–	11.0	–	–	[69]
BTDA-3MPDA linear (Type I)	30	–	5.00	–	5.0	–	–	[69]
BTDA-3MPDA (2h)	30	–	0.84	–	11.0	–	–	[69]

Table 1.4 *Gas permeability coefficients (Barrer) and selectivity of hyperbranched polyimides (Type III)*

Polymer	Temperature (°C)	Feed pressure (atm)	$P(O_2)$	$P(CO_2)$	$\alpha(O_2/N_2)$	$\alpha(CO_2/N_2)$	$\alpha(CO_2/CH_4)$	Reference
6FDA-TAPA (AM-T) EDGE 1.4	35	1	–	3.7	–	30.8	46	[20]
6FDA-TAPA (AM-T) EDGE 0.34	35	1	1.9	11	5.8	33.3	50	[20]
6FDA-TAPA (AM-T) EDGE 0.15	35	1	–	16	–	–	32	[20]
6FDA-TAPA (AM-T) TPA 0.25	35	1	11	65	5.1	30.1	41	[20]
6FDA-TAPA (AH-T) ODA 0.04	35	1	–	6.7	–	26.8	61	[20]
DSDA-TAPA (AH-T) TPA 0.05	35	1	–	4.0	–	43.5	–	[20]
DSDA-TAPA (AH-T) DDS 0.03	35	1	–	1.0	–	41.7	–	[20]
6FDA-TAPOB	25	1	2.25	12.80	6.2	35.3	–	[21]
6FDA-TAPOB (6FMA)	25	1	1.9	9.0	6.3	30.0	–	[22]
6FDA-TAPOB (p-3FMA)	25	1	1.5	7.2	6.5	31.3	–	[22]
6FDA-TAPOB	25	1	1.2	5.4	7.5	33.8	–	[22]
6FDA-TAPOB (m-3FMA)	25	1	1.3	6.3	6.8	33.2	–	[22]
HQDPA-TFAPOB(amine-terminated)	20	1	0.058	1.55	7.7	3.4	–	[24]
HQDPA-TFAPOB(CF3-terminated)	20	1	0.114	2.72	6.5	3.6	–	[24]
Matrimid 5218 (H40 0.0 wt%)	35	3.5	1.02	5.20	6.8	34.7	–	[25]
Matrimid 5218 (H40 1.0 wt%)	35	3.5	1.31	5.25	2.0	8.2	–	[25]
Matrimid 5218 (H40 5.0 wt%)	35	3.5	0.70	3.29	6.4	29.9	–	[25]
Matrimid 5218 (H40 10.0 wt%)	35	3.5	0.61	2.85	6.8	31.7	–	[25]
P84 (H40 0.0 wt%)	35	3.5	0.57	1.67	11.4	33.4	–	[25]
P84 (H40 1.0 wt%)	35	3.5	0.51	1.20	13	30.0	–	[25]
P84 (H40 5.0 wt%)	35	3.5	0.39	1.00	13	33.3	–	[25]
P84 (H40 10.0 wt%)	35	3.5	0.37	0.94	12	31.3	–	[25]
ODPA/TAP/ODA (AM-T) (1/1/0)	30	1.5	0.024	0.096	12	48.0	48	[26]
ODPA/TAP/ODA (AM-T) (1/0.95/0.05)	30	1.5	0.043	0.156	10.8	39.0	52	[26]
ODPA/TAP/ODA (AM-T) (1/0.75/0.25)	30	1.5	0.056	0.246	9.3	41.0	62	[26]
ODPA/TAP/ODA (AM-T) (1/0.5/0.5)	30	1.5	0.101	0.352	7.8	27.1	50	[26]
ODPA/TAP/ODA (AM-T) (1/0.25/0.75)	30	1.5	0.087	0.450	8.7	45.1	75	[26]
ODPA/TAP/ODA (AH-T) (2/1/0)	30	1.5	–	0.079	–	–	–	[26]
ODPA/TAP/ODA (AH-T) (2/0.95/0.05)	30	1.5	0.024	0.140	8.0	46.7	47	[26]
ODPA/TAP/ODA (AH-T) (2/0.75/0.25)	30	1.5	0.054	0.135	10.8	27.0	36	[26]
ODPA/TAP/ODA (AH-T) (2/0.5/0.5)	30	1.5	0.065	0.257	9.3	36.7	66	[26]
ODPA/TAP/ODA (AH-T) (2/0.25/0.75)	30	1.5	0.071	0.285	8.9	35.6	51	[26]

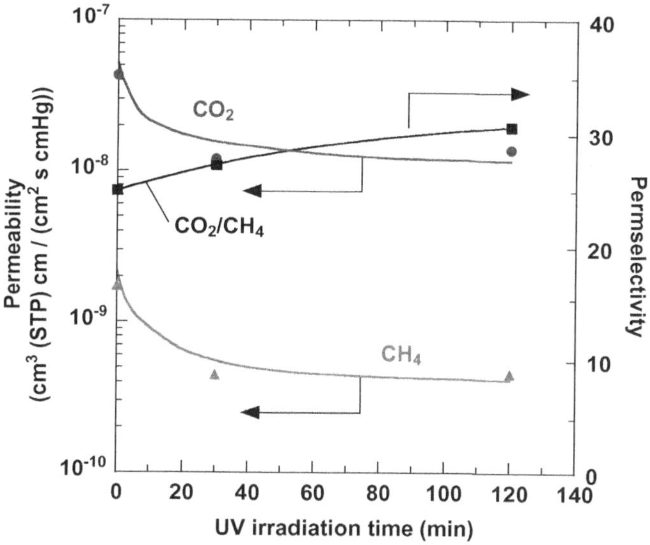

Figure 1.2 *Dependence of* P*(CO_2) (•),* P*(CH_4) (▲) and α(CO_2/CH_4) (■) on irradiation time in UV-cross-linked 6FDA-TeMPD-BEI membranes at 35 °C. Data from Reference [33]*

In the same manner the gas permeability in hyperbranched polyimide prepared from vinyl-terminated telechelic polyimide, 6FDA-TeMPD-PAS is reduced by UV irradiation [32]. The gas permeabilities of this polymer after 120 min of irradiation are as follows: 50 Barrer for O_2 and 345 Barrer for CO_2 at 30 °C and 1 atm. The selectivities are 3.5 for $\alpha(O_2/N_2)$ and 24 for $\alpha(CO_2/CH_4)$.

1.4.3 Selectivity and Permeability

Figures 1.3–1.5 present Robeson diagrams for different gas pairs, in various polyimide membranes. The first examination of these plots shows that the data points for Type II (cross-linked) and Type III (hyperbranched) structures can be found in the whole 'cloud' of the data points including those of linear structures (Type I). However, the data points for some hyperbranched and cross-linked polyimides are located near the upper bound for O_2/N_2 pair as shown in Figure 1.3. On the other hand, the majority of the hyperbranched polyimides tend to be located among the common linear polyimides, as well as that of cross-linked polyimides in the diagram for CO_2/N_2 pair relationship as shown in Figure 1.4. Finally, the data points for all three types of the structures are far from upper bound for the pair CO_2/CH_4, as seen in Figure 1.5. Apparently, more data are required in order to discuss more specifically the relationship between gas permeation properties and the structure of hyperbranched and cross-linked polymers.

1.5 Concluding Remarks

Initially, hyperbranched and cross-linked polymers were the subject of investigation in the field of polymer science (both polymer chemistry and physics). Gradually, since the

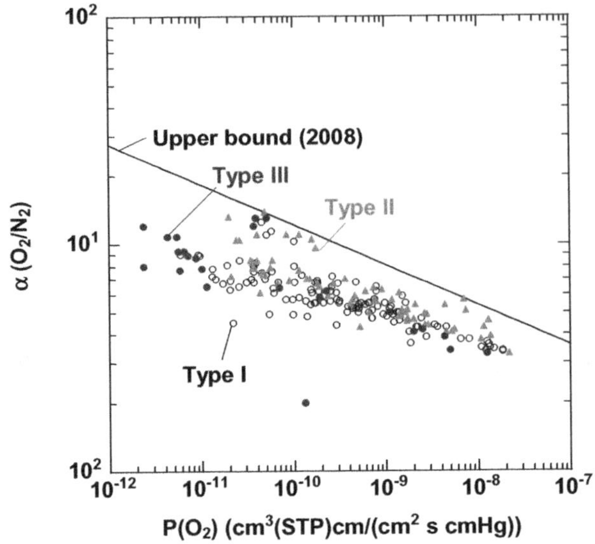

Figure 1.3 *Relationship between* $\alpha(O_2/N_2)$ *and* $P(O_2)$ *in polyimide membranes. Type I: linear (○), Type II: cross-linked (▲) and Type III: hyperbranched (•)*

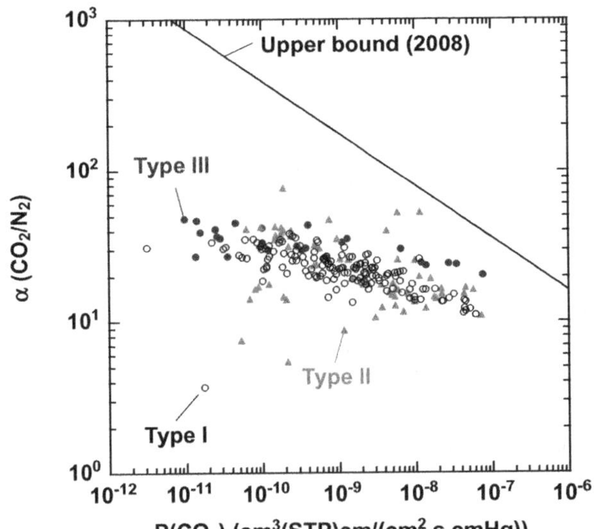

Figure 1.4 *Relationship between* $\alpha(CO_2/N_2)$ *and* $P(CO_2)$ *in polyimide membranes. Type I: linear (○), Type II: cross-linked (▲) and Type III: hyperbranched (•)*

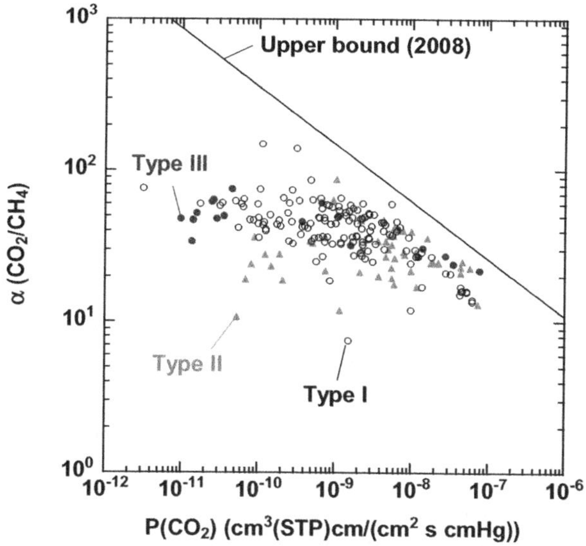

Figure 1.5 *Relationship between $\alpha(CO_2/CH_4)$ and $P(CO_2)$ in polyimide membranes. Type I: linear (○), Type II: cross-linked (▲) and Type III: hyperbranched (•)*

early 2000s, hyperbranched and cross-linked polyimides attracted interest as materials for gas separation membranes. Bearing in mind the general requirements for 'good' gas separation materials (high permeability, selectivity, physical stability, processability and low cost) they did not demonstrate so far (as the classes of polymer architecture) the desired combination of the properties. However, since some promising results have been obtained and experience has been accumulated, further investigations in the synthesis, modifications and membrane properties are important: these should show the real potential impact of these polymers in membrane science and technology.

References

(1) Y. H. Kim, Hyperbranched polymers 10 years after, *J. Polym. Sci., Part A: Polym. Chem.*, **36**, 1685–1698 (1998).

(2) B. Voit, New developments in hyperbranched polymers, *J. Polym. Sci., Part A: Polym. Chem.*, **38**, 2505–2525 (2000).

(3) M. Jikei, M. Kakimoto, Dendritic aromatic polyamides and polyimides, *J. Polym. Sci., Part A: Polym. Chem.*, **42**, 1293–1309 (2004).

(4) A. D. Jenkins, P. Kratochvìl, R. F. T. Stepto, U. W. Suter, Glossary of basic terms in polymer science, *Pure Appl. Chem.*, **68**, 2287–2311 (1996).

(5) J. M. J. Frechet, D. A. Tomalia, Dendrimers and other dendritic polymers, J. M. J. Frechet, D. A. Tomalia (eds). 2001, New York, United States: John Wiley & Sons.

(6) K. Yamanaka, M. Jikei, M. Kakimoto, Synthesis of hyperbranched aromatic polyimides via polyamic acid methyl ester precursor, *Macromolecules*, **33**, 1111–1114 (2000).

(7) D. S. Thompson, L. J. Markoski, J. S. Moore, Rapid synthesis of hyperbranched aromatic polyetherimides, *Macromolecules*, **32**, 4764–4768 (1999).

(8) P. J. Flory, Molecular size distribution in three dimensional polymers. VI. Branched polymers containing A-R-B, f-1, type units, *J. Am. Chem. Soc.*, **74**, 2718–2723 (1952).
(9) Y. H. Kim, O. W. Webster, Water-soluble hyperbranched polyphenylene: 'a unimolecular micelle?' *J. Am. Chem. Soc.*, **112**, 4592–4593 (1990).
(10) D. S. Thompson, L. J. Markoski, J. S. Moore, I. Sendijarevic, A. Lee, A. J. McHugh, Synthesis and characterization of hyperbranched aromatic poly(ether imide)s with varying degree of branching. *Macromolecules*, **33**, 6412–6415 (2000).
(11) J. Fang, H. Kita, K. Okamoto, Hyperbranched polyimides for gas separation applications. 1. Synthesis and characterization, *Macromolecules*, **33**, 4639–4646 (2000).
(12) J. Hao, M. Jikei, M. Kakimoto, Preparation of hyperbranched aromatic polyimides via A2 + B3 approach, *Macromolecules*, **35**, 5372–5381 (2002).
(13) G. Yang, M. Jikei, M. Kakimoto, Synthesis and properties of hyperbranched aromatic polyamide, *Macromolecules*, **32**, 2215–2220 (1999).
(14) M. Jikei, S.-H. Chon, M. Kakimoto, S. Kawauchi, T. Imase, J. Watanabe, Synthesis of hyperbranched aromatic polyamide from aromatic diamines and trimesic acid, *Macromolecules*, **32**, 2061–2064 (1999).
(15) G. Yang, M. Jikei, M. Kakimoto, Successful thermal self-polycondensation of AB2 monomer to form hyperbranched aromatic polyamide, *Macromolecules*, **31**, 5964–5966 (1998).
(16) M. Trollsas, D. Hedrick, R. Mecerreyes, R. Jerome, P. H. Dubois, Internal functionalization in hyperbranched polyesters, *J. Polym. Sci., Part A: Polym. Chem.*, **36**, 3187–3192 (1998).
(17) A. Morikawa, Preparation and properties of hyperbranched poly(ether ketones) with a various number of phenylene units, *Macromolecules*, **31**, 5999–6009 (1998).
(18) D. H. Bolton, K. L. Wooley, Synthesis and characterization of hyperbranched polycarbonates, *Macromolecules*, **30**, 1890–1896 (1997).
(19) K. Yamanaka, M. Jikei, M. Kakimoto, Preparation and properties of hyperbranched aromatic polyimides via polyamic acid methyl ester precursors, *Macromolecules*, **33**, 6937–6944 (2000).
(20) J. Fang, H. Kita, K. Okamoto, Gas permeation properties of hyperbranched polyimide membranes, *J. Membrane Sci.*, **182**, 245–256 (2001).
(21) T. Suzuki, Y. Yamada, Y. Tsujita, Gas transport properties of 6FDA-TAPOB hyperbranched polyimide membrane, *Polymer*, **45**, 7167–7171 (2004).
(22) Y. Yamada, J. Sakai, Siloxane-modified hyperbranched polyimide, 2006: WO 2006/082814.
(23) V. A. Bershtein, L. M. Egorova, P. N. Yakushev, P. Sysel, R. Hobzova, J. Kotek, P. Pissis, S. Kripotou, P. Maroulas, Hyperbranched polyimides crosslinked with ethylene glycol diglycidyl ether: Glass transition dynamics and permeability, *Polymer*, **47**, 6765–6772 (2006).
(24) H. Gao, D. Wang, W. Jiang, S. Guan, Z. Jiang, Gas permeability of fluorinated hyperbranched polyimide, *J. Appl. Polym. Sci.*, **109**, 2341–2346 (2008).
(25) D. M. Sterescu, D. F. Stamatialis, M. Wessling, Boltorn-modified polyimide gas separation membranes, *J. Membr. Sci.*, **310**, 512–521 (2008).
(26) J. Peter, A. Khalyavina, J. Kriz, M. Bleha, Synthesis and gas transport properties of ODPA-TAP-ODA hyperbranched polyimides with various comonomer ratios, *Eur. Polym. J.*, **45**, 1716–1727 (2009).
(27) D. R. Paul, Y. P. Yampol'skii (eds.), *Polymeric Gas Separation Membranes*. 1994, Boca Raton, FL, United States: CRC Press.
(28) R. W. Baker, Membrane Technology and Applications. 2000, New York: McGraw-Hill.
(29) M. Onda, S. Kanehashi, S. Kazama, K. Nagai, Department of novel networked polyimide membranes for gas separation at high temperature and pressure, *Polym. Prep. Jpn.*, **55**, 5026 (2006).
(30) K. Nagai, S. Kazama, Gas separation membranes, 2007: *JP Patent* 2007-044653.
(31) H. Shirokura, M. Onda, K. Imai, S. Ishimatsu, S. Kazama, K. Nagai, Gas permeability of hyperbranched polyimide membranes containing fluorine structure, *Polym. Prep. Jpn.*, **57**, 1610 (2008).
(32) H. Shirokura, K. Yoshida, S. Miyata, K. Nagai, Gas permeation properties of UV-cured telechelic polyimides, *Polym. Prep. Jpn.*, **57**, 4145 (2008).
(33) H. Shirokura, K. Yoshida, S. Miyata, K. Nagai, Gas separation properties of UV-crosslinked telechelic polyimide membranes, *Polym. Prep. Jpn.*, **58**, 1596 (2009).

(34) C. Staudt-Bickel, W. J. Koros, Improvement of CO_2/CH_4 separation characteristics of polyimides by chemical crosslinking, *J. Membr. Sci.*, **155**, 145–154 (1999).
(35) J. D. Wind, C. Staudt-Bickel, D. R. Paul, W. J. Koros, Solid-state covalent cross-linking of polyimide membranes for carbon dioxide plasticization reduction, *Macromolecules*, **36**, 1882–1888 (2003).
(36) A. Taubert, J. D. Wind, D. R. Paul, W. J. Koros, K. I. Winey, Novel polyimide ionomers: CO_2 plasticization, morphology, and ion distribution, *Polymer*, **44**, 1881–1892 (2003).
(37) J. D. Wind, C. Staudt-Bickel, D. R. Paul, W. J. Koros, The effects of crosslinking chemistry on CO_2 plasticization of polyimide gas separation membranes, *Ind. Eng. Chem. Res.*, **41**, 6139–6148 (2002).
(38) J. H. Kim, W. J. Koros, D. R. Paul, Effects of CO_2 exposure and physical aging on the gas permeability of thin 6FDA-based polyimide membranes: Part 2. with crosslinking, *J. Membr. Sci.*, **282**, 32–43 (2006).
(39) J. D. Wind, S. M. Sirard, D. R. Paul, P. F. Green, K. P. Johnston, W. J. Koros, Carbon dioxide-induced plasticization of polyimide membranes: Pseudo-equilibrium relationships of diffusion, sorption, and swelling, *Macromolecules*, **36**, 6433–6441 (2003).
(40) J. D. Wind, S. M. Sirard, D. R. Paul, P. F. Green, K. P. Johnston, W. J. Koros, Relaxation dynamics of CO_2 diffusion, sorption, and polymer swelling for plasticized polyimide membranes, *Macromolecules*, **36**, 6442–6448 (2003).
(41) J. D. Wind, D. R. Paul, W. J. Koros, Natural gas permeation in polyimide membranes, *J. Membr. Sci.*, **228**, 227–236 (2004).
(42) M. Alexis, W. Hillock, W. J. Koros, Cross-linkable polyimide membrane for natural gas purification and carbon dioxide plasticization reduction, *Macromolecules*, **40**, 583–587 (2007).
(43) M. Adam, M. Kratochvil, W. J. Koros, Decarboxylation-induced cross-linking of a polyimide for enhanced CO_2 plasticization resistance, *Macromolecules*, **41**, 7920–7927 (2008).
(44) R. A. Hayes, Amine-modified polyimide membranes, 1991: *US Patent* 4981497.
(45) P. S. Tin, T. S. Chung, Y. Liu, R. Wang, S. L. Liu, K. P. Pramoda, Effects of cross-linking modification on gas separation performance of Matrimid membranes, *J. Membr. Sci.*, **225**, 77–90 (2003).
(46) H.-Y. Zhao, Y.-M. Cao, X.-L. Ding, M.-Q. Zhou, J.-H. Liu, Q. Yuan, Poly(ethylene oxide) induced cross-linking modification of Matrimid membranes for selective separation of CO_2, *J. Membr. Sci.*, **320**, 179–184 (2008).
(47) H.-Y. Zhao, Y.-M. Cao, X.-L. Ding, M.-Q. Zhou, Q. Yuan, Effects of cross-linkers with different molecular weights in cross-linked Matrimid 5218 and test temperature on gas transport properties, *J. Membr. Sci.*, **323**, 176–184 (2008).
(48) Y. Liu, R. Wang, T.-S. Chung, Chemical cross-linking modification of polyimide membranes for gas separation, *J. Membr. Sci.*, **189**, 231–239 (2001).
(49) L. Shao, T.-S. Chung, S. H. Goh, K. P. Pramoda, The effects of 1,3-cyclohexanebis (methylamine) modification on gas transport and plasticization resistance of polyimide membranes, *J. Membr. Sci.*, **267**, 78–89 (2005).
(50) L. Shao, T.-S. Chung, S. H. Goh, K. P. Pramoda, Polyimide modification by a linear aliphatic diamine to enhance transport performance and plasticization resistance, *J. Membr. Sci.*, **256**, 46–56 (2005).
(51) C. E. Powell, X. J. Duthie, S. E. Kentish, G. G. Qiao, G. W. Stevens, Reversible diamine cross-linking of polyimide membranes, *J. Membr. Sci.*, **291**, 199–209 (2007).
(52) L. Shao, L. Liu, S.-X. Cheng, Y.-D. Huang, J. Ma, Comparison of diamino cross-linking in different polyimide solutions and membranes by precipitation observation and gas transport, *J. Membr. Sci.*, **312**, 174–185 (2008).
(53) T.-S. Chung, M. L. Chng, K. P. Pramoda, Y. Xiao, PAMAM dendrimer-induced cross-linking modification of polyimide membranes, *Langmuir*, **20**, 2966–2969 (2004).
(54) L. Shao, T.-S. Chung, S. H. Goh, K. P. Pramoda, Transport properties of cross-linked polyimide membranes induced by different generations of diaminobutane (DAB) dendrimers, *J. Membr. Sci.*, **238**, 153–163 (2004).

(55) Y. Xiao, T.-S. Chung, M. L. Chung, Surface characterization, modification chemistry, and separation performance of polyimide and polyamidoamine dendrimer composite films, *Langmuir*, **20**, 8230–8238 (2004).
(56) Y. Xiao, L. Shao, T.-S. Chung, D. A. Schiraldi, Effects of thermal treatments and dendrimers chemical structures on the properties of highly surface cross-linked polyimide films, *Ind. Eng. Chem. Res.*, **44**, 3059–3067 (2005).
(57) K. Nagai, N. Booker, A. Mau, J. Hodgkin, S. Kentish, G. Stevens, A. Geertsema, Gas permeation properties of polyimide/epoxy composite materials, *Polym. Mater. Sci. Eng.*, **85**, 91 (2001).
(58) X. J. Duthie, S. E. Kentish, C. E. Powell, G. G. Qiao, K. Nagai, G. W. Stevens, Plasticization suppression in grafted polyimide-epoxy network membranes, *Ind. Eng. Chem. Res.*, **46**, 8183–8192 (2007).
(59) N. J. Turro, Molecular photochemistry. 1965, New York: W. A. Benjamin.
(60) A. A. Lin, V. R. Sastri, G. Tesoro, A. Reiser, On the cross-linking mechanism of benzophenone-containing polyimides, *Macromolecules*, **21**, 1165–1169 (1988).
(61) S. Kuroda, I. Mita, Degradation of aromatic polymers II. The crosslinking during thermal and thermo-oxidative degradation of a polyimide, *Eur. Polym. J.*, **25**, 611–620 (1989).
(62) I. K. Meier, M. Langsam, H. C. Klotz, Selectivity enhancement via photooxidative surface modification of polyimide air separation membranes, *J. Membr. Sci.*, **94**, 195–212 (1994).
(63) H. Kita, T. Inada, K. Tanaka, K. Okamoto, Effect of photocrosslinking on permeability and permselectivity of gases through benzophenone-containing polyimide, *J. Membr. Sci.*, **87**, 139–147 (1994).
(64) S. Matsui, T. Ishiguro, A. Higuchi, T. Nakagawa, Effect of ultraviolet light irradiation on gas permeability in polyimide membranes. 1. Irradiation with low pressure mercury lamp on photosensitive and nonphotosensitive membranes, *J. Polym. Sci., Part B: Polym. Phys.*, **35**, 2259–2269 (1997).
(65) S. Matsui, H. Sato, T. Nakagawa, Effects of low molecular weight photosensitizer and UV irradiation on gas permeability and selectivity of polyimide membrane, *J. Membr. Sci.*, **141**, 31–43 (1998).
(66) S. Matsui, T. Nakagawa, Effect of ultraviolet light irradiation on gas permeability in polyimide membranes. II. Irradiation of membranes with high-pressure mercury lamp, *J. Appl. Polym. Sci.*, **67**, 49–60 (1998).
(67) R. A. Hayes, Polyimide gas separation membranes, 1988: *US Patent* 4717393.
(68) W. F. Burgoyne, M. Langsam, M. E. Ford, J. P. Casey, Membranes formed from unsaturated polyimides. 1990: *US Patent* 4931182.
(69) Y. Liu, C. Pan, M. Ding, J. Xu, Gas permeability and permselectivity of photochemically crosslinked copolyimides, *J. Appl. Polym. Sci.*, **73**, 521–526 (1999).
(70) N. Bilow, R. H. Boschan, A. L. Landis, Acetylene-substituted aromatic primary amines and the process of making them, 1975: *US Patent* 3928450.
(71) N. Bilow, A. L. Landis, L. J. Miller, Copolymer of polyimide oligomers and terephthalonitrile N,N-dioxide and their methods of preparation, 1975: *US Patent* 3864309.
(72) M. E. Rezac, E. T. Sorensen, H. W. Beckham, Transport properties of crosslinkable polyimide blends, *J. Membr. Sci.*, **136**, 249–259 (1997).
(73) M. E. Rezac, B. Schoberl, Transport and thermal properties of poly(ether imide)/acetylene-terminated monomer blends, *J. Membr. Sci.*, **156**, 211–222 (1999).
(74) A. Bos, I. G. M. Punt, M. Wessling, H. Strathmann, Suppression of CO_2-plasticization by semiinterpenetrating polymer network formation, *J. Polym. Sci., Part B: Polym. Phys.*, **36**, 1547–1556 (1998).
(75) Y. Xiao, T.-S. Chung, H. M. Guan, M. D. Guiver, Synthesis, cross-linking and carbonization of co-polyimides containing internal acetylene units for gas separation, *J. Membr. Sci.*, **302**, 254–264 (2007).
(76) K. Yamanaka, M. Jikei, M. Kakimoto, Preparation of hyperbranched aromatic polyimide without linear units by end-capping reaction, *Macromolecules*, **34**, 3910–3915 (2001).
(77) L. J. Markoski, D. S. Thompson, J. S. Moore, Synthesis and characterization of linear-dendritic aromatic etherimide copolymers: tuning molecular architecture to optimize properties and processability, *Macromolecules*, **33**, 5315–5317 (2000).

(78) J. Hao, M. Jikei, M. Kakimoto, Synthesis and comparison of hyperbranched aromatic polyimides having the same repeating unit by AB_2 self-polymerization and $A_2 + B_3$ polymerization, *Macromolecules*, **36**, 3519–3528 (2003).
(79) L. M. Robeson, Correlation of separation factor versus permeability for polymeric membranes, *J. Membr. Sci.*, **62**, 165–185 (1991).
(80) L. M. Robeson, The upper bound revisited, *J. Membr. Sci.*, **320**, 390–400 (2008).
(81) A. S. Kovvali, H. Chen, K. K. Sirkar, Dendrimer membranes: a CO_2-selective molecular gate, *J. Am. Chem. Soc.*, **122**, 7594–7595 (2000).

2

Gas Permeation Parameters and Other Physicochemical Properties of a Polymer of Intrinsic Microporosity (PIM-1)

Peter M. Budd[a], *Neil B. McKeown*[b], *Detlev Fritsch*[c], *Yuri Yampolskii*[d] *and Victor Shantarovich*[e]

[a] *Organic Materials Innovation Centre, School of Chemistry, University of Manchester, UK*

[b] *School of Chemistry, Cardiff University, UK*

[c] *Institute of Polymer Research, GKSS Research Centre Geesthacht GmbH, Geesthacht, Germany*

[d] *A. V. Topchiev Institute of Petrochemical Synthesis, Russian Academy of Sciences, Moscow, Russia*

[e] *N. N. Semenov Institute of Chemical Physics, Russian Academy of Sciences, Moscow, Russia*

2.1 Introduction

Membrane technologists are well aware that the most permeable glassy polymers are those which possess a very high free volume, where the term 'free volume' refers to the intermolecular voids within a material [1]. Scientists who work with molecular sieves, such as zeolites, commonly use the term 'microporous material' to describe those materials which contain pores or channels less than 2 nm in width, a definition that arises in the context of gas adsorption studies [2].

Two questions can thus be asked.

1 Can the macromolecular backbone of a polymer be designed to generate a large amount of interconnected free volume?

Membrane Gas Separation Edited by Yuri Yampolskii and Benny Freeman

2 If so, does the polymer then behave like a molecular sieve, specifically in terms of gas adsorption?

If the answer to both these questions is 'yes', then the polymer can be said to have 'intrinsic microporosity' [3,4]; 'intrinsic' because it is a consequence of the molecular structure and 'microporosity' because we are dealing with 'micropores' in the sense defined by IUPAC [2]. This chapter introduces a concept for creating just such polymer structures, outlines the properties of the archetypal 'polymer of intrinsic microporosity', PIM-1, and describes recent results for PIM-1 obtained in the course of a European collaborative research project [5].

2.2 The PIM Concept

Most polymer chains are flexible, in the sense that they can change conformation by rotation about backbone bonds. If sufficient free volume is available, it is possible for large-scale movements of flexible polymer chains to occur, giving a rubbery or fluid material. In order to trap additional free volume in the glassy state, chain mobility must be constrained. This can be done by creating a polymer backbone that cannot easily change conformation. The extreme example is a ladder polymer, for which rotation about the backbone axis requires a bond to be broken. However, rigid polymer chains, such as ladder polymers, tend to pack together effectively and interact strongly, giving dense materials that are difficult to process. Packing must be disrupted for a high free volume polymer to be achieved. One way to do this is to introduce 'sites of contortion', which force the backbone to twist and turn erratically. Thus, the two key features of a 'polymer of intrinsic microporosity' (PIM) are that the backbone incorporates (1) sequences that are 'rigid' in the sense that rotational movements are highly restricted (e.g. comprising fused rings giving a ladder structure) and (2) sites that introduce kinks into the backbone. A suitable 'site of contortion' is a spiro-centre where a single tetrahedral carbon atom is part of two five-membered rings. These features are clearly seen in the polymer referred to as PIM-1, which is most commonly prepared from 5,5′,6,6′-tetrahydroxy-3,3,3′,3′-tetramethyl-1,1′-spirobisindane and 1,4-dicyanotetrafluorobenzene (Figure 2.1) [6–10]. The highly efficient double aromatic nucleophilic substitution reaction gives a planar dibenzodioxane linkage, generating a fused ring (ladder) sequence. The spiro-centre provides a site of contortion that puts a bend in the polymer backbone. This design concept developed out of research originally aimed at producing catalytically active polymer networks with high surface area [11–14].

PIM-1 was first prepared by a lengthy polymerization (72 h) at modest temperature (65 °C) in dimethylformamide (DMF) [6,7]. Subsequently, a rapid, high temperature (155 °C) procedure has been developed [15]. The product, which precipitates during the polymerization, is a yellow, fluorescent polymer that is soluble in tetrahydrofuran, chloroform, dichloromethane, *o*-dichlorobenzene and acetophenone. It can be cast from solution to form free-standing membranes with a film density of about 1.1 $g\,cm^{-3}$. It is a glassy polymer with a high degradation temperature (above 350 °C) by thermogravimetric analysis. There is no evidence of a glass transition below the degradation temperature. As a membrane, in pervaporation it is selective for organics over water [7], and in gas separation it exhibits an outstanding combination of permeability and selectivity [5,16],

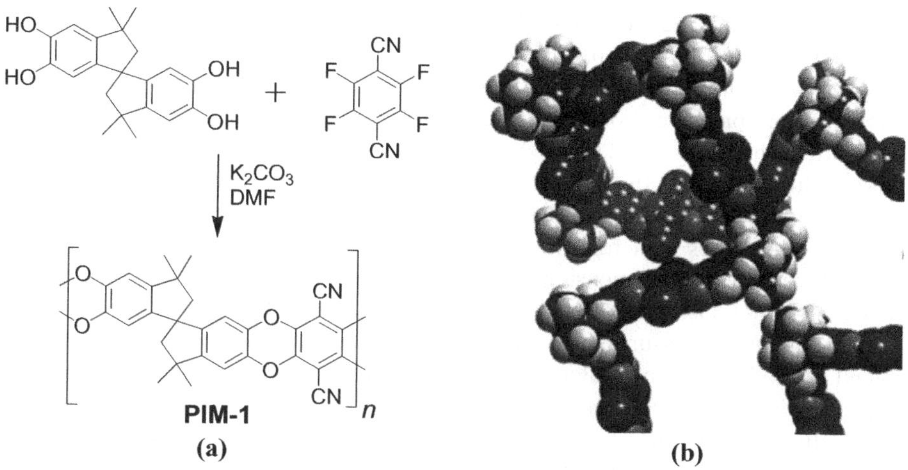

Figure 2.1 (a) Preparation of PIM-1. (b) Molecular model of a fragment of PIM-1 showing the contorted structure

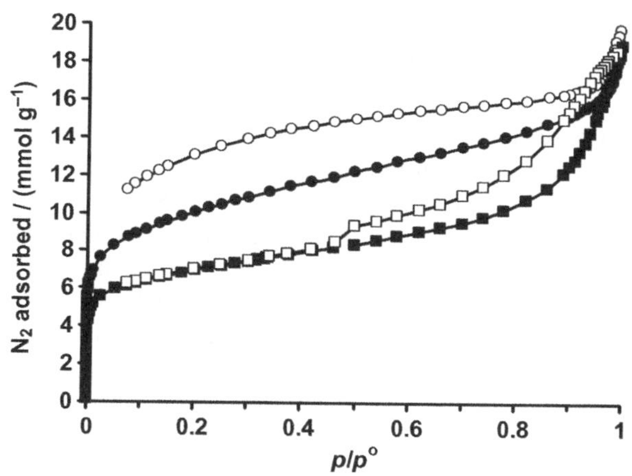

Figure 2.2 Nitrogen adsorption (filled symbols) and desorption (open symbols) isotherms at 77 K for PIM-1 (●, ○) and for Darco 20-40 mesh activated carbon (■, □)

as is discussed further below. The rationale for describing it as 'microporous' comes from gas adsorption studies and this is discussed first.

2.3 Gas Adsorption

The sorption of nitrogen at liquid nitrogen temperature (77 K) has long been used to characterize porous materials [17]. Figure 2.2 compares the low temperature N_2 adsorption/desorption behaviour of PIM-1 with that of a typical activated carbon [18]

(Darco 20–40 mesh). The data were obtained using a Micromeritics ASAP 2020 instrument. The amount of gas adsorbed is plotted against equilibrium relative pressure (p/p^0), where p^0 is the saturation pressure of the gas at the temperature of measurement, in this case atmospheric pressure. For both materials, high uptake is seen at very low relative pressure. It is the very smallest pores (micropores, dimensions <2 nm) that fill first, because multi-wall interactions give rise to strong adsorption. For the carbon, there is further adsorption associated with a hysteresis (the desorption curve lies above the adsorption curve) that closes at a relative pressure above 0.4. This is typical of a material that includes mesopores (dimensions 2–50 nm), which fill by a capillary condensation mechanism. PIM-1 also shows hysteresis, but it differs in that the hysteresis extends down to low relative pressures. This cannot be explained by mesoporosity, but it may be attributed to swelling of the material on adsorption, or possibly to the tortuosity of the micropores. Thus the carbon sample has both micropores and mesopores, whilst PIM-1 is essentially microporous. Superimposed on the 'microporous material' behaviour, however, is polymer behaviour, since rearrangement of polymer chains and swelling can occur, despite large-scale changes of polymer conformation being forbidden.

Brunauer, Emmet, Teller (BET) analysis of adsorption isotherms allows a surface area to be calculated [17,19]. The BET model is for multilayer adsorption on a flat surface, so when applied to a porous material the surface area obtained is an apparent value. Nevertheless, BET surface area provides a useful comparison between materials. PIM-1 shows a higher BET surface area than the activated carbon in Figure 2.2, a value of $780\,m^2\,g^{-1}$ for PIM-1 as compared to $545\ m^2\ g^{-1}$ for the carbon.

Since the smallest pores are filled first, the very low pressure region of the adsorption isotherm contains information about micropores. A method for determining the distribution of micropore size was developed by Horvath and Kawazoe [17,20]. The method assumes that all pores of a certain size will be filled at a particular relative pressure. The original Horvath–Kawazoe equation assumed the pores to be slit-shaped, but models for alternative pore geometries, such as cylinders [21], have been developed. A pore size distribution for PIM-1, calculated by the Horvath–Kawazoe method, can be seen in Figure 2.3. When interpreting data such as these, one should be aware of anomalies that arise because one is operating near the limits of the technique. The micropore distribution extends down to below the size corresponding to the lowest pressure that can be achieved experimentally. The data in Figure 2.3 are for a sample that was thoroughly outgassed at elevated temperature prior to measurement at liquid nitrogen temperature. The first few data points are therefore inflated by a contribution from smaller pores, giving an artificially sharp peak in the early part of the distribution. Despite this distortion of the distribution, low temperature N_2 adsorption data support the concept that PIM-1 behaves like a molecular sieve, with pore sizes in the micropore (<2 nm) region.

The Horvath–Kawazoe method can also be applied to adsorption of carbon dioxide at 273 K. This gives information about smaller micropores than does nitrogen adsorption. A partial micropore distribution for PIM-1 from CO_2 adsorption is compared with the result from N_2 adsorption in Figure 2.3. Extension of the distribution to higher pore widths requires higher pressures than were possible on the instrument used to obtain these data. The apparent distribution from CO_2 adsorption may be affected by the presence of specific adsorption sites. Nevertheless, the data support the idea that the material is essentially microporous.

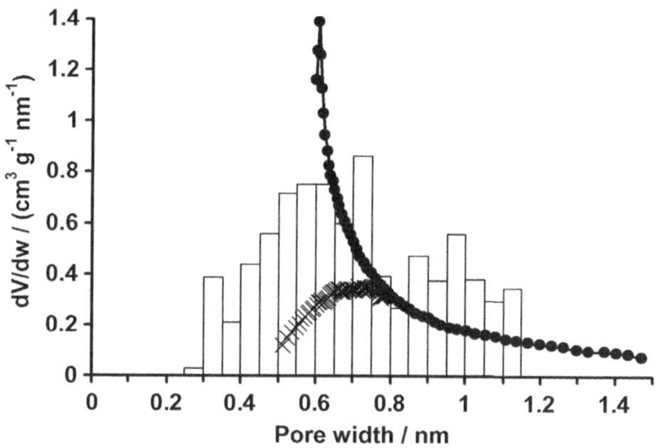

Figure 2.3 *Apparent distributions of pore size for PIM-1 derived from N_2 adsorption at 77 K (●), CO_2 adsorption at 273 K (×) and from atomistic computer simulation [22] (bars)*

Computer simulation may be used to visualize the microporosity of PIM-1. A pore size distribution derived by Heuchel *et al.* [22] from atomistic modelling of packed PIM-1 is included in Figure 2.3. This distribution was obtained using an approach that subdivides very elongated regions of free volume into smaller elements. The simulation, like the experimental gas adsorption results, indicates PIM-1 to contain pores or spaces with widths in the nanometre and sub-nanometre range, i.e. microporosity as defined by IUPAC. A simulated N_2 adsorption isotherm may be obtained from the computer model. Since the model is a purely static system, the predicted isotherm is what would be expected for a rigid microporous material, reaching a plateau at low relative pressure [22]. An apparent surface area of $435\,m^2\,g^{-1}$ may be calculated, lower than the experimental value of about $750\,m^2\,g^{-1}$. The additional adsorption and hysteresis observed experimentally may be attributed to the effects of polymer dynamics.

Further information about the sizes of free volume elements or micropores can be obtained from inverse gas chromatography and positron annihilation lifetime spectroscopy (PALS), as is discussed later.

2.4 Gas Permeation

It is well known that the gas permeation properties of glassy polymers tend to show 'trade-off' behaviour, with highly selective polymers exhibiting low permeability and vice versa. This is commonly represented in terms of a double logarithmic Robeson [23,24] plot of selectivity against permeability. Initial gas permeability measurements on PIM-1 membranes demonstrated that the polymer exhibits a high permeability (e.g. CO_2 permeability $P(CO_2) = 2300$ Barrer) [16], exceeded only by ultra-high free volume polymers such as poly(1-trimethylsilyl-1-propyne) (PTMSP) [25,26]. Of particular interest, however, is that this relatively high permeability is accompanied by a selectivity above Robeson's 1991 upper bound [23] for a number of important gas pairs, including O_2/N_2 and CO_2/CH_4. PIM-1 helps to define the recently revised upper bounds [24] for these gas pairs.

Table 2.1 *Permeability coefficients for poly(vinyltrimethylsilane) determined by the pressure increase method (GKSS, measurements made by Sergey Shishatskiy) and by the gas chromatographic method (TIPS)*

Gas	*P* / Barrer	
	GKSS	TIPS
H_2	215	220
He	178	160
O_2	45	44
N_2	11	11

Subsequent work has shown that even higher gas permeabilities can be achieved for PIM-1, depending on the history of the sample, with little loss of selectivity [5,27]. Comparative gas permeation experiments have been carried out using two different techniques in two different laboratories, using the same batch of PIM-1 [5]. In Germany, at the GKSS Research Centre Geesthacht GmbH (GKSS), a pressure increase time-lag method was used, operating with typical feed pressures of 200–300 mbar. This enabled calculation of diffusion coefficient, *D*, as well as permeability coefficient, *P*. In Moscow, at the A. V. Topchiev Institute of Petrochemical Synthesis of the Russian Academy of Sciences (TIPS), a gas chromatographic (GC) method was used. Pure penetrant gas at a pressure of 1 atm was passed through the upstream part of a permeation cell, while a carrier gas (He or Ar) was passed through the downstream part. GC was used to analyse the permeate and the flow of permeate was measured with a soap bubble flowmeter. In order to check that the two methods of measuring permeability gave comparable results, despite differences in the boundary conditions, a standard polymer, poly(vinyltrimethylsilane), was studied in both laboratories. Good agreement was obtained, as can be seen in Table 2.1.

A study was undertaken of the effects of membrane preparation protocol on the permeability of PIM-1. Membranes were cast from solutions of the polymer in tetrahydrofuran (THF) or chloroform onto a cellophane, glass or Teflon surface. Three 'states' of PIM-1 were identified.

1 'Water-treated' PIM-1, for which a flow of water was used to assist removal of the membrane from the surface onto which it was cast.
2 PIM-1 for which no water was used in membrane preparation, or else that was exhaustively dried at elevated temperature under vacuum.
3 'Methanol-treated' PIM-1, that was soaked in methanol for at least a day, to flush out any residual casting solvent and allow relaxation of chains in the swollen state, then dried under vacuum to constant weight.

Representative permeability data from both methods of measurement are given in Table 2.2 for PIM-1 in each of these three states. It can be seen that contact with water during membrane preparation can lead to a significant reduction in permeability. This is caused by trapped water inside the micropores and is hard to remove by vacuum and elevated temperature. Methanol treatment, however, substantially enhances the permeability. Indeed, methanol-treated PIM-1 is amongst the most permeable known polymers. Furthermore, the selectivities for some important gas pairs are significantly higher than

Table 2.2 Permeability coefficients at 30 °C for PIM-1 in various states (1 = water-treated; 2 = water-free; 3 = methanol-treated) determined by the pressure increase method (GKSS) and by the gas chromatographic method (TIPS)

State	$P(CO_2)$ / Barrer		$P(O_2)$ / Barrer		$P(N_2)$ / Barrer	
	GKSS	TIPS	GKSS	TIPS	GKSS	TIPS
1	950	1540	128	157	51	49
2	3700	4350	530	590	155	190
3	11 200	12 600*	1530	1610*	610	500*

*Temperature = 23 °C.

achieved for other polymers in the permeability range, as can be seen in the Robeson plots of Figure 2.4.

Many high permeability polymers exhibit a reduction in permeability over time (ageing) [28]. This effect was investigated for a methanol-treated (state 3) PIM-1 membrane. The sample was kept at ambient temperature in air and permeabilities measured at intervals by the gas chromatographic method. Results are shown in Figure 2.5. A reduction in permeability is seen, although changes are small after the first few days of measurement. Over 45 days of observation, $P(O_2)$ decreased by about 23%, with an accompanying increase in permselectivity $P(O_2)/P(N_2)$ from 3.3 to 4.2. The effects of ageing are less marked than have been observed for other high free volume polymers. For example, uncross-linked poly(4-methyl-2-pentyne) showed a decrease in N_2 permeability from about 1000 Barrer to 650 Barrer within 45 days [29].

The temperature dependence of the permeability of PIM-1 membranes was investigated in order to evaluate activation energies of permeation, E_p, for various gases. Values of E_p derived from pressure increase (barometric) and gas chromatographic methods for a water-free (state 2) PIM-1 membrane are given in Table 2.3. The activation energies are small, and for CO_2 are actually slightly negative.

The pressure increase time-lag method employed at GKSS enabled diffusion coefficient, D, as well as permeability coefficient, P, to be evaluated. Solubility coefficient, S, could then be calculated assuming a simple solution-diffusion model of permeability, $P = SD$. Values of D and S for PIM-1 membranes in various states are given in Table 2.4. The D values are smaller than obtained for other high permeability polymers, whereas S values are significantly higher (e.g. for PTMSP $D(CO_2)$ has been reported as 250×10^{-7} $cm^2 s^{-1}$ and $S(CO_2)$ as $76 \times 10^{-3} cm^3[STP] cm^{-3} cmHg^{-1}$) [30]. It is the high S values that give PIM-1 its remarkable combination of high permeability and good selectivity. The unusually high S values may be attributed to the microporous structure of PIM -1, coupled with the presence of polar nitrile (–C≡N) groups. To obtain further information about solubility coefficients and deeper insight of sorption thermodynamics in this polymer inverse gas chromatography was used.

2.5 Inverse Gas Chromatography

For inverse gas chromatography (IGC), the polymer of interest acts as the stationary phase and the retention time is measured for a range of solutes in a carrier gas. Solutes

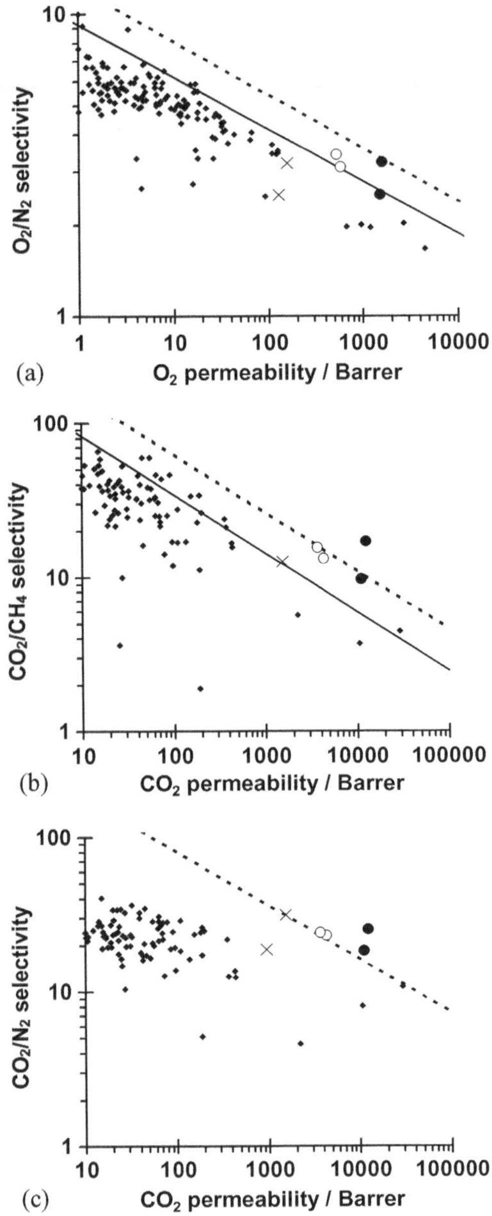

Figure 2.4 *Double logarithmic plots of selectivity against permeability for the gas pairs (a) O_2/N_2, (b) CO_2/CH_4 and (c) CO_2/N_2, showing the 1991 Robeson upper bound (solid line), [23] the 2008 Robeson upper bound (dashed line), [24]; literature data [16] for various polymers measured since 1991 (◆), and measurements for PIM-1 in state 1 (×), state 2 (○) and state 3 (●)*

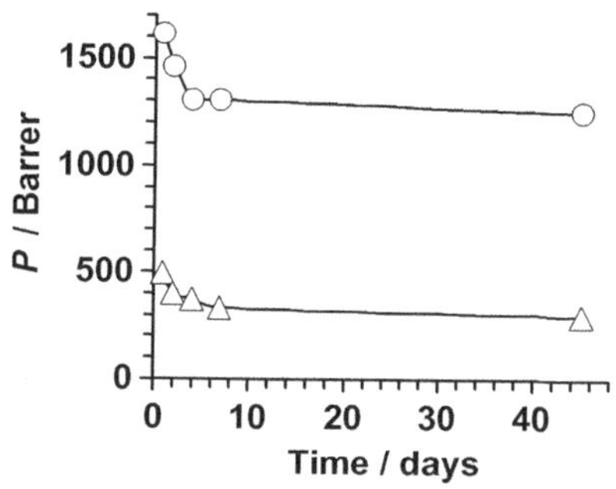

Figure 2.5 Dependence of oxygen permeability (○) and nitrogen permeability (△) on time for PIM-1 membrane (state 3, methanol-treated) kept at ambient temperature in air

Table 2.3 Activation energies of permeation, E_p, for water-free (state 2) PIM-1 membranes, determined by the pressure increase method (GKSS) and by the gas chromatographic method (TIPS)

Gas	$E_p/(kJ\ mol^{-1})$	
	GKSS	TIPS
CO_2	−4.5	−1.5
H_2	3.2	3.6
He	5.4	5.0
O_2	3.3	1.1
CH_4	10.9	10.9
N_2	10.5	7.5

Table 2.4 Values of diffusion coefficient, D, and solubility coefficient, S, at 30 °C for PIM-1 in various states (1 = water-treated; 2 = water-free; 3 = methanol-treated)

Gas	$D/(10^{-7}\ cm^2\ s^{-1})$			$S/(10^{-3}\ cm^3[STP]\ cm^{-3}\ cmHg^{-1})$		
	State 1	State 2	State 3	State 1	State 2	State 3
CO_2	3.5	4.5	16	270	770	700
H_2	100	290	500	2.7	6.9	6.6
He	190	360	680	0.82	2.6	1.9
O_2	9.3	15	39	14	40	39
CH_4	–	1.3	7.1	–	160	163
N_2	5.7	4.3	16	9.1	35	37

investigated in this work included *n*-alkanes (C_3–C_{10}), *n*-alcohols (C_1–C_4), acetone, chloroform, tetrachloromethane, tetrahydrofuran, cyclohexane, benzene, perfluorobenzene, toluene, perfluorotoluene, 4-fluorotoluene and 2,3,4,5,6-pentafluorotoluene. PIM-1 was coated from solution in THF onto a large pore, low surface area support (Inerton AW), which was packed into a stainless steel column (internal diameter = 3 mm; length = 1.5 m). The proportion of polymer in the dry packing material was determined by back-extracting

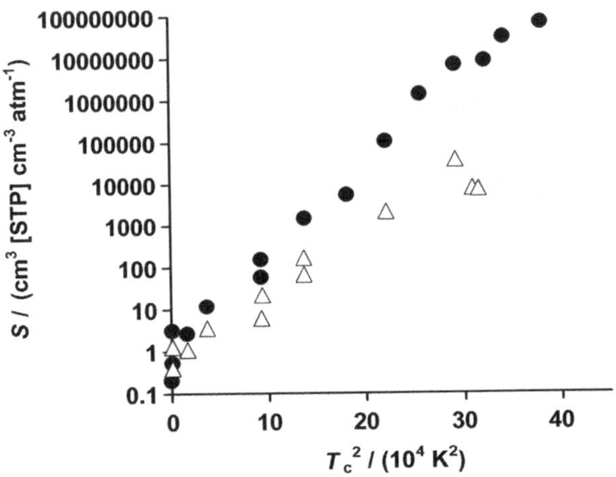

Figure 2.6 Dependence of solubility coefficient at 35 °C on T_c^2, where T_c is the critical temperature of the solute, for PIM-1 (●) and PTMSP [32] (△)

in a Soxhlet apparatus and found to be 8.3% by mass. Experiments were carried out using a LKhM-8MD chromatograph with thermal conductivity detector, over the temperature range 45–250 °C. Helium was used as the carrier gas. The retention time for an air peak was used to correct for the dead volume of the chromatograph. Knowing the flux of carrier gas, the pressure drop in the column and the mass of polymer in the column, a specific retention volume for each solute was calculated, from which the infinite dilution solubility coefficient, S, was determined. The temperature dependence of S allowed the enthalpy of sorption, ΔH_s to be determined, and hence the partial molar enthalpy of mixing, $\Delta H_m = \Delta H_s - \Delta H_c$, where ΔH_c is the enthalpy of condensation of the solute [5].

It has been found, for a wide range of solutes in glassy and rubbery polymers, that S varies with T_c^2, where T_c is the critical temperature of the solute [31]. This correlation was found to apply for PIM-1, as can be seen in Figure 2.6, which includes S values from IGC and S values derived from gas permeation experiments. Figure 2.6 also includes data for the previous 'champion' amongst membrane polymers, PTMSP [32], and the results show that PIM-1 exhibits the largest solubility coefficients of all polymers studied in this way. This confirms the exceptional affinity that PIM-1 has for small molecules.

Enthalpies of sorption and partial molar enthalpies and entropies of mixing for PIM-1 with a selection of solutes are listed in Table 2.5. Values of ΔH_s are more strongly negative than is generally observed for other glassy polymers, reflecting strongly negative ΔH_m values (exothermic mixing). Values of ΔS_m are also large and negative, showing that for the solute in the polymer there is considerable restriction in the degrees of freedom. The mixing process is thus entropically unfavourable, but driven by a large enthalpy term.

The size of free volume elements in a glassy polymer can be estimated from the dependence of ΔH_m on solute size [31]. A plot of ΔH_m against V_c, where V_c is the critical volume of the solute, generally passes through a minimum at a value of V_c that corresponds to a mean size of free volume elements. The data for n-alkanes in Table 2.5 show increasingly negative values of ΔH_m up to decane. Unfortunately, temperature limitations meant that

Table 2.5 Enthalpy of sorption, ΔH_s, partial molar enthalpy of mixing, ΔH_m, and partial molar entropy of mixing, ΔS_m, determined by IGC for a selection of solutes in PIM-1

Solute	ΔH_s/(kJ mol^{-1})	ΔH_m/(kJ mol^{-1})	ΔS_m/(J K^{-1} mol^{-1})
propane	−43	−24	−56
n-butane	−51	−21	−51
n-pentane	−61	−26	−58
n-hexane	−64	−28	−61
n-heptane	−69	−30	−62
n-octane	−80	−38	−77
n-nonane	−87	−40	−77
n-decane	−88	−40	−72
methanol	−43	−7	−32
ethanol	−51	−16	−44
n-propanol	−59	−21	−54
n-butanol	−64	−24	−56
acetone	−58	−27	−60
chloroform	−60	−25	−50
tetrachloromethane	−60	−26	−50
tetrahydrofuran	−60	−27	−56
benzene	−57	−25	−48
toluene	−71	−33	−66

n-alkanes larger than decane could not be studied, so a clear minimum is not observed for PIM-1. However, this suggests a minimum at a value of $V_c \geq 600\,cm^3\,mol^{-1}$, which translates into a sphere of radius ≥0.6 nm, or an equivalent pore width in the region of 1.2 nm. This supports the evidence from adsorption analysis and computer simulation, discussed earlier, that PIM-1 contains cavities of dimensions in the micropore region, as defined by IUPAC. Yet further evidence of this comes from PALS, as discussed below.

2.6 Positron Annihilation Lifetime Spectroscopy

Reaction of a positron with an electron gives a metastable positronium (Ps) particle, which may have antiparallel spins (*para*-positronium, *p*-Ps) or parallel spins (*ortho*-positronium, *o*-Ps). Within a polymer, the longer lifetimes of *o*-Ps may be related to the size, concentration and distribution of free volume elements. There have been a number of studies of PIM-1 by positron annihilation lifetime spectroscopy (PALS) [33–36].

This work investigated PIM-1 membranes in the three 'states' discussed above. Nickel-foil supported ^{22}NaCl was used as a positron source and stacks of film samples, each about 1 mm thick, were placed either side of the source. Annihilation lifetime decay curves were measured with an EG&G Ortec 'fast-fast' lifetime spectrometer. Measurements were made both in air and under an inert atmosphere (N_2). However, *o*-Ps lifetimes in air were reduced due to quenching by oxygen, so only results obtained under N_2 are discussed here. Results were analysed in terms of a four component lifetime distribution, which allowed obtaining better statistical fit. The two longest lifetimes, τ_3 and τ_4, for PIM-1 in

Table 2.6 *Long o-Ps lifetimes, τ_3 and τ_4, and corresponding radii of free volume elements, R_3 and R_4, for PIM-1 in various states (1 = water-treated; 2 = water-free; 3 = methanol-treated) under a nitrogen atmosphere*

State	τ_3/ns	τ_4/ns	R_3/nm	R_4/nm
1	1.4	5.6	0.22	0.51
2	2.1	5.8	0.30	0.52
3	1.8	7.1	0.27	0.58

each state are listed in Table 2.6, along with the corresponding radii of free volume elements, calculated according to the Tao-Eldrup model [37,38]. It can be seen that the longest lifetime suggests free volume elements with a radius of about 0.5 nm, or equivalent pore widths in the vicinity of 1 nm. This once again confirms the 'microporosity' of PIM-1. Furthermore, there is arguably a detectable increase in R_4 in going from the lower permeability 'water-treated' state 1 to the high permeability 'methanol-treated' state 3.

2.7 Conclusions

A variety of techniques, including gas adsorption, computer simulation, inverse gas chromatography and positron annihilation lifetime spectroscopy, all support the assertion that PIM-1 behaves like a molecular sieve, with effective pores of dimensions <2 nm (i.e. micropores as defined by IUPAC). This, combined with its chemical structure, which includes both aromatic rings and highly polar nitrile groups, gives rise to extraordinarily high solubility coefficients for a wide range of gases and vapours, which in turn results in membranes with an exceptional combination of permeability and permselectivity for a number of important gas separations. Strongly interacting solutes can profoundly affect transport properties, so that water-treated membranes have reduced permeability but a methanol treatment can substantially enhance permeability.

Acknowledgements

This collaboration was made possible by support from INTAS (Project 05-1000008-7862). The UK participants thank EPSRC for funding.

References

(1) P. M. Budd, N. B. McKeown, D. Fritsch, Free volume and intrinsic microporosity in polymers, *J. Mater. Chem.*, **15**, 1977–1986 (2005).
(2) D. H. Everett, Definitions, terminology and symbols in colloid and surface chemistry, Part I, *Pure Appl. Chem.*, **31**, 577–638 (1972).
(3) N. B. McKeown, P. M. Budd, K. J. Msayib, B. S. Ghanem, H. J. Kingston, C. E. Tattershall, S. Makhseed, K. J. Reynolds, D. Fritsch, Polymers of intrinsic microporosity (PIMs): Bridging the void between microporous and polymeric materials, *Chem. – Eur. J.*, **11**, 2610–2620 (2005).

(4) N. B. McKeown, P. M. Budd, Polymers of intrinsic microporosity (PIMs): organic materials for membrane separations, heterogeneous catalysis and hydrogen storage, *Chem. Soc. Rev.*, **35**, 675–683 (2006).
(5) P. M. Budd, N. B. McKeown, B. S. Ghanem, K. J. Msayib, D. Fritsch, L. Starannikova, N. Belov, O. Sanfirova, Y. Yampolskii, V. Shantarovich, Gas permeation parameters and other physicochemical properties of a polymer of intrinsic microporosity: Polybenzodioxane PIM-1, *J. Membr. Sci.*, **325**, 851–860 (2008).
(6) P. M. Budd, B. S. Ghanem, S. Makhseed, N. B. McKeown, K. J. Msayib, C. E. Tattershall, Polymers of intrinsic microporosity (PIMs): robust, solution-processable, organic nanoporous materials, *Chem. Commun.*, 230–231 (2004).
(7) P. M. Budd, E. S. Elabas, B. S. Ghanem, S. Makhseed, N. B. McKeown, K. J. Msayib, C. E. Tattershall, D. Wang, Solution-processed, organophilic membrane derived from a polymer of intrinsic microporosity, *Adv. Mater.*, **16**, 456–459 (2004).
(8) H. R. Kricheldorf, D. Fritsch, L. Vakhtangishvili, G. Schwarz, Cyclic ladder polymers by polycondensation of silylated tetrahydroxy-tetramethylspirobisindane with 1,4-dicyanotetrafluorobenzene, *Macromol. Chem. Phys.*, **206**, 2239–2247 (2005).
(9) H. R. Kricheldorf, N. Lomadze, D. Fritsch, G. Schwarz, Cyclic and telechelic ladder polymers derived from tetrahydroxytetramethylspirobisindane and 1,4-dicyanotetrafluorobenzene, *J. Polym. Sci., Part A: Polym. Chem.*, **44**, 5344–5352 (2006).
(10) J. Song, N. Du, Y. Dai, G. P. Robertson, M. D. Guiver, S. Thomas, I. Pinnau, Linear high molecular weight ladder polymers by optimized polycondensation of tetrahydroxytetramethylspirobisindane and 1,4-dicyanotetrafluorobenzene, *Macromolecules*, **41**, 7411–7417 (2008).
(11) N. B. McKeown, S. Makhseed, P. M. Budd, Phthalocyanine-based nanoporous network polymers, *Chem. Commun.*, 2780–2781 (2002).
(12) N. B. McKeown, S. Hanif, K. Msayib, C. E. Tattershall, P. M. Budd, Porphyrin-based nanoporous network polymers, *Chem. Commun.*, 2782–2783 (2002).
(13) P. M. Budd, B. Ghanem, K. Msayib, N. B. McKeown, C. Tattershall, A nanoporous network polymer derived from hexaazatrinaphthylene with potential as an adsorbent and catalyst support, *J. Mater. Chem.*, **13**, 2721–2726 (2003).
(14) H. J. Mackintosh, P. M. Budd, N. B. McKeown, Catalysis by microporous phthalocyanine and porphyrin network polymers, *J. Mater. Chem.*, **18**, 573–578 (2008).
(15) N. Du, J. Song, G. P. Robertson, I. Pinnau, M. D. Guiver, Linear high molecular weight ladder polymer via fast polycondensation of 5,5′,6,6′-tetrahydroxy-3,3,3′,3′-tetramethylspirobisindane with 1,4-dicyanotetrafluorobenzene, *Macromol. Rapid Commun.*, **29**, 783–788 (2008).
(16) P. M. Budd, K. J. Msayib, C. E. Tattershall, B. S. Ghanem, K. J. Reynolds, N. B. McKeown, D. Fritsch, Gas separation membranes from polymers of intrinsic microporosity, *J. Membr. Sci.*, **251**, 263–269 (2005).
(17) P. A. Webb, C. Orr, Analytical methods in fine particle technology, Micromeritics, Norcross, GA, 1997.
(18) A. V. Maffei, P. M. Budd, N. B. McKeown, Adsorption Studies of a Microporous Phthalocyanine Network Polymer, *Langmuir*, **22**, 4225–4229 (2006).
(19) S. Brunauer, P. H. Emmett, E. Teller, Adsorption of gases in multimolecular layers, *J. Am. Chem. Soc.*, **60**, 309–319 (1938).
(20) G. Horvath, K. Kawazoe, Method for the calculation of effective pore size distribution in molecular sieve carbon, *J. Chem. Eng. Japan*, **16**, 470–475 (1983).
(21) A. Saito, H. C. Foley, Curvature and parametric sensitivity in models for adsorption in micropores, *AIChE J.*, **37**, 429–436 (1991).
(22) M. Heuchel, D. Fritsch, P. M. Budd, N. B. McKeown, D. Hofmann, Atomistic packing model and free volume distribution of a polymer with intrinsic microporosity (PIM-1), *J. Membr. Sci.*, **318**, 84–99 (2008).
(23) L. M. Robeson, Correlation of separation factor versus permeability for polymeric membranes, *J. Membr. Sci.*, **62**, 165–185 (1991).
(24) L. M. Robeson, The upper bound revisited, *J. Membr. Sci.*, **320**, 390–400 (2008).
(25) T. Masuda, E. Isobe, T. Higashimura, K. Takada, Poly[1-(trimethylsilyl)-1-propyne]: a new high polymer synthesized with transition-metal catalysts and characterized by extremely high gas permeability, *J. Am. Chem. Soc.*, **105**, 7473–7474 (1983).

(26) K. Nagai, T. Masuda, T. Nakagawa, B. D. Freeman, I. Pinnau, Poly[1-(trimethylsilyl)-1-propyne] and related polymers: synthesis, properties and functions, *Progr. Polym. Sci.*, **26**, 721–798 (2001).
(27) S. Thomas, I. Pinnau, N. Du, M. D. Guiver, Pure- and mixed-gas permeation properties of a microporous spirobisindane-based ladder polymer (PIM-1), *J. Membr. Sci.*, **333**, 125–131 (2009).
(28) L. Starannikova, V. Khodzhaeva, Y. Yampolskii, Mechanism of aging of poly[1-(trimethylsilyl)-1-propyne] and its effect on gas permeability, *J. Membr. Sci.*, **244**, 183–191 (2004).
(29) L. Shao, J. Samseth, M.-B. Haegg, Crosslinking and stabilization of nanoparticle filled PMP nanocomposite membranes for gas separations, *J. Membr. Sci.*, **326**, 285–292 (2009).
(30) T. Nakagawa, S. Fujisaki, H. Nakano, A. Higuchi, Physical modification of poly [1-(trimethylsilyl)-1-propyne] membranes for gas separation, *J. Membr. Sci.*, **94**, 183–193 (1994).
(31) Y. P. Yampolskii, Methods for investigation of the free volume in polymers, *Russ. Chem. Rev.*, **76**, 59–78 (2007).
(32) T. C. Merkel, V. Bondar, K. Nagai, B. D. Freeman, Sorption and transport of hydrocarbon and perfluorocarbon gases in poly(1-trimethylsilyl-1-propyne), *J. Polym. Sci., Part B: Polym. Phys.*, **38**, 273–296 (2000).
(33) R. Lima de Miranda, J. Kruse, K. Raetzke, F. Faupel, D. Fritsch, V. Abetz, P. M. Budd, J. D. Selbie, N. B. McKeown, B. S. Ghanem, Unusual temperature dependence of the positron lifetime in a polymer of intrinsic microporosity, *Phys. Status Solidi RRL FIELD* (Full Journal Title: *Physica Status Solidi RRL: Rapid Research Letters*), **1**, 190–192 (2007).
(34) V. P. Shantarovich, T. Suzuki, Y. Ito, K. Kondo, R. S. Yu, P. M. Budd, Y. P. Yampolskii, S. S. Berdonosov, A. A. Eliseev, Structural heterogeneity in glassy polymeric materials revealed by positron annihilation and other supplementary techniques, *Phys. Status Solidi C*, **4**, 3776–3779 (2007).
(35) V. P. Shantarovich, T. Suzuki, Y. P. Yampol'skii, P. Budd, V. V. Gustov, I. B. Kevdina, A. V. Pastukhov, S. S. Berdonosov, V. E. Bozhevol'nov, Positronium annihilation data and actual free-volume distribution in polymers, *High Energy Chem.*, **41**, 370–380 (2007).
(36) C. L. Staiger, S. J. Pas, A. J. Hill, C. J. Cornelius, Gas separation, free volume distribution, and physical aging of a highly microporous spirobisindane polymer, *Chem. Mater.*, **20**, 2606–2608 (2008).
(37) S. J. Tao, Positronium annihilation in molecular substances, *J. Chem. Phys.*, **56**, 5499–5510 (1972).
(38) M. Eldrup, D. Lightbody, J. N. Sherwood, The temperature dependence of positron lifetimes in solid pivalic acid, *Chem. Phys.*, **63**, 51–58 (1981).

3

Addition-type Polynorbornene with $Si(CH_3)_3$ Side Groups: Detailed Study of Gas Permeation, Free Volume and Thermodynamic Properties

Yuri Yampolskii[a], *Ludmila Starannikova*[a], *Nikolai Belov*[a], *Maria Gringolts*[a], *Eugene Finkelshtein*[a] *and Victor Shantarovich*[b]

[a]*A. V. Topchiev Institute of Petrochemical Synthesis, Moscow, Russia*

[b]*N. N. Semenov Institute of Chemical Physics, Moscow, Russia*

3.1 Introduction

Norbornene and its numerous derivatives are an unusual case in polymer and membrane science: by selecting different catalysts for the polymerization process, materials with entirely different structures can be prepared. The norbornene bicyclic molecule is strongly strained, so ring opening metathesis polymerization (ROMP) is significantly favoured by thermodynamic factors. Because of this, numerous cyclolinear structures of ROMP polymers were prepared in the presence of the catalysts such as WCl_6, $RuCl_3$, Re_2O_7/Al_2O_3 or Grubbs Ru complexes and characterized during the last decades [1–6]. These polymers include as side groups the following substituents: $SiMe_3$, CN, Et, Bu, Cl, F, CF_3, OC_3F_7, etc., have modest permeability ($P(O_2) = 2$–100 Barrer) and behave like conventional glassy polymers.

Meanwhile, norbornene being a bicyclic olefin can be polymerized also via an addition scheme with opening of the double bonds. This reaction proceeds in the presence of Ni or Pd catalysts and results in polymers with entirely different structures and, hence, different properties (Scheme 3.1).

Membrane Gas Separation Edited by Yuri Yampolskii and Benny Freeman

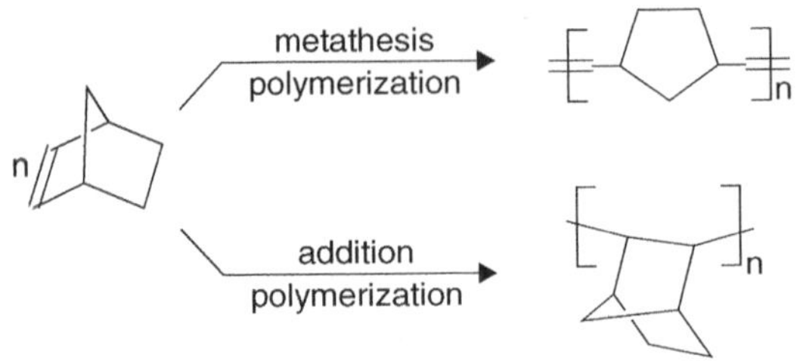

Scheme 3.1

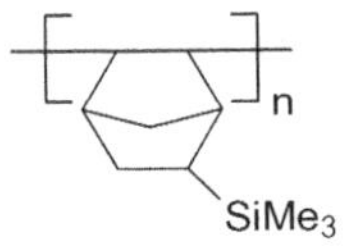

Structure 3.1

Table 3.1 *Effects of introduction of $Si(CH_3)_3$ groups in various main chains*

Polymer type[a]	$P(O_2)$, Barrer	Ref.
PS	2.9	[14]
$SiMe_3$-PS	56	[15]
PNB (ROMP)	2.6	[16]
$SiMe_3$-PNB (ROMP)	20.9	[17]
APN	6.9	[12]
$SiMe_3$-APN	840	[12]

[a] PS, polystyrene; $SiMe_3$-PS, poly(*p*-trimethylsilylstyrene); PNB, polynorbornene; $SiMe_3$-PNB: ROMP poly(trimethylsilylnorbornene); APN addition-type polynorbornene, $SiMe_3$-PNB addition-type poly (trimethylsilylnorbornene).

Several authors reported the transport properties of the addition-type polynorbornenes bearing such alkyl substituents as CH_3, n-C_4H_9, n-$C_{10}H_{21}$ [7–10]. These polymers are distinguished by very rigid structure (T_g at about 300 °C or higher) and permeability coefficients $P(O_2)$ in the range 20–90 Barrer, so not a big difference is observed in comparison with ROMP polynorbornenes studied. However, much higher permeability was reported for the product of addition-type polymerization of trimethylsilylnorbornene [11,12] (Structure 3.1).

This amorphous glassy polymer has the permeability coefficients $P(O_2)$ in the range 800–1000 Barrer (for different samples) [12,13]. Such behaviour is in line with well-documented effects of the introduction of bulky $Si(CH_3)_3$ group into various main chains; however, the effect here, bearing in mind very rigid structure of the backbone chain, is much larger (Table 3.1).

In this article we shall give a survey of the transport parameters, thermodynamic properties and free volume of addition-type poly(trimethylsilylnorbornene) (PTMSN).

3.2 Experimental

Synthesis of PTMSN is described in detail elsewhere [12]. The monomer was prepared from dicyclopentadiene and vinyltrimethylsilane by a Diels–Alder reaction. Polymerization was performed in toluene solution in the presence of nickel naphthenate [(Nph)$_2$Ni] as a catalyst and methylalumoxane (MAO) as co-catalyst. A vast surplus of MAO resulted in higher molecular mass of the polymer (up to 440 000 Da) which led to better film-forming properties and less brittleness of the films. The films of the polymers studied were cast from the 2–3 mass % solution in toluene (or cyclohexane in the case of non-substituted addition-type polynorbornene studied for comparison). Before testing, the films were kept in a vacuum until a constant weight is attained (for about 1 week). The thickness of the films was in the range 100–200 µm. Two methods were used in the determination of the transport parameters. Permeability (*P*) and diffusion (*D*) coefficients were determined using gas chromatographic and mass spectrometric methods. In the former technique the steady-state stream of penetrant gases under atmospheric pressure flew around the upstream part of the film, while the downstream part of it was moved around by the gas carrier – helium or nitrogen (the latter was used in the measurement of permeation rate of H_2 and He). The permeability coefficients were determined by measuring the penetrant concentration in the gas carrier and the total flow of this mixture. The determination of the diffusion coefficients *D* involved an adjustment of the experimental sigmoid-type curve of the attaining of the steady-state stream across the tested film and the theoretical one computed for the case of Fickian diffusion [18]. Measurements were performed in the range 20–60 °C. In the mass spectrometric set-up the permeability is measured by the slope of the increase in a certain ion current in the mass spectrum of the diffusing gas molecule after attaining steady-state conditions. The diffusion coefficient was determined using the Daynes–Barrer method. The measurements were carried out at 22 °C and pressure 0.1–1 atm over the film and about 10^{-4} mmHg below the film, i.e. in the receiving volume of the mass spectrometer. The details of the investigation of PTMSN using the positron annihilation lifetime spectroscopy (PALS) and inverse gas chromatography (IGC) can be found in Refs. 12 and 13.

3.3 Results and Discussion

3.3.1 Physical Properties

Some physical properties of PTMSN are presented in Table 3.2. This is an amorphous glassy polymer with very high T_g. Its glass transition is very close to the decomposition temperature and can be discerned only by accurate analysis of the TGA curves. Its mechanical properties are rather modest, though sufficient for preparation of stable films if the molecular mass of the sample is 400 000 Da or higher.

The density of PTMSN is relatively small, which results in a high value of the fractional free volume (FFV) according to Bondi, as will be discussed in more detail later.

3.3.2 Gas Permeability

A feature of PTMSN is a relatively high gas permeability coefficient *P*, higher than those of other addition-type polynorbornenes and comparable with permeability coefficients of

Table 3.2 Physical properties of PTMSN

Parameter	Method	Value
Crystallinity	WAXD (Rigaku)	None
Density ρ, g/cm^3	Geometric density	0.883 ± 0.001
FFV, %	Bondy method	27.5
T_g, °C	TMA-7 (Perkin-Elmer)	293–303
T_{dec}, °C (5%)	TGA-7 (Perkin-Elmer)	
In air		330
In Ar		420
Strain strength, MPa	INSTRON-2111	16.1
Elongation at break, %	INSTRON-2111	4.1

Table 3.3 Permeability coefficients (P, Barrer) of PTMSN and other addition-type polynorbornenes (Ni catalysts were used)

Side group R	He	H_2	O_2	N_2	CO_2	CH_4	Ref.
$Si(CH_3)_3$	840	1800	980	390	4480	990	This work
H	29.4	42	6.9	1.5	34	2.6	This work
CH_3	88.8	–	–	4.3	81.1	5.6	[9]
CH_3*	174	–	–	12.6	202.1	16.9	[9]

*Pd catalyst was used.

Table 3.4 Separation factors $\alpha(M_1/M_2) = P(M_1)/P(M_2)$ for some gas pairs

M_1/M_2	PTMSN	PTMSP [20]	AF2400 [21]
O_2/N_2	2.6	1.7	2.0
H_2/N_2	5.7	2.9	4.3
CO_2/N_2	14.6	10.5	4.7
CO_2/CH_4	5.5	4.4	6.0

PTMP, poly(trimethylsilylpropyne); AF2400, random copolymer of 87 mol% 2.2bis(trifluoromethyl)-4,5-difluoro-1,3-dioxole and 13 mol% tetrafluoroethylene.

polyacetylenes and amorphous Teflons AF [19], that is, it is a new member of the family of high permeability, high free volume glassy polymers. By the P values it is similar to the 'polymer of intrinsic microporosity', PIM-1 (see Chapter 2 by P. M. Budd *et al.* in this book). Table 3.3 presents a survey of permeability coefficients of PTMSN, where they are compared with those of other addition-type norbornene polymers.

The high permeability of PTMSN is due to the two design peculiarities: the bulkiness of the $Si(CH_3)_3$ group and the rigid character of the main chain (see Tables 3.1 and 3.3). It can also be concluded that PTMSN is much more permeable that its isomer, ROMP polymer containing $Si(CH_3)_3$ group [17]. It can be noted that these differences are much less marked for non-substituted polynorbornenes of addition and ROMP types.

Permselectivity or separation factors $\alpha(M_1/M_2) = P(M_1)/P(M_2)$ are compared in Table 3.4 for PTMSN and some other high permeability polymers. It is seen that this polymer is more permselective than PTMSP and AF2400 in the most cases. However, the observed

Table 3.5 *Permeability coefficients* P *(Barrer) of high free volume polymers with respect to hydrocarbons (at 1 atm)*

Polymer	CH_4	C_2H_6	C_3H_8	C_4H_{10}	Ref.
PTMSN	790	1430	1740	17500	This work
PTMSP	15000	31000	38000	–	[22]
PMP	2900	3700	7300	26000	[23]
AF2400	435	252	97	–	[20]

PMP, poly(methylpentyne).

separation factors of PTMSN are not high, hence common 'trade-off' behaviour takes place in this case.

Possibly, one of the most interesting peculiarities of PTMSN is the variation of the P values along the n-alkane series in this polymer. It is well known that conventional glassy polymers reveal size sieving (or mobility) selectivity as a function of molecular mass of penetrant: permeability coefficients correlate with the diffusion coefficients in such polymers. Opposite behaviour is typical for rubbery polymers, where the P values correlate with the solubility coefficients (S): solubility controlled permeation. However, some high free volume glassy materials, first and foremost PTMSP, show solubility controlled selectivity [19], so the separation factors reveal higher permeation rates of heavier components, e.g. $R(C_4H_{10}/CH_4) < 1$.

In this regard, PTMSN behaves like PTMSP and some other highly permeable polyacetylenes, as Table 3.5 indicates.

It is seen that the trend observed for PTMSN is similar to that typical for polyacetylenes, while perfluorinated material AF2400 having more or less the same permeability for light gases as PTMSN shows diffusivity controlled permeation; a decrease in P values with the size of the penetrant. However, bearing in mind the relatively high experimental pressure of 1 atm, which in the case of n-butane permeation corresponds to activity $p/p_s = 0.4$ (p_s is the saturated vapour pressure of n-butane at 22 °C), the observed high value $P(C_4H_{10})$ can be partly affected by the concentration dependence of P. Solubility controlled permeation here, as in other cases, can be ascribed to an open porosity kind of free volume typical for these polymers. For PTMSP, it was confirmed by a computer simulation study using molecular dynamics [24]. So it can be assumed that the nanostructure of PTMSN also includes large and partly opened pores. This subject will be considered in more detail later in this chapter. From the practical viewpoint, this result means that membrane based on PTMSN can be used to remove higher hydrocarbons from natural and associated petroleum gas.

Continuous decrease in permeability in time (so called ageing) is a well-known peculiarity of PTMSP and some other polyacetylenes [25]. Such ageing can be considered as a shortcoming of potential membrane materials, so it is quite desirable to test the time stability of the permeation parameters for any high free volume, high permeability polymers. Bearing this in mind, we undertook a study of the effects of ageing on permeability coefficients of PTMSN. The results are given in Figure 3.1. It can be concluded that some ageing takes place also in the case of PTMSN, though the rate of ageing is much slower that that of PTMSP.

A feature of highly permeable polyacetylenes is a low (sometimes negative) value of activation energy of permeation. Since

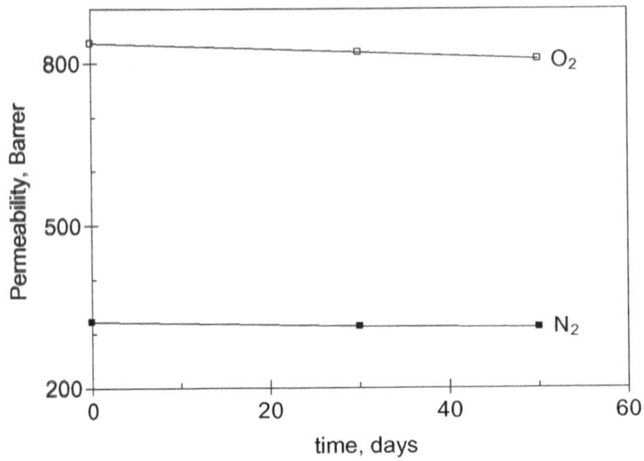

Figure 3.1 *Time dependence of the permeability coefficients of PTMSN (storing at room temperature)*

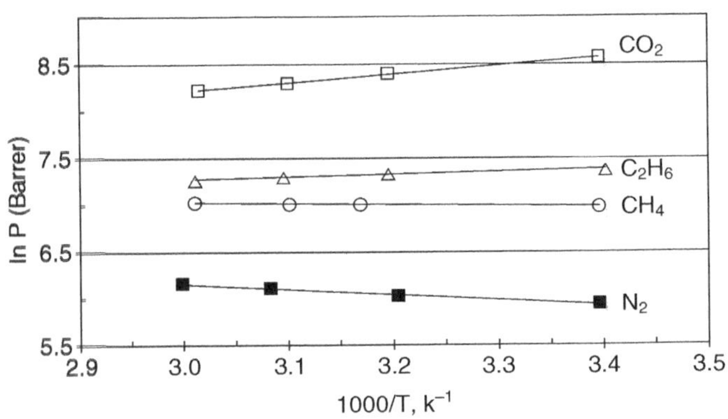

Figure 3.2 *Arrhenius dependence of permeability coefficients in PTMSN*

$$P = DS \tag{3.1}$$

$$P = P_0 \exp(-E_P/RT) = D_0 \exp(-E_D/RT) + S_0 \exp(-\Delta H_s/RT) \tag{3.2}$$

the activation energy of permeation can be presented as the sum:

$$E_P = E_D + \Delta H_s. \tag{3.3}$$

where E_D is the activation energy of diffusion and ΔH_s is enthalpy of sorption.

Since in such polymers activation energy barriers for diffusion are low, the prevailing effects are exerted by negative ΔH values. The temperature dependence of the permeability coefficients in PTMSN shows that such behaviour is typical also for this polymer (Figure 3.2).

Already a qualitative analysis of Figure 3.2 indicates that the activation energies of carbon dioxide and ethane are negative, while those of other gases have relatively small values (see also Table 3.6). This is one more feature that explains solubility controlled permeation in PTMSN and its large gas permeability.

Pressure also exerts effects on the permeability coefficients of gases in PTMSN. The strongest effects were observed for permeability of carbon dioxide, as Figure 3.3 indicates.

Such shape of the $P(p)$ curves is typical for glassy polymers. The left, low pressure branch of the curve is explained by semi-empirical dual-mode sorption and mobility model. The right branch that shows increases in permeability with increasing pressure is usually explained by the plasticization effects (see Ref. 19, p. 24). There is, however, one interesting peculiarity of PTMSN: in conventional glassy polymers such as polyimides, polycarbonates, etc. the pressure p_{min} where the curve passes through a minimum amount

Table 3.6 *Parameters of the Arrhenius dependence of permeability in PTMSN*

Gas	P_0 (Barrer)	E_P (kJ/mol)
H_2	6700	3.1
He	5100	4.2
O_2	1700	1.3
N_2	2500	4.6
CO_2	250	−7.4
CH_4	1390	0.63
C_2H_6	620	−2.3

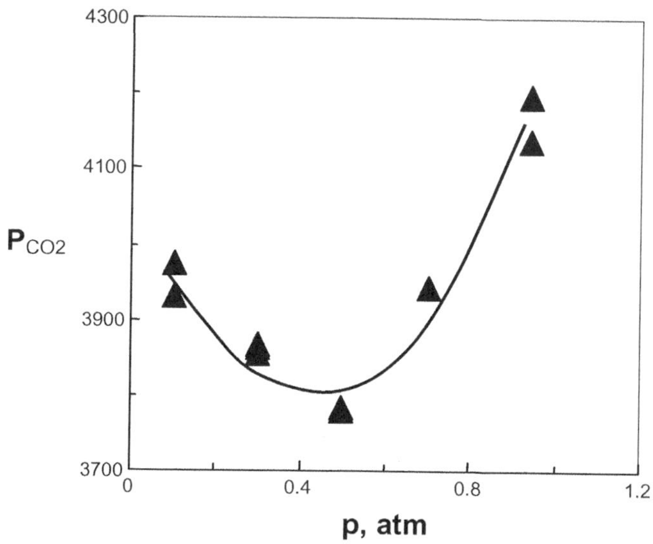

Figure 3.3 *Peremeability coefficients P(CO_2) (in Barrer) as a function of upstream pressure in the cell of mass spectrometric unit (downstream pressure is zero, T = 22°C)*

10–20 atm [26]. In the case of PTMSN, as is seen from Figure 3.3, p_{min} is approximately 0.5 atm. The reason for a strong tendency to plasticization of PTMSN by carbon dioxide is not completely clear. For its explanation, the whole shape of the sorption isotherm in a wide range of pressure should be compared, which is not yet available for PTMSN. The differences of solubility coefficients at infinite dilution (initial slopes of the sorption isotherms) are sufficiently large to explain such strong plasticization effects.

3.3.3 Sorption and Diffusion

Sorption thermodynamics were studied using the IGC method for a number of solutes (n-alkanes C_3–C_{16}, cyclohexane, methylcyclohexane) in the range 50–250 °C. For all the solutes, the linear dependencies of log S versus $1/T$ enabled an estimation of the enthalpies of sorption ΔH_s. It was shown that when the size of the solutes (e.g. their critical volume V_c) increases the negative enthalpies of sorption also increase. This dependence can be presented by the equation:

$$\Delta H_s = aV_c + b \tag{3.4}$$

where $a = -0.071 \pm 0.002$ and $b = -16.9 \pm 0.9$, correlation factor $R^2 = 0.993$. Using this correlation, enthalpies of sorption of light gases and, via Equation (3.3), the activation energies of diffusion of light gases were also estimated. The found E_D (kJ/mol) values are as follows: $O_2 = 23.5$; $N_2 = 27.9$; $CO_2 = 16.2$; $CH_4 = 24.6$. These values are not as small as those observed in PTMSP [20], however, they are relatively low as compared to the diffusion activation energies in conventional glassy polymers. So judging by this parameter too, PTMSN can be considered as intermediate between ultra-permeable polymers like PTMSP and conventional glassy polymers.

The IGC method also resulted in the determination of the solubility coefficients S in PTMSN. The S values were also determined as the ratio P/D for light gases. The combined results are presented in Figure 3.4 in the form of correlation of S versus squared critical temperature of solutes. The data for PTMSN obtained by the two methods are compared with the results of the investigation of PTMSP, the polymer having extremely high solubility coefficients.

It was shown that PTMSN is characterized by unusually great solubility coefficients. It is seen from Figure 3.4 that PTMSN has only marginally smaller S values as compared with PTMSP. A similar conclusion in comparison of these two polymers can be made based on comparison the S values obtained as the ratio $S = P/D$ (see Table 3.7). It can be assumed that rigid chains of PTMSN, as evidenced by its very high glass transition temperature and the presence of bulky $Si(CH_3)_3$ groups attached directly to the main chains, are the reasons for great sorption parameters. Table 3.7 also indicates that higher permeability of PTMSP is caused obviously by significantly larger values of its diffusion coefficients. The values of D and E_D in PTMSN imply that the structure of free volume in this polymers is not as opened as it is in PTMSP.

3.3.4 Free Volume

A description of high permeability polymers is not complete if the free volume of the material is not considered. In this work, several methods for evaluation of free volume

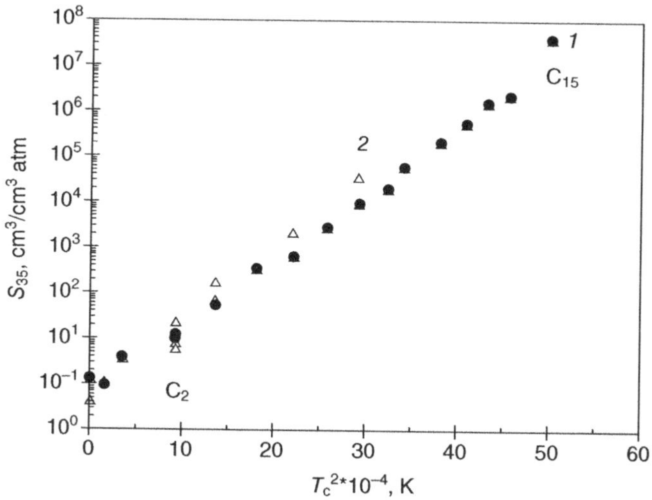

Figure 3.4 *Correlation of S (35 °C) versus T_c^2, where T_c is the critical temperature of solutes: (1) PTMSN; (2) PTMSP [22]*

Table 3.7 *Permeability, diffusion and solubility coefficients of gases in PTMSN and PTMSP[27][a]*

Gas	PTMSN[b]			PTMSP[c]		
	P	*D* /10^6	*S*	*P*	*D* /10^6	*S*
O_2	838	5.0	1.3	7730	35.8	1.6
N_2	321	2.7	0.91	4970	25.5	1.5
CO_2	4180	3.2	9.8	28000	21.7	9.8
CH_4	849	1.8	3.6	13000	22.7	4.4

[a] *P* in Barrer, *D* in cm^2 s^{-1}, *S* in cm^3(STP)/cm^3 atm.
[b] At 22 °C.
[c] At 30 °C.

are used. All of them show that the appearance of bulky $Si(CH_3)_3$ side groups results in a significant increase in free volume. The simplest and, maybe, the most traditional approach for estimation of free volume (FFV) is based on the Bondi method [28]:

$$\text{FFV} = V_f/V_{sp} = 1 - 1.3V_w/V_{sp} \tag{3.5}$$

where V_f (cm^3/g) is the free volume, V_{sp} (cm^3/g) is the specific volume of the polymer defined as $1/\rho$ where ρ (g/cm^3) is the density, and V_w is the van der Waals volume of the repeat unit. This is a very approximate approach, however, as it gives systematic scales of glassy polymers with various density of chain packing, and the FFV values usually correlate with the permeability and diffusion coefficients. It is seen from Table 3.8 that PTMSN has higher values of FFV than other addition-type polynorbornenes. The found

Table 3.8 Fractional free volume (FFV) and X-ray scattering data of addition-type polynorbornenes with different side groups R

R	FFV	WAXD				Ref.
		$(2\theta)_1$, degree	$(2\theta)_2$, degree	d_{B1}, Å	d_{B2}, Å	
$Si(CH_3)_3$	0.275	6.5	15.5	13.6	5.7	This work
CH_3	0.18	9.3	18.2	9.3	4.9	[9]
H	0.15	10	18.5	8.8	4.7	This work, [8]

Table 3.9 PALS data for PTMSN and other addition-type polynorbornenes

R	τ_3, ns	I_3, %	τ_4, ns	I_4, %	R_3/R_4, Å	Ref.
$Si(CH_3)_3$	3	10	7	30	3.7/5.7	This work
CH_3	3.15	33.3	–	–	3.34 (R_3)	[9]
H	2	25	3.6	8.9	2.9/4.1	This work

absolute FFV value is comparable to those earlier observed for other highly permeable polymers like e.g. perfluorinated copolymers AF [21].

X-ray wide angle scattering in glassy polymers usually shows one or two broad amorphous halos. Most high free volume polymers are characterized by two bands, while e.g. polyimides show a single halo. The coordinates of the maximum of 2θ allow an estimation of periodicity d_B (so-called d spacing) using Bragg's formula:

$$d_B = \lambda/(2\sin\theta) \tag{3.6}$$

The first value of d_B in PTMSN can be attributed to intersegmental spacing and is indicative of looser chain packing as compared with other presented norbornene polymers.

It is worthwhile to note that the conventional glassy polymers like polycarbonates and polysulfones are characterized by much smaller values of d spacing (4–6 Å); that is, they have more densely packed chains.

The generally recognized and the most reliable method for investigation of free volume in polymers is positron annihilation lifetime spectroscopy (PALS). It was applied for investigation of PTMSN and related polymers. This method is based on the measurement of lifetime spectra of positrons in polymers – lifetimes τ_i (ns) and corresponding intensities I_i (%). Longer lifetimes τ_i (or τ_3 and τ_4) (so-called o-orthopositronium lifetimes) can be related to the mean size of free volume R.

$$\tau_i = 1/2[1-(R_i/R_0)+(1/2)\pi\sin(2\pi R_i/R_0)]^{-1} \tag{3.7}$$

where $\tau_i = \tau_3$ or τ_4 are o-Ps lifetimes and $R_i = R_3$ or R_4 are the radii of free volume elements expressed in nanoseconds and angstroms, respectively; $R_0 = R_i + \Delta R$, where $\Delta R = 1.66$ Å is the fitted empirical parameter. Earlier studies [8,9] with other polynorbornenes with H and CH_3 substituents revealed relatively short lifetimes τ_3 and smaller sizes of free volume elements in these polymers. Table 3.9 presents the PALS data obtained for the PTMSN investigated. First, the treatment of the primary experimental results shows that much better statistical fit can be obtained for bimodal lifetime and size distribution of free volume for PTMSN. According to Equation (3.7), very large free volume elements are present in this polymer. The values $2R_4$ are consistent with the results of WAXD study (Table 3.8). Second, it should be reminded that bimodal lifetime distribution observed for PTMSN is a common feature of all polymers with great gas permeability [19].

An analysis of kinetics of o-positronium formation and decay in the polymer allows an estimation of the concentrations of free volume elements of the two kinds in PTMSN: $N_3 = 5.6 \times 10^{19}\,cm^{-3}$ and $N_4 = 9.8 \times 10^{19}\,cm^{-3}$. For comparison in PTMSP the corresponding values are somewhat higher: $N_3 = 5.6 \times 10^{19}\,cm^{-3}$, $N_4 = 26 \times 10^{19}\,cm^{-3}$. A big difference in the N_4 values can explain more opened porosity and other properties of PTMSP.

Temperature dependence of the PALS parameters in PTMSN is shown in Figure 3.5 for a wide range of temperature. Strong independence of the observed PALS parameters on temperature is rather unusual for common polymers, however, it has been observed for high free volume materials like PTMSP [29]. Usually temperature dependence of lifetimes and radii of free volume can be described by two linear dependencies with different slopes and a break in vicinity of the glass transition temperature [30]. In conventional glassy polymers, increases in the lifetimes or hole size with temperature are caused mainly by small scale movements of groups that form 'walls' of free volume elements in polymers. In extra high permeability polymers like PTMSN and PTMSP, the sizes of holes (and corresponding lifetimes) are determined mainly by loose packing of rigid chains, so this small scale mobility is masked by structure porosity frozen in polymer.

Inverse gas chromatography provides an additional possibility to determine the size of free volume elements in glassy polymers. For several highly permeable glassy polymers, it was shown that the dependencies of the partial molar enthalpies of mixing ΔH_m are strong functions of molecular sizes of solutes and pass through minima that correlate with the gas permeability [19]. So it is worthwhile to use the size of solutes, e.g. the critical volume V_c, as a scaling parameter and compare the coordinates of these minima for different glassy polymers. It can be assumed that the molecular size of the solute, which

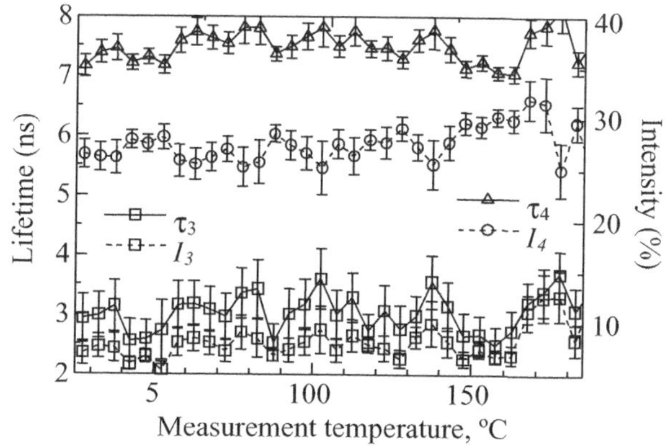

Figure 3.5 *Temperature dependence of the PALS parameters in PTMSN*

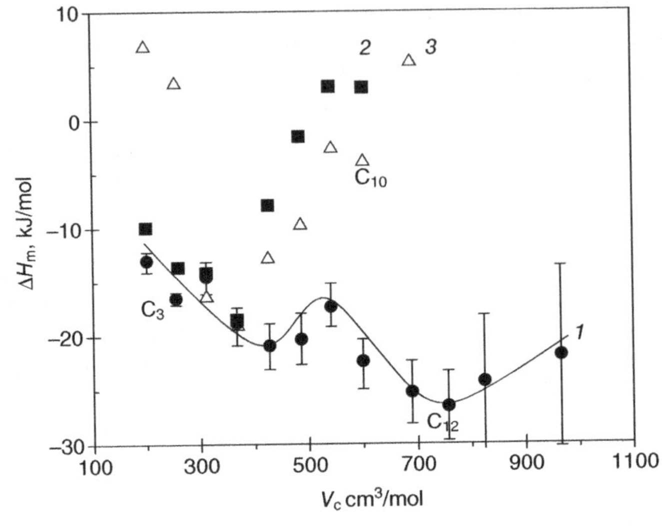

Figure 3.6 *Correlation of* ΔH_m *versus* V_c*: (1) PTMSN; (2) PVTMS [31]; (3) AF1600 [32]*

corresponds to the minimum at this dependence, is close to the mean size of free volume elements in the polymer. So it was interesting to check this for PTMSN and make some comparisons. Figure 3.6 shows the correlation of the ΔH_m values of PTMSN with the size of the solutes (*n*-alkanes) whose size is characterized by V_c.

One can discern for PTMSN the two broad minima in the curve $\Delta H_m(V_c)$ that correspond to the values of V_c of 426 cm^3/mol (n-C_7) and 754 cm^3/mol (n-C_{12}). The presence of the two minima is rather unexpected because in most cases only one minimum was observed in the curves $\Delta H_m(V_c)$. The only exception so far was amorphous Teflon AF2400, where also two minima, though flat and broad, could be noted [19]. Here we have apparently another manifestation of bimodal size distribution of free volume in glassy polymers that attracts now a keen interest [33]. The error bars shown in Figure 3.6 seem to support the assumption of bimodal size distribution in this polymer.

The task of the quantitative determination of the free volume size using the IGC method must depend on a choice of molecular volume of n-alkanes, the solutes used by us as the probes in free volume determination. Different scales can be employed as measures of molecular volumes: van der Waals volume V_w, molecular volume in liquid phase V_b (at corresponding boiling point T_b) and V_c. These quantities vary significantly and differ from each other; however, the general trends are the same for series of the solutes: for example for n-hexane the following values are reported(in Å^3/molecule): $V_w = 113$, $V_b = 235$ and $V_c = 611$ (the sources of these values is given in Ref. [34]). Since the density of the sorbed phase is unknown, the only criterion for selection the of best scaling parameter should be a comparison with the results of other probe methods for the investigation of free volume in polymers.

In Table 3.10, the sizes of free volume elements (V_f, Å^3) in PTMSN and other polymers are compared. They were found via PALS and IGC. The pairs of V_f values for PTMSN based on the IGC method correspond to the two minima shown in Figure 3.6.

Table 3.10 *Sizes of free volume elements in polymers as determined by the PALS and IGC methods [19,35]*[a]

Polymer	$P(O_2)$, Barrer	V_f (Å^3)		
		PALS	IGC	
			V_c	V_b[b]
PVTMS	44	345 ± 10	610	235
AF1600	170	490 ± 10	710	270
AF2400	1140	880 ± 20	710/1100	2700/455
PTMSN	800–1000	775	710/1260	270/500
PTMSP	7000–12 000	1320 ± 35	–	–

[a] AF1600 and AF2400 are random copolymers of 2,2-bis(trifluoromethyl)-4,5-difluoro-1,3-dioxole and tetrafluoroethylene with content of the latter of 35 and 13 mol%, respectively.
[b] V_b were calculated via Benson formula $V_b = V_c/(0.422 \log p_c + 1.981)$, where p_c is the critical pressure.

The PALS values correspond to the larger size of microcavities in the case of the bimodal size distribution. In addition, the permeability of the polymers in respect of oxygen is also shown. It is seen that the polymers with greater gas permeability are characterized by larger free volume elements independently of the method of the determination. However, the values obtained by the IGC method differ systematically from those found using the maybe more refined PALS method. In particular, V_f(PALS) are smaller than $V_f(V_c)$ but larger than $V_f(V_b)$ with approximately equal deviation of V_f(PALS) from the both quantities. There are several different reasons for these discrepancies. In the PALS method, the radii of free volume elements are estimated via *o*-positronium lifetimes. The size of free volume elements V_f is calculated then in assumption of its spherical symmetry, which is the subject of some doubts. In the case of IGC, the sizes of the free volume element are averaged over the temperature range, where ΔH_m values were determined, while the PALS method gives the information at specific temperatures. In addition, as has been mentioned, the main uncertainty in estimation of V_f in the IGC method is related to the density of sorbed phase and, consequently, the best choice for the scaling parameter – V_b or V_c. It can be speculated that the critical density, which defines V_c, is too small for the condensed phase. On the other hand, the conformations of the sorbed probe molecules (n-alkanes) can poorly correspond to irregular structure of free volume in polymers, so their accommodation can be rather imperfect. Since the deviations of $V_f(V_c)$ and $V_f(V_b)$ from V_f(PALS) are more or less the same for all the polymers considered, one can assume that the density of the sorbed phase is larger than the critical density, which defines the V_c values, by a factor about 2. This conclusion can be taken into account in further determination of the sizes of free volume elements using the IGC method.

3.4 Conclusions

PTMSN, described in this work, is another member of a family of high permeability, large free volume membrane materials. It is characterized by relatively high gas permeability caused mainly by great solubility coefficients, small or negative activation energies of permeation, solubility controlled permeation that is revealed by increase in permeability coefficients when the size of the penetrants increases, and great free volume as estimated by several independent methods. On the other hand, very slow ageing of this polymer was observed.

All these properties can be explained by the unusual design of the repeat unit of PTMSN – bulky $Si(CH_3)_3$ groups attached directly to the very stiff main chain.

PTMSN can be considered as a candidate material for membrane separation of hydrocarbons, and this research is now in progress.

References

(1) Y. Kawakami, H. Toda, M. Higashino, Y. Yamashita, *Polymer J.*, **20**, 285 (1988).
(2) E. Finkelshtein, N. Bespalova, E. Portnykh, K. Makovetskii, I. Ostrovskaya, S. Shishatskii, Yu. Yampolskii, N. Plate, N. Kaliuzhnyi, *Polymer Science*, A, **35**, 589 (1993).
(3) E. Finkelshtein, M. Gringolts, N. Ushakov, V. Lakhtin, S. Soloviev, Yu. Yampolskii, *Polymer*, **44**, 2843 (2003).
(4) E. Finkelshtein, E. Portnykh, N. Ushakov, M. Gringolts, Yu. Yampolskii, *Macromol. Chem. Phys.*, **198**, 1085 (1997).
(5) Yu. Yampolskii, N. Bespalova, E. Finkelshtein, V. Bondar, A. Popov, *Macromolecules*, **27**, 2872 (1994).
(6) M. Tlenkopachev, J. Vargas, M. A. Almareaz-Giron, M. Lopez-Gonzales, E. Riande, *Macromolecules*, **38**, 2696 (2005).
(7) K. D. Dorkenoo, P. H. Pfromm, M. E. Rezac, *J. Polym. Sci., Part B: Polym. Phys.*, **36**, 797 (1998).
(8) C. Zhao, M. do Rosario Ribeira, M. N. dePinho, V. S. Subrahamnyam, C. L. Gil, A. P. de Lima, *Polymer*, **42**, 2455 (2001).
(9) B. R. Wilks, W. J. Chung, P. J. Ludovice, M. E. Rezac, P. Meakin, A. J. Hill, *J. Polym. Sci., Part B: Polym. Phys.*, **41**, 2185 (2003).
(10) B. R. Wilks, W. J. Chung, P. J. Ludovice, M. E. Rezac, P. Meakin, A. J. Hill, *J. Polym. Sci., Part B: Polym. Phys.*, **44**, 215 (2006).
(11) E. Finkelshtein, K. Makovetskii, M. Gringolts, Yu. Rogan, T. Golenko, Yu. Yampolskii, L. Starannikova, N. Plate, *Patent RF*, 2296773 (2007).
(12) E. Finkelshtein, K. Makovetskii, M. Gringolts, Yu. Rogan, T. Golenko, L. Starannikova, Yu. Yampolskii, V. Shantarovich, T. Suzuki, *Macromolecules*, **39**, 7022 (2006).
(13) L. Starannikova, M. Pilipenko, N. Belov, Yu. Yampolskii, M. Gringolts, E. Finkelshtein, *J. Membr. Sci.*, **323**, 134 (2008).
(14) A. C. Puleo, N. Miruganandam, D. R. Paul, *J. Polym. Sci., Part B:Polym. Phys.*, **27**, 2385 (1989).
(15) V. S. Khotimskii, V. G. Filippova, I. S. Bryantseva, V. I. Bondar, V. P. Shantarovich, Yu. P. Yampolskii, J. *Appl. Polym. Sci.*, **78**, 1612 (2000).
(16) Yu. Yampolskii, E. Finkelshtein, K. Makovetskii, I. Ostrovskaya, E. Portnykh, M. Gringolts, Yu. Ishunina, I. Kevdina, V. Shantarovich, *Vysokomol. Soed. A*, **38**, 1480 (1996).
(17) Yu. Yampolskii, E. Finkelshtein, K. Makovetskii, V. Bondar, V. Shantarovich, *J. Appl. Polym. Sci.*, **62**, 349 (1996).
(18) A. Shvyryaev, I. Beckman, *Vestnik, Moscow State Univ., Ser. 2 (Chem.)*, **22**, 517 (1981).
(19) Yu. Yampolskii, I. Pinnau, B. D. Freeman, Materials Science of Membranes for Gas and Vapor Separation, John Wiley & Sons, Ltd., Chichester, 2006, pp. 231, 251.
(20) T. Masuda, Y. Iguchi, B.-Z. Tang, T. Higashimura, *Polymer*, **29**, 2041 (1988).
(21) A. Alentiev, Yu. Yampolskii, V. Shantarovich, S. Nemser, N. Plate, *J. Membr. Sci.*, **126**, 123 (1997).
(22) T. C. Merkel, V. Bondar, K. Nagai, B. D. Freeman, *J. Polym. Sci.: Part B: Polym. Phys.*, **38**, 273 (2000).

(23) A. Morisato, I. Pinnau, *J. Membr. Sci.*, **121**, 243 (1996).
(24) D. Hofmann, M. Entrialgo-Castano, A. Lerbret, M. Heuchel, Yu. Yampolskii, *Macromolecules*, **36**, 8528 (2003).
(25) K. Nagai, T. Masuda, T. Nakagawa, B. D. Freeman, I. Pinnau, *Progr. Polym. Sci.*, **26**, 721 (2001).
(26) A. Bos, High pressure CO_2/CH_4 separation with glassy polymer membranes, *PhD Thesis*, Twente, 1996.
(27) T. Nakagawa, T. Saito, S. Asakawa, Y. Saito, *Gas Separ. Purif.* **2**, no 3, 3 (1988).
(28) A. Bondi, Physical Properties of Molecular Crystals, Liquids and Glasses, John Wiley & Sons, Ltd., NY, 1968.
(29) G. Consolati, I. Genco, M. Pegoraro, L. Zanderghi, *J. Polym. Sci.: Prt B: Polym. Phys.*, **34**, 357 (1996).
(30) G. Dlubek, Positron annihilation lifetimes spectroscopy, Encyclopedia of Polymer Science and Technology, A. Seidel, Ed., John Wiley & Sons, Hoboken, NJ, 2008.
(31) Yu. Yampolskii, S. Durgaryan, Sorption thermodynamics in glassy polymers, in: R. Stryek, Yu. Yampolskii (Eds.), Chromatography and Thermodynamics, IChPh, PAN, Warsaw, 1986, p. 185.
(32) A. Yu. Alentiev, V. P. Shantarovich, T. C. Merkel, V. I. Bondar, B. D. Freeman, Yu. P. Yampolskii, *Macromolecules* **35**, 9513 (2002).
(33) M. Rudel, J. Kruse, K. Ratzke, F. Faupel, Yu. Yampolskii, V. Shantarovich, G. Dlubek, *Macromolecules*, **41**, 788 (2008).
(34) Yu. Yampolskii, S. Soloviev, M. Gringolts, *Polymer*, **45**, 6945 (2004).
(35) V. P. Shantarovich, I. B. Kevdina, Yu. P. Yampolskii, A. Yu. Alentiev, *Macromolecules*, **33**, 7453 (2000).

4

Amorphous Glassy Perfluoropolymer Membranes of Hyflon AD®: Free Volume Distribution by Photochromic Probing and Vapour Transport Properties

Johannes Carolus Jansen[a], *Karel Friess*[b], *Elena Tocci*[a], *Marialuigia Macchione*[c], *Luana De Lorenzo*[a], *Matthias Heuchel*[d], *Yuri P. Yampolskii*[e] *and Enrico Drioli*[a]

[a]*Institute of Membrane Technology, ITM-CNR, c/o University of Calabria, Rende, Italy*

[b]*Department of Physical Chemistry, Institute of Chemical Technology, Prague, Czech Republic*

[c]*University of Calabria, Rende, Italy*

[d]*GKSS Research Center, Institute of Chemistry, Teltow, Germany*

[e]*A. V. Topchiev Institute of Petrochemical Synthesis, Moscow, Russia*

4.1 Introduction and Scope

Fluorinated polymers have quite different properties compared to their hydrocarbon counterparts, due to the nature of the C–F bond. They often have a particularly high thermal and chemical stability, insolubility in common organic solvents and resistance to swelling by condensable gases or vapours. Such properties are an advantage for those membrane applications where the separation takes place under harsh conditions. Crystalline perfluoropolymers such as Teflon® are extremely stable but their crystalline nature makes them mostly suitable for porous membranes, either for filtration of liquid media, or for instance for membrane distillation processes which take advantage of the

Membrane Gas Separation Edited by Yuri Yampolskii and Benny Freeman

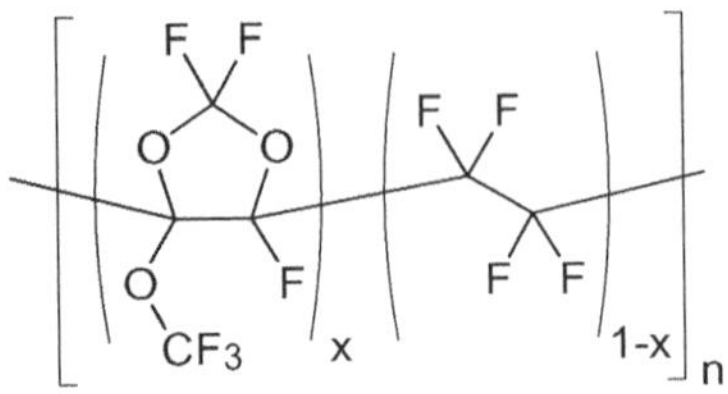

Figure 4.1 *Molecular structure of Hyflon® AD60X (x = 0.6) and Hyflon® AD80X (x = 0.8), poly(2,2,4-trifluoro-5-trifluoromethoxy-1,3-dioxole-co-tetrafluoroethylene)*

extremely high hydrophobicity of perfluorinated polymers. This prevents wetting of the pores by water and thus enables the transport through the vapour phase alone.

The increasing pressure on industry to develop more sustainable technologies is an additional driving force for radical changes in the materials choice. In this respect, in recent years, amorphous glassy perfluoropolymers (PFPs) such as Cytop®, Tefon AF® and Hyflon AD® are under the attention of the membrane community for their application in gas and/or vapour separation processes. The latter is possible due to the presence of bulky groups in the polymer chain which adds a relatively high permeability to their already outstanding set of properties. In theory this allows their application in fields with particularly harsh conditions where other polymers may fail due to chemical attack, thermal instability and plasticization by condensable species.

An important aspect for the understanding of the transport properties of such polymers is the study of their structure, down to the molecular level. Especially in the absence of swelling phenomena, when the penetrant solubility in the perfluoropolymer matrix is relatively low, as is usually the case for non-fluorinated hydrocarbon vapours, the main factor which determines the transport properties is the Free Volume (FV). Knowledge of the free volume is therefore often of great importance for the understanding of the transport properties of dense membrane materials.

The scope of this chapter is to determine the average free volume size and size distribution for Hyflon AD® (Figure 4.1) by the photochromic probe method and to study the transport of organic vapours in membranes of this polymer. The final aim is to correlate the transport data in the polymer with the free volume element (FVE) size distribution and to gain a better understanding of structure–property relationships in perfluoropolymers.

4.1.1 Free Volume and Free Volume Probing Methods for Polymers

The free volume in polymers is defined as the volume not directly occupied by the atoms constituting the polymer chains. It is a characteristic of all solids, but in the case of glassy polymers it is of particular interest, especially in relation to their use as membrane materials. Polymers below their T_g are in a non-equilibrium state. Depending on the sample history and on the specific material, the frozen non-equilibrium state, a result of the dramatically reduced mobility of the polymer chains in the glassy state, may give rise to time-dependent relaxation phenomena or physical ageing, leading to a gradual reduction of the free volume fraction and often to a shift in the FV distribution (FVD). For the understanding of mass transport in dense polymer membranes, insight in their free volume is essential.

By definition, the free volume cannot be measured directly: it is a vacuum which can neither be seen, nor sensed in any way. The only possible way for the determination of the free volume is by sensing of its surroundings, from which we can deduce the part of the volume that is not occupied. Several probe methods have been developed for probing the occupied volume in a polymeric sample, each of them based on different physical properties and on different measurement principles [1,2].

- Positron annihilation lifetime spectroscopy (PALS) studies the lifetime spectrum of ortho-positrons after being injected into the sample [3,4]. This lifetime depends on the probability of the ortho-positronium (o-Ps) particle (a hydrogen-like bound state formed by a positron–electron pair) to be quenched and annihilate. This probability is higher in condensed matter than in vacuum. Of all the probe methods PALS is nowadays probably the most versatile one and the most widely used. The o-Ps particle is the smallest probe available and can thus detect the smallest free volume elements; furthermore, the method furnishes information on the average free volume size and on the FV size distribution.
- Inverse gas chromatography, IGC [5]. In this method the sorption thermodynamics of a homologous series of usually alkanes in the polymeric matrix are determined. The partial molar enthalpy of mixing, ΔH_m, depends on the size of the sorbent and it was found that a plot of ΔH_m against probe size shows a minimum for the sorbent size which best corresponds to the average size of a free volume element.
- ^{129}Xe-NMR spectroscopy [6]. This method evaluates the chemical shift of xenon-129, dissolved in the polymer matrix. The chemical shift of the ^{129}Xe atoms depends on the interaction of the electron cloud with those of the atoms in the surrounding polymer. This interaction is stronger the smaller the size of the FVE in which the ^{129}Xe probe is located.
- Electrochromic probing [7] relies on the detection by UV spectroscopy of the reorientation of suitable electrochromic probes, usually azo or other dye molecules, dissolved in the polymer matrix and then oriented in an electric field. The rate and kinetics of the reorientation after release of the electric field is related to the probe molecule size and to the available free volume size.
- The spin probe method was one of the first methods used to evaluate the free volume in polymers [1,7]. It is based on the principle that the rotational frequency of spin probes, usually stable nitroxyl radicals such as TEMPO (2,2,6,6-tetramethylpiperidine-1-oxyl) is sensitive to the free volume. The relatively complex correlation between spectral data and FV makes this method more suitable for qualitative comparison of different polymers than for quantitative analysis of the FV [1,8].
- The photochromic probe method uses different probe molecules, usually substituted stilbenes and azobenzenes, and in this case the evaluation is based on the capacity of the probe molecules to undergo a *trans–cis* isomerization upon irradiation (Figure 4.2) [9]. This depends on the probe size and on the size of the FVE in which it is located. The probe molecules are dissolved in the polymer matrix, isomerization is induced by a strong UV lamp and the degree of isomerization is determined from the absorbance spectra before and after the isomerization.

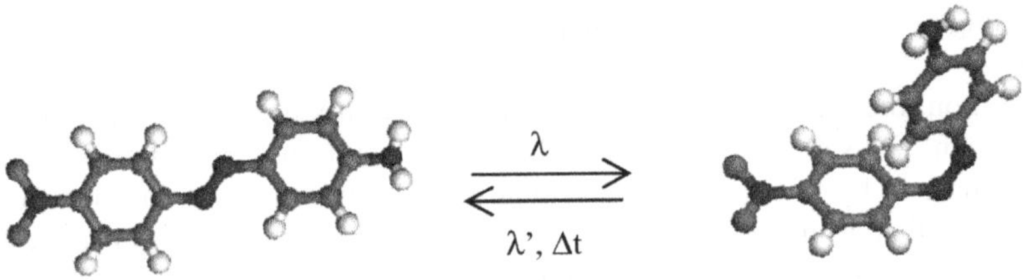

Figure 4.2 *Photo-induced* trans–cis *isomerization of 4-(4-nitrophenylazo)aniline (4-amino-4′-nitro-azobenzene, Disperse orange 3). The back reaction to the more stable trans isomer may be photo-induced or thermally induced*

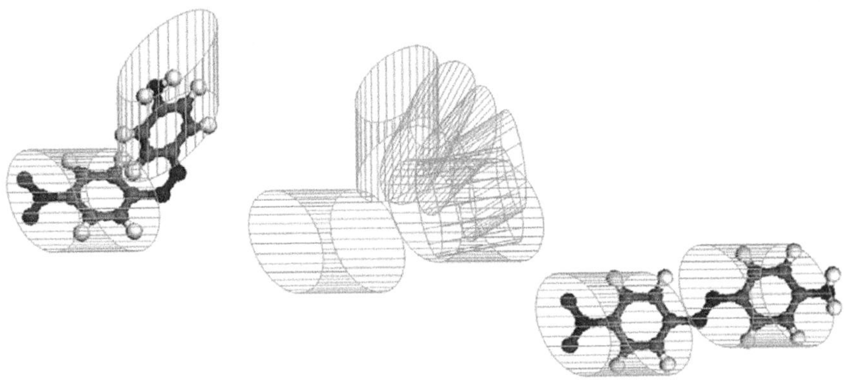

Figure 4.3 *Illustration of the total volume required for the isomerization of the photochromic probe 4-(4-nitro-phenylazo)aniline (4-amino-4′-nitro-azobenzene, Disperse orange 3)*

4.1.2 Details of the Photochromic Probe Method

As anticipated above, photochromism of molecular probes is a phenomenon which is sensitive to the distribution of local free volume in polymer glasses [9]. Photochromic molecular probes in glassy polymers were first used as far back as 1968 [10]. Since then they have been successfully used in evaluating the FV of various polymer matrices [11–14]. The principle of the method is based on the space required for photoisomerization of probe molecules. For this purpose the photochromic or photo-isomerizable molecules, often stilbenes and azobenzenes, are dispersed homogeneously into the polymer matrix. Upon irradiation with UV or visible light with the proper wavelength and/or upon heating, the photochromic probes may undergo *trans–cis* isomerization and back. The crucial hypothesis at the basis of this technique is that photoisomerization in the glassy state requires a minimum critical size of local free volume in the vicinity of the chromophore. This isomerization requires a certain amount of extra volume during the isomerization of the molecule from the *trans* to the *cis* configuration [9] not only because the molecular volume of the *cis* and the *trans* isomers may be different, but also because the structural rearrangement requires some freedom of motion. This is schematically displayed in Figure 4.3. Only if this volume is available, i.e. if the photochromic molecule is located in a sufficiently large free volume element, then photoisomerization will occur.

An important and generally accepted assumption is that the dye molecule does not influence its surroundings, i.e. it does not interact with the polymer matrix and it does not change the free volume itself. The polymer chain dynamics must be sufficiently slow that there is no significant relaxation of the free volume elements on the time scale required for the *trans–cis* transition. The latter makes this method only suitable for glassy polymers at temperatures sufficiently far below the glass transition.

In practice, the amount of probe photoisomerization in the polymer is measured relative to that in dilute solution in a non-viscous model solvent, where free volume is not a constraint to isomerization. The ratio between the two is an indication for the available local free volume of a given size and it is usually plotted as a function of the volume required for the photoisomerization of different probes. The *cis* fraction in solution represents the maximum possible degree of isomerization under the given conditions and the value in the polymer phase can range from 0% to 100% of the value in solution. For the same probe compound irradiated under the same conditions, a lower degree of isomerization in the polymer phase compared to the solution indicates that a fraction of the molecules is trapped in too small free volume elements to enable successful isomerization. This procedure gives a first qualitative impression of the distribution of local free volume if a series of photochromic molecules is used which require each a different volume for the isomerization. Under the proper conditions a *quantitative* free volume distribution can be obtained. If the concentration or number of probes (n_{probe}) is much lower than the number of FVEs (n_{hole}) in the polymer matrix, and if we further assume that the probes are homogeneously distributed in the polymer matrix, that is, without aggregation, then the relative degree of isomerization in film compared to solution is equal to the fraction of probe molecules located in a FVE larger than or equal to the minimum isomerization volume of the probe molecule. Since $n_{hole} >> n_{probe}$, we can invert this relation and conclude that the relative degree of isomerization is also equal to the fraction of FVEs larger than the minimum volume required for the isomerization reaction. This represents a point in the cumulative free volume distribution curve. Using a series of different probe molecules with a sufficiently wide range of isomerization volumes we will thus obtain several points, which together make up the entire cumulative distribution curve.

This method has been successfully applied to various glassy polymers. Torkelson and co-workers studied the local free volume distribution in polystyrene (PS) [9], poly(methyl methacrylate) (PMMA) [15] as well as bisphenol-A polycarbonate (PC) and poly(vinyl acetate) (PVAc) [16]. An interesting feature was that they were able to follow the process of physical ageing. The overall distribution is larger in PMMA than in PS and in both polymers physical ageing leads to a larger decrease in the fraction of large free volume elements than in the fraction of small free volume elements. In PS, on the other hand, a rearrangement of the entire FVD takes place and both the small and the large FVEs are affected. Results of PVAc demonstrated that the method cannot be successfully applied near the glass transition, where the typical chain relaxation times are of the same order of magnitude as the time required to reach the photostationary state of the probe molecules upon irradiation [16]. In this case, changes of the FVEs during the photoisomerization reaction compromise the interpretation of the data. In general the method can be applied to a wide range of systems. The matrix may be a thermoplastic polymer [11], as in the studies above, or a gel [13,14] and the probes may not only be dispersed in the polymer matrix but also chemically bound [12,14,17].

A fundamental requirement for the system to be suitable for the application of the photochromic probe method is that the polymer is sufficiently transparent to UV or visible light in the spectral range of the photochromic probe molecules, first of all to induce the photo-isomerization reaction upon irradiation of the sample, and secondly to allow the determination of the absorption spectrum of the isomers before and after the irradiation, and thus to calculate the degree of isomerization. In this respect, Hyflon AD is an excellent polymer because it is highly transparent in the visible spectrum and in most of the UV spectrum, the reason for its original application in optical devices.

4.1.3 Relevance of Free Volume for Mass Transport Properties

It is generally accepted that mass transport in dense polymer membranes takes place according to the well-known solution-diffusion mechanism [18,19]. For non-swelling and non-plasticizing species the amount of penetrant that can dissolve in the glassy polymer matrix depends on the available sorption sites and often a typical dual mode sorption behaviour is observed [20]. The number of Langmuir sorption sites is strongly related to the free volume distribution of the sample.

Also penetrant diffusion depends strongly on the free volume distributions, because it takes place through a kind of 'hopping' mechanism where the penetrant molecules move around in a free volume element until they find enough energy to exceed the energy barrier for a jump to the neighbouring FVE [21–23]. These energy barriers depend on many different factors, such as the rigidity of the polymer matrix, the size of the molecule and the degree of interconnectivity of the free volume. Therefore, knowledge of the free volume is of great help for a better understanding of the transport properties of dense polymeric membranes. The gas transport properties of solution-cast and melt-pressed Hyflon AD60X membranes have been published previously and the results show that the solution-diffusion model can be applied to this polymer [24].

4.2 Membrane Preparation

4.2.1 Materials

Two different grades of glassy amorphous perfluorpolymers were used, Hyflon® AD60X and Hyflon® AD80X (Solvay Solexis), having glass transition temperatures of 130 °C [24] and 134 °C [25], respectively. In this study 1-methoxy-nonafluorobutane (3M, trade name HFE 7100, b.p. 60 °C, mol. wt. 250 $g\,mol^{-1}$), was used as the solvent to prepare solution-cast Hyflon AD films. Photochromic molecules (Table 4.1) were purchased from Aldrich in their stable *trans* isomeric form and they were used without further purification. Dichloromethane (Carlo Erba Reagenti, DCM) was used as received, serving as the solvent to prepare the concentrated mother solution of the photochromic molecules.

4.2.2 Procedures

The polymer solution was prepared as a 5 wt.-% solution in HFE 7100 (0.25 g in 5.0 g of total solution). A $5.9 \times 10^{-3}\,mol\,L^{-1}$ mother solution of the *trans* isomer of the dye

Table 4.1 *Overview of all the probe molecule structures, their van der Waals structures in the* cis *and* trans *isomeric form and their isomerization volume*

Photochromic Probes Molecular structure	van der Waals structure *Cis* isomer	 *Trans* isomer	Total volume[a] (Å^3)	Ref.
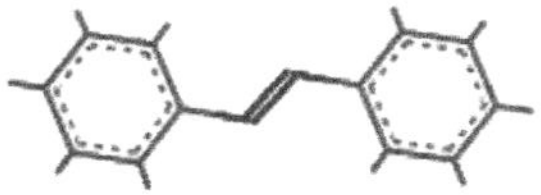 **Azobenzene**	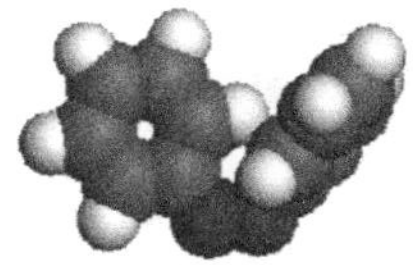	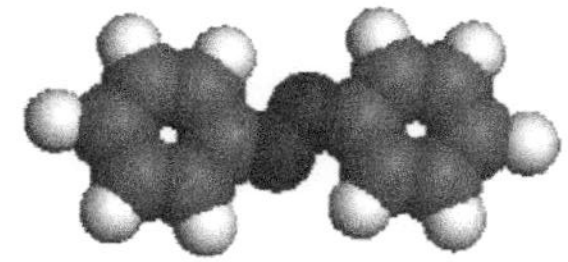	271	9
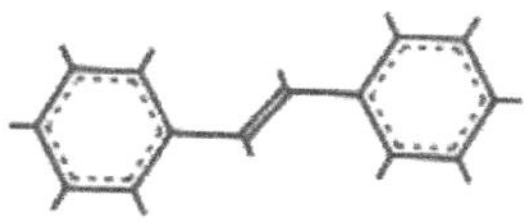 **Stilbene**	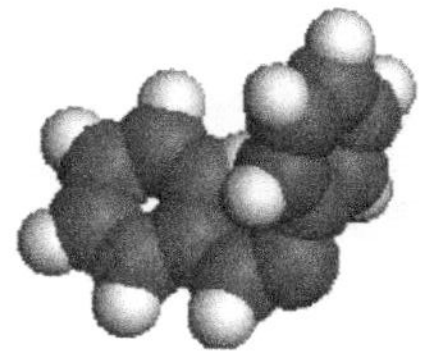	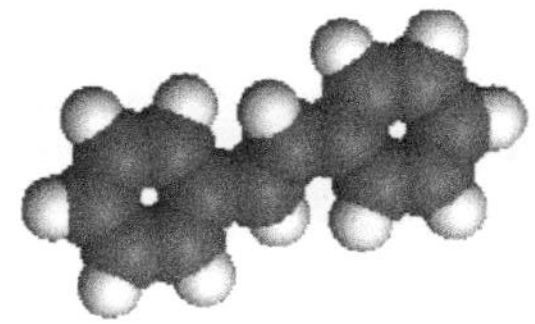	373	9
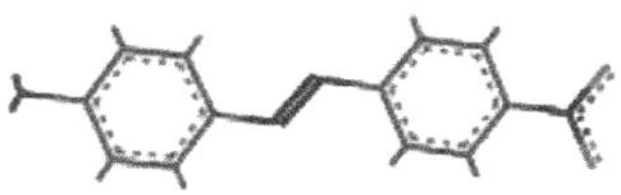 **Disperse orange 3** Azobenzene, R = NO_2, R′ = NH_2	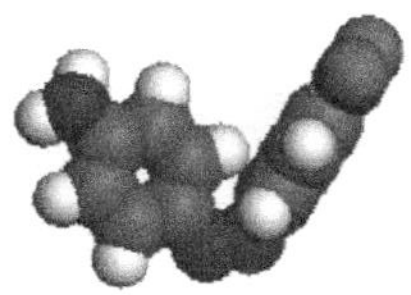	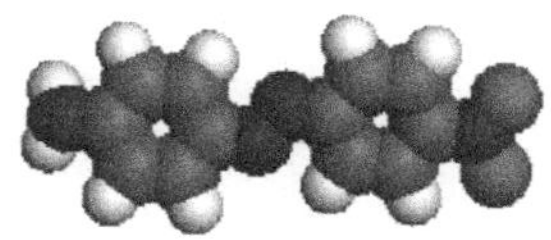	421	27

(continued overleaf)

Table 4.1 *(continued)*

Photochromic Probes Molecular structure	van der Waals structure *Cis* isomer	 *Trans* isomer	Total volume[a] (Å^3)	Ref.
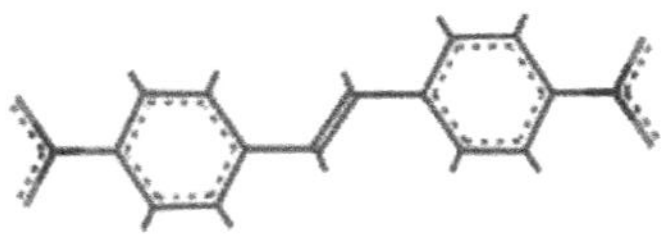 **4,4′-Dinitrostilbene**	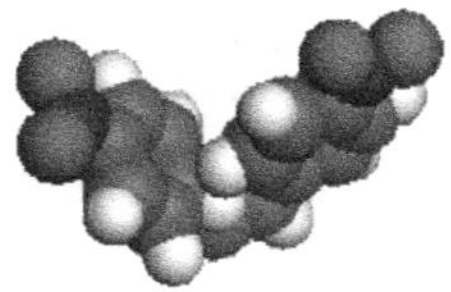	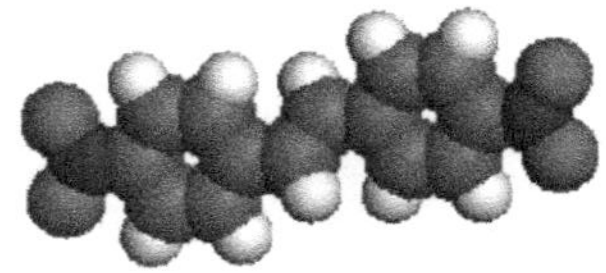	449	9
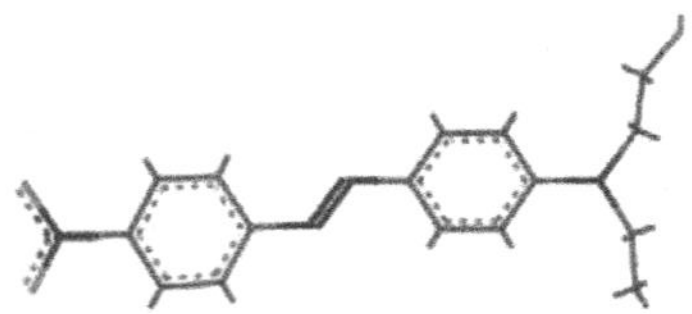 **Disperse red 1** Azobenzene, R = NO_2, R′ = $N(C_2H_5)C_2H_4OH$	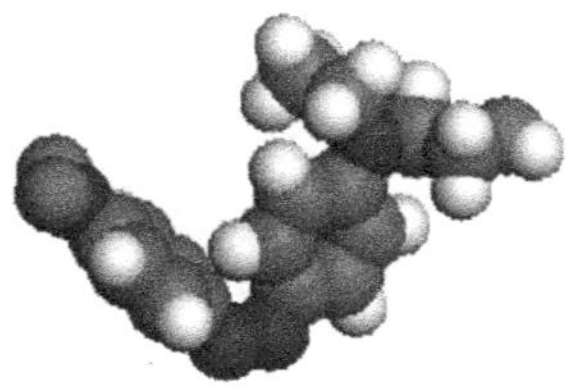	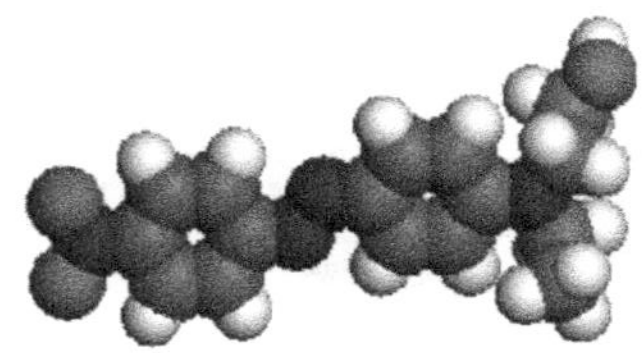	517	27
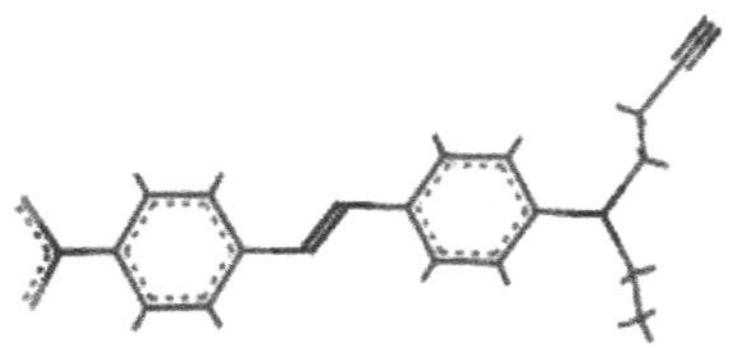 **Disperse orange 25** Azobenzene, R = NO_2, R′ = $N(C_2H_5)C_2H_4CN$	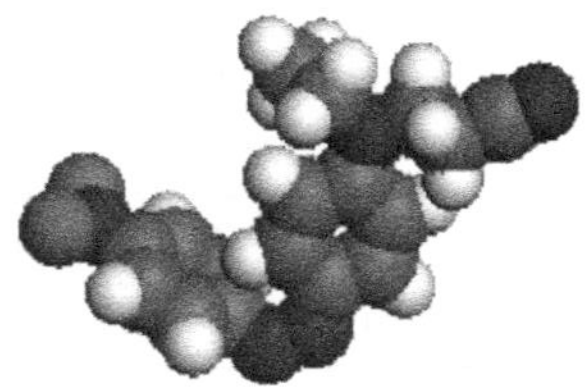	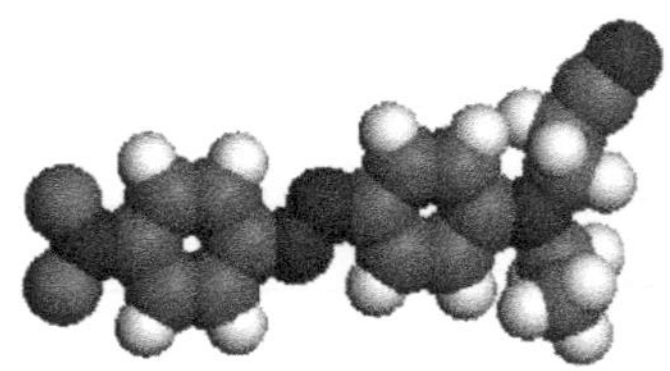	574	27

[a] Total volume needed for isomerization reaction.

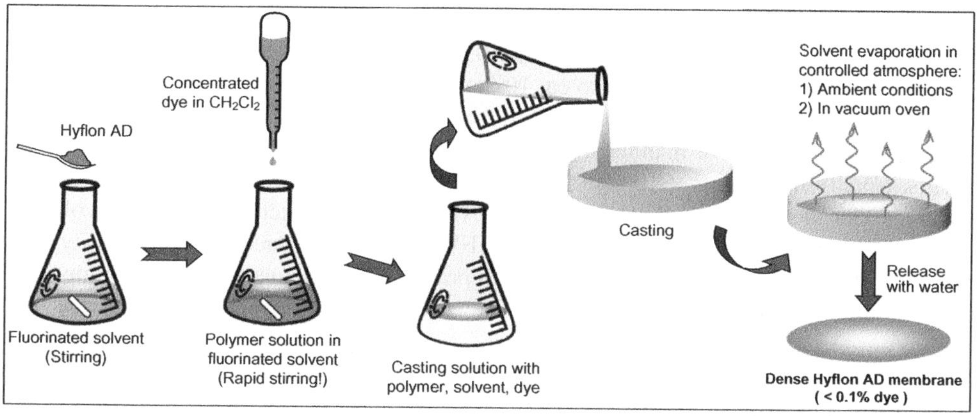

Figure 4.4 Membrane preparation procedure. All steps but the preparation of the polymer solution are carried out in the dark

molecules was prepared separately in DCM. The polymer solution containing the photochromic molecules was then prepared by addition of 0.05 g of the concentrated dye solution to the Hyflon solution under vigorous stirring. All photochromic solutions were prepared and stored in flasks wrapped with aluminium foil to avoid uncontrolled light exposure and premature photoisomerization. Finally, the photochromic dense membranes were prepared by the solvent evaporation method in a Petri dish. Evaporation of the solvent yielded slightly coloured or colourless completely transparent haze-free films. The procedure is schematically displayed in Figure 4.4.

Films were first dried for at least 48 h at room temperature and then in a vacuum oven at 50 °C for 6 h. In all operations the exposure of the dyes to ambient light was minimized in order to prevent uncontrolled photoisomerization. The resulting dye concentration in the membrane is approx. 0.06 wt.-%, based on the polymer dry weight. It is assumed that during evaporation of the solvent the rapidly increasing viscosity of the solution could immobilize the photochromic molecules, and that the dye concentration is sufficiently low to prevent noticeable aggregation or crystal nucleation of the dye. Since DCM is not a solvent for Hyflon, too high concentrations in the casting solution could cause precipitation of the polymer. Therefore the demixing behaviour of the ternary system Hyflon® AD60X / HFE 7100 / DCM was determined by drop-wise addition of DCM to a Hyflon solution in HFE, analogous to the procedure described previously for the determination of phase diagrams [26]. At all polymer concentrations from 1–10 wt.-% the maximum allowable amount of DCM in the solution without inducing phase separation was at least 18 wt.-%, well above the DCM concentration reached in the casting solution upon addition of the concentrated dye solution (ca. 1 wt.-%).

4.2.3 Membrane Properties

The membranes were prepared as a dense film of about 60 μm by solution casting and solvent evaporation, as described in the previous section. No photochromic dyes were used in the membranes for the transport measurements. For the FV analysis, six different probes were used (Table 4.1). Their final concentration in the dried membrane is approximately 1.7×10^{-6} mol cm^{-3}. This is less than 0.1% of the final weight and corresponds to a total number of molecules equal to ca. 10^{18} cm^{-3}. In terms of weight (or volume) this is practically negligible compared to the overall free volume in Hyflon AD of ca. 23 vol.-% [27]. Also in quantitative manner this is much lower than the concentration of FV elements, usually between 2×10^{20} cm^{-3} and 8×10^{20} cm^{-3} [1]. It may therefore indeed be assumed that the probes themselves do not have a notable influence on the free volume. The more than 200-fold excess of available FV holes also reduces the probability that two molecules might occupy the same FVE to practically zero, justifying the assumptions made in the introduction above.

4.3 Free Volume Analysis

4.3.1 Photoisomerization Procedures and UV-Visible Characterization

The *trans–cis* isomerization reaction (Figure 4.2) of all chromophores was carried out by irradiation with a UV-visible lamp (Helios Italquartz, model GRE 500W) selecting a wavelength of 350 nm with a quartz filter (OptoSigma). The back reaction was induced by irradiation at 440 nm, using a blue glass filter (OptoSigma). The lamp to film distance was 15 cm. The band pass filters were placed directly onto the sample surface to avoid irradiation of the films by unfiltered scattered light from the sides, which may compromise the efficiency of the photoisomerization. The UV-visible absorption spectra of the membranes before and after irradiation were recorded on Shimadzu UV 1601 spectrophotometer, interfaced to a personal computer for the data recording and elaboration.

4.3.2 Probes and Spectrophotometric Analysis

Preliminary studies of the *trans–cis* photisomerization kinetics recommended an irradiation time of at least 15 minutes to reach the photostationary state and to have the maximum photoconversion for all dyes. This time was therefore used in all photoisomerization experiments. Molar absorption coefficients, ε_{trans} and ε_{cis}, of the photochromic molecules dispersed in the polymer matrix were calculated from absorbance measurements, using the Lambert–Beer law:

$$A = C \cdot \varepsilon \cdot d \tag{4.1}$$

where A is the value of the absorbance peak at the maximum absorption wavelength of the dye, λ, while d is the path length of the UV beam in the sample, i.e. the membrane thickness, and C is the dye concentration, calculated from the mass ratio of the dye and the polymer in the original casting solution. All absorbance measurements were carried out at ambient temperature. The *cis* isomer fraction, Y, in the films is a measure of the *trans–cis* isomerization efficiency, and is given by:

$$Y = \frac{1 - A/A_{dark}}{1 - \varepsilon_{cis}/\varepsilon_{trans}} \tag{4.2}$$

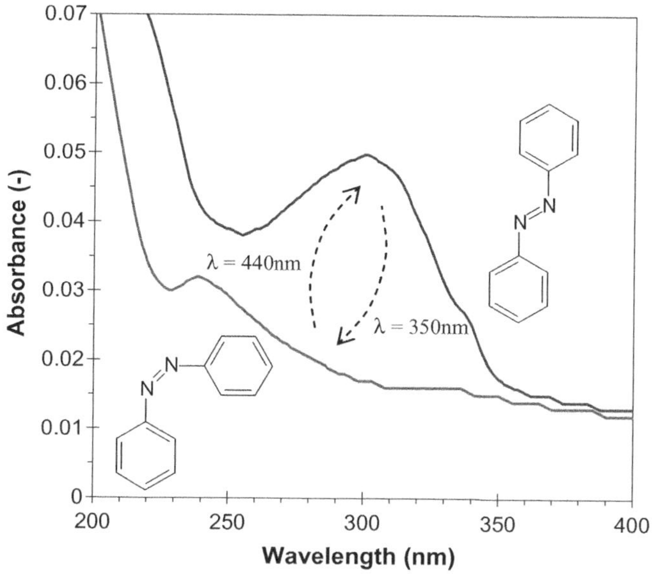

Figure 4.5 *Absorption spectrum of a Hyflon AD60X membrane doped with azobenzene, starting with the film as prepared, containing the* trans *isomer, and using a blank film without the dye as a reference. The isomerization is reversible over several cycles of alternating irradiation at 350 nm and 440 nm*

where A_{dark} is the initial peak absorbance with only *trans* isomer present, A is the actual peak absorbance in the photostationary state. In the present work a total of four azobenzene and two stilbene probe molecules with different substituents were used. Their molecular structures in the form of the *trans* isomer and the van der Waals structures of both isomers are shown in Table 4.1.

An example of the absorption spectra of a membrane containing unsubstituted azobenzene before and after isomerization is given in Figure 4.5. It must be noted that the isomerization is completely reversible over various cycles of isomerization with a wavelength of 350 nm and 440 nm, respectively.

The details of the spectrophotometric analysis and the interpretation of the results have been published recently [27]. The quantitative results of the cumulative free volume distribution, i.e. the relative *trans–cis* isomerization of the films compared to the solution are listed in Table 4.2 as a function of the probe size.

4.3.3 Free Volume Distribution

As discussed above, the ratio of the *cis* fraction in the film and the *cis* fraction in solution indicates which fraction of the probe molecules is located in a sufficiently large FVE necessary to undergo isomerization. This thus represents the cumulative pore size distribution and the data are given in the last two columns of Table 4.2. The experimental data are also plotted in Figure 4.6. A statistical distribution of the void sizes was assumed and the values of F were fitted with a cumulative normal distribution:

Table 4.2 Relative isomerization efficiency, F, for all probes used, defined as the ratio of the cis fraction, Y, in the film and in the solution [data from 27]

Probe	Size		$F = Y_{film}/Y_{solution}$	
	Probe volume[a] (Å³)	Eq. sphere radius (Å³)	Hyflon AD60X	Hyflon AD80X
Azobenzene	271	4.014	0.08	0.0123
Stilbene	373	4.466	0.20	0.0244
Disperse orange 3	421	4.649	0.50	0.249
4,4′-Dinitrostilbene	449	4.750	0.73	0.50
Disperse red 1	517	4.979	0.977	0.75
Disperse orange 25	574	5.156	0.991	0.92

[a]Total volume, including the van der Waals volume of the molecule [28] and the extra volume required for the photo-isomerization [9,29–31].

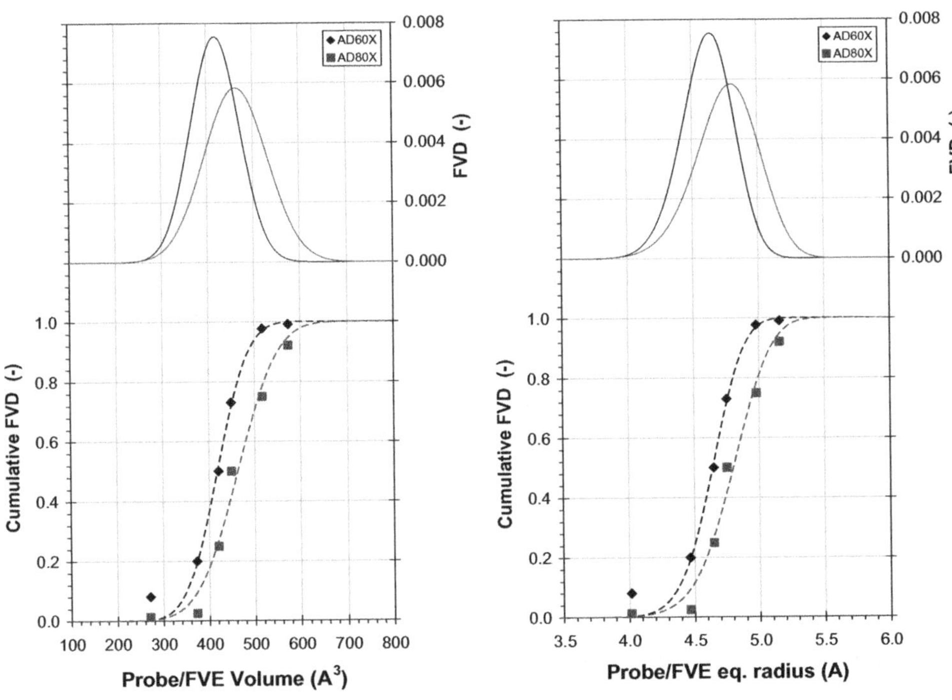

Figure 4.6 FVE distribution curve in terms of probe or FVE volume (left) and in terms of probe or FVE equivalent radius, r_{eq} (right). Fitting results: Hyflon AD60X: $\mu = 419$ and $\sigma = 53 Å^3$ ($r_{eq} = 4.64 Å$, $\sigma = 0.19 Å$); Hyflon AD80X: $\mu = 462$ and $\sigma = 68 Å^3$ ($r_{eq} = 4.79 Å$, $\sigma = 0.24 Å$). The values between parentheses are based on the fit of the radius

$$F = \frac{1}{2} + \frac{1}{2}\operatorname{erf}\left(\frac{V-\mu}{\sigma\sqrt{2}}\right) \tag{4.3}$$

where V is the probe size or corresponding FVE size, μ is the mean value and σ is the standard deviation, a measure of the width of the distribution. The void size distribution is then described by the normal probability function:

$$f = \frac{1}{\sigma\sqrt{2\pi}}\exp\left(-\frac{(V-\mu)^2}{2\sigma^2}\right) \tag{4.4}$$

The latter describes a peak with a maximum at the average void volume $V = \mu$. The two experimental data sets in the present work can be fitted quite satisfactorily with the normal distribution function. The resulting quantitative data of the fitting procedure are given in the caption of Figure 4.6.

The size of the free volume elements is in the range of about 250–560 Å^3 in Hyflon® AD60X and ca. 280–640 Å^3 in Hyflon® AD80X. This means that, if we assume a spherical shape of free volume elements, the equivalent radius is roughly between 4 Å and 5.4 Å, slightly lower in Hyflon AD60X than in Hyflon AD80X, with a maximum in the distribution of about 420 Å^3 and 460 Å^3 (r_{eq} = 4.6 Å and 4.8 Å), respectively. Recently, a comparative study of various experimental and computational free volume probing techniques on Hyflon AD was reported [27] and the results of photochromic probing were consistent with those of PALS, IGC and ^{129}Xe-NMR. Molecular dynamics (MD) simulations showed that the free volume distribution is much more complex than what can be described by a simple Gaussian distribution, due to a strongly interconnected three-dimensional void structure. Small discrepancies between the individual experimental probing techniques can all be traced back to this complex void structure and to the fundamentally different physicochemical interactions between the probes and the polymer matrix. For a qualitative description of the difference between the two grades of Hyflon the Gaussian distribution is more than satisfactory.

4.4 Molecular Dynamics Simulations

Molecular dynamics (MD) simulation studies were carried out to get deeper insight into the free volume structure and its spatial arrangement in the polymers, and into the correlation between the free volume and the transport properties of the materials. It is of crucial importance that MD simulations provide not only local characteristics of the free volume such as FVE size and FVE size distribution, but also information on its topology, enabling a complete visualization in 3D [1,27,32].

The construction of a correct model for the polymer structure is the prerequisite for obtaining accurate results and it is always the first step in the FV modelling. Nowadays several well established simulation methods exist for the preparation of atomistic packing models, both for high and for low free volume polymers [33]. In molecular modelling of amorphous polymers, usually a cubic characteristic volume element is filled with polymer chain segments with or without small permeant molecules in a way that represents most closely the behaviour to be expected in reality.

The analysis and visualization of the free volume of a polymeric model can be obtained in several ways, e.g. the Monte Carlo method [34], geometrical sizing methods [35] and energetic sizing method using the cavity energetic sizing algorithm (CESA) method [36]. The free volume in the packing models can be described qualitatively but for comparison with experimental methods a more quantitative evaluation is necessary.

A first evaluation is often made by a direct visualization of the packed polymer chains, cut into thin slices. The slice representations give important qualitative differences in the amount and distribution of free volume in the polymer. Generally, a homogeneous distribution of free volume is typical for 'normal' amorphous polymers with small and medium amounts of free volume [33], but polymers with high content of free volume such as PTMSP [32] and also glassy PFPs like Teflons AF [32] or Hyflons AD [27], behave differently. These materials contain regions of high segmental packing density where the free volume distribution resembles that in low and medium free volume polymers. The other regions of the materials show rather large voids that may extend in three dimensions to form a partly continuous hole phase. The shape of such free volume elements is highly irregular and non-spherical.

For the quantitative description of the free volume distributions, different methods have been developed, based on energetic [36] or geometric considerations [35]. In the first case the cavity size distribution is defined using a spherical volume with variable dimensions and a well-defined centre, located at the minimum in a repulsive particle energy field. In the second approach, used in the present work, the free volume is determined overlaying a tight three-dimensional grid on the polymer box and inserting a probe at every grid point. All grid points showing overlap with the van der Waals radius of the atoms in the polymer matrix are classified as occupied. Those without overlap are part of the free volume. All neighbouring free grid points are grouped into single elements that represent the individual holes. This approach identifies also cavities that are large and with an irregular shape.

The results of this free volume analysis are visualized in Figure 4.7 for a representative slice of Hyflon AD80X. Every free grid point is represented by a sphere of radius of 1.1 Å, corresponding to a positronium-sized test particle. The individual FVEs are represented by the collection of adjacent and partially overlapping spheres. In this representation several large FVEs seem to be present but in reality most of them belong to a single very large interconnected free volume element which extends over the neighbouring slices. Due to the periodic boundary conditions the FVEs which reach one edge of the slice continue at the opposite side.

For the representation of the complex geometry in a quantitative way different sets of shape parameters have been introduced, for instance based on the void surface area, equivalent sphere volume, radius of gyration. Each of these methods introduces an over-simplification of the very complex nature of the voids. For the same reason the photochromic probe method described above shows a much narrower distribution than one might expect on the basis of the clearly elongated voids displayed in Figure 4.7. This is a common limit of all experimental probing methods, none of which can probe the complex void structure to its full extent.

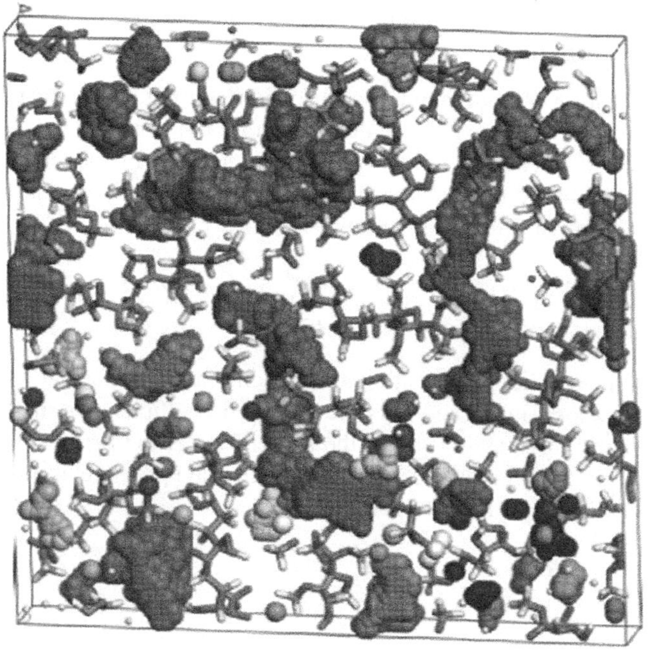

Figure 4.7 Representative slice of an atomistic packing model of Hyflon AD80X. The polymer chains are represented by the small sticks. The free volume elements are shown as the darker continuous regions, represented by a collection of neighbouring and overlapping spheres. Slice thickness ca. 5 Å

4.5 Transport Properties

4.5.1 Permeation Measurements

Vapour permeation experiments were carried out at 25 °C in a fixed volume–pressure increase instrument constructed by GKSS (Germany) according to the procedure described before [24]. Circular membranes with an effective exposed area of 11.3 cm^2 or 2.14 cm^2 were used. Depending on the specific vapour to be tested, the permeation measurements were performed at feed pressures ranging from 35 mbar to almost saturation. The membrane, mounted inside the permeation cell, was thoroughly evacuated for at least 1 h before the first measurement in order to remove all absorbed species. Evacuation was continued until the baseline drift was significantly below the expected steady state pressure increase rate of the species to be tested. Between two subsequent measurements the system was evacuated for a period of at least five times the time lag of the previous species in order to guarantee the complete removal of the penetrant from the membrane and from the rubber seals.

Measurements were carried out in the time lag mode, starting to record the permeate pressure as a function of time as soon as the membrane was exposed to the vapour.

4.5.2 Data Elaboration: Determination of the Time Lag and Steady State Permeability

For the gas permeation measurements the diffusion coefficient was determined by the well-known time lag procedure, based on the penetration theory. If a penetrant-free membrane is exposed to the penetrant at the feed side at $t = 0$ and the penetrant concentration is kept very low at the permeate side, then the total amount of penetrant, Q_t, passing through the membrane in time t is given by [37]:

$$\frac{Q_t}{l \cdot c_i} = \frac{D \cdot t}{l^2} - \frac{1}{6} - \frac{2}{\pi^2} \sum_1^{\infty} \frac{(-1)^n}{n^2} \exp\left(-\frac{D \cdot n^2 \cdot \pi^2 \cdot t}{l^2}\right) \tag{4.5}$$

in which c_i is the penetrant concentration at the membrane interface at the feed side, l is the membrane thickness (in m) and D is the diffusion coefficient (in $m^2 s^{-1}$).

For the barometric instrument used in the present work, the permeate volume is constant and permeation results in an increase of the permeate pressure as a function of time, which is described by a similar equation:

$$p_t = p_0 + (dp/dt)_0 \cdot t + \frac{RT \cdot A \cdot l}{V_P \cdot V_m} \cdot p_f \cdot S\left(\frac{D \cdot t}{l^2} - \frac{1}{6} - \frac{2}{\pi^2} \sum_1^{\infty} \frac{(-1)^n}{n^2} \exp\left(-\frac{D \cdot n^2 \cdot \pi^2 \cdot t}{l^2}\right)\right) \tag{4.6}$$

In which p_t is the permeate pressure (in bar) at time t (in s), p_0 is the starting pressure (in bar), R is the universal gas constant ($8.314 \times 10^{-5} m^3 bar mol^{-1} K^{-1}$), T is the absolute temperature (in K), A is the exposed membrane area (in m^2), V_P is the permeate volume (in m^3), $V$is the molar volume of a gas at standard temperature and pressure ($22.41 \times 10^{-3} m^3_{STP} mol^{-1}$ at 0 °C and 1 atm), p_f is the feed pressure (in bar) and S is the gas solubility coefficient ($m^3_{STP} m^{-3} bar^{-1}$). The term $(dp/dt)_0$ represents the baseline slope (in $bar s^{-1}$). Usually this should be negligible, but in the case of very slow permeating species it may be necessary to correct for this term. This is also the case if minor cracks are formed in these rather brittle perfluoropolymers under the pressure of the sealing rings in the membrane cell, which may give rise to Knudsen-type of diffusion and apparent baseline drift. At very long times the exponential term approaches zero and Equation (4.6) reduces to:

$$p_t = p_0 + (dp/dt)_0 \cdot t + \frac{RT \cdot A \cdot l}{V_P \cdot V_m} \cdot p_f \cdot S\left(\frac{D \cdot t}{l^2} - \frac{1}{6}\right) \tag{4.7}$$

or

$$p_t = p_0 + (dp/dt)_0 \cdot t + \frac{RT \cdot A}{V_P \cdot V_m} \cdot \frac{p_f \cdot S \cdot D}{l}\left(t - \frac{l^2}{6D}\right) \tag{4.8}$$

Thus, at long times a plot of p_t versus time describes a straight line which, upon extrapolation, intersects the baseline at $t = l^2/6D$, defined as the time lag, Θ (s)

$$\Theta = \frac{l^2}{6D} \tag{4.9}$$

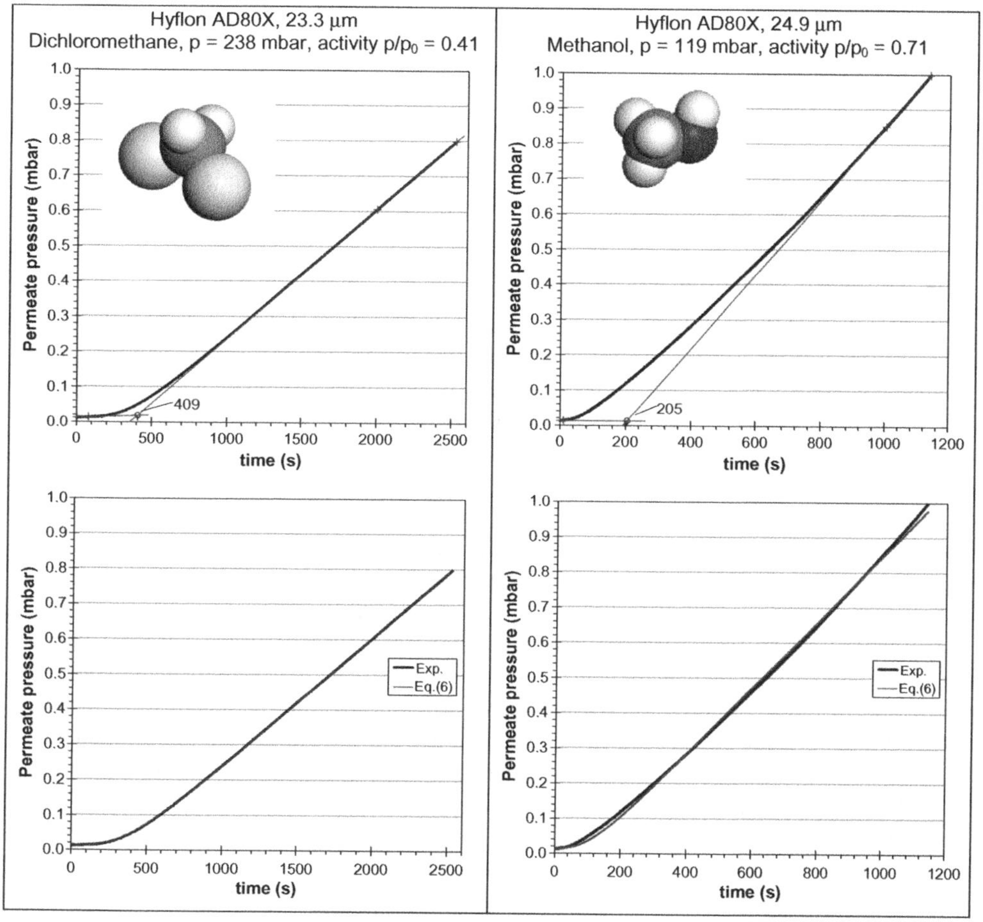

Figure 4.8 *Permeation curves of dichloromethane and methanol vapour in Hyflon AD80X membranes at 25 °C. Time lag calculation by the tangent method (top) and by a direct least squares fit of the entire permeation curve according to Eq. (4.6) (bottom). The thick dark line represents the experimental data; the thin brighter line gives the tangent (top) or the least squares fit. The experimental and fitted lines superimpose completely in the case of DCM, and only one curve can be distinguished, whereas MeOH gives a poor fit. See text for further explanation*

With this equation the diffusion coefficient can be obtained from the time lag if the membrane thickness is known. This way to calculate the diffusion coefficient will be further referred to as the '*tangent method*' and the procedure is schematically displayed in Figure 4.8, top section. The permeability follows from the steady state pressure increase rate and can be defined as:

$$P = \frac{V_P \cdot V_m \cdot l}{RT \cdot A \cdot p_f} \cdot [(dp/dt) - (dp/dt)_0] \tag{4.10}$$

If the simple solution-diffusion model applies, the penetrant solubility can be calculated from P and D from the simple equation:

$$P = D \cdot S \tag{4.11}$$

This procedure thus allows the determination of all three fundamental transport parameters, P, D and S.

4.5.3 Vapour Permeation Measurements

In this work preliminary vapour permeation measurements were carried out with two different species, the rather bulky dichloromethane (DCM) molecules and the much smaller methanol molecules. Two typical permeation curves are displayed in Figure 4.8. The transport parameters, determined on the basis of the tangent method and Equations (4.9)–(4.11), are listed in Table 4.3. It contains the parameters defined above as well as solubility C in the membrane in equilibrium with the feed pressure of penetrants.

At first sight the behaviour of both vapours is quite similar. Their permeability is of the same order of magnitude, with a somewhat higher value for methanol, as expected for the smaller molecule in the glassy polymer. The higher methanol permeability is indeed mainly due to the faster diffusion of methanol in the PFP matrix, about twice as fast as that of DCM. This confirms the size-sieving character of PFPs, already observed for permeation of permanent gases [24]. On the other hand, the solubility and solubility coefficient of DCM are higher, especially on a weight basis. At the actual feed pressure, for DCM the C values and solubility coefficient are more than 8 times higher than those of methanol, due to a combination of a higher solubility coefficient , a higher molar mass and a higher vapour pressure.

More careful analysis of the permeation curves show that the two vapours actually display completely different behaviour. The tangent method can be applied without problems to the DCM permeation curve, resulting in a time lag of ca. 400 s. In contrast, this method shows that methanol has an unusually wide transient period. While the extrapo-

Table 4.3 *Comparison of the transport properties of dichloromethane and methanol vapour in Hyflon AD80X at 25 °C*[a]

Vapour	DCM	MeOH	
p_{Feed}	238	119	mbar
p/p_0	0.41	0.71	–
P	3.35	4.85	$10^{-8}\,m^3_{STP}\,m\,m^{-2}\,h^{-1}\,bar^{-1}$
	12.4	17.9	Barrer
Θ	409	205	s
D	22.1	50.3	$10^{-10}\,cm^2\,s^{-1}$
S	42.0	26.7	$m^3_{STP}\,m^{-3}\,bar^{-1}$
	160	38.1	$10^{-3}\,g\,cm^{-3}\,bar^{-1}$
C	10.0	3.18	$cm^3_{STP}\,cm^{-3}$ at p_{Feed}
	38.0	4.53	$10^{-3}\,g\,cm^{-3}$ at p_{Feed}

[a] Effective values determined by the tangent method.

lated time lag is about 200 s, the onset of permeation is only a few seconds for methanol. Surprisingly it then still takes at least 20 minutes to reach steady state permeation, about as long as it takes for the much larger DCM molecules to reach steady state permeation.

The different behaviour between the two vapours is even more evident if we fit the entire experimental curve directly with Equation (4.6) after expansion into 10 terms (n = 1, 2, 3, ... 10). The DCM curve yields a nearly perfect fit (Figure 4.8, bottom), indicating that the DCM transport can be described well by the simple Fickian diffusion with a single diffusion constant, independent of time or concentration. In this case the values of D and S are directly obtained from the curve fit and they agree within an error of a few percent with the values in Table 4.3, obtained by the tangent method.

In contrast, the methanol permeation curve cannot be fitted satisfactorily. Nevertheless, if only the first 150 s are considered the fit is perfect, as shown in Figure 4.9B. In this short time interval we can assume normal Fickian behaviour and the value of D is directly obtained from the curve fit. For this case the corresponding time lag would be approx. 65 s. At longer times the fit (thin line delimiting area 1 in Figure 4.9A) underestimates the experimental data (thick solid line in Figure 4.9A). If we then take the difference between the first fit and the experimental curve (grey area 2), the shape is remarkably similar to a normal time lag curve (Figure 4.9C). Indeed, this curve can again be fitted satisfactorily with Equation (4.6) without the terms p_0 and $(dp/dt)_0 t$, which have been taken into account in the first step. The resulting time lag of the second step would approx. 336 s. Still, the sum of the two fits (thin line delimiting area 2 in Figure 4.9A) slightly underestimates the experimental curve. The same procedure can be repeated a third time (Figure 4.9D) and again a nearly perfect fit is obtained. The calculated time lag in the third step, approx. 1765 s but since it falls outside the window of the experimental data it should be considered with care. The sum of the three individual curves now coincides exactly with the experimental data.

This type of evaluation of the transient behaviour in the permeation of methanol vapour as the sum of three separate permeation curves suggest that methanol vapour transport is described by three independent diffusion phenomena. Given the polar character of methanol, the presence of OH groups and the highly hydrophobic character of the polymer, in combination with the relatively large free volume elements, it speaks for itself that one would hypothesize that the formation of clusters of two, three or more methanol molecules, held together by hydrogen bridges between the hydroxyl groups could be responsible for the observed anomalies in the transport. It is worthwhile to note that evidence of the cluster formation in the process of methanol diffusion was obtained in the study of another PFP (AF2400) [38]. Molecular dynamics (MD) simulation studies have revealed that diffusion proceeds via a 'hopping' process, in which the penetrant molecules move around in a free volume element until their energy is high enough to overcome the energy barrier for jumping to the adjacent free volume element [21,23]. The required energy rapidly increases with increasing dimensions of the molecule and as a consequence the diffusion coefficient of larger molecules decreases, as observed for a series of permanent gases with different kinetic diameter [24]. The same would be expected for clusters of more molecules in comparison with the single species.

In this light it is too simplistic to assume that the three time lags are related to the presence of single, dimeric and trimeric species, each with its own characteristic diffusion

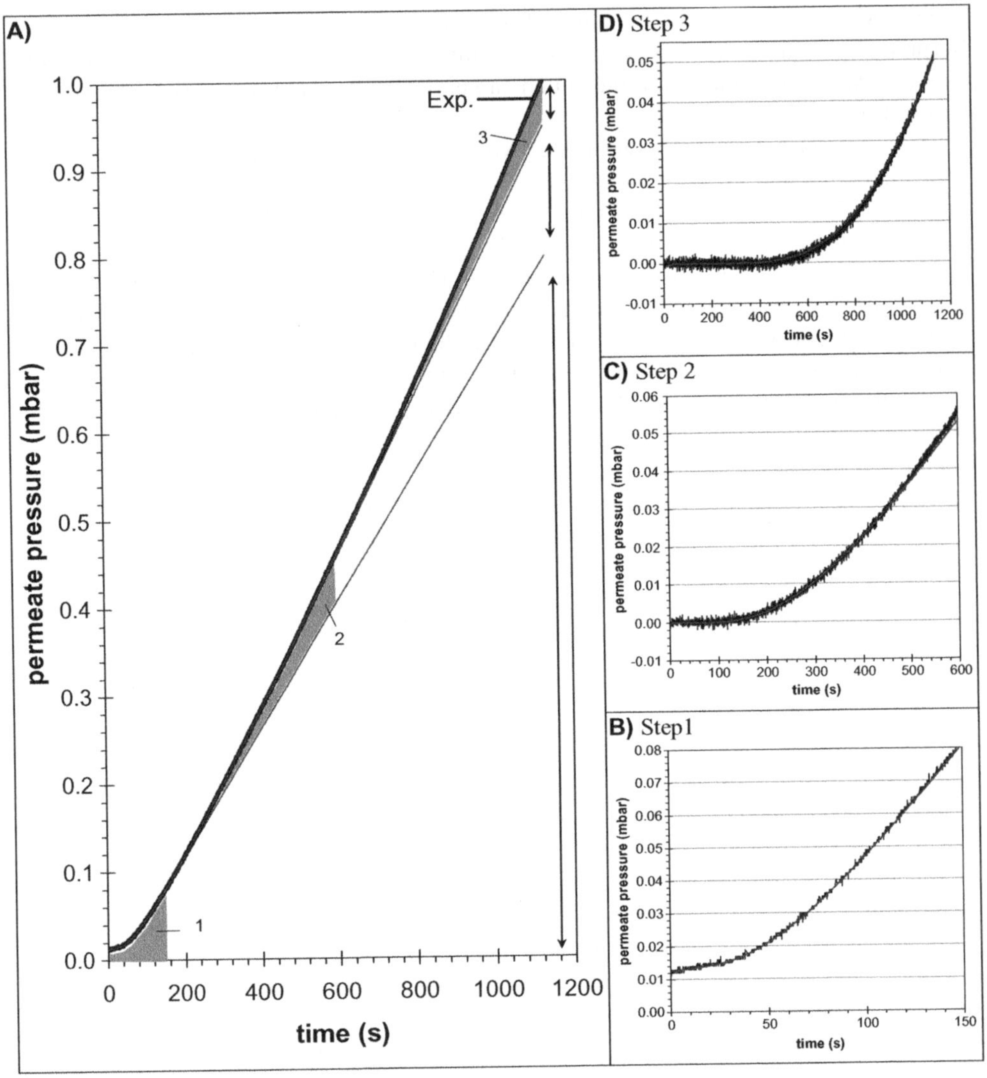

Figure 4.9 *Left side: Experimental permeation curve of methanol vapour in the Hyflon AD80X membrane of Figure 8 (thick dark line). The shaded areas and the thin brighter lines represent the different steps of the fitting procedure according to Eq. (4.6). Right side, B, C, D: experimental data or residual experimental data (noisy dark lines), with the corresponding fit of the individual steps in the three different time intervals (thin brighter lines). The sum of the three fits coincides perfectly with the experimental data. In most cases the fit is nearly perfect and only one curve can be distinguished. See further explanation in the text*

coefficient. It is reasonable to assume that the first part of the curve indeed corresponds to the diffusion of relatively fast, single methanol molecules which are not delayed by interaction with other molecules. The situation is more complex for the rest of the permeation curve. Since hydrogen bonds are relatively weak bonds, the formation of clusters is a dynamic process of continuous aggregation and dissociation. The clusters may therefore diffuse either as a whole or they may first dissociate and then diffuse as single molecules. If the average lifetime of a cluster is shorter than the corresponding jump frequency of the cluster, then it is more likely that methanol diffuses as the single molecules. In that case the observed 'slower' diffusion constants must be interpreted as a temporary 'immobilization' of the methanol molecules in larger clusters. The procedure described above then becomes a potential method to analyze the average cluster size. Further evaluation of this possibility is in progress [39].

4.6 Correlation of Transport and Free Volume

It is known that glassy polymer membranes can have a considerable size-sieving character, reflected mainly in the diffusive term of the transport equation. Many studies have therefore attempted to correlate the diffusion coefficient and the membrane permeability with the size of the penetrant molecules, for instance expressed in terms of the kinetic diameter, Lennard-Jones diameter or critical volume [40]. Since the transport takes place through the available free volume in the material, a correlation between the free volume fraction and transport properties should also exist. Through the years, authors have proposed different equations to correlate transport and FFV, starting with the historical model of Cohen and Turnbull for self diffusion [41], later adapted by Fujita for polymer systems [42]. Park and Paul adopted a somewhat simpler form of this equation to correlate the permeability coefficient with fractional free volume [43]:

$$P = A \cdot \exp(-B/FFV) \tag{4.12}$$

The latter was recently further elaborated by Rowe *et al.* in order to correlate permeability directly to o-Ps lifetime from PALS analysis [44]. Freeman discussed the physical basis for the Robeson upper bound, closely related to the fractional free volume and to penetrant size ratio [45].

In the present work the van der Waals volumes were calculated as a measure of the penetrant size. The values of the different vapour species are listed in Table 4.4. The

Table 4.4 *Van der Waals volume and (apparent) diffusion coefficient of dichloromethane and of hydrogen bonded methanol clusters of different size*[a]

Species	van der Waals volume $(\text{Å})^3$	D $(10^{-10}\,cm^2\,s^{-1})$	Method
MeOH	37.4	160	Fit with eq. (4.6)
$(MeOH)_2$	71.6	30.7	Fit
$(MeOH)_3$	105.8	5.9	Fit
MeOH Average	–	50.3	Tangent
CH_2Cl_2	56.9	22.1	Tangent

[a] Calculated with the *'Materials Studio'* software package, Accelrys, Inc. (San Diego, CA).

volume of a single methanol molecule is 37.4 Å^3 and that of the dimeric and trimeric species are nearly twice and three times larger, 71.6 Å^3 and 105.8 Å^3, respectively. These are all significantly smaller than the average FVE size determined with the photochromic probe method. MeOH is a nonsolvent for Hyflon AD, unable to plasticize or swell the polymer. The present results show that even without swelling of the polymer matrix sufficient space is available to host clusters of two or more molecules. Indeed, methanol has a strong preference to form clusters inside these free volume elements, so that the very polar hydroxyl groups are directed towards each other and they are shielded by the less polar methyl groups, which are oriented towards the hydrophobic perfluoropolymer matrix. Also the dichloromethane molecule, with a van der Waals volume of 56.9 Å^3 is significantly smaller than the available free volume elements (Figure 4.6). Lacking the possibility to form hydrogen bridges, the latter cannot form similar clusters as methanol and normal transport behaviour is therefore observed.

The solubility coefficients of DCM and methanol, calculated by Equation (4.11) from the permeability and the effective diffusion coefficient obtained by the tangent method, are 42.0 $m^3_{STP}m^{-3}bar^{-1}$ and 26.7 $m^3_{STP}m^{-3}bar^{-1}$, respectively. At the given feed pressures of 119 mbar for MeOH and 238 mbar for DCM this corresponds to approximately 4.5 mg cm^{-3} for MeOH and 38 mg cm^{-3} for DCM. Especially for MeOH this corresponds to a much lower volume than the total free volume fraction of 23% in the polymer. These results confirm that both in terms of FVE size and in terms of FFV the transport of methanol vapour, as single molecules or as clusters, and of DCM vapour, may take place through the available free volume without the necessity of swelling.

4.7 Conclusions

The photochromic probe method was successfully used to analyze the FVE size distribution in membranes of the glassy amorphous perfluoropolymer Hyflon AD. Two grades of Hyfon with a slightly different content of the rigid cyclic co-monomer showed a relatively narrow FV size distribution, and the entire distribution of Hyflon AD80X, with the highest cyclic co-monomer content is shifted to somewhat higher volumes compared to the AD60X grade. The results of the photochromic probe method, indicating a FVE size distribution ranging from ca. 250–560 Å^3 in Hyflon® AD60X and from ca. 280–640 Å^3 in Hyflon® AD80X, are in fair agreement with literature data from other experimental techniques like inverse gas chromatography, positron annihilation lifetime spectroscopy, ^{129}Xe-NMR spectroscopy. Molecular dynamics simulations show a combination of single small FVEs and large interconnected FVEs, extending in three dimensions.

Organic vapour transport reveals fundamental differences between a polar substance like methanol and the less polar dichloromethane. An extremely wide transient region for methanol vapour permeation may be related to clustering of the alcohol molecules through the formation of hydrogen bridges between the hydroxyl groups. This is supported by careful analysis of the time lag behaviour, which indicates a spectrum of diffusion constants, rather than a single one.

Based on their van der Waals volume, even large clusters of at least three methanol molecules are still considerably smaller than the FVE size observed experimentally. Their presence in the film should therefore not be hindered for sterical reasons. The total concentration of the vapours at the feed pressure used, calculated by the tangent method, is $10.0\,cm^3{}_{STP}cm^{-3}$ ($38\,mg\,cm^{-3}$) for DCM and $3.18\,cm^3{}_{STP}cm^{-3}$ ($4.5\,mg\,cm^{-3}$) for methanol. In the condensed (liquid) state this corresponds to a volume far below the available free volume of ca. 23%.

References

(1) Yu. P. Yampolskii, Methods for investigation of the free volume in polymers, *Russ. Chem. Rev.*, **76**, 59–78 (2007).

(2) Yu. Yampolskii, V. Shantarovich, Positron annihilation lifetime spectroscopy and other methods for free volume evaluation in polymers, in: Yu. Yampolskii, I. Pinnau, B. D. Freeman, (Eds.), Materials Science of Membranes for Gas and Vapor Separation, John Wiley & Sons, Chichester, England, 2006, Chapter 6, p. 191–210.

(3) D. M. Shrader, Y. C. Jean, Positron and positronium chemistry; Elsevier, Amsterdam, 1988.

(4) G. Dlubek, A. P. Clarke, H. M. Fretwell, S. B. Dugdale, M. A. Alam, Positron lifetime studies of free volume hole size distribution in glassy polycarbonate and polystyrene, *Phys. Stat. Sol. A*, **157**, 351 (1996).

(5) Yu. P. Yampolskii, N. E. Kaliuzhnyi, S. G. Durgaryan, Thermodynamics of sorption in glassy poly(vinyltrimethylsilane), *Macromolecules*, **19**, 846–850 (1986).

(6) G. Golemme, J. B. Nagy, A. Fonseca, C. Algieri, Y. Yampolskii, ^{129}Xe-NMR study of free volume in amorphous perfluorinated polymers: comparsion with other methods, *Polymer*, **44**, 5039–5045 (2003).

(7) Yu. P. Yampolskii, V. P. Shantarovich, F. P. Chernyakovskii, A. I. Kornilov, N. A. Plate, Estimation of free volume in poly(trimethylsilyl propyne) by positron annihilation and electrochromism methods, *J. Appl. Polym. Sci.*, **47**, 85–92 (1993).

(8) A. Kh. Vorobiev, V. S. Gurman, T. A. Klimenko, Rotational mobility of guest molecules studied by method of oriented spin probe, *Phys. Chem. Chem. Phys.*, **2**, 379–385 (2000).

(9) J. G. Victor, J. M. Torkelson, On measuring the distribution of local free volume in glassy polymers by photochromic and fluorescence techniques, *Macromolecules*, **20**, 2241–2250 (1987).

(10) Z. G. Gardlund, Effect of a polymer matrix on photochromism of some benzospirans, *J. Polym. Sci., Polym. Lett. Ed.*, **B6**, 57 (1968).

(11) I. Mita, K. Horie, K. Hirao, Photochemistry in polymer solids. 9. Photoisomerization of azobenzene in a polycarbonate film, *Macromolecules*, **22**, 558–563 (1989).

(12) C. S. P. Sung, I. R. Gould, N. J. Turro, Pulsed laser spectroscopic study of the photo-isomerization of azo labels at three different locations on a polystyrene chain, *Macromolecules*, **17**, 1447–1451 (1984).

(13) M. Ueda, H. B. Kim, T. Ikeda, K. Ichimura, Photoisomerization of an azobenzene in sol-gel glass-films, *Chem. Mater.*, **4**, 1229–1233 (1992).

(14) M. Ueda, H. B. Kim, T. Ikeda, K. Ichimura, Photoisomerizability of an azobenzene covalently attached to silica-gel matrix, *J. Non-Cryst. Solids*, **163**, 125–132 (1993).

(15) J. Scot Royal, J. G. Victor, J. M. Torkelson, Photochromic and fluorescent probe studies in glassy polymer matrices. 4. Effects of physical ageing on poly(methyl methacrylate) as sensed by a size distribution of photochromic probes, *Macromolecules*, **25**, 729–734 (1992).

(16) J. Scot Royal, J. G. Victor, J. M. Torkelson, Photochromic and fluorescent probe studies in glassy polymer matrices. 5. Effects of physical ageing on bisphenol-A polycarbonate and poly(vinyl acetate) as sensed by a size distribution of photochromic probes, *Macromolecules*, **25**, 4792–4796 (1992).

(17) J. Algers, P. Sperr, W. Egger, L. Liszkay, G. Kögel, J. de Baerdemaeker, F. H. J. Maurer, Free volume determination of azobenzene-PMMA copolymer by a pulsed low-energy positron lifetime beam with in-situ UV illumination, *Macromolecules*, **37**, 8035–8042 (2004).

(18) J. Wijmans, R. Baker, The solution-diffusion model: a review, *J. Membr. Sci.*, **107**, 1–21 (1995).

(19) J. G. Wijmans, R. W. Baker 'The solution-diffusion model: a unified approach', in: Yu. Yampolskii, I. Pinnau and B. D. Freeman (Eds.), Materials science of membranes for gas and vapor separation, John Wiley & Sons, Chichester, England, 2006 (chapter 5).
(20) R. M. Barrer, Diffusivities in glassy polymers for the dual model sorption model, *J. Membr. Sci.*, **18**, 25–35 (1984).
(21) H. Takeuchi, A jump motion of small molecules in glassy polymers: A molecular dynamics simulation, *J. Chem. Phys.*, **93**, 2062–2067 (1990).
(22) A. A. Gray-Weale, R. H. Henchman, R. G. Gilbert, M. L. Greenfield, D. N. Theodorou, Transition-state theory model for the diffusion coefficients small penetrants in glassy polymers, *Macromolecules*, **30**, 7296–7306 (1997).
(23) M. L. Greenfield, Sorption and diffusion of small molecules using the Transition-State theory, in: Simulation Methods for Polymers, M. J. Kotelyanskii and D. N. Theodorou (Eds.), Marcel Dekker, New York, 2004.
(24) M. Macchione, J. C. Jansen, G. De Luca, E. Tocci, M. Longeri, E. Drioli, Experimental analysis and simulation of the gas transport in dense Hyflon® AD60X membranes. Influence of residual solvent, *Polymer*, **48**, 2619–2635 (2007).
(25) R. S. Prabhakar, B. D. Freeman, I. Roman, Gas and vapor sorption and permeation in poly(2,2,4-trifluoromethoxy-1,3-dioxole-co-tetrafluoroethylene), *Macromolecules*, **37**, 7688–7697, (2004).
(26) J. C. Jansen, M. Macchione and E. Drioli, High flux asymmetric gas separation membranes of modified poly(ether ether ketone) prepared by the dry phase inversion technique, *J. Membrane Sci.*, **255**, 167–180 (2005).
(27) J. C. Jansen, M. Macchione, E. Tocci, L. De Lorenzo, Yu. P. Yampolskii, O. Sanfirova, V. P. Shantarovich, M. Heuchel, D. Hofmann, E. Drioli, Comparative study of different probing techniques for the analysis of the free volume distribution in amorphous glassy perfluoropolymers, *Macromolecules*, **42**, 7589–7604 (2009).
(28) A. Bondi, Van der Waals volumes and radii, *J. Phys. Chem.*, **68**, 441–451 (1964).
(29) M. Traetteberg, I. Hilmo, K. Hagen, A gas electron diffraction study of the molecular structure of trans-azobenzene, *J. Mol. Struct.*, **39**, 231–239, (1977).
(30) G. Orlandi, W. Siebrand, Model for the direct photo-isomerization of stilbene, *Chem Phys. Lett.*, **30**, 352–354 (1975).
(31) J. B. Birks, The photo-isomerization of stilbene, *Chem. Phys. Lett.*, **38**, 437–440 (1976).
(32) D. Hofmann, M. Entrialgo-Castano, A. Lerbret, M. Heuchel, Y. Yampolskii, Molecular modeling investigation of free volume distributions in stiff chain polymers with conventional and ultrahigh free volume: Comparison between molecular modeling and positron lifetime studies. *Macromolecules*, **36**, 8528–8538 (2003).
(33) D. Hofmann, L. Fritz, J. Ulbrich, C. Schepers, M. Boëhning, Detailed atomistic molecular modeling of small molecule diffusion and solution processes in polymeric membrane materials, *Macromol. Theory Simul.*, **9**, 293–327 (2000).
(34) M. L. Greenfield, D. N. Theodorou, Geometric analysis of diffusion pathways in glassy and melt atactic polypropylene. *Macromolecules*, **26**, 5461–5472 (1993).
(35) D. Hofmann, M. Heuchel, Y. Yampolskii, V. Khotimskii, V. Shantarovich, Free volume distributions in ultrahigh and lower free volume polymers: Comparison between molecular modeling and positron lifetime studies. *Macromolecules*, **35**, 2129–2140 (2002).
(36) P. J. in't Veld, M. T. Stone, T. M. Truskett, I. C. Sanchez, *J. Phys. Chem. B*, **104**, 12028–12034 (2000).
(37) J. Crank, G. S. Park, Diffusion in Polymers, Academic Press, London, 1986.
(38) A. Tokarev, K. Friess, J. Machkova, M. Sipek, Yu. Yampolskii, Sorption and diffusion of organic vapors in amorphous Teflon AF2400, *J. Polym. Sci.: Part B: Polym. Phys.*, **44**, 832–844 (2006).
(39) J. C. Jansen, K. Friess, E. Drioli, Organic vapour transport in glassy perfluoropolymer membranes. A simple semi-quantitative approach to analyze clustering phenomena by time lag measurements, *submitted.*
(40) S. Mateucci, Y. Yampolskii, B. D. Freeman, I. Pinnau, Transport of gases and vapours in glassy and rubbery polymers, in: Yu. Yampolskii, I. Pinnau, B. D. Freeman, (Eds.), Materials

Science of Membranes for gas and vapor separation, John Wiley & Sons, Chichester, England, 2006, Chapter 1, p. 1–47.
(41) M. H. Cohen, D. Turnbull, Molecular transport in liquid and glasses, *J. Phys. Chem.*, **31**, 1164–1169 (1959).
(42) H. Fujita, Diffusion in polymer-diluent systems, *Fortschr. Hochpolym.-Forsch.*, **3**, 1–47 (1961).
(43) J. Y. Park, D. R. Paul, Correlation and prediction of gas permeability in glassy polymer membrane materials via a modified free volume based group contribution method, *J. Membr. Sci.*, **125**, 23–39 (1997).
(44) B. W. Rowe, S. J. Pas, A. J. Hill, R. Suzuki, B. D. Freeman, D. R. Paul, A variable energy positron annihilation lifetime spectroscopy study of physical aging in thin glassy polymer films, *Polymer*, **50**, 6149–6156 (2009).
(45) B. D. Freeman, Basis of permeability/selectivity tradeoff relations in polymeric gas separation membranes, *Macromolecules*, **32**, 375–380 (1999).

5

Modelling Gas Separation in Porous Membranes

Aaron W. Thornton[a], *James M. Hill*[b] *and Anita J. Hill*[a]
[a]*CSIRO Materials Science and Engineering, Clayton, Victoria, Australia*
[b]*Nanomechanics Group, School of Mathematics and Applied Statistics, University of Wollongong, New South Wales, Australia*

5.1 Introduction

Instead of adopting a trial-and-error approach to membrane development, it is far more efficient to have a real understanding of the separation phenomena to guide membrane design. For example, relationships have been found between experimental separation results and pore sizes which have been determined from positron annihilation lifetime spectroscopy (PALS) [1–4] and small angle X-ray scattering [5–7], and these may be used to guide the tailoring of pore sizes to enhance separation performance. Similarly, methods such as Monte Carlo, molecular dynamics and other computational techniques [8–19] have improved our understanding of the relationships between membrane characteristics and separation properties. In addition to these inputs, it is also beneficial to have simple models and theories that give an overall insight into separation performance [20–33]. Recently, an increased interest can be noted in gas transport in porous (microporous or nano-porous) membranes, both inorganic and polymeric. In this chapter a brief review of the current models and conceptual frameworks of the transport in porous media are given. In addition, we present a new approach to describe the diffusion of gas molecules in small pores as outlined in more detail in ref. 23, which does indeed provide some new insight into the gas separation phenomena, and therefore facilitate the optimization of material design.

Membrane Gas Separation Edited by Yuri Yampolskii and Benny Freeman

5.2 Background

A quantitative measure of gas transport is the flux (or permeation rate), which is defined as the number of molecules that pass through a unit area per unit time [25]. It is believed that this molecular flux J follows Fick's first law. The flux is proportional to the concentration gradient through the membrane, i.e. there is a movement of gas from regions of high concentration to regions of low concentration, which may be expressed in the form

$$J = -D\frac{dc}{dx} \tag{5.1}$$

where D is the diffusivity (or diffusion coefficient), $c(x)$ is the concentration and x is the position across the membrane. By assuming a linear concentration gradient across the membrane, the flux can be approximated as

$$J = D\frac{C_2 - C_1}{L} \tag{5.2}$$

where $C_1 = c(0)$ and $C_2 = c(L)$ are the downstream and upstream concentrations (corresponding to the pressures p_1 and p_2 via sorption isotherm $c(p)$), respectively, and L is the membrane thickness, as labelled in Figure 5.1.

The membrane performance of various materials is commonly compared using the thickness-independent material property, namely the permeability P, which is related to the flux J in the following way

$$P = \frac{J\,L}{p_2 - p_1} = \left(\frac{C_2 - C_1}{p_2 - p_1}\right)D \tag{5.3}$$

In the case where the upstream pressure is much greater than the downstream pressure ($p_2 >> p_1$ and $C_2 >> C_1$) the permeability can be simplified to give

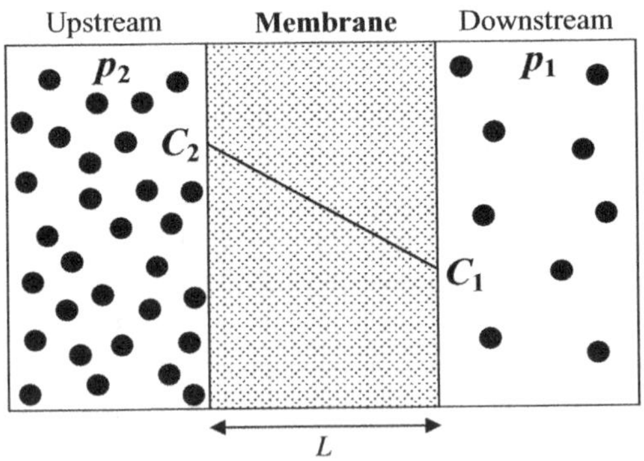

Figure 5.1 Gas separation membrane with a constant concentration gradient across membrane thickness L

$$P = \frac{C_2}{p_2} D \tag{5.4}$$

By introducing a solubility coefficient S, that is, the ratio of concentration over pressure C_2/p_2, when sorption isotherm can be represented by the Henry's law, the permeability coefficient may be expressed simply as

$$P = S\,D \tag{5.5}$$

This form is useful as it facilitates the understanding of this physical property by representing it in terms of two components:

- solubility, S, which is an equilibrium component describing the concentration of gas molecules within the membrane, that is the driving force, and
- diffusivity, D, which is a dynamic component describing the mobility of the gas molecules within the membrane.

The separation of a mixture of molecules A and B is characterized by the selectivity or ideal separation factor $\alpha_{A/B} = P(A)/P(B)$, i.e. the ratio of permeability of the molecule A over the permeability of the molecule B. According to Equation (5.5), it is possible to make separations by diffusivity selectivity $D(A)/D(B)$ or solubility selectivity $S(A)/S(B)$ [25,26]. This formalism is known in membrane science as the solution-diffusion mechanism. Since the limiting stage of the mass transfer is overcoming of the diffusion energy barrier, this mechanism implies the activated diffusion. Because of this, the temperature dependences of the diffusion coefficients and permeability coefficients are described by the Arrhenius equations.

Gas molecules that encounter geometric constrictions experience an energy barrier such that sufficient kinetic energy of the diffusing molecule, or the groups that form this barrier, in the membrane is required in order to overcome the barrier and make a successful diffusive jump. The common form of the Arrhenius dependence for the diffusion coefficient can be expressed as

$$D_A = D_A^* \exp(-\Delta E_a/RT) \tag{5.6}$$

For the solubility coefficient the van't Hoff equation holds

$$S_A = S_A^* \exp(-\Delta H_a/RT) \tag{5.7}$$

where $\Delta H_a < 0$ is the enthalpy of sorption.

Bearing in mind Equation (5.5)

$$P_A = P_A^* \exp(-\Delta E_P/RT) \tag{5.8}$$

where $\Delta E_p = \Delta E_a + \Delta H_a$

Gases are known to diffuse within non-porous or porous membranes according to various transport mechanisms [20,25,32,34]. Table 5.1 illustrates the mechanism of transport depending on the size of pores. For very narrow pores, size sieving mechanism

Table 5.1 Transport mechanisms

Mechanism	Schematic	Process
Activated diffusion	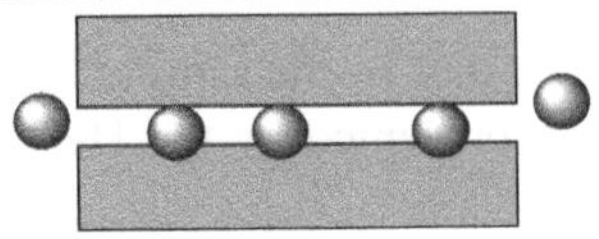	constriction energy barrier 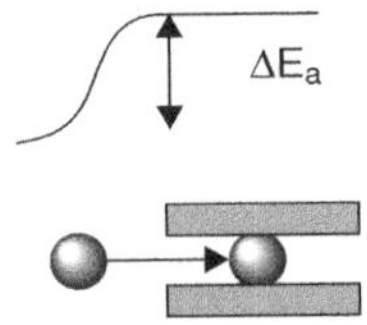
Surface diffusion	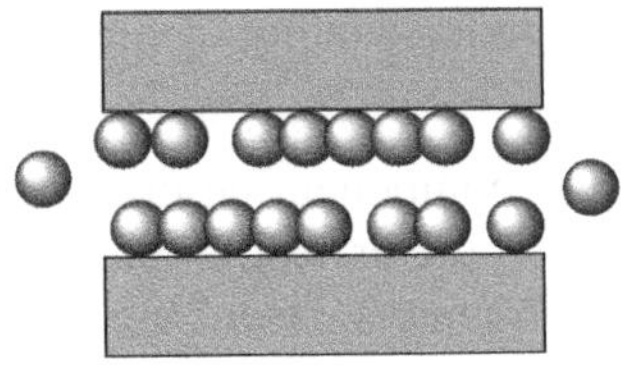	adsorption-site energy barrier 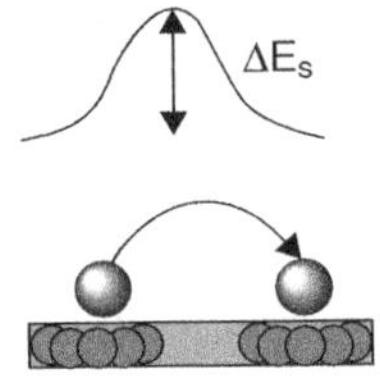
Knudsen diffusion	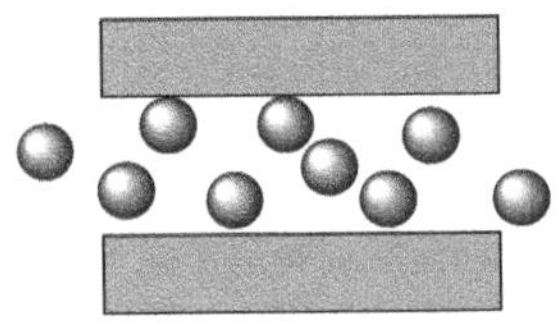	Direction and velocity 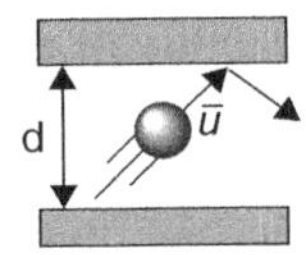

is realized that can be considered as a case of activated diffusion. This mechanism of diffusion is most common in the case of extensively studied non-porous polymeric membranes. For wider pores, the surface diffusion (also an activated diffusion process) and the Knudsen diffusion are observed.

5.3 Surface Diffusion

Surface diffusion is the diffusion mechanism which dominates in the pore size region between activation diffusion and Knudsen diffusion [29,30,34,35]. A model that well described the surface diffusion on the pore walls was proposed many years ago [36]. It was shown to be consistent with gas and vapour transport parameters in porous polymeric membranes (nucleopore). When the pore size decreased below a certain level, which depends on both membrane material and the permeating gas, the gas permeability coefficient exceeds the value for free molecular flow (Knudsen diffusion), especially in the case of organic vapours. Note that surface diffusion usually occurs simultaneously with Knudsen diffusion but it is the dominant mechanism within a certain pore size region discussed later in this chapter. Since surface diffusion is also a form of activated diffusion, the energy barrier ΔE_S is the energy required for the molecule to jump from one adsorption site to another across the surface of the pore. By allowing the energy barrier to be proportionate to the enthalpy of adsorption, Gilliland *et al.* [35] established an equation for the surface diffusion coefficient expressed here as

$$D_S = D_S^* \exp\left(\frac{-aq}{RT}\right) \tag{5.9}$$

where D_S^* is a pre-exponential factor depending on the frequency of vibration of the adsorbed molecule normal to the surface and the distance from one adsorption site to the next. The quantity q (>0) is the heat of adsorption and a is a proportionality constant ($0 < a < 1$) such that aq is the energy barrier which separates the adjacent adsorption sites. An important observation is that more strongly adsorbed molecules are less mobile than weakly adsorbed molecules [30].

In the case of surface diffusion, the gas concentration is well described by Henry's law $c = Kp$, where K is the temperature dependent Henry's law coefficient $K = K_0\exp(q/RT)$, where K_0 is a proportionality constant [29,30]. Since solubility is the ratio of the equilibrium concentration over pressure, the solubility is equivalent to the Henry's law coefficient,

$$S_s = K_0 \exp\left(\frac{q}{RT}\right) \tag{5.10}$$

which implies that solubility is a decreasing function of temperature.

The product of diffusivity and solubility gives

$$P_s = P_S^* \exp\left(\frac{(1-a)q}{RT}\right) \tag{5.11}$$

where P_S^* is a constant and since $0 < a < 1$ the total permeability will decrease with increased temperature meaning that any increase in the diffusivity is counteracted by a decrease in surface concentration [30].

5.4 Knudsen Diffusion

Knudsen diffusion [20,30,32,37–40] depending on gas pressure and mean free path in the gas phase applies to pores between 10 Å and 500 Å in size; however, there are examples in the literature where it was observed for much larger pores [41]. In this region, the mean free path of molecules in gas phase λ is much larger than the pore diameter d. It is common to use the so-called Knudsen number $K_n = \lambda/d$ to characterize the regime of permeation through pores. When $K_n << 1$, viscous (Poiseuille) flow is realized. The condition for Knudsen diffusion is $K_n >> 1$. An intermediate regime is realized when $K_n \approx 1$.

The Knudsen diffusion coefficient can be expressed in the following form

$$D_K = \frac{d}{3\tau}\bar{u} \tag{5.12}$$

where τ is the pore tortuosity and $\bar{u}$ is the average molecular speed. This expression shows that the separation outcome should depend on the differences in molecular speed (or molecular mass). The average molecular speed $\bar{u}$ is calculated using the Maxwell speed distribution,

$$\bar{u} = \sqrt{\frac{8RT}{\pi m}} \tag{5.13}$$

where m is the molecular mass and hence the diffusion coefficient can be presented as

$$D_K = (d/3\tau)(8RT/\pi m)^{1/2} \tag{5.14}$$

For the flux in the Knudsen regime the following equation holds [42,43]:

$$J = n\pi d^2 \Delta p D_K / 4RTL \tag{5.15}$$

where n is the surface concentration of the pores with diameter d, Δp is the pressure drop across the membrane and L is the membrane thickness. After substituting Equation (5.14) into Equation (5.15), one has the following expressions for the flux J and permeability coefficient P:

$$J = (n\pi^{1/2} d^3 \Delta p / 6\tau L)(2/mRT)^{1/2} \tag{5.16}$$

$$P = (n\pi^{1/2} d^3 / 6\tau)(2/mRT)^{1/2} \tag{5.17}$$

Two important conclusions can be made from analysis of Equations (5.16) and (5.17). First, selectivity of separation in Knudsen regime is characterized by the ratio $\alpha_{ij} = (M_j/M_i)^{1/2}$. It means that membranes where Knudsen diffusion predominates are poorly selective. For example, separation factor of separation of O_2/N_2 pair is 1.07. The highest gas separation selectivity can be observed in separation of lightest and heaviest gases, e.g. hydrogen and butane: in this case $\alpha = 5.4$. Another unusual feature of Knudsen diffusion is that increases in temperature result in slight decreases of the flux and permeability coefficient, as J and P depend on temperature as $T^{1/2}$. Numerous confirmation of this dependence can be found in the literature.

5.5 Membranes: Porous Structures?

The available range of membrane materials includes polymeric, carbon, silica, zeolite and other ceramics, as well as composites. Each type of membrane can have a different porous structure, as illustrated in Figure 5.2. Membranes can be thought of as having a fixed (immovable) network of pores in which the gas molecule travels, with the exception of most polymeric membranes [28,44]. Polymeric membranes are composed of an amorphous mix of polymer chains whose interactions involve mostly van der Waals forces. However, some polymers reveal a behaviour that is consistent with the idea of existence of opened pores within their matrix. This is especially true for high free volume, high

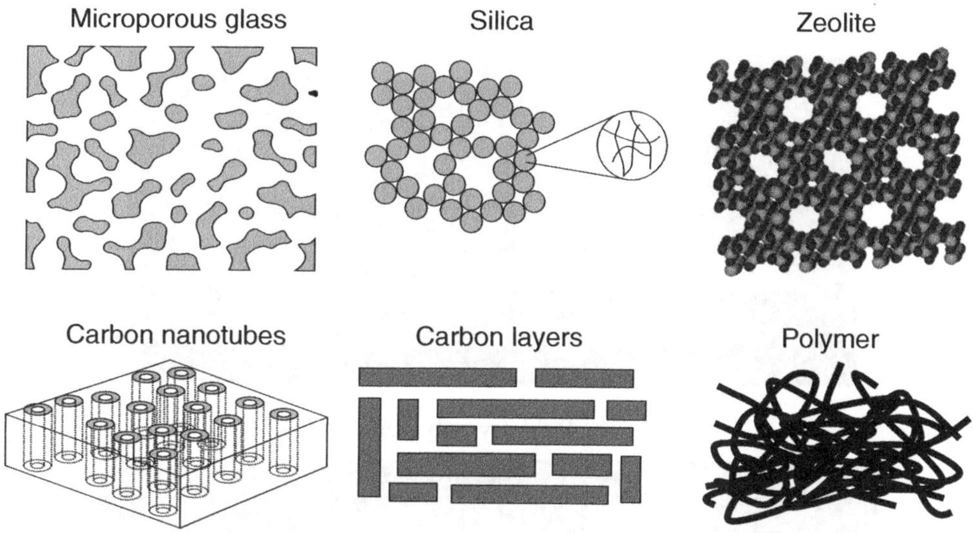

Figure 5.2 Porous structure within various types of membranes [3,22,37]. Microporous glass figure from [22], reprinted with permission of John Wiley & Sons, Inc. Silica figure from [3], reprinted with permission of Wiley-VCH Verlag GmbH & Co. KGaA. Carbon nanotubes figure reprinted with permission from Science, Aligned multiwalled carbon nanotube membranes, by B. J. Hinds, N. Chopra, T. Rantell, R. Andrews, V. Gavalas and L. G. Bachas, 303, 62–65. Copyright (2004) American Association for the Advancement of Science.

permeability polymers like poly(trimethylsilylpropyne), as has been proved by computer modelling, low activation energy of diffusion, negative activation energy of permeation, solubility controlled permeation in this and similar polyacetylenes [10,25] (see also Chapters 2 and 3 of this book). Although polymeric membranes have often been viewed as non-porous, in the modelling framework discussed here it is convenient to consider them nonetheless as porous. Glassy polymers have pores that can be considered as 'frozen' over short times scales, as demonstrated in Figure 5.3a, while rubbery polymers have dynamic fluctuating pores (or more correctly free volume elements) that move, shrink, expand and disappear, as illustrated in Figure 5.3b [14].

5.6 Transition State Theory (TST)

The diffusion of molecules within porous networks similar to that of microporous silica and non-porous glassy polymers can be modelled within the framework of the so-called transition state theory [17,22,28,33,45]. A gas molecule bounces around in a reactant cavity eventually bouncing towards the transition state by which it transports through to the product cavity and therefore successfully makes a diffusive jump, as demonstrated in Figure 5.4a. Within glassy polymers, see Figure 5.4b, the transition state is a dynamical section that becomes available through polymer chain motions. Within microporous silica, see Figure 5.4c, the transition state is a permanent pathway for the transport of the gas molecule. The transition state theory offers a method to express the rate of diffusion D (or diffusivity) within these porous networks in the following way:

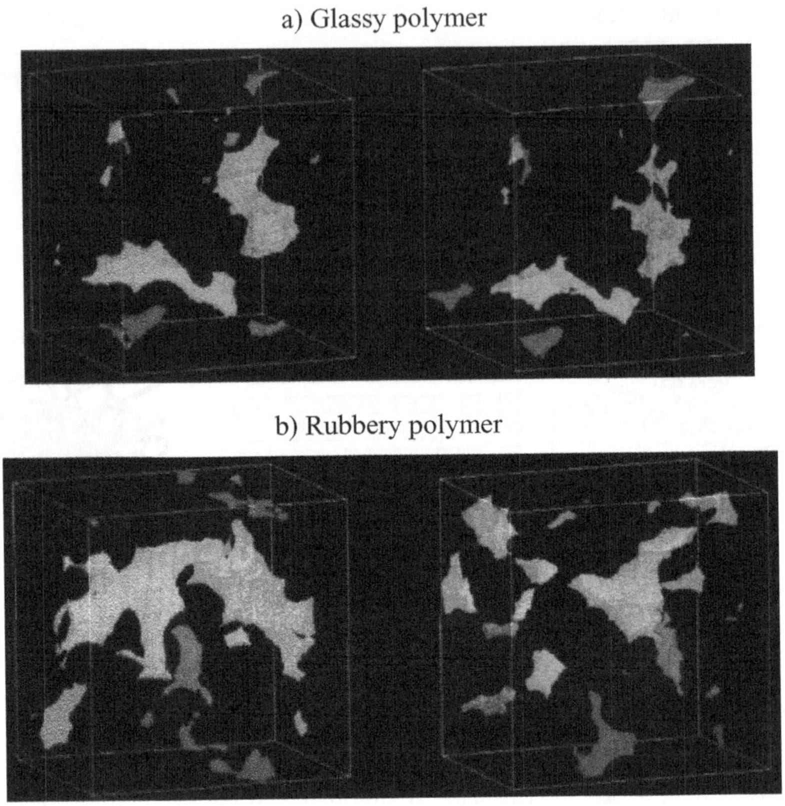

Figure 5.3 *Computer simulations performed by Greenfield and Theodorou [14] for free volume clusters before and after 10^7 Monte Carlo steps within (a) glassy polymer and (b) rubbery polymer. Reprinted with permission from Macromolecules, Geometric analysis of diffusion pathways in glassy and melt atactic polypropylene by M. L. Greenfield and D. N. Theodorou, 26, 5461–5472. Copyright (1993) American Chemical Society*

D = the probability that the molecule will travel towards a transition (ρ_g)
× the probability that the molecule will pass through the transition (ρ_E)
× the velocity of the molecule through the transition (u)
× the jump length from the reactant cavity to the product cavity (λ).

This formula

$$D = \rho_g \rho_E u \lambda$$

provides some insight into the major factors contributing to the separation of particular molecules. If the transition state has the form of a narrow constriction then the smaller molecules are more likely to pass through and therefore have a higher rate of diffusion than their larger counterparts. On the other hand, if the transition state is wide enough for both molecules to freely pass through, then the velocity at which they travel may be the

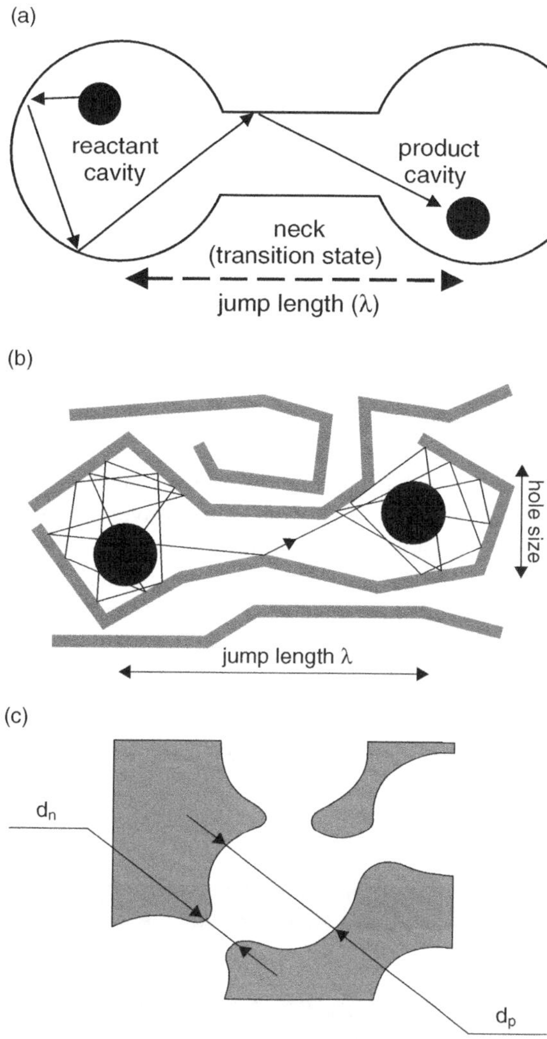

Figure 5.4 *Transition State Theory for diffusion in condensed media. (a) General representation of the transition state theory. (b) Diffusive jump in glassy polymer [17]. Reprinted from Journal of Membrane Science,* ***73****, E. Smit, M. H. V. Mulder, C. A. Smolders, H. Karrenbeld, J. van Eerden and D. Feil, Modelling of the diffusion of carbon dioxide in polyimide matrices by computer simulation, 247–257, Copyright (1992), with permission from Elsevier. (c) Diffusive jump in microporous silica, reprinted with permission from AIChE, Theory of gas diffusion and permeation in inorganic molecular-sieve membranes by A. B. Shelekhin, A. G. Dixon and Y. H. Ma, 41, 58–67, Copyright (1995) AIChE*

dominant factor in determining the rate of diffusion. Further, within glassy polymers the rate of diffusion could be dominated by the rate of polymer chain movements in the walls of free volume elements or closed pores which occasionally provide a transition pathway for the molecules.

5.7 Transport Models for Ordered Pore Networks

Membranes with ordered structures such as zeolites or nanotubes have considerable potential as gas separation membranes [46–48]. In addition to having thermal and chemical stability, the porosity of these structures is ordered, and therefore there is usually more control over the separation properties. The pores within these structures are such that gas transport can not be completely explained by the transition state theory. This is because, in nanotubes for example, there is only one transition, from outside of the tube to inside of the tube. Two alternative models are outlined here, the parallel transport model and the resistance in series transport model, which are illustrated in Figure 5.5, and they are explained in detail by the work of Gilron and Soffer [27].

5.7.1 Parallel Transport Model

The parallel transport model considers the total flux as the contribution from the molecules travelling via surface diffusion and from the molecules travelling via Knudsen diffusion [27,36,49,50]. This model does not consider transition stages and is applicable to pores that remain roughly the same size throughout the entire membrane such as nanotube-based membranes. Gilron and Soffer [27] presented the following expression,

$$P_{tot} = P_S + P_K \tag{5.18}$$

where P_S is the surface diffusion permeability and P_K is the Knudsen diffusion permeability as defined earlier.

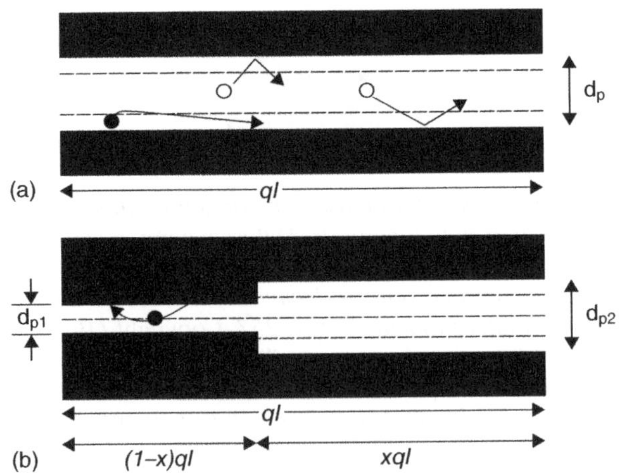

Figure 5.5 Schematic models for (a) parallel transport and (b) resistance in series transport [27]. Reprinted from Journal of Membrane Science, **209**, J. Gilron and A. Soffer, Knudsen diffusion in microporous carbon membranes with molecular sieving character, 339–352, Copyright (2002), with permission from Elsevier

5.7.2 Resistance in Series Transport Model

Resistance model for transport in composite hollow fibre membranes based on polysulfone with siloxane coating has been described in a classical work by Henis and Tripody [51]. The resistance in series model assumes that the gas molecules encounter constrictions at certain positions throughout the pore which control the rate of diffusion [20,27,33]. For this scenario the total permeability is inversely related to the total resistance, thus

$$R_{tot} = \frac{l}{P_{tot}} = \frac{x_K l_\tau}{P_K} + \frac{(1-x_K) l\tau}{P_A} \tag{5.19}$$

where x_K is the fraction of the total pore length l with pore size d_{p2} in which Knudsen diffusion dominates while $(1 - x_K)$ is the fraction of the total pore length l with pore size d_{p1} in which activated diffusion dominates and τ is the pore tortuosity (see Figure 5.5b). The total permeability simplifies to give

$$P_{tot} = \frac{P_A P_K}{\tau(x_K P_A + (1-x_K) P_K)} \tag{5.20}$$

where P_A is the activation diffusion permeability and P_K is the Knudsen diffusion permeability as defined earlier.

5.8 Pore Size, Shape and Composition

To understand gas transport phenomenon it is critical to consider the interactions between the gas molecules and the pore wall. The van der Waals interactions between particles are explained well by the Lennard-Jones function containing two parameters, the kinetic diameter σ (the distance where the potential energy between the particles is zero) and the well depth ε (the deepest potential minimum between the particles). These parameters were used, e.g. by Freeman [52], to establish a theoretical basis for the relationship between selectivity and permeability for a range of polymers and gases (so-called first upper bound empirically determined by Robeson in 1991 [53]; see also more recent paper by Robeson [54]). The rate of diffusion of a gas is dependent on its kinetic diameter while its solubility mainly depends on the condensability of the gas and consequently on the well depth for gas-gas interactions.

Gas–pore wall interactions have been considered to identify different pore size regimes for gas adsorption by Everett and Powl [31] and later modified to determine gas separation scenarios by de Lange *et al.* [30]. Figure 5.6 shows the potential energy within slit-shaped pores. A deep single minimum occurs within small pores and the shallower double minimum occurs in larger pores, as calculated by Everett and Powl [31]. This potential energy can be thought of as the adsorption energy which is enhanced at an optimal pore size, indicated by the peaks in Figure 5.7 for cylindrical and slit-shaped pores. Everett and Powl [31] used these calculations in order to further understand adsorption of the noble gases within microporous carbons. One of the key points outlined in Everett and Powl's work is that the separation outcomes may be predicted by comparing the potential energy curves of the particular gases.

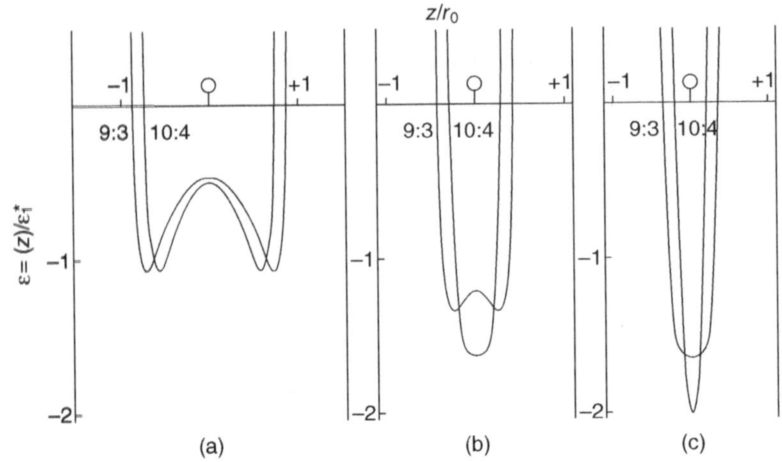

Figure 5.6 *Potential energy ε=(z) between two parallel planes (10:4) and two parallel slabs (9:3) at a distance apart of 2*d *for (a)* $d/r_0 = 1.60$*, (b)* $d/r_0 = 1.14$ *and (c)* $d/r_0 = 1.00$*, normalized by the energy minimum* ε_1^* *located at a distance of* r_0 *from a single slab. Reprinted with permission from Journal of the Chemical Society; Faraday Transactions 1, Adsorption in slit-like and cylindrical micropores in the Henry's law region. A model for the microporosity of carbons, by D. H. Everett and J. C. Powl, 72, 619–636, Copyright (1976) Royal Society of Chemistry*

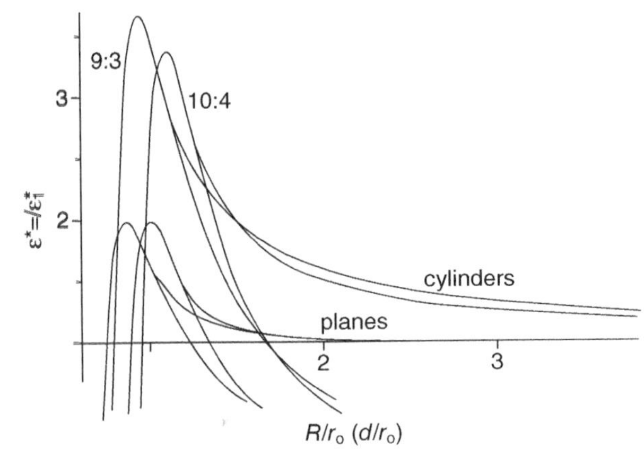

Figure 5.7 *Scaled potential energy minima* $\varepsilon_=^*/\varepsilon_1^*$ *within cylindrical and slit-shaped pores with varying radius* R *and slit size 2*d*, respectively, where* $\varepsilon_=^*$ *is the minimum potential within the pore and* ε_1^* *is the minimum potential with a single flat surface. Curves that go below the horizontal axis are the scaled potentials within the centre of the pore* $\varepsilon(0)/\varepsilon_1^*$ *where the potential in the centre ε(0) becomes less than the minimum potential with a single flat surface* ε_1^**, i.e.* $\varepsilon(0)/\varepsilon_1^* < 1$*, within larger pores. Reprinted with permission from Journal of the Chemical Society; Faraday Transactions 1, Adsorption in slit-like and cylindrical micropores in the Henry's law region. A model for the microporosity of carbons by D. H. Everett and J. C. Powl, 72, 619–636, Copyright (1976) Royal Society of Chemistry*

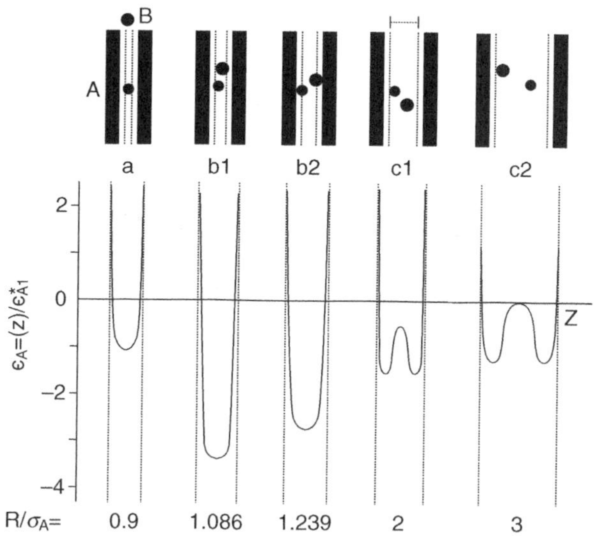

Figure 5.8 *Separation regimes determined by the potential energies within pores of different sizes. Potential energy* $\varepsilon_A^{=}(z)$ *for molecule A within cylindrical pores with radius* R, *scaled by the potential minimum* ε_{A1}^{*} *for molecule A with a single free surface and the Lennard-Jones kinetic diameter parameter* σ_A *[30]. Reprinted from Journal of Membrane Science,* ***104****, R. S. A. de Lange, K. Keizer and A. J. Burggraaf, Analysis and theory of gas transport in microporous sol-gel derived ceramic membranes, 81–100, Copyright (1995), with permission from Elsevier*

de Lange *et al.* [30] later extended the work of Everett and Powl [31] by relating transport mechanisms to potential energy calculations. Figure 5.8 demonstrates the separation scenarios within cylindrical shaped pores. Situation 'a' is where molecule A is accepted within the pore while larger molecule B is rejected by the repulsive forces experienced. This refers to true molecular sieving or size-sieving. Situations 'b1' and 'b2' are where both molecules are accepted within the pore but molecule A has a much deeper potential than molecule B. Since the pore is cylindrical, molecules may not pass each other and therefore the rate of diffusion is governed by the slowest component. Situations 'c1' and 'c2' are where molecules may pass each other and the potential energy becomes weaker having less influence on transport. In Ref. 30 these scenarios are combined with an extensive model that incorporates different stages of transport through the membrane and existing transport equations.

The extensive model considers the total flux as the sum of contributions of the flux at different stages, indicated schematically in Figure 5.9 and composed of the following.

1 Adsorption onto surface and flux from position $\theta_{0,\mathrm{surf}}$ to θ_0 at the pore entrance via surface diffusion (f2.J).
2 Adsorption directly at the pore entrance at position θ_0 (f1.J).
3 Flux directly to pore entrance with no adsorption taking place (F1.J).
4 Entrance of adsorbed molecules at position θ_0 to position θ_1 within the pore (F2.J).

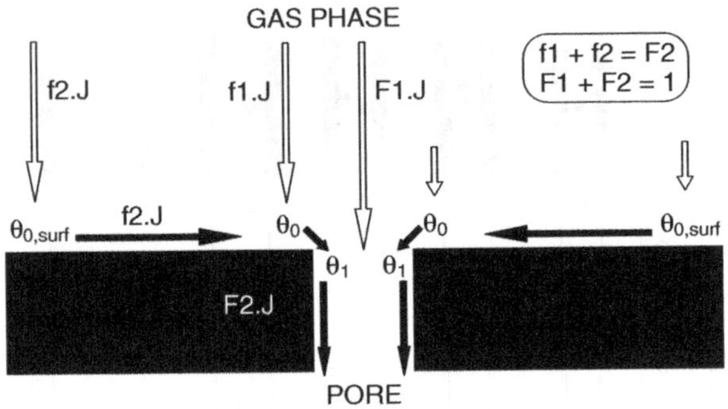

Figure 5.9 Schematic model of the total flux components through microporous membranes [30]. Reprinted from Journal of Membrane Science, **104**, R. S. A. de Lange, K. Keizer and A. J. Burggraaf, Analysis and theory of gas transport in microporous sol-gel derived ceramic membranes, 81–100, Copyright (1995), with permission from Elsevier

5 Micropore diffusion through the pores (J).
6 Desorption of the molecules from within the pore to the external surface or directly to the gas phase.
7 Desorption from the external surface to the gas phase.

5.9 The New Model

Recent work has combined existing transport theories to develop a model through which separation performance may be predicted as a function of pore size, shape and composition, and temperature [23]. As a molecule approaches a pore opening each atom in the gas molecule will interact with every atom making up the pore wall through van der Waals forces. The difference between the potential energy within the pore E_{in} and outside the pore E_{out} is useful for determining the dominant transport mechanism as well as the necessary driving force–energy barrier component. By assuming that the pore is cylindrically shaped the potential energy difference $W = E_{out} - E_{in}$ can be expressed as

$$W_{cyl} = E_{out} - \frac{48\varepsilon\pi^2\eta}{d^4}\left(\frac{42\sigma^{12}}{d^6} - \sigma^6\right) \tag{5.21}$$

where $\varepsilon = \sqrt{\varepsilon_g \varepsilon_s}$ is the well depth and $\sigma = (\sigma_g + \sigma_s)/2$ is the kinetic diameter that determines the forces between the gas molecule g and the pore surface atoms s, which is defined using the Lorentz–Berthelot mixing rules. The pore size d is defined here as the distance between the surface nuclei while characterization methods such as PALS use a pore size d^* which considers the electron cloud thickness δd surrounding the surface atoms such that, $d = d^* + \delta d$, where $\delta d = 3.32$ Å [55,56]. Similarly, by assuming a slit-shaped pore the potential difference can be expressed as

$$W_{slit} = E_{out} - 8\pi\eta\varepsilon\sigma^2\left[\frac{1}{5}\left(\frac{2\sigma}{d}\right)^{10} - \frac{1}{2}\left(\frac{2\sigma}{d}\right)^4\right] \tag{5.22}$$

Table 5.2 Lennard-Jones constants, molecular masses and average velocities at room temperature used throughout this chapter [57–59]

Gas / Pore atoms	σ (Å)	ε / k_B (K)	m (g/mol)	$\bar{v}$(m/s)
from UFF [59]				
C	3.43	53	12.01	–
H	2.57	22	1.01	–
O	3.12	30	16.00	–
N	3.26	35	14.01	–
Si	3.83	202	28.09	–
from Breck [57]				
He	2.60	10	4.00	1277
H_2	2.89	60	2.02	1800
CO_2	3.30	195	44.01	385
O_2	3.46	107	32.00	452
N_2	3.64	71	28.01	483
CH_4	3.87	149	16.04	638
from Poling [58]				
CO	3.69	92	25.01	476
Ar	3.54	93	39.95	399
n-C_5H_{12}	5.78	341	72.15	297
C_2H_6	4.44	216	30.07	455
SF_6	5.13	222	146.06	209

The parameters utilized for this approach are from Breck [57] for the gases He, H_2, CO_2, O_2, N_2 and CH_4, Poling [58] for the gases CO, Ar, C_2H_6, n-C_5H_{12} and SF_6, and the universal force field (UFF) values [59] are used for the surface atoms, as summarized in Table 5.2. This potential difference has been termed the suction energy since a positive W translates to a suction force of the molecule from the outside to the inside of the pore, while a negative W translates to a repulsive force directing the molecule away from the pore [23]. From this, a new transport mechanism is proposed in [23] as 'suction diffusion', where enhanced velocities are predicted as the gas molecules are sucked into the pore.

In Figure 5.10 an example of W is demonstrated for a single oxygen molecule entering a carbon cylindrical pore from outside where the potential energy is zero, i.e. $E_{out} = 0$. This single curve can be used to distinguish the pore size regions for which each transport mechanism dominates. Additionally, W can be used to estimate the energy barrier for activated diffusion ΔE_a (= $|W|$) and the energy barrier for surface diffusion ΔE_S (= $aq = aW$). The critical pore sizes d_{min} and d_K distinguish between the regions where three different diffusion mechanisms dominate the transport, namely, activated diffusion ($W < 0$), surface diffusion ($0 < W < RT$) and Knudsen diffusion ($W > RT$).

5.9.1 Enhanced Separation by Tailoring Pore Size

The three most common diffusion mechanisms known as activated diffusion, surface diffusion and Knudsen diffusion, usually dominate in small pores ($d^* < 3$ Å), medium pores (3 Å $< d^* <$ 10 Å) and large pores (10 Å $< d^* <$ 500 Å) for light gases, respectively. Separation by differences in diffusivity and/or differences in solubility can be enhanced

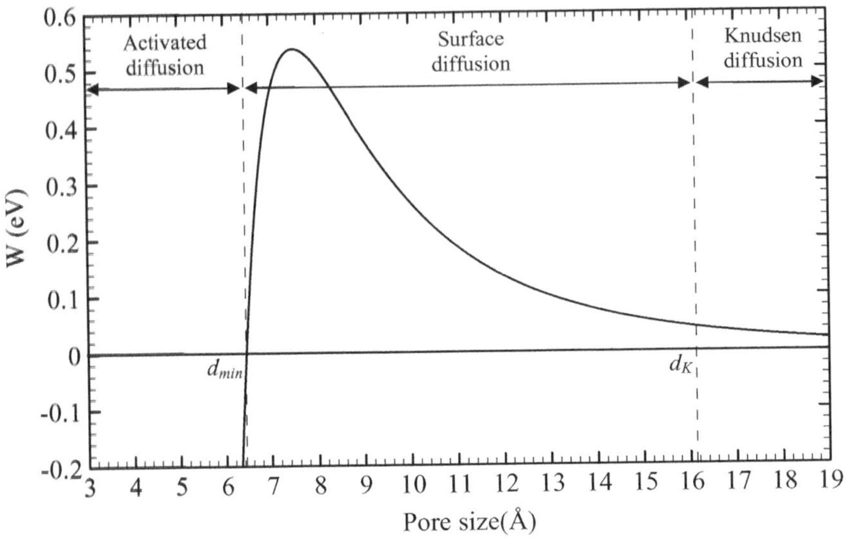

Figure 5.10 *Potential energy difference (W) for a single oxygen molecule at the entrance of a carbon cylindrical pore of diameter* d. *The pore regions where the diffusion mechanisms (activated diffusion, surface, and Knudsen flow) dominate are separated by the critical pore sizes* d_{min} *(where W = 0) and* d_K *(where W = 0.04 eV), indicated by dashed lines*

by tailoring the pore size such that the differences are maximized. The greatest separations are usually achieved when the competing gases are in different modes of transport. Thus it is important to know the critical pore sizes that distinguish the different diffusion mechanisms for each gas.

The critical pore sizes are summarized in Table 5.3 for the light gases He, H_2, O_2, N_2 CO_2, and CH_4, entering carbon and silica pores of cylindrical and slit shape. Additionally, Table 5.3 includes the results for carbon monoxide (a key component of synthesis gas), argon (an inert gas frequently used in industrial processes), ethane and n-pentane (hydrocarbons present in fossil fuels), and sulfur hexafluoride (the most potent greenhouse gas according to the Intergovernmental Panel on Climate Change [60]; in permeation studies SF_6 is often considered as a penetrant with an unusually large size). The results can be used as a guide for pore size design of a membrane according to the desired gas separation application. For example, if the application was natural gas purification (separation of CO_2 from CH_4) then the pore size range that allows CO_2 through while rejecting CH_4 can be found from Table 5.3 (carbon tube: 2.95–3.49 Å; silica tube: 3.17–3.69 Å; carbon slit: 2.46–2.95 Å; silica slit: 2.65–3.13 Å). Further, by using the transport equations (outlined earlier) it is possible to determine the pore size necessary to achieve a desired permeability and selectivity at the specified operating temperature, demonstrated later in this chapter.

The first observation to be made from the results in Table 5.3 is that the minimum pore sizes for barrier-free transport d_{min} of each gas are in the same order as the kinetic diameter with slightly different values because the model takes into account the interaction with

Table 5.3 *Minimum pore size for barrier-free transport (d_{min}) and minimum pore size for Knudsen diffusion (d_K) at room temperature (298 K). All pore sizes are given as the PALS pore size* d*

(A) Cylindrical pore

Gas	Carbon		Silica	
	d_{min} (Å)	d_K (Å)	d_{min} (Å)	d_K (Å)
He	2.30	5.69	2.53	6.37
H_2	2.57	8.92	2.79	9.79
CO_2	2.95	12.30	3.17	13.35
O_2	3.10	11.68	3.31	12.68
N_2	3.27	11.49	3.48	12.46
CH_4	3.49	13.74	3.69	14.84
CO	3.32	12.14	3.52	13.16
Ar	3.18	11.68	3.39	12.67
n-C_5H_{12}	5.27	23.51	5.45	25.04
C_2H_6	4.02	16.69	4.21	17.93
SF_6	4.66	19.46	4.85	20.81

(B) Slit-shaped pore

Gas	Carbon		Silica	
	d_{min} (Å)	d_K (Å)	d_{min} (Å)	d_K (Å)
He	1.86	2.19	2.06	4.35
H_2	2.10	6.52	2.31	7.22
CO_2	2.46	9.27	2.65	10.12
O_2	2.59	8.76	2.79	9.58
N_2	2.75	8.60	2.94	9.39
CH_4	2.95	10.43	3.13	11.32
CO	2.79	9.14	2.98	10.00
Ar	2.66	8.76	2.86	9.56
n-C_5H_{12}	4.59	18.33	4.75	19.57
C_2H_6	3.44	12.82	3.62	13.82
SF_6	4.03	15.06	4.20	16.15

the pore wall and not kinetic size only. This means that the model is a more accurate method for predicting whether a gas molecule will experience an energy barrier or not, consequently predicting the mode of transport. Another important observation is that the model predicts that Knudsen diffusion occurs in different pore size regions for each gas. For example, within a 12 Å sized pore, the model predicts that helium and hydrogen will be in Knudsen flow while all the other light gases will not. Excellent agreement between experimental separation results and the model predictions is found [23].

5.9.2 Determining Diffusion Regime from Experimental Flux

A widely used experimental approach that involves the determination of the mode of transport is performed by fitting permeation data to the Arrhenius equation for the flux or permeance $J = J_0\exp(-\Delta E/RT)$. As the flux J and permeability coefficient P are related as $J = P\Delta p/L$, where Δp is the pressure gradient and L is the membrane thickness, the above permeability expressions in the previous section provide the decision criteria. If ΔE is positive then the dominant transport mechanism is size-sieving activated diffusion with ΔE representing the energy barrier $|W|$ for $d^* < d_{min}$. If ΔE is negative then the dominant transport mechanism is surface diffusion with ΔE representing the weighted contribution of adsorption energy (or potential energy difference) for the energy barrier and surface concentration $(a - 1)W$. If ΔE is zero, i.e. no change in permeability with varying temperature, then either the size-sieving energy barrier and the enthalpy of adsorption are equal to zero ($W = 0$) or the surface diffusion energy barrier and the heat of adsorption are both equal ($a = 1$). If the Arrhenius equation does not fit the data then the mode of transport could be Knudsen flow, or a combination of all mechanisms since the dominant mode of transport can change as the temperature is varied. The above criteria will be a useful tool in understanding the mode of transport for each gas in a range of porous membrane materials.

5.9.3 Predictions of Gas Separation

In this section predictions are made using the model to demonstrate the different transport behaviours with varying pore size and temperature. Figure 5.11 predicts the permeability P as a function of pore size d for the different transport mechanisms, described earlier, with Equation (5.21) as the definition for W. For activated diffusion $\Delta E_a = |W|$ and for surface diffusion $\Delta E_S = aq = aW$. The results represent the permeability within a single pore. In reality there will be a distribution of pore sizes, and therefore the transition

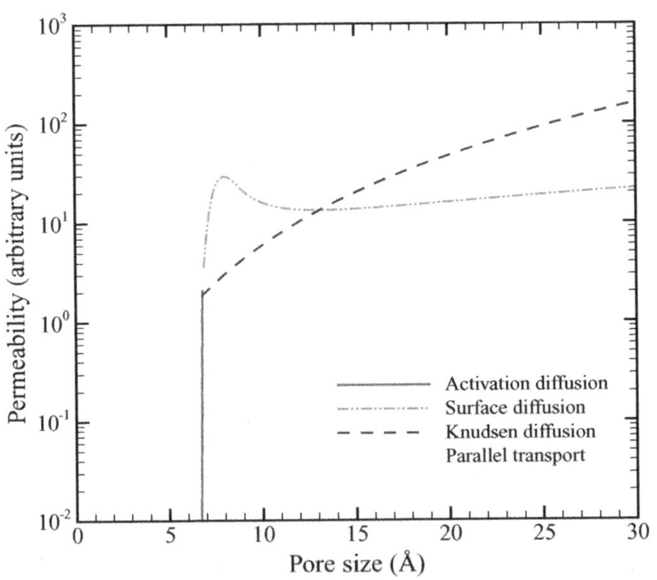

Figure 5.11 *Model prediction of normalized permeability* P *as a function of pore size (distance between surface nuclei,* d*). Modes of transport are indicated*

between activated diffusion and the other modes of transport will usually be smooth. Each mode of transport is scaled arbitrarily such that the trends may be clearly seen and therefore the magnitude is insignificant. The permeability for activated diffusion is a sharply increasing function of pore size as the energy barrier changes dramatically with pore size.

As explained earlier, the permeability for surface diffusion is dominated by the surface concentration and therefore the model predicts a peak at which the heat of adsorption is maximized. The permeability for Knudsen diffusion increases as a cubic function of pore size (see Equations 5.16 and 5.17) since the diffusivity depends linearly on the pore size d and the concentration depends on the volume of the pore d^2 (assuming cylindrical pores). The parallel transport model assumes that surface diffusion and Knudsen diffusion are occurring simultaneously such that the total permeability is given by Equation (5.18). This model is explained in further detail earlier in this chapter and has been used by various groups [27,49,50]. Parallel transport is initially dominated by surface diffusion within the smaller pores where the surface concentration is high while the mode of Knudsen diffusion dominates within the larger pores.

Permeability varies with temperature for each transport mechanism as demonstrated in Figure 5.12. The permeability for activation diffusion is the only increasing function with respect to increasing temperature. When gases are in the mode of surface diffusion, the surface concentration decreases more than the increase in surface mobility, resulting in an overall decrease in permeability with increasing temperature. Knudsen diffusion displays decreasing permeability with increasing temperature as the concentration loss dominates the increase in diffusivity. As temperature increases, the surface diffusion part

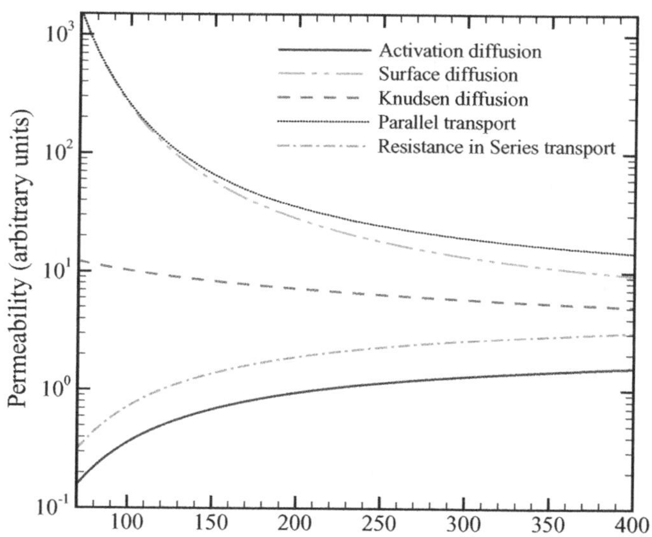

Figure 5.12 *Model prediction of permeability as a function of temperature. Modes of transport are indicated for the following pore sizes: activated diffusion (d = 6.8Å), surface diffusion (d = 10Å), Knudsen diffusion (d = 10Å), parallel transport (d = 10Å), and resistance in series transport (d_{small} = 6.8Å, d_{large} = 10Å, x_K = 0.8)*

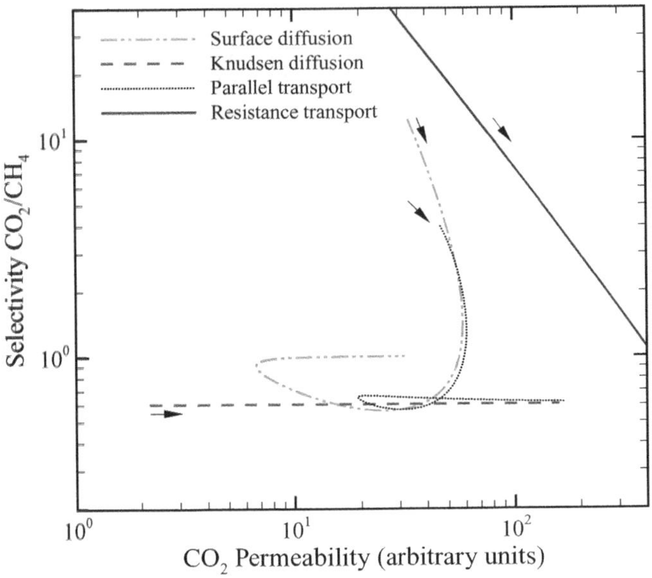

Figure 5.13 *Model predictions of CO_2/CH_4 selectivity versus CO_2 permeability for varying pore size* d. *Arrows indicate the direction of increasing pore size. The pore size range, 7.22 > d > 30 Å, was chosen for all the modes of transport, apart from the resistance in series transport where the constriction size varied while the large pore size remained constant (d_{small} = 5–7 Å, d_{large} = 10 Å, $x_K = 0.99$, Equation 5.20)*

of the parallel transport model has less influence causing the permeability to tend towards a Knudsen-type transport at high temperatures. The resistance in series transport model, detailed earlier in this chapter, assumes that the diffusing molecules travel through pores in the mode of Knudsen diffusion while occasionally encountering constrictions where activation diffusion occurs. The total permeability is therefore expressed by Equation (5.20), where x_K is the fraction of the pore length where Knudsen diffusion occurs. As shown in Figure 5.12, the total permeability predicted by the resistance in series model behaves mostly in accordance with the mode of activated diffusion even for a small fraction of constrictions ($1 - x_K$).

In the interest of gas separation, the model prediction for CO_2/CH_4 selectivity versus CO_2 permeability has been calculated with varying pore size and temperature, and the results are shown in Figures 5.13 and 5.14, respectively. Equation (5.21) is used for W_{cyl} with the parameter values taken from Table 5.2. Since the resistance in series transport behaves like activated diffusion, the predictions for activation diffusion have been omitted from the plot. The selectivity is high for small pores with surface diffusion as the transport mechanism where permeability is dominated by the concentration component for which CO_2 forms denser surface layers than CH_4. As pores become larger, the enthalpy of adsorption results in a maximum CO_2 permeability, followed by a decrease in the enthalpy of adsorption leading to a surface concentration loss. In this case, a single pore is considered and therefore the surface concentration eventually increases with increasing pore size according to the surface area of the cylindrical pore with the density of both gases tending toward that upon a flat surface. The Knudsen diffusion selectivity favours

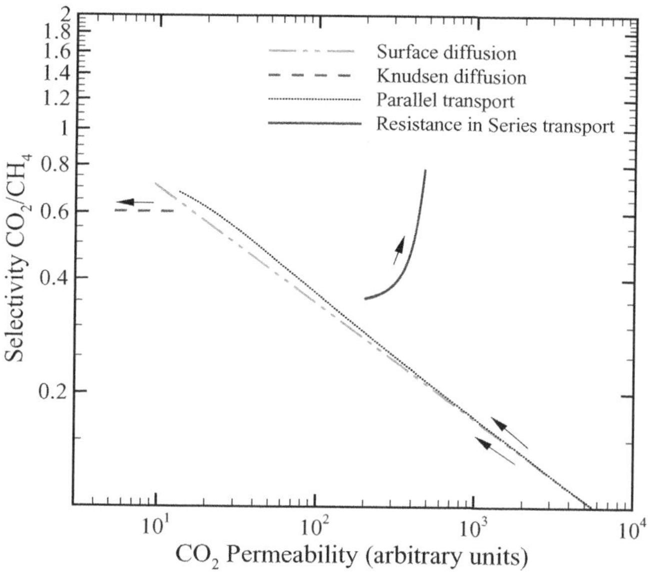

Figure 5.14 *Model predictions of CO_2/CH_4 selectivity versus CO_2 permeability for varying temperature T, within a pore of size d = 10 Å for all mechanisms expect for the resistance in series transport for which a pore size of d = 6.5 Å was chosen. Arrows indicate the direction of increasing temperature, from 70 to 500 K*

CH_4 because of its lighter mass resulting in a higher molecular velocity and does not change with pore size. Parallel transport follows the same trend as surface diffusion in small pores and tends toward Knudsen behaviour as the pore sizes increase. Finally, the resistance in series transport model predicts a decrease in selectivity as the permeability of CH_4 increases more rapidly than for CO_2 with increasing pore size.

As seen in Figure 5.14, the selectivity is predicted to slightly increase with increasing temperature when gases are in the mode of surface diffusion. This is due to the larger adsorption energy that CH_4 experiences over CO_2 for this particular pore size. Knudsen selectivity does not depend on temperature. Parallel transport is dominated by the surface diffusion component at low temperatures and gradually becomes more dependent on the Knudsen diffusion component at high temperatures. Note that the trends will be different depending on the pore size. For example, in the case of $d = 8$ Å, the selectivity is predicted to decrease as the temperature increases. The resistance in series transport demonstrates an increase in selectivity as the CO_2 permeability increases more than the CH_4 permeability with increasing temperature, as a consequence of the lower energy barrier that CO_2 experiences for this particular pore size ($d = 6.5$ Å).

5.10 Conclusion

Tailoring the pore size distribution within materials has been suggested as a means of fine tuning the transport properties in membranes [61,62]. To facilitate the development of gas separation membranes, the modelling of the transport of individual gas molecules through pores has been undertaken by various groups. The interactions between the gas molecules and the pore wall at the pore opening are considered to be of great importance and from which certain information about the gas kinetics can be obtained. A combination of theories provides the theoretical determination of the size of pores in which different modes of diffusion may dominate for each gas. Critical pore sizes d_{min} and d_K indicate the division of the three diffusion regimes, namely, activation diffusion, surface diffusion and Knudsen diffusion. By using this model one can predict the separation outcome for a variety of membranes in which the pore shape, size and composition are known, and conversely one can predict pore characteristics with known permeation rates. Further, one can also have a desired separation in mind and use this model to guide the design of the micro-structure of the membrane material.

List of symbols

J	flux or permeation rate
D	diffusivity
C	gas concentration
C_2 ,C_1	upstream and downstream gas concentration
p_2, p_1	upstream and downstream pressure
L	membrane thickness
P	permeability
S	solubility
η	atomic surface density
σ	kinetic diameter
ε	well depth
d	theoretical pore size (between surface nuclei)
d^*	experimental PALS pore size (d minus electron cloud thickness)
W	difference in the potential energies outside and inside of the pore
λ	mean free path
d_{min}	minimum pore size for barrier-free transport
d_K	minimum pore size for dominant Knudsen transport
ΔE_a	energy barrier for activation diffusion
ΔH_a	enthalpy of adsorption for activation diffusion
ΔE_S	energy barrier for surface diffusion
T	temperature
n	surface concentration of pore entrance sites
v_f	free volume
q	heat of adsorption
K	Henry's law coefficient, $K = K_0\exp(q/RT)$, where K_0 is a constant
m	mass of gas molecule
τ	pore tortuosity
$\bar{u}$	average molecular gas speed
ρ_g	probability that the molecule will travel towards a transition
ρ_E	probability that the molecule will pass through the transition

u	velocity of the molecule through the transition
λ	jump length from the reactant cavity to the product cavity
R	resistance (=L/P)
l	pore length
x_K	fraction of pore length in which Knudsen dominates
$\varepsilon_{=}(z)$	potential energy between two parallel planes

References

(1) F. Peng, L. Lu, H. Sun, Y. Wang, J. Liu, Z. Jiang, Hybrid organic-inorganic membrane: solving the tradeoff between permeability and selectivity, *Chem. Mater.*, **17**, 6790–6796 (2005).

(2) Y. Yampol'skii, N. B. Bespalova, E. S. Finkel'shtein, V. I. Bondar, A. V. Popov, Synthesis, gas permeability, and gas sorption properties of fluorine containing norbornene polymers, *Macromolecules*, **27**, 2872–2878 (1994).

(3) M. C. Duke, S. J. Pas, A. J. Hill, Y. S. Lin, J. C. Diniz da Costa, Exposing the molecular sieving architecture of amorphous silica using positron annihilation spectroscopy, *Adv. Funct. Mater.*, **18**, 1–9 (2008).

(4) F. T. Willmore, X. Wang, I. C. Sanchez, Free volume properties of model fluids and polymers: Shape and connectivity, *J. Polym. Sci., Part B: Polym. Phys.*, **44**, 1385–1393 (2006).

(5) J. A. Hedstrom, M. F. Toney, E. Huang, H. C. Kim, W. Volksen, T. Magbitang, R. D. Miller, Pore morphologies in disordered nanoporous thin films, *Langmuir*, **20**, 1535–1538 (2004).

(6) H. Jinnai, Y. Nishikawa, S.-H. Chen, S. Koizumi, T. Hashimoto, Morphological characterization of bicontinuous structures in polymer blends and microemulsions by the inverse-clipping method in the context of the clipped-random-wave model, *Phys. Rev. E*, **61**, 6773–6780 (2000).

(7) P. K. Pujari, D. Sen, G. Amarendra, S. Abhaya, A. K. Pandey, D. Dutta, S. Mazumder, Study of pore structure in grafted polymer membranes using slow positron beam and small-angle X-ray scattering techniques, *Nucl. Instrum. Meth. B.*, **254**, 278–282 (2007).

(8) M. L. Greenfield, D. N. Theodorou, Description of small penetrant jump motions in a polymer glass using transition-state theory, *Polym. Mater. Sci. Eng.*, **71**, 407–408 (1994).

(9) X. Y. Wang, K. M. Lee, Y. Lu, M. T. Stone, I. C. Sanchez, B. D. Freeman, Cavity size distributions in high free volume glassy polymers by molecular simulation, *Polymer*, **45**, 3907–3912 (2004).

(10) D. Hofmann, M. Entrialgo-Castano, A. Lerbret, M. Heuchel, Y. Yampolskii, Molecular modeling investigation of free volume distributions in stiff chain polymers with conventional and ultrahigh free volume: Comparison between molecular modeling and positron lifetime studies, *Macromolecules*, **36**, 8528–8538 (2003).

(11) A. I. Skoulidas, D. S. Sholl, Self-diffusion and transport diffusion of light gases in metal-organic framework materials assessed using molecular dynamics simulations, *J. Phys. Chem. B.*, **109**, 15760–15768 (2005).

(12) L. Xu, T. T. Tsotsis, M. Sahimi, Nonequilibrium molecular dynamics simulation of transport and separation of gases in carbon nanopores. I. Basic results, *J. Chem. Phys.*, **111**, 3252–3264 (1999).

(13) M. L. Greenfield, D. N. Theodorou, Coupling of penetrant and polymer motions during small-molecule diffusion in a glassy polymer, *Mol. Simulat.*, **19**, 329–361 (1997).

(14) M. L. Greenfield, D. N. Theodorou, Geometric analysis of diffusion pathways in glassy and melt atactic polypropylene, *Macromolecules*, **26**, 5461–5472 (1993).

(15) R. C. Sivakumar, D. S. Sholl, J. K. Johnson, Adsorption and separation of hydrogen isotopes in carbon nanotubes: Multicomponent grand canonical Monte Carlo simulations, *J. Chem. Phys.*, **116**, 814–824 (2002).

(16) D. Hofmann, L. Fritz, J. Ulbrich, C. Schepers, M. Bohning, Detailed-atomistic molecular modelling of small molecule diffusion and solution processes in polymeric membrane materials, *Macromol. Theory Simul.*, **9**, 293–327 (2000).
(17) E. Smit, M. H. V. Mulder, C. A. Smolders, H. Karrenbeld, J. van Eerden, D. Feil, Modelling of the diffusion of carbon dioxide in polyimide matrices by computer simulation, *J. Membr. Sci.*, **73**, 247–257 (1992).
(18) X.-Y. Wang, P. J. in 't Veld, Y. Lu, B. D. Freeman, I. C. Sanchez, A molecular simulation study of cavity size distributions and diffusion in para and meta isomers, *Polymer*, **46**, 9155–9161 (2005).
(19) J. H. Zhou, R. X. Zhu, J. M. Zhou, M. B. Chen, Molecular dynamics simulation of diffusion of gases in pure and silica-filled poly(1-trimethylsilyl-1-propyne) [PTMSP], *Polymer*, **47**, 5206–5212 (2006).
(20) D. J. Way, D. L. Roberts, Hollow fiber inorganic membranes for gas separations, *Sep. Sci. Technol.*, **27**, 29–41 (1992).
(21) A. B. Shelekhin, A. G. Dixon, Y. H. Ma, Adsorption, permeation, and diffusion of gases in microporous membranes. II. Permeation of gases in microporous glass membranes, *J. Membr. Sci.*, **75**, 233–244 (1992).
(22) A. B. Shelekhin, A. G. Dixon, Y. H. Ma, Theory of gas diffusion and permeation in inorganic molecular-sieve membranes, *AIChE*, **41**, 58–67 (1995).
(23) A. W. Thornton, T. Hilder, A. J. Hill, J. M. Hill, Predicting gas diffusion regime within pores of different size, shape and composition, *J. Membr. Sci.*, **336**, 101–108 (2009).
(24) R. Krishna, J. A. Wesselingh, The Maxwell-Stefan approach to mass transfer, *Chem. Eng. Sci.*, **52**, 861–911 (1997).
(25) Yu. Yampol'skii, I. Pinnau, B. D. Freeman, Materials Science of Membranes for Gas and Vapor Separation, John Wiley & Sons, Ltd, 2006.
(26) J. G. Wijmans, R. W. Baker, The solution-diffusion model: a review, *J. Membr. Sci.*, **107**, 1–21 (1995).
(27) J. Gilron, A. Soffer, Knudsen diffusion in microporous carbon membranes with molecular sieving character, *J. Membr. Sci.*, **209**, 339–352 (2002).
(28) A. A. Gray-Weale, R. H. Henchman, R. G. Gilbert, M. L. Greenfield, D. N. Theodorou, Transition-state theory model for the diffusion coefficients of small penetrants in glassy polymers, *Macromolecules*, **30**, 7296–7306 (1997).
(29) R. M. de Vos, H. Verweij, Improved performance of silica membranes for gas separation, *J. Membr. Sci.*, **143**, 37–51 (1998).
(30) R. S. A. de Lange, K. Keizer, A. J. Burggraaf, Analysis and theory of gas transport in microporous sol-gel derived ceramic membranes, *J. Membr. Sci.*, **104**, 81–100 (1995).
(31) D. H. Everett, J. C. Powl, Adsorption in slit-like and cylindrical micropores in the Henry's law region. A model for the microporosity of carbons, *J. Chem. Soc., Faraday Trans. 1*, **72**, 619–636 (1976).
(32) M. B. Rao, S. Sircar, Performance and pore characterization of nanoporous carbon membranes for gas separation, *J. Membr. Sci.*, **110**, 109–118 (1996).
(33) M. G. Baschetti, F. Doghieri, B. Freeman, G. C. Sarti, Transient and steady-state effective diffusivity in high free volume glassy polymers, *J. Membr. Sci.*, **344**, 144–154 (2009).
(34) T. C. Merkel, Z. He, I. Pinnau, B. D. Freeman, P. Meakin, A. J. Hill, Effect of nanoparticles on gas sorption and transport in poly(1-trimethylsilyl-1-propyne), *Macromolecules*, **36**, 6844–6855 (2003).
(35) E. R. Gilliland, R. F. Baddour, G. P. Perkinson, K. J. Sladek, Diffusion on surfaces. I. Effect of concentration on the diffusivity of physically adsorbed gases, *Ind. Eng. Chem. Fundam.*, **13**, 95–100 (1974).
(36) K. Kamide, S. Manabe, H. Makino, T. Nohmi, H. Narita, T. Kawai, Surface diffusion flow on the pore wall when gas permeates through a polymeric membrane, *Polymer J.*, **15**, 179–193 (1983).
(37) B. J. Hinds, N. Chopra, T. Rantell, R.,rews, V. Gavalas, L. G. Bachas, Aligned multiwalled carbon nanotube membranes, *Science*, **303**, 62–65 (2004).
(38) M. Knudsen, Die Gesetze der Molekularströmung und der inneren Reibungsströmung der Gase durch Röhren, *Ann. Phys.*, **28**, 75–130 (1909).

(39) S. T. Hwang, K. Kammermeyer, Membranes in Separations, Krieger, Malabar, Florida, 1984.
(40) J. K. Holt, H. G. Park, Y. Wang, M. Stadermann, A. B. Artyukhin, C. P. Grigoropoulos, A. Noy, O. Bakajin, Fast mass transport through sub-2-nanometer carbon nanotubes, *Science*, **312**, 1034–1037 (2006).
(41) D. I. Petukhov, D. A. Buldakov, R. V. Aziev, A. A. Eliseev, Gas selective membranes and membrane catalysis based on porous alumina films, *Rusnano*, 328–330 (2009).
(42) A. J. Burggraaaf, L. Cot, Fundamentals of Inorganic Membrane Science and Technology, Elsevier, Amsterdam, 1996.
(43) M. Mulder, Basic Principles of Membrane Technology, Kluwer Academic Publishers, Dordrecht, 1996.
(44) D. Rigby, R.-J. Roe, Molecular dynamics simulation of polymer liquid and glass. I. Glass transition, *J. Chem. Phys.*, **87**, 7285–7292 (1987).
(45) F. Jagodic, B. Borstnik, A. Azman, Model calculation of gas-diffusion through polymer bulk, *Makromol. Chem.*, **173**, 221–231 (1973).
(46) J. Caro, M. Noack, P. Kölsch, R. Schäfer, Zeolite membranes – state of their development and perspective, *Microporous Mesoporous Mater.*, **38**, 3–24 (2000).
(47) M. B. Rao, S. Sircar, Nanoporous carbon membranes for separation of gas mixtures by selective surface flow, *J. Membr. Sci.*, **85**, 253–264 (1993).
(48) M.-D. Jia, B. Chen, R. D. Noble, J. L. Falconer, Ceramic-zeolite composite membranes and their application for separation of vapor/gas mixtures, *J. Membr. Sci.*, **90**, 1–10 (1994).
(49) A. J. Burggraaf, Single gas permeation of thin zeolite (MFI) membranes: theory and analysis of experimental observations, *J. Membr. Sci.*, **155**, 45–65 (1999).
(50) D. Li, S. T. Hwang, Preparation and characterization of silicon based inorganic membrane for gas separation, *J. Membr. Sci.*, **59**, 331–352 (1991).
(51) J. M. S. Henis, M. Tripody, A novel approach to gas separation using composite hollow fiber membranes, *Sep. Sci. Technol.*, 15, 1059 (1980).
(52) B. D. Freeman, Basis of permeability/selectivity tradeoff relations in polymeric gas separation membranes, *Macromolecules*, **32**, 375–380 (1999).
(53) L. M. Robeson, Correlation of separation factor versus permeability for polymeric membranes, *J. Membr. Sci.*, **62**, 165–185 (1991).
(54) L. M. Robeson, The upper bound revisited, *J. Membr. Sci.*, **320**, 390–400 (2008).
(55) G. Dlubek, A. P. Clarke, H. M. Fretwell, S. B. Dugdale, M. A. Alam, Positron lifetime studies of free volume hole size distribution in glassy polycarbonate and polystyrene, *Phys. Status Solidi A*, **157**, 351–364 (1996).
(56) D. M. Schrader, Y. C. Jean, Positron and Positronium Chemistry, Studies in Physical and Theoretical Chemistry, Elsevier Sci. Publ. Co., Amsterdam, 1988.
(57) D. W. Breck, Zeolite Molecular Sieves: Structure, Chemistry, and Use, John Wiley & Sons, New York, 1973.
(58) B. E. Poling, J. M. Prausnitz, J. O. O'Connell, The Properties of Gases and Liquids, McGraw-Hill, New York, 2001.
(59) A. K. Rappe, C. J. Casewit, K. S. Colwell, W. A. Goddard, W. M. Skiff, Uff, a full periodic-table force-field for molecular mechanics and molecular-dynamics simulations, *J. Am. Chem. Soc.*, **114**, 10024–10035 (1992).
(60) T. Stocker, D. Qin, *The physical science basis of climate change*, WMO and UNEP, 2008.
(61) T. C. Merkel, B. D. Freeman, R. J. Spontak, Z. He, I. Pinnau, P. Meakin, A. J. Hill, Ultrapermeable, reverse-selective nanocomposite membranes, *Science*, **296**, 519–522 (2002).
(62) H. B. Park, C. H. Jung, Y. M. Lee, A. J. Hill, S. J. Pas, S. T. Mudie, E. van Wagner, B. D. Freeman, D. J. Cookson, Polymers with cavities tuned for fast selective transport of small molecules and ions, *Science*, **318**, 254–258 (2007).

[39] S.-T. Hwang, K. Kammermeyer, Membranes in Separations, Wiley, New York, 1975.
[40] J. K. Holt, H. G. Park, Y. Wang, M. Stadermann, A. B. Artyukhin, C. P. Grigoropoulos, A. Noy, O. Bakajin, Fast mass transport through sub-2 nanometer carbon nanotubes, Science, 312, 1034–1037 (2006).
[41] [illegible]
[42] [illegible]
[43] [illegible]
[44] [illegible]
[45] [illegible] Journal of Membrane Science, 175, [illegible]
[46] [illegible]

Part II

Nanocomposite (Mixed Matrix) Membranes

6

Glassy Perfluorolymer–Zeolite Hybrid Membranes for Gas Separations

Giovanni Golemme[a,b], *Johannes Carolus Jansen*[b], *Daniela Muoio*[a], *Andrea Bruno*[a], *Raffaella Manes*[a], *Maria Giovanna Buonomenna*[a], *Jungkyu Choi*[c] *and Michael Tsapatsis*[c]

[a]*Università della Calabria, Dipartimento di Ingegneria Chimica e dei Materiali, and INSTM Consortium, Rende, Italy*

[b]*Istituto per la Tecnologia delle Membrane, ITM-CNR, Rende, Italy*

[c]*University of Minnesota, Department of Chemical Engineering and Materials Science, Minneapolis, USA*

6.1 Introduction

Commercial polymeric membranes are cheap, but very often not sufficiently permselective. For any given gas pair, polymers typically show high selectivities with modest permeabilities, or high permeabilities coupled with reduced selectivities, so that in a selectivity vs. permeability log–log plot all polymers fall below a so-called upper bound line [1]. On the other hand, inorganic membranes are often very permselective, easy to clean, thermally and chemically resistant, but are usually expensive, brittle, difficult to be prepared in a reproducible way, and are typically characterized by a low surface-to-volume ratio in modules which, in turn, affects industrial applications (increased dead volume, need of larger compressors) and therefore translates into higher investment and running costs.

Mixed-matrix membranes (MMMs), also referred to as hybrid membranes, contain a separating layer made of a continuous phase (usually a polymer) embedding a second,

Membrane Gas Separation Edited by Yuri Yampolskii and Benny Freeman

dispersed phase, the chemical nature of which is different. They represent a viable opportunity for enhancing the separation capabilities of polymeric membranes for gas separation (GS). The combination of two materials with different gas diffusivity and solubility, in fact, allows a combination of high permselectivity of the filler (e.g. carbon molecular sieves, zeolites, inorganic particles) with an ease of production of polymeric membranes [2].

In general, the preparation of defect-free MMMs is not an easy task, especially dealing with rigid fillers in glassy polymers. In fact, the poor adhesion between the two phases can produce voids at the interface and preferential paths for the diffusing species, which spoil the selectivity of the composite membrane [3]. This problem is particularly severe for Hyflons AD and Teflons AF, high free volume, amorphous and glassy perfluoropolymers with interesting GS properties for hydrocarbon mixtures, for the sweetening of natural gas and for the N_2/CH_4 separation [4]: the well-known difficulty of sticking anything on the surface of perfluorinated polymers, such as polytetrafluoroethylene, can be rationalized by considering their very low solubility parameters [5]. In fact, in the literature the only composite membranes based on amorphous glassy perfluoropolymers are made of dispersions of non-porous nano-sized amorphous SiO_2 in Teflon AF 2400 [6] and AF 1600 [7]. SiO_2 nanoparticles in Teflons AF interfere with the packing ability of the polymer and increase the amount of free volume. The increase in the permeability of large, condensable molecules exceeds the one of small molecules, thereby decreasing the intrinsic size selectivity of the polymer: as a consequence, a selectivity reversal for the n-butane/CH_4 pair takes place above a 18 wt % SiO_2 content in Teflon AF 2400 [6].

The dispersion of suitable porous fillers inside such polymers offers a new degree of freedom for improvement of their permselectivity, by enhancing the solubility selectivity and/or the mobility selectivity or, in the limiting case, the sieving capabilities of the zeolites. The main goal of this work is to demonstrate that porous fillers can be effectively dispersed in amorphous glassy perfluoropolymers. Silicalite-1, the siliceous form of ZSM-5 (structure topology MFI) [8], has a three-dimensional channel structure with straight channels of 5.4–5.6 Å along the *b* axis and zigzag channels of 5.6 Å in the *ac* plane. The zigzag channels connect subsequent layers of straight channels. There were several reasons for choosing it as the porous filler: first, because this zeolite is hydrophobic and, unlike zeolites containing aluminium, its pores do not absorb significant amounts of moisture; secondly, because it can be easily prepared in a wide range of sizes and shapes; and finally, because the sorption and mobility data of several gases and vapours in it are already available in the literature.

Several mixed matrix membranes made of silicalite-1 crystals of different size and shape embedded in Teflon AF 1600, Teflon AF 2400 and Hyflon AD 60X have been prepared. The gas permeation properties of the membranes have been tested with pure gases, and some experiments with n-C_4H_{10}/CH_4 mixtures have also been carried out on silicalite-1/Teflon AF 2400 membranes.

6.2 Materials and Methods

The random co-polymers of tetrafluoroethylene (TFE) used in this work are amorphous and glassy. They are Teflon AF 1600 and 2400 (second monomer perfluoro-2,2-dimethyl-1,3-dioxole, DuPont) and Hyflon AD 60X (second monomer perfluoro-4-methoxy-

Table 6.1 *Copolymers used in this work and their relevant physical–chemical properties*

Name	Repeat units	T_g/°C	d/g cm^{-3}	FFV (%)	Reference
Teflon® AF 2400	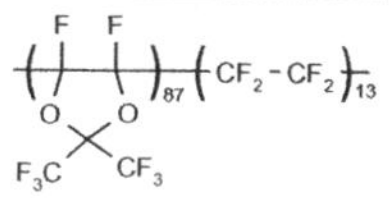	240	1.74	33	4
Teflon® AF 1600	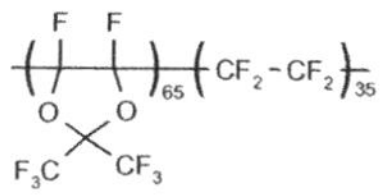	160	1.84	31	4
Hyflon® AD 60X	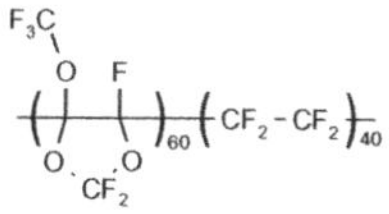	121	1.93	23	9

1,3-dioxole, courtesy of Dr. V. Arcella, Solvay Solexis). Their properties are listed in Table 6.1.

Silicalite-1 crystals of different size and habits have been prepared according to literature procedures [10,11]. In a typical synthesis tetrapropylammonium hydroxide (TPAOH) or bis-1,5-(tripropylammonium)pentane diiodide (dC5), sodium or potassium hydroxide, and tetraethylorthosilicate (TEOS) were stirred overnight to hydrolyze. The solution was filtered (qualitative paper) into polypropylene bottles (Nalgene) or Teflon-lined stainless steel autoclaves and heated statically or under rotation for the due time at the given temperature. After quenching in cold water, crystals were recovered by repeated centrifugation and washing with water until pH was 8, and dried at 90 °C overnight. The larger crystals were then calcined in air (2 °C/min) at 625 °C for 10 h; the sub-micron crystals instead were embedded in a polyacrylamide gel that was dried at 110 °C, heated under N_2 for 5 h at 550 °C (heating and cooling 1 °C/min) and finally calcined in air for 5 h at 550 °C (heating and cooling 1 °C/min) [12].

The MFI crystals were made fluorophilic by means of a surface chemical modification, the details of which will be given elsewhere, and then dispersed in highly fluorinated solvents (Galden HT 110 from Solvay Solexis, FC-72 from 3M, and 1-methoxyperfluorobutane from Alfa Aesar) by sonication for 1 h. The homogeneous polymer solutions were added to the MFI suspensions, and the mixtures were sonicated further for 1 h prior to pouring them into casting rings on glass or Teflon at room temperature. The rings were covered with a lid or a glass plate in order to reduce the evaporation rate of the solvent. In two or three days the 17–151 μm thick membranes became dry, and detached spontaneously from the surface in a few hours after filling the rings with water. The residual solvent was removed by heating (1–4 °C/min) in vacuum up to 200 °C (MMMs with Teflon AF 2400) or a few degrees beyond the T_g of the polymer [13]. The thickness of the membranes was determined with a Mahr micrometer as the average value of 8–20 measurements in different points.

The SEM observations were carried on different instruments (Jeol 6500 FEG, Cambridge Stereoscan 360, and FEI Quanta 200) after sputtering a conductive layer of C, Pt or Au on the samples. The samples for the TEM observation (Zeiss EM900) were included in an Epon Araldite matrix and cured at 60 °C for 3 days; the cross-sections were cut in 80 nm thick slices with a Leica Ultracut UCT ultramicrotome.

The pure gas (99.99% or higher purity) transport properties were tested with an instrument (GKSS, Geesthacht, Germany) with constant permeate volume described elsewhere [14]. The short response time of the instrument allows one to record transient permeation behaviours of less than 1 second. The pure gas permeability is the amount of gas permeating in the unit time, multiplied by the thickness of the membrane and normalized for the membrane surface and the pressure gradient. The recorded pressure vs. time plots were used to derive diffusion coefficients from the initial transient permeation, and permeability at the steady state. The time-lag θ is the intercept of the linear part of the pressure vs. time curve with the axis of time. In a homogeneous membrane in which the solubility of a gas obeys Henry's law, its diffusion coefficient D can be calculated from the ratio:

$$D = l^2/6\theta \tag{6.1}$$

where l represents the thickness of the membrane. On the GKSS instrument the value of D is obtained with an average uncertainty of less than ± 20%, and permeability within ±10%. When adsorptive filler are present, a modification of the simple Equation (6.1) is required [15]. In this work, however, the simplified Equation (6.1) is used throughout in order to draw qualitative conclusions. Before each measurements the system was evacuated. Feed pressure was 1 bar, the permeate pressure never reached 1 mbar. The separation factor is calculated as the ratio of the single gas permeability coefficients; the gas solubility coefficients (S) is found from the permeability (P) and the diffusion coefficients (D) through the well known relation $P = DS$ valid for membranes with a solution-diffusion transport mechanism [16].

Binary CH_4/n-butane permeation properties were determined using a continuous flow system [17]. Typically, a He sweep flow rate of 100 cm^3/min and an i-butane internal standard flow rate of 11 cm^3/min were used, and the equimolar mixed gas feed was 100 cm^3/min. At 25 °C the backflow of helium was negligible. The pressure on either side of the membrane was increased up to about two bars by using a back pressure regulator. Films were loaded into a Wicke–Kallenbach cell, the feed gas was introduced, and the permeating species across the membranes swept by the He stream were analyzed by GC (HP 6890). Three measurements were taken at 1 or 3 h intervals. The temperatures used was 25 °C; between each pressure increase/decrease, the membrane was allowed to equilibrate, typically for 3–22 h, before the next set of measurements was taken. The selectivity P_i/P_j was calculated according to the following expression:

$$P_i/P_j = C_i^p C_j^f / C_i^f C_j^p$$

where the superscript of the concentration C of the generic component refers to the feed or to the permeate streams.

6.3 Results and Discussion

The synthesis parameters of the four silicalite-1 crystals prepared in this work (Figure 6.1) are summarized in Table 6.2. The average length of MFI_80 and MFI_350 crystals

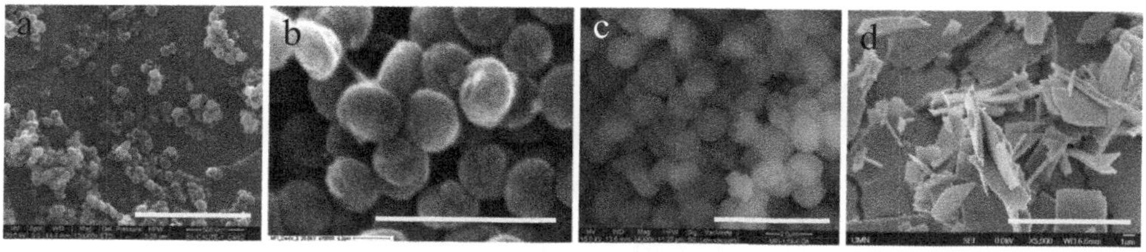

Figure 6.1 SEM pictures of the silicalite-1 samples used in this work: (a) MFI_80, (b) MFI_350, (c) MFI_1500, (d) MFI_dC5. Scale bar is 1 μm (a, b), 5 μm (c) or 10 μm (d)

Table 6.2 Synthesis conditions of the four silicalite-1 samples

Sample	Molar composition[a]	T^{b}/°C	t^{c}/h	Ref.
MFI_80	25 TEOS : 9 TPAOH : 0.1 Na_2O : 461 H_2O	96	72	10
MFI_350	25 TEOS : 5 TPAOH : 0.1 Na_2O : 1550 H_2O	98	60	10
MFI_1500 [d]	40 TEOS : 7.29 TPAOH : 3237 H_2O	130	12	11
MFI_dC5 [d]	40 TEOS : 7.49 dC5 : 12.5 K_2O : 18906 H_2O	175	120	11

[a] TEOS: tetraethoxysilane; TPAOH: tetrapropylammonium hydroxide; dC5: bis-1,5-(tripropylammonium)pentane diiodide.
[b] Synthesis temperature.
[c] Synthesis time.
[d] Autoclaves under rotation.

Figure 6.2 Droplet of water on a layer of fluorophilic 80 nm silicalite-1 crystals

is 80 and 350 nm respectively. MFI_1500 are slabs (010) with frequent twin intergrowths; the average dimensions are ca. 1.4 × 0.65 × 1.5 μm ($a \times b \times c$). The MFI_dC5 crystals are aggregates of highly intergrown thin sheets with the minimum length along the b axis of less than 400 nm, and dimensions along a and c up to about 5 and 10 μm, respectively. The surface treatment makes the zeolites fluorophilic, and this reflects into a sharp increase in the water contact angle from 89° (calcined zeolite) to 140–150° (surface modified zeolite, Figure 6.2). N_2 sorption experiments at 77 K indicate that the modified zeolites are permeable, with about the same sorption capacity as the pristine calcined samples.

The hybrid membranes are designated as follows: e.g. AD60_1500_42 indicates a membrane containing Hyflon AD 60 and 1500 nm silicalite-1 crystals, in which the mass of the zeolite is 42% of the total; AF24_dC5_40 indicates instead a membrane containing Teflon AF 2400 and MFI_dC5 crystals, in which the mass of the zeolite is 40% of the

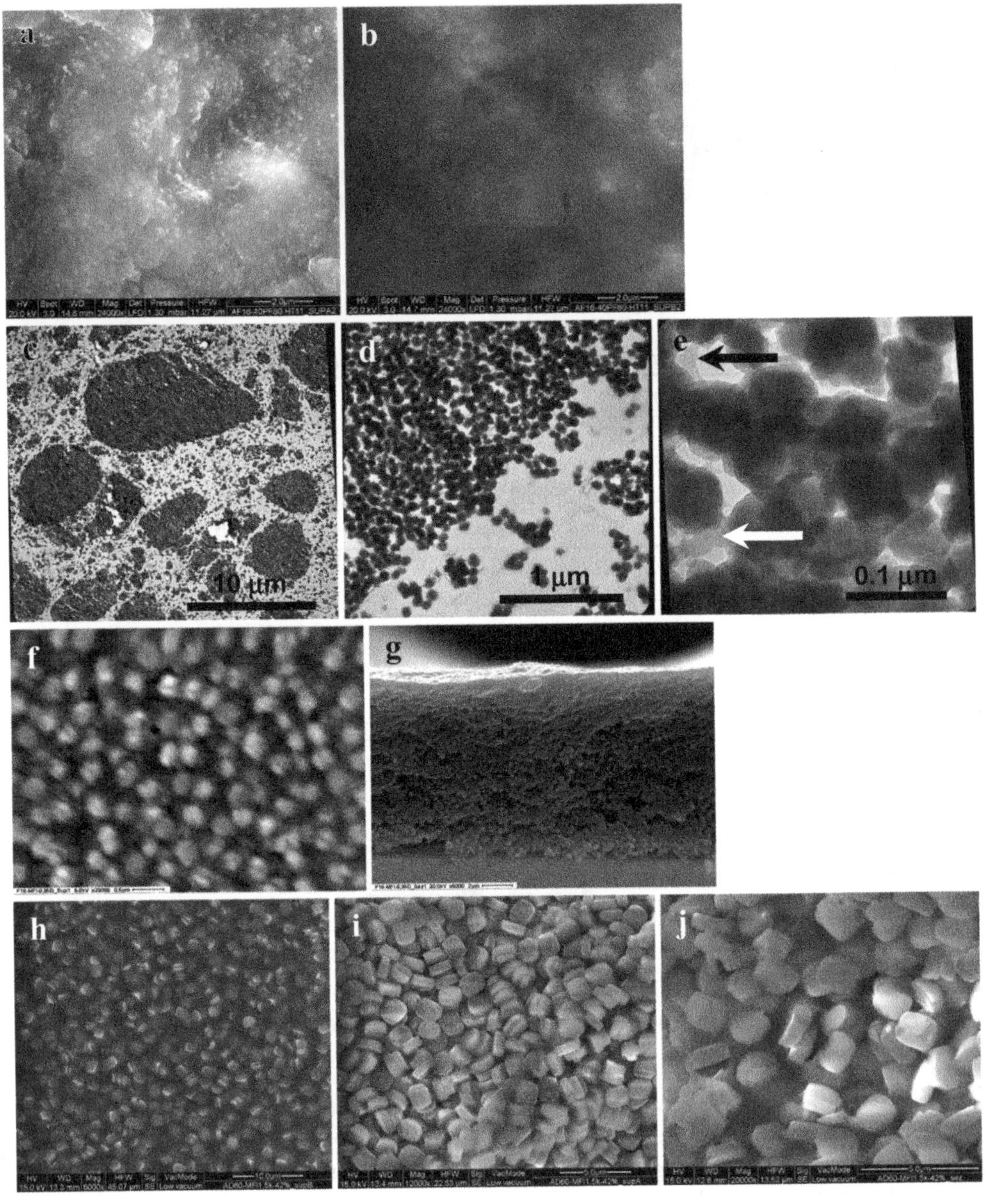

Figure 6.3 *(a) SEM pictures of top view and (b) bottom view of membrane AF16_80_40; (c), (d), (e) TEM pictures of the cross sections of membrane AF16_80_40 at different magnifications; SEM pictures of top view (f) and cross section (g) of membrane AF16_350_30; SEM pictures of bottom view (h), top view (i) and cross section (j) of membrane AD60_1500_42; SEM pictures of bottom view (k), top view (l) and cross section (m) of membrane AF24_1500_40; SEM pictures of bottom view (n), top view (o) and cross section (p) of membrane AF24_dC5_40*

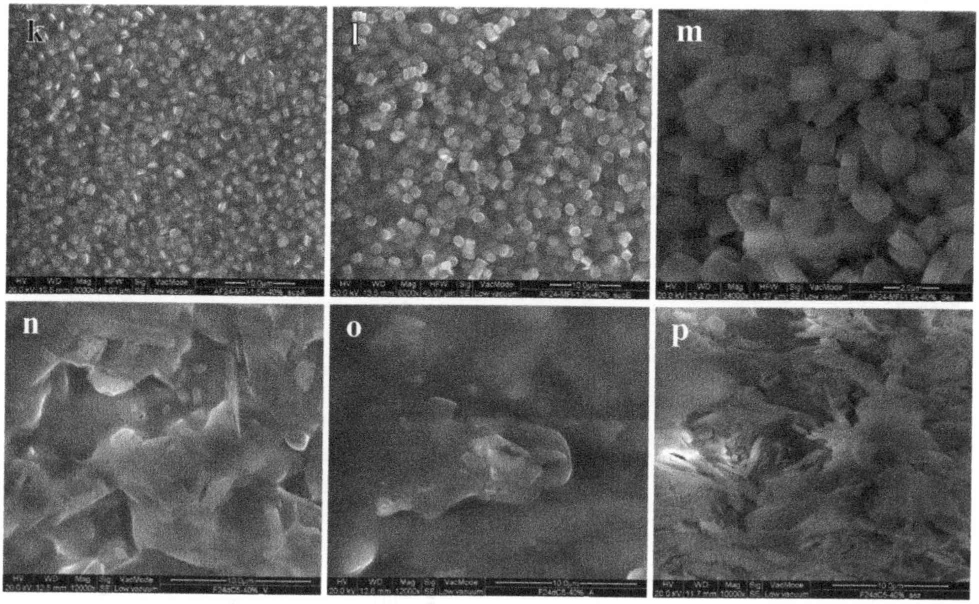

Figure 6.3 (*continued*)

total. The membranes are defect free and still flexible up to about 40 wt% of zeolite. The top and the bottom surfaces of membrane AF16_80_40 (Figures 6.3a and b) are smooth, although the dispersion of the small 80 nm MFI crystals in the polymer matrix is very poor even at a 15 wt% loading. Figures 6.3c–e show that the 80 nm crystals form mainly aggregates containing voids. The TEM picture at the largest magnification in Figure 6.3e, however, indicates the presence of polymer (highlighted by the arrows) around each single crystal in the aggregates, and a good wetting of zeolite surface by Teflon AF 1600. Zeolites of 350 nm or larger instead are well dispersed in Teflon AF 1600 (Figures 6.3f and 6.3g), Hyflon AD 60X (Figures 6.3h–j), Teflon AF 2400 (Figures 6.3k–p).

Before considering the permeability of the membranes, it is useful to evaluate the transport of gas in the silicalite-1 phase. If we assume a solution-diffusion transport mechanism inside a MFI crystal, then an estimate of the permeability can be obtained by multiplying gas solubility and diffusion coefficient at the specific temperature and pressures of our permeation experiments. The diffusion coefficients of CO_2 and CH_4 in silicalite-1, at 25 °C and infinite dilution, reported in Figure 6.4c, were derived from the data of Nijhuis *et al.* [18], because the pulse-response technique used in that work yields diffusivity values which are in good agreement with the self-diffusion coefficients determined by pulse-field-gradient NMR under equilibrium conditions [19]. In other words, the diffusion coefficients reported refer to the transport *inside* the crystal and do not consider the presence of barriers for the transport of matter at the crystal surface. The solubility of N_2, CH_4 and CO_2 in silicalite-1 (Figure 6.4d) was calculated from literature data [20,21] by considering an average value between 0 and 10^5 Pa at 25 °C. It can be seen from Figures 6.4c and d that both the solubility and the diffusion coefficients of N_2,

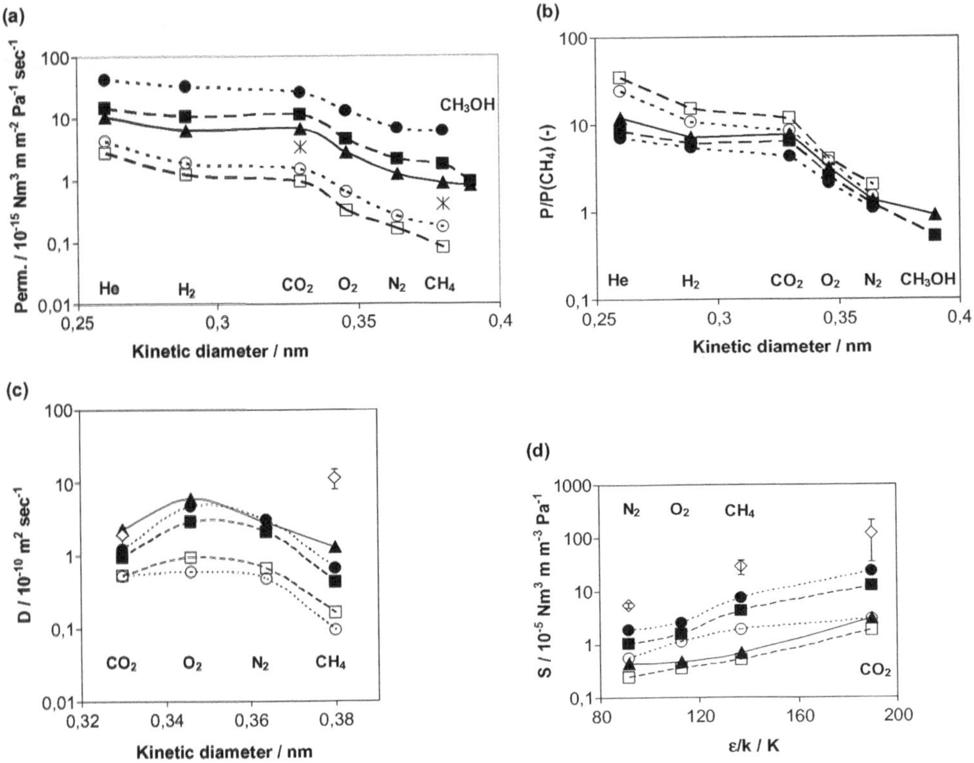

Figure 6.4 *Pure gas transport data at 25 °C of membranes AF1600 (○), AF16_350_30 (□), AF16_80_15 (▲), AF16_80_30 (■), AF16_80_40 (●), silicalite-1 (◇) as derived from literature data (see text), and predictions of the Maxwell model for a AF16/MFI 30% membrane (*); (a) Pure gas steady state permeability vs kinetic diameter of the permeating molecules; (b) gas/methane separation factor; (c) gas diffusion coefficients from time-lag experiments vs kinetic diameter; (d) gas solubility vs the ε/k Lennard-Jones parameter*

CH_4 and CO_2 in silicalite-1 are higher than in Teflon AF1600, and therefore, if no barrier to transport exists in the MFI/Teflon AF1600 membranes, their permeability should be higher than that of the polymer. The permeability to CO_2 and CH_4 of a Teflon AF1600/MFI 30% membrane, as calculated by means of the Maxwell model (random distributions of non interacting solid spheres in a continuous matrix) [22,23], is shown in Figure 6.4a: none of the two membranes containing 30 wt% MFI crystals is adequately described by this simplified model.

The pure gas permeabilities of a Teflon AF 1600 membrane and of four AF1600/MFI membranes (Figure 6.4a) indicate a good correlation with the kinetic diameter of the penetrant molecules. However, the membrane AF24_350_30, containing larger crystals, is less permeable and more selective than the others (Figure 6.4b). Therefore, we can conclude that a potential energy barrier to transport exists at or near the surface of the zeolite crystals. The other data in Figure 6.4 exclude that this barrier might be due to polymer densification at the interface. In fact, the permeability of the membranes containing 80 nm crystals is always larger, and the selectivity smaller, than in Teflon

AF1600: a fact that can be rationalized in terms of a loosened polymer packing at the interface with the inorganic phase. When the size of the crystal passes from 350 to 80 nm, the external surface of the inorganic phase increases of a factor of 19, and the volume of the poorly packed polymer increases proportionally. It is possible that such an increase of the volume of the more permeable 'shells' around the individual crystals might cause them to coalesce and to give rise to percolation paths of higher permeability and reduced selectivity. In the literature, in similar systems formed by nano-sized fumed silica dispersed in amorphous and glassy high free volume polymers, it was assumed that the fillers disrupt the chain packing at the interface [6,7]. The sustained diffusivity of the gases examined in the membranes containing 80 nm MFI crystals (Figure 6.4c) and the increase in the gas solubility they exhibit with the increase of the content of the filler (Figure 6.4d), are also consistent with this picture, by considering that a larger free volume can host more molecules.

It must be pointed out that the information on gas diffusivity and solubility reported in Figure 6.4 are qualitative, since they are derived from a simplified treatment of the transient permeation behaviour; equilibrium gas sorption experiments would be required to obtain detailed information on the different role of polymer and zeolites in these mixed matrix membranes.

The data regarding Teflon AF1600 mixed matrix membranes therefore seem to indicate that, when the zeolite particles are small, the gas flow runs mainly around them, whereas with larger particle sizes the overall resistance increases, and the gas molecules have a higher probability to cross the membrane surface and pass through the zeolites. For this reasons some membranes have been prepared with silicalite-1 crystals of about 1.5 μm and with dC5 MFI.

The properties of Hyflon AD 60X are clearly improved by the presence of silicalite-1. As can be seen in Figure 6.5a the CO_2/CH_4 separation factor gets closer to the Robeson's upper bound. Although other polymers have higher separation factors, the resistance to plasticization of medium free-volume perfluorinated polymers [4,9] makes this polymer/filler combination of some interest.

The best results of the AD60_1500_42 membrane is in the difficult N_2/CH_4 separation (Figure 6.5b), beyond the Robeson's upper bound.

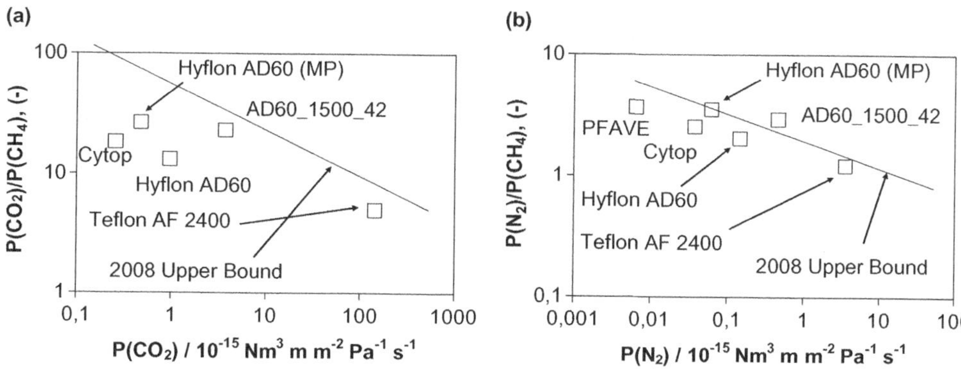

Figure 6.5 *Permeability vs. selectivity plots at 25 °C of perfluorinated polymers [4,13,14] and hybrid membranes for the CO_2/CH_4 pair (a) and for the N_2/CH_4 separation (b)*

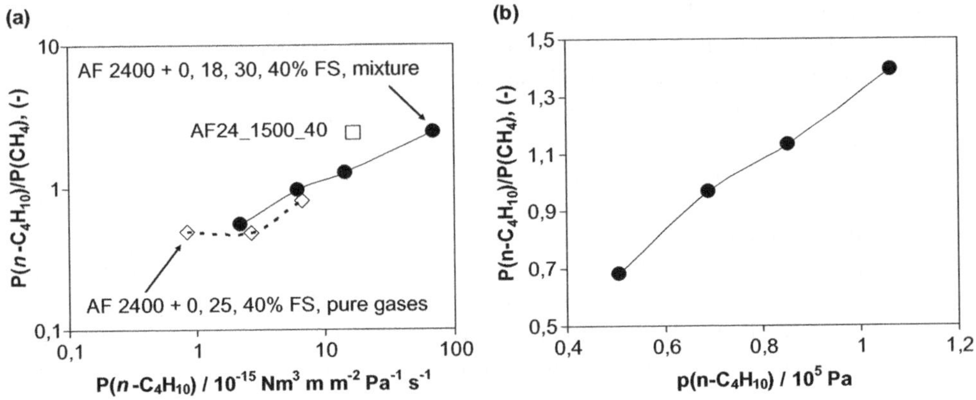

Figure 6.6 *(a) Permeability vs. selectivity plot of Teflon AF 2400 based hybrid membranes for the n-C_4H_{10}/CH_4 pair at 25 °C: pure gas data of fumed silica filled AF2400 (◇; 0, 25 and 40%), data from ref. [6]; pure gas data of membrane AF24_1500_40 (□); mixed gas data (2% n-butane feed, 11.2 bar) of fumed silica filled AF2400 (•; 0, 18, 30 and 40%), from ref. [6]; (b) butane/methane selectivity vs. butane partial pressure in the feed (C_4H_{10}/CH_4 50:50 molar ratio) of membrane AF24_1500_40*

A similar effect can be seen in Figure 6.6a for Teflon AF 2400: when compared to hybrid membranes containing fumed silica, the AF24_1500_40 membrane shows better butane permeability and butane/CH_4 separation factors in pure gas permeation experiments. Mixed gas permeation experiments at 25 °C with equimolar butane–methane mixtures demonstrate that the increase of the butane partial pressure from 50 to 106 kPa can revert the methane selective membrane to butane selective. In these experiments the n-butane permeability grows monotonically from 2.4 to $3.2 \times 10^{-15}\,m^3\,m\,m^{-2}\,Pa^{-1}\,s^{-1}$. The experiments with the membrane AF24_dC5_40, containing large and thin silicalite-1 crystals, the results of which are not shown here, always yield better butane/CH_4 selectivity and same permeability than with the AF24_1500_40 membrane, and demonstrate the importance of the shape of the filler on the performance of mixed matrix membranes [24].

6.4 Conclusions

Defect-free membranes comprising zeolites and amorphous glassy perfluoropolymers can be prepared by modifying the surface of the filler. The pure gas permeation experiments of a series of Teflon AF 1600 membranes with various amounts of 80 and 350 nm silicalite-1 crystals cannot be interpreted on the basis of the Maxwell model, but are compatible with a model in which a barrier to transport exists on the zeolite surface and a lower density polymer layer surrounds the crystals. With a small zeolite size (80 nm) the low density layers around the crystals may coalesce and form percolation paths of lesser resistance and less selectivity. Silicalite-1 crystals improve the CO_2/CH_4 selectivity of Hyflon AD60X, and drive the N_2/CH_4 selectivity beyond the Robeson's upper bound. It also turns out that the presence of silicalite-1 crystals, like fumed silica, promote the inversion of the methane/butane selectivity of Teflon AF2400 in mixed gas experiments.

Acknowledgements

We are grateful to E. Perrotta (TEM), M. Davoli (SEM), L. Pasqua (sorption isotherms) from the University of Calabria, to S. Maheshwari (SEM) of the University of Minnesota, and to S. Morelli (Contact Angle, ITM-CNR). AB and RM acknowledge the Italian Ministry of University and Research for two fellowships.

References

(1) (a) L. M. Robeson, Correlation of separation factor versus permeability for polymeric membranes, *J. Membr. Sci.*, **62**, 165 (1991); (b) L. M. Robeson, The upper bound revisited, *J. Membr. Sci.*, **320**, 390–400 (2008).
(2) T.-S. Chung, L. Y. Jiang, Y. Li, S. Kulprathipanja, Mixed matrix membranes (MMMs) comprising organic polymers with dispersed inorganic fillers for gas separation, *Prog. Polym. Sci.*, **32**, 483–507 (2007).
(3) T. T. Moore, W. J. Koros, Non-ideal effects in organic–inorganic materials for gas separation membranes, *J. Mol. Struct.*, **739**, 87–98 (2005).
(4) T. C. Merkel, I. Pinnau, R. Prabhakar, B. D. Freeman, Gas and vapor transport properties of perfluoropolymers, in Materials Science of Membranes for Gas and Vapor Separation, Y. Yampolskii, I. Pinnau, B. Freeman (Eds.), John Wiley & Sons, Ltd., Chichester, 2006, pp. 251–270.
(5) E. A. Grulke, Solubility Parameters, in Polymer Handbook, 3rd edn., J. Brandrup, E. H. Immergut (Eds.), Wiley-Interscience, New York, 1989.
(6) T. C. Merkel, Z. He, I. Pinnau, B.D. Freeman, P. Meakin, A. J. Hill, Sorption and transport in poly(2,2-bis(trifluoromethyl)-4,5-difluoro-1,3-dioxole-co-tetrafluoro-ethylene) containing nanoscale fumed silica, *Macromolecules*, **36**, 8406–8414 (2003).
(7) (a) J. Y. Zhong, W. Y. Wen, A. A. Jones, Enhancement of diffusion in a high-permeability polymer by the addition of nanoparticles, *Macromolecules*, **36**, 6430–6432 (2003). (b) J. Y. Zhong, G. X. Lin, W. Y. Wen, A. A. Jones, S. Kelman, B. D. Freeman, Translation and rotation of penetrants in ultrapermeable nanocomposite membrane of poly(2,2-bis(trifluoromethyl)-4,5-difluoro-1,3-dioxole-*co*-tetrafluoroethylene) and fumed silica, *Macromolecules*, **38**, 3754–3764 (2005).
(8) C. H. Baerlocher, W. M. Meier, D. H. Olson, Atlas of Zeolite Framework Types, Fifth revised edn., Elsevier, New York, 2001.
(9) I. Pinnau, Z. He, A. R. Da Costa, K. D. Amo, R. Daniels, Gas Separation using Organic-Vapor-Resistant Membranes, US Pat. 6361583 B1 (2002).
(10) A. E. Persson, B. J. Schoeman, J. Sterte, J. E. Otterstedt, The synthesis of discrete colloidal particles of TPA-silicalite-1, *Zeolites*, **14**, 557–567 (1994).
(11) M. Woo, J. Choi, M. Tsapatsis, Poly(1-trimethylsilyl-1-propyne)/MFI composite membranes for butane separations, *Micropor. Mesopor. Mater.*, **110**, 330–338 (2008).
(12) H. Wang, B. Holmberg, Y. Yan, Homogenous polymer-zeolite nanocomposite membranes by incorporating dispersible template-removed zeolite nanocrystals, *J. Mater. Chem.*, **12**, 3640–3643 (2002).
(13) J. C. Jansen, M. Macchione, E. Drioli, On the unusual solvent retention and the effect on the gas transport in perfluorinated Hyflon AD membranes, *J. Membr. Sci.*, **287**, 132–137 (2007).
(14) J. C. Jansen, M. Macchione, E. Drioli, High flux asymmetric gas separation membranes of modified poly(ether ether ketone) prepared by the dry phase inversion technique, *J. Membr. Sci.*, **255**, 167–180 (2005).
(15) D. R. Paul, D. R. Kemp, The diffusion time lag in polymer membranes containing adsorptive fillers, *J. Polym. Sci., Pt. C, Polym. Symp.*, **41**, 79–93 (1973).

(16) J. G. Wijmans, R. W. Baker, The solution-diffusion model: a unified approach to membrane permeation, in Materials Science of Membranes for Gas and Vapor Separation, Y. Yampolskii, I. Pinnau, B. Freeman (Eds.), John Wiley & Sons, Ltd., Chichester, 2006, pp. 159–189.
(17) J. Choi, S. Ghosh, L. King, M. Tsapatsis, MFI zeolite membranes from *a*- and randomly oriented monolayers, *Adsorption*, **12**, 339–360 (2006).
(18) J. Kärger, D. M. Ruthven, Diffusion in Zeolites and Other Microporous Solids, Wiley-Interscience, New York, 1992.
(19) T. A. Nijhuis, L. J. P. van den Broeke, M. J. G. Linders, J. M. van de Graaf, F. Kapteijn, M. Makkee, J. A. Moulijn, Measurement and modeling of the transient adsorption, desorption and diffusion processes in microporous materials; *Chem. Eng. Sci.*, **54**, 4423–4436 (1999).
(20) M. S. Sun, D. B. Shah, H. H. Xu, O. Talu, Adsorption equilibria of C1 to C4 alkanes, CO_2, and SF_6 on silicalite, *J. Phys. Chem. B*, **102**, 1466–1473 (1998).
(21) K. Makrodimitris, G. K. Papadopoulos, D. N. Theodorou, Prediction of permeation properties of CO_2 and N_2 through silicalite via molecular simulation, *J. Phys. Chem. B*, **105**, 777–88 (2001).
(22) C. Maxwell, Treatise on Electricity and Magnetism, vol. 1, Oxford University Press, London, 1973.
(23) E. E. Gonzo, M. L. Parentis, J. C. Gottifredi, Estimating models for predicting effective permeability of mixed matrix membranes, *J. Membr. Sci.*, **277**, 46–54 (2006).
(24) E. L. Cussler, Membranes containing selective flakes, *J. Membr. Sci.*, **52**, 275–288 (1990).

7

Vapor Sorption and Diffusion in Mixed Matrices Based on Teflon® AF 2400

Maria Chiara Ferrari, Michele Galizia, Maria Grazia De Angelis and Giulio Cesare Sarti
Dipartimento di Ingegneria Chimica, Mineraria e delle Tecnologie Ambientali (DICMA), Alma Mater Studiorum-Università di Bologna, Bologna, Italy

7.1 Introduction

Polymers are currently the dominant materials for gas separation membranes because of their processability that allows the economic production of modules for large scale separation. Their separation ability, however, may be low in comparison with more rigid molecular sieving media and this may limit their performance. A significant improvement can be achieved by dispersing an inorganic more selective phase in a polymeric processable phase, with the possibility of taking advantage of the peculiar characteristics of both materials. This class of composites is usually referred to as mixed matrix membranes (MMM) where the highly selective rigid phase is represented by zeolites, carbon molecular sieves or silica particles. The polymeric membranes employed are glassy polymers that already show good selective and/or gas permeation properties on their own, such as polyimides, polysulfone, amorphous Teflon®, PTMSP and PMP [1].

Measurements of gas transport in AF 2400 filled by non-porous hydrophobic fumed silica (FS) nanoparticles showed that the inorganic filler enhances gas permeability [2]; such an increase is more pronounced for larger penetrants and leads to a lower size selectivity of the mixed matrix membrane with respect to the unloaded polymer. The sorption

Membrane Gas Separation Edited by Yuri Yampolskii and Benny Freeman

isotherm of these materials, reported per unit mass of solid membrane, is not greatly affected by the inorganic filler: the latter adsorbs only marginal amounts of penetrants. Surprisingly, on the contrary, the diffusion coefficients are increased significantly by the addition of the nanoparticles, which are in fact obstacles to diffusion through the membrane, thus determining an important increase in permeability. Fumed silica is non-porous and therefore a penetrant is adsorbed only on its surface, and diffusion around it requires a more tortuous path. The observed sorption and diffusion behaviour can be explained by considering that the nano-filler changes the packing structure of the polymeric matrix increasing the overall fractional free volume (FFV) of the polymer phase and thus increasing its sorption capacity; that can compensate for the lower contribution to sorption offered by the particles. The additional free volume created affects diffusivity even more significantly, in spite of the presence of obstacles in the diffusive path. That physical interpretation is in agreement with PALS measurements that in the super-glassy polymers such as amorphous Teflons and PTMSP found two different populations of holes whose distribution is modified in MMM after the addition of FS, with an increase of the fraction of the larger FFV elements [3].

In a previous work [4] we already proposed a method to predict the enhancement in diffusivity due to the addition of fumed silica particles to high free volume matrices such as PTMSP and Teflon® AF 2400. The model was tested only on the bases of available literature data while in this work we performed a detailed characterization of Teflon® AF 2400 mixed matrices to further inspect and document the validity of the approach proposed.

7.2 Theoretical Background

In gas separation, the performance of a membrane is usually evaluated through the selectivity $\alpha_{i,j}$ between two penetrants i and j, that is expressed as:

$$\alpha_{i,j} = \frac{P_i}{P_j} = \alpha_D \cdot \alpha_S = \frac{D_i}{D_j} \cdot \frac{S_i}{S_j} \tag{7.1}$$

where P_i is the permeability of the ith penetrant and can be calculated as the product of D_i and S_i if the solution-diffusion model holds true and Fick's law is appropriate to represent the diffusive mass flux.

The addition of nanoparticles in a polymeric matrix has been considered with the expectation of taking advantage of the specific qualities of the two materials, obtaining in fact a variety of unexpected behaviours which apparently demand an extensive experimental analysis. Therefore, it would be highly desirable and useful to be able to predict the behaviour of a composite matrix from the properties of the pure constituents, glassy polymer and filler nanoparticles, while to our knowledge, there are no models available that can predict correctly the solubility and diffusivity behaviour of the mixed matrices under consideration. Only models for permeability are available and they all predict a decrease in permeability as rigid impermeable particles are added to the matrix, due to the increase in the tortuosity of the path that the penetrant molecules have to follow during diffusion through the membrane. One of the most commonly used is the Maxwell model [5], initially derived for the permittivity of a dielectric medium, that states:

$$P_i = P_{i,P}\left(\frac{1-\Phi_F}{1+\Phi_F/2}\right) \quad (7.2)$$

where Φ_F is the volume fraction of the particles loaded. Even if suitable for some MMM with non-porous fillers, this model is certainly not applicable to the case of amorphous Teflon® loaded with FS nanoparticles. Thus a new method has been proposed [4] to calculate solubility and transport properties in MMM, based on the NELF model for solubility, that can predict the behaviour of permeability through the separate calculation of penetrant solubility and diffusivity. The general features of the model are briefly revised hereafter, before presenting the results of the experimental analysis.

7.2.1 NELF Model

The NELF model [6–9] enables us to calculate the solubility isotherms in glassy polymers, on the bases of the results of non-equilibrium thermodynamics of glassy polymers [9]. It offers an explicit relationship between penetrant fugacity on one hand and penetrant solubility in the glass and glassy polymer density on the other hand. Thus it accounts for the strong and important relationship existing between solubility and fractional free volume in the glassy polymeric phase. In particular, the NELF model is an extension of the lattice fluid equation of state [10] to the non-equilibrium state that characterizes glassy polymers. The model uses the same characteristic parameters (p^*, ρ^*, T^*) of the Sanchez and Lacombe theory [11] to evaluate the properties of pure components, and the same mixing rules to estimate the mixture properties.

The characteristic parameters of the polymer can be calculated by best-fitting the LF equation of state to PVT data above T_G, while for the penetrant either PVT or vapour–liquid equilibrium data can be profitably used.

The number of lattice sites occupied by a molecule in its pure phase is given by:

$$r_i^0 = \frac{P_i^* M_i}{RT_i^* \rho_i^*} = \frac{M_i}{\rho_i^* v_i^*} \quad (7.3)$$

where M_i is the molar mass of component i and v_i^* is the volume occupied by a mole of lattice sites of pure substance. The parameter r_i^0 is usually set to infinity for the polymer species. Mixing rules with only one adjustable binary parameter, Ψ, are adopted to calculate the mixture properties. Ψ affects the binary characteristic pressure P_{12}^* that represents the energetic interactions per unit volume between gas and polymer molecules:

$$P_{12}^* = \Psi\sqrt{P_{11}^* \cdot P_{22}^*} \quad (7.4)$$

A first-order approximation is given by $\Psi = 1$, to use when no specific data for the mixture are experimentally available.

The pseudo-equilibrium state of the glassy mixture is described by the usual state variables (temperature, pressure and composition) plus the polymer density ρ_2 that accounts for the departure from equilibrium frozen into the glass.

The phase equilibrium between the external gas phase and the glassy polymer mixture is given by the usual relationship:

$$\mu_1^{\mathrm{NE(s)}}(T,p,\Omega_1^{\mathrm{PE}},\rho_2^{\mathrm{PE}})=\mu_1^{\mathrm{Eq(g)}}(T,p) \tag{7.5}$$

where the penetrant chemical potential in the solid non-equilibrium phase, $\mu_1^{\mathrm{NE(s)}}$, is defined as:

$$\mu_1^{\mathrm{NE(s)}}=\left.\frac{\partial G_{\mathrm{tot}}}{\partial n_1}\right|_{T,p,n_2,\rho_2} \tag{7.6}$$

and is given by an explicit expression [6–8] according to the Sanchez-Lacombe equation of state. Superscripts PE in Equation (7.5) label pseudo-equilibrium conditions.

In order to evaluate solubility isotherms through Equation (7.5) one thus requires the knowledge of the characteristic parameters of the pure components that may be found in specific collections, as well as of the actual density of the glassy polymer at the actual experimental conditions, that depends also on the thermo-mechanical history of the samples and can be determined from separate direct dilation experiments. For non-swelling penetrants, like many permanent gases such as O_2, N_2 and CH_4, one can use the pure unpenetrated polymer density ρ_2^0 for the entire isotherm. For swelling penetrants the value ρ_2^0 for the polymer density can be used only in the low pressure range [7], while for the entire isotherm one needs to account for the actual polymer density given by the dilation isotherm which parallels the penetrant sorption isotherm. It has been shown that in the cases experimentally tested [12–14] there is a linear dependence of the polymer density on the partial pressure of the vapour sorbed and thus we can introduce a swelling coefficient, k_{sw}, to account for such a linear relationship between polymer density and penetrant pressure during sorption:

$$\rho_2^{PE}(p)=\rho_2^0(1-k_{sw}p) \tag{7.7}$$

Since specific dilation isotherms are not usually available in open literature data, use of Equation (7.7) allows reasonable and simple procedures: in the absence of dilation isotherm data, one single solubility datum at high pressure can be used in Equations (7.5) and (7.7) to obtain the swelling coefficient k_{sw}, so that the swelling isotherm of the matrix at all other pressures can then be obtained through Equation (7.7).

Using Equations (7.5) and (7.7), the phase equilibrium condition becomes of the following type:

$$\mu_1^{NE(s)}(T,p,\Omega_1^{PE},\rho_2^0,k_{sw})=\mu_1^{Eq(g)}(T,p) \tag{7.8}$$

7.2.2 Modelling Gas Solubility Into Mixed Matrix Glassy Membranes

Frequently, gas solubility in a composite matrix is represented through a simple additive rule that considers the sorption capacity of the polymer and the filler unaffected by the mixed matrix conditions. Unfortunately this is not the case for mixed matrices obtained by loading impermeable nano-particles in high free volume glassy polymers, where the addition of the inorganic rigid phase affects essentially the actual density of the polymer and its swelling behaviour. The FFV increase is reasonably associated to the generation of defects at the polymer/filler interface, which on the average results in a decreased density of the polymer phase of the composite. Thus the presence of filler essentially affects only the polymeric phase, by adding free volume elements at the organic/inorganic interface, while the adsorption capacity on the filler surface can be considered practically unchanged, in view also of the fact that the adsorption on the filler surface is rather small compared to the total solubility.

In general we can write the total solubility of a penetrant as a sum of two contributions, one due to the polymeric phase and one due to the filler; per unit volume of the mixed matrix one has:

$$C_{i,M} = \frac{n_{i,F} + n_{i,P}}{V} = \Phi_F \cdot C_{i,F} + (1 - \Phi_F) \cdot C_{i,P} \quad (7.9)$$

where Φ_F is the volume fraction of the filler in the composite, $C_{i,M}$ is the gas solubility per unit volume of mixed matrix, $C_{i,F}$ and $C_{i,P}$ are the gas solubilities in the filler and in the polymer, per unit volume of filler and per unit volume of polymer, respectively. (Indeed, on the filler particles only surface adsorption takes place so that the filler contribution should be more properly written using the surface concentration γ_i and the area per unit volume of the particles, $A''' = 6/d_p$ for spheres of uniform diameter d_p, so that $C_{i,F} = \gamma_i A'''$. However, since the particles here used are reasonably monodisperse we can equivalently use $C_{i,F}$ in place of $\gamma_i A'''$.) Since the contribution due to adsorption is minor if not negligible, it appears reasonable to accept that the adsorption capacity of the filler in the composite is practically the same as onto the pure free particles, $C^0_{i,F}$; thus we now assume that:

$$C_{i,F} = C^0_{i,F}$$

so that Equation (7.9) becomes:

$$C_{i,M} = \Phi_F \cdot C^0_{i,F} + (1 - \Phi_F) \cdot C_{i,P} \quad (7.10)$$

Normally the major contribution to the solubility in the mixed matrix under consideration is given by the glassy polymer phase and thus by the second term in the right-hand side of Equation (7.10). Therefore, if one can measure reliably the density ρ^0_2 of the glassy polymer phase of the mixed matrix one can rely on the NELF model to calculate the solubility $C_{i,P}$ in the low pressure range or in the entire isotherm in the case of non-swelling penetrants. For swelling penetrants also the swelling coefficient k_{sw}, or the dilation isotherm of the polymer phase, would be required. Unfortunately neither ρ^0_2 nor k_{sw} values are commonly available for the polymer phase of a mixed matrix, and the limited existing data indicate that the polymer density in mixed matrices require techniques more delicate than usual to be determined, in the absence of which its value is obtained with an uncertainty too high to allow a direct use in the calculations of solubility as indicated above.

In the absence of a reliable value of the volumetric properties of the glassy polymer phase of the mixed matrix we can tackle the problem following a different procedure, according to which the required volumetric properties are obtained from the solubility isotherm of a penetrant, arbitrarily chosen as reference. We can thus select a convenient test penetrant and measure directly its adsorption onto the pure filler and its solubility isotherms in the mixed matrix as well as in the pure unloaded polymer. Then we can calculate the solubility isotherm in the polymer phase alone of the MMM, by means of Equation (7.10). Use of the NELF model allows one to calculate the unpenetrated polymer density ρ_2 at any pressure, considering for the binary parameter Ψ the same value valid for the unloaded polymer phase. From the dilation isotherm thus calculated for the polymer phase alone we obtain ρ_2^0 as well as the swelling coefficient k_{sw} in the mixed matrix. The value of ρ_2^0 represents the pure unpenetrated polymer density in the mixed matrix, which obviously holds true also when other penetrants are used in the same MMM. The value of k_{sw} accounts for the swelling effects of the specific penetrant on the polymer and thus it may vary for every polymer–gas couple. The value of ρ_2^0 thus obtained may then be used to calculate through the NELF model the solubility of all other penetrants, at least in the low pressure range in which swelling is negligible. In the presence of important swelling effects one needs to determine also the swelling coefficient k_{sw} of the polymer phase in the MMM, which requires the knowledge of one solubility value of that penetrant at relatively high pressures, according to the procedure illustrated in a previous work [7]. The determination of ρ_2^0 as described above allows us also to evaluate the fractional free volume in the unpenetrated polymer phase, commonly defined as:

$$FFV = \frac{V_2 - 1.3 \cdot V_W}{V_2} = \frac{\rho_2^W - 1.3 \cdot \rho_2^0}{\rho_2^W} \tag{7.11}$$

where V_W is the van der Waals volume of the repeating unit of the polymer, whose value can be calculated based on Bondi's group contribution method [15], and is already available for the matrices of interest in this study [4].

7.2.3 Modelling Gas Diffusivity and Permeability in Mixed Matrix Glassy Membranes

As far as diffusivity is concerned, it has been seen [4] that the influence of the penetrant concentration $C_{i,P}$ on the diffusion coefficient in composite membranes can be well represented with an exponential law:

$$D_{i,M} = D_{i,M}(0) \cdot \exp(\beta \cdot C_{i,P}) \tag{7.12}$$

where $D_{i,M}(0)$ is the infinite dilution apparent diffusion coefficient in the mixed matrix, $C_{i,P}$ the average penetrant concentration in the polymeric phase and β is a constant characteristic of the polymer–gas couple that depends on temperature.

If only the infinite dilution diffusivity $D_{i,M}(0)$ is considered, no swelling effects is present and we can consider that the diffusion coefficient depends only on the properties of the mixed matrix, namely tortuosity and FFV of the glassy polymeric phase.

A semi-empirical law, based on free volume theory [16], is usually considered to hold between the infinite dilution diffusion coefficient in the pure polymeric phase, $D_{i,P}(0)$, and its FFV:

$$\ln(D_{i,P}(0)) = A - \frac{B}{\mathrm{FFV}_M^0} \tag{7.13}$$

where A and B are parameters that depend on temperature and are specific to the polymer–gas pair under consideration. Indeed it is known that, although the parameters entering the original free volume expression [16] can be determined from pure component properties only, their values are practically unavailable or very difficult to determine for the majority of polymers. As a consequence, a more compact expression, represented by Equation (7.13), is used that maintains the functional form of the dependence between diffusivity and fractional free volume, but makes use of only two adjustable parameters that can be determined from experimental diffusion data.

The diffusion coefficient in the mixed matrix is affected by the presence of the filler not only through the effects on polymer FFV but also through changes in the diffusive path: the particles are impermeable and act as an obstacle in the path of the gas molecules through the membrane, increasing its tortuosity. The value of the infinite dilution apparent diffusion coefficient in the mixed matrix, $D_{i,M}(0)$, can then be related to $D_{i,P}(0)$ using a tortuosity factor τ, as usually made:

$$D_{i,M}(0) = \frac{1}{\tau} D_{i,P}(0) \tag{7.14}$$

The tortuosity factor is reasonably evaluated from the Maxwell model [5], actually specifically derived for the insertion of spherical particles as:

$$\tau = 1 + \frac{\Phi_F}{2} \tag{7.15}$$

Thus, in view of Equations (7.13) and (7.14) we have:

$$D_{i,M}(0) = \frac{1}{\tau} \exp\left(A - \frac{B}{\mathrm{FFV}_M^0} \right) \tag{7.16}$$

Equation (7.16) relates the apparent infinite dilution diffusion coefficient in the mixed matrix to the FFV_M^0 of the unpenetrated polymeric phase of the MMM.

The FFV value in the polymer phase of the mixed matrix can be calculated through Equation (7.11) based on the unpenetrated polymer density ρ_2^0 in the polymer phase of the MMM, obtained from the solubility data as indicated above. Since different filler loadings will induce different FFV in the polymer phase, one can use Equation (7.16) to obtain the parameters A and B virtually from as little as two different mixed matrices with two different filler loadings; then the same Equation (7.16) can be applied to calculate or correlate the infinite dilution apparent diffusivity $D_{i,M}$ for any other filler loading in the same glassy polymer.

If we then consider the ratio between the diffusivity in the composite and the diffusivity in the pure unloaded polymer, we obtain:

$$\frac{D_{i,M}(0)}{D_{i,P}^0(0)} = \frac{1}{\tau} \exp\left[B \left(\frac{1}{\mathrm{FFV}_P^0} - \frac{1}{\mathrm{FFV}_M^0} \right) \right] \tag{7.17}$$

Equation (7.17) contains only one adjustable parameter, B, together with FFV_P^0, which represents the fractional free volume of the pure unloaded glassy polymer.

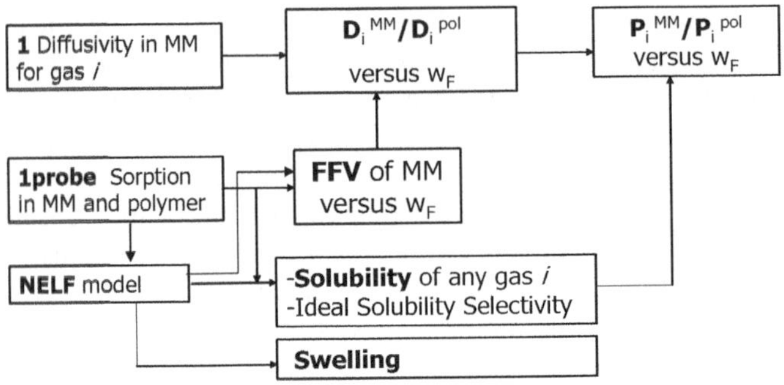

Figure 7.1 Scheme of the modelling approach used for the mixed matrices transport behaviour [4]

Of course, from the diffusivity and solubility values one can evaluate also the permeability ratio between composite and unloaded polymer, considering the solution-diffusion model:

$$\frac{P_{\mathrm{i,M}}(0)}{P_{\mathrm{i,P}}^{0}(0)}=\frac{D_{\mathrm{i,M}}(0)}{D_{\mathrm{i,P}}^{0}(0)}\cdot\frac{S_{\mathrm{i,M}}(0)}{S_{\mathrm{i,P}}^{0}(0)} \tag{7.18}$$

where the ratio between diffusivities can be calculated from Equation (7.17) and the solubility ratio comes from the calculations based on the NELF model.

The procedure proposed to estimate the solubility and transport properties in the mixed matrices is visually summarized in the scheme reported in Figure 7.1.

7.3 Experimental

7.3.1 Materials

The polymer matrix used is Teflon® AF 2400, a glassy perfluorinated copolymer obtained by random copolymerization of 2,2-bis(trifluoromethyl)-4,5-difluoro-1,3-dioxole (BDD) and tetrafluoroethylene (TFE) at mole fractions of 87% and 13% respectively, with glass transition temperature of 240 °C and density equal to 1.74 g/cm^3; the polymer has an excellent chemical resistance and high gas permeability. The filler used (non-porous fumed silica TS-530, Cabot Corporation) has been chemically treated with hexamethyldisilazane, in order to replace hydroxyl surface groups with hydrophobic trimethyl-silyl groups. The particles have an average diameter of 13 nm and a density of 2.2 g/cm^3. The structures of the repeating units of the copolymers investigated and of the inorganic filler used are shown in Figure 7.2.

Membranes were obtained by casting a solution of 1% wt of polymer in perfluoro-*N*-methyl-morpholine (PF5060, supplied by 3M) on a glass plate and allowing the solvent to evaporate, following the procedure proposed by Merkel *et al.* [2]. Fumed silica is added to the polymer solution in the case of mixed matrices, the mixture is blended for 2–3 min

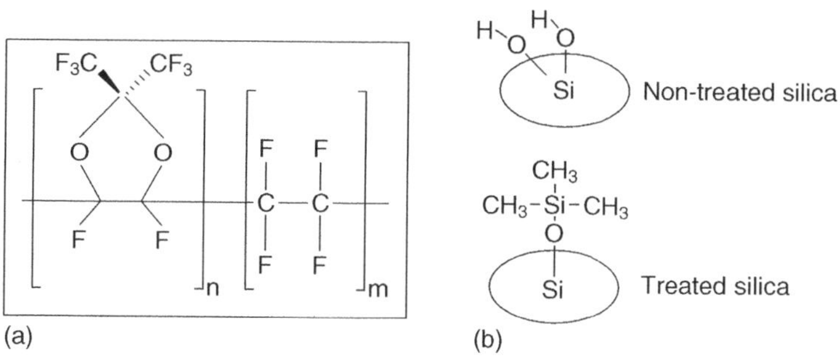

Figure 7.2 Molecular structure of (a) Teflon® AF and (b) non-porous fumed silica before and after treatment to have a hydrophobic surface

at 18000 rpm in a Waring two-speed laboratory blender and then cast onto a Petri dish and covered with an aluminium foil, allowing controlled solvent evaporation. It seems that solvent evaporation is the critical step to obtain good films, while the stirring time has no influence on the resulting membranes structure. The films were completely dried within about 48 h and were not treated further before sorption experiments. The final thickness of the films is in the range 60–140 µm. Mixed matrices were prepared with two different filler loadings, 25% and 40% wt of FS.

The vapours used in the experiments, n-butane and n-pentane, were supplied by Sigma-Aldrich, reactant grade, and were used as received.

7.3.2 Vapour Sorption

The determination of penetrant solubility and diffusivity was performed at pressures below 1 bar using a pressure decay apparatus. A known amount of vapour is fed into the sample chamber and the mass uptake is evaluated by measuring the pressure decrease of the gaseous phase versus time; the equilibrium solubility is equal to the final, asymptotic value of the mass uptake. Pressure is measured with an absolute capacitance manometer (FS value 1000 mbar, accuracy 0.15% of the reading), and vapour activity is calculated as the ratio between the equilibrium pressure and the penetrant vapour pressure at the temperature of the experiment. Subsequent sorption tests are performed by increasing the external pressure in a stepwise manner. The system is placed in an air-thermostated chamber where temperature is kept fixed to within ±0.1 °C. A scheme of the equipment is shown in Figure 7.3. The penetrant diffusivity in the film can be evaluated from the sorption kinetics, which follows Fickian behaviour, taking into account the variation of interfacial concentration during the experiment, due to the decrease of the gas pressure. The expression for the mass uptake in step (i) as a function of time, $M_t^{(i)}$, for the mass sorption from a limited volume where the variation of the interfacial concentration is due to mass sorption in the membrane, is given by the well-known expression [17]:

$$\frac{M_t^{(i)} - M_0^{(i)}}{M_\infty^{(i)} - M_0^{(i)}} = 1 - \sum_{n=1}^{\infty} \frac{2\alpha(1+\alpha)}{1+\alpha+\alpha^2 q_n^2} e^{-Dq_n^2 t/l^2} \tag{7.19}$$

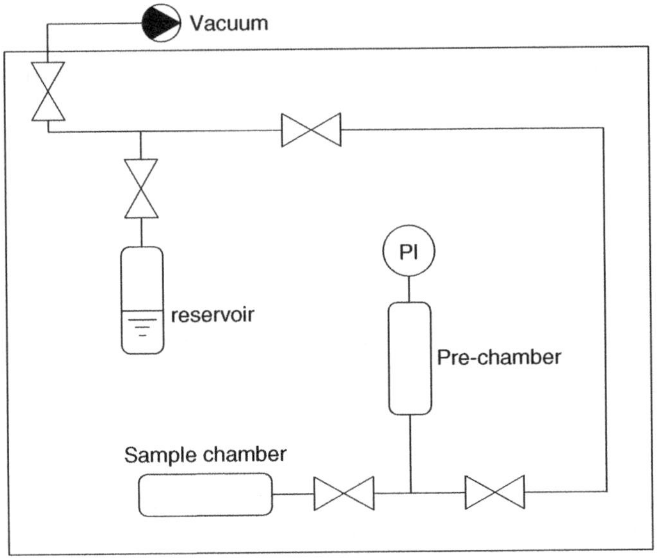

Figure 7.3 Pressure decay apparatus for vapour sorption

where $M_0^{(i)}$ and $M_t^{(i)}$ are the initial and final mass uptakes of step (i), respectively, α is the ratio between the volume of the chamber and that of the membrane, corrected for the partition coefficient of vapour between the gaseous phase and the polymer, l is the semi-thickness of the membrane, while q_n are the positive, non-zero solutions of the equation: $t_g(q_n) = -\alpha q_n$. By best fitting to Equation (7.19) the experimental data of mass uptake versus time, one obtains the average diffusivity value within the concentration interval inspected in the differential sorption step. This apparatus was used to perform experiments with pure polymer and mixed matrices based on Teflon® AF 2400 and also to measure adsorption onto pure fumed silica particles.

7.4 Results and Discussion

7.4.1 Vapour Sorption in Mixed Matrices Based on AF 2400

The solubility isotherms obtained for n-butane and n-pentane in mixed matrices of AF 2400 are shown in Figure 7.4, as grams of penetrant per grams of total solid, together with the solubility isotherms for the pure polymer and the mass adsorption onto the pure filler.

The addition of filler in AF 2400 apparently does not affect the solubility of n-C_4 per unit mass of solid (Figure 7.4a), in spite of the fact that fumed silica adsorbs much less n-C_4 than the pure polymer can sorb, and influences the mass uptake of n-C_5 (Figure 7.4b) only at high activity where the plasticization effect of the penetrant is significant, as it can be seen from the shape of the isotherm that shows a change in concavity.

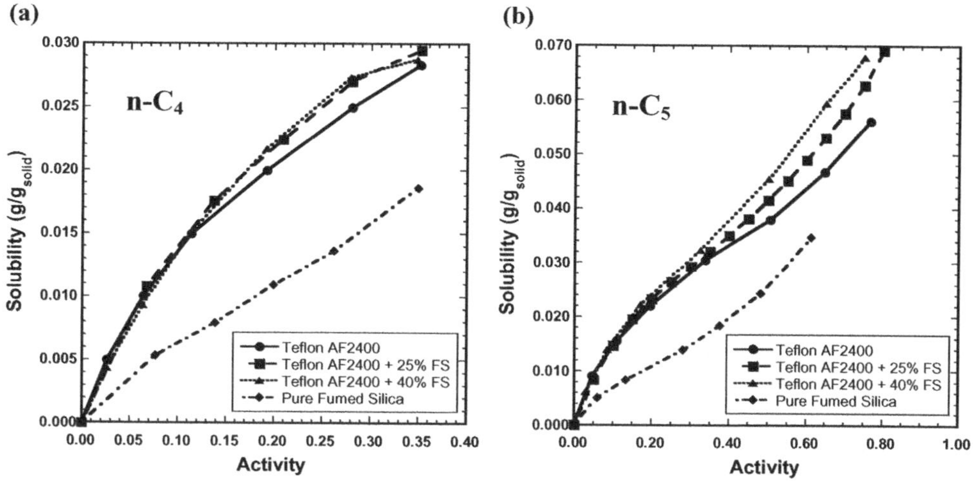

Figure 7.4 Solubility isotherms of (a) n-C_4 and (b) n-C_5 in Teflon AF 2400, AF 2400 + 25%wt FS and AF 2400 + 40%wt FS and adsorption onto pure silica at 25 °C

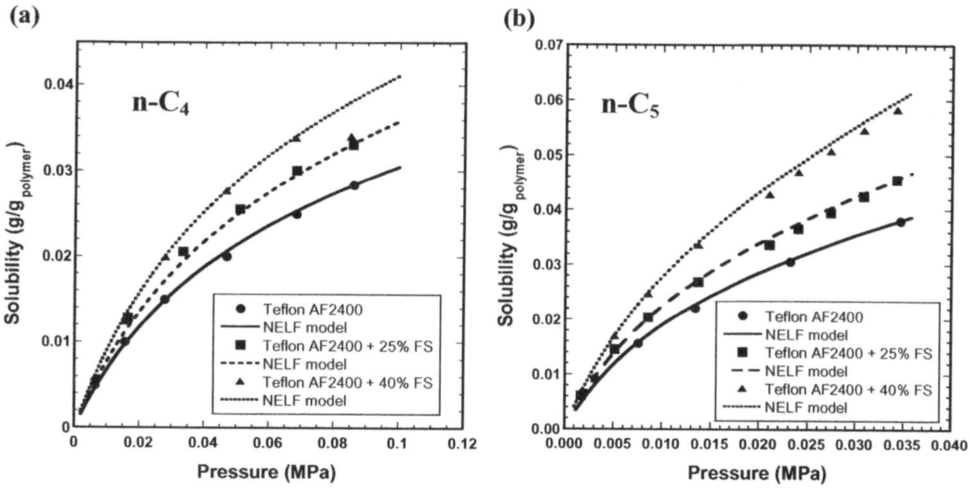

Figure 7.5 Experimental solubility of (a) n-C_4 and (b) n-C_5 in the polymeric phase of Teflon® AF 2400-based MMM at 25 °C and comparisons with NELF model

To understand better the underlying mechanism and to model the system behaviour it is more meaningful to report the solubility data in terms of mass of penetrant absorbed per unit mass of polymer phase alone, $C_{i,P}$; to that aim we apply the pseudo-additive approximation embodied by Equation (7.10) and report the values of $C_{i,P}$ thus obtained versus penetrant pressure. The results are shown in Figure 7.5, where they are also

Table 7.1 NELF model parameters for n-C_4, n-C_5, mixed matrices and their mixtures

Matrix	ρ^*(kg/L)	T* (K)	P* (MPa)	ρ_2^0 (g/cm^3)	*FFV* (Equation 7.11)
AF 2400 [18]	2.13	624	250	1.740[a]	0.320
AF 2400/25%wt FS				1.714[b]	0.330
AF 2400/40%wt FS				1.680[b]	0.344
Penetrant					
n-C_4	0.720	430	290		
n-C_5	0.749	451	305		
Penetrant	k_{ij}	k_{sw} (MPa^{-1})	**Matrix**		
n-C_4	0.14	0.13	AF 2400		
		0.20	AF 2400/25% wt FS		
		0.21	AF 2400/40% wt FS		
n-C_5	0.14	0.85	AF 2400		
		1.20	AF 2400/25% wt FS		
		2.00	AF 2400/40% wt FS		

[a] Experimental value.
[b] Value estimated based on the experimental solubility isotherms of *n*-C_4.

Table 7.2 Values of the coefficients A, B, E and F for the correlations of Equations (7.16) and (7.20)

	A	*B*	*E*	*F*
	(cm^2/s)		(g/g$_{pol}$)$^{-1}$	
n-C_4H_{10}	2.28×10^{16}	18.008	8.27×10^9	57.08
n-C_5H_{12}	5.06×10^{16}	18.355	3×10^{18}	118.93

compared with the simulations obtained by the NELF model using the parameter values reported in Table 7.1. The binary parameter used in the model calculations was adjusted on sorption data in the pure polymer and was kept constant even for composite materials, while the swelling coefficient was adjusted for each penetrant on the data of the isotherm at higher activities. In the absence of reliable experimental density data for the polymer phase alone of the mixed matrix, the polymer density value in the MMM was estimated by fitting the NELF model on the solubility isotherms of n-C_4. It can be seen from Table 7.2 that such values decrease with increasing FS content, and that the values thus obtained are good to calculate the solubility behaviour of n-C_5: this testifies that the density obtained in this way is a good representation of the actual polymer density and the corresponding FFV value will also be used to correlate the diffusivity data shown in the following.

7.4.2 Diffusivity

The diffusion coefficient was calculated in every sorption step of the pressure decay experiments, fitting Equation (7.19) to the transient data of mass uptake. Results for pure polymers and MMM are reported in Figure 7.6, where it can be seen that the addition of FS particles increases the diffusivity value, at fixed penetrant concentration in the polymeric phase. This is against the common thought that the addition of impermeable particles decreases the diffusion coefficient by increasing the tortuosity of the path that the penetrant molecules have to follow inside the membrane. However, the observed behaviour is completely consistent with previous results obtained for several gases and vapours in mixed matrices based on fumed silica and PTMSP, PMP and amorphous Teflon [2,3]. For those systems, the diffusivity enhancement was observed with every penetrant inspected, while the extent of the phenomenon varies with the specific penetrant. The same results are obtained considering diffusion coefficients obtained both from sorption and permeation experiments [2,3]. This trend has been considered as a direct consequence of the increase of the concentration of larger free volume elements with increasing filler content in Teflon AF 2400, which was documented with positron annihilation lifetime spectroscopy technique [3].

Also the variation of the apparent diffusivity with penetrant average concentration in the MMM shows a peculiar behaviour. Such a dependence is represented by the parameter β used in Equation (7.12), and we can notice that its value decreases with increasing filler content. An analogous phenomenon was already evidenced by Merkel *et al.*, although no clear explanation was provided [2,3]. It appears that the matrices characterized by high filler loadings have rather high infinite dilution diffusivities so that the favourable effect of swelling, which increases with penetrant concentration, results in smaller relative increases of diffusivity.

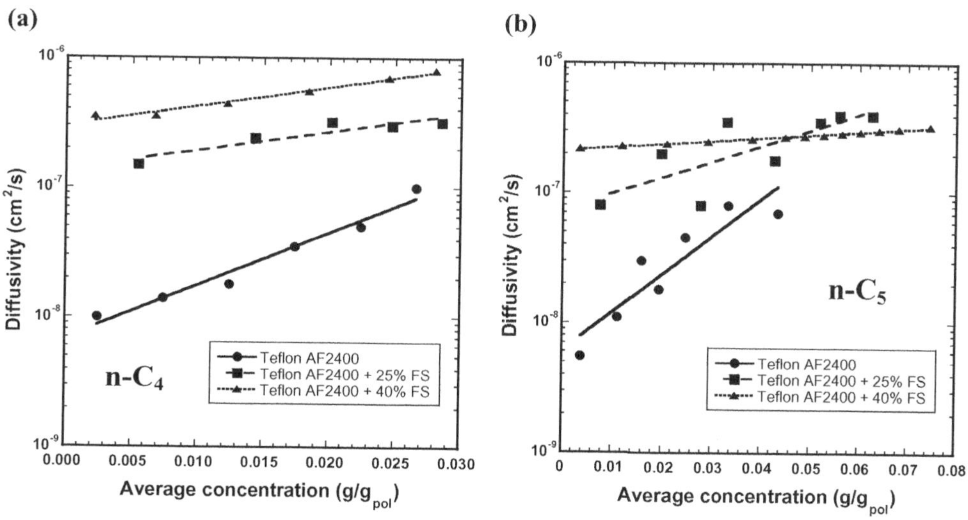

Figure 7.6 *Average diffusivity of (a) n-C_4 and (b) n-C_5 in Teflon AF 2400-based mixed matrices at 25 °C, versus average penetrant concentration*

In the following section, we will present quantitative correlations for both infinite dilution diffusion coefficients and β values as a function of the mixed matrix structural parameters.

7.4.3 Correlations for Diffusivity

The FFV in the unpenetrated polymer phase of the MMM calculated from the sorption isotherm of the test penetrant (n-C_4) through the NELF model can also be used to correlate the diffusivity data collected during sorption tests. The infinite dilution diffusion coefficient in the polymeric phase, $D_P(0)$ determined from the experimental diffusivity data can be related to the FFV initially present in the polymer matrix (FFV_M^0). As shown in Figure 7.7 for diffusion of n-butane in AF 2400, there is a correlation between the infinite dilution diffusion coefficient and $1/FFV_M^0$, that agrees with Equation (7.16). Similar behaviour is observed for n-C_5, with different values of the parameters A and B, as it can be seen from the results reported in Table 7.2. This is not surprising, because the values of these empirical parameters are expected to be a function of the penetrant nature and size.

Finally, it is found that the coefficient β appearing in the concentration dependence of diffusivity, Equation (7.12), is related to the FFV of the polymer phase of the MMM at infinite dilution (Figure 7.7c) through an exponential law:

$$\beta = E \cdot \exp(-F \cdot FFV_M^0) \tag{7.20}$$

The parameters E and F are both positive, vary with the penetrant (Figure 7.7d), and their values are reported in Table 7.2. The correlation expressed by Equation (7.20) actually states that the higher the initial free volume of the matrix, the lower the effect of concentration on diffusion, i.e. the lower the value of β.

7.4.4 Permeability and Selectivity

From the solubility and diffusivity data collected in the sorption experiments the value for the permeability of the two penetrants in the different mixed matrices was calculated. If the solution-diffusion model is considered to hold true, and Fick's law is suitable to represent the diffusive flux, the permeability can be calculated as the product of diffusivity and solubility coefficient. The calculated permeability is shown in Figures 7.8a and b, where it is clear that the addition of filler increases the permeability of the membrane for both penetrants.

In Figure 7.8c the ratio between the permeability in the composite and that in the pure polymer at infinite dilution for the two penetrants is reported as calculated from Equation (7.18), and clearly the addition of inorganic filler to the polymer enhances the permeability of the material.

From the knowledge of the permeability isotherms, the behaviour of the ideal selectivity of the composite can thus be calculated; as shown in Figure 7.9 there is an increase in the permeability of the larger penetrant as the initial FFV increases in the membrane. The case of 25% of filler in AF 2400 is peculiar, since a change in the selectivity is observed: the membrane is n-butane selective at low activity and becomes n-pentane selective at higher activity values.

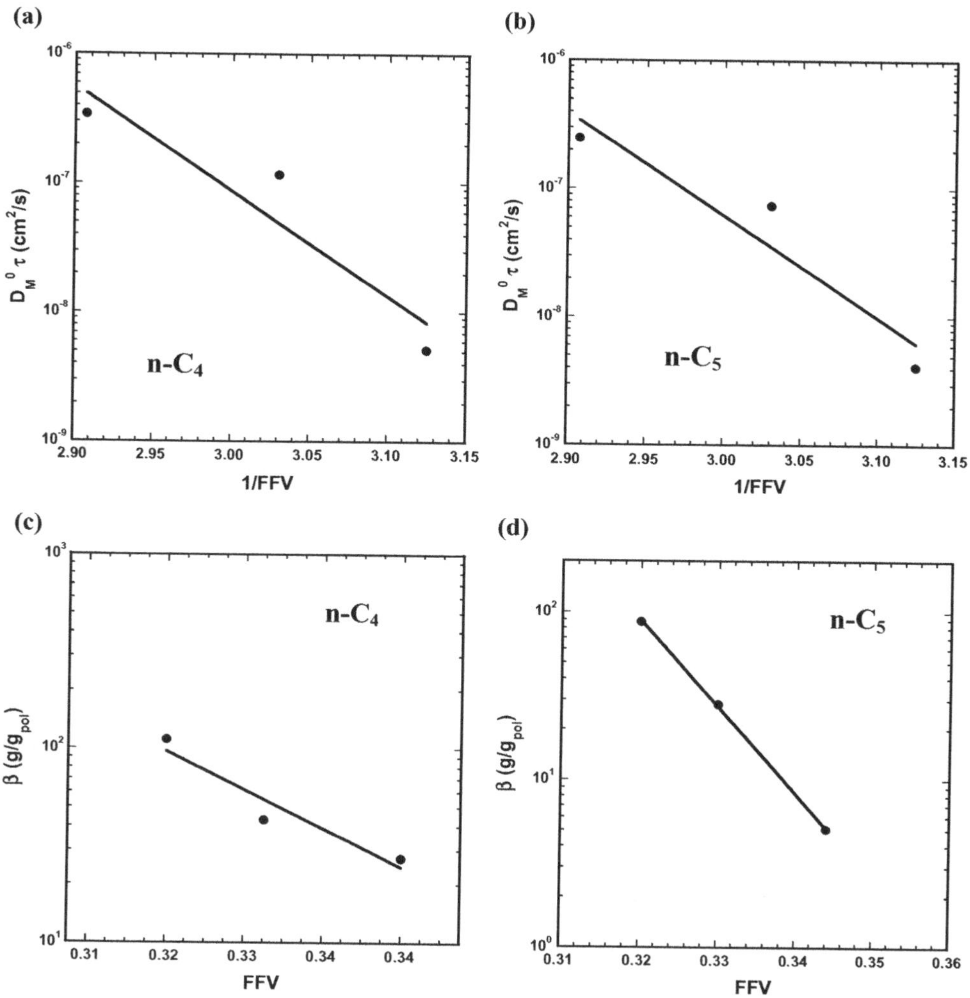

Figure 7.7 *Infinite dilution diffusion coefficient of (a) n-C_4 and (b) n-C_5 in the polymeric phase of mixed matrices based on AF 2400 as a function of FFV. β coefficient for diffusion of (c) n-C_4 and (d) n-C_5 in the mixed matrices based on AF 2400 as a function of FFV*

It is apparent that, for the case of the n-C_4/n-C_5 pair, the matrices considered are vapour-selective (solubility selective), meaning that they are more permeable to the more condensable penetrant, and that the selectivity increases with increasing penetrant pressure. It is also clear from Figure 7.9 that the n-C_5/n-C_4 selectivity of the matrices, at fixed permeability, decreases with increasing filler content. In any event, the behaviour of the matrices with the higher inorganic content would be much closer to a hypothetical trade-off curve traced for this particular mixture for vapour selective materials. Of course the ideal values of selectivity obtained above from pure vapour measurements could be different from the actual case of mixed gases in the feed, in which different interactions take place between the penetrants and between penetrants and polymer phase.

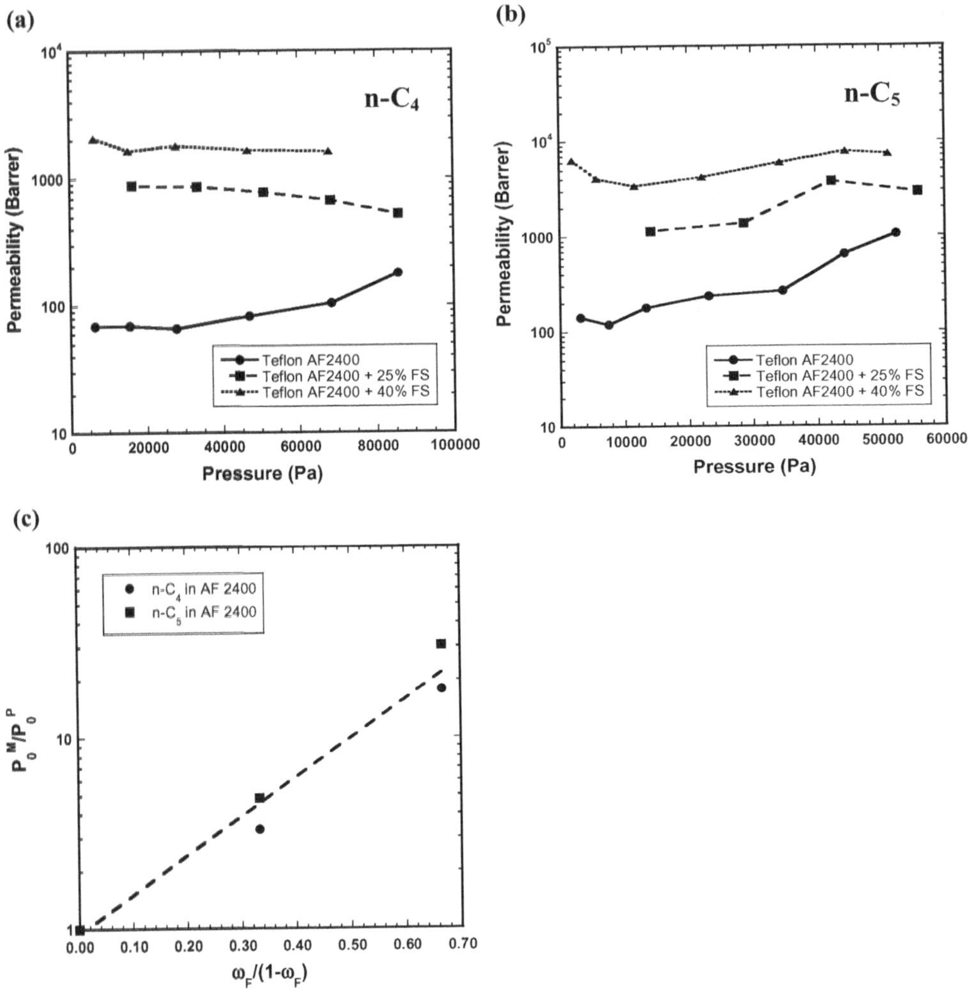

Figure 7.8 *Permeability of (a) n-C_4 and (b) n-C_5 in Teflon AF 2400-based mixed matrices at 25 °C, versus penetrant partial pressure. (c) Permeability ratio at infinite dilution*

7.5 Conclusions

Mixed matrix membranes based on fluorinated high free volume matrices loaded with fumed silica nanoparticles show attractive features, especially if compared to unloaded polymeric membranes. The addition of filler changes the FFV of the glassy polymeric matrix, thus positively affecting both solubility and diffusivity. A simple procedure has been proposed to calculate the relevant properties of the MMM, which is based on limited experimental data represented by the solubility isotherm of a test penetrant and few values of diffusion coefficients. The unexpected behaviour shown by the MMM obtained by using nano-filler loadings can be properly described by the method developed. The

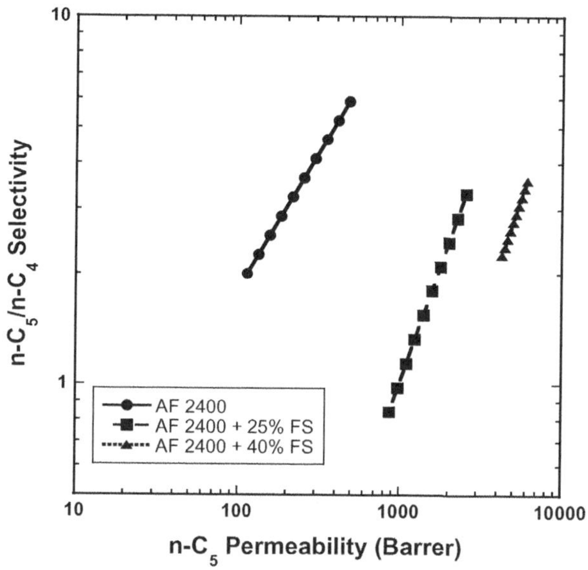

Figure 7.9 n-C_5/n-C_4 selectivity in Teflon AF 2400-based mixed matrices at 25 °C, versus n-C_5 permeability

method has proved to be useful also for the determination of the diffusion coefficient and for an estimation of the permeability in the composite materials. The evaluation of the contributions of the permeability can hence be used to determine the ideal selective behaviour of the composite membrane.

We believe that the procedure can be properly applied also for the case of MMM obtained by loading glassy matrices with other nano-fillers and can be a useful tool to develop better mixed matrix membranes.

Acknowledgements

The financial support of EU Strep NMP3-CT-2005-013644 (MultiMatDesign) is gratefully acknowledged. This work could not be possible without the experimental support of Lelio Basile. We also are grateful to Dr Tim Merkel of MTR for providing some of the samples characterized and for his precious advice.

References

(1) T.-S. Chung, L. Y. Jiang, Y. Li, S. Kulprathipanja, Mixed matrix membranes (MMMs) comprising organic polymers with dispersed inorganic fillers for gas separation, *Progress in Polymer Science*, **32**, 483 (2007).

(2) T. C. Merkel, B. D. Freeman, R. J. Spontak, Z. He, I. Pinnau, P. Meakin, Ultrapermeable, reverse-selective nanocomposite membranes. *Science*, **296**, 519–522 (2002).

(3) T. C. Merkel, Z. He, I. Pinnau, B. D. Freeman, P. Meakin, A. J. Hill, Sorption and transport in poly (2,2-bis(trifluoromethyl)-4,5-difluoro-1,3-dioxole-co-tetrafluoroethylene) containing nanoscale fumed silica, *Macromolecules*, **36**, 8406–8414 (2003).

(4) M. G. De Angelis, G. C. Sarti, Solubility and diffusivity of gases in mixed matrix membranes containing hydrophobic fumed silica: correlations and predictions based on the NELF model, *Ind. Eng. Chem. Res.*, **47**, 5214–5226 (2008).

(5) C. Maxwell, Treatise on Electricity and Magnetism. Vol. 1, Oxford University Press, London, 1873.

(6) M. Giacinti Baschetti, M. G. De Angelis, F. Doghieri, G. C. Sarti, Solubility of gases in polymeric membranes, in Chemical Engineering: Trends and Developments, M. A. Galan and E. M. d. Valle, Editors, John Wiley & Sons, Ltd., Chichester, 41–61, 2005.

(7) M. Giacinti Baschetti, F. Doghieri, G. C. Sarti, Solubility in glassy polymers: correlations through the non-equilibrium lattice fluid model, *Ind. Eng. Chem. Res.*, **40**, 3027–3037 (2001).

(8) G. C. Sarti, F. Doghieri, Predictions of the solubility of gases in glassy polymers based on the NELF model, *Chem. Eng. Sci.*, **19**, 3435–3447 (1998).

(9) F. Doghieri, M. Quinzi, D. G. Rethwisch, G. C. Sarti, Predicting gas solubility in glassy polymers through non-equilibrium EOS, in: Advanced Materials for Membrane Separations, ACS Symposium Series No. 876 Pinnau Freeman Eds., 2004, pp. 74–90.

(10) I. C. Sanchez, R. H. Lacombe, An elementary molecular theory of classical fluids. pure fluids, *J. Phys. Chem.*, **80**, 2352–2362 (1976).

(11) I. C. Sanchez, R. H. Lacombe, Statistical thermodynamics of polymer solutions, *Macromolecules*, **11**, 1145–1156 (1978).

(12) R. G. Wissinger, M. E. Paulaitis, Swelling and sorption in polymer-CO_2 mixtures at elevated pressures, *J. Polym. Sci., Part B: Polym. Phys.*, **25**, 2497–2510 (1987).

(13) S. Jordan, W. J. Koros, Free volume distribution model of gas sorption and dilation in glassy polymers, *Macromolecules*, **28**, 2228 (1995).

(14) G. K. Fleming, W. J. Koros, Carbon dioxide conditioning effects on sorption and volume dilation behaviour for bisphenol A-polycarbonate, *Macromolecules*, **23**, 1353 (1990).

(15) W. van Krevelen, Properties of Polymers, Elsevier, Amsterdam, 1990.

(16) M. H. Cohen, D. J. Turnbull, Molecular transport in liquids and glasses, *J. Chem. Phys.*, **31**, 1164–1169 (1959).

(17) J. Crank, The Mathematics of Diffusion, 2nd Edition, Oxford University Press, Oxford, 1975.

(18) M. G. De Angelis, T. C. Merkel, V. I. Bondar, B. D. Freeman, F. Doghieri, G. C. Sarti, Gas sorption and dilation in poly(2,2-bistrifluoromethyl-4,5-difluoro-1,3-dioxole-co-tetrafluoroethylene): comparison of experimental data with predictions of the non-equilibrium lattice fluid model, *Macromolecules*, **35**, 1276–1288 (2002).

8

Physical and Gas Transport Properties of Hyperbranched Polyimide–Silica Hybrid Membranes

Tomoyuki Suzuki, Yasuharu Yamada, Jun Sakai and Kumi Itahashi
Department of Biomolecular Engineering, Kyoto Institute of Technology, Matsugasaki, Kyoto, Japan

8.1 Introduction

Gas separation processes using polymeric membranes have greatly been developed during the last three decades. Especially, polyimides have been of great interest in gas separation membranes because of their high gas selectivity and excellent thermal and mechanical properties. In this regard, gas transport properties have been reported for a large number of polyimides [1–4]. In recent years, novel triamine-based hyperbranched polyimides (HBPIs) have been synthesized and characterized with the aim to assess their potential as high-performance gas separation materials. Fang *et al.* have synthesized HBPIs derived from a triamine, tris(4-aminophenyl)amine (TAPA), and commercially available dianhydrides and studied their physical and gas transport properties [5,6]. These studies demonstrated that HBPIs have good gas separation performance compared to linear-type polyimides. In our previous work, gas transport properties of the HBPI prepared by polycondensation of a triamine, 1,3,5-tris(4-aminophenoxy)benzene (TAPOB), and a dianhydride, 4,4′-(hexafluoroisopropylidene)diphthalic anhydride (6FDA), have been investigated, and it has been found that the 6FDA-TAPOB HBPI exhibits high gas permeability and selectivity arising from the characteristic hyperbranched structure [7].

Membrane Gas Separation Edited by Yuri Yampolskii and Benny Freeman

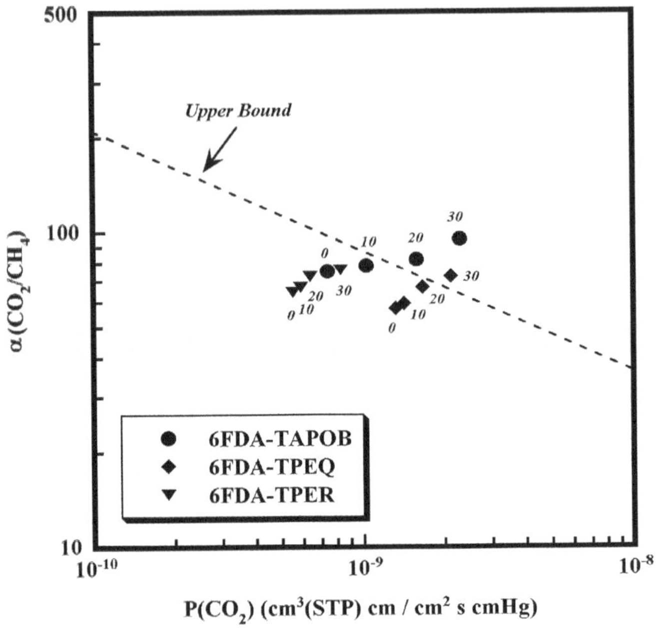

Figure 8.1 *Ideal CO_2/CH_4 selectivity ($\alpha(CO_2/CH_4)$) of 6FDA-based polyimide–silica hybrid membranes plotted against CO_2 permeability coefficient; attached number represents silica content (wt%) in the membrane*

Organic–inorganic hybrids (often called mixed matrix membranes) are attractive materials since they generally possess desirable properties (combination of high permeability and permselectivity). Hybridization with inorganic compounds has been focused on the modification of polyimides in order to improve their thermal, mechanical and gas transport properties [8–10]. We have also reported that gas transport properties of 6FDA-TAPOB HBPI–silica hybrid membranes prepared by sol–gel reaction using tetramethoxysilane (TMOS) have been investigated and compared with those of linear-type 6FDA-based polyimide–silica hybrid membranes synthesized with 1,4-bis(4-aminophenoxy)benzene (TPEQ) and 1,3-bis(4-aminophenoxy)benzene (TPER) as diamine monomers [11]. Figure 8.1 shows ideal CO_2/CH_4 selectivity ($\alpha(CO_2/CH_4)$) of 6FDA-based polyimide–silica hybrid membranes. It is found that gas permeability and $\alpha(CO_2/CH_4)$ of the HBPI–silica hybrid membranes significantly increase with increasing silica content, suggesting characteristic distribution and interconnectivity of free volume elements created by the incorporation of silica [12,13].

In this paper, physical and gas transport properties of HBPIs synthesized with various dianhydride monomers and their silica hybrid membranes prepared by sol–gel reaction with two kinds of alkoxysilanes, TMOS and methyltrimethoxysilane (MTMS), individually or simultaneously, are discussed. The methyl group of MTMS will prevent the formation of robust three-dimensional Si–O–Si network in the hybrid membranes. It is expected that the loose Si–O–Si network formed from MTMS induces unique physical and gas transport properties of the hybrid membranes.

8.2 Experimental

8.2.1 Materials

TAPOB was synthesized by reduction of 1,3,5-tris(4-nitrophenoxy)benzene with palladium carbon and hydrazine in methanol [14]. 6FDA was kindly supplied from Daikin Industries, Ltd. Pyromellitic dianhydride (PMDA) and 4,4′-oxidiphthalic anhydride (ODPA) were obtained from Daicel Chemical Industries, Ltd. and Manac Incorporated, respectively. TMOS, MTMS, 3-aminopropyltrimethoxysilane (APTrMOS), and *N,N*-dimethylacetamide (DMAc) were purchased from Aldrich. Chemical structures of monomers and alkoxysilanes are shown in Figure 8.2.

8.2.2 Polymerization

3 mmol of dianhydride (6FDA, PMDA, and ODPA) was dissolved in 40 mL of DMAc in a 100 mL three-neck flask under N_2 flow at room temperature. To this solution, 1.6 mmol of TAPOB in 20 mL of DMAc was added dropwise through a syringe with stirring for 3 h. After that 0.4 mmol of APTrMOS was added in the reaction mixture with stirring for 1 h to afford hyperbranched polyamic acid.

8.2.3 Membrane Formation

The HBPI–silica hybrid membranes were prepared by thermal imidization and sol–gel reaction with two kinds of alkoxysilanes, TMOS and/or MTMS. Appropriate amounts of TMOS and/or MTMS and ion exchange water were added in the DMAc solution of the hyperbranched polyamic acids. For TMOS/MTMS combined system, TMOS and MTMS were added to achieve equivalent weight fraction of silica components arising from TMOS and MTMS. The mixed solutions were stirred for 24 h and cast on PET films and

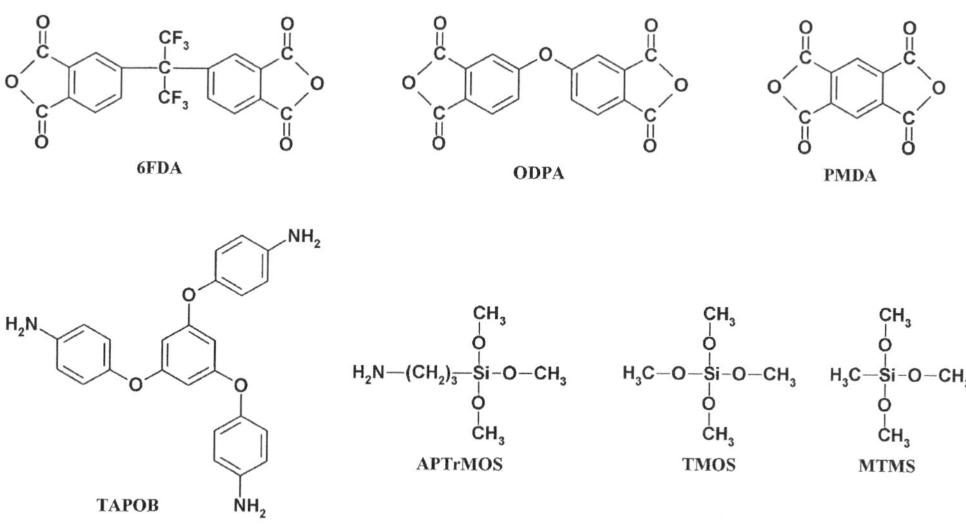

Figure 8.2 *Chemical structures of monomers and alkoxysilanes*

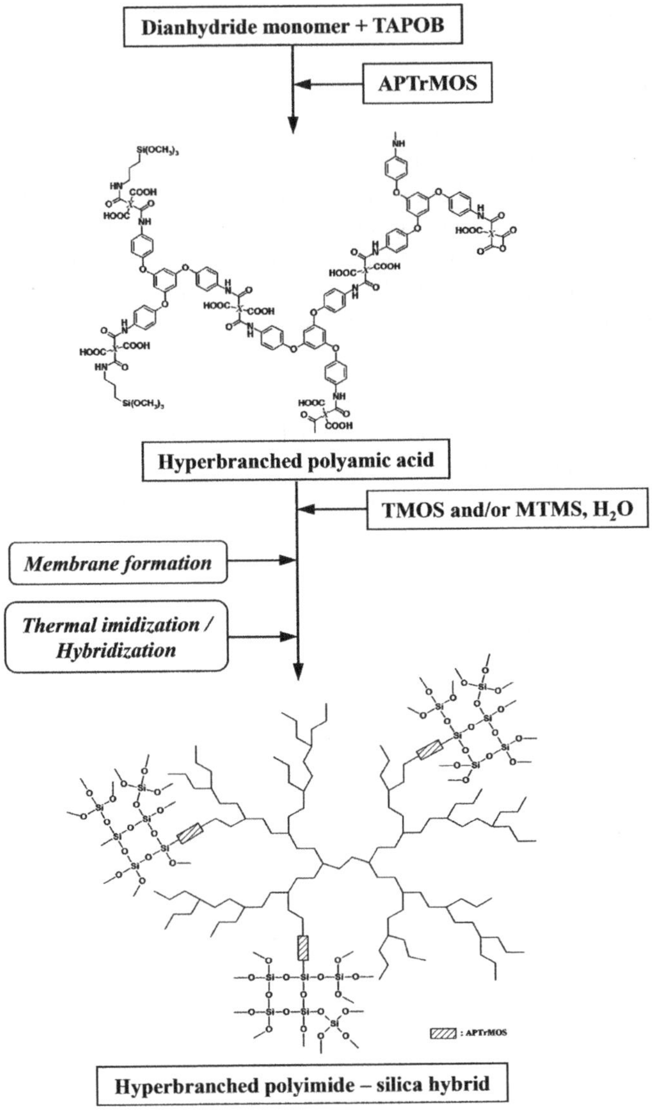

Scheme 8.1 *Preparation of hyperbranched polyimide–silica hybrid membrane*

dried at 85 °C for 3 h. The prepared membranes were peeled off and subsequently imidized at 100 °C for 1 h, 200 °C for 1 h, and 300 °C for 1 h in a heating oven under N_2 flow. The thickness of the HBPI–silica hybrid membranes was about 30 µm. A schematic diagram for the hybrid membrane preparation is shown in Scheme 8.1.

8.2.4 Measurements

Infrared (IR) spectra were recorded on a JASCO FT/IR-460 plus. UV/VIS optical transmittances were investigated by a JASCO V-530 UV/VIS spectrometer at a wavelength of 200–800 nm. Thermogravimetric–differential thermal analysis (TG-DTA) experiments were performed with a Seiko TG/DTA6300 at a heating rate of 10 °C/min under air flow. Thermal mechanical analysis (TMA) measurements were carried out using a Seiko TMA/SS6100 at a heating rate of 5 °C/min under N_2 flow. CO_2, O_2, N_2 and CH_4 permeation measurements were carried out by a constant volume/variable pressure apparatus at 76 cmHg and 25 °C. The permeability coefficient, P (cm^3(STP) cm/cm^2 s cmHg), was determined by the following equation [15];

$$P = \frac{22414}{A} \frac{L}{p} \frac{V}{RT} \frac{dp}{dt} \tag{8.1}$$

where A is the membrane area (cm^2), L is the membrane thickness (cm), p is the upstream pressure (cmHg), V is the downstream volume (cm^3), R is the universal gas constant (6236.56 cm^3 cmHg/mol K), T is the absolute temperature (K), and dp/dt is the permeation rate (cmHg/s). The gas permeability coefficient can be explained on the basis of the solution-diffusion mechanism, which is represented by the following equation [16,17];

$$P = D \times S \tag{8.2}$$

where D (cm^2/s) is the diffusion coefficient and S $(\mathrm{cm}^3(\mathrm{STP})/\mathrm{cm}^3_{\mathrm{polym}}\ \mathrm{cmHg})$ is the solubility coefficient. The diffusion coefficient was calculated by the time-lag method represented by the following equation [18];

$$D = \frac{L^2}{6\theta} \tag{8.3}$$

where θ (s) is the time-lag. Density of the pure HBPIs was measured by the floating method with bromoform and 2-propanol at 25 °C. According to the group contribution method, fractional free volume (FFV) of a polymer can be estimated by the following equation [19];

$$\mathrm{FFV} = \frac{\mathrm{V_{sp}} - 1.3\mathrm{V_w}}{\mathrm{V_{sp}}} \tag{8.4}$$

where $_{\mathrm{sp}}$ (cm^3/mol) is the specific molar volume which can be calculated from the experimental density, ρ (g/cm^3), and $\mathrm{V_w}$ (cm^3/mol) is the van der Waals volume of the repeat unit.

8.3 Results and Discussion

8.3.1 Polymer Characterization

Figure 8.3 shows FT-IR spectra of the 6FDA-TAPOB HBPI–silica hybrid membranes containing 30 wt% of silica prepared with various alkoxysilanes. The bands observed around 1784 cm^{-1} (C=O asymmetrical stretching), 1725 cm^{-1} (C=O symmetrical stretching), 1378 cm^{-1} (C–N stretching), and 723 cm^{-1} (C=O bending) are characteristic absorption bands of polyimides [6,20]. In contrast, the characteristic band of polyamic acids around 1680 cm^{-1} is not found. These results indicate that the prepared membranes are well imidized. For MTMS and TMOS/MTMS combined systems, the bands around

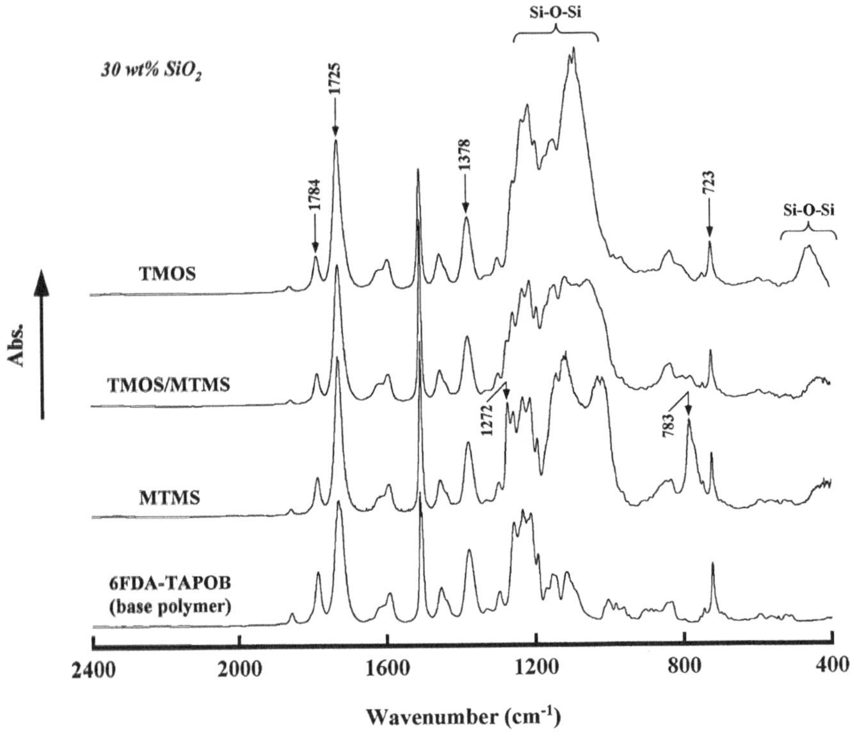

Figure 8.3 *FT-IR spectra of 6FDA-TAPOB HBPI–silica hybrid membranes (silica content = 30 wt%)*

1272 and 783 cm^{-1}, assigned to Si–CH$_3$ stretching and rocking, respectively [9], are also observed. The absorption bands at approximately 960–1280 and 410–480 cm^{-1}, identified as asymmetric stretching and rocking modes of the Si–O–Si group, respectively [9,21], are observed, indicating the formation of Si–O–Si network in the hybrid membranes. For the TMOS system, the strong absorption bands of the Si–O–Si asymmetric stretching appear around 1180 and 1090 cm^{-1}. On the other hand, for the MTMS system, they appear around 1120 and 1030 cm^{-1} with the red shift in turn to a lower wavenumber. The red shift of the absorption bands reveals decreasing bonding energy and increasing bond distance of the bond between Si and O atoms in the Si–O–Si network [21]. Therefore, for the MTMS system, the red shift of the Si–O–Si peaks suggests the formation of a looser silica network caused by the methyl group in MTMS which function as a terminating substituent in the formation of Si–O–Si bonds to perturb microstructure of resulting silica matrix. In addition, it is pointed out that the absorption bands of the Si–O–Si asymmetric stretching for the TMOS/MTMS combined system are the overlap of those for TMOS and MTMS systems, suggesting the formation of Si–O–Si network with an intermediate structure. Similar behaviours are observed for PMDA and ODPA systems.

Optical transmittances of the hybrid membranes are shown in Table 8.1. The hybrid membranes have high homogeneity and silica particles are finely dispersed in the HBPIs

Table 8.1 *Physical properties of HBPI–silica hybrid membranes*

Sample	Transmittance[1] (%)	T_g (°C)	T_d^5 (°C)	Silica content[2] (wt%)	CTE[3] (ppm/°C)
6FDA-TAPOB	88.9	282	457	0	54
TMOS, 10 wt% SiO_2	89.9	305	490	10	47
TMOS, 20 wt% SiO_2	90.2	311	496	20	38
TMOS, 30 wt% SiO_2	90.3	318	509	30	31
TMOS/MTMS, 10 wt% SiO_2	89.6	286	487	10	50
TMOS/MTMS, 20 wt% SiO_2	90.1	299	494	18	48
TMOS/MTMS, 30 wt% SiO_2	89.8	301	502	29	46
TMOS/MTMS, 40 wt% SiO_2	89.1	303	509	40	41
TMOS/MTMS, 50 wt% SiO_2	91.6	n.d.[4]	515	48	–
MTMS, 10 wt% SiO_2	90.2	278	471	10	59
MTMS, 20 wt% SiO_2	90.1	281	480	19	66
MTMS, 30 wt% SiO_2	91.4	287	496	30	73
MTMS, 40 wt% SiO_2	91.3	291	501	39	74
MTMS, 50 wt% SiO_2	91.7	n.d.[4]	505	48	–
ODPA-TAPOB	85.8	272	470	0	52
TMOS, 10 wt% SiO_2	87.0	276	494	10	45
TMOS, 20 wt% SiO_2	87.5	289	504	20	38
TMOS, 30 wt% SiO_2	88.2	291	506	30	32
TMOS/MTMS, 10 wt% SiO_2	86.6	272	484	12	49
TMOS/MTMS, 20 wt% SiO_2	88.3	279	491	21	47
TMOS/MTMS, 30 wt% SiO_2	90.5	281	501	31	41
MTMS, 10 wt% SiO_2	86.1	259	474	11	54
MTMS, 20 wt% SiO_2	87.5	263	484	21	61
MTMS, 30 wt% SiO_2	89.8	267	496	30	64
PMDA-TAPOB	85.2	326	464	0	35
TMOS, 10 wt% SiO_2	85.9	n.d.[4]	490	14	35
TMOS, 20 wt% SiO_2	85.1	n.d.[4]	497	22	32
TMOS, 30 wt% SiO_2	86.6	n.d.[4]	505	31	29
TMOS/MTMS, 10 wt% SiO_2	86.7	n.d.[4]	482	12	42
TMOS/MTMS, 20 wt% SiO_2	86.7	n.d.[4]	489	21	39
TMOS/MTMS, 30 wt% SiO_2	88.4	n.d.[4]	501	31	39
MTMS, 10 wt% SiO_2	86.5	n.d.[4]	475	13	42
MTMS, 20 wt% SiO_2	87.7	n.d.[4]	484	24	53
MTMS, 30 wt% SiO_2	88.9	n.d.[4]	500	33	56

1) Optical transmittance at 600 nm.
2) Determined from the residual at 800 °C.
3) CTE at 100–150 °C.
4) Not detected.

because of their good transparency, similar to corresponding pure HBPIs without silica. The high homogeneity is considered to be maintained not only by APTrMOS moiety which functions as covalent bond parts between organic and inorganic components but also by the characteristic hyperbranched structure of the molecular chains.

Thermal properties of the hybrid membranes were investigated by TG-DTA and TMA measurements. From T_g curves it was again demonstrated that the prepared membranes are well imidized because no weight-losses attributed to dehydration by imidization are observed all through the measurements. Glass transition temperature (T_g) determined from DTA curves and 5% weight-loss temperature (T_d^5) of the hybrid membranes are summarized in Table 8.1 in addition to the silica content determined from the residual at 800 °C. It is confirmed from the residual that all hybrid membranes contain appropriate amount of silica as expected. The T_g and T_d^5 values of the 6FDA-TAPOB HBPI–silica hybrid membranes are plotted against silica content in Figure 8.4. T_d^5 of the hybrid membranes increases with increasing silica content. This result indicates the increase in thermal stability of the HBPIs by hybridization with silica. The T_g of the TMOS system considerably increases with increasing silica content, suggesting the formation of robust three-dimensional Si–O–Si network. However, for the MTMS system, the T_g shows a minimum value at low silica content region, and the T_g of the TMOS/MTMS combined system is lower than that of the TMOS system. This fact suggests that the introduction of MTMS leads to an insufficient formation of three-dimensional Si–O–Si network, which brings about less constraint of molecular chains in the hybrid membranes compared to the hybrid membranes prepared solely with TMOS.

Coefficients of thermal expansion (CTEs) from 100 to 150 °C of the hybrid membranes are listed in Table 8.1. The CTE of the TMOS system greatly decreases with increasing silica content, indicating the enhancement of thermal mechanical stability of the HBPIs by the formation of robust three-dimensional Si–O–Si network. In contrast, the CTE of

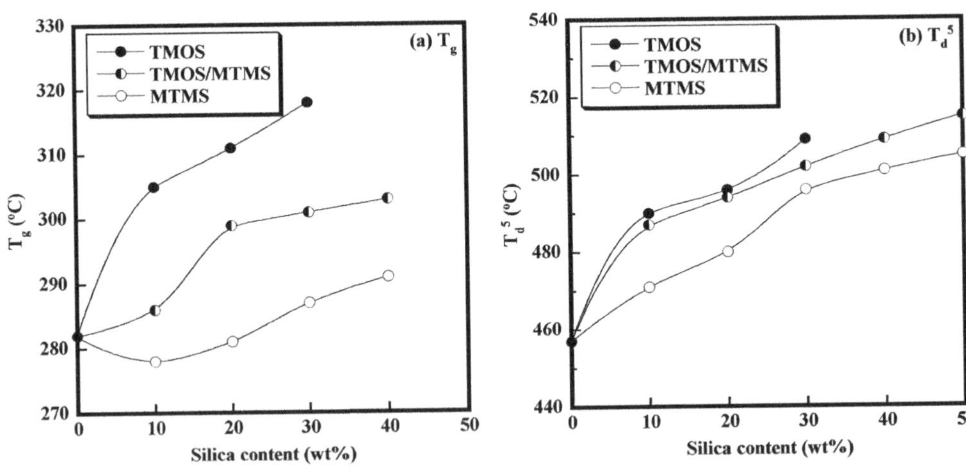

Figure 8.4 *Glass transition temperature (T_g) (a) and 5% weight-loss temperature (T_d^5) (b) of 6FDA-TAPOB HBIP–silica hybrid membranes plotted against silica content*

the MTMS system increases with increasing silica content. The increased CTE might be caused by a loose Si–O–Si network which induces an effective disruption of molecular chains packing and, as a result, provides a large amount of free volume elements [22]. The increased free volume decreases the hindrance of thermal expansion of molecular chains. The CTE of the TMOS/MTMS combined system shows the intermediate value between those of TMOS and MTMS systems.

8.3.2 Gas Transport Properties of HBPI–Silica Hybrid Membranes

Gas permeability, diffusion and solubility coefficients of the HBPI–silica hybrid membranes are summarized in Tables 8.2–8.4, respectively. Gas permeability of pure 6FDA-TAPOB HBPI is higher than that of pure PMDA- and ODPA-TAPOB HBPIs, owing to its higher diffusivity. It is well known that gas diffusivity of polymers strongly depends on the fractional free volume (FFV) [23,24]. Densities of pure 6FDA-, PMDA- and ODPA-TAPOB HBPIs are 1.433, 1.399 and 1.396 g/cm^3, respectively, and the FFVs of them calculated from Equation (8.4) are 0.165, 0.138 and 0.129, respectively. The highest diffusivity of the 6FDA-TAPOB HBPI is, therefore, attributed to the highest FFV. Figure 8.5 shows CO_2 permeability, diffusion and solubility coefficients of 6FDA-TAPOB HBPI–silica hybrid membranes plotted against silica content. Gas permeability coefficients of the hybrid membranes increase with increasing silica content. The increased permeability suggests additional formation of free volume holes effective for gas transport behaviour. Similar enhancement of free volume holes has been reported by Merkel *et al.* for high-free-volume, glassy polymer–nanosilica composites, and they have concluded that the nanosilica particles provide disruption of polymer chain packing and affect gas transport parameters [25,26]. He *et al.* have also reported that nano-sized inorganic fillers incorporated into high-free-volume, glassy polymers with rigid molecular chains can function as a spacer material which disrupts molecular chain packing, leading to increases in fractional free volume and gas permeability [27]. It is worth noting, for MTMS and TMOS/MTMS combined systems, significant improvements of gas permeability are mainly caused by the increased diffusivity observed. The enormous improvements of gas diffusivity suggest large enhancements of mean size and maybe total amount of free volume elements attributed to the loose Si–O–Si networks by the incorporation of methyl groups for MTMS and TMOS/MTMS combined systems.

8.3.3 O_2/N_2 and CO_2/CH_4 Selectivities of HBPI–Silica Hybrid Membranes

The ideal gas selectivity for the combination of gases A and B (α(A/B)) is defined by the following equation [28];

$$\alpha(\mathrm{A/B}) = \frac{P(\mathrm{A})}{P(\mathrm{B})} = \frac{D(\mathrm{A})}{D(\mathrm{B})} \times \frac{S(\mathrm{A})}{S(\mathrm{B})} = \alpha^{\mathrm{D}}(\mathrm{A/B}) \times \alpha^{\mathrm{S}}(\mathrm{A/B}) \tag{8.5}$$

where α^D(A/B) is the diffusivity selectivity and α^S(A/B) is the solubility selectivity. The O_2/N_2 and CO_2/CH_4 selectivities of the hybrid membranes are listed in Tables 8.2–8.4

Table 8.2 Permeability coefficients and ideal selectivity of HBPI–silica hybrid membranes at 76 cmHg and 25 °C

Sample	$P \times 10^{10}$, (cm^3(STP)cm/cm^2 s cmHg)				$\alpha(O_2/N_2)$	$\alpha(CO_2/CH_4)$
	CO_2	O_2	N_2	CH_4		
6FDA-TAPOB	7.4	1.5	0.23	0.098	6.8	75
TMOS, 10 wt% SiO_2	10	2.0	0.31	0.13	6.6	79
TMOS, 20 wt% SiO_2	13	2.1	0.32	0.16	6.7	82
TMOS, 30 wt% SiO_2	23	3.0	0.46	0.24	6.6	95
TMOS/MTMS, 10 wt% SiO_2	13	2.7	0.41	0.20	6.6	65
TMOS/MTMS, 20 wt% SiO_2	28	5.4	0.89	0.45	6.0	62
TMOS/MTMS, 30 wt% SiO_2	44	7.9	1.3	0.74	5.9	59
TMOS/MTMS, 40 wt% SiO_2	72	12	2.3	1.4	5.2	53
TMOS/MTMS, 50 wt% SiO_2	133	22	4.5	2.8	5.0	48
MTMS, 10 wt% SiO_2	23	4.4	0.75	0.45	5.9	51
MTMS, 20 wt% SiO_2	53	9.5	1.9	1.5	4.9	34
MTMS, 30 wt% SiO_2	99	18	4.1	4.1	4.3	24
MTMS, 40 wt% SiO_2	158	28	7.6	10	3.7	15
MTMS, 50 wt% SiO_2	251	46	14	20	3.4	12
ODPA-TAPOB	0.63	0.13	0.013	0.0064	10	98
TMOS, 10 wt% SiO_2	0.93	0.18	0.017	0.0094	11	98
TMOS, 20 wt% SiO_2	0.88	0.17	0.017	0.0078	10	112
TMOS, 30 wt% SiO_2	1.2	0.22	0.024	0.011	9.0	111
TMOS/MTMS, 10 wt% SiO_2	1.3	0.27	0.029	0.022	9.1	58
TMOS/MTMS, 20 wt% SiO_2	2.6	0.48	0.062	0.038	7.7	68
TMOS/MTMS, 30 wt% SiO_2	6.7	1.2	0.16	0.11	7.3	62
MTMS, 10 wt% SiO_2	1.8	0.37	0.052	0.036	7.2	51
MTMS, 20 wt% SiO_2	3.9	0.85	0.13	0.092	6.7	43
MTMS, 30 wt% SiO_2	17	3.1	0.61	0.64	5.1	26
PMDA-TAPOB	3.3	0.59	0.088	0.066	6.7	50
TMOS, 10 wt% SiO_2	4.1	0.69	0.11	0.069	6.4	60
TMOS, 20 wt% SiO_2	4.6	0.71	0.10	0.063	7.0	73
TMOS, 30 wt% SiO_2	7.1	1.0	0.15	0.073	6.8	97
TMOS/MTMS, 10 wt% SiO_2	4.8	0.76	0.12	0.081	6.5	59
TMOS/MTMS, 20 wt% SiO_2	13	1.9	0.31	0.22	6.1	58
TMOS/MTMS, 30 wt% SiO_2	33	4.9	0.89	0.68	5.5	48
MTMS, 10 wt% SiO_2	9.3	1.5	0.25	0.21	5.9	44
MTMS, 20 wt% SiO_2	30	4.8	1.0	1.0	4.6	30
MTMS, 30 wt% SiO_2	55	8.8	2.2	2.5	4.0	22

Table 8.3 *Diffusion coefficients and diffusivity selectivity of HBPI–silica hybrid membranes at 76 cmHg and 25 °C*

Sample	$D \times 10^8$ (cm^2/s)				$\alpha^D(O_2/N_2)$	$\alpha^D(CO_2/CH_4)$
	CO_2	O_2	N_2	CH_4		
6FDA-TAPOB	0.30	1.4	0.25	0.028	5.8	11
TMOS, 10 wt% SiO_2	0.35	1.5	0.29	0.026	5.2	13
TMOS, 20 wt% SiO_2	0.37	1.3	0.25	0.030	5.3	12
TMOS, 30 wt% SiO_2	0.57	1.7	0.29	0.040	5.8	14
TMOS/MTMS, 10 wt% SiO_2	0.59	2.1	0.46	0.060	4.7	9.9
TMOS/MTMS, 20 wt% SiO_2	1.2	4.0	0.90	0.090	4.4	13
TMOS/MTMS, 30 wt% SiO_2	1.7	5.9	1.2	0.17	5.1	10
TMOS/MTMS, 40 wt% SiO_2	2.6	7.9	1.8	0.25	4.4	11
TMOS/MTMS, 50 wt% SiO_2	5.2	14	3.7	0.63	3.7	8.2
MTMS, 10 wt% SiO_2	1.0	3.8	0.83	0.13	4.5	8.1
MTMS, 20 wt% SiO_2	2.6	8.7	2.2	0.42	4.0	6.0
MTMS, 30 wt% SiO_2	5.4	16	4.8	1.2	3.4	4.4
MTMS, 40 wt% SiO_2	9.9	28	9.3	3.7	3.0	2.7
MTMS, 50 wt% SiO_2	17	46	17	6.8	2.6	2.5
ODPA-TAPOB	0.030	0.16	0.021	0.0019	7.7	16
TMOS, 10 wt% SiO_2	0.039	0.19	0.025	0.0025	7.6	16
TMOS, 20 wt% SiO_2	0.033	0.17	0.022	0.0021	7.6	16
TMOS, 30 wt% SiO_2	0.040	0.19	0.024	0.0026	7.6	15
TMOS/MTMS, 10 wt% SiO_2	0.071	0.36	0.060	0.011	6.1	6.3
TMOS/MTMS, 20 wt% SiO_2	0.13	0.54	0.10	0.016	5.2	8.4
TMOS/MTMS, 30 wt% SiO_2	0.34	1.3	0.24	0.027	5.4	13
MTMS, 10 wt% SiO_2	0.12	0.54	0.11	0.012	5.0	9.4
MTMS, 20 wt% SiO_2	0.33	1.1	0.23	0.037	4.6	9.1
MTMS, 30 wt% SiO_2	1.1	3.6	0.89	0.25	4.1	4.3
PMDA-TAPOB	0.14	0.57	0.11	0.020	5.0	6.9
TMOS, 10 wt% SiO_2	0.16	0.64	0.11	0.023	5.9	7.1
TMOS, 20 wt% SiO_2	0.14	0.53	0.090	0.012	5.9	12
TMOS, 30 wt% SiO_2	0.19	0.61	0.11	0.019	5.3	10
TMOS/MTMS, 10 wt% SiO_2	0.17	0.70	0.13	0.019	5.4	9.4
TMOS/MTMS, 20 wt% SiO_2	0.45	1.6	0.34	0.071	4.6	6.4
TMOS/MTMS, 30 wt% SiO_2	1.1	3.2	0.77	0.13	4.1	8.5
MTMS, 10 wt% SiO_2	0.43	1.0	0.35	0.066	2.9	6.5
MTMS, 20 wt% SiO_2	1.5	4.8	1.5	0.40	3.2	3.8
MTMS, 30 wt% SiO_2	2.8	9.0	3.6	0.71	2.5	4.0

Table 8.4 *Solubility coefficients and solubility selectivity of HBPI–silica hybrid membranes at 76 cmHg and 25 °C*

Sample	$S \times 10^2$ ($cm^3(STP)/cm^3_{polym}$ cmHg)				$\alpha^S(O_2/N_2)$	$\alpha^S(CO_2/CH_4)$
	CO_2	O_2	N_2	CH_4		
6FDA-TAPOB	25	1.1	0.92	3.5	1.2	7.0
TMOS, 10 wt% SiO_2	30	1.4	1.1	5.0	1.3	5.9
TMOS, 20 wt% SiO_2	35	1.7	1.3	5.2	1.3	6.8
TMOS, 30 wt% SiO_2	41	1.8	1.6	6.0	1.1	6.7
TMOS/MTMS, 10 wt% SiO_2	22	1.3	0.90	3.3	1.4	6.6
TMOS/MTMS, 20 wt% SiO_2	24	1.4	1.0	4.9	1.4	4.9
TMOS/MTMS, 30 wt% SiO_2	26	1.3	1.2	4.4	1.2	5.9
TMOS/MTMS, 40 wt% SiO_2	27	1.5	1.3	5.5	1.2	5.0
TMOS/MTMS, 50 wt% SiO_2	26	1.6	1.2	4.4	1.3	5.8
MTMS, 10 wt% SiO_2	23	1.2	0.90	3.6	1.3	6.3
MTMS, 20 wt% SiO_2	21	1.1	0.90	3.7	1.2	5.6
MTMS, 30 wt% SiO_2	19	1.1	0.86	3.4	1.3	5.4
MTMS, 40 wt% SiO_2	16	1.0	0.82	2.7	1.2	5.8
MTMS, 50 wt% SiO_2	15	1.0	0.79	3.0	1.3	4.9
ODPA-TAPOB	21	0.81	0.60	3.3	1.3	6.3
TMOS, 10 wt% SiO_2	24	0.98	0.70	3.8	1.4	6.2
TMOS, 20 wt% SiO_2	26	1.0	0.78	3.6	1.3	7.2
TMOS, 30 wt% SiO_2	29	1.2	1.0	4.0	1.2	7.3
TMOS/MTMS, 10 wt% SiO_2	18	0.74	0.49	2.0	1.5	9.1
TMOS/MTMS, 20 wt% SiO_2	20	0.90	0.61	2.4	1.5	8.1
TMOS/MTMS, 30 wt% SiO_2	20	0.92	0.69	4.0	1.3	5.0
MTMS, 10 wt% SiO_2	16	0.69	0.48	2.9	1.4	5.4
MTMS, 20 wt% SiO_2	12	0.80	0.54	2.5	1.5	4.7
MTMS, 30 wt% SiO_2	16	0.85	0.69	2.6	1.2	5.9
PMDA-TAPOB	24	1.0	0.77	3.3	1.3	7.3
TMOS, 10 wt% SiO_2	26	1.1	1.0	3.1	1.1	8.5
TMOS, 20 wt% SiO_2	32	1.3	1.1	5.3	1.2	6.0
TMOS, 30 wt% SiO_2	38	1.7	1.3	4.0	1.3	9.7
TMOS/MTMS, 10 wt% SiO_2	27	1.1	0.89	4.4	1.2	6.3
TMOS/MTMS, 20 wt% SiO_2	29	1.2	0.93	3.1	1.3	9.2
TMOS/MTMS, 30 wt% SiO_2	29	1.5	1.1	5.1	1.3	5.7
MTMS, 10 wt% SiO_2	22	1.4	0.71	3.2	2.0	6.7
MTMS, 20 wt% SiO_2	20	1.0	0.69	2.5	1.5	7.9
MTMS, 30 wt% SiO_2	19	1.0	0.60	3.5	1.6	5.5

and $\alpha(O_2/N_2)$ and $\alpha(CO_2/CH_4)$ are plotted against O_2 and CO_2 permeability coefficients in Figures 8.6 and 8.7, respectively.

In Figure 8.6, $\alpha(O_2/N_2)$ values of the hybrid membranes slightly decrease with increasing O_2 permeability along with the upper bound trade-off line for O_2/N_2 separation demonstrated by Robeson [29]. Nevertheless, the hybrid membranes show relatively high $\alpha(O_2/N_2)$ values just below the upper bound.

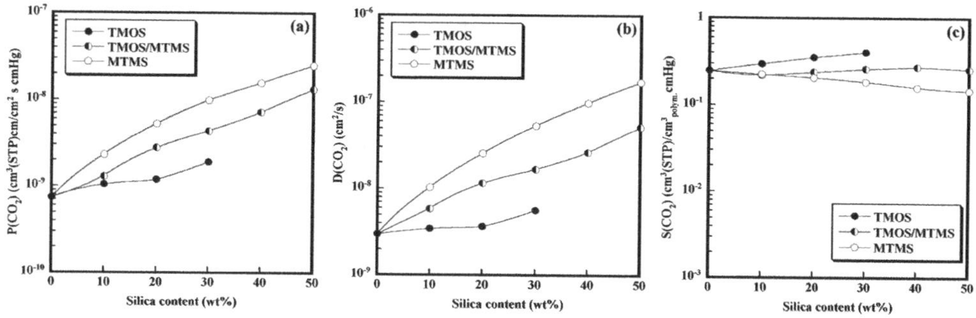

Figure 8.5 *CO_2 permeability (a), diffusion (b) and solubility (c) coefficients of 6FDA-TAPOB HBPI–silica hybrid membranes plotted against silica content*

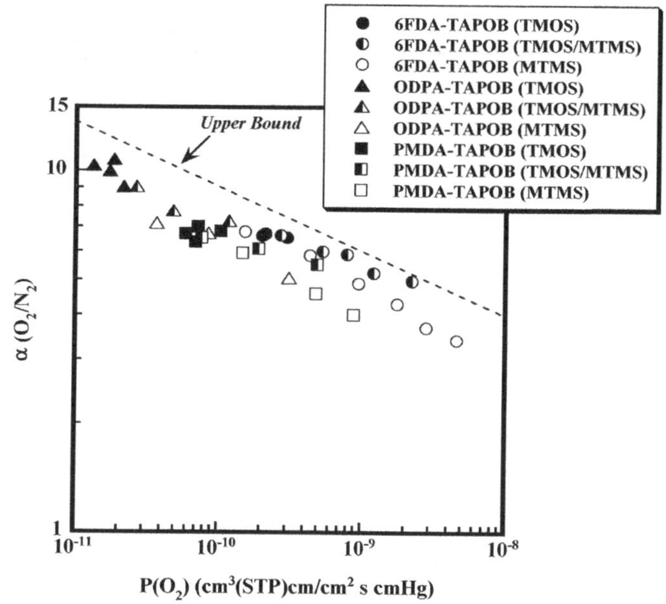

Figure 8.6 *Ideal O_2/N_2 selectivity ($\alpha(O_2/N_2)$) of HBPI–silica hybrid membranes plotted against O_2 permeability coefficient*

For CO_2/CH_4 separation, more attractive behaviour is observed (Figure 8.7). The values of $\alpha(CO_2/CH_4)$ of the hybrid membranes prepared solely with TMOS increase with increasing silica content in connection with increased CO_2 permeability. The remarkable CO_2/CH_4 separation behaviour of the HBPI–silica hybrid membranes prepared with TMOS is considered to be due to the characteristic distribution and interconnectivity of free volume elements created by the incorporation of silica, which provides a size-selective CO_2/CH_4 separation ability [11]. On the other hand, $\alpha(CO_2/CH_4)$ values of the

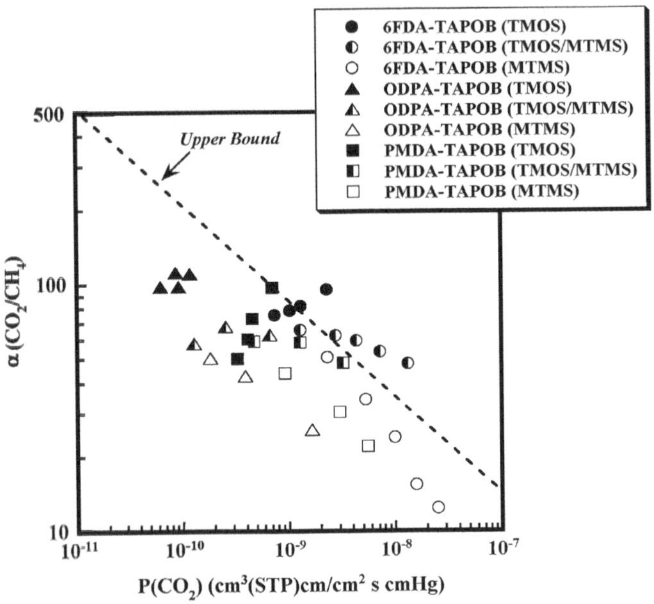

Figure 8.7 *Ideal CO_2/CH_4 selectivity ($\alpha(CO_2/CH_4)$) of HBPI–silica hybrid membranes plotted against CO_2 permeability coefficient*

MTMS systems decrease with increasing CO_2 permeability in connection with increased silica content as similar as the cases of conventional polymeric membranes. The distinctive difference between TMOS and MTMS systems is considered to be due to the differences in mean size and total amount of free volume elements created by the incorporation of silica. The size-selective CO_2/CH_4 separation ability of the hybrid membranes prepared with MTMS is sacrificed by excess enhancements of mean size and total amount of free volume elements although CO_2 permeability of the MTMS systems is markedly increased. It should be noted the CO_2/CH_4 separation ability of the TMOS/MTMS combined systems shows an intermediate behaviour between those of TMOS and MTMS systems. As shown in Figure 8.7, $\alpha(CO_2/CH_4)$ of the TMOS/MTMS combined systems have high values and tend to exceed the upper bound for CO_2/CH_4 separation [29] with increasing CO_2 permeability or silica content. Especially, 6FDA-TAPOB HBPI–silica hybrid membranes contain more than 20 wt% of silica show high $\alpha(CO_2/CH_4)$ values, exceeding the upper bound. Although a similar tendency of improvement of CO_2/CH_4 separation ability has been reported for some linear-type polyimide–silica hybrid membranes [9], such unique improvements of both CO_2 permeability and CO_2/CH_4 separation ability for the TMOS/MTMS combined system observed in this study have not been yet reported. This fact indicates that mean size and distribution of free volume elements in the TMOS/MTMS combined systems are successfully controlled for efficient CO_2/CH_4 separation and, as a result, the TMOS/MTMS combined systems show simultaneous large enhancements of CO_2 permeability and CO_2/CH_4 separation ability.

8.4 Conclusions

Physical and gas transport properties of HBPI–silica hybrid membranes prepared by the sol–gel method with two kinds of alkoxysilanes, TMOS and MTMS, were investigated. From the comparison of FTIR spectra for the TMOS and MTMS systems, the red shift of the Si–O–Si peaks is observed for the hybrid membranes prepared with MTMS, suggesting the formation of a looser Si–O–Si network caused by the methyl groups in MTMS, which induce large enhancements of mean size and total amount of free volume elements compared to those of the TMOS system. Thermal stabilities of the hybrid membranes are essentially improved by hybridization with silica. The T_g and T_d^5 of the hybrid membranes increase with increasing silica content, resulting from the formation of three-dimensional Si–O–Si network. However, the T_g values slightly decrease at low silica content region for MTMS and TMOS/MTMS combined systems due to the insufficient and loose formation of three-dimensional Si–O–Si network and the enhanced free volume holes. The CTE of the TMOS system greatly decreases with increasing silica content, and in contrast, the CTE of the MTMS system increases with increasing silica content. CO_2, O_2, N_2, and CH_4 permeability coefficients of the hybrid membranes increase with increasing silica content because of additional formation of free volume elements. In particular, MTMS and TMOS/MTMS combined systems show enormous improvements of gas diffusivity because of large enhancements of mean size and total amount of free volume elements attributed to loose Si–O–Si networks. The TMOS/MTMS combined systems also exhibit simultaneous large enhancements of CO_2 permeability and CO_2/CH_4 separation ability. This fact indicates the mean size and distribution of free volume elements in the TMOS/MTMS combined systems are successfully controlled to achieve the unique selective CO_2/CH_4 separation ability, and are expected to apply to high-performance gas separation membranes.

References

(1) G. F. Sykes, A. K. S. Clair, The effect of molecular structure on the gas transmission rates of aromatic polyimides, *J. Appl. Polym. Sci.*, **32**, 3725–3735 (1986).

(2) K. Okamoto, K. Tanaka, H. Kita, M. Ishida, M. Kakimoto, Y. Imai, Gas permeability and permselectivity of polyimides prepared from 4,4′-diaminotriphenylamine, *Polym. J.*, **24**, 451–457 (1992).

(3) M. Langsman, W. F. Burgoyne, Effect of diamine monomer structure on the gas permeability of polyimides. I. Bridged diamines, *J. Polym. Sci. Part A: Polym. Chem.*, **31**, 909–921 (1993).

(4) Y. Li, X. Wang, M. Ding, J. Xu, Effect of molecular structure on the permeability and permselectivity of aromatic polyimides, *J. Appl. Polym. Sci.*, **61**, 741–748 (1996).

(5) J. Fang, H. Kita, K. Okamoto, Hyperbranched polyimides for gas separation applications. 1. Synthesis and characterization, *Macromolecules*, **33**, 4639–4646 (2000).

(6) J. Fang, H. Kita, K. Okamoto, Gas permeation properties of hyperbranched polyimide membranes, *J. Membr. Sci.*, **182**, 245–256 (2001).

(7) T. Suzuki, Y. Yamada, Y. Tsujita, Gas transport properties of 6FDA-TAPOB hyperbranched polyimide membrane, *Polymer*, **45**, 7167–7171(2004).

(8) C. J. Cornelius, E. Marand, Hybrid silica–polyimide composite membranes: gas transport properties, *J. Membr. Sci.*, **202**, 97–118 (2002).

(9) C. Hibshman, C. J. Cornelius, E. Marand, The gas separation effects of annealing polyimide – organosilicate hybrid membranes, *J. Membr. Sci.*, **211**, 25–40 (2003).

(10) C. Hibshman, M. Mager, E. Marand, Effects of feed pressure on fluorinated polyimide – organosilicate hybrid membranes, *J. Membr. Sci.*, **229**, 73–80 (2004).
(11) T. Suzuki, Y. Yamada, Characterization of 6FDA-based hyperbranched and linear polyimide–silica hybrid membranes by gas permeation and ^{129}Xe NMR measurements, *J. Polym. Sci. Part B: Polym. Phys.*, **44**, 291–298 (2006).
(12) T. Suzuki, Y. Yamada, J. Sakai, Gas transport properties of ODPA-TAPOB hyperbranched polyimide–silica hybrid membranes, *High Perform. Polym.*, **18**, 655–664 (2006).
(13) T. Suzuki, Y. Yamada, Effect of end group modification on gas transport properties of 6FDA-TAPOB hyperbranched polyimide–silica hybrid membranes, *High Perform. Polym.*, **19**, 553–564 (2007).
(14) T. Takeichi, J. K. Stille, Star and linear imide oligomers containing reactive end caps: preparation and thermal properties, *Macromolecules*, **19**, 2093–2102 (1986).
(15) R. S. Prabhakar, B. D. Freeman, I. Roman, Gas and vapor sorption and permeation in poly(2,2,4-trifluoro-5-trifluoromethoxy-1,3-dioxole-co-tetrafluoroethylene), *Macromolecules*, **37**, 7688–7697 (2004).
(16) N. Muruganandam, W. J. Koros, D. R. Paul, Gas sorption and transport in substituted polycarbonates, *J. Polym. Sci. Part B: Polym. Phys.*, **25**, 1999–2026 (1987).
(17) A. Morisato, H. C. Shen, S. S. Sankar, B. D. Freeman, I. Pinnau, C. G. Casillas, Polymer characterization and gas permeability of poly(1-trimethylsilyl-1-propyne) [PTMSP], poly(1-phenyl-1-propyne) [PPP], and PTMSP/PPP blends, *J. Polym. Sci. Part B: Polym. Phys.*, **34**, 2209–2222 (1996).
(18) D. H. Weinkauf, H. D. Kim, D. R. Paul, Gas transport properties of liquid crystalline poly (p-phenyleneterephthalamide), *Macromolecules*, **25**, 788–796 (1992).
(19) D. W. van Krevelen, Properties of Polymers, Elsevier, Amsterdam, 1990.
(20) H. Chen, J. Yin, Synthesis and characterization of hyperbranched polyimides with good organosolubility and thermal properties based on a new triamine and conventional dianhydrides, *J. Polym. Sci. Part A: Polym. Chem.*, **40**, 3804–3814 (2002).
(21) H. Zhu, Y. Ma, Y. Fan, J. Shen, Fourier transform infrared spectroscopy and oxygen luminescence probing combined study of modified sol–gel derived film, *Thin Solid Films*, **397**, 95–101 (2001).
(22) M. Smaihi, T. Jermoumi, J. Marignan, R. D. Noble, Organic–inorganic gas separation membranes: preparation and characterization, *J. Membr. Sci.*, **116**, 211–220 (1996).
(23) J. Y. Park, D. R. Paul, Correlation and prediction of gas permeability in glassy polymer membrane materials via a modified free volume based group contribution method, *J. Membr. Sci.*, **125**, 23–39 (1997).
(24) A. Thran , G. Kroll, F. Faupel, Correlation between fractional free volume and diffusivity of gas molecules in glassy polymers, *J. Polym. Sci. Part B: Polym. Phys.*, **37**, 3344–3358 (1999).
(25) T. C. Merkel, B. D. Freeman, R. J. Spontak, Z. He, I. Pinnau, P. Meakin, A. J. Hill, Ultrapermeable, reverse-selective nanocomposite membranes, *Science*, **296**, 519–522 (2002).
(26) T. C. Merkel, L. G. Toy, A. L. Andrady, H. Gracz, E. O. Stejskal, Investigation of enhanced free volume in nanosilica-filled poly(1-trimethylsilyl-1-propyne) by ^{129}Xe NMR spectroscopy, *Macromolecules*, **36**, 353–358 (2003).
(27) Z. He, I. Pinnau, A. Morisato, Novel nanostructured polymer–inorganic hybrid membranes for vapor–gas separation, in Advanced Materials for Membrane Separation, I. Pinnau, B. D. Freeman (Eds.), ACS Symposium Series 876, American Chemical Society, Washington, DC, 2004.
(28) B. D. Freeman, Basis of permeability/selectivity tradeoff relations in polymeric gas separation membranes, *Macromolecules*, **32**, 375–380 (1999).
(29) L. M. Robeson, Correlation of separation factor versus permeability for polymeric membranes, *J. Membr. Sci.*, **62**, 165–185 (1991).

9

Air Enrichment by Polymeric Magnetic Membranes

Zbigniew J. Grzywna, Aleksandra Rybak and Anna Strzelewicz
Faculty of Chemistry, Silesian University of Technology, Gliwice, Poland

9.1 Introduction

Nowadays, obtaining cheap high purity gases or enriched gas mixtures (the air, in particular) is a very important problem in industry and medicine as well as in everyday life [1]. Methods for air separation or oxygen enrichment can be divided into two groups [2,3]: cryogenic and non-cryogenic. The gaseous oxygen and nitrogen market is dominated by cryogenic distillation of air, and vacuum swing adsorption [4]. Of the non-cryogenic methods, selective adsorption on zeolites and carbon adsorbents are available [5,6], and more and more attention is attracted to the membrane separation techniques [7,8]. The success of polymeric membranes has been largely based on their mechanical and thermal stability, along with good gas separation properties. The process of membrane separation is continuous, has a low capital cost, low power consumption, and the membranes, at least in gas separation, do not require regeneration [3]. In the case of gas separation two groups of factors have to be taken into account: solution-diffusion properties of membrane materials and molecular size differences of components. For similar-sized molecules such as oxygen and nitrogen, permselectivities $P(O_2)/P(N_2)$ are usually less than 10 in traditional polymer membranes, and it prevents the high degree of oxygen enrichment easily achieved in the cryogenic industry. In this situation a certain breakthrough was needed. Different approaches have been applied to improve performance of polymeric membranes, e.g. use of carbon molecular sieving (CMS) membranes or application of mixed matrix membranes [9–11]. Further development resulted in higher O_2/N_2 selectivity of

Membrane Gas Separation Edited by Yuri Yampolskii and Benny Freeman

membranes (up to 14 [12–14], i.e. about 80% of oxygen in the permeate or oxygen enriched stream). Efforts in various directions performed by another group [15–23] did not result in creation of very selective polymeric membranes.

The challenge with air, when separation by polymer membrane is concerned, is to provide conditions for differences (other than sorption and diffusion properties) in mass transport between oxygen and nitrogen. The idea of 'magnetic membranes' was introduced, and described briefly, only recently [24–26], based on the observation that oxygen and nitrogen have quite different magnetic properties, i.e. oxygen is paramagnetic whereas nitrogen diamagnetic, which gives a real chance for their separation. This idea can be utilized in at least two ways: magnets dispersed in a membrane ('magnetic membranes'), and a magnetic field applied on both sides of the permeation cell ('magnetic cell'). Of course, using both would be the most effective.

9.2 Formulation of the Problem

To separate gas mixtures one should take into account one of two (partly interrelated) factors by which gas components could differ, and which lead to an effective separation: solution-diffusion properties and molecular size differences of the components. Separation can be based on differences of molecular size, i.e. in the medium (a membrane) that behaves like a molecular sieve. In cases of similar sized molecules, like oxygen and nitrogen, the separation in polymer membranes can be also accomplished due to the differences thermodynamic properties of the penetrants. The gas mixture is made to pass through the membrane by applying a pressure difference on either side of the membrane (which causes the chemical potential gradient within the membrane). Differences in permeability result not only from diffusivity differences of the various gas species but also from differences in physico-chemical interactions of these species within the polymer [27]. A challenge with air, when separation by polymer membrane is concerned, is to provide conditions for differences (other than solution-diffusion) in mass transport between oxygen and nitrogen. The necessary differences can appear by introduction of a magnetic field which actively influences the oxygen transport, and is supposed to be neutral to nitrogen. A magnetic field gives rise to the 'drift term' in the overall oxygen transport, i.e. a deterministic term of a value dependent on the magnetic field induction. We consider a model of the mass transport system consisting of one or two gases, permeating through a dense membrane of thickness l. At the membrane boundaries for $x = 0$ and $x = l$, concentrations c_0 and 0 are posed, respectively (Figure 9.1).

The first approximation to the mathematical description reads

$$\begin{cases} \dfrac{\partial c_1}{\partial t} = D_{11}\dfrac{\partial^2 c_1}{\partial x^2} + D_{12}\dfrac{\partial^2 c_2}{\partial x^2} \pm w\dfrac{\partial c_1}{\partial x} \\ \dfrac{\partial c_2}{\partial t} = D_{21}\dfrac{\partial^2 c_1}{\partial x^2} + D_{22}\dfrac{\partial^2 c_2}{\partial x^2} \end{cases} \quad 0 < x < l, t \in R^+ \tag{9.1}$$

for initial and boundary conditions (indexes 1 and 2 refer to oxygen and nitrogen, respectively).

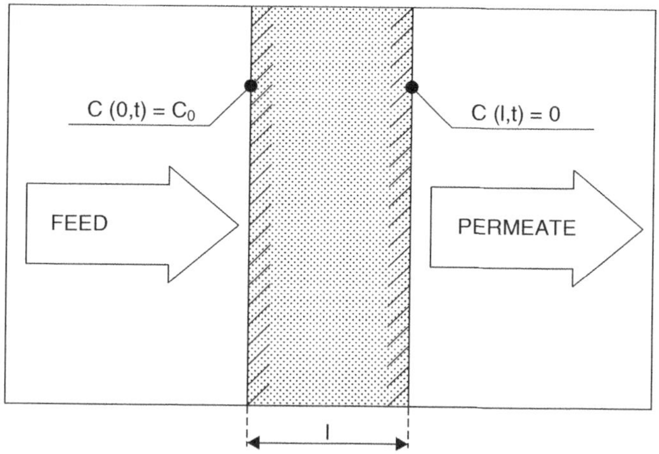

Figure 9.1 *Schematic membrane with boundary conditions and direction of flow*

$$\begin{array}{ll} c_1(x,0)=0 & c_2(x,0)=0 \\ c_1(0,t)=c_{01} & c_2(0,t)=c_{02} \\ c_1(l,t)=0 & c_2(l,t)=0 \end{array}$$

The experiment enabled the measurement of the permeation rates [25], whereas the composition of mixture was determined using a gas chromatograph.

9.2.1 One-component Permeation Process – Fick's and Smoluchowski Equation

The second of Fick's law (Equation 9.2) can be used as a sort of the reference point for describing a permeation process of one component gas (N_2, O_2) through a dense polymeric membrane with no external field. Functional dependence of the diffusion coefficient on coordinate, time or concentration can result from different types of circumstances, i.e. when the membrane is heterogeneous [28], is subjected to some relaxation phenomena [29,30], or membrane transport is accompanied by some other processes, like adsorption, etc. [9,12].

$$\frac{\partial c(x,t)}{\partial t} = div\,(D(\cdot)\;grad\;c) \tag{9.2}$$

where $D(\cdot)$ takes the $D(c)$, $D(x)$ or $D(t)$ form. To provide a coherent frame for this sort of approach we repeat several formulae for Fick's permeation.

The simplest case of the diffusion equation with constant diffusion coefficient D with initial and boundary conditions for permeation reads

$$\begin{cases} \dfrac{\partial c}{\partial t} = D\dfrac{\partial^2 c}{\partial x^2}, & x \in (0,l), \quad t \in R^+ \\ c(x,0)=0 \\ c(0,t)=c_0 \\ c(l,t)=0 \end{cases} \tag{9.3}$$

Solution of Equation (9.3) in the form of a Fourier series discussed by Crank [31] has a form

$$c(x,t) = c_0\left(1-\frac{x}{l}\right) - \frac{2c_0}{\pi}\sum_{n=1}^{\infty}\frac{1}{n}\sin\frac{n\pi x}{l}\exp\left(-\frac{n^2\pi^2}{l^2}Dt\right) \tag{9.4}$$

The diffusive flux $J(x, t)$ can be obtained from Equation (9.4) using the definition of flux:

$$J(x,t) = -D\frac{\partial c}{\partial x} \tag{9.5}$$

to get a form

$$J(x,t) = \frac{Dc_0}{l} + \frac{2Dc_0}{l}\sum_{n=1}^{\infty}\cos\frac{n\pi x}{l}\exp\left(-\frac{n^2\pi^2}{l^2}Dt\right) \tag{9.6}$$

In case of functional dependence of diffusion coefficients, the system (Equation 9.3) can be solved numerically if an analytical solution is not known. To start with, the diffusion equation is rewritten as a difference quotient [32]:

$$\frac{c_{i,j+1} - c_{i,j}}{\Delta t} = D\frac{c_{i+1,j} - 2c_{i,j} + c_{i-1,j}}{(\Delta x)^2} \tag{9.7}$$

Figure 9.2 shows the solution of Equation (9.3), i.e. a concentration profile as a function of x for some chosen values of t.

When other processes accompany diffusion (for example reaction) or when an external field is present, Equation (9.3) must be modified by adding the appropriate terms. If a potential field acts on a system (in our case magnetic), we add the 'drift term' to Equation (9.3), i.e. $w\frac{\partial c}{\partial x}$, to get finally the Smoluchowski equation [33]:

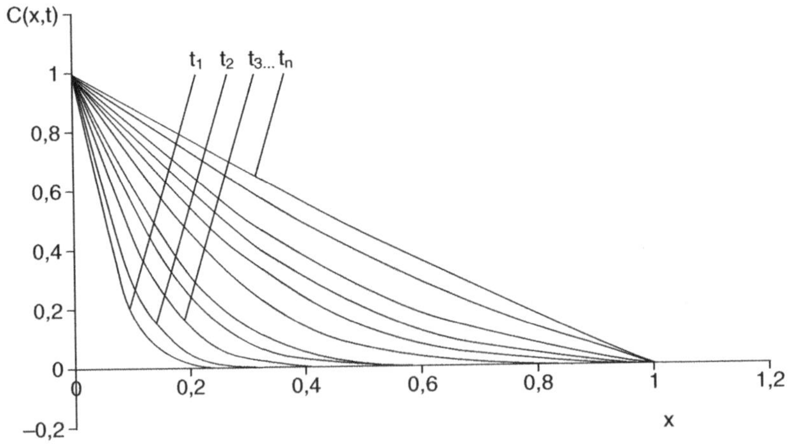

Figure 9.2 *Solution of Equation (9.3), i.e. concentration profiles, as a function of x and t. Reprinted with permission from Journal of Membrane Science, Studies on the air membrane separation in the presence of a magnetic field by Anna Strzelewicz and Zbigniew J. Grzywna, 294, 1–2, 60–67 Copyright (2007) Elsevier Ltd*

$$\frac{\partial c}{\partial t} = D\frac{\partial^2 c}{\partial x^2} - w\frac{\partial c}{\partial x} \tag{9.8}$$

where D is the constant diffusion coefficient, and w is the constant drift coefficient.

One of the ways to get an analytical solution of the Smoluchowski equation is to use some transformations, which reduce the Smoluchowski equation into Fick's equation [34,35]. The interesting fact is that for sufficiently large 'w', the solution of the Smoluchowski equation behaves like a 'travelling wave', i.e. it starts to behave like a solution of the following equation:

$$\frac{\partial c}{\partial t} = -w\frac{\partial c}{\partial x} \tag{9.9}$$

and represents a unidirectional wave motion with velocity w. Figure 9.3 shows the concentration profiles for steady state of Smoluchowski equation for different values of the drift coefficient.

For the steady state, Equation (9.8) takes the form

$$D\frac{d^2 c_s}{dx^2} - w\frac{dc_s}{dx} = 0 \tag{9.10}$$

which gives the formula for concentration profile $c_s(x)$

$$c_s(x) = c_0\left(\frac{e^{\frac{w}{D}x} - e^{\frac{wl}{D}}}{1 - e^{\frac{wl}{D}}}\right) \tag{9.11}$$

shown in Figure 9.3.

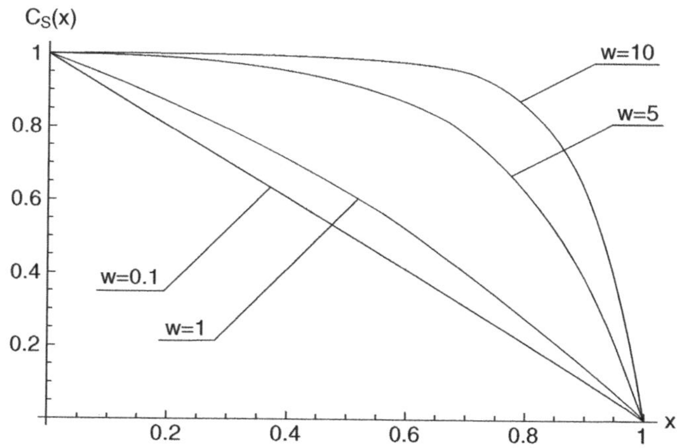

Figure 9.3 *Concentration profiles for steady state of Smoluchowski equation for different values of the drift coefficient:* w = *0.1,* w = *1,* w = *5 and* w = *10*

We might also consider the problem of diffusion with a chemical reaction. In a sufficiently strong magnetic field, molecular clusters can be formed. For the case of air, the clusters N_2-O_2-O_2 are preferable [36]. This can be regarded as a problem of diffusion with chemical reaction, which has described [31,37]

$$\frac{\partial c}{\partial t} = \frac{\partial}{\partial x}\left[D(\cdot)\frac{\partial c}{\partial x} - wc\right] - k_0 k(x) f(c) \tag{9.12}$$

where k_0 is the rate constant, $k(x)$ is the distribution function, and $f(c)$ is the reaction kinetic term, 'w', means the drift coefficient, as before. The term 'chemical reaction' has a rather formal meaning, and can represent also adsorption or 'trapping' processes.

9.2.2 Diffusion Equation for Two-component Gas Mixture (Without and With a Potential Field)

A diffusion process, in the simplest case described by the Fick's law, for the multicomponent diffusion is described by its generalization to an n-component system [38,39]. The flux has then a form:

$$-j_i = \sum_{j=1}^{n} D_{ij} \nabla c_j \tag{9.13}$$

In this chapter we consider the separation of a binary mixture by a dense membrane. The simplest case i.e. when D_{ij} is constant, is represented by a system (14):

$$\begin{cases} \dfrac{\partial c_1}{\partial t} = D_{11}\dfrac{\partial^2 c_1}{\partial x^2} + D_{12}\dfrac{\partial^2 c_2}{\partial x^2} \\ \dfrac{\partial c_2}{\partial t} = D_{21}\dfrac{\partial^2 c_1}{\partial x^2} + D_{22}\dfrac{\partial^2 c_2}{\partial x^2} \end{cases} \quad \mathrm{x} \in (0, l), \mathrm{t} \in \mathrm{R}^+ \tag{9.14}$$

with initial and boundary conditions:

$$\begin{aligned} c_1(x,0) &= 0 \quad & c_2(x,0) &= 0 \\ c_1(0,t) &= c_{01} \quad & c_2(0,t) &= c_{02} \\ c_1(l,t) &= 0 \quad & c_2(l,t) &= 0 \end{aligned} \tag{9.15}$$

In general, the coefficients are not symmetric, i.e. $D_{ij} \neq D_{ji}$. The diagonal terms are called the 'main term' coefficients, while off diagonal terms, i.e. $D_{ij,\, i \neq j}$, are called the 'cross-term' coefficients [38,39]. In fact, they represent the degree of coupling between the diffusing gas components. The set (Equation 9.14) can be solved numerically. The finite-difference method gives the following schemes:

$$\begin{cases} \dfrac{c_{1,i,j+1} - c_{1,i,j}}{\Delta t} = D_{11}\dfrac{c_{1,i+1,j} - 2c_{1,i,j} + c_{1,i-1,j}}{(\Delta x)^2} + D_{12}\dfrac{c_{2,i+1,j} - 2c_{2,i,j} + c_{2,i-1,j}}{(\Delta x)^2} \\ \dfrac{c_{2,i,j+1} - c_{2,i,j}}{\Delta t} = D_{21}\dfrac{c_{1,i+1,j} - 2c_{1,i,j} + c_{1,i-1,j}}{(\Delta x)^2} + D_{22}\dfrac{c_{2,i+1,j} - 2c_{2,i,j} + c_{2,i-1,j}}{(\Delta x)^2} \end{cases} \tag{9.16}$$

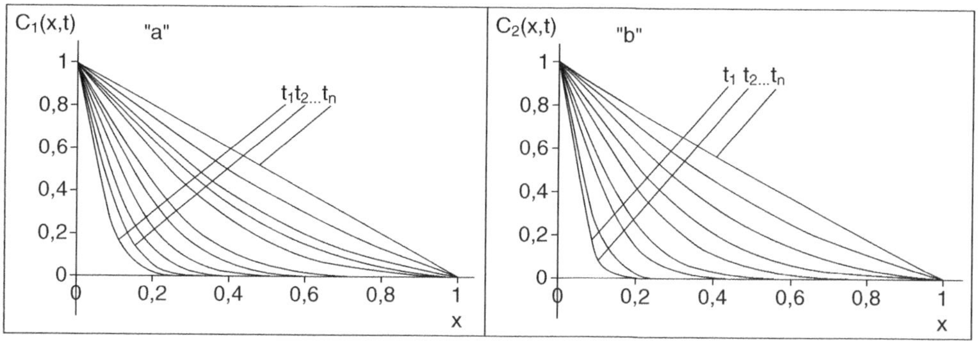

Figure 9.4 *Concentration profiles–solutions of Equation (9.14) for constant coefficients* $D_{11} = 1$, $D_{12} = 0.13$, $D_{21} = 0.14$, $D_{22} = 1.1$, *for different times: (a) concentration profiles for component 1 of the mixture; (b) concentration profiles for component 2 of the mixture. Reprinted with permission from Journal of Membrane Science, Studies on the air membrane separation in the presence of a magnetic field by Anna Strzelewicz and Zbigniew J. Grzywna, 294, 1–2, 60–67, Copyright (2007) Elsevier Ltd*

Figure 9.4 shows the concentration profiles of $c_1(x,t)$, and $c_2(x,t)$, i.e. solution of Equation (9.14) with conditions shown in Equation (9.15).

In the case of the presence of an additional force which acts on one component of a permeating gas mixture, the system shown in Equation (9.14) must be modified. We add the drift term, i.e. $w\frac{\partial c}{\partial x}$, to the first equation of Equation (9.14), and now it reads:

$$\begin{cases} \frac{\partial c_1}{\partial t} = D_{11}\frac{\partial^2 c_1}{\partial x^2} + D_{12}\frac{\partial^2 c_2}{\partial x^2} - w\frac{\partial c_1}{\partial x} \\ \frac{\partial c_2}{\partial t} = D_{21}\frac{\partial^2 c_1}{\partial x^2} + D_{22}\frac{\partial^2 c_2}{\partial x^2} \end{cases} \quad x \in (0, l), t \in R^+ \tag{9.17}$$

along with the initial and boundary conditions shown in Equation (9.15).

The resulting concentration profiles for large values of the drift term are presented in Figure 9.5.

Since our system (Equation 9.17) is weakly coupled, we can collect rough, but still essential information, about separation by analysis of just a single equation addressed to the particular air components, i.e. Fick's equation to the nitrogen, and the Smoluchowski equation to the oxygen, respectively.

9.3 Experimental

9.3.1 Membrane Preparation

Flat ethylcellulose membranes of thickness 28–80 μm and EC magnetic membranes of thickness 90–202 μm (depending on magnetic powder granulation and on the amount of added powder) were used. Both types of films were made by casting. The solution in

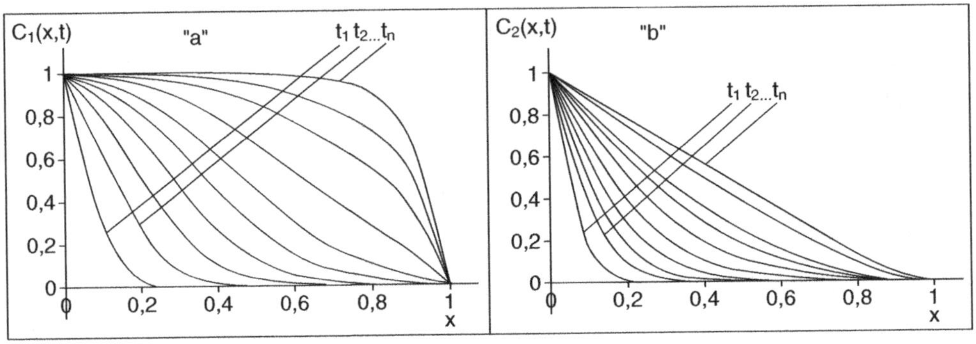

Figure 9.5 Concentration profiles–solutions of Equation (9.17) with conditions (Equation 9.15) for constant diffusion coefficients $D_{11} = 1$, $D_{12} = 0.13$, $D_{21} = 0.14$, $D_{22} = 1.1$, and drift coefficient $w = 10$, for different times: (a) concentration profiles for component 1 of the mixture; (b) concentration profiles for component 2 of the mixture. Reprinted with permission from Journal of Membrane Science, Studies on the air membrane separation in the presence of a magnetic field by Anna Strzelewicz and Zbigniew J. Grzywna, 294, 1–2, 60–67, Copyright (2007) Elsevier Ltd

40:60 ethanol/toluene with EC concentration of 3% was poured into a Petri dish, and evaporated at room temperature for 24 h. Magnetic membranes were made by pouring the 3% ethylcellulose solution with a dispersed magnetic neodymium powder (of appropriate granulation 20–50 μm) into a Petri dish, and then evaporated in external field of a coil (stable magnetic field with range of induction 0–40 mT) for 24 h. Magnetic membranes with 1.0–10.5% of neodymium powder content in polymer solution and 19.1–73.0% in dry membrane, were obtained [25,26].

The membranes were removed from the Petri dish (with some distilled water), and dried at 40 °C. For membranes with dispersed magnetic powder, the permeation measurements were carried out before (just powder) and after magnetization (magnetic membranes). Magnetic induction of this field was approximately of 2 T. Teslameter FH 54 was used for measurement of magnetic induction. The membranes were stored in an desiccator under vacuum conditions. We have also tested polyphenyleneoxide (PPO) membranes [40–44], courtesy of B. Kruczek (Faculty of Engineering, University of Ottawa, Canada).

9.3.2 Experimental Setup

The experimental setup APG-1 (Figure 9.6), described elsewhere [25], was used for measurement of the nitrogen, oxygen and air permeability. For pure gases and gas mixture permeation experiments, compressed gas cylinders of O_2, N_2, and air (21% O_2/79% N_2) were supplied by Praxair. This setup was furnished with a gas chromatograph HP 5890A, which let us measure the oxygen and nitrogen concentration in the permeate. The main part of this experimental setup was a diffusive chamber, where the membrane in the form of a disc was placed. The membrane effective area was 19.63 cm^2. This setup was used for a low-pressure (from 0.1 to 1.0 MPa) analysis of gas permeation. After installation of membrane in the diffusive chamber, we have rinsed this chamber with an appro-

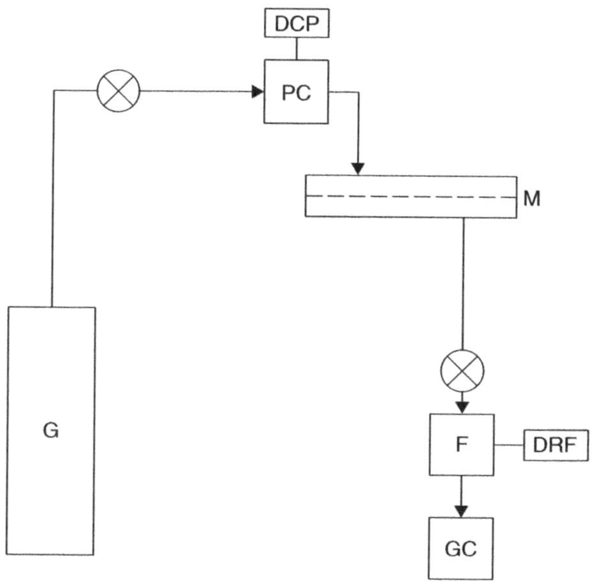

Figure 9.6 Scheme of the experimental setup APG-1. G, gas cylinder; F, flowmeter; DRF, digital registrar of flowmeter; PC, pressure controller; DCP, digital controller of pressure; M, diffusive chamber; GC, gas chromatograph

priate gas. Then, between the high- and low-pressure part of a chamber, a suitable pressure difference was established (0.3–0.7 MPa) and controlled by an electronic pressure meter and controller EL-press P-506C (the error of the measurement of step change in a feed pressure, about 0.2 s, is very low, and repeatable, for all the measurements). After that, the flow rate of the permeate was recorded using a Flow-Bus flowmeter (with a range 0–3 ml/min) with a computer data acquisition. All measurements were carried out at room temperature. Using flow rate data, air enrichment and transport coefficients were calculated.

9.3.3 Time Lag Method and D1-D8 System

In the experiments we were able to measure the flux and then, after integration with respect to time, we obtain a downstream permeation curve (shown in Figure 9.7) where the typical plot of $Q^a(l,t)$ versus time is presented.

The most important parameter is the average diffusion coefficient $\bar{D}$, calculated from the stationary permeation data. To calculate this coefficient, we have used experimental data (the thickness of membrane l, a stationary flux J_S and Δc_0, obtained from an intercept of the asymptote to the stationary permeation curve with the $Q^a(l,t)$ axis) (Figure 9.7).

$$\bar{D} = \frac{J_s \cdot l}{\Delta c_0} \text{ in cm}^2/\text{s} \tag{9.18}$$

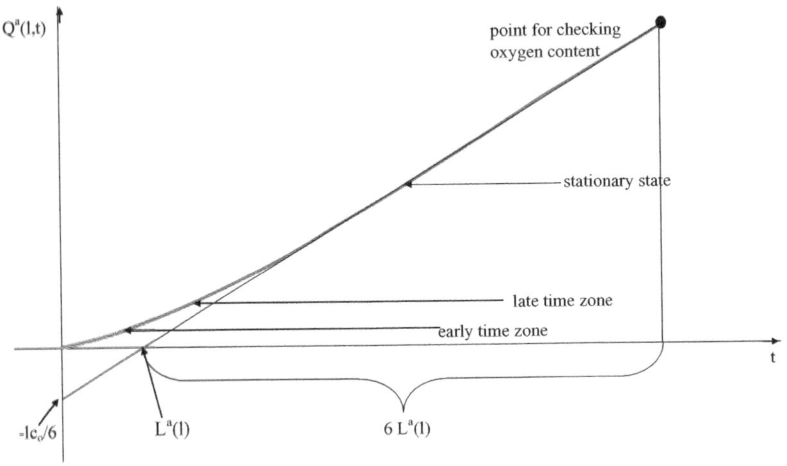

Figure 9.7 *Downstream absorption permeation curve – a schematic view. Reprinted with permission from Journal of Membrane Science, On the air enrichment by polymer magnetic membranes by A. Rybak, Z. J. Grzywna and W. Kaszuwara, 336, 1–2, 79–85, Copyright (2009) Elsevier Ltd*

We can get some insight into the nature of the transport process by comparing $\bar{D}$ with the value of diffusion coefficient D_L, calculated by the time lag method, obtained as an intercept of the asymptote to the stationary permeation curve with the time axis (Figure 9.7):

$$D_L = \frac{l^2}{6L^a(l)} \text{ in cm}^2/\text{s} \tag{9.19}$$

If $\bar{D} = D_L$, the diffusion is supposed to be 'ideal Fickian', unless there are some hidden processes, like a drift, which are not included in this simple protocol [31]. In fact, in case of the 'magnetic membranes' the presence of a drift is almost certain.

Having the drift and time lag values, the average diffusion coefficient can be calculated from a proper theory from which the following formula for time lag can be derived [37].

$$\hat{L}^a(l) = \frac{\bar{D}^2\left(2 - 2e^{-\frac{w}{\bar{D}}l} - 2\bar{D}wl + w^2l^2\right)}{2w^3l} \tag{9.20}$$

To understand properly air enrichment by the 'magnetic membranes' we had to estimate the values of Δc_0, i.e. an equilibrium concentration of the air components (no sorption experiments have been carried out). Since Δc_0 is not measured (in this chapter), we have to calculate it. The most realistic value of Δc_0 can be estimated by comparing its values obtained by the different methods.

The first method is based on the determination of the permeability coefficient:

$$P = D \cdot S = D \cdot \frac{\Delta c_0}{\Delta p} \tag{9.21}$$

We have to remember that the solubility coefficient for a pure EC is known. We do not know, however, its dependence on the parameters of the experiments with a magnetic field!

The subsequent three methods are based on analysis of $Q^a(l,t)$ (see Figure 9.7). If we are to deal with an 'ideal Fickian' description of a permeation process, the value $-\frac{lc_0}{6}$ can be pointed out as an intercept of the asymptote to the stationary permeation curve with $Q^a(l,t)$ axis.

The next two ways use a description of transient, early- and late-time zones.

For an early-time we can use a formula [45]

$$\ln\left[t^{\frac{1}{2}} \cdot J^a(l,t)\right] = \ln\left[2c_0\left(\frac{D_4}{\pi}\right)^{\frac{1}{2}}\right] - \frac{l^2}{4D_3 t} \tag{9.22}$$

which can be presented as a $\ln\left[t^{1/2} \cdot J^a(l,t)\right]$ versus $\frac{1}{t}$ graph (Figure 9.8). D_3 could be obtained from the slope of the asymptote to the early-time permeation curve, and D_4, along with c_0, can be obtained from the intercept of the asymptote with $\ln\left[t^{1/2} \cdot J^a(l,t)\right]$ axis.

While, for the late-time permeation, we get [45]

$$\ln\left[Q^a(l,t) - Q_s^a(l,t)\right] = \ln\frac{2lc_0}{\pi^2} - \frac{D_5\pi^2}{l^2}t \tag{9.23}$$

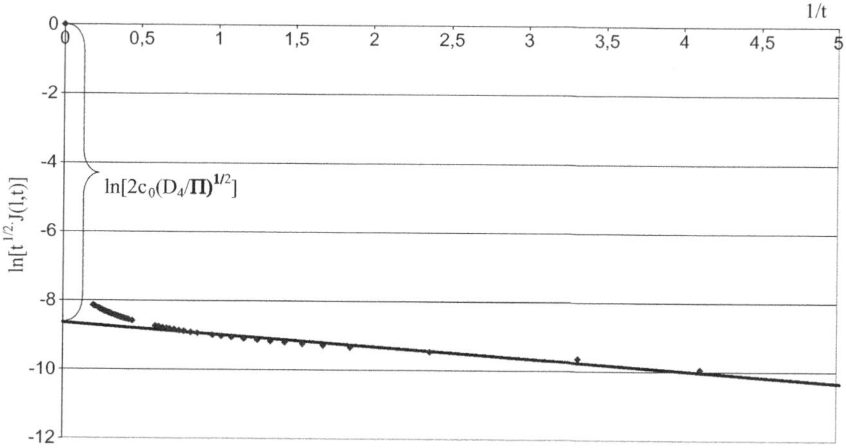

Figure 9.8 *An example of early-time permeation curve according to Equation (9.22). Reprinted with permission from Journal of Membrane Science, On the air enrichment by polymer magnetic membranes by A. Rybak, Z. J. Grzywna and W. Kaszuwara, 336, 1–2, 79–85, Copyright (2009) Elsevier Ltd*

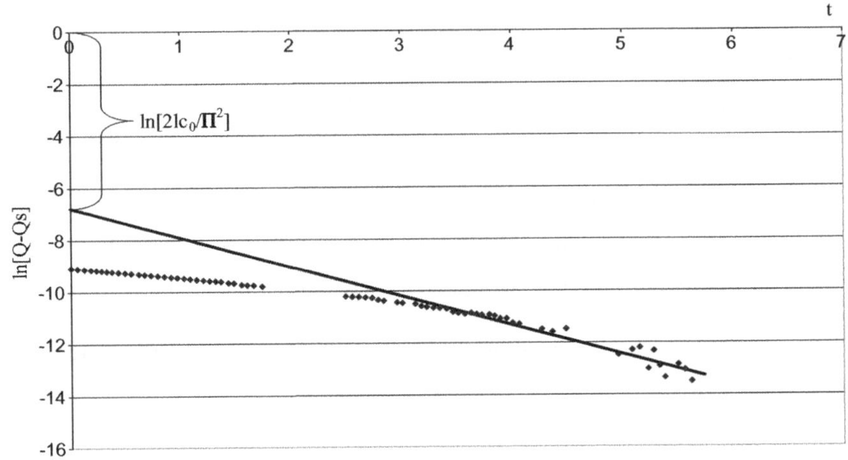

Figure 9.9 *An example of late-time permeation curve according to Equation (9.23). Reprinted with permission from Journal of Membrane Science, On the air enrichment by polymer magnetic membranes by A. Rybak, Z. J. Grzywna and W. Kaszuwara, 336, 1–2, 79–85, Copyright (2009) Elsevier Ltd*

and consequently $\ln\left[Q^a(l,t)-Q_s^a(l,t)\right]$ versus t graph (Figure 9.9). D_5 could be obtained from the slope of the asymptote to the late-time permeation curve, and c_0 from the intercept of the asymptote with $\ln\left[Q^a(l,t)-Q_s^a(l,t)\right]$ axis.

We have collected the data for the oxygen and nitrogen, both for individual gases and a mixture of the components of air, and analysed the system using formulas (18–23).

9.3.4 Permeation Data

The main permeation data are collected in Tables 9.1, 9.2 and Figure 9.10. The data presented are the mean values of six measurements with standard deviations.

Results 6 and 9 from Table 9.2 were obtained by using a newly designed permeation cell with strong magnets on both sides of a membrane. The percentage of air enrichment was directly measured, while the value of B in this case was just estimated. Values of calculated oxygen content in the permeate originate from theoretical consideration [24]. We can see that theoretical expectations are in really good agreement with experimental data. But, if we present the measured oxygen content in the permeate as a function of the magnetic induction (Figure 9.11) we can see the tendency of deviation from the linear relationship.

All that might be due to the influence of a strong magnetic field in which nitrogen and oxygen form aggregates [36] which carry more nitrogen, and spoil, in a sense, the separation process.

Table 9.1 *Mass transport coefficients for various membranes*

No.	Membrane	B (mT)	N_2, pure			N_2, in air		
			$P \times 10^7$	La(l)	Δc_0	$P \times 10^7$	La(l)	Δc_0
			$(cm^3_{STP}\ cm)/(cm^2\ s\ cmHg)$	(s)	(cm^3_{STP}/cm^3)	$(cm^3_{STP}\ cm)/(cm^2\ s\ cmHg)$	(s)	(cm^3_{STP}/cm^3)
1	Flat EC	0.00	0.19 ± 0.01	1.37 ± 0.10	0.71 ± 0.05	0.19 ± 0.01	1.21 ± 0.01	0.50 ± 0.04
2	EC + 1.30 g of Nd	0.00	0.22 ± 0.01	1.23 ± 0.10	0.73 ± 0.07	0.22 ± 0.01	1.31 ± 0.01	0.39 ± 0.04
3	EC + 1.23 g of Nd	0.50	0.31 ± 0.01	0.52 ± 0.04	0.71 ± 0.05	0.46 ± 0.01	0.47 ± 0.01	0.37 ± 0.03
4	EC + 1.38 g of Nd	0.79	0.34 ± 0.01	1.03 ± 0.10	0.73 ± 0.07	0.94 ± 0.01	0.35 ± 0.01	0.37 ± 0.03
5	EC + 1.49 g of Nd	1.25	0.42 ± 0.02	0.83 ± 0.07	0.72 ± 0.07	1.98 ± 0.02	0.30 ± 0.01	0.39 ± 0.04
No	**Membrane**	**B (mT)**	**O_2, pure**			**O_2, in air**		
			P·10^7	La(l)	Δc_0	P·10^7	La(l)	Δc_0
			$(cm^3_{STP}\ cm)/(cm^2\ s\ cmHg)$	(s)	(cm^3_{STP}/cm^3)	$(cm^3_{STP}\ cm)/(cm^2\ s\ cmHg)$	(s)	(cm^3_{STP}/cm^3)
1	Flat EC	0.00	0.23 ± 0.01	1.21 ± 0.10	0.63 ± 0.05	0.22 ± 0.01	1.17 ± 0.01	0.11 ± 0.01
2	EC + 1.30 g of Nd	0.00	0.24 ± 0.01	1.35 ± 0.10	0.60 ± 0.05	0.22 ± 0.01	1.26 ± 0.01	0.12 ± 0.01
3	EC + 1.23 g of Nd	0.50	0.52 ± 0.01	0.51 ± 0.05	0.62 ± 0.05	0.77 ± 0.01	0.45 ± 0.01	0.11 ± 0.01
4	EC + 1.38 g of Nd	0.79	0.63 ± 0.01	0.56 ± 0.06	0.63 ± 0.05	2.23 ± 0.01	0.32 ± 0.01	0.11 ± 0.01
5	EC + 1.49 g of Nd	1.25	1.18 ± 0.02	0.52 ± 0.05	0.64 ± 0.05	5.79 ± 0.01	0.23 ± 0.01	0.12 ± 0.01

Source: Reprinted with permission from Journal of Membrane Science, On the air enrichment by polymer magnetic membranes by A. Rybak, Z. J. Grzywna and W. Kaszuwara, 336, 1–2, 79–85 Copyright (2009) Elsevier Ltd

* Δc_0 was obtained using a late time permeation curve according to Equation (9.23). Other methods mentioned above could be used only in the case of the sufficient amount of data, and for an 'ideal Fickian' description of a permeation process.

Table 9.2 Comparison of measured and calculated air enrichment data for various membranes

No	Membrane	B (mT)	Measured oxygen content in permeate (%)	Predicted (calculated) oxygen content in permeate (%)
1	EC	0.00	23.8 ± 1.0	25.0
2	$EC_{0.0}$	0.00	22.1 ± 1.1	25.0
3	$EC_{0.50}$	0.50	30.7 ± 1.1	33.3
4	$EC_{0.79}$	0.79	40.7 ± 1.1	38.2
5	$EC_{1.25}$	1.25	43.8 ± 1.1	45.8
6	$EC_{2.25}$	2.25	55.6 ± 1.4	62.5
7	PPO	0.00	37.4 ± 1.2	–
8	$PPO_{1.70}$	1.70	54.1 ± 1.4	–
9	$PPO_{2.70}$	2.70	61.9 ± 1.5	–

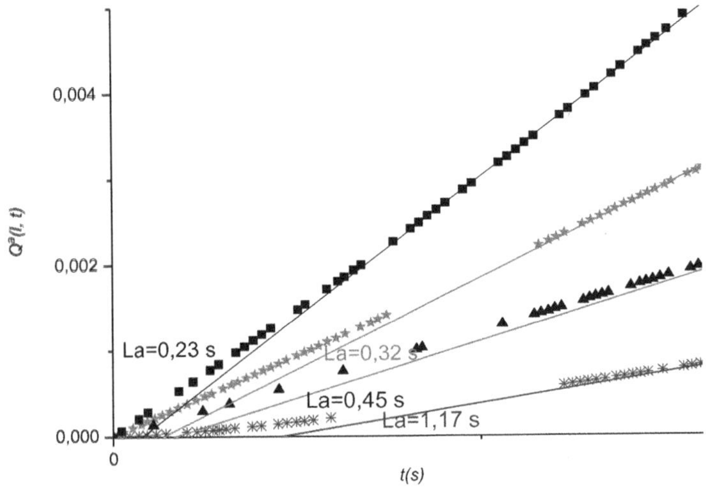

Figure 9.10 *Downstream absorption permeation curves for various polymer and magnetic membranes. Points marked as: squares, magnetic membrane with B = 1.25 mT; stars, magnetic membrane with B = 0.79 mT; triangles, magnetic membrane with B = 0.50 mT; asterisk, plane EC membrane*

9.4 Results and Discussion

We have found (Table 9.3) that EC films for pure and mixed components as well as EC + Nd/N_2 form ideal Fickian systems, while EC + Nd/N_2 air, EC + Nd/O_2 and EC + Nd/O_2 air, do not. Based on that we can calculate $\bar{D}$ and Δc_0 from the appropriate 'ideal' permeation curve to get a sort of a reference point for non-ideal cases.

We can separate the drift and diffusion in an overall flux assuming that the diffusional contribution to the overall flux is magnetic field independent.

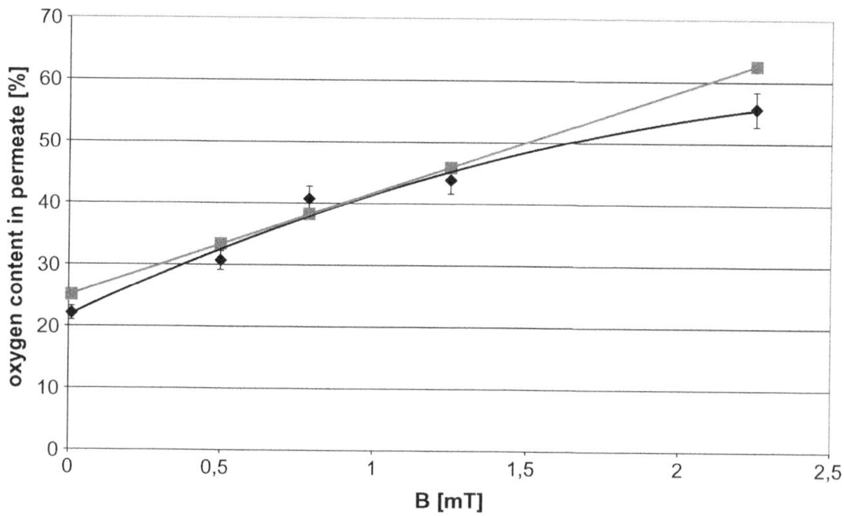

Figure 9.11 *Dependence of the oxygen content in permeate vs. magnetic field induction. Points marked as rhombus were obtained in experiment, points marked as squares were predicted by the theory*

$$J_{Bi} = \overline{D}\frac{\Delta c_0}{l} + w_{Bi}\Delta c_0 \qquad (9.24)$$

plane EC   drift

$$J_{\mathrm{Bi}} = J_0 + w_{\mathrm{Bi}}\Delta c_0 \qquad (9.25)$$

Here '$\mathrm{B_i}$' stands for magnetic induction of a membrane. The drift coefficient can be calculated for an imposed induction B from equation:

$$w_{\mathrm{Bi}} = \frac{J_{\mathrm{Bi}} - J_0}{\Delta c_0} \qquad (9.26)$$

and its values for different values of B are collected in Table 9.4.

As can be seen from Table 9.5 (pure gases) the influence of drift is meaningful only for oxygen, whose transport is strongly affected by the magnetic field.

Quite a different situation happens for air transport through magnetic membranes (i.e. with a nonzero induction), where much larger differences between the nitrogen and oxygen diffusion coefficients were observed (Table 9.6). Surprisingly, it was found that not only is oxygen transport affected by the magnetic field, but the nitrogen diffusion is as well.

Table 9.3 Diffusion coefficients for ideal, and non-ideal, Fickian systems

Membrane	Ideal Fickian system																			
	N_2 pure					O_2 pure					N_2 in air					O_2 in air				
	$\bar{D}$	D_L	D_3	D_4	D_5	$\bar{D}$	D_L	D_3	D_4	D_5	$\bar{D}$	D_L	D_3	D_4	D_5	$\bar{D}$	D_L	D_3	D_4	D_5
	10^5 (cm^2/s)					10^5 (cm^2/s)					10^5 (cm^2/s)					10^5 (cm^2/s)				
Flat EC	0.8 ± 0.2	0.8 ± 0.2	1.1 ± 0.2	0.9 ± 0.1	1.1 ± 0.2	1.1 ± 0.1	1.0 ± 0.1	1.1 ± 0.2	1.0 ± 0.1	1.2 ± 0.1	0.9 ± 0.2	0.9 ± 0.2	1.1 ± 0.4	0.9 ± 0.1	1.1 ± 0.2	0.9 ± 0.1	0.9 ± 0.1	1.4 ± 0.2	1.0 ± 0.1	1.2 ± 0.1
EC + 1.30 g Nd	1.3 ± 0.2	1.1 ± 0.2	1.4 ± 0.3	1.3 ± 0.2	0.9 ± 0.2	**out of the case**					**out of the case**					**out of the case**				
EC + 1.38 g Nd	1.4 ± 0.3	1.0 ± 0.2	1.3 ± 0.3	1.1 ± 0.2	0.9 ± 0.2	**out of the case**					**out of the case**					**out of the case**				

Membrane	Non-ideal Fickian system																			
	N_2 pure					O_2 pure					N_2 in air					O_2 in air				
	$\bar{D}$	D_L	D_3	D_4	D_5	$\bar{D}$	D_L	D_3	D_4	D_5	$\bar{D}$	D_L	D_3	D_4	D_5	$\bar{D}$	D_L	D_3	D_4	D_5
	10^5 (cm^2/s)					10^5 (cm^2/s)					10^5 (cm^2/s)					10^5 (cm^2/s)				
EC + 1.30 g Nd	**out of the case**					2.5 ± 0.2	6.1 ± 0.6	5.2 ± 0.5	0.8 ± 0.1	1.0 ± 0.1	2.2 ± 0.4	6.5 ± 1.2	5.6 ± 1.0	0.7 ± 0.1	0.4 ± 0.1	3.2 ± 0.3	6.8 ± 0.6	8.1 ± 0.3	1.0 ± 0.1	0.5 ± 0.05
EC + 1.38 g Nd	**out of the case**					3.0 ± 0.3	8.6 ± 0.8	7.2 ± 0.7	0.9 ± 0.1	1.5 ± 0.1	4.5 ± 0.9	11.7 ± 2.2	12.2 ± 2.4	1.2 ± 0.2	1.0 ± 0.2	9.4 ± 0.9	18.6 ± 1.2	25.5 ± 2.7	2.5 ± 0.2	1.8 ± 0.15

Source: Reprinted with permission from Journal of Membrane Science, On the air enrichment by polymer magnetic membranes by A. Rybak, Z. J. Grzywna and W. Kaszuwara, 336, 1–2, 79–85, Copyright (2009) Elsevier Ltd

Table 9.4 *Drift coefficients estimated via eq. 9.26 for different magnetic induction*

Membrane	B (mT)	w 10^3 (cm/s)
EC + 1.30 g Nd	0.00	0.01
EC + 1.23 g Nd	0.50	0.91
EC + 1.38 g Nd	0.79	3.10
EC + 1.49 g Nd	1.25	4.70

Source: Reprinted with permission from Journal of Membrane Science, On the air enrichment by polymer magnetic membranes by A. Rybak, Z. J. Grzywna and W. Kaszuwara, 336, 1–2, 79–85, Copyright (2009) Elsevier Ltd

From what we have shown it is clear that flat EC shows no selectivity. So it can be concluded that the magnetic field, and its strength, has to be a reason for the air enrichment reported in this work. The overall situation is, however, much more complicated. As was found by Tagirov *et al.* [36] the molecular clusters could be formed in sufficiently strong magnetic fields. The clusters N_2–O_2–O_2 are preferable for the case of air. So it means that a magnetic field, when sufficiently strong, can affect also the transport of N_2 by means of dragging its clusters along with O_2. This is exactly the situation that has been presented. As can be seen from Table 9.6, the diffusion coefficients increase both for O_2 and N_2 in air, and the ratio of their values is roughly 2:1, like the composition of a cluster. They have much larger values than for pure components. It could simply be explained by a larger magnetic dipole being developed on a cluster, and its interaction with the magnetic field. This leads to a conclusion that the mechanism of gas transport has no solely diffusion character, and for a stronger field drift dominates (see Equations 9.24 and 9.25), providing the basis for the two components to be separated.

As can be seen from Table 9.7, the flux of oxygen in air is smaller than that of the pure gas, contrary to nitrogen where the trend is the opposite. As we already have mentioned, this is probably due to the clusters forming [36]. Table 9.7 also shows some interesting, directional properties of a mass transport through 'magnetic membranes', i.e. the fluxes measured from two opposite sides are different. The reason for that is straightforward, and is simply a consequence of a neodymium powder sedimentation during the membrane casting process. The description, analysis, and practical aspects, of this phenomena is under our current investigation.

Additional interesting information is provided by Table 9.8, which presents some results for PPO magnetic membranes. The idea behind these experiments was to show the constructive role of a magnetic field in improving the already good separation properties of PPO membranes. As can be seen from Tables 9.2 and 9.8, it works in both directions, i.e. the air enrichment is remarkably high, and the oxygen flux is also bigger, as compared with plain PPO membrane [2,8,46,47]. It can be assumed that the tight structure of the PPO membrane cause the deaggregation of N_2-O_2-O_2 clusters, while the magnetic powder softens the structure a bit.

9.5 Conclusions

It is clear that the idea of 'magnetic membranes' works. There are, however, some obstacles that depend on the kind of membrane used, especially when the stronger magnetic

Table 9.5 *Diffusion coefficients for pure gases*

Membrane	B (mT)	N_2 pure					O_2 pure				
		$\bar{D}$	D_L	D_3	D_4	D_5	$\bar{D}$	D_L	D_3	D_4	D_5
		10^5 (cm^2/s)					10^5 (cm^2/s)				
Flat EC	0.00	0.8 ± 0.2	0.8 ± 0.2	1.1 ± 0.2	0.9 ± 0.1	1.1 ± 0.2	1.1 ± 0.1	1.0 ± 0.1	1.1 ± 0.2	1.0 ± 0.1	1.2 ± 0.1
EC + 1.30 g Nd	0.00	0.9 ± 0.2	1.3 ± 0.22	1.5 ± 0.3	1.1 ± 0.2	1.0 ± 0.2	1.2 ± 0.1	2.2 ± 0.2	3.4 ± 0.5	0.6 ± 0.1	0.9 ± 0.1
EC + 1.30 g Nd	0.50	1.3 ± 0.2	1.1 ± 0.2	1.4 ± 0.3	1.3 ± 0.2	0.9 ± 0.2	2.5 ± 0.2	6.1 ± 0.6	5.2 ± 0.5	0.8 ± 0.1	1.0 ± 0.1
EC + 1.38 g Nd	0.79	1.4 ± 0.3	1.0 ± 0.2	1.3 ± 0.3	1.1 ± 0.2	0.9 ± 0.2	3.0 ± 0.3	8.6 ± 0.8	7.2 ± 0.7	0.9 ± 0.1	1.5 ± 0.1
EC + 1.49 g Nd	1.25	1.8 ± 0.3	1.5 ± 0.3	1.6 ± 0.3	1.0 ± 0.2	1.1 ± 0.2	5.6 ± 0.5	14.3 ± 1.0	13.7 ± 1.1	1.4 ± 0.1	3.4 ± 0.3

Source: Reprinted with permission from Journal of Membrane Science, On the air enrichment by polymer magnetic membranes by A. Rybak, Z. J. Grzywna and W. Kaszuwara, 336, 1–2, 79–85 Copyright (2009) Elsevier Ltd

Table 9.6 *Diffusion coefficients for nitrogen and oxygen in air*

Membrane	B (mT)	N_2 in air					O_2 in air				
		$\bar{D}$	D_L	D_3	D_4	D_5	$\bar{D}$	D_L	D_3	D_4	D_5
		10^5 (cm^2/s)					10^5 (cm^2/s)				
Flat EC	0.00	0.9 ± 0.2	0.9 ± 0.2	1.1 ± 0.4	0.9 ± 0.1	1.1 ± 0.2	0.9 ± 0.1	0.9 ± 0.1	1.4 ± 0.2	1.0 ± 0.1	1.2 ± 0.1
EC + 1.30g Nd	0.00	1.0 ± 0.2	2.5 ± 0.4	4.1 ± 0.8	0.1 ± 0.05	0.8 ± 0.1	0.9 ± 0.1	2.8 ± 0.2	3.1 ± 0.3	0.05 ± 0.01	0.6 ± 0.1
EC + 1.30 g Nd	0.50	2.2 ± 0.4	6.5 ± 1.2	5.6 ± 1.0	0.7 ± 0.1	0.4 ± 0.1	3.2 ± 0.3	6.8 ± 0.6	8.1 ± 0.3	1.0 ± 0.1	0.5 ± 0.05
EC + 1.38 g Nd	0.79	4.5 ± 0.9	11.7 ± 2.2	12.2 ± 2.4	1.2 ± 0.2	1.0 ± 0.2	9.4 ± 0.9	18.6 ± 1.2	25.5 ± 2.7	2.5 ± 0.2	1.8 ± 0.15
EC + 1.49 g Nd	1.25	8.9 ± 1.8	24.2 ± 4.2	25.8 ± 4.4	2.0 ± 0.4	2.6 ± 0.4	22.5 ± 1.9	42.9 ± 3.8	51.7 ± 5.2	5.0 ± 0.4	4.9 ± 0.4

Table 9.7 Dependence of the flux vs. membrane side A, B for various membranes

No.	Membrane	B (mT)	N_2 pure		O_2 pure		N_2 in air		O_2 in air	
			J_A	J_B	J_A	J_B	J_A	J_B	J_A	J_B
			$10^4(cm^3_{STP}/cm^2\ s)$							
1	EC + 1.30 g of Nd	0.00	5.41	5.28	8.50	8.58	6.25	6.26	1.65	1.66
2	EC + 1.23 g of Nd	0.50	4.77	5.16	6.99	8.66	5.85	6.05	2.55	2.69
3	EC + 1.38 g of Nd	0.79	4.04	4.89	7.24	8.97	7.79	8.05	4.85	5.18
4	EC + 1.49 g of Nd	1.25	2.16	3.40	7.72	9.56	8.76	9.06	6.85	7.44

field acts. For example, the creation of N_2-O_2-O_2 clusters modifies the air enrichment leading to a less effective process in the case of EC membranes. A careful selection of the polymeric matrix can result in a 'magnetic membrane' with better properties. Formation of these aggregates can be the reason for the difference between theoretical expectations and experimental data for oxygen content in the permeate. It means that we have to take into account the influence of some kind of the 'reaction' between oxygen and nitrogen in a strong magnetic field, using for example the Smoluchowski equation with reaction term for theoretical predictions.

One more problem should be discussed in further investigations. We assumed in our preliminary mathematical description that membranes are homogeneous. However, the method of preparing membranes with magnetic powder by casting suggests that we deal with regular heterogeneous membranes. That is why a mathematical description in further studies should take into account the 'directional' mass transport properties. Diffusion of a penetrant in the membrane where diffusion properties are changing in the x-direction should be expressed in the transport equation by the functional form of the diffusion and drift coefficients, i.e. $D_{ij}(x)$ and $w(x)$ in Equation (9.1).

Acknowledgements

The authors would like to thank The Ministry of Science and Higher Education for providing financial support under the project N N508 409137.

List of Symbols

x	position (cm)
l	thickness of membrane (cm)
t	time (s)
D	diffusion coefficient for a one component system (cm²/s)

Table 9.8 Mass transport coefficients for membranes based on PPO matrix

Membrane	B (mT)	N_2 pure						O_2 pure					
		P	$\bar{D}$	D_L	D_3	D_4	D_5	P	$\bar{D}$	D_L	D_3	D_4	D_5
		$\cdot 10^{10}$ (cm^3_{STP} cm)/(cm^2 s cmHg)	10^7 (cm^2/s)					10^{10} (cm^3_{STP} cm)/(cm^2 s cmHg)	10^7 (cm^2/s)				
Plane PPO	0.00	2.70 ± 0.25	1.8 ± 0.4	1.6 ± 0.3	1.9 ± 0.4	1.5 ± 0.3	1.6 ± 0.3	10.01 ± 1.11	1.2 ± 0.2	1.3 ± 0.2	1.2 ± 0.3	1.4 ± 0.1	1.5 ± 0.2
PPO + 1.80 g Nd	1.70	32.50 ± 2.98	3.3 ± 0.6	2.9 ± 0.5	3.0 ± 0.6	2.9 ± 0.5	3.4 ± 0.7	144.0 ± 8.2	13.3 ± 1.1	17.9 ± 1.2	28.1 ± 2.3	5.3 ± 0.5	14.7 ± 1.2
Membrane	B (mT)	N_2 in air						O_2 in air					
		P	$\bar{D}$	D_L	D_3	D_4	D_5	P	$\bar{D}$	D_L	D_3	D_4	D_5
		$\cdot 10^{10}$ (cm^3_{STP} cm)/(cm^2 s cmHg)	$\cdot 10^7$ (cm^2/s)					$\cdot 10^{10}$ (cm^3_{STP} cm)/(cm^2 s cmHg)	10^7 (cm^2/s)				
Plane PPO	0.00	3.07 ± 0.34	1.8 ± 0.3	1.8 ± 0.3	1.9 ± 0.3	1.4 ± 0.1	1.4 ± 0.2	10.25 ± 1.12	1.4 ± 0.1	1.3 ± 0.1	1.6 ± 0.2	1.2 ± 0.1	1.2 ± 0.1
PPO + 1.80 g Nd	1.70	54.30 ± 5.01	22.7 ± 1.1	10.3 ± 0.9	51.6 ± 3.1	13.0 ± 1.0	3.0 ± 0.2	241.0 ± 15.0	73.1 ± 6.8	29.3 ± 2.2	169.0 ± 14.3	41.7 ± 3.0	8.8 ± 0.8

w	drift coefficient (cm/s)
$c_1(x,t)$	concentration at position x and time t for oxygen
$c_2(x,t)$	concentration at position x and time t for nitrogen
D_{ij}	diffusion coefficients for mixtures, $i,j = 1$ (oxygen), $= 2$ (nitrogen)
$\bar{D}$	mean diffusion coefficient (cm^2/s)
J_S	diffusive mass flux in stationary state ($cm^3_{STP}/cm^2\ s$)
$J_{A,B}$	diffusive mass flux in stationary state on various membrane sides A, B
$L^a(l)$	time lag (s)
D_L	diffusion coefficient calculated using time lag method (cm^2/s)
P	permeation coefficient ($cm^3_{STP}\ cm^2$)/($cm^3\,cmHg\,s$)
Δp	pressure difference (cmHg)
S	sorption coefficient (cm^3_{STP}/cm^3 cmHg)
c_0	initial concentration at position $x = 0$ (cm^3_{STP}/cm^3)
D_3	diffusion coefficient calculated using $D1$–$D8$ system (cm^2/s)
D_4	diffusion coefficient calculated using $D1$–$D8$ system (cm^2/s)
D_5	diffusion coefficient calculated using $D1$–$D8$ system (cm^2/s)
$J^a(l,t)$	diffusive mass flux at position $x = l$
$Q^a(l,t)$	penetrant mass which is flowing out from the membrane (cm^3_{STP}/cm^2)
$Q^a_S(l,t)$	penetrant mass which is flowing out from the membrane in stationary state (cm^3_{STP}/cm^2)
EC	ethylcellulose
PPO	poly(2,6–dimethyl-1,4-phenylene oxide)
Nd	neodymium powder
J_0	oxygen flux for pure EC membrane
J_{Bi}	oxygen flux for magnetic membrane with appropriate magnetic induction
w_{Bi}	drift coefficient
Δc_0	equilibrium concentration of oxygen
B_i	magnetic induction, $i = 0.50$, 0.79 and 1.25 mT, respectively
k_0	a reaction rate constant
$k(x)$	the distribution function of an active, reaction points
$f(c)$	the reaction kinetic term

References

(1) M. Bodzek, Membrane techniques in air cleaning, *Pol. J. Environm. Stud.*, **9**, 1 (2000).
(2) B. Freeman, Y. Yampolskii, I. Pinnau, Eds., Materials Science of Membranes for Gas and Vapor Separation, John Wiley and Sons, Ltd., Chichester, 2006.
(3) E. F. Vansant, R. Dewolfs, *Gas Separation Technology*, Elsevier, 1990.
(4) W. J. Koros, R. Mahajan, Pushing the limits on possibilities for large scale gas separation: which strategies? *J. Membr. Sci.*, **181**, 141 (2001).
(5) T. R. Gaffney, Porous solids for air separation, *Curr. Op. Solid State Mater. Sci.*, **1**, 69 (1996).
(6) L. Lin, J. T. Guthrie, Novel oxygen-enhanced membrane assemblies for biosensors, *J. Membr. Sci.*, **278**, 173 (2006).

(7) S. S. Dhingra, E. Marand, Mixed gas transport study through polymeric membranes, *J. Membr. Sci.*, **141**, 45 (1998).
(8) D. R. Paul, Y. P. Yampolskii, Eds., Polymeric Gas Separation Membranes, CRC Press, Boca Raton, 1993.
(9) A. Singh-Ghosal, W. J. Koros, Air separation properties of flat sheet homogeneous pyrolytic carbon membranes, *J. Membr. Sci.*, **174**, 177 (2000).
(10) D. Q. Vu, W. J. Koros, S. J. Miller, Mixed matrix membranes using carbon molecular sieves I. Preparation and experimental results, *J. Membr. Sci.*, **211**, 311 (2003).
(11) C. M. Zimmerman, A Singh, W. J. Koros, Tailioring mixed matrix composite membranes for gas separations, *J. Membr. Sci.*, **137**, 145 (1997).
(12) C. M. Zimmerman, W. J. Koros, Comparison of gas and sorption in the ladder polymer BBL and some semi-ladder polymers, *Polymer*, **40**, 5655 (1999).
(13) M. Moaddeb, W. J. Koros, Gas transport properties of thin polymeric membranes in the presence of silicon dioxide particles, *J. Membr. Sci.*, **125**, 143 (1997).
(14) M. Moaddeb, W. J. Koros, Effects of colloidal silica incorporation on oxygen/nitrogen separation properties of ceramic-supported 6FDA-IPDA thin films, *J. Membr. Sci.*, **111**, 283 (1996).
(15) X.-G. Li, M.-R. Huang, G. Lin, Temperature dependence and stability of oxygen enrichment through liquid crystalline triheptyl cellulose-containing membranes cast from three solvents, *J. Membr. Sci.*, **116**, 143 (1996).
(16) X.-G. Li, M.-R. Huang, Multilayer ultrathin-film composite membranes for oxygen enrichment, *J. Appl. Polym. Sci.*, **66**, 2139 (1997).
(17) X.-G. Li, M.-R. Huang, L. Hu, G. Lin, P.-C. Yang, Cellulose derivative and liquid crystal blend membranes for oxygen enrichment, *Eur. Polym. J.*, **35**, 157 (1999).
(18) M.-R. Huang, X.-G. Li, X.-L. Ji, W. Qiu, L.-X. Gu, Actual air separation across multilayer composite membranes, *J. Appl. Polym. Sci.*, **77**, 2396 (2000).
(19) X.-G. Li, M.-R. Huang, G.-F. Gu, W. Qiu, J.-Y. Lu, Actual air separation through poly(aniline-co-toluidine)/ethylcellulose blend thin-film composite membranes, *J. Appl. Polym. Sci.*, **75**, 458 (2000).
(20) X.-G. Li, M.-R. Huang, L.-H. Zhu, Y. Yang, Synthesis and air separation of soluble terpolymers from aniline, toluidine, and xylidine, *J. Appl. Polym. Sci.*, **82**, 790 (2001).
(21) M. R. Huang, S. Xi. Li, Z. Q. Dong, W. Feng, X. Y. Wang, S. Y. Gu, Y. H. Wu, X.-A. Huang, Oxygen enrichment from air through multilayer thin low-density polyethylene films, *J. Appl. Polym. Sci.*, **83**, 3013 (2002).
(22) M. R. Huang, Q. W. Xu, S. Y. Gu, Y. H. Wu, Z. Q. Dong, J. Wang, H. T. Shao, X. S. Wang, Oxygen enrichment across blend membranes of bipyridine and ethyl cellulose, *J. Appl. Polym. Sci.*, **87**, 1371 (2003).
(23) X. G. Li, H. J. Zhou, M. R. Huang, Synthesis and properties of processable conducting copolymers from N-ethylaniline with aniline, *J. Polym. Sci. A*, **42**, 6109 (2004).
(24) A. Strzelewicz, Z. J Grzywna, Studies on the air membrane separation in the presence of a magnetic field, *J. Membr. Sci.*, **294**, 60 (2007).
(25) A. Rybak, Z. J. Grzywna, W. Kaszuwara, On the air enrichment by polymer magnetic membranes, *J. Membr. Sci.*, **336**, 79 (2009).
(26) A. Rybak, M. Krasowska, A. Strzelewicz, Z. J. Grzywna, 'Smoluchowski type' equations for modelling of air separation by membranes with various structure, *Acta Phys. Pol. B*, **40**, 1001 (2009).
(27) K. Fritzsche, J. E. Kurz, Handbook of Industrial Membrane Technology, Ed. by M. C. Porter, Noyespublications, 1990.
(28) J. H. Petropoulos, Directional membrane permeability in polymer-vapor systems, *J. Polym. Sci.: Polym. Phys.*, **12**, 35 (1974).
(29) Z. J. Grzywna, J. Stolarczyk, Modelling of gold dispersion in glassy polymers by Smoluchowski equation, *Int. J. Modern Phys. C*, **13**, 9 (2002).
(30) Z. J. Grzywna, J. Stolarczyk, Diffusion in glassy polymers – from random walks to partial differential equations, *Acta Phys. Pol. B*, **36**, 1001 (2005).
(31) J. Crank, The Mathematics of Diffusion, Oxford University Press, 1975.

(32) G. D. Smith, Numerical Solution of Partial Differential Equations: Finite Difference Methods, Clarendon Press, Oxford, 2nd edn, 1978.
(33) H. Risken, The Fokker-Planck Equation, Springer-Verlag, 2nd edn, Berlin, Heidelberg, New York, 1996.
(34) N. G. van Kampen, Stochastic Processes in Physics and Chemistry, North-Holland, 1981.
(35) W. Jost, Diffusion in Solids, Liquids, Gases, Academic Press, 1960.
(36) M. S. Tagirov, R. M. Aminova, G. Frossati, V. N. Efimov, G. V. Mamin, V. V. Naletov, D. A. Tayurskii, A. N. Yudin, On the magnetism of liquid nitrogen-liquid oxygen mixture, *Phys. B*, **329–333**, 433 (2003).
(37) A. Strzelewicz, Z. J. Grzywna, On the permeation time lag for different transport equations by Frisch method, *J. Membr. Sci.*, **322**, 460 (2008).
(38) E. L. Cussler, Diffusion Mass Transfer in Fluid Systems, 2nd edn, Cambridge University Press, 1997.
(39) E. L. Cussler, Multicomponent Diffusion, Elsevier Scientific Publishing Company, Amsterdam, 1976.
(40) K. C. Khulbe, G. Chowdhury, B. Kruczek, R. Vujosevic, T. Matsuura, G. Lamarche, Characterization of the PPO dense membrane prepared at different temperatures by ESR, atomic force microscope and gas permeation, *J. Membr. Sci.*, **126**, 115 (1997).
(41) B. Kruczek, F. Shemshaki, S. Lashkari, R. Chapanian, H. L. Frisch, Effect of a resistance-free tank on the resistance to gas transport in high vacuum tube, *J. Membr. Sci.*, **280**, 29 (2006).
(42) S. Lashkari, B. Kruczek, Development of a fully automated soap flowmeter for micro flow measurements, *Flow Measure and Instrument*, **19**, 397 (2008).
(43) S. Lashkari, A. Tran, B. Kruczek, Effect of back diffusion and back permeation of air on membrane characterization in constant pressure system, *J. Membr. Sci.*, **324**, 162 (2008).
(44) A. Tran, B. Kruczek, Development and characterization of homopolymers and copolymers from the family of polyphenylene oxides, *J. Appl. Polym. Sci.*, **106**, 2140 (2007).
(45) Z. J. Grzywna, H. L. Frisch, An application of WKB approximation to transient diffusion in inhomogeneous membranes. Part 3: Permeation, *Pol. J. Chem.*, 1–3 (1984).
(46) G. Golemmea, J. B. Nagy, A. Fonseca, C. Algieri, Yu. Yampolskii, ^{129}Xe-NMR study of free volume in amorphous perfluorinated polymers: comparsion with other methods, *Polymer*, **44**, 5039 (2003).
(47) A. Alentiev, E. Drioli, M. Gokzhaev, G. Golemme, O. Ilinich, A. Lapkin, V. Volkov, Yu. Yampolskii, Gas permeation properties of phenylene oxide polymers, *J. Membr. Sci.*, **138**, 99 (1998).

Part III

Membrane Separation of CO_2 from Gas Streams

10

Ionic Liquid Membranes for Carbon Dioxide Separation

Christina R. Myers, David R. Luebke, Henry W. Pennline, Jeffery B. Ilconich and Shan Wickramanayake
P.O. Box 10940, Pittsburgh, PA 15236-0940, USA

10.1 Introduction

A new industrial revolution is occurring around the world. Rapid improvements in equipment, control, and process configuration are being made in an attempt to minimize raw material consumption and waste production while improving efficiency. 'Environmentally friendly' and 'green' have become the buzz words for the new millennium. This growing environmental awareness represents not just a shift in public opinion but a global recognition that sustainable practices have become imperative.

The focus of much of the new awareness has turned towards examining greenhouse gas emissions and their impact on global climate. Greenhouse gases include carbon dioxide (CO_2), methane (CH_4), nitrous oxide (N_2O), ozone (O_3), hydrofluorocarbons (HFCs), perfluorocarbons (PFCs), and sulfur hexafluoride (SF_6) [1]. Among these gases, CO_2, CH_4, and N_2O have become the focal point of efforts to limit climate change since these gases are released in large quantities and have significant global warming potential (GWP), a measure of the effect of a species on climate based on its ability to absorb infrared radiation, the particular location of that absorbance on the spectrum, and atmospheric lifetime. A breakdown of emissions based on their relative climate impacts is shown in Figure 10.1. Reducing emissions of CO_2 from energy production is particularly important because those emissions are the largest and fastest growing contributor to climate change. In the United States, emissions rose 21.8% during the period 1990–2007, and has likewise increased in other developing economies [1].

Membrane Gas Separation Edited by Yuri Yampolskii and Benny Freeman

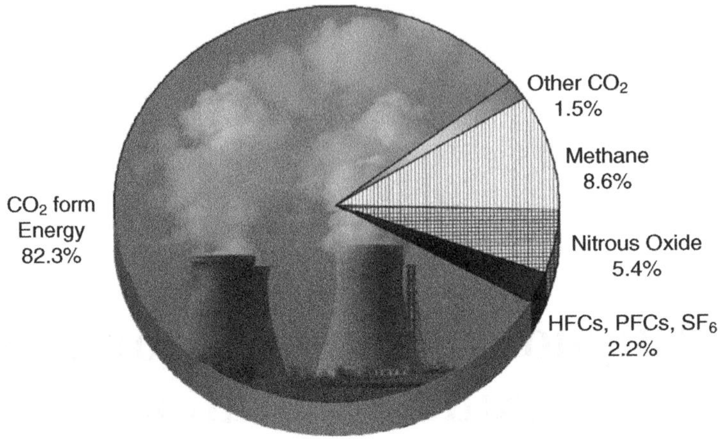

Figure 10.1 *United States greenhouse gas emissions (equivalent global warming basis). Reproduced with permission from the National Energy Technology Laboratory (NETL), US Department of Energy. Data taken from Energy Information Administration, Emissions of Greenhouse Gases in the United States 2006 (Washington D.C., November 2007)*

As the reality of future limits on CO_2 emissions and the low readiness level of renewable energy technologies have become apparent, many businesses, academics, and government agencies have taken an aggressive approach in developing technologies addressing CO_2 capture and sequestration. The majority of these technologies focus on fossil energy power generation systems since approximately 83% of CO_2 emissions are energy related [2] and capture from stationary point sources is a more attainable goal than capture from mobile sources.

The power generation systems currently of greatest research interest are those based on combustion and gasification of coal since energy from coal is responsible for more than half of electricity production in the United States [3,4]. Research issues associated with CO_2 capture from these systems include minimization of parasitic load, reduction in the cost of oxygen production, application of capture technology to existing system designs, integration with other power system components, scale-up, and reduction of capital costs [5].

Pulverized coal combustion (PCC) technology, which generates the majority of electricity in the U.S. [3], has been used for more than 100 years with steady improvement in efficiency and reduced pollution. Coal is combusted with air in a boiler to produce high-pressure steam. The steam drives a steam turbine coupled with a generator producing electricity. Particulates are captured from the flue gases exiting the boiler and subsequently the acid gases are removed by selective catalytic reduction and wet lime scrubbing. The flue gas, which is saturated with moisture after the flue gas desulfurization unit, contains approximately 10–15% CO_2 at atmospheric pressure that is then emitted through a stack. Capturing CO_2 from these systems presents a particular challenge. The low pressure, dilute concentration, and a high volume of gas lead to a low driving force for CO_2 separation. Compressing the large gas volume requires a large parasitic load, and trace impurities in the flue gas limit the capture technologies that may be used.

In the integrated gasification combined cycle (IGCC) power generation, oxygen and steam react with coal in a gasifier under pressure producing synthesis gas (syngas), a mixture containing CO, CO_2, H_2, H_2O and contaminants. The syngas exits the gasifier at high temperature and pressure, meaning that there is an innate advantage for CO_2 capture from IGCC processes compared to conventional pulverized coal combustion systems since compression requirements to reach sequestration pressures is appreciably less. Contaminants and particulates are removed from the syngas, and the gas is then fed to a combustion turbine/generator to produce electricity. The combination of a combustion turbine/generator, a heat recovery steam generator, and a steam turbine/generator, known as a combined cycle, is more efficient than conventional systems, allowing for higher efficiency and lower operating costs. Of the two energy production technologies, the IGCC process has the potential for higher efficiency but at increased capital costs.

Process efficiency can be improved in IGCC systems by taking advantage of potential benefits associated with integration of CO_2 capture and the water gas shift (WGS) reaction. The WGS reaction is exothermic and equilibrium limited under the process conditions.

$$CO + H_2O \leftrightarrow H_2 + CO_2;\ \Delta H_{298} = -41\ \text{kJ/mol}$$

Conventional WGS reactors take advantage of Le Chatelier's principle [6] by cooling the reacting gases to 533 K and adding steam to increase conversion and produce a shifted synthesis gas containing approximately 30% CO_2 and less than 1% CO.

Le Chatelier's principle is the basis for the WGS reaction's sensitivity to temperature and its tendency to shift towards the products side as the temperature decreases. The principle can alternatively be utilized by selectively removing one of the reaction products, CO_2 or H_2. Under these conditions only stoichiometric steam and cooling to the separation temperature would be required, allowing for an increase in process efficiency as depicted in Figure 10.2.

Currently CO_2 can be removed from the IGCC system at 313 K using Selexol [7] or at even lower temperatures with Rectisol [8]. Such substantial cooling to accommodate the low operating temperatures of conventional capture solvents followed by reheating to the

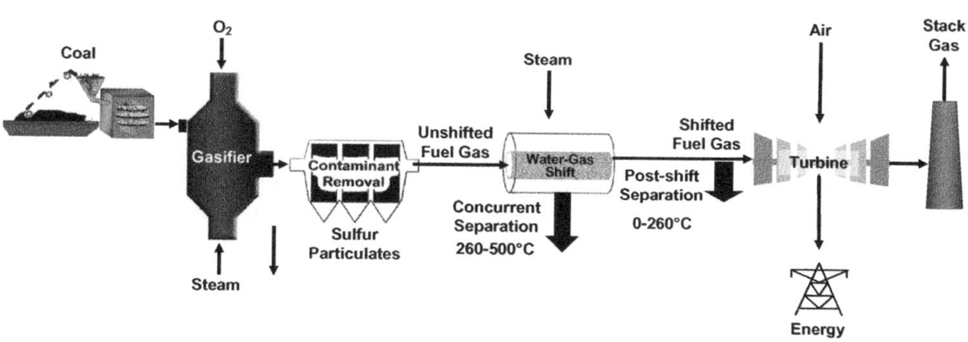

Figure 10.2 *The integrated gasification combined cycle (IGCC) with integral water-gas shift reaction incorporated*

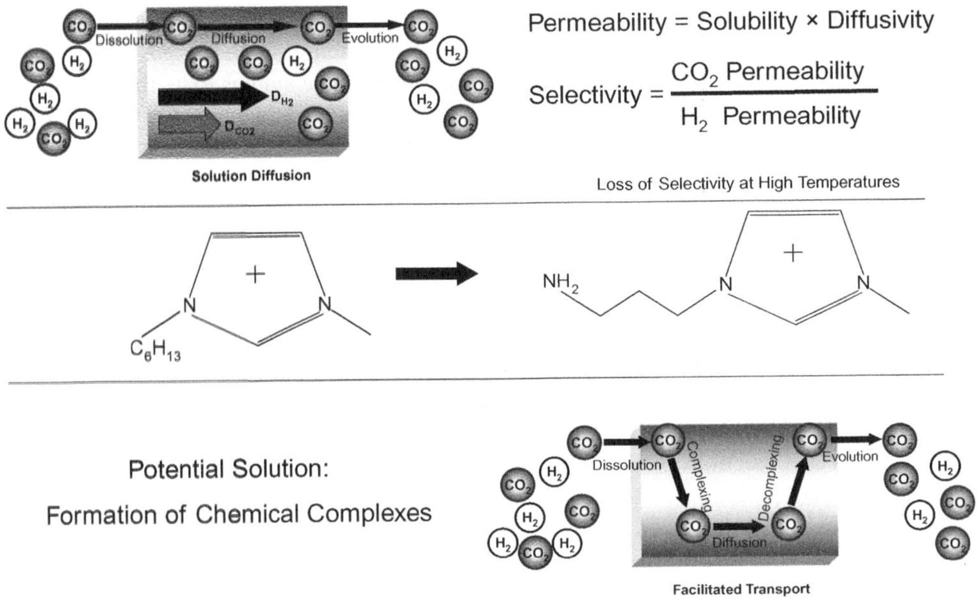

Figure 10.3 *Potential improvement of CO_2 permeability through facilitated transport*

combustion turbine inlet temperature contributes significantly to costs. Additionally, water is condensed from the gas stream during the cooling process and cannot then be expanded across the turbine to produce electricity.

Given the desired high temperature and pressure for CO_2 capture in IGCC systems and the current state of the art, major advances in separation technology will be necessary. Membranes are one technology with which such advancement might be possible. The National Energy Technology Laboratory (NETL) has pursued an approach based on supported liquid membranes (SLMs), which consist of a liquid transport medium and a solid support. The liquid allows higher diffusivity, leading to an increase in permeability over solid membranes and the potential to add chemical complexes which increase CO_2 solubility, as shown in Figure 10.3. There are two significant problems in the development of practical SLMs: evaporation of the liquid at high temperatures and forcing of the transport medium out of the support due to large pressure differentials. A possible solution to evaporative failure would be the use of ionic liquids (ILs). Ionic liquids are organic salts with melting temperatures below ambient due to their inability to easily form stable crystal lattices and associated delocalized charge [9]. As salts, these materials have negligible vapour pressure [10] and are also interesting for their considerable thermal stability and high CO_2 solubility relative to hydrogen (H_2), nitrogen (N_2), and methane (CH_4). Proposed explanations for this high solubility include the interaction of CO_2 and the anion occurring through a weak Lewis acid–base interaction and interactions of the polar ionic liquid ions with the large quadrupole moment of CO_2 [10–16].

The exceptional ability of ionic liquids to absorb CO_2 along with their other chemical and physical properties makes them potentially interesting materials for CO_2 separation

[10]. NETL has embarked on a programme to make use of these properties to develop technologies for cost effective and efficient capture and sequestration of the CO_2 emissions from power generation. One of the areas being investigated is the use of ionic liquid transport media in supported liquid membranes for high temperature capture applications such as the IGCC process.

10.2 Experimental

The ionic liquid 1-*n*-hexyl-3-methylimidazolium bis(trifluorosulfonyl)imide, [hmim][Tf_2N], was synthesized and characterized using standard procedures [17–19] at the University of Notre Dame. The ionic liquid 1-(3-aminopropyl)-3-methylimidazolium bis(trifluoromethylsulfonyl)imide, [$H_2NC_3H_6$mim][Tf_2N], was also synthesized and characterized at the University of Notre Dame [20].

The supported ionic liquid membranes (SILMs) were prepared by depositing the ionic liquids [hmim][Tf_2N] and [$H_2NC_3H_6$mim][Tf_2N] on top of cross-linked nylon support discs in a shallow glass container. A sufficient amount of ionic liquid to cover the membrane was used. The membrane was allowed to absorb the ionic liquid for at least 4 hours, and then the SILMs were removed from container and blotted dry. Further details concerning this procedure were published earlier [19].

Membrane performance testing was carried out using a constant pressure flow system equipped with a Clarus 500 gas chromatograph (GC) with twin thermal conductivity detector (TCD) and Alltech Hayesep D100/200 packed columns for measurement of the permeate and retentate gas compositions. A commercially available filter holder was used to mount the membrane discs for testing with gas mixtures of known concentration as previously described [19].

Membrane permeabilities for both CO_2 and H_2 and the CO_2/H_2 separation factor were evaluated in the presence of varying concentrations of CO (10, 100, 500 ppm) in mixtures containing ~20 mol% CO_2, ~20% H_2, and a balance of Ar. Performance was also evaluated in the presence of varying concentrations of H_2S (10, 100, 500 ppm).

Membrane performance was also tested in a simulated fuel gas (SFG) with the following gas composition: 30.1% CO_2, 40.1% H_2, 1.01% CO, 0.992% CH_4, 202 ppm H_2S, and balance Ar.

10.3 Results

Performance testing of supported [hmim][Tf_2N] membranes was originally carried out using gas mixtures containing only CO_2, H_2 and Ar [20]. The CO_2 permeability progressively increased over the temperature range from 310 K to 573 K from $3.75 \times 10^{-15}\,m^2\,s^{-1}\,Pa^{-1}$ to over $8.25 \times 10^{-15}\,m^2\,s^{-1}\,Pa^{-1}$ (490–1085 Barrer), while the selectivity steadily decreased with increasing temperature from 9.1 to 1.3. This behaviour is characteristic of solution-diffusion without facilitation.

CO_2 permeabilities were evaluated using gas mixtures with varying concentrations of CO (10 ppm, 100 ppm, 500 ppm, and ~5 mol%) on the supported [hmim][Tf_2N] membrane over the temperature range 310 K to 423 K, as shown in Figure 10.4. The CO_2 permeability

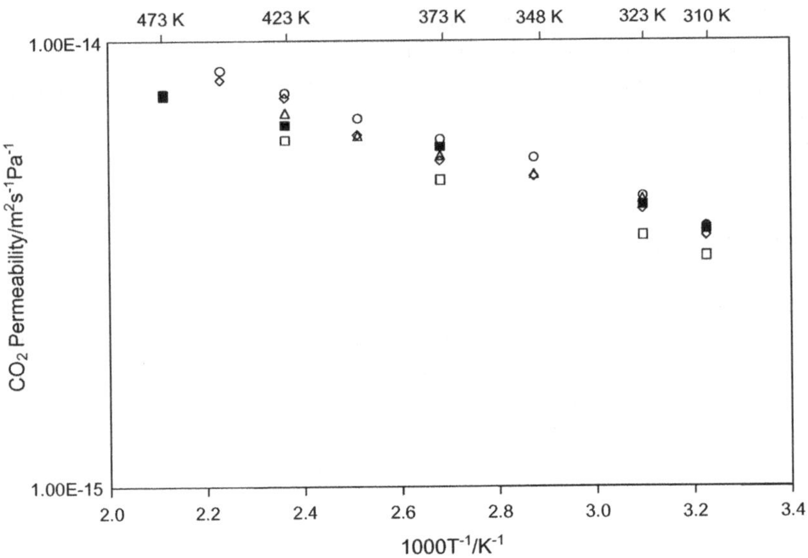

Figure 10.4 *Carbon dioxide permeability of [hmim][Tf$_2$N] on cross-linked nylon; ■ → without CO; □ → 5% CO; ◇ → 10 ppm CO; ○ → 100 ppm CO; Δ → 500 ppm CO*

in gas mixtures containing 10 ppm CO over the temperature range 310 K to 373 K ranged from 3.65×10^{-15} to $5.37 \times 10^{-15}\,m^2\,s^{-1}\,Pa^{-1}$, which is lower than the permeability measured for the mixture without CO. As the temperature increases the CO_2 permeability in the presence of 10 ppm CO increases relative to the permeability without CO. The permeabilities at 423 K and 448 K in the presence of 10 ppm CO are $7.44 \times 10^{-15}\,m^2\,s^{-1}\,Pa^{-1}$ and $8.16 \times 10^{-15}\,m^2\,s^{-1}\,Pa^{-1}$ as compared to $6.46 \times 10^{-15}\,m^2\,s^{-1}\,Pa^{-1}$ at 423 K and $7.57 \times 10^{-15}\,m^2\,s^{-1}\,Pa^{-1}$ at 473 K with no CO present.

In gas mixtures containing 100 ppm CO, CO_2 permeability is 3.82×10^{-15} and $8.58 \times 10^{-15}\,m^2\,s^{-1}\,Pa^{-1}$ at 310 K and 448 K, respectively. These permeabilities represent a slight increase in CO_2 permeability over the entire temperature range compared to mixtures not containing CO.

CO_2 permeabilities in the presence of 500 ppm CO are 3.84×10^{-15} and $6.87 \times 10^{-15}\,m^2\,s^{-1}\,Pa^{-1}$ at 310 K and 423 K, respectively. The observed permeabilities are quite similar to or slightly higher than those observed in the absence of CO.

Permeabilities for CO_2 in mixtures containing 5% CO show negligible change as compared to CO_2 permeabilities in mixtures without CO for temperatures between 373 K and 473 K. Below 373 K, the permeability for CO_2 in mixtures without CO was slightly higher (3.77×10^{-15}–$5.79 \times 10^{-15}\,m^2\,s^{-1}\,Pa^{-1}$) than in the mixture with 5% CO (3.28×10^{-15}–$4.87 \times 10^{-15}\,m^2\,s^{-1}\,Pa^{-1}$).

CO_2 permeability in nylon-supported [hmim][Tf$_2$N] displayed Arrhenius behaviour for all the CO concentrations tested with the following coefficients of determination: 0.987, 0.989 , 0.989, and 0.997, for the 10 ppm, 100 ppm, 500 ppm, and 5% CO, respectively, which is consistent with transport based on physical solubility.

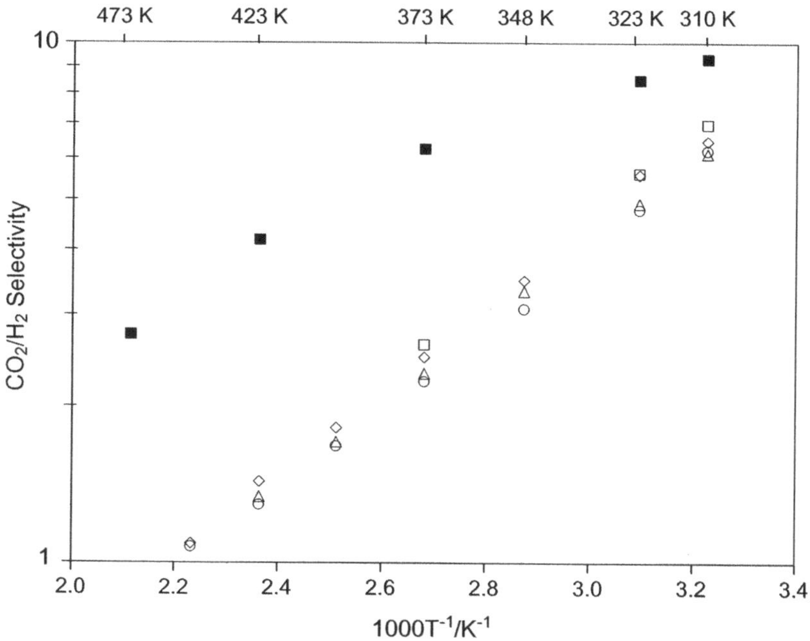

Figure 10.5 *Carbon dioxide selectivity of [hmim][Tf_2N] on cross-linked nylon; ■ → without CO; □ → 5% CO; ◇ → 10 ppm CO; ○ → 100 ppm CO; Δ → 500 ppm CO*

The CO_2/H_2 selectivities for supported [hmim][Tf_2N] decrease progressively with increasing temperature, as shown in Figure 10.5, as anticipated for a solution diffusion mechanism. Differences in the selectivity for varying CO concentrations are quite similar ranging from 6.46 to 1.44, 6.20 to 1.30, 6.13 to 1.34, and 4.70 to 1.01 in the temperature range 310 K to 423 K for the 10 ppm, 100 ppm, 500 ppm, and 5% CO, respectively. The trends are strongly Arrhenius over the entire temperature range with coefficients of determination of 0.998, 0.998, 0.999, and 0.999 for the 10 ppm, 100 ppm, 500 ppm, and 5% CO, respectively.

The membrane performance of nylon-supported [hmim][Tf_2N] was measured in gas mixtures with varying concentrations of H_2S, as shown in Figure 10.6. The CO_2 permeability for mixtures containing 10 ppm H_2S ranged from 3.33×10^{-15} to $6.04 \times 10^{-15}\,m^2\,s^{-1}\,Pa^{-1}$ in the temperature range from 310 K to 423 K, a slight decrease compared with mixtures not containing H_2S. At H_2S concentrations of 100 ppm, the CO_2 permeability increased from 3.62×10^{-15} to $5.58 \times 10^{-15}\,m^2\,s^{-1}\,Pa^{-1}$ in the temperature range from 310 K to 373 K, closely matching CO_2 permeability without H_2S. At 423 K, the CO_2 permeability in the presence of 100 ppm H_2S, $7.03 \times 10^{-15}\,m^2\,s^{-1}\,Pa^{-1}$, was increased as compared to the permeability measured in the absence of H_2S. In mixtures containing 0.9% H_2S, the CO_2 permeability increased from 3.00×10^{-15} to $5.40 \times 10^{-15}\,m^2\,s^{-1}\,Pa^{-1}$ over the temperature range 310 K to 423 K, a uniform decrease from the permeability in mixtures not containing H_2S.

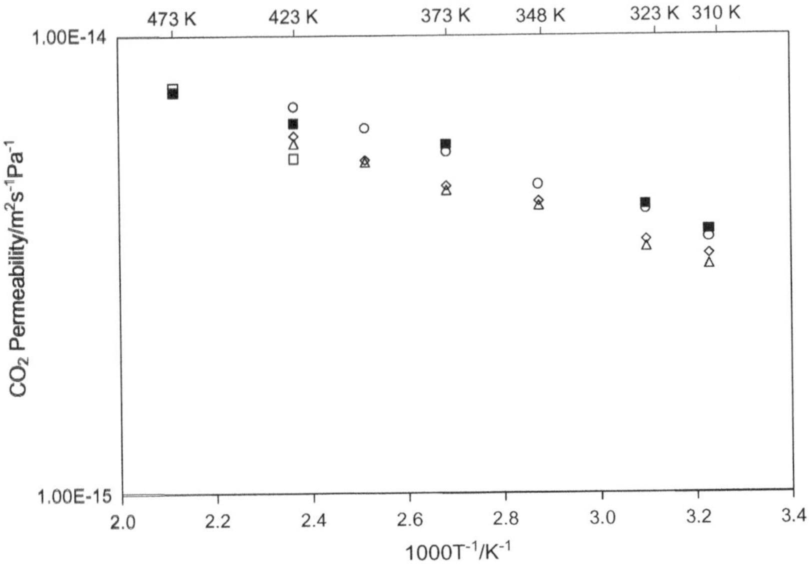

Figure 10.6 *Carbon dioxide permeability of [hmim][Tf_2N] on cross-linked nylon; ■ → without H_2S; □ → 0.9% H_2S; ◇ → 10 ppm H_2S; ○ → 100 ppm H_2S; Δ → 500 ppm H_2S*

The CO_2/H_2 selectivity for supported [hmim][Tf_2N], as shown in Figure 10.7, decreases with increasing temperature as anticipated for a solution diffusion mechanism. The CO_2/H_2 selectivities for varying H_2S concentrations are quite similar ranging from 6.53 to 1.37, 6.71 to 1.33, 6.82 to 1.42, and 7.24 to 1.57 in the temperature range 310 K to 423 K for H_2S concentrations of 10 ppm, 100 ppm, 500 ppm, and 0.9%, respectively. The trends are strongly Arrhenius over the entire temperature range with coefficients of determination of 0.999, 0.995, 0.996, and 0.999 for H_2S concentrations of 10 ppm, 100 ppm, 500 ppm, and 0.9%, respectively.

The membrane performance of [$H_2NC_3H_6$mim][Tf_2N], which contains a primary amine functionality, was also examined and the results shown in Figure 10.8. The CO_2 permeability exhibits Arrhenius behaviour with coefficient of determination 0.998 in the temperature range from 310 K to 368 K. Little change is observed in permeability from 368 K to 448 K. The CO_2/H_2 selectivity for supported [$H_2NC_3H_6$mim][Tf_2N] increases from 9.5 to 13.8 in the temperature range from 310 K to 348 K then decreases from 10.5 to 3.11 in the range from 373 K to 423 K, creating a selectivity maximum. Ionic liquids which contain primary amines such as [$H_2NC_3H_6$mim][Tf_2N] are thought to form chemical complexes with CO_2 [21]. It has been previously suggested that the observed trends in permeability and selectivity may be explained by a change in rate-limiting step from decomplexation of the CO_2 at low temperature to diffusion of complexed CO_2 at high temperatures [20].

The CO_2 permeability of supported [$H_2NC_3H_6$mim][Tf_2N] in the presence of 5% CO shows an Arrhenius behaviour as it increases from 2.84×10^{-15} to $5.85 \times 10^{-15}\,m^2 s^{-1} Pa^{-1}$ in the temperature range from 310 K to 423 K with a coefficient of determination of 0.971,

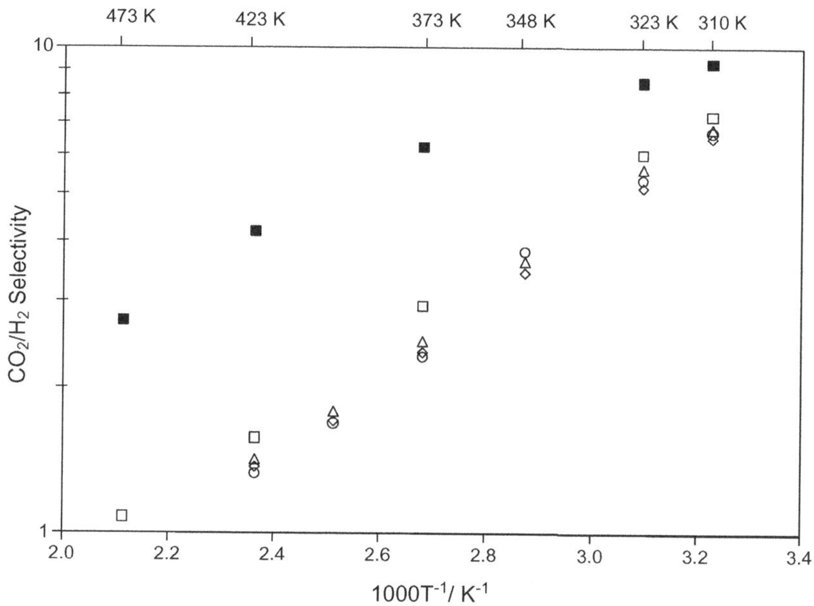

Figure 10.7 *Carbon dioxide selectivity of [hmim][Tf_2N] on cross-linked nylon; ■ → without H_2S; □ → 0.9% H_2S; ◇ → 10 ppm H_2S; ○ → 100 ppm H_2S; Δ → 500 ppm H_2S*

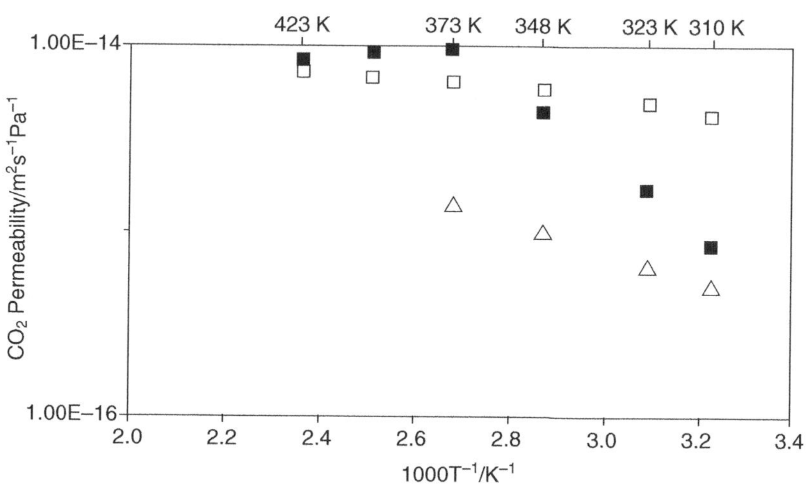

Figure 10.8 *Carbon dioxide permeability of [$H_2NC_3H_6$mim][Tf_2N] on cross-linked nylon; ■ → without contaminants; □ → 5% CO; Δ → 0.9 % H_2S*

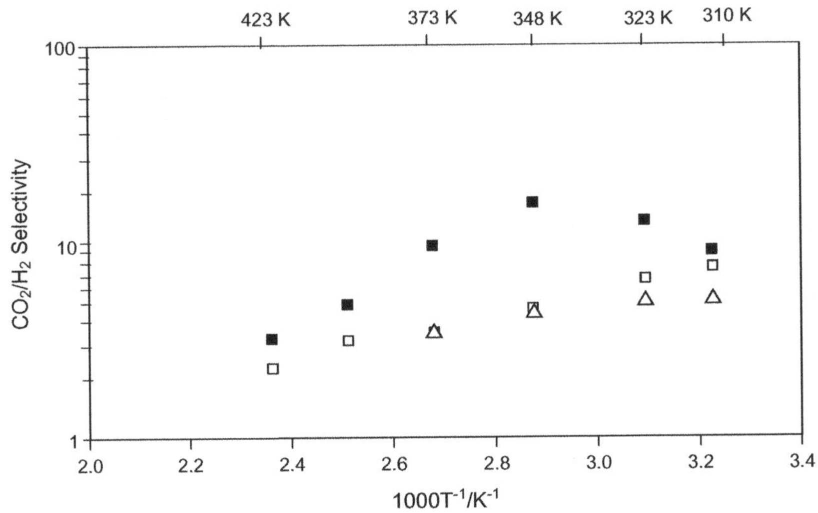

Figure 10.9 *Carbon dioxide selectivity of [$H_2NC_3H_6$mim][Tf_2N] on cross-linked nylon; ■ → without contaminants; □ → 5% CO; Δ → 0.9 % H_2S*

as shown in Figure 10.8. The CO_2/H_2 selectivity in the presence of 5% CO decreases from 7.82 to 2.64 over the temperature range from 310 K to 423 K and shows a rough Arrhenius dependence in Figure 10.9. In the presence of 0.9% H_2S, CO_2 permeability for supported [$H_2NC_3H_6$mim][Tf_2N] increases from 3.87×10^{-16} to $1.10 \times 10^{-15}\,m^2\,s^{-1}\,Pa^{-1}$ over the temperature range from 310 K to 373 K with a roughly Arrhenius dependence exhibited in Figure 10.8. The CO_2/H_2 selectivity decreases from 5.14 to 3.72 with a similarly rough Arrhenius dependence exhibited in Figure 10.9. Given the observed Arrhenius behaviours, which are consistent with solution-diffusion instead of facilitated transport, it is reasonable to suggest that both CO and H_2S are interfering with the ability of the ionic liquid to form complexes with CO_2.

The membrane performance of the supported ionic liquids was examined in a simulated fuel gas (SFG) containing both H_2S and CO. The CO_2 permeability for both supported ionic liquids in separations from SFG, as shown in Figure 10.10, differs in magnitude but not trend from that in mixtures not containing contaminants shown in Figure 10.8. While separation from the SFG shows more of an effect on [$H_2NC_3H_6$mim][Tf_2N] than on [hmim][Tf_2N] [20], the facilitated transport behaviour is only reduced but not eliminated as in the case of the higher concentrations of CO and H_2S. Selectivity results for [$H_2NC_3H_6$mim][Tf_2N] and [hmim][Tf_2N] with SFG, as shown in Figure 10.11, also follow similar trends to the [$H_2NC_3H_6$mim][Tf_2N] and [hmim][Tf_2N] mixtures without contaminants shown in Figure 10.9.

10.4 Discussion

In understanding the similar effects of CO and H_2S on the macroscopic membrane performance parameters of supported [hmim][Tf_2N], it is necessary to consider the molecu-

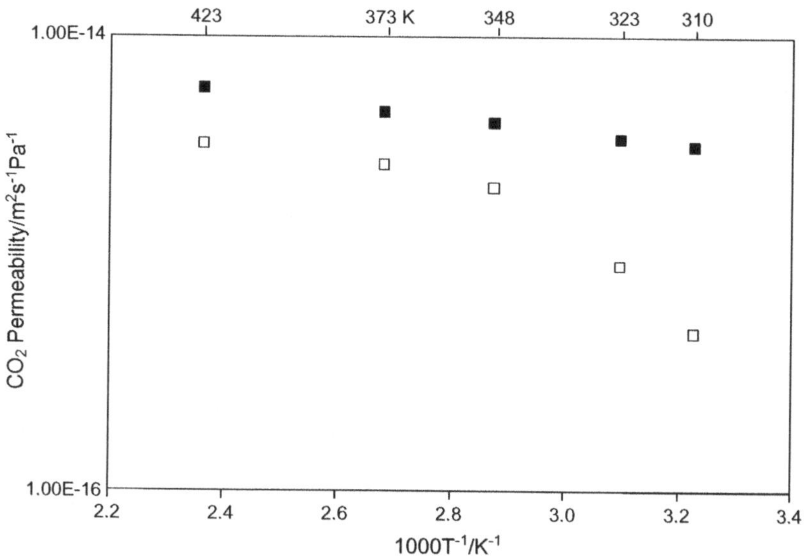

Figure 10.10 *Carbon dioxide permeability of [hmim][Tf$_2$N] ■ and [H$_2$NC$_3$H$_6$mim][Tf$_2$N] □ on cross-linked nylon with simulated fuel gas*

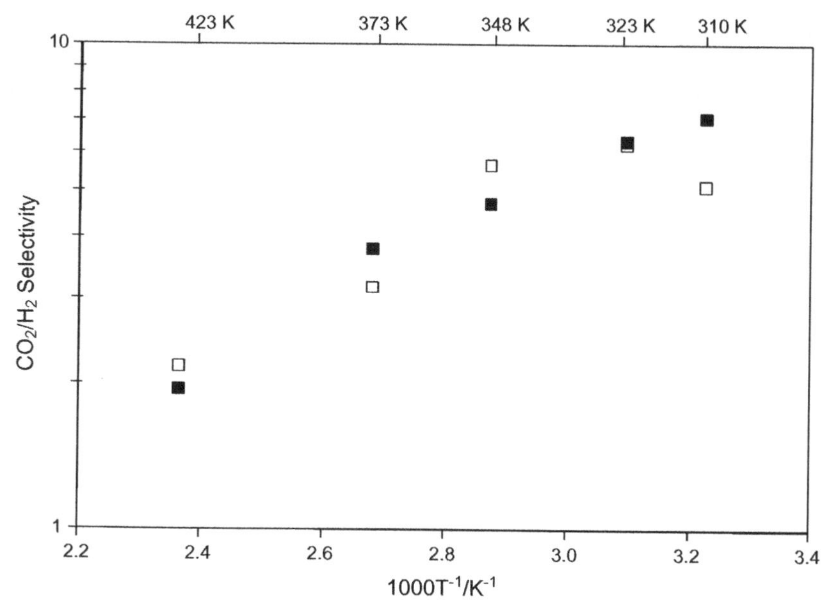

Figure 10.11 *Carbon dioxide selectivity of [hmim][Tf$_2$N] ■ and [H$_2$NC$_3$H$_6$mim][Tf$_2$N] □ on cross-linked nylon with simulated fuel gas*

lar-level interactions of the gases with the liquid ions. The results are initially puzzling since it appears that both CO and H_2S have very little effect on the CO_2 permeability, but a substantial effect is observable on CO_2/H_2 selectivity owing to an increase in H_2 permeability. H_2 permeability is a combination of diffusivity and solubility. Gas diffusivities are closely tied to viscosity in ionic liquids [22–24]. If the [hmim][Tf_2N] viscosity were being substantially affected by the presence of CO and H_2S, it is likely that there would be an observable change in CO_2 permeability as well. Since this is not the case, it is reasonable to believe that CO and H_2S are impacting H_2 solubility. Also puzzling is the lack of substantial differences in the magnitude of the H_2 solubility effect at different CO and H_2S concentrations. It appears that 10 ppm of CO or H_2S is a sufficient concentration to saturate the effect.

CO solubilities in ionic liquids were reported by Laurenczy *et al.* [25]. The researchers noted that the CO solubility is more dependent upon the nature of the ionic liquid than H_2 and attributed this to CO's strong dipole moment and greater polarizability. Also reported was the greater solubility of CO in ionic liquids made up of the 1-butyl-3-methylimidazolium cation and the [Tf_2N] anion than ionic liquids having the same cation and other anions. This result is consistent with CO having relatively strong interactions, with the [Tf_2N] anion. That anion is also known to have strong interactions with CO_2 associated with weak acid–base interactions [13,26]. Since H_2S is also an acid gas, similarly strong interactions are expected between that gas and [Tf_2N].

Given the strong associations of the polar CO and H_2S molecules with the ionic liquid, one explanation for the increase in H_2 solubility may be a net depolarization effect. By closely associating with the anion (and potentially also with the cation), even low concentrations of the relatively small, polar gas molecules may decrease the net strength of the interactions among the much larger anions and cations, increasing the molar volume of the ionic liquid [7], represented in Figure 10.12. The weak interaction of H_2 with the ionic liquid causes its solubility to be strongly tied to molar volume [27,28]. The increase in ionic liquid molar volume associated with the presence of CO and H_2S could result in an increased H_2 solubility and permeability which would in turn decrease the CO_2/H_2 selectivity.

A similar effect is observed on the membrane performance parameters of supported [$H_2NC_3H_6$mim][Tf_2N] in the presence of CO and H_2S. A decrease occurs in CO_2/H_2

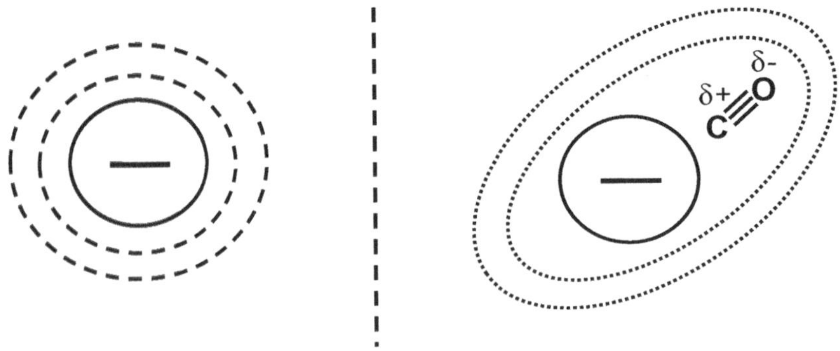

Figure 10.12 Ion depolarization effect in the presence of CO

selectivity associated with an increase in H_2 solubility. The behaviour of the membrane is complicated in this case by the contribution of the amine group to CO_2 solubility through a strong acid–base interaction [29,30] and the impact of CO and H_2S on that interaction. As shown in Figure 10.8, when CO and H_2S are not present, CO_2 permeability rises rapidly from 310 to 373 K then levels off. This temperature dependence has been previously attributed to a change in the rate-determining step from dissociation of the acid–base complex to CO_2 diffusion, as shown in Figure 10.3. This CO_2 permeability temperature dependence is tied to a peak in CO_2/H_2 selectivity at 348 K. In the presence of CO and H_2S, no change in slope occurs in CO_2 permeability and no peak is observed in CO_2/H_2 selectivity. As with [hmim][Tf_2N] and other physical solvents, CO_2 permeability increases and CO_2/H_2 selectivity decrease in an Arrhenius dependence with increasing temperature.

In the presence of H_2S, the behaviour is straightforward to explain. As the stronger acid gas, H_2S competitively interferes with CO_2 interaction with the amine site on the cation. This interference prevents the increase in CO_2 solubility which normally results from the presence of that amine site and causes the ionic liquid to behave as if the site were not present, i.e. as if it were [hmim][Tf_2N].

In the presence of CO, the rationale for the alteration in behaviour is less clear. Based on the similar trends, it would be reasonable to suggest that CO is also interacting with the amine site on the cation and interfering with the ability of CO_2 to do so. There are, however, no known strong associations between primary amines and CO. It is possible that CO forms such an association with another portion of the ionic liquid cation or with the anion. This adduct could then behave as an acid, interacting with the amine site to block it from forming complexes with CO_2.

In the simulated fuel gas studies where a gas mixture including approximately 10 000 ppm CO and 200 ppm H_2S was used, the results for supported [hmim][Tf_2N] were similar to those in the presence of either CO or H_2S alone. The result is consistent with the argument made earlier that very low concentrations of polar solute gases are sufficient to depolarize the ionic liquid slightly increasing molar volume and H_2 permeability. The membrane performance for supported [$H_2NC_3H_6$mim][Tf_2N] is less affected by the presence of lower concentrations of both CO and H_2S (10 000 ppm CO and 200 ppm H_2S, as shown in Figures 10.10 and 10.11) than by a higher concentration of either contaminant alone (50 000 ppm CO and 9000 ppm H_2S, respectively, as shown in Figures 10.8 and 10.9), i.e. the slope change in CO_2 permeability and the CO_2/H_2 selectivity peak are not eliminated in the simulated fuel gas studies. The result indicates that the lower concentrations of both contaminants are insufficient to competitively interfere with the CO_2–amine interaction. It also suggests that the direct H_2S interference with the interaction is more effective than the indirect interference of CO through adduct formation since 9000 ppm H_2S has a greater effect on the permeability and selectivity trends than the 10 000 ppm CO present in the simulated fuel gas.

10.5 Conclusions

The results presented demonstrate that multi-component gas interactions with supported ionic liquid membranes are more complex than might be at first suspected. The macroscopic membrane performance measurements shown here are sufficient to open a variety of new questions about the complicated nature of those interactions but do not allow confirmation of the molecular level explanations proposed. Further work will be required to fully understand the effects of common syngas constituents on ionic liquid membranes. That work will be critical to the successful application of these membranes to IGCC power systems and the efficient mitigation of greenhouse gas emissions from those power systems.

References

(1) Inventory of U.S. Greenhouse Gas Emissions and Sinks: 1990–2007, in *The US EPA, 2009 U.S. Greenhouse Gas Inventory Report*, April 2009.

(2) Emissions of Greenhouse Gases in the United States 2006, in The Energy Information Administration (EIA), Washington D.C., November 2007.

(3) Annual Energy Outlook 2006, in *The Energy Information Administration (EIA)*, www.eia.doe.gov/oiaf/aeo/.

(4) International Energy Outlook 2006, in *The Energy Information Administration (EIA)*, www.eia.doe.gov/oiaf/ieo/index.html.

(5) Cost and Performance Baseline for Fossil Energy Plants, *in the DOE/NETL-2007/1281; Volume 1: Bituminous Coal and Natural Gas to Electricity Final Report* (Original Issue Date, May 2007); Revision 1, August 2007.

(6) D. D. Ebbing, M. S. Wrighton (Eds), General Chemistry, 3rd edn., Houghton Mifflin Company, Boston, 1990.

(7) C. Higman, M. Van der Burgt, Auxiliary Technologies in Gasification, 2nd edn., Elsevier Inc., 2008.

(8) G. R. Maxwell, Hydrogen Production in Synthetic Nitrogen Products: A Practical Guide to the Products and Processes, Kluwer Academic/Plenum Publishers, New York, 2004.

(9) H. V. Spohr, G. N. Patey, Structural and dynamical properties of ionic liquids: The influence of charge location, *J. Chem. Phys.*, **130**, 1–11 (2009).

(10) A. Shariati, S. Raeissi, C. J. Peters, CO_2 solubility in alkylimidazolium-based ionic liquids, in Development and Applications in Solubility, T. M. Letcher (Ed.), The Royal Society of Chemistry, Cambridge, 2007.

(11) Ionic Liquids in Chemical Analysis, Mihkel Koel (Ed.), CRC Press, 2008.

(12) J. L. Anthony, J. L. Anderson, E. J. Maginn, J. F. Brennecke, Anion effects on gas solubility in ionic liquids, *J. Phys. Chem. B*, **109**, 6366–6374 (2005).

(13) C. Cadena, J. L. Anthony, J. K. Shah, T. I. Morrow, J. F. Brennecke, E. J. Maginn, Why is CO_2 so soluble in imidazolium-based ionic liquids? *J. Am. Chem. Soc.*, **126**, 5300–5308 (2004).

(14) M. C. Kroon, E. K. Karakatsani, I. G. Economou, G. J. Witkamp, C. J. Peters, Modeling of the carbon dioxide solubility in imidazolium-based ionic liquids with the tPC-PSAFT equation of state, *J. Phys. Chem. B*, **110**, 9262–9269 (2006).

(15) S. G. Kazarian, B. J. Briscoe, T. Welton, Combining ionic liquids and supercritical fluids: in situ ATR-IR study of CO_2 dissolved in two ionic liquids at high pressures, *Chem. Commun.*, **20**, 2047–2048 (2000).

(16) L. A. Blanchard, Z. Gu., J. F. Brennecke, High-pressure phase behavior of ionic liquid/CO_2 systems, *J. Phys. Chem. B*, **105**, 2437–2444 (2001).

(17) J. M. Crosthwaite, S. N. V. K. Aki, E. J. Maginn, J. F. Brennecke, Liquid phase behavior of immidazolium-based ionic liquids with alcohols, *J. Phys. Chem. B*, **108**, 5113–5119 (2004).

(18) J. M. Crosthwaite, S. N. V. K. Aki, E. J. Maginn, J. F. Brennecke, Liquid phase behavior of immidazolium-based ionic liquids with alcohols: effect of hydrogen bonding and non-polar interactions, *Fluid Phase Equilibria*, **228**, 303–309 (2005).

(19) J. Ilconich, C. Myers, H. Pennline, D. Luebke, Experimental investigation of the permeability and selectivity of supported liquid membranes for CO_2/He separation at temperatures up to 125°C, *J. Membr. Sci.*, **298**, 41–47 (2007).
(20) C. Myers, H. Pennline, D. Luebke, J. Ilconich, J. K. Dixon, E. J. Maginn, J. F. Brennecke, High temperature separation of carbon dioxide/hydrogen mixtures using facilitated supported ionic liquid membranes, *J. Membr. Sci.*, **322**, 28–31 (2008).
(21) E. D. Bates, R. D. Mayton, I. Ntai, J. H. Davis, Jr., CO_2 capture by a task-specific ionic liquid, *J. Am. Chem. Soc.*, **124**, 926–927 (2002).
(22) Y. Hou, R. E. Baltus, Experimental measurement of solubility and diffusivity of CO_2 in room-temperature ionic liquids using a transient thin-liquid film method, *Ind. Eng. Chem. Res.*, **46**, 8166–8175 (2007).
(23) D. Morgan, L. Ferguson, P. Scovazzo, Diffusivities of gases in room-temperature ionic liquids: data and correlations obtained using a lag-time technique, *Ind. Eng. Chem. Res.*, **44**, 4815–4823 (2005).
(24) L. Ferguson, P. Scovazzo, Solubility and permeability of gases in phosphonium-based room temperature ionic liquids: data and correlations, *Ind. Eng. Chem. Res.*, **46**, 1369–1374 (2007).
(25) C. A. Ohlin, P. J. Dyson, G. Laurenczy, Carbon monoxide solubility in ionic liquids: determination, prediction, and relevance to hydroformylation, *Chem. Commun.*, 1070–1071 (2004).
(26) M. J. Muldoon, S. N. V. K. Aki, J. L. Anderson, J. K. Dixon, J. F. Brennecke, Improving carbon dioxide solubility in ionic liquids, *J. Phys. Chem. B*, **111**, 9001–9009 (2007).
(27) G. Maurer, Gas solubility and related phenomena in systems with ionic liquids, Abstracts of Papers, 236th ACS National Meeting, Philadelphia, PA, U.S., August 17–21, 2008. IEC-008, CODEN: 69KXQ2 AN 2008:952394 CAPLUS.
(28) J. Kumelan, D. Tuma, G. Maurer, Partial molar volumes of selected gases in some ionic liquids, *Fluid Phase Equilibria*, **275**, 132–144 (2009).
(29) G. Gutowski, E. J. Maginn, Amine-functionalized task-specific ionic liquids: a mechanistic explanation for the dramatic increase in viscosity upon complexation with CO_2 from molecular simulation, *J. Am. Chem. Soc.*, **130**, 14690–14704 (2008).
(30) G. Yu, S. Zhang, G. Zhou, X. Liu, X. Chen, Structure, interaction and property of amino-functionalized imidazolium ILs by molecular dynamics simulation and Ab initio calculation, *AIChE J.*, **53**, 3210–3221 (2007).

11

The Effects of Minor Components on the Gas Separation Performance of Polymeric Membranes for Carbon Capture

Colin A. Scholes, Sandra E. Kentish and Geoff W. Stevens
Cooperative Research Centre for Greenhouse Gas Technologies (CO2CRC), University of Melbourne, Parkville, Victoria, Australia

11.1 Introduction

Gas separation polymeric membranes are a developing separation technology, which has the potential to efficiently and economically separate gases in a range of industrial processes. Already, non-porous polymer membranes are used commercially for the removal of carbon dioxide and hydrogen sulfide from natural gas, known as sweetening, small-scale oxygen enrichment of air, as well as hydrogen recovery in ammonia production. The technology has also been researched for dehydration, hydrocarbon recovery, syngas separation, as well as carbon dioxide separation from flue gas [1,2]. The importance of the last two examples in the membrane field has been developing over the past decade, as membrane technology has progressively demonstrated potential in carbon dioxide capture, part of the strategy of carbon capture and storage (CCS) to mitigate carbon emissions from a range of industrial processes [3]. There are three main strategies for CO_2 capture from fossil fuel based power plants [4]. Post-combustion is where CO_2 capture occurs after combustion from the exiting flue gas. Pre-combustion occurs where fossil fuels are reformed into syngas comprising primarily of hydrogen and carbon monoxide,

Membrane Gas Separation Edited by Yuri Yampolskii and Benny Freeman

carbon dioxide and water in a reducing environment. This is then further converted to more hydrogen through the water gas shift reaction and CO_2 is then separated from hydrogen before combustion. Finally, in oxy-fuel combustion air is replaced by oxygen in the combustion process which produces a flue gas of mainly H_2O and CO_2, which is readily captured. For all three strategies, membrane technology has sufficiently matured to become commercially competitive with other separation technologies in the coming decade.

Alongside the major gases, a range of minor gas components are present in all of these applications [5]. For example, in natural gas sweetening the main separation is of CO_2 and H_2S from CH_4 but water and a range of hydrocarbons are also present. In post-combustion capture the main separation is CO_2 from N_2; however, the process also experiences SOx (a combination of sulfur oxides), NOx (a combination of nitric oxides) and water. In pre-combustion capture, separation is of CO_2 from H_2 and N_2, with H_2S, CO, NH_3, water and hydrocarbons also present. In addition, O_2 is often present in many of these processes, along with argon if air is used. Depending on the fuel source, heavy metals, such as mercury and arsenic, may exist, as well as halogens, such as fluorine and chlorine, in their acidic form. Other processes than power generation with the possibility for carbon capture, such as cement production, refineries and blast furnaces, also have a range of minor components present. The concentration of these minor components varies considerably and is dependent on the process, fuel, temperature and pressure of reaction. Potential concentrations of some of these minor components are provided in Table 11.1. For membrane gas separation processes the effects of these minor components have to be considered on a case by case basis.

The impact of some of these minor components can be significantly reduced by a suitable pre-treatment process, such as a de-sulfurizer or an adsorbent guard bed. However, remaining concentrations of these minor components in the membrane gas separation process may compete with CO_2 (or other gas of interest) for separation and therefore decrease the observed permeability, as well as degrade the membrane structure, altering separation performance. Furthermore, the permeabilities of these minor components are of interest, because of the possibility of generating component-rich permeate or retentate streams, dependent on membrane selectivity. Thus for example, it may be possible to capture both acid gases (H_2S and CO_2) simultaneously from a pre-combustion stream.

Table 11.1 *Composition of minor components in a range of industrial processes [6–8]*

	Pre-combustion	Post-combustion	Iron blast furnace	Cement production
CO_2	10–35 mol%	12–14 vol%	20 vol%	14–33 vol%
SOx	–	1000–5000 ppm	–	0–150 ppm
NOx	–	100–500 ppm	–	0–150 ppm
CO	300–4000 ppm	~10 ppm	22 vol%	0–130 ppm
H_2S	500–1000 ppm	–	0–5000 ppm	–
NH_3	0–1500 ppm	–	–	–
H_2	~20 vol%	–	4 vol%	–
Water	Saturated	Saturated	4 vol%	12.8 vol%
Hydrocarbons	0–100 ppm	–	–	–

Hence, this provides the opportunity to reduce the number of separation processes the feed gas may go through.

Gas permeation through non-porous polymeric membranes is generally described by the solution-diffusion mechanism [2]. This is based on the solubility of specific gases within the membrane and their diffusion through the dense membrane matrix. In turn, the solubility of a specific gas component within a membrane is a function of its critical temperature, as this is a measure of the gas condensability. Critical temperatures for a range of gas components are provided in Table 11.2. Conversely, the diffusivity depends upon the molecular size, as generally indicated by the kinetic diameter. Indeed, Robeson *et al.* [9] have recently postulated that the relationship between the ideal permeability of one species P_i and that of another P_j are related by a simple function:

$$P_j = kP_i^n \tag{11.1}$$

Where:

$$n = \left(\frac{d_j}{d_i}\right)^2 \tag{11.2}$$

However, to achieve the best fit to this relationship they propose slight modifications to the kinetic diameters provided by Breck [10] which are more generally used. Both values are included in Table 11.2 where available.

Table 11.2 *Kinetic diameter and critical temperature of common gases*

Gas	Breck kinetic diameter (Å) [10]	Robeson kinetic diameter (Å) [9]	Critical temperature (K) [11]
H_2	2.89	2.875	33.2
N_2	3.64	3.568	126.2
CO	3.76		132.9
Ar	3.40		150.8
O_2	3.46	3.347	154.6
NO	3.17		180
CH_4	3.8	3.817	190.6
CO_2	3.3	3.325	304.2
HCl	3.2		324.6
C_3H_8	4.3		369.8
H_2S	3.6		373.2
NH_3	2.6		405.6
SO_2	3.6		430.8
NO_2			431.4
SO_3			491
C_6H_{14}			507.4
C_6H_6	5.85		561.9
H_2O	2.65		647.3
Hg			1750

11.2 Sorption Theory for Multiple Gas Components

11.2.1 Rubbery Polymeric Membranes

Rubbery membranes operate above the glass transition temperature, therefore polymer chains are able to rearrange on a meaningful timescale and are usually in thermodynamic equilibrium. Examples of rubbery membranes are polydimethylsiloxane, polyethylene glycol and silicone rubber. These membranes generally achieve their selectivity by differences in solubility [12]. Hence, they often allow larger gas molecules to permeate at a greater rate than smaller gases. A good example of this is hydrocarbon recovery from natural gas [13]. This is because the liquid-like polymer matrix has a poor ability to sieve molecules based on size because of chain motion. This is of considerable interest in syngas separation (pre-combustion capture), where it is desirable to keep hydrogen at high pressure and therefore allow the larger gases, such as CO_2, to transport through the membrane. This ensures hydrogen is at the pressure required for combustion as part of an integrated gasification and combined cycle process (IGCC).

For rubbery polymeric membranes, the process of mixing gas A into a polymer P can be described by Flory–Huggins regular solution theory for mixing components within polymer solutions [14]:

$$\mu_A - \mu_A^0 = \frac{\Delta G_0}{RT} = \ln\left(\frac{f_A}{f_{A0}}\right) = \ln\phi_A + \left(1 - \frac{V_A}{V_P}\right)\phi_P + \chi_A\phi_P^2 \quad (11.3)$$

Where f_A and f_{A0} are the fugacity of A and the saturated fugacity of A respectively, μ_A and μ_A^0 are the corresponding chemical potentials, ϕ_A and ϕ_P are the volume fraction of gas A and the polymer P respectively, V_A and V_P are the relevant molar volumes and χ_A the Flory–Huggins parameter of gas A and the polymer P.

The true concentration of the penetrant A expressed as the volume of gas (cm^3 at STP) dissolving in 1 cm^3 of polymer, can be related to the volume fraction in the above expression by [15]:

$$C_A = \frac{\phi_A V_G}{V_A} \quad (11.4)$$

where V_G is the molar volume of a gas at STP, which is 22400 cm^3/mol.

Recognising that:

$$\ln S_A = \ln(C_A) - \ln(f_A) = \ln(C_A) - \ln\frac{f_A}{f_{A0}} - \ln f_{A0} \quad (11.5)$$

Substitution of Equations (11.4) and (11.5) into Equation (11.3) gives:

$$\ln(S_A) = \ln\frac{\phi_A V_G}{V_A} - \ln\phi_A - \left(1 - \frac{V_A}{V_P}\right)\phi_P - \chi_A\phi_P^2 - \ln(f_{A0}) \quad (11.6)$$

Given $\phi_P = 1 - \phi_A$ and assuming that both $V_A << V_P$ and ϕ_A^2 is vanishingly small; Equation (11.6) becomes:

$$\ln(S_A) = \ln\frac{V_G}{V_A} - 1 - \chi_A + \frac{V_A}{V_G} C_A(1+2\chi_A) - \ln(f_{A0}) \tag{11.7}$$

The non-concentration dependent terms can be grouped together to yield:

$$\ln(S_A) = \ln(K_A) + \sigma_A C_A \tag{11.8}$$

where

$$\ln(K_A) = -(1+\chi_A) - \ln\frac{V_A}{V_G} - \ln(f_{A0}) \tag{11.9}$$

is the Henry's law constant of gas A in polymer P and

$$\sigma_A = \frac{(1+2\chi_A)V_A}{V_G} \tag{11.10}$$

Hence, the solubility of gas A within a rubbery polymeric membrane can be determined from the Flory–Huggins parameter.

Suwandi and Stern [16] point out that in many sorption studies the polymer swells as sorption occurs, but the concentration of A is calculated as the volume of gas dissolving in 1 cm^3 of unswollen polymer. That is, the weight change recorded by the sorption analyzer is converted to a concentration by division by the density of the unswollen system. In this case:

$$C' = \frac{V_G \phi_A}{V_A(1-\phi_A)} \text{ or } \phi_A = \frac{C'(V_A/V_G)}{C'(V_A/V_G)+1} \tag{11.11}$$

Mathematical manipulation of equations yields expressions identical to Equations (11.8) and (11.9), but importantly in this case:

$$\sigma'_A = \frac{2(1+\chi_A)V_A}{V_G} \tag{11.12}$$

For a tertiary system, the presence of gas B may influence the sorption of gas A in the polymer. Hence, the Flory–Huggins equation of gas A for a tertiary system is [14,17]:

$$\ln\left(\frac{f}{f_0}\right)_A = \ln\phi_A + (1-\phi_A) - \frac{\phi_B V_A}{V_B} - \phi_P\frac{V_A}{V_P} + (\chi_{AB}\phi_B + \chi_A\phi_P)(\phi_B+\phi_P) - \chi_B\phi_B\phi_P\frac{V_A}{V_B} \tag{11.13}$$

Here χ_{AB} characterizes the mixing between gas A and B, while the other parameters subscripts denote if it is for gas A, B or the polymer P. Again, the Flory–Huggins equation for a tertiary system can be simplified by the same assumptions for the binary system. That is, V_A and $V_B << V_P$, due to difference in gas volume compared to the polymer, and at low pressures the volume fraction of gas A and B is small, and $\phi_A^2 \to 0$ and $\phi_B^2 \to 0$. Further, $\phi_P = 1 - \phi_A - \phi_B$. Hence:

$$\ln\left(\frac{f}{f_0}\right)_A = \ln\phi_A + (1-\phi_A) - \frac{\phi_B V_A}{V_B} + (\chi_{AB}\phi_B + \chi_A(1-\phi_A-\phi_B))(1-\phi_A) - \chi_B\phi_B(1-\phi_A-\phi_B)\frac{V_A}{V_B} \quad (11.14)$$

rearranging assuming $\phi_A\phi_B \rightarrow 0$:

$$\ln\left(\frac{f}{f_0}\right)_A = \ln\phi_A - \left[\chi_A - \chi_{AB} + (1+\chi_B)\frac{V_A}{V_B}\right]\phi_B + 1 + \chi_A - \phi_A(1+2\chi_A) \quad (11.15)$$

Substitution of Equations analogous to Equations (11.4) and (11.5) yields:

$$\ln S_{AB} = \ln\frac{V_G}{V_A} - 1 - \chi_A + \frac{C_A V_A}{V_G}(1+2\chi_A) + \frac{C_B V_B}{V_G}\left(\chi_A - \chi_{AB} + (1+\chi_B)\frac{V_A}{V_B}\right) - \ln f_{A0} \quad (11.16)$$

where S_{AB} is the solubility of A in the polymer with gas B present. This can be simplified:

$$\begin{aligned}\ln S_{AB} &= \ln(K_A) + \frac{C_B V_B}{V_G}\left[\chi_A - \chi_{AB} + (1+\chi_B)\frac{V_A}{V_B}\right] + \sigma_A C_A \\ &= \ln(K_A) + \sigma_{AB} C_B + \sigma_A C_A \end{aligned} \quad (11.17)$$

where

$$\sigma_{AB} = \left(\chi_A - \chi_{AB} + (1+\chi_B)\frac{V_A}{V_B}\right)\frac{V_B}{V_G} \quad (11.18)$$

Therefore, the solubility of gas A in polymer P with gas B present can be expressed in terms of the solubility of gas A in the binary system (Equation 11.8):

$$\ln(S_{AB}) = \ln(S_A) + \sigma_{AB} C_B \quad (11.19)$$

If gas B concentration is low, then Henry's law allows Equation (11.19) to be approximated by:

$$\ln(S_{AB}) = \ln(S_A) + \sigma_{AB} K_B p_B \quad (11.20)$$

Where p_B is the pressure of gas B, and K_B the Henry's law constant. Hence, gas B may enhance or decrease the solubility of gas A within the rubbery polymeric membrane, depending on the relative strengths of the Flory–Huggins parameters for the gases between themselves and for the polymers.

For a quaternary system, the presence of gases B and C may influence the sorption of gas A into the polymer, P. The Flory–Huggins equation for such a system is [17]:

$$\ln\left(\frac{f}{f_0}\right)_A = \ln\phi_A + (1-\phi_A) - \frac{\phi_B V_A}{V_B} - \frac{\phi_C V_A}{V_C} - \frac{\phi_P V_A}{V_P}$$
$$+ \chi_{AB}\phi_B^2 + \chi_{AC}\phi_C^2 + \chi_A\phi_P^2 + \phi_B\phi_C(\chi_{AB} + \chi_{AC} - \chi_{CB})$$
$$+ \phi_B\phi_P\left(\chi_{AB} + \chi_A - \chi_B\frac{V_A}{V_B}\right) + \phi_C\phi_P\left(\chi_{AC} + \chi_A - \chi_C\frac{V_A}{V_C}\right) \quad (11.21)$$

Where χ_{ij} is the binary interaction parameter between gases i and j. For simplicity this model assumes no higher interaction parameters between the gases and polymer, e.g. $\chi_{ABC} = 0$. Again, Equation (11.21) can be simplified by the same assumptions used in the binary and tertiary systems. The solution, assuming gases B and C obey Henry's law, is:

$$\ln(S_{ABC}) = \ln(S_A) + \sigma_{AB}K_B p_B + \sigma_{AC}K_C p_C \quad (11.22)$$

where

$$\sigma_{AC} = \left(\chi_A - \chi_{AC} + (1+\chi_C)\frac{V_A}{V_C}\right)\frac{V_C}{V_G} \quad (11.23)$$

where K_C is the Henry's law constant of gas C. Hence, in a quaternary system the presence of gas B and C may enhance or decrease the solubility of gas A in the membrane, which can be determined from the various Flory–Huggins parameters.

Impurities can also affect the diffusivity of a gas through polymer swelling or dilation. These effects can be modelled using the well-known expression originally proposed by Fujita [18]

$$D = A\exp(-B/v_f) \quad (11.24)$$

Where v_f is the fractional free volume in the polymer. The effect of plasticization by penetrants A and B on this fractional free volume is given by [19]:

$$v_f = v_f^0 + \gamma_A\phi_A + \gamma_B\phi_B \quad (11.25)$$

Where v_f^0 is the fractional free volume in the polymer at the same temperature and pressure in the absence of plasticization and γ_A and γ_B are positive constants characteristic of the system.

11.2.2 Glassy Membranes

Glassy membranes operate below the glass transition temperature, and therefore polymer chain rearrangement is on an extraordinary long timescale, meaning the membrane never reaches thermodynamic equilibrium. The polymer chains are packed imperfectly, leading to excess free volume in the form of microscopic voids in the polymeric matrix. Within these voids Langmuir adsorption of gases occurs that increases the solubility. Examples of glassy membranes are polysulfone and Matrimid© polyimide.

The considerable free volume within glassy polymeric membranes, due to the presence of microvoids, generally means this class of membranes are diffusivity selectivity controlled [12]. That is, the membrane is selective towards the smaller gas molecules. Because of this, glassy polymeric membranes have been suggested for post-combustion capture as well as natural gas sweetening.

Gas concentration within glassy membranes consists of gas within the polymeric matrix as well as adsorbed in the microvoids. Therefore, the total concentration of absorbed gas within a glassy membrane (C) can be described by [15]:

$$C = C_{\mathrm{D}} + C_{\mathrm{H}} \tag{11.26}$$

where C_{H} is the standard Langmuir adsorption relationship

$$C_{\mathrm{H}} = \frac{C'_{\mathrm{H}} bf}{(1+bf)} \tag{11.27}$$

C'_{H} is the maximum adsorption capacity while b is the ratio of rate coefficients of adsorption and desorption, or the Langmuir affinity constant, defined as:

$$b = \frac{C_{\mathrm{H}}}{(C'_{\mathrm{H}} - C_{\mathrm{H}})f} \tag{11.28}$$

Hence, the dual-mode sorption for glassy membranes is written as:

$$C = K_{\mathrm{D}} f + \frac{C'_{\mathrm{H}} bf}{(1+bf)} \tag{11.29}$$

Petropoulos [20] and Paul and Koros [21] independently developed models where the diffusion of the gas species adsorbed in the Langmuir region is partially, or even totally, immobilized. In this case, a parameter F_{A} is introduced, defined either as the ratio of diffusion coefficients in the Langmuir (D_{H}) and Henry's Law region (D_{D}) or as the fraction of the Langmuir species that are fully mobile. In this case, the concentration of mobile species is given by:

$$C_{\mathrm{m}} = K_{\mathrm{D}} f \left[1 + \frac{F_{\mathrm{A}} C'_{\mathrm{H}} b}{K_{\mathrm{D}}(1+bf)}\right] \tag{11.30}$$

To determine the permeability across the membrane from this expression, it is necessary to account for the fugacity gradient by integration between the upstream and downstream values, f_0 and f_1 [20]:

$$\overline{P} = \frac{1}{f_0 - f_1} \int_{f_1}^{f_0} P(f)\,\mathrm{d}f \tag{11.31}$$

The result is the mean or integral permeability:

$$\overline{P} = K_{\mathrm{D}} D_{\mathrm{D}} + \frac{C'_{\mathrm{H}} F_{\mathrm{A}} D_{\mathrm{D}}}{(f_0 - f_1)} \ln\left(\frac{1+bf_0}{1+bf_1}\right) \tag{11.32}$$

When two gas species are present, competition can restrict both the solubility within the polymer matrix and the amount adsorbed in the Langmuir free volume. Competition in the former case is best modelled by adjustments to K_D based on Equations (11.20) or (11.22). To account for changes to the occupancy of the Langmuir sites for a binary mixture of gases A and B, the mobile concentration of gas A becomes [22]:

$$C_{mA} = K_{DA} f_A \left[1 + \frac{F_A C'_{HA} b_A}{K_{DA}(1 + b_A f_A + b_B f_B)} \right] \tag{11.33}$$

Similarly, the mobile concentration of gas B is:

$$C_{mB} = K_{DB} f_B \left[1 + \frac{F_B C'_{HB} b_B}{K_{DB}(1 + b_A f_A + b_B f_B)} \right] \tag{11.34}$$

Each parameter has the same definition as in the single gas case with the subscript denoting whether it is a property of gas A or B. The solubility of both gases A and B is reduced compared to the single gas case (Equation 11.30), and is heavily dependent on the relationship between b, the affinity constant, and fugacity.

When three components or more are present, the mobile concentration of gas A becomes:

$$C_{mA} = K_{DA} f_A \left[1 + \frac{F_A C'_{HA} b_A}{K_{DA}(1 + b_A f_A + b_B f_B + b_C f_C + \ldots)} \right] \tag{11.35}$$

The Langmuir affinity constant is generally proportional to the critical temperature of the gas, and Figure 11.1 demonstrates the relationship between Langmuir affinity constant and critical temperature [23]. For example, water has a very high critical temperature compared to N_2 and CO_2. This means water is more condensable within the free volume and correspondingly a higher Langmuir affinity constant is observed. Hence, the presence of water even in trace amounts may dominate observed gas permeabilities, because even though the partial pressure is low, water will successfully compete for sorption sites in the membrane.

It is common for gases at high pressure, especially acidic gases such as CO_2 and SO_2, to also plasticize glassy polymeric membranes. This results in an increase in the diffusivity of gases through the membrane due to disruption to the polymer chain–chain intermolecular bonding network, leading to an increase in the fractional of free volume between polymer chains. When the penetrant gas causes plasticization of a glassy membrane, the diffusion coefficient becomes concentration dependent, and a simple model based on Equations (11.24) and (11.25) can be written as [24]:

$$D(C) = D_0 \cdot \exp(\beta \cdot C) \tag{11.36}$$

where C is the concentration of the plasticizing gas, D_0 is the diffusion coefficient in the limit $C \to 0$, and β is an empirical constant, known as the plasticization potential, that depends on the nature of the gas–polymer system and the temperature. For the specific

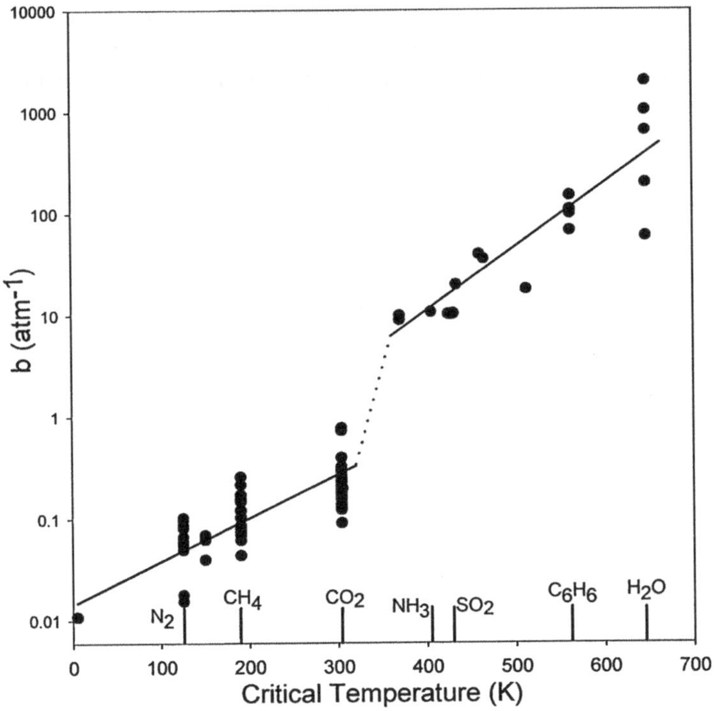

Figure 11.1 *Langmuir affinity constants (b) for a range of gases in various polymeric membranes relative to gas critical temperature. Reprinted with permission from Separation & Purification Reviews, Effects of minor components in carbon dioxide capture using polymeric gas separation membranes, by Scholes, C. A., S. E. Kentish, and G. W. Stevens, 38: 1–44, Copyright (2009) Taylor and Francis*

case when the permeate fugacity is zero, the mean or integral permeability for a ternary system can then be determined from:

$$\overline{P_{\mathrm{A}}} = \frac{D_{\mathrm{A}}(0)}{\beta_{\mathrm{A}} f_{\mathrm{A0}}}\left(\exp\left(\beta_{\mathrm{A}} \cdot K_{\mathrm{DA}} f_{\mathrm{A0}}\left(1+\frac{F_{\mathrm{A}} \cdot C'_{\mathrm{HA}} b_{\mathrm{A}}}{1+b_{\mathrm{A}} f_{\mathrm{A0}}+b_{\mathrm{B}} f_{\mathrm{B0}}+b_{\mathrm{C}} f_{\mathrm{C0}}}\right)\right)-1\right) \tag{11.37}$$

Hence, the effect of competitive sorption on glassy membranes is to reduce the permeability of all gases. This in turns alters the selectivity of the membrane. The relative magnitude of the Langmuir constants for each component dictates whether the selectivity decreases or increases.

11.3 Minor Components

11.3.1 Sulfur Oxides

Sulfur oxides, a combination of SO_2 and SO_3, are generated by combustion of sulfur-containing fuels such as coal [25]. Their emission is a major environmental problem,

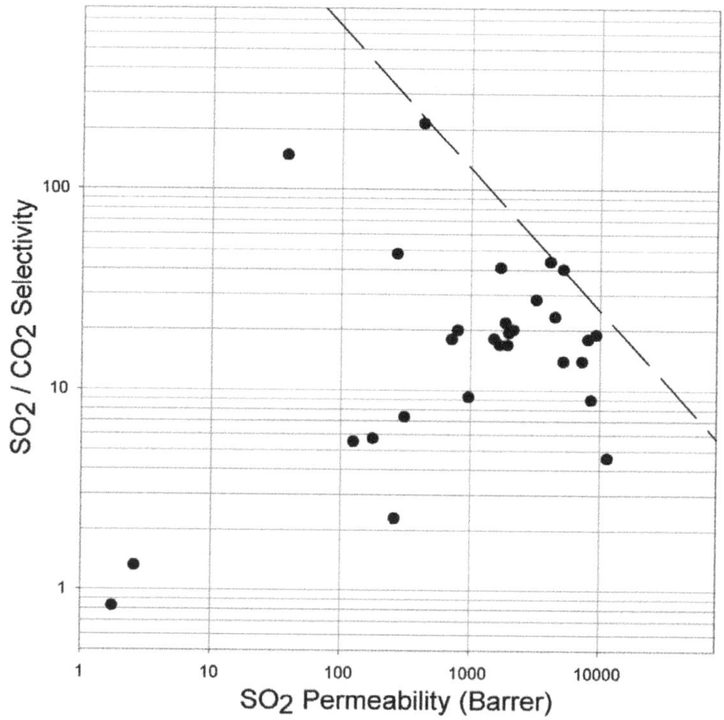

Figure 11.2 SO_2 permeability within polymeric membranes and selectivity relative to CO_2. Reprinted with permission from Separation & Purification Reviews, Effects of minor components in carbon dioxide capture using polymeric gas separation membranes, by Scholes, C. A., S. E. Kentish, and G. W. Stevens, 38: 1–44, Copyright (2009) Taylor and Francis

being the major cause of acid rain. A large range of non-porous polymeric membranes have been tested for SO_2 separation, with the permeability of SO_2 relative to SO_2/CO_2 selectivity for a range of membranes with potential in CO_2 capture shown in Figure 11.2. The selectivity almost always favours SO_2, implying that it permeates faster than CO_2 and therefore enriches the permeate stream. This is due to the higher condensability of SO_2 within the membrane, due to a larger critical temperature compared to CO_2, given that the kinetic diameter difference does not favour diffusion selectivity.

Sulfur dioxide has also been reported to plasticize polymeric membranes, which produces a more rubbery material and increases the diffusivity of penetrant gases [26–28]. Plasticization also reduces the mechanical integrity of the membrane, meaning it is more likely the membrane will rupture. However, plasticization is a strongly pressure dependent phenomenon, for example it has been reported in polyvinylidene membranes to occur at SO_2 pressures greater than 10 psi [29]. For many of the processes in carbon capture, such high partial pressures of SO_2 are not observed (Table 11.1), and therefore only minor plasticization by SO_2 is likely to occur.

A critical factor for membranes and SOx is the presence of high water vapour in many industrial processes. The mixture of SOx and water allows for the generation of sulfuric acid, which can chemically degrade the polymeric membrane, its support or the surrounding module structure. This has been observed for cellafan, a polysaccharide-based polymer, where SO_2 permeability experiences a 30-fold increase with water present, undoubtedly due to sulfuric acid degradation of the polymeric matrix [30].

11.3.2 Nitric Oxides

Combustion with air at high temperature results in the generation of nitrogen oxides, NOx, of which the dominant component is NO and to a lesser extent NO_2 [25]. There appears to be no information on the permeability of NO in polymeric membranes reported in the literature. Based on kinetic diameter (Table 11.1), NO should diffuse faster through polymeric membranes than CO_2; however, the larger difference in critical temperature suggests that CO_2 will sorb more strongly in the membrane than NO, and so this process should dominate in sorption selective membranes. For NO_2, only polytetrafluoroethylene has been documented, with a permeability of 15.9 Barrer and NO_2/CO_2 selectivity of 1.6 [31]. The plasticization effects of NOx are unknown, though in the presence of water NOx forms nitric acid that will degrade polymeric membranes, similar to that reported for SOx.

11.3.3 Hydrogen Sulfide

Hydrogen sulfide is present within natural gas and biogas, as well as being found in sufficient quantities in syngas to warrant removal [32,33]. The permeability of H_2S through a range of polymeric membranes and their selectivity to CO_2 is shown in Figure 11.3.

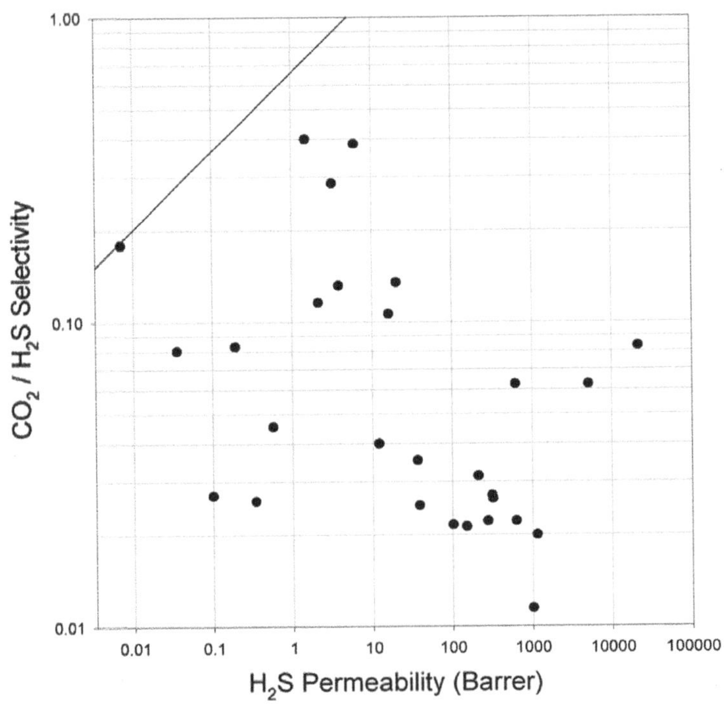

Figure 11.3 *H_2S permeability within polymeric membranes and selectivity relative to CO_2*

Many of these membranes are designed for sweetening of natural gas and therefore exhibit high permeabilities for H_2S and CO_2 with large selectivity relative to methane [34]. A large number of membranes demonstrate selectivity towards H_2S over CO_2, attributed to the higher condensability of H_2S (for critical temperature see Table 11.2) within the membrane, compared to CO_2. Similar to SO_2, H_2S has been reported to plasticize or swell polymeric membranes [35].

11.3.3.1 H_2S Competitive Sorption into Polydimethylsiloxane (PDMS)

In our recent work, a PDMS membrane (thickness of ~10 µm) was exposed to a 90% N_2–10% CO_2 mixture with 500 ppm H_2S present, to simulate syngas-like conditions [36]. The resulting permeability of CO_2 and N_2 through PDMS at different temperatures and pressures, with and without H_2S present, is shown in Figures 11.4 and 11.5 respectively.

In the presence of H_2S, CO_2 permeability through PDMS decreases by an average ~8% (Figure 11.4), implying that H_2S competes with CO_2 for transport through the membrane. Conversely, N_2 permeability in general increases slightly when H_2S is present (Figure 11.5). While this increase is often within the error of the measurement, it implies that H_2S enhances either the solubility or diffusion of N_2 within PDMS. Similar behaviour has been measured for H_2 permeability through PDMS in the presence of 3% H_2S [37]. This was attributed to plasticization (swelling) of the polymer matrix by H_2S reducing the resistance to gas transport leading to greater flux. However, the H_2S concentration here is in comparison quite low at 0.05%.

11.3.3.2 H_2S Competitive Sorption into Polysulfone and Matrimid 5218

The influence of H_2S on the CO_2 permeability in the glassy polymeric membranes polysulfone and Matrimid 5218 (a polyimide) are shown in Figures 11.6 and 11.7 respectively, where the membrane was exposed to 90% N_2–10% CO_2 gas mixture with H_2S at 500 ppm.

Two processes are occurring within these membranes upon exposure to H_2S. Firstly, competitive sorption between H_2S and CO_2 occurs within the microvoids, where H_2S will replace some adsorbed CO_2, and reduce the concentration of CO_2 within the membrane. This correlates to a reduction in the permeability of CO_2, which is evident in the time-resolved data for all three glassy membranes upon exposure to H_2S, at every temperature and partial pressures. Hence, H_2S strongly competes with CO_2 for space within the Langmuir voids, given that H_2S is present in trace amounts. The second process, plasticization of the membrane by H_2S, results in the increase in the measured permeability at a later time for both membranes under certain conditions, e.g. relative high temperatures and high H_2S partial pressures, as the diffusivity of gas increases. Plasticization is delayed because adsorbing H_2S requires time to build up the necessary concentration, due to competitive sorption, to have a significant plasticization influence on the polymeric matrix. The observed plasticization is attributed to the presence of H_2S only, and not CO_2. This assumption is supported by the partial pressure of CO_2 in all experiments being less than that require to plasticize each polymer, for polysulfone 34 bar CO_2 at 21 °C [39] is required and for Matrimid it is 14.8 bar CO_2 at 35 °C [40]. Secondly, each membrane is initially exposed to the CO_2/N_2 mixture, at the beginning of the experiment for 2 hours. Therefore, any plasticization by CO_2 should have occurred during this time period, and all later time effects are attributed to H_2S only. Hence, competitive sorption of H_2S occurs for both glassy membranes at the studied temperatures and pressures, but each polymer

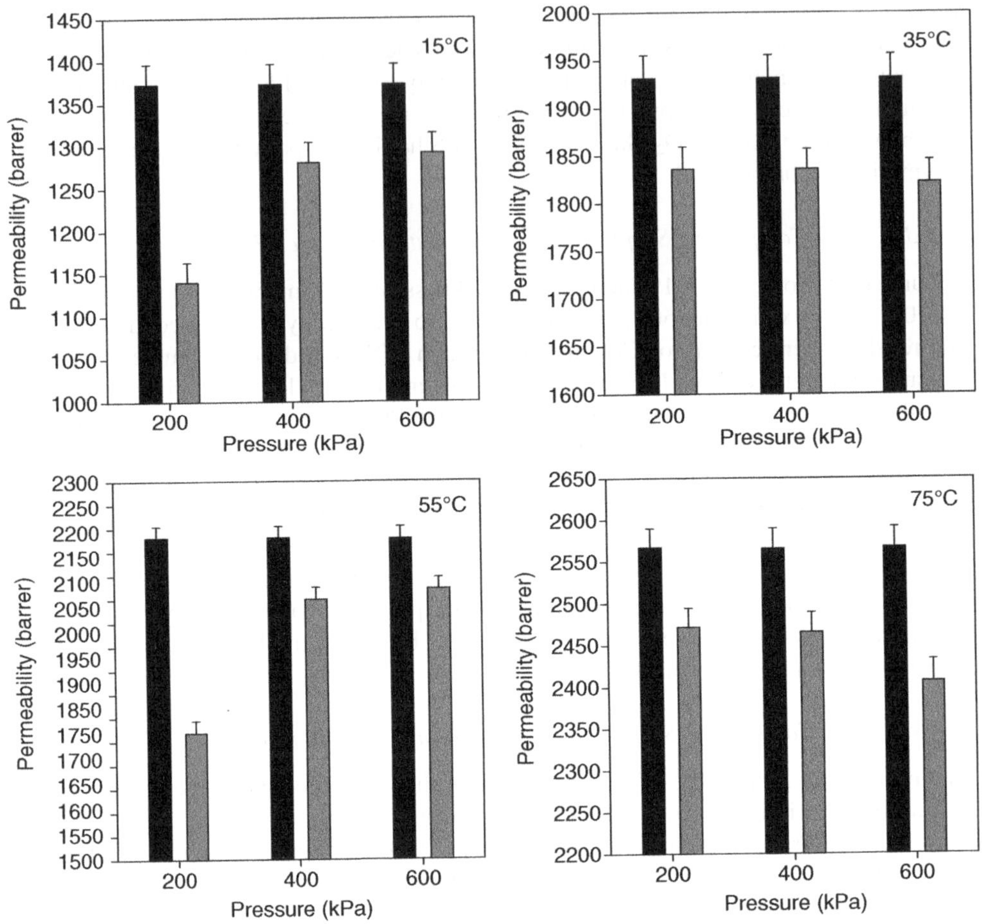

Figure 11.4 *CO_2 permeability in PDMS under mixed gas conditions with (grey) and without (black) 500 ppm or H_2S present. Reprinted with permission from Journal of Membrane Science, The effect of hydrogen sulfide, carbon monoxide and water on the performance of a PDMS membrane in carbon dioxide/nitrogen separation by Colin Scholes, Sandra Kentish and Geoff Stevens, 350, 1–2, 189–199, Copyright (2010) Elsevier Ltd*

is only susceptible to H_2S-induced plasticization at certain temperature and partial pressure conditions.

The increased occurrence of plasticization at high pressures and temperatures is attributed to the concentration of H_2S within the membrane, as well as the mobility of polymer chains. At high H_2S partial pressures, the concentration of H_2S within the glassy polymeric membrane reaches a critical concentration (the plasticization pressure) where the number of H_2S molecules per polymeric chain ratio is great enough to disrupt the polymeric chain–chain interactions network, allowing plasticization to occur. Bos *et al.* [39] postulates that this critical concentration is the same for polymeric membrane groups. Temperature also plays a critical role; polymer chains at higher temperature require less

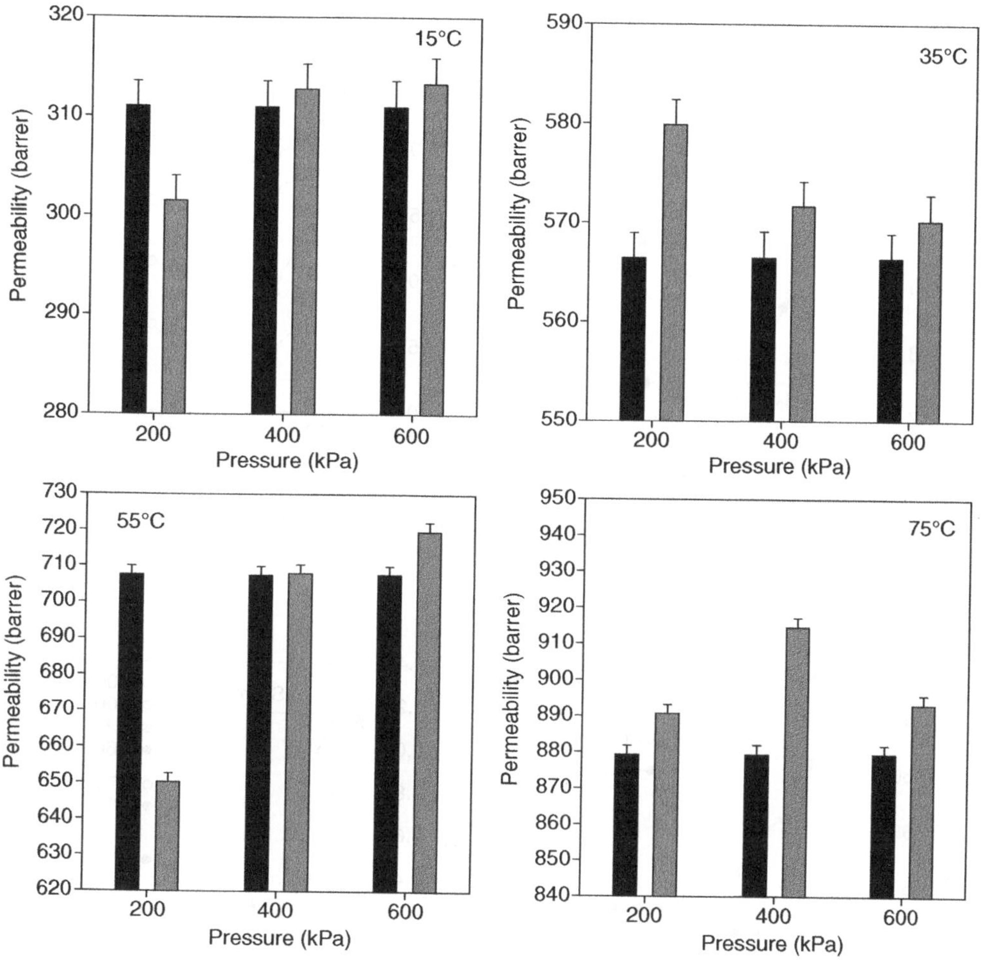

Figure 11.5 N_2 permeability in PDMS under mixed gas conditions with (grey) and without (black) 500 ppm H_2S present. Reprinted with permission from Journal of Membrane Science, The effect of hydrogen sulfide, carbon monoxide and water on the performance of a PDMS membrane in carbon dioxide/nitrogen separation by Colin Scholes, Sandra Kentish and Geoff Stevens, 350, 1–2, 189–199, Copyright (2010) Elsevier Ltd

energy to disrupt the chain–chain intermolecular bonding network, due to increased chain mobility. Therefore, it becomes easier for H_2S to disrupt this network at higher temperature and generate the plasticized state.

N_2 also experiences competitive sorption from H_2S, correspondingly the permeability of N_2 through the glassy membranes decreases. The reduction is not as pronounced as that reported above for CO_2, because the CO_2/N_2 selectivity during the competitive sorption event is reduced. An example measurement is provided in Figure 11.8. Hence, H_2S undergoes greater competitive sorption with CO_2, because both molecules are competing

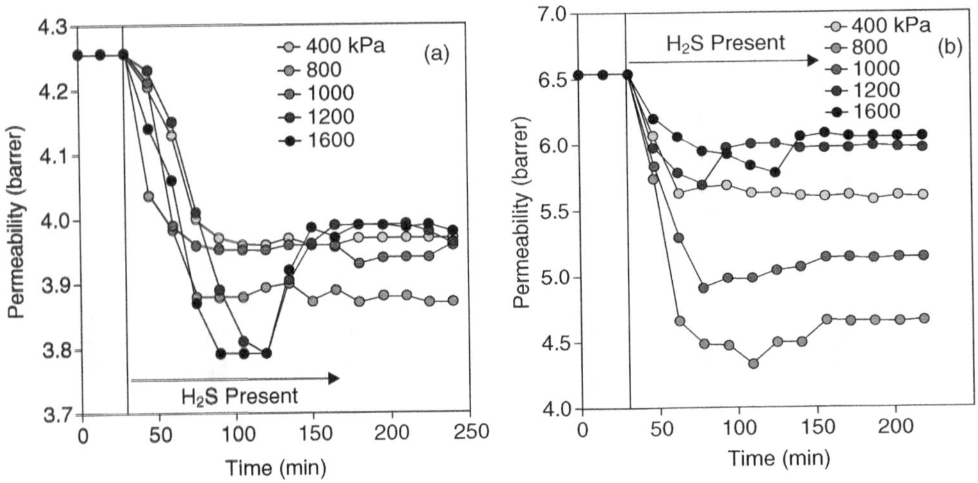

Figure 11.6 CO_2 *permeability through polysulfone at pressures between 400–1600 kPa (increasing greyscale); (a) 15 °C and (b) 55 °C [38]*

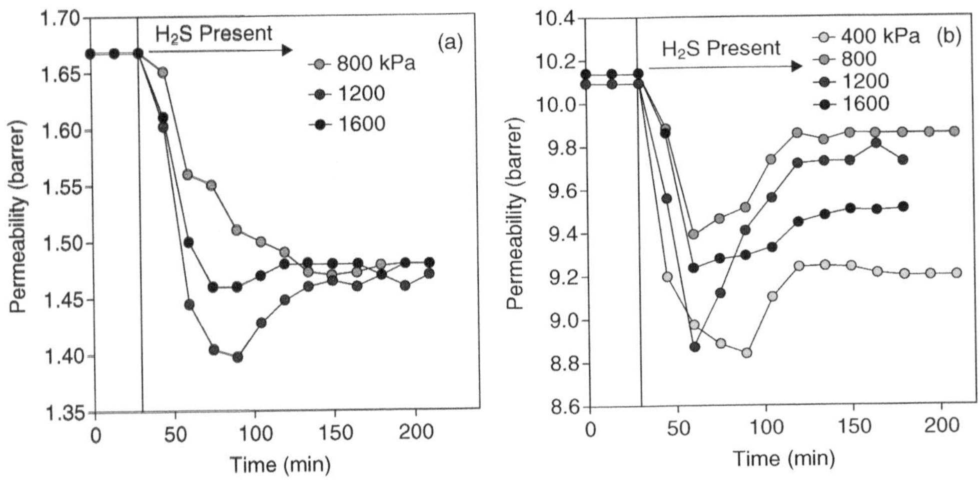

Figure 11.7 CO_2 *permeability through Matrimid 5218 at pressures between 400–1600 kPa (increasing grayscale); (a) 15 °C and (b) 55 °C [38]*

in the microvoids, while N_2 transport through these microvoids in comparison is already low. H_2S-induced plasticization also increases the permeability of N_2, because of the increase in fractional free volume. However, the CO_2/N_2 selectivity rises, revealing that while increased diffusivity occurs for both gases, it favours CO_2 because of it is high concentration within the microvoids, which benefits from the increase in pathways between them due to plasticization.

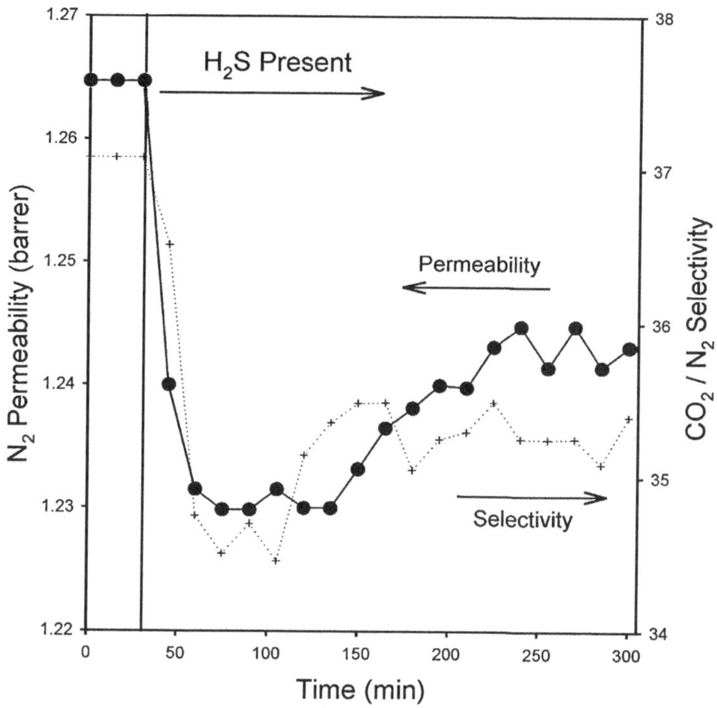

Figure 11.8 N_2 permeability through polysulfone, 55 °C 600 kPa, along with the CO_2/N_2 selectivity [38]

11.3.4 Carbon Monoxide

Carbon monoxide is present in post-combustion carbon capture as the product of partial oxidation due to reduced availability of oxygen, along with the reduced environment in iron smelting. Importantly, it is a major component of syngas and exists in significant quantities due to the equilibrium nature of the water gas shift reaction [41]. Hence, there is considerable research into membrane reactors that can be incorporated into the water gas shift reaction to drive hydrogen and carbon dioxide production. The performance of CO in polymeric membranes is less documented, due to their limitations in processing syngas at high temperatures. The majority of polymeric membranes will favour CO_2 over CO because the kinetic diameter of CO is greater [5], hence a sieving mechanism will favour CO_2, while the critical temperature of CO is lower than CO_2.

11.3.5 Water

In the majority of industrial processes where carbon dioxide capture can occur, the feed stream for separation is saturated with water vapour [42]. Therefore, competitive water sorption in the membrane, as well as plasticization and ageing effects, will have a much stronger influence on membrane performance compared to the previously mentioned

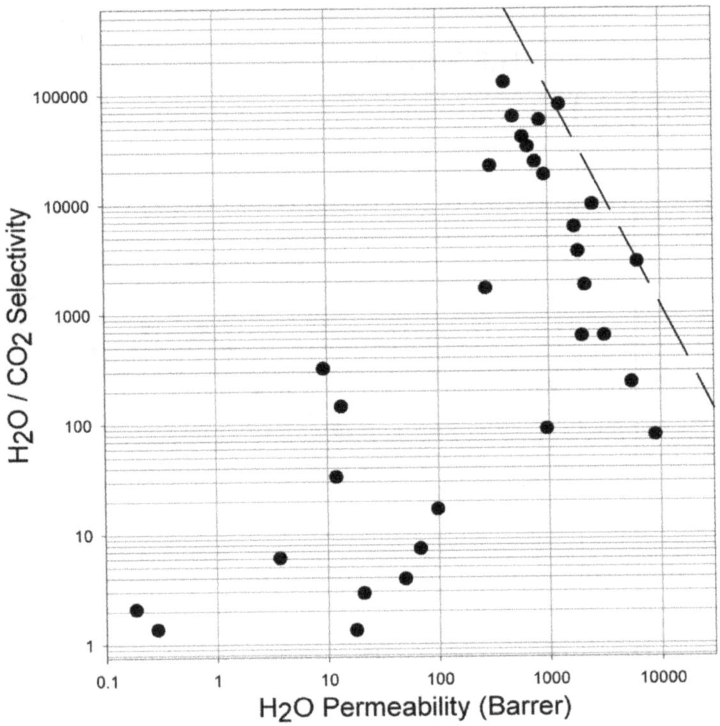

Figure 11.9 *Water permeability within polymeric membranes and selectivity relative to CO_2. Reprinted with permission from Separation & Purification Reviews, Effects of minor components in carbon dioxide capture using polymeric gas separation membranes by Scholes, C. A., S. E. Kentish, and G. W. Stevens, 38: 1–44, Copyright (2009) Taylor and Francis*

components. The permeability of water for a range of polymeric membranes and selectivity with respect to CO_2 is shown in Figure 11.9. Water permeability is almost always greater than CO_2, in many cases substantially, because water has both a smaller kinetic diameter than CO_2, therefore diffuses faster, and also has a higher critical temperature, meaning greater sorption in the membrane. Hence, for membrane CO_2 separation the permeate stream will generally have a higher water percentage than the feed. If the permeate temperature is below the dew point, i.e. supersaturated, there is the possibility of condensation formation, which will create an additional transfer layer for gases to cross, reducing performance. Membrane support structures can also be severely impacted by water condensation. Specifically, the capillary forces that act upon later evaporation of the water can cause porous supports to collapse irreversibly.

Plasticization due to water sorption has been observed, with a polyethersulfone membrane experiencing a dramatic flux increase of ~250% upon exposure to water, with a corresponding decrease in selectivity [43]. This is also of critical concern in membrane processing since it can permanently alter the membrane structure, meaning performance does not always return once the membrane is dried [44].

11.3.5.1 Water Competitive Sorption into Polydimethylsiloxane (PDMS)

Figures 11.10 and 11.11 show the results of our recent work where the permeability of the N_2/CO_2 mixture through PDMS was recorded upon exposure to different partial pressures of H_2O.

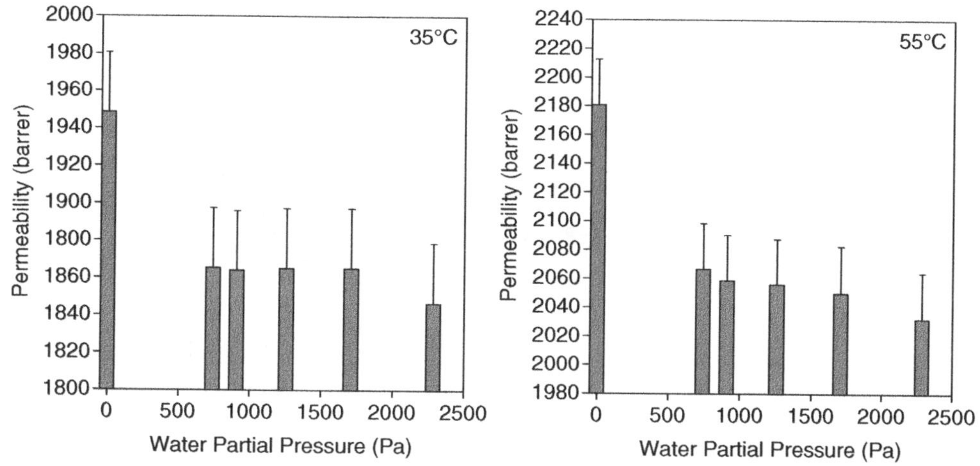

Figure 11.10 *PDMS permeability of CO_2 upon exposure to wet mixed gas at different water partial pressures. Reprinted with permission from Journal of Membrane Science, The effect of hydrogen sulfide, carbon monoxide and water on the performance of a PDMS membrane in carbon dioxide/nitrogen separation by Colin Scholes, Sandra Kentish and Geoff Stevens, 350, 1–2, 189–199, Copyright (2010) Elsevier Ltd*

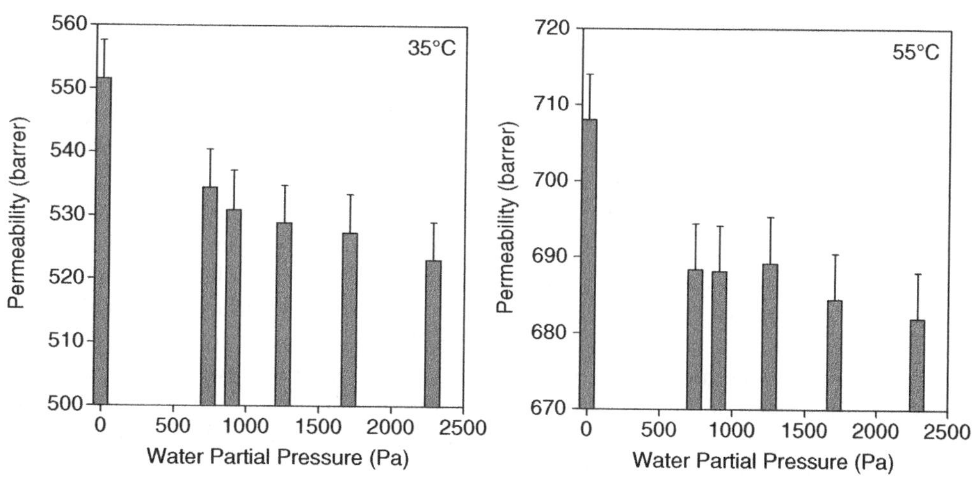

Figure 11.11 *PDMS permeability of N_2 upon exposure to wet mixed gas at different water partial pressures. Reprinted with permission from Journal of Membrane Science, The effect of hydrogen sulfide, carbon monoxide and water on the performance of a PDMS membrane in carbon dioxide/nitrogen separation by Colin Scholes, Sandra Kentish and Geoff Stevens, 350, 1–2, 189–199. Copyright (2010) Elsevier Ltd*

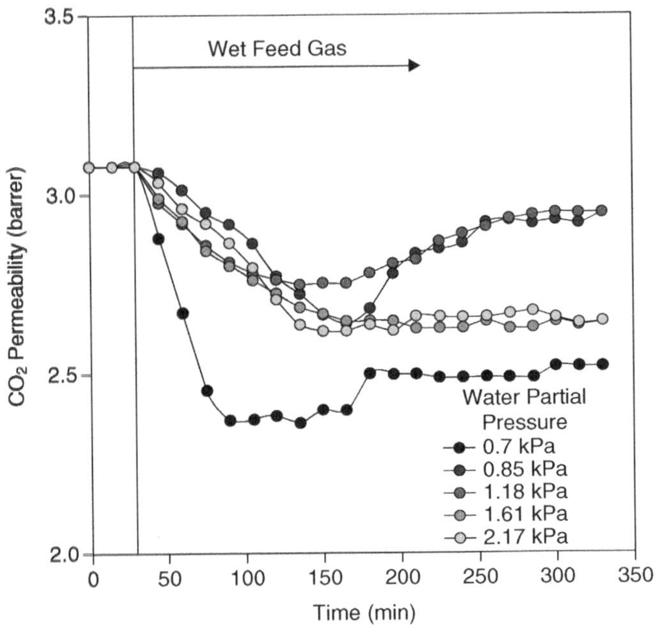

Figure 11.12 *Polysulfone CO_2 permeability from a 90% N_2–10% CO_2 gas mixture upon exposure to wet feed gas at 35 °C and 800 kPa [45]*

Upon exposure to the wet feed, the permeability of both CO_2 and N_2 reduces, due to competition from water, with a trend of low permeability with increased humidity of the feed.

11.3.5.2 Water Competitive Sorption into Polysulfone and Matrimid 5218

The influence of water on glassy polymeric membrane performance is important in many applications. The effect on the glassy polymeric membranes polysulfone and Matrimid 5218 is shown in Figures 11.12 and 11.13 respectively, taken from a 90% CH_4–10% CO_2 mixture, to simulate natural gas processing.

The change in CO_2 permeability upon initial exposure to wet feed gas for both polysulfone and Matrimid is indicative of competitive sorption of water within the glassy membrane reducing the flux of CO_2. For Matrimid this behaviour is clearly dependent on the relative humidity of the feed gas and indicates that the amount of water within the microvoids is dependent on the relative humidity. Polysulfone does not display the same behaviour, with CO_2 permeability reduction most significant for the lowest water pressure. However, it is clear from the increase in permeability at latter time for polysulfone that plasticization is occurring, and for some water pressures, the plasticization is of sufficient magnitude to overcome any loss in permeability due to competitive sorption. A similar effect is observed for CH_4 permeability (Figure 11.14), with the initial reduction in permeability due to competitive sorption, followed by plasticization. The correlation of selectivity demonstrates the significance of plasticization on performance. Initially, when only competitive sorption occurs, the selectivity falls; however, as plasticization of

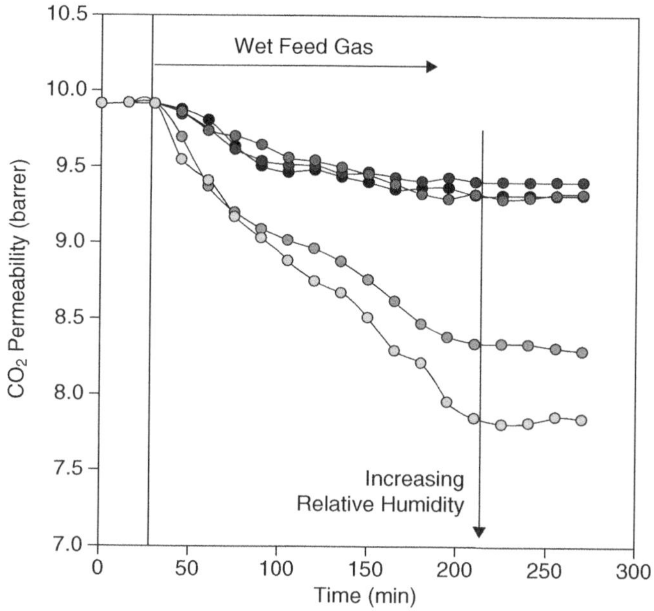

Figure 11.13 *Matrimid CO_2 permeability from a 90% N_2–10% CO_2 gas mixture upon exposure to wet feed gas at 35 °C and 800 kPa [45]*

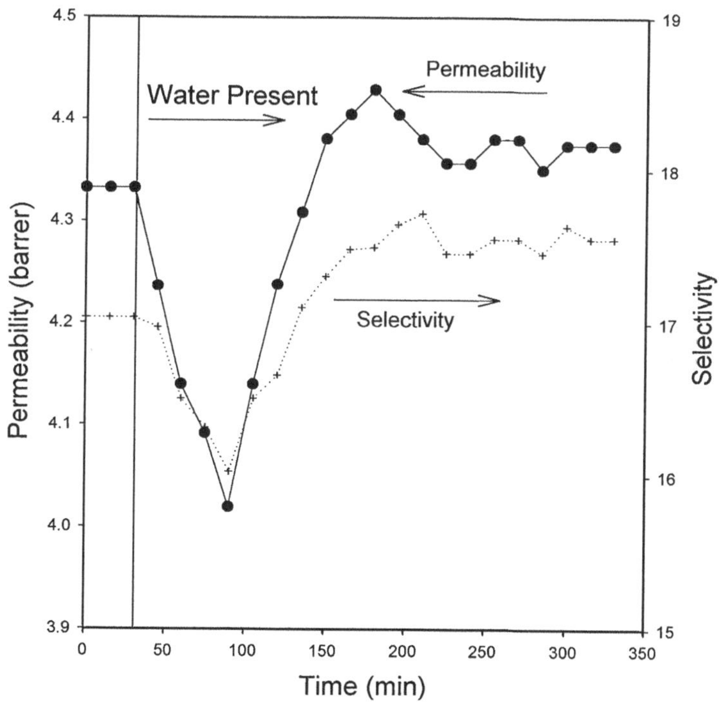

Figure 11.14 *Polysulfone CH_4 permeability with water present as a function of time. Data at 55 °C and 800 kPa [45]*

polysulfone occurs, the selectivity rises and becomes greater than the non-plasticized membrane.

In the case of water, the hydrogen bonding potential between molecules allows for multilayer adsorption to occur within the microvoids of the membrane and clustering of water molecules has been observed for a number of polymers. Hence, additional adsorption is occurring and the Langmuir model presented as Equations (11.29) to (11.37) may not be appropriate. Park [46] puts forward a simple model for the concentration of water within the glassy membrane when clustering occurs:

$$C = K_{\mathrm{D}} f + \frac{C'_{\mathrm{H}} b f}{1 + bf} + K_{\mathrm{C}} K_{\mathrm{D}}^{n} \frac{f^{n}}{n f_0} \tag{11.38}$$

where n is the number of water molecules within the cluster, K_{C} the equilibrium constant for the reaction between monomeric water and cluster water and f_0 the saturation fugacity. Alternatively, a multi-layer Brunauer–Emmett–Teller (BET) model may be a more appropriate isotherm to apply [47]:

$$C = K_{\mathrm{D}} f + \left(\frac{C'_{\mathrm{H}} b_{\mathrm{BET}} f}{1 - f} \right) \left(\frac{1 - (n+1) f^{n} + n f^{n+1}}{1 + (b_{\mathrm{BET}} - 1) f - b_{\mathrm{BET}} f^{n+1}} \right) \tag{11.39}$$

The parameter n can be correlated with the number of multilayers formed. If n is unity the isotherm converts to the Langmuir model, and if n is infinity it becomes the standard BET isotherm.

11.3.6 Hydrocarbons

In natural gas sweetening, hundreds of heavy hydrocarbons, both paraffinic and aromatic, are present in the feed gas [48] and can influence CO_2 separation through polymeric membranes with a detrimental loss in performance. It is alleged that low quantities of condensable hydrocarbons can also cause pre-mature ageing and membrane failure [49]. The competitive sorption effects of hexane on CO_2 permeability and selectivity against CH_4 in the glassy polyimide membrane poly (4,4′-hexafluoroisopropylidene diphthalic anhydride -2,3,5,6-tetramethyl-1,4-phenylenediamine) (6FDA-TMPDA, a polyimide) is shown in Figures 11.15 and 11.16 respectively [50].

There is a decrease in the permeability of both CO_2 and CH_4 as a result of exposure to hexane at different partial pressures (Figure 11.15) as the result of competitive sorption. The decrease is more pronounced for CO_2, because the solubility of this gas within the polymer is more strongly influenced by Langmuir adsorption. Indeed at the hexane concentrations used, almost all the carbon dioxide is displaced from the Langmuir voids by hexane with the permeability approaching that predicted by Henry's Law sorption alone. The greater reduction in CO_2 permeability results in a net loss of selectivity from 16 to around 10.

The use of Equation (11.37) allows for an estimation of the b value for hexane, based on known dual sorption parameters [50]. The resulting value of $240 \pm 100\,\mathrm{atm}^{-1}$ is two orders of magnitude greater than that available for CO_2 ($0.51\,\mathrm{atm}^{-1}$ [51]), and indicates that hexane has a much stronger affinity for the micro-voids than CO_2. This is consistent

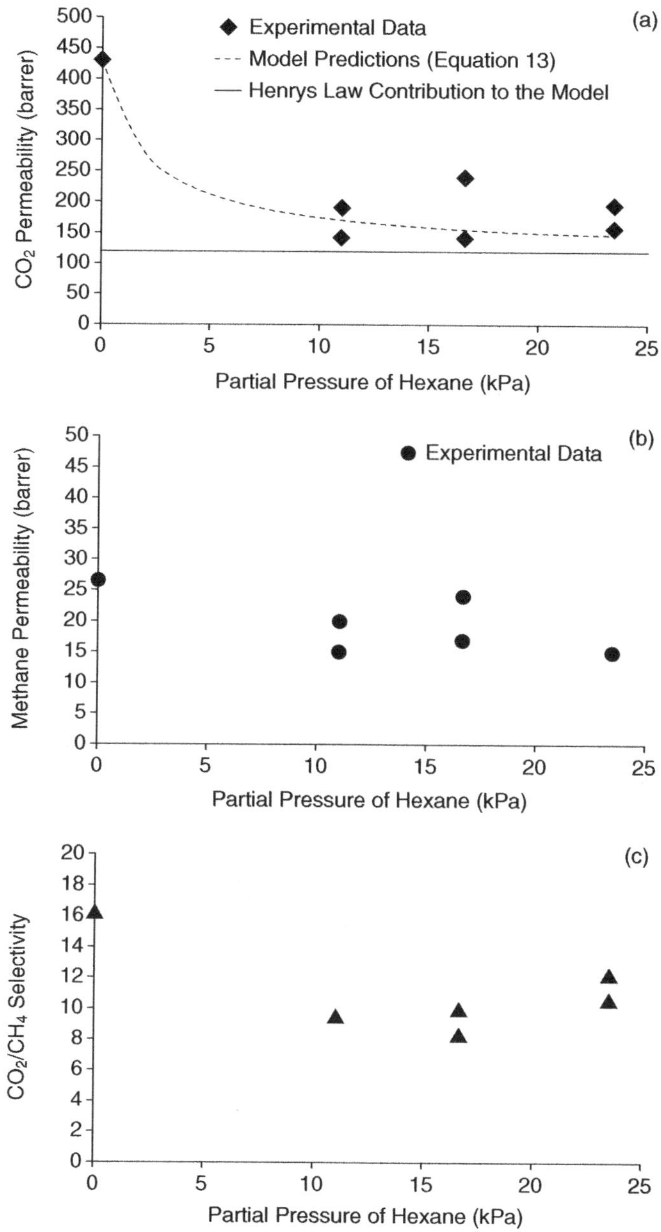

Figure 11.15 *The effect of an addition of hexane to a mixture of methane (90%) and carbon dioxide (10%) at a feed pressure of 1025 kPag. (a) CO_2 permeability, (b) CH_4 permeability, (c) CO_2/CH_4 selectivity. Reprinted with permission from Ind Eng Chem Res., The effect of hydrocarbons on the separation of carbon dioxide from methane through a polyimide gas separation membrane by Hasan, R., C. A. Scholes, G. W. Stevens and S. E. Kentish, 48, 5415–5419, Copyright (2009) American Chemical Society*

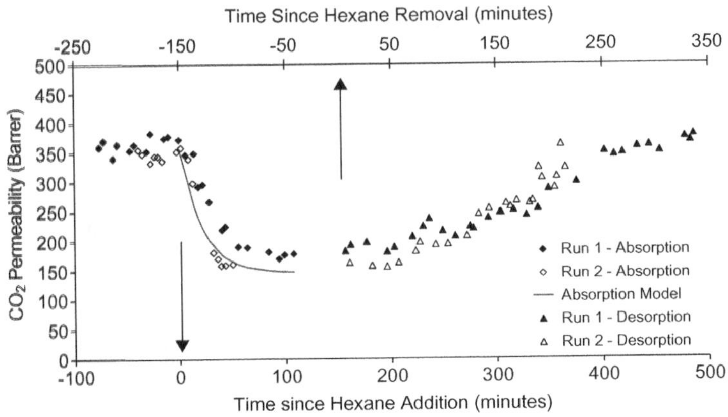

Figure 11.16 *The time dependent effect of addition of hexane on the permeability of CO_2 in 6FDA-TMPDA from a methane (90%) and carbon dioxide (10%) gas mixture. Reprinted with permission from Ind Eng Chem Res., The effect of hydrocarbons on the separation of carbon dioxide from methane through a polyimide gas separation membrane by Hasan, R., C. A. Scholes, G. W. Stevens and S. E. Kentish, 48, 5415–5419, Copyright (2009) American Chemical Society*

with the correlation between the critical temperature of a component and this Langmuir constant (Figure 11.1). The determined Langmuir constant is slightly higher than that reported for other species with similar critical temperatures and reflects the use of fugacity as the basis for calculations. As a fugacity is lower than the corresponding partial pressure, the corresponding b value will be higher.

Upon reaching steady-state conditions, the feed gas was returned to dry conditions and the permeability of both CO_2 and CH_4 was observed to recover (Figure 11.16). The permeabilities approach initial values after around 300 minutes, implying performance is recoverable and that the polymeric structure is not permanently altered by the presence of hexane.

11.4 Conclusions

This chapter has provided a range of mathematical models that can be used to characterize the effects of minor components on membrane performance. However, the quantity of good experimental data that can be used to fit these models remains quite limited. Good characterization of membrane performance in both pre-combustion and post-combustion flue gases will require accurate experimentally based determination of parameters such as Flory–Huggins interaction parameters, Langmuir constants and plasticization potentials.

Many membrane scientists working in this field focus only on the development of novel polymers for these separation processes, as this is of great appeal. Testing is restricted to pure gas samples only, or at most, binary mixtures. However, without the full characterization of each polymer with respect to true mixed gas streams the likelihood of commercial success will remain limited.

References

(1) W. J. Koros, G. K. Fleming, Membrane-based gas separation, *J. Membr. Sci.*, **83**, 1 (1993).
(2) D. R. Paul, Y. P. Yampol'skii, Gas Separation Membranes, Boca Raton, CRC Press (1994).
(3) C. E. Powell, G. G. Qiao, Polymeric CO_2/N_2 gas separation membranes for the capture of carbon dioxide from power plant flue gases, *J. Membr. Sci.*, **279**, 1–49 (2006).
(4) IPCC, Climate Change 2007: The Physical Science Basis. Contribution of Working Group I to the Fourth Assessment Report of the Intergovernmental Panel on Climate Change. Cambridge, United Kingdom and New York, USA: Cambridge University Press (2007).
(5) C. A. Scholes, S. E. Kentish, G. W. Stevens, Effects of minor components in carbon dioxide capture using polymeric gas separation membranes, *Sep. Purif. Rev.*, **38**, 1–44 (2009).
(6) K. Thambimuthu, M. Soltanieh, J. C. Abanades, *IPCC Special Report on Carbon Dioxide Capture and Storage*. Cambridge: Cambridge University Press (2005).
(7) F. Moavenzadeh, R. W. Cahn, M. B. Bever, Iron Production, in Encyclopedia of Building and Construction Materials, MIT Press: Boston. pp. 360–366 (1990).
(8) Emission estimation technique manual for cement manufacturing. Department of the Environment, Water, Heritage and the Arts (2008).
(9) L. M. Robeson, B. D. Freeman, D. R. Paul, B. W. Rowe, An empirical correlation of gas permeability and permselectivity in polymers and its theoretical basis, *J. Membr. Sci.*, **341**, 178–185 (2009).
(10) D. W. Breck, *Zeolite Molecular Sieves*. New York: John Wiley & Sons (1973).
(11) N. A. Lange, *Lange's Handbook of Chemistry*. New York: McGraw-Hill (1992).
(12) J. D. Perry, K. Nagai, W. J. Koros, Polymer membranes for hydrogen separation, *MRS Bulletin*, **31**, 745–749 (2006).
(13) R. W. Baker, *Membrane Technology and Applications*. New York: John Wiley & Sons (2004).
(14) P. Flory, *Principles of Polymer Chemistry*. Ithaca, New York: Cornell University Press (1953).
(15) J. H. Petropoulos, Mechanisms and Theories for Sorption and Diffusion of Gases in Polymers, in Polymeric Gas Separation Membranes, D. R. Paul, Y. P. Yampol'skii, Editors. CRC Press: Boca Raton. pp. 17–81 (1994).
(16) M. S. Suwandi, S. A. Stern, Transport of heavy organic vapors through silicone rubber, *J. Polym. Sci. Polym. Phys.*, **11**, 663–681 (1973).
(17) C.-S. Lu, Y.-L. Wang, Symmetrical form of the four-suffix margules equations for quaternary systems, *Ind. Eng. Chem. Fundamen.*, **3**, 271–272 (1964).
(18) H. Fujita, Diffusion in polymer-diluent systems, *Fortschr. Hochpolymer*, **3**, 1–47 (1961).
(19) S. M. Fang, S. A. Stern, H. L. Frisch, Free volume model of permeation of gas and liquid mixtures through polymeric membranes, *Chem. Eng. Sci.*, **30**, 773–780 (1975).
(20) J. H. Petropoulos, Quantitative analysis of gaseous diffusion in glassy polymers, *J. Polym. Sci. Polym. Phys. Ed.*, **8**, 1797–1801 (1970).
(21) D. R. Paul, W. J. Koros, Effect of partially immobilizing sorption on permeability and the diffusion time lag, *J. Polym. Sci. Polym. Phys. Ed.*, **14**, 675 (1976).
(22) W. J. Koros, R. T. Chern, V. Stannett, H. B. Hopfenberg, Model for permeation of mixed gases and vapors in glassy polymers, *J. Polym. Sci. Polym. Phys. Ed.*, **19**, 1513–1530 (1981).
(23) N. Plate, Y. P. Yampol'skii, Relationship between structure and transport properties for high free volume polymeric materials, Boca Raton: CRC Press (1994).
(24) C. E. Rogers, Solubility & Diffusivity, in Physics and Chemistry of the Organic Solid State, D. Fox, M. M. Labes, A. Weissberger, Editors. John Wiley & Sons: New York (1965).
(25) R. J. Heinsohnand, R. L. Kabel, *Sources and Control of Air Pollution*. Englewood Cliffs, NJ: Prentice Hall (1999).
(26) E. S. Sanders, Penetrant-induced plasticization and gas permeation in glassy polymers, *J. Membr. Sci.*, **37**, 63–80 (1988).
(27) Y. P. Yampol'skii, V. V. Volkov, Studies in gas permeability and membrane gas separation in the Soviet Union, *J. Membr. Sci.*, **64**, 191–228 (1991).
(28) S. I. Semenova, S. I. Smirnov, H. Ohya, Performance of glassy polymer membranes plasticized by interacting penetrants, *J. Membr. Sci.*, **172**, 75–89 (2000).
(29) R. Zavaleta, F. P. McCandless, Selective permeation through modified polyvinylidene fluoride membranes, *J. Membr. Sci.*, **1**, 333–353 (1976).

(30) J. Hanousek, L. Herynk, Determination of permeability of foils to sulfur dioxide, *Chem. Listy*, **56**, 376–382 (1962).

(31) R. A. Pasternak, M. V. Christenson, J. Heller, Diffusion and permeation of oxygen, nitrogen, carbon dioxide and nitrogen dioxide through polytetrafluoroethylene, *Macromolecules*, **3**, 366–371 (1970).

(32) J. Hao, P. A. Rice, S. A. Stern, Upgrading low-quality natural gas with H_2S- and CO_2-selective polymer membranes Part 1. Process design and economics of membrane stages without recycle streams, *J. Membr. Sci.*, **209**, 177–206 (2002).

(33) M. Harasimowicz, P. Orluk, G. Zakrzewska-Trznadel, A. G. Chmielewski, Application of polyimide membranes for biogas purification and enrichment, *J. Hazard Mater.*, **144**, 698–702 (2007).

(34) C. J. Orme, J. R. Klaehn, F. F. Stewart, Gas permeability and ideal selectivity of poly[bis-(phenoxy)phosphazene], poly[bis-(4-tert-butyl-phenoxyl)phosphazene] and poly[bis-(3,5-di-tert-butylphenoxy)1,2(chloro)0.8 phosphazene), *J. Membr. Sci.*, **238**, 47–55 (2004).

(35) C. Y. Pan, Gas separation by high-flux, asymmetric hollow-fibre membrane, *AIChE J.*, **32**, 2020–2027 (1986).

(36) C. A. Scholes, S. E. Kentish, G. W. Stevens, The effect of hydrogen sulfide, carbon monoxide and water on the performance of a PDMS membrane in carbon dioxide/nitrogen separation, *J. Membr. Sci.*, **350**, 189–199 (2010).

(37) B. Wilks, M. E. Rezac, Properties of rubbery polymers for the recovery of hydrogen sulfide from gasification gases, *J. Appl. Polym. Sci.*, **85**, 2436–2444 (2002).

(38) C. A. Scholes, S. E. Kentish, G. W. Stevens, Competitive sorption and plasticization of CO_2 selective glassy polymeric membranes by hydrogen sulfide. 2010. in preparation.

(39) A. Bos, I. G. M. Punt, M. Wessling, H. Strathmann, CO_2-induced plasticization phenomena in glassy polymers, *J. Membrane Sci.*, **155**, 67–78 (1999).

(40) P. S. Tin, Effects of cross-linking modification on gas separation performance of Matrimid membranes, *J. Membr. Sci.*, **225**, 77–90 (2003).

(41) C. A. Scholes, K. H. Smith, S. E. Kentish, G. W. Stevens, CO_2 capture from pre-combustion processes – strategies for membrane gas separation, *Int. J. Greenhouse Gas Control*, in press (2010).

(42) R. Steeneveldt, B. Berger, T. A. Torp, CO_2 capture and storage. Closing the knowing-doing gap, *Chem. Eng. Res. Des.*, **84**, 739–763 (2006).

(43) G. T. Paulson, A. B. Clinch, F. P. McCandless, The effects of water vapor on the separation of methane and carbon dioxide by gas permeation through polymeric membranes, *J. Membr. Sci.*, **14**, 129–137 (1983).

(44) D. G. Pye, H. H. Hoehn, M. Panar, Measurement of gas permeability of polymers. II. Apparatus for determination of permeabilities of mixed gases and vapors, *J. Appl. Polym. Sci.*, **20**, 287–301 (1976).

(45) C. A Scholes, S. E. Kentish, G. W. Stevens, Competitive sorption of water within glassy polymeric membranes, 2010: p. in preparation.

(46) G. S Park, Transport Principles – Solution, Diffusion and Permeation in Polymer Membranes, in Synthetic Membranes: Science, Engineering and Applications, P. M. Bungay, H. K. Lonsdale, M. N. Pinho, Editors. D. Reidel Publishing Company: Dordrecht. pp. 57–108 (1983).

(47) S. Brunauer, L. S. Deming, W. E. Deming, E. Teller, A theory of the van der Waals adsorption of gases, *J. Am. Chem. Soc.*, **62**, 1723–1732 (1940).

(48) T. W. Legatski, J. W. Tooke, L. A. Grundy, Approximating natural gas composition, *Petrol Refiner.*, **32**, 105 (1953).

(49) D. Q. Vu, W. J. Koros, S. J. Miller, Fouling of carbon molecular sieve hollow-fiber membranes by condensable impurities in carbon dioxide-methane separation, *Ind. Eng. Chem.*, **42**, 1064 (2003).

(50) R. Hasan, C. A. Scholes, G. W. Stevens, S. E. Kentish, The effect of hydrocarbons on the separation of carbon dioxide from methane through a polyimide gas separation membrane, *Ind. Eng. Chem. Res.*, **48**, 5415–5419 (2009).

(51) X. Duthie, S. Kentish, C. Powell, K. Nagai, G. Qiao, G. Stevens, Operating temperature effects on the plasticization of polyimide gas separation membranes, *J. Membr. Sci.*, **294**, 40–49 (2007).

12

Tailoring Polymeric Membrane Based on Segmented Block Copolymers for CO_2 Separation

Anja Car[a], Wilfredo Yave[a], Klaus-Viktor Peinemann[a] and Chrtomir Stropnik[b]
[a]*Institute of Material Research, GKSS-Research Centre Geesthacht GmbH, Geesthacht, Germany*
[b]*University of Maribor, Faculty of Chemistry and Chemical Engineering, Smetanova, Slovenia*

12.1 Introduction

The problems of global warming and climate change are today being popularly discussed in terms of CO_2 emission by human activities [1]. CO_2 is produced by different industrial processes. Finding a proper separation technology for the removal of CO_2 from different gas streams will drastically lead to the reduction of emissions. Although CO_2 production contributes to global warming, it is also considered as a raw material in many industries [2]. Global CO_2 consumption for example has reached approximately 20 million tons per year, which is strictly dedicated to the so-called merchant market, food, beverage, chemical and other industries [3]. Nevertheless, either for storage or for its utilization, CO_2 must be separated.

At present, the most widely used CO_2 separation processes consist of reversible chemical and physical absorption. Membrane processes are not used very much; however, they are attractive because of their simplicity and energy efficiency [4].

Actually, there are not yet suitable technologies or technical solutions which can effectively fight against CO_2 increase in the atmosphere. In the membrane community, it is believed that membrane technology capable of operating under a wide variety of conditions will have a considerable impact on the separation and purification of CO_2 [5–9].

Membrane Gas Separation Edited by Yuri Yampolskii and Benny Freeman

As known, membranes have become an established technology for CO_2 removal since 1981 (natural gas purification) [10]. The multiple benefits of membrane technology promised by early innovators have been proven in a wide variety of installations around the world, and one of the world's largest membrane systems for CO_2 removal from natural gas is the Grissik processing plant in South Sumatra, Indonesia [11]. The Grissik plant is a hybrid separation system (membrane plus amine treatment) which offers particularly attractive operational benefits. Thus, it was demonstrated that the most economical approach is to combine membranes with existing technologies.

Commercial membranes for CO_2 removal are polymer based, and the materials of choice are cellulose acetate, polyimides, polyamides, polysulfone, polycarbonates, and polyetherimide [12]. The most tested and used material is cellulose acetate, although polyimide has also some potential in certain CO_2 removal applications. The properties of polyimides and other polymers can be modified to enhance the performance of the membrane. For instance, polyimide membranes were initially used for hydrogen recovery, but they were then modified for CO_2 removal [13]. Cellulose acetate membranes were initially developed for reverse osmosis [14], and now they are the most popular CO_2 removal membrane. To overcome state-of-the-art membranes for CO_2 separation, new polymers, copolymers, block copolymers, blends and nanocomposites (mixed matrix membranes) have been developed [15–22]. However, many of them have failed during application because of different reasons (expensive materials, weak mechanical and chemical stability, etc.).

In order to design new membrane materials with superior separation performance, low cost and feasible for massive production, the old work of Kulprathipanja can be considered as a proper tool [23]. He dispersed polyethylene glycol (PEG) in PDMS and improved the CO_2 permeability. A similar idea led us to modify or blend existing commercial block polymers containing ethylene oxide (EO) units with poly(ethylene) glycol [24]; and thus by a blending process it was possible to produce high CO_2-selective membrane materials with improved permeability.

Permeability of polymers is an important feature for a broad range of applications including packaging, bio-materials (e.g. for controlled drug release or encapsulating membranes), barrier materials, high performance impermeable breathable clothing and membrane separation processes [25–27]. Copolymers were considered many years ago, starting with the most investigated polyurethanes and polyurethane-ureas followed by polyimides, polyamides and miscellaneous types of block copolymers (e.g. block copolymers containing siloxane segments, hydrocarbon block copolymers and related materials) [28,29]. Block copolymers offer a great structural versatility which is highly interesting for a fundamental analysis of permeation through polymeric materials. Systematic structure/property relationships allow one to design block copolymers with improved gas transport properties [30].

The simplest are diblock copolymers, where two different polymeric chains are bound together; and with an increase of block number, tri- or multiblocks with a variety of structures can be obtained [31,32]. Most block copolymers used today are prepared by living anionic polymerization, which is a feasible method to prepare block copolymers with controlled architecture. Different polymers do not mix well due to thermodynamic reasons [33], especially if their molecular mass is sufficiently high, they have a strong tendency to form separated phases. In block copolymers, this phase separation can occur only intermolecularly (micro- or nanophase separation) [34]. Those block copolymers based on poly(ethylene oxide) (PEO) exhibited excellent CO_2 permeability, and many research groups demonstrated that they can effectively separate CO_2 from other gases [35–42].

Poly(amide-b-ethylene oxide) copolymers were presented in 1990 as a promising membrane material [43]. These block copolymers were developed in 1972 but in 1981 began to be used for commercial purpose under the trade name Pebax®, produced by ATOCHEM [44] (now ARKEMA). Another important group of segmented poly(ester)s used for membranes are block copolymers based on PEO and PBT (poly(butylene terephthalate), known under commercial name of Polyactive® [45]. By changing the polyamide and polyether segment, molecular mass and the content of each block, the mechanical, chemical, and physical properties are nicely tuned as well [46].

There are many techniques to improve gas transport properties of the above-mentioned polymers, and the simplest way to do it is by physical incorporation of additives into the polymeric matrix (blending). Because the performance of the blend membranes is tightly related to the nature of additives (fillers) and polymer characteristics, they must be carefully selected in order to obtain a novel membrane material with superior properties. Generally, the presence of additives can have three effects on membrane transport: they can act as molecular sieves (zeolites, carbon molecular sieves), they can disrupt the ordering of polymer chains, which is reflected in a permeability increase (non-porous nanoparticles and plasticizer), and they can act as barrier reducing the permeability (e.g. clays) [47–50]. The incorporation of plasticizers in polymers appears to have a relatively complex influence on the permeability [51]. Plasticizers may operate as a spacer, separating the macromolecular chains (free volume increase), thus they lead to a decline of the intermolecular attractive forces increasing the mobility of them and facilitating the diffusion of the adsorbed molecules by lowering the activation energy. By inserting plasticizer between macromolecular chains, the glass transition temperature of the polymer is decreased due to the local segmental motions [52]; accordingly, the diffusion coefficients are changed. In the literature, there are many publications related to the effect of plasticizer on morphology and gas transport properties of polymer blends [53–55].

Immiscible or partially miscible polymers (blends) often form separated phases, leading to a complex morphology. In many cases, the miscibility in polymer blends [56] is a direct result of enthalpy-of-mixing (ΔH_{mix}) contributing to the overall free energy ΔG_{mix}, which is the fundamental thermodynamic quantity that controls the miscibility. To obtain homogeneous blend (single phase), the free Gibbs energy ΔG_{mix} must be <0. The free Gibbs energy for mixing per unit volume is given by the classical Flory–Huggins equation (Equation 12.1):

$$\frac{\Delta G_{mix}}{V} = R \cdot T \cdot \left[\frac{\rho_A \cdot \Phi_A}{M_A} \cdot \ln \Phi_A + \frac{\rho_B \cdot \Phi_B}{M_B} \cdot \ln \Phi_B \right] + \chi \cdot \Phi_A \cdot \Phi_B \tag{12.1}$$

where χ is the interaction parameter, R is gas constant, T is the absolute temperature and ρ, Φ and M are the density, volume fraction and molecular mass of the polymer, respectively. This equation includes implicitly enthalpy of mixing and entropy of mixing given by the Flory–Huggins model [57,58]. Although polymer blend phase behaviour is determined predominately by the nature of the molecular interactions, it may be influenced by many other factors, including molecular mass, copolymer composition, blend ratio, temperature and pressure. Morphology study is one of the most important research topics in polymer blends, and it has been shown that the surface and bulk chemical compositions of polymer blends are not the same even when they are miscible in the bulk [59]. However, a number of other factors such as solvent, specific interaction, crystallization, etc. can also affect the blend, and consequently its transport properties.

This chapter covers multi-step work of tailoring CO_2-selective membrane material, from new copolymers designs to tailor-made copolymer/PEG blends, moreover thin film composite membrane performance are also discussed. The relationships between gas transport properties, structure, morphology and physical properties are analyzed. The performance at different operating conditions and with mixed gases is monitored as well in order to have a guideline for scaling up the membranes. The benefits of these membranes are the simplicity of preparation, low cost and resistance toward acid gas treatment.

12.2 Tailoring Block Copolymers with Superior Performance

Figure 12.1 presents the chemical structure of poly(amide-b-ethylene oxide) (commercial name 'Pebax®') and poly(ethylene oxide)-poly(butylene terephthalate) (PEO-PBT) (commercially known as 'Polyactive®'). These copolymers with high content of PEO are hydrophilic and show excellent chemical resistance towards solvents. The solubility of these copolymers in different solvents is determined by the ratio of segmented blocks. Pebax® MH 1657 is soluble only in few solvents and generally the polymer solution is prepared under reflux at high temperature and low polymer concentration by using n-butanol or a mixture of n-butanol/1-propanol, after cooling down to room temperature,

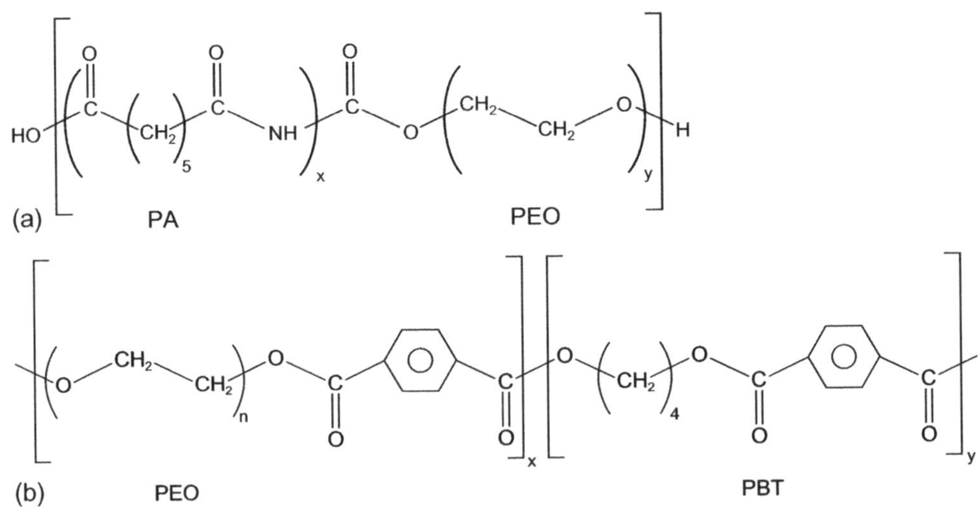

Figure 12.1 *Structure of (a) PA-b-PEO and (b) PEO-PBT, where x and y represent weight percent of each block and n gives information about the molecular weight (g/mol) of PEO in PEO-PBT*

however, the solution shows a strong tendency to form a gel [60]. During dissolution studies, surprisingly it was found that a simple binary mixture of ethanol/water (70/30 wt.%) can easily dissolve Pebax®. An advantage of this solvent mixture is the stability of the polymer solution at room temperature, which simplifies the membrane preparation [24]. Polyactives with relatively low molecular weight and high PEO fraction are soluble in tetrahydrofuran (THF), and those with high molecular weight are soluble in chloroform; the membrane preparation therefore is easy.

Pebax® and Polyactive are thermoplastic with elastomeric behaviour. PA and PBT rigid segments give strength to the polymeric material, it is able to crystallize and its glass transition is above room temperature. The rubbery PEO flexible segment (glass transition below room temperature) provides flexibility and hydrophilicity to the system. The gas transport mainly proceeds through the amorphous PEO soft phase [61]. As reported, the morphology of multi-block copolymers is very complex; these can comprise up to five different phases: amorphous and crystalline phase of both blocks and phases between them (interphases) [62]. Hence, the variation of both segments in content and molecular weight lead to materials with different properties, and increase of the molecular weight of blocks favour the micro-phase separation and crystallization [63]. Typical physical and gas transport properties of these copolymers are shown in Table 12.1.

From all the commercial copolymers presented in Table 12.1, Pebax® 1657, 1500PEO77PBT and 4000PEO55PBT (Polyactives) have high CO_2 selectivity over H_2 and N_2 and moderate CO_2 permeability, thus by adding PEG with low molecular weight, their CO_2 permeability can be improved without affecting the selectivity [24,67]. The other

Table 12.1 *Physical and gas transport properties of Pebax® and Polyactive® multi-block copolymers [24,64–67]*

Polymer	Flexible segment (wt.%)	Density (g/cm^3)	T_g of flexible segment (°C)	$P(CO_2)$ (Barrer)	$\alpha(CO_2/H_2)$	$\alpha(CO_2/N_2)$
[a,c] Pebax® 1657	60	1.14	−53	73	9.1	45
[a,d] Pebax® 1074	55	1.09	−55	110	8.0	46
[b,d] Pebax® 2533	80	1.01	−77	350	5.8	35
[b,d] Pebax® 3533	73	1.01	−72	230	4.9	–
[b,d] Pebax® 4033	53	1.01	−78	76	5.0	29
[b,d] Pebax® 5533	30	1.01	n.d	–	–	–
600PEO77PBT	77	1.20	−42	58	7.8	46
1000PEO80PBT	80	1.19	−45	82	8.9	44
1500PEO77PBT	77	1.19	−49	115	10.2	50
300PEO55PBT	55	1.28	−20	9	3.7	21
4000PEO55PBT	55	1.23	−49	95	10.9	44
1000PEO55PBT	52	–	−49	71	–	45
2000PEO55PBT	52	–	−53	116	–	–
3000PEO55PBT	52	–	−52	146	–	–

[a] Flexible segment PEO (Polyethylene oxide).
[b] Flexible segment PTMO (Polytetramethylene oxide).
[c] Rigid segment PA6 (Polyamide-6).
[d] Rigid segment PA12 (Polyamide-12).
[e] Rigid segment PBT (Polybutylene terephthalate).

copolymers can be also used for developing new membrane materials, but the selection of PEG or other plasticizer or fillers must be done very carefully in order to avoid phase separation. The lack of a commercial copolymer with higher performance than those presented in Table 12.1 led many researchers to investigate, chemically modify and synthesize new PEO-containing polymeric materials [68–72], and one of the best membrane materials with high CO_2 performance was reported by Freeman group [73], who cross-linked PEO copolymers by UV irradiation and the membranes exhibited promising properties. However, those membranes could be complex and expensive for mass production.

Therefore, research focused on the development of new membrane materials and up-scaling for production is still important. In that direction, and in order to understand better the membrane behaviour of PEO-PBT copolymers, the CO_2 permeability (Table 12.1) were fitted to polynomial mathematical models (Equation 12.2), and plotted versus the molecular weight (M_w) of the PEO block, as shown in Figure 12.2.

$$P(CO_2) = a + b \cdot (M_w) + c \cdot (M_w)^2 \tag{12.2}$$

This equation fitted to the experimental data for each family of copolymer (PEO55PBT45 and PEO77PBT23) gives an estimation of CO_2 permeability for the optimal membrane material. The triangle data points and solid line (Figure 12.2) are theoretical values for copolymers with apparent 80 wt.% of PEO (permeability data of copolymers containing 77 and 80 wt.% of PEO and a blend with 84 wt.% of PEO were used for fitting), and the fitted mathematical expression is as follows:

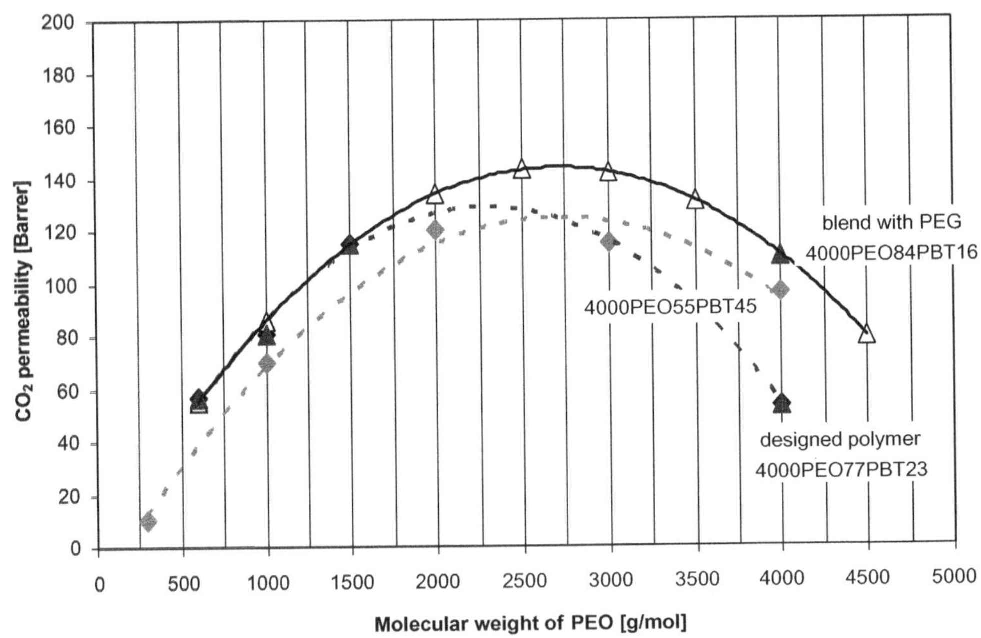

Figure 12.2 *Experimental and predicted data of CO_2 permeability (for different block copolymer) as a function of molecular weight of PEO block [modified plot from ref. 67]*

Table 12.2 *Thermal properties of PEO-PBT multi-block copolymers [Reprinted with permission 67]*

Polymer	PEO			PBT	
	T_g (°C)	T_m (°C)	X_c (%)	T_m (°C)	X_c (%)
600PEO77PBT23	−42	–	–	110	3
1000PEO80PBT20	−45	19.6	16	–	–
1500PEO77PBT23	−49	27.0	24	–	–
300PEO55PBT45	−20	–	–	152	6
4000PEO55PBT45	−49	40.5	19	213	9
4000PEO77PBT23*	−53	45.0	42	~180	–

Source: Reprinted with permission from Advanced Functional Materials, Tailor-made polymeric membranes based on segmented block copolymers for CO_2 separation, by A. Car, C. Stropnik, W. Yave, K.-V. Peinemann, 23, 2815–2823. Copyright (2008) Wiley-VCH.
*designed polymer.

$$P(CO_2) = -2.2746 + 0.1081 \cdot (M_w) - 2 \cdot 10^{-5} (M_w)^2 \tag{12.3}$$

From those points (theoretical data) one can conclude that the CO_2 permeability would have a maximum value for the copolymer containing PEO with a molecular weight between 2500 and 3000 g/mol. However, the commercially available polymer is that with 1500 g/mol (1500PEO77PBT). In order to validate the model and to have real information about the behaviour of these copolymers, the polymer 4000PEO77PBT23 was designed by us, and synthesized by PolyVation (Netherlands) [67]. The permeability of the synthesized polymer was lower than expected, as observed in Figure 12.2.

The low permeability in 4000PEO77PBT23 copolymer has been attributed to high crystallinity of PEO phase ($X_c = 0.42$), moreover the T_m was high (45 °C), which shows that at room temperature the crystallities are big. The data for commercial polymers are given for comparison in Table 12.2 [67]. Therefore, besides the PEO content in the polymer system, the amorphous/crystalline phase ratio of PEO is also very important for improving the CO_2 permeability, because an increase in the content of the crystalline phase leads to a strong decrease of solubility and diffusivity, and consequently the permeability drops. As mentioned above, one way to enhance the gas permeability is destroying the crystalline phase, which will improve the solubility as well as diffusivity. Addition of 50 wt.% of PEG (200 g/mol) reduced the crystallinity of the PEO phase (T_m = 38 °C and X_c = 0.23) in the polymer designed (4000PEO77PBT23), thus a membrane with higher permeability has been obtained. The high permeability observed in that membrane is also due to the increased chain motion (T_g decrease from −53 to −79°C).

Table 12.2 shows thermal properties (obtained from second heating) of different PEO-PBT copolymers; since the PEO content and molecular weight vary (300–4000 g/mol) different glass transition temperatures (T_g) are observed for each sample. It was not possible to detect by DSC method, the T_g of PBT blocks and according to literature it should be around 60 °C [62]. T_g in block copolymers with 77–80 wt.% of PEO content slightly decreases according to the PEO molecular weight increase, whereas T_m and degree of crystallinity (X_c) increase. Interpretation of T_g and T_m is consistent for copolymers containing 55 wt.% of PEO as well. For the PBT phase, an increase in PBT fraction seems to lead to higher values of T_m and X_c. T_m for crystalline PBT is between 110 and 213 °C [67,74]. The length of PBT chains (M_w) is unknown; hence their influence on the thermal properties can

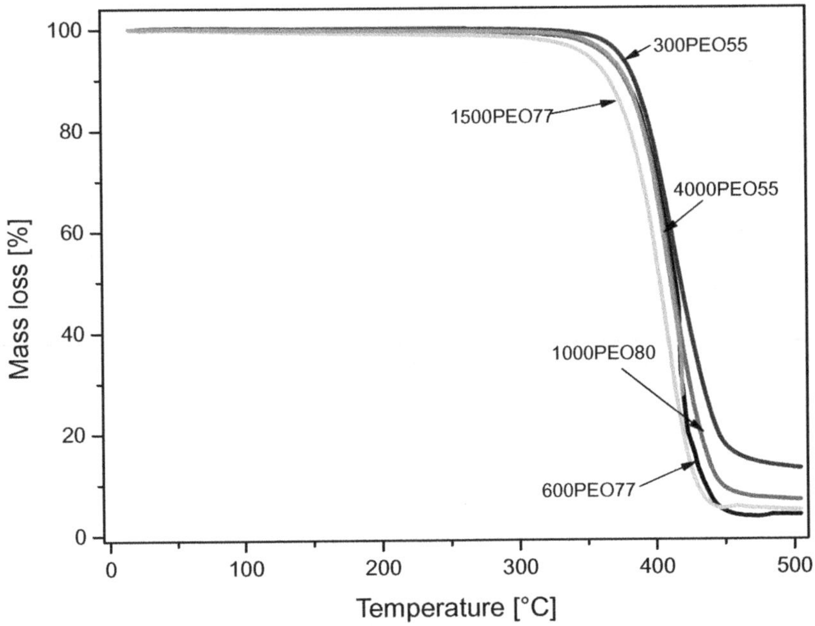

Figure 12.3 *Thermal gravimetric analyses of PEO-PBT multi-block copolymers. Reprinted with permission from Advanced Functional Materials, Tailor-made polymeric membranes based on segmented block copolymers for CO_2 separation, by A. Car, C. Stropnik, W. Yave, K.-V. Peinemann, 23, 2815–2823. Copyright (2008) Wiley-VCH*

not be discussed. Taking into account these results, block copolymers with high M_w, high fraction of PEO and low crystallinity could exhibit desired properties for gas separation.

Figure 12.3 illustrates thermogravimetric results of PEO-PBT copolymers performed under argon. The samples start to degrade at approximately 330 °C. There is no significant difference in temperature of degradation between different copolymers, but it is possible to distinguish that samples with high PEO content start to degrade at lower temperatures. All samples show good thermal stability, and thus, they are promising membrane materials, which can be used for gas separation at moderately high temperature (~100 °C). The temperature of exhaust gas in power plants for example is >55°C; hence the membranes developed from these copolymers have potential application in this scenario.

12.3 Block Copolymers and their Blends with Polyethylene Glycol

It has been observed [24] that for PEG (200 g/mol) modified Pebax® membrane for CO_2 separation the CO_2 permeability increased by a factor of about 2 (from 73 to 151 Barrer) and the separation factor CO_2/H_2 also increased by PEG addition (50 wt.%). This enhancement was attributed to the appearance of additional ethylene oxide (EO) units and free volume increase. Higher content of EO units results in an increase in the solubility of CO_2. Later, the total free volume increase and hence the increase of the permeability was demonstrated by measurements of density and by positron annihilation lifetime spectroscopy (PALS) analysis [75].

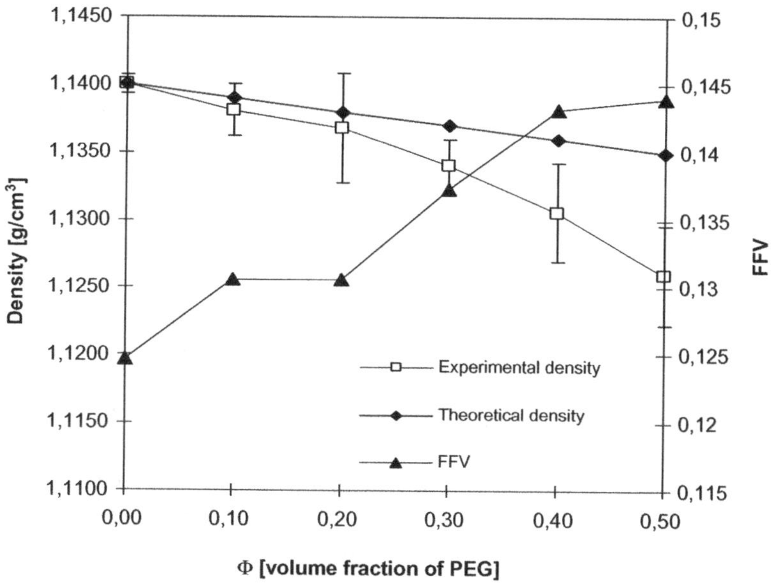

Figure 12.4 *Density and fractional free volume of Pebax®/PEG as function of volume fraction of PEG determined by additive model and buoyancy method [modified plot from ref. 24]*

Figure 12.4 presents the density decrease and the total fractional free volume (FFV) increase by addition of PEG into the Pebax®. The experimental values of density, which are lower than those calculated by the additive model, suggest that the total free volume is increased. The FFV was estimated from a definition of the volume fraction of the filler by using density values [76], and by solving the following equations:

$$\phi_F^N = \frac{(W_F/\rho_F)}{(W_P/\rho_P + W_F/\rho_F)} \tag{12.4}$$

$$\phi_F^T = \frac{\phi_F^N\left[\dfrac{W_P}{\rho_P}+\dfrac{W_F}{\rho_F}\right]}{\left[\dfrac{W_P}{\rho_P}+\dfrac{W_F}{\rho_P}+V_V\right]} = \frac{\left[\dfrac{W_F}{\rho_F}\right]}{\left[\dfrac{1}{\rho_C}\right]} = W_F\frac{\rho_C}{\rho_F} \tag{12.5}$$

$$FFV = FFV_P + FV_V \tag{12.6}$$

where ϕ_F^N and ϕ_F^T are the nominal and the true filler volume fraction respectively, W_F, W_P, ρ_F and ρ_P, are the weight percents and densities of filler and pure polymer respectively and, ρ_C, V_V, FFV_P and FV_V are the density of composite, the void volume, the fractional free volume of pure polymer and the void volume fraction. Note that the total FFV has been calculated by adding the void volume fraction created by PEG addition to the free volume fraction of pristine copolymer.

These results encouraged us to study in more detail this kind of blends, and the demonstration of free volume increase by using PALS was our goal. If we could demonstrate it for rubbery-like membranes, novel strategies for improving the membrane performance could be developed. The success of application of PALS to polymers is largely due to the so-called standard model developed by Tao and Eldrup which allows a quantitative description of the relation between free volume and ortho-positronium (o-Ps) lifetime ($\tau_{\text{o-Ps}}$) [77,78]. This quantum mechanical model is based on an assumption that o-Ps is confined in spherical holes (free volume elements) with infinitely high walls and gives a direct relationship between $\tau_{\text{o-Ps}}$ and the size of the free volume:

$$\tau_{\text{o-Ps}} = \frac{1}{2}\left[1-\left(\frac{R}{R+R_0}\right)+\frac{1}{2\pi}\sin\left(\frac{2\pi R}{R+R_0}\right)\right]^{-1} \tag{12.7}$$

where R is the average radius of holes (free volume element), $R_0 = R + \Delta R$ and $\Delta R = 0.166\,\text{nm}$.

PALS gives information about size and size distribution of free volume in amorphous polymers. In interpreting the results on free volume variation in rubbery polymers we use the concept of dynamic free volume [75,79]. The results show that in these membranes, the FFV varies by addition of PEG (Figure 12.4), but the average size of the free volume element does not change significantly at room temperature (Figure 12.5). As expected, this is due to the rubbery or apparent molten state of PEO. Although the size of these holes did not significantly increase, the dynamic free volume has been greatly enhanced because of plasticizer behaviour of PEG. Thus the total free volume was increased and consequently the observed CO_2 permeability was higher.

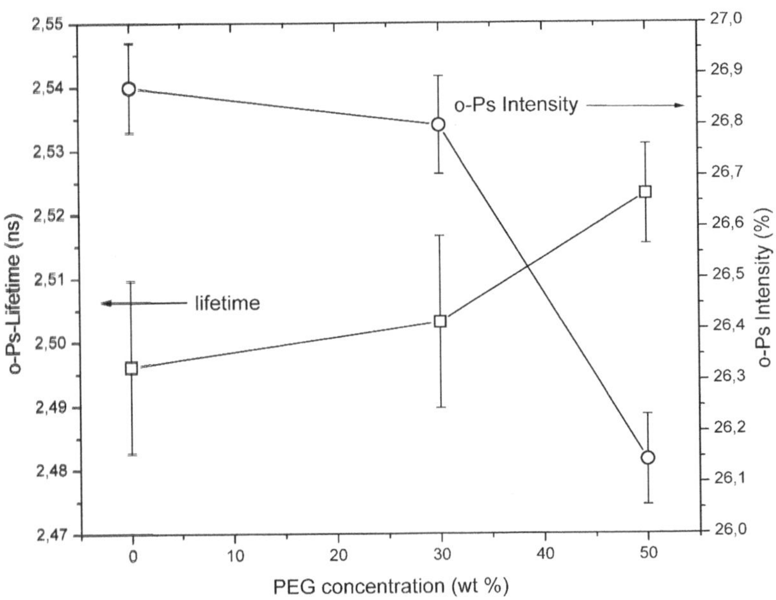

Figure 12.5 *o-Ps lifetime and o-Ps intensity as a function of PEG content [reprinted with permission from ref. 75]*

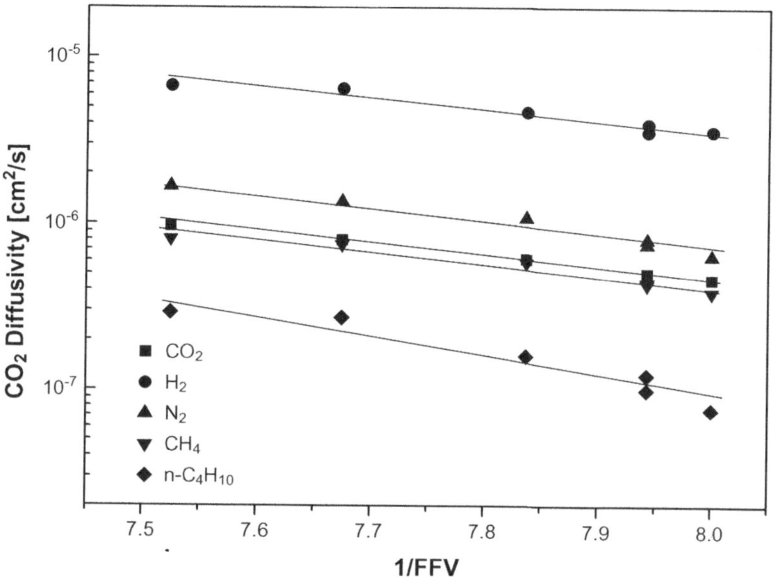

Figure 12.6 *CO_2 diffusivity as a function of 1/FFV in Pebax®/PEG membranes [modified plot from ref. 75]*

When gas diffusivity is plotted versus 1/FFV, a linear relationship is obtained (Figure 12.6). As is known, models based on free volume have been used to describe gas diffusion in polymers [80–84], and the correlation between the logarithm of diffusion coefficient and 1/FFV often gave a straight line. Because of FFV calculation from density, gas diffusivity and PALS measurements are completely independent; the good correlation between them supports the idea of increasing the total free volume in rubbery-like membranes, in our case increase of total free volume in Pebax® due to the PEG presence. Therefore, besides CO_2 solubility increase in Pebax®/PEG blend membranes, the total free volume increase due to the plasticizer is also considered as an important factor for improvement of gas permeability. In searching for the design of novel membrane rubbery-like materials it is desirable to develop strategies to enhance the chain flexibility and the total free volume in order to increase the gas permeability keeping the selectivity intact.

The two most promising Polyactive copolymers, i.e. 1500PEO77PBT23 and 4000PEO55PBT45, were also used for the preparation of blends with PEG. It was made in order to tailor a blend membrane material with improved CO_2 separation performance. Blend membranes prepared from both copolymers showed excellent miscibility with PEG200 up to 50 wt.%. Table 12.3 presents the permeabilities of different gases and selectivities for 1500PEO77PBT23/PEG200 blend membranes. CO_2 permeability slightly increases from 115 Barrer (pristine copolymer) to 134 Barrer (blend with 30 wt.% of PEG), then it decreased to 110 Barrer. As can be seen, all gases behave in similar way, i.e. first the permeability increases at lower content of PEG (<30 wt.%), and then at higher PEG content it decreases to values close to the pristine copolymer. This behaviour has been related to the crystallinity increase or chain ordering because of strong hydrogen

Table 12.3 Permeability and perm selectivity of 1500PEO77PBT23 blend membranes with PEG [Reprinted with permission 67]

Permeability, Barrer						
PEG (wt.%)	0	10	20	30	40	50
H_2	11.3	11.8	11.5	12.1	10.8	10.1
N_2	2.52	3.1	3.2	2.8	3.8	3
CH_4	6.76	7.63	8.3	8.9	7.6	7.5
CO_2	115	130	126	134	114	110
Permselectivity						
CO_2/H_2	10.2	11.0	11.0	11.1	10.6	10.9
CO_2/N_2	45.6	42.1	40.0	48.7	30.1	36.9
CO_2/CH_4	17.0	17.0	15.2	15.1	15.0	14.7

Source: Reprinted with permission from Advanced Functional Materials, Tailor-made polymeric membranes based on segmented block copolymers for CO_2 separation, by A. Car, C. Stropnik, W. Yave, K.-V. Peinemann, 23, 2815–2823, Copyright (2008) Wiley-VCH.

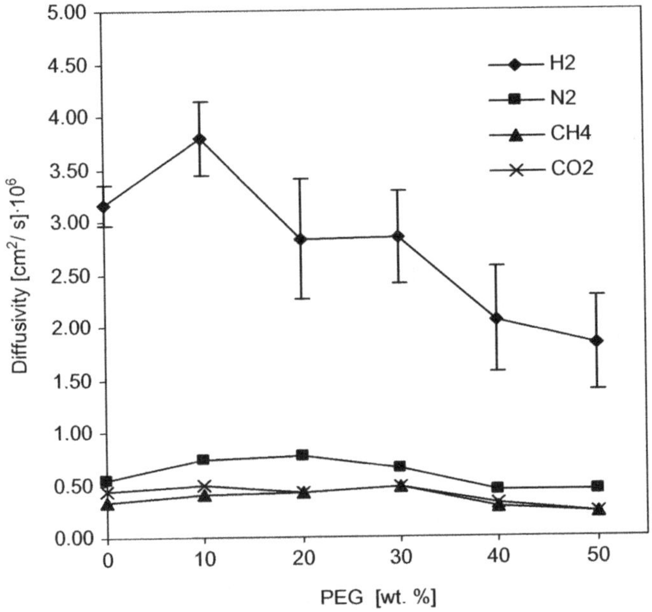

Figure 12.7 *Diffusivity of different gases (H_2, N_2, CH_4 and CO_2) in copolymer 1500PEO77PBT23. Reprinted with permission from Advanced Functional Materials, Tailor-made polymeric membranes based on segmented block copolymers for CO_2 separation, by A. Car, C. Stropnik, W. Yave, K.-V. Peinemann, 23, 2815–2823. Copyright (2008) Wiley-VCH*

bonding, since a new melting point was observed between 50 and 150 °C (discussed later). The selectivities are almost constant.

The diffusivity of gases plotted as a function of PEG content is shown in Figure 12.7, and the H_2 diffusivity decreases after the addition of 10 wt.% of PEG. The diffusivity of other gases has a similar trend. Therefore we concluded that the amorphous phase fraction decreased in the blends with higher PEG content. Blend membranes were optically homogeneous, since all samples were completely transparent. Thus, no phase separation is expected. Transport in blends depends upon composition, miscibility and phase morphology. In homogeneous blends, the diffusion process is influenced by the interaction between the components [85,86], while in heterogeneous blends the interfacial phenomena and the rubbery or glassy nature of phases are important [87]. Consequently, the permeability depends on heterogeneity of the system and the method of blend preparation [88]. A selection of polymer and additive with good compatibility are important in order to have a homogeneous system. However, strong interactions which can produce a perfectly homogeneous blend cannot be required, because hydrogen bonding can decrease the permeability due to the free volume decreases.

Copolymer 4000PEO55PBT45 has longer PEO segment than 1500PEO77PBT23, hence the addition of PEG with 200 g/mol effectively hindered the hydrogen bonding. It can be assumed that somehow the PEG chains are accommodated between polymer segments and create higher fractional free volume (more amorphous phase). The CO_2 permeability in pristine copolymer is 96 Barrer and it is increased two fold (~190 Barrer) by addition of 50 wt.% of PEG (as 77.5 wt.% calculated by Equation 12.8). Thus, the incorporation of EO units as PEG into the 4000PEO55PBT45 polymer matrix clearly led to a material with improved gas transport properties (Figure 12.8). This fact has been attributed to the structural (increase in the content of EO units) and morphological

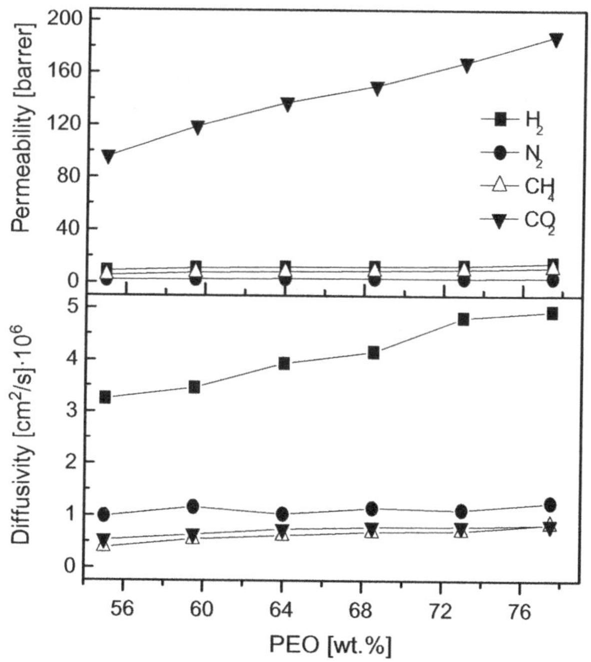

Figure 12.8 *Permeability and diffusivity of 4000PEO55PBT45 blends with PEG as function of total PEO. Reprinted with permission from Advanced Functional Materials, Tailor-made polymeric membranes based on segmented block copolymers for CO_2 separation, by A. Car, C. Stropnik, W. Yave, K.-V. Peinemann, 23, 2815–2823. Copyright (2008) Wiley-VCH*

(decrease in crystallinity) changes in the membrane As can be observed in Figure 12.8, the diffusivity of all gases is increased, as is the case with Pebax, and this is in agreement with the increases in the free volume.

It is worth to mention that permeabilities for 4000PEO55PBT45/PEG blend membranes shown in Figure 12.8 are according to recalculated values of total PEO content, which was done by using the following equation:

$$wt\cdot\%_{\mathrm{PEOtotal}} = 0.55\cdot m_{\mathrm{copolymer}} + m_{\mathrm{PEG200}} \tag{12.8}$$

where 0.55 represents the mass fraction of PEO in the copolymer, $m_{\mathrm{copolymer}}$ and m_{PEG200} are the mass of copolymer and PEG in 100 g of blend, respectively.

Figure 12.9 describes the strategy of preparation of a tailor-made blend membrane. By adding 50 wt.% of PEG as additive into the copolymers, membranes with improved properties can be produced. However, a limit may exist because of hydrogen bonding, which results in decreased chain mobility and CO_2 solubility. In order to develop novel membrane materials with better performance as compared to those developed here, other strategies must be studied i.e. strategies for a simultaneous increase in solubility and diffusivity (hindering the hydrogen bonding).

The enhancement of permselectivity, solubility selectivity and diffusivity selectivity of 4000PEO55PBT45/PEG blend membranes are presented in Table 12.4. The successful

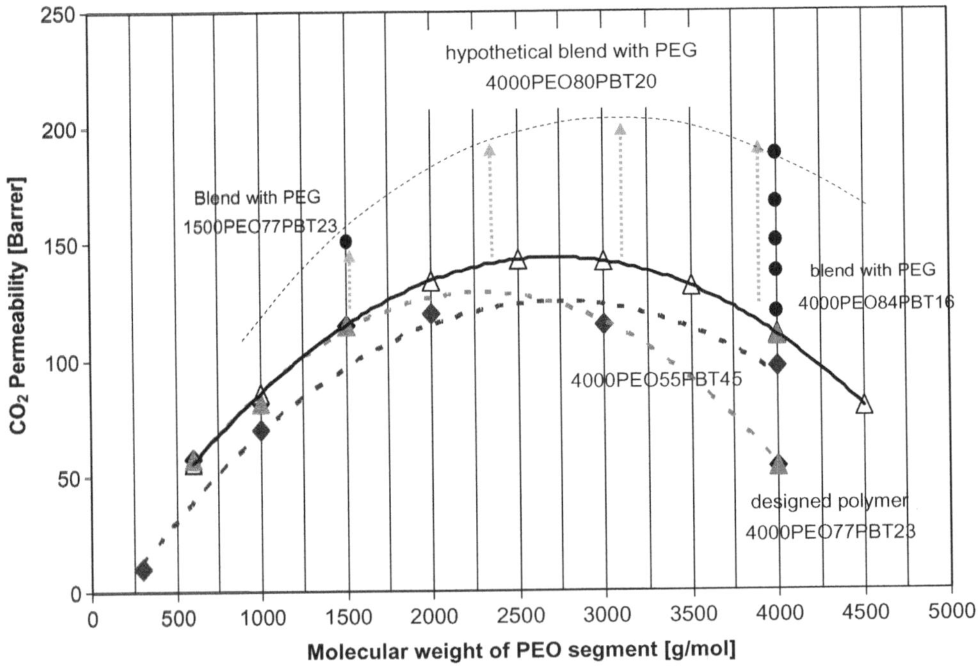

Figure 12.9 CO_2 *permeability of 1500PEO77PBT23 and 4000PEO55PBT45 blends with 50 wt.% of PEG as function of molecular mass of PEO segment (hypothetical blend, dash line) [modified plot from ref. 67]*

Table 12.4 *Permselectivity, solubility and diffusivity selectivity of 4000PEO55PBT45 blend membranes with PEG [Reprinted with permission 67]*

Permselectivity						
PEO [wt.%]	55.0	59.5	64.0	68.5	73.0	77.5
CO_2/H_2	10.8	10.6	11.7	12.3	12.4	13.2
CO_2/N_2	43.5	46.8	44.9	45.4	49.1	47.3
CO_2/CH_4	17.1	15.8	16.4	16.4	16.4	15.9
Solubility selectivity						
CO_2/H_2	67.5	58.8	61.7	64.9	72.9	77.8
CO_2/N_2	82.1	85.0	62.3	66.7	68.2	71.7
CO_2/CH_4	12.6	13.5	13.8	14.5	14.5	16.1
Diffusivity selectivity						
CO_2/H_2	0.16	0.18	0.19	0.19	0.17	0.17
CO_2/N_2	0.53	0.55	0.72	0.68	0.72	0.66
CO_2/CH_4	1.36	1.17	1.19	1.13	1.13	0.99

Source: Reprinted with permission from Advanced Functional Materials, Tailor-made polymeric membranes based on segmented block copolymers for CO_2 separation, by A. Car, C. Stropnik, W. Yave, K.-V. Peinemann, 23, 2815–2823. Copyright (2008) Wiley-VCH.

CO_2 separation by using a membrane prepared from a rubbery polymer is based on high solubility selectivity, while diffusivity selectivity should be maintained or even increased. The CO_2/H_2 permselectivity is increased (due to the increased EO units), which is a consequence of solubility selectivity increase. The strategy of increasing the solubility plays an important role, because of the unfavourable diffusivity selectivity. Incorporation of PEG increases the permselectivity of the pair CO_2/H_2 to 13. However, the CO_2/N_2 and CO_2/CH_4 permselectivities are maintained almost constant due to the similar penetrants size.

Numerous models have been developed to describe transport properties in heterogeneous polymer systems, but the one mostly used is the Maxwell approach [89]. The Maxwell equation has been used by Robeson *et al.* [90] to calculate the gas permeability of block copolymers. Other authors also applied it to mixed matrix membranes, and permeability has been calculated:

$$P = \left(\frac{P_d + 2 \cdot P_c - 2 \cdot \Phi \cdot (P_c - P_d)}{(P_d + 2 \cdot P_c + \Phi \cdot (P_c - P_d))} \right) \tag{12.9}$$

where P is the effective permeability, Φ is volume fraction of dispersed phase, and the subscripts d and c refer to dispersed and continuous phases, respectively.

In homogeneous systems (blends), the permeability can be calculated by simple additive model:

$$\ln P = \Phi_1 \ln P_1 + (1 - \Phi_1) \ln P_2 \tag{12.10}$$

where P, P_1 and P_2 are gas permeabilities in homogeneous blend, pristine polymer and dispersed polymer respectively, and Φ_1 is the volume fraction of polymer matrix. In Figure 12.10 the theoretical values and experimental data for three copolymers and their blends are presented, i.e. Pebax®, 1500PEO77PBT23, and 4000PEO55PBT45 with PEG.

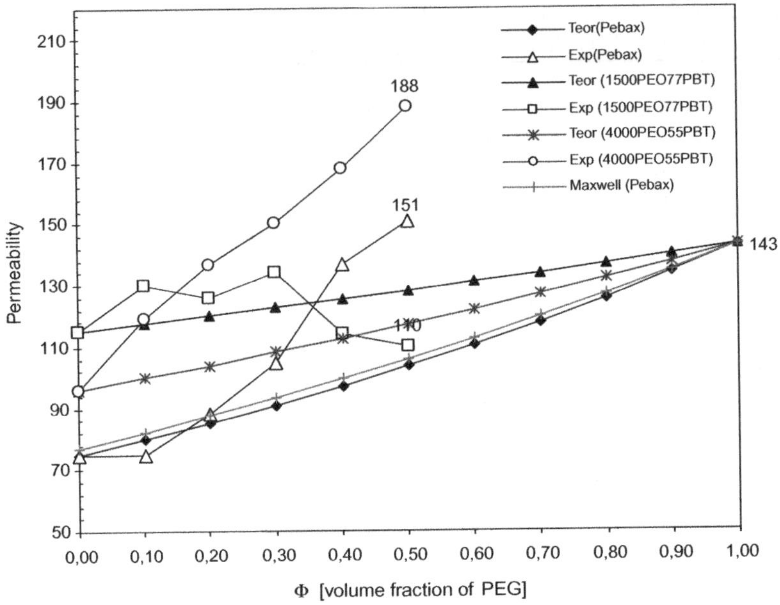

Figure 12.10 *Experimental and estimated values of permeability in Pebax®/PEG, 1500PEO77PBT /PEG and 40000PEO55PBT/PEG blend membranes [modified plot from ref. 24]*

The Maxwell model was calculated only for Pebax/PEG system in order to compare the data with an additive model.

Pebax®/PEG and 40000PEO55PBT/PEG exhibit similar trend (experimental data): the higher the PEG content, the higher the permeability. 1500PEO77PBT/PEG blends, however, are different: permeability increases and then is reduced below theoretical values calculated by the additive model. It can be observed that every copolymer shows different behaviour; hence, the selection of an appropriate additive for a specific copolymer is very important.

DSC curves obtained for the 1500PEO77PBT23 copolymer and its blends with PEG are presented in Figure 12.11. In the first heating (solid lines) and second heating (dashed lines) the melting temperatures of PEO phase are between 10 and 35 °C; for the second heating the melting temperatures are shifted to higher values. The pristine copolymer does not show the melting temperature of PBT, and the blends present peak around 140 °C in the first heating, which disappear in the second heating. These results show that an ordered structure is present in the as-cast membranes and therefore indicate that the microphase separation between PEO and PBT was induced, as revealed in endothermic peaks.

On the other hand, pristine 4000PEO55PBT45 copolymer showed two characteristic T_m values (40 °C for PEO and 213 for PBT, see Table 12.2). It is consistent with the microphase separated structure in block copolymers [91]. For the blend samples T_m for the PEO crystalline phase was shifted to higher values when PEG is added into the polymer matrix (Figure 12.12a). These results are in contradiction to those observed in Pebax/PEG system reported earlier [24]. High PEG content slightly increases the crystallinity: from 19% (pristine polymer) to 24% (blend with 50% of PEG). T_m values of PBT

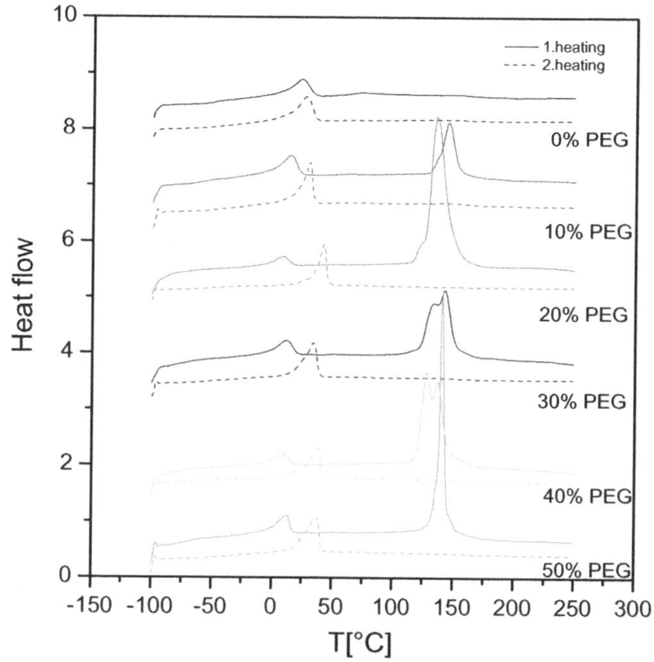

Figure 12.11 *DSC thermogram of copolymer 1500PEO77PBT23 and its blends with PEG200. The termograms have been displaced vertically for easier viewing. Reprinted with permission from Advanced Functional Materials, Tailor-made polymeric membranes based on segmented block copolymers for CO_2 separation, by A. Car, C. Stropnik, W. Yave, K.-V. Peinemann, 23, 2815–2823, Copyright (2008) Wiley-VCH*

in blend samples are not well defined (Figure 12.12b), they are shifted to lower values with PEG content, and blends with high PEG content presented only an amorphous phase. Thus, the decrease of crystallinity in the PBT phase led to greatly improved diffusivity for all gases (Figure 12.8), especially for H_2 and CO_2. Thermal properties and the crystallinity values are in a good agreement with permeability values: the higher the content of the amorphous phase, the higher is the rate of gas transport. The observed peak (~140 °C) in a previous blend based on 1500PEO77PBT23 copolymer was not detected in the present blend (4000PEO55PBT45/PEG). Therefore, the lower permeability obtained in 1500PEO77PBT23/PEG blends has been mainly attributed to the ordered structure found by DSC analysis. Although this ordered phase is not thermally stable, its presence in fresh films could have affected the gas permeability.

Some results of atomic force microscopy (AFM) of pristine copolymers and their blends are presented in Figure 12.13. The blends contain 40 wt.% of PEG and they are representative samples for discussion of morphology. Microphase separation can be observed in Polyactive®/PEG blend (Figure 12.13b) and not in pristine copolymer (Figure 12.13a). As observed in DSC analysis, the blends presented an ordered structure (Figure 12.11, peak around 140 °C), in good agreement with AFM images. The PEG in Polyactive® can induce microphase separation because of strong hydrogen bonding, and the PBT hard phase is surrounded by PEO and PEG phase, thus the CO_2 permeability is decreased.

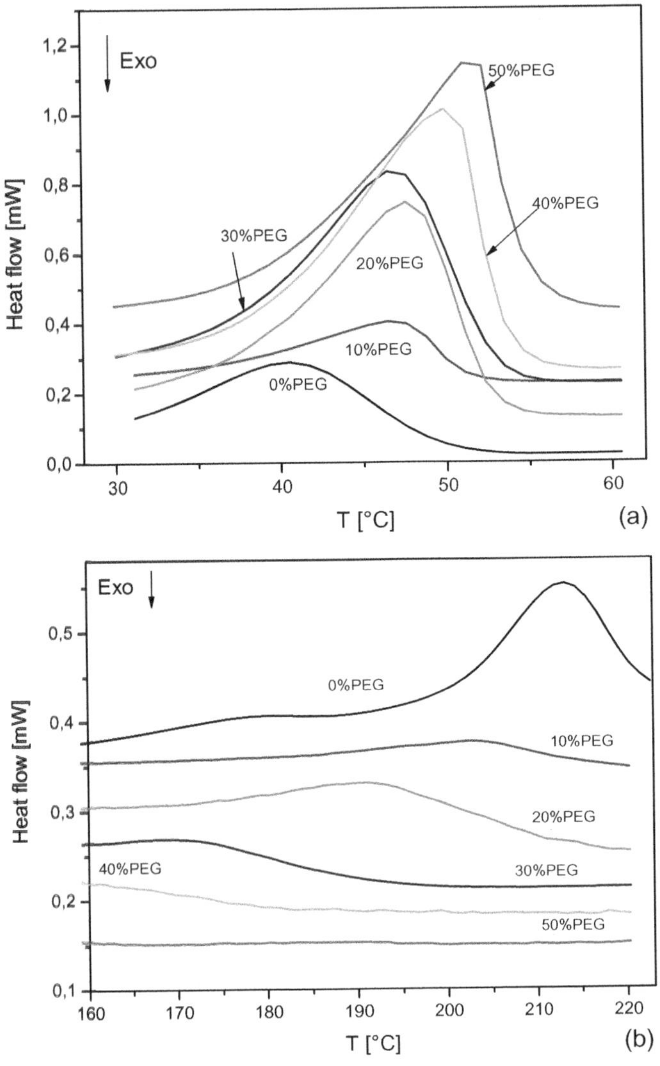

Figure 12.12 *DSC thermograms for (a) PEO and (b) PBT blocks in copolymer 4000PEO55PBT45. Reprinted with permission from Advanced Functional Materials, Tailor-made polymeric membranes based on segmented block copolymers for CO_2 separation, by A. Car, C. Stropnik, W. Yave, K.-V. Peinemann, 23, 2815–2823. Copyright (2008) Wiley-VCH*

The morphology of Pebax® samples (Figures 12.13c and 12.13d) are different in comparison with Polyactive® samples. The pristine Pebax® presents ordered structures as a result of microphase separation, and its blend with PEG does not exhibit any organized structure, hence a higher gas permeability is observed in these blend membranes. In Polyactive/PEG membranes the hydrogen bonding can be stronger than in Pebax®/PEG blend, this can induce an organization of phases and consequently the CO_2 permeability will be reduced.

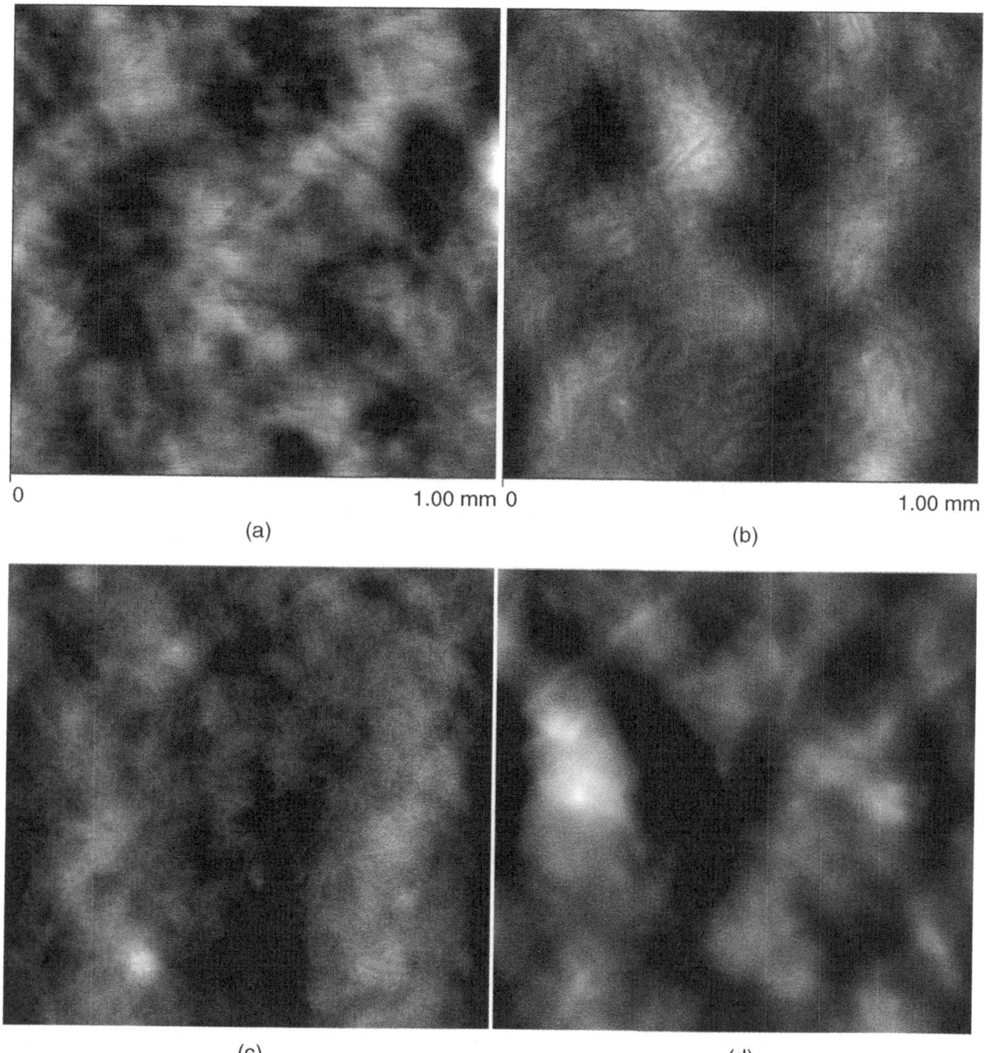

Figure 12.13 *AFM height images (surface) for: (a) pristine Polyactive®, (b) Polyactive®/PEG blend, (c) pristine Pebax® and (d) Pebax®/PEG blend [Reprinted with permission 24]*

12.4 Composite Membranes

The novel developed membrane materials should have potential applications on a large scale. The composite membrane preparation and its evaluation therefore must present promising results before testing the membranes in a real gas mixture. Single gas (CO_2, H_2, N_2 and CH_4) measurements of Pebax®/PEG blend membrane at high pressure (up to 20 bar) and at 293 K presented higher CO_2 fluxes than pristine copolymer (Figure 12.14).

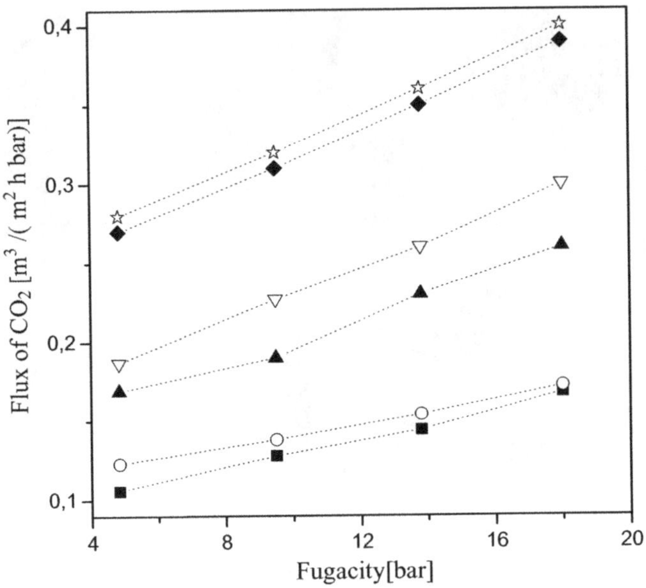

Figure 12.14 *Flux of CO_2 as a function of fugacity in single gas. (-■- 0% PEG, -○- 10% PEG, -▲- 20% PEG, -▽- 30% PEG, -◆- 40% PEG and -☆- 50% PEG [Reprinted with permission 93]*

This figure shows the CO_2 flux as a function of fugacity, and in all samples (Pebax® and blend membranes) the CO_2 flux increased when the fugacity was increased from 5 to 20 bar. This is expected because in rubbery polymers, the solubility coefficient is the determinant and it strongly depends on gas condensability, especially when strong permeant–polymer interactions exist.

After single gas measurements, mixed gas experiments were carried out with 50/50 vol.% in the feed for CO_2/H_2 and CO_2/CH_4, and 25/75 vol.% for CO_2/N_2 at 293 K and a total feed pressure up to 20 bar. The fluxes of CO_2 as a function of the total fugacity are plotted in Figure 12.15. The pressure increase enhances the CO_2 flux for the CO_2/H_2 and CO_2/CH_4 mixture, which is a consequence of membrane swelling. However, the flux of CO_2 in the CO_2/N_2 mixture slightly decreases with a pressure increase. This behaviour has been attributed to the gas composition in the feed, since high concentration of N_2 would decrease the CO_2 flux due to the plasticization effect, thus the N_2 flux would increase and simultaneously the CO_2/N_2 selectivity is reduced. When a membrane is plasticized by one component from a mixture, the permeability or flux of the other one is increased [92]; a stronger plasticization effect could be observed in CO_2/CH_4 mixture (30–50 wt.% of PEG) due to methane presence, since it has stronger sorption behaviour than hydrogen and nitrogen. However, all CO_2 fluxes of the membrane measured with mixed gases show good agreement with those obtained by single gas measurements. Although the CO_2 flux is not significantly increased with the feed pressure, the Pebax®/PEG membrane with 50 wt.% of PEG presented high CO_2 flux in the whole operating pressure range (>0.3 m^3/m^2 h bar for CO_2/N_2 mixture).

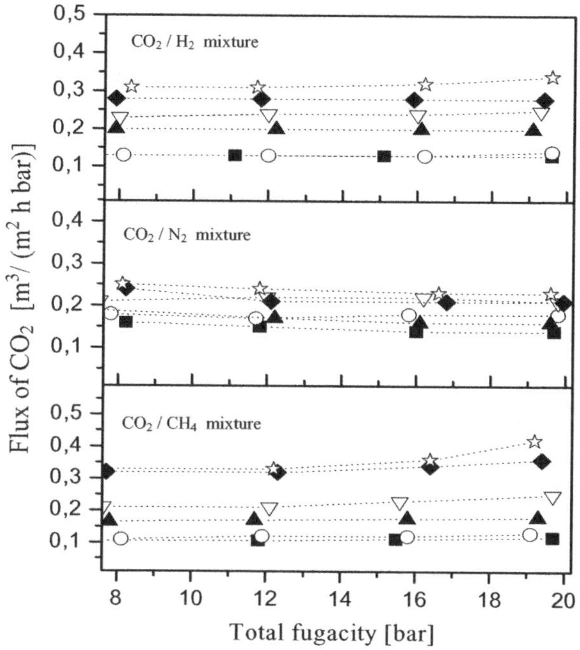

Figure 12.15 *Flux of CO_2 in mixed gas as a function of total fugacity. (-■- 0% PEG, -○- 10% PEG, -▲- 20% PEG, -▽- 30% PEG, -◆- 40% PEG and -☆- 50% PEG [Reprinted with permission 93]*

The selectivity of membranes for mixed gas has been different compared to the selectivity based on single gas. Although the data are not collected in the same operating pressure as in single gas experiments, all results are in the same operating range (see Figures 12.14 and 12.15). The observed CO_2/gas selectivities based on single gas measurement are higher than mixed gas selectivities due to the plasticization effect. Membranes with high PEG content presented better performance under all operating conditions, and the best improvement was observed for CO_2/H_2 mixed gas (selectivity >9.4 and CO_2 flux >0.31 m^3/m^2 h bar). Selectivity is constant for the whole pressure range. This result is in reasonable agreement with that reported earlier for homogeneous films [65]. The CO_2/N_2 selectivity is between 63 and 70 depending on the feed pressure, and CO_2 flux is higher than 0.23 m^3/m^2 h bar. The CO_2/CH_4 selectivity for Pebax®/PEG blend membranes is similar when compared with pristine polymer at 8 bar (around 17), but by increasing the pressure it dropped (~10), which is most evident in blends with higher PEG content (Pebax®/PEG with 50 wt.% of PEG). As expected and according to the sorption effects of CO_2 and CH_4, the high flux (>0.33 m^3/m^2 h bar) of CO_2 in this mixture is accompanied with a flux increase of CH_4, which leads to lower selectivity in mixed gas.

These membranes, evaluated at different temperatures, presented promising results [93]. They also presented good stability and a CO_2 flux higher than 0.5 m^3/m^2 h bar was obtained at 55 °C, which is approximately the flue gas temperature. This membrane was tested and used for simulation and optimization of multi-stage membrane systems used

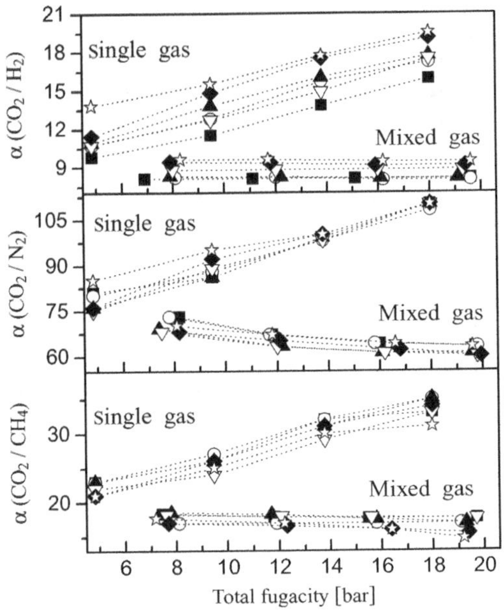

Figure 12.16 CO_2/gas selectivity in single gas and different mixed gas as a function of fugacity. (-■- 0% PEG, -○- 10% PEG, -▲- 20% PEG, -▽- 30% PEG, -◆- 40% PEG and -☆- 50% PEG. Reprinted with permission from Advanced Functional Materials, Tailor-made polymeric membranes based on segmented block copolymers for CO_2 separation, by A. Car, C. Stropnik, W. Yave, K.-V. Peinemann, 23, 2815–2823, Copyright (2008) Wiley-VCH

in a coal-fired power plant [94]. Based on ideal feed gas of 14% CO_2 and 86% N_2, a two-stage cascade membrane system presented the best results. For the developed membrane (Pebax®/PEG) a CO_2-capture cost of 57 Euro/tonne of separated CO_2 was calculated which is clearly higher than absorption process (30–50 Euro/tonne). However, an improvement of flux from 0.5 to 1.4 m^3(STP)/m^2 h bar can reduce the cost to 40 Euro/tonne, which demonstrates that membrane technology can compete with absorption process, although a membrane with higher flux is required.

12.5 Conclusions and Future Aspects

It is accepted that in the last 20 years there were no significant improvements in CO_2-separation membranes; rather, improvement or development of new membrane material was in laboratory scale but not on large scale. It seems that performance of polymeric membranes has reached a maximum level and it is difficult to achieve significant improvements. On the one side, the membrane technology (e.g. modules, engineering) is being daily improved, but on the other side, the lack of new materials with higher performance does not allow further progress. Actually, the cardo-polyimide membrane developed by RITE institute could be the best for CO_2 separation, but it is still in consideration for large-scale applications [95].

From results discussed above, one can see that an improvement of existing membrane material with high performance can be cheap and easy to scale-up.

Tailor-made polymeric membranes such as poly(ethylene(oxide)-poly(butylene terephthalate), a multi-block copolymer discussed in this chapter, is controlled by fraction of PEO phase and its molecular mass, thus allowing its structural manipulation. The addition of a plasticizer and filler which can simultaneously enhance the solubility and diffusivity selectivity of CO_2 over other gases will produce materials with the desired transport properties, thereby the study of block copolymers, its assembly and its nanostructures seem to be the way that will enable to find a superior membrane material. Following the proposed strategy (simultaneous increase of solubility and diffusivity by increasing the free volume in rubbery-like membranes), membranes with better performance were recently reported, hence we strongly recommend to the readers the references [96–98].

References

(1) Gas exchange: CO_2 emissions 1990–2006, *Nature*, **447**, 1038 (2007).

(2) Carbon Dioxide (CO_2) properties, uses, applications, CO_2 gas and liquid carbon dioxide, http://www.uigi.com/carbondioxide.html.

(3) S. R. Rushing, Opportunities for CO_2 coproduction, Ethanol producer magazine, June 2005.

(4) M. R. M. Abu-Zahra, P. H. M. Feron, P. J. Jansens, E. L.V. Goetheer, New process concepts for CO_2 post-combustion capture process integrated with co-production of hydrogen, *Inter. J. Hydro. Energy*, **34**, 3992–4004 (2009).

(5) M. Czyperek, P. Zapp, H. J. M. Bouwmeester, M. Modigell, K.-V. Peinemann, I. Voigt, W. A. Meulenberg, L. Singheiser, D. Stöver, MEM-BRAIN gas separation membranes for zero-emission fossil power plants, *Energy Procedia*, **1**, 303–310 (2009).

(6) L. Shao, B. T. Low, T.-S. Chung, A. R. Greenberg, Polymeric membranes for the hydrogen economy: Contemporary approaches and prospects for the future, *J. Memb. Sci.*, **327**, 18–31 (2009).

(7) R. W. Baker, K. Lokhandwala, Natural gas processing with membranes: An overview, *Ind. Eng. Chem. Res.*, **47**, 2109–2121 (2008).

(8) C. E. Powell, G. G. Qiao, Polymeric CO_2/N_2 gas separation membranes for the capture of carbon dioxide from power plant flue gases, *J. Membr. Sci.*, **279**, 1–49 (2006).

(9) J.-L. Li, B.-H. Chen, Review of CO_2 absorption using chemical solvents in hollow fiber membrane contactors, *Sep. Purif. Technol.*, **41**, 109–122 (2005).

(10) W. A. Bollinger, D. L. MacLean, R. S. Narayan, Separation systems for oil refining and production, *Chem. Eng. Prog.*, **1982**, 27–32.

(11) C. Anderson, A. Siahaan, Membrane CO_2 Removal from natural gas, Grissik gas plant Sumatra, Indonesia; Proceedings of the Laurance Reid gas conditioning conference, (2006) 71–84.

(12) S. P Nunes, K.-V. Peinemann (Eds.), Membrane Technology in the Chemical Industry, Second, Revised and Extended Edition, Wiley-VCH Verlag GmbH, Germany, 2006.

(13) R. W Baker, Membrane Technology and Applications; McGraw-Hill: New York, 2000.

(14) S. Loeb, S. Sourirajan, Sea-water demineralization by means of an osmotic membrane, *Adv. Chem. Ser.*, **38**, 117–132 (1962).

(15) S. Adhikari, S. Fernando, Hydrogen membrane separation techniques, *Ind. Eng. Chem. Res.*, **45**, 875–881 (2006).

(16) S. J. Miller, W. J. Koros, D. Q. Vu, Mixed matrix membrane technology: enhancing gas separations with polymer/molecular sieve composites, *Studies Surf. Sci. Catal.*, **170**, 1590–1596 (2007).

(17) H. Lin, B. D. Freeman, Materials selection guidelines for membranes that remove CO_2 from gas mixtures, *J. Molec. Struct.*, **739**, 57–74 (2005).
(18) Y. Xiao, B. T. Low, S. S. Hosseini, T. S. Chung, The strategies of molecular architecture and modification of polyimide-based membranes for CO_2 removal from natural gas – a review, *Prog. Polym. Sci.*, **34**, 561–580 (2009).
(19) D. Shekhawat, D. R. Luebke, H. W. Pennline, A review of carbon dioxide selective membranes, A Topical Report, DOE/NETL-2003/1200 December 1, 2003.
(20) Y. Hirayama, Y. Kase, N. Tanihara, Y. Sumiyama, Y. Kusuki, K. Haraya, Permeation properties of CO_2 and N_2 of poly(ethylene oxide)-containing and cross-linked polymer films, *J. Membr. Sci.*, **160**, 87–99 (1999).
(21) L. Deng, T.-J. Kim, M.-B. Hägg, Facilitated transport of CO_2 in novel PVAm/PVA blend membrane, *J. Membr. Sci.*, Accepted Manuscript, Available online 23 May 2009.
(22) J. Ahn, W.-J. Chung, I. Pinnau, M. D. Guiver, Polysulfone/silica nanoparticle mixed-matrix membranes for gas separation, *J. Membr. Sci.*, **314**, 123–133 (2008).
(23) S. Kulprathipanja, Separation of polar gases from nonpolar gases, *US Patent* 4606740 (1986).
(24) A. Car, C. Stropnik, W. Yave, K.-V. Peinemann, PEG modified poly(amide-b-ethylene oxide) membranes for CO_2 separation, *J. Membr. Sci.*, **307**, 88–95 (2008).
(25) G. A. Walton, R. M. Shane, G. T. Mildenhall, Styrenic block copolymer breathable elastomeric films, *US Patent* 6794024 (2004).
(26) K. Yagita, O. Yotsuo, Hollow fiber membrane formed from a block copolymer of aromatic polyester and polyoxyethylene for dialysis of blood, *US Patent* 4705632 (1987).
(27) S. M. Hauck, J. P. Viola, Refrigerator liner structures, *US Patent* 5340208 (1969).
(28) J. Kozakiewicz, Polysiloxaneurethanes: new polymers for potential coating applications, *Prog. Organ. Coat.*, **27**, 123–131 (1996).
(29) C. Park, J. Yoon, E. L. Thomas, Enabling nanotechnology with self assembled block copolymer patterns, *Polymer*, **44**, 6725–6760 (2003).
(30) A. Jonquières, R. Clément, P. Lochon, Permeability of block copolymers to vapors and liquids, *Prog. Polym. Sci.*, **27**, 1803–1877 (2002).
(31) L. Mingqi, C. K. Ober, Block copolymer patterns and templates, *Materials Today*, **9**, 30–39 (2006).
(32) S. Krishnamoorthy, C. Hinderling, H. Heinzelmann, Nanoscale patterning with block copolymers, *Materials Today*, **9**, 40–47 (2006).
(33) Van Krevelen, D. W. Properties of Polymers; Elsevier: Amsterdam, 1990.
(34) L. Leibler, Theory of microphase separation in block copolymers, *Macromolecules*, **13**, 1602–1617 (1980).
(35) L. Liu, A. Chakma, X. Feng, Preparation of hollow fiber poly(ether block amide)/polysulfone composite membranes for separation of carbon dioxide from nitrogen, *Chem. Eng. J.*, **105**, 43–51 (2004).
(36) L. Liu, A. Chakma, X. Feng, A novel method of preparing ultrathin poly(ether block amide) membranes, *J. Membr. Sci.*, **235**, 43–52 (2004).
(37) J. H. Kim, S. Y. Ha, Y. M. Lee, Gas permeation of poly(amide-6-b-ethylene oxide) copolymer, *J. Membr. Sci.*, **190**, 179–193 (2001).
(38) D. M. Muñoz, E. M. Maya, J. de Abajo, J. G. de la Campa, A. E. Lozano, Thermal treatment of poly(ethylene oxide)-segmented copolyimide based membranes: An effective way to improve the gas separation properties, *J. Membr. Sci.*, **323**, 53–59 (2008).
(39) E. Tocci, A. Gugliuzza, L. De Lorenzo, M. Macchione, G. De Luca, E. Drioli, Transport properties of a co-poly(amide-12-b-ethylene oxide) membrane: A comparative study between experimental and molecular modelling results, *J. Membr. Sci.*, **323**, 316–327 (2008).
(40) V. I. Bondar, B. D. Freeman, I. Pinnau, Gas transport properties of poly(ether-b-amide) segment block copolymers, *J. Polym. Sci.:Part B: Polym. Phys.*, **38**, 2051–2062 (2000).
(41) V. Barbi, S. S. Funari, R. Gehrke, N. Scharnagl, N. Stribeck, SAXS and gas transport in poly(ether-block-polyamide) copolymer membranes, *Macromolecules*, **36**, 749–758 (2003).
(42) M. Yoshino, K. Ito, H. Kita, K-I. Okamoto, Effect of hard-segment polymers on CO_2/N_2 gas separation properties of poly(ethylene oxide)-segmented copolymers, *J. Polym. Sci.: Part B: Polym. Phys.*, **38**, 1707–1715 (2000).

(43) I. Blume, I. Pinnau, Composite membrane, method of preparation and use, *US Patent* 4963165 (1990).
(44) L. A. Utracki, History of Commercial Polymer Alloys and Blends (From a Perspective of the Patent Literature), *Polym. Eng. Sci.*, **35**, 2–17 (1995).
(45) http://www.isotis.com/.
(46) A. A. Deschamps, D. W. Grijpma, J. Feijen, Poly(ethylene oxide)/poly(butylene terephthalate) segmented block copolymers: the effect of copolymer composition on physical properties and degradation behavior, *Polymer*, **42**, 9335–9345 (2001).
(47) T.-S. Chung, S. S. Chan, R. Wang, Z. Lu, C. He, Characterization of permeability and sorption in matrimid/C60 mixed matrix membranes, *J. Membr. Sci.*, **211**, 91 (2003).
(48) M. Moaddeb, W. J. Koros, Gas transport properties of thin polymeric membranes in the presence of silicon dioxide particles, *J. Membr. Sci.*, **125**, 143 (1997).
(49) M. Jia, K.-V. Peinemann, R.-D. Behling, Molecular sieving effect of the zeolite-filled silicone rubber membranes in gas permeation, *J. Membr. Sci.*, **57**, 289–292 (1991).
(50) J. M. Herrera-Alonso, E. Marand, J. C. Little, S. S. Cox, Transport properties in polyurethane/clay nanocomposites as barrier materials: Effect of processing conditions, *J. Membr. Sci.*, **337**, 208–214 (2009).
(51) J. Qiu, J.-M. Zheng, K.-V. Peinemann, Gas transport properties of poly(trimethylsilylpropyne) and ethylcellulose filled with different molecular weight trimethylsilylsaccharides: impact on fractional free volume and chain mobility, *Macromolecules*, **40**, 3213–3222 (2007).
(52) M. D. Sefcik, J. Schaefer, F. L. May, D. Raucher, S, M. Dub, Diffusivity of gases and main-chain cooperative motions in plasticized polyvinylchloride, *J. Polym. Sci. Polym. Phys. Ed.*, **21**, 1041–1054 (1983).
(53) A. Gonzalez, M. Iriarte, P. J. Iriondo, J. J. Iruin, Miscibility and carbon dioxide transport properties of blends of bacterial poly(3-hydroxybutyrate) and a poly(vinylidene chloride-co-acrylonitrile) copolymer, *Polymer*, **43**, 6205–6211 (2002).
(54) X.-G. Li, I. Kresse, J. Springer, J. Nissen, Y.-L. Yang, Morphology and gas permselectivity of blend membranes of polyvinylpyridine with ethylcellulose, *Polymer*, **42**, 6859–6869 (2001).
(55) K. Nagai, B. D. Freeman, A. Cannon, H. R. Allcock, Gas permeability of poly(bis-trifluoroethoxyphosphazene) and blends with adamantane amino/trifluoroethoxy (50/50) polyphosphazene, *J. Membr. Sci.*, **172**, 167–176 (2000).
(56) D. R. Paul, C. B. Bucknall, Polymer blends, Vol. **1**: Formulation, John Wiley & Sons (1999), New York.
(57) P. J. Flory, Thermodynamics of high polymer solution, *J. Chem. Phys.*, **10**, 51–61 (1942).
(58) M. L. Huggins, Theory of solutions of high polymers, *J. Chem. Soc.*, **64**, 1712–171 (1942).
(59) H. Al-Turaif, Surface coating properties of different shape and size pigment blends, *Prog. Organ. Coat.*, **65**, 322–327 (2009).
(60) B. C. Anderson, S. M. Cox, P. D. Bloom, V. V. Sheares, S. K. Mallapragada, Synthesis and characterization of diblock and gel-forming pentablock copolymers of tertiary amine methacrylates, poly(ethylene glycol), and poly(propylene glycol), *Macromolecules*, **36**, 1670–1676 (2003).
(61) D. R. Paul, Gas transport in homogeneous multicomponent polymers, *J. Membr. Sci.*, **18**, 75–86 (1984).
(62) S. Fakirov, T. Gogeva, Poly(ether/ester)s based on poly(butylene terephthalate) and poly(ethylene glycol), 1. Poly(ether/ester)s with various polyether: polyester ratios, *Makromol. Chem.*, **191**, 603–614 (1990).
(63) M. M. Sonpatki, K. Ravindranath, S. Ponrathnam, Random thermotropic elastomers. I: Effects of substitution and hard/soft segment lengths on properties *J. Polym. Sci.: Part A: Polym. Chem.*, **32**, 2999–3007 (1994).
(64) S. Metz, Water vapor and transport throught polymaric membranes, PhD thesis (2003) Netherlands.
(65) M. E. Rezac, T. John, P. Pfromm, Effect of copolymer composition on the solubility and diffusivity of water and methanol in a series of polyether amides, *J. Appl. Polym. Sci.*, **65**, 1983 (1997).

(66) V. Bondar, B. D. Freeman, I. Pinnau, Gas sorption and characterization of poly(ether-b-amide) segmented block copolymers, *J. Polym. Sci.: Part B: Polym. Phys.*, **37**, 2463–2475 (1999).
(67) A. Car, C. Stropnik, W. Yave, K.-V. Peinemann, Tailor-made polymeric membranes based on segmented block copolymers for CO_2 separation, *Adv. Funct. Mater.*, **23**, 2815–2823 (2008).
(68) K. Okamoto, N. Umeo, S. Okamyo, K. Tanaka, H. Kita, Selective permeation of carbon dioxide over nitrogen through polyethyleneoxide-containing polyimide membranes, *Chem. Lett.*, **2**, 225–228 (1993).
(69) M. Kawakami, H. Iwanaga, Y. Hara, M. Iwamoto, S. Kagawa, Gas permeabilities of cellulose nitrate/poly(ethylene glycol) blend membranes, *J. Appl. Polym. Sci.*, **27**, 2387–2393 (1982).
(70) J. H. Kim, S. Y. Ha, Y. M. Lee, Gas permeation of poly(amide-6-b-ethylene oxide) copolymer, *J. Membr. Sci.*, **190**, 179–193 (2001).
(71) N. P. Patel, R. J. Spontak, Mesoblends of polyether block copolymers with Poly(ethylene glycol), *Macromolecules*, **37**, 1394–1402 (2004).
(72) N. P. Patel, A. C. Miller, R. J. Spontak, Highly CO_2-permeable and selective polymer nanocomposite membranes, *Adv. Mater.*, **15**, 729–733 (2003).
(73) H. Lin, E. Wagner, B. D. Freeman, L. G. Toy, R. P. Gupta, Plasticization-enhanced hydrogen purification using polymeric membranes, *Science*, **311**, 639–642 (2006).
(74) S. J Metz, M. H. Mulder, V. M. Wessling, Gas-permeation properties of poly(ethylene oxide) poly(butylene terephthalate) block copolymers, *Macromolecules*, **37**, 4590–4597 (2004).
(75) W. Yave, A. Car, K.-V. Peinemann, M. Q. Shaikh, K. Rätzke, F. Faupel, Gas permeability and free volume in poly(amide-b-ethylene oxide)/polyethylene glycol blend membranes, *J. Membr. Sci.*, **339**, 177–183 (2009).
(76) W. Yave, S. Shishatskiy, V. Abetz, S. Matson, S. E. Litvinova, V. Khotimskiy, K.-V. Peinemann, A novel poly(4-methyl-2-pentyne)/TiO_2 hybrid nanocomposite membrane for natural gas conditioning: butane/methane separation, *Macromol. Chem. Phys.*, **208**, 2412 (2007).
(77) S. J. Tao, Positronium annihilation in molecular substances, *J. Chem. Phys.*, **56**, 5499 (1972).
(78) M. Eldrup, D. Lightbody, N. J. Sherwood, The temperature dependence of positron lifetimes in solid pivalic acid, *Chem. Phys.*, **63**, 51–58 (1981).
(79) M. Forsyth, P. Meakin, D. R. MacFarlane, A. J. Hill, Free volume and conductivity of plasticized polyether-urethane solid polymer electrolytes, *J. Phys. Condens. Matter*, **7**, 7601–7617 (1995).
(80) V. Shantarovich, I. Kevdina, Yu. Yampolskii, A. Alentiev, Positron annihilation lifetime study of high and low free volume glassy polymers: effects of free volume size on the permeability and permselectivity, *Macromolecules*, **33**, 7453–7466 (2000).
(81) K.-I. Okamoto, K. Tanaka, M. Katsube, H. Kita, O. Sueoka, Y. Ito, Correlation between positron annihilation and gas diffusion properties of various rubbery polymers, *Polym. J.*, **25**, 275–284 (1993).
(82) H. Li, B. D. Freeman, S. Kalakkunnat, D. S. Kalika, Effect of copolymer composition, temperature, and carbon dioxide fugacity on pure- and mixed-gas permeability in poly(ethylene glycol)-based materials: Free volume interpretation, *J. Membr. Sci.*, **291**, 131–139 (2007).
(83) M. E. Rezac, T. John, Correlation of penetrant transport with polymer free volume: Additional evidence from block copolymers, *Polymer*, **39**, 599–603 (1998).
(84) S.-M. Fang, S. A. Stern, H. L. Frisch, A 'free volume' model of permeation of gas and liquid mixtures through polymeric membranes, *Chem. Eng. Sci.*, **30**, 773 (1975).
(85) H. B. Hopfenberg, D. R. Paul, In: D. R. Paul, S. Newman, Eds. Polymer blends, New York: Academic Press (1978).
(86) A. Bos, I. Pünt, H. Strathmann, M. Wessling, Suppression of gas separation membrane plasticization by homogeneous polymer blending, *AIChE J.*, **47**, 1088–1093 (2004).
(87) H. Odani, M. Uchikura, Y. Ogino, M. Kurata, Diffusion and solution of methanol vapor in poly(2-vinylpyridine)-block-polyisoprene and poly(2-vinylpyridine)-block-polystyrene, *J. Membr. Sci.*, **15**, 193–208 (1983).

(88) L. Y. Lai, G. J. Wu, S. Shyu, TPX/Siloxane blend membrane for oxygen enrichment, *J. Appl. Polym. Sci.*, **34**, 559 (1987).
(89) J. C. Maxwell, A treatise on electricity and magnetism, Dover Publication, (1954) New York.
(90) L. M. Robeson, A. Noshay, M. Matzner, C. N. Merriam, Physical property characteristic of polysulfone/poly(dimethylsiloxane) block copolymer, *Angew. Macromol. Chem.*, **29/30**, 47 (1973).
(91) W. Li, X. Kong, E. Zhou, D. Ma, Isothermal crystallization kinetics of poly(ethylene terephthalate-poly(ethylene oxide) segmented copolymer with two crystallizing blocks, *Polymer*, **46**, 11655–11663 (2005).
(92) T. Visser, G. H. Koops, M. Wessling, On the subtle balance between competitive sorption and plasticization effects in asymmetric hollow fiber gas separation membranes, *J. Membr. Sci.*, **252**, 265–277 (2005).
(93) A. Car, C. Stropnik, W. Yave, K.-V. Peinemann, Pebax®/polyethylene glycol blend thin film composite membranes for CO_2 separation: Performance with mixed gases, *Sep. Purif. Tech.*, **62**, 110–117 (2008).
(94) L. Zhao, R. Menzer, E. Riensche, L. Blum, D. Stolten, Concepts and investment cost analyses of multi-stage membrane systems used in post-combustion processes, *Energ. Proced.*, **1**, 269–278 (2009).
(95) S. Kazama, S. Morimoto, S. Tanaka, H. Mano, T. Yashima, K. Yamada, K. Haraya, Cardo polyimide membranes for CO_2 capture from flue gases, in: E.S. Rubin, D.W. Keith, C.F. Gilboy (Eds.), Greenhouse Gas Control Technologies, Vol. I, Elsevier, Amsterdam 75 (2005).
(96) W. Yave, A. Car, K.-V. Peinemann, Nanostructured membrane materials designed for carbon dioxide separation, *J. Memb. Sci.*, **350**, 124–129 (2010).
(97) S. R. Reijerkerk, M. H. Knoef, K. Nijmeijer, M. Wessling, Poly(ethylene glycol) and poly(dimethyl siloxane): combining their advantages into efficient CO_2 gas separation membranes, *J. Memb. Sci.*, **352**, 126–135 (2010).
(98) W. Yave, A. Car, S. S. Funari, S. P. Nunes, K.-V. Peinemann, CO_2-philic polymer membrane with extremely high separation performance, *Macromolecules* **43**, 326–333 (2010).

13

CO_2 Permeation with Pebax®-based Membranes for Global Warming Reduction

Quang Trong Nguyen[a], *Julie Sublet*[b], *Dominique Langevin*[a], *Corinne Chappey*[a], *Stéphane Marais*[a], *Jean-Marc Valleton*[a] *and Fabienne Poncin-Epaillard*[c]

[a]*Laboratoire 'Polymères, Biopolymères, Surfaces', Université de Rouen-CNRS, Mon-Saint-Aignan Cedex, France*

[b]*Institut de recherches sur la catalyse et l'environnement de Lyon, CNRS-Université de Lyon 1, Villeurbanne Cedex, France*

[c]*Laboratoire 'Polymères, Colloïdes, Interfaces', CNRS-Université du Maine, Le Mans Cedex, France*

13.1 Introduction

There is a worldwide accepted viewpoint that global warming is real, and that anthropological carbon dioxide is a major cause of this. Adding more greenhouse gas would lead to a serious increase in the planet's surface temperature. Although nobody knows for sure what will happen, the resulting climate changes appear as the greatest threat to the planet. Therefore, reducing CO_2 emissions to the atmosphere is an important issue worldwide.

Major CO_2 emissions sources are generated by fossil fuel combustion within the power and industrial sectors. Storage of the greenhouse gas in empty gas fields and aquifers could be a solution. To store CO_2 underground it should be captured first. The most currently used post-combustion technology is absorption columns by means of an organic liquid. The flue gases are scrubbed in several reactors in which the CO_2 is bound by

Membrane Gas Separation Edited by Yuri Yampolskii and Benny Freeman

amines. This technology consumes a lot of energy and is therefore not very cost effective. As well as the energy consumption and pollution due to amine loss to the environment, huge installations are needed. The use of membranes to selectively remove CO_2 from gas mixtures would be of great interest due to the known advantages of the membrane processes [1]. However, the extraction of CO_2 from low concentration fumes is difficult. According to a study, it is necessary to have at least a CO_2 concentration of 10–15 vol.% (coal fumes, steel production fumes) to have a competitive membrane process [2].

The major problem for the use of membrane-based CO_2 separation is the lack of commercial membranes with both high permeability and high selectivity. Gas separation membranes based on polymer blends were shown to have improved mechanical properties, better membrane-forming ability and higher gas permeability [3]. By blending, a combination of useful properties of each polymer can be obtained without the tedious work required in the design of a new product. Compared with other modification technologies or the synthesis of entirely new materials, polymer blending has several advantages, such as simplicity, reproducibility and low fabrication cost, thus membrane fabrication is commercially feasible [4].

Most of the research work has been focused on polymer membrane materials involving a solution-diffusion mechanism. The performances of such materials generally fall within the trade-off relationship between permeability and selectivity suggested by Robeson [5], with an 'upper bound' limit for the membrane performances.

Kawakami *et al.* [6] in a study on PEG loaded in porous regenerated cellulose concluded that the acid–base reaction between the acidic CO_2 and the electron-rich ether oxygen atoms of the PEG molecules is probably responsible for the enhanced solubility of CO_2 in PEG.

The upper bound relationship between permeability and selectivity might also be overcome with facilitated transport membranes, because the latter have both a high permeability and a high selectivity via reversible reactions between reactive carriers and the target gas, carbon dioxide.

Recently, a new type of polymer material was reported to have a great potential for carbon dioxide separation from natural gas. It very selectively removed carbon dioxide by permeation through hourglass-shaped pores of molecular size, while impeding that of methane through these same pores [7].

Ward and Neulander [8] investigated the separations of SO_2/CO_2 mixtures by liquid membranes and found that polyethylene glycol (PEG) exhibited much higher solubility coefficients ($0.7\,mol\,L^{-1}\,atm^{-1}$) for SO_2 compared to that of CO_2 ($0.1\,mol\,L^{-1}\,atm^{-1}$) at 25 °C. Although the separation of CO_2/SO_2 was not effective, PEG was found to be an excellent solvent for polar gases.

Lin and Freeman [9] have reported an overview about material selection for membrane preparation for removal of CO_2 from gas mixtures. In that article, CO_2 solubility and CO_2/gas solubility selectivity in solvents and polymers containing different polar groups were discussed. They have concluded that ethylene oxide (EO) units in the polymer appear to be the most useful groups to achieve high CO_2 permeability and high CO_2/light gas selectivity. Although homo-poly(ethylene oxide) (PEO) consists of EO monomer units, the disadvantage of pure PEO membranes lies in their strong tendency to crystallize, leading to a low membrane permeability. They defined three strategies for incorporating high concentrations of PEO with low crystallization tendency into polymers:

(1) using low molecular weight liquid PEO or PEG
(2) designing phase-separated block copolymers with EO-rich but short segments
(3) building highly branched, cross-linked networks with high concentrations of PEG.

Block copolymers containing EO units like poly(amide-*b*-ether) have been shown as a good material for this purpose.

In this chapter, we followed the idea of Lin and Freeman of enhancing CO_2 separation properties. We investigated high selectivity membranes made of Pebax® block copolymers and their blends with free PEG (PEG is the standard abbreviation of polyethylene glycol, a short-chain polyethylene oxide polymer). These membranes were developed using a grant from the French Environment and Energy Management Agency (ADEME) and the Normandy Region from 2004 to 2007. The report on the results obtained with these membranes won an award in the innovative technologies for the environment at the Lyon Pollutec meeting in 2006.

It should be noted that Car *et al.* [10,11] (see also Chapter 12 of this volume) worked independently on similar blend membranes, which were made of Pebax® 1657, a grade of commercial poly(amide-*b*-ether) block copolymer with six polyamide blocks, and free PEG. Those membranes were also shown to exhibit high selectivity and permeability performances, which were attributed to changes in both the chemical composition (i.e. higher EO content) and the morphological structure (i.e. lower material crystallinity). On the contrary, Jaipurkar [12] observed CO_2 permeability and selectivity improvements for the blend of Pebax® 2533 with 25% of PEG 10000, but not for the blends with other PEG molecular weights or composition.

Those results motivated us, in the present work, to study the structure, physical characteristics and gas transport properties of membranes made of different grades of Pebax® block copolymers and their blends with PEG (Table 13.1).

Another type of membrane with high CO_2 separation performances is the facilitated transport membrane. Membranes of this type are those in which an amine-based liquid supported by a solid membrane promotes a selective CO_2 permeation. They exhibit generally a remarkably high selectivity and also high permeability at low CO_2 pressure in the feed gas. Unfortunately, they suffer from very poor stability. For the separation of CO_2, fixed carrier membranes were found to have high permselectivity without the drawback of poor stability, but with much lower permeability [13–15]. However, all membranes with amine-based carriers suffer from a strong reduction in performance with the reduction in the relative humidity in the gas mixtures, because the presence of water in the

Table 13.1 *General features of the different Pebax®*

Pebax® series	11	13	31				33
Pebax® grade	1657	1878	1074	3000	1041	6100	1205
PA/PE block nature	PA6/PEO	PA6/PTMO	PA12/PEO				PA12/PTMO
PA/PE wt.%	50/50	66,7/33,3	50/50		75/25		50/50
Density	1.14	1.09	1.07		1.04		1.01
Melting point (°C)	204	195	158		170		147

whole membrane is required for the CO_2–amine group interactions [16], thus affecting the rate of the CO_2 permeation through the membrane. We think that, if amine groups were grafted on the membrane surface facing the feed mixture, the CO_2 selective sorption into the membrane upstream face would be enhanced even at low humidity, leading to a general improvement of the membrane performance.

The plasma technique [17] either through surface modification or deposition of additional layers is often used to improve the transport properties of polymer films. For example, the CF_4 plasma treatment of the poly(trimethylsilylpropyne) membrane [18] induces an increase of its permselectivity towards oxygen and nitrogen compared to the virgin membrane. A composite polybutadiene/polycarbonate (PB/PC) [19] membrane treated in $CHCl_3$ plasma shows higher oxygen permeation with transport properties depending on the plasma parameters like duration and power. These alterations of transport properties are associated with a chemical modification of the membrane surface, i.e. densification and/or cross-linking.

When the plasma is created in gases such as N_2, NH_3, H_2/N_2, NO, NO_2 or their mixtures, the interactions between nitrogen, ammonia and mixture of hydrogen and nitrogen lead to attachment of different nitrogen-containing functions (amino, amide, nitrile, imide groups) [20,21]. An attraction of such modifications is to prepare surfaces with high levels of hydrophilicity and with attached sites having polar and basic characters that may be involved in interactions through Lewis acid–base interactions [22]. Good examples of increased interaction strength with nitrogen and ammonia from a plasma are those of treated polyaramid fibres and polyaramid reinforced resin-matrix [23]. Poly(p-phenylene terephthalamide) (PPTA) can be functionalized with N_2, NH_3 or methylamine plasma. Effects of nitrogen and nitrogen oxide plasmas on polyolefins and polyimides were also investigated and compared to those of oxygen and oxygen derivatives plasma treatments [23–26]. However, in NO-plasma, the main components are fragmented species from NO molecules and minor components are from O_2, N_2 molecules, oxygen and nitrogen atoms; therefore the surface-grafted groups are dominated by oxidized species with low interaction ability with carbonic acid gas.

In line with these observations, in the second part of the chapter, cold plasma in N_2 and hydrogen was used to graft amine groups onto a Pebax® membrane surface in order to enhance the overall transport through the surface-modified Pebax® membrane. We expect that the amine groups created on the polymer film in a N_2/H_2 cold plasma gas [27] promote enhanced CO_2 sorption at the upstream membrane surface, and therefore enhanced CO_2 selectivity in respect of nitrogen.

13.2 Experimental

13.2.1 Chemicals

PEG with an average molecular mass of $300\,g\,mol^{-1}$ and poly(ethylene oxide-co-epichlorhydrine) (64–69 wt.% of epichlorhydrine) were purchased from Aldrich and used as received without further purification (Figure 13.1). Pebax® samples (extruded films and granules) were given by Arkema. Pebax® are polyamide block polyether copolymers (cf. Figure 13.2) and their molecular weight (M_w) are between 30 000 and $50\,000\,g\,mol^{-1}$.

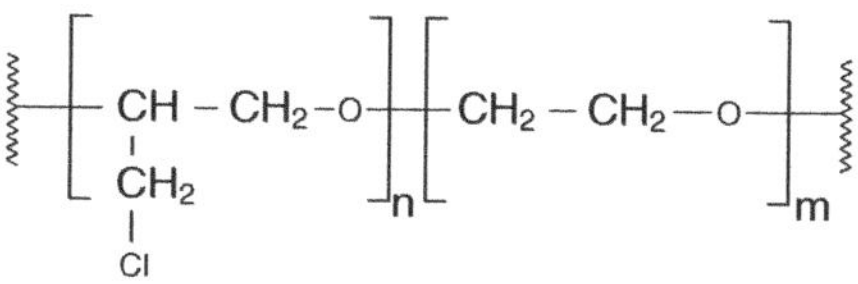

Figure 13.1 *Chemical structure of poly(ethylene oxide-co-epichlorhydrine)*

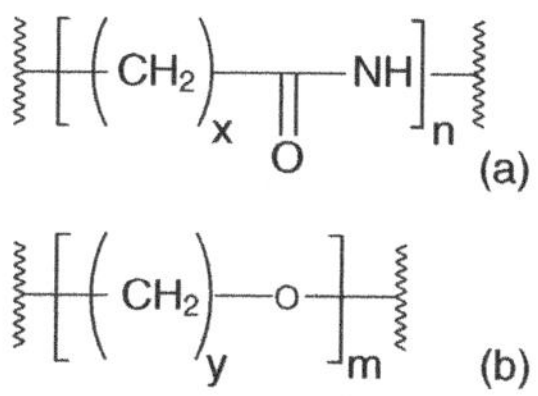

Figure 13.2 *Chemical structures of (a) polyamide and (b) polyether blocks; polyamide block could be PA6 (x = 5) or PA12 (x = 11) and polyether block could be poly(ethylene oxide) (PEO, y = 2) or poly(tetramethylene oxide) (PTMO, y = 4)*

13.2.2 Membranes

Extruded films provided by Arkema were directly used for the permeation studies of Pebax®.

Pebax® and PEG or PEGEPI blend membranes were prepared by solution casting and solvent evaporation techniques. The blend solutions were prepared by thoroughly mixing their individual solutions in different ratios. We only used limpid solutions, i.e. solutions without phase separation. The bubble-free blend solution was cast to the desired thickness on a glass plate pre-heated to 100 °C, and the solvent *N*-methyl pyrrolidone (NMP) was allowed to evaporate slowly at 100 °C for a period of 24 hrs. We consider NMP as the best solvent for Pebax® 1657, the main polymer used in the present study, since its Hildebrand solubility parameter value (22.9 $MPa^{1/2}$) coincides with that of the polymer (22 $MPa^{1/2}$). Due to the semi-crystalline nature of Pebax®, heating was still needed for thorough polymer dissolution. The membranes were designated by their relative weight percentage in the blend, with the numbers given in the same order as that of their abbreviated name.

The film thickness was determined by means of a Mitutoyo® micrometer. To obtain reliable values, the thickness was determined in nine places on each sample.

13.2.3 Gas Permeability Measurements

In order to assess the reproducibility of the permeation data obtained for each gas, five measurements were carried out per Pebax® sample. CO_2 and N_2 permeation properties of films were determined using the permeation apparatus previously described [28]. Before measurement, the air present in the permeation cell was completely evacuated by applying a vacuum on both sides of the film for one night. The pressure in the permeation cell had a constant value (<5 × 10^{-3} mbar) before starting the permeation experiment. Then the upstream side of the permeation cell was exposed to a fixed pressure of the gas under test (0.5 bar < p_1 <3 bars in different experiments).

The increase in the pressure p_2 was measured using a sensitive pressure gauge (0–10 mbar, Effa AW-10-T4) linked to a data acquisition system. The quantity $Q(t)$ of gas present in the receiving compartment was calculated using the perfect gas law:

$$Q = \frac{p_2 V_d}{RT}$$

where V_d is the calibrated downstream volume (98.13 cm^3), R is the gas constant and T is the measurement temperature.

The permeability coefficient P (cm^3(STP) cm cm^{-2} s^{-1} $cmHg^{-1}$) and the ideal selectivity α were determined by using the following equations (assuming $p_1 >> p_2$) [29]:

$$P = \frac{Q\,L}{A\,t\,p_1}$$

$$\alpha = \frac{P_{CO_2}}{P_{N_2}}$$

where Q is the quantity of STP gas permeated in a time interval t at the steady state of gas flow, A is the effective film area for gas permeation (11.64 cm^2), L is the sample thickness and P_{CO_2} and P_{N_2} are the permeability coefficients of pure CO_2 and N_2, respectively. The accuracy on P value is ca. 6%.

The sorption coefficient S was indirectly obtained, by dividing the permeability coefficient P by the diffusion coefficient D. The well-known relationship $P = DS$ with constant diffusivity D was assumed to be valid (which is generally the case for gas permeation in rubbery polymers at relatively low pressure). The diffusivity D is considered to be affected by the free volume, i.e. the unoccupied space in the polymer matrix, and the solubility S by physical and chemical interactions between gas and polymer.

13.2.4 Gas Sorption Measurements

The gas uptakes were measured with an electronic microbalance, IGA model 002, supplied by Hiden Analytical, Warrington (England). After transferring a flat piece of membrane to the sample pan, the system was evacuated for several days by turbo-molecular pump to evacuate all traces of volatile substances, until a stable weight was reached. The sample environment temperature was controlled by a sensor and a thermostatic water bath. A gas at a controlled pressure was fed to the sample chamber, and the sample weight was monitored as a function of time, until a stable weight was obtained. Afterward, several successive pressure levels can be applied and the successive equilibrium mass gains allow one to draw the gas sorption isotherm:

mass gain = f (pressure applied)

13.2.5 Thermal Behaviour Studies by DSC (Differential Scanning Calorimetry)

The thermal behaviour of samples was analyzed by using the conventional differential scanning calorimeter (DSC Q100 TA Instruments). Samples with a mass of about 10 mg were used. The heating step was done from −90 °C to 220 °C at a rate of 10 °C/min followed by an isothermal heating for 10 min, a first cooling from 220 °C to −90 °C, and a second heating in the same range. All DSC runs were carried out under nitrogen atmosphere to minimize the oxidative degradation. Before all DSC experiments, the baseline was calibrated using empty aluminium pans, and the DSC apparatus was calibrated using the melting temperature and enthalpy of a high-purity indium standard (156.6 °C and 28.45 J g^{-1}).

The degree of crystallinity (X_c) was determined with the following equation:

$$X_c(\%) = 100\left(\Delta H_m / \Delta H_m^0\right)$$

where ΔH_m is the melting enthalpy per gram of polyamide 12 phase in the membrane sample, ΔH_m^0 is the extrapolated value of the enthalpy corresponding to the melting of 100% crystalline ($\Delta H_m^0 = 246$ J g^{-1} for PA12 and 230 J g^{-1} for PA6) [30].

13.2.6 AFM Surface Imaging

AFM imaging in the contact mode and in the friction mode was done with a 'Nanoscope III A' from Digital Instruments (Santa Barbara, USA) using a 100 µm piezoelectric scanner. The cantilevers used were characterized by a spring constant of 0.16 N m^{-1}. A standard pyramidal tip of silicon nitride was used. The measurements were carried out in air and at a constant force in the 10^{-9} to 10^{-8} N range.

13.2.7 Surface Modification of Pebax® Membrane by Cold Plasma

The surface modification of the polymer film was carried out in a 13 L glass reactor equipped with a microwave generator working at 433 MHz. Nitrogen and hydrogen were fed at controlled flow rates on top of the reactor after a vacuum purge under 10^{-9} bar. The plasma gases were generated in the discharge cavity with the energy transferred from the generator via a surfatron waveguide. The polymer film reacted with the plasma gases that flowed on its surface under a pressure of 10^{-4} bar towards the reactor bottom. The exact conditions of the surface modification will be given in the results section.

13.3 Results and Discussions

13.3.1 Sorption and Permeation of CO_2 and N_2 in Extruded Pebax® Films

The CO_2 extraction from flue gases would be economically feasible if the membranes can be produced at a low price. Pebax® polymers can give an opportunity to produce low cost membranes, since they are industrially produced and are easily processable by extrusion into thin films of good quality according to the website of Arkema (www.pebax.com). Such films could be laminated onto a microporous support for a production of membranes in extrusion-lamination processes similar to those used for packaging films. Such an opportunity motivated us to study extruded Pebax® films in sorption and permeation.

The permeation, diffusion and sorption coefficients obtained from the 'time-lag' experiments are summarized in Table 13.2 for seven extruded-Pebax® grades. The values of these coefficients vary in a large range according to the chemical composition of the Pebax® films. The most significant features of the data are (1) the high sorption selectivity in favour of CO_2 gas (CO_2 sorption is up to 50 times higher than that of nitrogen); (2) the high permeability of polyether-rich Pebax ® grades, and (3) the low diffusion selectivity, which can be in favour of CO_2 or of N_2. These features are characteristic of a rubbery material. The sorption experiments carried out on the microbalance showed a Henry-type isotherm and confirmed the high CO_2 sorption calculated from the 'time-lag' data for all grades (Figure 13.3). The 1657, 6100 and 1205 grades showed the best CO_2 permeability (ca. 100 Barrer), and the 1657 grade the best ideal selectivity ($\alpha = 48$).

As Pebax® is a very complex material, more detailed studies on the physical structure and characteristics are needed to understand better the transport properties of Pebax® of different grades. The transport properties through Pebax® film would depend on the volume fraction of the PA and polyether blocks, the PA crystallinity, the density and availability for interactions of polar groups, the block nature, length and their organization in the films. Indeed, the two-step gas permeation mechanism requires first a sorption of gas molecules and then gas diffusion in the free volume of the entangled polymer chains. It should be noted that the close molecular sizes of the studied gases (kinetic diameter values of 0.33 and 0.36 nm for CO_2 and nitrogen, respectively) makes the diffusion selectivity close to unity for all membranes. The best films for the acid gas abatement would be those with high sorption coefficients for CO_2.

As the CO_2 molecule (polarizability value of $26.5 \times 10^{-25}\,cm^3$, compared with $17.6 \times 10^{-25}\,cm^3$ for nitrogen) is highly polarizable, it can interact with the polar groups in the membrane. It seems normal that Pebax® 1657 is the best membrane, due to its highest content in polar ether (EO) and amide groups. However, the situation was not quite clear for the other membranes, and a more detailed study on the fine membrane structure was required.

The studies of the films by DSC provide us with information about the phases and their physical properties. The DSC thermograms for the different Pebax® grades are shown in Figure 13.4. From the values of glass transition temperatures and melting points of the homopolymers, we were able to assign the measured values of the thermal characteristics to different phases. The glass transition temperature values reported in the literature are –55 °C for homopolymer PEO with a molecular mass of approximately 300 g mol^{-1}, 51 and 36 °C for high molecular weight PA6 and PA12 [31] (no value for PTMO is available since it has variable T_g values depending on its sequence length). The melting points for homopolymer standards ($M_n \approx 1000\,g\,mol^{-1}$) are 35, 15, 225 and 178 °C for PEO, PTMO, PA6 and PA12, respectively [31].

The detected temperature values for the first-order (endothermic peak) and the second-order transitions (heat capacity jump) (Table 13.3) were close to the values of the glass transition temperature and melting points of the pure homopolymers. The slightly smaller values of the PA and polyether melting points probably reflect the low molecular weight of the blocks and the less ordered crystallite structure. They can be therefore attributed

Table 13.2 *CO_2 and N_2 permeability (P), diffusion (D) and sorption (S) coefficients obtained for several Pebax® grades by using the gas permeation method in time-lag mode (under an upstream pressure of 4 bars and at 25 °C)*

Pebax® series	11	13	31				33
Pebax® grade	1657	1878	1074	3000**	1041	6100**	1205
Pa/PE block nature	PA6/PEO	PA6/PTMO	PA12/PEO				PA12/PTMO
PA/PE wt.%	50/50	66,7/33,3	50/50		75/25		50/50
PA/PE block M_w*	1.5PAr/1.5PEr	–	1.5PAr/1.5PEr		4.5PAr/1.5PEr		1PAr/1PEr
P_{CO2} (a)	97.9 ± 0.4	9.6 ± 0.1	25.5 ± 0.2	45.8 ± 0.3	23.3 ± 0.1	100 ± 1	93.0 ± 0.5
P_{N2} (a)	2.04 ± 0.03	2	0.583 ± 0.007	2.86 ± 0.09	0.7 ± 0.1	2.44 ± 0.07	3.43 ± 0.06
$\alpha_{CO2/N2}$	48.0	4.8	43.7	16.0	33.3	41	24.5
D_{CO2} (b)	5.0 ± 0.1	0.38 ± 0.01	1.74 ± 0.04	4.4 ± 0.4	1.7 ± 0.1	6.1 ± 0.1	6.32 ± 0.04
D_{N2} (b)	4 ± 3	0.24	0.4 ± 0.1	5 ± 1	0.11 ± 0.04	9 ± 6	13 ± 4
$\alpha_{DCO2/DN2}$	1.3	1.6	4.4	0.9	15.5	0.7	0.5
S_{CO2} (c)	197 ± 6	238 ± 7	147 ± 3	106 ± 11	139 ± 9	164 ± 3	147 ± 1
S_{N2} (c)	8 ± 6	81	16 ± 4	6 ± 2	72 ± 23	3 ± 2	2.9 ± 0.8
$\alpha_{SCO2/SN2}$	25	2.9	9.0	18	1.9	55	51

(a) Barrer or 10^{-10} cm^3(STP) cm cm^{-2} s^{-1} cmHg^{-1}.
(b) 10^{-7} cm^2 s^{-1}.
(c) 10^{-4} cm^3 (STP) cm^{-3} cmHg^{-1}.
* Pebax® are multiblock copolymers whose PA (polyamide) and PE (polyether) blocks alternate in the linear chain of the copolymers. The weight ratio of the PA to the PE phase of different grades was obtained by alternating blocks of pre-fixed molecular weights. The relative molecular weight of the PA (PE) block of a grade is expressed here as a multiple of a reference molecular weight of the polyamide, PAr (or polyether PEr) block.
** Pebax® 3000 and 6100 are obtained by incorporating a polymer additive to the 1074 and 1041 grades, respectively (undisclosed data).

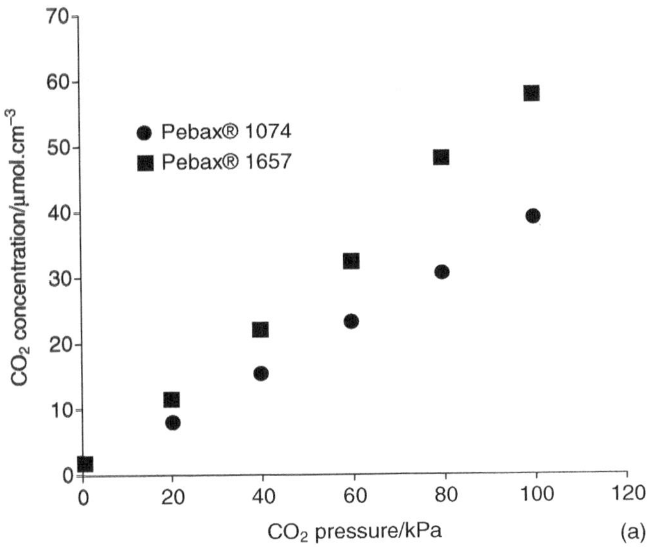

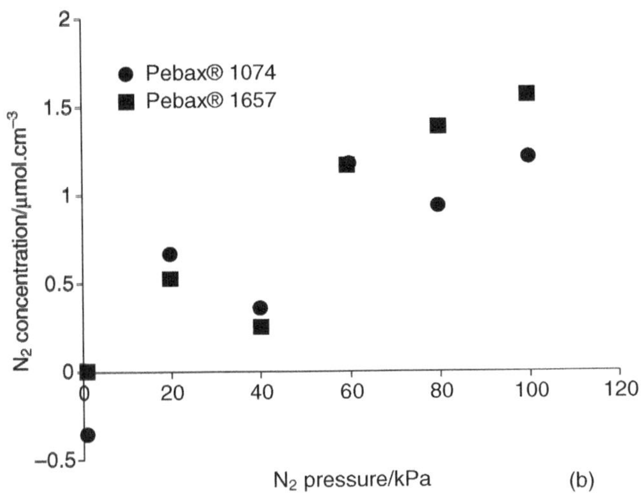

Figure 13.3 *Sorption isotherms for (a) CO_2 and (b) N_2 on Pebax® 1657 and 1074 at 25 °C*

to the quasi-pure polyether or polyamide phases. It should be noted that glass transition of the polyamide phase was not clearly detected in the thermograms, probably because of the weak change in heat capacity associated to the polyamide phases. The clear detection of the separated melting of the polyether and polyamide crystalline phases, on the one hand, and the absence of intermediate glass transition temperatures in-between those of the two polymers, on the other hand, indicate that the Pebax® films consist of two phases, polyamide and polyether phases, without any polyether-PA blend phase. Indeed, it is well known that, on the one hand, the crystalline phase of a polymer consists of only

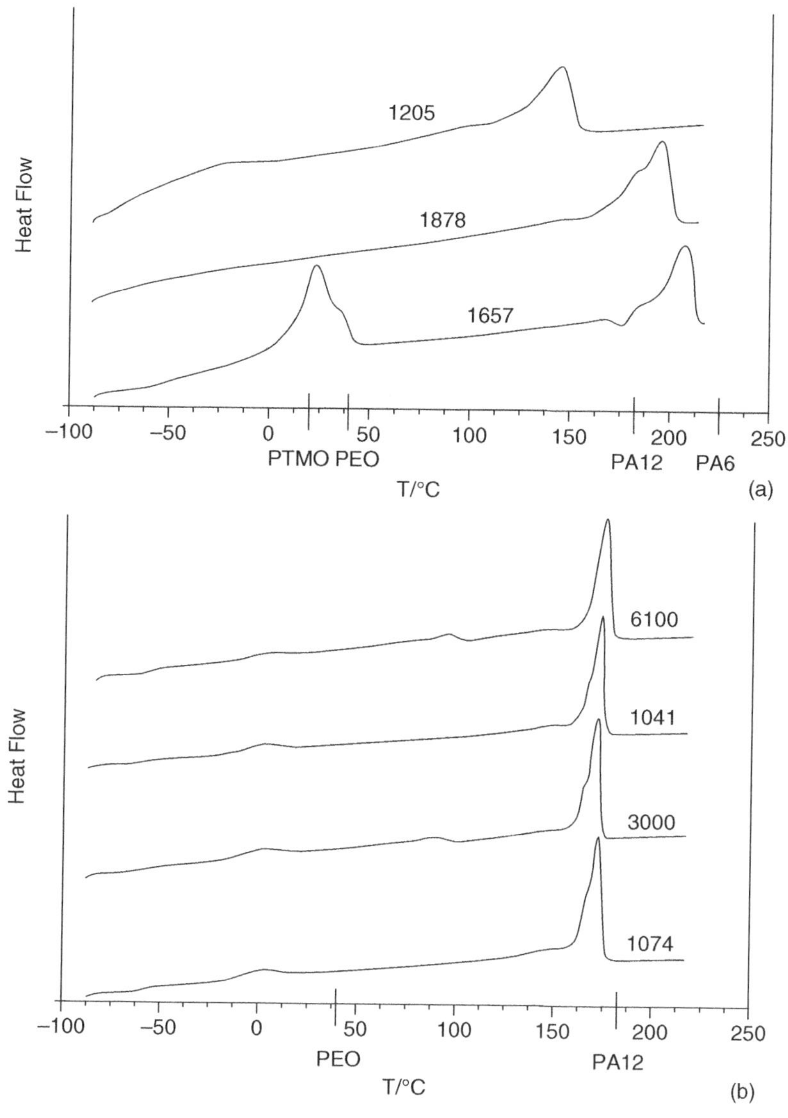

Figure 13.4 *DSC thermograms (second heating scans in temperature ramps from –90 to 220 °C) for (a) Pebax® 1657, 1878, 1205 and (b) Pebax® 1074, 3000, 1041, 6100 (curves from the bottom to the top in the graphs). The melting temperatures are approximately marked on the abscissa axis for comparison*

one type of segments, since a stereo-regularity of their repetitive unit is required for the segment folding into lamellar crystal structure, and on the other hand, the glass transition temperature of an amorphous phase of two miscible polymer segments would be intermediate between that of the two polymers.

The polymer crystallinity is an important parameter in permeation: the crystallites are not permeable to any species, and their distribution in the polymer materials defines the

Table 13.3 Values of the glass transition temperatures of PE blocks and the melting temperatures of PE and PA blocks for the different Pebax® grades, determined from DSC measurements (in the first heating scans). The PA crystallinity is obtained with the values found in the literature for the melting enthalpy of PA6 and PA12

	T_m polyamide block (°C)	T_m polyether block (°C)	T_g polyether block (°C)	wt.% PA crystallinity in Pebax®	wt.% PA crystallinity in PA block
1657	208	14	–55	20	40
1878	198	–	–	18	27
1074	173	6	–60	19	38
3000	171	5	–60	18	36
1041	172	5	–60	22	29
6100	160	3	–60	9	12
1205	147	–	–40	11	22

diffusion paths of the permeating molecules in the amorphous/molten polymer phases. The crystallinity of the PA phase can be calculated by assuming its melting enthalpy is identical to that of PA homopolymers. It falls within the 0.22–0.40 range for the different Pebax® grades (except the 6100 grade whose crystallinity is very low due to the presence of an undisclosed molecular additive). Rather low crystallinity of the PA phase in Pebax® in comparison with that of the homopolyamide (larger than 0.4) is also consistent with the low molecular weight of the blocks and the less ordered crystallite structure. The polyether phase, whose melting point is lower than room temperature, was in the molten state at room temperatures, i.e. in our permeation experiments.

AFM patterns revealed the complex nature of the Pebax® film surface. The phases whose stiffness was probed by the cantilever tip in friction or deflection mode appeared in better contrast than in contact mode. Figure 13.5 shows the soft (molten phase) polyether phase dispersed in a stiff polyamide matrix (crystallites intermingled with a glassy amorphous phase) for Pebax® 1657 membrane. The Pebax® 1657 grade appears also in optical microscopy images as a bi-continuous material with two phases (Figure 13.6). The polyamide phase consists of thread-like bundles (bright relief phase with high aspect ratio and no curvature, Figure 13.5) of folded lamellar PA blocks. In-between were the fragments of the amorphous phase of random morphology. There were phases with a difference in stiffness in the amorphous phase (which appear in different colours inside the plane phase). These two phases of the amorphous part must be the amorphous polyamide, and the molten polyether phase.

As the size and distribution of the three phases change with the nature and the volume ratio of the phases, the gas permeability could be quite different for materials of close chemical nature and crystallinity. Unfortunately, there is apparently no correlation between the gas permeability and the PA phase crystallinity. The overall PA crystalline phase fraction in the Pebax® films did not vary very much with the Pebax® grades (except for the 6100 and 1205 grades). The apparent absence of correlations between the gas permeability and the PA phase crystallinity could mean that the polyamide phases (intermingled amorphous and crystalline phases) did not play a crucial role in the permeation.

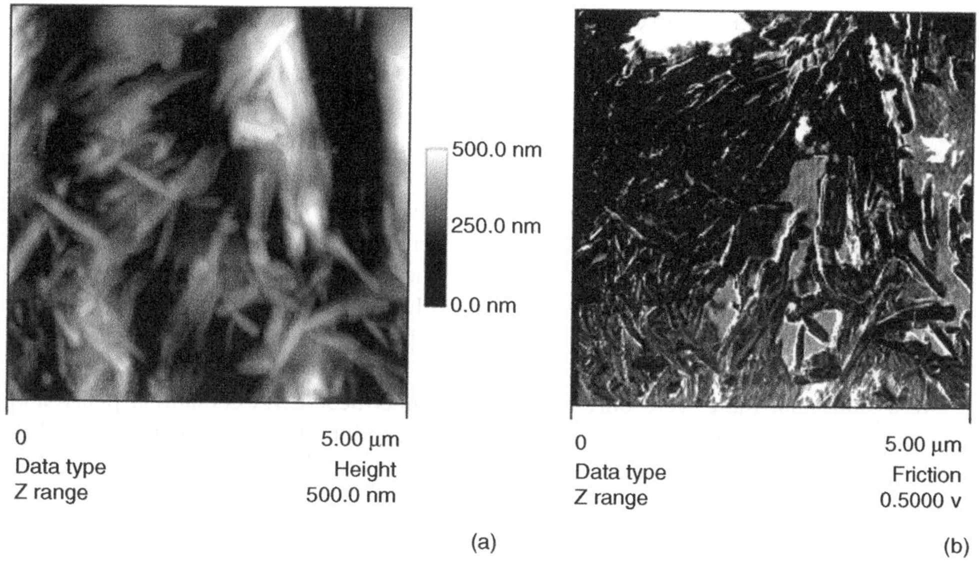

Figure 13.5 *Surface morphology of Pebax® 1657 membrane obtained by AFM in contact (a) and in friction (b) modes. The contact mode images give the surface topology*

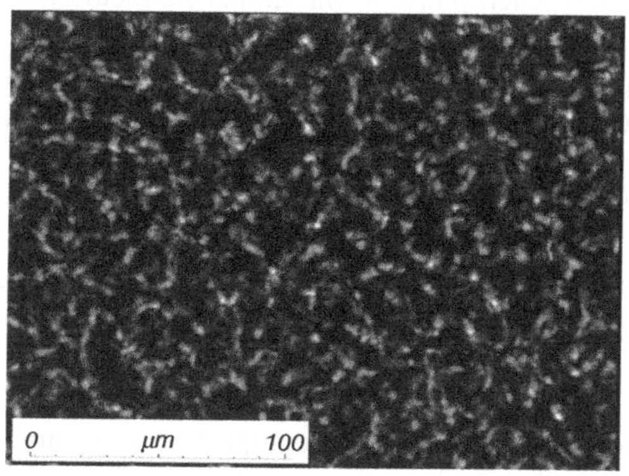

Figure 13.6 *Optical microscopy image of Pebax® 1657 membrane*

However, in addition to the three phases indicated above, there are interphase zones which may contribute significantly to the gas permeation. The interphases that connect the crystalline PA phase to the amorphous PA phase and to the molten polyether phase could behave as a quasi-crystalline phase, or as a liquid-like phase. In polyethylenes, the constraints due to the crystallites reduce the molecular diffusion, and make it more selective [32]. In a recent study, we pointed out the significant contribution of these constrained interphases to the permeation in nanocomposite materials with a semi-crystalline polyamide 12 matrix [33]. Although the existence of such an interphase can be hardly proven, the result analysis based on the idea of coexistence of two amorphous fractions ('real' and 'semi-ordered') may give a broader understanding of the relationship between polymer history and properties [34].

The permeation of gases in such a complex structure is very difficult to model due to the lack of information on the phase structures and properties, as well as the complexity of such modelling. Qualitatively, the reduced mobility and the chain orientation in semi-ordered interphases due to the stiff and ordered crystallites would make the permeability smaller. For the Pebax® grades with shorter polyether blocks and longer polyamide blocks, the tortuosity of the diffusion path will increase sharply when the polyether and amorphous PA phases become finely divided by the crystalline phase. Nevertheless, we tried to use the PA phase crystallinity to simulate the CO_2 and nitrogen permeabilities in Pebax® films with the simple 'resistance model' [35] to estimate the influence of the Pebax® structure on the permeability.

We assume that there were only two phases, impermeable PA crystalline phase and permeable molten polyether phase, which are both 'continuous' (percolated phases). The model in this case consists of two continuous phases in parallel paths of permeation. The overall permeability for such a case in the resistance model is given by:

$$P_{\mathrm{Pebax}} = \Phi_{\mathrm{PE}} P_{\mathrm{PE}} + \Phi_{\mathrm{PA}} P_{\mathrm{PA}}$$

where P_{Pebax} is the Pebax® permeability, Φ_i are the volume fractions and P_i are the permeability of the polyether (PE) and polyamide (PA) phases. This is a simple way to estimate the contribution of each phase to the global permeability in a block copolymer [36].

To simplify, we supposed that the CO_2 permeability in PA phase was at a constant value of ca. 0.5 Barrer (compilation of various literature data [37,38]). The CO_2 permeability in PE phase was estimated by extrapolating the CO_2 permeability obtained for different blends of PEG 300–Pebax® 1657 to 100% of PEG; we found for it a value of 275 Barrer, which is consistent with the literature [39].

The obtained simulation results are given in Table 13.4. The values of the permeability calculated in taking into account the polyether and PA volume fractions indicate that the 1657 and 1205 Pebax® grades had a permeability of the polyether phase 30% lower than that of pure PEO. Such reduction in permeability of the polyether phase can be attributed to its confinement in the polyamide phase. The higher permeability of the polyether phase in the 6100 grade compared with that of pure PEO is probably due to its additive. The lowest permeability of the polyether phase in the 1878 grade is easily interpreted by its highest content in polyamide, and its short polyether block. Its polyether phase is likely confined as very small fragments in the stiff semi-crystalline PA matrix, leading to a high

Table 13.4 *CO_2 permeability coefficient of different Pebax® grades and values of the CO_2 permeability coefficient for the polyether phase calculated with the resistance model*

	1657	1878	1074	3000	1041	6100	1205
P_{CO2} (Pebax®) (a)	97.9	9.6	25.5	45.8	23.3	100	93
P_{CO2} (polyether block) (a)	195	28	51	91	92	399	186

(a) 10^{-10} cm^3(STP) cm cm^{-2} s^{-1} cmHg^{-1}.

tortuosity of the diffusion path and a lowered mobility in 'semi-ordered' interphase zones. However, if the polyether block is longer, its fragments, and thus the permeability, would be larger: this was probably the case of the Pebax® 1041 grade, that was three times more permeable to CO_2 than the 1878 grade, in spite of the lower PA content of the latter. The Pebax® 1041 grade also had better diffusivity selectivity than the 1074 grade, owing to its higher PA content. It is difficult to discuss further the structure–property relationship. For more advanced simulations, a networked-resistance model or a molecular dynamic model could be used, but they required details on the morphological structure, as well as their permeation or molecular properties, which are not easy to determine.

The influence of temperature on the transport properties of CO_2 in Pebax® 1657 and 6100 extruded films was studied in the 20–40 °C range. Their negative sorption enthalpies (−25 and −5 kJ mol^{-1}, for 1657 and 6100, respectively) can be expected from fundamentals of sorption thermodynamics in polymers [40]. Negative sorption enthalpy reduces the membrane selectivity for gas treatments at higher temperature.

In summary, the gas transport in Pebax® membranes depends not only on the fractions and the nature of the constituting chemical blocks (i.e. the polar group content) but also on the length of the blocks. All these parameters are likely to control the phase structure and properties, such as the compactness (i.e. the free volume) of the permeating phase and the detailed organization of the polymer segments in this phase that governs the gas mobility. The polyamide crystalline phase affects these structural properties, and thus the gas permeability, via its interfacial actions on the polymer segment in the vicinity of the phase interfaces and via its distribution in the membrane. The Pebax® 1657 grade offered the best compromise in different structural factors for CO_2 extraction, with a CO_2 permeability of 100 Barrer and an ideal selectivity for carbon dioxide relative to nitrogen $\alpha_{CO2/N2}$ of 50.

13.3.2 Pebax® 1657-based Blend Membrane: Structure and Properties

As an acid–base reaction between the acidic CO_2 and the electron-rich ether oxygen atoms of the PEG molecules can impart high CO_2 permeability and selectivity [6,9], we looked at the use of membranes made by blending the best Pebax® copolymer (1657 grade) with polymers containing ethylene oxide units to improve further the membrane performance. To reduce the overall crystallinity of the membranes by an intimate mixing of the second polymer in the blends, PEG of low molecular weight and poly(ethylene oxide-co-epichlorhydrine) (PEGEPI) were chosen as the additive polymers for the blend membranes.

PEGEPI is a random copolymer which showed attractive performances in CO_2 removal [41]. As all the blend membranes were prepared by solvent casting from NMP solutions, we measured first the structure, morphology and gas permeation characteristics of the pure Pebax® 1657 membrane prepared by solvent casting in the same conditions as those for the blend membranes; they were found to be very similar to those of the extruded film. Melting points of PA and PE blocks and T_g of PE block were the same; crystallinity of the PA block in the solvent casting Pebax® 1657 membrane was slightly lower than that of 1657 extruded film. The permeation results were the same for both.

PEGEPI appeared to be miscible with the PEG phase of Pebax® 1657: in DSC, the PEGEPI- Pebax® 1657 blends showed a unique glass transition temperature (around

Table 13.5 *CO_2 and N_2 permeability (P), diffusion (D) and sorption (S) coefficients obtained for three Pebax® 1657–PEGEPI blends by using the gas permeation method in time-lag mode under an upstream pressure of 4 bars and at 25 °C*

	1657/10% PEGEPI	1657/20% PEGEPI	1657/50% PEGEPI
P_{CO2} (a)	104 ± 1	90 ± 10	30 ± 1
P_{N2} (a)	1.78 ± 0.02	1.8 ± 0.5	0.5
$\alpha_{CO2/N2}$	58.5	50	60
D_{CO2} (b)	4.8 ± 0.3	5.0 ± 0.6	4.4 ± 0.6
D_{N2} (b)	14	0.16 ± 0.05	0.08
$\alpha_{DCO2/DN2}$	0.3	31	55
S_{CO2} (c)	220 ± 13	180 ± 32	69 ± 7
S_{N2} (c)	1	129 ± 64	60
$\alpha_{SCO2/SN2}$	220	1.4	1.2

(a) Barrer or 10^{-10} cm^3(STP) cm cm^{-2} s^{-1} cmHg^{-1}.
(b) 10^{-7} cm^2 s^{-1}.
(c) 10^{-4} cm^3 (STP) cm^{-3} cmHg^{-1}.

–50 °C) whose value changed with the membrane composition according to the Fox equation. All the membranes of these series exhibited lower permeability than the pure Pebax® 1657 membrane with an improvement in selectivity (Table 13.5). The permeability reduction can be explained by a lower contribution to the permeability of PEGEPI than that of pure PEG in the membrane polyether phase. In fact, the high performances of PEGEPI copolymers compared with those of Pebax® 1657 were only found for cross-linked PEGEPI of very high PEG contents: $PCO_2 = 96$ Barrer and $\alpha_{CO2/N2} = 63$ for the copolymer with 90 wt.% PEG [41]. However, PEGEPI of high PEG content (much higher than that of the commercially available PEGEPI we used) could be attractive in blend membranes with Pebax® 1657, due to expected selectivity improvements. Pebax® 1657 would then impart its good mechanical properties to PEGEPI, without additional cross-linking processes.

The PEG-Pebax® blend membranes were prepared with different amounts of PEG of 300 g mol^{-1} molecular mass. A preliminary study of blends with PEG of higher molecular mass showed that the miscibility of the two polymers in solid blends decreased with increasing PEG molecular mass. For instance, the blend with only 1 wt.% of PEG of 3400 g mol^{-1} molecular mass exhibited a melting peak of PEG3400 crystallites in DSC thermograms, while a single melting peak in DSC was found for blends with PEG300 in blends up to 50 wt.%. The better miscibility of low molecular mass PEG is understandable, if one considers its favourable mixing entropy, in addition to the favourable mixing enthalpy caused by hydrogen bonds between PEG hydroxyl end-groups with ether groups in Pebax®. As non-miscible blends are known to be mechanically weak [42], we focused our further study on Pebax® 1657 blends with PEG300.

The permeation data (Figure 13.7) indicate first a reduction in CO_2 permeability from that of a pure Pebax® 1657 membrane, then a steady increase in CO_2 permeability. The highest performances were obtained with a blend containing 20 wt.% PEG300: the CO_2 permeability reached 128 Barrer and the CO_2/N_2 ideal selectivity was 80.

Car *et al.* [9,10] reported a two-fold increase in CO_2 permeability (up to 150 Barrer) without changes in CO_2/N_2 selectivity when the membrane material changes from pure

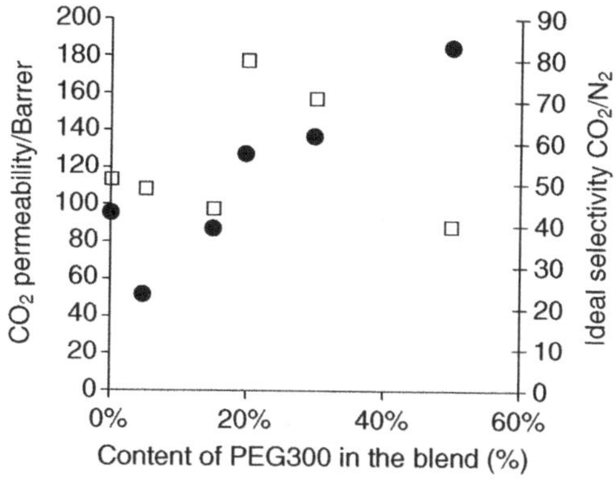

Figure 13.7 *CO_2 permeability (•) and CO_2/N_2 ideal selectivity coefficient (□) of the different PEG300-Pebax® 1657 blend membranes, 25 °C, 4 bars*

Pebax® 1657 to its 50:50 blend with PEG 200. The CO_2 permeability increase was attributed to a solubility increase with the increase in ethylene oxide (EO) units, but also to a diffusivity increase due to an increase in free volume (either total fractional free volume or individual free volume space): at high PEG content, there would be an increase of Pebax® segment motion, since low molecular weight PEG would act as a plasticizer in inserting its molecules between chain segments to screen out polymer/polymer interactions, and generating additional intermolecular space for small Brownian motion [43].

Our sorption and diffusion data indicated an apparently more complex behaviour of the blends. We did not observe a clear increase in CO_2 solubility coefficient in the membrane with increasing PEG content but an increase in the CO_2 diffusivity. As no free volume measurements were performed in the present study, we can only suggest a similar mechanism of diffusivity enhancement by PEG additive in Pebax® 1657 as that proposed by Yave *et al.* [43].

The influence of the nature of the polymer component on the permeability of blend membranes is worth discussing. Based on DSC and AFM data, Car *et al.* [10] reported a good miscibility of Pebax® 1657 with PEG200 in blends up to 50 wt.% PEG. For PEG300, we observed a similar behaviour of the blends in DSC, i.e. for the polyether phase, unique glass transition temperature and melting temperature, whose values decrease when the PEG content increases (Table 13.6). However, in optical microscopy we observed that at a low PEG content, the crystalline PA phase was finely dispersed in the blend, even more than in pure Pebax® 1657; at higher PEG300 contents, the dispersed crystalline phase became coarser. At 50 wt.% of PEG300, the phase was very coarse, and a slight exudation of PEG was observed on the membrane faces (Figure 13.8).

These features suggest a phase separation in the matrix, followed by a release of liquid PEG to the surface, although no PEG300 melting peak was observed in DSC trace of the 50% PEG membrane, probably because the expelled PEG was easily removed in handling operations. We assume that the phase separation limits the improvement in CO_2 diffusivity by a saturation of the polymer matrix, leading to a stagnancy of the blend membrane performances.

Table 13.6 *Values of the glass transition temperatures of PE and the melting temperature of PE and PA blocks for the different Pebax® grades, determined from DSC measurements (in the first heating scans). The PA crystallinity is obtained with the values found in the literature for the melting enthalpy of PA6*

	T_m polyamide block (°C)	T_m polyether block (°C)	T_g polyether block (°C)	wt.% PA crystallinity in Pebax®	wt. % PA crystallinity in PA block
1657	207	14	−55	15	30
1657/PEG300 5%	205	9	−56	12	24
1657/PEG300 15%	204	7	−65	–	–
1657/PEG300 20%	203	7	−65	–	–
1657/PEG300 30%	200	1	–	–	–
1657/PEG300 50%	196	0	–	–	–

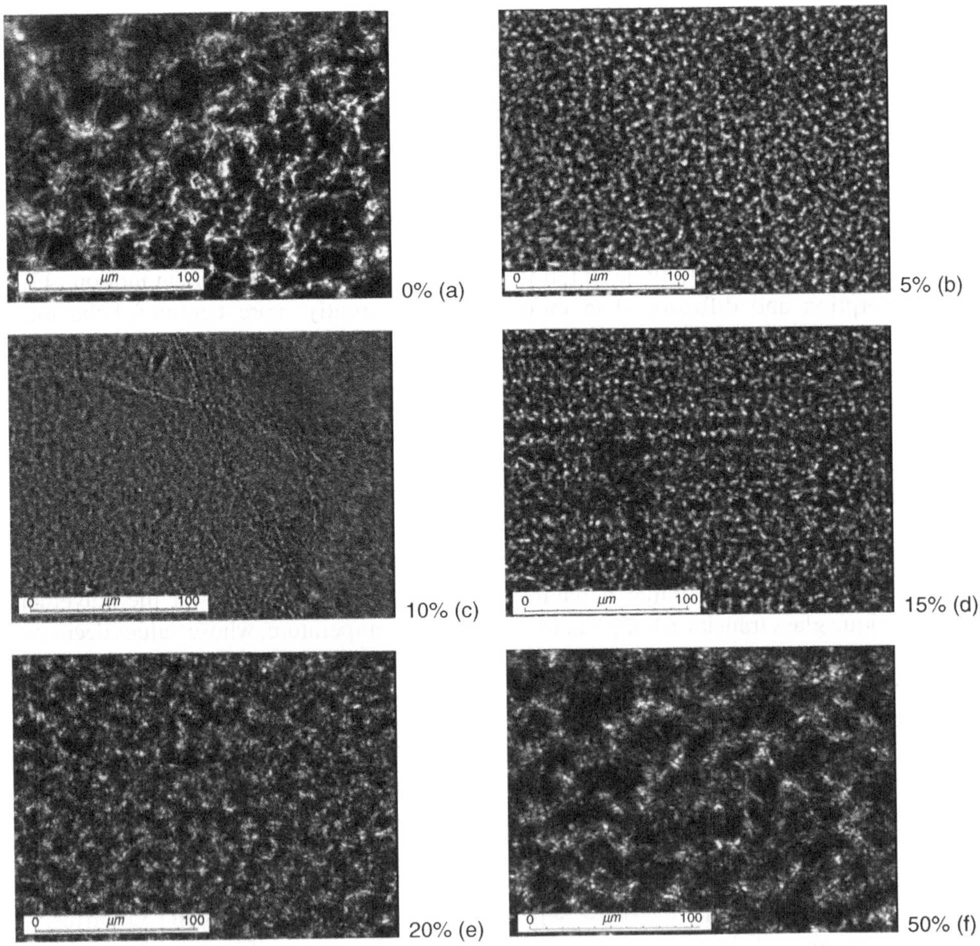

Figure 13.8 *Optical microscopy pictures of the different Pebax® 1657/PEG300 blends*

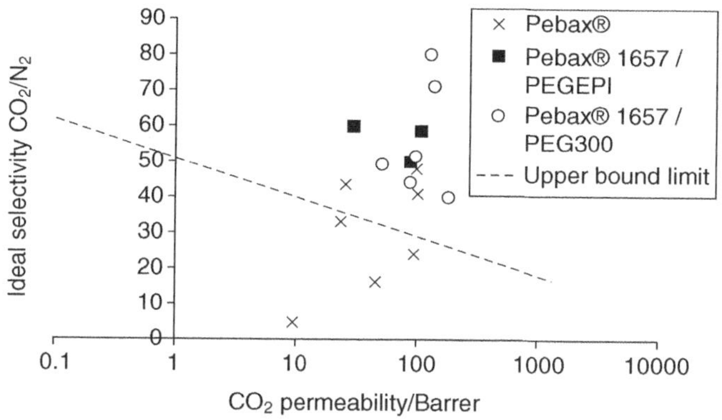

Figure 13.9 *Robeson's plot of (×) the different Pebax®, (○) the Pebax® 1657/PEGEPI blends and (■) the Pebax® 1657/PEG300 blends. (- - -) is the polymer upper bound*

The clearly lower miscibility of PEG300 compared with that PEG200 is attributed to the difference in their chemical structure: PEG200 has four EO units for two hydroxyl end groups, while PEG300 has six EO units. The gains in mixing entropy and hydrogen-bonding interactions should be much larger for PEG200 than for PEG300.

The larger increase in CO_2 permeability for PEG200 blend membranes compared with the PEG300 one can be explained by the larger positive contribution of PEG200 to the gas diffusion (larger free volume effect), and the higher PEG content obtainable in the blend membranes.

In summary, blending of Pebax® 1657 with PEG300 resulted in membranes of higher CO_2 permeability and selectivity due to enhanced CO_2 diffusivity. However, a phase separation that occurred at high PEG300 contents led to a stagnancy of performance. The performance of Pebax® membranes are compared with those obtained for the Pebax® blend membranes on a CO_2/N_2 Robeson's plot [5,44] (Figure 13.9).

13.3.3 Pebax® 1657 Membrane Modified by Cold Plasma

Although Pebax®–PEG blends can exhibit high performances in the CO_2 extraction from flue gases, a limitation in industrial operations can be foreseen: the long-term stability of the blend membranes, especially at high temperatures. In fact, a low molecular weight PEG is more a plasticizer than a true polymer in the blends; its chains are too short to be permanently blocked in a polymer network by simple entanglements. This was evidenced by the PEG migration towards the surface observed in the blends of high PEG300 contents. In order to obtain a more permanent modification of Pebax®, we propose the use of cold plasma to modify the Pebax® 1657 film surface with chemical groups capable of strong CO_2 absorption.

When a cold plasma in N_2, NH_3 or H_2/N_2 gas is put in contact with a membrane, the chemical modification by reactions of the plasma active species with the membrane surface results in alterations of transport properties associated with changes in the permeant sorption into that surface. Under the conditions where the modification occurs on the external surface, the underneath of the membrane would entirely control the diffusion of the penetrants towards the downstream side. Our aim in this work was to graft amine groups [45] onto the surface of a Pebax® 1657 film to enhance CO_2 sorption into the membrane while keeping unchanged the transport by diffusion through the Pebax® 1657, which is already quite good for mixtures of CO_2 with permanent gases. Several modification tests in N_2/H_2 cold plasma were performed on Pebax® 1657 films to determine the operating conditions for the largest change in film surface properties. The conditions for the best surface property modification (Table 13.7) were obtained by assessment tests with measurements of the contact angle of the modified surface with pure water. In fact, we considered that the contact angle with water is the lowest for a modification where hydrophilic amine groups are grafted to the film surface.

The friction mode AFM images (Figures 13.5 and 13.10) of the surface before and after plasma treatment under the conditions given in Table 13.7 show the difference in the surface properties. The uniform colour of the image of the surface-treated film is in big contrast with that of the pristine film, where crystalline zones, amorphous PA zones

Table 13.7 *Best conditions used for the surface modification by H_2/N_2 plasma*

Parameter	Value
Gas flow	10 STP $cm^3\,min^{-1}$
Power	45 W
Treatment time	3 min
Distance between the bottom of the excitation source and the sample	2 cm
Vol.% H_2	15

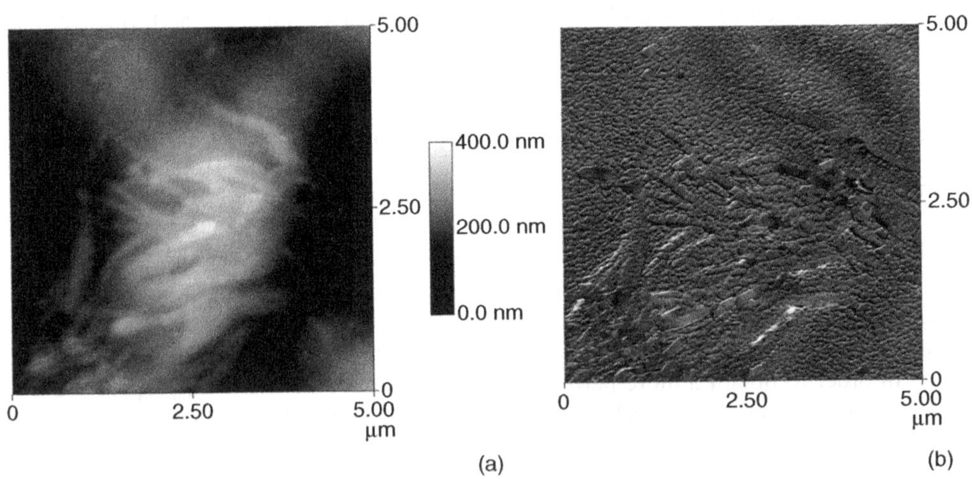

Figure 13.10 *Surface morphology of plasma treated Pebax® 1657 membrane obtained by AFM in contact (a) and in friction (b) modes. The contact mode image gives the surface topology*

and molten polyether zones can be clearly distinguished. We infer from the uniform friction characteristics of the plasma-modified surface that the grafted amine groups were fairly uniformly distributed on the surface.

However, the grafted layer is apparently extremely thin, since no changes were detected in DSC thermograms. The CO_2 permeation test indicated that the plasma-modified membrane exhibited a greatly enhanced permeability (144 Barrer at 25 °C).

We speculated that, in line with the mechanism proposed for facilitated transport of CO_2 with amine molecules, the CO_2 sorption was enhanced due to the following reversible complexation reaction [46,47], in addition to the normal gas sorption according to Henry's law into a rubbery/molten polymer phase:

$$RNH_2 + CO_2 \Leftrightarrow RNHCOO^- + H^+$$

$$RNH_2 + H^+ \Leftrightarrow RNH_3^+$$

In the case of our plasma-modified membrane where amine groups were only grafted on the surface, CO_2 facilitated transport cannot occur through the membranes. Instead, the high interfacial concentration in CO_2 served as entrance concentration for the classical diffusion through the unmodified Pebax® 1657 polymer, leading to an overall improvement of the CO_2 permeability coefficient.

13.4 Conclusions

Extruded films of different Pebax® grades and dense membranes made by solution blending of Pebax® 1657 grade and poly(ethylene glycol) (PEG) or poly(ethylene oxide-co-epichlorhydrine) (PEGEPI) were characterized by DSC, optical and atomic-force microscopies and gas permeation. The structure–property relationship of the films was analyzed. The membrane permeability appears to depend not only on the content of polar groups of the Pebax® grade, but also on the Pebax® phase structure, which is governed by the length of the polyamide and polyether blocks. The Pebax® 1657 grade offered the best compromise in CO_2 permeation properties, with a CO_2 permeability of 100 Barrer, and an ideal selectivity for CO_2 relative to nitrogen $\alpha_{CO2/N2}$ of 50. Our best membrane for global warming reduction was the one obtained by blending 20 wt.% of PEG 300 and Pebax® 1657 grade; it exhibited a CO_2 permeability of 128 Barrer, and an ideal selectivity $\alpha_{CO2/N2}$ of 80; these performances rank the material among the best membranes for CO_2 abatement from air sources like flue gases. Finally, a 40% enhancement in CO_2 permeability of a dense Pebax® 1657 membrane obtained by surface modification by cold plasma-activated nitrogen and hydrogen suggests that the plasma-assisted surface modification in order to graft CO_2 capturing amine groups is a promising technique to improve further the performances of a dense membrane.

References

(1) R. Baker, Future directions of membrane gas-separation technology, *Membr. Technol.*, **138**, 5–10 (2001).

(2) R. Bounaceur, N. Lape, D. Roizard, C. Vallieres, E. Favre, Membrane processes for post-combustion carbon dioxide capture: A parametric study, *Energy*, **31**, 2556–2570 (2006).
(3) X. G. Li, I. Kresse, J. Springer, J. Nissen, Y. L. Yang, Morphology and gas permselectivity of blend membranes of polyvinylpyridine with ethylcellulose, *Polymer*, **42**, 6859–6869 (2001).
(4) G. C. Kapantaidakis, G. H. Koops, High flux polyethersulfone–polyimide blend hollow fiber membranes for gas separation, *J. Membr. Sci.*, **204**, 153–171 (2002).
(5) L. M. Robeson, Polymer membranes for gas separation, *Curr. Opin. Solid. State Mater. Sci.*, **4**, 549–552 (1999).
(6) M. Kawakami, Y. Yamashita, M. Yamasaki, M. Iwamoto, S. Kagawa, Effects of dissolved inorganic salts on gas permeabilities of immobilized liquid polyethylene glycol membranes, *J. Polym. Sci., Polym. Lett. Ed.*, **20**, 251–257 (1982).
(7) H. B. Park, C. H. Jung, Y. M. Lee, A. J. Hill, S. J. Pas, S. T. Mudie, E. Van Wagner, B. D. Freeman, D. J. Cookson, Polymers with cavities tuned for fast selective transport of small molecules and ions, *Science*, **318**, 254–257 (2007).
(8) W. J. Ward, C. K. Neulander, Immobilized liquid membranes for sulfur dioxide separation, 65th Annu. Meet. AIChE, Cleveland, paper 7A (1969).
(9) H. Lin, B. D. Freeman, Materials selection guidelines for membranes that remove CO_2 from gas mixtures, *J. Molec. Struct.*, **739**, 57–74 (2005).
(10) A. Car, C. Stropnik, W. Yave, K. V. Peinemann, PEG modified poly (amide-b-ethylene oxide) membranes for CO_2 separation, *J. Membr. Sci.*, **307**, 88–95 (2008).
(11) A. Car, C. Stropnik, W. Yave, K. V. Peinemann, Pebax®/polyethylene glycol blend thin film composite membranes for CO_2 separation: Performance with mixed gases, *Sep. Pur. Technol.*, **62**, 110–117 (2008).
(12) K. M. Jaipurkar, S. R. Patwardhan, U. K. Kharul, Gas permeation using poly (ether-b-amide) segmented block copolymer, *Chem. Eng. World*, **42**, 73–78 (2007).
(13) Y. Zhang, Z. Wang, S. C. Wang, Synthesis and characteristics of novel fixed carrier membrane for CO_2 separation, *Chem. Lett.*, **31**, 430–431 (2002).
(14) Y. Zhang, Z. Wang, S. C. Wang, Selective permeation of CO_2 through new facilitated transport membranes, *Desalination*, **145**, 385–388 (2002).
(15) H. Matsuyama, A. Terada, T. Nakagawara, Y. Kitamura, M. Teramoto, Facilitated transport of CO_2 through polyethylenimine/poly(vinyl alcohol) blend membrane, *J. Membr. Sci.*, **163**, 221–227 (1999).
(16) Y. Cai, Z. Wang, C. Yi, Y. Bai, J. Wang, S. Wang, Gas transport property of polyallylamine-poly(vinyl alcohol)/polysulfone composite membranes, *J. Membr. Sci.*, **310**, 184–196 (2008).
(17) R. D'Agostino, Plasma deposition, treatment, and etching of polymers, Academic Press, New York, 1990.
(18) X. Lin, X. Qiu, G. Zheng, J. Xu, Gas permeabilities of poly(trimethylsilylpropyne) membranes surface modified with CF4 plasma, *J. Appl. Polym. Sci.*, **58**, 2137–2139 (1995).
(19) S. H. Chen, W. H. Chuang, A. A. Wang, R. C. Ruaan, J. Y. Lai, Oxygen/nitrogen separation by plasma chlorinated polybutadiene/polycarbonate composite membrane, *J. Membr. Sci.*, **124**, 273–281 (1997).
(20) R. Foerch, G. Beamson, D. Briggs, XPS valence band analysis of plasma-treated polymers, *Surf. Interface Anal.*, **17**, 842–846 (1991).
(21) P. Benoist, G. Legeay, R. Morello, F. Poncin-Epaillard, Preparation de surfaces de polymères par plasma froid pour métallisation par galvanochimie, *Eur. Polym. J.*, **28**, 1383–1393 (1992).
(22) J. R. Brown, P. J. C. Chappell, Z. Mathys, Plasma surface modification of advanced organic fibres, *J. Mater. Sci.*, **26**, 4172–4178 (1991).
(23) N. Inagaki, S. Tasaka, H. Kawai, Y. Kimura, Hydrophilic surface modification of polyethylene by NO-plasma treatment, *J. Adhesion Sci. Technol.*, **4**, 99–107 (1990).
(24) D. N. Hild, P. Schwartz, Plasma-treated ultra-high strength polyethylene fibers. I: Characterization by electron spectroscopy for chemical analysis, *J. Adhesion Sci. Technol.*, **6**, 879–896 (1992).
(25) N. Inagaki, S. Tasaka, K. Hibi, Surface modification of kapton film by plasma treatments, *J. Polym. Sci. Polym. Chem.*, **30**, 1425–1431 (1992).

(26) N. Inagaki, S. Tasaka, J. Ohkubo, H. Kawai, *J. Appl. Polym. Sci., Appl. Polym. Symp.*, **46**, 399 (1990).
(27) M. J. Wang, Y. I. Chang, F. Poncin-Epaillard, Effects of the addition of hydrogen in the nitrogen cold plasma: The surface modification of polystyrene, *Langmuir*, **19**, 8325–8330 (2003).
(28) C. Joly, D. Le Cerf, C. Chappey, D. Langevin, G. Muller, Residual solvent effect on the permeation properties of fluorinated polyimide films, *Sep. Purif. Technol.*, **16**, 47–54 (1999).
(29) F. P. Glatz, R. Mulhaupt, J. D. Schultze, J. Springer, Gas permeabilities and permselectivities of amorphous segmented 6F poly(arylene thioether imide)s and the corresponding poly(arylene sulfone imide)s, *J. Membr. Sci.*, **90**, 151–160 (1994).
(30) A. Xenopoulos, B. Wunderlich, Thermodynamic properties of liquid and semicrystalline linear aliphatic polyamides, *J. Polym. Sci. Part B: Polym. Phys.*, **28**, 2271–2290 (1990).
(31) V. I. Bondar, B. D. Freeman, I. Pinnau, Gas sorption and characterization of poly (ether-b-amide) segmented block copolymers, *J. Polym. Sci. Part B: Polym. Phys.*, **37**, 2463–2475 (1999).
(32) M. Hedenqvist, A. Angelstok, L. Edsberg, P. T. Larsson, U. W. Gedde, Diffusion of small-molecule penetrants in polyethylene: Free volume and morphology, *Polymer*, **37**, 2887–2902 (1996).
(33) B. Alexandre, D. Langevin, P. Médéric, T. Aubry, H. Couderc, Q. T. Nguyen, A. Saiter, S. Marais, Water barrier properties of polyamide 12/montmorillonite nanocomposite membranes: Structure and volume fraction effects, *J. Membr. Sci.*, **328**, 186–204 (2009).
(34) A. Danch, W. Osoba, D. Chrobak, G. Nowaczyk, S. Jurga, Relaxation of the constrained amorphous phase polyethylene-chalk composites, *J. Thermal Anal. Cal.*, **90**, 201–208 (2007).
(35) J. M. S. Henis, M. K. Tripodi, A novel approach to gas separations using composite hollow fiber membranes, *Sep. Sci. Technol.*, **15**, 1059–1068 (1980).
(36) J. H. Petropoulos, Mechanisms and theories for sorption and diffusion of gases in polymers, in Polymeric Gas Separation Membranes, D. R. Paul, Y. P. Yampolskii (Eds), CRC Press, Boca Raton, 1994.
(37) G. Härtel, T. Püschel, Permselectivity of a PA6 membrane for the separation of a compressed CO_2/H_2 gas mixture at elevated pressures, *J. Membr. Sci.*, **162**, 1–8 (1999).
(38) S. Jose, P. S. Thomas, S. Thomas, J. Karger-Kocsis, Thermal and crystallisation behaviours of blends of polyamide 12 with styrene-ethylene/butylene-styrene rubbers, *Polymer*, **47**, 6328–6336 (2006).
(39) H. Lin, E. V. Wagner, J. S. Swinnea, B. D. Freeman, S. J. Pas, A. J. Hill, S. Kalakkunnath, D. S. Kalika, Transport and structural characteristics of crosslinked poly (ethylene oxide) rubbers, *J. Membr. Sci.*, **276**, 145–161 (2006).
(40) J. H. Kim, S. Y. Ha, Y. M. Lee, Gas permeation of poly (amide-6-b-ethylene oxide) copolymer, *J. Membr. Sci.*, **190**, 179–193 (2001).
(41) C. Charmette, J. Sanchez, P. Gramain, A. Rudatsikira, Gas transport properties of poly (ethylene oxide-co-epichlorohydrin) membranes, *J. Membr. Sci.*, **230**, 161–169 (2004).
(42) Y. Gnanou, M. Fontanille, Organic and physical chemistry of polymers, Wiley-Interscience, 2008.
(43) W. Yave, A. Car, K. V. Peinemann, M. Q. Shaikh, K. Rätzke, F. Faupel, Gas permeability and free volume in poly (amide-b-ethylene oxide)/polyethylene glycol blend membranes, *J. Membr. Sci.*, **339**, 177–183 (2009).
(44) S. A. Stern, Polymers for gas separations: The next decade, *J. Membr. Sci.*, **94**, 1–66 (1994).
(45) D. Nicolas-Debarnot, M. Pascu, C. Vasile, F. Poncin-Epaillard, Influence of the polymer pre-treatment before its electroless metallization, *Surf. Coat. Technol.*, **200**, 4257–4265 (2006).
(46) T. Yamaguchi, L. M. Boetje, C. A. Koval, R. D. Noble, C. N. Bowman, Transport properties of carbon dioxide through amine functionalized carrier membranes, *Ind. Eng. Chem. Res.*, **34**, 4071–4077 (1995).
(47) J. Zou, W. S. W. Ho, CO_2-selective polymeric membranes containing amines in crosslinked poly (vinyl alcohol), *J. Membr. Sci.*, **286**, 310–321 (2006).

Part IV

Applied Aspects of Membrane Gas Separation

14

Membrane Engineering: Progress and Potentialities in Gas Separations

Adele Brunetti[a], *Paola Bernardo*[a], *Enrico Drioli*[a,b] *and Giuseppe Barbieri*[a]
[a] *National Research Council – Institute for Membrane Technology (ITM–CNR), The University of Calabria, Rende, Italy*
[b] *The University of Calabria – Department of Chemical Engineering and Materials, Rende, Italy*

14.1 Introduction

In the last few years the potentialities of membrane operations have been widely recognized. The attention to the process intensification (PI) strategy, as the best available approach for an appropriate sustainable industrial growth [2], confirmed that membrane engineering is a powerful tool to realize best this strategy. Today, membrane technology for gas separation (GS) is a well-consolidated technique, in various cases competitive with traditional operations. Separation of air components, H_2 from refinery industrial gases, natural gas dehumidification, separation and recovery of CO_2 from biogas and natural gas are some examples in which membrane technology is already applied at industrial level [3,4]. The separation of air components or oxygen enrichment has advanced substantially during the past 10 years. The oxygen-enriched air produced by membranes has been used in several fields, including chemical and related industries, the mechanical field, food packing, etc. In industrial furnaces and burners, for example, injection of oxygen-enriched air (30–35% of oxygen) leads to higher flame temperature and reduces the volume of parasite nitrogen to be heated; this means lower energy consumption. Mixtures containing more than 40% oxygen or 95% nitrogen can also be obtained. Today, membranes dominate the fraction of the nitrogen market for applications less than 50 tons/day and relatively low purity (95–99.5% nitrogen). On the contrary, oxygen separation with membranes is

Membrane Gas Separation Edited by Yuri Yampolskii and Benny Freeman

still underdeveloped. The main reason is that most of the industrial oxygen applications require purity higher than 90%, which is easily achieved by adsorption or cryogenic technologies, whereas it is difficult to obtain with a single membrane stage. The possibility of utilizing membrane technology in solving problems such as the greenhouse effect related to CO_2 production is also ongoing. Membranes able to separate CO_2 from flue gas streams emitted by power plant, kilns, steel mills, etc. with a high CO_2/N_2 selectivity, might be used at any large-scale industrial CO_2 source. The separation and recovery of organic solvents from gas streams is also rapidly growing at industrial level. Polymeric rubbery membranes that selectively permeate volatile organic compounds (VOC) and gasoline vapours in particular from air or nitrogen have been used. Such systems typically achieve greater than 99% removal of VOC from the feed gas and reduce the VOC content of the stream to 100 ppm or less.

Amorphous perfluoropolymers might be utilized for casting asymmetric composite membranes [5] with interesting selectivity and permeability for various low molecular species. Their cost is, however, a negative aspect. The possibility of also realizing new mass transport mechanisms as the ones characterizing the perovskite membranes is becoming of interest. The case of O_2 and H_2 transport in ion transport membranes might be extended to other species by realizing new specific materials. Molecular dynamics studies, fast growing in this area, might contribute to the design of these new inorganic materials or to the appropriate functionalization of existing polymeric membranes.

The significant positive results achieved in GS membrane systems are, however, still far from realizing the potentialities of this technology. Problems related to pre-treatment of the streams, to the membrane life time, to their selectivity and permeability still exist, slowing down the growth of large-scale industrial applications. Together with the investigation of new polymeric, inorganic and hybrid materials, the design and optimization of new membrane plant solutions, also integrated with the traditional operations, will lead to significant innovation toward the large-scale diffusion of the membranes for GS. With the introduction of process intensification strategy and of the methods related to it, large new areas will be open for GS membrane systems. Some of the drawbacks of the membrane operations such as the necessity of pre-treatment, might be solved by combining various membrane operations in the same industrial process, by developing better membranes (higher chemical stability, permeability and selectivity) and modules. In this respect, the role of chemical engineering is crucial.

14.2 Materials and Membranes Employed in GS

The criteria for selecting membranes for a given application are strongly related to several factors such as durability, mechanical integrity at the operating conditions, productivity and separation efficiency, etc. [6]. Different aspects must be deeply investigated for assuring a good membrane separation process:

- membrane transport properties (permeability, selectivity, permeance)
- membrane material and structure
- module and process design.

The membranes used in GS can be distinguished in two main categories, polymeric and inorganic. Polymeric membranes, specifically used for GS, are generally asymmetric or composite and based on a solution-diffusion transport mechanism. These membranes, made as flat sheet or hollow fibres, have a thin, dense skin layer on a micro-porous support that provides mechanical strength [7]. Typically, polymeric membranes show high, but finite, selectivities with respect to porous inorganic materials due to their low free-volume [8]. They generally undergo a trade-off limitation between permeability and selectivity [9]: as permeability increases, selectivity decreases, and vice-versa. Currently, however, only eight or nine polymer materials are used for preparing at least 90% of the membranes in use [10].

Polymeric membranes represent a good solution for the application in large-scale separations due to the low cost, ease of processing and high packing density. However, they cannot withstand high temperatures and aggressive chemical environments; moreover, when applied in petrochemical plants, refineries and natural gas treatment, heavy hydrocarbons in feed gas streams can be a problem, particularly in hollow fibre modules. Many polymers can be swollen or plasticized when exposed to hydrocarbons or CO_2 at high partial pressure; their separation capabilities can be dramatically reduced, or, the membranes irreparably damaged. Therefore, pre-treatment selection and condensate handling are critical decision factors for a proper operation of GS modules [11].

In order to enhance the properties of the polymeric membranes, new mixed matrix membranes consisting in nano- or micro- particles of inorganic material (metal, zeolite, carbon nanotubes, etc.) incorporated in the polymeric matrix, have been recently studied [12]. These membranes offer very interesting properties; however, their cost, difficulty of commercial scale manufacture and brittleness remain important challenges [10]. Facilitated transport membranes in which a carrier (typically, metal ions) with a special affinity toward a target gas molecule is mixed in the polymeric matrix showed also interesting results in GS [13]. There are several types of facilitated transport membranes; their drawbacks are related to the low fluxes. Hägg *et al.* [14] patented a facilitated transport membrane suitable for CO_2 capture. The membrane has a support coated with cross-linked polyvinylamine; a fixed carrier in the membrane helps so that the CO_2 molecules in combination with moisture form HCO_3^-, which is then quickly and selectively transported through the membrane [15]. Within a five-year period, these membranes are expected to be tested in four large power plants in Europe [16].

Although almost all industrial GS processes use polymeric membranes, a significant interest is focused on the development of inorganic membranes (metal, zeolite, ceramic, carbon, etc.), particularly for their use in high temperature separations and, also, in high temperature membrane reactors. Metal membranes are specifically used for the purification of specific gaseous streams, e.g. Pd-based membranes for hydrogen purification or silver membranes for pure oxygen production. Several metal membranes, including tantalum, niobium, vanadium, copper, gold, iron, cobalt and platinum, are used for hydrogen separation. These membranes are extraordinarily selective, since they are extremely permeable to hydrogen and essentially impermeable to all other gases. They are generally used for the production of a pure stream such in the case of the Pd-Ag membrane reactors widely studied [17] for the production of pure hydrogen reactions of high industrial interest (water gas shift, steam reforming of light hydrocarbons, dehydrogenation, etc.). However, they must be operated at high temperatures (>300 °C) to obtain useful permeation rates and to prevent embrittlement and cracking of the metal by sorbed gas. Additional drawbacks are the high cost, the low permeability in the case of highly selective dense membranes (e.g. metal oxides at temperatures below 400 °C) and the difficult sealing at high temperatures (greater than 600 °C).

Zeolite membranes show high thermal stability and chemical resistance compared with those of polymeric membranes. They are able to separate mixtures continuously on the basis of differences in the molecular size and shape [18], and/or on the basis of different adsorption properties [19], since their separation ability depends on the interplay of the mixture adsorption equilibrium and the mixture. Different types of zeolites have been studied (e.g. MFI, LTA, MOR, FAU) for the membrane separation. They are used still at laboratory level, also as catalytic membranes in membrane reactors (e.g. CO clean-up, water gas shift, methane reforming, etc.) [20,21]. The first commercial application is that of LTA zeolite membranes for solvent dehydration by pervaporation [22]. Some other pervaporation plants have been installed since 2001, but no industrial applications use zeolite membranes in the GS field [23]. The reason for this limited application in industry might be due to economical feasibility (development of higher flux membranes should reduce both costs of membranes and modules) and poor reproducibility.

Among the inorganic membranes, carbon molecular sieve membranes show good transport properties which allow their application in GS. They are obtained by the pyrolysis (at a high temperature in an inert atmosphere) of polymeric precursors already processed in the form of membranes [24]. These membranes combine good gas transport properties for light gases (gases of molecular sizes smaller than 4.0–4.5 Å) with thermal and chemical stability. However, the major disadvantages that hinder their commercialization are their brittleness, which means that they require careful handling [25–27] and, when compared to polymeric membranes, the cost of carbon-based membranes which is 1 to 3 orders higher per unit area.

Ion transport membranes are new dense inorganic membranes able to be permeated only by oxygen (or hydrogen). They show good performance in terms of permeability and selectivity at a very high temperature (>600 °C); however, the main problem is related to their durability. Once the time of operation is passed, the formation of micro-pinholes depletes the membrane properties, significantly reducing the selectivity [28].

14.2.1 Membrane Modules

GS membranes are formed in hollow-fibre or spiral-wound modules and, in some applications, also in plate and frame modules. A hollow-fibre module (Figure 14.1a) contains a large number of membrane fibres housed in a shell. Feed can be introduced on either the fibre or shell side. Permeate is usually withdrawn in a co-current or counter-current manner, with the latter being generally more effective. Spiral-wound modules (Figure 14.1b) are made from flat membrane envelopes, wrapped around a central tube. The feed passes along the length of the module and the permeate passes into a membrane envelope and then out via the central tube. Both the feed and the permeate are transported through the module in fluid-conductive spacer material. Modern modules tend to contain multiple membranes that are all attached to the same central tube. Usually in spiral-wound membrane modules, the size of the feed channel is greater than the active membrane area. In

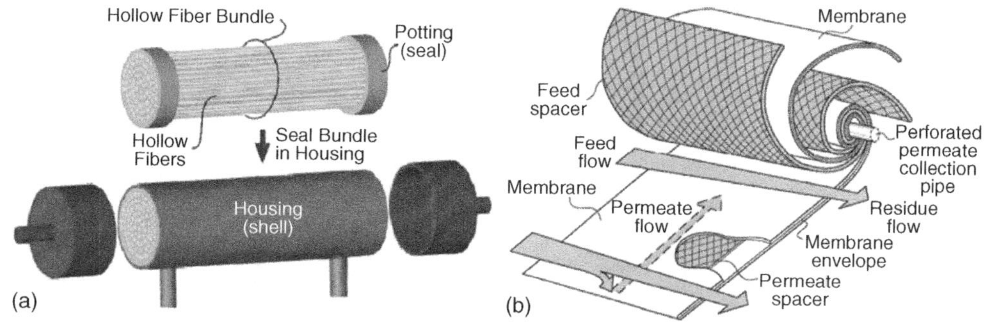

Figure 14.1 (a) Hollow fibre module (image courtesy of www.medarray.com, Copyright (2010) Medarray); (b) Spiral-wound module (image courtesy of www.mtrinc.com, Copyright (2010) mtrinc)

particular, the feed channel spacer is typically about 20% wider than the membrane envelope. This additional width may promote flow perpendicular to the bulk flow (along the spiral) [29]. Currently, the spiral wound modules contain around 1–2 m of rolled sheets, for 20–40 m^2 of membrane area.

Each membrane module configuration presents advantages and drawbacks. Hollow fibres are the cheapest on a per square metre basis (with the highest membrane area to module volume ratio); however, to make very thin selective layers in hollow-fibre form is harder than in flat sheet configuration. This implies that the permeance of hollow fibres is generally lower than that of a flat sheet membrane prepared with the same material. Therefore, larger membrane area is required for achieving the same separation by HF modules. Hollow fibre modules also require more pre-treatments of the feed than is usually required by spiral wound modules for removing particles, oil residue and other fouling components. These factors strongly affect the cost of the hollow fibre module design, therefore, currently, spiral wound modules are employed in several separations (e.g. in natural gas processing) particularly for those separations which cannot support the costs associated with the hollow fibre modules.

The size of the membrane modules for GS is rapidly increasing in order to follow the economies of scale. As reported by Baker [30], from 1980 up to now the size of Cynara's hollow-fibres modules, employed for natural gas treatment, increased by a factor of 6, while the spiral-wound modules are increasing, with the diameter passing from 20 cm to 30 cm (Figure 14.2). GS systems require membrane modules contained in high-pressure, code-stamped vessels. The costs of the vessels, frames and associated pipes, valves, etc. can be several times the cost of the membranes. Therefore, considerable reduction of costs can be achieved by packing larger membrane modules into fewer vessels [29]. This fact allows weight and footprint to be reduced. This is extremely important on offshore platforms and other space-constrained facilities.

A membrane GS production process can be realized assembling the membrane modules in several configurations, depending on the particular type of separation. The single-stage, double-stage, multi-stage with recycle constitute the main design solutions. Some examples are given in the following paragraphs relative to the industrial applications for the GS.

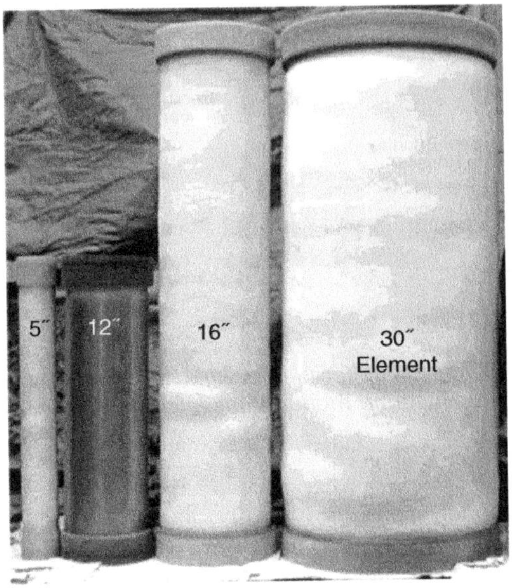

Figure 14.2 Photograph showing the development of hollow-fibre membrane modules at Cynara, from the first 5-inch modules of the 1980s to the 30-inch diameter currently being introduced [51] (Cynara Company now part of NATCO Group, Inc.). Reprinted with permission from Ind. Eng. Chem. Res., Natural gas processing with membranes: an overview, by R. W. Baker and K. Lokhandwala, 47, 7, 2109–2121, Copyright (2008) American Chemical Society

14.3 Membranes Applications in GS

The key factors to qualify the performance of a specific membrane material for GS applications are the permeability and the selectivity. It is generally recognized that there is a trade-off limitation for polymeric materials between these two parameters: as selectivity increases, permeability decreases and vice versa. Robeson showed such empirical upper-bound relationships for different gas separations, in 1991 [8]. In 2008, Robeson [30] published an upload of these trade-off diagrams. This diagram virtually summarized all the existing membrane materials, giving an indication of the best performance achievable (at present) by a polymeric membrane material for a specific gaseous mixture.

The upper bound correlation follows the relationship that is valid in double logarithmic scale:

$$Permeability = k\left(Selectivity_{i,j}\right)^{n}$$

where k is referred to as the 'front factor' and n is the slope of the log–log plot of the noted relationship [30]. Figure 14.3 shows a plot of selectivity versus permeability, the upper bound trade-off of the most important type of separations. With respect to the first Robeson's plots [8], a significant shift of the upper bound was obtained for several separations, due to the creation of novel high performance membrane materials; in many cases such materials are perfluorinated polymers. Furthermore, some new trade-off limits, like those relative to the CO_2/N_2 and N_2/CH_4 separations, were added. A more detailed analysis

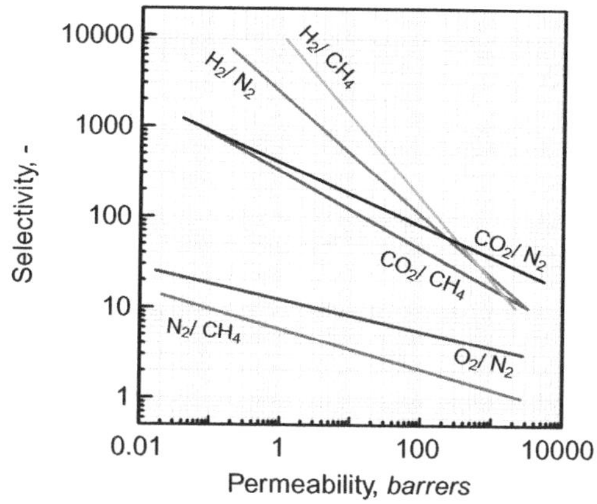

Figure 14.3 *Trade-off of Robeson's plot for different GS applications. Data elaborated from [30]*

on the state of the art of the membrane GS field in which many experimental results relative to different membrane materials are reported can be found in [10,31].

14.3.1 Current Applications of Membranes in GS

As early as in 1950, Weller and Steiner [32] considered membrane processes as feasible for the separation of hydrogen from hydrogenation tail gas, enrichment of refinery gas and air separation. However, commercial membranes capable of performing these separations economically only became available in 1970. In 1980, Permea, with its hydrogen separating Prism-membrane, launched the first large industrial application of gas separation membranes [9,33]. Since then, membrane-based operations, substituting or integrated with the traditional ones, had a rapid growth, with many companies, such as Cynara-Natco, Separex-UOP, GMS, Generon, Praxair, AirProducts, UBE are involved in this field [34,35].

A list of the main industrial applications of membrane technology in GS is given in Table 14.1, along with the membrane materials used and the status of the membrane technology.

14.3.2 Hydrogen Recovery

As previously mentioned, the first widespread commercial application of membranes in GS was the separation of hydrogen in the ammonia purge stream, by using Permea Prism™ systems. Hydrogen recovery is applicable to several processes, divided into three main categories:

- hydrogen recovery from ammonia purge streams
- syngas ratio adjustment
- hydrogen recovery in refineries.

In the ammonia process, the purge stream, almost clean and free of condensable vapours, consists of a mixture of hydrogen, nitrogen, methane and argon, delivered at a high pressure (136 bar). It is, thus, the ideal application for the membrane technology, since

Table 14.1 Main industrial applications of membrane technology for GS

Separation	Process	Traditional technology	Membrane material	Status of membrane technology application
H_2/N_2	Ammonia purge gas	PSA	Polysulfone	Plant installed (Prism by Permea)
			Pd-based	Lab scale
H_2/CO	Adjustment of H_2/CO ratio in syngas plant	PSA	Silicon rubber	Lab scale
			Polyimide	Plant installed (Separex)
			Pd-based	Lab scale
H_2/ hydrocarbons	Hydrogen recovery in refineries	PSA	Silicon rubber Polyimide	Plant installed (Sinopec Zhenhai (china) Prism by Permea Du Pont)
			Pd-based	Lab scale
H_2/light hydrocarbons	Ethylene cracker old trains	Cryogenic distillation	PTMSP; PMP	Pilot plant
			Pd-based	Lab scale
O_2/N_2	Air separation	Cryogenic distillation	Silicon rubber	Plant installed (Cynara; Separex; GMS; Air Products)
			Polysulfone	Plant installed (Permea)
			Polyimide	Plant installed (Medal; Dow-Generan; UBE)
			Polyphenilene	Plant installed (Aquilo)
			Ethyl Cellulose	Plant installed (Air Liquide)
			Ion transport (perovskite)	Lab scale
			Pd-based	Lab scale
N_2/CH_4	Nitrogen removal	Cryogenic distillation	Silicon rubber PMP Parel	Pilot plant
			PEBAX	Lab scale
H_2O/CH_4	Dehydration	Glycol absorption	Cellulose acetate Polyimide polyaramide	Plant installed [51]

Table 14.1 (*continued*)

Separation	Process	Traditional technology	Membrane material	Status of membrane technology application
CO_2/CH_4	Sweetening of natural gas	Amine absorption	Cellulose acetate	Offshore platforms In Thailand gulf. (Cynara-NATCO); Grace-Separex
			Polyaramide	Plant installed (Medal)
			Polyimide	Plant installed (Medal)
			Perfluoro-polymers	Plant installed (MTR)
CO_2/ Hydrocarbons	Natural gas liquid removal	Glycol absorption and cooling inn a propane refrigeration plant (−20 °C)	Cellulose acetate Polyimide polyaramide	Early commercial stage. A demonstration system installed
CO_2/N_2	CO_2 capture from flue gas streams	Amine absorption	Polyimide FSC; PEEKWC; Zeolite, silica based; carbon	Pilot plant Lab scale
VOCs/gas	Polyolefin plant resin degassing; Ethylene recovery	Adsorption, Refrigeration and turbo-expander plants	Silicone rubber PTMSP	Plant installed (MTR; OPW Vaporsaver™)

hydrogen is highly permeable with respect to the other gases and the stream already provides the necessary driving force for promoting the permeation. Actually, Permea, Inc. (now owned by Air Products and Chemicals, Inc.) designed a two-step membrane process for this separation (Figure 14.4).

Similar membrane systems are also already applied for syngas ratio adjustment (H_2/CO ratio) of the streams coming out by reformers. The process specifications are strictly related to the use of the stream. Generally, membranes are employed for stripping hydrogen out of the syngas in order to reduce the H_2/CO ratio. The stream is clean and already at a high pressure, thus ideal for use of a membrane. At the moment, several hundred hydrogen separation plants have been installed [36].

Figure 14.4 Two-stage H_2 membrane separation plant (from PRISM™ brochure [1]) in ammonia installation operating since 1979, which recycles 90% pure H_2 to the reactor. Reprinted with permission from Industrial & Engineering Chemical Research, Membrane gas separation: a review/state of the art, by P. Bernardo et al., 48, 10, Copyright (2009) American Chemical Society

The demand of hydrogen recovery in refineries is rapidly increasing also because of environmental regulations. The hydrogen content in the various refinery purges and off-gases ranges between 30–80%, mixed with light hydrocarbons (C_1–C_5); 90–95% hydrogen purity is required to recycle it to a process unit. A typical refinery operation is the separation of the hydrogen contained in the stream coming out from the hydrocracker. Actually, this process has the disadvantage of removing 4 mol of hydrogen for every mol of hydrocarbon removed. The membranes can be used alone or together with an absorber system, at a reduced capital cost and better process efficiency [32]. At the moment, the Prism™ system (using polysulfone hollow fibres with a thin silicone film on it) is dominant on the market for this kind of separation, showing interesting selectivities [37].

The main problem, related to an actual limited application of membrane technology, is membrane plasticization by strongly adsorbing components and the condensation of the light hydrocarbons on the membrane surface, which strongly affect the membrane performance and durability. The development of new membrane materials more resistant to high hydrocarbon partial pressure and the introduction of more efficient pre-treatment stage able to reduce pollutant content, will rapidly expand this technology [7].

14.3.3 Air Separation

A membrane process that has grown rapidly in the last few decades is the separation of the air into nitrogen- and oxygen-enriched streams. The main part of the membranes in use are oxygen selective, therefore the nitrogen-rich stream is recovered in the high pressure side, whereas an O_2-enriched stream is obtained as permeate at a low pressure.

Significant efforts have been made for increasing O_2/N_2 selectivity of the polymeric membranes; these two molecules have a very close kinetic diameter (nitrogen, 3.64 Å; oxygen, 3.46 Å), thus the use of a membrane whose selective properties are related only on the molecular size selection was very difficult. As reported by Baker [7], the first membranes used for this separation showed an O_2/N_2 selectivity of ca. 4. Significant improvement in membrane selectivity allowed values ranging from 8 to 12 to be reached, implying relevant reduction in compressor size and equipment costs.

Today, nitrogen separation by membrane systems is the largest GS process in use. Membrane selectivity does not need to be high in order to produce a relatively pure nitrogen stream, thus they became the dominant technology instead of pressure swing adsorption (PSA) or cryogenic distillation.

At the moment, thousands of compact on-site membrane systems generating nitrogen gas are installed in the offshore and petrochemical industry. Air Products Norway has delivered more than 670 PRISM® systems producing N_2 for different ship applications, and more than 160 PRISM® systems for offshore installations [38]. In December 2006, Air Products started with another PRISM® production plant in Missouri (USA) [39]. Another new air separation unit with a capacity of 550 ton/day of oxygen was installed by Air Liquide in Dalian (China) [40]. In Japan, Ube Industries [41] is increasing the production of polyimide hollow fibres for nitrogen separation to introduce a number of ethanol refining plants, mainly in the USA and Europe, driven by the rapid increase in the demand for bio-ethanol as an additive for oil products.

Concerning oxygen separation (Figure 14.4), great improvements in membrane performances are required to produce oxygen-enriched air and, moreover, pure oxygen to be used for chemical industries, electronic fields, medical fields, etc. The first application of membrane technology was carried out in 1980 using ethyl cellulose membranes, but the performance was not good enough to make the process competitive [7]. A compressor on the feed stream or a vacuum pump on the permeate side is required in order to guarantee the necessary driving force for the permeation of the oxygen through the membrane. Both these solutions are expensive, thus, for making this operation cheaper high flux membranes are required. Depending on the quality of the oxygen to be obtained, the separation process can be developed in one or two stages (Figure 14.5). Membranes are economically convenient when the final O_2 concentration is in the range 25–50% [42]; in this case, a single stage (Figure 14.5a) is the most economic configuration [43]. Pure oxygen can be produced in two stages (Figure 14.5b), as proposed by Baker [9]. However, many studies

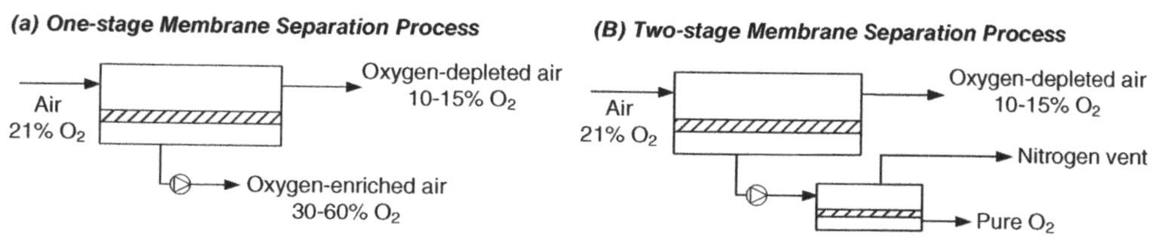

Figure 14.5 *Membrane process flow schematics for the production of (a) oxygen-enriched air and (b) pure oxygen. Reprinted with permission from Industrial & Engineering Chemistry Research, Future directions of membrane gas separation technology, by R. W. Baker, 41, 1393–1404, Copyright (2002) American Chemical Society*

are focused on new membrane materials exhibiting both high selectivity and high fluxes in order to develop the production of pure oxygen in only one stage.

Promising results have been obtained with the facilitated transport membranes in which an oxygen-complexing carrier compound acts like a shuttle to transport the oxygen selectively through the membrane [44].

Currently, dense inorganic membranes (ion transport membranes, ITM), able to be permeated only by oxygen (typically above 700 °C) are being developed [45,46]. A high-temperature air separation process could be integrated with power generation systems. Commercial-scale ion transport membrane oxygen modules have been fabricated by Air Products (0.5 ton/day of oxygen); this technology requires 35% less capital (much simpler flow sheet) and 35–60% less energy (less compression energy associated with oxygen separation) than cryogenic air separation [47].

14.3.4 Natural Gas Treatment

Raw natural gas composition changes depending on the source. In general, the CH_4 content is in the range of 75–90%, the rest being significant amounts of ethane, propane and butane, carbon dioxide and other impurities such as nitrogen and hydrogen sulfide. In order to meet the composition specifications for natural gas domestic use imposed by each country, natural gas requires some treatment before being delivered in the pipeline. The main treatments are related to the carbon dioxide, natural gas liquids, water and nitrogen removal.

Removal of carbon dioxide (natural gas sweetening) increases the calorific value and transportability of the natural gas stream. Carbon dioxide content in the natural gas obtained from the gas or oil well can vary from 4 to 50%. It has to be reduced down to ca. 2–5%. This goal is typically achieved by means of absorption with an aqueous alkanolamine solution. Amine treatment is a widely commercialized technology in which the hydrocarbon loss is almost negligible. However, this process has a tendency to corrode equipment and, over a short period of time, the amine solution loses viability through amine degradation and loss. However, the capital and operating cost shoots up very rapidly as the concentration of carbon dioxide in feed gas increases [48]. Membrane GS systems represent an alternative technology for separation of carbon dioxide from the natural gas, particularly for offshore applications [49].

Membrane systems can also be integrated with traditional units. The design of a hybrid membrane separation system depends on several aspects, such as membrane permeance and selectivity, CO_2 concentration of the inlet gas and the target required, the gas value (per ca. 30 Nm^3, the price of gas in 2007 was \$6–\$7 in the United States, whereas in Nigeria, which is far from being as well-developed a gas market, it may be as low as \$0.50 if the gas can be used at all) and the location of the plant (on an offshore platform, the weight, footprint, and simplicity of operation are critical; onshore, total cost is more significant) [51].

Figure 14.6 reports the block diagrams of two typical carbon dioxide membrane systems that treat natural gas with low CO_2 concentration, as proposed by Baker [51]. Both systems are designed to treat a feed stream with 10% of carbon dioxide. One-stage systems are preferred for very small gas flows. In such plants, methane loss to the permeate is often 10–15%. If there is no fuel use for this gas, it must be flared, which represents

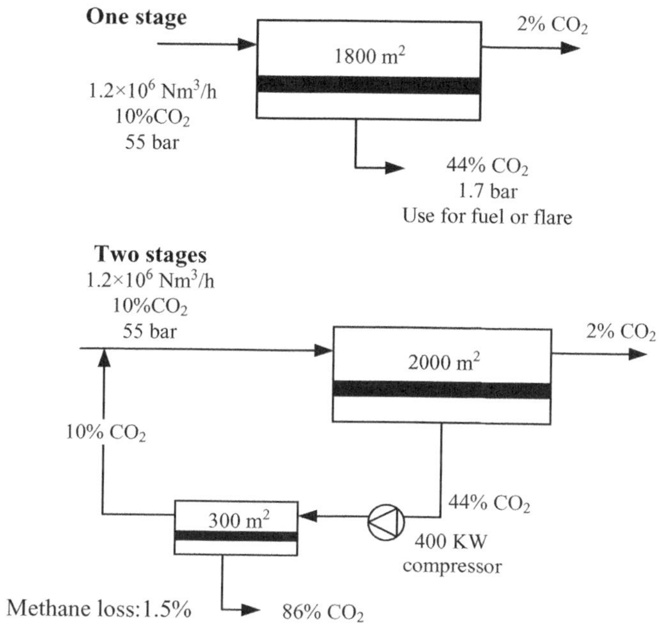

Figure 14.6 *Flow scheme of one-stage and two-stage membrane separation systems to remove CO_2 from natural gas. Reprinted with permission from Ind. Eng. Chem. Res., Natural gas processing with membranes: an overview, by R. W. Baker and K. Lokhandwala, 47, 7, 2109–2121, Copyright (2008) American Chemical Society*

a significant revenue loss. As the flow rate of the natural gas stream increases, the methane loss increases, therefore, usually, the permeate gas is recompressed and passed through a second membrane stage which reduces the methane loss to a few percent.

During the years 1980–1985, the first membrane plants using cellulose acetate membranes were installed [32]. Currently, several membrane systems are installed for small size applications (less than 6000 Nm^3/h), since amine processes are too complicated for small productions. Membrane and amine systems become competitive with respect to the traditional operations (PSA and cryogenic distillation) for size of 6000–50 000 Nm^3/h, while bigger plants are installed for offshore platforms or for enhanced oil recovery. Cellulose acetate is still the most widely used and tested material for natural gas sweetening as in UOP's membrane systems [50,51] (Figure 14.7). Recently, in an offshore platform located in the Gulf of Thailand (830 000 Nm^3/h) Cynara-NATCO [52] provided the biggest membrane system with 16-inch hollow fibre modules based on cellulose triacetate membranes for natural gas sweetening [53]. Polyimide membranes, originally developed for hydrogen separation, have been recently commercialized also for CO_2 separation.

Raw natural gas generally contain low concentrations of light hydrocarbon vapours. Their condensation must be prevented in order to avoid damages in the pipeline. So removal of these vapours is performed by glycol absorption and refrigeration at −20 °C in a propane atmosphere. Membrane use can make the process cheaper, drastically reducing the energy consumption due to refrigeration; however, it is still at an early commercial stage with only one demonstration plant installed in Pascagoula (USA), in 2002 [54].

Figure 14.7 Photographs of UOP membrane plants using spiral-wound modules of cellulose acetate membranes for CO_2 separation. (a) Plant compact enough to be moved when the gas field is exhausted after a few years of operation; capacity of 10 000 Nm^3/h membrane designed to reduce CO_2 gas from 6% to 2%. (b) A 600 000 sdNm^3/h unit designed to reduce CO_2 gas from 5.7% to 2%

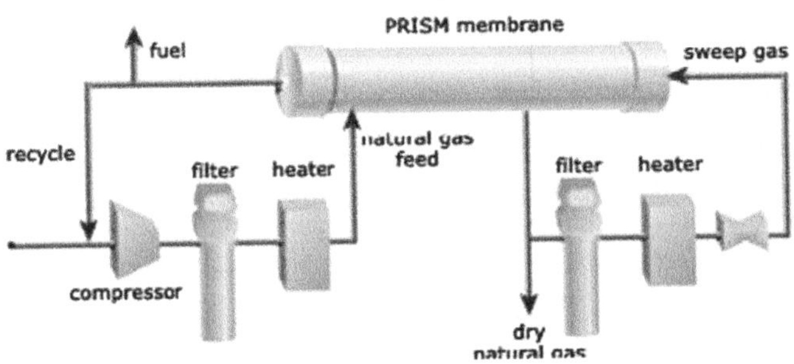

Figure 14.8 Membrane process flow schematics of a natural gas dehydration plant of PRISM

Water is a common impurity in natural gas that must be removed to prevent hydrate formation. This is a good opportunity for the application of membrane technology, but to be competitive, the membrane system must minimize the loss of methane with the permeate water [5]. This loss can be reduced, on the one hand, by choosing membranes with desired selectivity, but, on the other hand, by the process parameters, because this separation is pressure ratio-limited [5].

Currently, PRISM® membranes provide an attractive alternative to traditional glycol dehydration systems (Figure 14.8) based on simple process designs, lower costs, and the other benefits listed below. These benefits become even more pronounced as the industry produces natural gas from very remote locations [55]. An offshore membrane system for Shell Nigeria was designed to dry 600 000 Nm^3/h of natural gas from an inlet dew-point of 41 °C to an outlet dew-point of 0 °C at 38 bar. It was designed and built by Petreco, an Air Products PRISM Membranes licensed partner. Other plants were installed in Italy and Holland.

Several natural gas reserves are considered sub-quality because of the high nitrogen content. The gas pipeline specifications for inert gases, in fact, fixes the nitrogen content to a 4% limit [56]. Currently, cryogenic distillation is used for this separation; however,

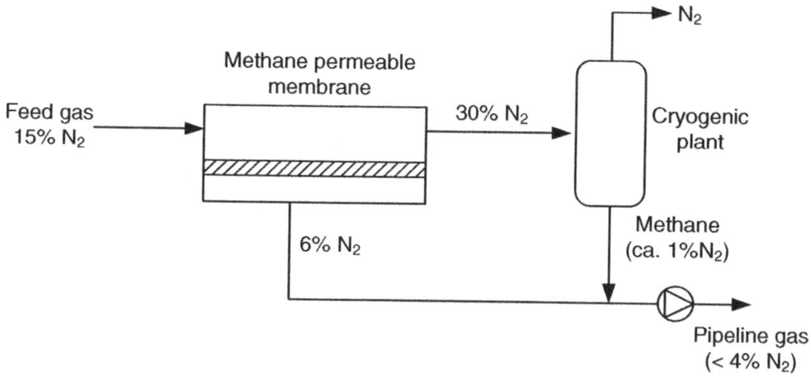

Figure 14.9 *Scheme of a hybrid membrane/cryogenic distillation plant for removal of nitrogen from natural gas. Reprinted with permission from Industrial & Engineering Chemistry Research, Future directions of membrane gas separation technology, by R. W. Baker, 41, 1393–1404, Copyright (2002) American Chemical Society*

membrane technology could be used here. The only challenge is the reduction of the methane loss in the permeate. However, methane-permeable membranes can be used conveniently in combination with a cryogenic plant (Figure 14.9). The feed gas, containing 15% nitrogen, is separated by a membrane into two streams: a residue stream (retentate) containing 30% nitrogen to be sent to the cryogenic plant and a permeate stream containing 6% nitrogen to be sent to the product pipeline gas. The membrane unit reduces the volume of the gas to be treated by the cryogenic unit by more than half. Simultaneously, the concentrations of water, C_{3+} hydrocarbons, and carbon dioxide are brought to very low levels, because these components also preferentially permeate the membrane. Removal of these components prior to cryogenic condensation is required to avoid freezing in the plant. The savings produced by using a smaller, simpler cryogenic plant more than offset the cost of the membrane unit [57].

14.3.5 CO_2 Capture

The recovery of carbon dioxide from large emission sources is a formidable technological and scientific challenge which has received considerable attention for several years [58–60]. In particular, the identification of a capture process which would fit the needs of target separation performances, together with a minimal energy penalty, is a key issue. Currently, the main strategies for the carbon dioxide capture in a fossil fuel combustion process are the following [61].

- Oxy-fuel combustion: this option consists in performing the oxygen/nitrogen separation on the feed stream, so that a CO_2/H_2O mixture is produced through the combustion process. The advantage of feeding an oxygen-enriched gas mixture (95% oxygen) instead of air, is the achievement of a purge stream rich in CO_2 and water with very low N_2 content, therefore the CO_2 can be easily recovered after the condensation of the water vapour.
- Pre-combustion capture: this solution is developed in two phases: (i) the conversion of the fuel to a mixture of H_2 and CO (syngas mixture) through, e.g. partial oxidation, steam reforming or auto-thermal reforming of hydrocarbons, followed by water-gas

shift; (ii) the separation of CO_2 (at 30–35%) from the H_2 that is then fed as clean fuel to the turbines. In these cases, the CO_2 separation could happen at very high pressures (up to 80 bar of pressure difference) and high temperatures (300–700 °C).

- Post-combustion capture: in this case, the CO_2 is separated from the flue gas emitted after the combustion of fossil fuels (from a standard gas turbine combined cycle, or a coal-fired steam power plant). CO_2 separation is realized at relatively low temperature, from a gaseous stream at atmospheric pressure and with low CO_2 concentration (ca. 5–15% if air is used during combustion). SO_2, NO_2 and O_2 may also be present in small amounts. This possibility is by far the most challenging since a diluted, low pressure, hot and wet CO_2/N_2 mixture has to be treated. Nevertheless, it also corresponds to the most widely applicable option in terms of industrial sectors (power, kiln and steel production for instance). Moreover, it shows the essential advantage of being compatible to a retrofit strategy (i.e. an already existing installation can be, in principle, subject to this type of adaptation) [60].

The conventional separation processes for CO_2 capture are: absorption (with amines), adsorption (with porous solids with high adsorbing capacity such as zeolites or active carbon) and cryogenic separation [62,63]. Even though amine absorption is the most common technology for post-combustion capture, its use is by far the best available technology owing to the high energetic cost (in the range 4–6 GJ/ton CO_2 recovered) related, in particular, to the significant energy consumption in the regeneration step. Furthermore, this option requires large-scale equipment for the CO_2 removal and chemicals handling. Membranes are most often listed as potential candidates for their application in post-combustion capture. However, the main problem related to their limited application is the low CO_2 concentration and pressure of the flue gas, which requires the use of membranes with high selectivities (ca. 100) for fitting the specification delivered by the International Energy Agency, i.e. a CO_2 recovery of 80% with a purity of at least 80% [64]. The commercial membranes (CO_2/CH_4 selectivity ca. 50), currently used to separate CO_2 from natural gas at high pressures are not suited for one-stage operation, implying a large membrane area and high compression costs [64].

Favre [48] provides a critical comparison of the application of polymeric dense membranes versus amine absorption in post-combustion capture. The energy requirement associated with membranes for post-combustion carbon capture for separating mixture containing 10% CO_2 is much larger than that of absorption. Even very selective membranes, showing selectivity above 120, require much more energy than absorption. However, for flue streams containing 20% or more CO_2 (steel or kiln productions), reasonable recoveries and permeate compositions can be attained with lower related cost than amine absorption. Scura *et al.* [65] proposed two different membrane system configurations specifically for applications in which the species to be recovered is at a low concentration (the CO_2 content in flue gas is around 10%) and low recoveries could be enough to meet process specifications. The flue gas stream compression (Figure 14.10a) and the vacuum on the permeate stream (Figure 14.10b), were compared for a permeate stream with the same CO_2 concentration and recovery. In particular, the vacuum option is a valuable alternative instead of the more expensive feed compression.

At a fixed pressure ratio, configuration (a) requires a lower installed membrane area but a higher investment and operating compression cost (the whole feed stream must be

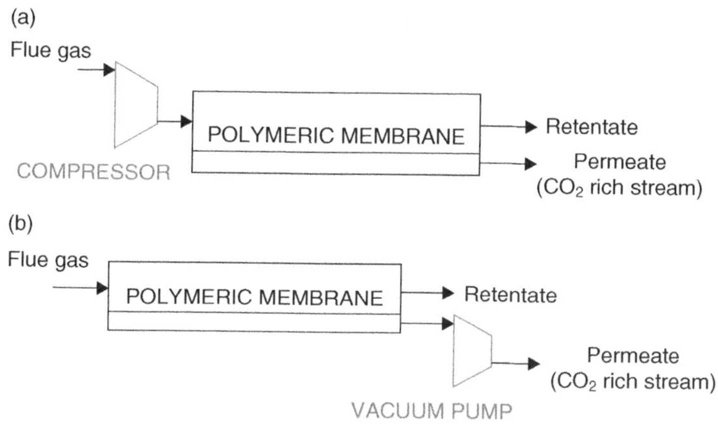

Figure 14.10 Configuration of the two membrane systems proposed for CO_2 capture

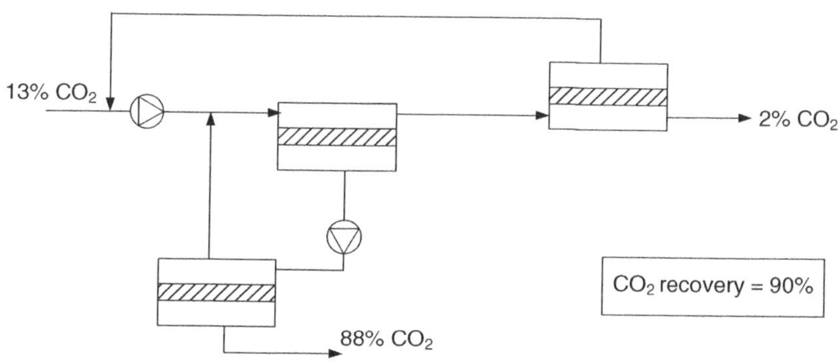

Figure 14.11 New applications: CO_2 from coal power plant flue gas. Adapted from the oral presentation of Reference 66

pressurized). On the contrary, in configuration (b), even if the total installed membrane area increases considerably, the compression cost strongly reduces (only the permeate stream must be sucked). The high size compressor in configuration (a) is substituted by a high number of modules, easy control, low investment and operating cost membrane modules. For a 20% recovery from a flue gas stream with the 13% of CO_2 and a pressure ratio of 15, the vacuum system (b) reduces the compression cost to less than 5% with respect to pressured system (a); in the meantime, the required membrane area in (b) increases up to more than 10 times.

In 2008, Baker [66] proposed the possibility of using membranes with CO_2/N_2 selectivity of ca. 50 (already commercial) as integrated multi-stage solutions. In this case, in fact, the appropriate choice of which kind of membrane can be used in each separation stage can make this application already feasible (Figure 14.11).

For the pre-combustion and oxyfuel capture processes, membranes based on alumina, zeolites, silica and carbon that show stability up to 300 °C are generally proposed [27].

Materials which conduct CO_3^{2-} ions (e.g. molten Li_2CO_3 formed from the reaction of Li_2ZrO_3 with CO_2) have also been studied for CO_2/CH_4 separation up to 600 °C [67,68]. These membranes may be economically efficient, stable and robust in applications where excess heat/energy is readily available to melt the carbonate. Also their use in high-temperature membrane reactors for integration in power generation cycles with CO_2 capture has been proposed [69]. However, significant design optimization would be required in order to identify efficient, feasible and environmentally sound technical solutions. In addition, further development and validation of performance of these membranes in real applications are needed.

14.3.6 Vapour/Gas Separation

The removal of organic solvents from gas streams has been widely developed at the industrial level. The main industrial applications of vapour/gas membrane separation are [70]:

- polyolefin plant resin degassing
- gasoline vapour recovery systems at large terminals
- polyvinyl chloride manufacturing vent gas
- ethylene recovery
- natural gas processing/fuel gas conditioning.

Baker [70] in 2006 reported the following data:

- the market for vapour separation systems is at least \$20–30 million per year
- more than 100 large systems, with a cost of \$1–5 million each, have been installed
- at least 500 small systems, with a cost of \$10,000–\$100,000 each, are operating to capture vapour emissions from retail gasoline stations, industrial refrigerator units and petrochemical process vents.

The recovery of hydrocarbon monomers from polyethylene and polypropylene plants is actually the largest application of vapour separation membranes (Figure 14.12). After the production of the polyolefin resin, there are unreacted monomer and hydrocarbon solvents, dissolved in the resin powder that must be separated in order to re-use the polymer. The traditional application involves stripping with hot nitrogen, in a column known as a 'degassing bin'. The value of nitrogen and monomer are both high, therefore the recovery and reuse of these components is of great interest. For this scope a membrane operation is profitably used. It consists of two membrane units in series where the off-gas from the 'bin' is compressed at 200 bar. The first membrane unit produces a permeate stream enriched in propylene and a purified residue stream containing ~97–98% nitrogen. The vapour-enriched permeate stream is recycled to the inlet of the compressor. The nitrogen-rich residue can often be recycled directly to the degassing bin without further treatment. The residue gas is passed to a second membrane unit to upgrade the nitrogen to better than 99% purity. During the last 10 years, almost 50 of these systems have been installed around the world.

In the polymerization of vinyl chloride, side reactions generate unwanted gas and some small amounts of air leak into the reactors. These impurities must be vented from the process. However, the vented gas stream, although small, may contain several hundred

Figure 14.12 Photograph of a membrane propylene recovery system installed at a polypropylene plant. This unit recovers approximately 450 kg/h of hydrocarbons. Image courtesy of www.mtrinc.com, Copyright 2010 mtrinc

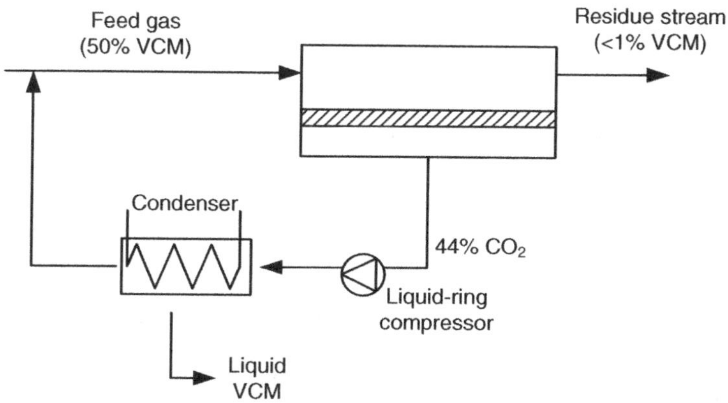

Figure 14.13 Membrane recovery of VCM monomer in a polyvinyl chloride plant. Image courtesy of www.mtrinc.com. Copyright 2010 mtrinc

thousand dollars of monomer values, owing to the high volatility of the vinyl chloride monomer (VCM). A typical flow scheme of a membrane process to recover VCM is shown in Figure 14.13. Feed gas containing VCM and air is sent to the membrane system. The VCM-enriched permeate from the membrane system is compressed in a liquid-ring compressor and cooled to liquefy the VCM. The non-condensable gases are mixed with the feed gas and returned to the membrane section. VCM recovery is more than 99%. The first unit of this type was installed by MTR in 1992. Since then, about 40 similar systems have been installed [71].

Gasoline vapour recovery has become an important field for membrane application in the last few years. Several hundred retail gasoline stations, in fact, have installed small membrane systems for the recovery of the hydrocarbon vapours during the transfer of hydrocarbons from tankers to holding tanks and then to trucks. GKSS licences have installed about 30 gasoline vapour recovery systems at fuel transfer terminals, mostly in Europe (www.gkss.de). MTR and OPW Fueling Components have developed a membrane vapour recovery system for the fuel storage tanks of retail gasoline stations. The OPW Vaporsaver™ system, fitted with MTR's membranes, reduces hydrocarbon emissions by 95–99% and pays for itself with the value of the recovered gasoline (www.mtrinc.com).

Ethylene oxide is produced through the catalytic oxidation of ethylene with 99.6% pure oxygen; carbon dioxide and water are by-products. The mixture of products is sent to a water-based scrubber to recover the ethylene oxide. Carbon dioxide is then removed by absorption with hot potassium carbonate, fresh ethylene and oxygen are added to the unreacted gases and the mixture is recycled back to the reactor. Owing to the presence of argon in the incoming oxygen and ethane in the incoming ethylene, a portion of the gases in the reactor loop must be purged to keep the concentration of these inerts under control. The purge gas for a typical ethylene oxide plant contains approximately 20–30% ethylene, 10–12% argon, 1–10% carbon dioxide, 1–3% ethane, 50% methane and 4–5% oxygen. A similar vent gas mixture is created in the production of vinyl acetate. This purge gas can be treated in a membrane-based recovery unit: ethylene preferentially permeates the membrane, producing an ethylene-enriched permeate stream and an argon-enriched residue stream.

14.4 New Metrics for Gas Separation Applications

For measuring progress towards sustainability, many efforts are being made for defining indicators of industrial process and, in particular their impact on three specific areas [72,73]:

- environmental indicators, which refer to the resource exploitation and the emission, effluents and waste related to the production
- economic indicators, which take into account the investments profits, values and taxes
- social indicators, that consider the employment situation, health and safety at work and the benefits/troubles for the community.

Most of these indicators are calculated in form of appropriate ratios. Ratio indicators, in fact, can provide a measure of impact independent of the scale of the operation, or to weigh cost against benefits and, in some cases, they can allow comparison between different operations [74]. Membrane operations are well-known for their modularity, compactness and flexibility, therefore they can be considered as new operations developed in the logic of PI. Recently, new metrics for comparing membrane performances with that of conventional units have been introduced [72,75]. With respect to the existing indicators, these new metrics take into account the size, the weight, the flexibility, the yield and the modularity of the plants. They are useful for having an immediate indication of the eventual gain that a membrane operation can offer with respect to a traditional one. In this sense, they are useful also for indicating which of a membrane or traditional operation is more suitable for a specific process.

14.4.1 Case Study: Hydrogen Separation in Refineries

The hydrogen upgrading in refineries is currently carried out by means of pressure swing adsorption (PSA) and cryogenic separation processes. However, the application of membrane systems for this type of separation is rapidly evolving to the commercial level, owing to the advantages related to the low capital costs, low energy requirements and modularity.

In the following, after a brief description of the conventional separation processes (PSA and cryogenic) a comparison among these processes and membrane systems for hydrogen separation is reported, also introducing some project considerations such as process flexibility, reliability, ease of response to the variations, expansion capability and versatility [76].

Pressure swing adsorption is a batch operation that uses multiple vessels to produce a constant product and off-gas flows. The PSA process is based on the capacity of some adsorbents (zeolites, molecular sieves, etc.) to adsorb more impurities at high gas-phase partial pressure. Proper selection of the adsorbents is critical to both the performance of the unit and adsorbent life. In the hydrogen separation, the impurities are adsorbed at higher partial pressure and then desorbed at lower partial pressure. The impurity partial pressure is lowered by 'swinging' the adsorber pressure from the feed pressure to the tail gas pressure, and by using a high-purity hydrogen purge. The hydrogen loss occurs through its use as a 'sweep gas', to carry away the contaminants, amounting to about 8–20% of the total hydrogen throughput [77]. Multiple adsorbers are used in order to provide constant feed, product and tail gas flows and each adsorber undergoes the same process steps in the same sequence. The separation is promoted by the impurity partial pressure difference between the feed and the tail gas. A minimum pressure ratio of approximately 4:1 between the feed and tail gas pressure is usually required for hydrogen separations. However, the absolute pressures of the feed and tail gas are also important, particularly to hydrogen recovery. The optimum feed pressure range for PSA units in refinery applications is 15–30 bar. The optimum tail gas pressure is as low as possible. Since vacuum is normally avoided, tail gas pressure between 1.1 and 1.4 bar are typically used when high recovery is needed. The product hydrogen from a PSA unit is available at essentially feed pressure.

Cryogenic separation is a low temperature separation process which uses the difference in boiling temperatures (relative volatilities) of the feed components to effect the separation. This process condenses the required amount of feed impurities by cooling the feed stream in multipass heat exchangers. The refrigeration required for the process is obtained by Joule–Thomson refrigeration derived from throttling the condensed liquid hydrocarbons. Additional refrigeration, if required, can be obtained by external refrigeration packages or by turboexpansion of the hydrogen product. One of the main advantage of the cryogenic process is the ability to produce separated hydrocarbon streams rich in C_{4+}, ethane/propane, etc. In order to choose the best technology, some parameters, such as those reported below, have to be taken into account.

14.4.1.1 Operating Flexibility

Operating flexibility is the ability to operate under variable feed quality conditions, either on a short-term or long-term basis.

The changes in feed composition occur very often in refinery applications, particularly when the source of the feed is a catalytic process or when the feedstock to the upstream unit changes.

In membrane processes, the increase in feed impurity concentrations tend to cause a decrease in product purity, which, however, can be maintained for small feed composition changes by adjusting the feed-to-permeate pressure ratio. In most refinery membrane applications, however, the major product impurity is methane, and this can be allowed to increase slightly in the product without major downstream impact. The response time of membrane systems is essentially instantaneous, and corrective action has immediate results. The start-up time required by the process is extremely short.

The PSA process shows a great ability to maintain hydrogen purity and recovery under changing conditions. The process is self-compensating and even relatively large changes in feed impurity concentrations have little impact on performance. As the concentration of a feed impurity increases, its partial pressure increases, increasing also the amount of the impurity which will be adsorbed. The purity of the hydrogen can be maintained constant by a simple cycle time adjustment. The response time to variations is rapid but not abrupt, generally requiring 5–15 minutes for responding to a step change in feed quality. The new steady-state upon restart following shutdown is reached in about 1 hour.

The cryogenic process has very low flexibility, because changes in the concentration of the lower boiling components of the feed affect the product purity directly. Recovery is not strongly affected. Response time is not as rapid as for PSA or membrane systems. Start-up is 8–24 hours, depending on the procedure used.

14.4.1.2 Turndown

Turndown is the capability of the system to operate at reduced capacity.

Membrane systems are highly capable of maintaining product purity even though the capacity is reduced down to 10% of the initial design. This is done by either reducing feed pressure, increasing permeate pressure, or by isolating modules from the system. The first two methods can be used for short-term operation, and the latter when operating at significantly reduced capacity for extended periods. No previsions are required in the design phase.

PSA units can maintain both recovery and product purity at throughputs of ca. 30–100% of design by adjustment of the cycle time. Between 0 and 30% of design product rates, purity can be maintained but recovery is reduced unless special provisions are made in the design.

The turndown capability of cryogenic systems strongly depends on the design. Partial condensation units can maintain product purity at slightly reduced recovery at flows down to 30–50% of design.

14.4.1.3 Reliability

Reliability takes into account the on-stream factors which can cause unscheduled shutdowns. It is an important project consideration, particularly if the process is a primary source of make-up hydrogen to a hydroprocessor or other mainstream refinery processes.

Membrane systems are extremely reliable with respect to the on-stream factor. The membrane separation process is continuous and has few control components which can cause a shutdown. Typically, the response to unscheduled shutdowns is rapid.

PSA systems are moderately reliable. The numerous valves associated with the process can cause unexpected shutdowns. The new PSA are designed with alternate modes of operation, in which 100% of design capacity can be achieved while bypassing any failed valve or instrument, with only a slight loss of recovery. Failures are automatically detected and bypassed by the microprocessor-based control system. However, stronger and periodic control cycles are required.

The cryogenic process is considered by refiners to be less reliable than the PSA or membrane processes; this is mainly due not to the process itself but to the need for feed pre-treatments. Failure of the pre-treatment system usually results in contaminants freezing in the cold box, leading to shutdown. In this case, a 'thaw' is necessary prior to restart. The pre-treatment system itself is often more complex than the cryogenic system.

14.4.1.4 Ease of Expansion

In some cases, future expansion is contemplated even during the initial phase of a project.

Expansion of membrane systems is very easy, since this only requires the addition of identical modules.

PSA systems can also be expanded, but it requires additional design considerations and adds cost in the initial phase of the project.

The cryogenic units cannot be expanded if it is not foreseen during the design phase. Generally they can be over-dimensioned and a capacity increase is often obtained without modification to the cold box itself through addition of a tail gas compressor.

14.4.1.5 By-product Value

By-product value is a measure of the capability of producing a high-value tail gas. In many refinery hydrogen upgrading processes, the impurities to be rejected include hydrocarbons which have value significantly above fuel value. This is particularly true for olefin-containing streams. The relative amounts of high-value hydrocarbons and the incremental cost of further separation determine whether by-product recovery is an important parameter to be considered.

The cryogenic process is best suited to applications involving by-product hydrocarbon recovery. It is possible to separate hydrocarbon streams containing C_2/C_3 and C_{4+} components, with recovery of these components generally quite high, with typical overall values of 90% for the C_2/C_3 components and up to 100% of the C_{4+}.

The membrane process is not capable of providing separate streams rich in specific hydrocarbon fractions; however, it is also used in these applications, since the hydrocarbon-rich stream is available at high pressure.

By-product hydrocarbon recovery is normally not economical when the PSA process is used, because the single hydrocarbon-rich tail gas is delivered at too low a pressure.

Combination of separation processes, such as membrane and cryogenic processes, may be applicable to these applications.

14.4.1.6 Comparison

Table 14.2 summarizes the comparison among the project parameters described above for the three operations considered in the case of H_2 separation.

The selection of the hydrogen purification process may be driven by specific considerations, strictly related to the output to be obtained. The feed composition (H_2%), the feed

Table 14.2 Comparison among some important project parameters [76]

	Membrane system	PSA	Cryogenic
Operating flexibility	Moderate	High	Low
Response to variations	Instantaneous	Rapid (5–15 minutes)	Slow
Start up after the variations	Extremely short (10 minutes)	1 h	8–24 h
Turndown	Down to 10%	Down to 30%	Down to 30–50%
Reliability	100%	95%	Limited
Control requirement	Low	High	High
Ease of expansion	Very high (modularity)	Moderate	Very low
By-product value	Moderate	Not economical	High

conditions (pressure and temperature), the product purity and the final destination of the product strongly affect the choice.

The composition of the feed and its variability has a large impact on the selection of a hydrogen separation process because it influences the performance, reliability and pre-treatment required by the three upgrading processes. Membrane systems are suitable for a wide range of feed compositions, owing to the possibility of driving the process acting on different parameters such as feed and permeate pressure, temperature, flow configuration, etc. Streams with 75–90 vol% hydrogen are most economically upgraded also by PSA with the selection being based on flow, pressure and pre-treatment requirements. Cryogenic upgrading is applicable to large streams with 30–75 vol% hydrogen.

Feed pressure and product flow rates are best considered together when selecting a hydrogen purification process because the three processes have drastically different economies of scale.

The membrane systems are the lowest capital cost alternative for small (less than 30 000 m^3/h product) flow rates, since the cost of a membrane system is proportional to the number of modules required. If the feed gas is already at high pressure or if there is downstream hydrocarbon recovery from the non-permeate, where the high pressure of this stream can be used to advantage, the membrane systems can be used also for larger flow rates. For small flow rates at high pressure, such as a purge stream from the high pressure separator of a hydroprocessor, membrane systems are the most economical.

PSA systems can be economical for flow rates from 10^3 to 10^5 m^3/h already compressed at 15–30 bar. Differently from membrane systems, PSA units cannot take advantage of high available feed pressures (50–70 bar) and the capital and operating costs associated with feed, product and/or tail gas compression are almost always a significant portion of the total separation costs and globally higher than that required by membrane systems. Cryogenic systems have high capital costs at low product rates, but have good economies of scale. This process is most applicable for feed pressures higher than 20 bar and for larger systems. However, there are few large, high pressure raw hydrogen streams available in most refineries, therefore the cryogenic system is currently limited. In general, when hydrogen upgrading alone is considered, small systems with high available feed pressure favour membrane systems, small systems with low feed pressure favour PSA and membrane systems, and large systems at high pressure favour cryogenic systems.

Table 14.3 Comparison among the three units as function of feed and product conditions

	Membrane system	PSA	Cryogenic
Feed composition, (H_2 %)	30–90%	75–90%	30–75%
Feed pressure and product flow	>10 bar <30 000 m^3/h	15–30 bar 1000–10^5 m^3/h	>20 bar >10^5 m^3/h
Product purity	90–98%	>99%	90–98%

Hydrogen product purity and the levels of specific product impurities are, of course, critical to process selection. Cryogenic and membrane processes normally produce hydrogen at 90–98 vol%, whereas the PSA process normally produces hydrogen at 99+ vol%. However, depending on the final destination of the upgraded hydrogen, it can be more conveniently applied to other operations (Table 14.3). If the upgraded hydrogen is used as a primary source of make-up hydrogen to a high pressure hydrotreater or hydrocracker, a high purity hydrogen is required and the PSA is often the best choice. If the upgraded hydrogen is only an incremental portion of the make-up hydrogen to a hydroprocessor, lower product purity is usually required and membrane systems are preferable, owing to the lower capital costs. If the feedstock to be upgraded contains only hydrogen and hydrocarbons, then the principal impurity in the hydrogen product will be methane. When 3–10 vol% methane can be tolerated in the hydrogen product, the membrane and cryogenic systems are the most economical. If the feed to the upgrading process contains substantial quantities of CO and CO_2, then PSA and the membrane units are most often selected, whereas cryogenic is not suitable.

Apart from the economic considerations, the choice among the three separation processes is generally made taking into account the specific process from which the hydrogen is produced and the specific requirements that the upgraded H_2 must have, also considering the final destination of the stream. Table 14.4 reports some selection guidelines of the most important processes of hydrogen production in refineries [77].

The largest source of easily recovered hydrogen in a refinery is the off-gas from catalytic reforming. This off-gas typically contains from 70–90+ vol% hydrogen with the balance being C_1 to C_{6+} hydrocarbons. If large quantities of reformer off-gas are to be upgraded for use as a primary source of make-up hydrogen for a hydrocracker or hydrotreater, the PSA process is normally used. The high hydrogen content makes the adsorbent requirements low, and the hydrocarbons in the tail gas sent to fuel are relatively small with respect to the feed gas. The aromatic and HCl content of the feed do not require special pre-treatment. In some cases, there is a requirement for relatively small quantities of hydrogen at 98+ vol% purity for uses such as catalyst regeneration. Membrane systems using feed compression may be most economical for these applications requiring less than about 1000 m^3/h of hydrogen product, owing to low capital costs. For large flows, cryogenic upgrading has sometimes been used. However, the feed hydrogen purity is often too high, and external refrigeration is required. Aromatic removal in the pre-treatment section is also an important design consideration. Unless recovered hydrocarbons are of high value, the cryogenic process is not normally used to process reformer off-gas alone.

The high pressure and low pressure purge gases from high pressure hydrocrackers and hydrotreaters are good candidates for hydrogen upgrading. The recovered hydrogen is

Table 14.4 Selection guidelines for specific application processes

Process application	Feed conditions	Membrane system	PSA	Cryogenic
Catalytic reformer off-gas	70–90%$_{vol}$ H_2 30–10% C_1 to C_6+ ppv of aromatics and HCl Pressure: 15–30 bar Room temperature	H_2 purity: 98%. Final destination: uses such as catalyst regeneration	H_2 purity: 99+% H_2 recovery: 82–90% Final destination: primary source of make up hydrogen for hydrocracker	Not used due to the necessity of pretreatment for aromatic removals
Hydroprocessor purge gases	High pressure purge gas: 75–90%$_{vol}$ H_2 Hydrocarbon balance Pressure: 50–200 bar Room temperature	H_2 purity: 92–98% H_2 pressure: 20–40 bar H_2 recovery: 85–95% Final destination: make-up hydrogen compressor	Not used because of the high feed pressure	Not used due to the low flow rates
	Low pressure purge gas: 50–75%$_{vol}$ H_2 Hydrocarbon balance Pressure: 5–20 bar Room temperature	Not used because of the low feed pressure	H_2 purity: 99% H_2 pressure: 1 bar H_2 recovery: 80–90% Final destination: primary source of make up hydrogen for hydrocracker	
FCC off-gas and other refinery purge streams	15–50% vol H_2 Hydrocarbon or olefins balance Pressure: 5–20 bar Room temperature	Feed compression: > 25 bar H_2 purity: 80–90% H_2 pressure: 5–20 bar H_2 recovery: 50–85% Final destination: low pressure hydrotreaters The tail gas is sent to downstream hydrogen recovery units	Not used because of the low H_2 concentration in the feed	H_2 purity: >95% H_2 pressure: 15–30 bar Recovery of a mixed stream containing >99+% of C_2+ components
Ethylene off-gas	80–90%vol H_2 CO, CO_2, CH_4 balance Pressure: 15–30 bar	H_2 purity: 95–97% H_2 recovery: 80–90% Final destination: make-up hydrogen compressor	H_2 purity: 99.5% H_2 recovery: 80–90% Final destination: primary source of make up hydrogen for hydrocracker	H_2 purity: 95% H_2 pressure: 15–30 bar Final destination: make-up hydrogen compressor

returned to the hydroprocessor with the make-up hydrogen. Many hydroprocessors have both high and low pressure purge streams. The high pressure purge streams are available at 50–200 bar and contain 75–90 vol% hydrogen, with the balance being hydrocarbons. Low pressure purge streams are available at much lower pressure, typically 5–20 bar and have hydrogen contents ranging from 50 to 75 vol%.

The membrane process is the most economical process for high pressure purge gas upgrading. The product delivery pressure is chosen to allow the product to enter one of the stages of the make-up hydrogen compressors. Low pressure purge gases are usually upgraded by the PSA process. The PSA process is better suited than the cryogenic process because the flow rates are relatively small and the stream composition can be highly variable. The combination of lower pressure and lower hydrogen content makes the PSA system less economical than the membrane system. The membrane system usually gives the highest rate of return on investment, the tail gas being at high pressure.

Hydrogen can be recovered from FCC off-gas and other low pressure refinery purge streams of low hydrogen content. Depending on flow rate, feed composition and variability, feed pressure and required hydrogen product purity, either the cryogenic or membrane process can be used. In almost all cases, the stream is not upgraded for its hydrogen content alone, but also the tail gas is of value. If there are valuable hydrocarbons, particularly olefins, which can be recovered in addition to hydrogen, or if hydrogen product purity in excess of 90 vol% is required, the cryogenic process is normally used. However, feed quality variations and contaminant levels are important considerations in determining whether the cryogenic process is appropriate. The membrane process can recover hydrogen efficiently from these streams at a hydrogen purity of 80–90 vol%. The low purity hydrogen product can be used effectively in some applications, such as low pressure hydrotreaters, and the tail gas can be sometimes sent to downstream hydrocarbon recovery units, being already compressed.

The ethylene production process produces large amounts of hydrogen which can be easily upgraded for refinery use if the ethylene plant is in close proximity to the refinery. The ethylene process uses a series of cryogenic units to separate the products from the ethylene furnace, and a high purity hydrogen stream is produced. Typically, only a small fraction of the available hydrogen is used in the ethylene plant (for acetylene hydrogenation) and the balance is sent to fuel unless it can be exported. The hydrogen-rich ethylene off-gas normally contains 80–90 vol% hydrogen, with CO, CH_4, ethylene and nitrogen as impurities. In few cases, the cryogenic system is employed, producing a hydrogen purity as high as 95 vol% with the same impurities through use of external refrigeration. Regardless of the hydrogen purity, the ethylene off-gas must be processed to remove CO to ppmv levels for general refinery use. For off-gas with 80–90 vol% hydrogen, the membrane systems are the most suitable for upgrading. The pressure and the high H_2 concentration, in fact, allow a high purity product to be produced, even though lower (95–97 vol%) than for the PSA system. Furthermore, the retentate gas stream is already compressed, on the contrary to the PSA systems that require tail gas compression, from 1.1–1.3 bar to fuel system pressure. The choice between the membrane and PSA processes will primarily depend upon the cost of compressing the membrane hydrogen product compared to the cost of compressing the PSA tail gas, assuming high hydrogen recovery is desired in both cases. The product purity from the membrane system will be lower (95–97 vol%) than for the PSA system.

For the process selection, general selection guidelines can be helpful, at least in eliminating an inappropriate process. In order to use such guidelines, feed characteristics, contaminant levels, required product purity, allowable product impurities, pressure levels and flow rates must be known. These parameters can be used in conjunction with experience-based, application-specific guidelines to select the optimum process. The membrane systems, owing to their high flexibility, reliability, modularity and ease of control, compete well with the other two consolidated technologies, in particular for some specific applications typical of refinery hydrogen upgrading.

In order to compare the three different systems, energy and mass intensity [78,79] were calculated as reported below, taking into account the power required by the system (Equation 14.1) and the steam and cooling water necessary (Equation 14.2). Moreover, a new indicator was defined as productivity to footprint ratio (Equation 14.3).

$$\text{energy intensity} = \frac{\text{power required}}{\text{H}_2 \text{ product flow rate}} \tag{14.1}$$

$$\text{mass intensity} = \frac{\text{steam and cooling water required}}{\text{H}_2 \text{ product flow rate}} \tag{14.2}$$

$$\text{productivity/footprint} = \frac{\text{H}_2 \text{ product flow rate}}{\text{footprint}} \tag{14.3}$$

Table 14.5 summarizes the data for H_2 recovery from refinery off-gas by means of three different separation technologies, as reported by Spillman [80], considering polyimide membranes. A comparison among them is provided by the calculation of the relative three indicators. A lower investment cost than pressure swing adsorption (PSA) or cryogenic separation was estimated for the H_2 recovery from refinery off-gas by polymeric membranes. This comparison is from 1989; however, since then polymeric membrane capital prices have dropped.

Table 14.5 *H_2 recovery from refinery off-gas (data adapted from Reference 80)*

	Membrane system (80 °C)	Membrane system (120 °C)	PSA	Cryogenic
H_2 recovery, %	87	91	73	90
H_2 purity, %	97	96	98	96
Product flow rate, m^3/h	3257	3375	2643	3375
Power, kW	220	220	370	390
Steam, kg/h	**230**	**400**	–	**60**
Cooling water, t/h	**38**	**38**	**64**	**99**
Investment, $ millions	**1.12**	**0.91**	**2.03**	**2.66**
Installation area, m^2	**8**	**5**	**60**	**120**
Metrics				
Energy intensity, kJ/kg H_2	2808	2738	5760	4853
Mass intensity, kg/kg H_2	136	133	277	342
Productivity/footprint, kg H_2 /(h m^2)	35	58	3.9	2.4

As can be seen, the energy intensity of the membranes system, for both the temperature values considered, is lower than that of the other two processes. This means that the energy required for producing a fixed amount of hydrogen is, in any case, significantly lower than that required by the other operations. Also the addition of the steam and cooling water is at least half of that necessary for the traditional systems. The productivity to footprint ratio calculated for the membrane technology is very interesting in the process intensification logic. Considering the same area occupied by the unit, the membrane systems show more than tenfold higher productivity.

References

(1) Advanced Prism® membrane systems for cost effective gas separations, http://www.airproducts.com/NR/rdonlyres/81FB384C-3DB5-4390-AED0-C2DB4EEDDB18/0/PrismPGS.pdf.

(2) F. Dautzenberg, M., Mukherjee, Process intensification using multifunctional reactors, *Chem. Eng. Sci.*, **56**, 251–267 (2001).

(3) A. Stankiewicz, J. A. Moulijn, Process intensification, *Ind. Eng. Chem. Res.*, **41** (8), 1920–1924 (2002).

(4) J. C. Charpentier, Process intensification, a path to the future, *Ing. Quim.* (Madrid), **38** (434), 16–24 (2006).

(5) R. W. Baker, Membranes for vapor/gas separation, (2006). http://www.mtrinc.com/publications/MT01%20Fane%20Memb%20for%20VaporGas_Sep%202006%20Book%20Ch.pdf.

(6) W. J. Koros, R. Mahajan, Pushing the limits on possibilities for large-scale gas separation: which strategies? *J. Membr. Sci.*, **175**, 181–196 (2000).

(7) W. J. Koros, I. Pinnau, Membrane formation for gas separation processes, in Polymeric gas separation membranes, D. R. Paul, Y. Yampolskii, (Eds.); CRC Press: Boca Raton, FL (1994).

(8) B. D. Freeman, Basis of permeability/selectivity trade-off relations in polymeric gas separation membranes, *Macromolecules*, **32**, 375 (1999).

(9) L. M. Robeson, Correlation of separation factor versus permeability for polymeric membranes. *J. Mem. Sci.*, **62**, 165–171 (1991).

(10) R. W. Baker, Future directions of membrane gas separation technology, *Ind. Eng. Chem. Res.*, **41**, 1393–1404 (2002).

(11) P. Bernardo, E. Drioli, Golemme G., Membrane gas separation. A review/state of the art, *Ind. Eng. Chem. Res.*, DOI: 10.1021/ie8019032 • Publication Date (Web): 22 April 2009.

(12) T. C. Merkel, B. D. Freeman, R. J. Spontak, Z. He, I. Pinnau, P. Meakin, A. J. Hill, Ultrapermeable, reverse-selective nanocomposite membranes. *Science*, **296**, 519–524 (2002).

(13) S. Hess, C. S. Bickel, R. N. Lichtenthaler, Propene/propane separation with copolyimide membranes containing silver ions., *J. Membr. Sci.*, **275**, 52–59 (2006).

(14) M. B. Hägg, T.-J. Kim, B. Li, Membrane for separating CO_2 and process for the production thereof, WO Patent 2005089907 (2005).

(15) T.-J. Kim, B. Li, M. B. Hägg, Novel fixed-site-carrier polyvinylamine membrane for carbon dioxide capture. *J. Polym. Sci., Part B: Polym. Phys.*, **42/43**, 4326–4332 (2004).

(16) www.nanoglowa.com.

(17) R. Dittmeyer, V. Höllein, K. Daub, Membrane reactors for hydrogenation and dehydrogenation processes based on supported palladium, *J. Mol. Cat. A: Chem.*, **173**, 135–184 (2001).

(18) J. Caro, M. Noack, P. Kolsch, R. Schafer, Zeolite membranes – state of their development and perspective. *Micropor. Mesopor. Mater.*, **38**, 3–16 (2000).

(19) M. Yang, B. D. Crittenden, S. P. Perera, H. Moueddeb, J. A. Dalmon, The hindering effect of adsorbed components on the permeation of a non-adsorbing component through a microporous silicalite membrane: the potential barrier theory. *J. Membr. Sci.*, **156**, 1–9 (1999).

(20) P. Bernardo, C. Algieri, G. Barbieri, E. Drioli, Hydrogen purification from carbon monoxide by means of selective oxidation using zeolite catalytic membranes, *Sep. Purif. Technol.*, **62**, 629–635 (2008).
(21) P. Bernardo, C. Algieri, G. Barbieri, E. Drioli, Catalytic (Pt–Y) membranes for the purification of H_2–rich streams, *Catalysis Today*, **118**, 90–97 (2006).
(22) Y. Morigami, M. Kondo, J. Abe, H. Kita, K. Okamoto, The first large-scale pervaporation plant using tubular-type module with zeolite NaA membrane. *Sep. Purif. Technol.*, **25**, 251–258 (2001).
(23) E. E. McLeary, J. C. Jansen, F. Kapteijn, Zeolite based films, membranes and membrane reactors: Progress and prospects. *Micropor. Mesopor. Mater.*, **90**, 198–210 (2006).
(24) P. S. Tin, Y. C. Xiao, T. S. Chung, Polyimide-carbonized membranes for gas separation: structural, composition and morphological control of precursors. *Sep. Purif. Rev.*, **35**, 285–292 (2006).
(25) S. M. Saufi, A. F. Ismail, Fabrication of carbon membranes for gas separation – a review, *Carbon*, **42**, 241–253 (2004).
(26) C. Liang, G. Sha, S. Guo, Carbon membrane for gas separation derived from coal tar pitch. *Carbon*, **37**, 1391–1396 (1999).
(27) J. E. Koresh, A. Soffer, The carbon molecular-sieve membranes – General properties and the permeability of CH_4/H_2 mixture. *Sep. Sci. Technol.*, **22**, 973–978 (1987).
(28) Y. S. Lin, Microporous and dense inorganic membranes: current status and prospective. *Sep. Purif. Technol.*, **25**, 39–54 (2001).
(29) J. Marriott, E. Sørensen, A general approach to modelling membrane modules, *Chem. Eng. Sci.*, **58**, 4975–4990 (2003).
(30) R. Baker, Membrane Gas-Separation: Applications. In membrane operations, E. Drioli, L. Giorno (Eds.), 2009, Wiley-VCH.
(31) L. M. Robeson, The upper bound revisited, *J. Membr. Sci.*, **320**, 390–400 (2008).
(32) S. M. Weller, W. A. Steiner, Separation of gases by fractional permeation through membranes. *J. Appl. Phys.*, **21**, 279 (1950).
(33) R. W. Spillman, Economic of gas separation by membranes, *Chem. Eng. Prog.*, **85**, 41–49 (1989).
(34) E. Sanders, D. O. Clark, J. A. Jensvold, H. N. Beck, G. G. Lipscomb, F. L. Coan, Process for preparing powadir membranes from Tetrahalobisphenol A polycarbonate, *U.S. Patent* 4772392 (1988).
(35) O. M. Ekiner, R. A. Hayes, P. Manos, Novel multicomponent fluid separation membranes, U.S. Patent 5085676 (Feb. 1992).
(36) N. W. Ockwig, T. M. Nenoff, Membranes for hydrogen separation, *Chem. Rev.*, **107**, 4078–4093 (2007).
(37) R. R. Zolandz, G. K. Fleming, Gas permeation applications, in W. S. W. Ho, K. K. Sirkar, (Eds.), Membrane handbook; Chapman &Hall: New York, pp.78–94, 1992.
(38) www.airproducts.no.
(39) www.airproducts.com.
(40) Air Liquide to install ASU for Dongbei Special Steel Group, *China Chemical Reporter*, May 2007.
(41) UBE expands gas separation membrane production, *Membrane Technology*, November 2006.
(42) S. L. Matson, W. J. Ward, S. G. Kimure, W. R. Browall, Membrane oxygen enrichment. II: Economic assessment. *J. Membr. Sci.*, **29**, 79–84 (1986).
(43) B. D. Bhide, S. A. Stern, A new evaluation of membrane processes for the oxygen enrichment of the air. Identification of optimal operating conditions and process configurations. *J. Membr. Sci.*, **62**, 13–21 (1991).
(44) A. Figoli, W. F. C. Sager, M. H. V. Mulder, Facilitated transport in liquid membranes: review and new concepts. *J. Membr. Sci.*, **97**, 181–194 (2001).
(45) S. Liu, G. Gavalas, Oxygen selective ceramic hollow fiber membranes, *J. Membr. Sci.*, **246**, 103–110 (2005).
(46) X. Tan, Z. Pang, K. Li, Oxygen production using La0.6Sr0.4Co0.2Fe0.8O3–α (LSCF) perovskite hollow fiber membrane modules. *J. Membr. Sci.*, **310**, 550–558 (2008).

(47) P. A. Armstrong, D. L. Bennett, E. P. Foster, E. E. Stein, ITM oxygen: the new oxygen supply for the new igcc market, presented at gasification technologies 2005, San Francisco, California, 9–12 October. 2005.
(48) R. Bounaceur, N. Lape, D. Roizard, C. Vallieres, E. Favre, Membrane processes for post-combustion carbon dioxide capture: A parametric study, *Energy*, **31**, 2556–2570 (2006).
(49) A. K. Datta, P.K. Sen, Optimization of membrane unit for removing carbon dioxide from natural gas. *J. Membr. Sci.*, **283**, 291–300 (2006).
(50) D. Dortmundt, K. Doshi, Recent developments in CO_2 removal membrane technology, UOP LLC, Des Plaines, Illinois (1999).
(51) R. W. Baker, K. Lokhandwala, Natural gas processing with membranes: an overview, *Ind. Eng. Chem. Res.*, **47**, 2109–2121 (2008).
(52) R. M. Kelly, Process for separating CO_2 from other gases, *US Patent* 4659343 Cynara, 1987.
(53) A. Callison, G. Davidson, Offshore processing uses membranes for CO_2 removal, *Oil & Gas Journal* (2007).
(54) R. W. Baker, T. Hofmann, J. Kaschemekat, K. A. Lokhandwala, Field demonstration of a membrane process to recover heavy hydrocarbons and to remove water from natural gas. Annual Report, Contract Number DE-FC26-99FT40723, Report Period Ending September 29, 2001.
(55) http://www.airproducts.com/Products/Equipment/PRISMMembranes/page05.htm.
(56) C. C. Tannerhill, Nitrogen removal requirement from natural gas; Gas Research Institute Topical Report GRI-99/0080; Gas Research Institute: Washington, D.C., May 1999.
(57) K. Lokhandwala, Membrane-Augmented Cryogenic Methane/Nitrogen Separation. *US Patent* 5647227 (1997).
(58) H. Herzog, What future for carbon capture and sequestration? *Environ. Sci. Technol.*, **35**, 148–153 (2001).
(59) C. M. White, Separation and capture of CO_2 from large stationary sources and sequestration in geological formations, *J. Air Waste Manage. Assoc.*, **53**, 645–715 (2003).
(60) E. Favre, Carbon dioxide recovery from post-combustion processes: Can gas permeation membranes compete with absorption? *J. Membr. Sci.*, **294**, 50–59 (2007).
(61) O. Davidson, B. Metz, Special report on carbon dioxide capture and storage, International Panel on Climate Change, Geneva, Switzerland, 2005. www.ipcc.ch.
(62) E. Drioli, M. Romano, Progress and new perspectives on integrated membrane operations for sustainable industrial growth, *Ind. Eng. Chem. Res.*, **40**, 1277–1300 (2001).
(63) M. Simmonds, P. Hurst, M. B. Wilkinson, C. Watt, C. A. Roberts, A study of very large scale post combustion CO_2 capture at a refining & petrochemical complex, 2003. www.co2captureproject.org.
(64) R. Bounaceur, N. Lape, D. Roizard, C. Vallieres, E. Favre, Membrane processes for post-combustion carbon dioxide capture: a parametric study, *Energy*, **31**, 2556–2570 (2006).
(65) F. Scura, G. Barbieri, E. Drioli, CO_2 removal from power plants flue gas: comparison of different membrane gas separation configurations, EPIC – European Process Intensification Conference, Copenhagen (Denmark), 19–20 September, 2007.
(66) R. W. Baker, Recent developments and future directions in membrane modules, proceedings of 12th Aachener membran kolloquim, October 29–30, 2008, Aachen (Germany).
(67) S. J. Chung, J. H. Park, D. Li, J. I. Ida, I. Kumakiri, J. Y. S. Lin, Dual-phase metal-carbonate membrane for high-temperature carbon dioxide separation. *Ind. Eng. Chem. Res.*, **44**, 7999–8005 (2005).
(68) H. Kawamura, T. Yamaguchi, B. N. Nair, K. Nakagawa, S. A. Nakao, Dual ion conducting membrane for high temperature CO_2 separation. *J. Chem. Eng. Jpn.*, **38**, 322–330 (2005).
(69) R. Bredesen, K. Jordal, O. Bolland, High-temperature membranes in power generation with CO_2 capture. *Chem. Eng. Process*, **43**, 1129–1137 (2004).
(70) R. W. Baker, Membranes for vapor/gas separation, 2006. http://www.mtrinc.com/publications/MT01%20Fane%20Memb%20for%20VaporGas_Sep%202006%20Book%20Ch.pdf.
(71) R. J. Lahiere, M. W. Hellums, J. G. Wijmans, J. Kaschemekat, Membrane vapor separation: recovery of vinyl chloride monomer from pvc reactor vents. *Ind. Eng. Chem. Res.*, **32**, 2236–2241 (1993).

(72) A. Criscuoli, E. Drioli, New metrics for evaluating the performance of membrane operations in the logic of process intensification. *Ind. Chem. Eng. Res.*, **46**, 2268–2271 (2007).
(73) S. K. Sikdar, Sustainable development and sustainability metrics, *AIChE J.*, **49**, 1928–1932 (2003).
(74) IChemE. Sustainable development progress metrics: recommended for use in the process industries; institution of chemical engineers: Rugby, U.K.; 2006, 1–28. http://www.icheme.org/sustainability/metrics.pdf.
(75) A. Brunetti, A. Caravella, G. Barbieri, E. Drioli, Simulation study of water gas shift in a membrane reactor, *J. Membr. Sci.*, **306**, 329–340 (2007).
(76) G. Q. Miller, J. Stöcker, Selection of a hydrogen separation process, 1989 NPRA Annual Meeting held March 19–21, 1989, San Francisco, California (USA).
(77) G. Q. Miller, J. Stöcker, Selection of a hydrogen separation process, 4th European Technical Seminar on Hydrogen Plants, Lisbon (Portugal), Oct 2003.
(78) A. D. Curzons, D. J. C. Constable, D. N. Mortimera, V. L. Cunningham, So you think your process is green, how do you know? – Using principles of sustainability to determine what is green – a corporate perspective, *Green Chem.*, **3**, 1 (2001).
(79) A. A. Martins, T. M. Mata, C. A. V. Costa, S. K. Sikdar, Framework for sustainability metrics, *Ind. Eng. Chem. Res.*, **46**, 2962 (2007).
(80) R. Spillman, Economics of gas separation membranes, *Chem. Eng. Progr.*, **41**, (1989).

15

Evolution of Natural Gas Treatment with Membrane Systems

Lloyd S. White
UOP LLC, Des Plaines, Illinois, USA

15.1 Introduction

The first commercial membrane system for carbon dioxide removal from natural gas was reported operational in 1983, utilizing cellulose acetate membranes [1,2]. Installations are now worldwide and systems continue to expand in size and scope. In keeping membrane treatment competitive with other separation technologies there have been incremental improvements in membranes, module designs and system performance. With the recent swings in energy prices the demand for natural gas treatment continues to grow. Although the membrane itself has attracted the most research efforts, there are many steps between the identification of a working membrane to the development of a commercially viable process. New generations of cellulose acetate membranes packaged as spiral wound modules are improving the economics and utility in natural gas treatment.

Membranes for natural gas treatment have been employed since the 1980s, with the earlier critical discoveries that cellulose acetate membranes for reverse osmosis (RO) could be transformed into gas separation membranes [3,4]. Since membranes for treatment of water have a high affinity for water transport, a molecule with unique polar properties, there is also an affinity for other polar molecules such as carbon dioxide and hydrogen sulfide. Water, CO_2 and H_2S are impurities in natural gas (methane) termed 'acid gases' that can promote corrosion of steel. Since pipelines are used in the transport

Membrane Gas Separation Edited by Yuri Yampolskii and Benny Freeman

of natural gas from the well-head to the consumer, the control of acid gases is critical. Typical specifications for pipeline quality natural gas call for amounts of less than 2% CO_2, 120 ppm water and 4 ppm H_2S [5].

Early formulations of reverse osmosis membranes called for mixtures of cellulose diacetate and/or cellulose triacetate that were prepared as flat sheet membranes [6,7]. Typically RO membranes are treated with a conditioning agent such as glycerol to allow membranes to be handled as dry sheets. These sheets are assembled into spiral-wound modules for commercial installation. Glycerol allows for easy rewetting with water and physically holds open the pores of the asymmetric microporous membrane structure. But glycerol also physically inhibits the passage of gases and therefore is not appropriate for use in gas separation membranes.

In order to convert RO membranes to gas separations, various techniques such as freeze drying and solvent exchange of water wet cellulose acetate (CA) membranes have been utilized [3,4,8]. When water is directly evaporated capillary forces collapse the microporous membrane structure leading to a denser active surface layer and loss of flux. Freeze drying is a technique that protects the microporous structure with cold temperature. Solvent exchange uses a polar solvent such as alcohol to first wash out and replace water, followed by a low surface tension solvent such as hydrocarbons to wash out the first solvent. This second solvent with low surface tension can then be evaporated while maintaining the pore structure. Another drying technique that has been reported is addition of a low surface tension hydrophobic organic compound such as n-octane that protects the pore walls from the interfacial tension of water and allows direct drying of the membrane structure [9]. n-Octane is preferably incorporated as part of the original dope formulation from which the membrane is formed by the phase inversion process.

Many of the design philosophies required for successful operation of a reverse osmosis system are transferable to natural gas treatment. Reverse osmosis systems can operate at high pressures, even above 172 bar (2500 psi), to overcome solutions with high osmotic forces. Natural gas streams can also be at high pressure, up to 138 bar (2000 psi), with increased productivity possible at higher pressures. In both applications better economics are achieved by employing modules with multi-year service life. Both water and natural gas are natural resources which can be contaminated with materials that can harm long-term membrane performance. Pre-treatment systems to control the quality of the feedstock are therefore a critical component for both in commercial applications.

Early designs for spiral wound modules or hollow fibre bundles came from the RO industry. Modules for treatment of natural gas have also borrowed from these technologies. A skid containing spiral wound modules built in the 1980s for gas treatment is shown in Figure 15.1. This early array had six high pressure housings while current installations for natural gas treatment can have hundreds of tubes.

This chapter will describe some of the market forces driving membrane growth in natural gas treatment, explore some of the competing technologies to CA membranes, examine membrane compaction and the implication to long-term performance, report both recent laboratory and field studies with CA membranes and comment on some of the technical challenges and trends existing today in natural gas treatment with membrane systems that would benefit from future research and development activities.

Figure 15.1 A gas treatment skid for hydrogen recovery from the 1980's containing cellulose acetate membranes as spiral wound modules. Image courtesy of W. R. Grace, Copyright (1988) W. R. Grace

15.2 Market for Natural Gas Treatment

Natural gas prices have tripled since 2002, which increases the incentive for treatment by producers to improve the quality and increase supply. The top 20 countries for proven gas reserves are listed in Table 15.1. This list is dominated by the top three of Russia, Iran and Qatar. The three top natural gas producing countries are the combined Russia and former Soviet Union states, the United States and Canada. Worldwide in 2007 there were 15 new gas plants, expansions and modifications scheduled and this number jumped to 24 for 2008 [10]. Opportunities for new membrane installations have been expanding.

Figure 15.2 breaks down the major technologies in use in 2001 for control of acid gases in natural gas. A good percentage of natural gases are not treated, but the dominant technology in use is amine treaters. In amine treaters aqueous solutions of various alkanolamines are contacted with natural gas in order to scrub out the acid gases. The 'other technology' includes various solid adsorbent systems and cryogenics. In 2001 only 1% by volume of natural gas was treated by membranes, indicating that substantial new opportunities for membranes are available.

Natural gas comes from a variety of sources which influences the composition of the gas streams.

- Natural gas from crude oil wells, termed 'associated gas', which has been in the presence of crude oil. Heavy hydrocarbons will likely be present.
- Natural gas from gas wells or condensate wells is termed 'unassociated gas'. Gas wells produce raw natural gas while the condensate wells also contain light hydrocarbons known as natural gas condensate or natural gasoline.
- Coalbed gas coming from methane deposits in coal seams.

Table 15.1 Top 20 of the major natural gas reserves worldwide in trillions of cubic feet (tcf, 35.3 ft^3 = 1 m^3), % share of the total volume and number of production units [10]

Rank	Country	Proven gas reserves (tcf)	% share	Number of plants
1	Russia	1680.0	27.2	24
2	Iran	948.2	15.3	22
3	Qatar	905.3	14.6	2
4	Saudi Arabia	252.6	4.1	10
5	USA	211.1	3.4	572
6	Abu Dhabi	198.5	3.2	
7	Nigeria	184.0	3.0	4
8	Venezuela	166.3	2.7	14
9	Algeria	159.0	2.6	4
10	Iraq	112.0	1.8	4
11	Kazakhstan	100.0	1.6	4
12	Turkmenistan	100.0	1.6	
13	Indonesia	94.0	1.5	13
14	Malaysia	83.0	1.3	5
15	China	80.0	1.3	
16	Norway	79.1	1.3	4
17	Uzbekistan	65.0	1.1	2
18	Egypt	58.5	0.9	19
19	Canada	58.2	0.9	967
20	Kuwait	55.5	0.9	4

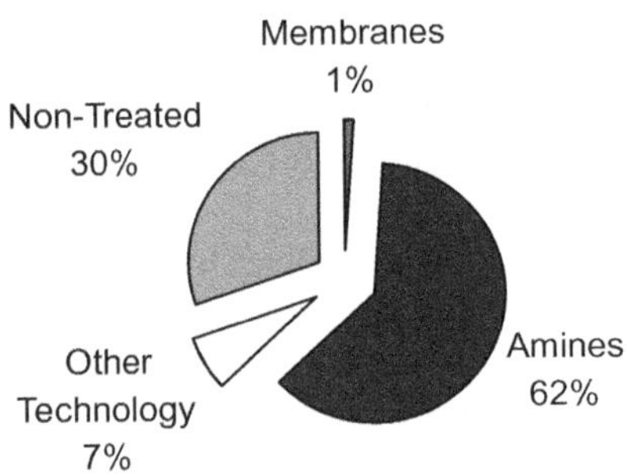

Figure 15.2 Estimate in 2001 of percentage of natural gas treated by various technologies to remove acid gases

- Town gas from gasification of coal.
- Biogas from the anaerobic decay of organic matter or biomass. Biogas is at low pressure and contains mostly methane and CO_2.
- Landfill gas from the decomposition of waste in landfills is also mostly methane and CO_2, again at low pressure.
- Enhanced oil recovery (EOR) where liquefied CO_2 is injected into the ground under high pressure to force out oil and natural gas. High CO_2 content is present in the natural

gas recovered and increases with the age of the field. In addition to methane recovery, the CO_2 can be recovered for recycling into the EOR operation.
- Off-shore platforms where space is a premium and the locations can be remote. Partial treatment of natural gas is used to protect pipelines to shore.
- On-shore treatment of natural gas produced off-shore.

Landfill and biogas are often small and localized operations where the methane is at low pressure and requires compression for membrane treatment. The methane gathered in these operations is often locally consumed. Both coal gasification and natural gas from Marcellus Shale on the East Coast have recently been championed in the United States as additional sources of fuels to help in the transition from a dependence on foreign oil to alternative energy sources.

Separex (now part of UOP) reported in 1985 an early membrane system as a trailer mounted skid. This skid treated 3.5 MMSCFD (105 000 m^3/day) at an EOR application [11]. The trailer allowed for moving this small system to other sites when a project was completed. MMSCFD are millions of standard cubic feet per day and are common engineering units used in this industry's literature.

A 1989 report by Grace [12] compared the economics of amine and membrane processes for natural gas treating for 3, 10, 30, 37, 60, 100, 104, 140 and 148 MMSCFD (90 000 to 4 440 000 m^3/day) scale systems. Costs for membrane/amine hybrid systems were also considered. The median system size in this collection of proposed applications is 60 MMSCFD (1 800 000 m^3/day).

UOP offers a report [13] that illustrates the expanding scale of membrane treatment systems. In 1994 a plant was installed in Michigan (USA) that processed 40 MMSCFD (1 200 000 m^3/day) to take CO_2 at 11% to less than 2% using spiral wound modules. In 1997 a membrane system was installed at an EOR facility in Mexico. The system processes 120 MMSCFD (3 600 000 m^3/day) of inlet gas containing 70% CO_2. The enriched CO_2 permeate stream at 93% CO_2 was reinjected into the EOR process, while the product methane stream was at 5% CO_2. A plant in Pakistan was started up in 1995 at a scale of 235 MMSCFD (7 050 000 m^3/day) of natural gas. An upgrade in 2003 increased plant capacity to 500 MMSCFD (15 000 000 m^3/day). Critical to these systems is pre-treatment to remove water, heavy hydrocarbons and other contaminants from the inlet natural gas.

NATCO reports the use of hollow fibre bundles of cellulose triacetate in an EOR application in Texas (USA) at a scale 100 MMSCFD (3 000 000 m^3/day) of gas since 1994 [14]. Pre-treatment includes chilling to remove heavy hydrocarbons and then dehydration by silica gel beds. An expansion of this facility to 200 MMSCFD (6 000 000 m^3/day) was planned in 2005.

Air Liquide and ConocoPhillips report that a membrane system in Indonesia installed in 1998 processed 310 MMSCFD (9 300 000 m^3/day) of natural gas, reducing CO_2 from 30% to 15% [15]. The polyimide hollow fibres are protected from heavy hydrocarbons through pre-treatment of the inlet gas with a thermal swing adsorption unit utilizing silica gel.

Another NATCO installation is an off-shore CO_2 removal facility in the Gulf of Thailand commissioned in December 2004 [16]. The feed gas volume is 700 MMSCFD (2 000 000 m^3/day) at a CO_2 concentration of 37% and pressure of 43.4 bar (630 psi). A plot of flux and separation factor from April 2006 through October 2006 shows stable operation with very little if any performance decline.

This history shows that there has been a steady progression to larger plant sizes. Current proposals for membrane treatment of natural gas exceed one billion SCFD (30 000 000 m^3/day). With the expanding scale of these projects, membrane penetration into natural gas in 2008 was approaching 5% of the volume treated. These global market forces have been favourable to increased module sales over recent years.

15.3 Amine Treaters

Amine treaters are in essence a small chemical plant operation where an aqueous amine solution is used to scrub out from natural gas the acid gases of CO_2 and H_2S with an absorber column. Typical amines in use are monoethanolamine (MEA), diethanolamine (DEA), methyldiethanolamine (MDEA), diisopropylamine (DIPA) and diglycolamine (DGA). The various amines are chosen on the basis of cost, capacity for acid gases, solubility, vapour pressure, regeneration conditions, chemical stability and corrosion resistance. The absorber column can be run up to 200 bar, requiring high pressure vessels. In order to release the acid gases and recovery the amine solution for recycle a regenerator tower is used. This regenerator runs at low pressure but is hot, in the temperature range of 115 to 126 °C.

One cost in running an amine system is the continuous slow loss of amine as natural gas is treated, requiring replacement of the chemicals. This is due to solubility of the amine in the product natural gas, the volatility of the amine, chemical degradation and operational problems such as foaming. Additional concerns in operating an amine plant are failure to meet product specification, corrosion, emulsions/carryovers, excessive amine solution losses, formation of heat-stable salts, degradation products and high filter replacement rates [17]. Spillage of amines is also an environmental concern.

As with membrane systems, installation of a good filtration system for pre-treatment of the inlet gas has become one of the key components of amine system design. They can remove particulates and take out heavy hydrocarbons that enter with the feed gas. The cleaner an amine system is the better it operates. This should be coupled with regular chemical analysis of the amine solution to ensure proper control of the plant operation and resolve operational problems before they impact performance [17].

Amine systems are established technology that is readily available with a proven track record. Incremental improvements over the years in the amine chemistry, design of the contactors and improved process schemes have raised the bar for membrane systems to compete. But higher CO_2 concentrations favour membrane systems since in amine treaters in order to remove more CO_2 one needs more amine solution. In contrast, when membranes see higher CO_2 contents they permeate higher CO_2 content. This joins with simpler (even unattended) operations, lower maintenance and operating costs, and smaller foot prints as factors in favour of membrane systems.

Membrane systems do not have to compete with amine treaters but instead can complement them by cutting high CO_2 levels down to levels more compatible with amine treaters. The long-term operations of combined membrane and amine systems by NATCO [14] and Air Liquide [15] are examples of this hybrid process.

15.4 Contaminants and Membrane Performance

The conundrum in designing membranes for high CO_2 permeability is that the separation is enhanced by the solubility of CO_2 in the polymer and that the polymer needs to be dissolved in solvents in order for the phase inversion process to occur and form the asymmetric structure. Properties that enhance CO_2 performance also can make the structure sensitive to contaminants that over the long term can hurt performance and in worst case scenarios completely shut down performance.

Liquid water collapses dried CA membrane structures, requiring in the field that natural gas streams be treated to reduce water content before they are introduced into the membrane. Simple droplet tests on the surface of flat sheet membranes also show that numerous solvents will also damage or dissolve these CA membranes. A partial listing includes acetone, 1,4-dioxane, acetonitrile, dichloromethane, chloroform, propyl acetate, ethyl acetate, 2-butanone, tetrahydrofuran and nitroethane.

Solubility parameters are a methodology that can be used to qualitatively characterize solvent and polymer interactions [18]. Hansen's 1971 Parameters provide tabulations of solvents broken into dispersive (d), polar (p) and hydrogen bonding (h) contributions. Following the conventions of Klein [19], plotting these p and h values for various compounds reported as solvents for cellulose diacetate and triacetate generates a solubility map (Figure 15.3). When a CA blend is used to make CA membranes the area covered in this map expands to an area larger than the contribution from a single polymer. A mixed CA structure may provide better performance but could also have a differing response to some contaminants.

In addition to the hydrocarbons that can be expected with a natural gas stream other organics can find their way into the system, both volatile and non-volatile. A leak in a

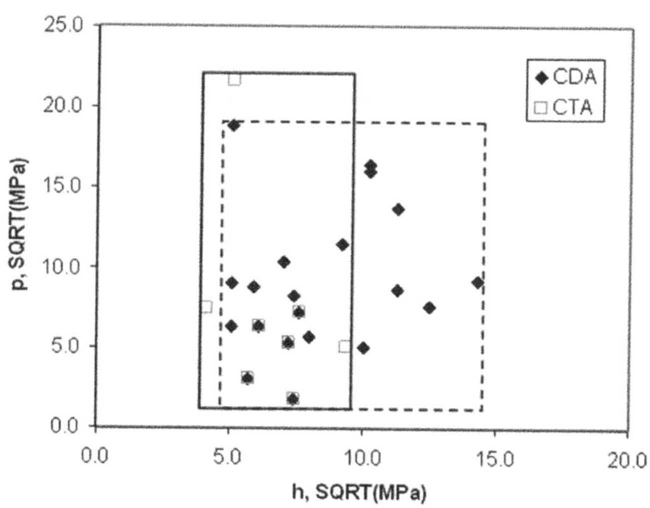

Figure 15.3 *Solubility map using Hansen's 1971 Parameters for reported solvents of cellulose diacetate (CDA) and/or cellulose triacetate (CTA). Image courtesy of W. R. Grace, Copyright (2008) W. R. Grace*

heat exchanger can introduce glycols or other heat exchange fluids. Gas compressors can leak oil. The high gas velocities that can be present can create and transport aerosols that can coat the membrane surface.

Glycol dehydrators are used to scrub water from natural gas to meet pipeline specifications but also pose a potential hazard to membrane systems. An upset in dehydrator operations can flood membrane modules with triethylene glycol (TEG), commonly employed as the working fluid.

Operations for oil and natural gas recovery can also introduce drilling additives, fracturing fluids and other processing liquids into the ground from which natural gas is extracted. These are low cost and sometimes proprietary mixtures for which the composition can be unknown and/or unreported [20]. There are many potential chemical hazards to a membrane operation so a proper pre-treatment system is recommended for successful long-term operations.

Funk *et al.* reported that high concentrations of water and H_2S can combine to deacetylate cellulose acetate membranes [21]. This report also shows that high relative humidity promoted increased thickness of the dense skin of the membrane with loss of flux. The conclusion was that water levels in the feed gas need to be controlled in these applications.

15.5 Cellulose Acetate versus Polyimide

Although CA is the focus of this study, there are other polymers available of which the most widely reported are the polyimides [22]. Reported values of even 100 for CO_2/CH_4 separation factors (α) are not uncommon with polyimides. This compares with reported pure gas dense film CA α values listed in Table 15.2 of about 32 depending on the acetyl content [23]. The higher acetyl content polymers have higher CO_2 permeability. A commercially available polyimide, Matrimid, has attracted much interest and had a pure gas α reported between 43 and 58 depending on how the membrane was prepared and tested [24]. Chemical structures of these two polymers are illustrated in Figure 15.4.

Table 15.2 *Permeability, P x 10^{10} cm^3(STP).cm/(sec.cm^2.cm(Hg)) and α for CA dense films 76–127 mμ (3–5 mils) thick prepared from Tennessee Eastman, Co. powders and tested at 1 atm (1.01 bar) and 35 °C (23). CDA corresponds to 39.8% acetyl content and CTA to 43.5%*

Eastman ID weight % acetyl	CA 320S 32.0	CA 398-30 39.8	CA 435-75 43.5
He	9.34	16.0	19.6
H_2	6.05	12.0	15.5
CO_2	1.84	4.75	6.56
O_2	0.32	0.82	1.46
N_2	0.057	0.15	0.23
CH_4	0.052	0.15	0.20
CO_2/CH_4, α	35.6	31.1	32.8
H_2/CH_4, α	116	80	78

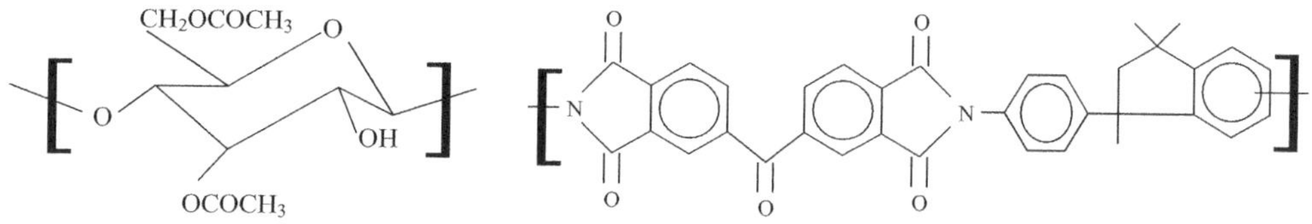

Figure 15.4 Chemical structures of two polymers investigated for use in natural gas separation membranes

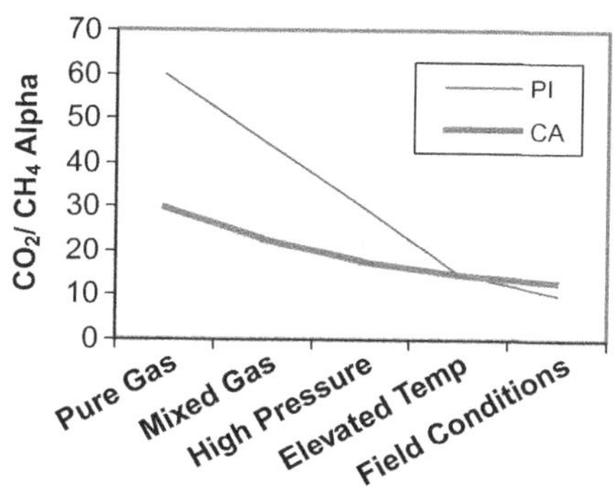

Figure 15.5 Illustration that polyimides and CA have changing α on going from pure gas measurements to field conditions

Both CA and polyimides are susceptible to plasticization by CO_2 [25–27]. When CO_2 solubilizes in the polymer matrix the polymer chains become more flexible and allow faster transport of CO_2 molecules. If other gas molecules such as methane or nitrogen are present they also increase their transport rate often even more so than CO_2. This means that mixed gas measurements often generate lower α than pure gas. Also, if pure gas measurements are done then methane is usually measured before CO_2 since the polymer 'remembers' exposure to CO_2 and needs time to relax the extra free volume that has been created. Various studies have documented the response of CA permeation rates to temperature, pressure and CO_2 concentration [28–31]. When pressure is increased more CO_2 solubilizes in the polymer matrix increasing this plasticization. Going up in temperature also softens the polymer and causes loss of α. In addition, common heavy hydrocarbons in gas fields such as hexane or toluene can negatively impact performance of polyimides in treating natural gas [32–34].

Polyimides would appear to have a clear advantage over CA but in practice other considerations need to be taken into account. Figure 15.5 illustrates how changing process conditions can impact polymer performance. Pure gas measurements are typically reported for dense films at 7 bar (100 psi) or less and 35 °C. Converting from thick dense films to asymmetric structures with thin active layers can introduce a loss of selectivity. Polyimides are also susceptible to plasticization by CO_2 so therefore on switching to mixed gas there is a drop in α. Increasing in pressure and temperature further plasticizes and softens the polymer structure, resulting in further loss of selectivity. Finally, in going to field conditions the trace contaminants that are very likely to be present can further soften the polymer. What initially looks like a promising and highly selective polyimide polymer has converted into a system with α in the single digits.

Why do polyimides appear to be more impacted than CA on going into field conditions? We have suggested [32] that the polyimides have a more ordered structure giving higher contributions to the diffusivity factor than CA. Therefore the polyimides are more susceptible to structural changes induced by plasticization by CO_2 or higher hydrocarbons.

Although CA was the original polymer system commercialized for natural gas treatment, CA still sets the bench mark against which other polymers can be compared. But if the natural gas stream is clean enough there are applications where the higher selectivities obtained by polyimides offer an advantage.

15.6 Compaction in Gas Separations

Compaction refers to a long-term densification of the active membrane layer which lowers flux over the commercial lifetime of a membrane. It is a difficult property to study experimentally since lab conditions are almost impossible to match with field conditions. Similar difficulties are found in reverse osmosis since it is a technical challenge to distinguish physical compaction from trace impurities in the water that promote fouling during long-term tests. In gas separation the treatment of natural gas is an exception to other gas separations in that CO_2 is a plasticizing agent especially for the more highly selective membrane materials.

There are relatively few reports of compaction of asymmetric gas separation membranes in the scientific literature. Reinsch *et al.* used ultrasonic time-domain reflectometry to look at the very first hour of compaction with a commercial CA membrane [35]. The 13.2% collapse of the membrane thickness in the first seconds probably comes from compression of the membrane sub-structure and/or the fabric support. The roughly 10% flux decline over this first hour was at a much slower rate, which may be an indication of the true compaction process.

In addition, the contaminants in natural gas present in the field can be poorly defined. If the pre-treatment of the feed stream is insufficient then even ppm levels of contaminants at these large gas volumes can collect over time on the membrane surface in large quantities and promote either surface fouling or enhanced compaction rates [36].

Another complication is that thin dense membrane films have been shown to decline in permeation properties over extended periods of time (10 000 hours) even without external stresses. The rate of ageing effect becomes greater the thinner the film [37]. This loss over extended time is presumably due to cast membrane films slowly reorienting polymer chains to achieve equilibrium properties with reduced free volume. The thin dense films used in these studies (0.4–25 μm) approach the same order of thickness as the active layer of cast asymmetric films at less than 2000 Å (0.2 μm). The property appears common in that it was reported in this paper with polysulfone, polyphenylene oxide and polyimide

(Matrimid) dense films. Therefore compaction may be driven not only by physical compressive forces but also by thermodynamic considerations.

15.7 Experimental

Data in this report are generated from both commercial and developmental flat-sheet CA membranes. CA membranes are prepared by dissolving commercial grades of CA polymers into a solvent/non-solvent mixture to give a highly viscous dope solution. After microfiltration a knife blade is used to spread the dope onto a woven nylon substrate. The commercial equipment utilized allows for a 1-m width to be cast. The thin dope film is quenched into a water bath to form the microporous structure by the phase inversion process. Membrane is washed with water and post-treated to give finished product in dry state as roll stock.

Figure 15.6 shows a typical scanning electron microscopy view of the finished membrane structure. The instrument used was a Hitachi 4500 Cold Field Emission Scanning Electron Microscope which allowed for sensitive structures to be examined at high magnification. Overall thickness of the CA layer is about 50 µm. The finely microporous structure helps resist the high pressures required in field applications. The active separation layer is at the very top of the structure. The 20 000 magnification is at the highest resolution that could be achieved with this instrument since the polymer structure distorts under the heat generated by the electron beam. Even at this high magnification the thickness of the active layer is not clear although it is believed to be well under 2000 Å.

Laboratory studies were done with flat-sheet membrane. Membrane coupons were mounted as disks in high pressure test cells. With safety considerations CO_2 in nitrogen (instead of methane) was used as feed in order to avoid the explosion potential with

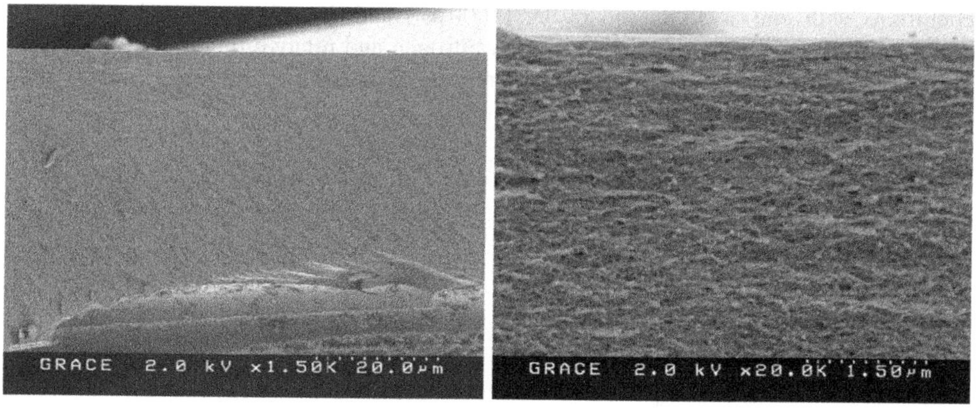

Figure 15.6 Scanning electron microscopy of the cross-section of a commercial CA membrane. The left view at 1500 times magnification is the overall cross-section with the backing layer removed; the right view at 20 000 times magnification includes the top active layer. The sample was prepared by freeze fracturing in liquid nitrogen and then electro sputtering with AuPd

methane. Nitrogen and methane have similar permeation rates: with these CA membranes methane averages about 10% higher rate than nitrogen. The feed manifold can be connected to as many as 20 cells in parallel. The feed gas was flowed over the surface of the membrane at a rate so that the retentate gas is at a 12:1 ratio to permeate flow. Flow rates were measured with an electronic bubble meter and gas chromatography was used to monitor CO_2 and N_2 levels in the feed, retentate and permeate. The test equipment could be left under pressure for months in order to follow long-term compaction rates. Flux is a measure of the productivity of an asymmetric membrane and is the polymer permeability (P) divided by the thickness of the active layer. Flux (J) is normalized to standard conditions for permeate volume, membrane area and applied pressure. Units for J are therefore volume/area/pressure for both CO_2 and N_2. The separation factor α is the ratio of CO_2 permeability (or flux) over methane or nitrogen permeability (or flux).

Field studies were done with commercial-scale spiral wound modules. Membrane, spacers and adhesives were wound around the central permeate pipe to form the module. Modules were both vacuum and high pressure tested to insure integrity.

15.8 Laboratory Tests of Cellulose Acetate Membranes

Varying the retentate to permeate balance (R:P ratio) impacts membrane performance because at low values the membrane surface becomes starved of the highly permeable CO_2. Figure 15.7 shows the average result from five coupons of a developmental CA membrane. At higher R:P ratio the % CO_2 in the permeate increases and the total feed flow to the coupon increases. This means smaller stage cut and higher CO_2 in the retentate. At the R:P ratio of 12:1 the mass balance calculates as a stage cut where 7.7 volume % of the feed is collected as permeate. Therefore the test results offer slightly conservative estimates of membrane performance.

A sample of commercial CA membrane was tested against two different sets of feed conditions with one at 38 °C and 69 bar (1000 psi) and the other at 54 °C and 90 bar (1300 psi). Both were with 10 mol.% CO_2 in nitrogen and relative fluxes are reported in Figure 15.8. The curve shapes are typical compaction curves where initial decline in flux

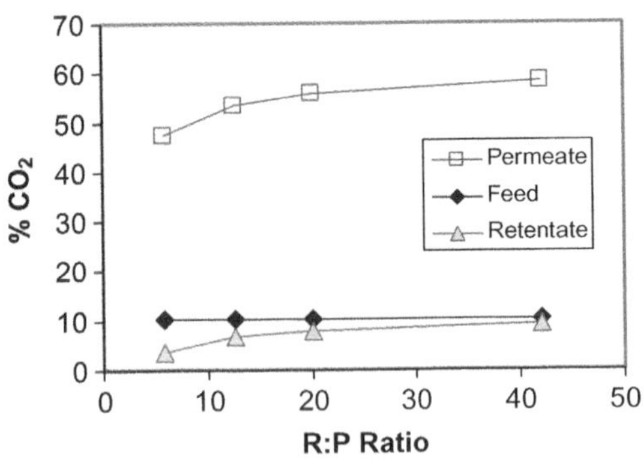

Figure 15.7 *Variation of retentate to permeate ratio for a coupon test of a developmental CA membrane under 10% CO_2 in nitrogen at 69 bar and 38 °C*

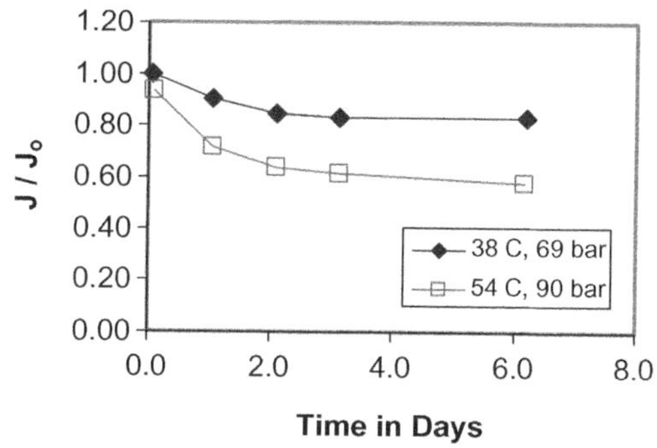

Figure 15.8 Relative CO_2 flux (J/J_0) (against the initial 38 °C reading for CA membrane) under 10% CO_2 in nitrogen for commercial CA membrane. Two separate trials are reported, one at each condition of temperature and pressure

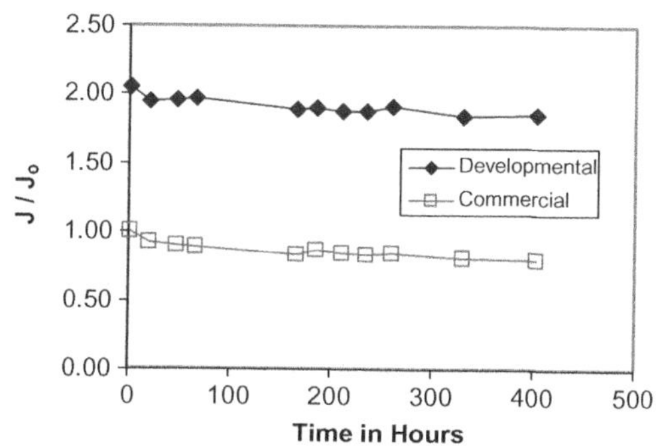

Figure 15.9 Relative CO_2 flux (J/J_0) (based on initial reading for commercial membrane) of two CA membranes under 6% CO_2 in nitrogen at 55 bar and 27 °C

is fairly rapid and then flux declines much more slowly over time. On day 6 the 38 °C trial was at 83% of the initial reading while the 54 °C trial was at only 62% of the initial reading and still visibly declining. Higher pressure and temperature have accelerated the compaction rate.

Literature [28–31] suggests that both the higher pressure and temperature in Figure 15.8 should have increased the initial CO_2 rate through the membrane at 54 °C and 90 bar (1300 psi) to be well above (perhaps 50%) of that for 38 °C and 69 bar (1000 psi). This was not observed, so during the initial few hours when the test equipment was being allowed to equilibrate the flux must have taken a substantial hit. This high decline rate would not be favourable to long-term applications.

Permeation tests are the primary method to evaluate alternative membrane formulations. Figure 15.9 shows a head-to-head test of commercial CA membrane coupons against a developmental CA membrane. Test conditions were set at 55 bar (800 psi) and 27 °C with 6 mol.% CO_2 in nitrogen. These conditions were chosen to simulate an upcoming field trial at a natural gas production facility. The lower operating pressure and temperature compared to Figure 15.8 results in reduced compaction rates and decline in flux over time. This developmental membrane is at almost twice the flux of the commercial CA membrane. The next question is whether these promising lab results would translate to real field conditions.

15.9 Field Trials of Cellulose Acetate Membranes

A large-scale natural gas producer made available a test skid that would allow for membrane products to be demonstrated under real field conditions. Commercial CA membranes from both Grace and another supplier and the developmental membrane were delivered to the field as full-scale 20 cm (8 inch) diameter spiral wound modules.

The feed gas contained about 6% CO_2 with about 50 ppm H_2S. The bulk of the gas was about 81% methane, with the major other components being about 12% nitrogen and 1% ethane. Target specifications for the sales gas (retentate) were to reduce CO_2 to less than 2% and H_2S content to less than 15 ppm. Feedstock conditions were relatively mild at 55 bar (800 psi) and 27 °C. The three trials were run sequentially over a period of months. One set of analysis was reported for each day of the trials.

The permeate fluxes for the three membrane systems over 30+ days are reported in Figure 15.10. CA membranes can be prepared at different points of the trade-off curve for flux and selectivity and these two commercial CA membranes are at slightly different flux. But the developmental membrane is substantially higher in permeate flow indicating a more productive membrane compared to the commercial CA membranes.

Although there is day-to-day variation as the feed flow rate to the test unit and gas composition varies slightly, none of the three products are showing discernible signs of compaction. This is consistent with the relatively mild process conditions employed in this trial. These results also confirm that laboratory tests – where a developmental membrane was also notably higher in flux than the commercial CA membrane (Figure 15.9) – can be predictive of field performance.

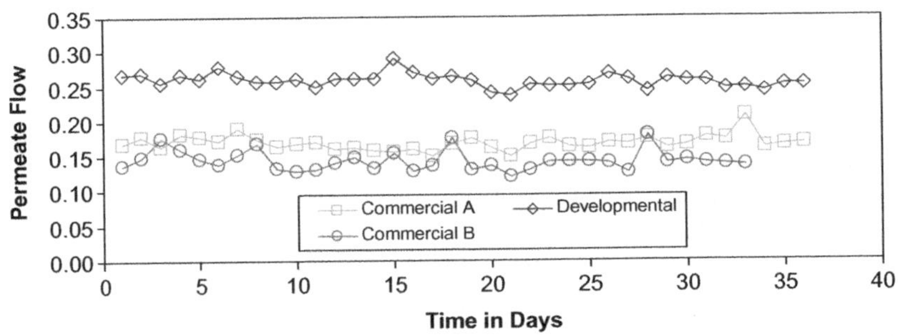

Figure 15.10 *Permeate flow for two commercial CA spiral wound modules and a developmental module during natural gas field trials. Process conditions are 6% CO_2 at 55 bar and 27 °C*

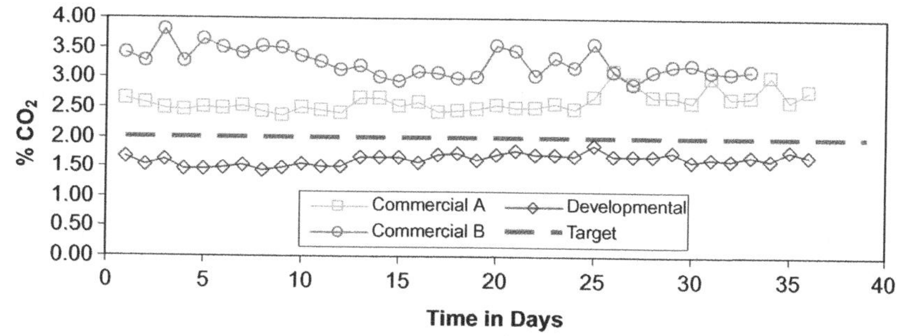

Figure 15.11 Amount of CO_2 in the retentate during the field trials of spiral wound modules at a natural gas plant. Target was less than 2% CO_2

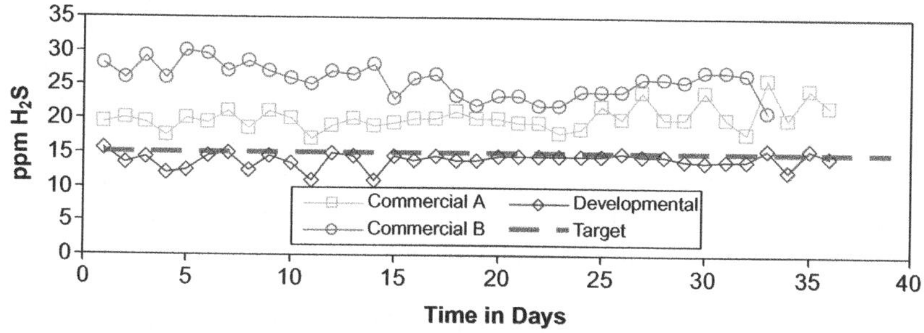

Figure 15.12 H_2S levels in the retentate during the field trials of spiral wound modules at a natural gas plant. Target was less than 15 ppm H_2S

Figures 15.11 and 15.12 report the CO_2 and H_2S levels of the retentate stream. All three products reduce the levels of these impurities in the product gas and are showing stable performance. But with the higher permeate flow from the developmental modules more CO_2 and H_2S are removed and therefore bring the retentate gas stream to the target levels of purity for both CO_2 and H_2S. The developmental membrane has demonstrated in the field substantially better performance and productivity than the two commercial CA membrane products that were also tested.

15.10 Strategies for Reduced Size of Large-scale Membrane Systems

Membrane, module and system performance are interdependent on each other. The membranes have to meet flux and selectivity requirements for a given application. There is a trade-off curve available between flux and selectivity so a preferred strategy is to supply a membrane that meets minimum selectivity and then boost the flux. But meeting the requirements in the lab for flux and selectivity are only the initial step since the membrane has to be designed and engineered into a working installation.

Only perhaps 20% of the cost of a membrane system is in the cost of the modules; since these are high-pressure systems with hazardous gases they require steel construction. Spiral wound modules are typically loaded end-to-end into housings made from piping 8 m in length. Standard industry size is 20 cm (8 inch) diameter modules and every high pressure flange or valve increases the system cost.

Reducing the number of modules required for a system therefore has a 4-fold multiplier in reducing the steel and construction costs going into an installation. Membrane area per spiral wound module has improved over the years with a more refined selection of feed and permeate spacers. This then allows for more functional membrane area to be packaged into every module. This reduces the total number of required modules in large-scale installations.

These techniques for improvements in membrane area have to be balanced against developing a pressure increase on the permeate side as the module attempts to evacuate the permeate gas and the pressure drop on the feed side from end-to-end of the modules as feed gas is forced through for treatment. Too high a permeate pressure increase will negatively impact both flux and selectivity. And since the membranes are highly selective to CO_2, the concentration of CO_2 can build up in the permeate channels and reverse permeate at the tail end of module leaves and contaminates the cleaned retentate gas. Too high a pressure drop on the feed side steals from the driving force for the separation and is exacerbated when modules are in multiple strings. Of concern is that a severe pressure drop can 'telescope' a module and cause complete failure. Anti-telescoping devices are built as end caps on spiral wound modules in order to provide protection against flow surges.

Another strategy to reduce the number of modules is to increase the dimension of the steel housing to handle larger modules. NATCO [5,14] has been successful in increasing the dimensions of hollow fibre bundles to greatly reduce system costs. The earlier modules were 13 cm (5 inch) diameter and 1 m (40 inch) length. This was boosted to 30 cm (12 inch) diameter, then to 41 cm (16 inch) and twice the length, and then to a mega-module at 76 cm (30 inch) diameter. These are potentially equivalent to 72 of the original modules. This represents a substantial decrease in foot print and required pressure housings for a large-scale installation.

For new system builds the same strategy can be done with spiral wound modules designed for natural gas treatment. Grace has constructed a 30-cm (12-inch) diameter module versus the standard 20-cm (8-inch) diameter module. Since membrane area is a squared function of diameter, the ratio of 144 to 64 means a 225% increase in membrane area per module. Although the larger diameter modules and pressure housings each cost more there is potential for reduction in total system cost.

And, of course, higher flux for a membrane would increase the volume of permeate output from the full-scale module. If selectivity is maintained than the total number of modules required is reduced.

The dependence on R:P ratio for coupon performance shown in Figure 15.7 also applies to module performance. In system design the module strings need to be supplied with a sufficient feed rate to promote top membrane performance. Supplying these high feed flow rates needs to be balanced against increasing end-to-end pressure drop. The design of an appropriate module can improve system performance or allow retrofit into existing systems. Figure 15.13 shows a variety of modules built for specific applications (some

Figure 15.13 A variety of module designs from lab-scale to commercial size for multiple applications including natural gas treatment. Courtesy of W.R. Grace

experimental) where different couplings, membrane area or overall dimensions have been incorporated. For example, reducing the diameter of a module increases feed velocity for the same given volume of feed and improves the R:P ratio.

15.11 Research Directions

Of high interest would seem to be improved flux and selectivity through a better polymer. Materials are reported every year with good permeability properties but many studies are carried out with relatively thick dense films on the order of 5 to 50 μm. In addition, with CO_2 any study with pure gas results has to be discounted until mixed gas results are reported because of CO_2 plasticization. Also, thin film permeation properties have been demonstrated as different than those of thick films [37].

For commercial membranes to be competitive in natural gas treatment they must be reliably manufactured with an active separation layer on the order of 0.1 μm. They also have to be resistant to warm and high pressure operating conditions and mechanically resistant to assembly into modules. These are all significant tasks to engineer and inherent permeability of the polymer is only one of the factors involved.

As gas fields age around the world they increase in contaminants that can potentially harm long-term membrane performance. Better chemical resistance for the membrane polymer would be beneficial. Whether highly selective polymers for CO_2 transport (see e.g. Chapters 10–13 of this book) that also have resistance to CO_2 plasticization can be developed remains an open question. Membrane structures with improved resistance to elevated process conditions of temperature and pressure are also desired. Converting new materials into working systems requires attention to preparing high productivity membranes with sub-micrometer active layers. These membranes need to be packaged into modules and then delivered as an engineered system that optimally enhances the original polymer performance.

The observations that very thin cast films lose permeability over time even without external stress is of interest since the active layers of commercial membranes are also thin films. Our commercial modules are pre-stressed with a pressure test to check integrity and can take months to get from the production floor to the customer. How does this thin dense film behaviour observed in the lab for other polymers relate to the asymmetric thin film structure of commercial CA membranes?

The material science of long-term membrane compaction in asymmetric membranes is not well reported. This phenomenon has been observed in gas separations including this report, but also in reverse osmosis [6] and organic solvent nanofiltration [38]. How to accurately predict multi-year performance in the field with short-term lab tests is a continuing challenge.

Since installed membrane systems for natural gas treatment are continuing to increase in scale, strategies that reduce the number of required modules and therefore total system cost continue to be of interest.

15.12 Summary

Natural gas separations are unusual in that not only is CO_2 highly permeable but that CO_2 and other contaminants can also plasticize the most selective polymer structures. Although cellulose acetate structures are sensitive to solvents and chemicals, effective pre-treatment schemes in the field can allow for stable long-term operations.

New installations in natural gas treatment are increasingly favourable to membrane systems as they expand in scale and treat streams higher in CO_2 content. In systems with large volumes of CO_2 a membrane system can be employed to make the bulk cut and then an amine system can provide the finishing treatment. This is especially applicable in fields employing CO_2 injection for enhanced oil recovery.

Long-term compaction and decline in flux with asymmetric membranes under gases that plasticize the polymer in use are rarely reported. These experiments confirm that higher pressure and temperature can increase compaction levels to unacceptable levels. Conversely, milder process conditions can allow for stable long-term performance in treatment of natural gas.

The developmental membranes reported here that have improved productivity and performance over conventional commercial cellulose acetate membranes indicate that there are avenues for improvement with new and future technologies.

Membrane materials that offer improved resistance to temperature, pressure, high CO_2 content, water and other chemical contaminants are all areas that offer gains in long-term performance desired in commercial applications.

Membrane systems for natural gas treatment are now approaching 30 years of service to the industry and continue to evolve toward better performance. Further advances in the technology should be expected.

Acknowledgements

The personnel at Grace Davison Membranes located in Littleton, Colorado (USA) delivered the membranes, modules and experimental results described in this report. This team has included Jerry T. Barnett, Anthony P. Lippl, Craig Paulaha, Samantha K. Stewart and Craig R. Wildemuth. In September 2009 this business was acquired by UOP LLC based in Des Plaines, Illinois (USA).

References

(1) J. Haggin, New generations of membranes developed for industrial separations, *Chem. & Eng. News*, **66**, 7–16 (June 6, 1988).

(2) W. J. Schell, C. G. Wensley, M. S. K. Chen, K. G. Venugopal, B. D. Miller, J. A. Stuart, Recent advances in cellulosic membranes for gas separation and pervaporation, *Gas Sep. & Pur.*, **3**, 162–169 (1989).

(3) U. Merten, S. Beach, P. K. Gantzel, Method and apparatus for gas separation by diffusion, *U.S. Patent* 3415038 (1968).

(4) W. MacDonald, C.-Y. Pan, Method for drying water-wet membranes, *U.S. Patent* 3842515 (1974).

(5) R. W. Baker, K. Lokhandwala, Natural gas processing with membranes: An overview, *Ind. Eng. Chem. Res.*, **47**, 2109–2121 (2008).

(6) W. M. King, D. L. Hoernschemeyer, C. W. Staltonstall, Jr., Cellulose acetate blend membranes, in Reverse Osmosis Membrane Research, H. K. Lonsdale, H. E. Podall (Eds), Plenum Press, New York, 1972.

(7) T. D. Nguyen, K. Chan, T. Matsuura, S. Sourirajan, Effect of shrinkage on pore size distribution of different cellulosic reverse osmosis membranes, *Ind. Eng. Chem. Prod. Res. Dev.*, **23**, 501–508 (1984).

(8) B. S. Minhas, T. Matsuura, S. Sourirajan, Formation of asymmetric cellulose acetate membranes for the separation of carbon dioxide–methane gas mixtures, *Ind. Eng. Chem. Res.*, **26**, 2344–2348 (1987).

(9) M.-W. Tang, W. M. King, C. G. Wensley, Air dried cellulose acetate membranes, *U.S. Patent* 4855048 (1989).

(10) W. R. True, Global processing capacity flat; US construction pace quickens, *Oil & Gas J.*, **106**, 50–57 (June 23, 2008).

(11) W. J. Schell, Commercial applications for gas permeation membrane systems, *J. Membr. Sci.*, **22**, 217–224 (1985).

(12) R. W. Spillman, Economics of gas separation membranes, *Chem. Eng. Prog.*, **85**, 41–62 (January 1989).

(13) D. R. Koch, W. R. Buchan, T. Cnop, W. I. Echt, Proper pretreatment systems reduce membrane replacement element costs and improve reliability, in Proceedings of the Presentations at Laurance Reid Gas Conditioning Conference, Norman, OK (2005).

(14) G. Blizzard, D. Parro, K. Hornback, CO_2 separation membranes a critical part of the Mallet CO_2 removal facility, in Proceedings of the Presentations at Laurance Reid Gas Conditioning Conference, Norman, OK (2005).

(15) C. L. Anderson, A. Siahaan, Case study: membrane CO_2 removal from natural gas, Grissik gas plant, Sumatra, Indonesia, in Proceedings of the Presentations at Laurance Reid Gas Conditioning Conference, Norman, OK (2005).

(16) G. Munoz, A. Callison, G. Davidson, Successful commissioning of CO_2 removal facility in Gulf of Thailand, in Proceedings of the Presentations at Laurance Reid Gas Conditioning Conference, Norman, OK (2007).

(17) R. Wagner, B. Judd, Fundamentals – gas sweetening, in Proceedings of the Presentations at Laurance Reid Gas Conditioning Conference, Norman, OK (2006).

(18) A. F. M. Barton, CRC Handbook of Solubility Parameters and Other Cohesion Parameters, CRC Press, Inc., Boca Raton, 1983.

(19) E. Klein, J. Eichelberger, C. Eyer, J. Smith, Evaluation of semi permeable membranes for determination of organic contaminants in drinking water, *Water Res.*, **9**, 807–811 (1975).

(20) B. Hileman, Balancing energy needs and safety, *Chem. Eng. News*, **84**, 38–42 (2008).

(21) E. W. Funk, S. S. Kulkarni, A. X. Swamikannu, Effect of impurities on cellulose acetate membrane performance, in *Recent Advances in Separation Techniques – III*, 82, AIChE Symposium Series No. 250 (1986).

(22) H. Ohya, V. V. Kudryavtsev, S. I. Semenova, Polyimide Membranes – Applications, Fabrications, and Properties, Gordon and Breach Publishers, Amsterdam, 1996.

(23) A. C. Puleo, D. R. Paul, S. S. Kelley, The effect of degree of acetylation on gas sorption and transport behavior in cellulose acetate, *J. Membr. Sci.*, **47**, 301–332 (1989).
(24) I.-f. Wang, B. S. Minhas, Asymmetric polyimide membranes, *U.S. Patent* 5067970 (1991).
(25) M. Wessling, M. Lidon Lopez, H. Strathmann, Accelerated plasticization of thin-film composite membranes, *Sep. Pur. Tech.*, **24**, 223–233 (2001).
(26) A. F. Ismail, W. Lorna, Penetrant-induced plasticization phenomenon in glassy polymers for gas separation membrane, *Sep. Pur. Tech.*, **27**, 173–194 (2002).
(27) T. Visser, N. Masetto, M. Wessling, Materials dependence of mixed gas plasticization behavior in asymmetric membranes, *J. Membr. Sci.*, **306**, 16–28 (2007).
(28) K. Haraya, K. Obata, T. Hakuta, H. Yoshitome, The permeation of gases through asymmetric cellulose acetate membranes, *J. Chem. Eng. Jpn.*, **15**, 431–436 (1986).
(29) E. Sada, H. Kumazawa, Y. Yoshio, S.-T. Wang, P. Xu, Permeation of carbon dioxide through homogeneous dense and asymmetric cellulose acetate membranes, *J. Polym. Sci.: Part B: Polym. Phys.*, **26**, 1035–1048 (1988).
(30) M. D. Donahue, B. S. Minhas, S. Y. Lee, Permeation behavior of carbon dioxide–methane mixtures in cellulose acetate membranes, *J. Membr. Sci.*, **42**, 197–214 (1989).
(31) E. Sada, H. Kumazawa, P. Xu, S.-T. Wang, Permeation of pure carbon dioxide and methane and binary mixtures through cellulose acetate membranes, *J. Polym. Sci.: Part B: Polym. Phys.*, **28**, 113–125 (1990).
(32) L. S. White, T. A. Blinka, H. A. Kloczewski, I.-f. Wang, Properties of a polyimide gas separation membrane in natural gas streams, *J. Membr. Sci.*, **103**, 73–82 (1995).
(33) J. D. Wind, D. R. Paul, W. J. Koros, Natural gas permeation in polyimide membranes, *J. Membr. Sci.*, **228**, 227–236 (2004).
(34) M. Al-Juaied, W. J. Koros, Performance of natural gas membranes in the presence of heavy hydrocarbons, *J. Membr. Sci.*, **274**, 227–243 (2006).
(35) V. E. Reinsch, A. R. Greenberg, S. S. Kelley, R. Peterson, L. J. Bond, A new technique for the simultaneous, real-time measurement of membrane compaction and performance during exposure to high-pressure gas, *J. Membr. Sci.*, **171**, 217–228 (2000).
(36) J. M. S. Henis, Commercial and practical aspects of gas separation membranes, in Polymeric Gas Separation Membranes, D. R. Paul, Y. P. Yampolskii (Eds), CRC Press, Boca Raton, 1994.
(37) Y. Huang, D. R. Paul, Effect of film thickness on the gas-permeation characteristics of glassy polymer films, *Ind. Eng. Chem. Res.*, **46**, 2342–2347 (2007).
(38) L. S. White, C. R. Wildemuth, Aromatics enrichment in refinery streams using hyperfiltration, *Ind. Eng. Chem. Res.*, **45**, 9136–9143 (2006).

16

The Effect of Sweep Uniformity on Gas Dehydration Module Performance

Pingjiao Hao and G. Glenn Lipscomb
Chemical and Environmental Engineering Department, University of Toledo, Toledo, Ohio, USA

16.1 Introduction

Membrane gas separation for air dehydration (AD) differs from other commercial applications in several ways. First, the component to be removed (water) possesses a permeability that may be more than three orders of magnitude greater than the other components in the feed (oxygen and nitrogen). Second, the feed concentration is small, less than 1% on a molar basis. Third, the product concentration may be two orders of magnitude less than the feed concentration.

Membranes in the form of hollow fibre modules are used commonly for gas dehydration. In comparison to spiral wound modules made from flat sheet membranes, hollow fibre membrane modules contain more membrane surface area per unit volume thereby reducing the size of the module. Additionally, hollow fibre module manufacturing costs are lower [1] and hollow fibre designs easily permit permeate sweep.

A hollow fibre module can be operated in three different modes: co-current, cross, or counter-current flow. In co-current flow, the permeate flows in the same direction as the feed and retentate. In cross-flow, the permeate flows perpendicularly to the feed and retentate while in countercurrent flow the permeate flows in the opposite direction. The countercurrent flow pattern gives the best performance as the driving force for transport is maximized along the module length. One can produce an arbitrarily high purity

Membrane Gas Separation Edited by Yuri Yampolskii and Benny Freeman

retentate product with the countercurrent design but the maximum permeate purity is limited by the intrinsic separation properties of the membrane [2,3]. Production of a high-purity permeate requires module staging. Thus, the production of a high-purity product is easier when the product is the retentate.

Wang *et al.* [4,5] demonstrate that for a countercurrent hollow fibre module the membrane resistance to water transport is negligible – the overall mass transfer coefficient is controlled by lumen and shell-side concentration boundary layers. Returning part of the retentate product as a permeate sweep increases the rate of water removal. The required membrane area decreases dramatically as the fraction of the retentate used as sweep is increased (sweep fraction). However, for sweep fractions greater than ~0.1, the increase in productivity is offset by an increase in consumption of product gas as sweep.

The patent literature describes three primary methods for introducing sweep from the retentate product into the shell of a hollow fibre module.

First, Skarstrom and Kertzman [6] teach the use of conduits and valves external to the module to return a portion of the retentate product to the shell as sweep. The sweep may be introduced either through an external port on a case enclosing the fibre bundle or through a tube that extends from outside the module through the tube sheet into the shell. Similarly, Makino and Nakagawa [7] teach the use of valves and conduits to feed the sweep to the fibre lumens in a shell-fed module. Friesen *et al.* [8,9] teach the use of the sweep stream that is mixed with the water-containing permeate at a point generally opposite the feed to the module, preferably through a port near the retentate product (i.e. the dehydrated air product) end of the module.

Second, Stookey [10] teaches the use of fibres that possess reduced selectivity and increased permeance near the retentate product end. If the fibres used in the module are composite membranes, one can change their transport properties simply by removing the discriminating coating in a region near the tube sheet. This allows introduction of the sweep from the fibres themselves but does not allow control of the sweep rate.

Third, Morgan *et al.* [11] teach the use of a plurality of passages (fibres, tubes, or other conduits) embedded in the retentate end tube sheet that allow fluid communication between the retentate header and the shell. The pressure difference between the header and shell drives a portion of the retentate product back into the shell. The sweep flow rate is determined by the number and size of the passages and cannot be regulated externally. This 'internal sweep' design is used extensively. Burban *et al.* [12] describe a modification in which a diffuser is used to distribute the sweep more uniformly in the shell. A channel or conduit extends from the retentate header through the tube sheet into the fibre bundle. The channel end in the header is left as an open orifice while the channel end in the fibre bundle is capped by a porous diffuser.

The patent literature also teaches the importance of countercurrent contacting [13] and recycle configurations to improve process performance [14]. Auvil *et al.* [14] describe two configurations in which the wet permeate is recycled to eliminate feed air losses. If the feed is at ambient pressure, the permeate is sent to the inlet of the feed compressor. If the feed is already at pressure, a recycle compressor is used to increase the permeate pressure to the feed pressure. In both configurations, water is removed from the process in the chiller/condenser that follows the compressor.

More recent literature addresses other issues associated with membrane dehydration. Vallieres and Favre [15] demonstrate that use of a permeate vacuum may be preferable to use of a product sweep in some cases. However, the changes in performance are modest and may not justify the purchase and maintenance of a vacuum pump. Metz *et al.* [16] report measurements of the dependence of water permeability on water concentration. This dependence is important if boundary layer resistances to mass transfer are comparable to or less than the membrane resistance.

Pan [17] describes one of the first attempts to evaluate the performance of a gas separation module for the separation of multicomponent mixtures. Pan's analysis assumes the local permeate composition, that determines the partial pressure driving force for permeation, is not equal to the bulk concentration of the permeate due to the membrane support resistance. Instead the composition within the membrane support is taken as the composition produced by cross-flow of the permeate through the support. Additionally, permeate pressure drop within the support is neglected. Kovvali *et al.* [18] propose a simplification to Pan's solution procedure by assuming a linear relationship between permeate and retentate mole fractions for sufficiently small composition ranges. Chowdhury *et al.* [19] obtain more accurate solutions by using a backward difference Adams-Moulton or Gear approximation for the governing differential equations. The resulting non-linear algebraic equations are solved with a modified Powell hybrid algorithm that utilizes a finite difference approximation for the Jacobian.

Coker *et al.* [20] avoid the assumption of a local permeate composition equal to the cross-flow composition by dividing a module into a series of N well-mixed stages and solving the Weller-Steiner equations [21] for each stage. If the stages are numbered from 1 to N from the feed to the retentate product end, the retentate stream exiting stage i is the feed to stage $i + 1$. The permeate from stage $i + 1$ is sent to stage i as a sweep for countercurrent modules; for co-current modules, the permeate from stage $i - 1$ is used. This simplification is equivalent to the use of first-order finite differences to approximate the derivatives in the governing mass balances. Kaldis *et al.* [22] describe an alternative solution procedure in which the governing differential equations are approximated using an orthogonal collocation algorithm.

Lemanski and Lipscomb [23,24] describe solutions of the governing conservation of mass equations for flows throughout the lumen and shell regions of a module. The solution assumes the flows are equivalent to flow through an effective porous media and explicitly accounts for the influence of inlet and outlet port placement on performance. Marriott *et al.* [25–27] also describe solutions to the governing conservation of mass, momentum, and energy equations for the flows within the lumen and shell regions. For countercurrent and co-current flows, the rigorous axis symmetric conservation equations are solved within the lumen of an individual fibre – all fibres are assumed identical. The lumen equations are coupled to the shell equations assuming uniform axial plug flow in the shell; fluid distribution to and from lumen distribution manifolds is not considered.

This work reports simulations of sweep distribution within the shell and its effect on module performance. Two types of simulations are considered: (1) simulations that assume the sweep flow around each fibre is distributed in a Gaussian manner and (2) simulations that explicitly predict flow fields within the shell based on how the sweep gas is introduced.

Predictions based on explicit calculation of the shell flow are in good agreement with those based a Gaussian sweep distribution using a standard deviation in sweep flow equal to ~20% of the average sweep flow rate. Surprisingly, sweep distribution has little effect on performance in both cases.

16.2 Theory

16.2.1 Gaussian Distribution of Sweep

The simulations assuming a Gaussian distribution of sweep flow are based on mass balances for each fibre in the module. Key assumptions in the theoretical analysis are as follows.

1 The retentate flows in the lumen and the permeate in the shell.
2 Gas mixtures behave ideally.
3 The shell pressure is constant.
4 The lumen pressure drop is described by the Hagen-Poiseuille equation for compressible flows.
5 Mass transfer by axial diffusion is small relative to axial convection.
6 The process is isothermal and steady-state.
7 The properties of a given fibre do not vary along its length. However, properties may vary from fibre to fibre.
8 Lumen-side mass transfer coefficients are calculated using the Leveque correlation and shell-side mass transfer coefficients are calculated using the Donohue correlation.

The Gaussian sweep distribution is given by Equation (16.1):

$$g(\phi) = \frac{1}{\sigma\sqrt{2\pi}} \exp\left(-\frac{(\phi - \bar{\phi})^2}{2\sigma^2} \right) \tag{16.1}$$

where ϕ is the sweep flow rate around a fibre, $\bar{\phi}$ the mean sweep flow rate per fibre, and σ the standard deviation. The fraction of fibres for which the sweep flow rate falls in the interval $[\phi, \phi + d\phi]$ is equal to $g(\phi)d\phi$. For the bundle of fibres, the average flow rate per fibre is given by Equation (16.2):

$$\bar{f} = \int_{\phi\min}^{\phi\max} f(\phi) g(\phi) d\phi \tag{16.2}$$

where f may be the retentate or permeate flow. Gauss-Hermite quadrature [28] is used to evaluate Equation (16.2) and determine the overall performance of the fibre bundle. The number of quadrature points is determined by increasing the number until the results change by less than the desired solution accuracy.

The lumen and shell mass balances for individual fibres in the bundle are solved by dividing each fibre into N Weller-Steiner Case I stages along its length, as illustrated in Figure 16.1.

For stage j and component i, the mass balance equations and permeation relationships are given by Equations (16.3) and (16.4), respectively:

$$R_{j-1}x_{i,j-1} + P_{j+1}y_{i,j+1} = R_j x_{i,j} + P_j y_{i,j} \tag{16.3}$$

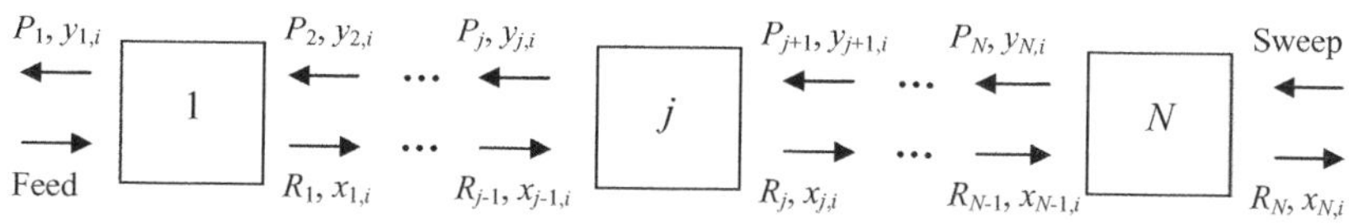

Figure 16.1 *Countercurrent module divided into N stages*

$$P_j y_{i,j} - P_{j+1} y_{i,j+1} = A_j J_{i,j} = A_j k_i (p_{\mathrm{h},j} x_{i,j} - p_1 y_{i,j}) \quad (16.4)$$

where R is retentate molar flow rate, P permeate molar flow rate, x retentate mole fraction, y permeate mole fraction, A membrane area, p pressure, J permeation flux, and k the overall mass transfer coefficient. The subscripts h and l denote the high pressure retentate and low pressure permeate, respectively. The number of stages N is varied until the results obtained by increasing N change by less than 1%.

Summing the mass balances for each of the n components, Equation (16.5), gives the following expressions for the change in retentate and permeate flow rate.

$$P_j - P_{j+1} = \sum_{i=1}^{n} A_j k_i (p_{\mathrm{h},j} x_{i,j} - p_1 y_{i,j}) \quad (16.5)$$

$$R_{j-1} - R_j = \sum_{i=1}^{n} A_j k_i (p_{\mathrm{h},j} x_{i,j} - p_1 y_{i,j}) \quad (16.6)$$

Pressure drops in the lumen are calculated using the Hagen-Poiseuille equation [29] modified by substituting the product of molar flow rate and molar density for the volumetric flow rate and using the ideal gas law to calculate molar density:

$$\frac{dp_{h,j}^2}{dz} = -\frac{16\eta R_g T}{\pi r_i^4 N_f} R_j \quad (16.7)$$

where z is distance along the module, η the gas viscosity, R_g the ideal gas constant, T temperature, r_i the fibre inner radius, and N_f the number of fibres. The integral of this equation for stage j is given by:

$$p_{h,j} = \sqrt{p_{h,j-1}^2 - \frac{16\eta R_g T}{\pi r_i^4 N_f}\left(\frac{R_{jm1} + R_j}{2}\right)\Delta L} \quad (16.8)$$

where ΔL is the length of the stage (i.e. (active fibre length)/N_f). Equation (16.8) is used to evaluate the pressure in each stage given the pressure of the feed gas to stage 1.

Equations (16.3)–(16.7) are solved with an iterative solution algorithm. Using the crossflow solution as an initial guess, improved estimates for $x_{i,j}$ and $y_{i,j}$ are obtained from Equations (16.9) and (16.10), respectively, which follow from Equations (16.3) and (16.4):

$$x_{i,j} = \frac{R_{j-1} x_{i,j-1} + A_j k_i p_l y_{i,j}}{R_j + A_j k_i p_{h,j}} \quad (16.9)$$

$$y_{i,j} = \frac{P_{j+1} y_{i,j+1} + A_j k_i p_{h,j} x_{i,j}}{P_j + A_j k_i p_l} \tag{16.10}$$

Improved estimates for P_j, R_j, and $p_{h,j}$ are obtained using Equations (16.5), (16.6), and (16.8), respectively. The iterative procedure is stopped when the variables change by less than 1% from one iteration to the next. This direct substitution iterative algorithm proved to be robust and converged quickly for all cases considered. The converged solution for each fibre is used to evaluate overall module performance by using Equation (16.2) to determine average flow rates.

The overall mass transfer coefficient for species i is calculated by summing the lumen-side, membrane, and shell-side mass transfer resistances as given by Equation (16.11):

$$\frac{1}{k_i} = \frac{r_o}{k_{l,i} r_i} + \frac{1}{Q_i} + \frac{1}{k_{s,i}} \tag{16.11}$$

where r_o is the fibre outer radius and Q the membrane permeance based on the fibre outer radius. As defined in Equation (16.11), the overall mass transfer coefficient is based on the fibre outer radius.

The shell and lumen side boundary resistances are calculated using the Donohue and Leveque equations, Equations (16.12) and (16.13) respectively, as suggested by Wang [4,5].

$$Sh = \frac{k_l L}{D} = 1.62 \left[Re\ Sc \left(\frac{2 r_i}{L} \right) \right]^{0.33} \tag{16.12}$$

$$Sh = \frac{k_s L}{D} = 0.0732 D_h^{0.6} Re^{0.6} Sc^{0.33} \tag{16.13}$$

where k_l is the lumen-side mass transfer coefficient, k_s the shell-side mass transfer coefficient, L the length of the fibres, D the diffusion coefficient, Re the Reynolds number, Sc the Schmidt number, and D_h the hydraulic diameter given by:

$$D_h = \frac{(2 r_m)^2 - N_f (2 r_i)^2}{2 r_m + 2 N_f r_i} \tag{16.14}$$

where r_m is the module radius. Note the Reynolds number is calculated based on the fibre inner radius for Equation (16.12) and the fibre outer radius for Equation (16.13).

Results obtained for a Gaussian sweep distribution are compared to results obtained for a Gaussian fibre inner radius distribution. Previous work [30,31] has shown that variation in fibre inner radius has the largest impact on module performance when the variability in fibre properties is included in module performance simulations.

16.2.2 Explicit Sweep Distribution Simulations

To explicitly calculate the shell and lumen flow distributions, the shell-side and lumen-side spaces are treated as bi-continuous porous media. Volume averaging the conservation equations for the lumen and shell spaces yields conservation equations in terms of volumetric average values of the field variables (velocity, pressure, and concentration) where the averaging volume is small compared to the macroscopic dimensions of the module but larger than fibre dimensions.

For low Reynolds number flows, volume average of the conservation of mass equations yields Darcy's law as the relationship between superficial velocity and pressure:

$$u = -\frac{K}{\eta}\nabla p \tag{16.15}$$

where K denotes the hydraulic permeability of the porous medium and u is the volume average superficial velocity. Note that for anisotropic porous media K is a tensor. However, previous work [23] shows that for sufficiently high module aspect ratios (i.e. module length to diameter ratios) the effect of anisotropy on module performance is negligible so one may assume the porous media are isotropic.

The steady-state volume average conservation of mass equation for component i is given by Equation (16.16)

$$\nabla \cdot (\rho_i u) = \pm J_i \tag{16.16}$$

where ρ is the molar fluid density and J the permeation flux defined in Equation (16.4). For a lumen-fed module, the positive sign is used for the shell flow and the negative for the lumen flow. Note that mass transfer due to molecular diffusion and Taylor dispersion is neglected relative to convection, as suggested in the literature [24].

Summing Equation (16.16) over all of the components yields the continuity equation for the porous media.

$$\nabla \cdot (\rho u) = \pm \sum_{i=1}^{n} J_i \tag{16.17}$$

The shell and lumen velocity fields are obtained by substituting Darcy's law, Equation (16.15) for the velocity and applying appropriate boundary conditions. An appropriate equation of state also is required to calculate density from pressure. The ideal gas law is used for the relatively low pressure dehydration process considered here.

Shell and lumen boundary conditions for external sweep ports are illustrated in Figure 16.2 for an axis symmetric module cross-section. Symmetry is applied along the module centreline while the radial velocity and radial mass fluxes are set to zero along the external case. The pressures along the inlet and outlet are specified for the lumen and shell regions. The lumen and shell flow rates are controlled by the magnitude of the pressure drop between inlet and outlet. Note that specification of uniform pressures along the inlet and outlet boundaries assumes uniform gas distribution in the header that connects these regions to the external plumbing that delivers/removes the gas flow.

Equations (16.14)–(16.16) are solved using COMSOL Multiphysics® [32]. This simulation environment solves the governing conservation equations using the finite element method [33]. It also readily accommodates introduction of the appropriate form for the permeation flux and its dependence on gas partial pressure.

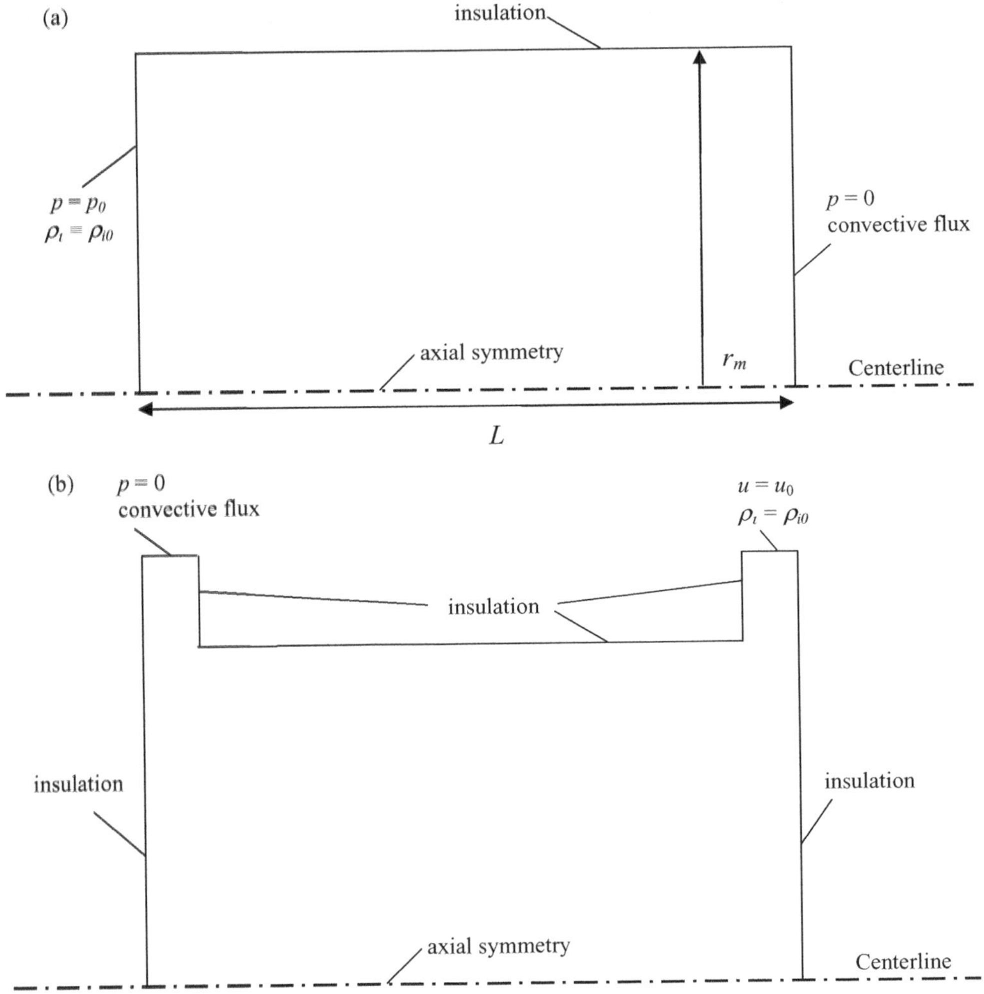

Figure 16.2 *Boundary conditions used in explicit simulations of sweep distribution to evaluate module performance: (a) lumen-side boundary conditions and (b) shell-side boundary conditions. Note that the boundary conditions for the shell correspond to one of the configurations used to simulate shell flows. Similar boundary conditions apply for the others. The shell extensions allow establishment of uniform velocity and concentration fields as described previously [24]*

16.3 Results and Discussion

The results presented here assume the selectivity for water permeation relative to nitrogen is 1000 and that for oxygen to nitrogen is 8 which are representative of current commercial air dehydration modules. Other fibre properties and operating conditions are tabulated in Table 16.1 and are representative of commercial hollow fibre dehydration modules.

Table 16.1 *Fibre and module properties and operating conditions. Literature correlations are used to convert dew points to water vapour concentrations [38,39]*

Fibre number	10,000
Fibre length (m)	0.1
Fibre inner diameter (μm)	200
Fibre outer diameter (μm)	150
Fibre packing fraction	0.5
H_2O/N_2 selectivity	1000
O_2/N_2 selectivity	8
N_2 permeance (GPU)	1
Feed pressure (psia)	147
Shell-side pressure (psia)	14.7
Feed dew point @ 14.7 (°F)	45
Operating temperature (°F)	115

The simulations assuming a Gaussian sweep distribution were performed using three quadrature points based on previous work demonstrating that use of five quadrature points gave equivalent results to within less than 1% [30,31]. Simulations with varying number of stages indicated that the results did not change by more than 5% if 200 or more stages were used. Thus, all simulations were performed with 200 stages.

The explicit sweep distribution simulations were performed with increasingly refined meshes until the results did not change by more than 5%. Typical meshes contained 5000 elements.

In the following discussion, percent variation prefers to the ratio of the standard deviation to the mean value of a property distribution expressed as a percentage ($100 * \sigma/\bar{\phi}$). Module performance refers to the dry gas (retentate) flow rate and recovery for a given purity.

16.3.1 Module Performance with Ideal Sweep Distribution

Figures 16.3 and 16.4 illustrate the effect of sweep fraction (i.e. fraction of the dry product returned as sweep to the shell) on module performance assuming uniform, ideal sweep distribution. For all sweep fractions, the product gas flow rate and recovery decrease as the dew point decreases since increased water removal is accompanied by increased loss of oxygen and nitrogen.

Increasing the sweep fraction dramatically increases the product gas flow rate. For example, Figure 16.3 indicates the dry gas flow rate increases by a factor of nearly 20 at a dew point of 0 °F if the sweep fraction is increased to 0.2. This increase in flow rate comes at the cost of reduced recovery. For high selectivities, recovery is reduced by an amount approximately equal to the sweep fraction. Figure 16.4 indicates recovery decreases from 0.88 to 0.79 at a dew point of 0 °F as the sweep fraction is increased to 0.2.

For the higher dew points, flow rates are high and significant pressure drops can occur. For the results in Figures 16.3 and 16.4, the highest pressure drops were ~8 kPa. Large pressure drops reduce the partial pressure driving force for water transport and consequently are detrimental to recovery.

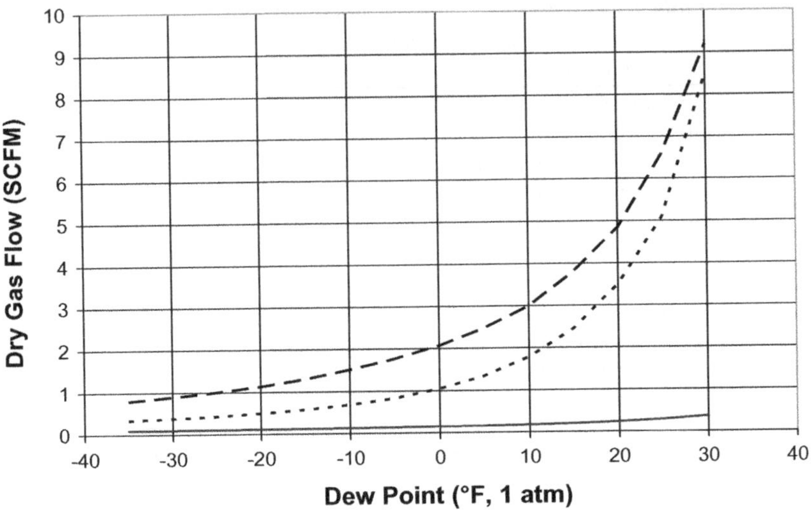

Figure 16.3 *Dry gas flow rate as a function of dew point for various sweep fractions: solid – 0; short dash – 0.1; long dash – 0.2*

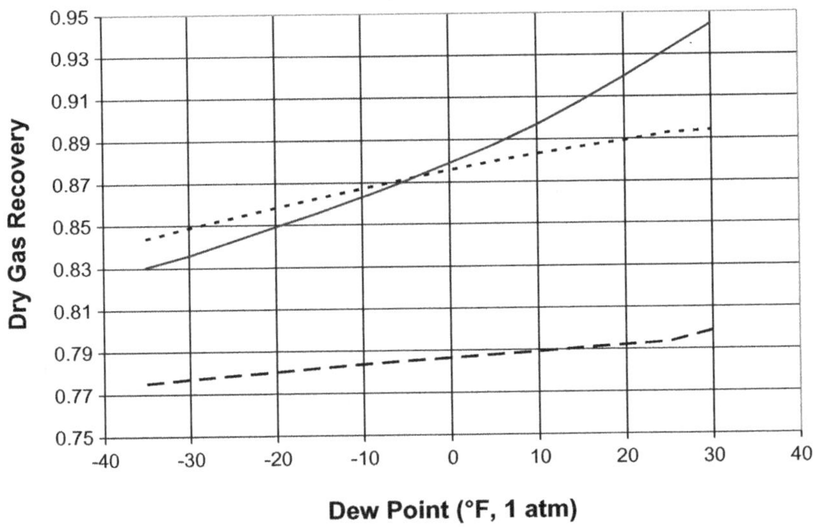

Figure 16.4 *Dry gas recovery as a function of dew point for various sweep fractions: sweep fractions: solid – 0; short dash – 0.1; long dash – 0.2*

The changes in performance for a lower water/nitrogen selectivity of 100 are illustrated in Figures 16.5 and 16.6. Relative to a selectivity of 1000, dry gas recovery decreases slightly but dry gas flow rates are significantly lower. A decrease in recovery is expected with a decrease in selectivity since more dry air will permeate along with the water. For a dew point of 0 °F and sweep fraction of 0.2, the recovery decreases from 0.79 to 0.75.

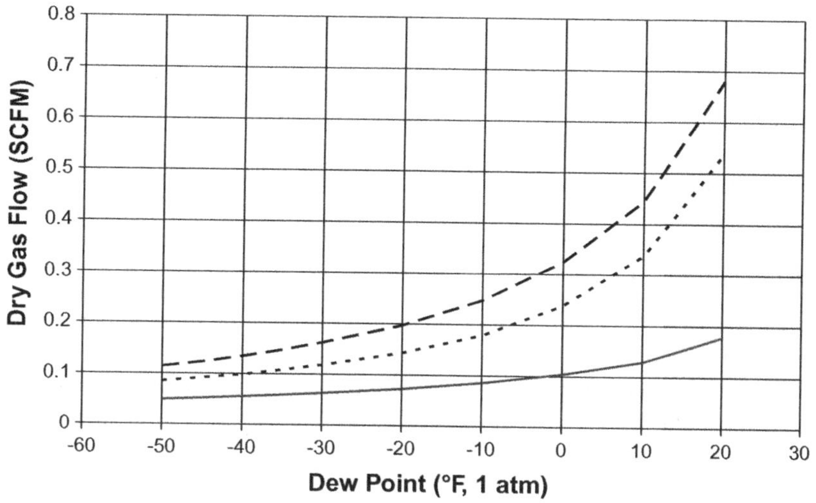

Figure 16.5 Dry gas flow rate as a function of dew point for a water/nitrogen selectivity of 100 and various sweep fractions: solid – 0; short dash – 0.1; long dash – 0.2

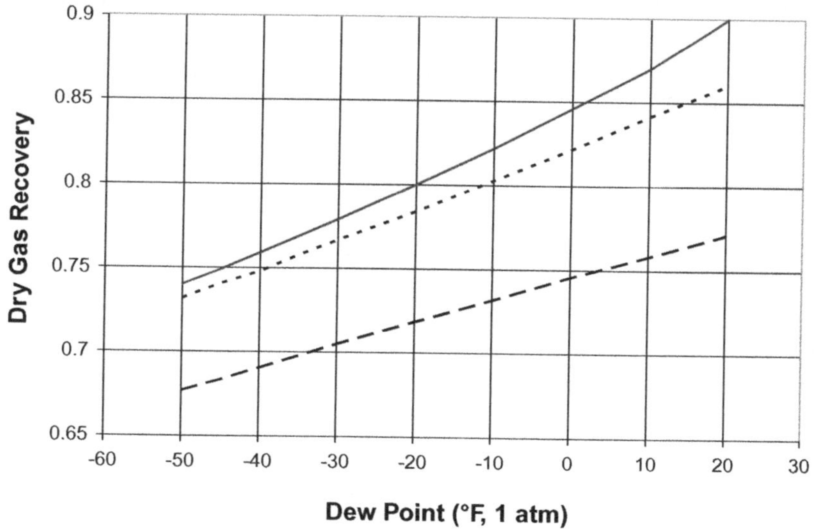

Figure 16.6 Dry gas recovery as a function of dew point for a water/nitrogen selectivity of 100 and various sweep fractions: solid – 0; short dash – 0.1; long dash – 0.2

The large drop in dry gas flow rate is due to the change in water permeance with selectivity. For the calculations, the nitrogen permeance was held constant so the decrease in selectivity led to an equal decrease in water permeance. However, the flow rate change is less than the permeance change; one expects the changes to be comparable if pressure drops in the fibre lumen are small and the permeate concentration does not change. For a dew point of 0 °F and sweep fraction of 0.2, the dry gas flow decreases by a factor of ~6 (from 2 to 0.3 SCFM) as the selectivity and water permeance decrease by a factor of 10. The decrease in flow rate is less than the decrease in permeance because more dry air permeates with the water. This reduces the water concentration in the permeate which in turn increases the water partial pressure difference that drives transport across the membrane. Consequently, the rate of water removal and the amount of humid gas that can be dried increase. Similar changes with selectivity have been reported previously [4,5].

16.3.2 Effect of Sweep and ID Variation

Figures 16.7 and 16.8 illustrate the effect of sweep variation on module performance. An average sweep fraction of 0.1 was used for the calculations.

Variability in the sweep flow around individual fibres has little effect on module performance. Only slight decreases in dry gas recovery and flow rate occur as the variability in sweep flow increases. The dry gas flow rate drops by less than 10% over the range of dew points considered as the variability in sweep flow rate increases to 20%. The dry gas recovery changes by less than 1%.

Figures 16.9 and 16.10 compare the effects of inner diameter and sweep variation on module performance. Like sweep variability, inner diameter variability is detrimental to performance – dry gas recovery and flow rate decrease as the variability increases. However, the effect of inner diameter variability is significantly larger. The dry gas flow rate decreases by a factor of two with 20% inner diameter variation for the lower dry gas dew points. Changes in dry gas recovery are smaller and do not exceed 5% at the lowest dew points considered.

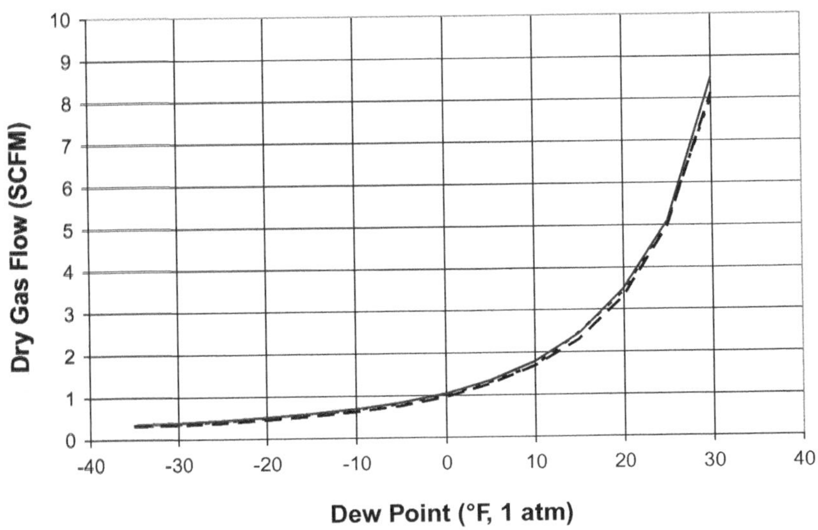

Figure 16.7 *Effect of sweep variation on dry gas flow rate as a function of dew point: solid – sweep variation = 0, short dash – sweep variation = 0.1, long dash – sweep variation = 0.2*

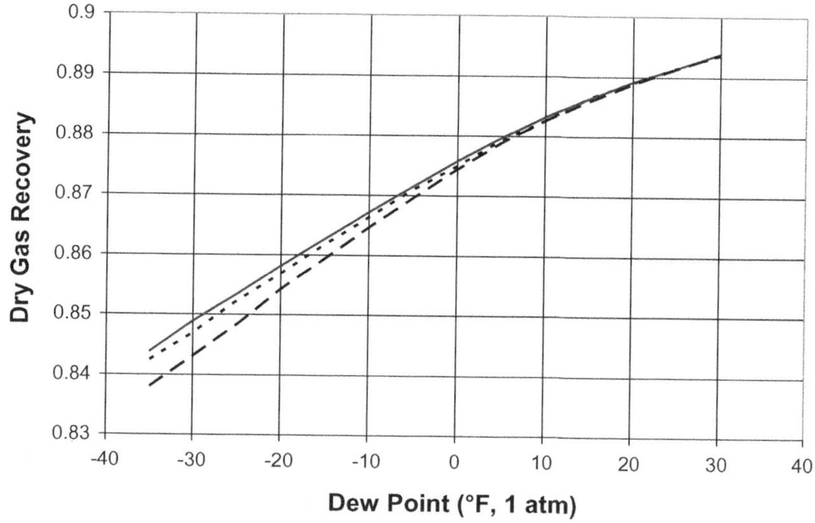

Figure 16.8 *Effect of sweep variation on dry gas recovery as a function of dew point: solid – sweep variation = 0, short dash – sweep variation = 0.1, long dash – sweep variation = 0.2*

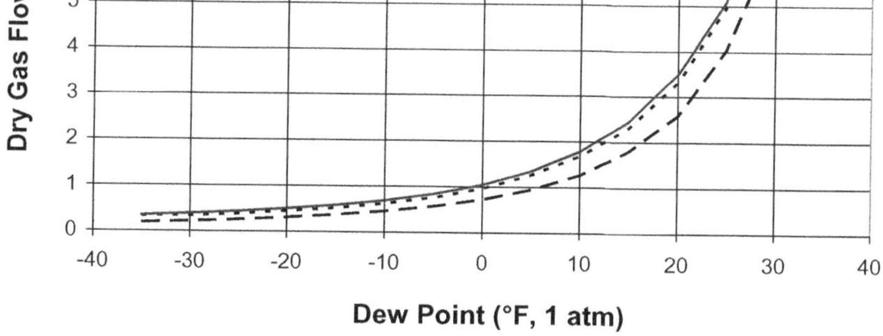

Figure 16.9 *Effect of sweep and fibre inner diameter variation on dry gas flow rate as a function of dew point: solid – no variation, short dash – sweep variation = 0.2, long dash – inner diameter variation = 0.2*

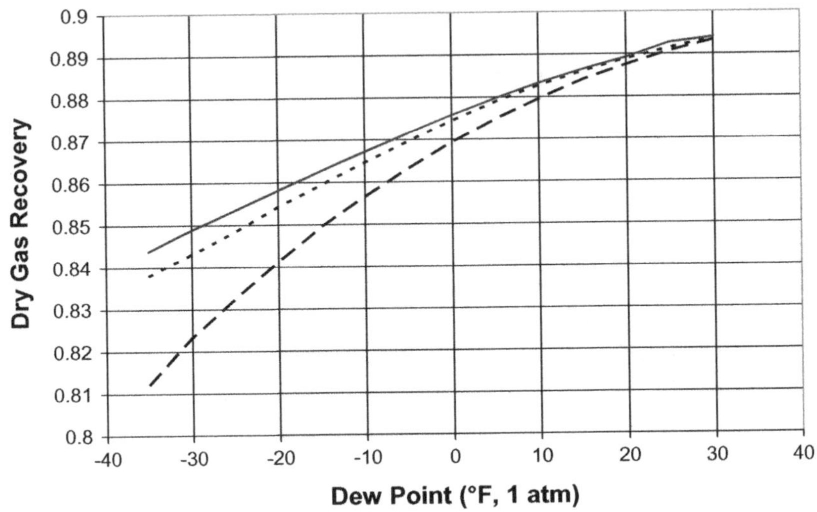

Figure 16.10 Effect of sweep and fibre inner diameter variation on dry gas recovery as a function of dew point: solid – no variation, short dash – sweep variation = 0.2, long dash – inner diameter variation = 0.2

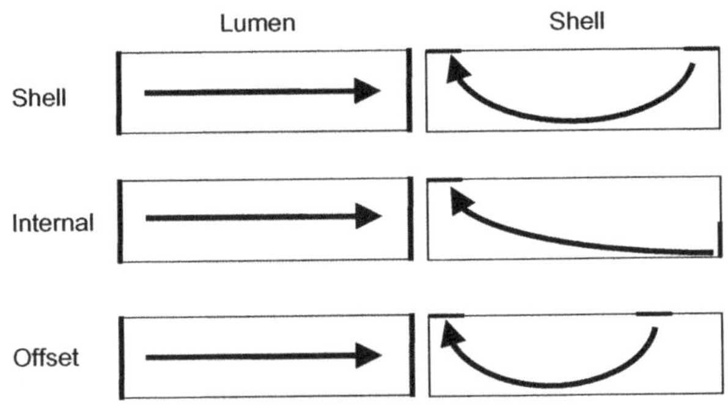

Figure 16.11 Three different sweep configurations used in explicit sweep distribution calculations to determine module performance. The arrows indicate the macroscopic flow direction. The thick solid lines indicate the location of the inlet and outlet

The effect of sweep variability in air dehydration is comparable to that observed previously for fibre variability in nitrogen production from air [30,31]. Significant changes in module performance are found only for inner diameter variability.

16.3.3 Effect of Sweep Distribution

Explicit sweep distribution calculations were performed for three different configurations as illustrated in Figure 16.11. The sweep is introduced either (1) on the periphery of the fibre bundle adjacent to the tube sheet, (2) on the periphery offset partially from the tube sheet, or (3) internally from within fibre bundle.

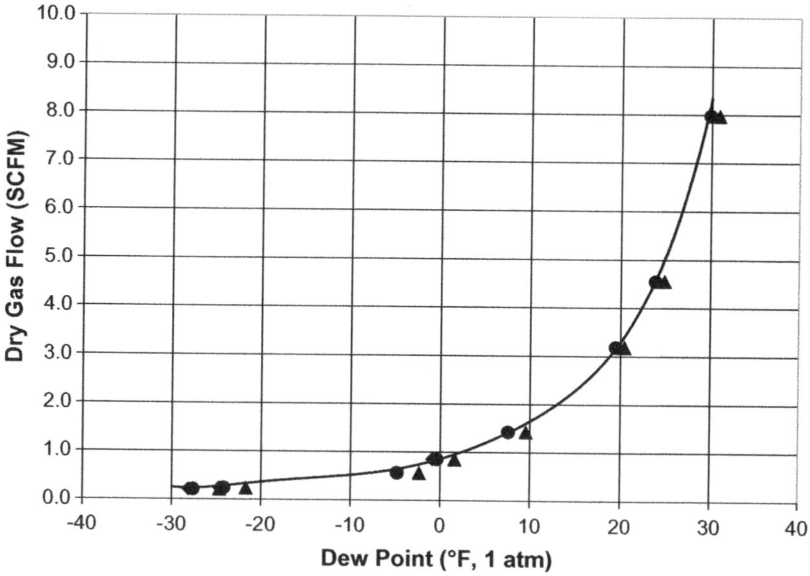

Figure 16.12 Effect of sweep configuration on dry gas flow rate as a function of dew point: diamond – internal, circle – shell, triangle – offset. The solid line corresponds to uniform sweep distribution. Note the diamond and circle symbols overlap

Figures 16.12 and 16.13 illustrate the performance predictions for the different configurations. The locations of the inlet and outlet regions for the sweep have little effect on performance. Dry gas flow rate and recovery decrease by less than 5% over the dew point range considered. The change in performance increases as dew point increases. Such results are consistent with the results obtained assuming a Gaussian distribution of the sweep around each fibre – variations in sweep flow rate do not significantly affect module performance.

The simulation results provide detailed concentration profiles within the module. Figure 16.14 illustrates a typical radial concentration profile for a module cross-section along the tube sheet at the retentate outlet (i.e. the dry gas product end). Clearly radial variations in concentration exist. The shell concentration increases by 600% from the periphery of the fibre bundle to the centreline while the lumen concentration increases by 60%. The difference in lumen and shell concentrations decreases by 400%.

Surprisingly, the large variations in concentration that exist do not impact overall module performance. The mixing cup average concentrations calculated from the concentration and velocity fields are nearly identical to the concentrations calculated assuming uniform sweep distribution and no radial variation in the concentration or velocity fields. This fortuitous result implies the inlet and outlet locations for the sweep are not a critical design variable.

16.3.4 Effect of Fibre Packing Variation Along Case

The literature indicates that large variations in packing of spheres [34] or fabrics [35] can occur at the boundary between the packing and a solid case enclosing the packing. At the

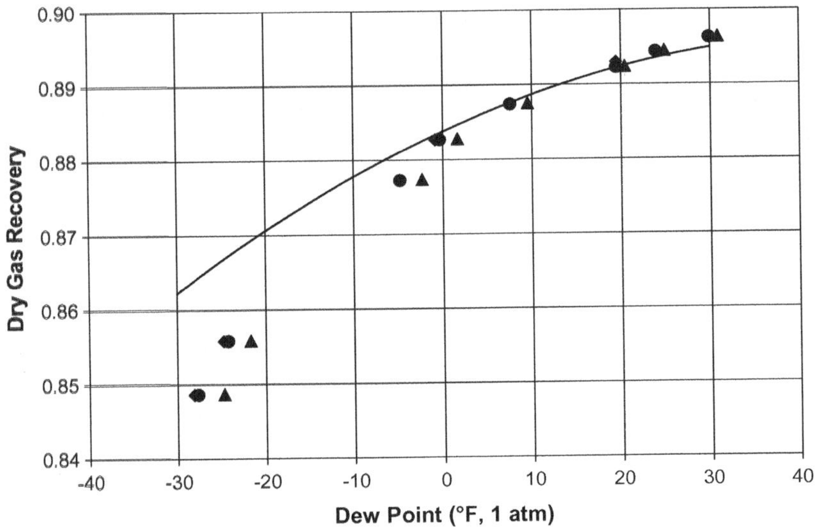

Figure 16.13 Effect of sweep configuration on dry gas recovery as a function of dew point: diamond – internal, circle – shell, triangle – offset. The solid line corresponds to uniform sweep distribution. Note the diamond and circle symbols overlap

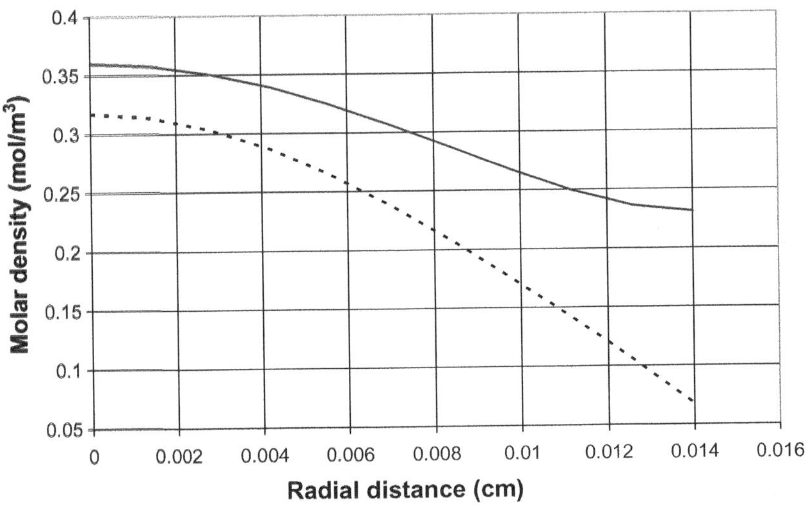

Figure 16.14 Variation in water concentration in the radial direction along the retentate outlet (i.e. the dry gas product end): solid – lumen and short dash – shell. The results are for operating conditions that give a dry gas dew point of −30 °F

packing/case interface, the packing fraction can decrease (i.e. the void volume increase) relative to the bulk packing. Increased void volume will lead to significant increases in hydraulic permeability and preferential flow at the packing/case interface. Such preferential flow is referred to as race tracking since fluid flows are much higher than the superficial flow within the packing and can impact heat and mass transfer [36,37].

The observation of large shell concentration gradients at the periphery of the fibre bundle suggests race tracking phenomena may have a significant impact on module performance. To test this hypothesis, the hydraulic permeability for the shell side flow was assumed to have the following dependence on the radial coordinate:

$$k = k_0 \exp(\alpha r) \tag{16.18}$$

where k_0 is the hydraulic permeability of the fibre bundle and α characterizes the extent and magnitude of the variation – as α increases the extent and magnitude increase. Equation (16.18) introduces an exponentially increasing hydraulic permeability with r. The maximum value occurs at the bundle/case boundary which will lead to race tracking for sufficiently large α.

Figures 16.15 and 16.16 illustrate the effect of the variable hydraulic permeability on module performance. For a fixed dew point, dry gas flow rate and recovery decrease as α increases. The flow rate decreases by up to 50% for the largest value of α but recovery decreases by less than 10%.

Reduced fibre packing at the bundle/case boundary is detrimental to performance. However, characterization of fibre packing in actual modules is required to determine the appropriate value for α and the impact of race tracking on performance.

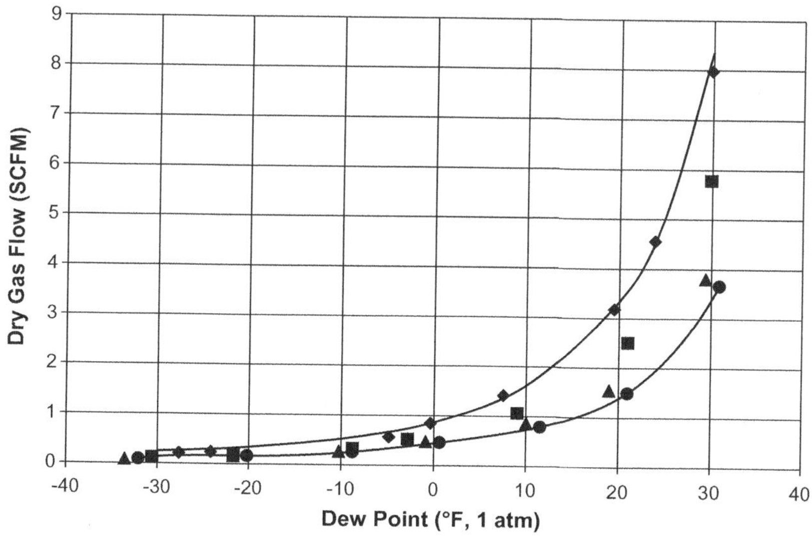

Figure 16.15 Effect of variable fibre packing near the module case on dry gas flow rate as a function of dew point. The symbols correspond to different values of α: diamond – 0, square – 200, triangle – 400, circle – 600. The upper solid line corresponds to uniform sweep distribution. The lower solid line is provided as a guide for the reader

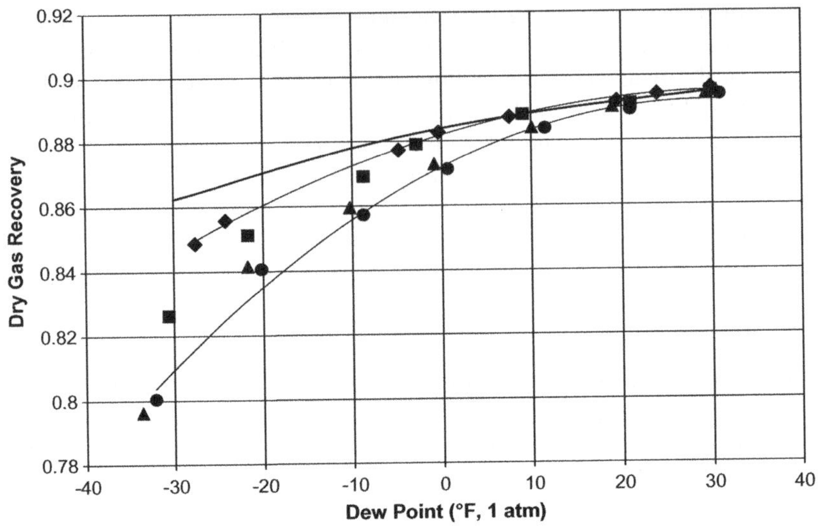

Figure 16.16 *Effect of variable fibre packing near the module case on dry gas recovery as a function of dew point. The symbols correspond to different values of α: diamond – 0, square – 200, triangle – 400, circle – 600. The upper solid line corresponds to uniform sweep distribution. The lower solid lines are provided as guides for the reader*

16.4 Conclusion

An analysis of sweep uniformity on the performance of hollow fibre gas dehydration modules is presented. The analysis is based on the governing conservation of mass and momentum equations and appropriate expressions for the overall mass transfer coefficient.

In all cases, for a given product dew point, the dry gas flow rate increases as the fraction of the product used as permeate sweep increases. However, the dry gas recovery (the fraction of the wet feed recovered as product) simultaneously decreases. The optimal sweep fraction will depend on the trade-off between increased productivity and fractional loss of the dry, high pressure product.

The effect of non-uniform sweep distribution is examined by: (1) assuming a Gaussian variation in the sweep flow around each fibre and (2) explicitly calculating the sweep distribution within the bundle for specified sweep inlet and outlet locations. In both cases, non-uniform sweep flows has little effect on module performance. The explicit sweep distribution calculations indicate large radial concentration gradients are present in the module. Surprisingly, these concentration gradients are not detrimental to performance.

If fibre packing is allowed to decrease at the interface between the fibre bundle and the enclosing case, preferential flow will occur along the interface relative to the centre of the bundle. Such flow is detrimental to performance for sufficiently large variation in hydraulic permeability. To determine the magnitude of this effect in commercial modules, fibre packing variations need to be characterized.

We believe the analysis presented here will be useful in evaluating alternative methods for introducing sweep in membrane modules for both gas and liquid separations. Although uniform sweep distribution appears not to be critical for air dehydration, other applications may be more sensitive.

List of Symbols

A	active membrane area
D	diffusivity
D_h	hydraulic diameter
f	flow rate
g	distribution function
J	permeation flux
k	mass transfer coefficient
K	porous medium hydraulic permeability
L	module or stage length
n	number of gas components
N	number of stages
N_f	number of fibres
p	pressure
P	permeate flow rate
Q	gas permeance
r	radius
R	retentate flow rate
R_g	ideal gas constant
Re	Reynolds number, $(2r)v\rho/\eta$
Sc	Schmidt number, $\eta/(\rho D)$
Sh	Sherwood number, kL/D
T	temperature
u	superficial velocity
v	bulk velocity
x	retentate mole fraction
y	permeate mole fraction
z	coordinate along flow direction

Greek Letters

η	gas viscosity
ϕ	sweep flow rate around a fibre
ρ	gas density
σ	standard deviation of sweep distribution

Subscripts and Superscripts

h	high pressure
i	gas component number

i	fibre inner diameter
j	stage number
l	low pressure or lumen side
m	module
max	maximum value
min	minimum value
o	fibre outer diameter
s	shell side

References

(1) R. W. Baker, Membrane Technology and Applications, John Wiley & Sons, Ltd., Chichester, 2004.
(2) M. Mulder, Basic Principles of Membrane Technology, 2nd Edition, Kluwer Academic, Dordrecht, 2000.
(3) S. A. Sonalkar, Membrane Gas Separation Processes for Permeate Purification, Doctoral Dissertation, University of Toledo, 2005.
(4) K. L. Wang, Improving Performance of Hollow Fiber Separators, Doctoral Dissertation, University of Minnesota, 1992.
(5) K. L. Wang, S. H. McCray, D. D. Newbold, E. L. Cussler, Hollow fiber air drying, *J. Membr. Sci.*, **72**, 231–244 (1992).
(6) C. W. Skarstrom, J. Kertzman, Process for separating fluids and apparatus, *U.S. Patent* 3735558 (1973).
(7) H. Makino, K. Nakagawa, Method for moving water vapor from water vapor-contacting gas, *U.S. Patent* 4718921 (1987).
(8) D. T. Friesen, R. J. Ray, D. D. Newbold, Countercurrent dehydration by hollow fibers, *U.S. Patent* 5002590 (1991).
(9) D. T. Friesen, R. J. Ray, D. D. Newbold, Countercurrent dehydration by hollow fibers, *U.S. Patent* 5108464 (1992).
(10) D. J. Stookey, Fluid separation membranes, *U.S. Patent* 4687869 (1987).
(11) W. H. Morgan, L. K. Bleikamp, D. G. Kalthod, Hollow fiber membrane dryer with internal sweep, *U.S. Patent* 5525143 (1996).
(12) J. H. Burban, R. O. Crowder, M. R. Spearman, K. A. Roberts, Membrane air dryer with integral diffuser and method of manufacture thereof, *U.S. Patent* 6585808 (2002).
(13) D. T. Friesen, R. J. Ray, D. D. Newbold, S. B. McCray, Counter-current dehydration by hollow fibers, *U.S. Patent* 5108464 (1992).
(14) S. R. Auvil, J. S. Choe, L. J. Kellogg, Jr., Use of membrane separation to dry gas streams containing water vapor, *U.S. Patent* 5259869 (1993).
(15) C. Vallieres, E. Favre, Vacuum versus sweeping gas operation for binary mixtures separation by dense membrane processes, *J. Membr. Sci.*, **244**, 17–23 (2004).
(16) S. J. Metz, W. J. C. Van de Ven, J. Potreck, M. H. V. Mulder, M. Wessling, Transport of water vapor and inert gas mixtures through highly selective and highly permeable polymer membranes, *J. Membr. Sci.*, **251**, 29–41 (2005).
(17) C. Y. Pan, Gas separation by permeators with high-flux asymmetric membranes, *AIChE J.*, **32**, 2020–2027 (1986).
(18) A. S. Kovvali, S. Vemury, W. Admassu, Modeling of Multicomponent Countercurrent Gas Permeators, *Ind. Eng. Chem. Res.*, **33**, 896–903 (1994).
(19) M. H. Chowdhury, X. Feng, P. Douglas, E. Croiset, A new numerical approach for a detailed multicomponent gas separation membrane model and AspenPlus simulation, *Chem. Eng. Technol.*, **28**, 773–782 (2008).
(20) D. T. Coker, B. D. Freeman, G. K. Fleming, Modeling multicomponent gas separation using hollow-fiber membrane contactors, *AIChE J.*, **44**, 1289–1302 (1998).

(21) S. Weller, W. A. Steiner, Engineering aspects of separation of gases. Fractional permeation through membranes, *Chem. Eng. Prog.*, **46**, 585–590 (1950).
(22) S. P. Kaldis, G. C. Kapantaidakis, G. P. Sakellaropoulos, Simulation of multicomponent gas separation in hollow fiber membrane by orthogonal collocation – hydrogen recovery from refinery gases, *J. Membr. Sci.*, **173**, 61–71 (2000).
(23) J. Lemanski, G. G. Lipscomb, Effect of shell-side flows on hollow-fiber membrane device performance, *AIChE J.*, **41**, 2322–2326 (1995).
(24) J. Lemanski, G. G. Lipscomb, Effect of shell-side flows on hollow-fiber gas separation modules, *J. Membr. Sci.*, **195**, 215–228 (2001).
(25) J. I. Marriott, E. Sorensen, I. D. L. Bogle, Detailed mathematical modeling of membrane modules, *Comp. Chem. Eng.*, **25**, 693–700 (2001).
(26) J. I. Marriott, E. Sorensen, A general approach to modeling membrane modules, *Chem. Eng. Sci.*, **58**, 4975–4990 (2003).
(27) J. I. Marriott, E. Sorensen, The optimal design of membrane systems, *Chem. Eng. Sci.*, **58**, 4991–5004 (2003).
(28) B. Carnahan, H. A. Luther, J. O. Wilkes, Applied Numerical Methods, Krieger, Malabar, FL, 1990.
(29) R. B. Bird, W. E. Stewart, E. N. Lightfoot, Transport Phenomena, Revised Second Edition, John Wiley & Sons, New York, 2007.
(30) J. Lemanski, B. Liu, G. G. Lipscomb, Effect of fiber variation on the performance of cross-flow hollow fiber gas separation modules, *J. Membr. Sci.*, **153**, 33–43 (1999).
(31) J. Lemanski, G. G. Lipscomb, Effect of fiber variation on the performance of countercurrent hollow fiber gas separation modules, *J. Membr. Sci.*, **167**, 241–252 (2000).
(32) http://www.comsol.com/.
(33) J. N. Ready, D. K. Gartling, The Finite Element Method in Heat Transfer and Fluid Dynamics, CRC Press, Boca Raton, FL, 1994.
(34) J. J. Lerou. G. F. Froment, Velocity, temperature and conversion profiles in fixed bed catalytic reactors, *Chem. Eng. Sci.*, **32**, 853–861 (1977).
(35) M. Devillard, K.-T. Hsiao, A. Gokce, S. G. Advani, On-line characterization of bulk permeability and race-tracking during the filling stage in resin transfer molding process, *J. Comp. Mat.*, **37**, 1525–1541 (2003).
(36) J. G. H. Borkink, K. R. Westererp, Significant of the radial porosity profile for the description of heat transport in wall-cooled packed beds, *Chem. Eng. Sci.*, **49**, 863–876 (1994).
(37) A. Grah, U. Nowak, M. Schreier, R. Adler, Radial heat transfer in fixed-bed packing with small tube/particle diameter ratios, *Heat Mass Transfer*, **45**, 417–425 (2009).
(38) J. A. Goff, S. Gratch, Low-pressure properties of water from −160° to 212 °F, *Trans. Amer. Soc. Heat Vent. Eng.*, **52**, 95–121 (1946).
(39) A. L. Buck, New equations for computing vapor pressure and enhancement factor, *J. Appl. Meteor.*, **20**, 1527–1532 (1981).

Index

Note: References in *italics* indicate a figure and references in **bold** indicate a table.

Membrane Gas Separation Edited by Yuri Yampolskii and Benny Freeman

Printed and bound by CPI Group (UK) Ltd, Croydon, CR0 4YY

Printed and bound by CPI Group (UK) Ltd, Croydon, CR0 4YY
16/09/2026
14963473-0001

Atlas of

Clinical Vascular Medicine

Atlas of Clinical Vascular Medicine

EDITED BY

Jessica Mintz
Cleveland Clinic Heart and Vascular Institute
Cleveland Clinic Lerner College of Medicine
Case Western Reserve University
Cleveland, OH, USA

Bruce L. Mintz
Clinical Associate Professor Internal Medicine
Director Vascular Technical Training Program
University of Medicine and Dentistry of New Jersey
Newark, NJ;
Attending Gagnon Heart Center
Director Anticoagulation Service
Morristown Medical Center
Morristown, NJ, USA

Michael R. Jaff
Associate Professor of Medicine
Harvard Medical School;
Chair, MGH Institute for Heart, Vascular and Stroke Care
Massachusetts General Hospital
Boston, MA, USA

WILEY-BLACKWELL
A John Wiley & Sons, Ltd., Publication

This edition first published 2013. © 2013 by John Wiley & Sons, Ltd

Wiley-Blackwell is an imprint of John Wiley & Sons, formed by the merger of Wiley's global Scientific, Technical and Medical business with Blackwell Publishing.

Registered office: John Wiley & Sons, Ltd, The Atrium, Southern Gate, Chichester, West Sussex, PO19 8SQ, UK

Editorial offices: 9600 Garsington Road, Oxford, OX4 2DQ, UK
The Atrium, Southern Gate, Chichester, West Sussex, PO19 8SQ, UK
111 River Street, Hoboken, NJ 07030-5774, USA

For details of our global editorial offices, for customer services and for information about how to apply for permission to reuse the copyright material in this book please see our website at www.wiley.com/wiley-blackwell.

Library of Congress Cataloging-in-Publication Data

Mintz, Jessica Mintz.
p. ; cm.
Includes bibliographical references and index.
ISBN 978-0-470-65809-3 (hardback : alk. paper)
I. Mintz, Jessica. II. Mintz, Bruce L. III. Jaff, Michael R.
[DNLM: 1. Vascular Diseases–diagnosis–Atlases. 2. Vascular Diseases–diagnosis–Case Reports.
3. Vascular Diseases–therapy–Atlases. 4. Vascular Diseases–therapy–Case Reports. WG 17]
616.8'1–dc23

2012044507

A catalogue record for this book is available from the British Library.

Wiley also publishes its books in a variety of electronic formats. Some content that appears in print may not be available in electronic books.

Set in 8/12pt Stone Serif by Toppan Best-set Premedia Limited, Hong Kong
Printed and bound in Singapore by Markono Print Media Pte Ltd

1 2013

Contents

Preface and Dedication

For those who have committed their professional careers to the diagnosis and management of vascular diseases, one of the true joys of this specialty is the broad and myriad presentations patients demonstrate. One can never get bored, nor can one assume that they have "seen it all".

Vascular medicine specialists must have a keen ability to identify subtle physical diagnostic findings, and link them to often difficult historical presentations and objective imaging tests. There are many tomes on vascular medicine; however, we felt that an easy to reference pictorial atlas would serve as a useful ready reference for practicing clinicians. We have asked renowned experts in the field to help us put together a comprehensive, organized pictorial with brief descriptions of what you see in each figure. We hope you find this a useful addition to your clinical libraries.

Each of us dedicates this *Atlas* to our families, who tolerate our hectic lifestyles. Still, the true dedication of this *Atlas* is to Jess R. Young, MD, Emeritus Chair of Vascular Medicine at the Cleveland Clinic. For those of us who were privileged to train under Dr. Young (B.L.M., M.R.J.), we have never encountered his equal. Dr. Young combined a mastery of clinical medicine with an encyclopedic fund of knowledge and a unique method of approaching patients. We use Dr. Young's approach to vascular medicine every day in our clinical practice, and attempt to pass on his acumen and diagnostic strategies to our trainees. Thank you, Dr. Young, on behalf of all of the physicians you have trained and all the patients you directly, and indirectly, touched and healed.

Contributor List

Jason Q. Alexander, MD
Assistant Professor of Clinical Surgery, University of Minnesota
Minneapolis Heart Institute, Minneapolis,
MN, USA

J. Michael Bacharach, MD, MPH, FACC, FSCAI
Section Head, Vascular Medicine and Peripheral Vascular Intervention
North Central Heart Institute
Sioux Falls, SD, USA

John R. Bartholomew, MD
Section Head of Vascular Medicine
Cleveland Clinic Heart and Vascular Institute; Cleveland Clinic Lerner College of Medicine of Case Western Reserve University,
Cleveland, OH, USA

Joshua A. Beckman, MD, MS
Harvard Medical School
Cardiovascular Division, Brigham and Women's Hospital
Boston, MA, USA

Stephen L. Chastain, MD
Assistant Professor of Medicine
University of South Carolina School of Medicine/Greenville;
Director of Inpatient Vascular Medicine Services
Greenville Health System, Greenville, SC, USA

Shane Clark, MD
Chief Dermatology Resident
Wexner Medical Center at the Ohio State University
Columbus, OH, USA

Steven M. Dean, DO, FACP, RPVI
Program Director, Vascular Medicine
Associate Professor of Medicine,
Department of Cardiovascular Medicine,
Wexner Medical Center at the Ohio State University
Columbus, OH, USA

Taki Galanis, MD
Assistant Professor of Medicine
Department of Surgery
Director of Hyperbaric Therapy and Wound Care
Jefferson Vascular Center
Thomas Jefferson University Hospitals
Jefferson Medical College
Philadelphia, PA, USA

Michael Gironta, DO
Eastern Vascular Associates
Denville, NJ, USA

Heather L. Gornik, MD, MHS
Medical Director, Non-Invasive Vascular Laboratory
Cleveland Clinic Heart and Vascular Institute
Assistant Professor Medicine
Cleveland Clinic Lerner College of Medicine of Case Western Reserve University
Cleveland, OH, USA

Bruce H. Gray, DO
Professor of Surgery/Vascular Medicine
University of South Carolina School of Medicine/Greenville;
Director of Endovascular Services
Greenville Health System, Greenville, SC, USA

Peter C. Grayson, MD, MSc
Assistant Professor of Medicine – Rheumatology
Vasculitis Center, Section of Rheumatology
Boston University Medical Center, Boston, MA, USA

Paul B. Haser, MD
Associate Professor of Surgery
Robert Wood Johnson Medical School
University of Medicine & Dentistry of New Jersey
New Brunswick, NJ, USA

Raghu Kolluri, MD, RVT, FACC, FACP, FSVM
Medical Director
Vascular Laboratories of Prairie Vascular Institute
Springfield, IL, USA

Peter A. Merkel, MD, MPH
Chief, Division of Rheumatology
Professor of Medicine and Epidemiology
University of Pennsylvania, Philadelphia, PA, USA

Geno J. Merli, MD, FACP, FHM, FSVM
Professor of Medicine
Co-Director, Jefferson Vascular Center,
Jefferson Medical School
Thomas Jefferson University Hospitals
Philadelphia, PA, USA

Ari J. Mintz, DO
Department of Internal Medicine
Lahey Clinic Medical Center
Burlington, MA, USA

Stephen C. Nelson, MD
Director, Hemoglobinopathy Program
Co-Director, Vascular Anomalies Clinic
Adjunct Assistant Professor
Department of Pediatrics
University of Minnesota
Children's Hospitals and Clinics of Minnesota
Minneapolis, MN, USA

Cory Perugino, DO
Cleveland Clinic Foundation
Cleveland, OH, USA

Georgann A. Poulos, MD
Wexner Medical Center at the Ohio State University
Division of Dermatology
Columbus, OH, USA

Robert J. Rosen, MD
Co-Director of Endovascular Surgery
Lenox Hill Heart and Vascular Institute
New York, NY, USA

Robert M. Schainfeld, DO
Lecturer on Medicine
Harvard Medical School
Section of Vascular Medicine
Massachusetts General Hospital, Boston,
MA, USA

Timothy Sullivan, MD
Minneapolis Heart Institute
Minneapolis, MN, USA

Sandra Yesenko, BS
Lead Technologist for Education
Non-invasive Vascular Laboratory
Cleveland Clinic
Cleveland, OH, USA

Ido Weinberg, MD MSc MHA
Vascular Medicine Section
Massachusetts General Hospital
Boston, MA, USA

Katherine P. Zimmerman
Student Research Intern
Minneapolis Heart Institute,
Minneapolis, MN, USA

About the Companion Web Site

There is a companion web site for this book at
http://www.wiley.com/go/Jaff/Vascular

On this site you will find
- A complete set of figures from the book

Atlas of

Clinical Vascular Medicine

Peripheral Artery Disease

Joshua A. Beckman
Cardiovascular Division, Brigham and Women's Hospital, and Harvard Medical School, Boston, MA, USA

BRIEF CASE PRESENTATION

A 74-year-old man presents with a chief complaint of limb discomfort with exercise which resolves with rest.

CHARACTERISTICS

- Peripheral artery disease (PAD) is a common manifestation of systemic atherosclerosis
- Patients with PAD present in myriad ways. They may have no symptoms, report muscular pain with ambulation, claudication, or present with critical limb ischemia. Critical limb ischemia is represented by pain at rest, a non-healing ulcer or gangrene
- The diagnosis is made through history, careful physical examination of the lower extremities, pulse evaluation and performance of the ankle-brachial index (ABI). An ABI less than 0.91 is diagnostic of PAD
- Several other non-invasive tests are used for the diagnosis of PAD including exercise stress testing, segmental blood pressure, and pulse volume detection. Duplex ultrasonography can be utilized to pinpoint the specific morphology of the vascular lesion
- Pulse volume recordings (PVR) are especially useful in diabetic patients with non-compressible arteries as it is less effected by medial calcinosis than segmental blood pressure recordings
- Patients with PAD have a similar risk of myocardial infarction and stroke as patients with coronary heart disease, yet PAD patients are typically less intensively treated with anti-platelet therapy and risk reduction than patients with coronary disease

TREATMENT

Cardiovascular risk reduction therapies including tobacco cessation, blood pressure control, cholesterol management, tight glycemic control and aspirin therapy are all indicated to reduce myocardial infarction, stroke, and death. Management of intermittent claudication include risk factor intervention and a supervised exercise walking program. Revascularization is considered when medical management and maximal exercise therapy have failed to improve the patients physical functioning. In the setting of critical limb ischemia, lower extremity revascularization may be required to avoid amputation

REFERENCES

Rooke TW, Hirsch AT, Misra S, et al. ACCF/AHA 2005 Focused Update of the Guideline for the management of patients with peripheral artery disease (updating the 2005 guideline). J Am Coll Cardiol. 2011; 58:19.

Hirsch AT, Criqui MH, Treat-Jacobson D, et al. Peripheral arterial disease detection, awareness, and treatment in primary care. JAMA. 2001; 286(11):1317–24.

Diehm C, Allenberg JR, Pittrow D, et al. Mortality and vascular morbidity in older adults with asymptomatic versus symptomatic peripheral artery disease. Circulation. 2009; 120:2053–61.

Atlas of Clinical Vascular Medicine, First Edition. Edited by Jessica Mintz, Bruce L. Mintz, and Michael R. Jaff.

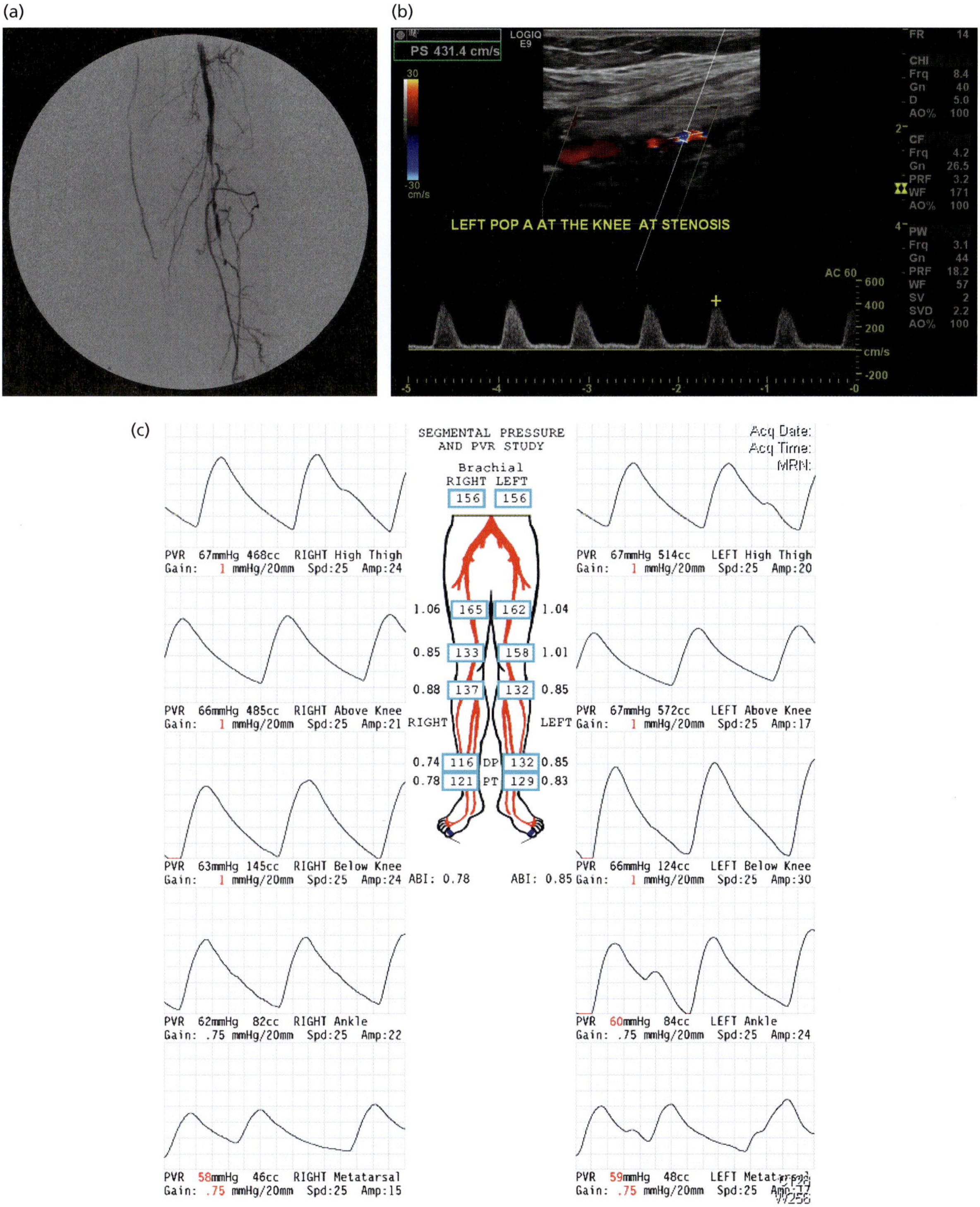

Peripheral Artery Disease Imaging. (a) Angiography of atherosclerotic popliteal artery stenosis. (b) Duplex ultrasonography of popliteal artery stenosis. (c) Pulse volume recordings demonstrating bilateral superficial femoral artery and popliteal artery stenoses.

Elevation Pallor and Dependent Rubor

Ido Weinberg and Michael R. Jaff
Massachusetts General Hospital, Boston, MA, USA

BRIEF CASE PRESENTATION

An 82-year-old woman with a history of extensive tobacco abuse and type 2 diabetes mellitus for 25 years developed foot pain during sleep. She attained relief by dangling her foot over the bed. Physical examination revealed a diminished femoral pulse and absent popliteal and pedal pulses on the right. In the dependent position, the right foot became deeply erythematous, while pallor of the forefoot developed rapidly with elevation of the foot 60 degrees above the level of the heart.

CHARACTERISTICS

- Dependent rubor and elevation pallor are elicited by the elevation-dependency test and denote severe peripheral artery disease
- Dependent rubor results from the dilation of dermal arterioles and capillaries in the presence of increased hydrostatic pressure
- Elevation pallor results from inadequate blood flow to reach the foot without the assistance of gravity

TREATMENT

This is a physical finding. Treatment must be directed towards the underlying peripheral artery disease.

REFERENCES

Creager MA, Libby P. Peripheral arterial disease. In: Libby P, Bonow RO, Mann DL, Zipes DP. Braunwald's Heart Disease: A Textbook of Cardiovascular Medicine. 8th ed. Philadelphia, Pa: Saunders; 2007:chap 57.

Rain, FA, Cooper LT. Peripheral arterial disease: diagnosis and management. Mayo Clinic Proceedings. 2008; 83:944–50.

Spittell JA Jr. Occlusive peripheral arterial disease. Guidelines for office management. Postgrad Med. 1982 Feb;71(2):137–42, 146, 150–1. PubMed PMID: 6460230.

Atlas of Clinical Vascular Medicine, First Edition. Edited by Jessica Mintz, Bruce L. Mintz, and Michael R. Jaff.

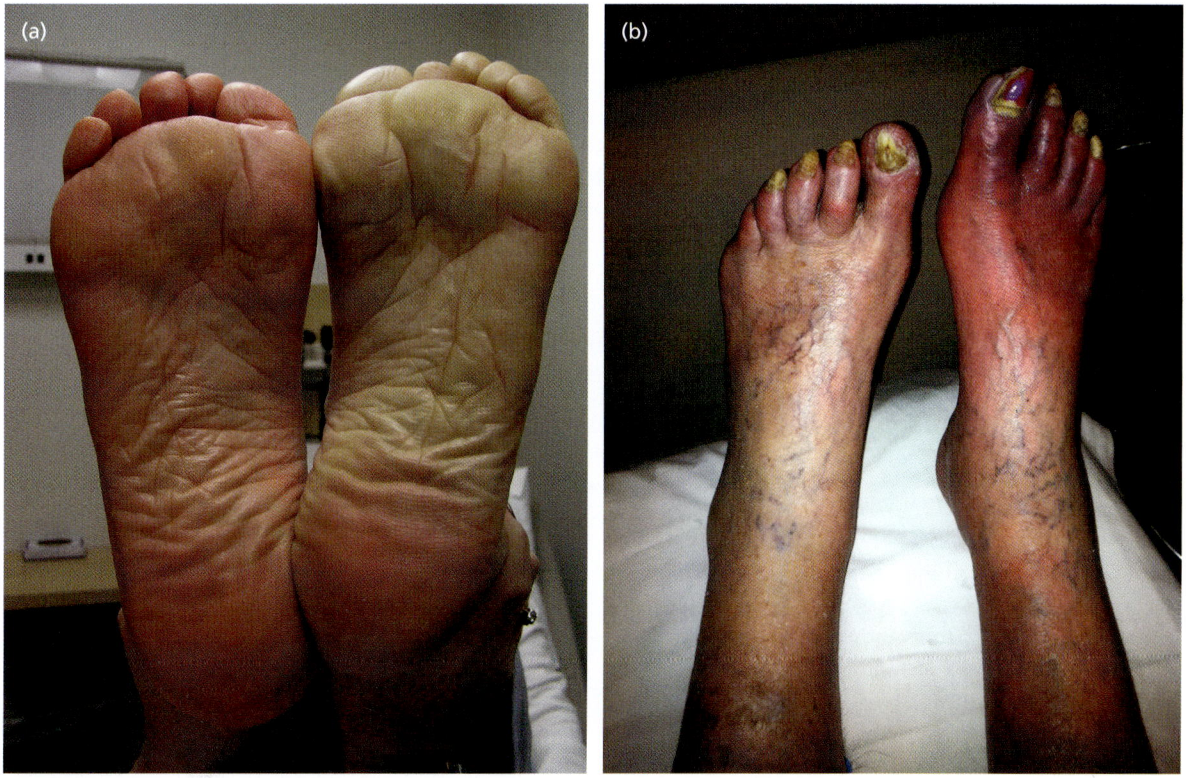

Peripheral artery examination. (a) Lower extremity demonstrating pallor on elevation. (b) Rubor on dependency.

Ischemic Toes

Ido Weinberg and Michael R. Jaff
Massachusetts General Hospital, Boston, MA, USA

BRIEF CASE PRESENTATION

A 75-year-old man presents with painful, discrete purple lesions on his toes bilaterally. Past medical history includes atrial fibrillation, hypercholesterolemia, hypertension and cigarette smoking.

CHARACTERISTICS

- Emboli to the toes cause local ischemic phenomena. They are typically intensely painful and can be located anywhere
- A bilateral lower extremity presentation suggests a central etiology such as an abdominal aortic aneurysm or cardiogenic emboli. Unilateral ischemic digits suggest a source in the extremity such as popliteal aneurysm
- Ischemic limbs can arise spontaneously or as a result of catheter manipulation
- There are multiple conditions that can mimic ischemic toes including, but not limited to, vasospastic disorders, thromboangiitis obliterans and vasculitis

TREATMENT

The specific treatment for ischemic toes includes management of pain; skin, toe, limb protection; and assessment of large artery circulation. Treatment should be directed at the source and associated atherosclerotic risk factors. Evaluation for emboli in other vascular beds should be considered (renal, mesenteric). Systemic anti-atherosclerosis therapy should be provided.

REFERENCES

Banerkee S, Thatthi T, Badhey N, Brilakis ES. Atheroemboli during Peripheral arterial interventions. J Inv Cardiol. 2009; 21(12):632–3.

O'Keeffe ST, Woods BO, Breslin DJ, Tsapatsaris NP. Blue toe syndrome. Causes and management. Arch Intern Med. 1992; 152(11):2197–202.

Richter JG, Sander O, Schneider M, Klein-Weigel P. Diagnostic algorithm for Raynaud's phenomenon and vascular skin lesions in systemic lupus erythematosus. Lupus. 2010; 19(9): 1087–95. Review. PubMed PMID: 20693202.

Atlas of Clinical Vascular Medicine, First Edition. Edited by Jessica Mintz, Bruce L. Mintz, and Michael R. Jaff.

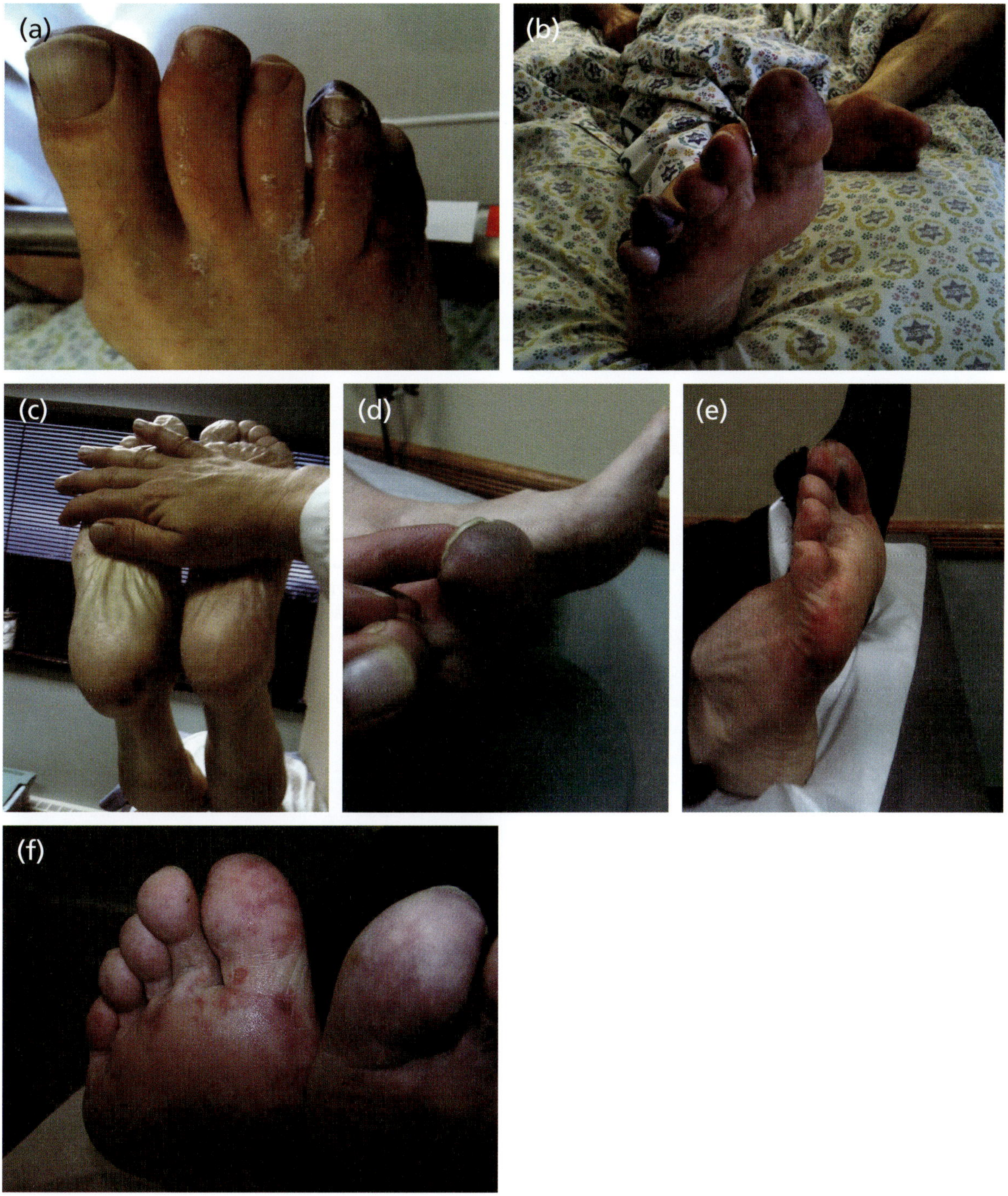

Limb Ischemia. (a) Thromboemboli, note plethoric changes in toes along with rubor and multiple circumferential cutaneous lesions on forefoot; (b) Livedo reticularis of the great toe. Irreversible ischemia of the 4th and 5th toes. (c) Atheromatous emboli to foot with evidence of heel emboli. (d) Emboli to great toe with impending tissue loss. (e) Livedo reticularis with tissue loss of the lateral forefoot and great toe. (f) Livedo reticularis of bilateral feet. Image (c) is courtesy of Bruce L. Mintz.

Severe Limb Ischemia Related to Systemic Vasopressor Use

Bruce L. Mintz[1] and Michael R. Jaff[2]
[1]Morristown Medical Center, Morristown, NJ, USA
[2]Massachusetts General Hospital, Boston, MA, USA

BRIEF CASE PRESENTATION

A 68-year-old man presented to an emergency department in cardiogenic shock. He suffered a massive anterior wall myocardial infarction and required emergency coronary artery bypass graft surgery. Due to hemodynamic instability, he required several days of vasopressor administration.

CHARACTERISTICS

- Prolonged systemic vasopressor use may result in peripheral vasoconstriction of all limbs
- Patients may develop necrosis of the tips of all digits
- These digital lesions are intensely painful and yet extremity pulses are commonly palpable
- If associated with pre-existing atherosclerosis more extensive and proximal ischemia may occur

TREATMENT

Cessation of vasopressor use must occur at the earliest possible moment. Analgesics for pain control as well as protection of necrotic tissue from further trauma or infection is of paramount importance. Autoamputation is preferable, but in the case of intractable pain or infection, digital sympathectomy, botulinum toxin injection, or amputation may be considered.

REFERENCES

Borrego R, Lopez-Herce J, Mencia S, et al. Severe ischemia of the lower limb and of the intestine associated with systemic vasoconstrictor therapy and femoral arterial catheterization. Pediatr Crit Care Med. 2006; 7:267–9.

Marine L, Castro P, Enriquez A, et al. Four-limb acute ischemia induced by ergotamine in an AIDS patient treated with protease inhibitors. Circulation. 2011; 124:1395–7.

Atlas of Clinical Vascular Medicine, First Edition. Edited by Jessica Mintz, Bruce L. Mintz, and Michael R. Jaff.

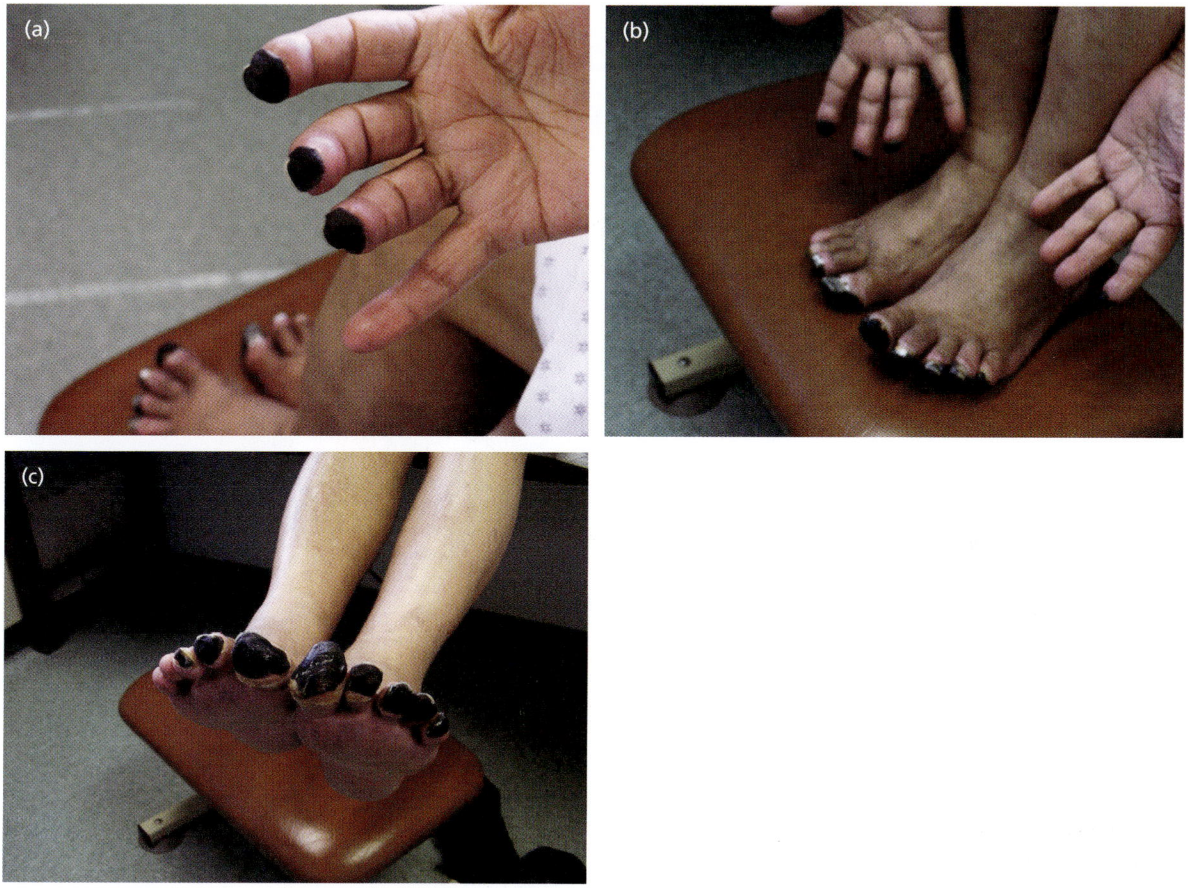

Vasopressor-induced limb ischemia. (a–c) Ischemic digits secondary to vasopressor agents. Images are courtesy of Dr Paul Haser.

Renal Artery Stenosis: Atherosclerotic

J. Michael Bacharach

Vascular Medicine and Peripheral Vascular Intervention, North Central Heart Institute, Sioux Falls, SD, USA

BRIEF CASE PRESENTATION

A 61-year-old female with a history of hypertension and stage II chronic kidney disease presented for evaluation of uncontrolled hypertension.

Digital subtraction angiography of the renal arteries demonstrate severe ostial narrowing of the left and right renal arteries.

CHARACTERISTICS

- Renal artery stenosis (RAS) secondary to atherosclerosis is the most common cause of renovascular hypertension and is a cause of chronic renal insufficiency
- RAS is strongly associated with generalized atherosclerosis. Dyslipidemia, hypertension, cigarette smoking and diabetes mellitus are all risk factors that contribute to the pathogenesis. RAS commonly presents with several comorbidities including coronary artery disease, peripheral artery disease and cerebrovascular disease
- Most cases are asymptomatic in the setting of uncontrolled hypertension
- Acute kidney injury or deterioration in renal function with initiation of angiotensin-converting enzyme inhibitors is suggestive of bilateral RAS
- Diagnosis is made by renal artery duplex ultrasonography documenting a renal artery peak systolic velocity greater than 200 cm/s, renal aortic peak systolic velocity ratio >3.5, with post-stenotic turbulence.

TREATMENT

Risk factor modification, including blood pressure control and lipid management is key in the treatment of atherosclerotic RAS. Patients with significant stenosis refractory to medical management are considered for revascularization.

REFERENCES

Kumbhani DJ, Bavry AA, Harvey JE, et al. Clinical outcomes after percutaneous revascularization versus medical management in patients with significant renal artery stenosis. Am Heart J. 2011; 161(3):622–30.

Forbes TL, Ricco JB. Role of interventions for atherosclerotic renal artery stenoses. J Vasc Surg. 2011; 54:2.

Rosenfield K, Jaff MJ. An 82-year- old woman with worsening Hypertension: review of renal artery stenosis. JAMA. 2008; 300(17):2036–44.

Atlas of Clinical Vascular Medicine, First Edition. Edited by Jessica Mintz, Bruce L. Mintz, and Michael R. Jaff.

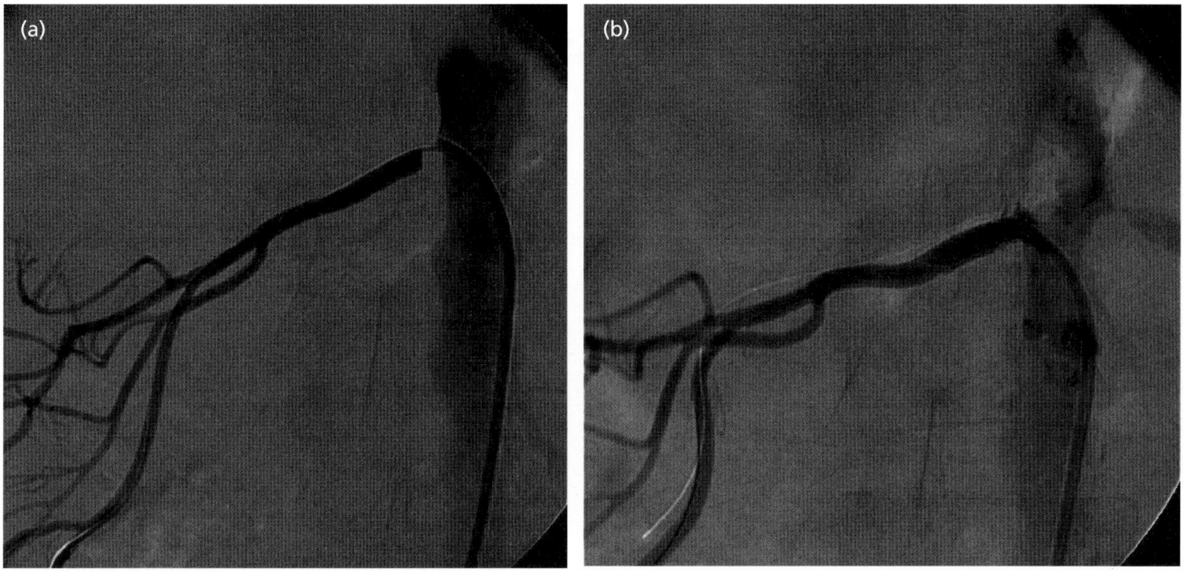

Atherosclerotic renovascular disease. (a) Angiography of the right renal artery demonstrating ostial stenosis. (b) Percutaneous transluminal renal angioplasty (PTRA) and stent deployment with post-revascularization angiography.

Renal Artery Stenosis: Fibromuscular Dysplasia

J. Michael Bacharach
Vascular Medicine and Peripheral Vascular Intervention, North Central Heart Institute, Sioux Falls, SD, USA

BRIEF CASE PRESENTATION

A 54-year-old female presents with labile and difficult to control hypertension. Selective digital subtraction angiography of the renal arteries demonstrated web-like narrowing with a "string-of-beads" pattern bilaterally. Angiographic findings were consistent with the medial fibroplasias variant of fibromuscular dysplasia.

CHARACTERISTICS

- Fibromuscular dysplasia (FMD) is a non-inflammatory, non-atherosclerotic vascular disease which is poorly understood and often under diagnosed. Seen most frequently in renal and carotid arteries, renal artery stenosis secondary to FMD manifests as difficult to control hypertension
- Primarily affects women from ages 20 to 60, but has been seen in men and women in all age groups
- FMD is categorized into five histological types: medial fibroplasia, perimedial fibroplasias, medial hyperplasia, intimal fibroplasia and adventitial fibroplasia
- Clues to the diagnosis include: onset of hypertension before age 35; systolic/diastolic epigastric bruits; and discrepancy in kidney size in females at low risk for atherosclerosis
- Diagnosis is made initially with a high index of suspicion, followed by either renal artery duplex ultrasonography, computerized tomographic angiography, magnetic resonance angiography, or contrast arteriography. Catheter angiography however remains the gold standard for diagnosis

TREATMENT

Asymptomatic patients are treated with antiplatelet therapy. Percutaneous transluminal angioplasty without a stent is indicated for patients with FMD and recent onset hypertension, resistant hypertension, or (rarely) ischemic nephropathy. Patients with renal artery aneurysms should be treated with a covered stent or surgical repair.

REFERENCES

Olin JW, Sealove BA. Diagnosis, management and future developments of fibromuscular dysplasia. J Vasc Surg. 2011; 53(3):826–36.

Slovut DP, Olin JW. Fibromuscular dysplasia. NEJM. 2004; 350(18):1862–71.

Mousa AY. Campbell JE, Stone PA, Broce M, Bates MC, AbuRahma AF. Short- and long- term outcomes of percutaneous transluminal angioplasty/stenting of renal fibromuscular dysplasia over a ten-year period. J Vasc Surg. 2012; 55(2):421–7.

Atlas of Clinical Vascular Medicine, First Edition. Edited by Jessica Mintz, Bruce L. Mintz, and Michael R. Jaff.

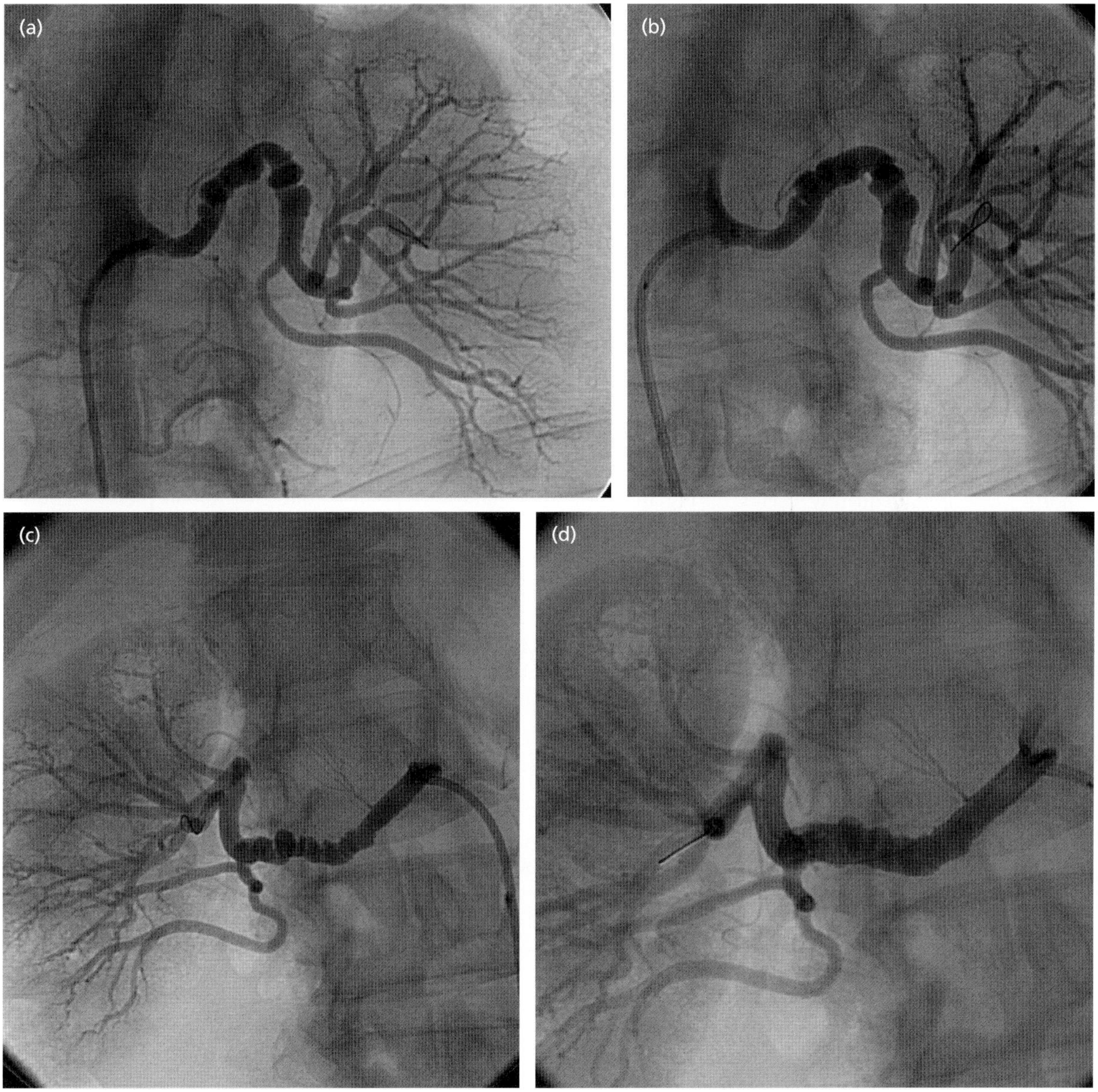

Renal artery fibromuscular dysplasia (FMD). (a) Angiography of the left renal artery demonstrating the "string-of-beads pattern" of medial-type FMD. (b) Angiography of the left renal artery following percutaneous transluminal renal angioplasty (PTA). (c) Angiography of the right renal artery demonstrating the medial fibroplasia variant of FMD. (d) Angiography of the right renal artery following PTA. The patient underwent PTA of the right and left renal arteries.

Aortic Dissection

Paul B. Haser

Robert Wood Johnson Medical School, University of Medicine & Dentistry of New Jersey, New Brunswick, NJ, USA

BRIEF CASE PRESENTATION

A 57-year-old male with hypertension, hyperlipidemia and depression presents to the emergency room with severe chest pain, thought initially to represent a myocardial infarction.

CHARACTERISTICS

- Aortic dissection frequently occurs during the fifth and sixth decade of life, usually associated with hypertension
- Chest pain is the most common presenting symptom, but the initial diagnosis can be quite difficult. Symptoms may occur anywhere based on the segment of the aortic or branch artery involvement
- Dynamic CT angiogram and/or transesophageal echocardiogram provide the most rapid means of diagnosis

TREATMENT

- Proximal dissections, (DeBakey types I & II or Stanford type A), require emergent surgical repair. Distal dissection beyond the subclavian artery (DeBakey type III or Stanford type B) require close blood pressure control and selective intervention for branch vessel obstruction, refractory pain related to the dissection, or aneurysmal degeneration of the aorta.

REFERENCES

DeBakey ME, McCollum CH, Crawford ES, et al. Dissection and dissecting aneurysms of the aorta: twenty-year follow-up of five hundred twenty-seven patients treated surgically. Surgery. 1982; 92:1118–34.

Easo J, Weigang E, Holzl PP, et al. Influence of operative strategy for the aortic arch in DeBakey type I aortic dissection: Analysis of the German Registry for Acute Aortic Dissection Type A. J Thorac Cardiovasc Surg. 2012; 144(3):617–23.

Wittels K. Aortic emergencies. Emerg Med Clin North Am. 2011; 29:789–800.

Atlas of Clinical Vascular Medicine, First Edition. Edited by Jessica Mintz, Bruce L. Mintz, and Michael R. Jaff.

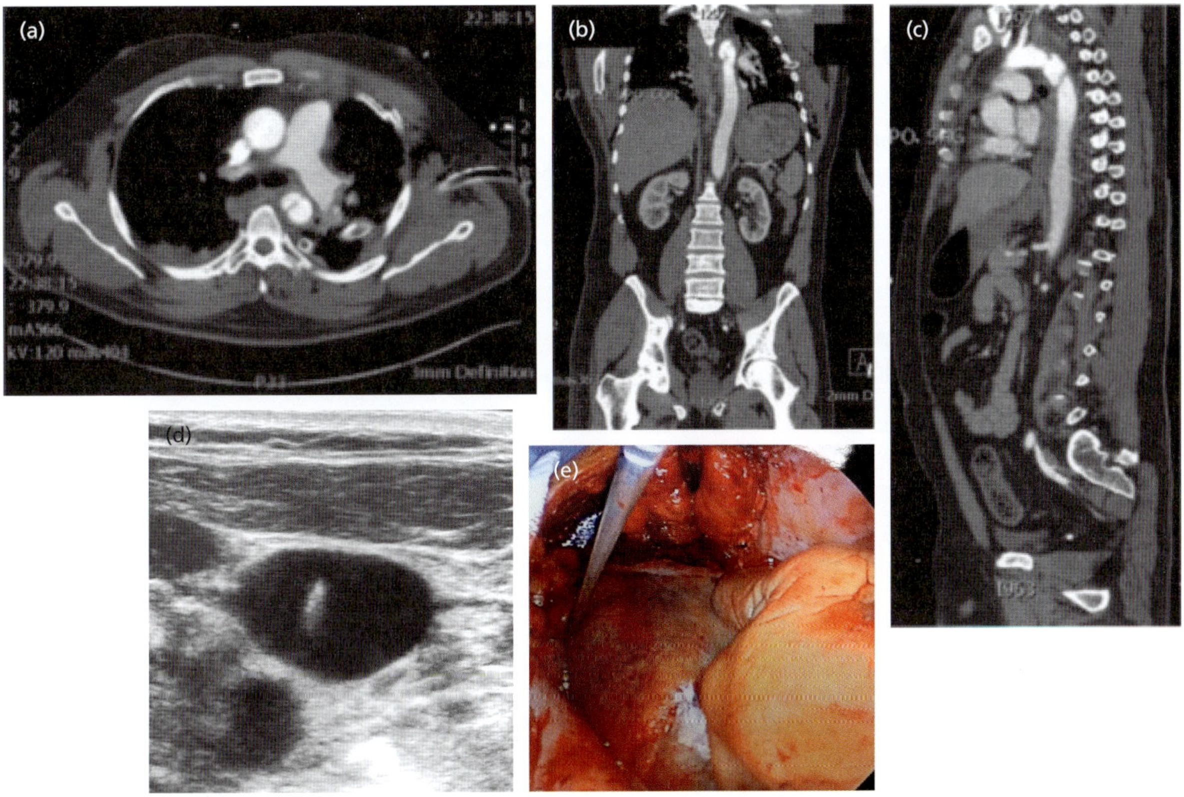

Aortic dissection. (a–c) Computed tomographic angiography of an aortic dissection. (d) Transthoracic echocardiogram demonstrating an aortic dissection flap. (e) The forceps are touching the hematoma of the dissection seen in the descending aorta as visualised via open thorocotomy.

Carotid Artery Stenosis

Paul B. Haser

Robert Wood Johnson Medical School, University of Medicine & Dentistry of New Jersey, New Brunswick, NJ, USA

BRIEF CASE PRESENTATION

A 54-year-old male presents with three episodes of right eye amaurosis fugax and one 5-minute period of left hand sensory and motor loss.

CHARACTERISTICS

- Carotid artery stenosis (CAS) is a narrowing of the carotid artery most commonly due to atherosclerosis
- Carotid stenosis at the bulb and bifurcation accounts for ~10–15% of transient ischemic attacks, (e.g., amaurosis fugax) and permanent neurologic events (e.g., stroke or loss of function of the ophthalmic nerve)
- Diagnosis is made by physical exam noting for carotid bruits, and non-invasive imaging including carotid duplex ultrasonography, computerized tomographic angiography and magnetic resonance angiography

TREATMENT

Medical management including antihypertensive therapy, antiplatelet agents, and statin therapy; cessation of all tobacco use should be offered to all patients with CAS. Treatment with carotid endarterectomy and carotid artery stenting are both options.

REFERENCES

Brott TG, Hobson RW, Howard G, et al. Stenting versus endarterectomy for treatment of carotid-artery stenosis. NEJM. 2010; 363:11–23.

Brott TG, Haperin JL, Abbara S, et al. 2011 ASA/ACCF/AHA/AANN/AANS/ACR/ASNRCNS/SAIP/SCAI/SIR/SVM.SVS Guideline on the MAnagement of Patients with Extracranial Carotid and Vertebral Artery Disase:Executive Summary. JACC. 2011; 57(8):1002–44.

Hobson RW, 2nd, Goldstein JE, Jamil Z, et al. Carotid restenosis: operative and endovascular management. J Vasc Surg. 1999; 29:228–35; discussion 35–8.

Ricotta JJ, Aburahma A, Ascher E, Eskandari M, Faries P, Lal BK. Updated Society for Vascular Surgery guidelines for management of extracranial carotid disease: executive summary. J Vasc Surg. 2011; 54:832–6.

Atlas of Clinical Vascular Medicine, First Edition. Edited by Jessica Mintz, Bruce L. Mintz, and Michael R. Jaff.

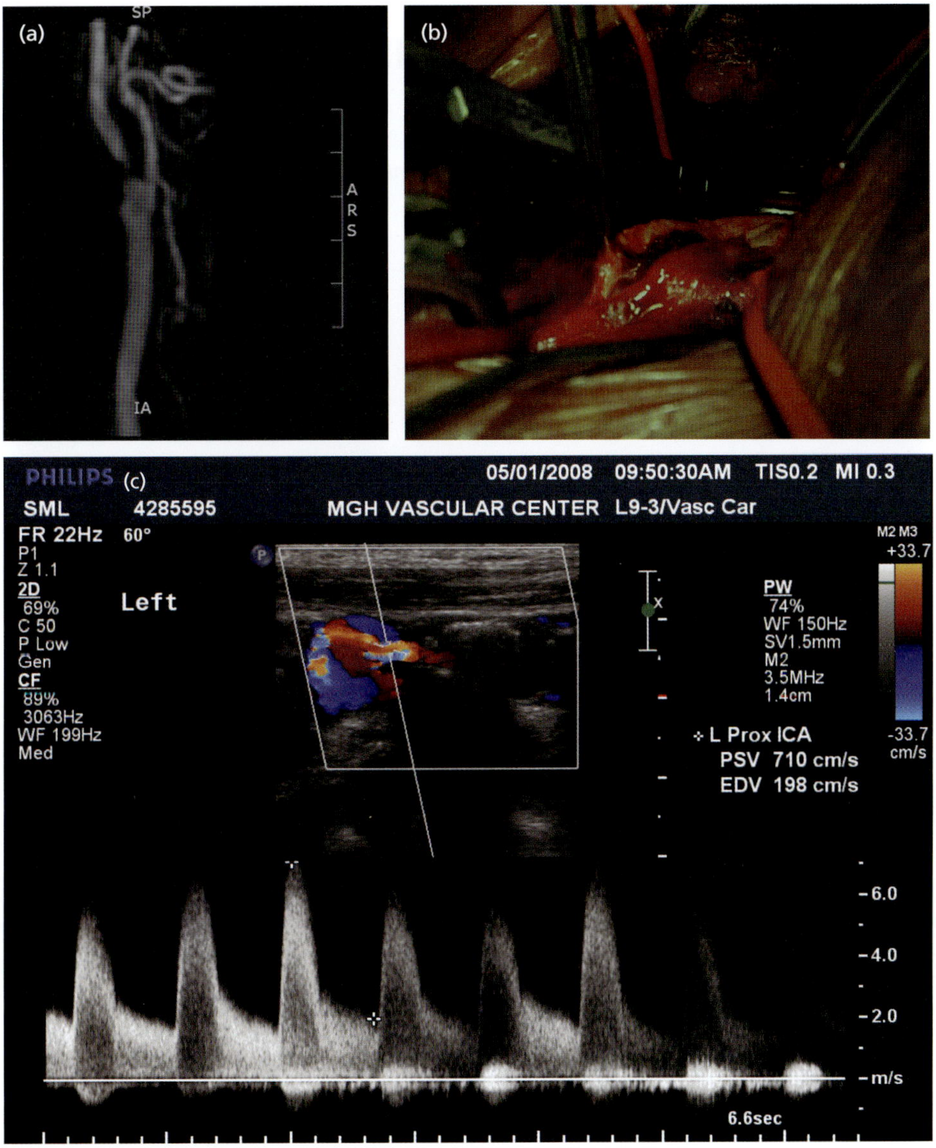

Atherosclerotic carotid artery stenosis. (a) Magnetic resonance angiography demonstrating significant extracranial right internal carotid artery stenosis. (b) Intraoperative image of carotid artery plaque. (c) Duplex ultrasonography demonstrating severe internal carotid artery stenosis.

Carotid Fibromuscular Dysplasia

Heather L. Gornik and Sandra Yesenko

Cleveland Clinic Heart and Vascular Institute and Cleveland Clinic Lerner College of Medicine, Case Western Reserve University, Cleveland, OH, USA

> **BRIEF CASE PRESENTATION**
>
> A 43-year-old woman presents with a longstanding history of migraine headaches, cervical bruits are discovered on physical examination. She has never had a stroke or transient ischemic attack.

CHARACTERISTICS

- Carotid fibromuscular dysplasia (FMD) can present as an asymptomatic cervical bruit or with symptoms such as pulsatile tinnitus or dizziness
- Many carotid FMD patients suffer from headache. Patients with carotid FMD may present with carotid (or vertebral) artery dissection, transient ischemic attack, or stroke

TREATMENT

Antiplatelet therapy is generally recommended for carotid FMD patients.

Evaluation for evidence of FMD in other vascular territories (e.g., vertebral, intracranial, renal) with a targeted vascular history, physical examination, and appropriate imaging, is undertaken. Asymptomatic brain aneurysm has been associated with carotid FMD, and thus a screening imaging study of the intracranial arteries is generally recommended, even for the asymptomatic patient, with carotid FMD.

REFERENCES

Kim Y, Schulman S. Cervical artery dissection: Pathology, epidemiology and management. Thromb Res. 2009; 123:810–21.

Olin JW, Sealove BA. Diagnosis, management, and future developments of fibromuscular dysplasia. J Vasc Surg. 2011; 53: 826.

Olin JW. Recognizing and managing fibromuscular dysplasia. Cleve Clin J Med. 2007;273.

Atlas of Clinical Vascular Medicine, First Edition. Edited by Jessica Mintz, Bruce L. Mintz, and Michael R. Jaff.

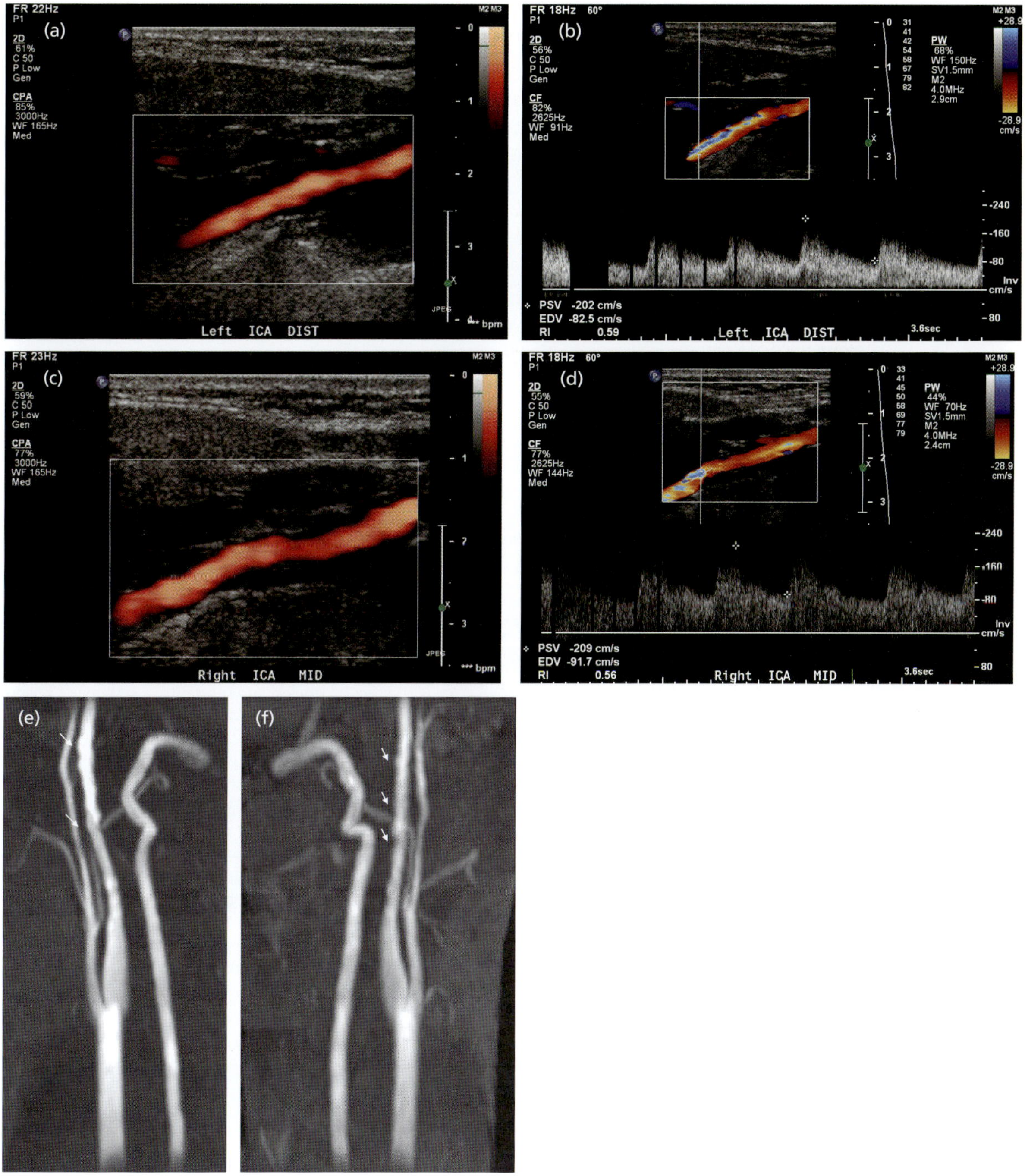

Carotid artery fibromuscular dysplasia. (a–d) Duplex ultrasonography and reconstructed magnetic resonance angiogram images (e,f) of a patient with known multifocal, medial fibroplasia type fibromuscular dysplasia. Images demonstrate turbulent flow and velocity shifts in both the distal internal carotid arteries with a "beaded appearance" of the vessels bilaterally seen on color power angiography. (a) Left internal carotid artery duplex ultrasonography with color power angiography. (b) Spectral Doppler of the distal left internal carotid artery with elevated velocity and turbulent flow. (c) Right internal carotid artery color power angiography. (d) Spectral Doppler of the distal right internal carotid artery with elevated velocity and turbulent flow. (e,f) Magnetic resonance angiography with beading in bilateral internal carotid arteries (arrows).

Subclavian Artery Stenosis

J. Michael Bacharach
Vascular Medicine and Peripheral Vascular Intervention, North Central Heart Institute, Sioux Falls, SD, USA

BRIEF CASE PRESENTATION

A 67-year-old right-handed male presents with right arm discomfort associated with activity. His symptoms have progressively worsened and now limit his activities.

CHARACTERISTICS

- Subclavian artery stenosis occurs in the proximal subclavian artery. Stenosis involving the left subclavian artery is three times more common than the right subclavian artery.
- Subclavian steal syndrome represents a proximal subclavian stenosis or occlusion. The vertebral artery provides collateral flow into the distal subclavian artery by reversing its' normal flow to the brain. This retrograde flow from the vertebral artery at the expense of the vertebral circulation is the result of the subclavian stenosis or occlusion
- Signs and symptoms include those of posterior circulation ischemia (diplopia, dysarthria, dysequilibrium, "drop attacks"), significant discrepancy in upper extremity blood pressure readings and arm claudication
- The most common cause of subclavian artery stenosis is atherosclerosis and thus, the prevalence is much higher in patients with a history of tobacco use, hypertension, diabetes mellitus and dyslipidemia

TREATMENT

In symptomatic patients only, revascularization with endovascular strategies including percutaneous transluminal angioplasty (PTA) and stent placement is indicated.

REFERENCES

Clark CE, Taylor RS, Shore AC, Ukoumunne OC, Campbell JL. Association of a difference in systolic blood pressure between arms with vascular disease and mortality: a systematic review and meta-analysis. Lancet. 2012 Published Online January 30, 2012 DOI:10.1016/S0140-6736(11)61710-8.

DeVries JP, Jager LC, Van den Berg JC, et al. Durability of percutaneous transluminal angioplasty for obstructive lesions of the proximal subclavian artery: long-term results. J Vasc Surg. 2005; 41(1):19–23.

Gutierrez GR, Mahrer P, Aharonian V, Mansukhani P, Bruss J. Prevalence of subclavian artery stenosis in patients with peripheral vascular disease. Angiology. 2001; 52:189–94.

Van Noord BA, Lin AH, Cavendish JJ. Rates of symptom recurrence after endovascular therapy in subclavian artery stenosis prior to coronary bypass grafting. Vasc Health Risk Manag. 2007; 3(5):759–62.

Atlas of Clinical Vascular Medicine, First Edition. Edited by Jessica Mintz, Bruce L. Mintz, and Michael R. Jaff.

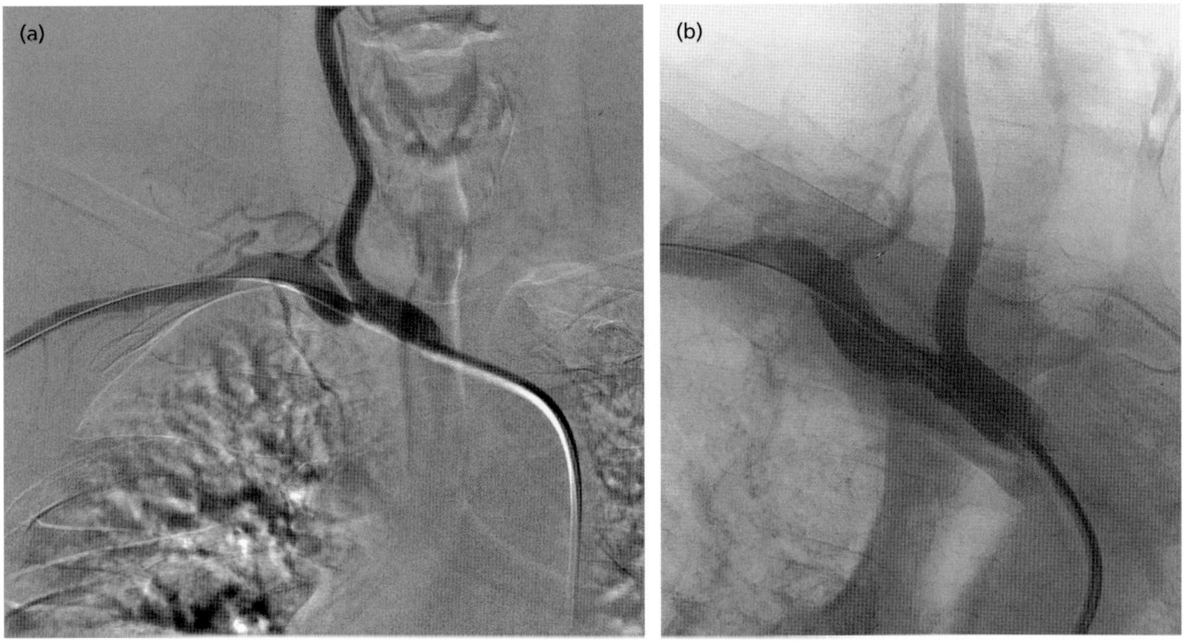

Subclavian artery stenosis. (a) Digital subtraction angiography demonstrating a severe focal stenosis at the origin of the right subclavian artery. (b) Angiography following percutaneous transluminal angioplasty and stenting of the subclavian artery.

Abdominal Aortic Aneurysm

J. Michael Bacharach

Vascular Medicine and Peripheral Vascular Intervention, North Central Heart Institute, Sioux Falls, SD, USA

BRIEF CASE PRESENTATION

A 66-year-old male presented with a 1-week history of progressive back pain. Initial ultrasound evaluation in the emergency department demonstrated a large infrarenal abdominal aortic aneurysm. Subsequent computerized tomography (CT) confirmed a 10 cm aneurysm. No overt extravasation of contrast was seen; however, fluid was present in the cul-de-sac suggesting the possibility of a contained leak. Emergency angiography was performed with subsequent endovascular repair.

CHARACTERISTICS

- The majority of aneurysms of the abdominal aorta are asymptomatic; however, when occuring acutely, they can present with back, lower extremity, or abdominal pain
- Abdominal aneurysms are most common in men over the age of 60 with comorbidities including hypertension, hyperlipidemia, history of tobacco abuse, and obesity
- Men over the age of 65 who have ever smoked should have a screening duplex ultrasound examination for abdominal aortic aneurysms
- Leaking but contained aneurysms may be initially stable but represent potential for rupture and require emergent repair
- Endograft repair has emerged as a viable and advantageous method of repair in these patients

TREATMENT

For asymptomatic abdominal aortic aneurysms less than 5.5 cm in diameter, conservative management is the standard of care. This includes tobacco cessation, blood pressure control, and aneurysm surveillance. When the aneurysm is larger than 5.5 cm, repair should be considered. Open and endovascular repair are options for intervention. Long-term mortality is similar in both groups. The endovascular approach has lower perioperative mortality however results in an increased rate of re-interventions.

REFERENCES

Starnes BW, Quiroga E, Hutter C, et al. Management of ruptured abdominal aortic aneurysm in the endovascular era. J Vasc Surg. 2010; 51:9–18.

Moore R, Nutley M, Cina CS, et al. Improved survival after introduction of an emergency endovascular therapy protocol for ruptured abdominal aortic aneurysms. J Vasc Surg. 2007; 45: 443–9.

Mehta M, Taggert J, Darling RC, et al. Establishing a protocol for endovascular treatment of ruptured abdominal aortic aneurysms: outcomes of a prospective analysis. J Vasc Surg. 2006; 44:1–8.

Atlas of Clinical Vascular Medicine, First Edition. Edited by Jessica Mintz, Bruce L. Mintz, and Michael R. Jaff.

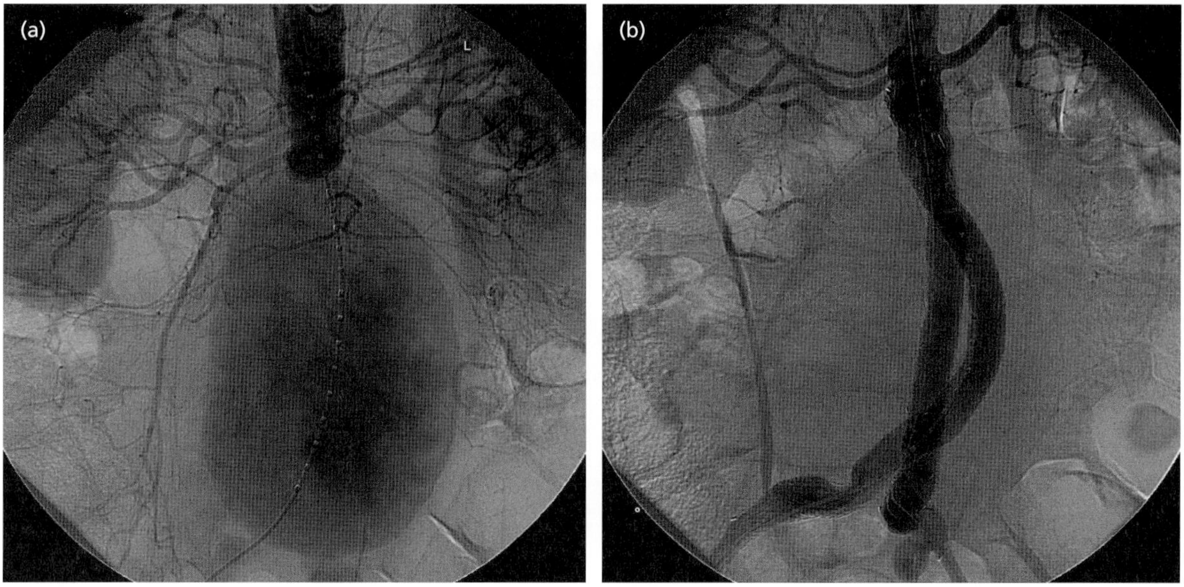

Abdominal aortic aneurysm. (a) Angiography demonstrating a large contained rupture of the abdominal aorta. The patient was treated with emergency endovascular repair with an Excluder endograft (W.L. Gore, Flagstaff, Arizona). (b) Angiography following endovascular repair of the abdominal aorta.

Popliteal Artery Aneurysm

Jason Q. Alexander and Timothy Sullivan
Minneapolis Heart Institute, Minneapolis, MN, USA

> **BRIEF CASE PRESENTATION**
> A 73-year-old male with a history of abdominal aortic aneurysm repair presents with increasing pain with walking and pulsatile mass behind the left knee.

CHARACTERISTICS

- Popliteal artery aneurysm is a dilatation of the popliteal artery at least 1.5 times the normal caliber
- Untreated, popliteal artery aneurysms will increase in size (aneurysmal degeneration), potentially compressing surrounding tissues (popliteal vein and nerve)
- As aneurysmal degeneration occurs, normal laminar flow of blood in the vessel is lost. Loss of normal flow leads to thrombus formation within the aneurysm sac. As thrombus collects, flow may dislodge emboli into the distal leg leading to limb threatening ischemia. Furthermore, the aneurysm itself may occlude leading to loss of limb
- Popliteal aneurysms are usually associated with atherosclerotic stigmata (hypertension, hyperlipidemia, diabetes mellitus, tobacco exposure) but may also be secondary to trauma
- Popliteal artery aneurysms are associated with aneurysm development in other anatomic distributions (most commonly contralateral popliteal artery and infrarenal abdominal aorta)

TREATMENT

- Both open surgical and endovascular approaches should be considered for popliteal artery aneurysm repair
- Open surgical approach includes arterial bypass around the aneurysm with ligation of the aneurysm above and below its origin to prevent both continued growth and dislodging of thrombus into the distal limb
- Endovascular treatment requires the placement of a covered stent across the area of aneurysmal degeneration, effectively excluding the aneurysm from normal arterial flow

REFERENCES

DeWeese JA, Shortell C, Green R, Ouriel K. Operative repair of popliteal aneurysm: Twenty-five years experience. In: Yao JS, Pearce WH (eds) Long-Term Results in Vascular Surgery. Norwalk, CN: Appleton & Lange, 1993, pp. 287–93.

Jung E, Jim J, Rubin BG, Sanchez LA, Choi ET, Sicard GA, Geraghty PJ. Long-term outcome of endovascular popliteal artery aneurysm repair. Ann Vasc Surg. 2010; 24(7):871–5.

Lovegrove RE, Javid M, Magee TR, Galland RB. Endovascular and open approaches to non-thrombosed popliteal aneurysm repair: a meta-analysis. Eur J Vasc Endovasc Surg. 2008; 36: 96–100.

Atlas of Clinical Vascular Medicine, First Edition. Edited by Jessica Mintz, Bruce L. Mintz, and Michael R. Jaff.

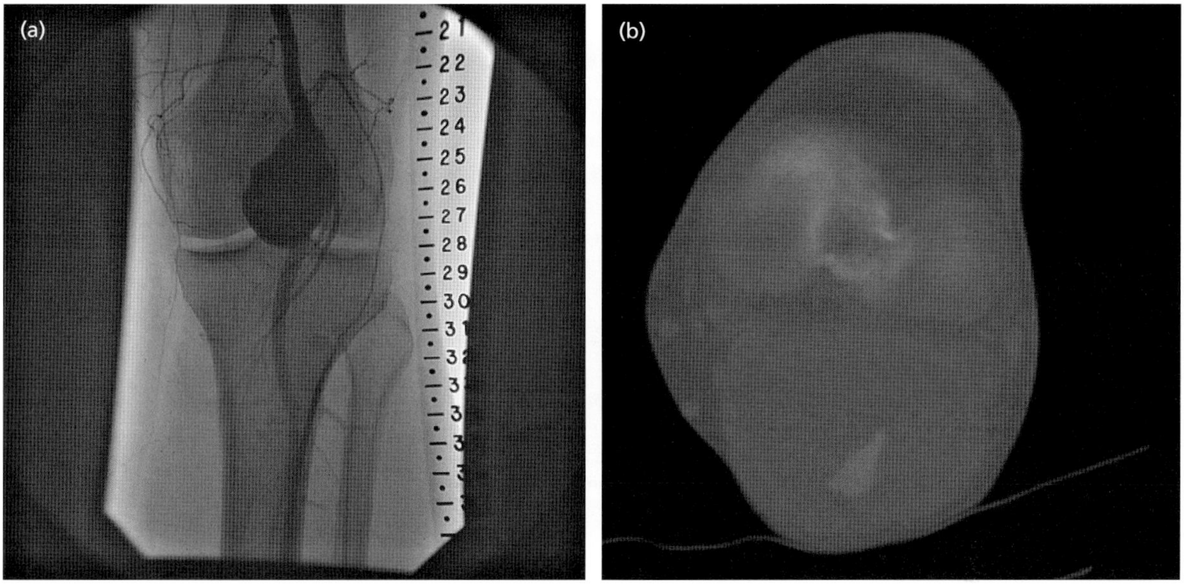

Popliteal artery aneurysm. (a) Angiography demonstrating a popliteal artery aneurysm. (b) Magnetic resonance demonstrating a popliteal artery aneurysm.

Post-Traumatic Compartment Syndrome

Ari J. Mintz[1] and Bruce L. Mintz[2]

[1]Lahey Clinic Medical Center, Burlington, MA, USA
[2]Morristown Medical Center, Morristown, NJ, USA

BRIEF CASE PRESENTATION

This 51-year-old male with a history of seizure disorder presents after being found post-ictal, clinching his laptop over his forearm. The heat of his laptop had caused partial thickness burns. He was brought to the hospital where he was observed with elevation and Silvadene. Over the next 24 hours his arm became progressively swollen, tender, and cool to the touch with loss of both radial and ulnar pulses.

BRIEF DESCRIPTION

- Compartment syndrome is a potentially limb-threatening condition associated with increased pressure in a confined space of a muscular and fascial compartment. The ensuing mechanical impingement leads to muscle and nerve injury as well as occlusion of blood flow
- Most commonly seen subsequent to traumatic injury such as motor vehicle accidents. Alcoholics, patients with seizure disorders, and individuals who do not respond to the early warning signs of the condition such as long distance runners are also at risk
- Compartment syndrome occurs in areas of the body where anatomically confined spaces restrict the expansion of the inflammatory process. The forearm and lower leg are where compartment syndrome is usually found, however this condition has been described in the hand, foot and in restricted areas of the abdomen
- Refractory pain resistant to analgesia, lack of improvement with raising the affected area, pallor upon elevation, rubor upon dependency, loss of thermoregulation, progressive weakness, neurosensory changes and ultimately analgesia are hallmarks of compartment syndrome
- Objective diagnosis is required and involves the insertion of a needle attached to a manometric device

TREATMENT

Surgical intervention with generous open incisions to relieve the pressure is the standard of care. The wounds are often left open for several days and closed when the edema diminishes. Skin grafting is commonly used and recovery is dependent upon attention to detail and prompt and aggressive decompression with fasciotomy. Permanent nerve injury and loss of muscle function can result if the diagnosis is delayed greater than 12–24 hours after the start of the insult.

REFERENCES

Geiderman JM, Katz D. General principles of orthopedic injuries. In: Marx J, (ed.) Rosen's Emergency Medicine: Concepts and Clinical Practice, 7th ed. Philadelphia, PA: Mosby Elsevier, 2009, Chap 46.

Hobson RW, Wright JG, Fox D, Kerr JC. Heparinization reduces endothelial permeability and hydrogen ion accumulation in canine skeletal muscle ischemia-reperfusion model. J Vasc Surg. 1988; 7:585–91.

Von Volkmann R. Veilletzungen und Krankenheiten der Berwegungsorgane. In: von Pithe F, Billroth T (eds) Handbuch der Allgemeinen und Speziellen Chirurgs. Stuttgart: Verlag von Ferdinand Enke Zweiter Band, Zweiter Abteilung, Abschmitt V, Ersted haft, 1882, pp. 234–920.

Atlas of Clinical Vascular Medicine, First Edition. Edited by Jessica Mintz, Bruce L. Mintz, and Michael R. Jaff.

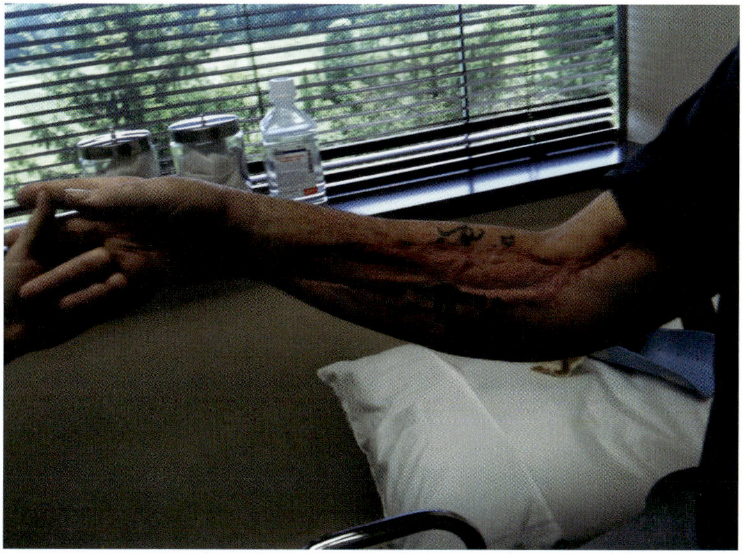

Compartment syndrome. Right upper extremity forearm status post fasciotomy from a compartment syndrome.

Livedoid Vasculopathy

Steven M. Dean
Department of Cardiovascular Medicine, Wexner Medical Center at the Ohio State University, Columbus, OH, USA

BRIEF CASE PRESENTATION

A 38-year-old woman presents with a 2-year history of relapsing, painful, bilateral lower extremity purplish discoloration and ulcerations.

CHARACTERISTICS

- Livedoid vasculopathy initially begins with clusters of painful purpuric macules or papules along the malleoli, dorsal feet, and distal calves in the setting of concurrent livedo reticularis
- With time, punched out ulcerations often evolve that ultimately heal with characteristic stellate porcelain-white scars, also known as "atrophie blanche"
- Occurs more frequently in middle-aged women and involves the lower extremities in approximately 80% of cases
- Thrombophilia appears to play a role in the pathogenesis
- Biopsy illustrates abundant luminal and vessel wall fibrin deposition consistent with a thrombotic occlusive vasculopathy (*not* a vasculitis)

TREATMENT

A variety of medications are sometimes successful and include pentoxifylline, antiplatelet agents (e.g. aspirin, dipyridamole), anticoagulants (e.g., subcutaneous unfractioned or low molecular weight heparin injections, warfarin), fibrinolytic agents (e.g., danazol, tissue plasminogen activator), anti-inflammatories (e.g., prednisone, intravenous immunoglobulin), and vasodilators (e.g., nifedipine, nicotinic acid). Supportive care including avoidance of trauma, control of edema, and antibiotic therapy are sometimes useful. Hyperbaric oxygen has also been used with some success.

REFERENCES

Criado PR, Rivitti EA, Sotto MN, de Carvalho JF. Livedoid vasculopathy as a coagulation disorder. Autoimmun Rev. 2011; 10(6):353–60.

Di Giacomo TB, Hussein TP, Souza DG, Criado PR. Frequency of thrombophilia determinant factors in patients with livedoid vasculopathy and treatment with anticoagulant drugs-a prospective study. J Eur Acad Dermatol Venereol. 2010; 24(11): 1340–6.

Hairston BR, Davis MD, Pittelkow MR, Ahmed I. Livedoid vasculopathy: further evidence for procoagulant pathogenesis. Arch Dermatol. 2006; 145(11):1413–8.

Atlas of Clinical Vascular Medicine, First Edition. Edited by Jessica Mintz, Bruce L. Mintz, and Michael R. Jaff.

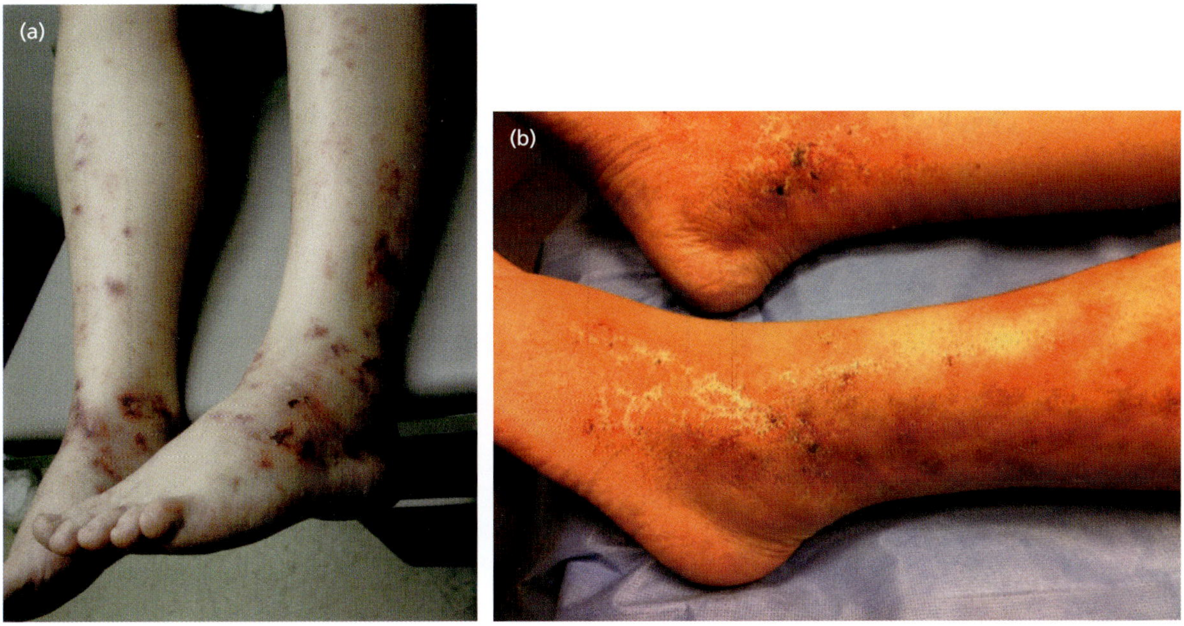

Livedoid vasculopathy. (a) Bilateral lower extremity purpuric macules and ulceration characteristic of livedoid vasculopathy. (b) Atrophie blanche demonstrated by white stellate scars surrounding the medial and lateral malleoli of bilateral lower extremities.

Digital Ischemia in Systemic Sclerosis

Peter C. Grayson[1] and Peter A. Merkel[2]
[1]Boston University Vasculitis Center, Boston, MA, USA
[2]Division of Rheumatology, University of Pennsylvania, Philadelphia, PA, USA

BRIEF CASE PRESENTATION

A 45-year-old woman with systemic sclerosis (scleroderma) and a longstanding history of Raynaud's phenomenon presents for evaluation of left foot pain. On physical examination, there is fixed cyanosis involving the toes of the left foot with gangrenous changes. Examination of the hands reveals several, digital pitting scars involving the fingertips.

CHARACTERISTICS

- Scleroderma is a widespread connective tissue disorder characterized by fibrosis involving the skin, blood vessels, joints and internal organs
- Reversible vasospasm (Raynaud's phenomenon) and progressive fixed structural microvascular changes are both associated with scleroderma. These processes can combine to result in ischemia and infarction, manifested as cyanosis, digital ulcers, and gangrene
- It is important to differentiate digital ulcers and subsequent pitting scars located on the pulp of the fingers secondary to ischemia from skin abrasions found along the lateral aspects of the fingers or over the bony prominences secondary to direct trauma and fragile skin

TREATMENT

It is important to protect the hands and feet from excessive cold exposure as well as use of skin moisturizers and avoidance of trauma to the digits. Pharmacologic therapies for ischemic digital ulcers include calcium channel blockers and phosphodiesterase type 5 inhibitors (i.e., sildenafil). In cases of severe digital ischemia (i.e. fixed cyanosis or gangrenous changes), therapy with endothelin receptor antagonists (e.g., bosentan) or prostacyclin analogs (e.g., epoprostenol) should be considered.

REFERENCES

Amanzi L, Braschi F, Fiori G, et al. Digital ulcers in scleroderma: staging, characteristics and sub-setting through observation of 1614 digital lesions. Rheumatology (Oxford). 2010; 49(7): 1374–82.

Herrick A. Diagnosis and management of scleroderma peripheral vascular disease. Rheum Dis Clin North Am. 2008; 34(1):89–114; vii.

Matucci-Cerinic M, Seibold JR. Digital ulcers and outcomes assessment in scleroderma. Rheumatology (Oxford). 2008; 47 (Suppl 5):v46–7.

Atlas of Clinical Vascular Medicine, First Edition. Edited by Jessica Mintz, Bruce L. Mintz, and Michael R. Jaff.

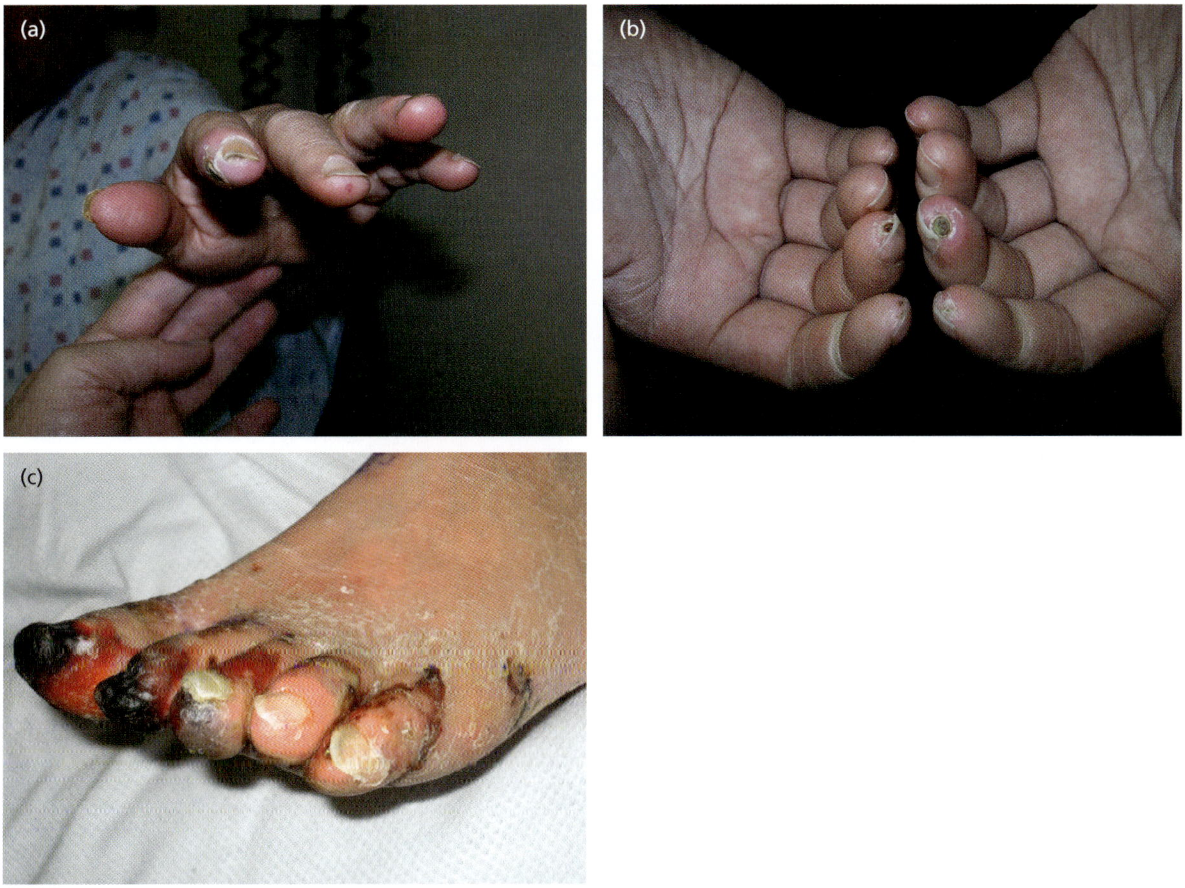

Systemic sclerosis. (a) Digital ulceration with telangiectasias at the tip of the 2nd digit. (b) Bilateral pitting scars at the tip of the 3rd and 4th digits demonstrating healed remnants of ischemia secondary to systemic sclerosis. (c) Gangrenous ulceration of the 1st through 5th digits of the left foot secondary to systemic sclerosis.

Digital Arterial Thrombosis Secondary to Polycythemia Vera

Jessica Mintz, John R. Bartholomew and Heather L. Gornik

Cleveland Clinic Heart and Vascular Institute and Cleveland Clinic Lerner College of Medicine, Case Western Reserve University, Cleveland, OH, USA

BRIEF CASE PRESENTATION

Patient presents with purple, painful second digit in the setting of polycythemia vera.

CHARACTERISTICS

- Polycythemia vera is a myeloproliferative disorder characterized by overproduction of red blood cells, high hematocrit, and a resulting increase in blood viscosity
- Associated with hypercoagulability, which may lead to both venous and arterial thrombosis
- Small vessel thrombosis can lead to digital ischemia, as occurred in this patient
- Polycythemia vera, along with other myeloproliferative disorders is associated with mutations of the Janus kinase 2 (JAK2) gene

TREATMENT

Phlebotomy is the first line treatment for polycythemia vera along with antiplatelet therapy, however myelosuppressive agents such as hydroxyurea may also be used. Use of JAK inhibitors are under investigation. Patients with arterial and/or venous thrombosis related to polycythemia vera are generally treated with systemic anticoagulation.

REFERENCES

De Stefano V, Za T, Rossi E et al. Recurrent thrombosis in patients with polycythemia vera and essential thrombocythemia: incidence, risk factors and effects of treatments. Haematologica. 2008; 93(3):372–80.

Kralovics R, Passamonti F, Buser AS, et al. A gain-of-function mutation of JAK2 in myeloproliferative disorders. NEJM. 2005; 28:1779–90.

Osler W. Chronic cyanosis, with polycythemia and enlarged spleen: a new clinical entity. Am J Med Sci. 1903; 126: 176–201.

Atlas of Clinical Vascular Medicine, First Edition. Edited by Jessica Mintz, Bruce L. Mintz, and Michael R. Jaff.

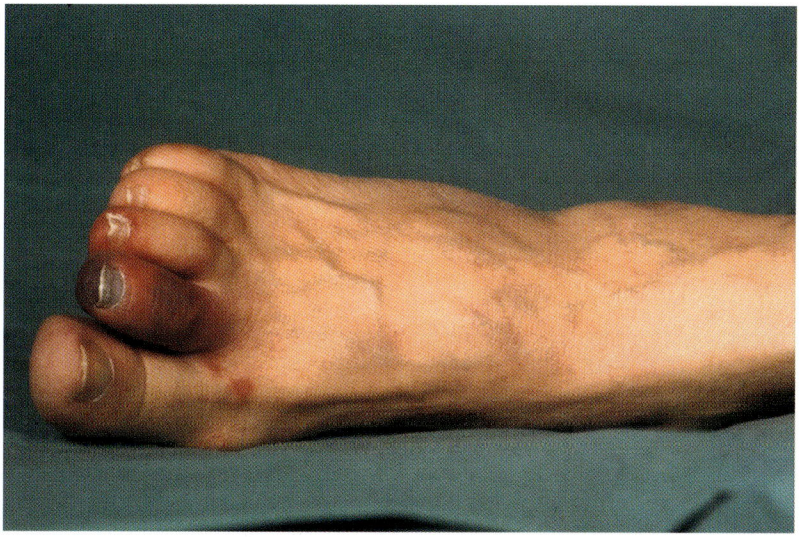

Polycythemia vera. Second toe discoloration from arterial occlusion secondary to polycythemia vera. Image courtesy of the Cleveland Clinic Foundation Department of Vascular Medicine archives.

Raynaud's Phenomenon

Peter C. Grayson[1] and Peter A. Merkel[2]

[1]Boston University Vasculitis Center, Boston, MA, USA
[2]Division of Rheumatology, University of Pennsylvania, PA, USA

BRIEF CASE PRESENTATION

A 50-year-old woman presents for evaluation of painful fingers for the last year. She describes color change in the fingers with cold exposure. She also notes thickening of the skin, heartburn, arthralgias, and dysphagia with solid foods. Physical examination is notable for sclerodactyly with mild contractures of the fingers. The fingers blanch white with cold exposure. Nailfold capillaroscopy reveals dilated capillary loops and capillary drop-out. She is diagnosed with systemic sclerosis (scleroderma) with secondary Raynaud's phenomenon.

CHARACTERISTICS

- Raynaud's phenomenon is an exaggerated vasospastic response to cold stimuli or stress
- Vasoconstriction of the digital arteries and cutaneous arterioles leads to a characteristic triphasic color change in the fingers – white/purple (vasoconstriction), and red (vasodilatation upon rewarming)
- Primary Raynaud's is not associated with underlying disease and tends to occur in women between 15 and 30 years of age. Secondary Raynaud's has been associated with certain connective tissue diseases (e.g., scleroderma, systemic lupus erythematosus, idiopathic inflammatory myopathies), medications (e.g., sympathomimetics), and hematologic diseases (e.g., cold agglutinin disease, paraproteinemia)
- In association with Raynaud's phenomenon, dilated nailfold capillary loops or capillary drop-out seen on capillaroscopy and a positive antinuclear antibody (ANA) increases the likelihood of an underlying connective tissue disease

TREATMENT

Lifestyle modification, including avoidance of sudden cold exposure, keeping body core and extremities warm, and smoking cessation can reduce frequency and severity of attacks. Pharmacologic therapies include calcium channel blockers and phospodiesterase type 5 inhibitors (i.e., Sildenafil).

REFERENCES

Block JA, Sequeira W. Raynaud's phenomenon. Lancet. 2001; 357(9273):2042–8.

Ingegnoli F, Boracchi P, Gualtierotti R, et al. Prognostic model based on nailfold capillaroscopy for identifying Raynaud's phenomenon patients at high risk for the development of a scleroderma spectrum disorder: PRINCE (prognostic index for nailfold capillaroscopic examination). Arthritis Rheum. 2008; 58(7):2174–82.

Wigley F. Raynauds phenomenon. NEJM. 2002; 347:1001–8.

Atlas of Clinical Vascular Medicine, First Edition. Edited by Jessica Mintz, Bruce L. Mintz, and Michael R. Jaff.

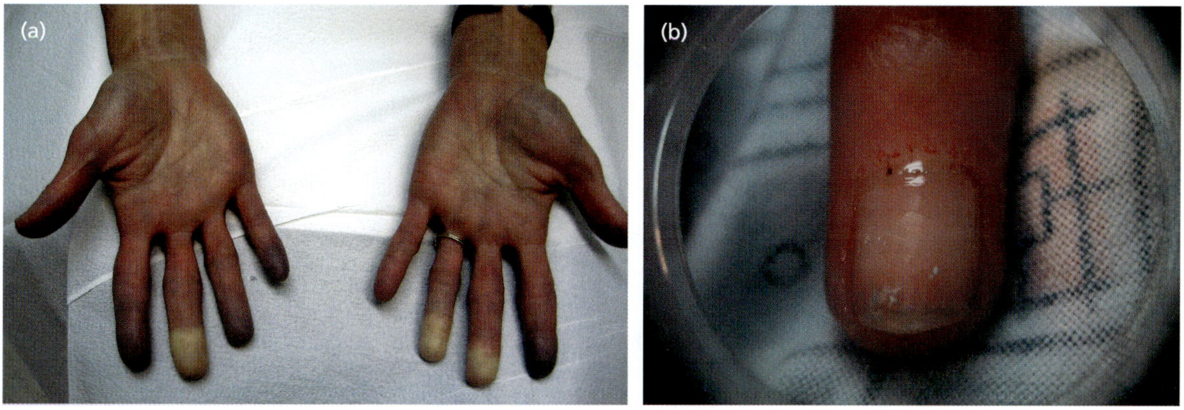

Raynaud's disease. (a) Bilateral digital discoloration with associated blanching seen secondary to cold exposure. Image courtesy of Michael R. Jaff. (b) Nail fold changes of a patient with Raynaud's phenomenon, note the dilated nail fold bed capillaries.

Acrocyanosis

Ido Weinberg and Michael R. Jaff
Massachusetts General Hospital, Boston, MA, USA

> **BRIEF CASE PRESENTATION**
> An 81-year-old woman presents with chronic, symmetric, non-painful, bluish discoloration of her hands and feet. This has been present for "as long as she can remember". It exacerbates upon exposure to cold and when she is excited and improves with warming. There are no ulcerations.

CHARACTERISTICS

- Acrocyanosis is symmetric bluish discoloration of the hands and less commonly the feet and face
- Acrocyanosis can be primary and benign or secondary, which may be asymmetric and can result in tissue loss
- Secondary acrocyanosis has been associated with hypoxemia, systemic infections, connective tissue disease, anorexia nervosa, malignancy, drugs, bipolar disease and heritable conditions such as Ehlers-Danlos syndrome

TREATMENT

Acrocyanosis is usually a benign disorder that does not mandate specific treatment. In these cases, warming the extremities promotes resolution of the discoloration. Management is directed at the underlying condition in cases of secondary acrocyanosis. Pharmacological treatment has been described for primary acrocyanosis with reserpine, minoxidil and bromocriptine but treatment with calcium channel blockers has not been found to be effective.

REFERENCES

Kurklinsky AK, Miller VM, Rooke TW. Acrocyanosis: the flying dutchman. Vasc Med. 2011 16(4): 288–301. Epub 2011 Mar 22.

Morrish DW, Crockford PM. Acrocyanosis treated with bromocriptine. Lancet. 1976; 2(7990):851.

Nousari HC, Kimyai-Asadi A, Anhalt GJ. Chronic idiopathic acrocyanosis. J Am Acad Dermatol. 2001; 45(6 Suppl):S207–8.

Atlas of Clinical Vascular Medicine, First Edition. Edited by Jessica Mintz, Bruce L. Mintz, and Michael R. Jaff.

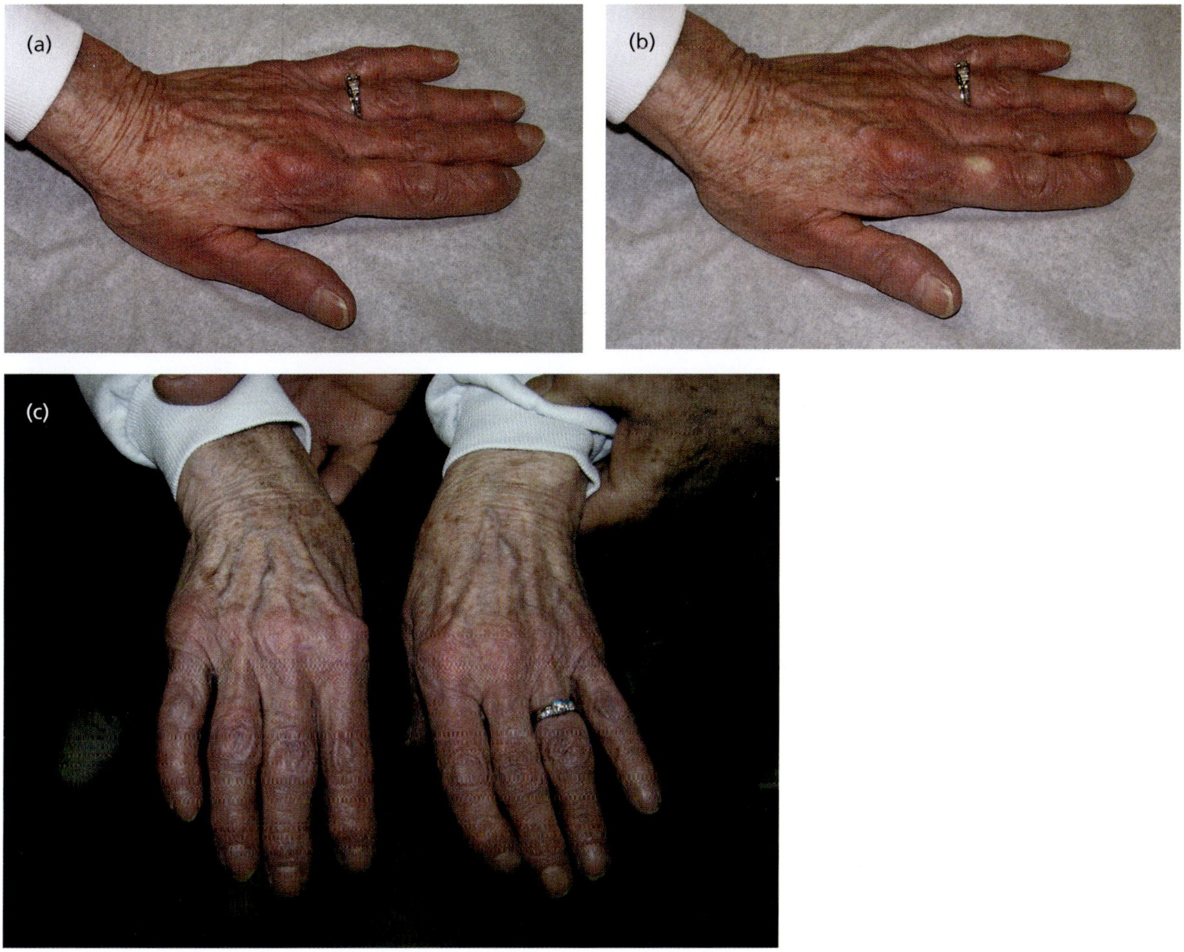

Acrocyanosis. (a) Erythematous distal digits prior to manual pressure applied to the hand. (b) Note blanching of 2nd digit with surrounding erythematous skin after manual pressure is applied. (c) Complete resolution of bilateral hand discoloration with warming.

Livedo Reticularis

Steven M. Dean
Department of Cardiovascular Medicine, Wexner Medical Center at the Ohio State University, Columbus, OH, USA

BRIEF CASE PRESENTATION

A 42-year-old healthy female presents for evaluation of an episodic asymptomatic mottled discoloration of the legs. It appears upon cold exposure and dissipates in warm environments.

CHARACTERISTICS

- Livedo reticularis is an ischemic dermopathy characterized by a violaceous reticular mottling of the skin. The symmetric "fishnet like" discoloration surrounds a pallorous conical core and arises from either deoxygenation or venodilatation within the dermal venous plexus
- Livedo reticularis is most often a primary disorder that affects young to middle aged females [20 to 50 years of age] that are otherwise healthy
- Livedo reticularis is often associated with vasospastic digits and/or acrocyanosis.
- It is imperative that a clinician differentiates the symmetrical, regular cones of benign livedo reticularis from the asymmetric, irregular, and "broken" pattern of the pathological variant (livedo racemosa)

TREATMENT

Other than cold avoidance, medical treatment for primary livedo reticularis is typically unwarranted. In the rare instance that the patient is socially inhibited by the cosmetic appearance of livedo or experiences associated uncomfortable numbness and/or tingling, a trial of vasodilator therapy can be attempted. No evidence-based medicine is available to guide therapy.

REFERENCES

Dean SM. Livedo reticularis and associated disorders. Curr Treat OptCardiovasc Med. 2011; 13(2):179–91.

Herrero C, Guilabert A, Mascaro-Galy JM. Diagnosis and treatment of livedo reticularis on the legs. Actas Dermosifiliogr 2008; 99:598–607.

Kauke T. Images in clinical medicine. Livedo reticularis and cold agglutinins. NEJM. 2007; 18;356(3):284.

Atlas of Clinical Vascular Medicine, First Edition. Edited by Jessica Mintz, Bruce L. Mintz, and Michael R. Jaff.

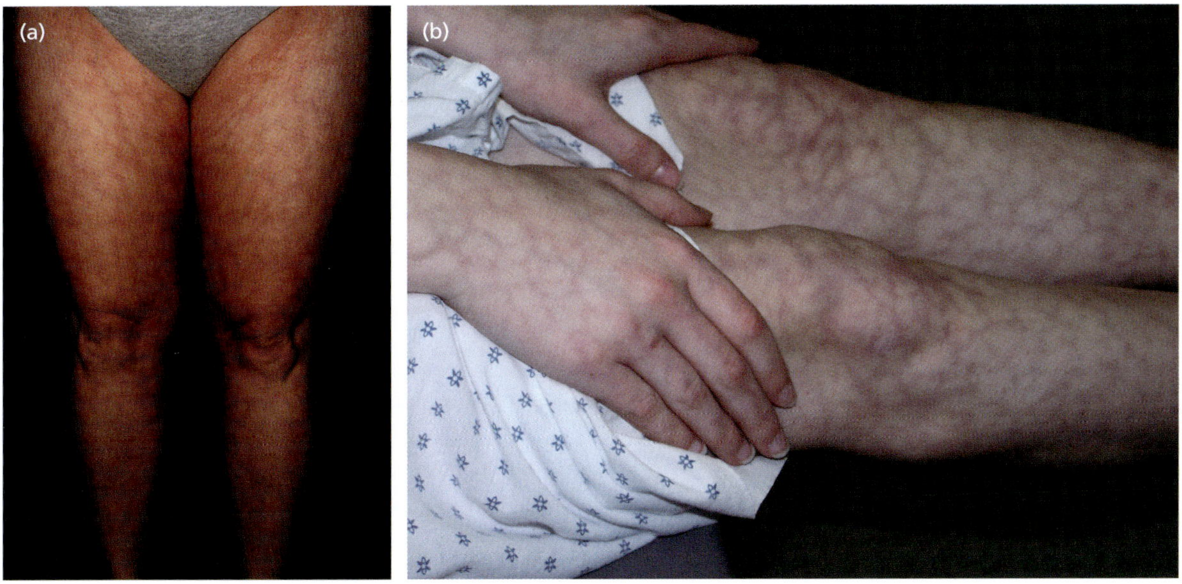

Livedo reticularis. (a) Bilateral lower extremity lace-like pattern of livedo. (b) Upper and lower extremity livedo.

Livedo Racemosa

Steven M. Dean
Department of Cardiovascular Medicine, Wexner Medical Center at the Ohio State University, Columbus, OH, USA

BRIEF CASE PRESENTATION

A 27-year-old male presents with a painful ischemic appearing leg ulceration and cognitive impairment. Examination showed widespread irregular areas of livedo involving the trunk as well as the upper and lower extremities. Serology was remarkable for both lupus anticoagulant and high titer IgG anticardiolipin positivity, and an MRI of his brain documented multiple, small remote cortical strokes. He was subsequently diagnosed with the antiphospholipid antibody syndrome and Sneddon syndrome (livedo racemosa with cerebral vascular accidents).

CHARACTERISTICS

- The ischemic discoloration of livedo reticularis is usually limited to the legs and arms. However, livedo racemosa tends to be more diffuse than livedo reticularis and may involve the trunk, buttocks, hands, feet, and even the face
- Livedo racemosa is described as "broken", irregular erythematous rings of livedo which is distinguished from the "unbroken", uniform discolored rings of livedo reticularis. In contrast to livedo reticularis, livedo racemosa does not resolve with warming or limb elevation
- Whereas livedo reticularis is often benign, livedo racemosa is always pathologic and a marker of an underlying disorder
- When livedo racemosa is identified, antiphospholipid antibodies should be obtained

TREATMENT

There is no specific treatment for livedo racemosa. The underlying pathology for livedo racemosa must be sought. Anticoagulation is recommended for treatment of antiphospholipid antibody syndrome with cortical strokes.

REFERENCES

Gibbs MB, English JC 3rd, Zirwas MJ. Livedo reticularis: an update. J Am Acad Dermatol. 2005; 52:1009–19.

Sneddon IB. Cerebro-vascular lesions and livedo reticularis. Br J Dermatol. 1965; 77:180–5.

Toubi E, Krause I, Fraser A, et al. Livedo reticularis is a marker for predicting multi-system thrombosis in antiphospholipid syndrome. Clin Exp Rheumatol. 2005; 23:499–504.

Atlas of Clinical Vascular Medicine, First Edition. Edited by Jessica Mintz, Bruce L. Mintz, and Michael R. Jaff.

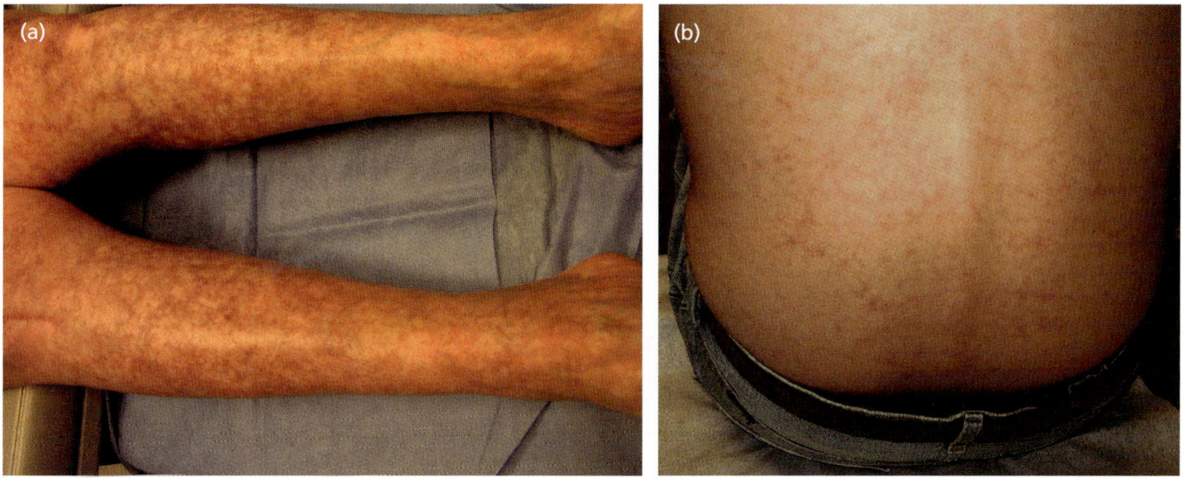

Livedo racemosa. (a) Bilateral lower extremity fishnet discoloration. Note the irregularity and broken pattern of the livedo. (b) Irregular, lace-like pattern over the lower back.

Ergotism

Bruce L. Mintz
Morristown Medical Center, Morristown, NJ, USA

BRIEF CASE PRESENTATION

This 44-year-old white female presents with lower extremity pain with ambulation, a burning sensation in her legs, digital pain in both hands, and abdominal discomfort. On physical exam she had bilateral digital discoloration and lower extremity livedo racemosa. History reveals that after a 25 year history of refractory headaches with multiple treatment failures, the patient's husband (who is a dentist) began injecting her with daily doses of subcutaneous ergotamine.

CHARACTERISTICS

- Ergotism is a rare phenomenon in current medical practice. Newer medications with more benign side effect profiles have made ergots an uncommon therapy choice
- Ergotism may be seen in refractory migraine sufferers if exposed to an ergot derivative and is more common in female patients
- Ergot use produces profound vasospasm in large blood vessels and digits which can cause claudication, and in severe cases, gangrene. Small vessel "pseudo-vasculitis" may affect both the upper or lower extremities and can produce a burning pain which has been referred to as "St. Anthony's fire"
- Classically, symptoms occur over the course of years. However, symtoms may also mainfest after the first dose of the drug
- The symptoms are not limited to peripheral arteries and may affect even the mesenteric vasculature

TREATMENT

The primary treatment consists of ergot cessation and avoidance of secondary trauma. There is no specific antidote for ergot toxicity. Though vasodilators, anticoagulants, dextrans and even sympathetic nerve blockade have been used, no reliable data have demonstrated benefit.

REFERENCES

Hans-Christoph D, Volker L. Medication-overuse headache: a worldwide problem. Lancet Neurology. 2005; 4(1):11–12.

Caporael, L. Ergotism: The satan loosed in Salem. Science. 1976; 192.

Wells KE, Steed DL, Zajko AB, Webster MW. Recognition and treatment of arterial insufficiency from Cafergot. J Vasc Surg. 1986; 4(1):8–15.

Atlas of Clinical Vascular Medicine, First Edition. Edited by Jessica Mintz, Bruce L. Mintz, and Michael R. Jaff.

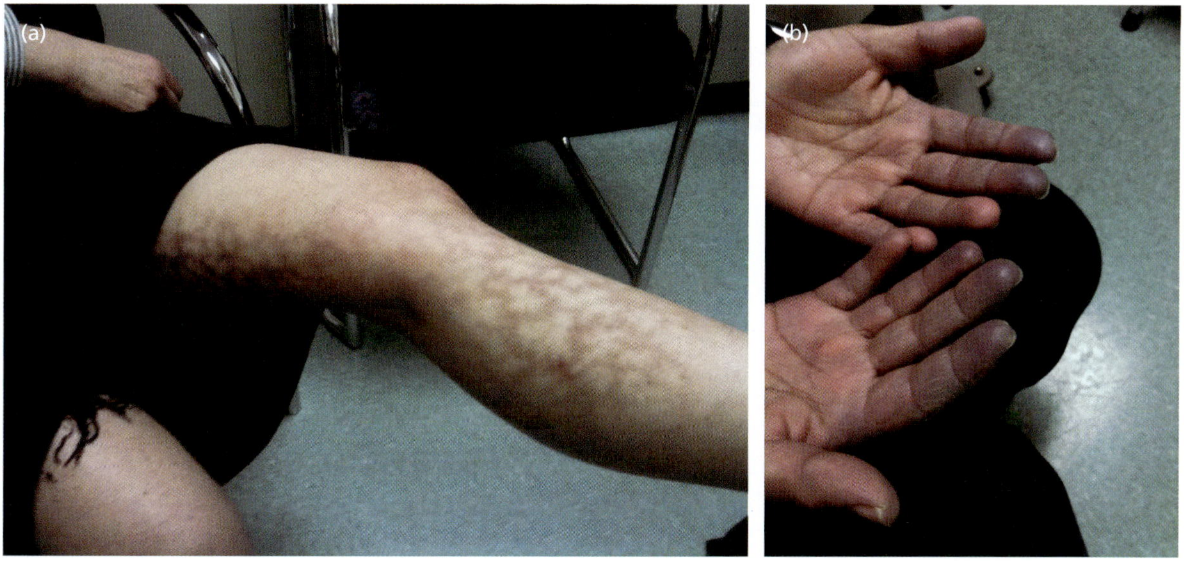

Ergotism. (a) Livedo racemosa associated with ergotamine use. (b) Bilateral distal digital cyanosis associated with ergotamine use.

Giant Cell Arteritis (Temporal Arteritis)

Bruce L. Mintz
Morristown Medical Center, Morristown, NJ, USA

BRIEF CASE PRESENTATION

This 73-year-old white female presents with a 3–4 week history of left-sided headache associated with generalized malaise, muscle aching, depression, weight loss and progressive jaw claudication. She denied visual changes but an erythrocyte sedimentation rate performed in her primary care physicians office was 104 mm/h.

CHARACTERISTICS

- Giant cell arteritis is a large or medium vessel vasculitis preferentially involving the aorta and its branches, most notably the temporal artery
- The American College of Rheumatology (ACR) classification criteria include ESR > 50, age > 50, painful or pulseless temporal artery, and new onset localized headache, which need not be temporal. The ACR criteria however were not designed for diagnostic purposes and carry 91% specificity
- Temporal artery biopsy is the gold standard for diagnosis yet provides only 85–90% sensitivity. Clinical suspicion is key to making the diagnosis

TREATMENT

Giant cell arteritis is a medical emergency. To avoid vision loss, therapy should be started prior to biopsy results if there is substantial clinical suspicion. Initial high dose prednisone therapy and early ophthalmology consultation are the standards of care.

REFERENCES

Ghosh P, Borg FA, Dasgupta B. Current understanding and management of giant cell arteritis and polymyalgia rheumatica. Expert Rev Clin Immunol. 2010; 6(6):913–28.

Gonzalez-Gay MA, Martinez-Dubois C, Agudo M, Pompei O, Blanco R, Llorca J. Giant cell arteritis: epidemiology, diagnosis and management. J. Curr Rheumatol Rep. 2010; 12(6): 436–42.

Maksimowicz- McKinnon K, Hoffman GS. Large vessel vasculitis. Clin Exp Rheumatol. 2007; 25(1Suppl 44):S58–9.

Atlas of Clinical Vascular Medicine, First Edition. Edited by Jessica Mintz, Bruce L. Mintz, and Michael R. Jaff.

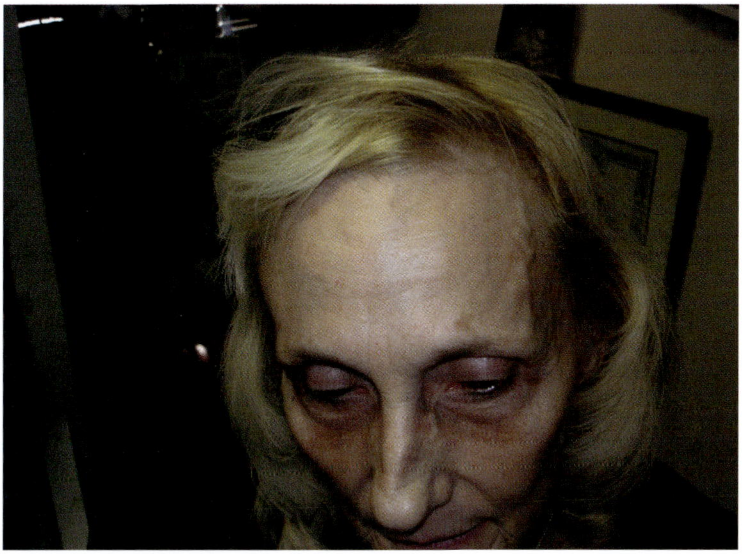

Temporal arteritis. Left temporal artery with branch enlargement.

Polyarteritis Nodosa

Peter C. Grayson[1] and Peter A. Merkel[2]
[1]Boston University Vasculitis Center, Boston, MA, USA
[2]Division of Rheumatology, University of Pennsylvania, PA, USA

BRIEF CASE PRESENTATION

A 64-year-old previously healthy male was hospitalized for severe, acute-onset abdominal pain. Angiography demonstrated a proximal splenic artery aneurysm and a ruptured, distal splenic artery aneurysm. The distal aneurysm was surgically repaired. One week later, repeat angiography demonstrated multiple, new arterial aneurysms along the hepatic artery. Inflammatory markers were elevated. An extensive evaluation for infectious diseases, including hepatitis B, was unremarkable.

CHARACTERISTICS

- Polyarteritis nodosa (PAN) is a systemic necrotizing vasculitis that typically affects medium-sized muscular arteries (e.g. renal, hepatic, coronary, and mesenteric arteries) and can result in aneurysm formation
- Symptoms include weight loss, rash (livedo reticularis, ulcers), testicular pain, myalgias, arthralgias, neuropathy, hypertension, and renal failure
- Most cases are idiopathic, although hepatitis B infection has been implicated as a causal agent in some cases

TREATMENT

Glucocorticoids, either alone (mild disease – constitutional, skin, musculoskeletal involvement) or in combination with other immunosuppressant medications (severe disease – kidney, gastrointestinal, cardiac, or neurologic involvement) are recommended. Aneurysms may require surgical/endovascular repair

REFERENCES

Guillevin L, Cohen P, Mahr A, et al. Treatment of polyarteritis nodosa and microscopic polyangiitis with poor prognosis factors: a prospective trial comparing glucocorticoids and six or twelve cyclophosphamide pulses in sixty-five patients. Arthritis Rheum. 2003; 49(1):93–100.

Lightfoot RW, Jr., Michel BA, Bloch DA, et al. The American College of Rheumatology 1990 criteria for the classification of polyarteritis nodosa. Arthritis Rheum. 1990; 33(8):1088–93.

Pagnoux C, Seror R, Henegar C, et al. Clinical features and outcomes in 348 patients with polyarteritis nodosa: a systematic retrospective study of patients diagnosed between 1963 and 2005 and entered into the French Vasculitis Study Group Database. Arthritis Rheum. 2010; 62(2):616–26.

Atlas of Clinical Vascular Medicine, First Edition. Edited by Jessica Mintz, Bruce L. Mintz, and Michael R. Jaff.

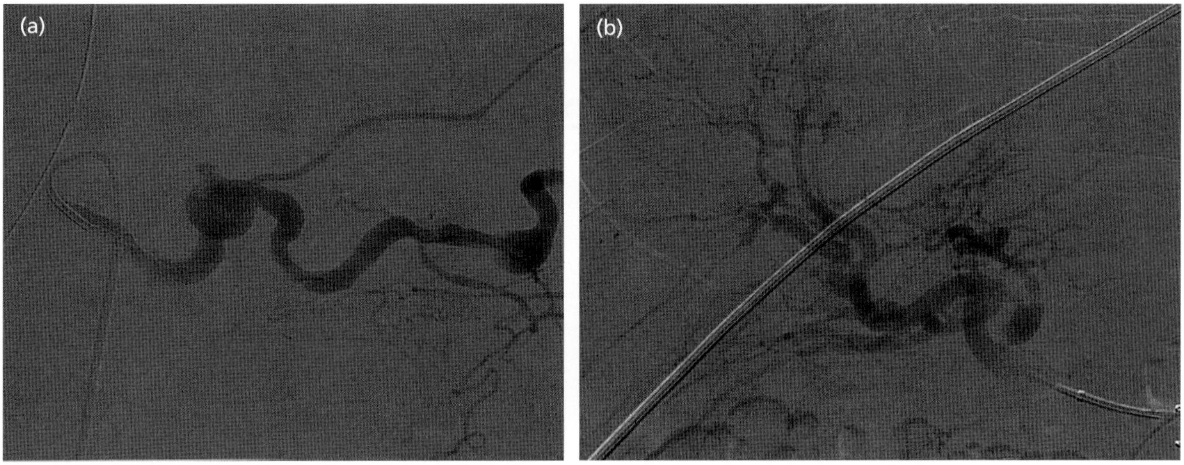

Polyarteritis nodosa. (a) Angiography demonstrating a proximal splenic artery aneurysm and ruptured distal splenic artery aneurysm. (b) Repeat angiogram with multiple new arterial aneurysms of the hepatic artery.

Thromboangiitis Obliterans (Buerger's Disease)

Raghu Kolluri
Prairie Vascular Institute, Springfield, IL, USA

BRIEF CASE PRESENTATION

A patient with a long-standing history of smoking presents with ischemia and ulceration of the right second toe.

CHARACTERISTICS

- Thromboangiitis obliterans is a non-atherosclerotic inflammatory disease of small and medium sized arteries
- It is most typical in young males under the age of 40 who have a history of active tobacco use
- Exposure to tobacco is universal to patients with Buerger's disease
- Highly cellular, inflammatory thrombi occur in the arterial and venous systems causing ischemia, intermittent claudication, and ultimately ischemic rest pain and painful ischemic ulcerations
- It is common for patients with thromboangiitis obliterans to have involvement in more than one limb. Therefore, it is imperative to assess all limbs in patients in which thromboangiitis obliterans is suspected
- Angiographic features include segmental occlusive disease in distal arteries with normal proximal arteries, no evidence of atherosclerosis and collateral arteries appearing in a corkscrew pattern
- To confirm the diagnosis of thromboangiitis obliterans, emboli, autoimmune disease and hypercoagulable states must be excluded

TREATMENT

While pain control, prevention of infection, and antiplatelet therapy are important, only complete cessation of tobacco use has been proven to prevent progression to amputation.

REFERENCES

Olin JW. Thromboangiitis obliterans (Buergers disease). NEJM. 2000; 343:864–9.

Olin JW, Shih A. Thromboangiitis obliterans (Buergers disease). Curr Opin Rheumatol. 2006; 18(10):18–24.

Puechal X, Fiessinger J.-N. Thromboangiitis obliterans or Buerger's disease: challenges for the rheumatologist. Rheumatology. 2007; 46:192–9.

Atlas of Clinical Vascular Medicine, First Edition. Edited by Jessica Mintz, Bruce L. Mintz, and Michael R. Jaff.

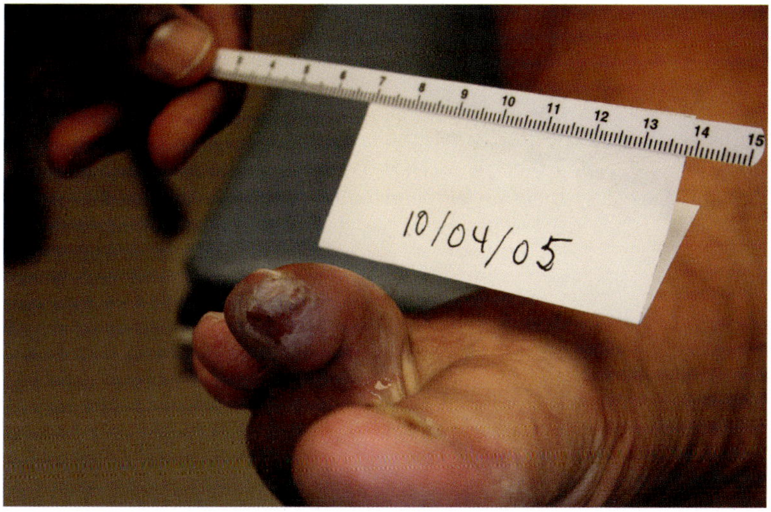

Buerger's disease (thromboangiitis obliterans). Ulceration of the second right toe in a smoker with thromboangiitis obliterans.

Small Vessel Vasculitis

Peter C. Grayson[1] and Peter A. Merkel[2]

[1]Boston University Vasculitis Center, Boston, MA, USA
[2]Division of Rheumatology, University of Pennsylvania, PA, USA

BRIEF CASE PRESENTATION

A 43-year-old man presents with 1 week of a palpable rash on the lower extremities. Review of systems is notable for fatigue, recurrent epistaxis, sinus pain, hearing loss, and arthralgias. Antineutrophil cytoplasmic antibody testing is positive in a cytoplasmic pattern by immunofluorescence with antibodies to proteinase 3 by ELISA. The patient is diagnosed with granulomatosis with polyangiitis (formerly known as Wegener's granulomatosis).

CHARACTERISTICS

- Small vessel vasculitis refers to necrotizing inflammation involving capillaries, venules, arterioles, or small arteries
- The eyes, lungs, skin, and kidneys are commonly affected
- Extravasation of red blood cells secondary to immune-mediated damage to vessel walls results in palpable purpura
- The most common types of small vessel vasculitis are the antineutrophil-cytoplasmic antibody associated vasculitides [granulomatosis with polyangiitis (Wegener's); microscopic polyangiitis; Churg–Strauss syndrome], cryoglobulinemic vasculitis, Henoch–Schönlein purpura, cutaneous leukocytoclastic angiitis, and drug-induced vasculitis

TREATMENT

Specific treatment is based upon identification of the type of small vessel vasculitis. Glucocorticoids and other immunosuppressant medications are often indicated.

REFERENCES

Falk RJ, Hoffman GS. Controversies in small vessel vasulitis – comparing the rheumatology and nephrology views. Curr Opin Rheumatol. 2007; 19(1):1–9.

Jennette JC, Falk RJ, Andrassy K, et al. Nomenclature of systemic vasculitides. Proposal of an international consensus conference. Arthritis Rheum. 1994; 37(2):187–92.

Langford CA. Small-vessel vasculitis: therapeutic management. Curr Rheumatol Rep. 2007; 9(4):328–35.

Atlas of Clinical Vascular Medicine, First Edition. Edited by Jessica Mintz, Bruce L. Mintz, and Michael R. Jaff.

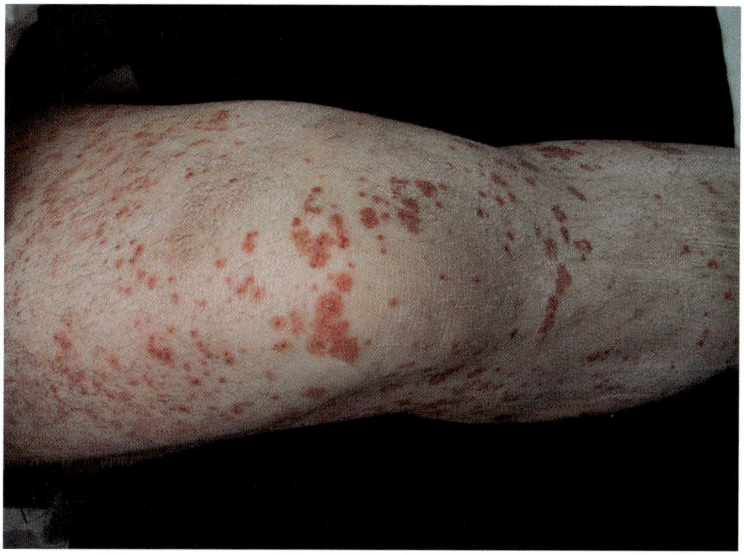

Small vessel vasculitis (leukocytoclastic vasculitis). Lower extremity palpable purpura.

Henoch–Schönlein Purpura

Shane Clark

Wexner Medical Center at the Ohio State University, Columbus, OH, USA

BRIEF CASE PRESENTATION

A 60-year-old male presents with palpable purpura of his lower extremities associated with melena and joint pain 1 week after recovering from an upper respiratory tract infection. A urinalysis demonstrates hematuria.

CHARACTERISTICS

- Henoch–Schönlein purpura (HSP) is a small-vessel vasculitis which occurs most commonly in children <10 and typically affects the GI tract, kidneys, joints and skin. It is the most common vasculitis of childhood
- Common signs and symptoms include a purpuric rash, abdominal symptoms (pain, nausea, melena), and arthralgias. One-third of cases follow upper respiratory tract infection
- No specific test is available for the diagnosis of HSP. However, lab tests often demonstrate hematuria and proteinuria, leukocytosis, an elevated ESR, and a positive stool guaiac
- A skin biopsy of the purpuric lesions is useful and will typically demonstrate leukocytoclastic vasculitis (LCV) of the small vessels. A direct immunofluorescence reveals perivascular IgA deposition in almost all patients
- Most features are self-limiting, but renal disease can be recurrent and result in significant morbidity

TREATMENT

HSP is typically self-limited and resolves in 4 weeks. Corticosteroids effectively treat the GI symptoms and arthropathy but have not been shown to aid in the resolution of purpura or prevent the progression of nephritis. Nephropathy of HSP is typically treated supportively however other immunosuppressive agents such as azathioprine, cyclosporine, and mycophenolate mofetil may be used to treat nephritis with some success. Children should be followed for at least 6 months after presentation or until urinalysis normalizes.

REFERENCES

Carder KR. Hypersensitivity reactions in neonates and infants. Dermatol Ther. 2005; 18(2):160–75.

McCarthy HJ, Tizard EJ. Clinical practice: Diagnosis and management of Henoch-Schönlein purpura. Eur J Pediatr. 2010; 169(6):643–50.

Saulsbury FT. Henoch–Schönlein purpura. Curr Opin Rheumatol. 2010; 22(5):598–602.

Atlas of Clinical Vascular Medicine, First Edition. Edited by Jessica Mintz, Bruce L. Mintz, and Michael R. Jaff.

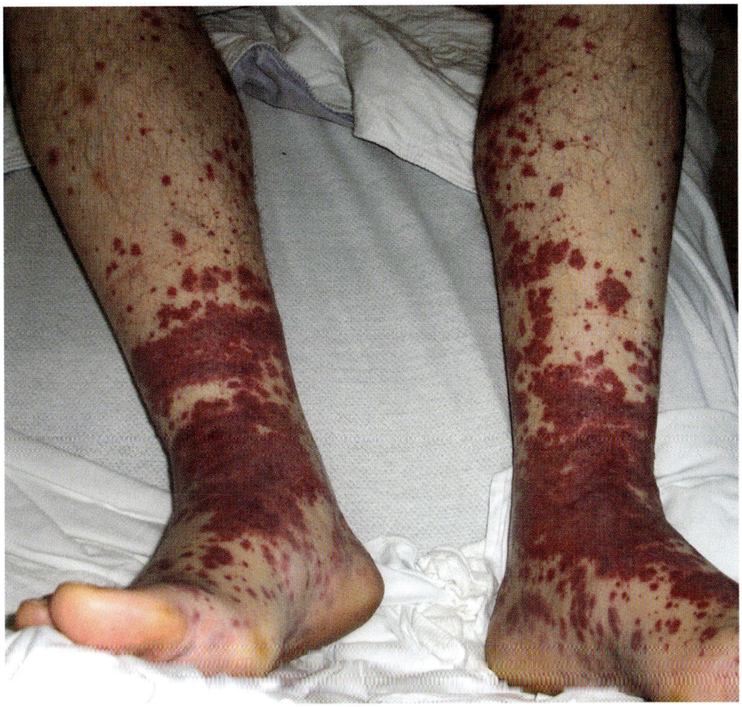

Henoch–Schönlein purpura. Bilateral lower extremity palpable purpura. Courtesy of Dr. Matthew Zirwas.

Behçet's Disease

Peter C. Grayson[1] and Peter A. Merkel[2]

[1]Boston University Vasculitis Center, Boston, MA, USA

[2]Division of Rheumatology, University of Pennsylvania, PA, USA

BRIEF CASE PRESENTATION

A 19 year-old man presents with several days of posterior left knee pain. Physical examination reveals oral and genital ulcers and a palpable, tender mass in the left popliteal fossa. Magnetic resonance angiography shown here demonstrates a partially thrombosed left popliteal artery pseudoaneurysm measuring up to 8 cm in diameter.

CHARACTERISTICS

- Behçet's disease is a multisystem inflammatory disease often classified as a form of vasculitis and associated with HLA-B51
- Behçet's disease is more common along the ancient Silk Road (eastern Asia to the Mediterranean), and is most prevalent in Turkey
- Typical manifestations include recurrent oral and genital ulcers, skin lesions, ocular disease, gastrointestinal disease, neurologic disease, and arthritis
- Vasculitis can involve small, medium, and large arteries and veins. Vascular inflammation may lead to arterial hemorrhage, stenosis, aneurysm, and thrombus formation
- The aorta and carotid, pulmonary, iliofemoral, and popliteal arteries are most commonly involved

TREATMENT

Colchicine, glucocorticoids, and other immunosuppressant agents are considered the pharmacologic treatments of choice, although surgical repair of a large artery aneurysm or pseudoaneurysm may be necessary. Venous thrombosis occurs secondary to endothelial inflammation or extrinsic compression by an adjacent aneurysm or pseudoaneurysm, and treatment is aimed at controlling underlying systemic inflammation. Use of standard anticoagulation therapy is controversial.

REFERENCES

Yazici Y, Yazici H (eds). Behcet's Syndrome. June 2010. Springer Science and Business Media, LLC.

Ahn JK, Lee YS, Jeon CH, Koh EM, Cha HS. Treatment of venous thrombosis associated with Behcet's disease: immunosuppressive therapy alone versus immunosuppressive therapy plus anticoagulation. Clin Rheumatol. 2008; 27(2):201–5.

Koksoy C, Gyedu A, Alacayir I, Bengisun U, Uncu H, Anadol E. Surgical treatment of peripheral aneurysms in patients with Behcet's disease. Eur J Vasc Endovasc Surg. 2011; 42:525–30.

Atlas of Clinical Vascular Medicine, First Edition. Edited by Jessica Mintz, Bruce L. Mintz, and Michael R. Jaff.

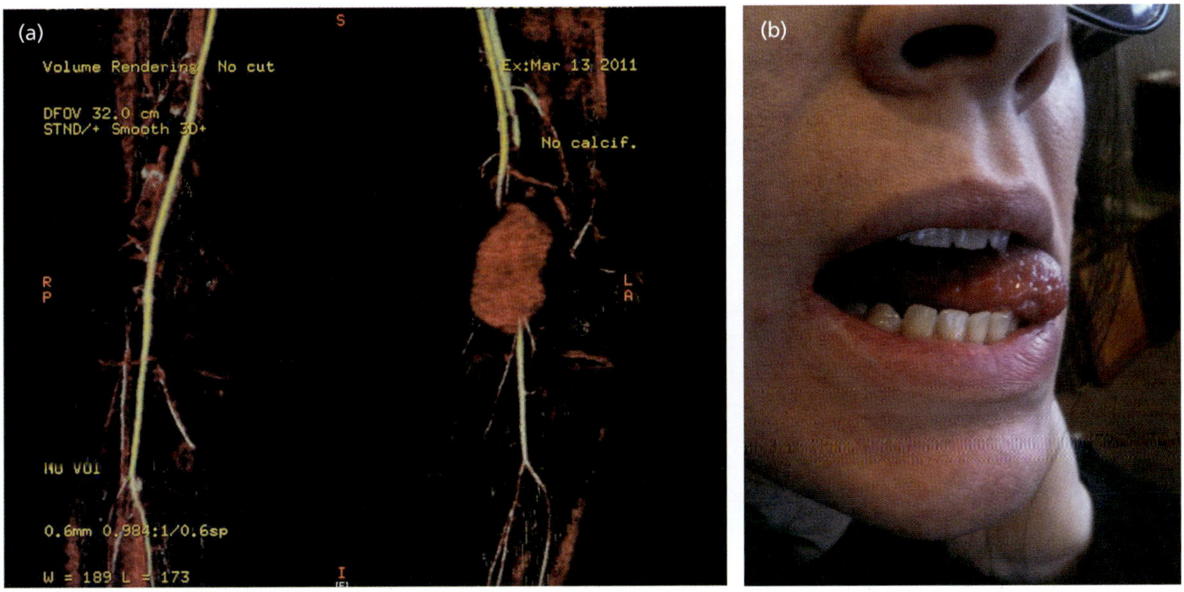

Behçet's disease. (a) Magnetic resonance angiography of a popliteal pseudoaneurysm. (b) Lateral tongue ulceration in a patient with Behçet's disease. Part b courtesy of Michael R Jaff.

Takayasu Arteritis

Ido Weinberg and Michael R. Jaff
Massachusetts General Hospital, Boston, MA, USA

BRIEF CASE PRESENTATION

A 35-year-old Asian-American female presented with new-onset visual disturbances in her right eye. Physical examination revealed a bruit over her right neck. Her erythrocyte sedimentation rate was 60 mm/h and the C-reactive protein level was 22.4 mg/L (normal is <8 mg/L). Duplex ultrasound revealed a long stenosis of the right common carotid artery. Computed tomography of the chest demonstrated aortic root dilatation.

CHARACTERICS

- Takayasu vasculitis is a large vessel vasculitis
- The inflammation in Takayasu arteritis typically affects the aorta and its main branches. It may be localized primarily to the chest or abdominal aorta
- Takayasu arteritis typically, though not exclusively, presents in young women of Asian descent
- Diagnosis involves a combination of age < 40 years, extremity claudication, decreased brachial pulses, interbrachial blood pressure difference >10 mmHg, bruit over the subclavian artery or aorta
- Other supportive findings include constitutional signs and symptoms and marked elevation of acute phase reactants
- Imaging findings include long diffuse tapering stenosis with segments of normal artery
- More recently, elevated interleukin 6 offers great specificity for disease activity

TREATMENT

Immunosuppression is typically achieved with corticosteroids. Long-term treatment may necessitate steroid-sparing medication. Symptomatic arterial manifestations (i.e. intermittent claudication, abdominal coarctation with hypertension) are treated by endovascular techniques or surgery. Non-arterial complications such as aortic valve insufficiency and congestive heart failure should be anticipated and treated appropriately. Interleukin 6 antagonists are a novel treatment modality.

REFERENCES

Arend WP, Michel BA, Bloch DA, et al. The American College of Rheumatology 1990 criteria for the classification of Takayasu arteritis. Arthritis Rheum. 1990; 33(8):1129–34.

Ogino H, Matsuda H, Minatoya K, et al. Overview of late outcome of medical and surgical treatment of Takayasu arteritis. Circulation. 2008; 118(25):2738–47.

Noris M, Daina E, Gamba S, Bonazzola S, Remuzzi G. Interleukin-6 and RANTES in Takayasu arteritis: a guide for therapeutic decisions? Circulation. 1999; 100:55–60.

Saadoun D, Lambert M, Mirault T, et al. Retrospective analysis of surgery versus endovascular intervention in Takayasu arteritis: A multicenter Experience. Circulation. 2010; 125:813–19.

Salvarani C, Magnanu L, Catanoso M, et al. Tocilizumab: a novel therapy for patients with large-vessel vasculitis. Rheumatology. 2012; 51:151–6.

Atlas of Clinical Vascular Medicine, First Edition. Edited by Jessica Mintz, Bruce L. Mintz, and Michael R. Jaff.

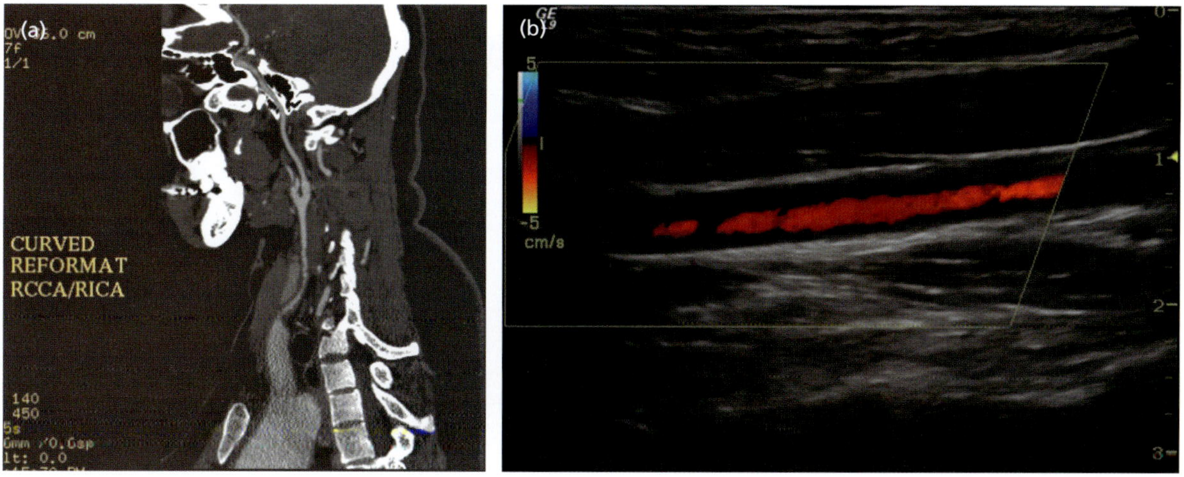

Takayasu arteritis. (a) Carotid computed tomography with aortic root dilatation and long tapering stenosis of the right common carotid artery. (b) Carotid Duplex ultrasonography with stenosis of the right common carotid artery.

Marfan's Syndrome

Ido Weinberg and Michael R. Jaff
Massachusetts General Hospital, Boston, MA, USA

BRIEF CASE PRESENTATION

A 52-year-old female nurse presents for follow-up after repair of an incidentally discovered rapidly expanding ascending thoracic aortic aneurysm. Her brother had an aneurysm repair at the age of 47. She describes hyperelastic joints. She is 6′3″ (1.90 m) and weighs 123 lbs (55 kg).

CHARACTERISTICS

- An autosomal dominant multisystem collagen disorder affecting at least 1 in 5000 of the population. May represent de-novo mutations
- The diagnosis is made by the Ghent criteria. The most important criteria are ectopia lentis and aortic root dilatation, a family history, or a proven fibrillin (FBN1) gene mutation. Other criteria include skeletal, skin and pulmonary signs and symptoms
- Positive wrist and thumb signs are major criteria, while joint hypermobility is a minor criterion
- The most common causes of death for patients with Marfan's syndrome are cardiovascular, including aortic dissection, aortic regurgitation and congestive heart failure

TREATMENT

Patients should refrain from isometric exercise and from competitive or highly strenuous dynamic exercise. All patients with Marfan's syndrome should be given beta-blockers at maximal tolerated doses, titrated to heart rate and blood pressure. An angiotensin converting enzyme inhibitor or angiotensin receptor blocker should be added when further blood pressure control is needed, such as in patients with arterial dissection. Indications for aortic root graft replacement are aortic dissection, aortic dilatation >50 mm and severe aortic regurgitation.

REFERENCES

Keane MG, Pyeritz RE. Medical management of Marfan syndrome. Circulation. 2008; 117(21):2802–13. Review.

Loeys BL, Dietz HC, Braverman AC, et al. The revised Ghent nosology for the Marfan syndrome. J Med Genet. 2010; 47(7):476–85.

Shores J, Berger KR, Murphy EA, Pyeritz RE. Progression of aortic dilatation and the benefit of long-term beta-adrenergic blockade in Marfan's syndrome. NEJM. 1994; 330(19):1335–41.

Atlas of Clinical Vascular Medicine, First Edition. Edited by Jessica Mintz, Bruce L. Mintz, and Michael R. Jaff.

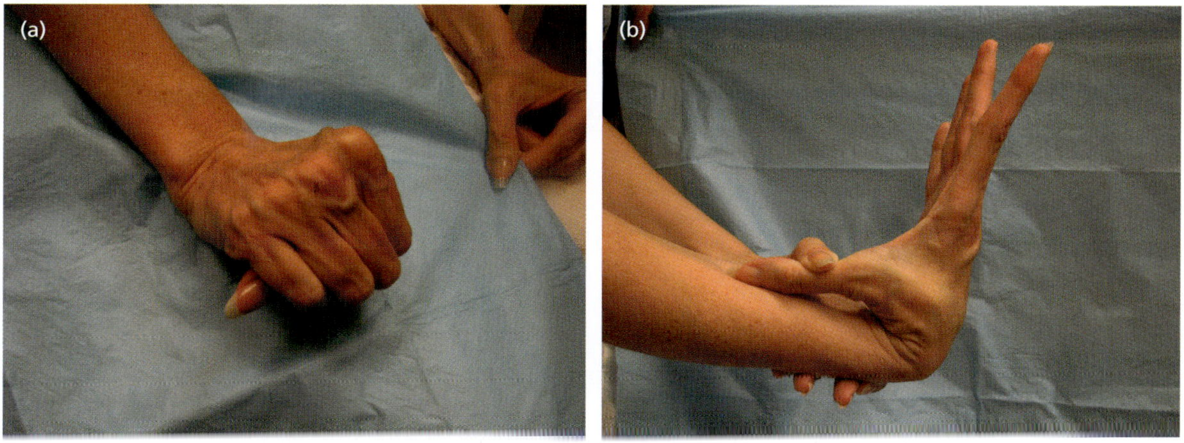

Marfan's syndrome. (a) Right hand demonstrating the "thumb sign". The thumb protrudes medially from a clenched fist in a patient with Marfan's syndrome. (b) Right hand with thumb hyperflexing to touch the forearm (Steinberg sign).

Ehlers–Danlos Type IV (Vascular Type)

Bruce L. Mintz
Morristown Medical Center, Morristown, NJ, USA

> **BRIEF CASE PRESENTATION**
> This 17-year-old male presents 3 days after starting a job as a waiter. He reports no significant trauma, but began to note sudden right-sided arm pain associated with generalized edema. Venous duplex ultrasonography revealed no deep vein thrombosis but marked muscular edema was noted in the area of the brachialis muscle.

CHARACTERISTICS

- Type IV Ehlers–Danlos is a genetic mutation of type III procollagen leading to fragility in arteries and all connective tissue
- Clinical diagnosis is made by four criteria: bruisability, visible veins secondary to thinned skin, arterial rupture and characteristic facial features. Uterine and digestive complications are also common in the vascular type of Ehlers–Danlos
- Hypermobility of joints and hyperextension which are classic for the other forms of Ehlers–Danlos are far less common in the vascular type
- 25% of all complications affect the gastrointestinal tract and prompt surgical intervention
- Diagnosis is made by clinical signs, non-invasive imaging and demonstration of the mutation of the *COL3A1* gene

TREATMENT

No therapy delays onset of complications in type IV Ehlers–Danlos, although knowledge of the disease may affect the way that patients are managed surgically.

REFERENCES

Parapia LA, Jackson C. Ehlers–Danlos syndrome – a historical review. Br J Haematol. 2008; 141(1):32–5.

Pepin M, Schwarze U, Superti-Furga A, Byers P. Clinical and genetic features of Ehlers–Danlos syndrome type IV, the vascular type. NEJM. 2000; 342:673–80.

Lum YW, Brooke BS, Black JH. Contemporary management of vascular Ehlers–Danlos syndrome. Curr Opin Cardiol. 2011; 26(6):494–501.

Atlas of Clinical Vascular Medicine, First Edition. Edited by Jessica Mintz, Bruce L. Mintz, and Michael R. Jaff.

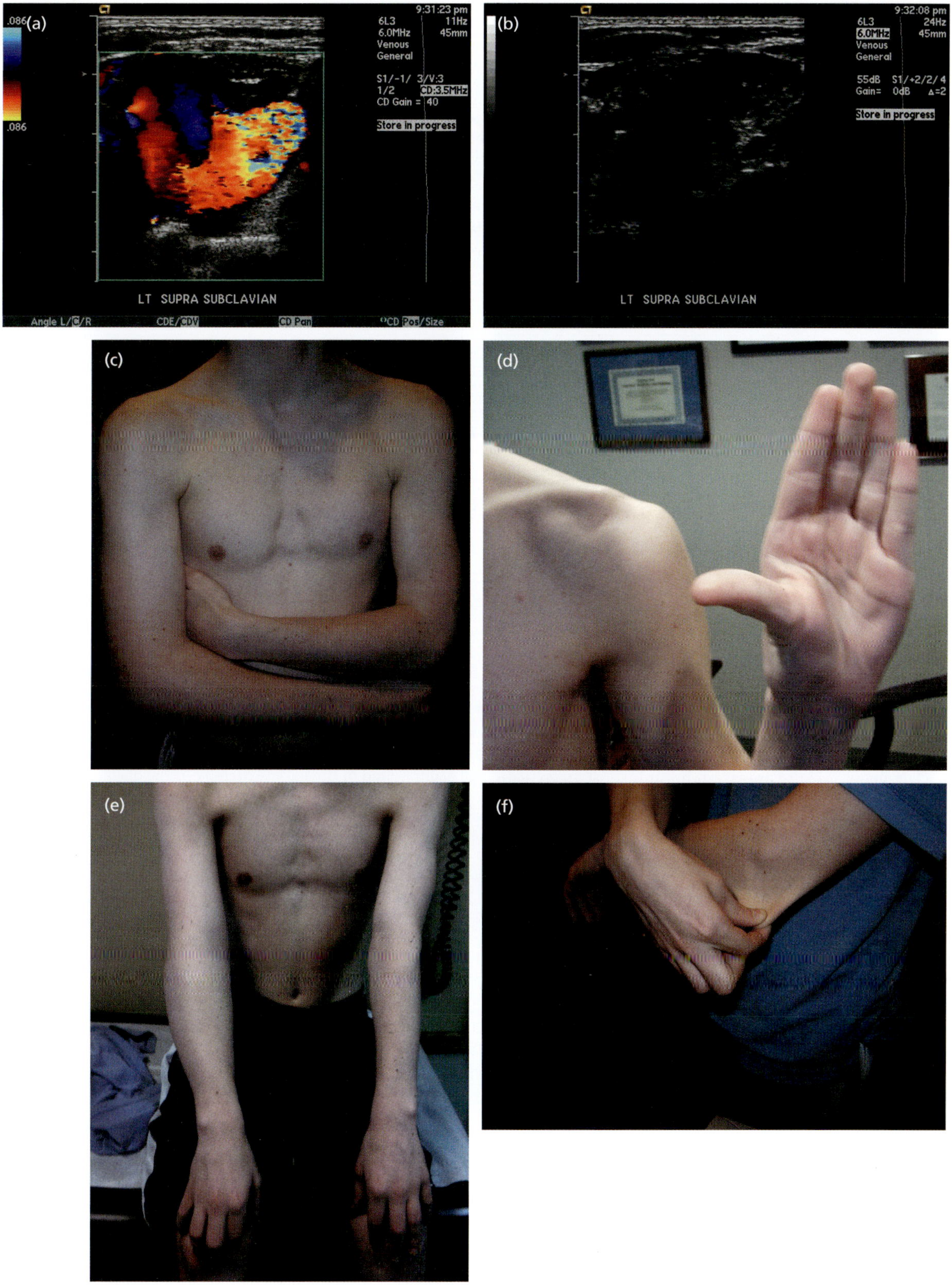

Ehlers–Danlos syndrome. (a) Subclavian artery aneurysm in a patient with Ehlers–Danlos demonstrated on color flow duplex ultrasonography. (b) Subclavian aneurysm in a patient with Ehlers–Danlos demonstrated on gray scale duplex ultrasonography. (c) Pectus excavatum in a patient with Ehlers–Danlos syndrome. (d) Left hand with signs of hypermobility and dislocated thumb. (e) Ruptured right brachialis muscle after minor stress. (f) Skin demonstrating hyperelasticity.

Deep Vein Thrombosis

Ari J. Mintz[1] **and Bruce L. Mintz**[2]
[1]Lahey Clinic Medical Center, Burlington, MA, USA
[2]Morristown Medical Center, Morristown, NJ, USA

BRIEF CASE PRESENTATION

This 50-year-old white female presents with an acutely swollen right lower extremity after a 7 hour car ride.

CHARACTERISTICS

- Thrombosis of the deep veins (DVT) most commonly occurs in the lower extremity. Other less common sites include the upper extremities, mesenteric, hepatic, renal and cerebral veins
- While DVT may be asymptomatic, the classic presentation includes a painful, swollen and discolored extremity
- The pathogenesis of thrombosis is described by Virchow's triad consisting of hypercoagulability, stasis, and endothelial injury
- Calf veins are the most common site of deep vein thrombosis. Thrombi limited to calf veins often spontaneously lyse without progression while 25% will extend to more proximal veins if left untreated. From 30–50% of untreated proximal DVTs will progress to a pulmonary embolism
- Rarely, thrombosis is massive causing vascular compromise of the lower extremity presenting as phlegmasia cerulea dolens
- The gold standard for diagnosis has become compression duplex ultrasonography. Characteristic lack of compression by duplex provides the diagnosis of deep vein thrombosis

TREATMENT

The cornerstone of therapy is antithrombotic therapy with medications such as vitamin K antagonists (warfarin) after a bridging period with a Xa inhibitor (enoxaparin, fondaparinux) or unfractionated heparin.

REFERENCES

Mintz BL, Clifford A, Kritharis A, Hobson RW II. Venous Duplex ultrasound of the lower extremity in the diagnosis of venous thrombosis. In: AbuRahma Ali F, Bergan JJ (eds) Noninvasive Vascular Diagnosis, 2nd edn. London: Springer, 2007, pp. 385–93.

Virchow RLK. Thrombose und Embolie. Gefässentzündung und septische Infektion. Gesammelte Abhandlungen zur wissenschaftlichen Medicin. Frankfurt: Von Meidinger & Sohn1856, pp. 219–732.

O'Dea, JA, Kiernan TJ, Jaff MR. Deep venous thrombosis and pulmonary embolism.In: Jaff MR, White CJ (eds) Vascular Disease: Diagnostic and Therapeutic Approaches. Minneapolis: Cardiotext, 2011, pp. 407–26.

Atlas of Clinical Vascular Medicine, First Edition. Edited by Jessica Mintz, Bruce L. Mintz, and Michael R. Jaff.

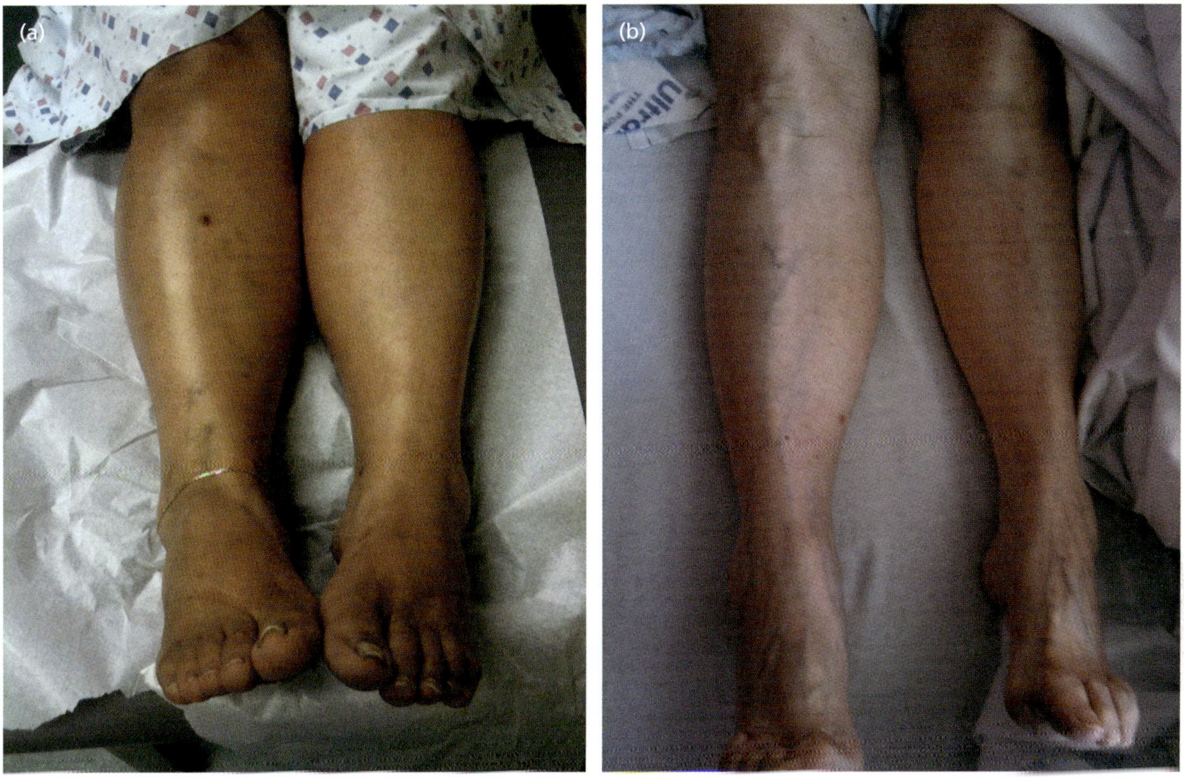

Lower extremity deep vein thrombosis. (a) Right lower extremity new dilated superficial vein in the setting of an acute deep vein thrombosis ("[illegible] sign"). Note the collateral vein over the anterior shin. (b) Right calf swelling with dilated superficial compensatory collateral veins in a patient with a deep vein thrombosis.

Upper Extremity Deep Vein Thrombosis

Bruce L. Mintz
Morristown Medical Center, Morristown, NJ, USA

BRIEF CASE PRESENTATION

This 40-year-old white female presented to the outpatient clinic with a chief complaint of right arm pain and swelling associated with shortness of breath. On physical exam there are prominent upper extremity collateral veins visible on the right side of the patients' chest.

CHARACTERISTICS

- Upper extremity deep venous thrombosis (DVT) is characterized by pain and edema and is commonly associated with new onset asymmetric veins on the chest. 50% of patients will be asymptomatic
- The most common cause of upper extremity DVT in the hospital is iatrogenic secondary to indwelling catheter placement. Outpatient upper extremity DVT is commonly due to thoracic outlet obstruction
- Inherited thrombophilia may be associated with an increased risk of upper extremity DVT. However, between 20 and 30% of upper extremity DVT's are de novo and not associated with any underlying etiology
- Objective diagnosis is mandatory prior to initiating anticoagulant treatment. Diagnosis is most commonly made by duplex ultrasonography, which is both sensitive and specific. Contrast venography may be necessary to evaluate thrombosis in the face of bony anatomic interference but is most commonly performed during planned intervention.
- Pulmonary embolism complicates upper extremity deep vein thrombosis in up to 36% of patients. The clinical course is complicated by recurrent thromboembolism and post-thrombotic swelling of the limb

TREATMENT

Traditional therapy includes 3–6 months of therapeutic anticoagulation; however, options have expanded to include catheter-directed thrombolysis, percutaneous transluminal angioplasty with stenting, and pharmacomechanical thrombolysis. First rib resection remains the standard treatment for mechanical obstruction.

REFERENCES

Kucher N. Deep vein thrombosis of the upper extremities. NEJM. 2011; 364:861–9.

Mustafa, S, Stein PD, Patel KC, Otten, TR, Holmes R, Silbergleit A. Upper extremity deep venous thrombosis. Chest. 2003; 123 (6):1953–6.

Martinelli I, Battaglioli T, Bucciarelli P, Passamonti SM, Mannuccio PM. Risk factors and recurrence rate of primary deep vein thrombosis of the upper extremities. Circulation. 2004; 110;566–70.

Seeger M, Bewig B. Paget Schroetter syndrome. NEJM. 2010; 363:3.

Atlas of Clinical Vascular Medicine, First Edition. Edited by Jessica Mintz, Bruce L. Mintz, and Michael R. Jaff.

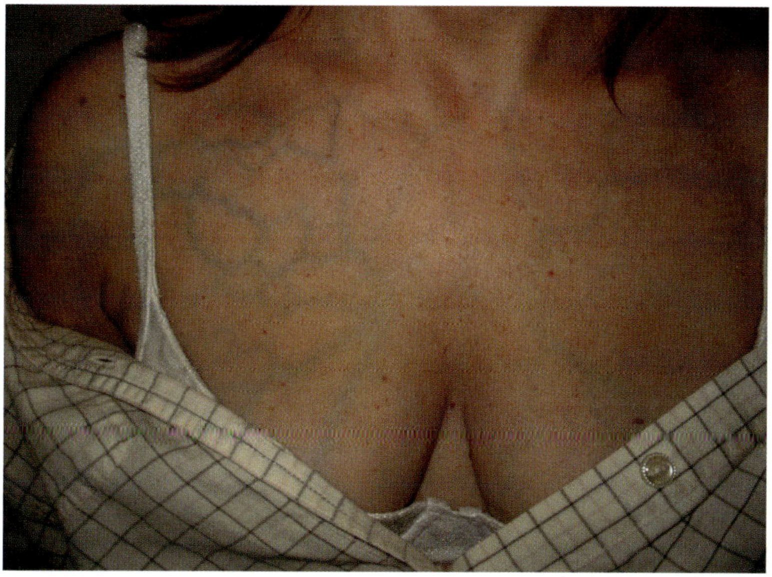

Upper extremity deep vein thrombosis. Prominent chest wall veins representing venous collaterals in a patient with an axillary-subclavian deep vein thrombosis.

May–Thurner Syndrome

Robert M. Schainfeld

Vascular Medicine Section, Massachusetts General Hospital, Boston, MA, USA

BRIEF CASE PRESENTATION

A 38-year-old female presents with acute onset of left leg swelling and erythema 2 days following arthroscopic right knee surgery. Past medical history is notable for recurrent retinal hemorrhages and peripheral artery disease, onset during early 30's, manifested by intermittent claudication.

CHARACTERISTICS

- Left iliac vein compression from the right common iliac artery (May–Thurner syndrome, Cockett's or iliac vein compression syndrome) or non-occlusive iliac vein lesion (NIVL) is estimated to be present in 49–62% of cases of left lower extremity deep vein thrombosis (DVT).
- Those who experience DVT frequently have anatomically abnormal veins with venous spur formation and are at high risk of developing recurrent deep vein thrombosis and post-thrombotic syndrome

TREATMENT

Endovascular treatment with venoplasty and stenting of iliac vein obstruction secondary to May–Thurner syndrome has shown promising initial results, with high technical success and excellent primary-assisted and secondary patency rates. At 72 months follow-up, 67% primary, 89% primary-assisted and 93% secondary patency rates are reported. Adequate coverage of the entire diseased segment as guided by intravascular ultrasound seems to provide best long-term results. This frequently necessitates placing the stent into the inferior vena cava to ensure full lesion coverage and on occasion extending the stents into the common femoral vein.

REFERENCES

Kibbe MR, Ujiki M, Goodwin AL, Eskandari M, Yao J, Matsumura J. Iliac vein compression in an asymptomatic patient population. J Vasc Surg. 2004; 39(5):937–43.

Neglén, P, Hollis, KC, Olivier, J, Raju, S. Stenting of the venous outflow in chronic venous disease: Long-term stent-related outcome, clinical, and hemodynamic result. J Vasc Surg. 2007; 46:979–90.

Atlas of Clinical Vascular Medicine, First Edition. Edited by Jessica Mintz, Bruce L. Mintz, and Michael R. Jaff.

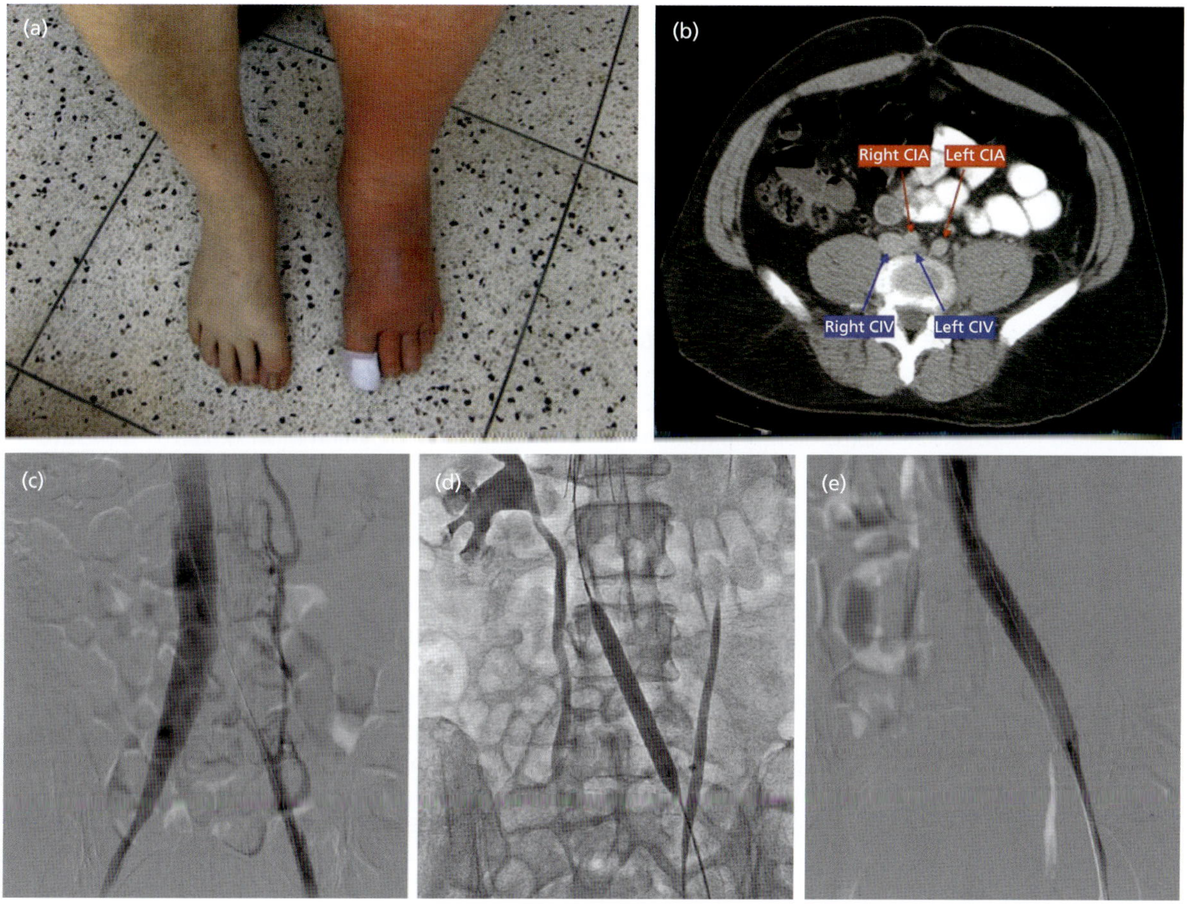

May–Thurner syndrome. (a) Left lower extremity swelling and erythema. (b) Computed tomography demonstrating severe compression of the left common iliac vein by the overlying right common iliac artery against the 5th lumbar vertebrae. (c) Iliocaval venogram after catheter-directed thrombolysis and adjunctive percutaneous mechanical thrombectomy of the iliofemoral veins. (d) Venoplasty of an occluded left common iliac vein presumed to be due to iliac vein compression (May–Thurner). Note the inferior vena cava filter deployed after retinal hemorrhage during tissue plasminogen activator infusion. (e) A Wallstent (Boston Scientific, Natick, MA, USA) deployed in the inferior vena cava and common iliac vein.

Phlegmasia Cerulea Dolens

Steven M. Dean
Department of Cardiovascular Medicine, Wexner Medical Center at the Ohio State University, Columbus, OH, USA

BRIEF CASE PRESENTATION

A 51-year-old male presents with exquisite pain, marked swelling, and discoloration throughout the left lower extremity. A venous duplex ultrasound examination documents acute occlusive deep venous thrombosis extending from the tibioperoneal veins to the external iliac vein with concurrent superficial venous thrombosis within the great and small saphenous veins.

CHARACTERISTICS

- Phlegmasia cerulea dolens (PCD) is a turgid, painful, blue swollen leg. This is a limb and life threatening emergency secondary to massive occlusive venous thrombosis throughout the deep, superficial, and collateral veins
- It most commonly affects the left lower extremity with a 3:1 to 4:1 predominance. Less than 5% of cases involve the upper extremities
- Physical findings include massive limb edema, petechia, bulla and absent distal arterial pulses
- Complications include pulmonary emboli, compartment syndrome, hypovolemic shock, venous gangrene, and death

TREATMENT

No standardized therapy exists. At a minimum, full dose heparin and extreme limb elevation are warranted. Aggressive therapy includes catheter-directed pharmacologic and/or pharmacomechanical thrombolysis, surgical or catheter-based aspiration thrombectomy, iliac vein stenting, or various combinations of these modalities. A fasciotomy may be required. Irreversible cases necessitate limb amputation.

REFERENCES

Chinsakchai K, Ten Duis K, Moll FL, de Borst GJ. Trends in management of phlegmasia cerulea dolens. Vasc Endovascular Surg. 2011; 45(1):5–14.

Vysetti S, Shinde S, Chaudhry S, Subramoney K. Phlegmasia cerulea dolens – a rare, life-threatening condition. Sci World J. 2009; 9:1105–6.

Oguzkurt L, Tercan F, Ozkan U. Manual aspiration thrombectomy with stent placement: rapid and effective treatment for phlegmasia cerulea dolens with impending venous gangrene. Cardiovasc Intervent Radiol. 2008; 31(1):205–8.

Atlas of Clinical Vascular Medicine, First Edition. Edited by Jessica Mintz, Bruce L. Mintz, and Michael R. Jaff.

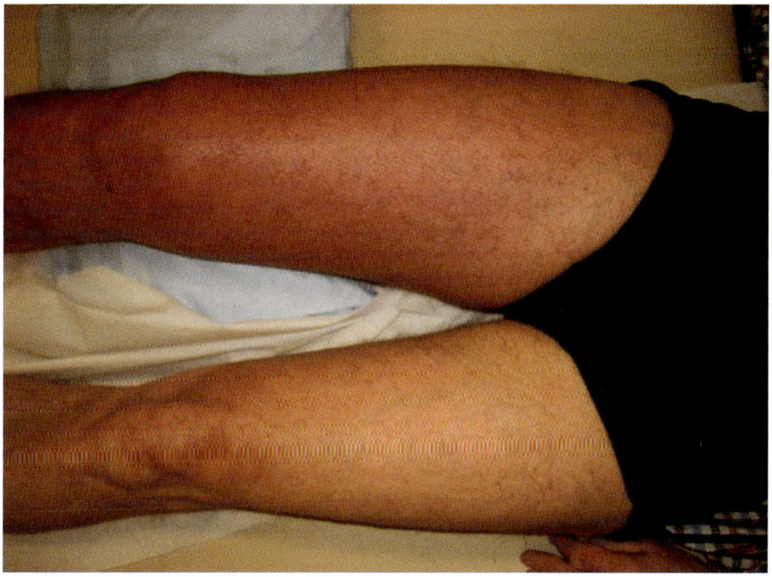

Phlegmasia cerulea dolens. Massive right limb swelling and erythrocyanosis with ipsilateral multisegmental deep and superficial venous thrombosis.

Heparin-induced Thrombocytopenia

Jessica Mintz, John R. Bartholomew and Heather L. Gornik
Cleveland Clinic Heart and Vascular Institute and Cleveland Clinic Lerner College of Medicine, Case Western Reserve University, Cleveland, OH, USA

BRIEF CASE PRESENTATION

A hospitalized patient being treated with IV unfractionated heparin develops arterial thrombosis associated with right lower extremity discoloration from mid calf extending into all five digits of the right foot.

CHARACTERISTICS

- Heparin-induced thrombocytopenia (HIT) is a life-threatening, hematologic syndrome associated with the use of unfractionated, or much less frequently, low-molecular-weight heparin
- There are two types of HIT
  - Type 1 presents after 2 days of heparin exposure. Platelets normalize with continued heparin therapy. Type 1 HIT is *not* an autoimmune phenomenon
  - Type 2 HIT is immune-mediated and typically occurs after 4–10 days of heparin therapy. This is a life threatening, prothrombotic condition affecting both arterial and venous vascular beds
- HIT is caused by antigen–antibody complexes. Antibodies are formed against heparin bound to platelet factor 4 (PF4) on platelet surfaces
- Thrombosis associated with HIT can be arterial, venous or both. Limb ischemia, deep vein thrombosis, myocardial infarction, and strokes are all complications of HIT. Thrombotic risk is 30 times that of control populations, with risk of thrombosis remaining high even after discontinuation of heparin
- Diagnosis of HIT is made by clinical suspicion as well as laboratory findings for confirmation. There are currently two types of laboratory tests available: functional tests including serotonin release assay (SRA) and antigen (immunoassays) including PF4.

TREATMENT

The most important treatment is to discontinue heparin (and low-molecular-weight-heparin products) and start an alternative anticoagulant such as argatroban, lepirudin or bivalirudin, as these are direct thrombin inhibitors which are not implicated in HIT. It is imperative to abstain from platelet transfusions when a patient is suspected to have HIT.

REFERENCES

Arepally, GM, Ortel TL. Heparin-induced thrombocytopenia. NEJM. 2006; 355:809–17.

Warkentin TE. Heparin-induced thrombocytopenia: diagnosis and management. Circulation. 2004; 110(18):454–8.

Bartholonew JR, Begelman SM, Almahameed A. Heparin-induced thrombocytopenia: principles for early recognition and management. Cleveland Clin J Med. 2005; 72(1):31–6.

Atlas of Clinical Vascular Medicine, First Edition. Edited by Jessica Mintz, Bruce L. Mintz, and Michael R. Jaff.

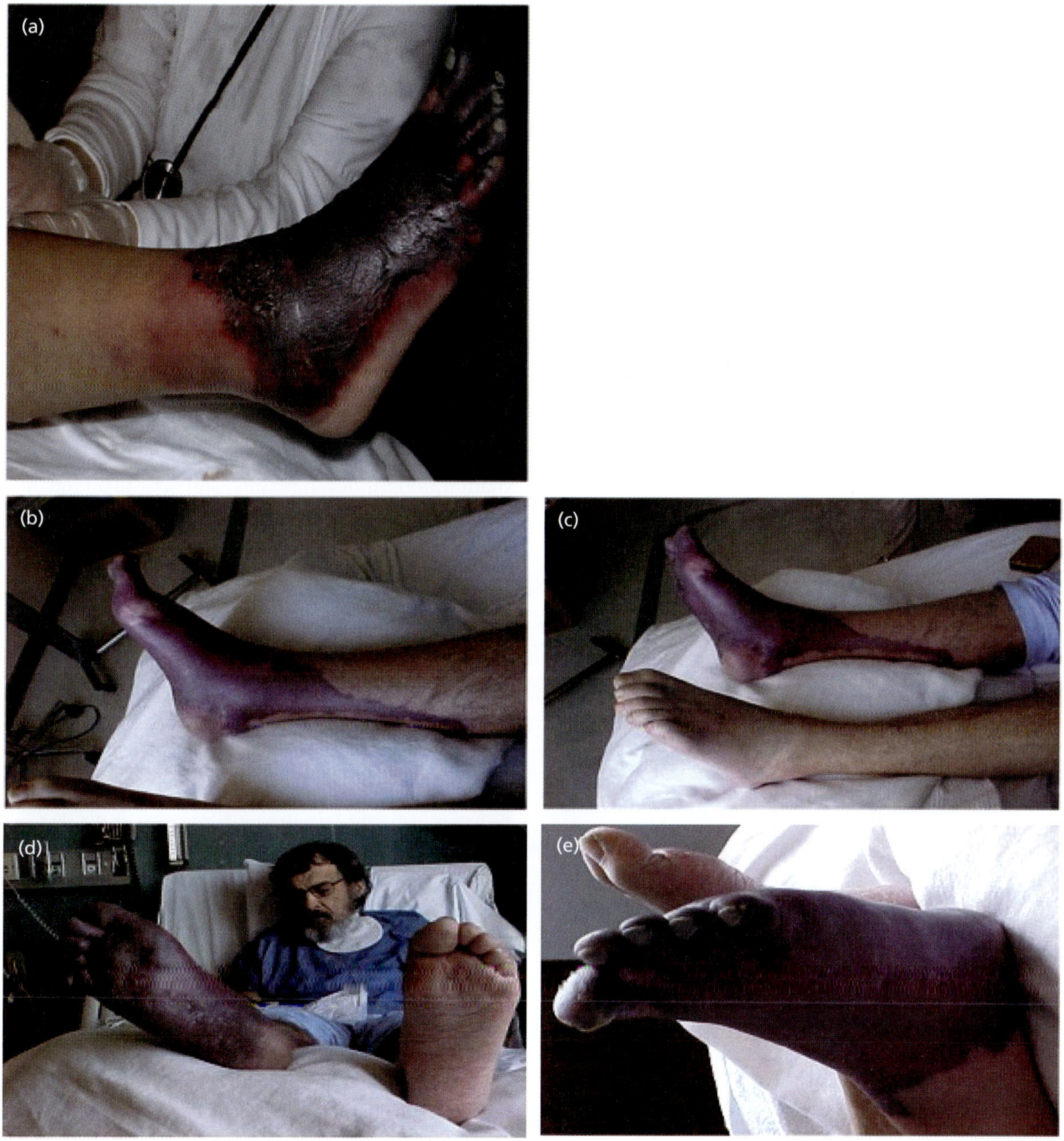

(a) Progressive area of necrosis with associated blisteriung occurring on the 8th day of heparin therapy. (b–e) Well delineated area of purpuric necrosis overlying the right distal lower extremity in a patient with new onset thrombocytopenia while exposed to heparin therapy.

Warfarin-induced Skin Necrosis

Bruce L. Mintz
Morristown Medical Center, Morristown, NJ, USA

BRIEF CASE PRESENTATION

A 55-year-old female with a history of breast cancer presents with an exquisitely painful progressive necrotic lesion overlying her right thigh while being treated for deep vein thrombosis with warfarin alone.

CHARACTERISTICS

- Warfarin-induced skin necrosis is a rare but potentially fatal complication of treatment with warfarin
- Skin necrosis often occurs between the third and tenth days of warfarin therapy
- Patients who are deficient in protein C and S are at most risk for this complication. Because routine screening for protein C and S deficiency is not part of the standard of care for warfarin therapy, detection of this subset of patients is often not possible
- Cutaneous findings include exquisitely painful erythema, ecchymosis and hemorrhage over subcutaneous fatty areas including breast, thigh, abdomen, buttocks, and penis

TREATMENT

Withdrawal of warfarin and initiating unfractionated heparin or low molecular weight heparin along with vitamin K, fresh frozen plasma, and activated protein C are all standard treatments for warfarin-induced skin necrosis. Wide excision, debridement, and skin grafting may be necessary.

REFERENCES

Flood EP, Redish MH, Bociek S, et al. Thrombophlebitis migrans disseminata: Report of a case in which gangrene of breast occurred. NY State J Med. 1943; 43:1121–4.

Srinivasan AF, Rice L, Bartholomew JR et al. Warfarin-induced skin necrosis and venous limb gangrene in the setting of heparin induced thrombocytopenia. Arch Intern Med. 2004; 164:66–70.

Verhagen H. Local haemorrhage and necrosis of the skin and underlying tissues, during anti-coagulant therapy with dicumarol or dicumacyl. Acta Med Scand. 1954; 148 (6):453–67.

Atlas of Clinical Vascular Medicine, First Edition. Edited by Jessica Mintz, Bruce L. Mintz, and Michael R. Jaff.

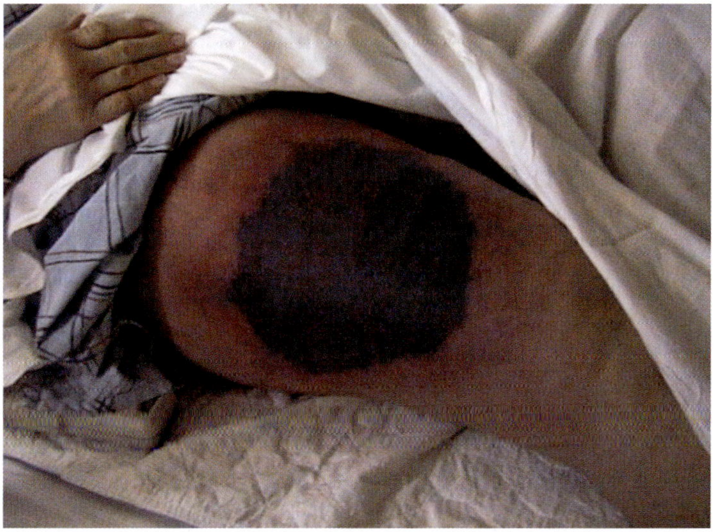

Exquisitely tender, circumscribed area of necrosis over the fatty tissue of the thigh in a patient exposed to warfarin.

Telangiectasias

Robert M. Schainfeld
Vascular Medicine Section, Massachusetts General Hospital, Boston, MA, USA

BRIEF CASE PRESENTATION

A 27-year-old female presents with scattered small, red lesions involving both thighs and calves, which she has had since 18 years of age. She feels these lesions have increased in number over past few years associated with symptoms of heaviness, burning, fatigue and exacerbation of pain during menses.

CHARACTERISTICS

- Chronic venous disorders (CVD) include a spectrum of clinical presentations ranging from uncomplicated telangiectasias and varicose veins to venous ulceration. Chronic venous insufficiency (CVI) usually refers more specifically to the spectrum of skin changes associated with sustained venous hypertension
- In order to develop a classification system for venous disease, a consensus document (CEAP) was formulated by members of the American Venous Forum. The classification was based on clinical manifestations (C), etiologic factors (E), anatomic distribution of disease (A) and the underlying pathophysiologic findings (P)
- Telangiectasias are a confluence of dilated intradermal venules of less than 1 mm in diameter. A synonym for telangiectasias is spider veins
- According to the CEAP classification, telangiectasias would be C1

TREATMENT

Compression stockings are mandated as first-line therapy for all chronic venous disorders. In those individuals who fail conservative therapy, options include injection sclerotherapy, laser and intense pulse light.

REFERENCES

Duffy, DM. Small vessel sclerotherapy: An overview. Adv Dermatol. 1988; 3:221–42.

Eklof, B, Rutherford, RB, Bergan JJ, et al., for the American Venous Forum International Ad Hoc Committee for Revision of the CEAP classification. Revision of the CEAP classification for chronic venous disorders: Consensus statement. J Vasc. Surg. 2004; 40:1248–52.

Atlas of Clinical Vascular Medicine, First Edition. Edited by Jessica Mintz, Bruce L. Mintz, and Michael R. Jaff.

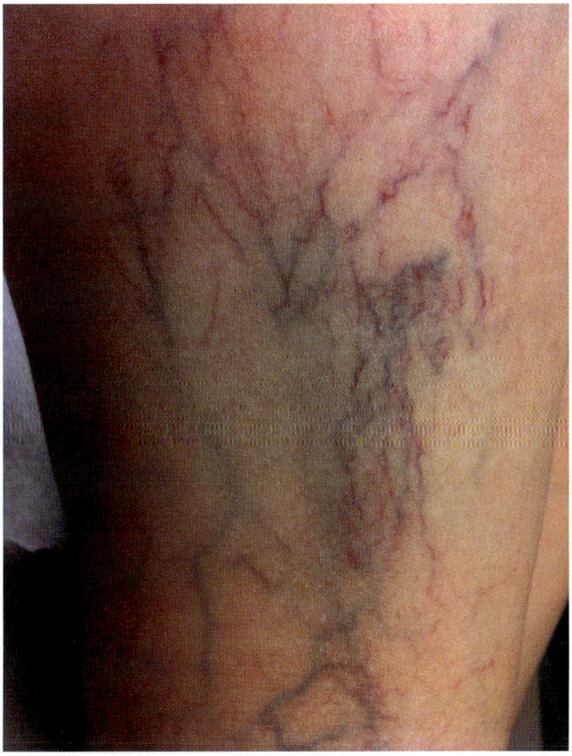

Photograph of a 21-year-old with blue veins, measuring 2–3 mm in diameter coursing proximally from her right lateral calf to thigh. Note in contrast, associated superficial red (spider) veins in adjacent distribution.

Reticular Veins

Robert M. Schainfeld
Vascular Medicine Section, Massachusetts General Hospital, Boston, MA, USA

BRIEF CASE PRESENTATION

A 21-year-old woman presents to office with complaints of pain along the thighs and calves bilaterally. On exam, blue dilated veins were observed on both lateral thighs and calves feeding telangiectatic webs.

CHARACTERISTICS

- Reticular veins are characterized by dilated bluish subdermal veins usually ranging in diameter from 1–3 mm. They tend to be tortuous in their course
- Similar to telangiectasias, reticular veins are classified by CEAP as C1
- Telangiectasias can result from refluxing reticular veins. When such reticular veins are associated with telangiectasias, they are called "feeder veins"

TREATMENT

The basic principle of interventional venous management is to treat the largest vein first and the smallest telangiectasia last. If sclerotherapy is chosen as the treatment for a particular patient, the reticular veins should be injected first, then telangiectatic webs or networks and finally isolated telangiectasias. Several solutions are reasonable sclerosant choices, including 23% normal saline and Polidocanol.

REFERENCE

Mundy L, Merlin TL, Fitridge RA, Hiller JE. Systemic review of endovenous laser treatment for varicose veins. Br J Surg. 2005; 92:1189–94.

Atlas of Clinical Vascular Medicine, First Edition. Edited by Jessica Mintz, Bruce L. Mintz, and Michael R. Jaff.

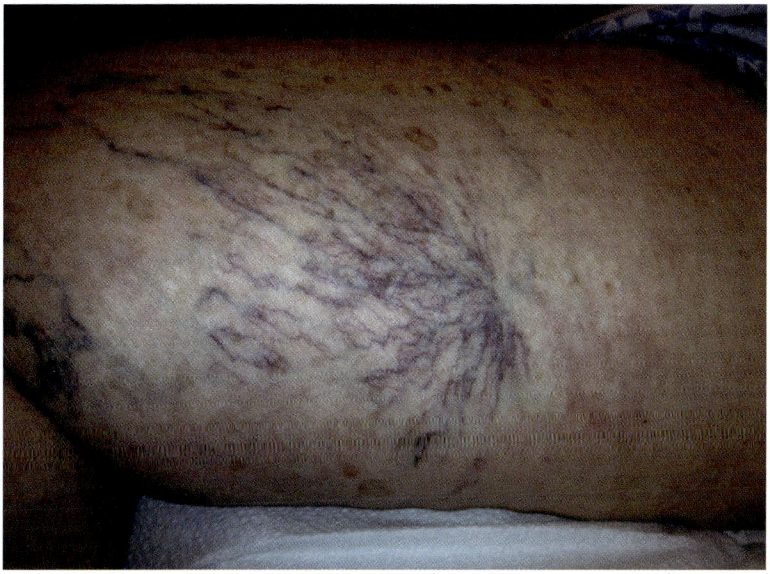

Right inner thigh dilated subdermal veins.

Venous Varicosities

Robert M. Schainfeld

Vascular Medicine Section, Massachusetts General Hospital, Boston, MA, USA

BRIEF CASE PRESENTATION

A 26-year-old female with history of left leg discomfort presents following delivery of her third child with large, bulging, painful veins exacerbated by standing and/or sitting for lengthy periods. In addition, she complains of fatigue and ankle swelling at the end of a day. Modest improvement was noted with limb elevation and graduated compression stockings.

CHARACTERISTICS

- Varicose veins are subcutaneous dilated veins measuring 3 mm in diameter or larger, obtained while in the upright position
- May involve saphenous or non-saphenous veins, lie subcutaneously and are palpable

TREATMENT

Traditional approach to treatment has been removal of the great saphenous vein after ligation and division of the saphenofemoral junction. Complications include bruising, hematoma, and pain with one-third of patients developing recurrence or neovascularization. Surgical methods of removing the great saphenous vein include proximal ligation alone, proximal ligation with ankle to groin stripping, and proximal ligation with knee to groin stripping. Endovenous ablation of the saphenous vein has revolutionized the treatment for varicose veins. Currently available methods to achieve ablation of diseased veins are: (1) Closure procedure using a radiofrequency (RF) catheter and generator (VNUS Medical Technologies Inc., Sunnyvale, CA); (2) endovenous laser ablation procedure using a laser fiber and generation (Covidien Inc., Manchester, MA); (3) endovenous chemical ablation with ultrasound-guided foam sclerotherapy (either injected or catheter-directed). The first two methods use electromagnetic energy to destroy the vein in situ. The latter method utilizes a foamed chemical detergent (Polidocanol or sodium tetradecylsulfate). As with stripping, portions of the great saphenous and/or small saphenous vein, perforator veins, and varicose tributaries remaining after these endovenous procedures must be treated with either injection sclerotherapy or phlebectomy.

REFERENCES

Barrett JM, Allen B, Ockelford A, Goldman MP. Microfoam ultrasound-guided sclerotherapy treatment for varicose veins in a subgroup with diameters at the junction of 10 mm or greater compared with a subgroup of less than 10 mm. Dermatol Surg. 2004; 30(11):1386–90.

Campbell B. Varicose veins and their management. BMJ. 2006; 333:286–92.

Mundy L, Merlin TL, Fitridge RA, Hiller JE. Systematic review of endovenous laser treatment for varicose veins. Br J Surg. 2005; 92:1189–94.

Shepherd AC, Gohel MS, Brown LC, Metcalfe MJ, Hamish M, Davis AH. Randomized clinical trial of VNUS ClosureFAST radiofrequency ablation versus laser for varicose veins. B J Surg. 2010; 97:810–18.

Atlas of Clinical Vascular Medicine, First Edition. Edited by Jessica Mintz, Bruce L. Mintz, and Michael R. Jaff.

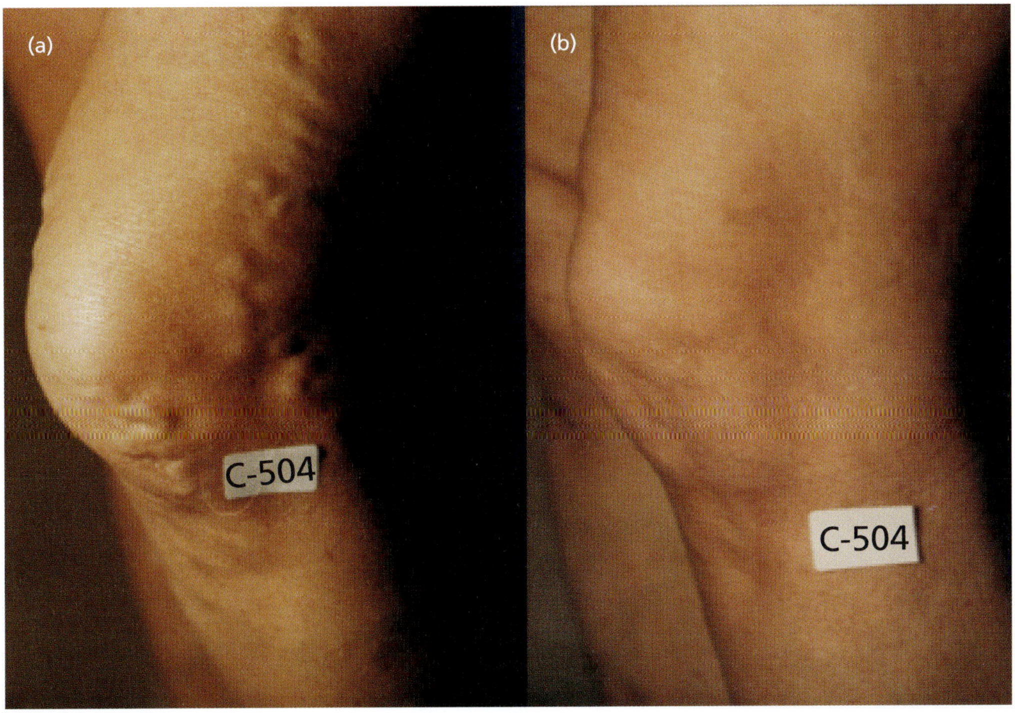

Venous varicosities. (a) Left lower extremity venous varicosities following delivery of patient's third child. (b) Clinical results following endovenous ablation (laser) of incompetent left great and posterior medial saphenous veins with adjunctive ambulatory phlebectomies.

Superficial Thrombophlebitis

Robert M. Schainfeld
Vascular Medicine Section, Massachusetts General Hospital, Boston, MA, USA

BRIEF CASE PRESENTATION

A 66-year-old male was seen in office with 7-day history of an exquisitely painful, red, warm area overlying a large varicose vein present for over 40 years. He injured his left medial thigh resulting in this clinical picture. Past medical history is significant for deep vein thrombosis (DVT) at age 57 associated with a pulmonary embolus, treated with anticoagulation for 6 months.

CHARACTERISTICS

- Superficial thrombophlebitis (ST) is a relatively common disorder which most often occurs in the veins of the lower extremities. Uncommonly, this may occur on the arms or other areas as well
- ST may affect both varicose and non-varicose veins, with the great saphenous vein and its tributaries being the most commonly involved veins
- Clinical presentation of patients with superficial thrombophlebitis is pain, erythema, and swelling overlying a superficial vein which when palpated feels firm (cord-like) and tender to the patient
- Predisposing risk factors include the presence of varicose veins, age >60 years, obesity, tobacco use, and prior history of either DVT or ST
- Estimates of the percentage of patients who have concomitant DVT or pulmonary embolism (PE) vary between 6% and 53% and symptomatic PE has been reported in 0% to 10% of patients. As a result, all patients with ST should undergo venous duplex ultrasound examinations to ensure no deep venous system involvement and assess the extent of superficial thrombosis

TREATMENT

Conservative treatment consists of compression stockings, ambulation, warm compresses, and nonsteroidal anti-inflammatory agents. Resolution of symptoms can be expected in 10 to 14 days. When conservative therapy fails or there is evidence of propagation of thrombophlebitis, anticoagulation is indicated. When the thrombosis involves either the saphenofemoral or saphenopopliteal junction without extension into the deep vein, saphenous vein ligation and anticoagulation are reasonable treatment strategies. Standard anticoagulation is warranted in cases if propagation occurs into the deep venous system.

REFERENCES

Decousus H, Epinat M, Guillot K, Quenet S, Boissier C, Trady B. Superficial vein thrombosis: risk factors, diagnosis, and treatment. Curr Opin Pulm Med. 2003; 9:393–7.

Decousus H, Quere I, Presles E, Leizorovicz A, for the POST (Prospective Observational Superficial Thrombophlebitis) Study Group. Ann Intern Med. 2010; 152:218–24.

Atlas of Clinical Vascular Medicine, First Edition. Edited by Jessica Mintz, Bruce L. Mintz, and Michael R. Jaff.

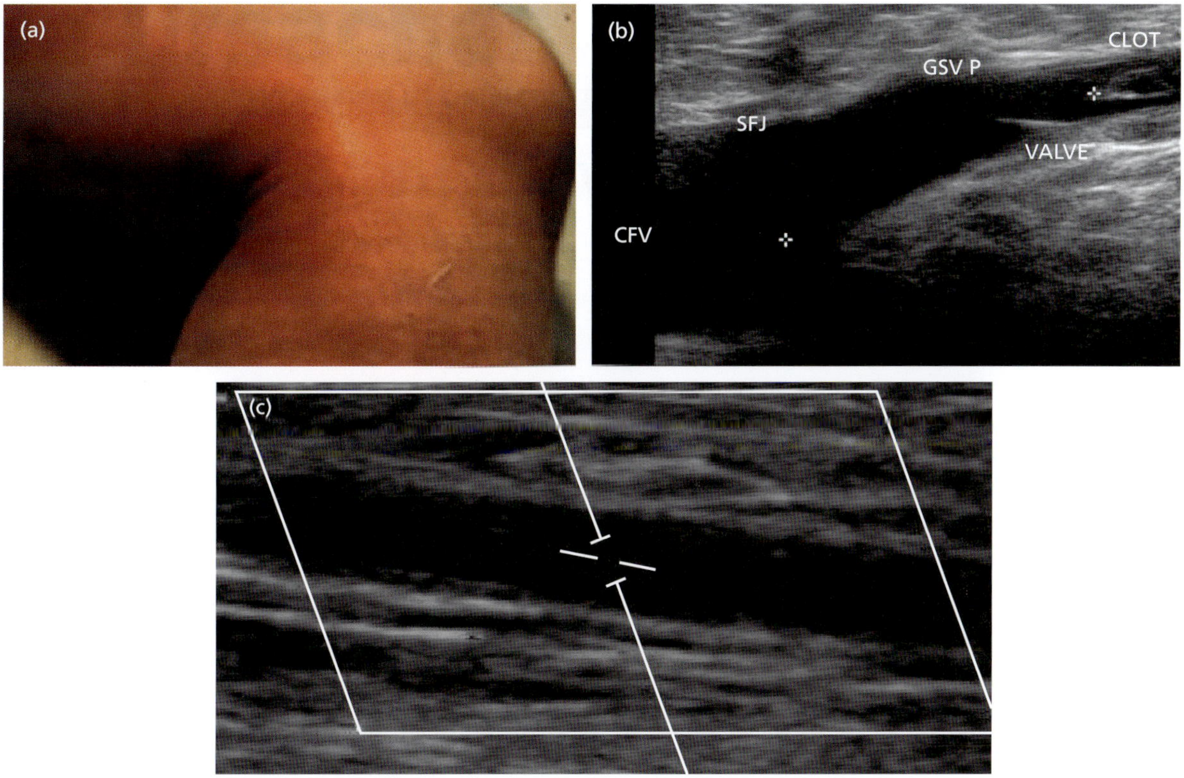

Superficial thrombophlebitis. (a) Left lower extremity erythema and swelling with superficial thrombophlebitis. (b) Venous duplex ultrasonography demonstrating an acute superficial venous thrombosis of the proximal great saphenous vein approximately 2 cm from the saphenofemoral junction. Note that the common femoral vein remains patent. (c) The great saphenous vein with lack of compression and a hypoechoic appearance is suggestive of an acute superficial venous thrombosis.

Spontaneous Venous Hemorrhage (Benign Blue Finger Syndrome)

Bruce L. Mintz
Morristown Medical Center, Morristown, NJ, USA

BRIEF CASE PRESENTATION

This 43-year old white female with no previous medical history presents with a 3 week history of a "black finger" involving the third digit of her right hand. Onset was sudden and painless.

CHARACTERISTICS

- Spontaneous venous hemorrhage, as the name implies, appear spontaneously without trauma or an underlying medical condition
- This is a benign condition most commonly seen in women
- The fingertip is classically spared of the discoloration
- Spontaneous resolution is seen over days in variable phases of blood resorption, although this entity can recur
- Allen's test is normal and capillary filling is unremarkable
- The diagnosis of spontaneous venous bleeding (benign blue finger syndrome) requires both high clinical suspicion along with a thorough history and physical examination to exclude other causes of digital discoloration such as thromboembolic disease and systemic autoimmune diseases

TREATMENT

No treatment is necessary, although clinician follow up to ensure resolution is important. Avoidance of unnecessary imaging, lab tests and anticoagulation is key in this clinical entity.

REFERENCES

Cowen R, Richards T, Dharmadasa A, Handa A, Perkins JMT. The acute blue finger: management and outcome. Ann R Coll Surg Engl. 2008; 90(7):557–60.

Robertson A, Liddington MI, Kay SP. Paroxysmal finger haematomas (Achenbach's syndrome) with angiographic abnormalities. J Hand Surg [Br]. 2002; 27:391–3.

Weinberg I, Jaff MJ. Spontaneous blue finger syndrome: a benign process. Am J Med. 2012; 125(1):e1–2.

Atlas of Clinical Vascular Medicine, First Edition. Edited by Jessica Mintz, Bruce L. Mintz, and Michael R. Jaff.

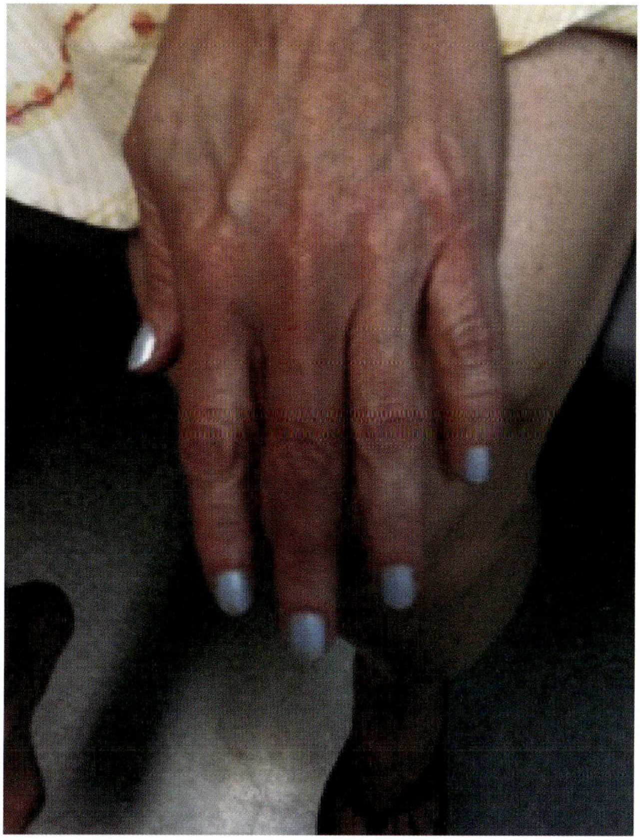

Spontaneous venous hemorrhage. Discoloration of the proximal phalanx of the left third digit.

Venous Aneurysm

Bruce L. Mintz
Morristown Medical Center, Morristown, NJ, USA

BRIEF CASE PRESENTATION

This 45-year-old male with no significant past medical history presents with profound shortness of breath and subsequent hemoptysis. A nuclear ventilation perfusion lung scan revealed multiple mismatched defects in a lobar distribution and systemic anticoagulation was started. He became hemodynamically unstable and a transthoracic echocardiogram demonstrated right ventricular dilatation. Systemic thrombolytic therapy was begun with marked hemodynamic improvement. Twenty-four hours later, venous duplex ultrasonography showed a thrombus free venous aneurysm in the popliteal space. A subsequent venogram clearly outlines the aneurysm.

CHARACTERISTICS

- Venous aneurysms are unusual vascular defects that are seen at any age, with no predilection for gender. By definition, venous aneurysms must be an area of venous dilatation not within a varicose vein, with no concomitant pseudoaneurysm or other arterial malformation. These lesions are often misdiagnosed as soft tissue masses or hernias
- The incidence of venous aneurysms is somewhat unclear, but it appears to be less than 0.5%
- Commonly, patients present with a throbbing mass in the distribution of the pathologic venous segment. However, patients can remain asymptomatic until an embolic event occurs. Venous aneurysms may initially present as a deep vein thrombosis or pulmonary embolism making incidental detection of this entity with duplex ultrasonography helpful in diagnosis and prevention of morbidity related to this condition
- Duplex ultrasonography not only defines the aneurysm origin and extent, but also delivers critical information about concurrent and confounding diagnoses such as popliteal cystic disease and other vascular abnormalities

TREATMENT

The management of venous aneurysm is predominantly surgical.

REFERENCES

Berggvist D, Bjorck M, Ljungman C. Popliteal venous aneurysm-a systematic review. World J Surg. 2006; 30(3):273–9.

Gillespie DL, Villavicencio JL, Gallagher C, et al. Presentation and management of venous aneurysms. J Vasc Surg. 1997; 26(5):845–52.

Hobson RW, Mintz BL, Jamil Z, Breitbart GB. Diagnosis of acute deep venous thrombosis. Surg Clin N Am. 1990; 70:143–57.

Atlas of Clinical Vascular Medicine, First Edition. Edited by Jessica Mintz, Bruce L. Mintz, and Michael R. Jaff.

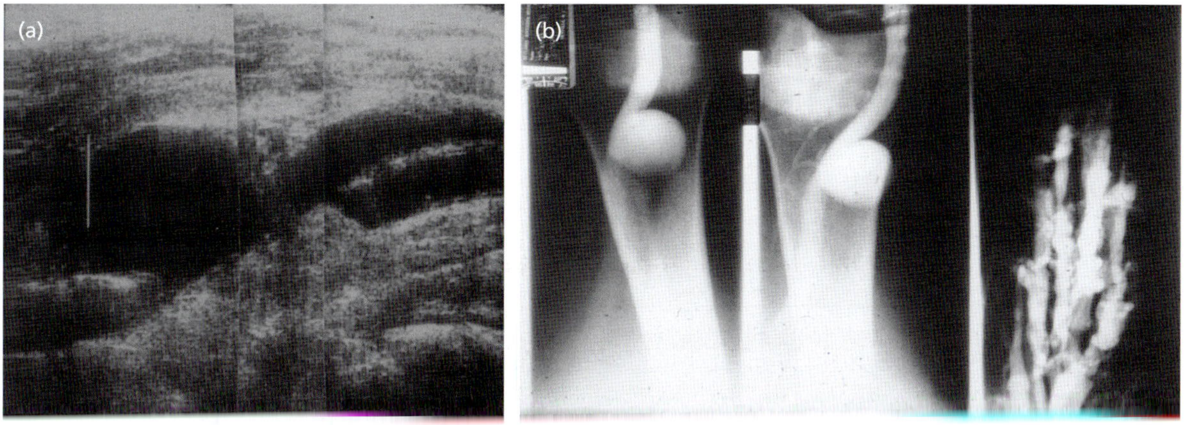

Venous aneurysm. (a) Duplex ultrasonography of a venous aneurysm. (b) Venography of a venous aneurysm. Reproduced from Hobson et al. Surg Clin N Am. 1990;70:143–57, with permission from Elsevier.

Lymphedema

Bruce L. Mintz
Morristown Medical Center, Morristown, NJ, USA

> **BRIEF CASE PRESENTATION**
> A 32-year-old women presents with unilateral left lower extremity swelling which first became manifest at menarche.

CHARACTERISTICS

- Lymphedema is defined as fluid accumulation secondary to impaired lymphatic drainage
- Lymphedema can be characterized as primary or secondary
- Primary lymphedema can be present at birth (congenital or hereditary lymphedema) or appear later in life. When not apparent at birth, it most commonly occurs at the time of menarche and is referred to as lymphedema praecox. Occasionally, true primary lymphedema can occur after age 35 and is then referred to as lymphedema tarde
- Secondary lymphedema may occur as a consequence of trauma, infections, radiation, or malignant invasion of the lymphatics. Secondary lymphedema should be considered in an abrupt onset of unilateral swelling with the absence of clear inflammation in a patient over the age of 40
- Characteristically, lymphedema patients exhibit none of the features of venous insufficiency evidenced by stasis pigmentation or venous varicosities and their swelling is typically painless in nature. On physical exam lymphedema is more resistant to pitting than edema secondary to venous insufficiency or volume overload states. One can often distinguish lymphedema from other causes of lower extremity swelling by examining the feet. Lymphedema often exhibits square toes and a positive Stemmers sign (inability to grasp the skin of the second toe)
- Lymphedema is a chronic condition that if left untreated often will be complicated by recurrent lymphangitis or cellulitis. In rare cases, lymphangiosarcoma can complicate long standing lymphedema

TREATMENT

Lymphedema is a chronic condition requiring lifelong management. The combination of physical therapy including manual lymphatic drainage, exercise, non-elastic wrappings and compression garments should be considered the standard of care. Intermittent pneumatic compression can be used as an adjuvant for excessive fluid removal. As warm serous fluid is a welcoming nidus of infection, aggressive management of any secondary bacterial disease should be undertaken to avert any potential further lymphatic compromise.

REFERENCES

Szuba A, Rockson SG. Lymphedema: classification, diagnosis and therapy. Vasc Med. 1998; 3:145–56.

Tiwari A, Cheng KS, Button M, Myint F, Hamilton G. Differential diagnosis, investigation, and current treatment of lower limb lymphedema. Arch Aurg. 2003; 138(2):152–61.

Warren AG, Brorson H, Borud LJ, Slavin SA. Lymphedema: a comprehensive review. Ann Plast Surg. 2007; 59:464–72.

Atlas of Clinical Vascular Medicine, First Edition. Edited by Jessica Mintz, Bruce L. Mintz, and Michael R. Jaff.

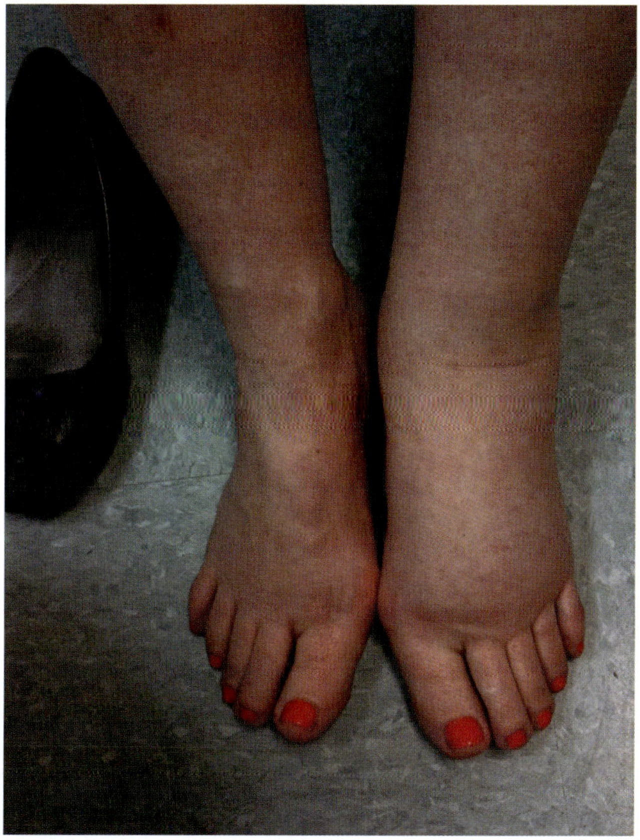

Lymphedema. Left lower extremity edema due to disruption of the lymphatic channels. Courtesy of Michael R Jaff.

Milroy's Disease

Jessica Mintz[1], John R. Bartholomew[1], Steven M. Dean[2] and Heather L. Gornik[1]

[1]Cleveland Clinic Heart and Vascular Institute and Cleveland Clinic Lerner College of Medicine, Case Western Reserve University, Cleveland, OH, USA

[2]Department of Cardiovascular Medicine, The Ohio State University, Columbus, OH, USA

BRIEF CASE PRESENTATION

Patient presents with lifelong history of bilateral edema of upper and lower extremities.

CHARACTERISTICS

- A congenital hereditary lymphedema with onset at birth
- Occurring most commonly in females, it is due to congenital malformation in the lymphatic system
- Painless, non-pitting edema in one or multiple extremities involving the dorsum of the hands and feet
- Associated with mutations in the *FLT4* gene, encoding vascular endothelial growth factor receptor-3
- Primarily autosomal dominant inheritance pattern with environmental contributors, but cases of recessive inheritance of mutation in the *FLT4* gene have been reported

TREATMENT

Manual lymphatic drainage, complete decongestive physiotherapy and compression garments are mainstays of treatment.

REFERENCES

Brice G, Child AH, Evans A, et al. Milroy disease and the VEGFR-3 mutation phenotype. J Med Genet. 2005; 42:98–10.

Butler MG, Dagenais SL, Rockson SG, Glover TW. A novel VEGFR3 mutation caused Milroy disease. Am J Med Genet Part A. 2007; 143(11):1212–7.

Milroy WF. An underdescribed variety of hereditary oedema. N York Med J. 1892; 56:505–8.

Atlas of Clinical Vascular Medicine, First Edition. Edited by Jessica Mintz, Bruce L. Mintz, and Michael R. Jaff.

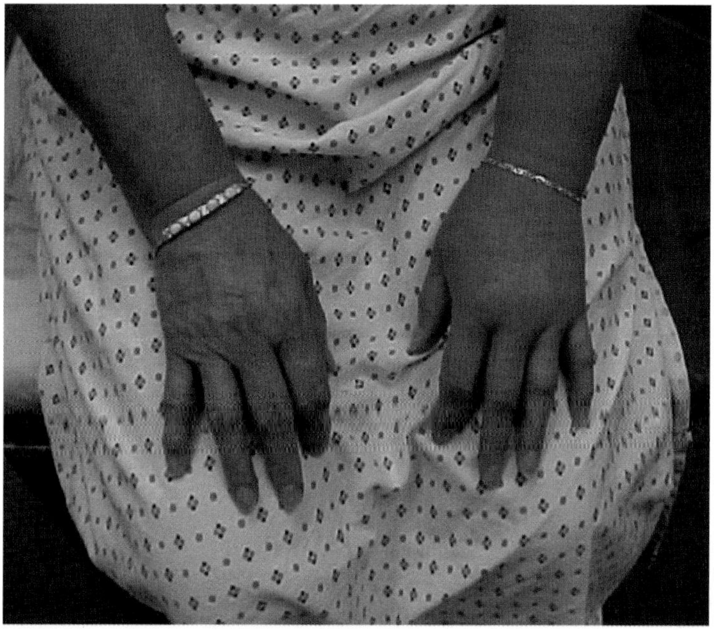

Milroy's disease. Upper extremity lymphedema. Image courtesy of the Cleveland Clinic Foundation Department of Vascular Medicine archives.

Filariasis

Jessica Mintz, John R. Bartholomew and Heather L. Gornik

Cleveland Clinic Heart and Vascular Institute and Cleveland Clinic Lerner College of Medicine, Case Western Reserve University, Cleveland, OH, USA

BRIEF CASE PRESENTATION

This man has severe swelling of the right lower extremity with classic involvement of the foot and toes, consistent with lymphedema. There is also marked swelling of the scrotum. The left leg is spared.

CHARACTERISTICS

- Lymphedema with thickening of the skin characteristically in the lower extremities, including involvement of the feet and toes
- Lymphedema is due to infection which occurs by transmission of the filarial parasite via mosquitoes
- Parasites are deposited in the skin, migrate to the lymphatic vessels and develop into adult worms. Inflammatory response to the worms leads to lymphatic disruption
- Chronic lymphatic obstruction can involve the genital region leading to hydrocele with subsequent scrotal enlargement as seen in this patient
- The most common cause of lymphedema in the developing world

TREATMENT

Anti-helminth therapies including albendazole, ivermectin and diethylcarbamazine are considered the treatment of choice. Once lymphedema is established, graduated elastic compression stockings and/or compressive wraps, leg elevation and routine exercise of the effected limb is key. If available, manual lymphatic drainage and complete decongestive physiotherapy can significantly reduce limb swelling.

REFERENCES

Anitha K, Shenoy RK. Treatment of lymphatic filariasis: Current trends. Ind J Dermatol Venereol Leprol. 2001; 67(2):60–5.

Dreyer G, Addiss D, Gadelha P, Lapa E, Williamson J, Dreyer A. Interdigital skin lesion of the lower limbs among patients with lymphoedema in an area endemic for bancroftian filariasis. Trop Med Int Health. 2006; 11(9):1475–81.

Taylor MJ, Hoerauf A, Bockarie M. Lymphatic filariasis and onchocerciasis. Lancet, 2010; 376(9747):1175–85.

Atlas of Clinical Vascular Medicine, First Edition. Edited by Jessica Mintz, Bruce L. Mintz, and Michael R. Jaff.

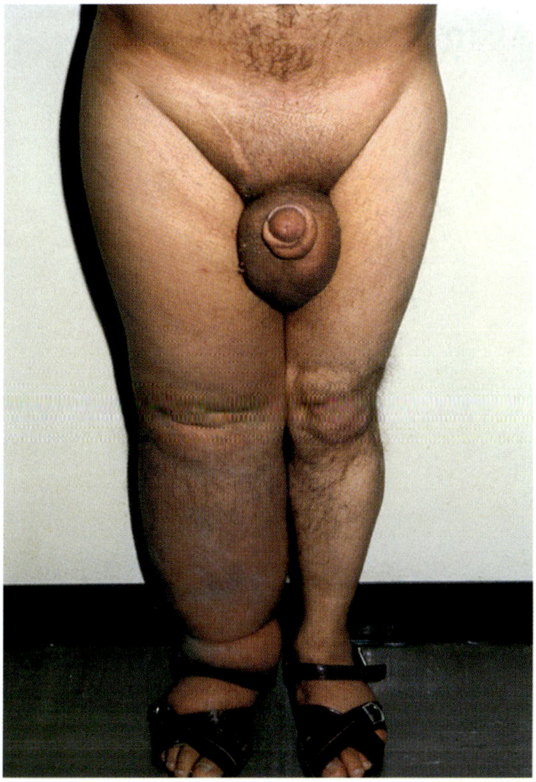

Filariasis. Left leg lymphedema extending from the foot to the groin. Note the associated scrotal edema.

Yellow Nail Syndrome

Bruce L. Mintz
Morristown Medical Center, Morristown, NJ, USA

BRIEF CASE PRESENTATION

A 71-year-old white male presented for an evaluation of shortness of breath. He was noted to have slightly labored breathing, bilateral lower extremity edema, and discolored, malformed nails. His transthoracic echocardiogram was unremarkable; however, a chest radiograph was notable for a pleural effusion.

CHARACTERISTICS

- Yellow nail syndrome classically consists of a triad of dystrophic yellow nails, pitting and non-pitting lymphedema, and respiratory tract illness. It may be associated with a chylous pleural effusion. Pleural effusions are usually a late manifestation and rarely regress spontaneously
- Occurs slightly more commonly in women than men. The onset of symptoms varies from as early as birth to age 60, with the median age of 40
- Symptoms seldom appear simultaneously and the full triad is not always present
- The etiology of this condition results from an irregularity of the lymphatic vessel architecture resulting in impaired drainage and lymphatic congestion
- Lymphangiography reveals patterned hypoplastic or dilated lymphatics consistent with incomplete development

TREATMENT

Therapy is generally symptomatic. Therapeutic thoracentesis has been utilized, however effusions typically recur in the following few days to months. Chemical pleurodesis and pleurectomy may prevent fluid reaccumulation in some patients. Spontaneous partial or complete resolution of nail abnormalities occurs in 30% of patients.

REFERENCES

Beer DJ, Pereira W Jr, Snider GL. Pleural effusion associated with primary lymphedema: a perspective on the yellow nail syndrome. Am Rev Respir Dis. 1978; 117(3):595.

Hiller E, Rosenow EC 3rd, Olsen AM. Pulmonary manifestations of the yellow nail syndrome. Chest. 1972; 61(5):452.

Samman PD, White WF. The "yellow nail" syndrome. Br J Dermatol. 1964; 76:153–7.

Atlas of Clinical Vascular Medicine, First Edition. Edited by Jessica Mintz, Bruce L. Mintz, and Michael R. Jaff.

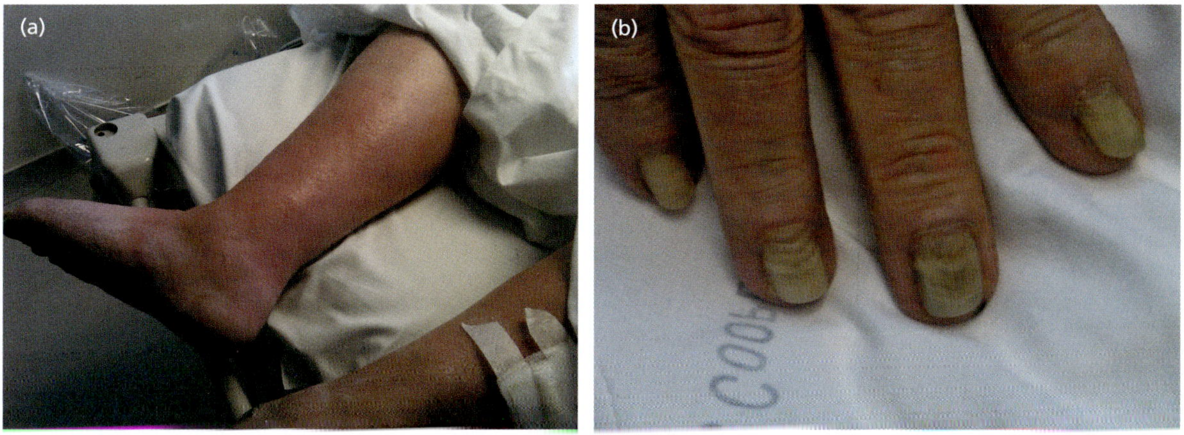

Yellow nail syndrome. (a) Lower extremity lymphedema associated with yellow nail syndrome. (b) The yellow hue of fingernails in a patient with yellow nail syndrome.

Elephantiasis Nostras Verrucosa

Stephen L. Chastain and Bruce H. Gray
Greenville Hospital System, Greenville, SC, USA

BRIEF CASE PRESENTATION

A 73-year-old obese woman presented complaining of skin infection over the left lower leg. Findings of severe chronic lymphedema with marked squaring of the toes and positive Stemmer's sign were noted bilaterally.

CHARACTERISTICS

- Elephantiasis nostras verrucosa (ENV) is a result of untreated chronic nonfilarial lymphedema. This subtype of acquired elephantiasis is seen outside the tropics as the limb becomes enlarged and then disfigured. Obesity and concurrent chronic venous insufficiency (CVI) have been reported in a majority of patients
- ENV is characterized by extensive fibrosis of the dermis and subcutaneous tissue associated with non-pitting edema that does not resolve on elevation. Hyperkeratosis of the dermis with nodule formation results in a cobblestone appearance. With time, the skin develops papillomatous lesions or verrucous areas described as "mossy" in texture. The lesions become malodorous due to chronic colonization with bacteria and fungi
- In a recent review of 21 cases of ENV, all patients were noted to be obese with a mean body mass index over 50. Distribution of ENV was predominantly bilateral and involved the calves. Concurrent CVI was noted in 71% of patients indicating this as a possible risk factor in the genesis of ENV. Lipodermatosclerosis, a sequela of CVI resulting in dermal and subcutaneous fibrosis, was found in 43% of patients with ENV. There was also noted to be an association between soft-tissue infection and ENV

TREATMENT

Current management includes techniques for reduction of lymphostasis and treatment of recurrent bacterial and fungal infections. Anecdotal reports using oral or topical retinoids offer a promising but expensive option for future treatment. Surgical debridement has been reported but is rarely performed.

REFERENCES

Sisto K, Khachemoune A. Elephantiasis nostras verrucosa. Am J Clin Dermatol. 2008; 9(3):131–46.

Dean SM, Zirwas MJ, Vander Horst A. Elephantiasis nostras verrucosa: an institutional analysis of 21 cases. J Am Acad Dermatol. 2011; 64(6):1104–10.

Atlas of Clinical Vascular Medicine, First Edition. Edited by Jessica Mintz, Bruce L. Mintz, and Michael R. Jaff.

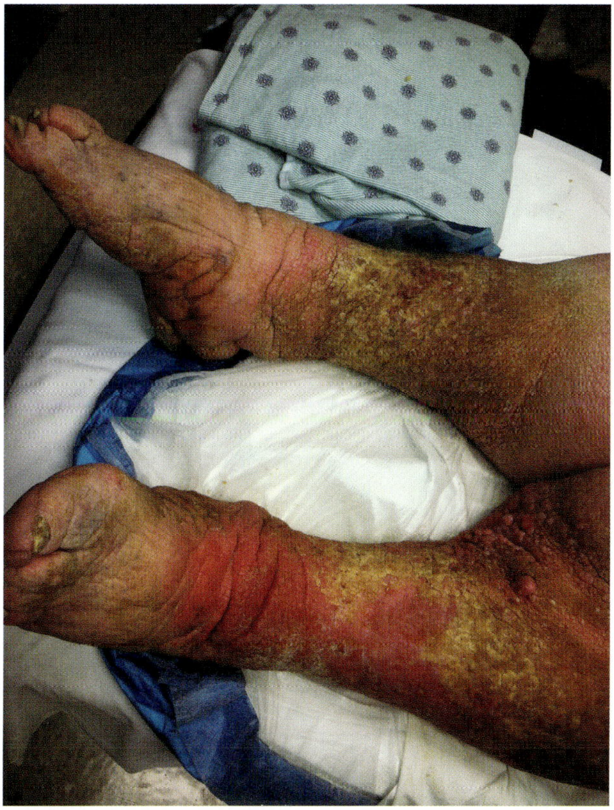

Elephantiasis nostras verrucosa. Bilateral skin changes including hyperkeratosis and dermal fibrosis with crowded wart like growths in the setting of chronic lymphedema.

Lipedema

Bruce L. Mintz
Morristown Medical Center, Morristown, NJ, USA

BRIEF CASE PRESENTATION

This 32-year-old female presents with gradually progressive dysmorphic leg swelling which spares the dorsum of her feet. It is associated with increasing pain without typical stasis skin changes.

CHARACTERISTICS

- Lipedema is symmetrical bilateral swelling of the lower extremities extending from hips to ankles classically sparing the dorsum of the feet, appearing like a band above the ankles
- Lipedema is encountered almost exclusively in women regardless of body habitus. It is most commonly activated by puberty, pregnancy or menopause
- When present in a male patient it is important to investigate hormonal abnormalities such as Cushing's disease or growth hormone abnormalities
- Lipedema predominantly occurs in the hips and lower legs and weight loss is ineffective as treatment for the fat deposition
- Lipedema may be inherited and involves a surplus of localized fatty deposition. Distinct from other fatty deposition, it cannot be altered by diet
- Chronic fatty redistribution may be associated with secondary lymphedema

TREATMENT

Early diagnosis may moderate progression of fat deposition with a combination of compression garments, and lymphatic massage. Nutritional support to avoid weight gain can be helpful. Surgical intervention and liposuction remain controversial and in many circumstances exacerbate the condition.

REFERENCES

Allen EV, Hines EA. Lipedema of the legs: A syndrome characterized by fat legs and orthostatic edema. Mayo Clin Proc. 1940; 15:184–7.

Child AH, Gordon KD, Sharpe P, et al. Lipedema: An inherited condition. AJMG. 2010; 152(4):970–6.

Fonder MA, Loveless JW, Lazarus GS. Lipedema, a frequently unrecognized problem. J Am Acad Dermatol. 2007; 57:S1–S3.

Atlas of Clinical Vascular Medicine, First Edition. Edited by Jessica Mintz, Bruce L. Mintz, and Michael R. Jaff.

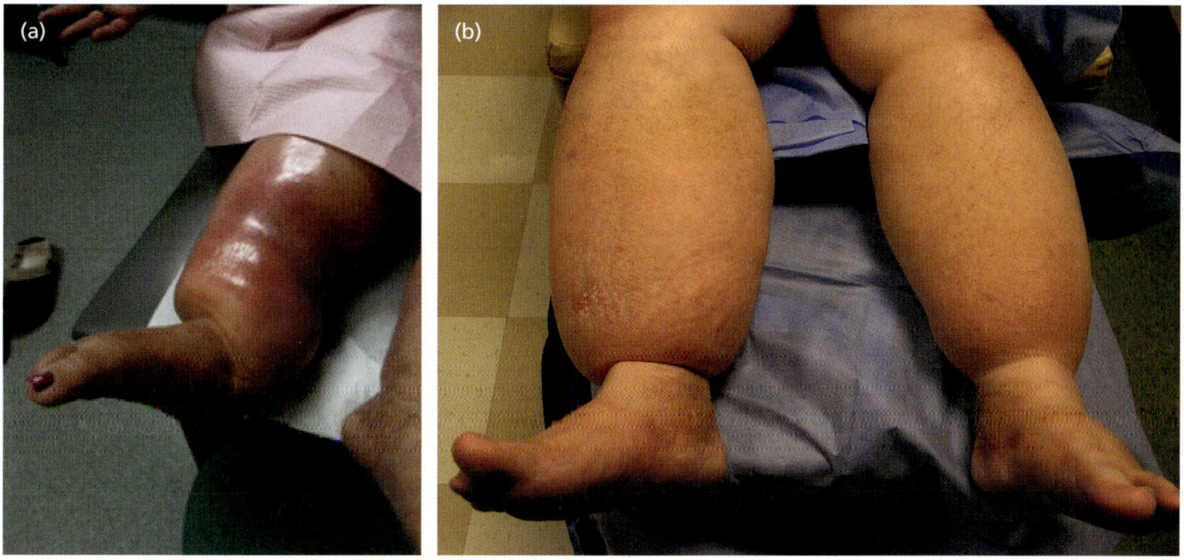

Lipedema. (a, b) Symmetric, bilateral, lower extremity edema sparing the dorsum of the feet. Part (b) courtesy of Michael R Jaff.

Pretibial Myxedema (Thyroid Dermopathy)

Steven M. Dean

Department of Cardiovascular Medicine, Wexner Medical Center of the Ohio State University, Columbus, OH, USA

BRIEF CASE PRESENTATION

A 52-year-old female with a history of Graves' disease and associated ophthalmopathy presents with bilateral lower extremity acral lymphedema with copious, well-demarcated reddish-orange non-pitting edema along the anterolateral calves.

CHARACTERISTICS

- Pretibial myxedema complicates 0.5–4% of Graves' patients, most commonly in females
- The distribution of the edema is pretibial often involving the toes, although rarely it can occur in other sites (<1%)
- 90% of pretibial myxedema occurs in hyperthyroidism with the remaining 10% hypothyroid/euthyroid patients
- Attendant ophthalmopathy (99%) appears first; dermopathy appears >1 year later (12–24 months)
- There are three distinct clinical variants:
  1. Non-pitting edema with hyperpigmentation (pink, yellow, orange, brownish-red) – 45% (as illustrated in photograph)
  2. Plaques – separate or confluent – 30%
  3. Nodules – 20%

TREATMENT

No treatment may be needed as the disorder is often asymptomatic and mostly cosmetic. When therapy is needed, topical corticosteroids [sometimes under occlusion] are considered the standard of care, with compressive therapy (gradient compression stockings/lymphatic pumps/manual lymphatic drainage) as a useful therapeutic adjunct. Systemic immunomodulation: oral corticosteroids; plasmapheresis; intravenous immunoglobulin as well as octreotide; pentoxifylline have been attempted with some success.

REFERENCES

Fatourechi V. Pretibial myxedema: pathophysiology and treatment options. Am J Clin Dermatol. 2005; 6(5):295–309.

Schwartz KM, Fatourechi V, Ahmed DD, Pond GR. Dermopathy of Graves' disease (pretibial myxedema): long-term outcome. J Clin Endocrinol Metab. 2002; 87(2):438–46.

Weetman A. Graves disease. NEJM. 2000; 343:1236–48.

Atlas of Clinical Vascular Medicine, First Edition. Edited by Jessica Mintz, Bruce L. Mintz, and Michael R. Jaff.

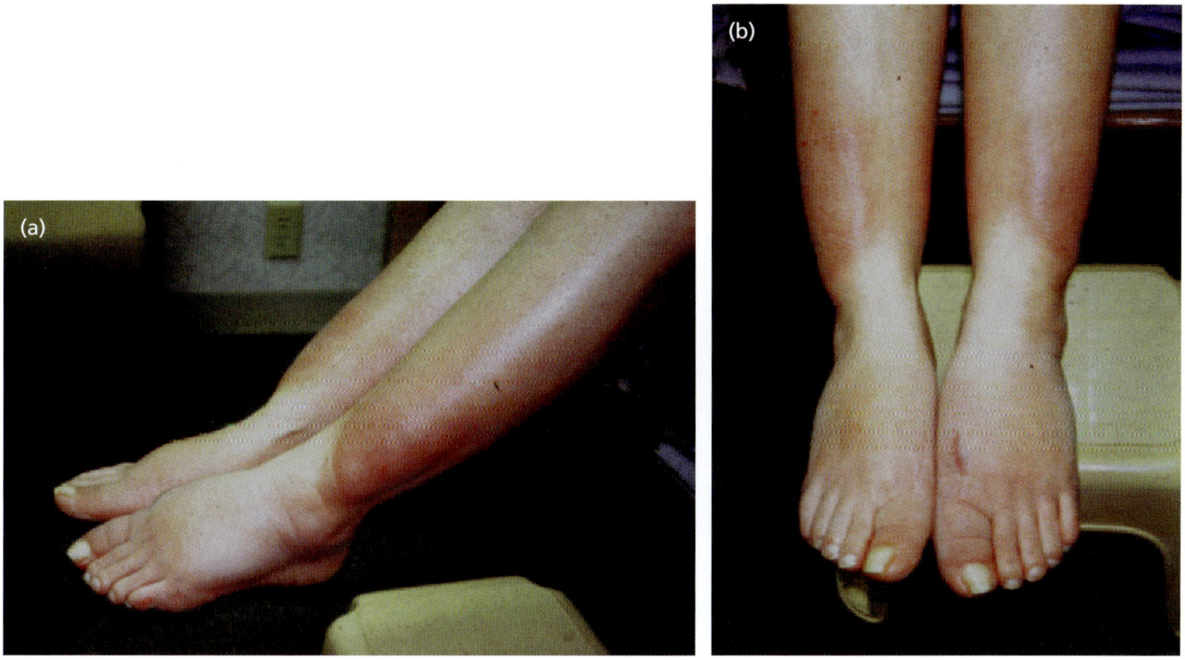

Pretibial myxedema. Bilateral, lower extremity, non-pitting discolored lymphedema along the anterolateral calves as well as marked swelling along the dorsal feet and toes.

Popliteal Cyst (Baker's Cyst)

Bruce L. Mintz
Morristown Medical Center, Morristown, NJ, USA

BRIEF CASE PRESENTATION

This 60-year-old female with a history of degenerative arthritis and bilateral knee pain presents with a progressive swelling of her right knee followed by a sudden tearing feeling and subsequent swelling of her calf.

CHARACTERISTICS

- Popliteal cysts consist of a synovial fluid filled mass in the popliteal fossa. The presence of popliteal cysts usually denotes increased intra-articular pressure due to underlying joint disease
- The vast majority of popliteal cysts in adults are secondary to meniscal tears, anterior cruciate ligament insufficiency, and degenerative cartilage. In contrast, the majority of popliteal cysts in children are primary and usually do not communicate with the knee itself
- Complications of popliteal cysts include cyst rupture into the posterior calf muscles and pseudothrombophlebitis. Pseudo-thrombophlebitis consists of edema exclusively below the knee and the appearance of a small hematoma descending inferiorly down to the medial malleolus. This clinical finding is known as the "crescent sign" or "meniscus sign"
- Diagnosis is commonly made by physical examination. Visualization of the cyst is enhanced with knees fully extended with the patient standing or when viewed from behind with the knee partially flexed

TREATMENT

Treatment is generally intended to address the underlying etiology. Conservative treatment is commonly preferred, however ultrasound guided aspiration may be needed in some more symptomatic cases. Surgical excision may be performed in rare, recalcitrant cases but even then recidivism is not uncommon.

REFERENCES

Baker WM. The formation of abnormal synovial cysts in connection with the joints. II. Saint Bartholomew's Hospital Reports, London. 1885; 21:177–90.

Fritschy D, Fasel J, Imbert JC, Bianchi S, Verdonk R, Wirth CJ. The popliteal cyst. Knee Surg Sports Traumatol Arthrosc. 2006; 14(7):623–8.

Kraag G, Thevathasan EM, Gordon DA, Walker IH. The hemorrhagic crescent sign of acute synovial rupture. Ann Intern Med. 1976; 85(4):477–8.

Atlas of Clinical Vascular Medicine, First Edition. Edited by Jessica Mintz, Bruce L. Mintz, and Michael R. Jaff.

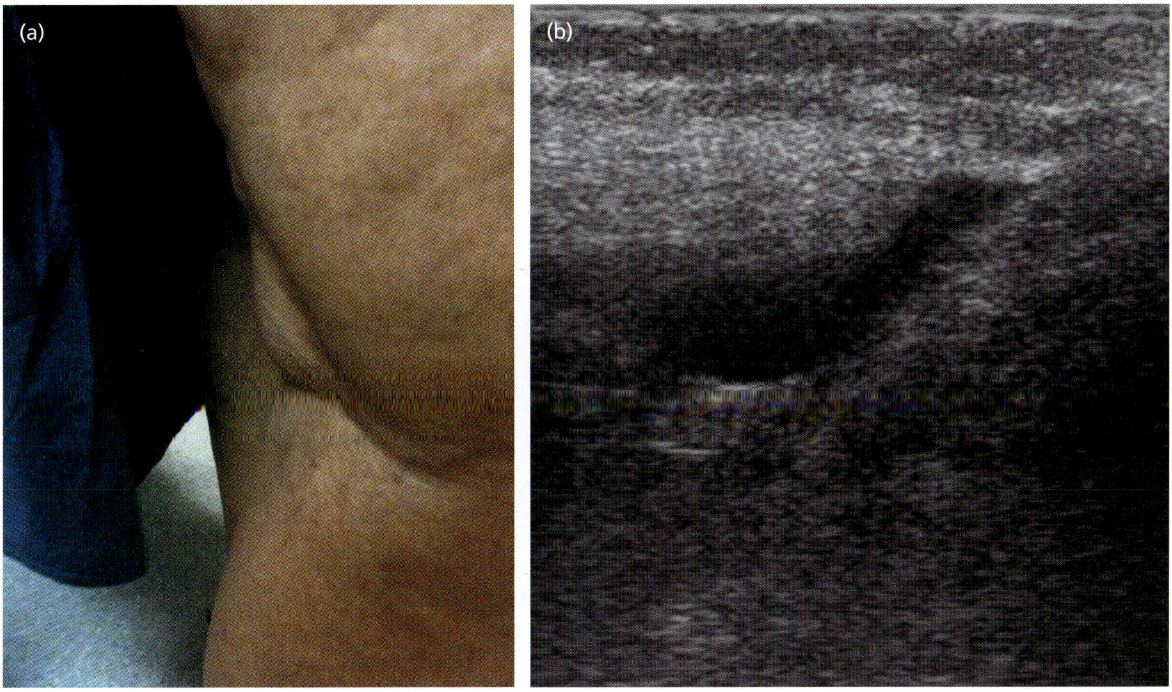

Baker's cyst. (a) Right-sided knee swelling with a mass shown in the popliteal fossa. (b) Duplex ultrasonography demonstrating a fluid filled sac in a non-vascular distribution.

Bacterial Cellulitis

Jessica Mintz, John R. Bartholomew and Heather L. Gornik

Cleveland Clinic Heart and Vascular Institute and Cleveland Clinic Lerner College of Medicine, Case Western Reserve University, Cleveland, OH, USA

BRIEF CASE PRESENTATION

This patient presents with an erythematous, warm and swollen leg with areas of normal skin. A demarcating margin of erythema is noted above the ankle. This patient also has evidence of concomitant chronic venous insufficiency with prominent reticular veins and some varicosities noted at the ankle and foot.

CHARACTERISTICS

- The most common infectious cause of the swollen limb
- Patients present with acute swelling, localized erythema, tenderness, and enlarged regional lymph nodes. The erythema is classically patchy and not confluent
- Frequently associated with systemic signs of infection including fevers, chills, nausea, and rigors
- Can occur chronically through an established or recurrent portal of entry, such as tinea pedis, ulceration, or sites of intravenous drug injection
- Recognition of recurrent cellulitis and eliminating the portal of entry can prevent secondary lymphedema, which occurs due to inflammatory destruction of lymphatic vessels

TREATMENT

Antibiotic therapy with adequate coverage of skin flora organisms as well as recognition and elimination of the portal of entry. Elimination of edema and meticulous skin care are also key in treatment.

REFERENCES

Stevens D, Bisno A, Chamber H, et al. Practice guidelines for the diagnosis and management of skin and soft tissue infections. Clin Infect Dis. 2005; 41(10):1373–406.

Swartz M. Cellulitis. N Engl J Med. 2004; 350:904–12.

Woo PCY, Lum PNL, Wong SSY, Cheng VC, Yen KY. Cellulitis complicating lymphoedema. Eur J Clin Microbiol Infect Dis. 2000; 19:294–7.

Atlas of Clinical Vascular Medicine, First Edition. Edited by Jessica Mintz, Bruce L. Mintz, and Michael R. Jaff.

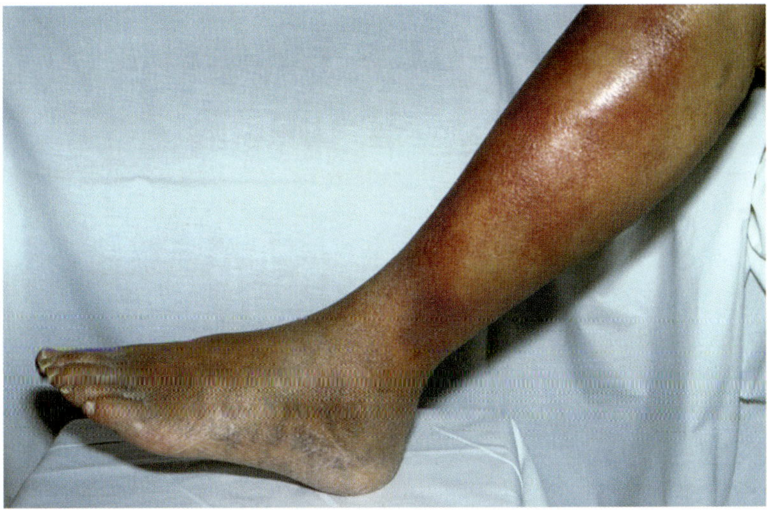

Cellulitis. Left lower extremity with edema and well-demarcated erythema. Image courtesy of the Cleveland Clinic department of Vascular Medicine archives.

Gastrocnemius Muscle Rupture

Bruce L. Mintz
Morristown Medical Center, Morristown, NJ, USA

BRIEF CASE PRESENTATION

This 32-year-old long distance runner presents with sudden onset of a "popping sound" in his calf associated with severe calf pain while running. He noted subsequent ecchymosis hours later.

CHARACTERISTICS

- Gastrocnemius muscle rupture may be partial or complete and can occur secondary to injury or overexertion
- First-degree strain involves less than 5% of the muscle. This causes a mild discomfort without significant sequelae
- Second-degree strain is an incomplete muscle tear. Defect of the muscle causes either a mass or an indentation. Patients with second-degree strains are unable to walk or stand without pain or a limp
- Third-degree rupture is a complete tear across the width of the muscle leading to an inability to contract the muscle. This is associated with a large protuberance under the skin associated with subcutaneous bleeding. Within days following the injury, an area of ecchymosis develops often with a meniscus sign over the medial, plantar-aspect of the foot

TREATMENT

Management depends upon the degree of muscle strain. Conservative management is appropriate for first-degree tears; however, surgical intervention may be necessary for complete disruption. Prompt recognition and avoidance of presumptive diagnosis of deep venous thrombosis is critical, as anticoagulation will exacerbate the clinical scenario.

REFERENCES

van Mechelen W. Running injuries. A review of the epidemiological literature. Sports Med. 1992; 14(5):320–35.

Macera CA, Pate RR, Powell KE, Jackson KL, Kendrick JS, Craven TE. Predicting lower-extremity injuries amoung habitual runners. Arch Intern Med. 1989; 149(11):2565–8.

Hough T. Ergographic studies in muscular soreness. J Boston Soc Med Sci. 1900; 5(3):81–92.

Atlas of Clinical Vascular Medicine, First Edition. Edited by Jessica Mintz, Bruce L. Mintz, and Michael R. Jaff.

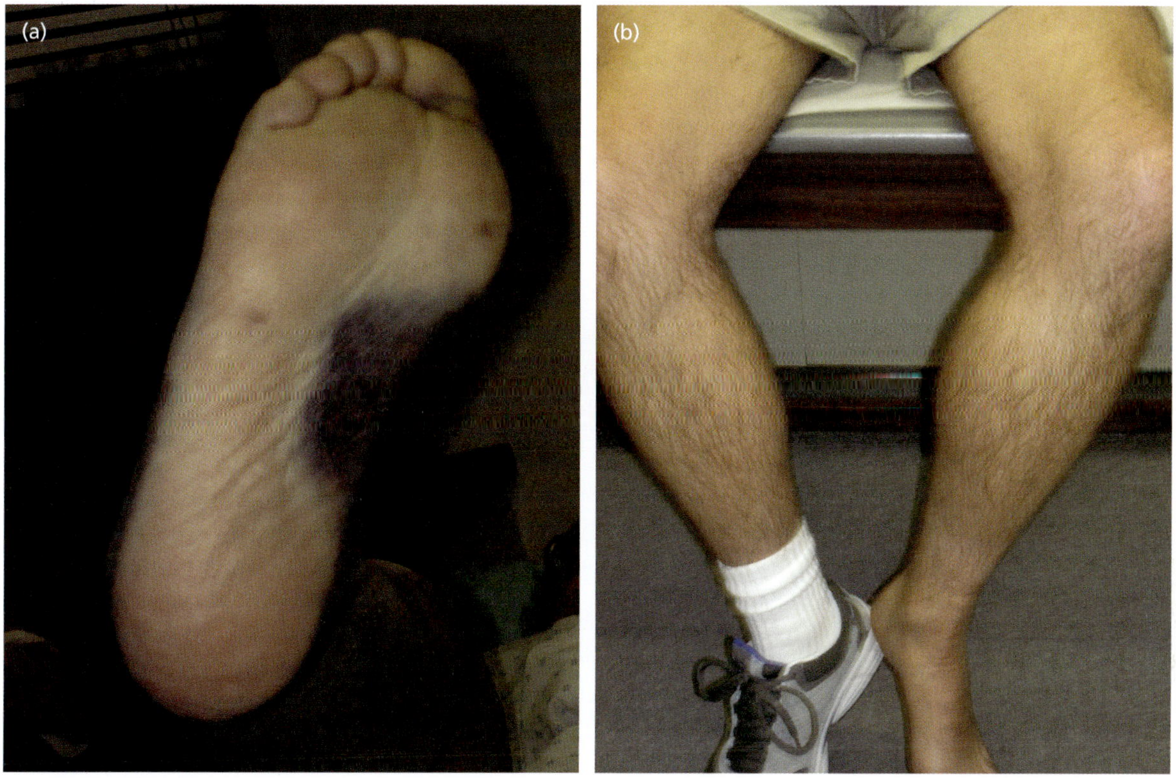

Gastrocnemius rupture. (a) Left medial plantar arch ecchymosis in a meniscus pattern (Scimitar sign). (b) Left leg swelling associated with gastrocnemius rupture.

Venous Stasis Ulcer

Geno J. Merli and Taki Galanis
Jefferson Vascular Center, Thomas Jefferson University Hospitals, Jefferson Medical College, Philadelphia, PA, USA

BRIEF CASE PRESENTATION

This 71-year-old female with a remote history of breast cancer complicated by a post-operative deep vein thrombosis, presents with a recent ulcer on her left medial malleolus.

CHARACTERISTICS

- Venous stasis ulceration is the most serious consequence of chronic venous insufficiency. These ulcers are a consequence of venous valvular dysfunction often due to remote venous thrombosis
- The ulcer is commonly seen on the medial malleolus and may be associated with hyperpigmentation of the surrounding skin, edema and poor healing
- It is important to monitor patients with venous insufficiency for skin changes as stasis ulceration may progress and become severe and recalcitrant to therapy
- Prevalence of venous stasis ulcers ranges from 0.06% to 2% and exceeds 4% in the population over age 75
- Risk factors identified for recurrence of venous stasis ulcers are severe lipodermatosclerosis, deep venous thrombosis, previous venous stasis ulcer history and time since first ulcer episode >2 years

TREATMENT

Compression therapy with multi-layer compression bandaging is the mainstay of acute treatment, while gradient elastic stockings are important for prevention. Education regarding the underlying causes and limb elevation are key in the treatment. Despite optimal therapy, ulcers may take up to twelve weeks to completely heal. Adjuvants for the treatment of venous stasis ulcer may include emollients, absorbents and zinc oxide.

REFERENCES

Abbade L, Lastoria S, Rollo H. Venous ulcer: clinical characteristics and risk factors. Int J Dermatol. 2011; 50:405–11.

Henke P, Kistner B, Wakefield TW, Eklof B, Lurie F. reducing venous stasis ulcers by fifty percent in 10 years: The next steps. J Vasc Surg. 2010; 52:37S–8S.

Vowden K, Vowden P. Preventing venous ulcer recurrence: a review. Int Wound J. 2006; 3:11–21.

Word R. Medical and surgical therapy for advanced chronic venous insufficiency. Surg Clin N Am. 2010; 90:1195–214.

Atlas of Clinical Vascular Medicine, First Edition. Edited by Jessica Mintz, Bruce L. Mintz, and Michael R. Jaff.

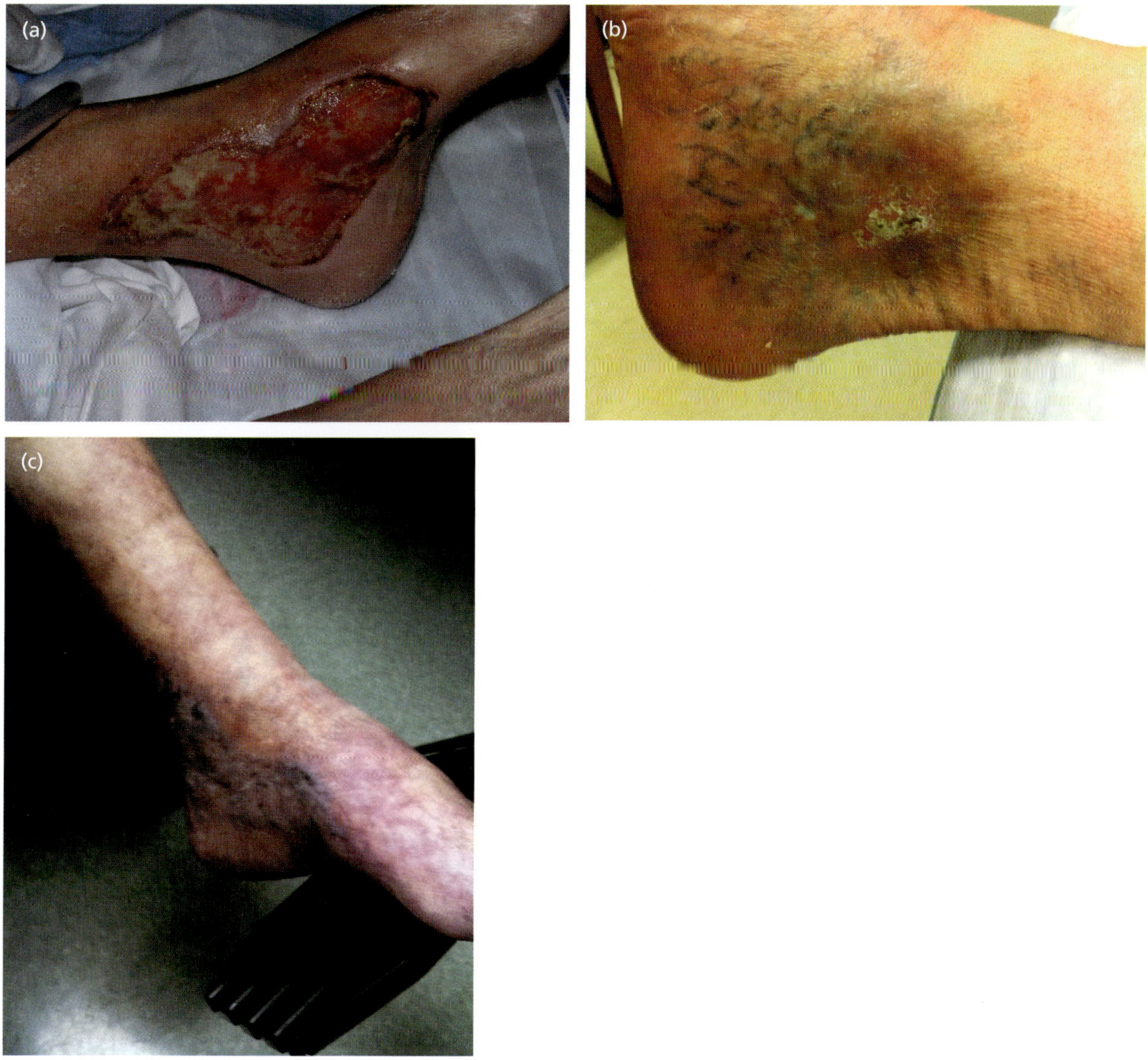

Venous stasis ulcer. (a) Stasis ulcer over the medial malleolus. (b) The early stages of a venous stasis ulcer with associated venous insufficiency. (c) Healing venous stasis ulcer. (Images (a) and (c) are courtesy of Bruce Mintz).

Arteriosclerosis Obliterans Ulcer (Arterial Ulcer)

Ari J. Mintz[1] **and Bruce L. Mintz**[2]
[1]Lahey Clinic Medical Center, Burlington, MA, USA
[2]Morristown Medical Center, Morristwon, NJ, USA

BRIEF CASE PRESENTATION

This 67-year-old female with a 10-year history of diabetes mellitus and lower extremity claudication presents with sudden onset of toe and forefoot pain. The pain was dull throbbing at first; however, it quickly transformed to searing discomfort.

CHARACTERISTICS

- Arterial insufficiency ulcers tend to occur on the lateral aspects of the foot or ankle with a predilection for the distal digits
- The pathophysiology stems from the decrease in perfused tissue which does not allow for repair to damaged skin
- Arteriosclerosis obliterans (ASO) ulcerations are typically very painful and appear punched out with a regular, circumferential border
- Risk factors for the development of ASO ulcerations are age, peripheral artery disease, hypertension, hyperlipidemia, and tobacco use
- Diagnosis is made clinically based on appearance, risk factors and history, as well as objectively with ankle-brachial index and pulse volume recordings
- Definitive diagnosis with angiography should be done if invasive treatment is required to localize the lesions

TREATMENT

Aggressive reduction of risk factors is key in the prophylaxis against ASO ulcerations as well in its treatment. Regular walking programs along with medical management including aspirin, statin therapy and treatment of underlying diabetes and hypertension has been shown to improve outcomes. For non-healing arterial ulcerations in the face of ankle brachial indices less than 0.5, tissue reperfusion may be required.

REFERENCES

Aboyans V, Criqui MH, Denenberg JO, et al. Risk factors for progression of peripheral arterial disease in large and small vessels. Circulation. 2006; 113:2623.

McDermott MM, Liu K, Greenland P, et al. Functional decline in peripheral arterial disease: associations with the ankle brachial index and leg symptoms. JAMA. 2004; 292:453.

Murabito JM, Evans JC, Nieto K, et al. Prevalence and clinical correlates of peripheral arterial disease in the Framingham Offspring Study. Am Heart J. 2002; 143:961.

Atlas of Clinical Vascular Medicine, First Edition. Edited by Jessica Mintz, Bruce L. Mintz, and Michael R. Jaff.

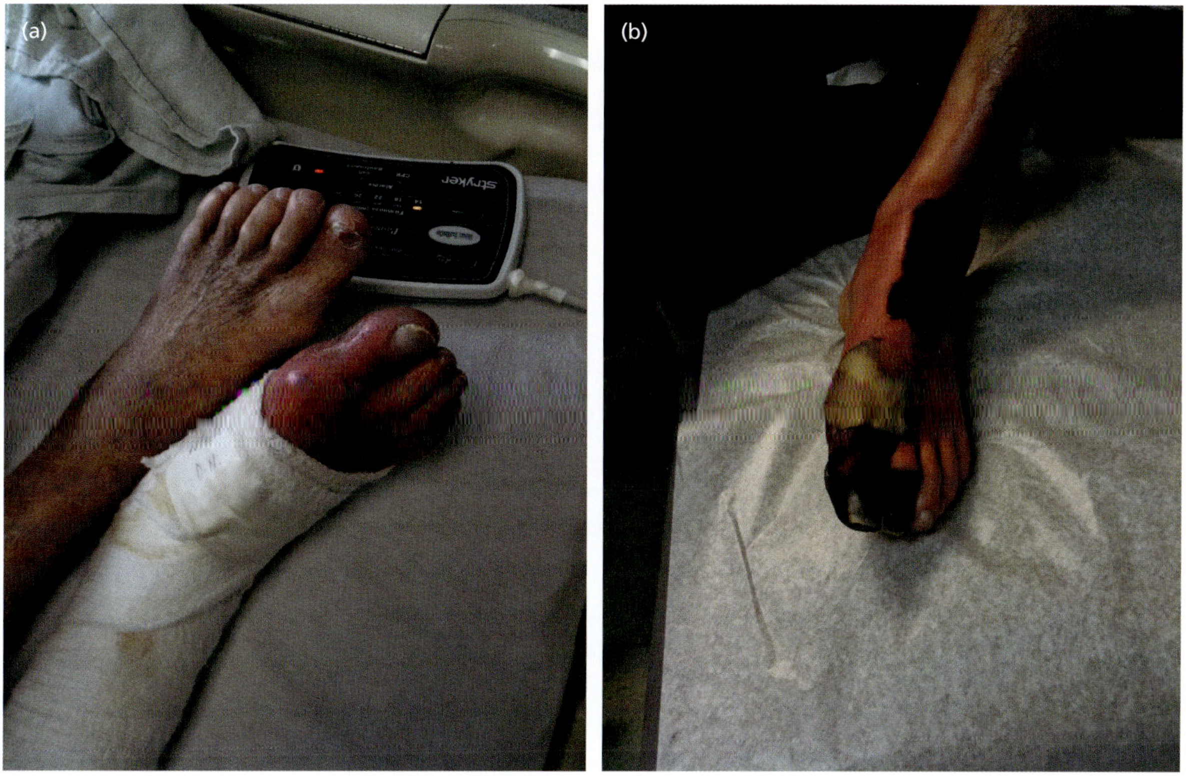

(a) Distal erythema of the right foot when compared to the unaffected left foot. (b) Ischemic digit with a traumatic non-healing ulcer overlying the anterior ankle.

Antiphospholipid Antibody Syndrome-induced Ulceration

Raghu Kolluri
Prairie Vascular Institute, Springfield, IL, USA

BRIEF CASE PRESENTATION

A patient with known antiphospholipid antibody syndrome presents with a painful ulceration with surrounding erythema over the anterior tibial region of her left lower extremity

CHARACTERISTICS

- Antiphospholipid antibody (APLA) syndrome is a hypercoagulable state in association with antiphospholipid antibodies. These antibodies are a heterogenous group, which prolong phospholipid dependent coagulation (lupus anticoagulant and anticardiolipin antibodies). APLA can occur alone (primary APLA) or in association with other autoimmune conditions such as systemic lupus erythematosus
- APLA syndrome consists of arterial and venous thrombosis, as well as small vessel thrombosis and pregnancy associated spontaneous abortion
- Up to 40% of patients with APLA have dermatologic manifestations, most commonly present in the form of livedo racemosa. Other presentations include painful ulceration secondary to ischemia, digital gangrene, superficial venous thrombosis, and subungual splinter hemorrhages
- Laboratory criteria include IgG/IgM anticardiolipin, lupus anticoagulant and IgG/IgM anti-beta 2 glycoprotein I antibodies. These antibodies must be present on two separate occasions at least 12 weeks or more apart

TREATMENT

Treatment of APLA-associated ischemic ulcerations is not well defined although when in association with cutaneous skin necrosis, full anticoagulation is suggested.

REFERENCES

Dean SM. Atypical ischemic lower extremity ulcerations: a differential diagnosis. Vasc Med. 2008; 13:47.

DiFrancesco LM, Burkart P, Hoehn JG. A cutaneous manifestation of antiphospholipid antibody syndrome. Ann Plast Surg. 2003; 51(5):517–22.

Asherson RA, Frances C et al. The antiphospholipid antibody syndrome: diagnosis, skin manifestations and current therapy. Clin Exp Rheumatol. 2006; 24(1Suppl 40):S46–51.

Atlas of Clinical Vascular Medicine, First Edition. Edited by Jessica Mintz, Bruce L. Mintz, and Michael R. Jaff.

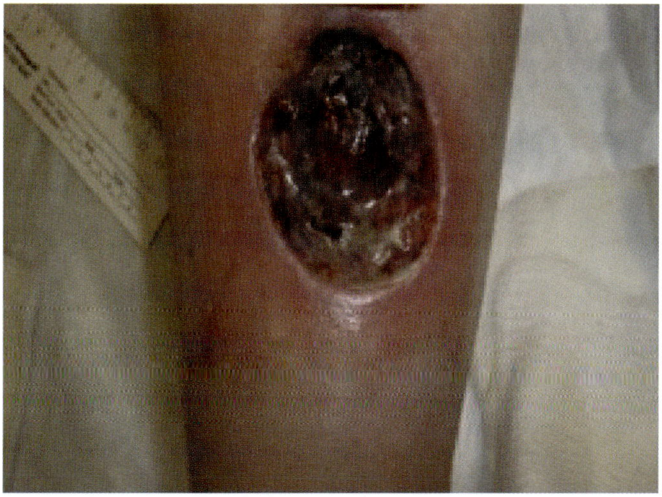

Antiphospholipid antibody ulceration. A pretibial ischemic ulceration with surrounding erythema in a patient with known antiphospholipid antibody syndrome.

Rheumatoid Vasculitis

Cory Perugino[1] and Raghu Kolluri[2]
[1]Cleveland Clinic Foundation, Cleveland, OH, USA
[2]Prairie Vascular Institute, Springfield, IL, USA

BRIEF CASE PRESENTATION

A 52-year-old female with rheumatoid arthritis diagnosed 8 years ago and maintained on low-dose prednisone and methotrexate presents to clinic with a 1 week history of subjective fevers, generalized fatigue, 6 lb (2.7 kg) weight loss, diffuse joint aches, and new onset leg ulceration. Lab work is revealing for an anti-CCP titer of 110 IU/ml, ESR of 65 mm/h, and CRP 11 mg/dL.

CHARACTERISTICS

- Rheumatoid vasculitis is nearly exclusive to seropositive rheumatoid patients with other risk factors including smoking, male gender, other extraarticular manifestations, and later age of disease onset
- Most commonly affects the skin and peripheral nervous system and can manifest as petechiae, palpable purpura, livedo racemosa, superficial ulcers (typically of the dorsum of the foot), digital necrosis, distal symmetric sensory polyneuropathy or mononeuritis multiplex
- Visceral manifestations include scleromalacia, stroke, transient ischemic attack, seizure, myocardial infarction, fibrosing alveolitis, alveolar hemorrhage, interstitial nephritis, glomerulonephritis, and ischemic colitis
- Diagnosis relies on pathologic verification of transmural infiltration or fibrinoid necrosis of small or medium-sized vessels to distinguish this entity from other causes of vascular insufficiency
- Highest diagnostic yield is with biopsy of skin lesions (up to 75%). Some evidence suggests the utility of high-titer anti-CCP antibodies for its specificity and negative predictive value for rheumatoid vasculitis

TREATMENT

- Management depends on severity of vasculitic flare
- Cutaneous involvement is generally treated with moderate-dose corticosteroids with the addition of either methotrexate or azathioprine
- Visceral or peripheral nervous involvement is managed with high-dose corticosteroids with the addition of either a tumor necrosis factor inhibitor or, more recently, rituximab
- Monitoring of the disease course is done both clinically and by trending of C3, C4, ESR, & CRP

REFERENCES

Bartels CM, Bridges AJ. Rheumatoid vasculitis: vanishing menace or target for new treatments? Curr Rheumatol Rep. 2010; 12(6):414–19. [PubMed: 2082467]

Laskaria K, Ahmadi-Simab K, Lamken M, et al. Are anti-cyclic citrullinated peptide autoantibodies seromarkers for rheumatoid vasculitis in a cohort of patients with systemic vasculitis? Ann Rheum Dis. 2010; 69:469–71. [PubMed: 20107034]

Scott DG, Bacon PA, Tribe CR. Systemic rheumatoid vasculitis: a clinical and laboratory study of 50 cases. Medicine. 1981; 60:288–97. [PubMed: 7242322]

Voskuyl AE, van Duinen SG, Zwinderman AH, et al. The diagnostic value of perivascular infiltrates in muscle biopsy specimens for the assessment of rheumatoid vasculitis. Ann Rheum Dis 1998; 57:114–17. [PubMed: 9613342]

Atlas of Clinical Vascular Medicine, First Edition. Edited by Jessica Mintz, Bruce L. Mintz, and Michael R. Jaff.

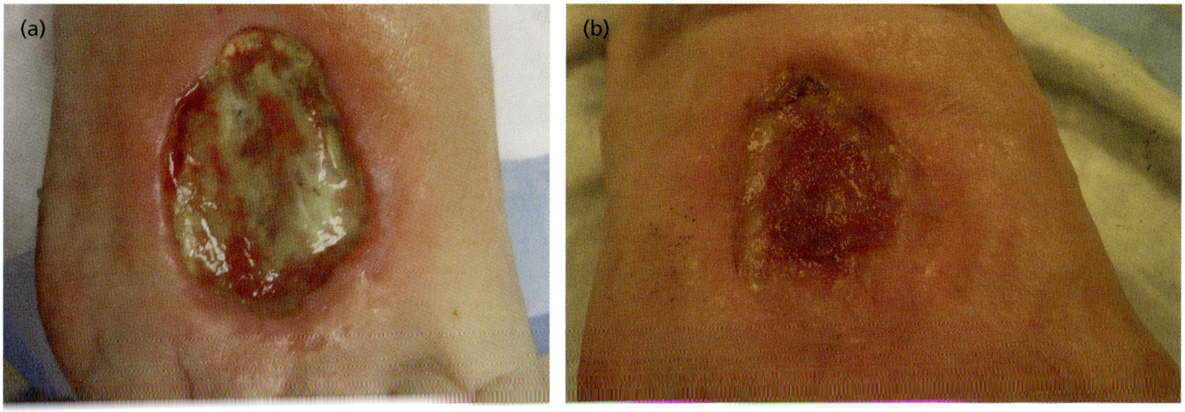

Rheumatoid vasculitis ulceration. Ischemic ulcer of the dorsum of the right foot in a patient with known rheumatoid arthritis.

Calciphylaxis

Jessica Mintz

Cleveland Clinic Heart and Vascular Institute and Cleveland Clinic Lerner College of Medicine, Case Western Reserve University, Cleveland, OH, USA

BRIEF CASE PRESENTATION

This patient with a history of end-stage renal disease on hemodialysis presents with lower extremity wounds with superficial ulceration and a black eschar overlying necrotic appearing tissue

CHARACTERISTICS

- Calciphylaxis is a condition resulting from a disorder in calcium phosphate metabolism, which can result in arterial calcification and organ calcium deposition. It is characterized by systemic medial calcification of arterioles ultimately leading to tissue ischemia
- Calciphylaxis is most commonly associated with end-stage renal disease (ESRD) on hemodialysis, although the condition can occur in patients without renal dysfunction, notably those patients with hyperparathyroidism
- Prolonged hyperphosphatemia, elevated calcium and phosphorus products, protein malnutrition, warfarin use and hypercoagulable states have been associated with calciphylaxis
- Mortality rate ranges from 60 to 80%. This poor prognosis is often secondary to sepsis, organ failure and wound infections. Poor wound healing and limb loss are both common phenomena
- Diagnosis is made by high clinical index of suspicion, serum levels of parathyroid hormone, calcium, and phosphate. Skin biopsy is often performed. Although biopsy can support the diagnosis, there is no pathognomonic finding for this entity
- Exclusion of underlying peripheral artery disease is important in the diagnosis as this is a common comorbidity and cause of ulceration in the ESRD population. Distinguishing calciphylaxis from ischemic ulcers can often present a challenge to the proper diagnosis

TREATMENT

Early recognition is key to treatment. Management of calciphylaxis consists of wound care and dialysis with calcium and phosphorus binding agents. Debridement of necrotic wounds and vascular surgery is often necessary with arterial revascularization and parathyroidectomy in the appropriate patient.

REFERENCES

Arseculeratne G, Evans AT, Morley SM. Calciphylaxis – a topical overview. J Eur Acad Dermatol Venereol. 2006; 20(5): 493–502.

Milas M, Bush R, Lin P, Brown K, MacKay G, et al. Calciphylaxis and nonhealing wounds: the role of the vascular surgeon in a multidisciplinary treatment. J Vasc Surg. 2002; 37(3):501–7.

Wilmer WA, Magro CM. Calciphylaxis: emerging concepts in prevention, diagnosis, and treatment. Semin Dial. 2002; 15(3):172–86.

Atlas of Clinical Vascular Medicine, First Edition. Edited by Jessica Mintz, Bruce L. Mintz, and Michael R. Jaff.

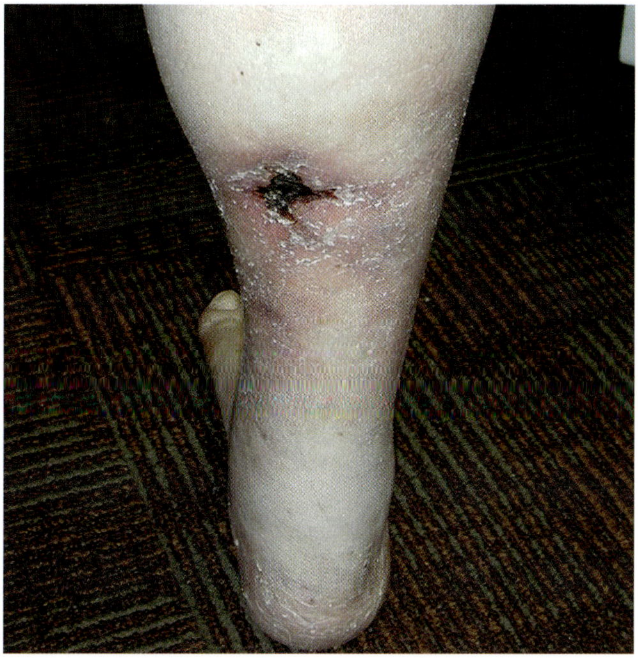

Calciphylaxis. Posterior calf ulceration with a central black eschar in a patient with end-stage renal disease. Courtesy of Steven Dean.

Hypertensive Ulcer

Jessica Mintz, John R. Bartholomew and Heather L. Gornik

Cleveland Clinic Heart and Vascular Institute and Cleveland Clinic Lerner College of Medicine, Case Western Reserve University, Cleveland, OH, USA

BRIEF CASE PRESENTATION

A non-diabetic, known hypertensive patient presents with exquisitely painful ulceration around the lateral malleolus.

CHARACTERISTICS

- Hypertensive ulcers are a rare and controversial diagnosis
- Hypertensive ulcers are superficial ulcerations most commonly located on the dorsolateral lower leg (i.e., especially lateral ankle)
- These are a result of ischemia caused by arteriolar lesions
- Associated with hypertension, particularly if poorly controlled and long-standing
- In order to diagnose hypertensive ulceration, there must be no significant arterial occlusive disease (palpable pulse in arteries of the lower limb and normal arterial physiological testing/ankle-brachial indices)
- The centrally necrotic ulceration typically is surrounded by a cyanotic halo due to ischemia within the tissue not yet ulcerated

TREATMENT

Standards of care for this condition include analgesics, antihypertensive therapies, and antibiotics if bacterial superinfection is present. Debridement and local wound care is often required.

REFERENCES

Graves JW, Morris JC, Sheps SG. Martorell's hypertensive leg ulcer: case report and concise review of the literature. J Hum Hypertens. 2001; 15(4):279–83.

Hafner J, Nobbe S, Partsch H, et al. Mortorell hypertensive ischemic leg ulcer: a model of ischemic subcutaneous arteriosclerosis. Arch Dermatol. 2010; 146(9):961–8.

Martorell F. Las ulceras supermaleolares por arteriolitis de las grandes hipertensas. Actas (Reun Cientif Cuerpo Facul) Inst Policlinico Barcelona. 1945; 1(1):6–9.

Roenigk H, Young JR. Leg ulcers. In: Young JR, Olin JW, Bartholomew JR (eds) Peripheral Vascular Diseases, 2nd Edition. St Louis: Mosby, 1996, pp. 624–6.

Atlas of Clinical Vascular Medicine, First Edition. Edited by Jessica Mintz, Bruce L. Mintz, and Michael R. Jaff.

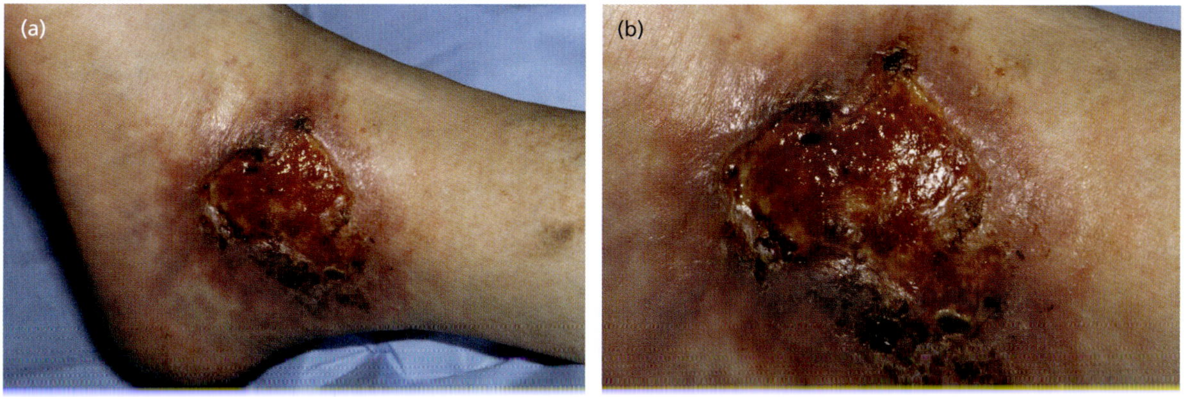

Hypertensive Ulcer. Superficial ulceration of the lateral malleolus with the classic cyanotic halo surrounding central necrosis. Images courtesy of the Cleveland Clinic Foundation Department of Vascular Medicine archives.

Neurotrophic Ulcer

Bruce L. Mintz
Morristown Medical Center, Morristown, NJ, USA

BRIEF CASE PRESENTATION

A diabetic patient presents with a well-circumscribed ulcer on the inferior heel. Patient's shoes noted to be a half a size too small.

CHARACTERISTICS

- Neurotrophic ulcers are usually located over increased pressure points of the foot. These ulcers may appear red/pink or more dark brown/black. They primarily occur in diabetics, although they may be found in any patient with neuropathy and impaired sensation
- Motor neuropathy causing muscular atrophy with resulting mechanical changes of the foot, sensory neuropathy leading to loss of sensation and repetitive trauma, and autonomic neuropathy altering skin turgor all contribute to accelerated damage to skin and ulcer formation
- The combination of neuropathy with impaired wound healing often leads to non-healing ulcers in diabetics

TREATMENT

Annual foot exams are recommended for all diabetics to identify and prevent high-risk conditions, which are known to lead to amputations. These include neuropathy, peripheral arterial disease and structural foot deformities. Debridement is the first step in wound care.

REFERENCES

Boulton AJ. The diabetic foot – an update. Foot Ankle Surg. 2008; 14(3):120–4.

Molines, D, Darmon P, Raccah D. Charcot's foot: newest findings on pathophysiology, diagnosis and treatment. Diabetes Metab. 2010; 36(4):251–5.

Simmons Z, Feldman EL. Update on diabetic neuropathy. Curr Opin Neurol. 2002; 15(5):595–603.

Atlas of Clinical Vascular Medicine, First Edition. Edited by Jessica Mintz, Bruce L. Mintz, and Michael R. Jaff.

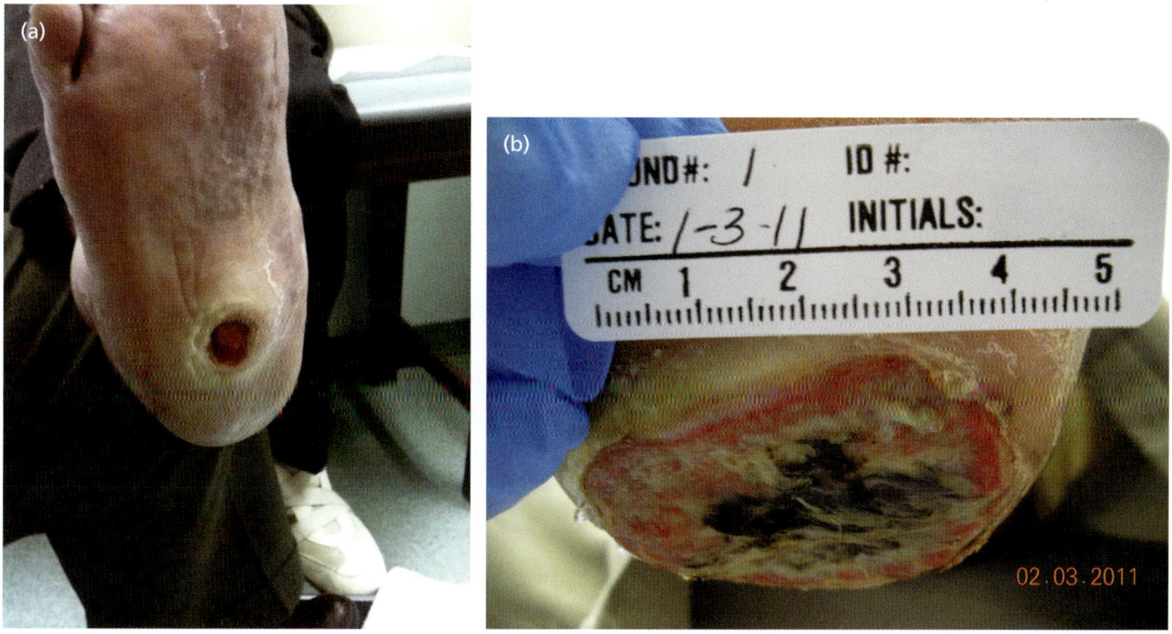

Neurotrophic ulcer. (a) Well-circumscribed ulceration of the inferior heel with surrounding callous formation. (b) Superimposed infection overlying a neurotrophic ulcer. Note the necrosis and lateral granulation tissue. (a) Courtesy of Bruce L. Mintz and (b) courtesy of Raghu Kolluri.

Necrobiosis Lipoidica Diabeticorum

Raghu Kolluri
Prairie Vascular Institute, Springfield, IL, USA

> **BRIEF CASE PRESENTATION**
> A 54-year-old female with a history of diabetes mellitus presents with several years of asymptomatic yellow-brown plaques on the anterior shin of her right lower extremity.

CHARACTERISTICS

- Necrobiosis lipoidica diabeticorum (NLD) is a rare and idiopathic dermatologic condition strongly associated with diabetes mellitus. NLD is also seen in patients without glucose intolerance. These lesions affect less than 1% of patients with diabetes mellitus and do not correlate with extent of glycemic control
- NLD is more commonly seen in women with an average age of 30–40 years
- Typically manifest as asymptomatic waxy appearing plaques over the pretibial regions. The lesion begins as violaceous oval patch advancing to a central yellow-brown lesion with a red border. Telangiectasias are commonly found coursing through the lesions

TREATMENT

Currently no evidence-based guidelines exist for the treatment of NLD. Therapies include topical and subcutaneous corticosteroids, photo chemotherapy, and new treatment options including infliximab, cyclosporine or tacrolimus, all with varying degrees of success. Flares of NLD are frequent and there has been no completely effective treatment found to eradicate these lesions.

REFERENCES

Klaber R. Necrobiosis lipoidica diabeticorum (Urbach-Oppenheim). Proc R Soc Med. 1934; 27(6):713–14.

Lowitt MH, Dover JS. Necrobiosis lipoidica. J Am Acad Dermatol. 1991; 25(5):735–48.

Atlas of Clinical Vascular Medicine, First Edition. Edited by Jessica Mintz, Bruce L. Mintz, and Michael R. Jaff.

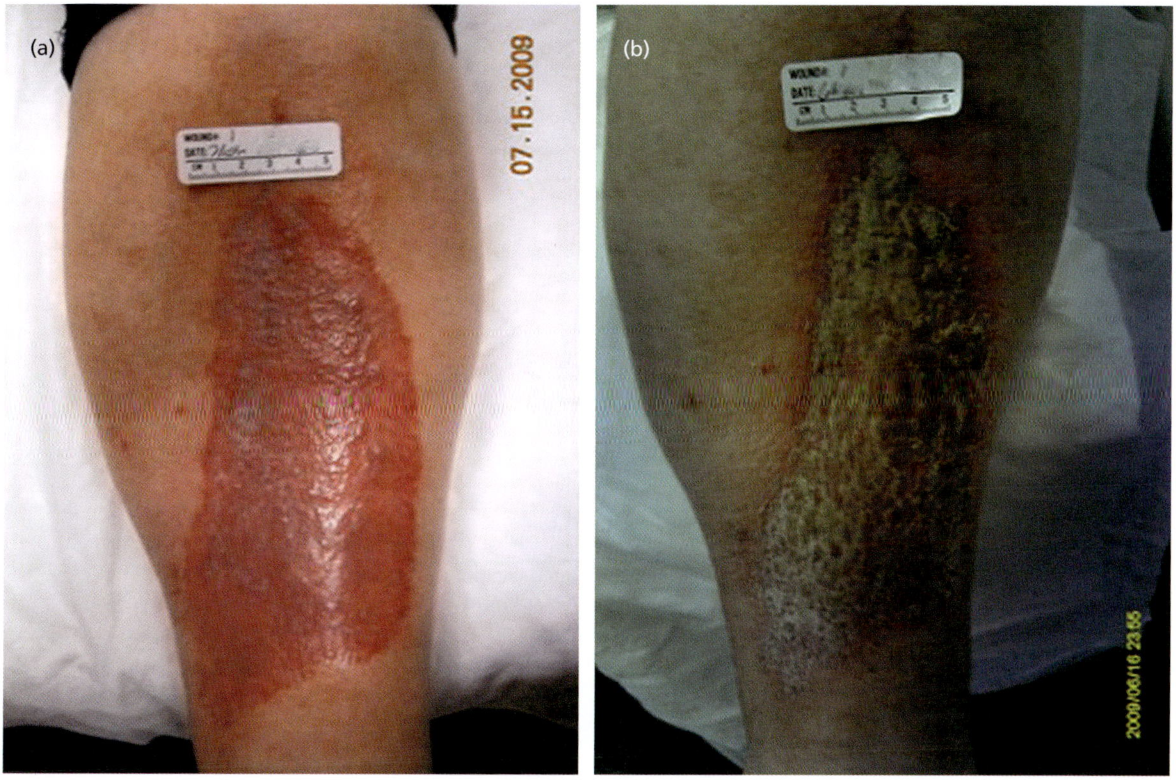

Necrobiosis lipoidica diabeticorum. (a) Waxy plaque of the pretibial region with a violaceous red appearance and granulation tissue. Courtesy of Michael R Jaff.(b) A pretibial lesion with advanced yellow-brown lesions.

Deep Foot Infection

Raghu Kolluri
Prairie Vascular Institute, Springfield, IL, USA

BRIEF CASE PRESENTATION

A diabetic patient status post trans-metatarsal amputation presents with purulent discharge over amputation site.

CHARACTERISTICS

- Deep foot infections are the most common reason for lower extremity amputations in diabetic patients
- Diabetics are particularly susceptible to deep foot infections due to neuropathy, peripheral artery disease (PAD), and altered immune function
- Neuropathy can lead to foot deformities. These deformities cause altered mechanics and pressure points that can cause ulcerations. Ulcerations of the skin can cause direct exposure of underlying tissue to pathogens and ultimately infection
- Most common pathogens in superficial wounds are *Staphylococcus* and *Streptococcus* species. These superficial wounds progress due to delay in treatment, as well as poor wound healing from PAD. Deep wounds tend to be caused by a mixture of gram positive, Gram-negative and anaerobic organisms

TREATMENT

All wounds should be appropriately debrided with removal of all foreign bodies, as well as off loading of all local pressure. Depending upon the severity of the infection, patients should be started on antibiotics with deep wound cultures. If infection is severe, surgical debridement is critical.

REFERENCES

Boulton AJ, Vileikyte L, Ragnarson-Tennvall G, Apelqvist J. The global burden of diabetic foot disease. Lancet. 2005; 366(9498):1719–24.

Lipsky BA, Berendt A, Deery HG, et al. Diagnosis and treatment of diabetic foot infections. Clin Infect Dis. 2004; 39(7): 885–910.

Williams DT, Hilton JR, Harding KG. Diagnosing foot infections in diabetes. Clin Infect Dis. 2004; 39(suppl 2):S83–6.

Atlas of Clinical Vascular Medicine, First Edition. Edited by Jessica Mintz, Bruce L. Mintz, and Michael R. Jaff.

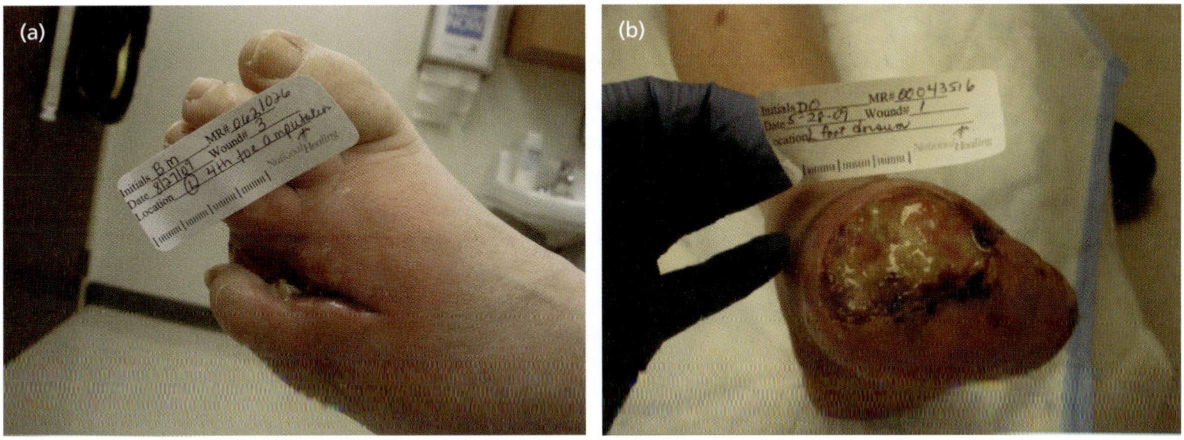

Deep foot infection. (a) Patient with a fourth metatarsal amputation and a deep foot infection overlying the amputation stub. (b) Lower extremity amputation complicated by a deep wound infection of the surgical site.

Sickle Cell Induced Leg Ulceration

Jessica Mintz[1], John R. Bartholomew[1], Michael Gironta[2] and Heather L. Gornik[1]

[1]Cleveland Clinic Heart and Vascular Institute and Cleveland Clinic Lerner College of Medicine, Case Western Reserve University, Cleveland, OH, USA

[2]Eastern Vascular Associates, Denville, NJ, USA

BRIEF CASE PRESENTATION

Two African American patients with known sickle cell anemia present with throbbing ankle ulcerations.

CHARACTERISTICS

- Sickle cell anemia is a hemoglobinopathy resulting in poor microvascular flow leading to tissue ischemia
- Sickle cell anemia patients may often present with skin ulcerations
- These ulcers are well-demarcated lesions occurring in the lower extremities at the malleoli. Lesions may be very painful
- Granulation tissue is commonly found at the base of the ulcer

TREATMENT

Local wound care, including debridement if needed comprise the standard of care for the lesions. In this condition, maintenance of hemoglobin above 10 g/dL with exchange transfusions as well as treatment of sickle cell disease with hydroxyurea have been suggested to promote healing.

REFERENCES

Halabi-Tawil M, Lionnet F, Bachmeyer C Levy PP, Aractingi S. Sicle cell leg ulcers: a frequently disabling complication and marker of severity. Br J Dermatol. 2008; 158:339–44.

Minniti CP, Eckman J, Sebasstiani P, Steinberg MH, Ballas SK. Leg ulcers in sickle cell disease. Am J Hematol. 2010; 85:831–3.

Roehigk HH, Young JR. Leg ulcers. In: Young JR Olin JW, Bartholomew JR (eds). Peripheral Vascular Diseases, 2nd edition. St Louis: Mosby, 1996, pp. 624–6.

Atlas of Clinical Vascular Medicine, First Edition. Edited by Jessica Mintz, Bruce L. Mintz, and Michael R. Jaff.

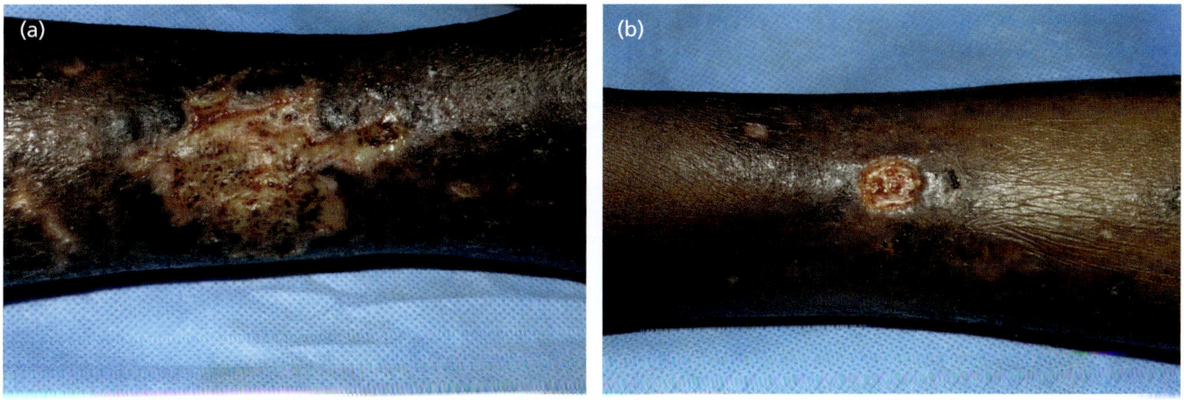

Sickle cell induced ulceration. Well-demarcated recurrent painful lesions on lower extremities with associated granulation tissue in a patient with known sickle cell anemia. Image courtesy of The Cleveland Clinic Foundation Department of Vascular Medicine archives.

Thalassemia Intermedia Induced Limb Ulceration

Jessica Mintz, John R. Bartholomew and Heather L. Gornik

Cleveland Clinic Heart and Vascular Institute and Cleveland Clinic Lerner College of Medicine, Case Western Reserve University, Cleveland, OH, USA

BRIEF CASE PRESENTATION

Patient with known thalassemia presents with a 2–3 cm circumferential, well-demarcated leg ulceration with surrounding fibrosis.

CHARACTERISTICS

- Thalassemia is an autosomal recessive genetic disorder resulting in mutations or deletions in the globin chains that make up hemoglobin
- Leg ulcerations secondary to thalassemia are a common complication of thalassemia intermedia
- Thalassemia intermedia presents later in life than thalassemia major with moderate anemia usually not requiring transfusions
- Ulcers are believed to occur as a result of local tissue hypoxia from anemia. However, hypercoagulability may also be a contributor to the pathogenesis of thalassemia induced ulceration
- Ulcers are well demarcated and typically occur over the medial or lateral malleolus
- These ulcers heal very slowly and often recur causing chronic pain

TREATMENT

Proper thalassemia management is critical as ulceration occurs more commonly in poorly managed individuals. Maintenance of a clean and moist environment, debridement of devitalized tissue, and proper dressings are also beneficial.

REFERENCES

Gamberini MR, Fortini M, DeSanctis V. Healing of leg ulcers in thalassemia intermedia patients with associated endocrine complications. Pediatr Endocrinol Rev. 2004; 2(Suppl 2): 319–22.

Levin C, Koren A. Healing of refractory leg ulcer in a patient with thalessemia intermedia and hypercoagulability after 14 years of unresponsive therapy. IMAJ. 2011; 13:316–18.

Stevens DM, Shupack JL, Javid J, Silber R. Ulcers of the leg in thalassemia. Arch Dermatol. 1977; 113(11):1558–60.

Atlas of Clinical Vascular Medicine, First Edition. Edited by Jessica Mintz, Bruce L. Mintz, and Michael R. Jaff.

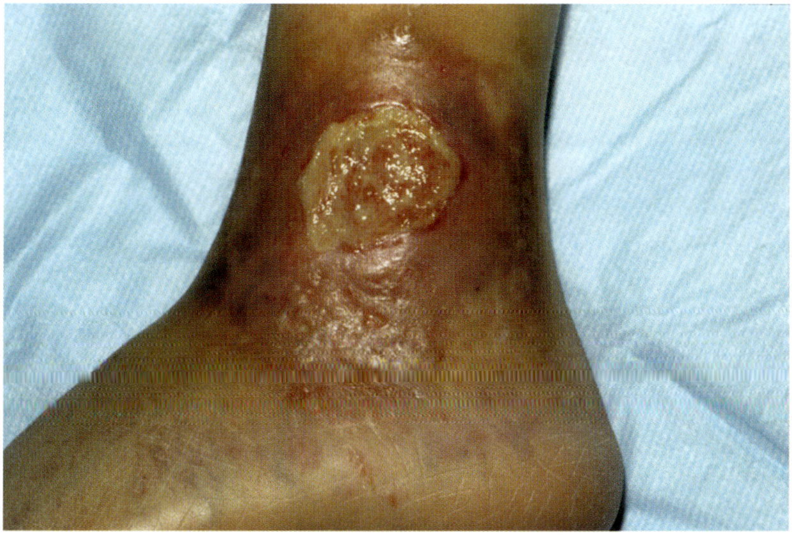

Thalassemia induced limb ulceration. Well-demarcated, circumferential ulceration over the lateral malleolus with surrounding fibrosis in a patient with known thalassemia.

Pyoderma Gangrenosum

Ido Weinberg and Michael R. Jaff
Massachusetts General Hospital, Boston, MA, USA

BRIEF CASE PRESENTATION

A 34-year-old woman with known inflammatory bowel disease presents with rapidly expanding, painful, ulcerations on both lower extremities.

CHARACTERISTICS

- Pyoderma gangrenosum represents markedly painful, violaceous ulcers with irregular borders, that are often rapidly progressive and proceeded by minor trauma to the skin
- Pyoderma gangrenosum can appear on any skin surface but is most common on the legs
- The most common association is with inflammatory bowel disease, more often Crohn's disease than ulcerative colitis. The activity of inflammatory bowel disease does not always coincide with the presence of the ulcerations. Other causes include hematologic malignancies and seronegative spondyloarthritides. However, it remains idiopathic in 50% of cases
- Pyoderma gangrenosum is a diagnosis of exclusion. Mimicking conditions include venous stasis ulcers, ulcers associated with antiphospholipid antibodies, cutaneous malignancy, and infection

TREATMENT

First line therapy includes topical and systemic corticosteroids. Cyclosporine has been reported to be successful in the treatment of pyoderma gangrenosum. Other immunosuppressive agents including anti-tumor necrosis factor-α agents such as infliximab and etanercept have been tried.

REFERENCES

Binus AM, Qureshi AA, Li VW, Winterfield LS. Pyoderma gangrenosum: a retrospective review of patient characteristics, comorbidities, and therapy in 103 patients. Br J Dermatol. 2011; doi: 10.1111/j.1365-2133.2011.10565.x. [Epub ahead of print] PubMed PMID: 21824126

Brooklyn T, Dunnill G, Probert C. Diagnosis and treatment of pyoderma gangrenosum. BMJ. 2006; 333(7560):181–4. Review.

Weenig RH, Davis MD, Dahl PR, Su WP. Skin ulcers misdiagnosed as pyoderma gangrenosum. NEJM. 2002; 347(18):1412–8. Review.

Atlas of Clinical Vascular Medicine, First Edition. Edited by Jessica Mintz, Bruce L. Mintz, and Michael R. Jaff.

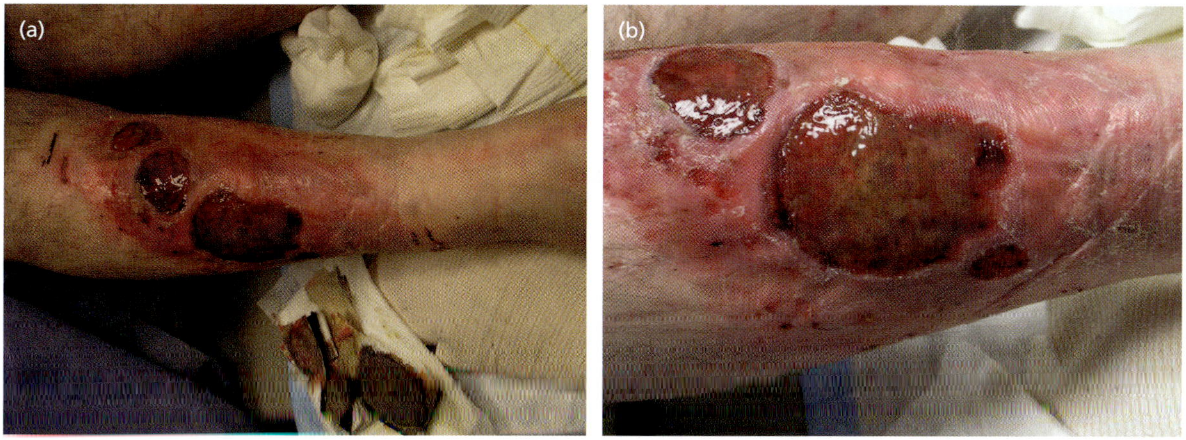

Pyoderma gangrenosum. Lower extremity violaceous ulceration with classic irregular borders.

Cocaine (Levamisole) Induced Vasculitis

Peter C. Grayson[1] **and Peter A. Merkel**[2]
[1]Boston University Vasculitis Center, Boston, MA, USA
[2]Division of Rheumatology, University of Pennsylvania, Philadelphia, PA, USA

BRIEF CASE PRESENTATION

This 48-year-old woman presents with extensive skin ulceration on the legs, arms, and earlobes. She reports recent intranasal cocaine use and urine toxicology confirms the presence of levamisole. Antibody testing was positive for anti cytoplasmic neutrophil antibodies (ANCA), antineutrophil antibodies (ANA), and lupus anticoagulant.

CHARACTERISTICS

- In the United States, cocaine is commonly adulterated with levamisole, an antihelminth medication known to have immunostimulating effects
- Levamisole can produce a syndrome that mimics vasculitis and thrombotic disease
- Purpuric lesions with predilection for earlobes can occur and often ulcerate
- Skin biopsy typically demonstrates a mixed pattern of leukocytoclastic vasculitis and microthrombus
- Positive tests for ANCA, ANA, antiphospholipid antibodies, and neutropenia are associated with this syndrome
- Levamisole, and not cocaine, is the suspected causative agent for vasculitis in such cases

TREATMENT

Cessation of cocaine use with local wound care.

REFERENCES

Bradford M, Rosenberg B, Moreno J, Dumyati G. Bilateral necrosis of earlobes and cheeks: another complication of cocaine contaminated with levamisole. Ann Intern Med. 2010; 152(11):758–9.

Ullrich K, Koval R, Koval E, Bapoje S, Hirsh JM. Five consecutive cases of a cutaneous vasculopathy in users of levamisole-adulterated cocaine. J Clin Rheumatol. 2011; 17(4):193–6.

Walsh NM, Green PJ, Burlingame RW, Pasternak S, Hanly JG. Cocaine-related retiform purpura: evidence to incriminate the adulterant, levamisole. J Cutan Pathol. 2010; 37(12): 1212–19.

Atlas of Clinical Vascular Medicine, First Edition. Edited by Jessica Mintz, Bruce L. Mintz, and Michael R. Jaff.

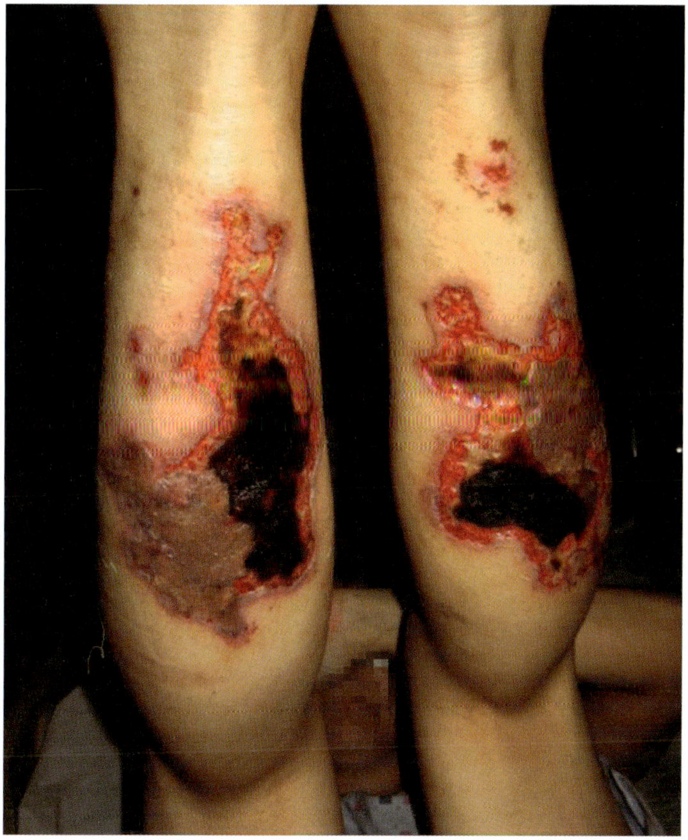

Cocaine (levamisole) induced vasculitis. Extensive lower extremity skin ulceration with irregular borders and marked necrosis.

Lichen Planus

Georgann A. Poulos[1] and Steven M. Dean[2]

Department of Dermatology[1] and Cardiovascular Medicine[2], Wexner Medical Center at the Ohio State University, Columbus, OH, USA

BRIEF CASE PRESENTATION

A 52-year-old African-American female presents for evaluation of multiple, pruritic, hyperkeratotic papules and plaques of the anterior legs. A skin biopsy revealed a lichenoid interface dermatitis with direct immunofluorescence (DIF) most consistent with lichen planus (LP).

CHARACTERISTICS

- LP is an inflammatory dermatosis that can affect not only skin, but also nails, hair, and mucosal surfaces
- In its classic form, LP is characterized by violaceous, "flat-topped, polygonal papules." A reticulated white scale, known as Wickham's striae, can often be appreciated
- Although 20 different LP variants have been described, the hypertophic variant is most classically found on the anterior legs
- LP typically affects middle aged adults (5th to 6th decade). The prevalence is estimated to be less than 1% of the population. There may be a slight female predominance
- It is imperative that the clinician confirm the diagnosis of hypertrophic LP via skin biopsy given clinical overlap with hypertrophic lupus. Several authors have also noted an increased development of squamous cell carcinoma within hypertrophic LP lesions further supporting the need for histologic confirmation

TREATMENT

Topical remedies in the treatment of LP include superpotent topical corticosteroids or intralesional corticosteroids. Systemic therapy with an oral retinoid (acitretin) is particularly helpful for severe hypertrophic LP. Phototherapy, oral corticosteroids, antimalarials, biologics, and oral antibiotics have also been studied with more limited evidence-based data.

REFERENCES

Asch S, Goldenberg G. Systemic treatment of cutaneous lichen planus: an update. Cutis. 2011; 87:129–34.

Bezanis G, Wang TS, Johnson TM, Lowe L. Verrucous plaques on the leg. Arch Dermatol. 2003; 139(7):933–8.

Boyd A, Nelder K. Lichen Planus. Journal of the American Academy of Dermatology. 1991; 25(4):593–613.

James WD, Berger TG, Elston DM. Andrews' Diseases of the Skin: Clinical Dermatology, 10th edn. Philadelphia, PA: WB Saunders, 2005.

Atlas of Clinical Vascular Medicine, First Edition. Edited by Jessica Mintz, Bruce L. Mintz, and Michael R. Jaff.

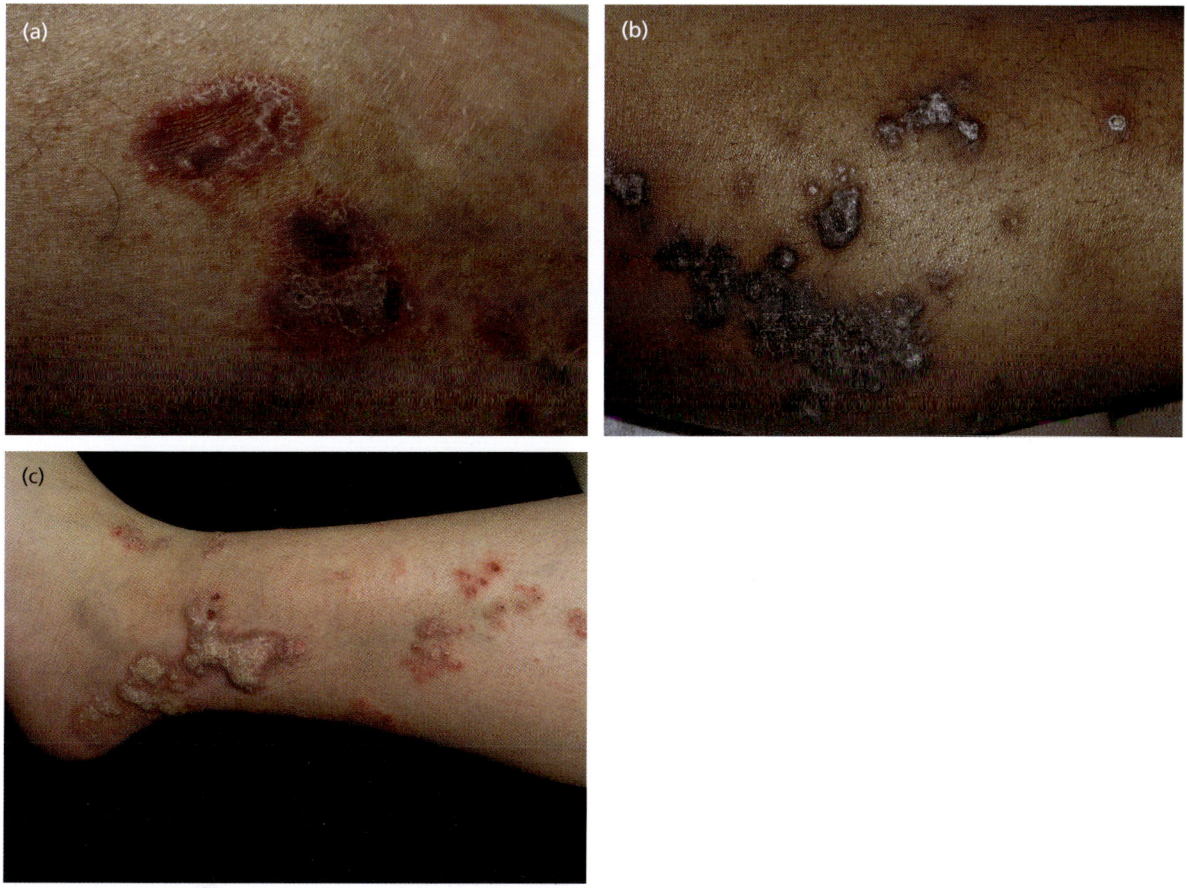

Lichen planus. (a, b) Flattened papules with white scaling consistent with lichen planus. (c) Hypertrophic variant of lichen planus of the lower leg. Photographs courtesy of Dr. Matt Zirwas.

Brown Recluse Spider Bite

Bruce L. Mintz
Morristown Medical Center, Morristown, NJ, USA

BRIEF CASE PRESENTATION

This 50-year-old diabetic male presents with a painful papule on dorsum of the foot noted upon returning from his tool shed. The wound became progressively necrotic with subsequent purulent drainage which cultured methicillin resistant *Staphylococcus aureus*.

CHARACTERISTICS

- The most common geographic regions of brown recluse spider bites are in the southeastern United States
- Most bites are asymptomatic or cause a constellation of symptoms suggestive of a flu-like illness. However, a condition characterized by necrotic skin ulcers (necrotic arachnidism) may also occur. These ulcers are often recalcitrant to heal and may necessitate skin grafting or local amputation of the affected limb
- Although brown recluse spider venom is poisonous and may result in local cell damage with disruption of skin, fat, and blood vessels, the relative paucity of venom in the inoculum makes mortality exceedingly rare
- Venom provokes both an immediate and delayed immune response. In rare circumstances, the consequence of this immune response may cause thrombocytopenia, anemia, acute kidney injury, deep vein thrombosis and/or coma.

TREATMENT

Treatment by and large consists of cleansing the wound and applying emollient barrier antibiotic ointments. Medical attention should be sought if constitutional symptoms such as nausea, vomiting, fever or a rash occur. There is no specific anti-venom available but close attention should be paid to the occurrence of secondary infection. Non-healing wounds may become deep and infected requiring surgical debridement.

REFERENCES

Ashurst J, Sexton J, Cook M. Approach and management of spider bites for the primary care physician. Osteopathic Family Physician. 2011; 3(4):149–53.

Rees R, Altenbern D, Lynch J, King L. Brown recluse spider bites. A comparison of early surgical excision versus dapsone and delayed surgical excision. Ann Surg. 1985; 202(5):659–63.

Swanson D, Vetter R. Bites of brown recluse spiders and suspected necrotic arachnidism. NEJM. 2005; 352 (7):700–7.

Atlas of Clinical Vascular Medicine, First Edition. Edited by Jessica Mintz, Bruce L. Mintz, and Michael R. Jaff.

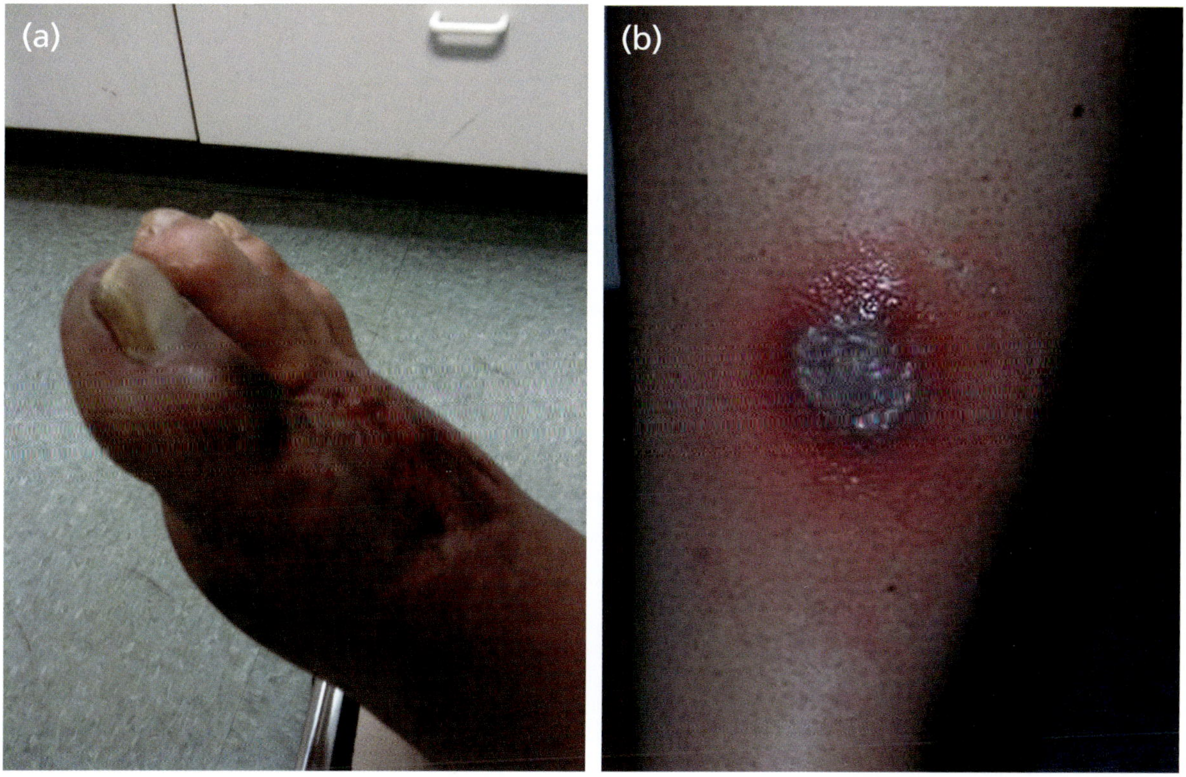

Brown recluse spider bite. (a) Papule of the dorsum of the foot with associated necrosis and purulent discharge. (b) Tarantula bite.

Frostbite

Ido Weinberg[1], Bruce L. Mintz[2] and Michael R. Jaff[1]
[1]Massachusetts General Hospital, Boston, MA, USA
[2]Morristown Medical Center, Morristown, NJ, USA

BRIEF CASE PRESENTATION

A 41-year-old homeless Caucasian female presents in mid December after a light wet snow with initially numb and then painful blistering of both feet. Her feet had been protected from the snow through wet canvas sneakers.

CHARACTERISTICS

- Frostbite is a condition characterized by irreversible cold-induced cell death which occurs with exposure of tissue to below freezing temperatures most commonly on the hands
- Patients experience pain followed by loss of sensation and loss of color to the exposed area
- The mechanism of cell death may be immediate with intracellular ice crystal formation or a more gradual development of a localized inflammatory process and tissue ischemia
- With reperfusion secondary to re-warming, patients develop localized edema and blister formation, which are clear or dark in color depending on the severity of frostbite
- Traditional classification system divides frostbite into four degrees of severity:
  - First degree: hyperemia and edema
  - Second degree: large clear blisters
  - Third degree: vesicles which may become hemorrhagic
  - Fourth degree: complete necrosis of the involved skin with gangrene

TREATMENT

The treatment of frostbite depends on the severity and time from insult. Pre-hospital care consists of gradual re-warming. Recurrent cold exposure, rapid re-warming or tissue friction (such as rubbing) can all be detrimental to healing and should be avoided. All patients with frostbite should be given tetanus toxoid and broad-spectrum antibiotics along with adequate analgesics.

Aspirin should be avoided as it may reduce synthesis of vasodilating prostaglandins. In recent onset, severe frostbite, arterial thrombolysis has been tried with promising preliminary results. While tissue necrosis in frostbite is progressive, it may also recede spontaneously. Therefore, amputation should be delayed as long as possible.

REFERENCES

Cauchy E, Cheguillaume B, Chetaille E. A controlled trial of a prostacyclin and rt-PA in the treatment of severe frostbite. NEJM. 2011; 364(2):189–90.

Biem J, Koehncke N, Classen D, Dosman J. Out of the cold: management of hypothermia and frostbite. CMAJ. 2003; 168:305–11.

Larrey DJ. Memoirs of Military Surgery and Campaigns of the French Armies. Baltimore: Cushing, 1814, pp. 156–64.

Atlas of Clinical Vascular Medicine, First Edition. Edited by Jessica Mintz, Bruce L. Mintz, and Michael R. Jaff.

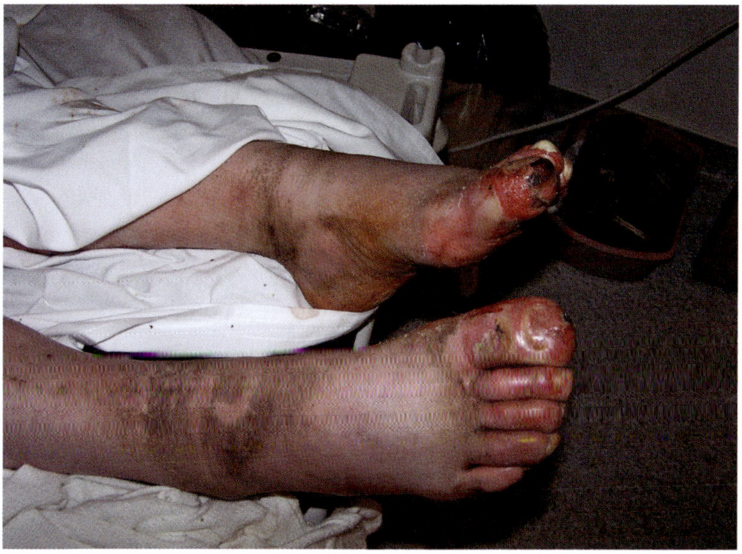

Frostbite. Bilateral lower extremity blistering and excoriation associated with cold exposure.

Pernio (Chilblains)

Bruce L. Mintz
Morristown Medical Center, Morristown, NJ, USA

BRIEF CASE PRESENTATION

A 41-year-old white female presents with a 3-day history of burning discomfort in both feet after a hike in the mountains on a cold spring day. Her feet were initially pruritic but progressed to discomfort and ultimately ulceration.

CHARACTERISTICS

- Pernio is a vasospastic disorder affecting skin upon cold exposure. The early findings consist of digital numbness and pruritus. Progression over hours to days includes pain, erythema, blisters and even ulceration of a digit. These ulcers likely represent localized capillaritis
- Pernio classically occurs in spring or fall in wet environments that commonly exclude subzero temperatures
- This disease is a clinical diagnosis and must be distinguished from all other causes of extremity discoloration by history and physical exam. A careful history may avoid unnecessary angiographic diagnostic testing
- Clinical findings suggestive of pernio in situations other than in cold climates should heighten concerns for secondary pernio. Laboratory examination for underlying collagen vascular diseases may be warranted

TREATMENT

Most lesions resolve of their own volition within days to months. Vasodilatation, steroid creams and antihistamines have been used; however, their effectiveness is controversial. Patients must be instructed to dress appropriately in cold weather and attempt to preserve body temperature with protective outdoor wear such as gloves and proper shoes.

REFERENCES

Ingram, JT. Chilblains. BMJ. 1949; 2(4639):1284–6.

Isenberg, DA, Black, C. ABC of rheumatology: Raynaud's phenomenon, scleroderma, and overlap syndromes. BMJ. 1995; 310(6982):795–8.

Patel, GK. ABC of wound healing: Uncommon causes of ulceration. BMJ. 2006; 332(7541):594–6.

Atlas of Clinical Vascular Medicine, First Edition. Edited by Jessica Mintz, Bruce L. Mintz, and Michael R. Jaff.

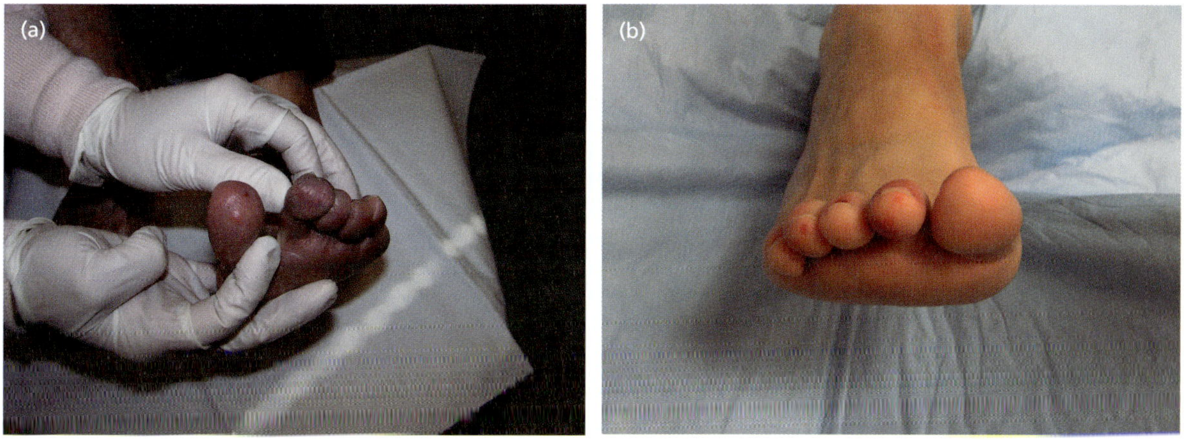

Pernio (chilblains). (a) Left foot with pruritic, erythematous, and violaceous papules overlying the digits. Note the blanching of the second toe. (b) Blanching erythema of the second, third, and fourth toe in a patient with pernio (courtesy of Michael R. Jaff).

Trench Foot

Bruce L. Mintz[1] and Jessica Mintz[2]

[1]Morristown Medical Center, Morristown, NJ, USA

[2]Cleveland Clinic Heart and Vascular Institute and Cleveland Clinic Lerner College of Medicine, Case Western Reserve University, Cleveland, OH, USA

BRIEF CASE PRESENTATION

A 55-year-old female presents with tingling, cold, edematous feet with necrotic lesions bilaterally. She had been outdoors for long periods of time raking leaves in the late fall wearing damp rain boots

CHARACTERISTICS

- Trench foot is an environmental immersion injury, which is associated with prolonged exposure to cold, non-freezing damp settings as well as immobility
- First descriptions of the condition were during the Napoleonic wars; however, it was during World War I that this condition got its name from soldiers spending prolonged periods in cold damp trenches with army boots
- Majority of cases have occurred in wartime conditions although there have been cases of patients following wet and cold exposures during prolonged athletic events in the cold
- Trench foot has three clinical phases: vasospastic, hyperemic phase occurring after removal from environment, and a late ischemic phase
- Pathology leading to gangrene is related to tissue disruption and not arterial occlusion
- It is important to distinguish this condition from environmental injury in freezing temperatures such as frostbite

TREATMENT

Most important is avoidance of wet and damp environments as well as frequent sock changes. Treatment of trench foot includes removal of wet or damp footwear with return to a warmer environment. If the hyperemic phase occurs, prompt elevation of the feet as well as return to a cooler environment for the injury to completely demarcate are important measures.

REFERENCES

Young JR, Olin JW, Bartholomew JR. Peripheral Vascular Diseases, 2nd edition. St Louis: Mosby, 1996.

Frost HM. Trench foot. Boston Med Surg J. 1917; 301–4.

Redisch W, Brandman O, Rainone S. Chronic trench foot: A study of 100 cases. Ann Intern Med. 1951; 34:1163–8.

Atlas of Clinical Vascular Medicine, First Edition. Edited by Jessica Mintz, Bruce L. Mintz, and Michael R. Jaff.

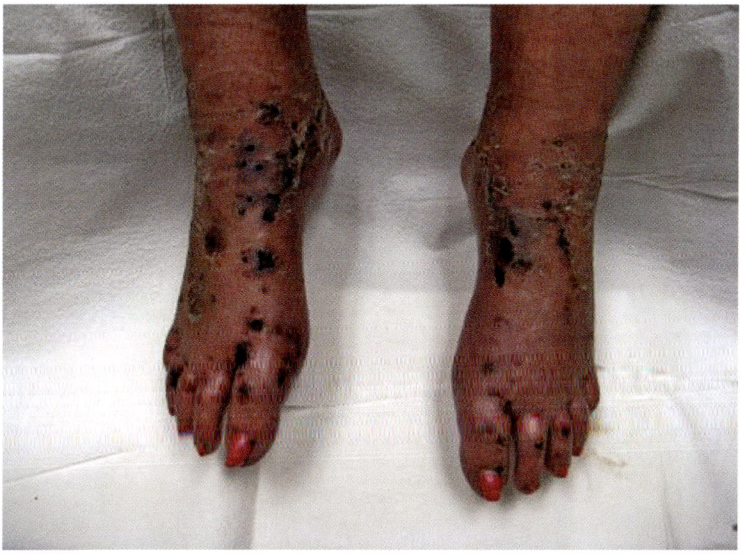

Trench foot. Bilateral edematous feet with advanced necrotic lesions. Courtesy of Steven Dean.

Erythromelalgia

Bruce L. Mintz
Morristown Medical Center, Morristown, NJ, USA

BRIEF CASE PRESENTATION

This 68-year-old female presents with longstanding burning pain, swelling and discoloration of her extremities. Her pain was profoundly worsened by warmth in summer months and seemingly improved by exposure to the cold. She denied any other medical problems

CHARACTERISTICS

- Erythromelalgia is a rare condition in which increase in temperature induces *erythros* (redness), *melos* (extremity) and *algia* (pain). The heat-induced erythema and burning pain are symmetric, involving the lower extremities and less frequently the upper extremities
- Erythromelalgia generally affects middle-aged women and young adults. The symptoms are often intermittent and occur with exposure to heat or with exercise
- The condition is classified as either primary or secondary. Primary is generally idiopathic, believed to be due to a small fiber neuropathy. Secondary erythromelalgia is associated with myeloproliferative disorders such as polycythemia vera
- Diagnosis is made by history and physical examination with no objective laboratory or other confirmatory tests
- Over time symptoms can worsen, improve, or stay the same

TREATMENT

Aspirin and non-steroidal anti-inflammatory agents are useful therapies for patients with an underlying myeloproliferative disease; however, they are less helpful in primary disease. Gabapentin, tricyclic antidepressants and vasoactive medications have all been tried with variable results. Symptoms are improved with exposure to cool temperatures and patients are encouraged to avoid heat.

REFERENCES

Buttaci CJ. Erythromelalgia: a case report and literature review. Pain Med. 2006; 7(6):534–8.

Davis MD, Rooke T. Erythromelalgia. Curr Treat Options Cardiovasc Med. 2006; 8(2):153–65.

Davis MD, O'Fallon WM, Rogers RS III, Rooke TW. Natural history of erythromelalgia: presentation and outcome in 168 patients. Arch Dermatol. 2000; 136(3):330–6.

Atlas of Clinical Vascular Medicine, First Edition. Edited by Jessica Mintz, Bruce L. Mintz, and Michael R. Jaff.

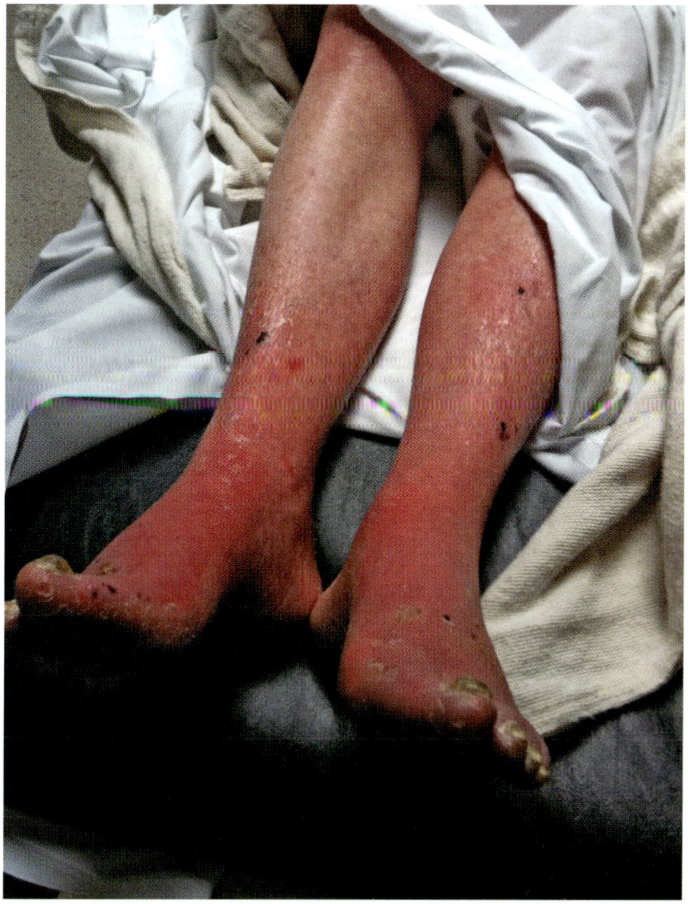

Erythromelalgia. Exquisitely painful edema and erythema of bilateral lower extremities occurring in response to warmth.

Heat-induced (Cholinergic) Urticaria

Joshua A. Beckman
Cardiovascular Division, Brigham and Women's Hospital, Boston, MA, USA

BRIEF CASE PRESENTATION
A 27-year-old man presents with hives on anterior shins after heavy exertion.

CHARACTERISTICS

- Heat-induced (cholinergic) urticaria is a rare form of urticaria characterized by the development of wheals limited to the areas of heat contact. It most commonly occurs after heavy sweating induced by exercise, warm environment, or sexual activities
- The lesions are typically, small, erythematous lesions with a halo 1–3 mm in diameter, although they may coalesce
- Cholinergic stimulation can elicit similar skin lesions, suggesting an allergic association
- It is most prevalent in the second and third decades of life

TREATMENT

Primary therapy involves treatment with antihistamines, although severe cases are commonly resistant to therapy. Many second line agents have been tried (danazol, beta-blockers, and scopolamine), but to little effect. Anti-IgE therapy has been shown effective in a single patient and may be a promising therapy in the future.

REFERENCES

Chang A, Zic JA. Localized heat uticaria. J Am Acad Dermatol. 1999; 41:354–6.
Horikawa, T., Fukunaga A, Nishigori C, et al. New concepts of hive formation in cholinergic urticaria. Curr Allergy Asthma Rep. 2009; 9(4):273–9.
Kozel MM, Sabroe RA. Chronic uticaria:aetiology, management and current future options. Drugs. 2004; 64(22):2515–36.

Atlas of Clinical Vascular Medicine, First Edition. Edited by Jessica Mintz, Bruce L. Mintz, and Michael R. Jaff.

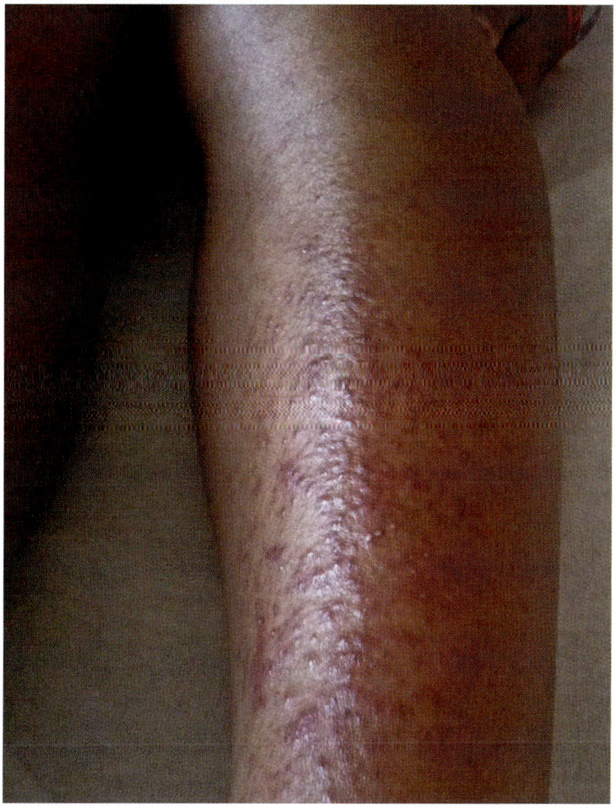

Heat-induced urticaria. Urticaria overlying the left pretibial region.
Courtesy Bruce L Mintz.

Erythema Ab Igne

Ido Weinberg and Michael R. Jaff
Massachusetts General Hospital, Boston, MA, USA

BRIEF CASE PRESENTATION

A 63-year-old female with chronic lower back pain uses a topical heating pad to manage the discomfort. On examination, there is a reticular, reddish-brown, pruritic, nontender, macular, nonblanching rash that has appeared on her lower back.

CHARACTERISTICS

- Erythema ab igne consists of local areas of erythema caused by chronic exposure to heat or infrared radiation. The skin manifestations are due to damage to the dermis and venous plexus systems
- Initially, erythema ab igne manifests as mild erythema, but with repeated exposures a darker hyperpigmentation may appear.
- Hot water bottles and topical heating pads are a common cause, although it was historically noted in women using open fires for cooking
- A modern manifestation of erythema ab igne has been described on the anterior thighs of people who use laptop computers

TREATMENT

Discontinued use of the heat source will often lead to gradual resolution of the discoloration.

REFERENCES

Arnold AW, Itin PH. Laptop computer-induced erythema ab igne in a child and review of the literature. Pediatrics. 2010; 126(5): e1227–30. Epub 2010 Oct 4. Review.

Beleznay K, Humphrey S, Au S. Erythema ab igne. CMAJ. 2010; 182(5):E228. Epub 2010 Feb 16.

Riahi RR, Cohen PR, Robinson FW, Gray JM. Erythema ab igne mimicking livedo reticularis. Int J Dermatol. 2010; 49(11): 1314–17.

Atlas of Clinical Vascular Medicine, First Edition. Edited by Jessica Mintz, Bruce L. Mintz, and Michael R. Jaff.

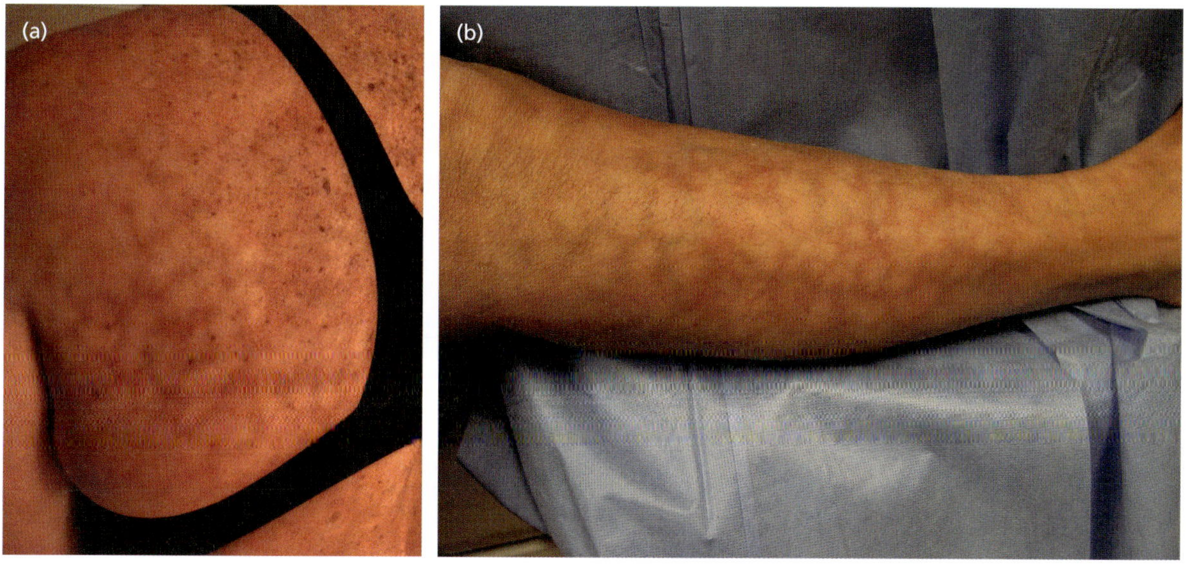

Reticular, non-blanching macular rash of erythema ab igne referred to as "toasted skin." Image (a) is courtesy of Bruce L. Mintz.

Hypothenar Hammer Syndrome

Jessica Mintz, Heather L. Gornik and John R. Bartholomew

Cleveland Clinic Heart and Vascular Institute and Cleveland Clinic Lerner College of Medicine, Case Western Reserve University, Cleveland, OH, USA

BRIEF CASE PRESENTATION

A patient presents for evaluation of cold, painful blanching of right fifth finger. There is an area of discoloration under the nail, which may be consistent with digital infarction vs. hemorrhage. There is blanching along the lateral aspect of the hypothenar eminence and a callous at an area which likely corresponds to the area of repetitive trauma in this patient.

CHARACTERISTICS

- Hypothenar hammer syndrome is an uncommon overuse syndrome secondary to repetitive trauma from hammering with the palm of ones hand. This action can result in ulnar artery occlusion
- Typically, it occurs in men involving the dominant hand in the setting of specific occupations requiring the use of the hypothenar portion of the hand (i.e., habitual hammering). Patients at highest risk are those who work as machinists, butchers, carpenters, and mechanics
- The cause of injury is due to the vulnerable anatomy of the ulnar artery after emerging from Guyon's canal. The superficial branch of the ulnar artery has very minimal protection. Palmar trauma can compress the artery against the hamate bone causing ulnar artery intimal damage and eventual thrombus formation
- Microemboli can occlude digital arteries distal to the trauma leading to ischemia
- Due to variable anatomy of the palmar circulation this condition can have protean manifestations
- Angiographic findings are typically "corkscrew" configurations suggesting that hypothenar hammer syndrome may have intrinsically abnormal arteries predisposing them to trauma
- Calluses over the hypothenar eminences, ischemia of the digits and a positive Allen's test are suggestive of this entity

TREATMENT

Conservative therapy most often leads to full recovery. Smoking cessation is recommended. Sympathectomy or surgical procedures have been tried with some success.

REFERENCES

Conn J, Bergan JJ, Bell JL. Hypothenar hammer syndrome: post traumatic digital ischemia. Surgery. 1970; 68(6):1122–8.

Cooke RA. Hypothenar hammer syndrome: a discrete syndrome to be distinguished from hand-arm vibration syndrome. Occupational Med. 2003; 53:320–4.

McCready RA, Bryant MA, Divelbiss JL. Combined thenar and hypothenar hammer syndromes: case report and review of the literature. J Vasc Surg. 2008; 48(3):741–4.

Atlas of Clinical Vascular Medicine, First Edition. Edited by Jessica Mintz, Bruce L. Mintz, and Michael R. Jaff.

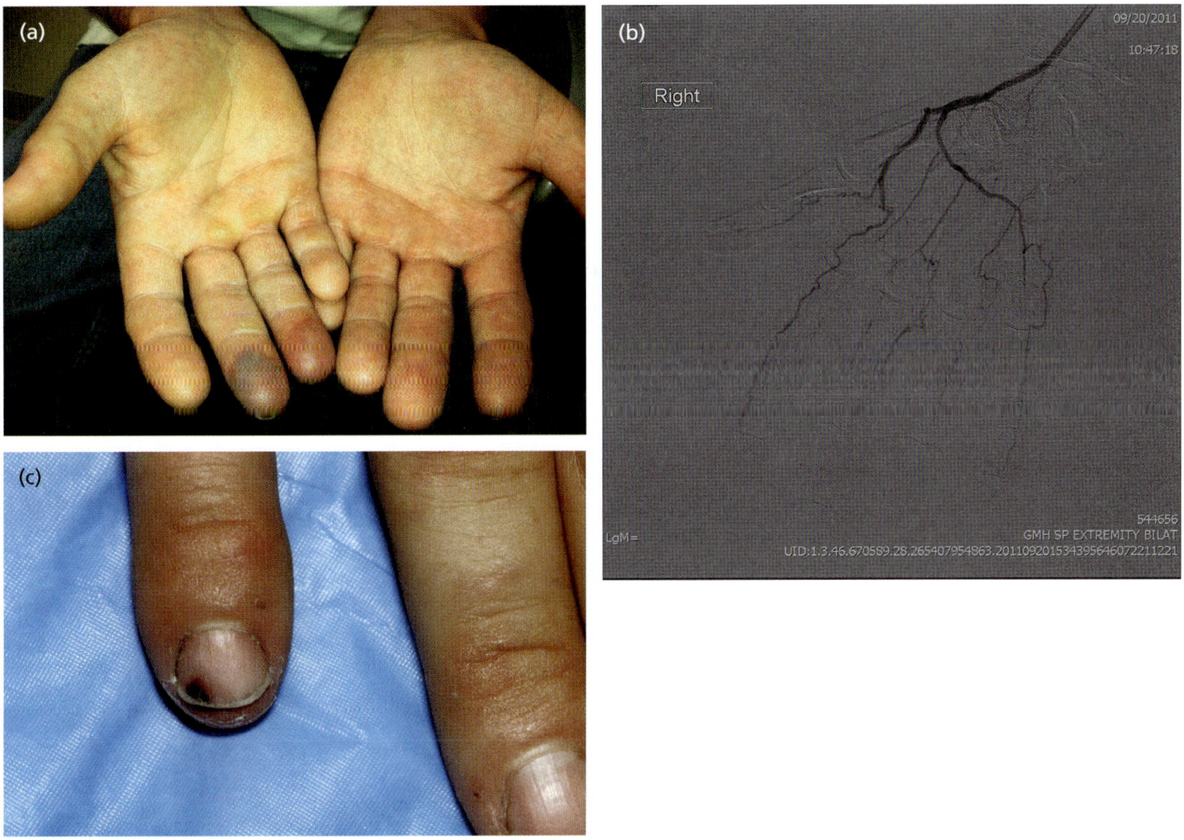

Hypothenar hammer syndrome. (a) Distal digital discoloration secondary to ulnar occlusion. Note bilateral calluses. (b) Angiogram with ulnar artery occlusion. (a and b) Courtesy of Bruce Gray and Stephen L Chastain. (c) Fifth digital ischemia consistent with digital infarction. Courtesy Cleveland Clinic Foundation Department of Vascular Medicine archives.

Arteriovenous Malformations

Robert J. Rosen
Department of Interventional Cardiology, Lenox Hill Hospital, New York, NY, USA

BRIEF CASE PRESENTATION

This 33-year-old man with known high flow arteriovenous malformations (AVM) of the mid-foot presents with venous hypertension and ischemic ulceration of the toes. Embolization of the proximal shunt with microspheres reduced both the steal and venous hypertension with rapid healing of the ulcerations.

CHARACTERISTICS

- AVM's arise from defects in vascular tissue during embryonic development, representing a direct connection between the arterial and venous systems
- An AVM with arterial components is considered a high-flow vascular malformation
- Complications of high-flow AVM's typically include swelling, pulsation, dilated veins, thickening of the skin due to venous hypertension, ulceration and bleeding due to ischemia from the proximal "steal"
- Non-invasive methods to define these lesions include duplex ultrasonography and magnetic resonance imaging, although transarterial catheter angiography is required to visualize the AVM and surrounding structure

TREATMENT

The mainstay of treatment lies in embolization. Vascular malformations of the hand are among the most difficult to treat. The vascular interventionalist must determine which arterial branches can be safely occluded without causing ischemia.

REFERENCES

Donnelly LF, Adams DM, Bisset GS III. Vascular malformations and hemangiomas: a practical approach in a multidisciplinary clinic. Am J Roentgenol. 2000; 174:597–608.

Herborn CU, Goyen M, Lauenstein TC, et al. Comprehensive time-resolved MRI of peripheral vascular malformations. Am J Roentgenol. 2003; 181(3):729–35.

Atlas of Clinical Vascular Medicine, First Edition. Edited by Jessica Mintz, Bruce L. Mintz, and Michael R. Jaff.

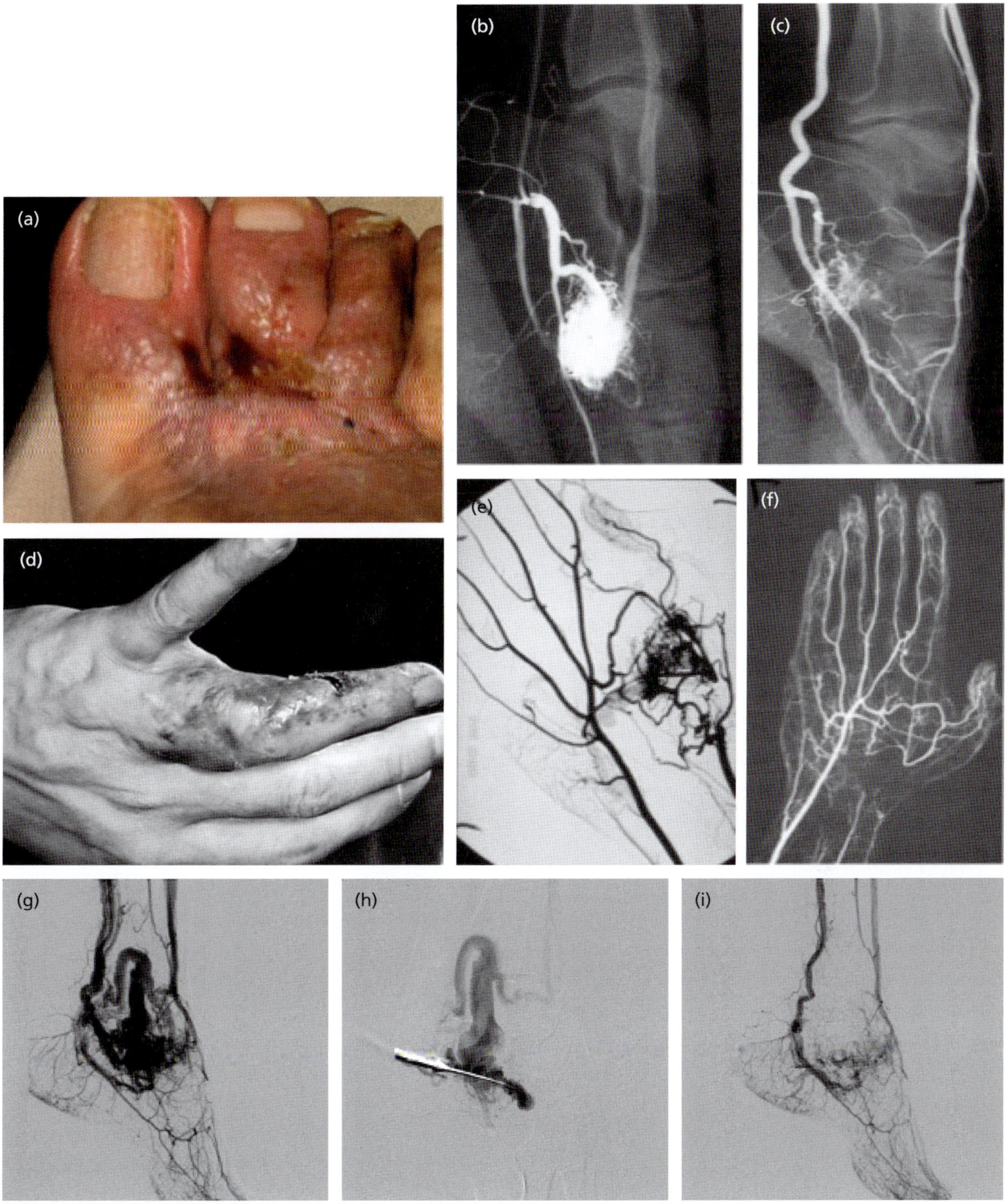

Arteriovenous malformation. (a) High-flow arteriovenous malformation in mid-foot causing venous hypertension and ischemic ulceration of toes. (b, c) Embolization of the proximal shunt with microspheres reduced both the steal and venous hypertension with rapid healing. (d) Complications of arteriovenous malformation in the hand including swelling, dilated veins, skin thickening, and ulceration. (e) High-flow hand lesion in a patient who has undergone prior ligation of the radial artery with recurrence of arteriovenous malformation. (f) Excellent angiographic result after microcatheter embolization using nBCA adhesive with preservation of the digital flow. (g) High-flow arteriovenous malformation in the ankle with multiple feeders. (h, i) Embolization using a direct puncture permitting occlusion of the nidus while not putting normal artery branches at risk.

Arteriovenous Fistula

Robert J. Rosen
Department of Interventional Cardiology, Lenox Hill Hospital, New York, NY, USA

BRIEF CASE PRESENTATION

A patient being treated in the intensive care unit following central venous catheter placement develops a loud bruit over the right infraclavicular space. An angiogram was performed and demonstrated a fistula between a branch of the subclavian artery and innominate vein.

CHARACTERISTICS

- An arteriovenous fistula is a direct communication between an artery and vein
- Most are acquired and are either traumatic or iatrogenic. If untreated, these fistulae may enlarge over time and can lead to high output congestive heart failure
- Failure to occlude the connection between artery and vein will result in collateral recruitment
- Diagnosis is confirmed by contrast angiography

TREATMENT

Successful treatment requires occlusion of the communication either by surgical ligation, covered stent grafts, or transarterial embolization. In many situations, the latter is the preferred treatment.

REFERENCES

Finlay DJ, Sanchez LA, Sicard GA. Subclavian artery injury, vertebral artery dissection, and arteriovenous fistulae following attempt at central line placement. Ann Vasc Surg. 2002; 16(6):774–8.

Halabi AR, Kandzari DE. The thrill is gone: Catheter-based exclusion of a posttraumatic arteriovenous fistula with a covered stent graft. Catheterization and Cardiovascular Interventions. 2005; 66(1):27–33.

Ziporin SJ, Ifune CK, MacConmara MP, Geraghty PJ, Choi ET. A case of external iliac arteriovenous fistula and high-output cardiac failure after endovenous laser treatment of great saphenous vein. J Vasc Surg. 2010; 51(3):715–9.

Atlas of Clinical Vascular Medicine, First Edition. Edited by Jessica Mintz, Bruce L. Mintz, and Michael R. Jaff.

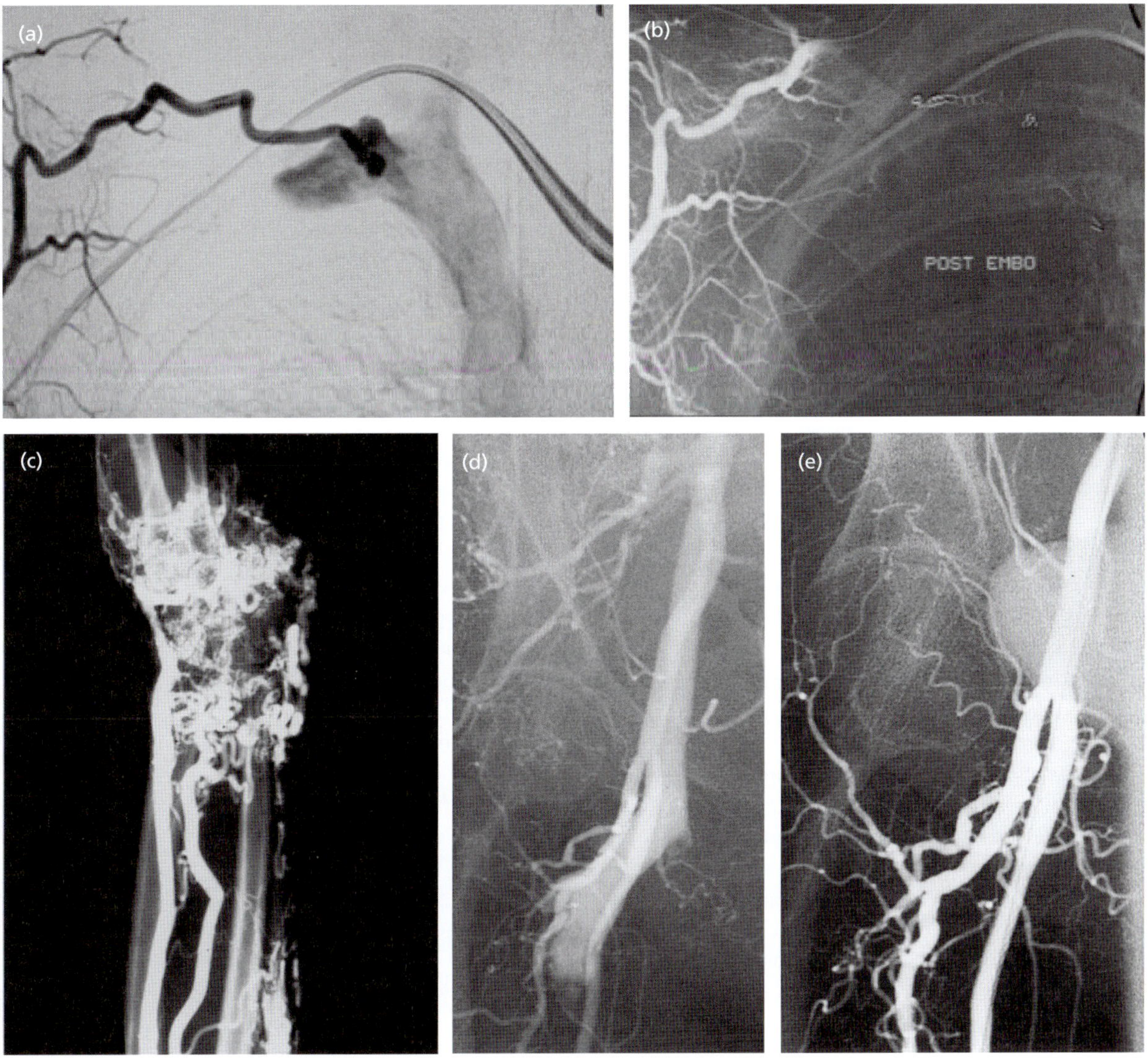

Arteriovenous fistula. (a) Angiogram of fistula between subclavian artery and brachiocephalic vein after central line placement. (b) Post-procedural angiogram after coil embolization. (c) This patient had penetrating trauma to the hand resulting in an arteriovenous fistula. Numerous surgical ligations were performed with failure to occlude actual site of AV communication. The resulting lesion is as complex as a congenital arteriovenous malformation. Note amputations performed in an effort to control the fistula.
(d & e): Post-catheterization side-to-side fistula (d) from right SFA treated with placement of covered stent (e).

Pelvic Arteriovenous Malformations

Robert J. Rosen
Department of Interventional Cardiology, Lenox Hill Hospital, New York, NY, USA

BRIEF CASE PRESENTATION

This young woman with pelvic arteriovenous malformation (AVM) underwent a 14-hour operation to ligate all feeding branches ("skeletonization") with apparent excellent immediate angiographic result. She then presented with recurrent symptoms of pelvic pain 4 months later, at which time a repeat angiography (selective study of the middle sacral artery) demonstrated recurrence of the lesion supplied by numerous collaterals.

CHARACTERISTICS

- The pelvis is a relatively common location for congenital arteriovenous malformation and occurs in both males and females
- Clinically, pelvic AVM's may be asymptomatic or may present with pain, gastrointestinal or urologic symptoms, bleeding, or rarely high-output states
- The hypogastric artery is the most common feeding vessel, but in some cases the supply can be quite complex, including both hypogastric arteries, the inferior mesenteric artery, and multiple other regional branches

TREATMENT

Embolization is the mainstay of treatment, the goal being penetration and eradication of the central nidus. Asymptomatic pelvic AVMs may be observed without intervention. Absolute indications for treatment include pain, functional impairment, high output heart failure, or hemorrhage.

REFERENCES

Calligari KD, Sedlacek TV, Savarese RP, et al. Congenital pelvic arteriovenous malformations: long term follow up care in two cases and review of the literature. J Vasc Surg. 1992; 16(1): 100–8.

Mallios A, Laurian C, Houbballah R, Gigou F, Marteau. Curative treatment of pelvic arteriovenous malformation – an alternative strategy:transvenous intra-operative embolisation. Eur J Vasc Endovasc Surg. 2011; 41(4);548–53.

Atlas of Clinical Vascular Medicine, First Edition. Edited by Jessica Mintz, Bruce L. Mintz, and Michael R. Jaff.

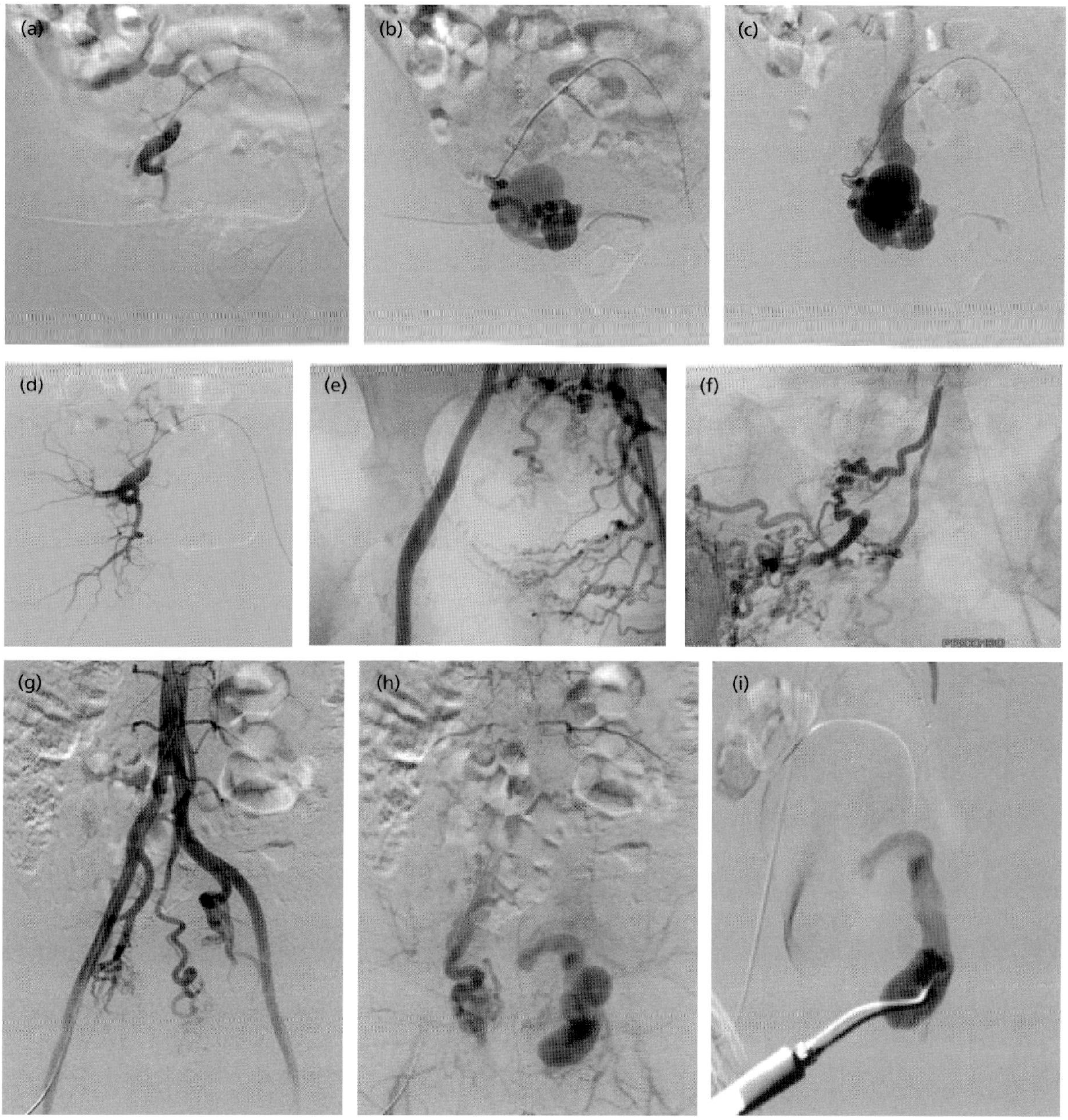

Pelvic arteriovenous malformation. (a–d) Images of right pelvic arteriovenous malformations supplied by the right hypogastric (a) artery with aneurysmal dilatation of a draining vein (b, c). In this type of lesion with simple architecture, super selective embolization using liquid agents is usually curative (d). (e) Patient with pelvic arteriovenous malformation following surgical ligation of all feeding branches. Initial angiographic result appears good. (f) Repeat angiograph 4 months later (after symptoms recurred) demonstrates resupply of lesion via numerous collaterals. (g) Complex pelvic arteriovenous malformation with multiple feeder vessels. (h) Later phase image demonstrates filling of aneurysmal hypogastric veins. (i) Treatment may be carried out by direct percutaneous puncture and embolization of the draining veins.

Pulmonary Arteriovenous Malformations

Robert J. Rosen

Department of Interventional Cardiology, Lenox Hill Hospital, New York, NY, USA

BRIEF CASE PRESENTATION

A patient with known hereditary hemorrhagic telangiectasia presents with brain absess.

CHARACTERISTICS

- Pulmonary arteriovenous malformations can be seen as an isolated anomaly or as part of hereditary hemorrhagic telangiectasia (Osler–Weber–Rendu Syndrome), in which case they tend to be multiple
- Anatomically, they are usually simple fistula-like arteriovenous connections with a single feeding pulmonary artery branch and a single draining pulmonary vein
- Clinical presentation can include stroke or brain absess (due to paradoxical embolization), arterial desaturation, and rarely, hemoptysis

TREATMENT

These malformations are usually curable with embolization, using coils, vascular plugs, or detachable balloons.

REFERENCES

Cottin V, Chinet T, Lavole A, et al. Pulmonary arteriovenous malformations in hereditary hemorrhagic telangiectasia: a series of 126 patients. Medicine (Baltimore). 2007; 86(1): 1–17.

Faughnan ME, Lui YW, Wirth JA, et al. Diffuse pulmonary arteriovenous malformations: characteristics and prognosis. Chest. 2000; 117(1):31–8.

Pollak JS, Saluja S, Thabet A, et al. Clinical and anatomic outcomes after embolotherapy of pulmonary arteriovenous malformations. J Vasc Interv Radiol. 2006; 17(1):35–45.

Trembath RC, Thomson JR, Machado RD, et al. Clinical and molecular genetic features of pulmonary hypertension in patients with hereditary hemorrhagic telangiectasia. NEJM. 2001; 345(5):325–34.

Atlas of Clinical Vascular Medicine, First Edition. Edited by Jessica Mintz, Bruce L. Mintz, and Michael R. Jaff.

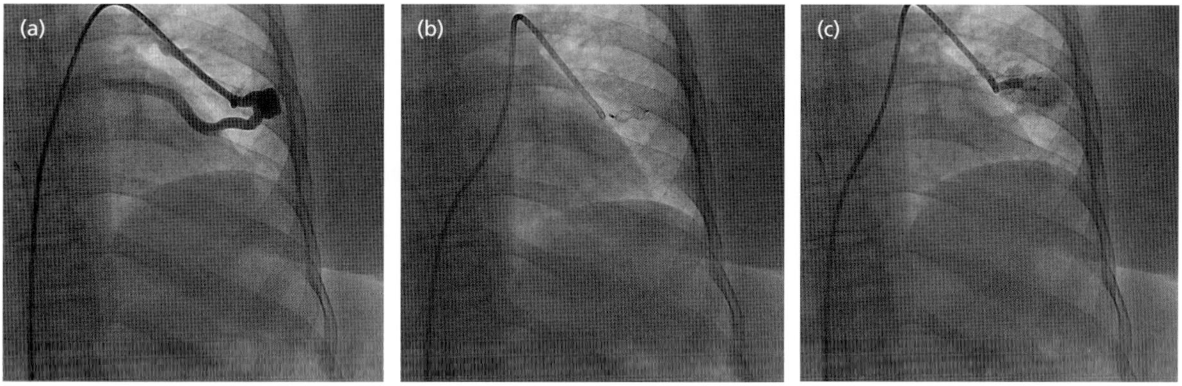

Pulmonary arteriovenous malformation (a), occluded with Amplatzer Plug (b) with no further filling seen on post embolization angiogram (c).

Parkes–Weber Syndrome

Robert J. Rosen
Department of Interventional Cardiology, Lenox Hill Hospital, New York, NY, USA

> **BRIEF CASE PRESENTATION**
> An 8-year-old female presents with multiple venous varicosities, port wine stains, and dyspnea. She was initially diagnosed with Klippel–Trenaunay syndrome (KTS); however, during this admission, echocardiography demonstrated pulmonary hypertension.

CHARACTERISTICS

- Parkes–Weber syndrome (PWS) is characterized by KTS abnormalities (hemi-hypertrophy, port wine stain and varicose veins) along with arteriovenous fistulae resulting in dyspnea and congestive heart failure
- Some cases of PWS are associated with *RASA1* gene
- Cases uniformly presents in childhood

TREATMENT

Supportive care includes graduated compression stockings to reduce edema and adequate pain management. Care must be taken when considering intervention for superficial venous varicosities, particularly with congenital absence of the deep venous system. Port wine stains are of cosmetic concern and can lead to psychosocial morbidity. Special attention should be paid to the psychological impact that is often associated with this disorder. If high output congestive heart failure develops, embolization of the arteriovenous fistulae must be considered.

REFERENCES

Garzon MC, Huang JT, Enjoiras O, Frieden BJ. Vascular malformations Part II: associated syndromes. J Am Acad Dermatol. 2007; 56(4):541–60.

Revencu N, Boon LM, Mulliken JB, et al. Parkes–Weber syndrome, vein of Galen aneurysmal malformation, and other fast-flow vascular anomalies are caused by *RASA1* mutations. Hum Mutat. 2008; 29(7):959–65.

Atlas of Clinical Vascular Medicine, First Edition. Edited by Jessica Mintz, Bruce L. Mintz, and Michael R. Jaff.

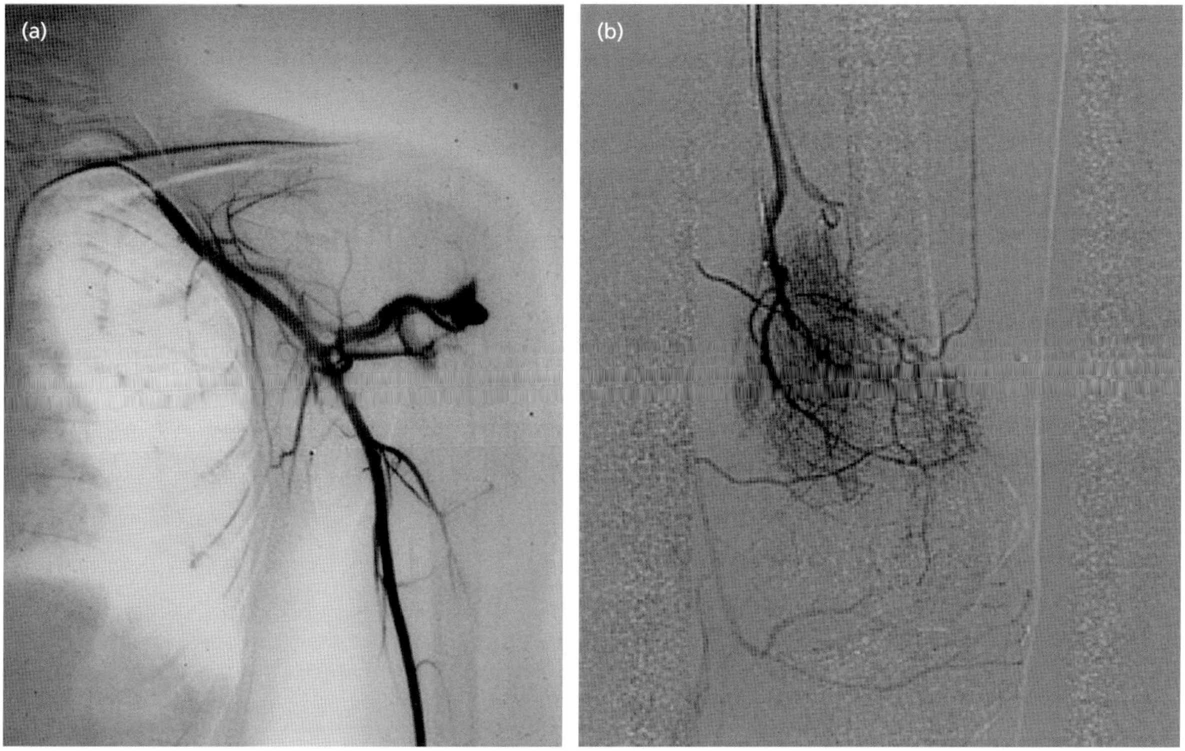

Parkes–Weber syndrome. Both images demonstrate a patient with venous appearing malformations found to have an arterial component on angiography.

Klippel–Trenaunay Syndrome

Bruce L. Mintz
Morristown Medical Center, Morristown, NJ, USA

BRIEF CASE PRESENTATION

This 55-year-old female presents for evaluation of marked venous varicosities associated with hemi-hypertrophy of her lower extremity and port wine stains which have been present since birth.

CHARACTERISTICS

- Klippel–Trenaunay syndrome (KTS) consists of the classic triad of muscular hemi-hypertrophy, varicose veins and port wine skin lesion which are present from birth
- KTS has been considered a sporadic mutation, although some literature demonstrates that it may occur with an autosomal dominant inheritance pattern
- Port wine stains are often unilateral and do not cross the midline. These lesions can become progressively darker with age. Although varicosities are often found at birth they can proliferate at puberty and may be associated with arteriovenous malformations

TREATMENT

Treatment should be tailored to the patient. Though surgical intervention may be helpful, a conservative approach including compression wraps and manual therapy are usually most appropriate.

REFERENCES

Garzon M, Huang J, Enjolras O, Frieden I. Vascular malformations. Part II: associated syndromes. J Am Acad Dermatol. 2007; 56(4):541–64.

Jacob AG, Driscoll DJ, Shaughnessy WJ, Stanson AW, Clay RP, Gloviczki P. Klippel–Trénaunay syndrome: spectrum and management. Mayo Clin Proc. 1998; 73(1):28–36.

Weber FP. Angioma-formation in connection with hypertrophy of limbs and hemi-hypertrophy. Br J Dermatol 1907; 19: 231–5.

Atlas of Clinical Vascular Medicine, First Edition. Edited by Jessica Mintz, Bruce L. Mintz, and Michael R. Jaff.

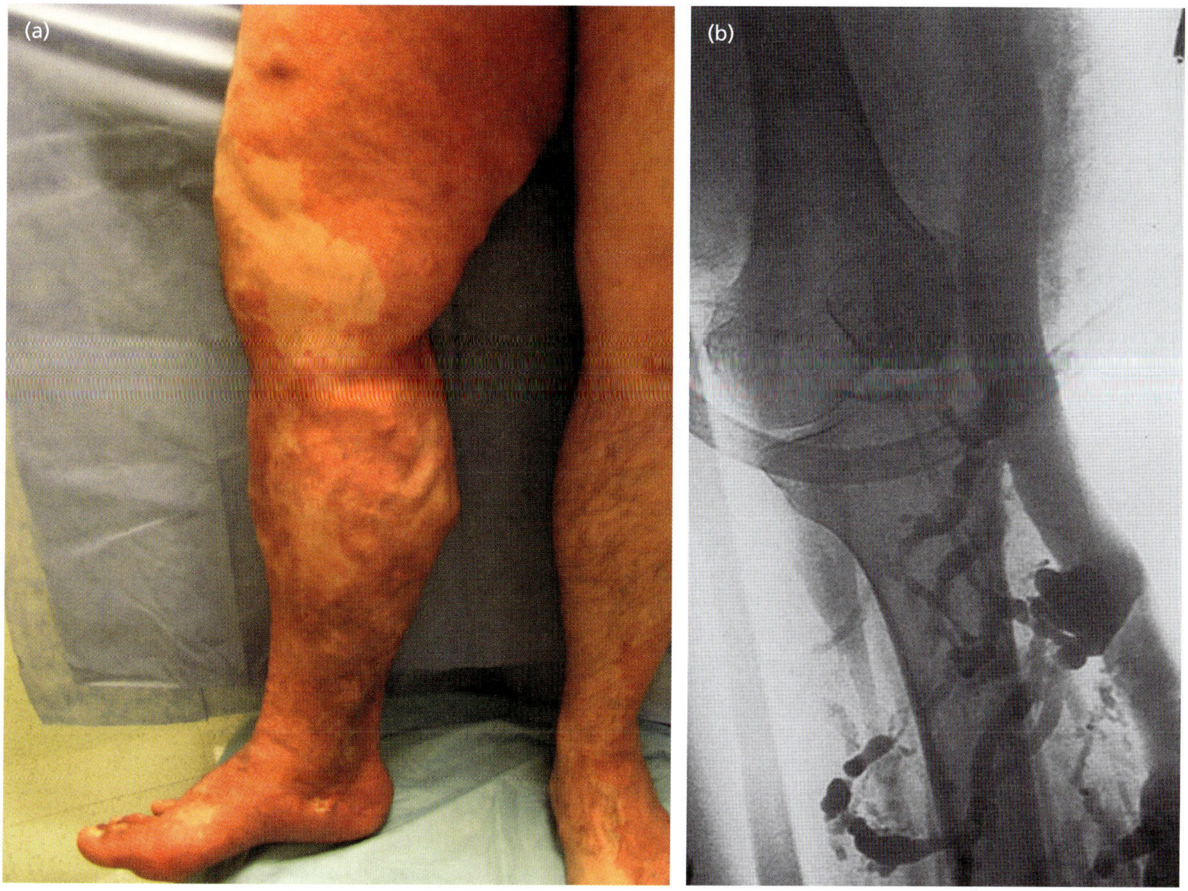

Klippel–Trenaunay syndrome. (a) Port wine stains, hemi-hypertrophy, and venous varicosities of the right lower extremity; all classic for Klippel–Trenaunay syndrome. Courtesy of Michael R. Jaff. (b) Contrast venography showing a large anomalous vein. Courtesy of Robert Rosen.

Lymphatic Malformations

Robert J. Rosen
Department of Interventional Cardiology, Lenox Hill Hospital, New York, NY, USA

> **BRIEF CASE PRESENTATION**
> A patient presents with right-sided upper extremity swelling with an associated non-pulsatile mass which has been present since birth.

CHARACTERISTICS

- Lymphatic malformations (LM) are congenital lesions, clinically categorized as macrocystic (>2 cm) or microcystic (<2 cm)
- LM are a collection of fluid filled spaces lined by one layer of endothelium. In some cases they are mixed with cavernous venous components
- Head and neck lesions are most common, followed by trunk, extremities and visceral lesions
- These malformations typically grow in proportion to the patients growth and have equal predilection for both sexes
- Infection, inflammation, and hemorrhage into cysts are the most common complications

TREATMENT

Treatment is multifaceted and includes ultrasound or fluorscopic-guided direct puncture followed by injection of different agents including doxycyline, bleomycin and sodium tetradecyl sulfate, absolute ethanol, or OK 432. Laser and surgical treatment modalities have also been described.

REFERENCES

Blei F. Cogenital lymphatic malformations. Ann NY Acad Sci. 2008; 1131:185–94.

Garzon MC, Huang JT, Enjoiras O, Frieden IJ. Vascular malformations Part II: associated syndromes. J Am Acad Dermatol. 2007; 56(4):541–64.

Atlas of Clinical Vascular Medicine, First Edition. Edited by Jessica Mintz, Bruce L. Mintz, and Michael R. Jaff.

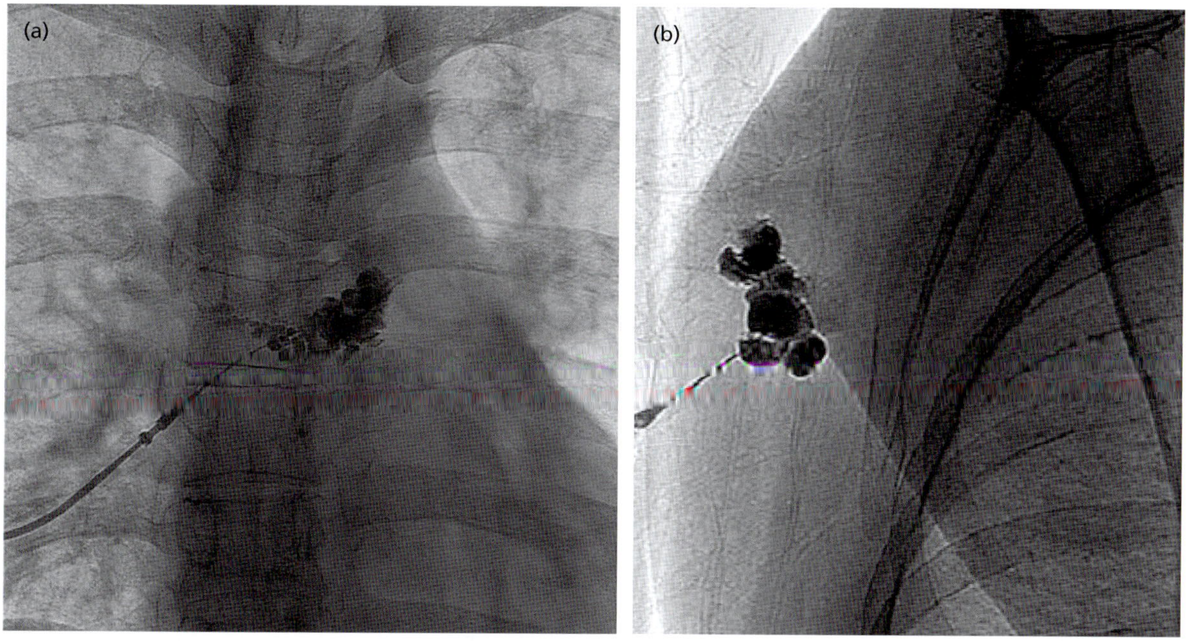

Lymphatic Malformations. Direct puncture study of subcutaneous chest wall lymphatic malformation demonstrates multicystic space from which clear lymph fluid was aspirated. Sclerotherapy was then carried out using doxycycline (a, b).

Hemangioma of Infancy

Jason Q. Alexander, Kate Zimmerman, Stephen C. Nelson and Timothy Sullivan
Minneapolis Heart Institute, Minneapolis, MN, USA

BRIEF CASE PRESENTATION

A 12-week-old infant presents with an expanding lesion over the eye. The parents initially describe the lesion as a "scratch" that was not present at birth but developed after the first few weeks. The lesion began to rapidly grow and impinge upon the infant's right eye.

CHARACTERISTICS

- Hemangioma of infancy is a vascular tumor which arises due to endothelial hyperplasia
- Most hemangiomas of infancy go through a period of nascent growth, followed by proliferation, followed by involution
- Fifty percent of hemangiomas of infancy will involute by 5 years and 90% by 9 years
- Certain hemangiomas of infancy mandate treatment. Hemangiomas of infancy may intrude upon the eye, preventing normal visual development, and may impact the lips preventing normal attachment for feeding
- Hemangiomas may also obstruct critical structures such as the trachea during the proliferative phase leading to life threatening complications. Finally, hemangiomas of infancy may bleed either internally (e.g., hemangiomas of the liver) or superficially from ulceration
- When a hemangioma of infancy involutes it will often leave a skin tag

TREATMENT

Traditional treatment of symptomatic hemangioma of infancy involved combination of systemic steroids, injectable steroids and/or chemotherapy as well as surgical resection. These treatments demonstrated poor results and were traumatic for the infant. In 2008, propranolol was discovered to have dramatic impacts on the resolution of hemangioma of infancy. Oral propanolol leads to clinical evidence of resolution within days of starting treatment. Often hemangioma of infancy will demonstrate complete resolution after a few months of treatment.

REFERENCES

Leaute-Labreze C, Dumas De La Roque E, Hubiche T, Boralevi F, Thambo JB, Taieb A. Propranolol for severe hemangiomas of infancy. NEJM. 2008; 358:2649–51.

Mulliken JB, Glowacki J. Hemangiomas and vascular malformations in infants and children: A classification based on endothelial characteristics. Plast Reconstr Surg. 1982; 69:412.

Nelson SC, Alexander JQ, Sidman JD, Zimmerman KP. Outpatient propranolol is safe and effective in infants with hemangioma: single center experience and case report of successful use in a seven-year-old. Poster session presented at: American Society of Pediatric Hematology and Oncology 24th Annual Meeting, Baltimore, Maryland, April 13–16, 2011.

Atlas of Clinical Vascular Medicine, First Edition. Edited by Jessica Mintz, Bruce L. Mintz, and Michael R. Jaff.

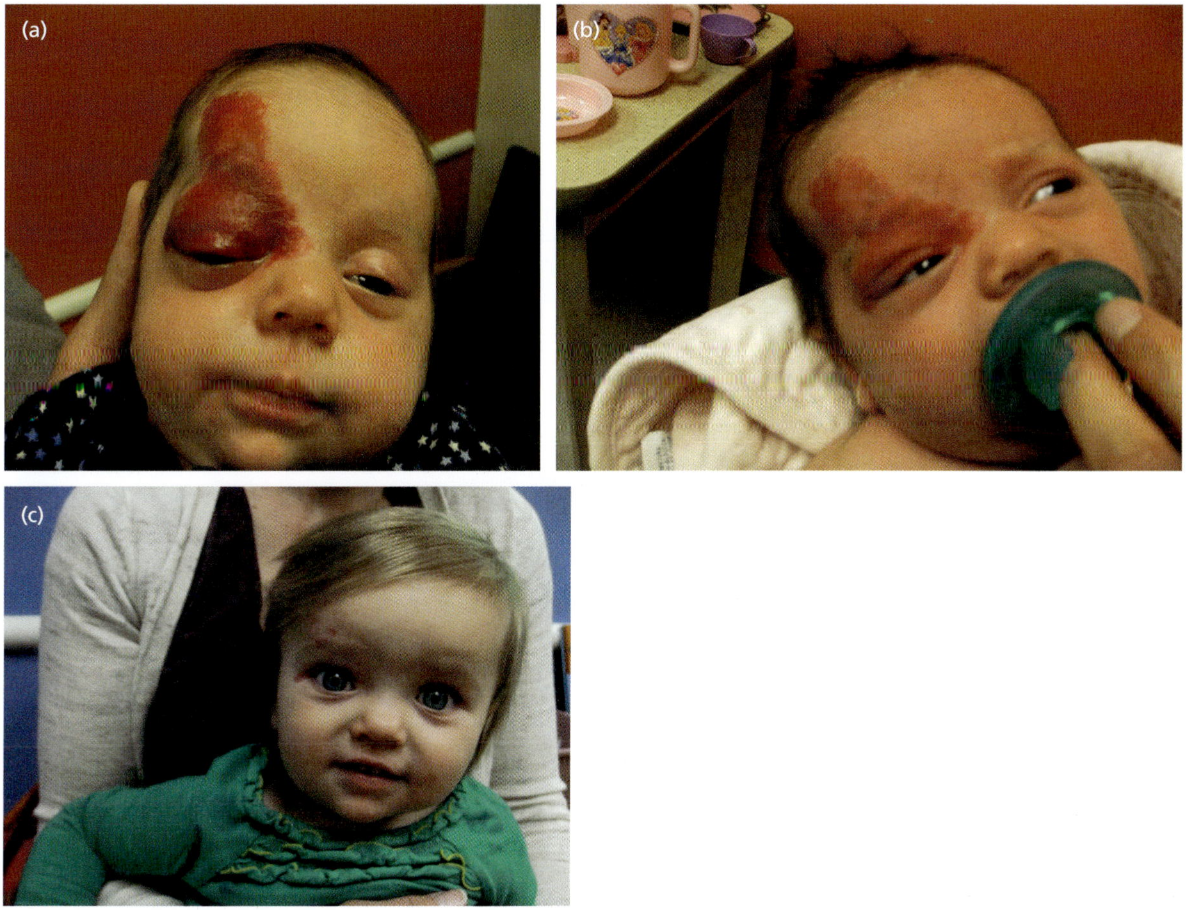

Infantile hemangioma. (a) Proliferative vascular tumor surrounding an infant's right eye prior to treatment with propranolol. (b) Patient able to open eyelid within days of propranolol therapy. (c) Infantile hemangioma after treatment with propranolol.

Kaposiform Hemangioendothelioma

Robert J. Rosen
Department of Interventional Cardiology, Lenox Hill Hospital, New York, NY, USA

BRIEF CASE PRESENTATION
An infant presents with a purpuric indurated lesion over his left thigh, which rapidly increased in size to include the entire length of the leg.

CHARACTERISTICS

- Kaposiform hemangioendothelioma (KHE) is a rare vascular tumor, which presents in infants from birth or in the first few months of life
- This lesion of infancy behaves much more aggressively than an infantile hemangioma. The neoplasm is generally cutaneous in origin, however it often infiltrates to deeper issue
- In severe cases it may be associated with marked shunting, resulting in high-output cardiac failure
- Lesions are ill-defined often occurring in truncal locations with local invasion. However, they are not known to cause distant metastasis
- It may also cause severe hematologic abnormalities, specifically, thrombocytopenia (Kasabach–Merritt Syndrome)
- KHE has a significant mortality rate related both to the coagulopathy as well as local tumor infiltration

TREATMENT

Treatment is primarily pharmacologic with corticosteroids, and more recently, with propranolol. Embolization has been successfully employed in some cases. Careful monitoring for coagulopathy is essential.

REFERENCES

Hermans DJ, van Beynum IM, van der Vijver RJ, Kool LJ, de Blaauw I, van der Vleuten CJ. Kaposiform hemangioendothelioma with Kasabach–Merritt syndrome: a new indication for propranolol treatment. J Pediatr Hematol Oncol. 2011; 33(4):171–3.

Lyons LL, North PE, Mac-Mourne Lai F, Stoler MH, Folpe AL, Weiss SW. Kaposiform hemangioendothelioma: A study of 33 cases, emphasizing its pathologic, immunophenotypic, and biologic uniqueness from juvenile hemangioma. Am J Surg Pathol. 2004; 28(5):559–68.

Wassef F. Vascular tumors and pseudotumors. Kaposiform hemangioendothelioma. Ann Pathol. 2011; 31(4):260–5.

Atlas of Clinical Vascular Medicine, First Edition. Edited by Jessica Mintz, Bruce L. Mintz, and Michael R. Jaff.

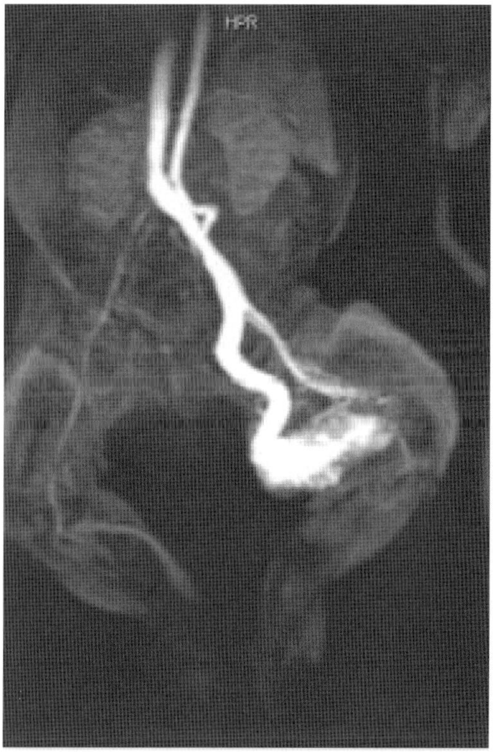

Kaposiform hemangioendothelioma (KHE). Angiogram of an infant with Kaposiform angioendothelioma. Note the left-sided prominent vascular lesion.

Glomus Tumor

Robert J. Rosen
Department of Interventional Cardiology, Lenox Hill Hospital, New York, NY, USA

> **BRIEF CASE PRESENTATION**
> A patient with a known glomus tumor presents for a second opinion regarding intervention.

CHARACTERISTICS

- Glomus tumors belong to a broad category of rare, mostly benign, slow growing, hypervascular neoplasms known as paragangliomas
- These locally invasive tumors have been reported in multiple locations, most commonly the middle ear (glomus tympanicum), the carotid body (carotid body tumor), and the cervical vagus nerve (glomus vagale)
- Glomus tumors commonly affect middle-age women and are most often left sided lesions originating from cells which develop in the neural crest

TREATMENT

Surgery and/or catheter-directed embolization is the treatment of choice for glomus tumors. Preoperative embolization can facilitate surgical excision and minimize intraoperative blood loss.

REFERENCES

Fletcher CDM, Unni KK, Mertens F (eds): Glomus tumors. In: World Health Organization Classification of Tumors. Pathology and Genetics of Tumors of Soft Tissue and Bone. IARC Press: Lyon, 2002, p. 137.

Luna-Ortiz K, Rascon-Ortiz M, Villavicencio-Valencia V, Granados-Garcia M, Herrera-Gomez A. Carotid body tumors: a review of 20 years of experience. Oral Oncol. 2005; 41(1): 56–61.

Sajid MS, Hamilton G, Baker DM. A multicenter review of carotid body tumor management. Eur J Vasc Endovasc Surg. 2007; 34(2):127–30.

Atlas of Clinical Vascular Medicine, First Edition. Edited by Jessica Mintz, Bruce L. Mintz, and Michael R. Jaff.

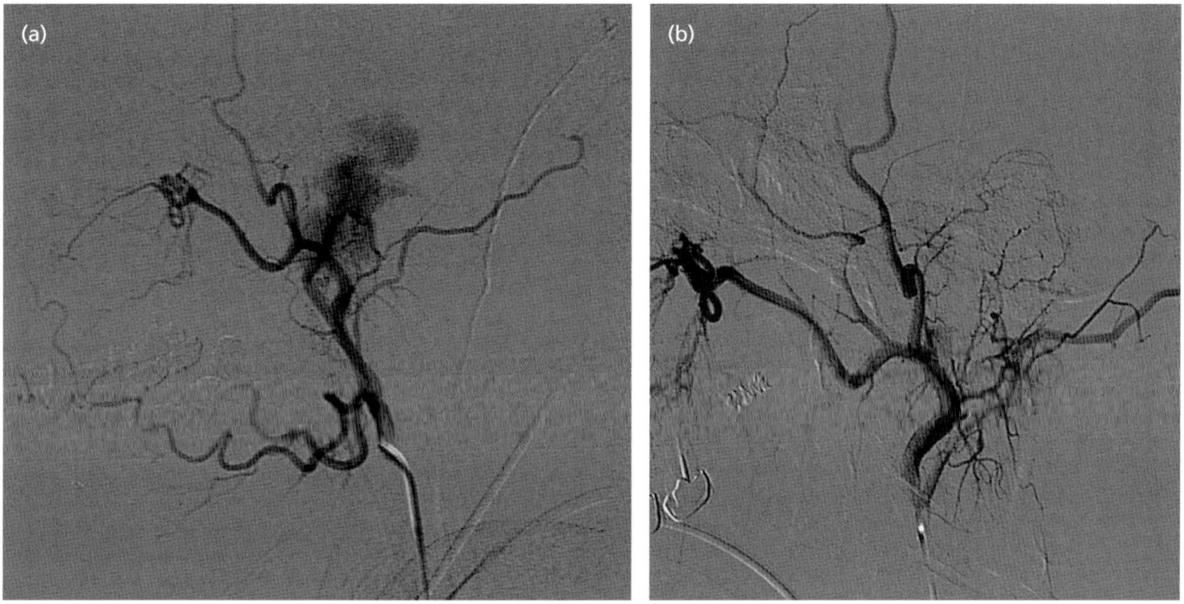

Glomus tumors. (a) Lateral external carotid artery angiogram revealing a cervical hypervascular mass embedded within the temporal bone and supplied by the ascending pharyngeal artery. (b) Post-embolization angiogram confirming a successful devascularization and preservation of flow to the uninvolved branches.

Angiomyolipoma

Robert J. Rosen
Department of Interventional Cardiology, Lenox Hill Hospital, New York, NY, USA

> **BRIEF CASE PRESENTATION**
> A 55-year-old man presents with right flank discomfort. He has a history of hypertension. A renal ultrasound suggests a tumor in the superior pole of the right kidney.

CHARACTERISTICS

- Angiomyolipomas are benign hypervascular tumors containing adipose and smooth muscle tissues
- They are associated with tuberous sclerosis but can be isolated lesions with the kidney being the organ most commonly involved
- Secondary hemorrhage from focal aneurysms within the mass is the most frequent complication; however, symptoms can also be caused by a tumor mass effect
- Tumor blood supply is usually derived from a single intrarenal branch artery. This facilitates super selective angiography and transcatheter embolization

TREATMENT

Recommendations for treatment are usually based upon size of the lesion and/or patient symptoms. Management options include observation, embolization, surgical resection of mass or nephrectomy. In the case presented above a selective renal angiogram demonstrates a hypervascular mass. Following embolization with nBCA adhesive, renal angiogram shows de-vascularization of the tumor with preservation of normal renal parenchyma.

REFERENCES

Hadley DA, Bryant LJ, Ruckle HC. Conservative treatment of renal angiomyolipomas in patients with tuberous sclerosis. Clin Nephrol. 2006; 65(1):22–7.

Schwartz MJ, Smith EB, Trost DW, Vaughan ED Jr. Renal artery embolization: clinical indications and experience from over 100 cases. BJU Int. 2007; 99(4):881–6.

Atlas of Clinical Vascular Medicine, First Edition. Edited by Jessica Mintz, Bruce L. Mintz, and Michael R. Jaff.

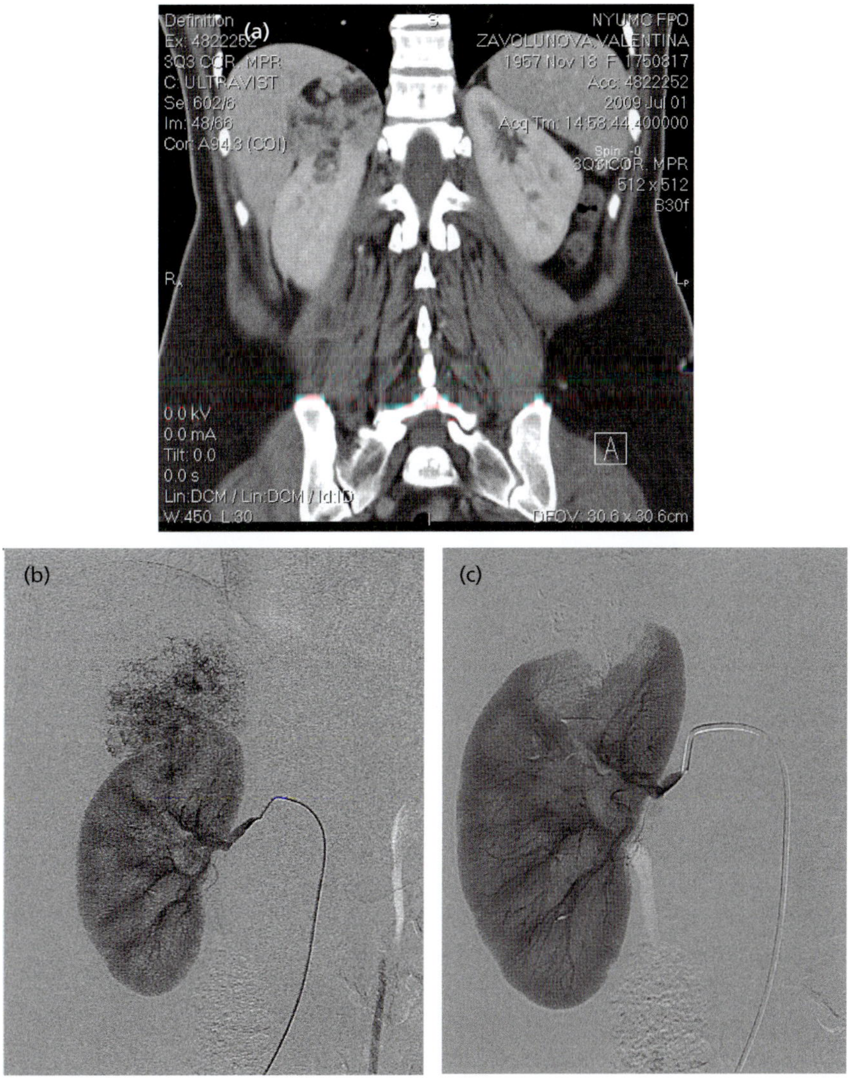

(a) Abdominal CT scan shows mass in upper pole of right kidney. Presence of fat is pathognomonic for angiomyolipoma. (b) selective right renal angiogram shows hypervascular mass containing pseudoaneurysms (c) following superselective embolization of feeding branch with nBCA, mass is devascularized with preservation of normal renal parenchyma.

Juvenile Nasopharyngeal Angiofibroma

Robert J. Rosen
Department of Interventional Cardiology, Lenox Hill Hospital, New York, NY, USA

BRIEF CASE PRESENTATION

A prepubescent male presents with refractory recurrent epistaxis. Patient was found to have an angiofibroma with the internal maxillary artery as the feeding vessel.

CHARACTERISTICS

- Juvenile nasopharyngeal angiofibroma is a benign hypervascular locally invasive tumor which typically presents as refractory epistaxis and recurrent infections in pre-pubertal and pubertal males
- Average age of onset is 15 years old
- It is frequently located in the posterior aspect of the nasal turbinates deriving blood supply from distal branches of internal maxillary artery
- These tumors are smooth masses in the nasopharynx that can be pale, purple or beefy red in appearance
- Staging of these tumors is made based upon invasion into local structures including the paranasal sinuses, orbital bones, skull base, and cavernous sinuses
- Diagnosis is made by history and physical as well as imaging including a CT scan, MRI and angiogram to evaluate feeding blood vessels

TREATMENT

The gold standard of treatment is surgical excision. Preoperative angiography can identify contributing arterial anatomy and allow for preoperative embolization to minimize bleeding complications.

REFERENCES

Cansiz H, Guvenc MG, Sekecioglu N. Surgical approaches to juvenile nasopharyngeal angiofobroma. J Craniomaxillofac Surg. 2006; 34(1):3–8.

Pryor SG, Moore El, Kasperbauer JL. Endoscopic versus traditional approaches for excision of juvenile nasopharyngeal angiofibroma. Laryngoscope. 2005; 115(7):1201–20.

Radkowski D, McGill T, Healy GB, et al. Angiofibroma. Arch Otolaryngol. 1996; 122(2):122–9.

Atlas of Clinical Vascular Medicine, First Edition. Edited by Jessica Mintz, Bruce L. Mintz, and Michael R. Jaff.

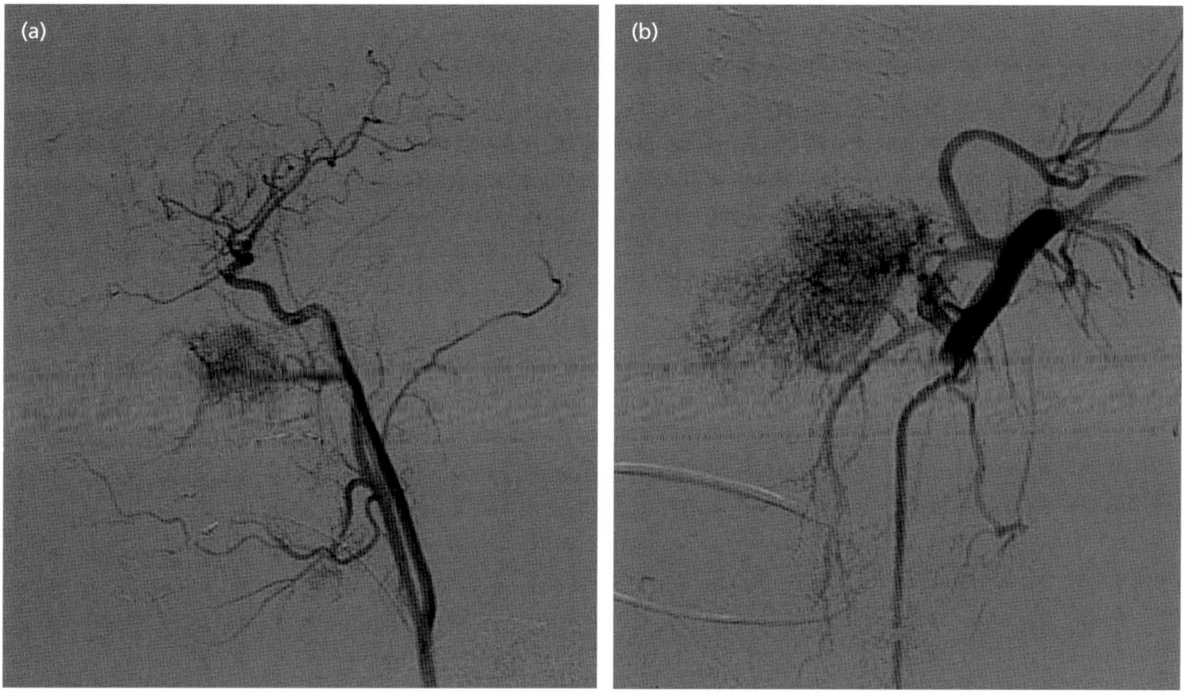

Juvenile nasopharyngeal angiofibroma. (a) Angiogram demonstrating an angiofibroma of the posterior aspect of the nasal turbinate (lateral projection). (b) Anterior – posterior projection.

Maffucci's Syndrome

Jessica Mintz, John R. Bartholomew and Heather L. Gornik

Cleveland Clinic Heart and Vascular Institute and Cleveland Clinic Lerner College of Medicine, Case Western Reserve University, Cleveland, OH, USA

BRIEF CASE PRESENTATION

A patient presents with a swollen leg and foot; she has no previous medical history.

CHARACTERISTICS

- Maffucci's syndrome is characterized by multiple soft tissue hemangiomias and enchondromas. Shortening of long bones occurs commonly
- Hemangiomas characteristically are blue subcutaneous nodules appearing most often in the hands, feet, arms, legs, and trunk
- Although hemangiomas and enchondromas of Maffucci's are not overtly malignant, the syndrome is associated with neoplastic changes

TREATMENT

No treatment is necessary for asymptomatic patients. Follow-up care and proper sub-specialty consultation for bone and skin lesions is required.

REFERENCES

Falk A, Allali F, El Hassani S, Hajjaj-Hassouni N. Maffucci's syndrome: a case report. Clin Rheumatol. 2005; 25:88–91.

Kaplan RP, Wang JT, Amron DM, Kaplan L. Maffucci's syndrome: two case reports with a literature review. J Am Acad Dermatol. 1993; 29:894.

Jermann M, Eid K, Pfammatter T, Stahel R. Maffucci's syndrome. Circulation. 2001; 104:1693.

Atlas of Clinical Vascular Medicine, First Edition. Edited by Jessica Mintz, Bruce L. Mintz, and Michael R. Jaff.

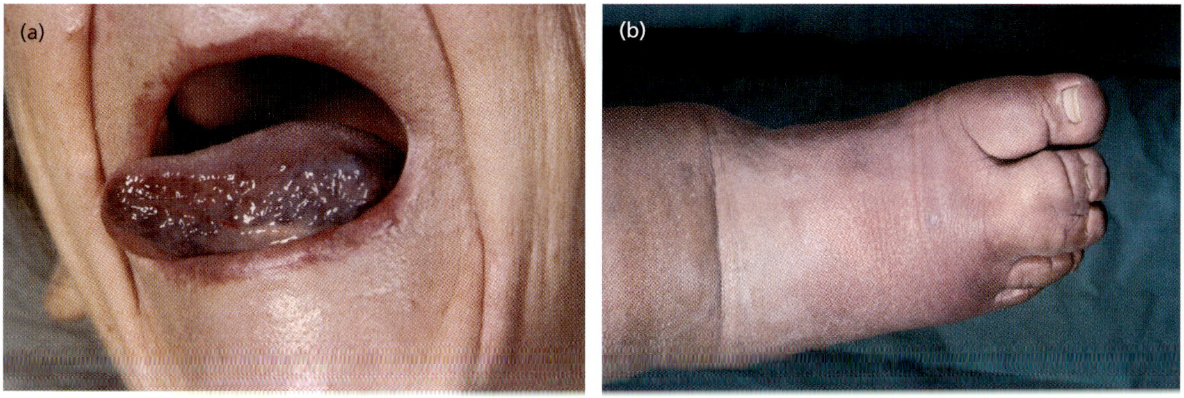

Maffucci's syndrome. (a) Tongue lesion in a patient with Maffucci's syndrome. (b) Lower extremity lymphedema, malformation of the toes and subtle dilated subcutaneous veins vs hemangiomas on the medial aspect of the foot and the metatarsal region in a patient with known Maffucci's syndrome. Image courtesy of the Cleveland Clinic Foundation Department of Vascular Medicine archives.

Osler–Weber–Rendu Syndrome

Jessica Mintz, John R. Bartholomew and Heather L. Gornik

Cleveland Clinic Heart and Vascular Institute and Cleveland Clinic Lerner College of Medicine, Case Western Reserve University, Cleveland, OH, USA

BRIEF CASE PRESENTATION

A patient presents with recurrent spontaneous epistaxis and multiple dilated capillaries which surround the oral mucosa.

CHARACTERISTICS

- An autosomal dominant condition which is also referred to as hereditary hemorrhagic telangiectasia (HHT). HHT type 1 has been associated with genetic mutations on chromosome 9q34
- Abnormal blood vessel formation leads to a constellation of symptoms depending on its organ involvement
- HHT is characterized by cutaneous telangiectasias of the lips, nose, and oral mucosa, recurrent epistaxis, and visceral arteriovenous malformations (see chapter 78) of the pulmonary, hepatic, cerebral or spinal vasculature

TREATMENT

Treatment options vary depending on the specific pattern of organ involvement. Nasal packing, humidification, cauterization and laser ablation are all treatments for the epistaxis, which occurs in the vast majority of patients. Topical agents may be used for skin involvement, whereas surgical and endovascular therapies may be necessary for visceral malformations.

REFERENCES

Fuchizaki U, Miyamori H, Kitagawa S, et al. Hereditary haemorrhagic telangiectasia (Rendu–Osler–Weber disease). Lancet. 2003; 362:1490–4.

Shovlin CL. Hereditary haemorrhagic telangiectasia: pathophysiology, diagnosis and treatment. Blood Reviews. 2010; 24(6): 203–19.

Online Mendelian Inheritance in Man [OMIM]. Reference #187300. www.omim.org.

Atlas of Clinical Vascular Medicine, First Edition. Edited by Jessica Mintz, Bruce L. Mintz, and Michael R. Jaff.

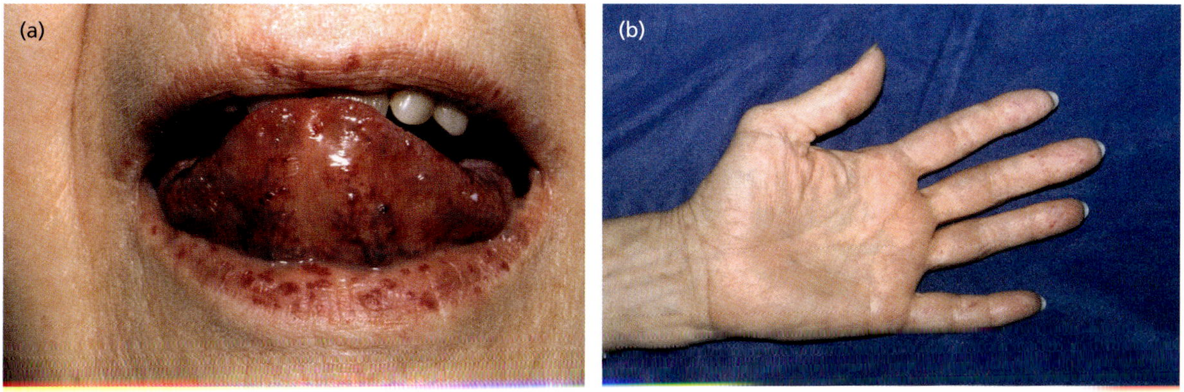

Osler–Weber–Rendu syndrome. (a) Oral mucosal capillary dilatation and telangiectasias in a patient with Osler–Weber–Rendu syndrome. (b) Cutaneous digital telangiectasias in a patient with Osler–Weber–Rendu syndrome. Images courtesy of Cleveland Clinic Foundation Department of Vascular Medicine archives.

Lymphangiosarcoma

Jessica Mintz, John R. Bartholomew and Heather L. Gornik

Cleveland Clinic Heart and Vascular Institute and Cleveland Clinic Lerner College of Medicine, Case Western Reserve University, Cleveland, OH, USA

BRIEF CASE PRESENTATION

Patient with longstanding lower extremity lymphedema presents with a tender, purple nodules of the leg, some with associated ulceration.

CHARACTERISTICS

- Lymphangiosarcoma is a malignancy of the endothelium and a rare complication of chronic lymphedema
- Lesions appear as purple and blue discoloration overlying the site of chronic lymphedema. These lesions may enlarge and ulcerate leading to skin necrosis and generally carry a poor prognosis given their propensity for rapid metastasis
- The majority of lymphangiosarcomas occur in patients multiple years following mastectomy complicated by chronic lymphedema
- Stewart–Treves syndrome is the eponym used to describe post-mastectomy lymphangiosarcoma

TREATMENT

Chemotherapy and radiation in conjunction with surgical management represent the mainstays of treatment. Treatment may also include amputation of the effected limb.

REFERENCES

Alessi E, Sala F, Berti E. Angiosarcoma in lymphedematous limbs. Am J Dermatopathol. 1986; 8:371.

Komorowski AL, Wysocki WM, Mitus J. Angiosarcoma in a chronically lymphedematous leg: an unusual presentation of Stewart–Treves syndrome. South Med J. 2003; 96(8):807–8.

Ruocco V, Schwartz RA, Ruocco E. Lymphedema: an immunologically vulnerable site for development of neoplasms. J Am Acad Dermatol. 2002; 47(1):124–7.

Stewart FW, Treves N. Lymphangiosarcoma in postmastectomy lymphedema: A report of six cases in elephantiasis chirurgica. Cancer. 1948; 1(1):64–81.

Atlas of Clinical Vascular Medicine, First Edition. Edited by Jessica Mintz, Bruce L. Mintz, and Michael R. Jaff.

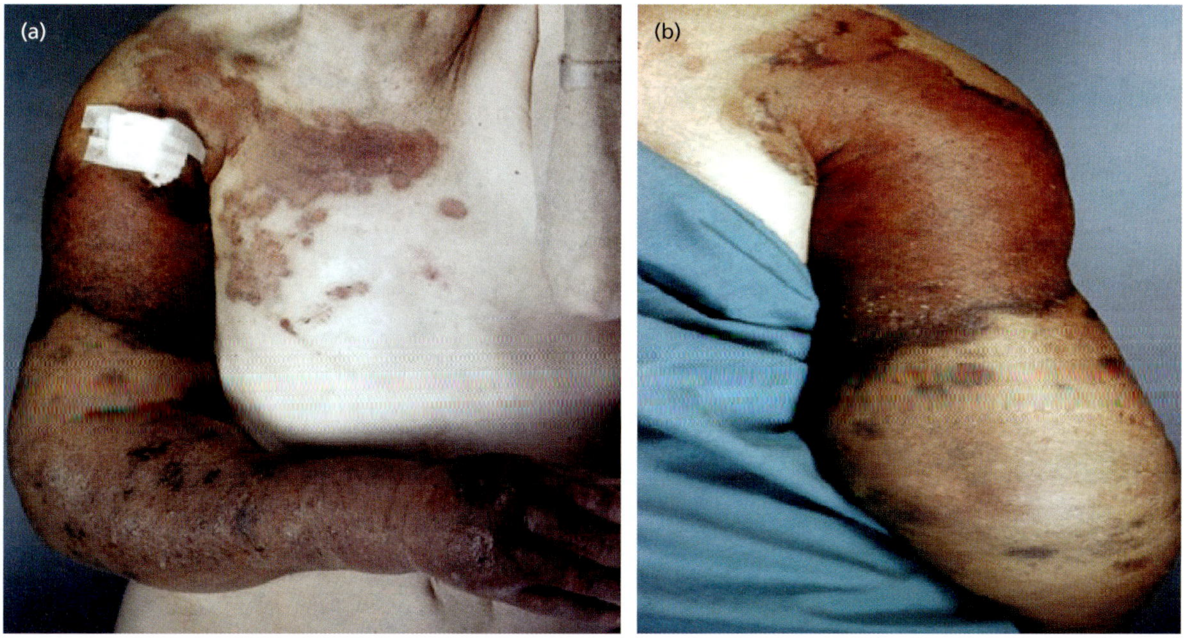

Lymphangiosarcoma. (a) Anterior view of right upper extremity purple lesions and nodules overlying edema in a patient post lymph node dissection. (b) Posterior image of lymphangiosarcoma of the right upper extremity.

Limb Ulceration Due to Malignant Melanoma

Jessica Mintz, John R. Bartholomew and Heather L. Gornik
Cleveland Clinic Heart and Vascular Institute and Cleveland Clinic Lerner College of Medicine, Case Western Reserve University, Cleveland, OH, USA

BRIEF CASE PRESENTATION

Patient presents with a non-healing ulceration on the lateral aspect of the right foot.

CHARACTERISTICS

- Dermatologic malignancy that usually originates as a nevus
- Classically described as a sudden and rapidly enlarging lesion with associated asymmetry, changes in color, and irregular borders
- Lesions often ulcerate or bleed
- Ulceration of a malignant melanoma may masquerade as a non-healing venous ulcer or hypertrophic granulation tissue. Despite proper wound care, the lesion may increase in size rather than heal

TREATMENT

Failure of a wound to show signs of progressive healing should prompt skin biopsy. All lesions suspicious for melanoma, especially pigmented lesions or any atypical-appearing wound, should be biopsied early in the treatment course. Diagnosis of melanoma warrants immediate referral to an experienced medical oncologist for further staging and discussion of treatment options.

REFERENCES

Koch SE, Lange JR. Amelanotic melanoma: the great masquerader. J Am Acad Dermatol. 2000; 42(5 Pt 1):731–4.

Roehigk HH, Young JR. Leg ulcers. In: Young JR, Olin JW, Bartholomew JR (eds). *Peripheral vascular diseases*, 2nd edition. St Louis: Mosby, 1996, Chapter 35, p. 632.

Thomas JM, Newton-Bishop J, A'Herm R, et al. Excision margins in high-risk malignant melanoma. NEJM. 2004; 350(8): 757–66.

Atlas of Clinical Vascular Medicine, First Edition. Edited by Jessica Mintz, Bruce L. Mintz, and Michael R. Jaff.

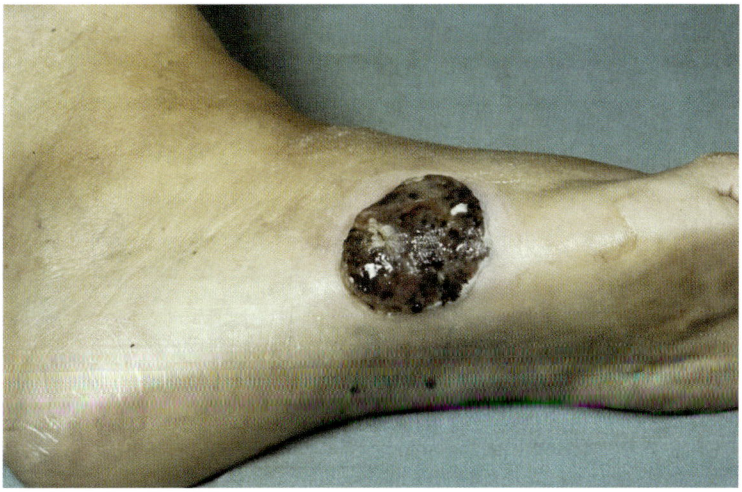

Malignant melanoma. Well-demarcated, circumferential, non-healing ulceration on the lateral aspect of the right foot. Image courtesy of Cleveland Clinic Foundation Department of Medicine archives.

Cutaneous Waldenström's Macroglobulinemia

Jessica Mintz, John R. Bartholomew and Heather L. Gornik
Cleveland Clinic Heart and Vascular Institute and Cleveland Clinic Lerner College of Medicine, Case Western Reserve University, Cleveland, OH, USA

BRIEF CASE PRESENTATION

Patient presents with bilateral nodular-pigmented purple papules on the surface of the anterior knees extending from the thigh to the calf.

CHARACTERISTICS

- Waldenström's macroglobulinemia is a lymphoplasmacytic disorder characterized by overproduction of monoclonal IgM antibodies
- Diagnosis is made by skin biopsy demonstrating cutaneous infiltration of a monoclonal lymphoplasmacytoid B-cell population. Patients often present with cytopenias or symptoms of hyperviscosity
- Cutaneous manifestations of Waldenström's macroglobulinemia are uncommon. Although rare, it must be considered and included in the differential diagnosis of ulcerative skin lesions

TREATMENT

High-dose corticosteroids and alkylating agents are traditionally used in treatment of Waldenström's macroglobulinemia with varying success. Cutaneous Waldenström's macroglobulinemia generally responds poorly to traditional systemic chemotherapies. However, cyclophosphamide, alone or in conjuction with vincristine has been used with some success.

REFERENCES

Libow LF, Mawhinney JP, Bessinger GT. Cutaneous Waldenström's macroglobulinemia: report of a case and overview of the spectrum of cutaneous disease. J Am Acad Dermatol. 2001; 45(6 Suppl):S202–6.

Mantouz F, Castanet J, Perrin C, Fuzibet JG, Lacour JP, Ortonne JP. Cutaneous Waldenstroms macroglobulinemia. Ann Dermatol Venereol. 2002; 129:53–5.

Orengo IF, Kettler AH, Bruce S, Levy ML, Rosen T, Tschen J. Cutaneous Waldenström's macroglobulinemia. A report of a case successfully treated with radiotherapy. Cancer. 1987; 60(6):1341–5.

Atlas of Clinical Vascular Medicine, First Edition. Edited by Jessica Mintz, Bruce L. Mintz, and Michael R. Jaff.

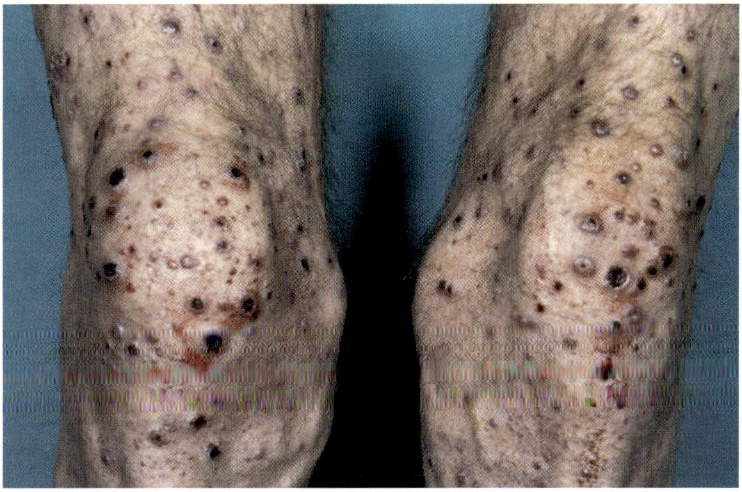

Waldenström's macroglobulinemia. Bilateral, nodular, purple, pigmented lesions encompassing the lower extremities. Image courtesy of the Cleveland Clinic Foundation Department of Vascular Medicine archives.

Verrucous Carcinoma

Ido Weinberg and Michael R. Jaff
Massachusetts General Hospital, Boston, MA, USA

BRIEF CASE PRESENTATION

A 75-year-old man with known severe peripheral artery disease has a non-healing ulcer on his right great toe for the past 4 months. He definitely remembers trauma to that area. On examination his right foot is warm; however, he does not have palpable pedal pulses.

CHARACTERISTICS

- Verrucous carcinoma is a low-grade, slow-growing, well-differentiated variant of squamous cell carcinoma that affects the skin and mucosa. It has a predilection for the foot
- Diagnosis is achieved by repeated and deep biopsies
- Metastasis is rare; however, local invasion is typical

TREATMENT

Medical treatment is ineffective, including electro-dissection, cryotherapy, and laser surgery. Treatment is surgical removal. Sometimes local amputation is needed.

REFERENCES

Ho J, Diven DG, Butler PJ, Tyring SK. An ulcerating verrucous plaque on the foot. Verrucous carcinoma (epithelioma cuniculatum). Arch Dermatol. 2000; 136(4):547–8, 550–1. Review.

Koch H, Kowatsch E, Hodl S, et al. Verrucous carcinoma of the skin: long-term follow-up results following surgical therapy. Dermatol Surg. 2004; 30:1124–30.

Mekkes JR, Loots MA, Van Der Wal AC, Bos JD. Causes, investigation and treatment of leg ulceration. Br J Dermatol. 2003; 148(3):388–401. Review.

Atlas of Clinical Vascular Medicine, First Edition. Edited by Jessica Mintz, Bruce L. Mintz, and Michael R. Jaff.

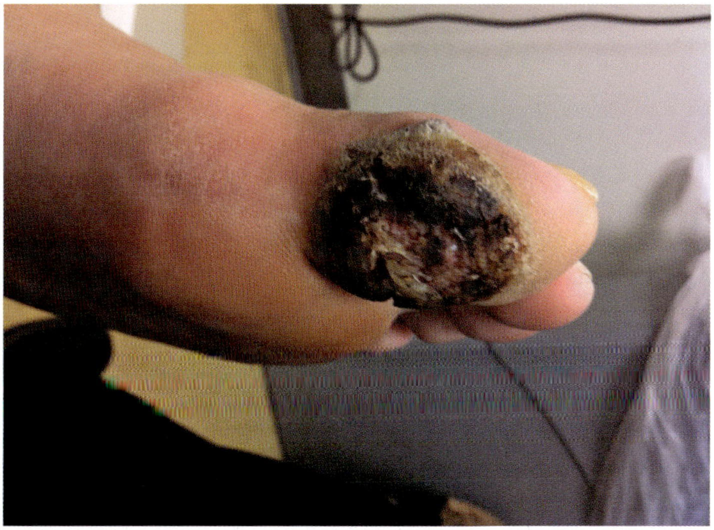

Verrucous carcinoma. Ulcerated lesion of the medial left great toe.

Onychomycosis (Dermatophytic Onychomycosis)

Bruce L. Mintz
Morristown Medical Center, Morristown, NJ, USA

BRIEF CASE PRESENTATION

This 70-year-old white female with long-standing brittle nails, recent loosening of the base of the nail bed and cracks in her skin, presents with fever, chills, and erythema.

CHARACTERISTICS

- Onychomycosis is a long-standing, largely asymptomatic fungal infection of the nails
- Onychomycosis is common in patients with peripheral artery disease and in the immunocompromised, including patients with diabetes mellitus
- Prevalence in the western adult population ranges from 2–14% and increases with age
- Other risk factors include trauma, occlusive footwear, and locker room exposure
- Onychomycosis is often associated with tinea pedis and cracking of the skin in the soles of the feet. Any break in the skin may serve as a portal of entry for bacteria resulting in consequent cellulitis

TREATMENT

Topical treatments are ineffective. Ketoconazole, as a systemic antifungal, also has relatively poor efficacy and is associated with significant adverse effects, including elevated liver enzymes and fulimant hepatic failure and is consequently not recommended. Other antifungals, such as terbinafine and itraconazole, have better penetrance into the nail bed and are more effective than ketoconazole. Treatment should also include good foot hygiene with regular inspection and address any uncontrolled underlying medical conditions that increase the risk of infection.

REFERENCES

Baum, Meissner. Treating onychomycosis, 1853 ('Arch. f. phys. Ileilk,' xii B., p. 193).

de Berker D. Fungal nail disease. NEJM. 2009; 360:2108–16.

Haugh M, Helou S, Boissel JP, Cribier BJ. Terbinafine in fungal infections of the nails: a meta-analysis of randomized clinical trials. Br J Dermatol. 2002; 147(1):118–21.

Atlas of Clinical Vascular Medicine, First Edition. Edited by Jessica Mintz, Bruce L. Mintz, and Michael R. Jaff.

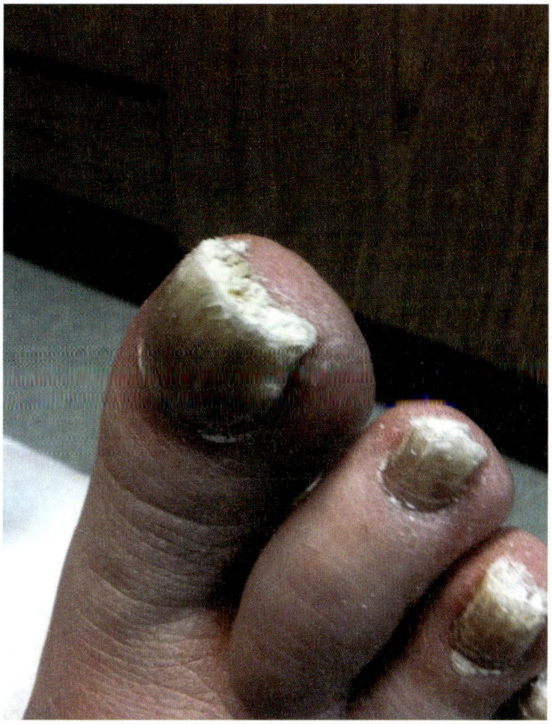

Onychomycosis. Thickened and brittle fungal nails in a patient with peripheral arterial disease.

Index

Atlas of Clinical Vascular Medicine, First Edition. Edited by Jessica Mintz, Bruce L. Mintz, and Michael R. Jaff.

Extending H^{∞} Control to Nonlinear Systems

Advances in Design and Control

SIAM's Advances in Design and Control series consists of texts and monographs dealing with all areas of design and control and their applications. Topics of interest include shape optimization, multidisciplinary design, trajectory optimization, feedback, and optimal control. The series focuses on the mathematical and computational aspects of engineering design and control that are usable in a wide variety of scientific and engineering disciplines.

Series Volumes

Helton, J. William and James, Matthew R., *Extending H^∞ Control to Nonlinear Systems: Control of Nonlinear Systems to Achieve Performance Objectives*

Extending H^∞ Control to Nonlinear Systems

Control of Nonlinear Systems to Achieve Performance Objectives

J. William Helton
University of California, San Diego
San Diego, California

Matthew R. James
Australian National University
Canberra, Australia

siam

Society for Industrial and Applied Mathematics
Philadelphia

10 9 8 7 6 5 4 3 2 1

Library of Congress Cataloging-in-Publication Data

Helton, J. William, 1944–
Extending H^∞ control to nonlinear systems : control of nonlinear systems to achieve performance objectives / J. William Helton, Matthew R. James.
p. cm. – (Advances in design and control)
Includes bibliographical references and index.
ISBN 0-89871-440-0 (pbk.)
1. H^∞ control. 2. Nonlinear control theory.
I. James, Matthew R. (Matthew Ronald) II. Title. III. Series.
QA402.35.H45 1999
629.8'312 -- dc21 99-35569

Contents

Preface

H^∞ control originated in an effort to codify classical control methods where one shapes frequency response functions to meet certain objectives. These techniques have dominated industrial design and commonly involved trial and error. H^∞ control underwent a tremendous development in the 1980s, and arguably this made a considerable step toward systematizing classical control. The next major issue, how this extends to nonlinear systems, is what this book addresses. What we present is an elegant general theory and corresponding formulas.

At the core of nonlinear control theory lie two partial differential equations (PDEs). One is a first-order evolution equation called the *information state* equation, and, as we shall see, it constitutes the dynamics of the controller. One can view the information state equation as a *nonlinear dynamical system* and much of this book is concerned with properties of this system, such as the nature of trajectories, stability, and, most importantly, how it leads to a general solution of the nonlinear H^∞ control problem.

In addition to the information state PDE discussed above, there is a second partial differential inequality (PDI) which is defined on the space of possible information states (which is an infinite-dimensional space). While the information state PDE determines the dynamics of the controller this second PDI determines the output of the controller. As it happens, this is a new type of PDI (and associated PDE), which is now being studied for its own sake. In this book we explore the system theoretic significance of this equation and present its gross structure (which is reasonably complete for smooth solutions) and ways to actually solve it in particular circumstances.

Many challenges are encountered, such as dealing with *singular* information states (functions that may assume the value $-\infty$ on nontrivial sets). These occur naturally, especially in linear systems, and though many technical issues concerning them are not resolved, they offer an enormous practical benefit. Namely, it is often possible to vastly reduce the dimension of the space on which the information state PDE must be solved, and thus vastly reduce the (online) computation required to implement the information state controller. The paradigm problem of classical control, which in the H^∞ context is called the mixed sensitivity problem, is one example where singular states are fruitful. This is because it is very common that the system to be controlled has a small number of unstable modes and this number is the dimension of the (reduced) space on which the singular information state PDE must be solved. While it is far from being proven in great generality, if a solution to the control problem exists, one based on singular solutions exists, and this controller not only solves the H^∞ problem, but the online part could in some cases be implemented.

The book presents a general structure, examples, and proofs at various levels of generality and in various stages of completeness. Thus we guide the reader to an area of vigorous research.

How to Read this Book

This book is divided into two parts, following the Introduction. Part I contains the basic problem definition and information state solution. The key formulas and results are presented. We have attempted to minimize the technical complexity of these first chapters. Part II contains further results and, in particular, some of the more detailed technical results. A number of appendices are provided for the reader's convenience. For readers interested primarily in using the results, we recommend studying the Introduction and Part I, in particular Chapters 2, 3, and 4. For readers wishing to learn the full details and to contribute to the subject, we recommend reading Part II after the main ideas of Part I have been digested.

	Chapter 1	**Introduction.** A quick, light presentation of the main ideas in this book, plus sections on history and classical control.
Part I	Chapter 2	**The H^∞ Control Problem.** The problem is specified and terminology is introduced.
	Chapter 3	**Information States.** The problem is expressed as a minimax game, and the information state is defined and studied.
	Chapter 4	**Information State Control.** Information state controllers are defined and used to solve the H^∞ problem using the dynamic programming PDE and dissipation PDIs.
	Chapter 5	**State Feedback H^∞ Control.** Ideas from the theory of dissipative systems, such as the bounded real lemma, stability, and the state feedback case are reviewed.
	Chapter 6	**Storage Functions.** The closed-loop plant and information state controller system are discussed in the context of storage functions on the product space.
	Chapter 7	**Special Cases.** Bilinear and linear systems and the certainty equivalence cases are presented.
	Chapter 8	**Factorization.** A general theory of factorization is developed and applied to the H^∞ problem.
	Chapter 9	**The Mixed Sensitivity Problem.** This paradigm control problem is solved.
Part II	Chapter 10	**Singular Information States.** These are important for both practical and theoretical reasons, and this chapter presents some detailed results.
	Chapter 11	**Stability of the Information State System.** The information state defines a nonlinear infinite-dimensional dynamical system, and this chapter analyzes its stability behavior.
	Chapter 12	**Time Varying Systems.** A brief discussion of the time varying case.
	Appendix A	**Differential Equations and Stability.** A brief review of some basic facts is presented.

Appendix B	**Nonlinear PDE and Riccati Equations.** Some key ideas are summarized.
Appendix C	**Max-Plus Convergence.** Self-contained background information is given.

Acknowledgments

We wish to thank many colleagues and friends for their support, technical advice, and comments on drafts of this book; in particular, thanks to John Baras, Peter Dower, Paul Dupuis, Wendell Fleming, Bill McEneaney, and Ian Petersen. Additional thanks to Pier Cardiglet, Eduardo Gallestey, Mike Hardt, Davor Hrovat, and Dell Kronewitter for comments and corrections. A special thanks to our family members, Ruth, Maxene, and Jocelyn, for their loving support during the writing of this book.

The writing of this book was supported with funds from Air Force Office of Scientific Research, National Science Foundation, Ford Motor Company, Civilian Research Defense Foundation, Australian Research Council, and Cooperative Research Center for Robust and Adaptive Systems.

Notation

Plant and controller

P	plant for classical mixed sensitivity, page 3
(G, K)	closed-loop plant and controller, page 1
(G, K^{u})	closed loop plant and information state controller, page 81
G	plant in generalized H^∞ control problem, page 45
K	measurement feedback controller, page 47
x, u, z, y	plant state, control, performance output, and measurement output
A, B_1, B_2, C_1, C_2, D_{12}, D_{21}	plant model data (nonlinear functions), page 45
E_1	$D_{12}'D_{12}$, page 46
E_2	$D_{21}D_{21}'$, page 46
$\tilde{G}$	reversed-arrow system, page 46
ξ	state of reversed-arrow system, page 46, or plant state in some formulas, page 61
$A^\times$	$A - B_1 D_{21}' E_2^{-1} C_2$, vector field in reverse-arrow dynamics, page 46
$\Phi_{s,t}^{u,y}(x)$	transition operator for state dynamics, page 52
(A, B, C)	system defined by vector field A and control matrix B, with output C, sometimes abbreviated, page 56
$\mathcal{W}_1$, $\mathcal{W}_2$	weights for the mixed sensitivity problem, page 203

Spaces, norms, etc.

L_2	$L_2[0, \infty)$
$L_{2,T}$	$L_2[0, T]$
$L_{2,loc,T}$	$\cup_{0 \le t < T} L_{2,t}$
$L_{2,loc}$	$L_{2,loc,\infty}$
$C(X)$	real-valued continuous functions on a topological space X
$C_b(X)$	bounded real-valued continuous functions on a topological space X
$C^1(\mathbf{R}^n)$	continuously differentiable functions
$\mathcal{X}$	$C(\mathbf{R}^n)$, page 49

$\mathcal{X}_l$	subspace of $\mathcal{X}$ of functions with at most linear growth, page 49
$\|\cdot\|_l$	norm of $\mathcal{X}_l$, page 49
$\mathcal{X}_q$	subspace of $\mathcal{X}$ of functions with at most quadratic growth page 49
$\mathcal{X}_{q,2,b}$	subspace of $\mathcal{X}$ of functions with at most quadratic growth and continuous, bounded second derivatives, page 51
$\|\cdot\|_q$	norm of $\mathcal{X}_q$, page 49
$\mathbf{R}_e$	$\mathbf{R} \cup \{-\infty\}$, page 49
$\mathcal{X}_e$	upper semicontinuous $\mathbf{R}_e$-valued functions that are bounded above, page 49
$\mathbf{USC}_e(\mathbf{R}^n)$	bounded above upper semicontinuous functions, page 302
$\mathcal{X}_{et} = \mathbf{CLS}_e(\mathbf{R}^n)$	bounded above functions with compact level sets, page 304
$\mathcal{X}_{et,1}$	normalized bounded above functions with compact level sets, page 304
$\mathbf{F}(\mathcal{X}_e)$	real valued functions on $\mathcal{X}_e$, page 216
$\mathbf{B}_e(\mathcal{X}_e)$	$\mathbf{R}_e$ valued functions on $\mathcal{X}_e$ bounded above, page 216
$\langle \cdot \rangle$	max-plus "sup-pairing" or inner "product," page 49
δ_M	max-plus "delta" function, supported on set $M \subset \mathbf{R}^n$, page 49
$\Longrightarrow$	max-plus "weak convergence," page 50
$\oplus$	max-plus "addition," $a \oplus b = \max(a, b)$, page 301
$\odot$	max-plus "multiplication," $a \odot b = a + b$, page 301
$\mathbf{0} = -\infty, \mathbf{1} = 0$	max-plus additive and multiplicative identities, page 301
$\mathcal{B}$	function space for biases, page 47
$q.i.\,D$	quadratic interior of a set $D \subset \mathcal{X}_e$, page 51
$\mathcal{G}, \mathcal{G}_b, \mathbf{F}(\mathcal{X}_e)$	spaces associated with information state transition operator, page 216
$\hat{\mathcal{G}}_b$	space associated with generator, page 218
$\mathcal{M}$	set of all k-dimensional submanifolds of $\mathbf{R}^n$, page 211
$\mathcal{F}(\mathcal{M})$	fiber bundle over $\mathcal{M}$ with fibers $C(M)$, $M \in \mathcal{M}$, page 211

Basic H^∞ and information state notation

γ	dissipation L_2 gain, page 47
β	bias, page 47
β_K	minimal bias, page 48
p	generic information state, a function of x, $p \in \mathcal{X}_e$, page 49
$p = \delta_M + \check{p}$	singular information state, page 49
support p	support of p, page 62

p_0	initial information state, page 59
p_t	information state at time t, page 61
$\dot{p} = F(p, u, y)$	information state dynamics, page 62
$S_t^{u,y}$	transition operator for information state, page 64
$\mathcal{S}_t^{u,y}$	function space transition operator associated with the information state, page 216
$\mathcal{L}^{u,y}$	"generator" of $\mathcal{S}_t^{u,y}$, page 218
p_∞	any equilibrium information state, $F(p_\infty, 0, 0) = 0$, page 69
p_e	a particular equilibrium information state, page 70, control attractor, page 73
A^{p_e}	$A^\times - \gamma^{-2}\tilde{B}_1\tilde{B}_1'\nabla_x p_e'$, page 70
p_{se}	infosubequilibrium, page 71
q_{tT}	adjoint information state, page 75
q_e	a particular equilibrium adjoint information state, page 71
$\mathcal{D}^0_{attr}(p_e)$, $\mathcal{D}_{attr}(p_e)$	domain of (control) attraction, page 73
$r_t(x)$	$p_t(x) + V(x)$, alternative certainty equivalence information state, page 162
$\dot{r} = G(r, u, y)$	dynamics for alternative certainty equivalence information state, page 163
$p(f)$	$\sup_x\{p(x) + f(x)\}$, page 304
$p(A)$	$\sup_{x\in A}\{p(x)\}$, page 304
$[[p + f]]$	$\text{argmax}_x\{p(x) + f(x)\}$, page 212
$\bar{x}$	an element of $[[p + f]]$, page 214

Information state control

$J_{p_0}(K)$	minimax cost functional, page 59
$\bar{J}_{p_0}(K)$	minimax cost functional, information state representation, page 66
$\mathbf{u}(p)$	information state control function, page 80
dom $\mathbf{u}(p)$	domain of information state control function, page 80
$K^{\mathbf{u}}_{p_0}$	information state controller, page 77
$W(p)$	value function, page 78
dom $\hat{W}$	domain of $\hat{W}$, page 51
$W(p) \geq \langle p \rangle$	domination, page 82
$W(p_1) \geq W(p_2)$	if $p_1 \geq p_2$, monotonicity, page 82
$W(p + c) = W(p) + c$	additive homogeneity, page 82
$\nabla_p W(p)[h]$	Frechet derivative, page 86
$\nabla_p^+ \Psi(p)[h]$	generalized directional derivative, page 89
$\mathbf{u}^*(p)$, $\mathbf{y}^*(p)$	optimal control and observation, page 87

K_{po}^*	optimal information state controller obtained from value function W, page 78
$K_{p_e}^*$	central controller, page 106
$\hat{W}(p)$	function of p, often solution of dissipation PDI, page 88
$\mathrm{dom}_{\mathrm{smooth}}\hat{W}$	domain on which $\hat{W}$ is smooth, page 94
$\mathrm{dom}_{\mathrm{pd}}\hat{W}$	set on which $\hat{W}$ solves PDE or PDI, page 95
τ_{esc}	escape time, page 93
$\mathbf{u}_{\hat{W}}^*(p)$, $\mathbf{y}_{\hat{W}}^*(p)$	optimal control and observation depending on $\hat{W}$, page 94
$K_{po}^{\mathbf{u}_{\hat{W}}^*}$	information state controller obtained from a solution $\hat{W}$ of the dissipation PDI, page 78
$\overline{W}(p)$	information state value function, page 101
$\bar{\mathbf{u}}_{po}^*(p)$	optimal control via direct minimization, page 101
$K_{po}^{\bar{\mathbf{u}}^*}$	information state controller obtained from direct minimization of $\overline{J}$, page 101
$\mathrm{dom}_{\mathrm{pd}_s}\hat{W}$, $\mathrm{dom}_{\mathrm{pd}_{cs}}\hat{W}$	stabilizing sets, page 122
$\mathcal{ZD}^0$, $\mathcal{ZD}$	zero dynamics sets, page 123
$V^{\hat{W}}(x,p)$	storage function for information state closed loop, page 147
$V_a^{\hat{W}}(x,p)$, $V_r^{\hat{W}}(x,p)$	available storage and required supply, page 147
$e^{\hat{W}}(x,p)$	$-p(x)+\hat{W}(p)$, an explicit storage function, page 150
$\mathbf{w}_{e,\hat{W}}^*(x,p)$	optimal disturbance for $e^{\hat{W}}(x,p)$, page 152
$W(x,p)$	modified value function, page 153
$\bar{x}(t)$, $\bar{x}(p)$	minimum stress estimate, page 161
$u^{ce}(t)$	certainty equivalence control, page 161
$\mathbf{y}_{state}^*(x)$	certainty equivalence optimal measurement, page 162

Dissipation and state feedback control

Σ	nonlinear system, page 127
$V_a(x)$	available storage, page 128
$V_r(x)$	required supply, page 128
$V(x)$, $\hat{V}(x)$	storage function, page 128, state feedback value function, page 140
$\mathbf{w}_V^*(x)$	$\gamma^{-2}B(x)'\nabla_x V(x)'$, optimal disturbance, page 130
A^*	$A+\gamma^{-2}BB'\nabla_x V'$, optimal vector field, BRL, page 134
$\mathbf{w}^*(x)$	$B'\nabla_x V'$, optimal disturbance, BRL, page 136
$K_{state}(x)$	state feedback controller, page 139
$\mathbf{u}_{state}^*(x)$	optimal state feedback control, page 142

$\mathbf{w}^*_{state}(x)$	optimal state feedback disturbance, page 142
K^*_{state}	optimal state feedback controller, page 142
$\mathbf{u}^*_{state,\,\hat{V}}(x)$	optimal state feedback control, depending on $\hat{V}$, page 143
$\mathbf{w}^*_{state,\hat{V}}(x)$	optimal state feedback disturbance, depending on $\hat{V}$, page 143
$K^{\mathbf{u}^*_{\hat{V}}}_{state}$	optimal state feedback controller depending on $\hat{V}$, page 143
$A^*_{state,\,\hat{V}}$	optimal state feedback-disturbance vector field, page 144

H_2 filtering and control

$V_{H_2}(x)$	H_2 state feedback value function, page 139
$A^*_{H_2}$	$A - B_2E_1^{-1}D_{12}C_1 - B_2E_1^{-1}B_2'\nabla_x V_{H_2}'$ optimal H_2 vector field, page 140
r_t	H_2 information state at time t, page 246
r_e	equilibrium H_2 information state, page 243
A^{r_e}	$A - B_1D_{21}'E_2^{-1}C_2 + \tilde{B}_1\tilde{B}_1'\nabla_x r_e'$, optimal H_2 filtering vector field, page 242

Linear/bilinear systems

P'	transpose of a matrix P
P^{-L}	left inverse of a matrix P
$P \geq 0\ (P > 0)$	symmetric positive definite matrix (strict), page 133
$G_{bilinear}$	bilinear system, page 155
G_{linear}	linear system, page 158
$A,\ B_1,\ B_2,\ C_1,$ $C_2,\ D_{12},\ D_{21}$	plant model data (matrices), page 158
B_3	bilinear product term B_3ux, page 155
$A(u)$	$A + B_3u$ bilinear controlled vector field, page 155
$\lVert\cdot\rVert_{H^\infty}$	H^∞-norm, page 4
$\hat{x}$	state estimate term, bilinear page 156, linear page 158
Y	Y-Riccati term, bilinear page 156, linear page 158
ϕ	constant term, bilinear page 156, linear page 158
$\tilde{W}(\hat{x}, Y, \phi)$	value function for bilinear/linear systems, page 156
$\hat{W}(\hat{x}, Y, \phi)$	dynamic programming PDE solution for linear systems, page 158
X_e, Y_e	stabilizing Riccati solutions, page 159
Σ_{lin}	system obtained by linearization, page 132
$A_{lin},\ B_{lin},\ C_{lin}$	matrices obtained by linearization, page 132

Miscellaneous

∇_x	gradient with respect to the variable x, page 52
∇_x^2	Hessian with respect to the variable x
M_s, M_{as}	stable and antistable manifolds for a hyperbolic vector field, page 55
$\mathcal{S}_{cs}$, $\mathcal{S}_{cas}$	stabilizable and antistabilizable sets for a system (A, B), page 55
$\xi_0(x_0)$	tracking point, page 56

Constants

The symbols b, c, C, k, etc., are used to denote various constants. Subscripts are used to emphasize particular meanings. Often in calculating estimates in proofs the same symbol is used repeatedly, with the understanding that the actual value of the symbol can change as the calculation progresses. This is used where the actual value is immaterial and is done to reduce the complexity of the notation.

Chapter 1

Introduction

The goal of this introduction is to give a quick, light presentation of the main ideas developed in this book. Thus we focus on laying out the structure for a basic nonlinear control problem, which has the remarkably jargony name, the "two block problem." This is a nice level of generality for an introduction, because it contains most of the ideas behind the more general four block problem (which is treated in detail in the book), the formulas are simpler, and the linear two block problem corresponds to the paradigm problem of classical control. This paradigm problem is often called the "mixed sensitivity problem," and Chapter 9 is devoted to its solution. We turn now to our overview of the theory. We postpone historical discussion and motivation to §1.11. Linear systems are discussed in §§1.2, 1.3.6.

1.1 The Standard Problem of Nonlinear H^∞ Control

Now we introduce a *special case of the standard problem* of H^∞ control. It entails a description of the plant and controller models and definitions of the control objectives. This is motivated in §1.8 and actually done carefully in Chapter 9. The standard control problem corresponds to Figure 1.1, which we now explain.

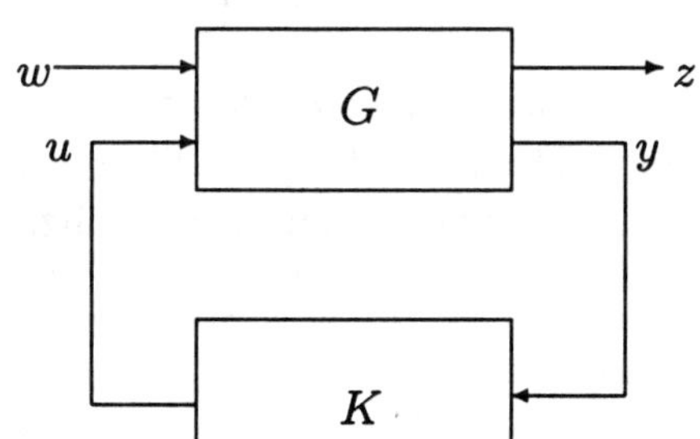

Figure 1.1: The closed-loop system (G, K).

1.1.1 The Plant

Let us consider nonlinear plants G with a two block structure

$$G: \begin{cases} \dot{x} = A(x) + B_1(x)w + B_2(x)u, \\ z = C_1(x) + D_{12}(x)u, \\ y = C_2(x) + w. \end{cases} \tag{1.1}$$

Here, $x(t) \in \mathbf{R}^n$ denotes the state of the system and is not in general directly measurable; instead an output $y(t) \in \mathbf{R}^p$ is observed. The additional output quantity $z(t) \in \mathbf{R}^r$ is a performance measure, depending on the particular problem at hand. The control input is $u(t) \in \mathbf{R}^m$, while $w(t) \in \mathbf{R}^s$ is regarded as an opposing disturbance input. Detailed assumptions concerning the functions appearing in (1.1) will be given in Chapter 2; however, we mention here that the origin is an equilibrium and the two block structure requires

$$E_1(x) \triangleq D_{12}(x)' D_{12}(x) > 0. \tag{2.4}$$

(More generally, in the four block case the coefficient of w in the third equation of (1.1) is a matrix D_{21}, which will satisfy condition (2.3) given in Chapter 2.)

1.1.2 The Class of Controllers

The plant G was described by an explicit state space model and is assumed given. However, in the spirit of optimal control theory, we do *not* prescribe a state space model for the controller K, since it is an unknown to be determined from the control objectives. Rather, we simply stipulate some basic input–output properties required of any *admissible controller*, namely, that the controller must be a causal function of the output

$$K : y(\cdot) \mapsto u(\cdot)$$

and the resulting closed-loop system be well defined in the sense that trajectories and signals exist and are unique. The controller K will be said to be *null initialized* if $K(0) = 0$, regardless of whether or not a state space realization of K is given.

1.1.3 Control Objectives

The H^∞ control problem is commonly thought of as having two objectives: find a controller K such that the closed-loop system (G, K) is

(i) dissipative and

(ii) stable.

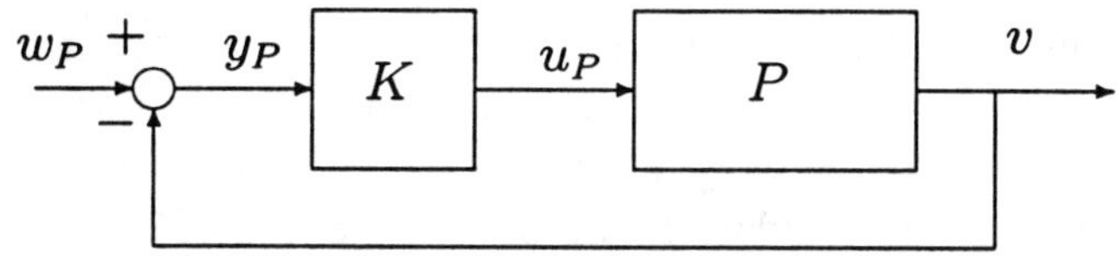

Figure 1.2: Mixed sensitivity setup.

In §1.2 we define what is meant by these terms in the case of linear systems. We now describe their meanings for nonlinear systems; this gives an extension of H^∞ control to nonlinear systems.[1]

The closed-loop system (G, K) is *γ-dissipative* if there exist $\gamma > 0$ and a function $\beta(x_0) \geq 0$ $\beta(0) = 0$, such that

$$\begin{cases} \frac{1}{2}\int_0^T |z(s)|^2\,ds \;\leq\; \gamma^2\frac{1}{2}\int_0^T |w(s)|^2\,ds + \beta(x_0) \\ \text{for all } \; w \in L_{2,T} \text{ and all } T \geq 0. \end{cases} \tag{1.2}$$

This definition is saying that the nonlinear input–output map $(G, K) : w \mapsto z$ defined by the closed-loop system has finite L_2 gain with a bias term due to the initial state x_0 of the plant G.

While dissipation captures the notion of performance of a control system, another issue with H^∞ control is stability of the system. The closed-loop system will be called *weakly internally stable* provided that if G is initialized at any x_0, then if $w(\cdot) \in L_2[0,\infty)$, all signals $u(\cdot), y(\cdot), z(\cdot)$ in the loops as well as $x(\cdot)$ converge to 0 as $t \to \infty$. By *internal stability* we mean that the closed-loop is weakly internally stable and in addition if the controller has a state space realization, then the controller state will converge to an equilibrium as $t \to \infty$.

Dissipation and stability are closely related; see, e.g., [Wil72], [HM76], [HM77], [vdS96]. Indeed, dissipative systems that enjoy a detectability or observability property also enjoy a stability property. In our context, suppose the system (G, K) is *z-detectable*; that is, $w(\cdot)$ and $z(\cdot) \in L^2[0,\infty)$ imply $x(\cdot) \in L^2[0,\infty)$ and $x(t) \to 0$ as $t \to \infty$. By *z-observable* we mean that if $w(\cdot) = 0$, $z(\cdot) = 0$, then $x(\cdot) = 0$. *If (G, K) is γ-dissipative and z-detectable, then (G, K) is weakly internally stable (see Theorem* 2.1.3*).*

1.1.4 A Classic Example

The book is written for readers with many different interests, so it is worth emphasizing for the reader with a classical control bent that the following problem is of the type we just introduced:

> Given plant P, find a controller K achieving a given H^∞ performance specification (see Figure 1.2).

[1]The term "nonlinear H^∞ control" has no precise mathematical meaning, but it has come into common use in the control engineering community and refers to nonlinear generalizations of H^∞ control (which has precise meaning for linear systems).

In linear H^∞ control the designer selects certain weights and optimizes a worst case frequency domain performance. This is called the *mixed sensitivity* problem of H^∞ control; see Chapter 9. If the weights are chosen correctly for a mixed sensitivity problem, then one gets the standard two block problem of H^∞ control that we just presented. Choice of weights is a serious business in practice, and some serious investigation of how this should be done for nonlinear systems is in its infancy. In §1.8 and Chapter 9 we describe some basic considerations in selecting weights.

1.2 The Solution for Linear Systems

The H^∞ problem is well understood when the systems are linear. The plant is linear provided

$$A(x) = Ax, \quad C_1(x) = C_1 x, \quad C_2(x) = C_2 x,$$

$$B_1(x) = B_1, \quad B_2(x) = B_2, \quad D_{12}(x) = D_{12}$$

where

$$A, \quad C_1, \, C_2, \quad B_1, \, B_2, \, D_{12}$$

are matrices of appropriate dimension. We recall here the well-known solution to the H^∞ control problem for the two block linear systems; see [DGKF89], [PAJ91], [GL95], etc. (these references also contain the "standard assumptions").

1.2.1 Problem Formulation

The class of admissible controllers K are those with finite-dimensional linear state space realizations

$$K : \begin{cases} \dot{\eta} = A_K \eta + B_{1K} y + B_{2K} u, \\ u = C_K \eta + D_K y. \end{cases}$$

Given $\gamma > 0$, the H^∞ control problem for G is to find, if possible, a compensator K such that the resulting closed-loop system $(G, K) : w \mapsto z$ satisfies the following:

(i) *Dissipation.* The required dissipation property is expressed in the frequency domain in terms of the H^∞-norm of the closed-loop transfer function $(G, K)(s)$ as follows:

$$\| (G, K) \|_{H^\infty} = \sup_{\omega \in \mathbf{R}} \sigma_{max}[(G, K)(j\omega)] < \gamma.$$

(ii) *Stability.* We require that the closed-loop system

$$(G, K) \text{ is } \textit{internally stable}.$$

Some discussion of the classical transfer function (classical loop shaping) pictures versus the state space picture of control is found in §1.8.

1.2.2 Background on Riccati Equations

Recall a few facts about Riccati equations. An algebraic Riccati equation

$$\Sigma A + A\Sigma + \Sigma R \Sigma + Q = 0 \tag{1.3}$$

with real matrix entries A, R, Q and R, Q self-adjoint, meeting suitable positivity and technical conditions (see, e.g., [ZDG96, Chapter 13]), has upper and lower solutions Σ_a, Σ_r, so that any other self-adjoint solution Σ lies between them

$$\Sigma_a \leq \Sigma \leq \Sigma_r.$$

The bottom solution is called the *stabilizing solution* because it has and is characterized by the property

$$A + R\Sigma_a \tag{1.4}$$

and is asymptotically stable. Likewise Σ_r is *antistabilizing* in that

$$-(A + R\Sigma_r) \tag{1.5}$$

is asymptotically stable.

1.2.3 Standard Assumptions

There are a number of "standard assumptions" that are needed for the necessity and sufficiency theorems about H^∞ control. These can be expressed in various ways and here we follow [PAJ91].

We have already seen the first condition, viz., rank condition (2.5). The D_{12} rank condition ensures that the cost term $|z|^2$ is strictly positive definite in the control u (while the more general four block condition (2.3), Chapter 2, relates to the solvability for w given x, y in the output equation $y = C_2x + D_{21}w$).

Next are two important technical conditions that take the form

$$\text{rank} \begin{bmatrix} A - j\omega I & B_2 \\ C_1 & D_{12} \end{bmatrix} = n + m \quad \text{for all } \omega \geq 0, \tag{1.6}$$

and

$$\text{rank} \begin{bmatrix} A - j\omega I & B_1 \\ C_2 & D_{21} \end{bmatrix} = n + l \quad \text{for all } \omega \geq 0. \tag{1.7}$$

The condition (1.6) can be replaced by a stronger H^2 state feedback assumption (Chapter 5), while (1.7) can be replaced by a stronger H^2 filtering assumption (Chapter 11). These two conditions are commonly used in H^2 control and filtering and concern the controllability and observability of underlying systems.

1.2.4 Problem Solution

The necessary and sufficient conditions for solvability of the H^∞ problem under the standard assumptions follow.

Condition 1. *State feedback control.* There exists $X_e \geq 0$ solving the control-type Riccati equation

$$\begin{aligned}\mathcal{R}_{state}(X_e) &\triangleq (A - B_2E_1^{-1}D_{12}'C_1)'X_e + X_e(A - B_2E_1^{-1}D_{12}'C_1) \\ &+X_e(\tfrac{1}{\gamma^2}B_1B_1' - B_2E_1^{-1}B_2')X_e + C_1'(I - D_{12}E_1^{-1}D_{12}')C_1 = 0,\end{aligned} \tag{1.8}$$

which is stabilizing; i.e.,

$$A - B_2E_1^{-1}D_{12}'C_1 + (\gamma^{-2}B_1B_1' - B_2E_1^{-1}B_2')X_e \quad \text{is asymptotically stable.} \tag{1.9}$$

Condition 2. *State estimation.* There exists $Y_e \geq 0$ solving the filter-type Riccati equation

$$\begin{aligned}&\mathcal{R}_{est}(Y_e) \\ &\triangleq (A - B_1C_2)Y_e + Y_e(A - B_1C_2)' + Y_e(\gamma^{-2}C_1'C_1 - C_2'C_2)Y_e = 0,\end{aligned} \tag{1.10}$$

which is stabilizing; i.e.,

$$A - B_1C_2 + Y_e(\gamma^{-2}C_1'C_1 - C_2'C_2) \text{ is asymptotically stable.} \tag{1.11}$$

Condition 3. *Coupling.* The matrix X_eY_e has spectral radius strictly less than γ.

THEOREM I.2.1 ([DGKF89], [PAJ91], [GL95]). *The H^∞ control problem for G, meeting certain technical conditions, is solvable if and only if the above three conditions are satisfied. If these conditions are met, one controller, called the central controller, is given by*

$$K^* : \begin{cases} \dot{\hat{x}} = (A - B_1C_2 + Y_e(\gamma^{-2}C_1'C_1 - C_2'C_2))\,\hat{x} \\ \qquad +(B_1 + Y_eC_2')y + (B_2 + \gamma^{-2}Y_eC_1'D_{12})u, \\ u = -E_1^{-1}[D_{12}'C_1 + B_2'X_e](I - \gamma^{-2}Y_eX_e)^{-1}\hat{x}. \end{cases} \tag{1.12}$$

We sometimes refer to K^* as the *"DGKF" central controller*, after its discoverers J. Doyle, K. Glover, P. Kargonekar, and B. Francis [DGKF89].

1.2.5 The Linear Solution from a Nonlinear Viewpoint

Of course the solution to the nonlinear control problem that we present in this book when specialized to linear systems solves the linear H^∞ control problem. The solution looks a bit different from the classical one we just saw. The linear solution has been put in coordinates that make degenerate cases appear unpathological. However, it is not easy to change coordinates in nonlinear solutions, so what we get is forced upon us. Let us see what the linear specialization of the nonlinear solution looks like.

If $Y_e > 0$, and hence invertible, the coupling condition is equivalent to

$$-\gamma^2 Y_e^{-1} + X_e < 0.$$

This foreshadows the nonlinear theory in that it focuses on the *inverse* of Y_e. Moreover, we shall see that one does not actually need the stabilizing properties of X_e and Y_e; positive definite inequalities will do. Indeed if we take the main results of this book given in Chapter 4, §4.10, and specialize them to the linear case we get Theorem 1.2.2.

THEOREM 1.2.2 *A solution to the linear H^∞ control problem exists (and there are formulas for producing it) if there exists solutions $X \geq 0$ and $Y > 0$ to the DGKF Riccati equations that satisfy strict coupling*

$$-\gamma^2 Y^{-1} + X < 0. \tag{1.13}$$

Conversely, if a solution to the linear H^∞ control problem exists, the stabilizing solutions X_e and Y_e to the Riccati equations are nonnegative definite and if $Y_e > 0$, we have

$$-\gamma^2 Y_e^{-1} + X_e < 0. \tag{1.14}$$

Note that the lower bounding properties

$$X_e \leq X \quad \text{and} \quad Y_e \leq Y$$

of X_e, Y_e imply

$$-\gamma^2 Y_e^{-1} + X_e \leq -\gamma^2 Y^{-1} + X < 0.$$

So the DGKF Theorem 1.2.1 has for simplicity presented the extreme case of the possible solutions.

As we soon see this funny way of writing the X, Y coupling condition is exactly the way it presents itself for general nonlinear systems. Also we have only discussed $Y > 0$. Actually, for the theory to hold Y need not be invertible. This may sound like a fine point, but a rank-one or -two Y contains much less information than a rank-17 Y, and such economies of information translate into major computational savings in the nonlinear case. Thus in the book we give considerable attention to the "singular cases," that is, where Y^{-1} is "not finite."

1.3 The Idea of the Nonlinear Solution

This section is the heart of the introductory outline of the book and contains a discussion of the main ideas of the solution to the nonlinear H^∞ control problem defined above.

State feedback control. The nature of the information available to the controller has a very significant bearing on the complexity of the problem and of the resulting controller. Accordingly, we begin with an easier problem in which the controller is allowed to read the state of the plant. This simpler problem is known as the state feedback H^∞ control problem (essentially one with full information) and is well understood in the literature.

Estimation—The information state. Next we turn to the general output feedback problem. Here the state is not known perfectly, and so we must estimate it. This estimation is done with something called the information state, a function on the state space of the plant G that satisfies a PDE. Thus the information state is produced by an infinite-dimensional controlled dynamical system. Much of this book is concerned with properties of this dynamical system and how it can be used to solve the H^∞ control problem.

Coupling—Information state feedback. Using the information state, the output feedback problem is converted to state feedback problem for a new system. This new system uses the information state as its state variable, and the solution of the new state feedback problem leads to the solution of the output feedback H^∞ control problem. This is a coupling of control and estimation.

This indicates the layout of the remainder of the Introduction.

1.3.1 The State Feedback Control Problem

The state feedback H^∞ has been extensively studied in the literature and is well understood; see [vdS96] and the references contained therein.

1.3.1.1 Problem Statement

A block diagram illustrating the state feedback H^∞ problem is given in Figure 1.3. The state space model for the plant is

$$G: \begin{cases} \dot{x} = A(x) + B_1(x)w + B_2(x)u, \\ z = C_1(x) + D_{12}(x)u. \end{cases} \tag{1.15}$$

The controller can read measurements of the plant state x, so that

$$u = K(x).$$

(For simplicity we only consider static state feedback controllers. Alternatively, one could work with full information controllers, where K is a causal function of the disturbance, and this would yield the same optimal controller under appropriate regularity assumptions.)

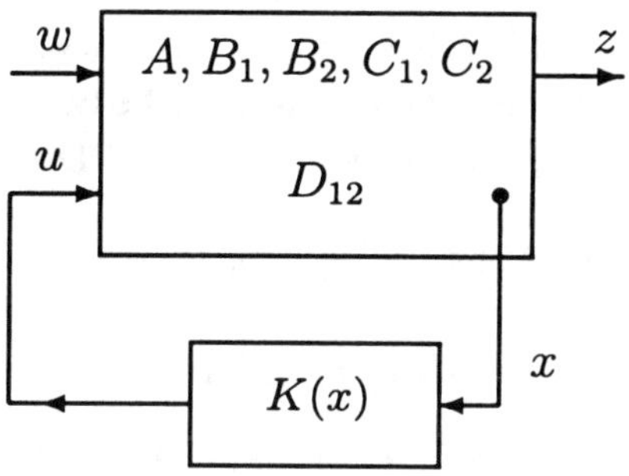

Figure 1.3: The state feedback closed-loop system (G, K).

The state feedback H^∞ control problem is to find a controller $u = K(x)$ which is dissipative in the sense of §1.1 and stable in the sense that the vector field

$$A + B_2K \text{ is \textit{asymptotically stable}.}$$

1.3.1.2 Problem Solution

The solution is determined by the *state feedback Hamilton–Jacobi–Bellman–Isaacs PDE* (HJBI PDE)

$$\begin{aligned}\nabla_x V \cdot (A - B_2E_1^{-1}D_{12}'C_1) + \tfrac{1}{2}\nabla_x V \cdot (\gamma^{-2}B_1B_1' - B_2E_1^{-1}B_2') \cdot \nabla_x V' \\ + \tfrac{1}{2}C_1'(I - D_{12}E_1^{-1}D_{12}')C_1 = 0.\end{aligned} \tag{1.16}$$

If one can find a strictly positive proper smooth solution V ($V(x) > 0$ if $x \neq 0$, $V(0) = 0$) that makes the vector field

$$A - B_2E_1^{-1}D_{12}'C_1 + (\gamma^{-2}B_1B_1' - B_2E_1^{-1}B_2')\nabla_x V' \text{ asymptotically stable}$$

then a solution to the state feedback problem follows.

1.3.1.3 The State Feedback Central Controller

$$K^*(x) = \mathbf{u}^*_{state}(x) = -E_1(x)^{-1}[D_{12}(x)'C_1(x) + B_2(x)'\nabla_x V(x)'].$$

Using this controller, the closed-loop system (G, K_s^*) becomes

$$\begin{cases}\dot{x} = A(x) + B_1(x)w - B_2(x)E_1(x)^{-1}[D_{12}(x)'C_1(x) + B_2(x)'\nabla_x V(x)'], \\ z = C_1(x) - D_{12}(x)E_1(x)^{-1}[D_{12}(x)'C_1(x) + B_2(x)'\nabla_x V(x)'],\end{cases}$$

and integration of the PDE (1.16) yields the dissipation inequality

$$V(x(t)) + \tfrac{1}{2}\int_0^T |z(t)|^2 dt \le \gamma^2 \tfrac{1}{2}\int_0^T |w(t)|^2 dt + V(x(0)).$$

The desired dissipation property (1.2) follows from this on setting $\beta = V$ since $V \geq 0$. Also, stability of the vector field $A + B_2K^*$ follows (see, e.g., [vdS96]).

It is important to note for practical reasons that the designer can solve for V *offline* (i.e., not in real time). This requires the solution of a PDE in n-dimensions.

EXAMPLE 1.3.1 For linear systems (cf. §1.2), the state feedback HJBI PDE has a quadratic solution

$$V(x) = \tfrac{1}{2}x'Xx$$

if it has any solution at all. One can substitute $\nabla_x V(x) = x'X$ into the HJBI, which we illustrate when $D_{12} = I = D_{21}$ and get

$$\begin{aligned} 0 &= \tfrac{1}{2}x'(A - B_2C_1)'\nabla V(x)' + \tfrac{1}{2}\nabla V(x)(A - B_2C_1)x \\ &\quad + \tfrac{1}{2}\nabla V(x)(\gamma^{-2}B_1B_1' - B_2B_2')\nabla V(x)' \\ &= \tfrac{1}{2}[\, x'(A - B_2C_1)'Xx + x'X(A - B_2C_1)x \\ &\quad + x'X(\gamma^{-2}B_1B_1' - B_2B_2')Xx \,]; \\ \text{i.e., } 0 &= \tfrac{1}{2}x'\mathcal{R}_{state}(X)x, \end{aligned}$$

which is the DGKF state matrix Riccati equation $\mathcal{R}_{state}(X) = 0$. Take $X_e \geq 0$ to be the stabilizing solution of this Riccati equation to get the optimal state feedback controller

$$K^*(x) = \mathbf{u}^*_{state}(x) = -[C_1 + B_2'X_e]x.$$

∇

REMARK 1.3.2 Actually to solve the state feedback H^∞ problem, it is enough to find a function $V(x) \geq 0$, $V(0) = 0$ satisfying the HJBI PDE (1.16) plus a detectability assumption. For example, if the closed-loop system (G, K_V^*) is detectable (here $K_V^*(x)$ is determined by the solution $V(x)$: $K_V^*(x) = -E_1(x)^{-1}[D_{12}(x)'C_1(x) + B_2(x)\nabla_x V(x)]$), then one can obtain stability of $A + B_2K_V^*$ from the dissipation inequality analogously to what is done in Theorem 2.1.3. This approach will be used frequently in the sequel. Note, however, that it is in general difficult to check detectability; however, the generic system is detectable (which of course does not imply that a system derived from some generic optimization process is detectable). Another addition to solutions of the HJBI that produce solutions to the state feedback control problem is the class of strict positive V with $V(0) = 0$ solving the strict HJBI inequality

$$\begin{aligned} &\nabla_x V \cdot (A - B_2E_1^{-1}D_{12}'C_1) + \tfrac{1}{2}\nabla_x V \cdot (\gamma^{-2}B_1B_1' - B_2E_1^{-1}B_2') \cdot \nabla_x V' \\ &+ \tfrac{1}{2}C_1'(I - D_{12}E_1^{-1}D_{12}')C_1 < 0. \end{aligned}$$

∇

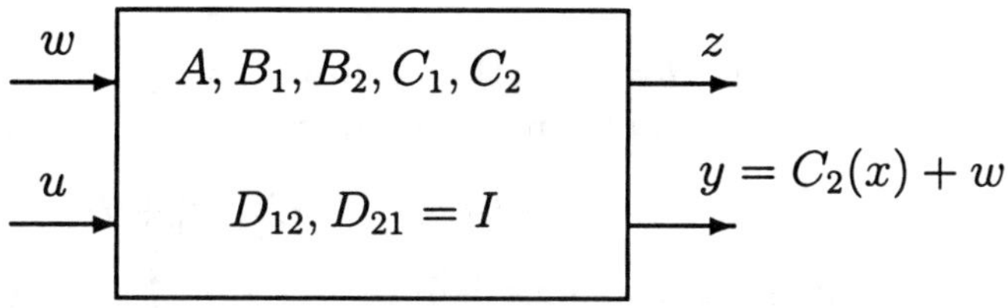

(a) The Original System G.

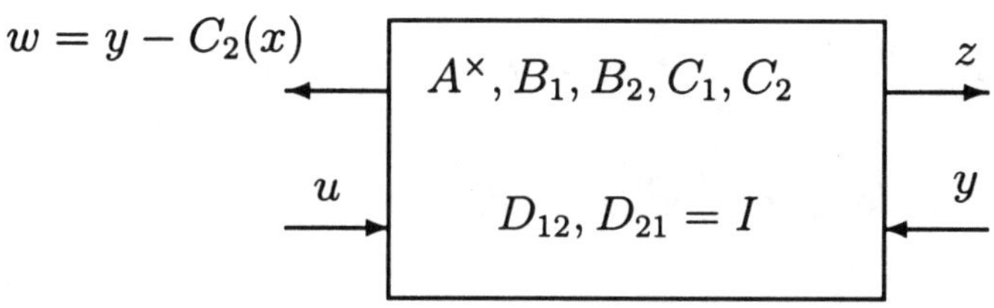

(b) The Reverse-Arrow System $\tilde{G}$.

Figure 1.4: Reversing arrows.

1.3.2 The Information State

We return to the output feedback problem. To solve it, we use an information state. This converts the output feedback problem to a new state feedback problem with a new state, namely, the information state. (This methodology is an old one from stochastic optimal control.) We now give definitions that lead directly to the construction of the controller dynamics.

1.3.2.1 Reversing Arrows

We start by defining the *reverse-arrow system.* It is a new system $\tilde{G}$ which, in the two block case, is obtained from G by reversing the w and y arrows. While the definition is algebraic, pictures help a lot; see Figures 1.4(a) and (b).

The *reverse-arrow system* is defined by

$$\tilde{G}: \begin{cases} \dot{\xi} = A(\xi)^{\times} + B_1(\xi)y + B_2(\xi)u, \\ z = C_1(\xi) + D_{12}(\xi)u, \\ w = -C_2(\xi) + y, \end{cases} \qquad (1.17)$$

with $A^\times$ defined by

$$A^\times \triangleq A - B_1 C_2.$$

Note that G and $\tilde{G}$ have the same state space. Clearly this is derived by substituting $w = y - C_2(x)$ into the G dynamics to produce

$$\dot{x} = A(x) + B_1(x)[y - C_2(x)] + B_2(x)u,$$

which is the same as the dynamics defined above in (1.17) for $\tilde{G}$.

1.3.2.2 Definition

Given time $t \geq 0$, past measurement $y \in L_2[0,t]$ and past control signal $u \in L_2[0,t]$ introduce a function $p_t(x) = p(x,t)$ on the states x of the plant G by

$$\begin{aligned} p_t(x) &= p_0(\xi(0)) + \tfrac{1}{2}\int_0^t [|C_1(\xi(s)) + D_{12}(\xi(s))u(s)|^2 - \gamma^2|y(s) - C_2(\xi(s))|^2]ds \\ &= p_0(\xi(0)) \ + \tfrac{1}{2}\int_0^t \ [\text{output}^2 - \gamma^2\text{input}^2] \ ds \\ &\qquad \text{given } u, y \text{ over } [0,t] \text{ and given } \xi(t) = x. \end{aligned} \tag{1.18}$$

Here ξ follows the state trajectory from 0 to t of the reverse-arrow system (1.17) with *final* state $\xi(t) = x$. This is the tricky part. Given x to define $p_t(x)$ we must run the $\tilde{G}$ system backward for t time units, using the given u, y. We see how much energy $E_{[0,t]}$ was consumed by the system and what state $\xi(0)$ the trajectory hits, with energy $p_0(\xi(0))$. Then

$$p_t(x) = p_0(\xi(0)) + E_{[0,t]},$$

the sum of two cost terms. This function is called the *information state* for the H^∞ control problem.

The information state plays the role of a "sufficient statistic" [JBE94], [JB95]. In [BB91], this function is introduced for linear systems and called the "cost to come." For $p_t = p(\cdot, t)$ to be defined everywhere we must assume that for each $u, y \in L_2[0,t]$ the differential equation (1.1) has trajectories whose endpoints at time t sweep out the whole state space provided the endpoints at time 0 do. (See Theorem 3.1.8.) Figure 1.5 illustrates some of the common shapes of information states; generally they are bounded above, point downward, and may take the value $-\infty$.

If the information state $p_t(x)$ is smooth, then it satisfies the *information state PDE*

$$\begin{aligned} \frac{\partial p_t(x)}{\partial t} &= F(p_t, u(t), y(t)) \triangleq -\nabla_x p_t(x) \cdot [A^\times(x) + B_2(x)u(t) + B_1(x)y(t)] \\ &\qquad + \tfrac{1}{2}|C_1(x) + D_{12}(x)u(t)|^2 - \gamma^2 \tfrac{1}{2}|y(t) - C_2(x)|^2, \end{aligned} \tag{1.19}$$

which is readily obtained by differentiating (1.18). Often we write this differential equation even when p is not smooth, but one should interpret it as the integral

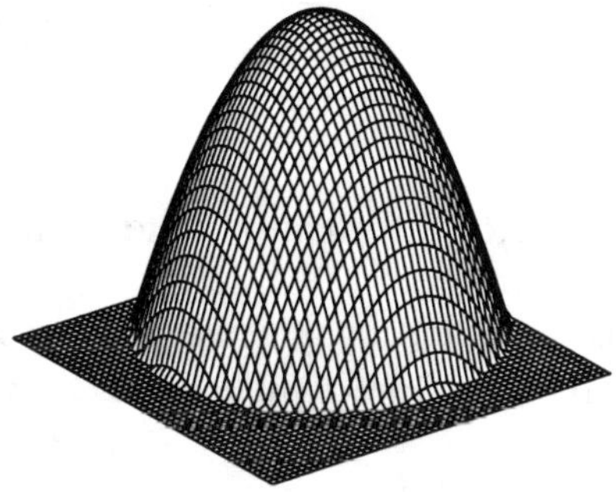

(a) Nonsingular (everywhere finite).

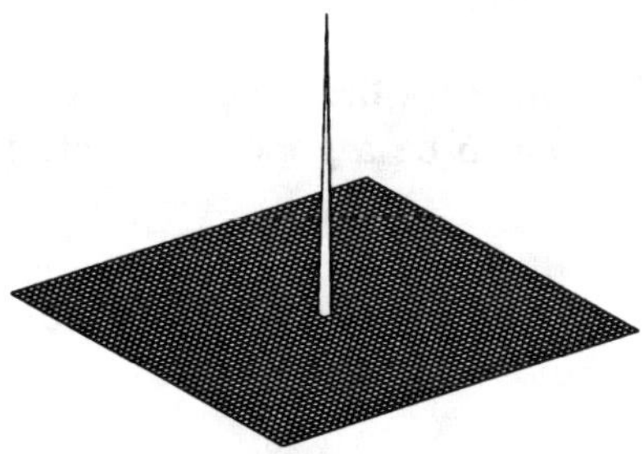

(b) Purely Singular (equal to $-\infty$ everywhere except at $x = 0$).

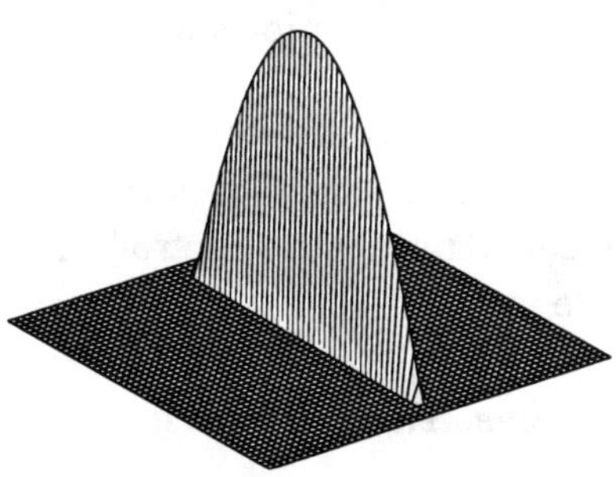

(c) Mixed Singular (finite on a subset of $\mathbf{R}^n$ and equal to $-\infty$ elsewhere).

Figure 1.5: Common information states. (The information state is a function $p(x)$ defined on the plant state space, the horizontal plane in the figure, with coordinates $x = (x_1, \ldots, x_n)$.)

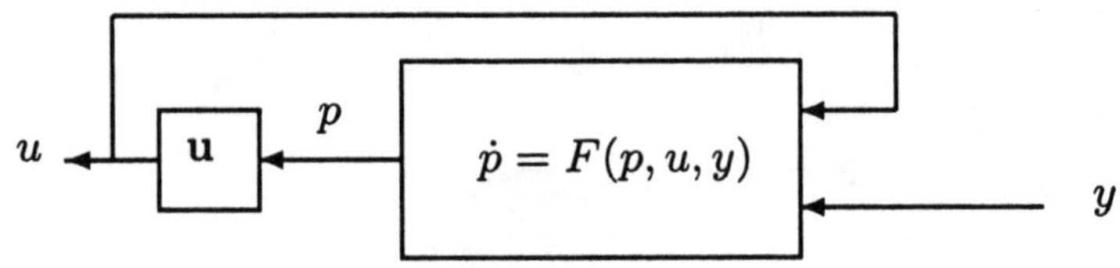

Figure 1.6: Information state controller.

equation (1.18) or perhaps in the viscosity sense (see Appendix B). We can think of the PDE (1.19) as describing an *infinite-dimensional dynamical system*, which can be written in shorthand form

$$\frac{dp}{dt} = F(p, u, y).$$

This system has a "state" p belonging to an infinite-dimensional function space, and it is driven by input signals u and y. The solution of the H^∞ problem depends on properties of this system. Some references include [JB95], [BB91], [DBB93].

1.3.3 The Central Controller

We now give a high-level formula for the structure and dynamics of a controller that (as the book unfolds) turns out to be a good candidate for the solution of the H^∞ problem.

State Space is a space $\mathcal{X}_e$ of functions $p = p(x)$ on the state space $\mathbf{R}^n$ of the plant G.

Dynamics are the PDE

$$\frac{dp}{dt} = F(p, u, y).$$

Output u is function

$$u = \mathbf{u}(p)$$

defined (on a subset of) the information state space $\mathcal{X}_e$.

We call this the *information state controller*, illustrated in Figure 1.6.
In §1.3.5 we show how $\mathbf{u}$ is constructed. Ultimately we shall focus on a particular information state controller called the central controller; it is obtained by optimization (yielding $\mathbf{u}^*(p)$) and suitable initialization.

An important point is that for this controller to be implementable *one must solve the information state PDE online*. This is a PDE in n-dimensions.

1.3.4 Equilibrium Information States

Our definition of the controller dynamics is not complete because in order to define its dynamics we must specify an initial information state p_0. As we shall see, careful choice of this initial state p_0 makes a big difference in the implementability of the controller and strongly affects the dynamical behavior. Thus we devote substantial effort and several subsections to the following question: *Which initial state p_0 do we use?*

An obvious requirement of p_0 stemming from the null initializing property $K(0) = 0$ is

$$\mathbf{u}(p_t) = 0 \text{ for all } t \text{ when } p_t \text{ solves } \tfrac{dp}{dt} = F(p, 0, 0) \text{ initialized at } p_0.$$

However, a stronger highly desirable condition is

$$0 = F(p_0, 0, 0) \text{ and } \mathbf{u}(p_0) = 0.$$

That is, p_0 is said to be an *equilibrium solution* p_e to the information state PDE. This is the correct initialization of the central controller: $p_0 = p_e$. (Below we discuss convergence of p_t to p_e; that is, the stability of the information state.)

As we shall see, the equilibria for two block information states have a surprising form. It is surprising enough that we had better retreat to an example before describing it; this is done in §1.3.6. In the meantime, we consider the problem of choosing the controller output function $\mathbf{u}(p)$.

1.3.5 Finding u* and Validating the Controller

We give now some details on the construction of the function $\mathbf{u}(p)$, which is a key component of the information state and central controllers. This is chosen optimally as follows (so that we will take $\mathbf{u} = \mathbf{u}^*$): solve an infinite-dimensional state feedback control problem. The HJBI PDE for this problem is

$$\inf_{u \in \mathbf{R}^m} \sup_{y \in \mathbf{R}^p} \nabla_p W(p)[F(p, u, y)] = 0 \text{ in dom } W. \tag{1.20}$$

Here, $\nabla_p W(p)$ is interpreted as a Frechet derivative (more general interpretations are discussed in Chapters 4 and 10). One attempts to solve this PDE for a smooth function $W(p)$ defined on a domain dom W, a subset of the state space, and satisfying auxiliary conditions such as $W(p) \geq \sup_x\{p(x)\}$, and $W(p_0) = 0$ for some $p_0 \in \text{dom } W$. The function $W(p)$ is called the *value function* for the H^∞ control problem and can be regarded as an analogue of the state feedback value function $V(x)$ (see §1.3) for the information state system. The information state feedback function $\mathbf{u}^*(p)$ is obtained by

$$\begin{aligned}\mathbf{u}^*(p) &= \operatorname*{argmin}_{u \in \mathbf{R}^m}\{\sup_{y \in \mathbf{R}^p} \nabla_p W(p)[F(p, u, y)]\} \\ &= (\nabla_p W(p)[E_1])^{-1} \nabla_p W(p)[-D_{12}' C_1 + B_2' \nabla_x p'].\end{aligned}$$

Necessary and sufficient conditions for the solvability of the H^∞ control problem can be expressed in terms of the function $W(p)$ and the PDE (1.20). The following "metatheorem" states the main idea without the clutter of technical details:

RESULT I.3.3 *If there exists some controller that solves the H^∞ problem, then there exists a function $W(p)$ solving the PDE (1.20) (in some sense) as well as auxiliary technical conditions. If the function $W(p)$ is smooth, then the central controller $K^*_{p_e}$ obtained from $W(p)$ solves the H^∞ problem. Key to this is the "coupling condition" ensuring that the controller is well defined for all time and along trajectories of the closed-loop system,*

$$p_t(x(t)) \leq \sup_x\{p_t(x)\} \leq W(p_t) \leq W(p_e) < +\infty, \quad (1.21)$$

where $u(t) = \mathbf{u}^*(p_t)$. Conversely, if one can solve the PDE (1.20) for a smooth function $W(p)$ satisfying some auxiliary technical conditions, then the central controller $K^*_{p_e}$ obtained from $W(p)$ solves the H^∞ problem.

The major objective of this book is to present intuition and theory for results of this type.

1.3.5.1 Construction of the Central Controller

Now we summarize the procedure for building the central controller:

(i) Obtain a function $W(p)$ and $\mathbf{u}^*(p)$ solving the PDE (1.20) and the coupling (1.21).

(ii) Compute p_e and check $\mathbf{u}^*(p_e) = 0$.

(iii) Use $\mathbf{u}^*$ as the output term of the central controller.

(iv) The information state PDE (1.19) initialized at $p_0 = p_e$ gives the dynamics of the controller $K^*_{p_e}$.

1.3.5.2 Validating the Controller

We review the context in which we sit. Let $w(\cdot) \in L_2$ and $x_0 \in \mathbf{R}^n$ be given. These determine signals $y(\cdot)$, $z(\cdot)$, and $u(\cdot)$ and trajectories $x(\cdot)$, $p.$ from the dynamics of the closed loop $(G, K^*_{p_e})$ with $p_0 = p_e$, $u(\cdot) = \mathbf{u}^*(p.)$. The idea behind confirming dissipativity of the closed-loop system follows.

(i) Integrate the PDE (1.20) along the trajectory p_t:

$$W(p_t) \leq W(p_0) = W(p_e) < +\infty.$$

Then use the property

$$p_t(x(t)) \leq \sup_x\{p_t(x)\} \leq W(p_t) \leq W(p_e) = 0 \quad (1.22)$$

and the definition of the information state to obtain

$$p_e(\xi(0)) + \tfrac{1}{2}\int_0^t [|C_1(\xi(s)) + D_{12}(\xi(s))u(s)|^2 - \gamma^2|y(s) - C_2(\xi(s))|^2]\, ds$$

$$= p_t(x) \leq W(p_e),$$

where $\xi(\cdot)$ is the solution of (1.17) with $\xi(t) = x$. Now if $w(\cdot)$ is input to the plant G with initial state x_0, we obtain signals $u(\cdot) = K^*_{p_e}(y(\cdot))$ and state $x(\cdot)$ in closed loop, and so if we set $x = x(t)$ we have $\xi(\cdot) = x(\cdot)$ and so

$$\tfrac{1}{2}\int_0^t [|C_1(x(s)) + D_{12}(x(s))u(s)|^2 - \gamma^2|w(s)|^2]ds \leq W(p_e) - p_e(x(0))$$

$$= -p_e(x_0),$$

which is the dissipation inequality (1.2) with $\beta = -p_e$.

(ii) If p_e is nonsingular and if $(G, K^*_{p_e})$ is detectable, then $K^*_{p_e}$ solves the H^∞ control problem. If p_e is singular, then with extra work and stronger conditions it is possible to prove that $K^*_{p_e}$ solves the H^∞ control problem. See Chapter 4.

(iii) The stability results discussed in §1.5 below for the information state system are used to deduce the asymptotic behavior of the information state in closed-loop (Chapter 4).

1.3.5.3 Storage Functions

Associated with a dissipative system are functions $e(x,p)$ on its state space called *storage functions*. Of course we are interested in the closed-loop system $(G, K^*_{p_e})$ and a storage function e for it is defined to be nonnegative and satisfy the "dissipation inequality":

$$e(x(t), p(t)) + \tfrac{1}{2}\int_0^t |z(s)|^2\, ds \leq e(x_0, p_e) + \gamma^2 \tfrac{1}{2}\int_0^t |w(s)|^2\, ds \tag{1.23}$$

for all $t \geq 0$ and all $w \in L_2[0,t]$. It is fairly remarkable that there is a storage function $e(x,p)$ for the closed-loop system $(G, K^*_{p_0})$, which has a very simple and explicit formula:

$$e(x,p) = -p(x) + W(p).$$

It is interesting to note that the content of (1.23) is the same as that of (1.22) as can be verified by adding minus the information state equation that p_t satisfies to (1.21). Also compare (1.23) with the dissipation inequality (1.2) of §1.1 (note $\beta(x) = e(x, p_e) = -p_e(x)$). This storage function gives a handy tool for validating that $K^*_{p_0}$ is γ-dissipative provided $p_0(x_0)$ is finite.

1.3.6 Example: Linear Systems

1.3.6.1 $W(p)$ for Linear Systems

The information state for linear systems is quadratic and will be described immediately below. For now we discuss the form of W. One has

$$W(p) = \max_x \{p(x) + \tfrac{1}{2}x'X_e x\}.$$

Thus we have the following:

(i) The integrated form of (1.20)

$$\begin{aligned} \max_x\{p_t(x)\} \le W(p_t) &= \max_x\{p_t(x) + \tfrac{1}{2}x'X_e x\} \\ \le W(p_0) &= \max_x\{p_0(x) + \tfrac{1}{2}x'X_e x\} \end{aligned}$$

is equivalent to $W(p)$ being finite and X_e being positive semidefinite.

(ii) The equilibrium information state is $p_e = -\gamma^2 \frac{1}{2}x'Y_e^{-1}x$, where Y_e solves the DGKF Y equation. $W(p_e)$ finite is equivalent to the matrix $-\gamma^2 Y_e^{-1} + X_e$ being negative semidefinite. If γ is suboptimal, then this is negative definite since small perturbations of this will be negative.

Thus we have that the DGKF conditions 1, 2, and 3 (except for the strictness) of §1.2.4 are implied by the existence of (finite) W and the existence of p_e solving (1.20) and (1.21). The converse is true and can be checked with a little effort.

1.3.6.2 The Information State

For linear systems, one can check that if $Y(t)$ is invertible, then solutions to the information state equation have the form

$$p_t(x) = -\gamma^2 \tfrac{1}{2}[x - \hat{x}(t)]'Y(t)^{-1}[x - \hat{x}(t)] + \phi(t)$$

whenever p_0 has this form, where

$$\begin{aligned} \dot{\hat{x}} &= [A^\times + Y(\gamma^{-2}C_1'C_1 - C_2'C_2)]\hat{x} \\ &\quad + [B_1 + YC_2']y + [B_2 + \gamma^{-2}YC_1'D_{12}]u, \\ \dot{Y} &= A^\times Y + YA^{\times\prime} + Y(\gamma^{-2}C_1'C_1 - C_2'C_2)Y. \end{aligned}$$

Now we compare this with the dynamics of the DGKF central controller (1.12) to the linear H^∞ problem. The $\hat{x}$ equation is exactly (1.12) if we take $Y(t)$ equal to Y_e, the stabilizing solution to the DGKF Y equation (1.10). The above Riccati differential equation for $Y(t)$ can be initialized in many ways which lead to a solution to the H^∞ control problem. However, the equilibrium solution has the great advantage that $dY_e/dt = 0$, so we have no Y differential equation to solve in real time (since $Y(t) = Y_e$ for all $t \geq 0$).

Now comes a crucial pair of exercises. They are so crucial that the reader should think for a minute and not race to the answers.

Exercise 1. Suppose $A^{\times}$ is a stable matrix. What is Y_e?

Answer: $Y_e = 0$. The reason is that the DGKF Y equation is homogeneous in Y, so $Y_e = 0$ certainly satisfies it. But is it stabilizing? Well yes since $A^{\times}$ is stable even without perturbing it.

Exercise 2. $A^{\times}$ has no pure imaginary eigenvalues. What can we say about Y_e?

Answer: Y_e is 0 on the stable eigenspace of $A^{\times}$.

In the first exercise the DGKF Y equation disappears when $Y_0 = Y_e$ since the stabilizing solution Y_e is 0, so the controller formulas will only involve the DGKF X equation. In the second exercise Y_e is usually low rank, so maybe *the controller will have a low dimension* (in some sense) if we initialize $Y_0 = Y_e$. For the nonlinear case this suggests a big simplification since Y determines the state estimator (the online part of the computation).

We return to the equilibrium information state p_e, which in the linear case is formally of the form

$$p_e(x) = -\gamma^2 \tfrac{1}{2} x' Y_e^{-1} x$$

and immediately worry because Y_e is typically not invertible. Indeed if $Y_e = 0$ we suspect that $p_e(x) = -\infty$. While this is close to correct it is not quite and so we now embark on definitions and a discussion of singular functions. Later we give precise formulas for singular information states and resulting controllers.

1.4 Singular Functions

When Y_e is not of full rank, the function $p_e(x) = -\gamma^2 \frac{1}{2} x' Y_e^{-1} x$ is interpreted as a singular function. In the first exercise $A^{\times}$ is stable and corresponds to $Y_e = 0$, so we define then

$$p_e = \delta_0,$$

where

$$\delta_0(x) = \begin{cases} 0 & \text{if } x = 0, \\ -\infty & \text{if } x \neq 0; \end{cases}$$

See Figure 1.7.

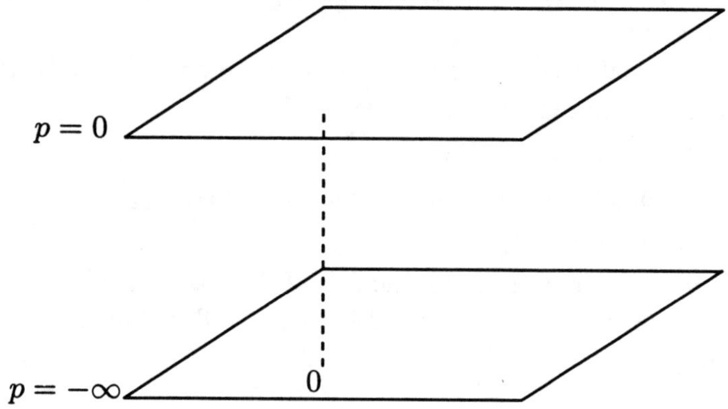

Figure 1.7: The singular function δ_0.

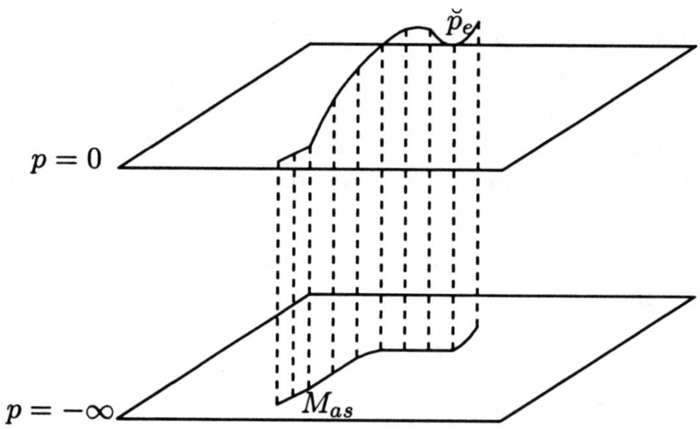

Figure 1.8: The singular function $p_e = \delta_{M_{as}} + \breve{p}_e$.

In the second exercise

$$p_e = \delta_{M_{as}} + \breve{p}_e.$$

Here

$$\delta_{M_{as}}(x) = \begin{cases} 0 & \text{if } x \in M_{as}, \\ -\infty & \text{if } x \notin M_{as}, \end{cases}$$

(see Figure 1.8) and M_{as} is the antistable subspace of $A^\times$ and $\breve{p}_e$ is a quadratic form on M_{as} (the analogous notation $p = \delta_M + \breve{p}$ will be used frequently where $M \subset \mathbf{R}^n$ and $\breve{p}$ is a function defined on M).

We emphasize again that in mixed sensitivity control applications M_{as} *is usually low dimensional!* Thus $p_0 = p_e$ is supported on a very thin set.

1.4.1 Singular Equilibrium Information States

For the nonlinear two block problem we shall assume that

$$A^\times \text{ is a hyperbolic vector field}$$

with global stable M_s and antistable M_{as} submanifolds. As we shall see the equilibrium information state p_e is given by

$$p_e = \delta_{M_{as}} + \breve{p}_e,$$

where M_{as} is the *antistable submanifold* of $A^\times$ and $\breve{p}_e$ is a smooth function on M_{as}. There are two important special cases:

(i) when $A^\times$ is stable, $p_e = \delta_0$ is a purely singular function, and

(ii) when $A^\times$ is antistable, p_e is a finite, smooth, nonsingular function.

In any case once M_{as} is computed $\breve{p}_e$ can be determined by computing for each $x \in M_{as}$ the integral

$$\breve{p}_e(x) = \tfrac{1}{2}\int_{-\infty}^{0} [|C_1(\xi(s))|^2 - \gamma^2|C_2(\xi(s))|^2]ds,$$

where $\xi(\cdot)$ is the solution in backward time to

$$\dot{\xi} = A^\times(\xi), \qquad \xi(0) = x \in M_{as}.$$

See §3.2 for a derivation.

1.4.2 The Central Controller Dynamics

The *central controller* is obtained by initializing the optimal information state controller ($\mathbf{u} = \mathbf{u}^*$) at equilibrium, $p_0 = p_e$, and is denoted $K^*_{p_e}$. If p_e is singular, the resulting information state dynamics is still quite concrete to write down, manipulate, and compute numerically.

The formula for the dynamics. Suppose $p_0 = p_e = \delta_{M_{as}} + \breve{p}_e$. Then for any $u, y \in L_2$, p_t is of the form

$$\delta_{M_t} + \breve{p}(\cdot, t),$$

where

$$M_t \triangleq \{\xi(t) \ : \ \dot{\xi} = A^\times(\xi) + B_1(\xi)y + B_2(\xi)u, \ 0 \le s \le t, \xi(0) \in M_{as}\},$$

and also $\breve{p}(\cdot, t)$ is the function on M_{as} given by

$$\begin{aligned}\breve{p}(x,t) &= \breve{p}_e(\xi(0)) + \frac{1}{2}\int_0^t [|C_1(\xi(s)) + D_{12}(\xi(s))u(s)|^2 - \gamma^2|y(s) - C_2(\xi(s))|^2]ds \\ &= \text{initial state energy} + \text{the energy it takes to get from } \xi(0) \text{ to } \xi(t) = x,\end{aligned}$$

where $\xi(s)$, $0 \le s \le t$, is given by the reversed system dynamics (1.17) with $\xi(t) = x \in M_t$.

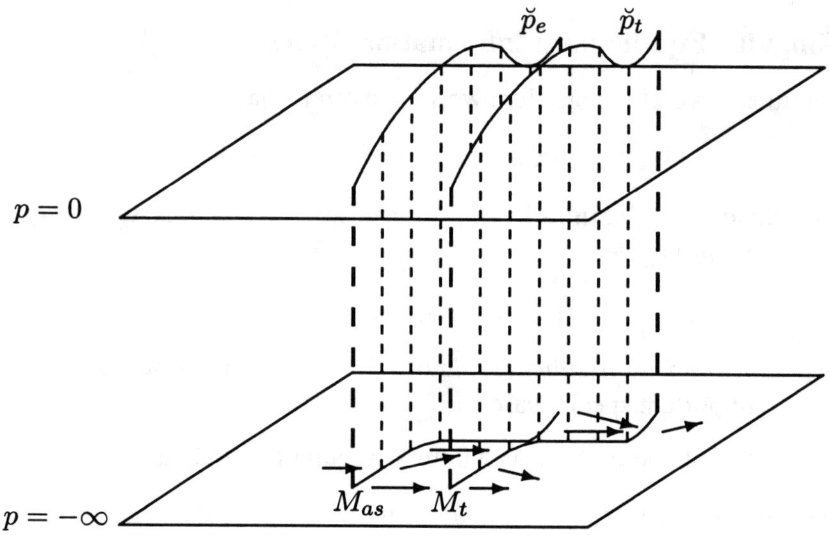

Figure 1.9: Flow of singular information states.

This evolution of functions $\breve{p}_t$ on the $M \equiv \mathbf{R}^k$ together with the evolution (2.6) constitutes a "reduced-dimensional" picture of the compensator dynamics (to be discussed further in Chapter 10); see Figure 1.9.

1.4.2.1 Computational Requirements

The amount of computation required for a singular function indeed is less than for a nonsingular function. This is because one need only solve a $k << n$-dimensional PDE (in real time). Intuitively, the singularity of information states reflects a degree of knowledge concerning the state trajectory, and this means that less computational effort is required.

To be more specific, suppose one wishes to approximately compute the compensator state $(M_t, \breve{p}_t)$ by numerically solving the ordinary differential equation (ODE)

$$\dot{x} = A^{\times}(x) + B_1(x)y + B_2(x)u,$$

which propagates M_t. Then $\breve{p}_t$ is computed by evaluating the integral recursively; i.e., one only needs to update the integral at each time step. One begins such a numerical computation by laying out a grid on M_0. For example, if $M_0 = M_{as}$ is k dimensional and one chooses N equally spaced grid points in each dimension, then if M_0 is k dimensional choosing N^k grid points would be natural. One initializes the ODE at each grid point x_g and solves the ODE numerically as the values of $u(t)$ and $y(t)$ become known. Since the ODE is an $n \times n$ system at any time its solution is an n-dimensional vector, and we get one of these per grid point. Thus the memory and operation requirements scale like

$$n\mathrm{O}(N^k).$$

This is a striking improvement over the $\mathrm{O}(N^n)$ required to solve the PDE (3.9) for a smooth solution in $\mathbf{R}^n$.

1.5 Attractors for the Information State

A very important issue is whether or not the equilibrium p_e is an attractor for the information state dynamics (1.19). By this we mean roughly that there exists a set $\mathcal{D}_{attr}(p_e)$ of initial states p_0 and a nonempty set of signals u, y in L_2 for which the resulting information state trajectory p_t converges to p_e (in an appropriate sense). We call an equilibrium p_e a *control attractor* if for all u, y in L_2

$$p_t \to p_\infty = p_e + c \quad \text{as } t \to \infty$$

for some constant $c \in \mathbf{R}$ (depending on u, y, p_0) for $p_0 \in \mathcal{D}_{attr}(p_e)$.

For the nonlinear two block problem with $A^\times$ hyperbolic we shall prove the following roughly:

> *Suppose $A^\times = A - B_1C_2$ is hyperbolic, that the initial function p_0 satisfies $F(p_0, 0, 0) < 0$, and that certain assumptions hold. If u, y are supported on $[0, T]$, then the solution p_t to*
>
> $$\dot{p} = F(p, u, y)$$
>
> *with smooth initial condition p_0 has the stability property*
>
> $$p_t \to p_e + c \stackrel{\triangle}{=} \delta_{M_{as}} + \breve{p}_e + c \;\; as \; t \to \infty,$$
>
> *where M_{as} is the antistable manifold of $A^\times$, and $\breve{p}_e$ is a function on M_{as} and c is a real number depending on u, y, and p_0.*

It is important to note that the special case $A^\times$ antistable leads to nonsingular equilibria p_e, whereas when $A^\times$ is stable $p_e = \delta_0$ is purely singular. Moreover, stronger results are proven (Chapter 11):

- With restrictions on p_0 it is possible to prove convergence for arbitrary u, y in L_2.
- Initial states p_0 in $\mathcal{D}_{attr}(p_e)$ are characterized.

Various notions of convergence are required, but these are mainly confined to the more theoretical sections of the book. In particular, we make use of the max-plus notion of weak convergence (analogous to weak convergence of probability measures) and hypoconvergence (from optimization); see Appendix C.

1.6 Solving the PDE and Obtaining u*

The value function $W(p)$ and the infinite-dimensional PDE (1.20) offer a high-level framework for solving nonlinear H^∞ control problems. The function PDE (1.20) is to be solved *offline* for $W(p)$, and hence $\mathbf{u}^*(p)$ can be constructed offline. As mentioned, the only online part of the central controller $K^*_{p_e}$ is the information state dynamics (1.19). We now discuss three situations in which $W(p)$ vastly simplifies.

The significance of these results is that the infinite-dimensional PDE (1.20) can be solved in terms of a PDE on a finite-dimensional space (i.e., one on $\mathbf{R}^n$). Solving such a PDE is a traditional pursuit of mathematics and engineering, and it bears directly on the (offline) construction of the central controller. Solving these PDEs gives formulas for $\mathbf{u}^*$ in terms of the optimal state feedback $\mathbf{u}_s^*$ control law applied to carefully selected states.

1.6.1 Certainty Equivalence

Under the certainty equivalence assumption (Whittle [Whi81], Basar and Bernhard [BB91]), it is possible to use the function

$$\hat{W}(p) = \sup_x \{p(x) + V(x)\}$$

as a value function for the H^∞ control problem. Here, $V(x)$ is the state feedback value function of §1.3, which determines the state feedback controller $u^*_{state}(x)$. The *certainty equivalence assumption* requires that the *minimum stress estimate*

$$\bar{x}(t) = \underset{x\in\mathbf{R}^n}{\operatorname{argmax}}\{p_t(x) + V(x)\}$$

is unique. If this assumption holds, then the certainty equivalence controller

$$u^{ce}(t) = \mathbf{u}^*_{state}(\bar{x}(t))$$

coincides with the central controller described above. Further, the function $\hat{W}(p)$ solves the PDE (1.20).

The certainty equivalence controller has dynamics

$$\begin{aligned}\dot{\bar{x}} &= A - B_2E_1^{-1}D_{21}'C_1 + (\gamma^{-2}B_1B_1' - B_2E_1^{-1}B_2')\nabla_x V' \\ &+(B_1D_{21}' - \gamma^2(\nabla_x^2 r_t)^{-1}\nabla_x y_s^{*\prime})(y - \mathbf{y}^*_{state}) \\ &+(B_2 + (\nabla_x^2 r_t)^{-1}\nabla_x u_s^{*\prime}E_1)(u - \mathbf{u}^*_{state}),\end{aligned} \tag{1.24}$$

where the RHS is evaluated at $x = \bar{x}(t)$, and $r = r(x,t)$ solves a PDE obtained by combining the PDEs for $p_t(x)$ and $V(x)$ (see Chapter 7). This yields the concrete formula for the u^*:

$$\mathbf{u}^*(p) = \mathbf{u}^*_{state}(\bar{x})$$

since $\bar{x} = \bar{x}(p)$.

REMARK 1.6.1 A generalization of certainty equivalence to cases of multiple maxima has been considered in [HV95]. ∇

1.6.2 Bilinear Systems

There are some classes of systems for which the information state is finite dimensional. Two such classes are those consisting of bilinear and linear systems. The plant is *bilinear* provided

$$A(x) = Ax, \quad C_1(x) = C_1 x, \quad C_2(x) = C_2 x,$$

$$B_1(x) = B_1, \quad B_2(x) = B_2 + B_3 x, \quad D_{12}(x) = D_{12},$$

where

$$A, \quad C_1, \, C_2, \quad B_1, \, B_2, \, B_3, \, D_{12},$$

are matrices of appropriate dimension, and we assume for simplicity that u is one dimensional ($m = 1$).

If $p_0(x) = -\gamma^2 \frac{1}{2} x' Y_0^{-1} x$ with $Y_0 > 0$, then the information state is given explicitly by

$$p_t(x) = -\gamma^2 \tfrac{1}{2}(x - \hat{x}(t))' Y(t)^{-1}(x - \hat{x}(t)) + \phi(t), \tag{1.25}$$

where

$$\begin{aligned} \dot{\hat{x}} &= (A(u) - B_1 C_2 + Y(\gamma^{-2} C_1' C_1 - C_2' C_2))\, \hat{x} \\ &\quad + (B_1 + Y C_2') y + (B_2 + \gamma^{-2} Y C_1' D_{12}) u, \end{aligned} \tag{1.26}$$

$$\begin{aligned} \dot{Y} &= (A(u) - B_1 C_2) Y + Y (A(u) - B_1 C_2)' \\ &\quad + Y(\gamma^{-2} C_1' C_1 - C_2' C_2) Y, \end{aligned} \tag{1.27}$$

$$\dot{\phi} = \tfrac{1}{2} |C_1 \hat{x} + D_{12} u|^2 - \gamma^2 \tfrac{1}{2} |y - C_2 \hat{x}|^2. \tag{1.28}$$

Here, we have written $A(u) = A + B_2 u$. Thus the information state p_t projects to a finite-dimensional triple $(\hat{x}, Y, \phi)$. Consequently, *the online computation of the information state is drastically simplified and feasible.*

If the value function is defined for the triple $(\hat{x}, Y, \phi)$, call it $\hat{W}(\hat{x}, Y, \phi)$; the corresponding PDE for $\hat{W}(\hat{x}, Y, \phi)$ is defined on a finite-dimensional space $\mathbf{R}^n \times \mathbf{R}^{n^2} \times \mathbf{R}$ and has the form

$$\inf_{u \in \mathbf{R}^m} \sup_{y \in \mathbf{R}^p} \nabla_{(\hat{x}, Y, \phi)} \hat{W}(\hat{x}, Y, \phi)[\hat{F}(\hat{x}, Y, \phi, u, y)] = 0, \tag{1.29}$$

where $\hat{F}(\hat{x}, Y, \phi, u, y)$ denotes the dynamics defined by (1.26), (1.27), (1.28). Evaluating the infimum in the RHS of the PDE (1.29) yields the central controller function

$$\begin{aligned} \mathbf{u}^*(\hat{x}, Y, \phi) &= \operatorname*{argmin}_u \{ \sup_y \nabla_{(\hat{x}, Y, \phi)} \hat{W}(\hat{x}, Y, \phi)[\hat{F}(\hat{x}, Y, \phi, u, y)] \} \\ &= -E_1^{-1} [D_{12}' C_1 \hat{x} + (B_2' + \gamma^{-2} D_{12}' C_1 Y + \hat{x}' B_3') \nabla_{\hat{x}} \hat{W}(\hat{x}, Y, \phi)' \\ &\quad + \nabla_Y \hat{W}(\hat{x}, Y, \phi) \cdot (B_3 Y + Y B_3')]. \end{aligned}$$

The PDE (1.29) for a function $\hat{W}(\hat{x}, Y, \phi)$ can seldom be solved explicitly and so approximations and numerical methods must be used. However, *it is important to note that this is already feasible in applications where the state space is very low dimensional.* In general, it is not possible to solve this explicitly.

REMARK 1.6.2 For linear systems ($B_3 = 0$), the value function is given explicitly by

$$\hat{W}(\hat{x}, Y, \phi) = \tfrac{1}{2}\hat{x}' X_e (I - \gamma^{-2} Y X_e)^{-1} \hat{x} + \phi,$$

where $X_e \geq 0$ is a solution of the X Riccati equation (1.8). ∇

1.7 Factorization

While engineers have a deep love for feedback diagrams like Figure 1.1 this is not familiar to the average mathematician. Most mathematicians are, however, quite fond of factoring. They will try to factor numbers or mappings on most objects you put in front of them. Fortunately, the H^∞ control problem for the plant G in (1.1) is equivalent under various hypotheses to a type of factorization problem for the reversed arrow system $\widetilde{G}$ in (1.1) or more accurately because of possible degeneracies to what we call a decomposition of $\widetilde{G}$. To be more specific we start with a given system $\Sigma = \widetilde{G}$ and seek another system Σ^0 so that the composition $\Sigma^I = \Sigma \circ \Sigma^0$ is dissipative with respect to a certain signed bilinear form and so that Σ^0 satisfies a fairly weak partial left invertibility type of assumption. If Σ^0 is invertible, this is equivalent to Σ having the factorization $\Sigma = \Sigma^I \circ (\Sigma^0)^{-1}$.

Notice that if Σ is a system whose input space is $\mathbf{R}^{p+m}$, then the output space of Σ^0 is constrained to be $\mathbf{R}^{m+p}$, but its input space can be of any dimension. Traditionally, investigators found factors whose input space is $\mathbf{R}^{m+p}$, which if Σ^0 is linear means that its transfer function has values that are square matrices. The square case does not correspond precisely to the H^∞ control problem in Figure 1.1, but it can be used to parameterize many solutions to the problem; thus having a good square factoring is more than is needed to solve the control problem. The bulk of Chapter 8 treats square factorization. Actually *equivalent to the control problem* is having a good factor Σ^0 whose input space is $\mathbf{R}^p$. This is described in §8.10.

Factoring of various types as a subject independent of control is presented in the first and last parts of Chapter 8. The middle part of the chapter treats the connection between factoring and control. A mathematician with little interest in control could skip directly to the factoring chapter after reading the introduction. Much of it is self-contained with only a few (key) proofs requiring machinery from the first part of the book.

1.8 A Classical Perspective on H^∞ Control

Most people who learn H^∞ control these days for linear systems see state space problems and state space theory. In fact the subject began as a purely input–output frequency domain theory; H^∞ engineering began with amplifier design and later came

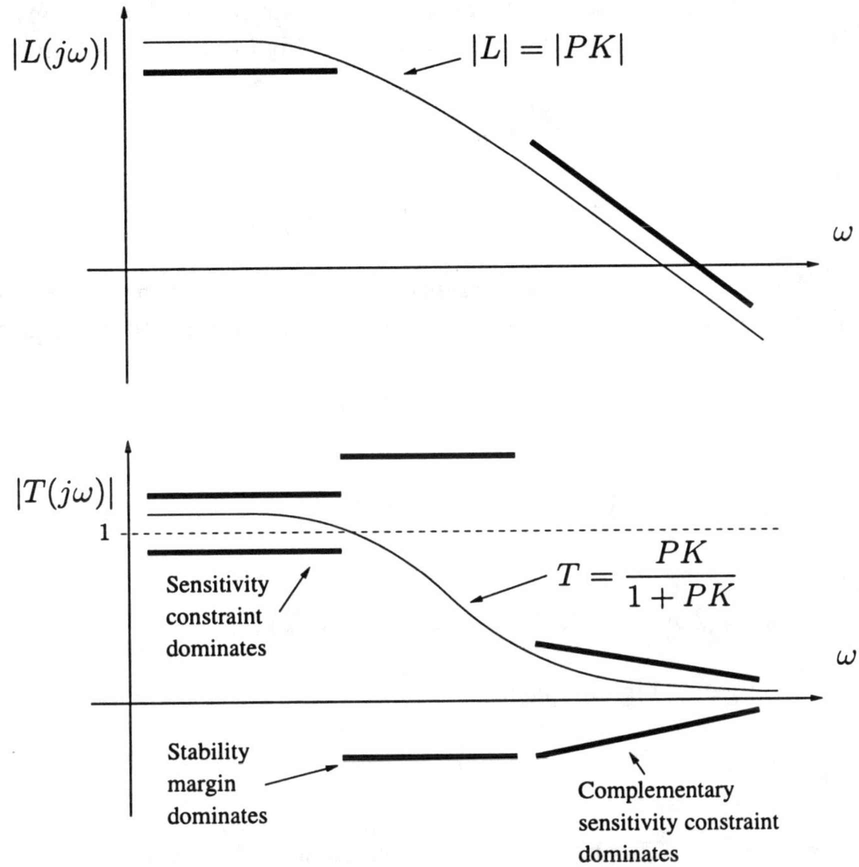

Figure 1.10: Bode diagram (transfer function magnitude versus angular frequency).

into control and gained prominence there; see §1.11. In this section we sketch some of these ideas. We start with H^∞ control and then mention a few ideas and connections with broadband impedance matching, which is an ingredient of classical amplifier design.

1.8.1 Control

One is given a system (plant) P and wishes to find a controller K so that the closed-loop transfer function

$$T(j\omega) \triangleq P(j\omega)K(j\omega)(1 + P(j\omega)K(j\omega))^{-1}$$

of the system in Figure 1.2 has a certain "shape." The desired shape corresponds to the specs laid out in the control problem. A typical situation is illustrated by the Bode plot in Figure 1.10. It contains two plots which contain equivalent information but in different coordinates.

You see in the top picture of Figure 1.10 that the absolute value of $L \triangleq PK$ must be bigger than the heavy line at low frequency and below the other heavy line at high frequency. At midrange frequencies there is a bit of flexibility, so precise constraints are typically not drawn in. Algebraically the low and high frequency constraints are written as

$$|PK| \geq \gamma_\ell >> 1 \quad \text{low } \omega,$$

$$|PK| \leq \gamma_h << 1 \quad \text{high } \omega,$$

where γ_ℓ and γ_h are given. The bottom figure contains the same information as the top figure but in terms of T, which we now see using simple algebra. At low frequency

$$|1 + PK| \geq \gamma_\ell - 1,$$

$$|(1 + PK)^{-1}| \leq (\gamma_\ell - 1)^{-1},$$

so

$$|T - 1| = |PK(1 + PK)^{-1} - 1| = |(1 + PK)^{-1}| \leq (\gamma_\ell - 1)^{-1}$$

is small if γ_ℓ is large. At high frequency

$$|T| \leq \gamma_h(1 + \gamma_h)^{-1}$$

is small if γ_h is near 0. This high frequency constraint is often called the *rolloff* or bandwidth constraint.

We rephrase the constraint on T in the form

$$|T - \kappa|^2 W_\ell + |T|^2 W_h \leq R, \tag{1.30}$$

where W_ℓ, W_h are positive weight functions and κ is a function which is 1 at low frequencies, 0 at high frequencies and interpolates smoothly in between. Note that (1.30) contains a constraint on frequencies at midrange and the Bode plot above does not. Actually (1.30) constitutes a well-posed problem, while the Bode plot constraints do not. Adding midrange (like stability margin) constraints to the Bode plot gives a well-posed problem. Note $W_h(j\omega) \to \infty$ as $\omega \to \infty$ to force the envelope containing T to pinch to 0 at ∞.

We would like to show how the problem of finding a stable closed-loop system meeting the constraint (1.30) translates to a familiar state space H^∞ problem. Actually there is a subtle issue on what we mean by the closed-loop system being stable. Certainly we want T to have no poles in the closed right half-plane (RHP), but we need in addition that small perturbations of P and K also have this property. This is one version of *internal stability*. We will not belabor this viewpoint because that would be time consuming and because internal stability corresponds directly *to stability of the state space equations for the closed-loop system* as was previously defined.

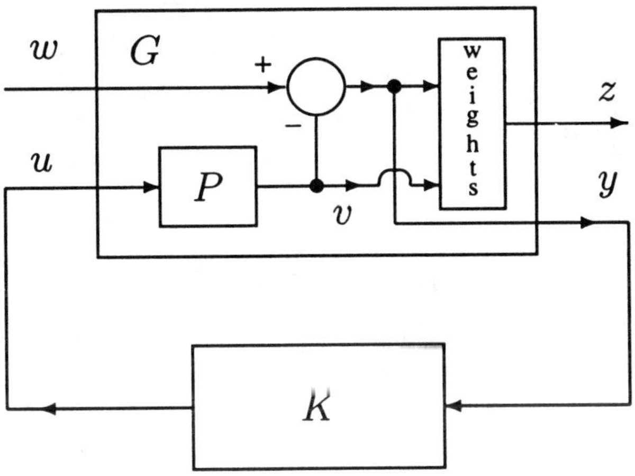

Figure 1.11: Mixed sensitivity embeds in the standard problem.

The next step in conversion of the H^∞ problem to state space form is embedding our H^∞ control problem in *the standard problem* described in §1.8. Figure 1.11 indicates how this is done. The transfer function $G(s)$ incorporates all information in the weights W_1, W_2, and plant P. One can read off the precise formula for G from Figure 1.11 and there is no reason to record it here, since we explicitly give the state space version of the formula in Chapter 9. Thus we have shown that our classical H^∞ problem is equivalent to finding K, which makes the closed-loop system in Figure 1.11 γ-dissipative or, equivalently, internally stable with $w \to z$ transfer function having sup-norm less than γ. Now Figure 1.11 has the form of Figure 1.1 and we see that the classical H^∞ control problem has the form of the standard problem of H^∞ control in §1.1.

1.8.2 Broadband Impedance Matching

A basic problem in classical circuit theory is, *given an amplifying device, connect it with a passive circuit that produces a total (closed-loop) amplifier which maximizes the worst gain over all frequencies.* An easier problem, which often bears heavily on the amplifier problem, is the broadband impedance matching problem: *Transfer as much power as possible from a given source to a passive (dissipative) load.*

This problem is illustrated by Figure 1.12. The top picture shows an amplifier gain maximization problem. The middle picture illustrates the impedance matching problem associated with the amplifier problem. The last picture draws Figure 1.2 and the classical control problem we discussed in §1.8.1 in a way which looks much like the middle picture of Figure 1.12.

One thing to mention is that the key tradeoff in impedance matching goes under the name *gain-bandwidth limitations*; to wit huge gain over high frequency bands is impossible. They have been studied (under this name for decades), and the basic "rule

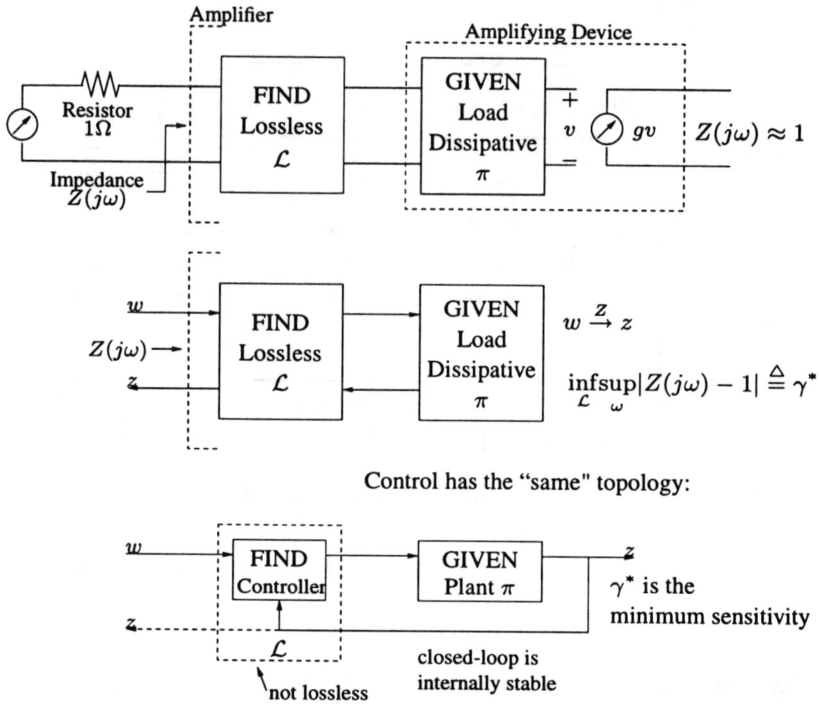

Figure 1.12: Amplifier.

of thumb" is the Bode–Fano integral constraints (an analogue of the Freudenberg–Looze constraints of control theory). Gain–bandwidth limitations are quite literally analogues of performance–rolloff constraints in control.

1.9 Nonlinear "Loop Shaping"

As mentioned in §1.8, in classical linear control the main objectives (in order of importance) are to make the controlled system

(i) stable,

(ii) have prescribed rolloff, and

(iii) achieve high performance at low frequencies.

A metaphor for their implications is that if we design an airplane that fails to be stable, it will crash immediately; if rolloff is poor then it will probably crash eventually; and if performance is mediocre, the plane will waste something, maybe fuel or maybe a passenger's lunch.

Controller design classically often consisted of choosing a candidate controller, and then checking the closed-loop transfer function to see if it met given performance and rolloff specs, hence the term *loop shaping*. H^∞ control originated with the goal of making loop shaping more systematic. The H^∞ formalism involves weight selection, which is reasonably intuitive. Once sensible weights are picked, solutions to the H^∞ problem often are not so far from desired that a few natural iterations give a solution.

Of course there are serious tradeoffs between stability, rolloff, and performance constraints. While frequency and hence rolloff have no meaning for nonlinear systems, it is hard to believe that when systems are nonlinear that these tradeoffs disappear. They must be important in some form.

What is nonlinear loop shaping? This is the subject of much current research and discussion, although the word loop shaping is not used. Indeed the issue is enough in flux that we do not presume to say anything definitive here. Our goal in this section is just to introduce a few issues. The main issues actually emerge in the state feedback problem, say, for a system of the form

$$\begin{aligned} \dot{x} &= A(x) + B_2(x)u, \\ z &= C_1(x) + D_{12}(x)u, \end{aligned}$$

so we focus on state feedback in this presentation rather than on the more complicated measurement feedback problem (the next section considers this briefly).

Much attention goes to stabilizing a system, and stability might be viewed as a type of performance constraint. This can be facilitated by solving a control Lyapunov inequality,

$$(A(x) + B_2(x)u) \cdot \nabla_x V(x) + \Omega(x) \leq 0, \tag{1.31}$$

where $\Omega \geq 0$. Given V and Ω this can be done explicitly. For example, for single input systems ($\dim u = 1$),

$$u(x) = \frac{-1}{B_2(x) \cdot \nabla_x V(x)} (A(x) \cdot \nabla_x V(x) + \Omega(x))$$

for $x \neq 0$.

More challenging is to decipher the analogue or function in nonlinear control of rolloff constraints. Mathematically, rolloff constraints for a linear system look something like

$$\sup_{s=j\omega} \left| \frac{1}{(1+s)^n} T(s) \right| \leq k_\infty$$

or

$$\int_{\substack{j\omega \\ axis}} \left| \frac{1}{(1+s)^n} T(s)u(s) \right|^2 \leq k_2.$$

In time domain terms these inequalities punish the size of

$$\frac{d^n}{dt^n} [\widehat{Tu}](t);$$

that is, one has rate bounds on the output of the system. The output of the system is $C_1(x(t)) + D_{12}u(t)$ and a rate bound is implied by a rate bound on $u(t)$ and $x(t)$ separately. Thus it suffices to impose constraints of the form

$$|u| \leq k_1, \quad |\dot{u}| \leq k_2, \quad \text{and } x \in \mathcal{R},$$

where $\mathcal{R}$ is a carefully chosen region in state and input space.

In our discussion we shall focus on bounding $\dot{u}$, since this is an actuator rate bound and these are very common; so set $C_1 = 0$, $D_{12} = 1$, and $n = 1$. We begin by considering a rate saturation constraint $|\dot{u}| \leq 1$ and use the standard trick of making u a state and adding an input v to get

$$\dot{u} = v$$

with v meeting the saturation constraint $|v| \leq 1$. Incorporate this into the control Lyapunov inequality (1.31) to get

$$(A(x) + B_2(x)u) \cdot \nabla_x V(x,u) + v\nabla_u V(x,u) + \Omega(x) \leq 0.$$

Next we use the saturation constraint on v and the inequality to get that it is sufficient that V satisfy the first-order PDI

$$(A(x) + B_2(x)u) \cdot \nabla_x V(x,u) - |\nabla_u V(x,u)| + \Omega(x) \leq 0, \qquad (1.32)$$

with

$$v^*(x,u) = -\text{sign}\,\{\nabla_u V(x,u)\}.$$

The corresponding controller satisfies

$$u(t) = u(0) + \int_0^t v^*(x(s), u(s))ds.$$

We emphasize that we are just giving sufficient conditions for the solution.

In fact many more formulas much more thoughtfully crafted than this can be written out in a large variety of circumstances. This is an effort pioneered by E. Sontag and collaborators; see [LS91], [LS95] for cases like those we have just treated. Other very direct approaches to finding control Lyapunov functions for systems with special structures have been developed by many authors. Particularly extensive developments are due to Kokotovic and collaborators and are reported in books such as [KKK95].

Now we describe another approach to imposing constraints on u. Rather than directly impose a hard rate bound $|\dot{u}| \leq 1$ we just punish "large" rates in some way, for example, we combine a cost of the form

$$\gamma^2 \tfrac{1}{2} \int_0^\infty |\dot{u}|^2 \, dt \qquad (1.33)$$

with the control Lyapunov inequality (1.31). This gives

$$(A + Bu) \cdot \nabla_x V + v\nabla_u V + \gamma^2 \tfrac{1}{2} v^2 + \Omega \leq 0$$

with $\dot{u} = v$. Just as before we optimize over v, but with this approach there is no constraint on v, so the maximizing v is

$$v^*(x) = -\gamma^{-2}\nabla_u V(x,u)$$

and the PDI becomes

$$(A + Bu) \cdot \nabla_x V - \frac{1}{2\gamma^2}(\nabla_u V)^2 + \Omega \leq 0, \qquad (1.34)$$

which is similar to a nonlinear Riccati inequality.

Similarly, we could be more cautious and treat disturbances w entering the system

$$\dot{x} = A(x) + B_1(x)w + B_2(x)u,$$

$$z = C_1(x) + D_{12}(x)u,$$

thereby getting H^∞-type inequalities

$$(A + B_1 w + B_2 u) \cdot \nabla_x V + v\nabla_u V + \gamma_1^2 \tfrac{1}{2} v^2 - \gamma_2^2 \tfrac{1}{2} w^2 + \Omega \leq 0$$

maximizing over w and minimizing over v to get

$$v^*(x) = -\gamma_1^{-2}\nabla_u V(x, u) \quad \text{and} \quad w^*(x) = \gamma_2^{-2} B_1 \cdot \nabla_x V(x, u)$$

and the PDI

$$(A + B_2 u) \cdot \nabla_x V(x, u) - \tfrac{1}{2\gamma_1^2}(\nabla_u V(x, u))^2$$

$$+ \tfrac{1}{2\gamma_2^2}\nabla_x V(x, u) B_1 B_1' \nabla_x V(x, u) + \Omega(x) \leq 0.$$

We summarize all of this by saying that the solution of the stabilization problem together with a rate saturation constraint amounts to solving a particular first-order PDI. The bottom line is that to handle constraints on u it is very likely that some first-order PDI or comparably difficulty problem must be addressed. What we have done here is done with the intention of provoking thought and is hardly conclusive.

While this book treats HJBI inequalities, much of what is done applies to large classes of first-order PDIs. The most extreme example is Bellman inequalities, since they are just the special case where $B_1 = 0$. The next section expands on this theme.

1.10 Other Performance Functions

A wide range of problems can be cast into a form that involves the use of optimization techniques, such as optimal control, game theory, and in particular, the dynamic programming method. In this book we emphasize measurement feedback problems, solved using the information state framework. This framework applies to a range of stochastic (or H_2) problems and, as we discuss in detail in this book, deterministic minimax problems.

The integrand $\frac{1}{2}|z|^2 - \gamma^2 \frac{1}{2}|w|^2$ in the cost functional used in this book (see (3.1)) has special meaning due to the L_2 dissipation inequalities and connection to the H^∞ norm (a frequency domain concept) in the case of linear systems. Any integrand $L(x, u, w)$ could be substituted in principle for $\frac{1}{2}|z|^2 - \gamma^2 \frac{1}{2}|w|^2$, and the corresponding solution could be derived using similar methods. In particular, a suitable information state can be defined:

$$p_t(x) = p_0(\xi(0)) + \tfrac{1}{2}\int_0^t L(\xi(s), u(s), y(s) - C_2(\xi(s)))\, ds$$

$$\text{given } u, y \text{ over } [0,t] \text{ and given } \xi(t) = x,$$

where as discussed above the trajectory $\xi(\cdot)$ is a solution of the reversed-arrow dynamics (1.17).

Further, measurement feedback versions of stabilization and loop shaping can also be developed. To illustrate, consider a robust version of the hard-constrained rate saturation example discussed above, where $|\dot{u}| \leq 1$, with plant model

$$\dot{x} = A(x) + B_1(x)w + B_2(x)u,$$

$$y = C(x) + w,$$

and Lyapunov integrand $\Omega(x)$. The minimax cost associated with this problem for an output feedback controller $K : y(\cdot) \mapsto v(\cdot)$ is

$$J_{p_0}(K) = \sup_{T\geq 0,\, w(\cdot)} \left\{ p_0(x_0) + \int_0^T [\Omega(x(s)) - \gamma^2 \tfrac{1}{2}|w(s)|^2] ds \right\}$$

with information state (a function of (x, u)) defined for $v(\cdot), y(\cdot)$

$$p_t(x, u) = p_0(\xi(0), u(0)) + \int_0^t [\Omega(\xi(s)) - \gamma^2 \tfrac{1}{2}|y(s) - C(\xi(s))|^2] ds,$$

where

$$\dot{\xi} = A(\xi) + B_1(\xi)w + B_2(\xi)u,$$

$$\dot{u} = v,$$

for $0 \leq s \leq t$ and $\xi(t) = x$, $u(t) = u$. The PDE for the information state is

$$\dot{p} = -\nabla_x p(x,u) \cdot (A(x) + B_1(x)w + B_2(x)u) - \nabla_u p(x,u) \cdot v$$
$$+\Omega(x) - \gamma^2|y - C(x)|^2.$$

The dynamic programming PDE is

$$\inf_{|v|\leq 1} \sup_y \{\nabla_p W(p)[-\nabla_x p \cdot (A + B_1(y - C) + B_2 \cdot) - \nabla_u p \cdot v$$
$$+\Omega - \gamma^2|y - C|^2]\} = 0.$$

If we optimize over $|v| \leq 1$ we get

$$\sup_y \{\nabla_p W(p)[-\nabla_x p \cdot (A + B_1(y - C) + B_2 \cdot) - \nabla_u p \cdot v$$
$$+\Omega - \gamma^2|y - C|^2] - |\nabla_p W(p)[-\nabla_u p]|\} = 0,$$

with optimizer

$$v^*(p) = -\text{sign}\{\nabla_p W(p)[-\nabla_u p]\}.$$

This gives the controller

$$u(t) = u(0) + \int_0^t v^*(p_s)ds.$$

The integral-constrained control rate example (1.33) can be handled in the same way.

1.11 History

The objective of this section is to give a history of developments preceding this book. Initially the account follows that given in [HM98a]. We have attempted to mention the main developments, and we apologize in advance if we have missed some references.

1.11.1 Linear Frequency Domain Engineering

In commonplace language H^∞ engineering amounts to achieving prescribed worst case frequency domain specs. Optimizing worst case error in the frequency domain along its present lines started not with control but with passive circuits. One issue was to design amplifiers with maximum gain over a given frequency band. Another was the design of circuits with minimum broadband power loss. Indeed H^∞ control is a subset of a broader subject, H^∞ *engineering*, which focuses on worst case design in the frequency domain. In paradigm engineering problems this produces what a mathematician calls an "interpolation problem" for analytic functions. These can be solved by Nevanlinna–Pick techniques. The techniques of Nevanlinna–Pick interpolation had their first serious introduction into engineering in a single-input single-output (SISO) circuits paper of Youla and Saito [YS67] in the middle 1960s. Further development waited until the mid-1970s, when Helton [Hel76], [Hel78], [Hel81] applied interpolation and more general techniques from operator theory to amplifier problems. Here the methods of commutant lifting [And63], [SNF70], [Sar67] and of Adamjam–Arov–Krein (AAK) [AAK68], [AAK72], [AAK78] were used to solve many-input many-output (MIMO) optimization problems.

In the late 1970s G. Zames [Zam79] began to marshall arguments indicating that H^∞ rather than H_2 was the physically proper setting for control. Zames suggested on several occasions that these methods were the appropriate ones for codifying classical control. These efforts yielded a mathematical problem which Helton identified as an interpolation problem solvable by existing means (see [ZF81]). Zames and Francis [ZF83] used this to solve the resulting SISO problem, and Chang–Pearson [CJ84] and Francis–Helton–Zames [FHZ84] solved it for the MIMO system.

The pioneering work of Zames and Francis treated only sensitivity optimization. In 1983 three independent efforts emphasized bandwidth constraints, formulated the problem as a precise mathematics problem, and indicated effective numerical methods for their solution: Doyle [Doy83], Helton [Hel83], and Kwakernaak [Kwa83]. All of these papers described quantitative methods which were soon implemented on computers. It was these papers that actually laid out precisely the tradeoff in control between performance at low frequency and rolloff at higher frequency and how one solves the resulting mathematics problem. This is in perfect analogy with amplifier design where one wants large gain over as wide a band as possible, producing the famous gain bandwidth tradeoff.

Another independent development was Tannenbaum's [Tan80] very clever use of Nevanlinna–Pick interpolation in a control problem in 1980. Also early on the H^∞ stage was Kwakernaak's polynomial theory [Kwa86]. Another major development that dovetailed closely with the invention of H^∞ control was a tractable theory of plant uncertainty. A good historical treatment appears in [DFT92]. Another application of these techniques is to robust stabilization of systems by H. Kimura [Kim84]. An early book on H^∞ control was [Fra84].

1.11.2 Linear State Space Theory

To describe the origins of state space H^∞ engineering we must back up a bit. Once the power of the commutant lifting AAK techniques was demonstrated on engineering problems, P. de Wilde played a valuable role by introducing them to signal processing applications (see [DVK78]) and to others in engineering. The state space solutions of H^∞ optimization problems originated not in H^∞ control, but in the area of model reduction. The AAK work with a shift of language is a paper on model reduction (although not in state space coordinates). This was recognized by Bettayeb–Safanov–Silverman [BSS80], which gives a state space viewpoint for SISO systems. Subsequently Glover [Glo84] gave the MIMO state space theory of AAK-type model reduction. Since the H^∞ control problem was already known to be solvable by AAK, this quickly gave state space solutions to the H^∞ control problem. These state space solutions were described first in 1984 by Doyle in a report [Doy84], which although never published was extremely influential. Earlier in his thesis (unpublished) he had given state space H^∞ solutions based on converting the geometric (now it would be called behavioral by engineers) version of commutant lifting AAK due to Ball and Helton to state space.

There was a vast effort on state space H^∞ control by many engineers and mathematicians. We mention now only a few major developments. In the beginning there were only crude bounds on the dimension of the state space of the controller, and numerical recipes for the controller relied on substantial cancellation, which of course is bad. It was discovered by [LH88] that the dimension of an H^∞ optimal controller equals that of the plant G. Next came the famous paper [DGKF89], which gave an elegant cancellation-free formula for the controller (as discussed in §1.2). The formulas in this paper have become standard. Other closely related results also appeared around this time or a little later; see [Tad90], [Sto92]. An excellent presentation is given in [GL95].

1.11.3 Factorization

It might be mentioned that factorization (the subject of Chapter 8) was known from early on to yield all controllers producing a certain performance, as well as solving other problems; cf. [HBJP87]. These methods were developed by Ball–Helton and H. Kimura and coworkers in many papers during the 1980s and 1990s (see [BHV91], [Kim97] and the references therein). This leads to an elegant proof of the original Glover state space AAK model reduction formulas as well as the first discrete time version by Ball–Ran [BR87]. A J-spectral factorization approach was presented in [GGLD90], [Gre92].

1.11.4 Game Theory

It was observed in [Pet87], [DGKF89] (and elsewhere) that there are close connections between H^∞ control and differential games. Basically, the two quite distinct problems can be solved using the *same* Riccati equations. These connections were pursued in depth by a number of researchers; see, e.g., [LAKG92], [BB91] (updated in 1995). The game theory view of H^∞ control is as a minimax game, where the disturbance or uncertainty is modelled by a malicious opponent, and the aim of the controller is to minimize the worst performance under these circumstances. This time domain formulation is very important for nonlinear systems.

1.11.5 Nonlinear H^∞ Control and Dissipative Systems

The efforts to extend H^∞ control to nonlinear systems began in the mid-1980s by mathematicians versed in linear commutant lifting and AAK techniques. Ball et al. formulated the nonlinear problem and showed that power series (Volterra) expansions lead to reasonable approximate solutions [BFHT87b], [BFHT87a]. This effort has continued to produce impressive results [FGT95], [FGT96], [FGT98]. Ball–Helton pursued several different approaches. One was what would today be described in terms of behaviors or games in extensive form [BH88c]. Another was in state space form [BH92a], [BH92b], [BH88a], [BH88b]. This reduced the solution of the measurement feedback discrete time nonlinear problem for a "strongly" stable plant P or more generally $\tilde{G}$ to solution of an HJBI equation.

For continuous time state feedback, basic work was done by van der Schaft [vdS91], [vdS92], [vdS96]. He reduced the solution of the state feedback problem for a nonlinear plant G to solution of an HJBI equation. This work was influenced by Willems' theory of dissipative systems and the works of Hill–Moylan [Wil72], [HM76], [HM77], etc. Indeed, van der Schaft emphasizes L_2-gain terminology and the bounded real lemma [AV73]. This is a powerful and natural formulation. Indeed, it is the L_2-gain inequality (which we refer to as the dissipation inequality in this book) that makes sense for nonlinear systems, whereas the frequency domain concept of H^∞ norm does not apply to nonlinear systems.

1.11.6 Filtering and Measurement Feedback Control

Classical control problems, as discussed earlier, are formulated in the frequency domain and are naturally measurement feedback problems. This is reflected in the papers of the 1980s; cf. [DFT92], [HM98a]. Optimal control with measurement feedback is difficult, and this explains in part the length of time it took to obtain a nice state space solution to the linear H^∞ control problem (most of a decade). The issue is how to represent and use the information contained in the measurements.

Much of optimal control theory (including games) is concerned with state feedback problems. This is natural, since the state of a system is a summary of its status and together with the current input values can be used to determine future behavior. Engineers are interested in feedback controllers, and solutions to state feedback optimal control problems lead to state feedback solutions (via, say, dynamic programming). However, given that the original problem of interest is a measurement feedback one, there is the difficulty of what to do with the lack of full state information. A common, but often suboptimal, approach is to design a state estimator (or observer) and plug the state estimate into the optimal state feedback controller. This is called certainty equivalence. The solution of the linear quadratic Gaussian (LQG) problem is an optimal certainty equivalence controller [Won68]. First, an optimal state feedback controller is designed and then coupled with the output of the optimal state estimator, i.e., the Kalman–Bucy filter [Kal60], [KB60]. The certainty equivalence approach is not optimal for the deterministic linear quadratic regulator (LQR) problem. Deterministic LQR designs may employ a Luenberger observer [Lue66].

The LQG problem is a stochastic optimal control problem. What is happening in Kalman's solution is that the optimal state estimate, the conditional mean, becomes the state of a new system, and the optimal controller for this new system turns out to coincide with the optimal state feedback controller for the original system. Actually, the optimal LQG controller feeds back the conditional probability distribution, which being a Gaussian distribution is completely determined by the conditional mean and covariance (finite parameters). For nonlinear optimal stochastic control problems analogous to LQG, the optimal controller is a function of the conditional distribution. Thus the conditional distribution serves as an "information state" for these optimal control problems. The measurement feedback optimal control problem is transformed into a new state feedback optimal control problem, with the information state serving as the state variable. The evolution of the conditional distribution is described by a stochastic PDE, called the Kushner–Stratonovich equation [Kus64], [Str68], or in unnormalized form, the Duncan–Mortensen–Zakai equation [Dun67], [Mor66], [Zak69]. These are the stochastic PDEs of nonlinear filtering and are the nonlinear counterparts to the Kalman filter equations. Thus nonlinear filtering is infinite dimensional, and measurement feedback optimal stochastic control involves the optimal state feedback control of an infinite-dimensional system.

The information state approach has been well known since at least the 1960s, both in the West and East. A nice explanation of these ideas is given in [KV86]. Of the many publications devoted to this problem, we mention only [Str65], [Nis76], [Ell82], [FP82], [Fle82], [Hij90], [EAM95]. It is still a difficult mathematical problem and presents challenging implementation issues.

For nonlinear problems analogous to the deterministic LQR problem, there is no information state solution, and one typically uses a suboptimal certainty equivalence design as discussed above. A key difficulty here is the design of the state estimator or observer. This is a major problem in nonlinear control [KET75], [HK77], [KR85], etc.

In contrast, it is relatively straightforward to write down a nonlinear filter, although one is faced with computational difficulties for implementation. In 1968, R.E. Mortensen derived a deterministic approach to nonlinear filtering, called minimum energy estimation [Mor68]. This is essentially a least squares approach and leads to a filter, which is a first-order nonlinear PDE. An interesting study of this filter was conducted in 1980 by O. Hijab [Hij80]. These deterministic filters are related to the stochastic filters via small noise limits. These limits are examples of the type which occur in the theory of large deviations. J.S. Baras was intrigued by these filters and their connections, and in [BK82] proposed using these methods as the basis of a design procedure for nonlinear observers [BBJ88], [JB88], [Jam91].

1.11.7 H^∞ Control, Dynamic Games, and Risk-Sensitive Control

In 1973, D.H. Jacobson [Jac73], introduced a new type of stochastic optimal control problem with an exponential cost function, which today is often called the risk-sensitive problem. He solved a linear exponential quadratic Gaussian (LEQG) problem with full state feedback and observed that his solution is the same as the solution for a related dynamic game (the same Riccati equation). It took until 1981 for the corresponding linear measurement feedback problem to be solved by Whittle [Whi81]. The structure of the controller is again of the certainty equivalence type, although the Kalman filter estimate is not used. Instead, the Kalman filter is modified with terms coming from the control objective. Whittle's solution was very interesting, since the conditional distribution is not used as the information state. Later, connections with H^∞ control were discovered [GD88], [DGKF89]. Thus H^∞ control, dynamic games, and risk-sensitive control are all related.

In the late 1980s and early 1990s Basar–Bernhard and coworkers developed the certainty equivalence principle for deterministic minimax games and H^∞ control. The key reference here is the 1989 monograph [BB91] (revised in 1995), as well as the papers [Ber91], [DBB93], [BR95]. The book [BB91] contains an excellent account of the minimax game approach and certainty equivalence mainly in the linear context, with some nonlinear results in the second edition. The certainty equivalence solution is very closely related to the solution of Whittle and is the basis of an important approach to measurement feedback nonlinear H^∞ control.

In the early 1990s a number of researchers began exploring the connections between H^∞ control, dynamic games, and risk-sensitive control in the nonlinear context, beginning with Whittle [Whi90a], [Whi90b], [Whi91]. The connections made use of small noise limits. This work inspired Fleming–McEneaney, leading to the papers [FM92], [FM95], and also to papers studying viscosity solutions of the H^∞ PDEs and PDIs [BH96], [Jam93], [McE95b], [McE95a], [Sor96]. Independently, J.S. Baras suggested investigating the risk-sensitive problem using small noise methods, in conjunction with earlier work on nonlinear filters. This led to the papers [Jam92], [JBE94], [JB95], [JB96], [BJ97]. The paper [JBE94] solved the nonlinear measurement feedback (discrete time) stochastic risk-sensitive problem, solved a nonlinear measurement feedback deterministic minimax game, and established connections between them via small noise limits. An information state was used for both problems, and in the risk-sensitive case, the information state was not the conditional probability distribution. The information state definition was inspired by the paper [BvS85], which used a method that generalizes to nonlinear systems. In the minimax case, the information state coincides with Basar–Bernhard's cost-to-come method and is related to the risk-sensitive information state in a manner analogous to the link between Mortensen's minimum energy estimator and stochastic nonlinear filters discussed above. See also the publications [KS89], [Ber96]. A large number of papers have since been written concerning various aspects of risk-sensitive control, filtering, games, and their connections: [PMR96], [CE95], [CH95], [FHH97], [FHH98], [Nag96], [RS91], [Run91], etc.

1.11.8 Nonlinear Measurement Feedback H^∞ Control

While stable plant problems had been known since the late 1980s to convert to HJBI inequalities, the unstable measurement feedback problem remained intractable. A substantial number of papers have been written, including: Isidori–Astolfi–Kang [IA92a], [IA92b], [Isi94], [IK95]; Ball–Helton–Walker [BHW93]; Didinsky–Basar–Bernhard [DBB93]; Krener [Kre94]; Lin–Byrnes [LB95]; Lu–Doyle [LD94]; Maas [Maa96]; Nguang [Ngu96]. These results illuminated various aspects of the measurement feedback problem, and indeed the results all specialized to the well-known DGKF solution when applied to linear systems. The results were generally of a sufficient nature, so that if certain PDEs or PDIs could be solved, then a solution to the nonlinear H^∞ control problem would be produced. However, in general these results are far from being necessary: H^∞ controllers could exist but not be of the form given in these papers. This is because nonlinear filtering, and hence optimal measurement feedback control, is intrinsically infinite dimensional.

Information state controllers for nonlinear H^∞ control were obtained by a number of authors in the early 1990s. van der Schaft [vdS96] identified some of the key measurement feedback equations, including the coupling condition, and obtained information state controller formulas assuming certainty equivalence. Didinsky–Basar–Bernhard [DBB93] obtained information state controllers assuming certainty equivalence and generalized certainty equivalence. The first general solution to the nonlinear H^∞ problem was given in [JB95] (see also [JBE94]). The information state was employed to give an intrinsically infinite-dimensional solution, complete with a clean set of basic necessity and sufficiency theorems. Also an independent effort of Chichka and Speyer [CS94] discovered the "general" information state in work on adaptive control. A number of related papers have appeared since then, e.g., [Teo94], [TYJB94], [JY95], [Yul96]. In 1994 Helton–James realized that the information state framework could be used for J inner–outer factorization, and preliminary results and formulas were published in [HJ94]. This initiated a detailed investigation and development of the information state solution, leading to the papers [HJ95], [HJ96b], [HJ96a], and, ultimately, to this book.

1.11.9 Prehistory

Now we turn back to sketch the origins of the HJBI equations that play such a big role in this book. This is an extensive subject which is well described in many places, so we give little account of the history and just list some references. Thus we urge the curious to read [Bel57], [Isa65], [You69], [FR75], [FS93], [BO95].

1.12 Comments Concerning PDEs and Smoothness

In this book we make extensive use of optimal control methods and nonlinear PDEs (Hamilton–Jacobi type). In general, solutions to such PDEs are *not globally smooth*, and in Appendix B we discuss these equations and their solutions, in particular, the concept of *viscosity solution*.

We have attempted to minimize technical issues arising because of lack of smoothness and to keep the focus of the book on control-theoretic ideas. In many places we use PDEs on finite-dimensional spaces (such as the PDE giving the dynamics of the information state) and use integrated (i.e., dynamic programming) representations that are meaningful without smoothness. In some results we assume smoothness to help keep statements clear (and readily connected to the familiar linear case) and to simplify proofs. However, readers should be aware that such results remain valid without the smoothness assumptions, with appropriate interpretations and proofs.

PDEs on infinite-dimensional spaces play a major role in this book. There are many unresolved purely mathematical issues concerning these PDEs. We have not attempted to describe in detail issues concerning the concept of solution for such equations (this is still an open question). Instead, we have stated a number of results that have no need of smoothness (these make use of the integrated dynamic programming equation). However, when one uses the dynamic programming PDE to obtain an optimal feedback controller (such as our construction of the central controller) some form of smoothness is required, so we formalize what we need and assume this in order to develop the control-theoretic ideas. We have tried to make clear where smoothness is or is not assumed.

We remark that the results in this book have discrete time analogues (see [JB95]), and differentiability is irrelevant in discrete time. Thus discrete time controllers can be obtained directly from discrete time analogues of the dynamic programming PDE without the need for the value function to be differentiable.

Part I

Basic Theory for Nonlinear H^{∞} Control

Chapter 2

The H^∞ Control Problem

In this chapter the nonlinear H^∞ problem is carefully posed. In addition some technical assumptions are made and various definitions are given.

2.1 Problem Formulation

A general state space model for the plant G might be

$$G : \begin{cases} \dot{x} = f(x, w, u), \\ z = g(x, w, u), \\ y = h(x, w, u). \end{cases} \tag{2.1}$$

Here, $x(t) \in \mathbf{R}^n$ denotes the state of the system and is not in general directly measurable; instead an output $y(t) \in \mathbf{R}^p$ is observed. The additional output quantity $z(t) \in \mathbf{R}^r$ is a performance measure, depending on the particular problem at hand. The control input is $u(t) \in \mathbf{R}^m$, while $w(t) \in \mathbf{R}^s$ is regarded as an opposing disturbance input.

However, to minimize obfuscating technical issues and to make a clear link to existing linear theory, we consider nonlinear plants G of the form

$$G : \begin{cases} \dot{x} = A(x) + B_1(x)w + B_2(x)u, \\ z = C_1(x) + D_{12}(x)u, \\ y = C_2(x) + D_{21}(x)w. \end{cases} \tag{2.2}$$

We also make some technical assumptions concerning the plant model data in the following assumption.

> ASSUMPTION 2.1.1 We assume that the functions appearing in (2.2) are smooth with bounded first- and second-order partial derivatives, that B_1, B_2, D_{12}, and D_{21} are bounded, and that zero is an equilibrium: $A(0) = 0$, $C_1(0) = 0$, and $C_2(0) = 0$.

These assumptions include the case of linear systems and are designed for an L_2 theory for nonlinear systems, which extends the H^∞ theory for linear systems. It is of course possible to develop a theory with weaker technical conditions, but the basic structure of the theory will be the same as presented in this book.

In H_∞ control the extent to which D_{12} and D_{21} are invertible has a major effect on the complexity of the solution of the control problem, enough so that varying amounts of invertibility have standard names: the "one, two, and four block" problems. In any case one assumes

$$D_{12}(x)'D_{12}(x) \triangleq E_1(x) > 0 \quad \text{and } D_{21}(x)D_{21}(x)' \triangleq E_2(x) > 0 \tag{2.3}$$

are invertible. This is called the *regular case* and with no further assumptions is the four block problem. The two block problems are characterized by

$$D_{12}(x)'D_{12}(x) = E_1(x) > 0 \quad \text{and} \quad D_{21}(x) = I_p \quad (s = p), \tag{2.4}$$

and

$$D_{12}(x) = I_m \quad (r = m) \quad \text{and } D_{21}(x)D_{21}(x)' = E_2(x) > 0. \tag{2.5}$$

The one block problem is defined by

$$D_{12}(x) = I_m \quad (r = m) \quad \text{and } D_{21}(x) = I_p \quad (s = p).$$

In the nonlinear case (as in the linear case) no new ideas are required to go from the 1 to the 2A block solution. Recall that for linear systems the 2A block problem is the mixed sensitivity problem that underlies classical control. Four blocks are only required for μ synthesis.

The following signal spaces will be used:

$$L_2 = L_2([0,\infty), \mathbf{R}^q), L_{2,T} = L_2([0,T], \mathbf{R}^q), \quad \text{and } L_{2,loc,T} = \cup_{0 \le t < T} L_{2,t},$$

where the dimension of the range space will not be stated explicitly but inferred from the context. We write $L_{2,loc} = L_{2,loc,+\infty}$.

A system $\tilde{G}$ is obtained from G by reversing the w and y arrows; the equations defining $\tilde{G}$ are as follows:

$$\tilde{G} : \begin{cases} \dot{\xi} = A^\times(\xi) + B_1(\xi)D_{21}(\xi)'E_2^{-1}(\xi)y + B_2(\xi)u + \tilde{B}_1(\xi)v, \\ z = C_1(\xi) + D_{12}(\xi)u, \\ w = [I - D_{21}'E_2^{-1}D_{21}]v + D_{21}'E_2^{-1}(y - C_2(\xi)). \end{cases} \tag{2.6}$$

Here u, y, and an auxiliary signal v are regarded as inputs to this system

$$\tilde{B}_1 = B_1[I - D_{21}'E_2^{-1}D_{21}] \tag{2.7}$$

and

$$A^\times \triangleq A - B_1D_{21}'E_2^{-1}C_2. \tag{2.8}$$

The system (2.6) will play a role in representing the information state (Chapter 3) and will be used for J-inner–outer factorization (Chapter 8). In the 1 and 2A block cases, $\tilde{B}_1 = 0$ and so the auxiliary input plays no role, and $A^\times$ simplifies to $A^\times = A - B_1 C_2$.

Under our assumptions on the plant data, the state equation in (2.2) has a unique solution on $[0, \infty)$ for all initial conditions x_0 and all inputs $u, w \in L_{2,loc}$. Similarly, the state equation in (2.6) has a unique solution on $[0, \infty)$ for all initial conditions x_0 and all inputs $u, y, v \in L_{2,loc}$. By Gronwall's inequality (Appendix A), the state solving (2.2) and the state satisfying (2.6) satisfy the respective bounds

$$\begin{aligned} |x(s)| &\leq C_a e^{c_a t}(1 + |x(0)|), \quad 0 \leq s \leq t, \\ |\xi(s)| &\leq C_b e^{c_b t}(1 + |\xi(t)|), \quad 0 \leq s \leq t, \end{aligned} \tag{2.9}$$

for $u, y, v \in L_{2,loc}$, for suitable constants $C_a, c_a, C_b, c_b \geq 0$ (which may depend on $u, y, v \in L_{2,loc}$).

A *controller* K is a causal mapping $K : L_{2,loc} \to L_{2,loc}$ taking outputs y to inputs u. Such a controller will be termed *admissible* if the following conditions are satisfied:

(i) *Causality*. If $y_1(s) = y_2(s), 0 \leq s \leq t$, then $K(y_1)(t) = K(y_2)(t)$.

(ii) The closed-loop equations for G (2.2) with $u = K(y)$ and any $x_0 \in \mathbf{R}^n$, $w \in L_{2,loc}$ are well defined in the sense that unique solutions $x(\cdot)$ exist with $u(\cdot), y(\cdot)$ in $L_{2,loc}$ and satisfy the first part of (2.9).

The controller K will be said to be *null initialized* if $K(0) = 0$, regardless of whether or not a state space realization of K is given. Note that if K is null initialized and $x_0 = 0$, $w(t) = 0$ for all $t \geq 0$, then $u(t) = 0$, $y(t) = 0$, and $x(t) = 0$ for all $t \geq 0$ in closed loop (G, K).

Let $\mathcal{B}$ denote the class of nonnegative real-valued functions satisfying $\beta(0) = 0$. A function $\beta \in \mathcal{B}$ will be called a *bias*. A controller K is said to solve the *dissipative control problem* provided the closed-loop system (G, K) is γ*-dissipative*, in the sense that there exist a *gain* $\gamma > 0$ and a bias $\beta \in \mathcal{B}$ such that

$$\begin{cases} \frac{1}{2}\int_0^T |z(s)|^2\, ds \;\leq\; \gamma^2 \frac{1}{2}\int_0^T |w(s)|^2\, ds + \beta(x_0) \\ \\ \text{for all } \; w \in L_{2,T} \text{ and all } T \geq 0. \end{cases} \tag{2.10}$$

Here, the integrand is evaluated along the trajectory of (2.2) with disturbance input w, controller $u = K(y)$, and initial plant state x_0. Note that $\beta(x) \geq 0$ for all x is automatically true, which is proved by placing $w(s) \equiv 0$ in (2.10). The function β is called a *coercive bias* if there exists a constant $c > 0$ such that $\beta(x) \geq c|x|^2$.

REMARK 2.1.2 The L_2-gain or H^∞-norm of (G, K) is the smallest γ for which (G, K) is γ-dissipative. ▽

In general, a bias β will depend on the controller K. Let $\mathcal{B}_K$ denote the class of all biases β for which (2.10) holds and $\beta(0) = 0$. The smallest of these, the *minimal bias*, is denoted β_K and is given by

$$\beta_K(x_0) \triangleq \sup_{T\geq 0} \sup_{w\in L_{2,T}} \left\{ \tfrac{1}{2}\int_0^T [|z(s)|^2 - \gamma^2 |w(s)|^2]\, ds \; : \; x_0, K \right\}. \tag{2.11}$$

Thus (G, K) is γ-dissipative if and only if $\beta_K(x_0) < +\infty$ for all x_0.

Sometimes *strict* γ-dissipativity is required. The closed-loop system is called *strictly γ-dissipative* if there exists $\varepsilon_\gamma > 0$ such that the *strict γ-dissipation inequality* holds for any $0 < \varepsilon \leq \varepsilon_\gamma$:

$$\begin{cases} \tfrac{1}{2}\int_0^T |z(s)|^2\, ds \;\leq\; (\gamma^2 - \varepsilon^2)\tfrac{1}{2}\int_0^T |w(s)|^2\, ds + \beta(x_0) \\ \\ \text{for all } \; w \in L_{2,T} \text{ and all } T \geq 0. \end{cases} \tag{2.12}$$

For a given plant G, the number γ^* denotes the smallest value of $\gamma > 0$ for which there exists a controller K (output feedback) with (G, K) γ-dissipative.

While dissipation captures the notion of performance of a control system, another issue with H^∞ control is stability of the system. We now give some definitions associated with this.

(i) The closed-loop system will be called *weakly internally stable* provided that if G is initialized at any x_0, and if $w(\cdot) \in L_2[0,\infty)$, then in the closed-loop defined by $u = K(y)$ the signals $u(\cdot), y(\cdot), z(\cdot)$ belong to L_2 and the plant state $x(t)$ converges to 0 as $t \to \infty$.

(ii) *Internal stability* will mean, in addition to weak internal stability, that if the controller K has a state space realization, then its internal state will converge to an equilibrium as $t \to \infty$ when in closed-loop $u = K(y)$ for any initial plant state x_0 and plant input $w \in L_2[0,\infty)$.

(iii) The system (G, K) is *z-detectable* if $w(\cdot)$ and $z(\cdot) \in L_2[0,\infty)$ implies $x(\cdot) \in L_2[0,\infty)$.

THEOREM 2.1.3 *If (G, K) is γ-dissipative and z-detectable, then (G, K) is weakly internally stable.*

Proof. The inequality

$$\frac{1}{2}\int_0^\infty |z(t)|^2 dt \leq \gamma^2 \frac{1}{2}\int_0^\infty |w(t)|^2 dt + \beta_K(x_0)$$

holds along the state trajectory starting at x_0, with K null initialized. Thus $z(\cdot)$ is in $L^2[0,\infty)$; so is $x(\cdot) \in L_2$ by z-detectability. This implies, using the growth rate of Assumption 2.1.1, that $y = C_2(x) + D_{21}(x)w \in L_2$ and $u = E_1(x)^{-1}D_{12}(x)'[z - C_1(x)] \in L_2$. Also, $\dot{x} = A(x) + B_1(x)w + B_2(x)u \in L_2$. Therefore, $x(t) \to 0$ as $t \to \infty$. □

Thus if we solve the dissipative control problem we also solve the (weak) internal stability problem and these together are what is usually thought of as the "H^∞ control" problem.

2.2 Appendix: Some Technical Definitions

This section provides most technical definitions that will be used in this book. The reader may refer to them at his or her convenience, and it is not necessary to read through this material on a first reading of the book.

2.2.1 Spaces, Convergence

In this section we define the function spaces that occur in this book. Fairly complicated ones are actually essential since they are the most natural state spaces for the information state controller.

2.2.1.1 Singular and Nonsingular Functions and Convergence

(i) The space $C(\mathbf{R}^n)$ of real-valued continuous functions on $\mathbf{R}^n$ is often denoted

$$\mathcal{X} = C(\mathbf{R}^n).$$

(ii) The space

$$\mathcal{X}_e$$

is defined to be the space of all upper semicontinuous (u.s.c.) $\mathbf{R}_e = \mathbf{R} \cup \{-\infty\}$-valued functions that are bounded above, so that the *max-plus norm*

$$\langle \phi \rangle \stackrel{\triangle}{=} \sup_{x \in \mathbf{R}^n} \{\phi(x)\} \tag{2.13}$$

is finite for all $\phi \in \mathcal{X}_e$ (see Appendix C).

The space $\mathcal{X}_e$ is an "extension" of the space $\mathcal{X}$, which includes extended real-valued *singular* functions such as

$$\delta_M(x) = \begin{cases} 0 & \text{if } x \in M, \\ -\infty & \text{if } x \notin M, \end{cases} \tag{2.14}$$

where $M \subset \mathbf{R}^n$ (e.g., a submanifold), or functions of the type

$$p(x) = \delta_M(x) + \breve{p}(x) = \begin{cases} \breve{p}(x) & \text{if } x \in M, \\ -\infty & \text{if } x \notin M, \end{cases} \tag{2.15}$$

where $\breve{p}$ is a real-valued function defined on M. These extended real-valued functions are of vital importance, and they will be used frequently in later chapters. Indeed, $\mathcal{X}_e$ will be the state space of the central controller in the general case, and, in particular, in singular cases.

(iii) Let $\{p_n\}_{n\geq 0}$ be a sequence in $\mathcal{X}_e$, and $p_\infty \in \mathcal{X}_e$. Weak convergence, p_n *converges weakly* to p_∞, denoted

$$p_n \Longrightarrow p_\infty \text{ as } n \to \infty,$$

is characterized by

$$\langle p_n + f \rangle \to \langle p_\infty + f \rangle \text{ as } n \to \infty$$

for all $f \in C_b(\mathbf{R}^n)$ (continuous, bounded functions).[2]

A word on terminology: the space $\mathcal{X}$ consists of *nonsingular* functions, meaning everywhere finite functions (not taking the values $-\infty$ or $+\infty$).

Two special functions are

$$\mathbf{0}(x) = -\infty \text{ for all } x \in \mathbf{R}^n,$$

and

$$\mathbf{1}(x) = 0 \text{ for all } x \in \mathbf{R}^n$$

(see Appendix C).

For $p : \mathbf{R}^n \to \mathbf{R}_e$, we use the notation

$$\text{support } p$$

to denote the subset of $x \in \mathbf{R}^n$ on which $p(x) > -\infty$.

2.2.1.2 Growth at Infinity

Under Assumption 2.1.1, many of the derived functions in this book will grow spatially at most linearly or quadratically; accordingly we introduce function spaces and norms to accommodate them. These growth rates are compatible with an L_2 theory.

(i) The subset of $\mathcal{X}$ consisting of continuous functions with at most linear growth is denoted

$$\mathcal{X}_l,$$

with norm

$$\| \phi \|_l = \sup_{x\in\mathbf{R}^n} \frac{|\phi(x)|}{1+|x|}.$$

(ii) The subset of $\mathcal{X}$ consisting of continuous functions with at most quadratic growth is denoted

$$\mathcal{X}_q,$$

with norm

$$\| \phi \|_q = \sup_{x\in\mathbf{R}^n} \frac{|\phi(x)|}{1+|x|^2}.$$

[2]This is one of several equivalent characterizations of this mode of convergence; see Appendix C for the definition.

(iii) The subset of $\mathcal{X}_q$ of C^2 functions satisfying the growth conditions

$$|\nabla_x p(x)| \leq b_1|x| + b_2, \quad |\nabla_x^2 p(x)| \leq b_3$$

(for suitable constants $b_1 > 0$, $b_2 > 0$, $b_3 > 0$) for all $x \in \mathbf{R}^n$ is denoted

$$\mathcal{X}_{q,2,b}.$$

(iv) Let D be a subset of $\mathcal{X}_e$. A point $p \in D$ is said to belong to the *quadratic interior* of D ($q.i.D$) if there exists $\varepsilon_0 > 0$ such that

$$p + \varepsilon|\cdot|^2 \in D \text{ for all } 0 \leq \varepsilon < \varepsilon_0.$$

2.2.2 Some Basic Properties of Functions

2.2.2.1 Domain

Consider a function $\hat{W} : \mathcal{X}_e \backslash \{0\} \to \mathbf{R} \cup \{+\infty\}$. The *domain* of $\hat{W}$, denoted

$$\text{dom}\, \hat{W},$$

is the largest subset of $p \in \mathcal{X}_e$ for which $\hat{W}(p)$ is *finite*.

2.2.2.2 Structure

For a function $\hat{W} : \mathcal{X}_e \backslash \{0\} \to \mathbf{R} \cup \{+\infty\}$, the following structural conditions will often be required to hold in dom $\hat{W}$:

(i) *Domination.* $\hat{W}(p) \geq \langle p \rangle$ for all $p \in \text{dom}\, \hat{W}$.

(ii) *Monotonicity.* If $p_1 \in \text{dom}\, \hat{W}$, $p_2 \in \mathcal{X}_e$, with $p_1 \geq p_2$, then $\hat{W}(p_1) \geq \hat{W}(p_2)$ and $p_2 \in \text{dom}\, \hat{W}$.

(iii) *Additive homogeneity.* If $p \in \text{dom}\, \hat{W}$ and $c \in \mathbf{R}$, then $p + c \in \text{dom}\, \hat{W}$ and $\hat{W}(p + c) = \hat{W}(p) + c$ for all constants $c \in \mathbf{R}$.

2.2.3 Differentiation

We turn next to differentiation of functions defined on function spaces. Let X be a Banach space with norm $\|\cdot\|_X$ (e.g., $X = \mathcal{X}_q$, $\|\cdot\|_X = \|\cdot\|_q$). The *dual space* X^* is the Banach space of all bounded linear functionals $L : X \to \mathbf{R}$. Bounded means $\| L \|_{X^*} = \sup_{x \in X : \|x\|_X \leq 1} |L(x)| < +\infty$, and the norm on X^* is $\|\cdot\|_{X^*}$. The value of the linear functional at x is denoted $L(x)$, $L[x]$ or simply Lx.

The *Frechet derivative* $f'(x_0) \in X^*$ of a function $f : X \to \mathbf{R}$ at the point $x_0 \in X$ is defined by

$$f(x_0 + h) = f(x_0) + f'(x_0)h + o(\| h \|_X)$$

as $h \to 0$ in X.

Figure 2.1: Forward flow of the differential equation.

Figure 2.2: Backward flow of the differential equation.

The *Gateaux derivative* $\nabla f(x_0) \in X^*$ at a point $x_0 \in X$ in the direction $h \in X$ is defined by

$$\lim_{t\to 0} \frac{f(x_0+th)-f(x_0)}{t} = \nabla f(x_0)[h].$$

Clearly, if $f'(x_0)$ exists, then $\nabla f(x_0)$ exists; further, if $\nabla f(x)$ exists for x belonging to a neighborhood of x_0 and is continuous at x_0, then $\nabla f(x_0) = f'(x_0)$ (see, e.g., [Dei85]).

When f is continuously Frechet differentiable, we will simply use the notation $f'(x_0) = \nabla f(x_0)$, and when we wish to emphasize the variable of differentiation we will write $\nabla_x f(x_0)$.

The *generalized directional derivative* $\nabla_x^+ f(x_0)$ in the direction $h \in X$ is defined by (cf. [Cla83])

$$\nabla_x^+ f(x_0)[h] = \limsup_{t\downarrow 0,\, x'\to x_0} \frac{f(x'+th)-f(x')}{t}. \tag{2.16}$$

2.2.4 Transition Operators and Generators

Consider the nonlinear control system

$$\dot{x} = A(x) + B(x)u \tag{2.17}$$

with drift vector field A and control matrix B.

Let $\Phi^u_{t,0}(x)$ denote the solution of (2.17) with initial state $x(0) = x$ and input $u \in L_{2,loc}$. $\Phi^u_{t,0}$ is called the *transition operator* of the controlled differential equation (2.17). This is illustrated in Figure 2.1.

Now $\Phi^u_{0,t}(x)$ denotes the solution at time 0 starting from x at time t; see Figure 2.2.

The transition operator enjoys the *semigroup property*, which reads, in the case u constant,

$$\Phi^u_{0,t} = \Phi^u_{0,s}\Phi^u_{s,t};$$

see Figure 2.3.

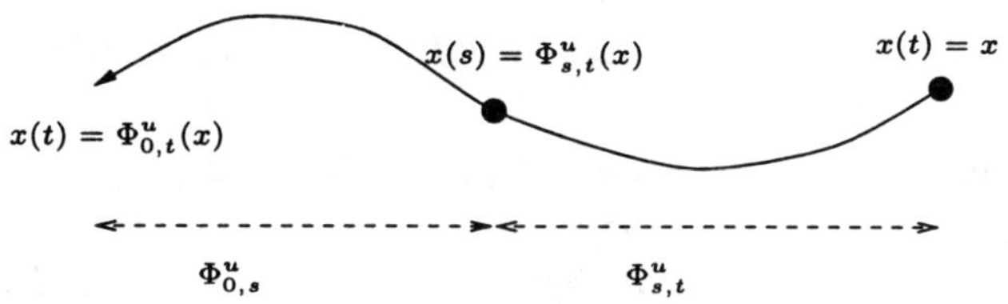

Figure 2.3: Semigroup property.

Here, $\Phi^u_{s,t}(x)$ denotes the solution at time s starting from x at time t. Other terminology for the transition operator is also used, e.g., *propagator*, *semigroup*, *flow*, etc.

The transition operator $\Phi^u_{t,s}$ can be used to define various other transition operators in function spaces. For example, consider the space $C(\mathbf{R}^n)$ of continuous functions defined on $\mathbf{R}^n$, and let $\phi \in C(\mathbf{R}^n)$. Define a transition operator $\mathcal{T}^u_t$ (u is fixed here) by

$$\mathcal{T}^u_t \phi(x_0) = \phi(\Phi^u_{t,0}(x_0))$$

for all $x_0 \in \mathbf{R}^n$. Under reasonable regularity assumptions, $\mathcal{T}^u_t \phi \in C(\mathbf{R}^n)$, so that for each t, $\mathcal{T}^u_t : C(\mathbf{R}^n) \to C(\mathbf{R}^n)$. This transition operator enjoys the semigroup property

$$\mathcal{T}_{t+s} = \mathcal{T}_t \mathcal{T}_s.$$

The *generator* of the transition operator $\mathcal{T}^u_t$ is defined by

$$L^u \phi(x_0) = \lim_{t \downarrow 0} \frac{\mathcal{T}^u_t \phi(x_0) - \phi(x_0)}{t}$$

when the limit exists (and is finite). The *domain* $\operatorname{dom} L^u$ of the generator L^u is the subset of $\phi \in C(\mathbf{R}^n)$ for which this limit exists for all $x_0 \in \mathbf{R}^n$. The space of continuously differentiable functions is contained in this domain: $C^1(\mathbf{R}^n) \subset \operatorname{dom} L^u$, and for $\phi \in C^1(\mathbf{R}^n)$ we can evaluate $L^u \phi$ explicitly as

$$L^u \phi(x_0) = \nabla_x \phi(x_0)(A(x_0) + B(x_0)u)$$

using the chain rule.

2.2.5 Stability

The definitions listed below refer to the system (A, B):

$$\dot{x} = A(x) + B(x)u. \tag{2.17}$$

The state $x = 0$ is an *equilibrium* of the vector field A if $A(0) = 0$. This will be assumed throughout. Also A and B are globally Lipschitz continuous and B is bounded.

The vector field A is *stable* if for all $\varepsilon > 0$ there exists $\eta > 0$ such that $|x_0| < \eta$ implies $|x(t)| < \varepsilon$ for all $t \geq 0$.

The vector field A is *asymptotically stable* if it is stable and if for all $\varepsilon > 0$ there exists $\eta > 0$ such that $|x_0| < \eta$ implies $\lim_{t\to\infty} x(t) = 0$.

The vector field A is *asymptotically L_2 stable* if it is asymptotically stable and the state trajectories $x(\cdot)$ are in $L_2[0,\infty)$.

From Lyapunov theory, we have that A is *globally exponentially stable* if and only if there exists a *Lyapunov function* $U(x) \geq 0$ such that

$$c_1|x|^2 \leq U(x) \leq c_2|x|^2$$

and

$$\frac{d}{dt}U(x(t)) = \nabla_x U(x(t)) \cdot A(x(t)) \leq -c_3|x|^2,$$

where $c_i > 0$, $i = 1,2,3$, are suitable constants. Thus

$$|x(t)| \leq C|x_0|e^{-c_3 t/2} \text{ for all } t \geq 0$$

for some constant $C \geq 0$.

By *local exponential stability* we mean that this definition is valid at least for x in a neighborhood of $x = 0$.

The vector field A is called *monotone stable* if

$$\xi' \nabla_x A(x)\xi \leq -c|\xi|^2 \quad (\text{some } c > 0), \tag{2.18}$$

or, equivalently,

$$(\xi - x)'(A(\xi) - A(x)) \leq -c|\xi - x|^2 \text{ for all } x, \xi. \tag{2.19}$$

A Lyapunov function for such a vector field is $U(x) = \frac{1}{2}|x|^2$.

The system (A, B) is *incrementally L_2 exponentially stable* if

$$\begin{aligned}|\Phi^{v_1}_{t,s}(x_1) - \Phi^{v_2}_{t,s}(x_2)| &\leq b|x_1 - x_2|e^{-c(t-s)/2} \\ &\quad + b(\textstyle\int_s^t e^{-c(t-r)}|v_1(r) - v_2(r)|^2 dr)^{1/2},\end{aligned} \tag{2.20}$$

where $t \geq s$, for suitable constants $b \geq 0$, $c > 0$.

By applying this definition to $(-A, B)$ we have that the system (A, B) is *incrementally L_2 exponentially antistable* if

$$\begin{aligned}|\Phi^{v_1}_{s,t}(x_1) - \Phi^{v_2}_{s,t}(x_2)| &\leq b|x_1 - x_2|e^{-c(t-s)/2} \\ &\quad + b(\textstyle\int_s^t e^{-c(r-s)}|v_1(r) - v_2(r)|^2 dr)^{1/2}.\end{aligned} \tag{2.21}$$

Recall that here $\Phi^v_{s,t}(x)$ denotes the transition operator associated with the system (A, B), and $\Phi^v_{t,t}(x) = x$.

We say that the system (A, B) is *L_2 stable* if

$$\int_s^t |\Phi^v_{r,s}(x)|^2 dr \leq b|x|^2 + b\int_s^t |v(r)|^2 dr, \tag{2.22}$$

for some $b \geq 0$, or L_2 *antistable* if

$$\int_s^t |\Phi_{r,t}^v(x)|^2 dr \leq b|x|^2 + b\int_s^t |v(r)|^2 dr. \qquad (2.23)$$

The system (A, B) is L_2 *exponentially stable* if

$$|\Phi_{t,s}^v(x)| \leq b|x|e^{-c(t-s)/2} + b\left(\int_s^t e^{-c(t-r)}|v(r)|^2 dr\right)^{\frac{1}{2}}, \qquad (2.24)$$

for some $b \geq 0$, or L_2 *exponentially antistable* if

$$|\Phi_{s,t}^v(x)| \leq b|x|e^{-c(t-s)/2} + b\left(\int_s^t e^{-c(r-s)}|v(r)|^2 dr\right)^{\frac{1}{2}}. \qquad (2.25)$$

Note that incremental L_2 exponential antistability implies (by setting $x_1 = 0$, $v_1 = 0$) L_2 exponential antistability, which implies L_2 antistability (by integration).

An operator $\Sigma : L_{2,loc} \to L_{2,loc}$ is called L_2 stable if $\Sigma(L_2) \subset L_2$.

2.2.6 Stabilizability

The following definitions apply to system (2.17) as above.

The set of L_2 *stabilizable states* $\mathcal{S}_{cs}$ is defined by $x \in \mathcal{S}_{cs}$ if and only if there exists $u \in L_2[0, \infty)$ such that $t \mapsto \Phi_{t,0}^u(x)$ is in $L_2[0, \infty)$.

(A, B) is called L_2 *stabilizable* if $\mathcal{S}_{cs} = \mathbf{R}^n$.

The set of L_2 *antistabilizable states* $\mathcal{S}_{cas}$ is defined by $x \in \mathcal{S}_{cas}$ if and only if there exists $u \in L_2(-\infty, 0]$ such that $t \mapsto \Phi_{0,t}^u(x)$ is in $L_2(-\infty, 0]$.

(A, B) is called L_2 *antistabilizable* if $\mathcal{S}_{cas} = \mathbf{R}^n$.

2.2.7 Hyperbolicity

The vector field A is called *exponentially hyperbolic* provided there exist sets M_s and M_{as} such that

(i) $x \in M_s$ implies $|\Phi_{t,0}(x)| \leq b_1(x)e^{-ct}$ for $t \geq 0$.

(ii) $x \in M_{as}$ implies $|\Phi_{t,0}(x)| \leq b_2(x)e^{ct}$ for $t \leq 0$.

(iii) $x \in \mathbf{R}^n \backslash M_s$ implies $|\Phi_{t,0}(x)| \geq b_3(x)e^{ct}$ for $t \geq 0$.

(iv) $x \in \mathbf{R}^n \backslash M_{as}$ implies $|\Phi_{t,0}(x)| \geq b_4(x)e^{-ct}$ for $t \leq 0$.

Here, $c > 0$ is a constant, and $b_i(x) \geq 0$ depend on x, $i = 1, \ldots, 4$.

This definition is motivated by the familiar concept of hyperbolic vector field; see Figure 2.4.

The system (A, B) is called *incrementally hyperbolic* if for all $x_0 \in \mathbf{R}^n$ there exists $\xi_0(x_0) \in M_{as}$ such that for all $v \in L_{2,loc}[0, \tau)$ (some $0 < \tau \leq +\infty$) if x and ξ are trajectories satisfying

$$\dot{x} = A(x) + B(x)v, \; t \geq 0, \; x(0) = x_0,$$

$$\dot{\xi} = A(\xi) + B(\xi)v, \; t \geq 0, \; \xi(0) = \xi_0(x_0)$$

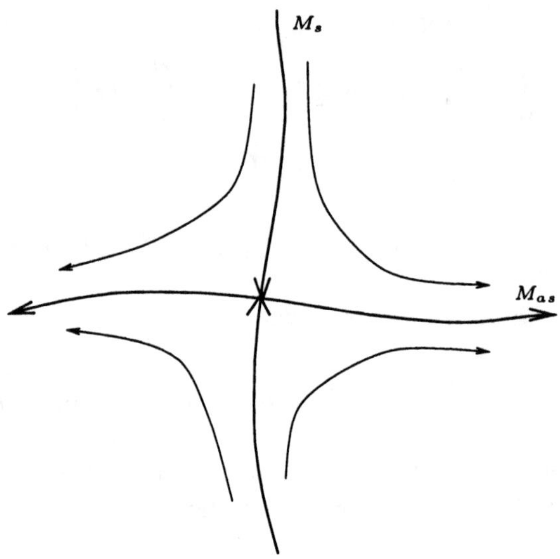

Figure 2.4: Hyperbolic flow.

for $0 \le t < \tau$ then

$$|x(t) - \xi(t)| \le b(x_0, \xi_0(x_0))e^{-ct}, \quad 0 \le t < \tau, \tag{2.26}$$

where $b \ge 0$, $c > 0$ are constants independent of the $L_{2,loc}$ signal v. Call $\xi_0(x_0)$ a *tracking point* for x_0.

2.2.8 Observability/Detectability

The following definitions refer to the system (A, B, C):

$$\begin{aligned} \dot{x} &= A(x) + B(x)u, \\ z &= C(x). \end{aligned} \tag{2.17)'$$

The pair (A, C) is *zero-state observable* if the output $z(\cdot) \equiv 0$ implies that the trajectory $x(\cdot) \equiv 0$ (here $B = 0$ or $u \equiv 0$).

The pair (A, C) is *zero-state detectable* if for $u \equiv 0$ the output $z(\cdot) \equiv 0$ implies that the state $x(t) \to 0$ as $t \to \infty$.

A state x_0 *appears* L_2 *stable from the output* of the system (A, C), provided $\int_0^\infty |C(x(t))|^2 dt$ is finite on any trajectory of an A initialized at x_0 with $u \equiv 0$.

(A, C) is called L_2*-detectable* if when any state x_0 *appears* L_2 *stable from the output*, then the corresponding trajectory $x(\cdot)$ is in L_2.

If the system is (A, B, C) and has input function u, then similar definitions apply to trajectories of (2.17). For this we use the phrase that the state x_0 *driven by input* w *appears* L_2 *stable from the output* of the system.

The system (A, C) is *strongly zero-state observable* if $w(t) = 0$, for all $t \geq 0$, implies there exists $T_1 > 0$, $c_1 > 0$ such that

$$\int_0^T |C(x(t))|^2 \, dt \geq c_1 |x_0|^2 \quad \text{for all } T \geq T_1. \tag{2.27}$$

Refer now to the closed-loop system $(G, K) : w \mapsto z$ (recall (2.2)).

The system (G, K) is *z-detectable* if $w(\cdot)$ and $z(\cdot) \in L^2[0, \infty)$ imply $x(\cdot) \in L^2[0, \infty)$. This is a type of L_2 detectability.

By *z-observable* we mean that $z(\cdot) = 0$, $w(\cdot) = 0$ implies $x(\cdot) = 0$.

The closed-loop system (G, K) is *L_2-observable* if

$$\int_0^t [|x(s)|^2 + |u(s)|^2 + |y(s)|^2] ds \leq C \int_0^t |z(s)|^2 ds \tag{2.28}$$

for all $t \geq 0$ for which the closed-loop signals are defined and all $w \in L_{2,loc}$.

2.3 Notes

(i) The notation used for the plant model (2.2) was used in [DGKF89] and has become quite standard. It is not the most general model that could be used for nonlinear systems; however, we have used it for several reasons. The information state framework developed in this book can handle more general models (e.g., [JB95]), but most of the essential ideas come out using the nonlinear DGKF model. Also, readers familiar with linear H^∞ control can readily interpret the formulas in this book since the notation is familiar.

(ii) For perspective on the effects of invertibility of D_{ij}, we point out that for the 1 and 2A block problem we can obviously reverse arrows as described in Chapter 1, but for the 2B and 4 block problems we cannot. For a pseudoreversal for 2B and 4 block systems, see (2.6). Cases where the rank conditions fail are important; see, e.g., the "cheap sensor" problem [HJM98] and Baramov [Bar98a], [Bar98b].

(iii) The dissipation inequality (2.10) corresponds to finite L_2 gain [Wil72], [HM76], [vdS96]. This is the most convenient and natural way of extending the linear H^∞-norm objective to nonlinear systems.

(iv) For a general discussion of the connections between dissipation and stability, see and the papers [Wil72], [HM76], and Chapter 6.

Chapter 3

Information States

In this chapter the information state is defined, and basic properties of the information state are investigated. In particular, it is shown how the output feedback H^∞ problem can be expressed in terms of the information state. The information state will be used in Chapter 4 to obtain a solution to the H^∞ control problem.

3.1 Differential Games and Information States

Game theoretic methods have been used by a number of authors to approach the H^∞ control problem; see [BB91], for example. In this book we follow the game approach developed in [JBE94], [JB95], [JB96] to solve the H^∞ problem, and, in particular, we make use of the information state solution for this problem. This section formulates the differential game and defines the information state.

3.1.1 Cost Function

We turn now to the definition of the game-theoretic cost function. For $p_0 \in \mathcal{X}_e$, controller K, time horizon $T \geq 0$, disturbance input w, and initial plant state $x(0)$, define the cost function

$$J_{p_0}(K; T, w, x(0)) = p_0(x(0)) + \tfrac{1}{2} \int_0^T \left[|z(s)|^2 - \gamma^2 |w(s)|^2 \right] ds.$$

Next define the minimax cost functional

$$J_{p_0}(K) = \sup_{T \geq 0} \sup_{w \in L_{2,T}} \sup_{x(0) \in \mathbf{R}^n} \left\{ J_{p_0}(K; T, w, x(0)) \right\}, \tag{3.1}$$

where the integrand is evaluated along trajectories of (2.2) with controller $u = K(y)$, initial plant state $x(0)$, and disturbance input w (in Chapter 4 we will minimize over all admissible output feedback controllers K achieving dissipation and weak internal stability). This functional characterizes the dissipation property, as shown in Lemma 3.1.2.

We will often use the notation

$$\langle p+q\rangle \stackrel{\triangle}{=} \sup_{x\in\mathbf{R}^n} \{p(x)+q(x)\} \tag{3.2}$$

for the "sup-pairing" [JBE94], and $\langle p\rangle = \sup_{x\in\mathbf{R}^n}\{p(x)\}$.

REMARK 3.1.1 The pairing (3.2) is actually an inner product in the max-plus or idempotent algebra; see, e.g., [LM95] and [Qua92]. This and related matters are discussed in Appendix C. ▽

LEMMA 3.1.2 *The closed-loop system* (G, K) *is* γ*-dissipative if and only if*

$$J_{-\beta}(K) \leq 0 \textit{ for some } \beta \in \mathcal{B}. \tag{3.3}$$

Moreover, if (G, K) *is dissipative, then*

$$J_{p_0}(K) \leq \langle p_0 + \beta\rangle \textit{ for some } \beta \in \mathcal{B}, \tag{3.4}$$

and

$$\beta_K(x_0) = J_{\delta_{x_0}}(K). \tag{3.5}$$

Lemma 3.1.2 is a version of [JB95, Lemma 4.2].

Proof. Assume that (G, K) is dissipative. Then (2.10) holds, for some finite bias $\beta \in \mathcal{B}_K$, so that for all $T \geq 0$ and all $w \in L_{2,T}$,

$$\tfrac{1}{2}\int_0^T [|z(s)|^2 - \gamma^2|w(s)|^2]ds \leq \beta(x_0).$$

Adding $p_0(x_0)$ to both sides gives

$$p_0(x_0)+\tfrac{1}{2}\int_0^T \left[|z(s)|^2-\gamma^2|w(s)|^2\right] ds \leq p_0(x_0)+\beta(x_0) \leq \sup_{x_0}\{p_0(x_0)+\beta(x_0)\}.$$

Taking the supremum over all $T \geq 0$ and all $w \in L_{2,T}$ gives

$$J_{p_0}(K) \leq \langle p_0+\beta\rangle,$$

proving (3.4), and setting $p_0 = -\beta$ gives (3.3).

Conversely, if (3.3) holds for some $\beta \in \mathcal{B}$, then we have immediately

$$-\beta(x_0) + \tfrac{1}{2}\int_0^T [|z(s)|^2 - \gamma^2|w(s)|^2]\, ds \leq 0$$

for all $T \geq 0$ and all $w \in L_{2,T}$, which implies the dissipation inequality (2.10).

The identity (3.5) follows from the definitions, i.e.,

$$\begin{aligned} J_{\delta_{x_0}}(K) &= \sup_{T\geq 0}\sup_{w\in L_{2,T}}\sup_{x(0)\in\mathbf{R}^n}\left\{J_{\delta_{x_0}}(K;T,w,x(0))\right\} \\ &= \sup_{T\geq 0,\, w\in L_{2,T}}\left\{\tfrac{1}{2}\int_0^T [|z(s)|^2-\gamma^2|w(s)|^2]\, ds \;:\; x_0, K\right\} \\ &= \beta_K(x_0) \end{aligned}$$

(recall (2.11)). □

Next, we see that $J_p(K)$ enjoys certain structural properties.

LEMMA 3.1.3 *The function* $p \mapsto J_p(K)$ *satisfies the following structural properties*

(i) Domination. $J_p(K) \geq \langle p \rangle$.

(ii) Monotonicity. $p_1 \geq p_2$ *implies* $J_{p_1}(K) \geq J_{p_2}(K)$.

(iii) Additive homogeneity. *For any constant* $c \in \mathbf{R}$, $J_{p+c}(K) = J_p(K) + c$.

Proof. Now $J_p(K; 0, 0, x_0) = p(x_0)$, and so

$$J_{p_0}(K) \geq p(x_0)$$

for any $x_0 \in \mathbf{R}^n$. Taking the supremum over all x_0 gives the domination property.

Next, if $p_1 \geq p_2$, then $J_{p_1}(K; T, w, x_0) \geq J_{p_2}(K; T, w, x_0)$, and on taking the supremum over all $x_0 \in \mathbf{R}^n$, $w \in L_{2,T}$, and $T \geq 0$ we obtain the monotonicity property.

Finally, $J_{p+c}(K; T, w, x_0) = J_p(K; T, w, x_0) + c$ for any constant $c \in \mathbf{R}$, and so taking the supremum over all $x_0 \in \mathbf{R}^n$, $w \in L_{2,T}$, and $T \geq 0$ gives the additive homogeneity property. □

From the above inequalities we see that the functional $J_{p_0}(K)$ need not necessarily be finite for all p_0 even if (G, K) is γ-dissipative. However, it follows that

$$\langle p_0 \rangle \leq J_{p_0}(K) \leq \langle p_0 + \beta \rangle \text{ for all } p_0 \in \mathcal{X}_e,\ \beta \in \mathcal{B}_K \tag{3.6}$$

(cf. [JB95, Lemma 4.3]) with equality in the RHS of (3.6) for the minimal $\beta = \beta_K \in \mathcal{B}_K$ (when finite):

$$J_{p_0}(K) = \langle p_0 + \beta_K \rangle.$$

Thus

$$J_{p_0}(K) \text{ finite iff } \langle p_0 + \beta_K \rangle \text{ finite.}$$

The *domain* of $J_.(K)$ is the set of p's for which $J_p(K)$ is finite. Note that $\operatorname{dom} J_.(K)$ is precisely the set of p's for which $\langle p + \beta_K \rangle$ is finite (if $\langle p + \beta_K \rangle$ is finite, then $\langle p \rangle$ is finite). The domain can contain both singular and nonsingular functions. By definition, δ_x and $-\beta$ ($\beta \in \mathcal{B}_K$) belong to $\operatorname{dom} J_.(K)$.

3.1.2 The Information State

Consider the information available to a controller K, viz., the measurement signal y. Since $u = K(y)$, the control signal u is also known. We represent this information as follows. For fixed $u, y \in L_{2,loc}$, the *information state* p_t is defined by

$$p_t(x) = \sup_{\tilde{w}(\cdot)} \Big\{ p_0(\xi(0)) + \tfrac{1}{2} \int_0^t [|\tilde{z}(s)|^2 - \gamma^2 |\tilde{w}(s)|^2]\, ds$$

$$: \xi(t) = x, y(s) = C_2(\xi(s)) + D_{21}(\xi(s))\tilde{w}(s),\ 0 \leq s \leq t \Big\}, \tag{3.7}$$

where $\xi(\cdot)$ satisfies the state equation in (2.2)

$$\dot{\xi} = A(\xi) + B_1(\xi)\tilde{w} + B_2(\xi)u, \tag{3.8}$$

and $\tilde{z} = C_1(\xi) + D_{12}(\xi)u$. The quantity p_t is interpreted as the worst value of the minimax cost that is consistent with the measurement information up to time t, plant dynamics (2.2), and terminal plant state x. The significance of this definition will be seen in what follows. The definition of p_t is as a reverse-time optimal control problem and involves a *constraint*. An important question is whether or not $p_t(x)$ is finite. A priori, there is no reason to expect that, for arbitrary $y(\cdot)$, the constraint can be achieved (if it cannot be achieved, the value is $-\infty$), nor is it apparent, even if the constraint can be achieved, that the supremum does not equal $+\infty$. This question is answered by Lemma 3.1.10 below.

The *support* of an information state p is the subset of $x \in \mathbf{R}^n$ on which $p(x)$ is finite. It is denoted

$$\text{support } p.$$

For the purpose of control, it is convenient to have a dynamical description of the information state. The dynamics for p_t is a PDE: for fixed $u \in L_{2,loc}$ and $y \in L_{2,loc}$ we have

$$\dot{p}_t = F(p_t, u(t), y(t)), \tag{3.9}$$

where $\dot{p} = \frac{dp}{dt}$ and $F(p, u, y)$ is the nonlinear first-order differential operator

$$\begin{aligned} F(p,u,y) = & \sup_{w\in\mathbf{R}^s} \{-\nabla_x p \cdot (A + B_1 w + B_2 u) + \tfrac{1}{2}|C_1 + D_{12}u|^2 - \gamma^2 \tfrac{1}{2}|w|^2 \\ & \quad : \; y = C_2 + D_{21}w\} \\ = & -\nabla_x p \cdot (A + B_1 D_{21}' E_2^{-1}(y - C_2) + B_2 u) \\ & +\tfrac{1}{2\gamma^2}\nabla_x p B_1 [I - D_{21}' E_2^{-1} D_{21}] B_1' \nabla_x p' \\ & +\tfrac{1}{2}|C_1 + D_{12}u|^2 - \gamma^2 \tfrac{1}{2}(y - C_2)' E_2^{-1}(y - C_2). \end{aligned} \tag{3.10}$$

The maximizing w here is

$$w^*(x) = -\gamma^{-2}[I - D_{21}' E_2^{-1} D_{21}] B_1' \nabla_x p' + D_{21}' E_2^{-1}(y - C_2).$$

Equation (3.9) is a Hamilton–Jacobi–Bellman (HJB) equation corresponding to the reverse-time optimal control problem defining the information state (3.7).

The function p_t is called the *information state* of the original system (2.2), and equation (3.9) is called the *information state equation*. When p_t is not smooth, it is typically interpreted in the sense of (3.7) (see also below).

An alternative representation for $p_t(x)$ is useful. Indeed, we can write

$$F(p,u,y) = \sup_{v\in\mathbf{R}^s} \{-\nabla_x p \cdot (A + B_1 D_{21}' E_2^{-1}(y - C_2) + B_2 u + \tilde{B}_1 v) - \gamma^2 \tfrac{1}{2}|v|^2\}$$

$$+\tfrac{1}{2}|C_1 + D_{12}u|^2 - \gamma^2\tfrac{1}{2}(y - C_2)'E_2^{-1}(y - C_2), \tag{3.11}$$

where

$$\tilde{B}_1 = B_1[I - D_{21}'E_2^{-1}D_{21}]. \tag{2.7}$$

Here v is an auxiliary input arising from the identity

$$\sup_v\{-\nabla_x p\cdot \tilde{B}_1 v - \tfrac{1}{2}\gamma^2|v|^2\} = \tfrac{1}{2}\gamma^{-2}\nabla_x p\tilde{B}_1\tilde{B}_1'\nabla_x p'$$

with optimal value $v^*(x) = -\gamma^{-2}\tilde{B}_1(x)'\nabla_x p(x)'$.

Lemma 3.1.4 *The information state is given by the alternative formula*

$$p_t(x) = \sup_{v(\cdot)\in L_2[0,t]} \Big\{p_0(\xi(0)) + \tfrac{1}{2}\int_0^t [|C_1(\xi(s)) + D_{12}(\xi(s))u(s)|^2 - \gamma^2(|v(s)|^2$$

$$+(y(s) - C_2(\xi(s)))'E_2^{-1}(\xi(s))(y(s) - C_2(\xi(s))))]\,ds\ :\ \xi(t) = x\Big\}, \tag{3.12}$$

where $\xi(s)$ satisfies

$$\dot{\xi} = Ax(\xi) + B_1(\xi)D_{21}'(\xi)'E_2^{-1}(\xi)(y - C_2(\xi)) + B_2(\xi)u + \tilde{B}_1(\xi)v \tag{2.6$'$}$$

for $0 \le s \le t$ with $\xi(t) = x$. (This is an unconstrained optimization.)

Proof. Let x, t, u, y be given. We prove (3.12) using two inequalities.

Choose $\tilde{w}$ such that the conditions in (3.7) are satisfied, with plant trajectory $\xi(\cdot)$ thus determined via (2.2). Now define v by

$$v = [I - D_{21}(\xi)'E_2^{-1}(\xi)D_{21}(\xi)]^{-L}\,(\tilde{w} - D_{21}(\xi)'E_2(\xi)^{-1}(\xi)(y - C_2(\xi)))$$

$$= [I - D_{21}(\xi)'E_2^{-1}(\xi)D_{21}(\xi)]\tilde{w} \tag{3.13}$$

using the left inverse and the fact that $y - C_2(\xi) = D_{21}(\xi)\tilde{w}$. Then the RHS of (3.12) is bounded below by

$$\ge p_0(\xi(0)) + \tfrac{1}{2}\int_0^t [|C_1(\xi(s)) + D_{12}(\xi(s))u(s)|^2 - \gamma^2(|v(s)|^2$$

$$+(y(s) - C_2(\xi(s)))'E_2^{-1}(\xi(s))(y(s) - C_2(\xi(s))))]\,ds$$

$$= p_0(\xi(0)) + \tfrac{1}{2}\int_0^t [|\tilde{z}(s)|^2 - \gamma^2|\tilde{w}(s)|^2]ds,$$

where $\tilde{z} = C_1(\xi) + D_{12}(\xi)u$. Taking the supremum over all such $\tilde{w}$ we obtain that the RHS of (3.12) is bounded below by $p_t(x)$.

Next, choose any v, and solve (2.6) to obtain $\xi(\cdot)$ with $\xi(t) = x$. Define $\tilde{w}$ by the formula

$$\tilde{w} = [I - D_{21}' E_2^{-1} D_{21}]v + D_{21}' E_2^{-1}(y - C_2). \tag{3.14}$$

Now $C_2(\xi) + D_{21}(\xi)\tilde{w} = y$, and $\tilde{z} = C_1(\xi) + D_{12}(\xi)u$, so by (3.7),

$$\begin{aligned}
p_t(x) &\geq p_0(\xi(0)) + \tfrac{1}{2}\int_0^t [|\tilde{z}(s)|^2 - \gamma^2|\tilde{w}(s)|^2]\,ds \\
&= p_0(\xi(0)) + \tfrac{1}{2}\int_0^t [|C_1(\xi(s)) + D_{12}(\xi(s))u(s)|^2 \\
&\qquad -\gamma^2(v(s)'[I - D_{21}(\xi(s))'E_2(\xi(s))^{-1}D_{21}(\xi(s))]v(s) \\
&\qquad +(y(s) - C_2(\xi(s)))'E_2^{-1}(\xi(s))(y(s) - C_2(\xi(s))))]\,ds \\
&\geq p_0(\xi(0)) + \tfrac{1}{2}\int_0^t [|C_1(\xi(s)) + D_{12}(\xi(s))u(s)|^2 - \gamma^2(|v(s)|^2 \\
&\qquad +(y(s) - C_2(\xi(s)))'E_2^{-1}(\xi(s))(y(s) - C_2(\xi(s))))]\,ds.
\end{aligned}$$

The supremum over v of the RHS of this inequality equals the RHS of (3.12), proving that $p_t(x)$ is an upper bound for the RHS of (3.12). This completes the proof of equation (3.12). □

REMARK 3.1.5 The representation for p_t given in Lemma 3.1.4 involves the reverse-arrow system $\tilde{G}$ introduced in §2.1; see display (2.6). ▽

The next lemma states properties of the information state concerning the dependence of the information state p_t on the initial state p_0. These can be elegantly interpreted in the max-plus algebra. In concrete terms, the first property listed says that if the initial condition p_0 produces the information state p_t at time $t \geq 0$, then the initial condition $p_0 + c$, where $c \in \mathbf{R}$ is a constant, produces the information state $p_t + c$ at time $t \geq 0$. The second property says that if two initial conditions p_0^1 and p_0^2 give information states p_t^1 and p_t^2, respectively, at time $t \geq 0$, then the initial condition $\max(p_0^1, p_0^2)$ produces the information state $\max(p_t^1, p_t^2)$ at time $t \geq 0$. A convenient way of expressing these properties is to use the solution or transition operator $S_t^{u,y}$ for the information state; so

$$p_t = S_t^{u,y} p_0 \tag{3.15}$$

is the information state at time $t \geq 0$ corresponding to the inputs u, y and initial condition p_0.

LEMMA 3.1.6 *The information state depends on its initial value in a max-plus linear manner. In particular,*

$$S_t^{u,y}(p_0 + c) = S_t^{u,y}(p_0) + c \text{ for any constant } c \tag{3.16}$$

and

$$S_t^{u,y}(\max(p_0^1, p_0^2)) = \max(S_t^{u,y}(p_0^1), S_t^{u,y}(p_0^2)). \tag{3.17}$$

Also, the transition operator is monotone*:*

$$p_0^1 \geq p_0^2 \text{ implies } S_t^{u,y}(p_0^1) \geq S_t^{u,y}(p_0^2). \tag{3.18}$$

Proof. Let $c \in \mathbf{R}$. Write

$$I_t(x,v) = \tfrac{1}{2}\int_0^t [|C_1(\xi(s)) + D_{12}(\xi(s))u(s)|^2 - \gamma^2(|v(s)|^2$$
$$+(y(s) - C_2(\xi(s)))' E_2^{-1}(\xi(s))(y(s) - C_2(\xi(s))))]\, ds,$$

where $\xi(\cdot)$ is the solution of (2.6) with $\xi(t) = x$. Then

$$\begin{aligned} S_t^{u,y}(p_0 + c) &= \sup_{v(\cdot)}\{p_0(\xi(0)) + c + I_t(x,v)\} \\ &= \sup_{v(\cdot)}\{p_0(\xi(0)) + I_t(x,v)\} + c \\ &= S_t^{u,y}(p_0) + c. \end{aligned}$$

Next,

$$\begin{aligned} S_t^{u,y}(\max(p_0^1, p_0^2)) &= \sup_{v(\cdot)}\{\max(p_0^1(\xi(0)), p_0^2(\xi(0))) + I_t(x,v)\} \\ &\geq \sup_{v(\cdot)}\{p_0^i(\xi(0)) + I_t(x,v)\} \end{aligned}$$

for $i = 1, 2$. Therefore,

$$S_t^{u,y}(\max(p_0^1, p_0^2)) \geq \max(S_t^{u,y}(p_0^1), S_t^{u,y}(p_0^2)).$$

Now for $i = 1, 2$, and any $v(\cdot)$,

$$S_t^{u,y}(p_0^i) \geq p_0^i(\xi(0)) + I_t(x,v)$$

and so

$$\max(S_t^{u,y}(p_0^1), S_t^{u,y}(p_0^2)) \geq \max(p_0^1(\xi(0)), p_0^2(\xi(0))) + I_t(x,v)$$

for any $v(\cdot)$. Taking the supremum over $v(\cdot)$ gives

$$\begin{aligned} \max(S_t^{u,y}(p_0^1), S_t^{u,y}(p_0^2)) &\geq \sup_{v(\cdot)}\{\max(p_0^1(\xi(0)), p_0^2(\xi(0))) + I_t(x,v)\} \\ &= S_t^{u,y}(\max(p_0^1(\xi(0)), p_0^2(\xi(0)))). \end{aligned}$$

These inequalities prove equation (3.17). The monotonicity assertion (3.18) follows similarly. □

REMARK 3.1.7 Using the notation from Appendix C, max-plus linearity of the transition operator $S_t^{u,y}$ reads

$$S_t^{u,y}(c \odot p_0) = c \odot S_t^{u,y}(p_0) \text{ for any constant } c \tag{3.16$'$}$$

and

$$S_t^{u,y}(p_0^1 \oplus p_0^2) = S_t^{u,y}(p_0^1) \oplus S_t^{u,y}(p_0^2). \tag{3.17$'$}$$

The monotonicity property (3.18) relates to the *comparison principle* in the theory of PDEs.

3.1.3 Information States and Closed-Loop Dissipation

The dissipative property can be represented in terms of the information state p_t via the function

$$\overline{J}_{p_0}(K) = \sup_{T\geq 0} \sup_{y\in L_{2,T}} \{\langle p_T \rangle \;:\; \text{given } p_0\}. \tag{3.19}$$

THEOREM 3.1.8 *We have*

$$J_{p_0}(K) = \overline{J}_{p_0}(K) \tag{3.20}$$

for all admissible controllers K.

Proof. We prove (3.20) by checking two inequalities. Let K be an admissible controller, and $p_0 \in \mathcal{X}_e$ be given.

Fix any $T \geq 0$, $w \in L_{2,T}$, and $x_0 \in \mathbf{R}^n$. The closed-loop system (G, K) with this input and $x(0) = x_0$ will produce trajectories $u(\cdot) = K(y(\cdot))$, $y(\cdot)$, $z(\cdot)$ and state $x(\cdot)$. The information state $p_t(x)$ is determined, for all $x \in \mathbf{R}^n$, $0 \leq t \leq T$ by (3.7) or (3.12). From (3.19) we have, using (3.7),

$$\begin{aligned}\overline{J}_{p_0}(K) &\geq \langle p_T \rangle \\ &\geq p_T(x(T)) \\ &\geq p_0(x(0)) + \tfrac{1}{2}\int_0^T [|z(s)|^2 - \gamma^2|w(s)|^2]\, ds.\end{aligned}$$

Since T, w and $x(0) = x_0$ were arbitrary, we get $\overline{J}_{p_0}(K) \geq J_{p_0}(K)$ on taking the supremum.

Next, fix any $T \geq 0$, $y, v \in L_{2,T}$, and $x \in \mathbf{R}^n$. Let $u = K(y) \in L_{2,T}$. Using (2.6), determine $\xi(\cdot)$ satisfying $\xi(t) = x$, and $\tilde{z} = C_1(\xi) + D_{12}(\xi)u$. Let $\tilde{w}$ be determined by (3.14), so $C_1(\xi) + D_{12}(\xi)\tilde{w} = y$. Then from (3.1) we have

$$J_{p_0}(K) \geq p_0(\xi(0)) + \tfrac{1}{2}\int_0^T [|\tilde{z}(s)|^2 - \gamma^2|\tilde{w}(s)|^2]\, ds$$

$$= p_0(\xi(0)) + \tfrac{1}{2}\int_0^T [|C_1(\xi(s)) + D_{12}(\xi(s))u(s)|^2$$

$$-\gamma^2(v(s)'[I - D_{21}(\xi(s))'E_2(\xi(s))^{-1}D_{21}(\xi(s))]v(s)$$

$$+(y(s) - C_2(\xi(s)))'E_2^{-1}(\xi(s))(y(s) - C_2(\xi(s)))))]\, ds$$

$$\geq p_0(\xi(0)) + \tfrac{1}{2}\int_0^T [|C_1(\xi(s)) + D_{12}(\xi(s))u(s)|^2 - \gamma^2(|v(s)|^2$$

$$+(y(s) - C_2(\xi(s)))'E_2^{-1}(\xi(s))(y(s) - C_2(\xi(s)))))]\, ds.$$

Since T, y, v, and x were arbitrary, we get $J_{p_0}(K) \geq \overline{J}_{p_0}(K)$ on taking the supremum. This completes the proof. □

An important consequence of this representation result is the following characterization of the dissipation property.

COROLLARY 3.1.9 *The closed-loop system* (G, K) *is* γ*-dissipative if and only if*

$$\sup_{T\geq 0}\sup_{y\in L_{2,T}} \{\langle p_T\rangle\} \leq 0, \tag{3.21}$$

where $p_0 = -\beta$ *for some* $\beta \in \mathcal{B}$ *(i.e.,* $p_0(x) \leq 0$, $p_0(0) = 0$*).*

Proof. The result follows from Lemma 3.1.2 and Theorem 3.1.8. □

The minimax game associated with the H^∞ control problem can be solved by minimizing the RHS of the representation formula (3.20). This is done in Chapter 4.

The following lemma concerns finiteness and regularity of the information state.

LEMMA 3.1.10 (i) *Assume there exists a controller* K_0 *that yields a* γ*-dissipative closed loop. Then with* $p_0 \in \mathcal{X}_e$, $u = K_0(y)$, *the resulting information state satisfies*

$$p_t(x) \leq \langle p_0 + \beta_{K_0}\rangle, \tag{3.22}$$

for all $t \geq 0$ *and all* $x \in \mathbf{R}^n$, *and so* $p_t(x)$ *is bounded from above whenever* $\langle p_0 + \beta_{K_0}\rangle < +\infty$.

(ii) *Let* $p_0 \in \mathcal{X}_e$ *be everywhere finite, and let* u, y *in* $L_{2,loc}$ *be arbitrary. Then*

$$-\infty < p_t(x) \tag{3.23}$$

for all $t \geq 0$ *and all* $x \in \mathbf{R}^n$, *and so* $p_t(x)$ *is finite from below.*

(iii) *Let* $p_0 \in \mathcal{X}_e$ *satisfy* $\langle p_0 \rangle$ *finite (so there exists at least one* x *for which* $p_0(x)$ *is finite), and let* u, y *in* $L_{2,loc}$ *be arbitrary. Then*

$$-\infty < \langle p_t \rangle \tag{3.24}$$

for all $t \geq 0$.

(iv) *Assume there exists a controller* K_0 *that yields a* γ*-dissipative closed-loop* (G, K_0) *which is* z*-detectable. Let* $p_0(x_0)$ *be finite, and* $w \in L_2$*. Then there exists a constant* $C \geq 0$ *such that*

$$-C \leq \langle p_t \rangle \tag{3.25}$$

for all $t \geq 0$*, where* p_t *is the information state determined by the closed-loop signals* $u = K_0(y)$.

(v) *Let* $p_0 \in \mathcal{X}_q$*, and suppose for some* $u, y \in L_{2,loc}$

$$\langle p_t \rangle \leq C \tag{3.26}$$

for some $C \geq 0$, $0 \leq t \leq T$*. Then*

$$p_t \in \mathcal{X}_q, \quad 0 \leq t \leq T. \tag{3.27}$$

Proof. From (3.6) and (3.19) we have immediately

$$p_t(x) \leq \langle p_t \rangle \leq \overline{J}_{p_0}(K_0) \leq \langle p_0 + \beta_{K_0} \rangle,$$

proving part (i).

For part (ii), we note that given any y, any $\tilde{w}$ of the form (3.14) satisfies the constraint in the information state definition (3.7) (this depends on x), with trajectory satisfying (2.6) and terminal condition $\xi(t) = x$. Therefore, the set over which the supremum is taken in the definition of $p_t(x)$ is nonempty, and so $p_t(x) > -\infty$ as asserted.

To prove part (iii), we need only show that for each t, there exists an x with $p_t(x) > -\infty$. To this end, select x_0 with $p_0(x_0) > -\infty$. Let $v = 0$, and integrate (2.6) from 0 to t, and set $x = \xi(t)$. The resulting integral in the alternative expression (3.12) is finite, and so $p_t(\xi(t)) > -\infty$.

Part (iv) is similar to part (iii), since by z-detectability $z, y, u = K_0(y) \in L_2$ and $x(\cdot) \in L_2$, which means that the integral lower bounding $\langle p_t \rangle$ converges to a finite number.

To prove part (v), we need a lower estimate of the form

$$-k_T(1 + |x|^2) \leq p_t(x), \quad 0 \leq t \leq T.$$

However, such an estimate can be obtained by setting $v = 0$ in (3.12) and making use of the linear growth Assumptions 2.1.1 on the plant data and (2.9). Continuity can be proven using standard methods, e.g., [FS93], [McE95b]. □

Thus the finiteness of the information state depends on the solvability of the dissipative control problem and on the nature of the initial information state. In particular, if the information state is initially nonsingular and if it is driven by signals u, y in a dissipative closed loop, then the information state is nonsingular for all time. Interestingly, as we shall see in Chapter 11, even if $p_t \in \mathcal{X}$ is finite for all $t \geq 0$, it can happen that $\lim_{t\to\infty} p_t \in \mathcal{X}_e \backslash \mathcal{X}$; i.e., the limit can be singular.

When finite, the smoothness of $p_t(x)$ and consequently the sense in which (3.9) is to be understood depends on the smoothness of the initial data p_0 (the other data are assumed smooth), the regularity of $u(\cdot)$ and $y(\cdot)$, and on the structure of $F(p, u, y)$. In general p_t is *not* smooth; however, in the 1 and 2A block cases it will be smooth if p_0 is; see [JB96]. In general, (3.9) can be interpreted in the viscosity sense; see [JB96], [FS93], [BCD98] (the viscosity interpretation applies when p_t is not necessarily differentiable, and extensions of it apply when p_t is singular).

> *It is important to note that, even if* p_t *is not smooth or perhaps not everywhere finite, the information state dynamics* (3.9) *can always be understood in the integrated form* (3.7).

The state space of the central controller we will construct (see below) is in general a subset of the space $\mathcal{X}_e$. The nature of this subset depends on the particular initial state p_0. The particular choice of p_0 is a very important issue. In general, the information state solution to the (nonlinear) H^∞ control problem is *infinite dimensional*, since in general it is not possible to compute $p_t(x)$ using a finite set of ODEs. Thus, in general, p_t evolves in the infinite-dimensional space $\mathcal{X}_e$. We note, however, that for some poor choices of p_0, p_t can escape $\mathcal{X}_e$ for some t. Also, even in typically infinite-dimensional cases, it can happen that for certain choices of initial state p_0, the resulting trajectory p_t stays in a finite-dimensional subset of $\mathcal{X}_e$, whereas for other choices of p_0 it does not. Consequently, the dimension and structure of the set of states reachable from p_0 can vary dramatically; see Chapter 10.

We conclude this section with a short discussion of some additional issues concerning the information state. The basic state space we are using for the information state is the space $\mathcal{X}_e$ of functions that are bounded above and are upper semicontinuous. It is possible that under certain circumstances p_t may fail to belong to $\mathcal{X}_e$ for some $t > 0$, even though $p_0 \in \mathcal{X}_e$. However, this does not happen in a properly running closed-loop system. Results related to this issue are given in Chapters 10 and 11. If $p_0 \in \mathcal{X}_q$ is finite, smooth, and of at most quadratic growth, it is possible to show that $p_t \in \mathcal{X}_q$. The results of [McE95b] can be applied to show this, at least for γ sufficiently large.

3.2 Equilibrium Information States

It will be important in what follows to consider equilibrium information states that solve a steady state version of the information state equation (3.9), i.e., solutions p_∞ of the *equilibrium* (or *steady state*) information state equation

$$0 = F(p_\infty, 0, 0); \tag{3.28}$$

i.e.,

$$-\nabla_x p_\infty \cdot (A - B_1 D_{21}' E_2^{-1} C_2) + \tfrac{1}{2\gamma^2} \nabla_x p_\infty B_1 [I - D_{21}' E_2^{-1} D_{21}] B_1' \nabla_x p_\infty'$$
$$+\tfrac{1}{2}|C_1|^2 - \gamma^2 \tfrac{1}{2} C_2' E_2^{-1} C_2 = 0.$$

This equation is interpreted in the usual sense if p_∞ is smooth; else one can use the viscosity interpretation or the equivalent integral form

$$p_\infty(x) = \sup_{v(\cdot)} \Big\{ p_\infty(x(0)) + \tfrac{1}{2} \int_0^t [|C_1(\xi(s))|^2 - \gamma^2(|v(s)|^2$$
$$+C_2(\xi(s))' E_2^{-1}(\xi(s)) C_2(\xi(s)))] \, ds \, : \, \xi(t) = x \Big\} \tag{3.29}$$

for all $t \geq 0$, where $\xi(s)$ satisfies (2.6) with $u = 0$, $y = 0$:

$$\dot{\xi}(s) = A^\times(\xi(s)) + \tilde{B}_1(\xi(s)) v(s). \tag{2.6$'$}$$

This is a steady state version of the information state definition in alternative form (3.12).

Of particular interest will be a special solution p_e of (3.28), which is defined by the formula

$$p_e(x) = \sup_{v(\cdot)} \Big\{ \lim_{T \to \infty} \tfrac{1}{2} \int_{-T}^0 [|C_1(\xi(s))|^2 - \gamma^2(|v(s)|^2$$
$$+C_2(\xi(s))' E_2^{-1}(\xi(s)) C_2(\xi(s)))] \, ds \tag{3.30}$$
$$: \lim_{T \to \infty} \xi(-T) = 0, \ \xi(0) = x \Big\},$$

where ξ solves (2.6) with $u = 0$, $y = 0$. Indeed, p_e will turn out to be a stable equilibrium for the information state system (3.9), as described in the following section (§3.3).

The nature of the equilibrium p_e has profound implications, and in particular singularity or otherwise has a significant bearing on the computational complexity of resulting central controllers. The critical factor that determines the support of p_e is the stabilizability of the system $(A^\times, \tilde{B}_1)$ (recall the definition in §2.2). In general, it turns out that when the H^∞ control problem is solvable, p_e has the form

$$p_e = \delta_{\mathcal{S}_{cas}} + \breve{p}_e, \tag{3.31}$$

where $\mathcal{S}_{cas}$ is the antistabilizable set for the system $(A^\times, [B_1 D_{21}' E_2^{-1} \, : \, B_2 \, : \, \tilde{B}_1)]$, and $\breve{p}_e \in C(\mathcal{S}_{cas})$ (see Chapter 11 for details). In particular, p_e is nonsingular if $(A^\times, [B_1 D_{21}' E_2^{-1} \, : \, B_2 \, : \, \tilde{B}_1])$ is antistabilizable, since $\mathcal{S}_{cas} = \mathbf{R}^n$ (see Lemma 3.2.1 to follow). We shall require in some of the results to follow in this book that the closed-loop vector field

$$A^{p_e} = A^\times - \gamma^{-2} \tilde{B}_1 \tilde{B}_1' \nabla_x p_e' \tag{3.32}$$

be incrementally L_2 exponentially antistable.

A second particular solution of (3.28) is the function $-q_e(x)$, defined by

$$q_e(x) = -\inf_{v(\cdot)}\Big\{ - \lim_{T\to\infty} \tfrac{1}{2}\int_0^T [|C_1(\xi(s))|^2 - \gamma^2(|v(s)|^2 + C_2(\xi(s))' E_2^{-1}(\xi(s)) C_2(\xi(s)))]\, ds \quad : \lim_{T\to\infty} \xi(T) = 0,\ \xi(0) = x\Big\}, \tag{3.33}$$

where ξ solves (2.6) with $u = 0$, $y = 0$. The function turns out to be the equilibrium solution to an adjoint information state PDE (3.47) below.

Also important are solutions p_{se} of the PDI

$$F(p_{se}, 0, 0) \leq 0, \tag{3.34}$$

which we call *infosubequilibriums* (see also Appendix B).

The following lemmas provide further information about the functions p_e and q_e.

LEMMA 3.2.1 (i) *The function p_e is defined by* (3.30) *and satisfies*

$$-\infty < p_e(x) \leq p_{se}(x) \tag{3.35}$$

for any L_2 antistabilizable state $x \in \mathcal{S}_{cas}$, where p_{se} is any continuous infosubequilibrium.

(ii) *The function q_e defined by* (3.33) *satisfies*

$$p_{se}(x) \leq -q_e(x) \tag{3.36}$$

for any L_2 stabilizable state $x \in \mathcal{S}_{cs}$, where p_{se} is any continuous infosubequilibrium.

(iii) $p_e(0) \geq 0$ *and* $q_e(0) \leq 0$. *If there exists a continuous infosubequilibrium, then* $p_e(0) = q_e(0) = 0$.

(iv) *Any finite, smooth L_2 antistabilizing solution p_e^- of* (3.28) *satisfying* $p_e^-(0) = 0$ *must equal p_e.*

(v) *Any finite, smooth L_2 stabilizing solution $-q_e^+$ of* (3.28) *satisfying* $q_e^+(0) = 0$ *must equal $-q_e$.*

Proof. The results follow from Theorem B.2.2. □

3.2.1 Quadratic Upper Limiting

We say that an equilibrium p_e is *quadratically upper limiting* for the information state system (3.9) if there exists a constant $\bar{\alpha}_0 >$ such that if $u, y \in L_{2,loc}$, $c_0 \geq 0$, and

$$p_0 \leq p_e(x) + \alpha_0|x|^2 + c_0 \tag{3.37}$$

for $0 \leq \alpha_0 \leq \bar{\alpha}_0$, we have for all $0 < \eta < 1$,

$$p_t(x) \leq p_e(x) + \frac{c_1(t)}{\eta} + (\eta + c_1(t)\alpha_0)k|x|^2 \text{ for all } t \geq 0, \tag{3.38}$$

where $k \geq 0$ and $c_1(t) = c_1(\| u \|_{L_{2,t}}, \| y \|_{L_{2,t}}) \geq 0$ are constants with $c_1(\infty)$ finite whenever $u, y \in L_2$.

3.3 Information State Dynamics and Attractors

Stability is of course a fundamental concept for dynamical systems and is central to the H^∞ control problem. This section and Chapter 11 concern the asymptotics (as $t \to \infty$) of the information state p_t (governed by equation (3.9)), viz.,

$$\dot{p} = F(p, u, y), \tag{3.9$'$}$$

for u and y in $L_2[0, \infty)$ and properties of the limiting function p_∞,

$$p_t \to p_\infty.$$

The limit is a stationary solution to the information state equation, i.e., a solution of the *equilibrium information state equation* (3.28)

$$0 = F(p_\infty, 0, 0). \tag{3.28}$$

We shall see that the limits p_∞ are often functions of the form

$$p_\infty = p_e + c,$$

where p_e is the particular equilibrium information state described in §3.2 and c is a constant (depending on the inputs u, y).

An equilibrium p_e is typically called a *local attractor or attractive equilibrium* by those who study dynamical systems if it is a limit of p_t for inputs $u = 0$, $y = 0$ (i.e., uncontrolled system) and for all initial states p_0 belonging to some open subset of $\mathcal{X}$ (or $\mathcal{X}_e$) called the *domain of attraction.* The attractor is called *global* if its domain of attraction is the whole space. The concept of domain of attraction is illustrated in Figure 3.1.

We shall introduce notation for domains of attraction, since the equilibrium p_e need not be globally attractive, and because the effect of inputs must be taken into account. The set $\mathcal{D}^0_{attr}(p_e)$ consists of all initial states p_0 for which p_e is the limit of p_t for inputs $u = 0$, $y = 0$, and it is called the *domain of attraction* for the equilibrium

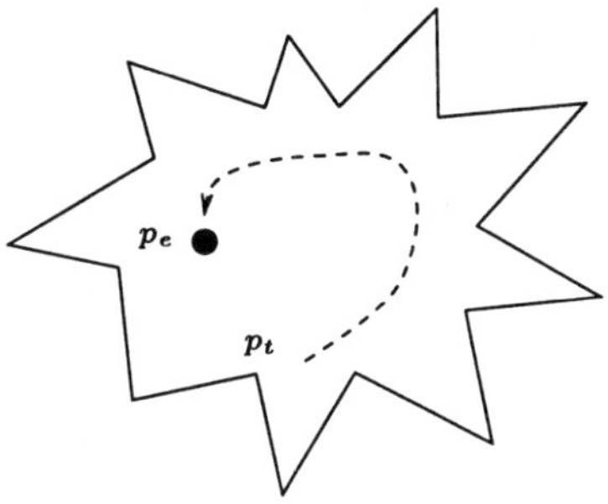

Figure 3.1: The domain of attraction of the equilibrium p_e.

p_e. Also important in control is the set of p which are driven to $p_\infty = p_e + c$ by all inputs L_2, inputs u, y (or at least reasonably decaying u, y). We call this the *domain of control attraction* and denote it $\mathcal{D}_{attr}(p_e)$. If $\mathcal{D}_{attr}(p_e) \neq \emptyset$, we say that p_e is a *control attractor*.

Upon examination we find that the information state system is highly unreachable with the munificent side effect that in many situations the information state equation has a "large" domain of control attraction. That is, it is common when we start at a certain equilibrium p_e and drive the system with L^2 inputs u, y that p_t never leaves (a subset of) $\mathcal{D}^0_{attr}(p_e)$. It should be noted that relative to the topology of $\mathcal{X}_e$ the domain of controlled attraction can be very thin.

We shall see that in nonsingular cases the attracting property of the equilibrium p_e is closely related to the condition that the vector field A^{p_e} (recall (3.32)) is strongly antistable. To help explain how the antistabilizing property of $(A^\times, \tilde{B}_1)$ is related to the attracting nature of p_e, consider the linearization of the uncontrolled information state system

$$\dot{p} = F(p, 0, 0), \tag{3.39}$$

viz.,

$$\begin{aligned}
\dot{\phi} &= \nabla_p F(p_e, 0, 0)[\phi] \\
&= \lim_{h \to 0} \frac{F(p_e + h\phi, 0, 0) - F(p_e, 0, 0)}{h} \\
&= -\nabla_x \phi \cdot (A - B_1 D'_{21} E_2^{-1} C_2 - \gamma^{-2} B_1 [I - D'_{21} E_2^{-1} D_{21}] B'_1 \nabla_x p'_e).
\end{aligned} \tag{3.40}$$

That is, $\dot{\phi} = -\nabla_x \phi \cdot A^{p_e}$. The linear transition operator S_t^{lin} for this linearized system is given by

$$\begin{aligned}
\phi_t(x) &= (S_t^{lin} \phi)(x) \\
&= \phi(\Phi_{0,t}(x))
\end{aligned}$$

($\phi_t(x)$ is the solution of (3.40) with initial condition $\phi_0 = \phi$), where $\Phi_{0,t}(x)$ is the flow for $\dot{\xi} = A^{p_e}(\xi)$, $0 \leq s \leq t$, $\Phi_{t,t}(x) = x$, i.e., $\xi(s) = \Phi_{s,t}(x)$. Now the antistability of

A^{p_e} means that $\Phi_{0,t}(x) \to 0$ as $t \to \infty$, and so if ϕ is continuous and without loss of generality $\phi(0) = 0$ we have

$$S_t^{lin}\phi(x) \to 0 \text{ as } t \to \infty \tag{3.41}$$

for all $x \in \mathbf{R}^n$. This calculation shows that the transition operator of the linearized system (3.40) is globally attracted to the zero function and suggests that the information state system (3.9) should enjoy at least a local form of stability. In fact, we shall see that global stability results can be obtained, since it turns out that we can explicitly represent and estimate the linearization error (Chapter 11).

The key stability properties of the equilibrium p_e are

(i) for $u = 0$, $y = 0$, and $p_0 = p_e$ we have $p_t = p_e$ for all $t \geq 0$ (equilibrium property);

(ii) the domain of attraction is a nonempty subset of $\mathcal{X}_e$, so that for $u = 0$, $y = 0$, and $p_0 \in \mathcal{D}^0_{attr}(p_e)$ we have $p_t \Longrightarrow p_e$ as $t \to \infty$;

(iii) the domain of control attraction $\mathcal{D}_{attr}(p_e)$ is a nonempty subset of $\mathcal{X}_e$, so that for $p_0 \in \mathcal{D}_{attr}(p_e)$ and $u, y \in L_2$ we have $p_t \Longrightarrow p_e + c(u, y, p_0)$ as $t \to \infty$, with $c(0, 0, p_e) = 0$.

Here p_t is the solution of (3.9). The equilibrium information state p_e is a *stable equilibrium for the information state dynamics* (3.9); i.e., p_e is a *control attractor.* The sense of convergence is in general weak convergence (see Appendix C), or in nonsingular cases it can be uniform convergence on compact sets.

Notice that the limit information states p_∞ are of the form

$$p_\infty = p_e + c,$$

where $c \in \mathbf{R}$ is a constant. This says that all limits are equilibrium solutions, i.e., solve (3.28). To obtain a *unique* limit, we can normalize or form a quotient space

$$\mathcal{X}_e/\mathbf{R} = \{[p] \;:\; p \in \mathcal{X}_e\},$$

where the equivalence class $[p]$ is defined by

$$[p] = \{p' \in \mathcal{X}_e \;:\; p' = p + c, \text{ for some } c \in \mathbf{R}\}.$$

In fact, we have an explicit formula

$$[p](x) = p(x) - \langle p \rangle.$$

Thus for $p_0 \in \mathcal{D}_{attr}(p_e)$ we have

$$\lim_{t\to\infty} [p_t] = [p_e].$$

We defer to Chapter 11 for detailed results concerning stability of the information state system.

3.4 Adjoint Information State

We conclude our introduction to information states by defining an adjoint information state q_{tT}. The state q_{tT} is *adjoint* to p_t in the sense that

$$\langle q_{tT} + p_t \rangle \text{ is a constant independent of } t. \tag{3.42}$$

Apart from being of independent interest, the adjoint information state will be used below in Chapter 11 for stability analysis.

The *adjoint information state* q_{tT} is defined by

$$q_{tT}(x) = \sup_{v(\cdot)} \Big\{ f(\xi(T)) + \tfrac{1}{2} \int_t^T [|C_1(\xi(s)) + D_{12}(\xi(s))u(s)|^2 - \gamma^2(|v(s)|^2$$
$$+ (y(s) - C_2(\xi(s)))' E_2^{-1}(\xi(s))(y(s) - C_2(\xi(s))))]\, ds : \ \xi(t) = x \Big\}, \tag{3.43}$$

where $x(s)$ satisfies (2.6), with initial condition $x(t) = x$, and $f \in C_b(\mathbf{R}^n)$. The dynamics for q_{tT} is the PDE (for fixed $u \in L_{2,loc}$ and $y \in L_{2,loc}$)

$$\frac{dq_{tT}}{dt} = -F(-q_{tT}, u(t), y(t)), \tag{3.44}$$

for $0 \leq t \leq T$, where $F(p,u,y)$ is the nonlinear differential operator given by (3.10) or (3.11), and the terminal condition is

$$q_{TT} = f. \tag{3.45}$$

LEMMA 3.4.1 *For any $u \in L_{2,loc}$, $y \in L_{2,loc}$ we have*

$$\langle p_0 + q_{0,T} \rangle = \langle p_t + q_{tT} \rangle = \langle p_T + f \rangle \ \textit{for all } 0 \leq t \leq T. \tag{3.46}$$

Proof. Following [JBE94], [JB96], the assertion can be proven by combining the definitions (3.12) and (3.43). However, let us explain this in the case where p_t, $q_t = q_{tT}$ are smooth and $\bar{x}(t) = \operatorname{argmax}_x \{p_t(x) + q_t(x)\}$ are unique. Write $\psi(t) = p_t(\bar{x}(t)) + q_t(\bar{x}(t)) = \langle p_t + q_t \rangle$, and note that $\nabla_x p_t(\bar{x}(t)) + \nabla_x q_t(\bar{x}(t)) = 0$. Then, using Danskin's theorem (see, e.g., [BB91, Appendix B])

$$\begin{aligned} \frac{d\psi(t)}{dt} &= \frac{\partial p_t}{\partial t}(\bar{x}(t)) + \frac{\partial q_t}{\partial t}(\bar{x}(t)) \\ &= F(p_t, u, y)(\bar{x}(t)) - F(-q_t, u, y)(\bar{x}(t)) \\ &= 0 \end{aligned}$$

and so $\psi(t) = \psi(0) = \langle p_0 + q_{0T} \rangle = \psi(T) = \langle p_T + f \rangle$. □

The steady state version of the PDE (3.44) for q_{tT} is the PDE

$$0 = -F(-q_\infty, 0, 0), \tag{3.47}$$

i.e.,

$$-\nabla_x q_\infty \cdot (A - B_1 D_{21}' E_2^{-1} C_2) - \tfrac{1}{2\gamma^2} \nabla_x q_\infty B_1 [I - D_{21}' E_2^{-1} D_{21}] B_1' \nabla_x q_\infty$$

$$-\tfrac{1}{2}|C_1|^2 + \gamma^2 \tfrac{1}{2} C_2' E_2^{-1} C_2 = 0.$$

By this we mean, in integrated form,

$$q_\infty(x) = \sup_{v(\cdot)} \Big\{ q_\infty(\xi(T)) + \tfrac{1}{2} \int_0^T [|C_1(\xi(s))|^2 - \gamma^2(|v(s)|^2 \\ + C_2(\xi(s))' E_2^{-1}(\xi(s)) C_2(\xi(s)))] \, ds \;:\; \xi(t) = x \Big\} \tag{3.48}$$

for all $T > 0$, where $x(s)$ satisfies (2.6), with initial condition $x(0) = x$ and $u = 0$, $y = 0$. The function q_e defined by (3.33) is a particular solution of this adjoint PDE.

3.5 Notes

(i) The information state is called the *cost to come* in [BB91] and is referred to as a *conditional storage function* in [Kre94]. Our terminology follows that used in stochastic control; see [KV86]. The definition here follows [JBE94], [JB95], [JB96].

(ii) Properties of the information state transition operator will be further explored in Chapter 10.

(iii) The quadratic upper limiting property is important and will be used in Chapter 4 to show that coupling with the controller is maintained. Results showing that there exists an equilibrium with this property are given in Chapter 11.

(iv) Normalization of the information state is used extensively in [FHH97], [FHH98].

(v) The adjoint information was defined in [JBE94], [JB96].

Chapter 4

Information State Control

4.1 Introduction

In this chapter we make use of the information state introduced in Chapter 3 to solve the nonlinear H^∞ control problem. We continue with the use of optimization techniques, specifically game-theoretic methods.

Theorem 3.1.8 shows how the measurement feedback minimax cost functional $J_{p_0}(K)$ can be expressed in terms of the information state, defined by (3.7) or (3.12). This results in the full information minimax game corresponding to the minimization of the cost function

$$\overline{J}_{p_0}(K) = \sup_{T \geq 0} \sup_{y \in L_{2,T}} \{ (p_T) \ : \ \text{given } p_0 \} \tag{3.19}$$

over suitable classes of admissible controllers. Since this cost functional is expressed in terms of the information state p, it is natural to consider controllers $K^{\mathbf{u}}_{p_0}$ that are functions of this information state, defined via functions $\mathbf{u}(p)$; see Figure 4.1. The structure of the controller $K^{\mathbf{u}}_{p_0}$ illustrated in Figure 4.1 constitutes a *separation principle*. These controllers, called *information state controllers*, are the subject of §4.2. A key issue here is whether or not a given information state controller is well defined for all time, and this leads to a discussion of *admissibility* and *coupling*.

We are particularly interested in *optimal information state controllers* and controllers obtained from solutions of *dissipation PDIs*. A traditional method for obtaining optimal controllers is dynamic programming (some relevant aspects of this technique are summarized in Appendix B). This method is used in this chapter, beginning in

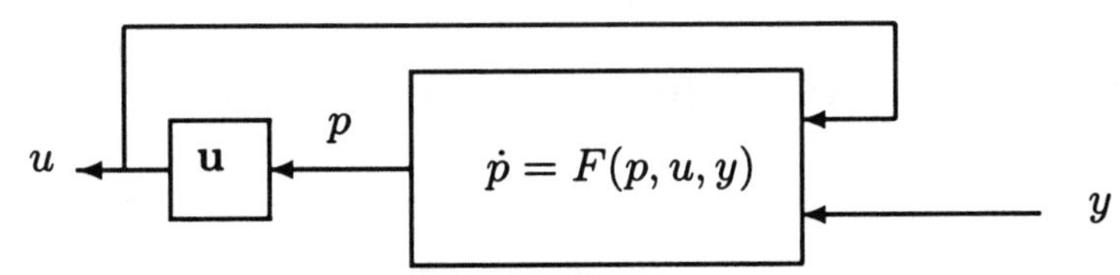

Figure 4.1: Information state controller $K^{\mathbf{u}}_{p_0}$.

§4.3. The dynamic programming *value function* $W(p)$ relevant to our minimax problem is defined by the formula

$$W(p) = \inf_K \overline{J}_p(K), \tag{4.4}$$

where the minimization ranges over the class of all admissible controllers (including information state controllers). Since this function need not be finite for all p, we denote by dom W its domain of finiteness. The value function enjoys some natural structural properties (domination, monotonicity, and additive homogeneity; see §2.2.2.2), and satisfies the dynamic programming principle: for any $t \geq 0$, $p \in \text{dom}\, W$,

$$W(p) = \inf_K \sup_{y \in L_{2,t}} \{W(p_t) : p_0 = p\}. \tag{4.8}$$

This identity is fundamental and is called the *dynamic programming equation.*

The differential version of the dynamic programming equation is the *dynamic programming PDE*:

$$\inf_{u \in \mathbf{R}^m} \sup_{y \in \mathbf{R}^p} \nabla_p W(p)[F(p,u,y)] = 0. \tag{4.16}$$

This PDE will make sense if the value function W is sufficiently smooth, and this is unlikely in general. However, if W is smooth, say, Frechet differentiable on $\mathcal{X}_q$, then the infimum and supremum in (4.16) can be evaluated explicitly, giving the important formulas

$$\begin{aligned} \mathbf{u}^*(p) &= \nabla_p W(p)[E_1^{-1}(-D_{12}'C_1 + B_2'\nabla_x p')], \\ \mathbf{y}^*(p) &= \nabla_p W(p)[C_2 - \gamma^{-2} D_{21} B_1' \nabla_x p'] \end{aligned} \tag{4.23$'$}$$

in the case where D_{12} and D_{21} are independent of x. The optimal function $\mathbf{u}^*(p)$ is used to obtain the optimal information state controller $K_{p_0}^* = K_{p_0}^{\mathbf{u}^*}$. In general, information state controllers $K_{p_0}^{\mathbf{u}^*_{\hat{W}}}$ can be obtained from solutions $\hat{W}$ of the PDI

$$\inf_{u \in \mathbf{R}^m} \sup_{y \in \mathbf{R}^p} \nabla_p \hat{W}(p)[F(p,u,y)] \leq 0 \tag{4.26}$$

via the minimizing function $\mathbf{u}^*_{\hat{W}}(p)$ (the value function W solves this PDI, if smooth).

What is important for H^∞ control is dissipation, and the important *dissipation inequality* which in this language is just the integrated version of the PDI (4.26), is

$$\hat{W}(p_t) \leq \hat{W}(p_0) \tag{4.9$'$}$$

for all $t \geq 0$ along trajectories of the closed-loop system $(G, K_{p_0}^{\mathbf{u}^*_{\hat{W}}})$, since inequality (4.9) yields

$$\begin{cases} \frac{1}{2}\int_0^T |z(s)|^2\, ds \leq \gamma^2 \frac{1}{2}\int_0^T |w(s)|^2\, ds - p_0(x_0) + \hat{W}(p_0) \\ \text{for all } w \in L_{2,T} \text{ and all } T \geq 0, \end{cases} \tag{4.10$'$}$$

as explained in Chapter 1.

The goals of this chapter are to present detailed results concerning the construction of information state controllers from solutions of the PDI (4.26) (sufficiency), and to establish necessity results in terms of the value function W defined by (4.4). The main technical complications that arise are as follows:

(i) *Smoothness.* We spend considerable effort on this issue and discuss generalized directional derivatives and other ways to interpret the PDE (4.16) and PDI (4.26). We define a concept of smoothness, called p-smooth, which is general enough to handle singular information states, and use this to define admissible information state controllers $K_{p_0}^{\mathbf{u}^*_{\hat{W}}}$. This amounts to ensuring that the PDE can be integrated along closed-loop trajectories that are well defined, at least for small time (*preadmissibility*).

(ii) *Coupling.* Given a preadmissible information state controller $K_{p_0}^{\mathbf{u}^*_{\hat{W}}}$, is it well defined for all time (*admissibility*)? At least one requires $\mathbf{u}^*_{\hat{W}}(p_t)$ to be well defined for all $t \geq 0$. Conditions are given that ensure this, expressed in terms of suitable initial information states p_0 and the *quadratic upper limiting* property.

(iii) *Singular information states.* From a practical as well as theoretical point of view it is important to consider singular information states. However, this complicates matters and, in particular, the term $-p_0(x_0)$ appearing in the dissipation inequality (4.10). If $p_0(x_0)$ is not finite, one must use additional hypotheses to obtain dissipation for such initial plant states x_0, viz., *incremental hyperbolicity.*

The following theorem summarizes the necessity Theorems 4.3.1, 4.7.1, and 4.10.1.

THEOREM 4.1.1 *Assume that the γ-dissipative control problem is solved by an admissible controller K_0. Then the following hold:*

(i) *The value function W defined by* (4.4) *satisfies the dynamic programming equation* (4.8) *and the structural conditions (domination, monotonicity, and additive homogeneity)* [*Theorem* 4.3.1].

(ii) *If the value function is p-smooth, then W solves the dynamic programming PDE* (4.16) [*Theorem* 4.7.1].

(iii) *If the optimal information state controller $K^*_{p_0} = K^{\mathbf{u}^*_W}$ constructed from W (via* (4.23)) *exists for all $t \geq 0$, then the closed-loop system $(G, K^*_{p_0})$ is dissipative, i.e.,* (4.10) *holds (with $\hat{W} = W$). Stability can be obtained with z-detectability (as in Theorem* 2.1.3) [*Theorem* 4.10.1].

This theorem essentially says (modulo technicalities) that if it is possible to solve the H^∞ problem, then there is always an information state controller that can solve it.

The next theorem is a sufficiency result summarizing Theorem 4.10.3. It says when controllers can be constructed from solutions of the PDI (4.26) that solve the H^∞ control problem.

THEOREM 4.1.2 *Assume that a p-smooth function* $\hat{W}$ *satisfies the structural conditions (domination, monotonicity, and additive homogeneity) and solves the PDI* (4.26). *If the information state controller* $K_{p_0}^{\mathbf{u}^*_{\hat{W}}}$ *constructed from* $\hat{W}$ *(via* (4.23), *with* $\hat{W}$ *replacing* W*) exists for all* $t \geq 0$, *then the closed-loop system* $(G, K_{p_0}^{\mathbf{u}^*_{\hat{W}}})$ *is dissipative; i.e.,* (4.10) *holds. Stability can be obtained with z-detectability* (*Theorem* 2.1.3).

Because of the complexities of some of the results in this chapter, due to the technical difficulties outlined above, subsequent chapters in Part I will not explicitly use the detailed results to follow, but rather rely on the abbreviated statements embodied in Theorems 4.1.1 and 4.1.2. **Readers are encouraged to skip to the next chapter and continue with the rest of Part I. Come back to the remainder of this chapter at a later time.**

To prove some of the results in this chapter, we will make use of the following assumption.

> ASSUMPTION 4.1.3 The matrix functions $D_{12}(x) = D_{12}$ and $D_{21}(x) = D_{21}$ are independent of $x \in \mathbf{R}^n$.

4.2 Information State Controllers

We define information state controllers as follows. Let $\mathbf{u}$ be a function

$$\mathbf{u} \,:\, \mathcal{X}_e \to \mathbf{R}^m$$

so that for each $p \in \mathcal{X}_e$ (or in a subset dom $\mathbf{u}$ of $\mathcal{X}_e$), $\mathbf{u}(p)$ is a control value in $\mathbf{R}^m$. The set

$$\mathrm{dom}\, \mathbf{u}$$

is called the *domain* of the function $\mathbf{u}$ and consists of all information states $p \in \mathcal{X}_e$ for which $\mathbf{u}(p)$ is defined and finite. Let $p_0 \in \mathcal{X}_e$ be an initial information state and $y \in L_{2,loc}$, and consider the information state trajectory p_t, determined by *information state feedback*

$$u(t) = \mathbf{u}(p_t), \; t \geq 0 \tag{4.1}$$

(the control at time t is a function of the current value p_t of the information state). Define the output feedback controller $K_{p_0}^{\mathbf{u}}$ by

$$K_{p_0}^{\mathbf{u}}[y](t) = \mathbf{u}(p_t[y]). \tag{4.2}$$

Then $K_{p_0}^{\mathbf{u}}$ is called an *information state controller*. Note that we have emphasized in the notation the role of the function $\mathbf{u}$ and the initial state p_0. This controller has state space realization (3.9) with initial state p_0. This is illustrated in Figure 4.1.

The plant G can be combined in the obvious way to form the closed-loop system $(G, K^{\mathbf{u}})$. To be explicit, the closed-loop system $(G, K^{\mathbf{u}})$ has state space $\mathbf{R}^n \times \mathcal{X}_e$

and the state space realization

$$(G, K^{\mathbf{u}}) : \begin{cases} \dot{x} = A(x) + B_1(x)w + B_2(x)u, \\ \dot{p} = F(p, u, y), \\ z = C_1(x) + D_{12}(x)u, \\ y = C_2(x) + D_{21}(x)w, \\ u = \mathbf{u}(p). \end{cases} \quad (4.3)$$

The initial state is denoted (x_0, p_0). Often, we will fix the initial information state p_0 and denote the resulting closed-loop system by

$$(G, K^{\mathbf{u}}_{p_0}) : (x_0, w) \mapsto z.$$

As we shall see, a critical issue is whether or not p_t stays inside dom $\mathbf{u}$. If p_t failed to stay in dom $\mathbf{u}$, the control $u(t) = \mathbf{u}(p_t)$ would not be defined. The information state p_t may remain in dom $\mathbf{u}$ up until a time τ_{adm} at which p_t escapes from dom $\mathbf{u}$. This is an issue of *coupling* between the information state and controller, and these considerations lead to the following definitions.

The information state controller is *preadmissible* if $K^{\mathbf{u}}_{p_0}$ is a controller for which if $p_0 \in q.i.\text{dom}\,\mathbf{u}$, then there exists

$$0 < \tau_{adm} = \tau_{adm}(K^{\mathbf{u}}_{p_0}) \le +\infty$$

such that

(i) the information state trajectory $p_t \in \mathcal{X}_e$ is defined for all $0 \le t < \tau_{adm}(K^{\mathbf{u}}_{p_0})$, $y \in L_{2,loc}[0, \infty)$, $u = K^{\mathbf{u}}_{p_0}(y)$, with $u \in L_{2,loc}[0, \tau_{adm})$, $p_t \in \text{dom}\,\mathbf{u}$ for all $0 \le t < \tau_{adm}(K^{\mathbf{u}}_{p_0})$; and

(ii) the trajectories of the closed-loop system $(G, K^{\mathbf{u}}_{p_0})$ are defined for all $0 \le t < \tau_{adm}(K^{\mathbf{u}}_{p_0})$; and $u, y \in L_{2,loc}[0, \tau_{adm}(K^{\mathbf{u}}_{p_0}))$ for any disturbance input $w \in L_{2,loc}$ and initial plant state x_0.

We say that $K^{\mathbf{u}}_{p_0}$ is *admissible* if always $\tau_{adm}(K^{\mathbf{u}}_{p_0}) = +\infty$. We take p_0 in the interior of dom $\mathbf{u}$ since by preadmissibility we want to be able to solve the closed-loop information state equation without immediately escaping the domain; and the use of quadratic interior is convenient given the growth assumptions used in this book.

Our main task is to find information state controllers that solve the H^∞ control problem; this is carried out using a function $\mathbf{u}^*$, which optimizes a performance measure defined on information states, viz., (3.19). Optimal information state controllers, and in particular the central controller, can be constructed from $\mathbf{u}^*$. To this end, we use dynamic programming (§4.3); this is complicated because of technical problems associated with smoothness, and we discuss in some detail the issues involved. We also consider direct minimization of the cost function (3.19); this may be possible in cases when the dynamic programming PDE fails to have a sufficiently smooth solution (§4.4), although it may not be possible to provide explicit formulas for the optimizers.

4.3 Dynamic Programming

In this and subsequent sections we apply the dynamic programming method. This allows us to determine, under regularity assumptions, optimal information state controllers and their properties. We define the *value function* for the problem of minimizing the cost function (3.19) over the class of all admissible controllers. In §4.4 we present the fundamental dynamic programming PDE and study it subsequently.

The *value function* $W(p)$ is defined by the formula

$$W(p) = \inf_{K} \overline{J}_p(K), \tag{4.4}$$

where the minimization ranges over the class of all admissible controllers achieving dissipation and weak internal stability. Here, p denotes the initial information state ($p_0 = p$), so the value function is a function of this initial state. It is an infinite horizon optimization problem, and so the value function is independent of time t. The value function need not be finite for all $p \in \mathcal{X}_e$, and so we denote by

$$\operatorname{dom} W$$

the subset of $\mathcal{X}_e$ for which $W(p)$ is finite.

The following theorem is an improvement of Theorem 4.23 of [JB95]. It proves some important properties enjoyed by the value function, in particular, certain structural properties and the dynamic programming principle. No assumptions about smoothness of the value function are made. It is also shown that if an optimal admissible controller exists, then it achieves a dissipative closed loop with internal stability.

THEOREM 4.3.1 *Assume that the γ-dissipative control problem for G is solved by an admissible controller K_0. Then the value function $W(p)$ defined on a set dom$W \subset \mathcal{X}_e$ by* (4.4) *enjoys the following properties:*

(i) dom W *is nonempty, and so W is finite on a nonempty set* dom W.

(ii) *Structural properties.*

(iia) W dominates $\langle \cdot \rangle$: $W(p) \geq \langle p \rangle$ *for all $p \in \mathcal{X}_e$ and, in particular, for all* $p \in \operatorname{dom} W$.

(iib) W *is* monotone: *if p_1 is in* dom W, *and if $p_1 \geq p_2$, then p_2 is in* dom W *and*

$$W(p_1) \geq W(p_2).$$

(iic) W *is* additive homogeneous*: for any constant $c \in \mathbf{R}$, we have*

$$W(p \pm c) = W(p) \pm c. \tag{4.5}$$

(iii) $-\mathcal{B}_{K_0} \subset \operatorname{dom} W$ *and $W(-\beta) = 0$ for all $\beta \in \mathcal{B}_{K_0}$.*

(iv) *Fix $p \in \mathcal{X}_e$ and assume that $\overline{J}_p(K_0)$ is finite. Then*

$$\begin{array}{c}\overline{J}_p(K_0) \geq W(p_t) \text{ and } p_t \in \operatorname{dom} W \text{ for all } t \geq 0, y \in L_2[0,t], \\ \text{where } u = K_0(y), p_0 = p.\end{array} \tag{4.6}$$

(v) *Fix* $p \in \operatorname{dom} W$. *Then if* K_ε *is an* ε*-optimal (i.e.,* $\overline{J}_p(K_\varepsilon) \leq W(p) + \varepsilon$*) admissible controller,*

$$W(p) + \varepsilon \geq W(p_t) \text{ and } p_t \in \operatorname{dom} W \text{ for all } t \geq 0, y \in L_2[0,t], \quad \text{where } u = K_\varepsilon(y), p_0 = p. \tag{4.7}$$

Thus any almost-optimal controller results in an information state trajectory along which $W(p_t)$ *is almost decreasing.*

(vi) *The dynamic programming principle holds: for any* $t \geq 0$, $p \in \operatorname{dom} W$,

$$W(p) = \inf_K \sup_{y \in L_{2,t}} \{W(p_t) : p_0 = p\}. \tag{4.8}$$

(Identity (4.8) *is called the* dynamic programming equation.*)*

(vii) *If for* $p \in \operatorname{dom} W$ *an admissible optimal controller* $\tilde{K}_p^* \in \operatorname{argmin}_K \overline{J}_p(K)$ *exists (so that* $W(p) = \overline{J}_p(\tilde{K}_p^*)$*), then* $W(p_t)$ *decreases along any optimal closed-loop trajectory determined by* $w \in L_{2,loc}$, $x_0 \in \mathbf{R}^n$*:*

$$W(p) \geq W(p_t), \tag{4.9}$$

where $u(t) = \tilde{K}_p^*(y)(t)$, $p_0 = p$, *and* $p_t \in \operatorname{dom} W$ *for all* $t \geq 0$.

Moreover, the closed-loop system $(G, \tilde{K}_p^*)$ *is dissipative for all* $x_0 \in \operatorname{support} p$

$$\begin{cases} \frac{1}{2}\int_0^T |z(s)|^2\,ds \leq \gamma^2 \frac{1}{2}\int_0^T |w(s)|^2\,ds - p(x_0) + W(p) \\ \text{for all } w \in L_{2,T} \text{ and all } T \geq 0, \end{cases} \tag{4.10}$$

and if $(G, \tilde{K}_p^*)$ *is* z*-detectable, the closed-loop system is weakly internally stable provided* $x_0 \in \operatorname{support} p$.

Proof. Part (i). From the bounds (3.6) and the definition (4.4), it follows that

$$\langle p \rangle \leq W(p) \leq J_p(K_0) = \langle p + \beta_{K_0} \rangle \leq \langle p + \beta \rangle \tag{4.11}$$

for all $p \in \mathcal{X}_e$, $\beta \in \mathcal{B}_{K_0}$. This implies $\operatorname{dom} \overline{J}_\cdot(K_0) \subset \operatorname{dom} W$, proving (i).

Part (ii). The domination property follows from (4.11). Monotonicity and additive homogeneity follow by inspecting the definition (4.4) of W (see Lemma 3.1.3).

Part (iii). Let $\beta \in \mathcal{B}_{K_0}$. Then by (4.11),

$$0 = \langle -\beta \rangle \leq W(-\beta) \leq \langle -\beta + \beta \rangle = 0,$$

which shows that $W(-\beta) = 0$ and hence $-\beta \in \operatorname{dom} W$.

Part (iv). Inequality (4.6) holds for $t = 0$ by definition of $W(p)$. Next assume $t > 0$ is fixed. Fix $y \in L_2[0,t]$, and for any $y^0 \in L_{2,loc}[t,\infty)$ define $\tilde{y} \in L_{2,loc}[0,\infty)$ by

$$\tilde{y}(s) = \begin{cases} y(s), & 0 \leq s \leq t, \\ y^0(s), & s \geq t. \end{cases}$$

Let $u(s) = K_0(\tilde{y})(s)$, so that $u \in L_{2,loc}$ since K_0 is admissible. Given p_0, this allows us to define p_s, $s \geq 0$. In particular, p_t exists (it is determined by $p_0 = p$, $u(s)$ and $y(s)$ on $[0, t]$) and $\langle p_t \rangle$ is finite, by Lemma 3.1.10. Let $0 \leq t \leq T$. Then

$$\begin{aligned} \overline{J}_p(K_0) &= \sup_{T \geq 0} \sup_{\tilde{y} \in L_2[0,T]} \langle p_T \rangle \\ &\geq \sup_{T \geq t} \sup_{y^0 \in L_2[t,T]} \langle p_T \rangle \\ &= \overline{J}_{p_t}(K_0^t) \geq \langle p_t \rangle, \end{aligned} \tag{4.12}$$

where $K_0^t(y^0)(s) = K_0(\tilde{y})(s)$, $s \geq t$. Since p_t is independent of $K_0^t(y^0)$ on $[t, \infty)$, we can minimize the last line of (4.12) over K_0^t to get

$$\overline{J}_p(K_0) \geq W(p_t) \geq \langle p_t \rangle. \tag{4.13}$$

This proves inequality (4.6) and in particular that $W(p_t)$ is finite (it is bounded from above and below).

Part (v). Apply part (iv) with $K_0 = K_\varepsilon$.

Part (vi). The proof is standard dynamic programming [ES84], [BB91], [JB95, Theorem 4.23], [JB96, Theorem 4.4]. Let $\varepsilon > 0$. From part (v), we know that there exists a controller K_ε such that for all $y \in L_{2,t}$,

$$W(p) + \varepsilon \geq W(p_t).$$

This implies

$$W(p) + \varepsilon \geq \inf_K \sup_{y \in L_{2,t}} \{W(p_t) : p_0 = p\}.$$

Therefore, since $\varepsilon > 0$ was arbitrary,

$$W(p) \geq \inf_K \sup_{y \in L_{2,t}} \{W(p_t) : p_0 = p\}. \tag{4.14}$$

This proves half of (4.8).

To prove the opposite inequality, define

$$R(p) = \inf_K \sup_{y \in L_{2,t}} \{W(p_t) : p_0 = p\}.$$

Let $\varepsilon > 0$. Choose K^1 admissible such that

$$R(p) \geq \sup_{y \in L_{2,t}} \{W(p_t) : p_0 = p\} - \varepsilon.$$

For all $q \in \operatorname{dom} W$, there exists K_q^2 admissible such that

$$W(q) \geq \overline{J}_q(K_q^2) - \varepsilon.$$

Define K^3 by

$$K^3(y)(s) = \begin{cases} K^1(y)(s), & 0 \le s \le t, \\ K^2_{p_t}(y)(s), & s \ge t. \end{cases}$$

Then K^3 is an admissible output feedback controller. Let $T \ge t > 0$, $y \in L_{2,loc}$. Then

$$R(p) \ge W(p_t) - \varepsilon \ge \overline{J}_{p_t}(K^2_{p_t}) - 2\varepsilon \ge \langle p_T \rangle - 2\varepsilon,$$

where $u = K^3(y)$, $p_0 = p$. Thus

$$W(p) \le \overline{J}_p(K^3) \le R(p) + 2\varepsilon,$$

which implies, since $\varepsilon > 0$ was arbitrary,

$$W(p) \le \inf_K \sup_{y \in L_{2,t}} \{W(p_t) : p_0 = p\}. \tag{4.15}$$

Combining the inequalities (4.14) and (4.15), we obtain the desired dynamic programming equation (4.8).

Part (vii). Inequality (4.9) follows from part (v) with $\varepsilon = 0$.

The dissipation property (4.10) is proven as follows. Let $w \in L_{2,loc}$ and $x_0 \in$ support p be given. These determine signals $y(\cdot)$, $z(\cdot)$, and $u(\cdot) = \tilde{K}^*_p(y(\cdot))$ arising in the closed-loop system $(G, \tilde{K}^*_p)$ and a corresponding plant state trajectory $x(\cdot)$.

Consider now the information state p_t driven by the signals u, y with initial condition $p_0 = p$ (we do not assert anything about any controller state here; in particular, it is not asserted at this stage that p_t is the state of the controller $\tilde{K}^*_p$). Inequality (4.9) implies

$$p_t(x) \le \langle p_t \rangle \le W(p_t) \le W(p) < +\infty$$

for all $x \in \mathbf{R}^n$. Set $x = x(t)$. By definition (3.7), we have

$$p_0(\xi(0)) + \tfrac{1}{2}\int_0^t [|\tilde{z}(s)|^2 - \gamma^2|\tilde{w}(s)|^2]\,ds \le p_t(x) \le \langle p_t \rangle \le W(p_t) \le W(p) < +\infty$$

for any $\tilde{w}$ with $y = C_2(\xi) + D_{21}(\xi)\tilde{w}$, and $\tilde{z} = C_1(\xi) + D_{12}(\xi)u$ with ξ solving the plant dynamics (3.8) with $\xi(t) = x$.

Setting $\tilde{w} = w$ implies $\xi(\cdot) = x(\cdot)$, and $\tilde{z}(\cdot) = z(\cdot)$; consequently,

$$p_0(x(0)) + \tfrac{1}{2}\int_0^t [|z(s)|^2 - \gamma^2|w(s)|^2]\,ds \le W(p) < +\infty,$$

which implies (4.10) when $p(x(0))$ is finite. □

4.4 The Dynamic Programming PDE

In this section we present and discuss the dynamic programming PDE for the value function W (equation (4.16) below) if it is smooth. (The results about the value function in the previous section do not involve smoothness.) We shall ultimately (§4.5) see how one obtains an optimal controller from a smooth solution to the PDE. In general, it is not likely to be possible to prove smoothness of the value function W in a global Frechet C^1 sense (this is a common problem in dynamic programming). Because additional technical issues arise when considering singular information states, we begin the discusion by focusing on smooth nonsingular information states (recall from §2.2.1 that nonsingular information states are everywhere finite) and defer to later sections our treatment of the dynamic programming PDE for singular information states.

4.4.1 Smooth Nonsingular Information States and Frechet Derivatives

The value function W satisfies, at least formally, the PDE

$$\inf_{u\in\mathbf{R}^m}\sup_{y\in\mathbf{R}^p} \nabla_p W(p)[F(p,u,y)] = 0 \text{ in } \operatorname{dom} W \cap \mathcal{X}_q, \tag{4.16}$$

which we call the *dynamic programming PDE*. To see this the PDE (4.16) is obtained from the dynamic programming equation (4.8) by rearranging

$$0 = \inf_{K}\sup_{y\in L_{2,t}} \frac{W(p_t) - W(p)}{t},$$

and (4.16) follows, since on sending $t \downarrow 0$, the chain rule says

$$\frac{(W(p_t) - W(p))}{t} \to \nabla_p W(p)\left[\frac{dp}{dt}\right] = \nabla_p W(p)[F(p,u,y)].$$

Because $F(p,u,y)$ is quadratic in u and y (see (4.24) below), the order of the inf and sup is immaterial; i.e., the Isaacs condition holds.

Note that p must be differentiable for $F(p,u,y)$ to make sense in the usual manner (since it is a function of the gradient $\nabla_x p$). Also the (infinite-dimensional) PDE (4.16) involves both a function W *and* a set $\operatorname{dom} W$ on which it is finite. In (4.16), $\nabla_p W(p)$ denotes a one-sided derivative of W with respect to p. For fixed p it is a function that acts on functions h and take values denoted $\nabla_p W(p)[h]$. If W is Gateaux differentiable at p (recall Chapter 2), it is a linear functional of h. In general, the PDE will actually only make sense for p belonging to a subset of $\operatorname{dom} W$ (this is discussed further below).

The following properties hold for any function ψ that is monotone, additively homogeneous, and Frechet (and hence Gateaux) differentiable (the following are to be interpreted for $p \in \operatorname{dom}\psi$ whenever the indicated derivatives exist and are finite).

(i) If $h_1 \geq h_2$ then the gradient is monotone:

$$\nabla_p\psi(p)[h_1] \geq \nabla_p\psi(p)[h_2]. \tag{4.17}$$

(ii) For any constant $c \in \mathbf{R}$, we have the identity relation

$$\nabla_p\psi(p)[c] = c. \tag{4.18}$$

(iii) For any constant $c \in \mathbf{R}$ and function h we have multiplicative homogeneity:

$$\nabla_p \psi(p)[ch] = c \nabla_p \psi(p)[h]. \tag{4.19}$$

(iv) For any pair of functions h_1, h_2, we have additivity

$$\nabla_p \psi(p)[h_1 + h_2] = \nabla_p \psi(p)[h_1] + \nabla_p \psi(p)[h_2]. \tag{4.20}$$

(Therefore the map $h \mapsto \nabla_p \psi(p)[h]$ is linear.)

(v) If $h \leq c$, for some constant $c \in \mathbf{R}$, then we have the upper bound

$$\nabla_p \psi(p)[h] \leq c. \tag{4.21}$$

(vi) If $c \leq h$, for some constant $c \in \mathbf{R}$, then we have the lower bound

$$c \leq \nabla_p \psi(p)[h]. \tag{4.22}$$

(See also Proposition 4.4.2 below.)

If the value function W is Frechet differentiable, then the Frechet derivative $\nabla_p W$ enjoys these properties, since we know from Theorem 4.3.1 that W is monotone and additively homogeneous.

Further, if the value function W is Frechet differentiable, and if the gradient $\nabla_x p$ exists, then it is possible to evaluate directly the inf and sup in (4.16) (these are *unique*); indeed, one has

$$\begin{aligned} \mathbf{u}^*(p) &= (\nabla_p W(p)[E_1])^{-1} \nabla_p W(p)[-D'_{12} C_1 + B'_2 \nabla_x p'], \\ \mathbf{y}^*(p) &= (\nabla_p W(p)[E_2^{-1}])^{-1} \nabla_p W(p)[E_2^{-1}(C_2 - \gamma^{-2} D_{21} B'_1 \nabla_x p')]. \end{aligned} \tag{4.23}$$

To see this, we note that the Frechet derivative defines a linear operator $h \mapsto \nabla_p W(p)[h]$, and so using the formula (3.10) for $F(p, u, y)$ we have

$$\begin{aligned} \nabla_p W(p)[F(p,u,y)] &= \nabla_p W(p)[-\nabla_x p \cdot (A - B_1 D'_{21} E_2^{-1} C_2) \\ &\qquad + \tfrac{1}{2\gamma^2} \nabla_x p B_1 [I - D'_{21} E_2^{-1} D_{21}] B'_1 \nabla_x p' \\ &\qquad + \tfrac{1}{2} |C_1|^2 - \gamma^2 \tfrac{1}{2} C'_2 E_2^{-1} C_2] \\ &\quad - \nabla_p W(p)[\nabla_x p C'_1 D_{12} - \nabla_x p \cdot B_2] u \\ &\quad + \nabla_p W(p)[(B_1 D'_{21} + \gamma^2 C'_2) E_2^{-1}] y \\ &\quad + \tfrac{1}{2} u' \nabla_p W(p)[E_1] u - \gamma^2 \tfrac{1}{2} y' \nabla_p W(p)[E_2^{-1}] y. \end{aligned} \tag{4.24}$$

This function is clearly strictly convex in u and strictly concave in y (in fact quadratic), and the evaluation of the saddle point $(\mathbf{u}^*(p), \mathbf{y}^*(p))$ is immediate.

If E_1, E_2 are independent of x, then

$$\begin{aligned}\mathbf{u}^*(p) &= \nabla_p W(p)[E_1^{-1}(-D_{12}'C_1 + B_2'\nabla_x p')],\\ \mathbf{y}^*(p) &= \nabla_p W(p)[C_2 - \gamma^{-2}D_{21}B_1'\nabla_x p'].\end{aligned} \tag{4.23)'$$

If E_1, E_2 do depend on x, we need to ensure that the matrices

$$\tilde{E}_1(p) = \nabla_p W(p)[E_1], \quad \tilde{E}_2^I(p) = \nabla_p W(p)[E_2^{-1}]$$

are invertible. However, this follows from the linearity, monotonicity (4.17), and identity (4.18) properties discussed above (using $\phi(x) = \xi' E_1(x)\xi > 0$ and $\phi(x) = \xi' E_2(x)^{-1}\xi > 0$ for $\xi \in \mathbf{R}^n, \xi \neq 0$). Indeed, $\tilde{E}_1(p) > 0$ and $\tilde{E}_2^I(p) > 0$.

We now make a point that while seemingly banal is used frequently. One can express the content of the dynamic programming PDE (4.16) as follows:

For all $y \in \mathbf{R}^p$, we have

$$\nabla_p W(p)[F(p, u^*(p), y)] \leq 0 \text{ in dom } W \cap \mathcal{X}_q. \tag{4.25}$$

Moreover, equality holds if and only if $y = \mathbf{y}^*(p)$.

This looks stronger than the dynamic programming PDE at first glance; however, the fact that F is strictly concave in y tells us that $\nabla_p W(p)[F(p, u^*(p), y)]$ is concave in y (see (4.24) above), so any critical point $\mathbf{y}^*(p)$ is a maximum.

What is important for dissipation is the dissipation inequality. Indeed, we will make use of the *dissipation PDI*

$$\inf_{u \in \mathbf{R}^m} \sup_{y \in \mathbf{R}^p} \nabla_p \hat{W}(p)[F(p, u, y)] \leq 0 \text{ in dom } W \cap \mathcal{X}_q, \tag{4.26}$$

which of course is (formally) satisfied by the value function W.

As we shall see, an issue of fundamental importance is whether or not $p_t \in \text{dom } W$, and furthermore if $\mathbf{u}^*(p_t)$ is well defined. This latter requirement necessitates p_t belonging to a domain of smoothness of W, a special type of coupling.

REMARK 4.4.1 The value function W is uniquely defined by (4.4) and is a particular solution to the dynamic programming PDE (4.16) (in a sense to be described below). Even for linear systems solutions to the dynamic programming PDE are not unique: one may produce an optimal $\mathbf{y}^*$ for which $t \mapsto \mathbf{y}^*(p_t)$ is a function in L_2; the other optimal $\mathbf{y}^*$ do not. This parallels the fact that Riccati equations can have stabilizing and antistabilizing solutions, as well as other possible solutions. This will be discussed in more detail in §4.5 and in Chapter 6. ▽

4.4.2 Directional Derivatives

The use of Frechet derivatives in the previous section illustrated the basic form of the dynamic programming PDE. However, Frechet differentiability is too strong, and in fact the value function W will rarely be Frechet differentiable in the sense we have described. A substantial amount of research has gone into weakening the notion of differentiability in a manner that is suitable for dynamic programming. For instance, the concept of viscosity solution has been very useful in providing an existence and uniqueness theory for certain types of nonlinear PDE (however, such a theory for the particular PDE (4.16) is not available at present, although a start has been made in [JB96]). What is most useful for dynamic programming is a synthesis theory that permits optimal controllers to be constructed from the dynamic programming PDE. Techniques from nonsmooth analysis have been employed with success in many applications; see, e.g., [Cla83]. Notions of generalized directional derivatives and gradients are used.

In the previous section we saw that to write the formal dynamic programming PDE (4.16) we needed only directional derivatives of W but at a higher level of generality. In this section we use a definition of generalized directional derivative that may be of use for a synthesis theory for the dynamic programming equation (4.16). We sketch here only a few basic ideas and properties of such a generalized directional derivative when it exists. In §§4.4.3–4.5.4 we provide a more detailed generalization of (4.16) based on the idea of directional derivatives along information state trajectories.

Following Clarke [Cla83], we define the *generalized directional derivative* $\nabla_p^+ \psi$ of a function $\psi : \mathcal{X}_e \to \mathbf{R} \cup \{+\infty\}$ in the direction of a function $h : \mathcal{X}_e \to \mathbf{R}$ by

$$\nabla_p^+ \psi(p)[h] = \limsup_{t \downarrow 0,\, p' \Longrightarrow p} \frac{\psi(p' + th) - \psi(p')}{t}. \tag{4.27}$$

Notice the use of lim sup and the base point moves $p' \Longrightarrow p$ in the max-plus weak sense[3] (see Appendix C) and that it is one-sided ($t \downarrow 0$). The generalized gradient will always exist, although it may not be finite.

PROPOSITION 4.4.2 *The following properties hold for any function ψ that is monotone and additively homogeneous (the following are to be interpreted for $p \in \operatorname{dom}\psi$ whenever the indicated derivatives exist and are finite):*

(i) *If $h_1 \geq h_2$ then the gradient is monotone:*

$$\nabla_p^+ \psi(p)[h_1] \geq \nabla_p^+ \psi(p)[h_2]. \tag{4.28}$$

(ii) *For any constant $c \in \mathbf{R}$, we have the identity relation*

$$\nabla_p^+ \psi(p)[c] = c. \tag{4.29}$$

(iii) *For any constant $c \geq 0$ and function h we have positive multiplicative homogeneity:*

$$\nabla_p^+ \psi(p)[ch] = c\nabla_p^+ \psi(p)[h]. \tag{4.30}$$

[3]Max-plus weak convergence turns out to be very natural and is even essential for singular information states. As the name implies, it is weaker than, say, uniform convergence on compact subsets, and there are situations where singular states do not converge in this stronger topology.

(iv) *For any pair of functions* h_1, h_2, *we have subadditivity*

$$\nabla_p^+\psi(p)[h_1 + h_2] \leq \nabla_p^+\psi(p)[h_1] + \nabla_p^+\psi(p)[h_2]. \quad (4.31)$$

(v) *For any pair of functions* h_1, h_2 *and constant* $0 \leq c \leq 1$ *we have convexity:*

$$\nabla_p^+\psi(p)[ch_1 + (1-c)h_2] \leq c\nabla_p^+\psi(p)[h_1] + (1-c)\nabla_p^+\psi(p)[h_2]. \quad (4.32)$$

(vi) *For any constant* $c \geq 0$ *and function* h *we have the additive relation*

$$\nabla_p^+\psi(p)[h + c] = \nabla_p^+\psi(p)[h] + c. \quad (4.33)$$

(vii) *If* $h \leq c$, *for some constant* $c \in \mathbf{R}$, *then we have the upper bound*

$$\nabla_p^+\psi(p)[h] \leq c. \quad (4.34)$$

(viii) *If* $c \leq h$, *for some constant* $c \in \mathbf{R}$, *then we have the lower bound*

$$c \leq \nabla_p^+\psi(p)[h]. \quad (4.35)$$

(ix) $\nabla_p^+\psi(p)[h]$ *is u.s.c. in* p.

Proof. All of these properties can be verified using the definition (4.27).
By monotonicity, if $h_1 \geq h_2$,

$$\frac{\psi(p' + th_1) - \psi(p')}{t} \geq \frac{\psi(p' + th_2) - \psi(p')}{t},$$

from which (4.28) follows, proving (i). Similarly, by additive homogeneity, for $c \in \mathbf{R}$,

$$\frac{\psi(p' + tc) - \psi(p')}{t} = \frac{\psi(p') - \psi(p')}{t} + c = c$$

giving (4.29), proving (ii). If $c \geq 0$ is a constant,

$$\begin{aligned}
\nabla_p^+\psi(p)[ch] &= \limsup_{t\downarrow 0,\, p'\Longrightarrow p} \frac{\psi(p' + tch) - \psi(p')}{t} \\
&= \limsup_{s\downarrow 0,\, p'\Longrightarrow p} \frac{\psi(p' + sh) - \psi(p')}{s/h} \\
&= c\nabla_p^+\psi(p)[h].
\end{aligned}$$

This proves (iii).

For any h_1, h_2,

$$\begin{aligned}
&\nabla_p^+\psi(p)[h_1 + h_2] \\
&= \limsup_{t\downarrow 0,\, p'\Longrightarrow p} \left\{ \frac{\psi(p' + t(h_1 + h_2)) - \psi(p' + th_2)}{t} + \frac{\psi(p' + th_2) - \psi(p')}{t} \right\} \\
&\leq \nabla_p^+\psi(p)[h_1] + \nabla_p^+\psi(p)[h_2];
\end{aligned}$$

hence (iv) follows.

Part (v) follows from (iii) and (iv).

For $c \in \mathbf{R}$,

$$\frac{\psi(p' + t(h+c)) - \psi(p')}{t} = \frac{\psi(p' + th) - \psi(p')}{t} + c,$$

from which (vi) follows.

If h is bounded above by a constant c,

$$\begin{aligned}\frac{\psi(p' + th) - \psi(p')}{t} &\le \frac{\psi(p' + tc) - \psi(p')}{t} \\ &\le \frac{\psi(p') - \psi(p')}{t} + c = c,\end{aligned}$$

which implies (vii), and (viii) is similar. □

REMARK 4.4.3 The fact that the base point slides in the definition (4.27) is important for part (ix) of Proposition 4.4.2, but it is irrelevant for properties (i)–(viii). ∇

COROLLARY 4.4.4 *Since the value function W (defined by (4.4)) is monotone and additively homogeneous (by Theorem 4.3.1), the generalized gradient $\nabla_p^+ W$ will enjoy the properties listed in Proposition 4.4.2 (at least when it is finite).*

One possible generalization of the dynamic programming PDE (4.16) is

$$\inf_{u \in \mathbf{R}^m} \sup_{y \in \mathbf{R}^p} \nabla_p^+ W(p)[F(p,u,y)] = 0 \text{ in } \operatorname{dom} W \cap \mathcal{X}_q, \tag{4.36}$$

where the Frechet derivative is replaced by the generalized directional derivative. In such an event the optimal control and observation functions are defined by

$$\begin{aligned}\mathbf{u}^*(p) &= \operatorname*{argmin}_{u \in \mathbf{R}^m} \left\{ \sup_{y \in \mathbf{R}^p} \nabla_p^+ W(p)[F(p,u,y)] \right\}, \\ \mathbf{y}^*(p) &\in \operatorname*{argmax}_{y \in \mathbf{R}^p} \left\{ \nabla_p^+ W(p)[F(p,\mathbf{u}^*(p),y)] \right\},\end{aligned} \tag{4.37}$$

and we note that it is not in general possible to directly evaluate them. When W is Frechet differentiable these formulas reduce to the concrete formulas (4.23).

However, it can be seen that a unique minimizer and maximizers will exist. To see this, let us assume that D_{12} and D_{21} are independent of x. Indeed, due to the convexity of the generalized directional derivative and the form (3.10) of $F(p,u,y)$ it is evident that for each y the map $u \mapsto \nabla_p^+ W(p)[F(p,u,y)]$ is strictly convex, and also the map $u \mapsto \sup_y \{\nabla_p^+ W(p)[F(p,u,y)]\}$ is strictly convex (the fact that $\nabla_p^+ W(p)[F(p,u,y)]$ is of the form $\nabla_p^+ W(p)[F_0(p,u,y)] + \frac{1}{2}|u|^2$, where the term $F_0(p,u,y)$ is linear in u, is used to verify this). This implies that $\mathbf{u}^*(p)$ exists and is unique. The map $y \mapsto \nabla_p^+ W(p)[F(p,u,y)]$ need not be concave, but we have the estimates

$$\nabla_p^+ W(p)[F_1^\eta(p)] + \tfrac{1}{2}(1-\eta)|u|^2 - \tfrac{1}{2}\gamma^2(1+\eta)|y|^2 \leq \nabla_p^+ W(p)[F(p,u,y)]$$
$$\leq \nabla_p^+ W(p)[F_2^\eta(p)] + \tfrac{1}{2}(1+\eta)|u|^2 - \tfrac{1}{2}\gamma^2(1-\eta)|y|^2 \tag{4.38}$$

for all $\eta > 0$. Here, $F_1^\eta(p)$ and $F_2^\eta(p)$ are functions not depending on u, y. We deduce that $\mathbf{y}^*(p)$ exists due to the coercive (in y) upper bound for $\nabla_p^+ W(p)[F(p,u,y)]$, but it is not necessarily unique.

The dissipation PDI expressed in terms of the directional derivative operator reads

$$\inf_{u\in\mathbf{R}^m} \sup_{y\in\mathbf{R}^p} \nabla_p^+ W(p)[F(p,u,y)] = 0 \ \text{ in dom } W \cap \mathcal{X}_q. \tag{4.39}$$

4.4.3 General Information States

The preceding sections considered derivatives of the value function W in directions $F(p,u,y)$ in the dynamic programming PDE and in the formulas for the optimal control and observation. When p is singular, as is often the case in important examples, alternative formulas are needed. Since these are important we go to some effort in Chapter 10 (see (10.15)) to define an operator $\mathcal{L}^{u,y}$ that affects an extension of directional derivative to more general situations. You do not need to know this construction until Chapter 10, but we mention it now because it is good to know that singular situations are not insurmountable.

In terms of the operator $\mathcal{L}^{u,y}$ the dynamic programming PDE takes the form

$$\inf_{u\in\mathbf{R}^m} \sup_{y\in\mathbf{R}^p} \{\mathcal{L}^{u,y} W(p)\} = 0 \ \text{ in dom}_{\text{pd}} W, \tag{4.40}$$

and the dissipation PDI becomes

$$\inf_{u\in\mathbf{R}^m} \sup_{y\in\mathbf{R}^p} \{\mathcal{L}^{u,y} W(p)\} \leq 0 \ \text{ in dom}_{\text{pd}} W. \tag{4.41}$$

Here $\text{dom}_{\text{pd}} \subset \text{dom } W$ is a set to be described shortly. If W is continuously Frechet differentiable at a smooth $p \in \text{dom } W$, we expect that

$$\mathcal{L}^{u,y} W(p) = \nabla_p W(p)[F(p,u,y)], \tag{4.42}$$

so (4.40) reduces to the earlier dynamic programming PDE (4.16) for W. However, if p is singular the RHS of formula (4.42) is not well defined, since, e.g., $F(p,u,y)$ is not well defined, which is what motivates us to use $\mathcal{L}^{u,y}$ and ignore the RHS.

The optimal control and observation functions are defined (for $p \in \text{dom}_{\text{smooth}} W$) by

$$\mathbf{u}^*(p) \in \underset{u\in\mathbf{R}^m}{\text{argmin}}\{\sup_y \mathcal{L}^{u,y} W(p)\},$$
$$\mathbf{y}^*(p) \in \underset{y\in\mathbf{R}^p}{\text{argmax}}\{\mathcal{L}^{\mathbf{u}^*(p),y} W(p)\}. \tag{4.43}$$

These formulas can be interpreted in both nonsingular and singular cases (see Chapter 10 for further details).

4.5 Solving the Dynamic Programming PDE and Dissipation PDI

In general, the dynamic programming PDE (4.40) and dissipation PDE (4.41) can have many solutions (as we shall see, the value function W is one of them). In this general situation we will need definitions of smoothness, etc., appropriate for functions $\hat{W}$ to be a solution of the dynamic programming PDE

$$\inf_{u \in \mathbf{R}^m} \sup_{y \in \mathbf{R}^p} \{\mathcal{L}^{u,y}\hat{W}(p)\} = 0 \qquad (4.40)'$$

or a solution of the dissipation PDI

$$\inf_{u \in \mathbf{R}^m} \sup_{y \in \mathbf{R}^p} \{\mathcal{L}^{u,y}\hat{W}(p)\} \leq 0. \qquad (4.41)'$$

The definitions relate to the following conditions:

(i) Structural conditions such as finiteness, domination, monotonicity, and additive homogeneity (defined in §2.2.2).

(ii) Smoothness conditions ensuring that the LHS of (4.40) or (4.41) is well defined and that the PDE or PDI can be integrated (in a manner to be specified below).

(iii) Admissibility conditions ensure that the controller obtained from $\hat{W}$ is sufficiently regular so that the information state and control signals are well defined.

(iv) Solvability conditions specify at which points p the function actually solves the PDE or PDI, i.e., points at which the LHS of (4.40) equals zero or at which the LHS of (4.41) is less than or equal to zero ($p \in \mathrm{dom}_{\mathrm{pd}}\hat{W}$).

REMARK 4.5.1 Our definition of smoothness will allow us to explain how information state controllers can be produced from smooth solutions of the PDE (4.40) or PDI (4.41). This material also serves the purpose of bringing out some of the technical issues involved in dealing with PDEs and PDIs at this level of generality. It is hoped that future research will improve on our present understanding of the complex and largely unresolved issues concerning (4.40) and (4.41).

Given a set $D \subset \mathcal{X}_e$, we write

$$\tau_{esc}(D) = \sup\{t \geq 0 \,:\, p_s \in D \text{ for all } 0 \leq s \leq t\}$$

for the time of first escape of p_t from D (when driven by inputs $u, y \in L_{2,loc}$).

4.5.1 Smoothness

We say that $\hat{W}$ is *p-smooth* provided

(i) $q.i.\mathrm{dom}\,\hat{W}$ is a nonempty subset of $\mathcal{X}_e$.

(ii) $\hat{W}$ is continuous in the interior of dom $\hat{W}$ with respect to the topology of $\mathcal{X}_e$.

(iii) There exists a nonempty subset $\text{dom}_{\text{smooth}}\hat{W} \subset q.i.\text{dom}\,\hat{W}$ satisfying the properties listed below.

(iv) The expression $\mathcal{L}^{u,y}\hat{W}(p)$ associates a real number to each $u \in \mathbf{R}^m$, $y \in \mathbf{R}^p$, and $p \in \text{dom}_{\text{smooth}}\hat{W}$. The set $\text{dom}_{\text{smooth}}\hat{W}$ is the largest subset of dom $\hat{W}$ for which $\mathcal{L}^{u,y}\hat{W}(p)$ is well defined for all $u \in \mathbf{R}^m$, $y \in \mathbf{R}^p$.

(v) For all $u \in \mathbf{R}^m$, $y \in \mathbf{R}^p$, $0 < \eta < 1$, there exists operators $\mathcal{L}_i^\eta$ $(i = 1, 2)$ such that

$$\begin{aligned}&\mathcal{L}_1^\eta W(p) + \tfrac{1}{2}(1-\eta)|u|^2 - \tfrac{1}{2}\gamma^2(1+\eta)|y|^2 \leq \mathcal{L}^{u,y}W(p)\\ &\leq \mathcal{L}_2^\eta W(p) + \tfrac{1}{2}(1+\eta)|u|^2 - \tfrac{1}{2}\gamma^2(1-\eta)|y|^2\end{aligned} \tag{4.44}$$

for all $p \in \text{dom}_{\text{smooth}}\hat{W}$.

(vi) For each $p_0 \in \text{dom}_{\text{smooth}}\hat{W}$ there exists $t_0 > 0$ such that $p_t \in \text{dom}_{\text{smooth}}\hat{W}$ for all $0 \leq t \leq t_0$ and all $u, y \in L_2[0, t_0]$, and

(a) the map $s \mapsto \mathcal{L}^{u(s),y(s)}W(p_s)$ is continuous on $[0, t_0]$ uniformly for u, y in $L_2[0, t_0]$-bounded subsets of $C[0, t_0]$;

(b) for all $0 \leq \bar{t}_0 \leq t_0$, the map $u, y \mapsto \int_0^{\bar{t}_0} \mathcal{L}^{u(s),y(s)}W(p_s)ds$ from $L_2[0, \bar{t}_0]$ to $\mathbf{R}$ is continuous;

(c) for all $0 \leq \bar{t}_0 \leq t_0$, $u, y \in L_2[0, \bar{t}_0]$,

$$\left|\int_0^{\bar{t}_0} \mathcal{L}_i^\eta W(p_s)ds\right| \leq C < \infty \tag{4.45}$$

$(i = 1, 2)$.

(Here, p_s denotes the information state at time s driven by inputs u, y with initial condition $p_0 \in \text{dom}_{\text{smooth}}\hat{W}$.)

(vii) The functions

$$\begin{aligned}\mathbf{u}_{\hat{W}}^*(p) &\in \underset{u \in \mathbf{R}^m}{\text{argmin}}\{\sup_y \mathcal{L}^{u,y}\hat{W}(p)\},\\ \mathbf{y}_{\hat{W}}^*(p) &\in \underset{y \in \mathbf{R}^p}{\text{argmax}}\{\mathcal{L}^{\mathbf{u}^*(p),y}\hat{W}(p)\}\end{aligned} \tag{4.46}$$

are well defined for all $p \in \text{dom}_{\text{smooth}}\hat{W}$, so that

$$\text{dom}\,\mathbf{u}_{\hat{W}}^* = \text{dom}_{\text{smooth}}\hat{W},$$

$$\text{dom}\,\mathbf{y}_{\hat{W}}^* = \text{dom}_{\text{smooth}}\hat{W}.$$

We say that the integral version of the *fundamental theorem of calculus* holds for the function $\hat{W}$ relative to the operator $\mathcal{L}^{u,y}$ if for all $t \geq 0$, $p_0 \in \text{dom}_{\text{smooth}}\hat{W}$, $u, y \in L_{2,loc}$:

$$\hat{W}(p_t) = \hat{W}(p_0) + \int_0^t \mathcal{L}^{u(s),y(s)}\hat{W}(p_s)\, ds \tag{4.47}$$

whenever $p_s \in \text{dom}_{\text{smooth}}\hat{W}$, $0 \leq s \leq t$, where $\dot{p}_s = F(p_s, u(s), y(s))$, $0 \leq s \leq t$, i.e., for all $0 \leq t < \tau_{esc}(\text{dom}_{\text{smooth}}\hat{W})$.

The smooth domain $\text{dom}_{\text{smooth}}\hat{W}$ is the subset of the domain $\text{dom}\,\hat{W}$ on which the LHS of (4.40) is defined; this set may be larger than the set $\text{dom}_{\text{pd}}\hat{W}$ on which the LHS equals zero.

4.5.2 Admissibility

We now require that the controller defined via (4.46) from a solution of the PDE (4.40) or PDI (4.41) is at least preadmissible (recall §4.2).

> ASSUMPTION 4.5.2 For each $p_0 \in q.i.\text{dom}_{\text{smooth}}\hat{W}$, the controller $K_{p_0}^{\mathbf{u}^*_{\hat{W}}}$ is preadmissible, with $p_t \in \text{dom}_{\text{smooth}}\hat{W}$ for all $0 \leq t \leq \tau_{adm}(K_{p_0}^{\mathbf{u}^*_{\hat{W}}})$.

Note that, for $p_0 \in \text{dom}_{\text{smooth}}\hat{W}$,

$$0 < \tau_{adm}(K_{p_0}^{\mathbf{u}^*_{\hat{W}}}) = \tau_{esc}(\text{dom}_{\text{smooth}}\hat{W}).$$

4.5.3 Solutions of the Dynamic Programming PDE and Dissipation PDI

Now that we have introduced a weak type of smoothness for functions on the space $\mathcal{X}_e$, and related definitions, we show how this meshes with solving the dynamic programming PDE (4.40) and dissipation PDI (4.41).

We will define the solvability of the PDE (4.40) or PDI (4.41) on a set

$$\text{dom}_{\text{pd}}\hat{W} \subset \text{dom}_{\text{smooth}}\hat{W}.$$

The reason for this is that the LHS of (4.40) or (4.41) may be defined for a larger set than that on which the PDE or PDI holds, and further the controller $K_{p_0}^{\mathbf{u}^*_{\hat{W}}}$ may be well defined irrespective of the PDE or PDI holding. [4]

A p-smooth function $\hat{W} : \mathcal{X}_e \to \mathbf{R}$ is *a p-smooth solution of the dynamic programming PDE* (4.40) on the set $\text{dom}_{\text{pd}}\hat{W}$ if the information state controller $K_{p_0}^{\mathbf{u}^*_{\hat{W}}}$ satisfies for all $p_0 \in \text{dom}_{\text{pd}}\hat{W}$ the following "integrated" inequality and equality:

(i) For all y in $L_{2,loc}$ we have

$$\hat{W}(p_t) \leq \hat{W}(p_0) \tag{4.48}$$

for all $0 \leq t < \tau_{esc}(\text{dom}_{\text{pd}}\hat{W})$.

---

[4] As an analogy, consider the classical PDE $\frac{d}{dx}\phi = 0$ in $\Omega = (-1, 1) \subset \mathbf{R}$. The C^1 function $\phi(x) = 0$ if $x \in \Omega$, $\phi(x) = \frac{1}{2}x^2 + x - \frac{3}{2}$ if $x < -1$, $\phi(x) = \frac{1}{2}x^2 - x + \frac{1}{2}$ if $x > 1$ solves this PDE in $\bar{\Omega}$ and is also well defined outside $\bar{\Omega}$. Now $\text{dom}_{\text{smooth}}\phi = \mathbf{R}$ (here smooth means continuously differentiable in the usual sense), and $\frac{d}{dx}\phi(x)$ is well defined for all $x \in \mathbf{R}$, but $\text{dom}_{\text{pd}}\phi = \bar{\Omega}$.

(ii) For $y(s) = \mathbf{y}^*_{\hat{W}}(p_s)$, $0 \le s \le t$, we have

$$\hat{W}(p_t) = \hat{W}(p_0) \tag{4.49}$$

for all $0 \le t < \tau_{esc}(\mathrm{dom}_{\mathrm{pd}}\hat{W})$.

A p-smooth function $\hat{W} : \mathcal{X}_e \to \mathbf{R}$ is *a p-smooth solution of the dissipation PDI* (4.41) on the set $\mathrm{dom}_{\mathrm{pd}}\hat{W}$ if the information state controller $K_{p_0}^{\mathbf{u}^*_{\hat{W}}}$ satisfies for all $p_0 \in \mathrm{dom}_{\mathrm{pd}}\hat{W}$ the following "integrated" inequality:

(i) For all y in $L_{2,loc}$ we have

$$\hat{W}(p_t) \le \hat{W}(p_0) \tag{4.48}$$

for all $0 \le t < \tau_{esc}(\mathrm{dom}_{\mathrm{pd}}\hat{W})$.

A function $\hat{W}$ is called a *good solution* of the dynamic programming PDE (4.40) or dissipation PDI (4.41) provided $\hat{W}$

(i) is a p-smooth solution,

(ii) satisfies the structural conditions (domination, monotonicity, additive homogeneity).

4.5.4 The Value Function Solves the Dynamic Programming PDE

We now have enough technical assemblage to state and prove the desirable result that the value function W defined by (4.4) solves the dynamic programming PDE (4.40) and in fact is the minimal good solution.

THEOREM 4.5.3 *Assume that the value function W is p-smooth and the fundamental theorem of calculus holds for W relative to the operator $\mathcal{L}^{u,y}$ in* $\mathrm{dom}_{\mathrm{smooth}}W$, *and the* Isaacs condition *is satisfied:*

$$\inf_{u\in\mathbf{R}^m}\sup_{y\in\mathbf{R}^p}\{\mathcal{L}^{u,y}W(p_0)\} = \sup_{y\in\mathbf{R}^p}\inf_{u\in\mathbf{R}^m}\{\mathcal{L}^{u,y}W(p_0)\} \tag{4.50}$$

for all $p_0 \in \mathrm{dom}_{\mathrm{smooth}}W$. Then the following conditions hold:

(i) *For each $p \in \mathrm{dom}_{\mathrm{smooth}}W$, we have*

$$\inf_{u\in\mathbf{R}^m}\sup_{y\in\mathbf{R}^p}\{\mathcal{L}^{u,y}W(p)\} = 0 \tag{4.40$'$}$$

so $\mathrm{dom}_{\mathrm{pd}}W = \mathrm{dom}_{\mathrm{smooth}}W$.

(ii) *W is a good solution of the dynamic programming PDE* (4.40), *and hence also of the dissipation PDI* (4.41).

(iii) *Let $\hat{W}$ be any good solution of the dissipation PDI* (4.41). *If* $p_0 \in \mathrm{dom}_{\mathrm{pd}}\hat{W}$ *and* $K_{p_0}^{\mathbf{u}^*_{\hat{W}}}$ *is admissible with* $p_t \in \mathrm{dom}_{\mathrm{pd}}\hat{W}$ *for all* $t \geq 0$ *(i.e.,* $\tau_{adm}(K_{p_0}^{\mathbf{u}^*_{\hat{W}}}) = +\infty$*),* $y \in L_{2,loc}$*, then*

$$W(p_0) \leq \hat{W}(p_0). \tag{4.51}$$

That is, W is the minimal good solution of the PDI (4.41).

Proof. To prove that W satisfies (4.40) for $p \in \mathrm{dom}_{\mathrm{pd}}W$, we shall need two facts whose proofs are deferred to the end of the proof of this theorem.

Fact 1. Assume W is p-smooth, $p_0 \in \mathrm{dom}_{\mathrm{smooth}}W$, and

$$\inf_{u\in\mathbf{R}^m}\sup_{y\in\mathbf{R}^p}\{\mathcal{L}^{u,y}W(p_0)\} \leq -\theta, \tag{4.52}$$

where $\theta > 0$. Then there exists $t_1 > 0$ and an admissible controller K_1 such that for all $0 \leq t \leq t_1$, $y \in L_{2,t_1}$,

$$\int_0^t \mathcal{L}^{K_1[y](s),y(s)}W(p_s)ds \leq -\theta t/2. \tag{4.53}$$

Fact 2. Assume W is p-smooth, $p_0 \in \mathrm{dom}_{\mathrm{smooth}}W$, and

$$\inf_{u\in\mathbf{R}^m}\sup_{y\in\mathbf{R}^p}\{\mathcal{L}^{u,y}W(p_0)\} \geq \theta, \tag{4.54}$$

where $\theta > 0$. Then there exists $t_2 > 0$ and a $y_2 \in L_{2,t_2}$ such that for all $0 \leq t \leq t_2$ and all admissible controllers K,

$$\int_0^t \mathcal{L}^{K[y_2](s),y_2(s)}W(p_s)ds \geq \theta t/2. \tag{4.55}$$

Continuing with the proof, let us suppose, to the contrary, that for some $p_0 \in \mathrm{dom}_{\mathrm{smooth}}W$,

$$\inf_{u\in\mathbf{R}^m}\sup_{y\in\mathbf{R}^p}\{\mathcal{L}^{u,y}W(p_0)\} \neq 0.$$

Then either (4.52) or (4.54) holds for some $\theta > 0$.

By the dynamic programming principle (4.8), we have for all $t \geq 0$,

$$\inf_{K}\sup_{y\in L_{2,t}}\{W(p_t) - W(p_0)\} = 0,$$

and by the fundamental theorem of calculus, this implies

$$\inf_{K}\sup_{y\in L_{2,t}}\left\{\int_0^t \mathcal{L}^{K[y](s),y(s)}W(p_s)\right\} = 0 \tag{4.56}$$

for all $0 \leq t \leq t_0$. However, this contradicts (4.53) and (4.55). Indeed, if (4.52) holds, then (4.56) implies

$$\sup_{y\in L_{2,t}}\left\{\int_0^t \mathcal{L}^{K_1[y](s),y(s)}W(p_s)\right\} \geq 0,$$

which contradicts (4.53) by Fact 1. Similarly, if (4.54) holds, then (4.56) implies

$$\inf_K \left\{ \int_0^t \mathcal{L}^{K[y_2](s), y_2(s)} W(p_s) \right\} \leq 0,$$

which contradicts (4.55) by Fact 2.

We conclude that (4.40) holds for $p_0 \in \text{dom}_{\text{smooth}} W$.

Making use of the fundamental theorem of calculus, we can integrate (4.40) to yield the integrated form of the PDE as described above.

Next, if $\hat{W}$ is a good solution of the PDI (4.41) with the stated additional assumptions, then

$$\langle p_t \rangle \leq \hat{W}(p_t) \leq \hat{W}(p_0) < \infty \text{ for all } t \geq 0.$$

(That this holds for all $t \geq 0$ follows from the hypothesis $\tau_{adm} = +\infty$.) Here, $u(t) = \mathbf{u}^*_{\hat{W}}(p_t)$ and $y \in L_{2,loc}$. This implies

$$\overline{J}_{p_0}(K_{p_0}^{\mathbf{u}^*_{\hat{W}}}) \leq \hat{W}(p_0),$$

and hence, on minimizing over the class of all admissible controllers, we get

$$W(p_0) \leq \overline{J}_{p_0}(K_{p_0}^{\mathbf{u}^*_{\hat{W}}}) \leq \hat{W}(p_0)$$

as required.

Proof of Fact 1.

Claim 1a. Fix $u_1 \in \mathbf{R}^m$. There exists a bounded set $B_{u_1} \subset L_2[0, t_0] \cap C[0, t_0]$ such that

$$\sup_{y \in B_{u_1}} \int_0^{\bar{t}_0} \mathcal{L}^{u_1, y} W(p_s) ds = \sup_{y \in L_2[0, \bar{t}_0]} \int_0^{\bar{t}_0} \mathcal{L}^{u_1, y} W(p_s) ds$$

and

$$\sup_{y \in B_{u_1}} \int_0^{t_0} |y(s)|^2 ds \leq C_{u_1}$$

for all $0 \leq \bar{t}_0 \leq t_0$ for some $C_{u_1} \geq 0$. (Here, $u \equiv u_1$ is constant, and p_s denotes the information state at time s driven by inputs u_1, y with initial condition $p_0 \in \text{dom}_{\text{smooth}} \hat{W}$.)

Proof of Claim 1a. Let $\varepsilon > 0$ and choose any $y^\varepsilon \in L_2[0, \bar{t}_0]$ such that

$$\int_0^{\bar{t}_0} \mathcal{L}^{u_1, y^\varepsilon} W(p_s) ds \geq \sup_{y \in L_2[0, \bar{t}_0]} \int_0^{\bar{t}_0} \mathcal{L}^{u_1, y} W(p_s) ds - \varepsilon.$$

Using (4.44) we have

$$\int_0^{\bar{t}_0} \mathcal{L}_2^\eta W(p_s) + \tfrac{1}{2}(1+\eta)|u_1|^2 - \tfrac{1}{2}\gamma^2(1-\eta)|y^\varepsilon(s)|^2 ds \geq \int_0^{\bar{t}_0} \mathcal{L}^{u_1, y^\varepsilon(s)} W(p_s) ds.$$

Combining these inequalities and rearranging we have

$$(1-\eta)\frac{\gamma^2}{2} \int_0^{\bar{t}_0} |y^\varepsilon(s)|^2 ds \leq -\sup_{y \in L_2[0, \bar{t}_0]} \int_0^{\bar{t}_0} \mathcal{L}^{u_1, y} W(p_s) ds + \varepsilon$$

$$+ \int_0^{\bar{t}_0} \mathcal{L}_2^\eta W(p_s) ds + \tfrac{1}{2}(1+\eta)|u_1|^2 \bar{t}_0,$$

and by (4.45) this implies

$$\int_0^{\bar{t}_0} |y^\varepsilon(s)|^2 ds \leq \tilde{C}_{u_1}$$

for some constant $\tilde{C}_{u_1}$ independent of $0 \leq \bar{t}_0 \leq t_0$, $0 < \varepsilon \leq 1$ (but depending on γ, η). Now set

$$\tilde{B}_{u_1,\bar{t}_0} = \left\{ y \in L_2[0, \bar{t}_0] \; : \; \int_0^{\bar{t}_0} |y(s)|^2 ds \leq \tilde{C}_{u_1} \right\}.$$

Clearly

$$\sup_{y \in \tilde{B}_{u_1,\bar{t}_0}} \int_0^{\bar{t}_0} \mathcal{L}^{u_1,y} W(p_s) ds \geq \sup_{y \in L_2[0,\bar{t}_0]} \int_0^{\bar{t}_0} \mathcal{L}^{u_1,y} W(p_s) ds - \varepsilon.$$

Since this inequality holds for all $0 < \varepsilon \leq 1$, we have

$$\sup_{y \in \tilde{B}_{u_1,\bar{t}_0}} \int_0^{\bar{t}_0} \mathcal{L}^{u_1,y} W(p_s) ds = \sup_{y \in L_2[0,\bar{t}_0]} \int_0^{\bar{t}_0} \mathcal{L}^{u_1,y} W(p_s) ds.$$

Now let $B_{u_1,\bar{t}_0} = \tilde{B}_{u_1,\bar{t}_0} \cap C[0, \bar{t}_0]$, and in view of p-smooth property vi(b),

$$\sup_{y \in B_{u_1,\bar{t}_0}} \int_0^{\bar{t}_0} \mathcal{L}^{u_1,y} W(p_s) ds = \sup_{y \in L_2[0,\bar{t}_0]} \int_0^{\bar{t}_0} \mathcal{L}^{u_1,y} W(p_s) ds.$$

Any element of $B_{u_1,\bar{t}_0}$ can be extended to an element in $L_2[0, t_0] \cap C[0, t_0]$, with perhaps a slight increase in norm, say, no more than 1. Let B_{u_1} be the set of all such elements as $\bar{t}_0$ ranges over $[0, t_0]$, and let $C_{u_1} = \tilde{C}_{u_1} + 1$. This completes the proof of Claim 1a.

To prove Fact 1, (4.52) implies there exists $u_1 \in \mathbf{R}^m$ such that

$$\mathcal{L}^{u_1,y} W(p_0) \leq -\frac{3\theta}{4} \quad \text{for all } y \in \mathbf{R}^p.$$

By Claim 1a, there is a bounded set B_{u_1} such that

$$\sup_{y \in L_2[0,t]} \int_0^t \mathcal{L}^{u_1,y(s)} W(p_s) ds = \sup_{y \in B_{u_1}} \int_0^t \mathcal{L}^{u_1,y(s)} W(p_s) ds.$$

By the p-smooth continuity property vi(a), there exists $0 \leq t_1 \leq t_0$ such that

$$\mathcal{L}^{u_1,y(0)} W(p_0) \geq \mathcal{L}^{u_1,y(s)} W(p_s) - \theta/4$$

for all $0 \leq s \leq t_1$, $y \in B_{u_1}$. Therefore, for all $y \in B_{u_1}$, $0 \leq t \leq t_1$,

$$\int_0^t \mathcal{L}^{u_1,y(s)} W(p_s) ds \leq -\frac{\theta t}{2}.$$

Hence using Claim 1a,

$$\sup_{y\in L_2[0,t]} \int_0^t \mathcal{L}^{u_1,y(s)} W(p_s)ds = \sup_{y\in B_{u_1}} \int_0^t \mathcal{L}^{u_1,y(s)} W(p_s)ds$$

$$\leq -\frac{\theta t}{2}.$$

Since we can define an admissible controller K_1 by $K_1(y) = u_1$ for all $y \in L_{2,loc}$, this inequality completes the proof of Fact 1.

Proof of Fact 2. The proof of Fact 2 is similar to that of Fact 1, and the details are omitted. The proof begins with the observation that (4.54) implies there exists $y_2 \in \mathbf{R}^p$ such that

$$\mathcal{L}^{u,y_2} W(p_0) \geq \frac{3\theta}{4} \quad \text{for all } u \in \mathbf{R}^m.$$

This makes use of the Isaacs condition (4.50). □

REMARK 4.5.4 Since in general $\hat{W}$ need not equal the value function W, the minimal solution, the controller $K_{p_0}^{\mathbf{u}^*_{\hat{W}}}$ is not in general optimal (in the sense of minimizing the cost functional (3.19)). ▽

4.5.5 Dissipation

In this subsection we explain that the controller obtained from a good solution to the dissipation PDI (4.41) (or of the dynamic programming PDE (4.40)) achieves predissipation. This is a step toward a more detailed sufficiency result Theorem 4.10.3.

THEOREM 4.5.5 *Let $\hat{W}$ be a good solution of the dissipation PDI* (4.41), *and consider the controller $K_{p_0}^{\mathbf{u}^*_{\hat{W}}}$, where $p_0 \in \mathrm{dom}_{\mathrm{pd}}\hat{W}$. Let $x_0 \in \mathrm{support}\, p_0$. Then the closed-loop system $(G, K_{p_0}^{\mathbf{u}^*_{\hat{W}}})$ is* predissipative*:*

$$\begin{cases} \frac{1}{2}\int_0^T |z(s)|^2\, ds \leq \gamma^2 \frac{1}{2}\int_0^T |w(s)|^2\, ds - p(x_0) + \hat{W}(p) \\ \\ \textit{for all } w \in L_{2,T} \textit{ and all } 0 \leq T < \tau_{esc}(\mathrm{dom}_{\mathrm{pd}}\hat{W}). \end{cases} \tag{4.57}$$

Proof. The predissipation property is proven as follows (similar to part (vii) of Theorem 4.3.1). Let $w \in L_{2,loc}$ and $x_0 \in \mathrm{support}\, p_0$ be given. These determine signals $y(\cdot)$, $z(\cdot)$, and $u(\cdot) = K_{p_0}^{\mathbf{u}^*_{\hat{W}}}(y(\cdot))$ arising in the closed-loop system $(G, K_{p_0}^{\mathbf{u}^*_{\hat{W}}})$, and a corresponding plant state trajectory $x(\cdot)$; all these trajectories are defined on the time interval $0 \leq t < \tau_{esc}(\mathrm{dom}_{\mathrm{pd}}\hat{W})$.

Consider now the information state p_t driven by the signals with initial condition $p_0 \in \mathrm{dom}_{\mathrm{pd}}\hat{W}$. Since $\hat{W}$ is a smooth solution of the PDI (4.41), we have (recall (4.48))

$$\hat{W}(p_t) \leq \hat{W}(p_0)$$

for all $0 \leq t < \tau_{esc}(\text{dom}_{\text{pd}}\hat{W})$. Since $\hat{W}$ is dominating (as it is a good solution), this implies

$$p_t(x) \leq \langle p_t \rangle \leq \hat{W}(p_t) \leq \hat{W}(p_0) < +\infty$$

for all $x \in \mathbf{R}^n$, $0 \leq t < \tau_{esc}(\text{dom}_{\text{pd}}\hat{W})$. Set $x = x(t)$. By definition (3.7), we have

$$p_0(\xi(0)) + \tfrac{1}{2}\int_0^t [|\tilde{z}(s)|^2 - \gamma^2|\tilde{w}(s)|^2]\, ds \leq p_t(x) \leq \langle p_t \rangle \leq \hat{W}(p_t) \leq \hat{W}(p_0) < +\infty$$

for any $\tilde{w}$ with $y = C_2(\xi) + D_{21}(\xi)\tilde{w}$, and $\tilde{z} = C_1(\xi) + D_{12}(\xi)u$ with ξ solving the plant dynamics (3.8) with $\xi(t) = x$, provided $0 \leq t < \tau_{esc}(\text{dom}_{\text{pd}}\hat{W})$.

Setting $\tilde{w} = w$ implies $\xi(\cdot) = x(\cdot)$ and $\tilde{z}(\cdot) = z(\cdot)$; consequently,

$$p_0(x(0)) + \tfrac{1}{2}\int_0^t [|z(s)|^2 - \gamma^2|w(s)|^2]\, ds \leq \hat{W}(p_0) < +\infty,$$

for all $0 \leq t < \tau_{esc}(\text{dom}_{\text{pd}}\hat{W})$, which implies (4.57). □

4.6 Optimal Information State Controllers

As we have seen, synthesizing controllers from the dynamic programming PDE (4.40) involves substantial technical issues. Accordingly, we consider the possibility, in §4.6.1 to follow, of direct minimization of the cost functional (3.19). We do not prove that such direct optimization is possible, but rather we define the optimal information state controller in an appropriate manner. Study of this controller, when it exists, is simplified because the smoothness issues discussed at length above do not arise. We also show, under suitable hypotheses, that the controller obtained from the value function W when it is a good solution to the dynamic programming PDE is optimal, and in §4.7 it is shown, in essence (modulo smoothness issues), that the existence of this controller is necessary. The central controller is defined in §4.8.

4.6.1 Direct Minimization and Dynamic Programming

Consider now the information state controller $K = K^{\mathbf{u}}_{p_0}$ defined by (4.2) for the information system (3.9) with initial state p_0, and associated cost

$$\overline{J}_{p_0}(K^{\mathbf{u}}_{p_0}) = \sup_{T\geq 0}\ \sup_{y\in L_{2,T}} \{\langle p_T \rangle\ :\ \text{given } p_0\}. \tag{3.19$'$}$$

Let us minimize this cost over the class of admissible static information state feedback controllers $K^{\mathbf{u}}_p$:

$$\overline{W}(p) = \inf_{K^{\mathbf{u}}_p} \overline{J}_p(K^{\mathbf{u}}_p) = \inf_{\mathbf{u}} \overline{J}_p(K^{\mathbf{u}}_p) \tag{4.58}$$

to obtain $\overline{W}$, which we call the *information state value function*. Associated with $\overline{W}$ are optimizers $\bar{\mathbf{u}}^*_{p_0}(\cdot) : \mathcal{X}_e \to \mathbf{R}^m$:

$$\bar{\mathbf{u}}^*_{p_0} \in \operatorname*{argmin}_{\mathbf{u}} \overline{J}_{p_0}(K^{\mathbf{u}}_{p_0}). \tag{4.59}$$

We will assume that

$$\text{dom}\,\bar{\mathbf{u}}^*_{p_0} = \text{dom}\,\overline{W}.$$

We shall use these optimizers to define *optimal information state controllers* $K^*_{p_0} = K^{\bar{\mathbf{u}}^*_{p_0}}_{p_0}$. Of course this means the controller $K^*_{p_0}$ satisfies

$$K^*_{p_0}[y](t) = \bar{\mathbf{u}}^*_{p_0}(p_t[y]), \tag{4.60}$$

where p_0 is the initial information state and p_t is the information state at time $t \geq 0$ (if it exists) corresponding to any $y \in L_{2,loc}$ and information state feedback

$$u(t) = \bar{\mathbf{u}}^*_{p_0}(p_t),\ t \geq 0. \tag{4.61}$$

REMARK 4.6.1 Recall that in order for the optimal information state controller to be defined, the coupling condition $p_t \in \text{dom}\,\bar{\mathbf{u}}^*_{p_0}$ must be satisfied. ▽

The information state value function $\overline{W}$ enjoys properties analogous to the value function W described in Theorem 4.3.1.

THEOREM 4.6.2 *Assume that the γ-dissipative control problem for G is solved by an admissible information state feedback controller $K_0 = K^{\bar{\mathbf{u}}_0}_{p_0}$. Then the list of properties of W in Theorem 4.3.1 also hold for $\overline{W}$. In particular, the value function $\overline{W}(p)$ defined (and finite) on a set* $\text{dom}\,\overline{W} \subset \mathcal{X}_e$ *by (4.58) enjoys the following properties:*

(i) $\text{dom}\,\overline{W}$ *is nonempty.*

(ii) *Structural properties.*

(a) $\overline{W}$ *dominates* $\langle \cdot \rangle$: $\overline{W}(p) \geq \langle p \rangle$ *for all* $p \in \mathcal{X}_e$, *and in particular for all* $p \in \text{dom}\,W$.

(b) $\overline{W}$ *is monotone: if* p_1 *is in* $\text{dom}\,\overline{W}$, *and if* $p_1 \geq p_2$, *then* p_2 *is in* $\text{dom}\,\overline{W}$ *and*

$$\overline{W}(p_1) \geq \overline{W}(p_2).$$

(c) $\overline{W}$ *is additive homogeneous: for any constant* $c \in \mathbf{R}$, *we have*

$$\overline{W}(p \pm c) = \overline{W}(p) \pm c. \tag{4.62}$$

(iii) $-\mathcal{B}_{K_0} \subset \text{dom}\,\overline{W}$ *and* $\overline{W}(-\beta) = 0$ *for all* $\beta \in \mathcal{B}_{K_0}$.

(iv) *Fix* $p \in \mathcal{X}_e$ *and assume that* $\overline{J}_p(K_0)$ *is finite. Then*

$$\overline{J}_p(K_0) \geq \overline{W}(p_t) \text{ and } p_t \in \text{dom}\,\overline{W} \text{ for all } t \geq 0, y \in L_2[0,t], \quad \text{where } u = K_0(y), p_0 = p. \tag{4.63}$$

(v) *Fix $p \in \operatorname{dom} \overline{W}$. Then if K_ε is an ε-optimal (i.e., $\overline{J}_p(K_\varepsilon) \le \overline{W}(p) + \varepsilon$) admissible information state controller,*

$$\overline{W}(p) + \varepsilon \ge \overline{W}(p_t) \text{ and } p_t \in \operatorname{dom} \overline{W} \text{ for all } t \ge 0, y \in L_2[0,t], \quad \text{where } u = K_\varepsilon(y), p_0 = p. \tag{4.64}$$

Thus any almost-optimal controller results in an information state trajectory along which $\overline{W}(p_t)$ is almost decreasing.

(vi) *The dynamic programming principle holds: for any $t \ge 0$, $p \in \operatorname{dom} \overline{W}$,*

$$\overline{W}(p) = \inf_{\mathbf{u}} \sup_{y \in L_{2,t}} \{\overline{W}(p_t) : p_0 = p\}. \tag{4.65}$$

(vii) *If for $p \in \operatorname{dom} \overline{W}$ a preadmissible optimal information state controller $K_p^{\bar{\mathbf{u}}^*} \in \operatorname{argmin}_K \overline{J}_p(K)$ exists, then $\overline{W}(p_t)$ decreases along any optimal closed-loop trajectory determined by $w \in L_{2,loc}$, $x_0 \in \mathbf{R}^n$:*

$$\overline{W}(p) \ge \overline{W}(p_t), \text{ where } u(t) = K_p^{\bar{\mathbf{u}}^*}(y)(t), \; p_0 = p, \tag{4.66}$$

and $p_t \in \operatorname{dom} \overline{W}$ for all $0 \le t < \tau_{adm}(K_p^{\bar{\mathbf{u}}^*}, y)$.

Moreover, the closed-loop system $(G, K_p^{\bar{\mathbf{u}}^*})$ is pre-γ-dissipative for all $x_0 \in$ support p_0

$$\begin{cases} \frac{1}{2}\int_0^T |z(s)|^2\, ds \;\le\; \gamma^2 \frac{1}{2}\int_0^T |w(s)|^2\, ds - p_0(x_0) + W(p_0) \\ \text{for all } \; w \in L_{2,T} \text{ and all } 0 \le T < \tau_{adm}(K_p^{\bar{\mathbf{u}}^*}, y), \end{cases} \tag{4.67}$$

and if $(G, K_p^{\bar{\mathbf{u}}^*})$ is z-detectable, admissible ($\tau_{adm} = +\infty$), the closed-loop system is weakly internally stable provided $x_0 \in$ support p_0.

Proof. The proof is similar to the proof of Theorem 4.3.1 and is omitted. □

The definition of optimal information state controller given here is a direct, self-contained one and does not assume smoothness of any solution to the dynamic programming PDE. This is the reason we introduced it. However, the downside is that we do not give a prescription for finding it. The next theorem compares the two value functions W and $\overline{W}$ and states that when the value function W is smooth, an optimal information state controller can be obtained from it (via (4.43)).

THEOREM 4.6.3 *The value function W is less than or equal to the information state value function $\overline{W}$:*

$$W(p) \le \overline{W}(p) \text{ for all } p \in \mathcal{X}_e.$$

If

(i) *the value function* W *is a good solution to the dynamic programming PDE* (4.40), *and*

(ii) *if* $p_0 \in \text{dom}_{\text{pd}}W$ *and* $K_{p_0}^{\mathbf{u}^*}$ $(\mathbf{u}^* = \mathbf{u}_W^*)$ *is admissible with* $p_t \in \text{dom}_{\text{pd}}W$ *for all* $t \geq 0$, $y \in L_{2,loc}$,

then W *equals the information state value function* $\overline{W}$*:*

$$\overline{W}(p) = W(p) \;\; \textit{for } p \in \text{dom}_{\text{pd}}W,$$

and formula (4.43) *defines an optimal information state controller* $K_{p_0}^{\mathbf{u}^*}$. *Note that* $W(p) = \overline{J}_p(K_p^{\mathbf{u}^*})$ *for* $p \in \text{dom}_{\text{pd}}W$.

Proof. Since the class of information state feedback controllers is contained in a larger class of output feedback controllers, we have immediately $W \leq \overline{W}$.

Let us explain why the opposite inequality $\overline{W} \leq W$ holds when W is a good solution of (4.40). Let $\mathbf{u}^*(\cdot)$ be defined from the value function W via (4.43). Since W is a good solution, for any $y \in L_{2,loc}$ and $p_0 \in \text{dom}_{\text{pd}}W$,

$$W(p_t) \leq W(p_0).$$

This inequality holds for all $t \geq 0$ by hypothesis (ii). We next have

$$\langle p_t \rangle \leq W(p_t) \leq W(p_0),$$

and hence

$$\overline{W}(p_0) = \sup_{t \geq 0, y \in L_{2,t}} \{\langle p_t \rangle\} \leq W(p_t) \leq W(p_0).$$

□

REMARK 4.6.4 For discrete time systems *all optimal* H^∞ *controllers are optimal information state controllers.* This is because issues concerning the differentiability of the value function do not rise. See, e.g., [JB95]. ▽

REMARK 4.6.5 *Question:* When does $\overline{W}$ equal W? We do not know any examples where this fails. In other words, *all known optimal* H^∞ *controllers are optimal information state controllers.* ▽

We say that an optimal information state controller $K_{p_0}^{\bar{\mathbf{u}}^*}$ is *pre-*p_0*-regular* if it exists and enjoys the following properties:

(i) $K_{p_0}^{\bar{\mathbf{u}}^*}$ is defined from a function $\bar{\mathbf{u}}^*(\cdot) : \mathcal{X}_e \to \mathbf{R}^m$ defined for all $p \in \text{dom}\,\overline{W}$ which is independent of p_0.

(ii) $K_{p_0}^{\bar{\mathbf{u}}^*}$ is preadmissible.

If $K_{p_0}^{\bar{\mathbf{u}}^*}$ is pre-p_0-regular for all $p_0 \in q.i.\text{dom}\,\overline{W}$, we say that $\mathbf{u}^*$ is *preregular.* p_0-regular means $\tau_{adm} = +\infty$.

LEMMA 4.6.6 *For any constant* $c \in \mathbf{R}$,

$$\begin{aligned} \bar{\mathbf{u}}^*(p \pm c) &= \bar{\mathbf{u}}^*(p), \\ \bar{\mathbf{y}}^*(p \pm c) &= \bar{\mathbf{y}}^*(p). \end{aligned} \tag{4.68}$$

Proof. A constant added to a function affects the value of its minimum (maximum) but not the optimizer itself. Thus the lemma follows directly from (4.59), (4.58), and the additive homogeneity property (Theorem 4.6.2(ii)). □

4.7 Necessity of an Optimal Information State Solution

Our first necessity result was Theorem 4.3.1, which proved key results about the value function W and dynamic programming (assuming a controller K solving the dissipative control problem exists); in particular, part (vii) asserted that if the optimal output feedback controller $\tilde{K}_p^*$ is admissible (recall also $W(p) = \overline{J}_p(\tilde{K}_p^*)$), then the controller $\tilde{K}_p^*$ solved the H^∞ problem in the sense that the closed-loop system is dissipative and weakly internally stable. We now show that under additional assumptions the optimal output feedback controller $\tilde{K}_p^*$ can be taken to be the optimal information state controller $K_p^{\mathbf{u}^*}$ obtained from the dynamic programming PDE (4.40), and consequently *it is necessarily the case* (modulo the additional assumptions, which do not arise in discrete time) *that an optimal information state controller solves the H^∞ control problem.*

THEOREM 4.7.1 *Assume*

(i) *the γ-dissipative control problem for G is solved by an admissible controller K_0;*

(ii) *the value function W defined by* (4.4) *is p-smooth and the fundamental theorem of calculus applies (relative to the operator $\mathcal{L}^{u,y}$);*

(iii) *if $p_0 \in \mathrm{dom}_{\mathrm{pd}} W$ and $K_{p_0}^{\mathbf{u}^*}$ ($\mathbf{u}^* = \mathbf{u}_W^*$) is admissible with $p_t \in \mathrm{dom}_{\mathrm{pd}} W$ for all $t \geq 0$, $y \in L_{2,loc}$.*

Then for $p \in \mathrm{dom}_{\mathrm{pd}} W$, the controller $K_p^{\mathbf{u}^*}$ is optimal, $K_p^{\mathbf{u}^*} \in \mathrm{argmin}_K \overline{J}_p(K)$ (so that in the notation of Theorem 4.3.1 *we can take $\tilde{K}_p^* = K_p^{\mathbf{u}^*}$), and if $p_0(x_0)$ is finite, the closed-loop system $(G, K_p^{\mathbf{u}^*})$ is γ-dissipative:*

$$\begin{cases} \frac{1}{2}\int_0^T |z(s)|^2\,ds \leq \gamma^2 \frac{1}{2}\int_0^T |w(s)|^2\,ds - p(x_0) + W(p) \\ \textit{for all } w \in L_{2,T} \textit{ and all } T \geq 0, \end{cases}$$

and if $(G, K_p^{\mathbf{u}^*})$ is z-detectable, the closed-loop system is weakly internally stable.

Proof. Assumption (i) ensures that the value function W is well defined on a nonempty domain $\operatorname{dom} W$, and thanks to Theorem 4.3.1 W satisfies the dynamic programming principle (4.8). Next, assumption (ii) implies, by Theorem 4.5.3, that W is a good solution to the dynamic programming PDE (4.40), and this defines the information state controller $K_p^{\mathbf{u}^*}$, which is optimal by Theorem 4.6.3, using assumption (iii).

Now the fact that W is a good solution implies that $W(p_t)$ is decreasing along closed-loop trajectories (provided $p_0 = p \in \operatorname{dom}_{pd} W$), and the dissipation and stability conclusions follow as in Theorem 4.3.1. □

REMARK 4.7.2 If the controller $K_{p_0}^{\mathbf{u}^*}$ is preadmissible, then the closed-loop system is dissipative on the interval $0 \leq T < \tau_{adm}(K_{p_0}^{\mathbf{u}^*})$. ▽

4.8 Definition of Central Controller

In general, there are many controllers that solve the H^∞ problem (provided it is at all solvable). Controllers can be obtained from solutions of the dynamic programming PDE, from solutions of the dissipation PDI, or by direct minimization (as in §4.6). In linear systems, there is a special controller called the central controller. It is the controller constructed from stabilizing (hence minimal) solutions $X_e \geq 0$, $Y_e \geq 0$, of the DGKF Riccati equations. The analogous controller in the nonlinear context is the information state controller constructed from the value function W (minimal solution of the dissipation PDI) initialized at the control attractor p_e. In view of our discussion of optimal information state controllers, this leads to the following definition:

> A *central controller* is an optimal information state controller initialized at the control attractor p_e and is denoted $K_{p_e}^*$.

When the value function W is a good solution, the formulas (4.23) or (4.43) may be used to define the central controller, since then $W(p) = \overline{W}(p)$. In this case since the value function W is unique, the central controller is unique.

4.9 Initialization of Information State Controllers

Now that the necessity of information state controllers has been established, we turn to the sufficiency side, i.e., to the construction of the solution to the H^∞ control problem.

Initialization of information state controllers is a very important issue. Initialization affects both the stability of the information state and its coupling with the controller. Results relating to coupling are given in §4.9.1, where it is shown that correct initialization (together with other assumptions) ensures that coupling is achieved and the controller is admissible.

The concept of a null-initialized controller was defined in Chapter 2 to be a controller K that maps the zero measurement $y(\cdot) = 0$ to the zero control $u(\cdot) = 0$: $K(0) = 0$. In §4.9.2 we present conditions under which information state controllers are null initialized.

4.9.1 Coupling

As has been remarked, it is vital that the coupling conditions connecting information states to controllers be satisfied, since otherwise the information state controllers would not be well defined. In this section we present sufficient conditions that guarantee that the information state p_t remains inside the domain $\mathrm{dom}_{\mathrm{pd}}\hat{W}$ of a good solution $\hat{W}$ of the dynamic programming PDE (4.40). This type of coupling means that $p_t \in \mathrm{dom}_{\mathrm{pd}}\hat{W}$ for all $t \geq 0$. *The sufficient conditions are that the equilibrium p_e is quadratically tight and has the property of quadratic upper limiting introduced in Chapter 3, that p_e belongs to the quadratic interior of* $\mathrm{dom}\,\hat{W}$, *that p_0 be "quadratically close" to p_e, and a technical condition concerning the set* $\mathrm{dom}_{\mathrm{pd}}\hat{W}$. Before presenting this result, we examine some less complicated coupling questions and remark that in §4.11 it is shown that p_e in fact belongs to the quadratic interior of the domain of the value function W, and §11.4 provides results that establish the quadratic upper limiting property in some circumstances.

We recall from Chapter 3 that an equilibrium p_e is *quadratically upper limiting* for the information state system (3.9) if there exists a constant $\bar{\alpha}_0 > 0$ such that if $u, y \in L_{2,loc}$, $c_0 \geq 0$, and

$$p_0 \leq p_e(x) + \alpha_0|x|^2 + c_0 \tag{3.37}$$

for $0 \leq \alpha_0 \leq \bar{\alpha}_0$, we have for all $0 < \eta < 1$:

$$p_t(x) \leq p_e(x) + \frac{c_1(t)}{\eta} + (\eta + c_1(t)\alpha_0)k|x|^2 \text{ for all } t \geq 0, \tag{3.38}$$

where $k \geq 0$ and $c_1(t) = c_1(\| u \|_{L_{2,t}}, \| y \|_{L_{2,t}}) \geq 0$ are constants with $c_1(\infty)$ finite whenever $u, y \in L_2$.

The following lemma considers the information state system driven by open-loop signals u, y and provides conditions under which a coupling condition holds.

LEMMA 4.9.1 *Let $\hat{W} : \mathcal{X}_e \to \mathbf{R}\cup\{+\infty\}$ be a function that is monotone and additively homogeneous. Assume that p_e is quadratically upper limiting and $p_e \in q.i.\mathrm{dom}\,\hat{W}$. If p_0 satisfies $p_0 \leq p_e$, then for any $u, y \in L_{2,loc}$ there exists $\eta > 0$ such that*

$$\hat{W}(p_t + \eta|\cdot|^2) \leq k_1 + \frac{C(t)}{\eta} < +\infty \text{ for all } t \geq 0, \tag{4.69}$$

where $k_1 \geq 0$, $c_1(t) \geq 0$, and therefore

$$p_t \in q.i.\mathrm{dom}\,\hat{W} \text{ for all } t \geq 0. \tag{4.70}$$

More generally, there exists $\alpha_0 > 0$ such that if p_0 satisfies (3.37)*, then for any $u, y \in L_2$* (4.69) *and* (4.70) *hold with $c_1(t) \leq k_2$ for some $k_2 \geq 0$.*

Proof. Since p_e belongs to the quadratic interior of $\mathrm{dom}\,\hat{W}$ we have

$$p_e + \varepsilon|\cdot|^2 \in \mathrm{dom}\,\hat{W}$$

for some $\varepsilon > 0$. If $p_0 \leq p_e$, we take $\alpha_0 = 0$ and so (3.38) implies

$$p_t(x) \leq p_e(x) + \frac{c_1(t)}{\eta} + k\eta|x|^2 \text{ for all } t \geq 0$$

for any $0 < \eta < 1$. Here, $k \geq 0$ and $c_1(t) \geq 0$. Pick η so that $(k+1)\eta < \varepsilon$; then we get

$$p_t(x) + \eta|x|^2 \leq p_e(x) + \frac{c_1(t)}{\eta} + (k+1)\eta|x|^2 < p_e(x) + \frac{c_1(t)}{\eta} + \varepsilon|x|^2.$$

Then by the structural monotonicity and additive homogeneity properties

$$\hat{W}(p_t + \eta|\cdot|^2) \leq \hat{W}(p_e + \varepsilon|\cdot|^2) + \frac{c_1(t)}{\eta} < +\infty.$$

This proves (4.69) with $k_1 = \hat{W}(p_e + \varepsilon|\cdot|^2)$.

The proof for $\alpha_0 \neq 0$ is similar, with the key being to choose $\alpha_0 \neq 0$ small enough to make (recall (3.38))

$$(\eta + c_1(t)\alpha_0)k + \eta < \varepsilon.$$

This is always possible when $u, y \in L_2$ since $c_1(t)$ is bounded: $c_1(t) \leq k_2$ for all $t \geq 0$ for some $k_2 \geq 0$. □

We now analyze the coupling condition in the closed-loop system. The next theorem considers the optimal information state controller obtained by direct minimization.

Recall from Chapter 2 that the closed-loop system (G, K) is *L_2-observable* if

$$\int_0^t [|x(s)|^2 + |u(s)|^2 + |y(s)|^2]ds \leq C \int_0^t |z(s)|^2 ds \quad (2.28)$$

for all $t \geq 0$ for which the closed-loop signals are defined and all $w \in L_{2,loc}$.

THEOREM 4.9.2 *Assume the following conditions hold:*

(i) *The information state value function* $\overline{W}$ *defined by* (4.58) *has nonempty domain* dom $\overline{W}$.

(ii) *There exists an equilibrium* p_e *that is quadratically tight: for some* $c_{p_e} > 0$, $p_e \leq -c_{p_e}|\cdot|^2$.

(iii) $p_e \in q.i.\text{dom}\,\overline{W}$.

(iv) p_e *is quadratically upper limiting.*

(v) $p_0 \in \mathcal{X}_e$, $p_0 \leq p_e$ *and* $x_0 \in \text{support}\, p_0$.

(vi) $w \in L_2[0, \infty)$.

(vii) *The optimal information state controller* $K_{p_0}^{\bar{\mathbf{u}}^*}$ *is pre-p_0-regular for all* $p_0 \in q.i.\text{dom}\,\overline{W}$ *with* $p_0 \leq p_e$.

(viii) *The closed-loop system* $(G, K_{p_0}^{\bar{\mathbf{u}}^*})$ *is* L_2*-observable.*

Then the closed-loop system $(G, K_{p_0}^{\bar{\mathbf{u}}^*})$ *is well defined for all* $t \geq 0$ $(\tau_{adm}(K_{p_0}^{\bar{\mathbf{u}}^*}) = +\infty)$*,* p_t *satisfies the coupling condition*

$$p_t \in q.i.\mathrm{dom}\,\overline{W} \ \ \textit{for all}\ t \geq 0,$$

and $u, y \in L_2[0, \infty)$*.*

Proof. Assume that $\tau_{esc}(\mathrm{dom}\,\overline{W}) < +\infty$.
Since $p_e \in q.i.\mathrm{dom}\,\overline{W}$, there exists $\varepsilon > 0$ such that

$$p_e + \varepsilon|\cdot|^2 \in \mathrm{dom}\,\overline{W}.$$

Since $p_0 \in \mathrm{dom}\,\overline{W}$, we have the dissipation inequality for the closed-loop system

$$\tfrac{1}{2}\int_0^t |z(s)|^2 ds \leq \gamma^2 \tfrac{1}{2}\int_0^t |w(s)|^2 ds - p_0(x_0) + \overline{W}(p_0) \tag{4.71}$$

for $0 \leq t < \tau_{esc}(\mathrm{dom}\,\overline{W})$, and by the quadratic upper limiting hypothesis,

$$p_t + \eta|\cdot|^2 \leq p_e + (k+1)\eta|\cdot|^2 + \frac{c_1(t)}{\eta} \tag{4.72}$$

(for all $0 < \eta \leq 1$) for $0 \leq t < \tau_{esc}(\mathrm{dom}\,\overline{W})$. Here, $k \geq 0$ and $c_1(t) \geq 0$.

Since $p_0(x_0)$ and $\overline{W}(p_0)$ are finite, and $w \in L_2$, it follows by combining (4.71) and (2.28) that

$$\int_0^t [|x(s)|^2 + |u(s)|^2 + |y(s)|^2] ds \leq C\left(\gamma^2 \tfrac{1}{2}\int_0^t |w(s)|^2 ds - p_0(x_0) + \overline{W}(p_0)\right)$$

for all $0 \leq t < \tau_{esc}(\mathrm{dom}\,\overline{W})$. Further, we have

$$\begin{aligned} &\int_0^{\tau_{esc}(\mathrm{dom}\,\overline{W})} [|x(s)|^2 + |u(s)|^2 + |y(s)|^2] ds \\ &\quad \leq C\left(\gamma^2 \tfrac{1}{2}\int_0^{\infty} |w(s)|^2 ds - p_0(x_0) + \overline{W}(p_0)\right), \end{aligned} \tag{4.73}$$

so that $u, y \in L_2[0, \tau_{esc}(\mathrm{dom}\,\overline{W})]$. Hence the constant $c_1(t)$ in (4.72) is bounded above uniformly in t (for some $k_2 \geq 0$):

$$c_1(t) \leq k_2 \ \ \text{for all} \ \ 0 \leq t < \tau_{esc}(\mathrm{dom}\,\overline{W}). \tag{4.74}$$

Inequality (4.74) implies that $p_t \in q.i.\mathrm{dom}\,\overline{W}$ for all $0 \leq t < \tau_{esc}(\mathrm{dom}\,\overline{W})$ and is uniformly quadratically tight; in fact,

$$\begin{aligned} p_t + \eta|\cdot|^2 &\leq p_e + (k+1)\eta|\cdot|^2 + \frac{k_2}{\eta} \\ &\leq -\frac{c_{p_e}}{2}|\cdot|^2 + \frac{k_2}{\eta} \end{aligned} \tag{4.75}$$

for $0 \leq t < \tau_{esc}(\mathrm{dom}\,\overline{W})$, where η is chosen so that $(k+1)\eta < \min(c_{p_e}/2, \varepsilon)$.

Since $u, y \in L_2[0, \tau_{esc}(\mathrm{dom}\,\overline{W})]$ and (4.75) holds, it now follows from the definition of the information state (3.12) that p_t is well defined for all $0 \leq t \leq \tau_{esc}(\mathrm{dom}\,\overline{W})$. Indeed $\bar{p} = p_{\tau_{esc}(\mathrm{dom}\,\overline{W})}$ satisfies

$$\bar{p} + \eta|\cdot|^2 \leq p_e + \varepsilon|\cdot|^2 + \frac{k_2}{\eta}. \tag{4.76}$$

Since $\overline{W}$ is monotone and additive homogeneous, we have

$$\overline{W}(\bar{p} + \eta|\cdot|^2) \leq \overline{W}(p_e + \varepsilon|\cdot|^2) + \frac{k_2}{\eta}. \tag{4.77}$$

Therefore, $\bar{p} \in q.i.\mathrm{dom}\,\overline{W}$.

Now the controller $K_{\bar{p}}^{\bar{\mathbf{u}}^*}$ is pre-$\bar{p}$-regular, and so we can restart the controller at time $\tau_{esc}(\mathrm{dom}\,\overline{W})$ and run it for at least a short time $\bar{\tau} > 0$. So the controller $K_{p_0}^{\bar{\mathbf{u}}^*}$ is well defined for $0 \leq t \leq \tau_{esc}(\mathrm{dom}\,\overline{W}) + \bar{\tau}$, and $p_t \in q.i.\mathrm{dom}\,\overline{W}$; this contradicts the definition of $\tau_{esc}(\mathrm{dom}\,\overline{W})$ if finite.

The remaining assertions now follow easily. □

The case of the closed-loop corresponding to a controller obtained from a good solution $\hat{W}$ of the dynamic programming PDE (4.40) is more complicated because of issues of smoothness. An assumption is needed to guarantee smooth-PDE coupling.

THEOREM 4.9.3 *Assume the following conditions hold:*

(i) $\hat{W}$ *is a good solution to the dissipation PDI* (4.41).

(ii) *There exists an equilibrium* p_e *that is quadratically tight: for some* $c_{p_e} > 0$, $p_e \leq -c_{p_e}|\cdot|^2$.

(iii) $p_e \in q.i.\mathrm{dom}\,\hat{W}$.

(iv) p_e *is quadratically upper limiting.*

(v) $p_0 \in \mathcal{X}_e$, $p_0 \leq p_e$ *and* $x_0 \in \mathrm{support}\, p_0$.

(vi) $w \in L_2[0, \infty)$.

(vii) *Every limit point of* $\{p_t,\ 0 \leq t < \tau_{esc}(\mathrm{dom}_{\,pde}\hat{W}, K_{p_0}^{\mathbf{u}^*_{\hat{W}}})\}$ *belongs to* $\mathrm{dom}_{\mathrm{pd}}\hat{W}$.

(viii) *The closed-loop system* $(G, K_{p_0}^{\mathbf{u}^*_{\hat{W}}})$ *is* L_2*-observable.*

Then for the closed-loop system $(G, K_{p_0}^{\mathbf{u}^*_{\hat{W}}})$, $\tau_{esc}(\mathrm{dom}_{\mathrm{pd}}\hat{W}, K_{p_0}^{\mathbf{u}^*_{\hat{W}}}) = +\infty$, p_t *satisfies the coupling condition*

$$p_t \in \mathrm{dom}_{\mathrm{pd}}\hat{W} \quad \textit{for all } t \geq 0,$$

and $u, y \in L_2[0, \infty)$; *so* $K_{p_0}^{\mathbf{u}^*_{\hat{W}}}$ *is admissible for all* $p_0 \leq p_e \in \mathrm{dom}_{\mathrm{pd}}\hat{W}$.

Proof. The proof is similar to that of Theorem 4.9.2 and the details are omitted. The key difference is that without hypothesis (vii) we have no way of knowing if limit points $\bar{p} = p_{\tau_{sec}(\mathrm{dom}\,\hat{W})}$ belong to $q.i.\mathrm{dom}_{\mathrm{pd}}\hat{W}$, since it is not enough that $\bar{p} \in q.i.\mathrm{dom}\,\hat{W}$. □

REMARK 4.9.4 A sufficient condition (strong) implying hypothesis (vii) is the following: there exists $\varepsilon_{pde} > 0$ such that

$$\{p \in \mathcal{X}_e \;:\; p \leq p_e + \varepsilon_{pde}|\cdot|^2\} \subset \mathrm{dom}_{\mathrm{pd}}\hat{W}. \tag{4.78}$$

∇

COROLLARY 4.9.5 *If the value function* W *(defined by* (4.4)*) is a good solution of the dynamic programming PDE* (4.40)*, then the results of Theorem* 4.9.3 *apply with* W *replacing* $\hat{W}$.

The preceding three results have assumed that $p_0(x_0)$ is finite. This assumption was needed to obtain a finite upper bound in the proofs. This assumption can be removed if we assume that the plant is incrementally hyperbolic.

THEOREM 4.9.6 *Assume the following conditions hold:*

(i) *The information state value function* $\overline{W}$ *defined by* (4.58) *has nonempty domain* $\mathrm{dom}\,\overline{W}$.

(ii) *There exists an equilibrium* p_e *that is quadratically tight: for some* $c_{p_e} > 0$, $p_e \leq -c_{p_e}|\cdot|^2$.

(iii) $p_e \in q.i.\mathrm{dom}\,\overline{W}$.

(iv) p_e *is quadratically upper limiting.*

(v) $p_0 \in \mathcal{X}_e$, $p_0 \leq p_e$, and $x_0 \in \mathbf{R}^n$.

(vi) $w \in L_2[0,\infty)$.

(vii) *The optimal information state controller* $K_{p_0}^{\bar{\mathbf{u}}^*}$ *is pre-*p_0*-regular for all* $p_0 \in q.i.\mathrm{dom}\,\overline{W}$ *with* $p_0 \leq p_e$.

(viii) *The closed-loop system* $(G, K_{p_0}^{\bar{\mathbf{u}}^*})$ *is* L_2*-observable.*

(ix) *The system* $(A^\times, [B_1 D_{21}' E_2^{-1} \;:\; B_2 \;:\; \tilde{B}_1])$ *is incrementally hyperbolic with* $M_{as} \subset \mathrm{support}\, p_0$.

(x) *Assumption* 4.1.3 *holds.*

Then the closed-loop system $(G, K_{p_0}^{\bar{\mathbf{u}}^*})$ *is well defined for all* $t \geq 0$ *(*$\tau_{adm}(K_{p_0}^{\bar{\mathbf{u}}^*}) = +\infty$*),* p_t *satisfies the coupling condition*

$$p_t \in q.i.\mathrm{dom}\,\overline{W} \;\; \textit{for all}\; t \geq 0,$$

and $u, y \in L_2[0,\infty)$. *That is, the controller* $K_{p_0}^{\bar{\mathbf{u}}^*}$ *is* p_0*-regular for all* $p_0 \in q.i.\mathrm{dom}\,\overline{W}$.

Proof. In view of Theorem 4.9.2, we need only consider $x_0 \notin \operatorname{support} p_0$. Initialize at (x_0, p_0) and input $w \in L_2[0, \infty)$ and denote by $x(\cdot)$, p_t, $u(\cdot)$, $y(\cdot)$, and $z(\cdot)$ the resulting closed-loop trajectories and signals, all defined on the time interval $[0, \tau)$, where $\tau = \tau_{esc}(\operatorname{dom} \overline{W})$. Define $v \in L_{2,loc}[0, \tau)$ by

$$w(t) = [I - D_{21}' E_2^{-1} D_{21}] v(t) + D_{21}' E_2^{-1} (y - C_2(x(t)));$$

i.e., in terms of the left inverse,

$$v(t) = [I - D_{21}' E_2^{-1} D_{21}]^{-L} (w(t) - D_{21}' E_2^{-1} (y - C_2(x(t)))).$$

Let ξ_0 denote a tracking point $\xi_0(x_0)$, with trajectory $\xi(\cdot)$ determined by the signals $u, y, v \in L_{2,loc}[0, \tau)$ via

$$\dot{\xi} = A^{\times}(\xi) + B_1(\xi) D_{21}(\xi)' E_2^{-1}(\xi) y + B_2(\xi) u + \tilde{B}_1(\xi) v. \qquad (2.6)'$$

Now define

$$\begin{aligned} \tilde{w}(t) &= [I - D_{21}' E_2^{-1} D_{21}] v(t) + D_{21}' E_2^{-1} (y - C_2(\xi(t))), \\ \tilde{z}(t) &= C_1(\xi(t)) + D_{12} u(t), \\ \tilde{y}(t) &= C_1(\xi(t)) + D_{21} \tilde{w}(t) = y(t), \\ \tilde{u}(t) &= K_{p_0}^{\bar{\mathbf{u}}^*}(\tilde{y})(t) = u(t). \end{aligned}$$

These signals are well defined on $[0, \tau)$, and $\tilde{z}$, $\tilde{y} = y$ are the outputs of the plant (2.2) when driven by input $\tilde{w}$ with control $\tilde{u} = u$ and initial state ξ_0.

We would now like to follow the method of proving Theorem 4.9.2, but applied to the tilde signals, since we have the predissipation inequality

$$\tfrac{1}{2} \int_0^t |\tilde{z}(s)|^2 ds \le \tfrac{1}{2} \gamma^2 \int_0^t |\tilde{w}(s)|^2 ds - p_0(\xi_0) + \overline{W}(p_0) < +\infty$$

for all $0 \le t < \tau$, since $p_0(\xi_0)$ is finite. In order to do so, we need to establish that

$$\int_0^\tau |\tilde{w}(s)|^2 ds \le C < +\infty, \qquad (4.79)$$

for some $C \ge 0$. In fact, it follows from the proof of Theorem 4.10.1 below that

$$\int_0^t |\tilde{w}(s)|^2 ds \le \frac{1+\eta}{1-\eta} \gamma^2 \int_0^t |w(s)|^2 ds + C \left(-p_0(\xi_0) + \overline{W}(p_0) + \int_0^t |x(s) - \xi(s)|^2 ds \right)$$

for some $C \ge 0$, and for all $0 < \eta < 1$, $0 \le t < \tau$. By incremental hyperbolicity (2.26), we have $|x(s) - \xi(s)|^2 \le b e^{-ct}$ $(b \ge 0, c > 0)$, for $0 \le t < \tau$, which implies

$$\int_0^\tau |x(s) - \xi(s)|^2 ds \le b(1 - e^{-c\tau}) < +\infty$$

for some $b \ge 0$. This proves (4.79).

By L_2-observability (2.28), and predissipation (above), we have

$$\frac{1}{2}\int_0^\tau [|\xi(s)|^2 + |\tilde{u}(s)|^2 + |\tilde{y}(s)|^2]ds$$

$$\leq C\left(\gamma^2 \frac{1}{2}\int_0^\tau |\tilde{w}(s)|^2 ds - p_0(x_0) + \overline{W}(p_0)\right) < +\infty.$$

This implies $\tilde{u} = u, \tilde{y} = y \in L_2[0,\tau]$, and from above we know $\tilde{w} \in L_2[0,\tau]$.

Extend $\tilde{w}$ to a function in $L_{2,loc}[0,\infty)$ by setting $\tilde{w}(t) = 0$ (say) for $t \geq \tau$. Now input this to the tilde closed-loop system and consider the signals $\tilde{u} = K_{p_0}^{\bar{\mathbf{u}}^*}(\tilde{y}), \tilde{y}, \tilde{z}$. The escape time for this system is also equal to τ. Since $p_0(\xi_0)$ is finite, we can apply the proof of Theorem 4.9.2 to conclude that $\tau_{esc} = +\infty$. This completes the proof. □

THEOREM 4.9.7 *Assume the following conditions hold:*

(i) $\hat{W}$ *is a good solution to the dissipation PDI* (4.41).

(ii) *There exists an equilibrium* p_e *that is quadratically tight: for some* $c_{p_e} > 0$, $p_e \leq -c_{p_e}|\cdot|^2$.

(iii) $p_e \in q.i.\text{dom}\,\hat{W}$.

(iv) p_e *is quadratically upper limiting.*

(v) $p_0 \in \mathcal{X}_e$, $p_0 \leq p_e$ *and* $x_0 \in \mathbf{R}^n$.

(vi) $w \in L_2[0,\infty)$.

(vii) *Every limit point of* $\{p_t,\ 0 \leq t < \tau_{esc}(\text{dom}_{pde}\hat{W}, K_{p_0}^{\mathbf{u}^*_{\hat{W}}})\}$ *belongs to* $\text{dom}_{\text{pd}}\hat{W}$.

(viii) *The closed-loop system* $(G, K_{p_0}^{\mathbf{u}^*_{\hat{W}}})$ *is* L_2*-observable.*

(ix) *The system* $(A^\times, [B_1 D_{21}' E_2^{-1} : B_2 : \tilde{B}_1])$ *is incrementally hyperbolic.*

(x) *Assumption* 4.1.3 *holds.*

Then for the closed-loop system $(G, K_{p_0}^{\mathbf{u}^*_{\hat{W}}})$, $\tau_{esc}(\text{dom}_{pde}\hat{W}, K_{p_0}^{\mathbf{u}^*_{\hat{W}}}) = +\infty$, p_t *satisfies the coupling condition*

$$p_t \in \text{dom}_{\text{pd}}\hat{W} \ \textit{for all}\ t \geq 0,$$

and $u, y \in L_2[0,\infty)$; *so* $K_{p_0}^{\mathbf{u}^*_{\hat{W}}}$ *is admissible for all* $p_0 \leq p_e \in \text{dom}_{\text{pd}}\hat{W}$.

Proof. The proof is similar to that of Theorem 4.9.6 and is omitted. □

COROLLARY 4.9.8 *If the value function* W *(defined by* (4.4)) *is a good solution to the dynamic programming PDE* (4.40), *then the results of Theorem* 4.9.7 *apply with* W *replacing* $\hat{W}$.

4.9.2 Null Initialization

To prove null initialization we will need to assume a simple condition connecting the information state with the cost function.

THEOREM 4.9.9 *Assume that the information state controller $K^{\mathbf{u}}_{p_0}$ is p_0-regular. Then*

$$\overline{J}_{p_0}(K^{\mathbf{u}}_{p_0}) = p_0(0) \tag{4.80}$$

implies that $K^{\mathbf{u}}_{p_0}$ is null initialized: $K^{\mathbf{u}}_{p_0}[0] = 0$.

Proof. Consider the closed-loop system $(G, K^{\mathbf{u}}_{p_0})$ and set $x_0 = 0$, $w(\cdot) = 0$. By Theorem 3.1.8, we get

$$J_{p_0}(K^{\mathbf{u}}_{p_0}) = \overline{J}_{p_0}(K^{\mathbf{u}}_{p_0}),$$

and by the definition of J we have

$$p_0(0) + \tfrac{1}{2}\int_0^t |z(s)|^2 ds \leq \overline{J}_{p_0}(K^{\mathbf{u}}_{p_0})$$

for all $t \geq 0$ (since $w = 0$). This says the hypothesis (4.80) implies $z(\cdot) = 0$. Consequently, using (2.2), we have $u(\cdot) = -E_1^{-1}D_{12}'C_1(x(\cdot))$, and so the plant state equation becomes

$$\dot{x} = A(x) - E_1^{-1}D_{12}'C_1(x),$$

with initial condition $x_0 = 0$. However, by uniqueness of solutions to ODEs, $x(\cdot) = 0$ ($x = 0$ is an equilibrium since $A(0) = 0$, $C_1(0) = 0$). Then $y(\cdot) = C_2(x(\cdot)) + D_{21}(x(\cdot))w(\cdot) = D_{21}(0)w(\cdot) = 0$, which implies $w(\cdot) = 0$. Therefore, we have that $0 = K^{\mathbf{u}}_{p_0}[0]$. □

The next corollary establishes null initialization for the information state controllers we have constructed.

COROLLARY 4.9.10 (i) *If the optimal information state controller $K^{\bar{\mathbf{u}}^*}_{p_0}$ is p_0-regular, $p_0 \in \mathrm{dom}\,\overline{W}$, and $p_0(0) = \overline{W}(p_0)$, then $K^{\bar{\mathbf{u}}^*}_{p_0}[0] = 0$.*

(ii) *If the optimal information state controller $K^{\mathbf{u}^*}_{p_0}$ is admissible, $p_0 \in \mathrm{dom}_{\mathrm{pd}} W$, and $p_0(0) = W(p_0)$, then $K^{\mathbf{u}^*}_{p_0}[0] = 0$.*

(iii) *If the information state controller $K^{\mathbf{u}^*_{\hat{W}}}_{p_0}$ is admissible, $p_0 \in \mathrm{dom}_{\mathrm{pd}} \hat{W}$, and $p_0(0) = \hat{W}(p_0)$, then $K^{\mathbf{u}^*_{\hat{W}}}_{p_0}[0] = 0$.*

REMARK 4.9.11 The control attractor p_e vanishes at the origin, and it will be shown in Theorem 4.11.1 below that if $p_e(0) = 0$ and $p_e(x) \leq 0$, then $W(p_e) = 0$. It then follows from Corollary 4.9.10 that the central controller $K^{\mathbf{u}}_{p_e}$ is null initialized. ∇

4.10 Solution of the H^∞ Control Problem

In this section we tie together the results obtained above and present the solution to the H^∞ control problem. We first state our main necessity result, which shows that if the H^∞ control problem for a plant G is solvable by some controller K, then (modulo technical issues) it is also solvable by the controller $K^{\mathbf{u}^*}_{p_0}$ obtained from the value function W (assumed to be a good solution of the dynamic programming PDE (4.40)), and in particular by the central controller $K^{\mathbf{u}^*}_{p_e}$. The second theorem can be regarded as a sufficiency result concerning controllers $K^{\mathbf{u}^*_{\hat{W}}}_{p_0}$ obtained from good solutions $\hat{W}$ of the dynamic programming PDE (4.40), asserting that such controllers solve the H^∞ control problem. Finally, we show that the controller $K^{\bar{\mathbf{u}}^*}_{p_0}$ obtained by directly minimizing the cost functional (3.19), when it exists, also solves the H^∞ control problem.

Our main necessity result is the following.

THEOREM 4.10.1 *Assume the following conditions hold:*

(i) *The γ-dissipative control problem for G is solved by an admissible controller K_0.*

(ii) *The value function W (defined by (4.4)) is a p-smooth solution of the dynamic programming PDE (4.40), and the fundamental theorem of calculus holds relative to the operator $\mathcal{L}^{u,y}$.*

(iii) *$p_0 \in \mathrm{dom}_{\mathrm{smooth}} W$ satisfies $p_0 \leq p_e$, where*

  (a) *there exists an equilibrium p_e that is quadratically tight: for some $c_{p_e} > 0$, $p_e \leq -c_{p_e}|\cdot|^2$;*

  (b) *p_e is quadratically upper limiting;*

  (c) *$p_e \in \mathrm{dom}_{\mathrm{smooth}} W$.*

(iv) *Every limit point of $\{p_t,\ 0 \leq t < \tau_{esc}(\mathrm{dom}_{\mathrm{smooth}} W, K^{\mathbf{u}^*}_{p_0})\}$ belongs to $\mathrm{dom}_{\mathrm{smooth}} W$.*

(v) *The closed-loop system $(G, K^{\mathbf{u}^*}_{p_0})$ is L_2-observable.*

(vi) *The system $(A^\times, [B_1 D_{21}' E_2^{-1} \ : \ B_2 \ : \ \tilde{B}_1])$ is incrementally hyperbolic with $M_{as} \subset \mathrm{support}\, p_0$.*

(vii) *Assumption 4.1.3 holds.*

Then we have the following:

(i) *The value function W exists on a nonempty domain $\mathrm{dom}\, W$ and W is a good solution of the dynamic programming PDE (4.40) with $\mathrm{dom}_{\mathrm{pd}} W = \mathrm{dom}_{\mathrm{smooth}} W$.*

(ii) *The coupling condition holds: $p_t \in \mathrm{dom}_{\mathrm{smooth}} W$ for all $t \geq 0$, and the controller $K^{\mathbf{u}^*}_{p_0}$ is admissible.*

(iii) *The closed-loop system* $(G, K_{p_0}^{\mathbf{u}^*})$ *is* $\tilde{\gamma}$*-dissipative for all* $\tilde{\gamma} > \gamma$*:*

$$\begin{cases} \frac{1}{2}\int_0^T |z(s)|^2\, ds \;\le\; \tilde{\gamma}^2 \frac{1}{2}\int_0^T |w(s)|^2\, ds + \beta(x_0) \\ \\ \text{for all } w \in L_{2,T} \text{ and all } T \ge 0. \end{cases} \tag{4.81}$$

(iv) *If also* $p_0 \in -\mathcal{B}$ *and* $W(p_0) = 0$, (4.81) *is the dissipation inequality* (2.10) for some $\beta = -\mathcal{B}$, *and* $K_{p_0}^{\mathbf{u}^*}$ *is null initialized.*

(v) *If* $x_0 \in \operatorname{support} p_0$*, then we have* $\tilde{\gamma} = \gamma$ *in the dissipation inequality* (4.81), *and we can take* $\beta = -p_0$.

(vi) *If* $(G, K_{p_0}^{\mathbf{u}^*})$ *is z-detectable, then the closed-loop system is weakly internally stable.*

(vii) *If* $(G, K_{p_0}^{\mathbf{u}^*})$ *is z-detectable, and if* p_e *is a control attractor,* $p_0 \in \mathcal{D}_{attr}(p_e)$*, then the closed-loop system is internally stable, and in particular*

$$x(t) \to 0 \;\; and \;\; p_t \Longrightarrow p_e + c \;\; as\; t \to \infty.$$

Proof. Hypothesis (i) implies, thanks to Theorem 4.3.1, that dom W is nonempty and that W satisfies the structural properties (domination, monotonicity, and additive homogeneity). Then hypothesis (ii) and Theorem 4.5.3 imply that W is a good solution of the dynamic programming PDE (4.40), with $\mathrm{dom}_{\mathrm{pd}} W = \mathrm{dom}_{\mathrm{smooth}} W \subset \mathrm{dom}\, W$.

Next, since $p_0 \in q.i.\mathrm{dom}\, W$, Theorem 4.5.5 implies that the closed-loop system $(G, K_{p_0}^{\mathbf{u}^*})$ is pre-γ-dissipative; i.e., (4.57) holds (setting $\hat{W} = W$), provided $p_0(x_0)$ is finite. By Theorem 4.9.6, we see that the coupling condition holds for all $t \ge 0$ and that the controller $K_{p_0}^{\mathbf{u}^*}$ is admissible, for any $x_0 \in \mathbf{R}^n$. If $p_0(x_0)$ is finite, we see that assertion (v) holds. We will prove assertion (iii), that $(G, K_{p_0}^{\mathbf{u}^*})$ is $\tilde{\gamma}$-dissipative, i.e., (4.81) holds below. Assuming this, we verify the remaining assertions (iv), (vi), (vii).

Assertion (iv) follows from Corollary 4.9.10, and assertion (vi) is a consequence of Theorem 2.1.3. The hypotheses require that $p_0 \in \mathcal{D}_{attr}(p_e)$, the domain of control attraction. By weak internal stability, we have $u, y \in L_2[0,\infty)$, and so assertion (vii) follows.

Proof of assertion (iii). The technicalities arise when M_{cas} is a strict subset of $\mathbf{R}^n$, with p_0 singular, as described above. We now treat this case. We initialize at an arbitrary $x_0 \in \mathbf{R}^n$ (note that $p_0(x_0)$ need not be finite). Before giving details we sketch the idea. The incremental hyperbolicity assumption tells us that there is ξ_0 in M_{cas} so that the trajectory $x(t)$ originating at x_0 rapidly gets close to the corresponding trajectory $\xi(t)$ originating at ξ_0. Inequality (4.81) above says that dissipation holds along $\xi(t)$, while rapid convergence guarantees the difference in energy and output "from $x(t)$ and $\xi(t)$" is finite. Thus a finite energy input signal w produces a finite energy signal out. Indeed the more refined argument, which we now give, implies dissipation.

Now we turn to the details. Initialize at (x_0, p_0) and let ξ_0 denote a tracking point $\xi_0(x_0)$. Input $w \in L_2[0,\infty)$ and denote $x(\cdot)$, p_t, $u(\cdot)$, $y(\cdot)$, and $z(\cdot)$ the resulting closed-loop trajectories and signals, all defined on the time interval $[0, \infty)$. Define $v \in L_{2,loc}[0, \infty)$ by

$$w = [I - D'_{21}E_2^{-1}D_{21}]v + D'_{21}E_2^{-1}(y - C_2(x))$$

i.e., in terms of the left inverse,

$$v = [I - D'_{21}E_2^{-1}D_{21}]^{-L}(w - D'_{21}E_2^{-1}(y - C_2(x))).$$

Let ξ_0 denote a tracking point $\xi_0(x_0)$, with trajectory $\xi(\cdot)$ determined by the signals $u, y, v \in L_{2,loc}[0, \infty)$ via (2.6). Now define

$$\begin{aligned}
\tilde{w} &= [I - D'_{21}E_2^{-1}D_{21}]v + D'_{21}E_2^{-1}(y - C_2(\xi)),\\
\tilde{z} &= C_1(\xi) + D_{12}u,\\
\tilde{y} &= C_1(\xi) + D_{21}\tilde{w} = y,\\
\tilde{u} &= K_{p_0}^{\mathbf{u}^*}(\tilde{y}) = u.
\end{aligned}$$

These signals are well defined on $[0, \infty)$, and $\tilde{z}$, $\tilde{y} = y$ are the outputs of the plant (2.2) when driven by input $\tilde{w}$ with control $\tilde{u} = u$ and initial state ξ_0.

Our next goal is to show that z is close to $\tilde{z}$, and that w is close to $\tilde{w}$.

We wish to bound

$$\begin{aligned}
\int_0^t |z(s)|^2 ds &= \int_0^t |C_1(x(s)) + D_{12}u(s)|^2 ds\\
&= \int_0^t |C_1(\xi(s)) + D_{12}u(s)|^2 ds \qquad (4.82)\\
&\quad + \int_0^t |C_1(x(s)) + D_{12}u(s)|^2 - |C_1(\xi(s)) + D_{12}u(s)|^2 ds.
\end{aligned}$$

Inequality (4.81) above tells us we can dominate the first term of (4.82), which is the energy of the output $\tilde{z}(s) = C_1(\xi(s)) + D_{12}u(s)$ of the ξ system by

$$\int_0^t |\tilde{z}(s)|^2 ds \leq \gamma^2 \int_0^t |\tilde{w}(s)|^2 ds - 2p_0(\xi_0) + 2W(p_0)$$

(we can take $\tilde{\gamma} = \gamma$ for $p_0(\xi_0)$ finite). The second term of (4.82) can be written as a sum

$$\int_0^t \tilde{z}(s)[C_1(x(s)) - C_1(\xi(s))]ds + \int_0^t z(s)[C_1(x(s)) - C_1(\xi(s))]ds,$$

the first term of which, because of the above energy bound for $\tilde{z}$ and Schwarz's inequality, is dominated by

$$\leq \left\{ \left[\gamma^2 \int_0^t |\tilde{w}(s)|^2 ds - 2p_0(\xi_0) + 2W(p_0)\right]^{1/2} + \left[\int_0^t |z(s)|^2\right]^{1/2} \right\} k(x_0, \xi_0)^{1/2},$$

where $k(x_0, \xi_0, t) = \int_0^t |C_1(x(s)) - C_1(\xi(s))|^2 ds$ is implied bounded in $t \geq 0$ by the incremental hyperbolic assumption. The age-old geometric versus arithmetic mean inequality $ab \leq \eta a^2 + \frac{1}{4\eta} b^2$ yields the estimate

$$\leq \eta \int_0^t |z(s)|^2 ds + \eta\gamma^2 \int_0^t |\tilde{w}(s)|^2 ds - 2\eta p(\xi_0) + 2W(p_0) + \frac{2}{4\eta} K(x_0, \xi_0)$$

for any $\eta > 0$ for the second term. Put these estimates together to get

$$\int_0^t |z(s)|^2 ds \leq \frac{1+\eta}{1-\eta}\gamma^2 \int_0^t |\tilde{w}(s)|^2 ds + k_\eta(x_0, \xi_0(x_0))$$

for any $0 < \eta < 1$ and some k_η.

A similar calculation gives the estimate

$$\int_0^t |\tilde{w}(s)|^2 ds \leq \frac{1+\eta}{1-\eta} \int_0^t |w(s)|^2 ds + \tilde{k}_\eta(x_0, \xi_0(x_0))$$

for any $0 < \eta < 1$ and some $\tilde{k}_\eta$.

These inequalities with $\tilde{\gamma}^2 = \left(\frac{1+\eta}{1-\eta}\right)^2 \gamma^2$ are the $\tilde{\gamma}$-dissipation inequality (4.81) as required to prove the theorem. If $W(p_0) = p_0(0)$, we set

$$\beta(x_0) = k_\eta(x_0, \xi_0(x_0)) + \frac{1+\eta}{1-\eta}\tilde{k}_\eta(x_0, \xi_0(x_0)),$$

and note $\beta \geq 0$, $\beta(0) = 0$ (since when $x_0 = 0$ we can take $\xi_0(0) = 0 \in M_{cas}$). □

COROLLARY 4.10.2 *Under the assumptions of Theorem 4.10.1, with $p_0 = p_e$, the central controller $K_{p_e}^{\mathbf{u}^*}$ solves the H^∞ control problem.*

We turn now to the following sufficiency result.

THEOREM 4.10.3 *Assume the following conditions hold:*

(i) *There exists a good solution $\hat{W}$ to the dissipation PDI* (4.41).

(ii) $p_0 \in \mathrm{dom}_{\mathrm{pd}}\hat{W}$ *satisfies* $p_0 \leq p_e$, *where*

  (a) *there exists an equilibrium p_e that is quadratically tight: for some $c_{p_e} > 0$, $p_e \leq -c_{p_e}|\cdot|^2$;*

  (b) *p_e is quadratically upper limiting;*

  (c) $p_e \in \mathrm{dom}_{\mathrm{pd}}\hat{W}$.

(iii) *Every limit point of* $\{p_t, 0 \leq t < \tau_{esc}(\mathrm{dom}_{pde}\hat{W}, K_{p_0}^{\mathbf{u}^*_{\hat{W}}})\}$ *belongs to* $\mathrm{dom}_{\mathrm{pd}}\hat{W}$.

(iv) *The closed-loop system $(G, K_{p_0}^{\mathbf{u}^*_{\hat{W}}})$ is L_2-observable.*

(v) *The system $(A^\times, [B_1 D_{21}' E_2^{-1} : B_2 : \tilde{B}_1])$ is incrementally hyperbolic with $M_{as} \subset$ support p_0.*

(vi) *Assumption* 4.1.3 *holds.*

Then the following conditions hold:

(i) *The coupling condition holds:* $p_t \in q.i.\text{dom}\,\hat{W}$ *for all* $t \geq 0$*, and the controller* $K_{p_0}^{\mathbf{u}^*_{\hat{W}}}$ *is admissible.*

(ii) *The closed-loop system* $(G, K_{p_0}^{\mathbf{u}^*_{\hat{W}}})$ *is* $\tilde{\gamma}$*-dissipative for all* $\tilde{\gamma} > \gamma$*:*

$$\begin{cases} \frac{1}{2}\int_0^T |z(s)|^2\,ds \;\leq\; \tilde{\gamma}^2\frac{1}{2}\int_0^T |w(s)|^2\,ds + \beta(x_0) \\ \text{for all } w \in L_{2,T} \text{ and all } T > 0. \end{cases} \tag{4.83}$$

(iii) *If* $p_0 \in -\mathcal{B}$ *and* $W(p_0) = 0$*,* (4.83) *is the dissipation inequality* (2.10) *for some* $\beta = -\mathcal{B}$*, and* $K_{p_0}^{\mathbf{u}^*_{\hat{W}}}$ *is null initialized.*

(iv) *If* $x_0 \in \text{support}\,p_0$*, then we have* $\tilde{\gamma} = \gamma$ *in the dissipation inequality* (4.83)*, and we can take* $\beta = -p_0$*.*

(v) *If* $(G, K_{p_0}^{\mathbf{u}^*_{\hat{W}}})$ *is z-detectable, then the closed-loop system is weakly internally stable.*

(vi) *If* $(G, K_{p_0}^{\mathbf{u}^*_{\hat{W}}})$ *is z-detectable, and if* p_e *is a control attractor,* $p_0 \in \mathcal{D}_{attr}(p_e)$*, then the closed-loop system is internally stable and, in particular,*

$$x(t) \to 0 \;\; \text{and} \;\; p_t \Longrightarrow p_e + c \;\; \text{as } t \to \infty.$$

Proof. Assertion (i) is a consequence of Theorem 4.9.7. From Theorem 4.5.5, we see that the closed-loop system $(G, K_{p_0}^{\mathbf{u}^*_{\hat{W}}})$ is γ-dissipative; i.e., (4.57) holds (on the time interval $[0,\infty)$), provided $p_0(x_0)$ is finite. If $p_0(x_0)$ is not finite, the proof of Theorem 4.10.1 shows that $(G, K_{p_0}^{\mathbf{u}^*_{\hat{W}}})$ is $\tilde{\gamma}$-dissipative for all $\tilde{\gamma} > \gamma$. The proof is completed as for Theorem 4.10.1. □

The last main result of this section applies to the controller $K_{p_0}^{\bar{\mathbf{u}}^*}$ obtained via direct minimization of the cost functional (3.19) and is useful in that it does not make use of the dynamic programming PDE (4.40) and avoids complications due to smoothness.

THEOREM 4.10.4 *Assume the following conditions hold:*

(i) *The* γ*-dissipative control problem for G is solved by an admissible information state controller* $K_0 = K_{p_0}^{\bar{\mathbf{u}}_0}$*.*

(ii) *The optimal information state controller* $K_{p_0}^{\bar{\mathbf{u}}^*}$ *exists and is* p_0*-regular.*

(iii) $p_0 \in \mathcal{X}_e$ *satisfies* $p_0 \leq p_e$*, where*

(a) *there exists an equilibrium* p_e *that is quadratically tight: for some* $c_{p_e} > 0$*,* $p_e \leq -c_{p_e}|\cdot|^2$*;*

(b) p_e *is quadratically upper limiting;*

(c) $p_e \in q.i.\mathrm{dom}\, \overline{W}$.

(iv) *Every limit point of* $\{p_t,\ 0 \le t < \tau_{esc}(\mathrm{dom}\, \overline{W}, K^{\mathbf{u}^*}_{p_0})\}$ *belongs to* $q.i.\mathrm{dom}\, \overline{W}$.

(v) *The closed-loop system* $(G, K^{\bar{\mathbf{u}}^*}_{p_0})$ *is* L_2*-observable.*

(vi) *The system* $(A^\times, [B_1 D'_{21} E_2^{-1} \ :\ B_2 \ :\ \tilde{B}_1])$ *is incrementally hyperbolic with* $M_{as} \subset \mathrm{support}\, p_0$.

(vii) *Assumption* 4.1.3 *holds.*

Then the following conditions hold:

(i) *The coupling condition holds:* $p_t \in q.i.\mathrm{dom}\, \overline{W}$ *for all* $t \ge 0$, *and the controller* $K^{\bar{\mathbf{u}}^*}_{p_0}$ *is admissible.*

(ii) *The closed-loop system* $(G, K^{\bar{\mathbf{u}}^*}_{p_0})$ *is* $\tilde{\gamma}$*-dissipative for all* $\tilde{\gamma} > \gamma$*:*

$$\begin{cases} \frac{1}{2}\int_0^T |z(s)|^2\, ds \ \le\ \tilde{\gamma}^2 \frac{1}{2}\int_0^T |w(s)|^2\, ds + \beta(x_0) \\ \\ \textit{for all}\ \ w \in L_{2,T}\ \textit{and all}\ T \ge 0. \end{cases} \tag{4.84}$$

(iii) *If* $p_0 \in -\mathcal{B}$ *and* $W(p_0) = 0$, (4.84) *is the dissipation inequality* (2.10) *for some* $\beta = -\mathcal{B}$, *and* $K^{\bar{\mathbf{u}}^*}_{p_0}$ *is null initialized.*

(iv) *If* $x_0 \in \mathrm{support}\, p_0$, *then we have* $\tilde{\gamma} = \gamma$ *in the dissipation inequality* (4.84), *and we can take* $\beta = -p_0$.

(v) *If* $(G, K^{\bar{\mathbf{u}}^*}_{p_0})$ *is z-detectable, then the closed-loop system is weakly internally stable.*

(vi) *If* $(G, K^{\bar{\mathbf{u}}^*}_{p_0})$ *is z-detectable, and if* p_e *is a control attractor,* $p_0 \in \mathcal{D}_{attr}(p_e)$, *then the closed-loop system is internally stable and, in particular,*

$$x(t) \to 0 \ \textit{and}\ p_t \Longrightarrow p_e + c \ \textit{as}\ t \to \infty.$$

Proof. Assertion (i) is a consequence of Theorem 4.9.2. From Theorem 4.6.2, we see that the closed-loop system $(G, K^{\bar{\mathbf{u}}^*}_{p_0})$ is γ-dissipative; i.e., (4.67) holds (on the time interval $[0, \infty)$), provided $p_0(x_0)$ is finite. If $p_0(x_0)$ is not finite, the proof of Theorem 4.10.1 shows that $(G, K^{\bar{\mathbf{u}}^*}_{p_0})$ is $\tilde{\gamma}$-dissipative for all $\tilde{\gamma} > \gamma$. The proof is completed as for Theorem 4.10.1. □

4.11 Further Necessity Results

The central controller $K_{p_e}^{\bar{\mathbf{u}}^*}$ (§4.8) is defined to be initialized at the control attractor: $p_0 = p_e$. For this to make sense, we need $p_e \in \operatorname{dom} \overline{W}$. Indeed, any information state controller $K_{p_e}^{\mathbf{u}^*_{\hat{W}}}$ obtained from a solution of the dynamic programming PDE (4.40) or dissipation PDI (4.41) and initialized at p_e requires $p_e \in \operatorname{dom} \hat{W}$. In this section we show that $p_e \in q.i.\operatorname{dom} W$ (analogous results for $\overline{W}$ or functions $\hat{W}$ can also be obtained). This result depends on other results that appear in Chapter 11.

THEOREM 4.11.1 *Make all the assumptions of Theorem* 11.4.4. *Then*

(i) *the control attractor p_e belongs to the quadratic interior of the domain of the value function W:*

$$p_e \in q.i.\operatorname{dom} W, \tag{4.85}$$

and

(ii) *if in addition W is lower semicontinuous (with respect to max-plus weak convergence), and $p_e(0) = 0$, $p_e(x) \leq 0$ for all $x \in \mathbf{R}^n$, then*

$$W(p_e) = 0. \tag{4.86}$$

Proof. Consider the closed-loop system (G, K_0) and set $w \in L_2$, $x_0 \in \mathbf{R}^n$. This produces L_2 signals u, y, z. By item (iii) of Theorem 11.4.4, $p_t \to p_e + c$ as $t \to \infty$ uniformly on compact sets, and in fact $\| p_t - p_e \|_q \to 0$. For any $\varepsilon > 0$, there exists $t_\varepsilon > 0$ and $c \in \mathbf{R}$ such that

$$p_t(x) \geq p_e(x) + c - \varepsilon(1 + |x|^2) \tag{4.87}$$

for all $t \geq t_\varepsilon$. By (4.6),

$$+\infty > \overline{J}_{p_0}(K_0) \geq W(p_t)$$

for all $t \geq 0$, and hence by the monotonicity of W and (4.87),

$$\overline{J}_{p_0}(K_0) \geq W(p_e - \varepsilon|\cdot|^2) + c - \varepsilon. \tag{4.88}$$

This implies $p_e - \varepsilon|\cdot|^2 \in \operatorname{dom} W$ for all $\varepsilon > 0$.

By the strict γ-dissipation hypothesis of Theorem 11.4.4, the above argument applies for some $\bar{\gamma} < \gamma$. Thus

$$\overline{J}^{\bar{\gamma}}_{p_0}(K_0) \geq W^{\bar{\gamma}}(p_e^{\bar{\gamma}} - \varepsilon|\cdot|^2) + c - \varepsilon, \tag{4.89}$$

where we now emphasize the value of γ in the notation. By definition (4.4),

$$W^{\bar{\gamma}}(p) \geq W^{\gamma}(p) \tag{4.90}$$

for any $p \in \mathcal{X}_e$, and

$$p_e^{\bar{\gamma}}(x) \geq p_e^{\gamma}(x) + \varepsilon_1^2 r_e(x),$$

where r_e is the equilibrium H_2 information state (§11.4.3) and $\bar{\gamma}^2 = \gamma^2 - \varepsilon_1^2$. Now r_e is coercive, and so

$$p_e^{\bar{\gamma}}(x) \geq p_e^{\gamma}(x) + \varepsilon_1^2 c_{r_e}|x|^2. \tag{4.91}$$

Choose $\varepsilon = \frac{1}{2}\varepsilon_1^2 c_{r_e}$. Then combining (4.91), (4.90), and (4.89), we get

$$\begin{aligned} W^{\gamma}(p_e^{\gamma} + \tfrac{1}{2}\varepsilon_1^2 c_{r_e}|\cdot|^2) &\leq W^{\bar{\gamma}}(p_e^{\gamma} + \tfrac{1}{2}\varepsilon_1^2 c_{r_e}|\cdot|^2) \\ &\leq W^{\bar{\gamma}}(p_e^{\bar{\gamma}} - \tfrac{1}{2}\varepsilon_1^2 c_{r_e}|\cdot|^2) \\ &\leq \bar{J}_{p_0}^{\bar{\gamma}}(K_0) - c + \tfrac{1}{2}\varepsilon_1^2 c_{r_e} < +\infty. \end{aligned} \tag{4.92}$$

This proves $p_e^{\gamma} \in q.i.\text{dom}\, W^{\gamma}$.

Now let $x_0 = 0$ and $w(\cdot) = 0$, and consider the resulting trajectories $y(\cdot)$, $u(\cdot)$ and p. (γ is assumed in what follows.) Then with $p_0 = -(1+\varepsilon_0^2)\beta_{K_0}$,

$$p_t(x) \leq \langle p_t \rangle \leq W(p_t) \leq \bar{J}_{p_0}(K_0) = 0$$

along the information state trajectory, which implies

$$\tfrac{1}{2}\int_0^t |z(s)|^2\, ds \leq p_t(x(t)) \leq \langle p_t \rangle \leq W(p_t) \leq 0$$

for all $t \geq 0$. Thus $z(\cdot) = 0$, and as in the proof of Theorem 4.9.9, $x(\cdot) = 0$, which in turn implies $y(\cdot) = 0$ and $u(\cdot) = 0$. Sending $t \to \infty$ gives

$$0 \leq \langle p_e \rangle + c \leq W(p_e) + c \leq 0.$$

Now since $\langle p_e \rangle = 0$, we have $c = 0$ and therefore $W(p_e) = 0$. This completes the proof. □

4.12 Optimal Control and Observation

4.12.1 Stabilizing Property

We next define the stabilizing property for solutions $\hat{W}$ of the dynamic programming PDE (4.40). The definitions refer to the optimal information state system

$$\dot{p} = F(p, \mathbf{u}_{\hat{W}}^*(p), \mathbf{y}_{\hat{W}}^*(p)) \tag{4.93}$$

and the following perturbation of it:

$$\dot{p} = F(p, \mathbf{u}_{\hat{W}}^*(p) + v_2, \mathbf{y}_{\hat{W}}^*(p) + v_1). \tag{4.94}$$

Define the sets

$$\begin{aligned} \text{dom}_{\text{pd}_s}\hat{W} = \{p_0 \in \text{dom}_{\text{pd}}\hat{W} \ &: \ \text{the solution } p_t \text{ of (4.93)} \\ &\text{is well defined with } p_t \in \text{dom}_{\text{pd}}\hat{W} \text{ and} \\ &t \mapsto \mathbf{u}_{\hat{W}}^*(p_t), \mathbf{y}_{\hat{W}}^*(p_t) \in L_2[0,\infty)\} \end{aligned} \tag{4.95}$$

and

$$\begin{aligned}\mathrm{dom}_{\mathrm{pd}_{cs}}\hat{W} = \{p_0 \in \mathrm{dom}_{\mathrm{pd}}\hat{W} \ &: \ \text{the solution } p_t \text{ of (4.93)} \\ &\text{is well defined with } p_t \in \mathrm{dom}_{\mathrm{pd}}\hat{W} \text{ and} \\ &t \mapsto \mathbf{u}^*_{\hat{W}}(p_t), \mathbf{y}^*_{\hat{W}}(p_t) \in L_2[0,\infty) \text{ for all } v_1, v_2 \in L_2[0,\infty)\}.\end{aligned} \tag{4.96}$$

Clearly these sets are invariant under the information state dynamics and $\mathrm{dom}_{\mathrm{pd}_{cs}}\hat{W} \subset \mathrm{dom}_{\mathrm{pd}_s}\hat{W} \subset \mathrm{dom}_{\mathrm{pd}}\hat{W}$. We say that W is *stabilizing* if $\mathrm{dom}_{\mathrm{pd}_s}\hat{W}$ is nonempty and *strongly stabilizing* if $\mathrm{dom}_{\mathrm{pd}_{cs}}\hat{W}$ is non-empty.

CONJECTURE 4.12.1 *Let $\hat{W}$ be a stabilizing solution to the dynamic programming PDE (4.40) and $p \in \mathcal{D}_{attr}(p_e) \cap \mathrm{dom}_{\mathrm{pd}}\hat{W}$. Then*

$$\hat{W}(p) = W(p). \tag{4.97}$$

4.12.2 Zero Dynamics

The *zero dynamics* $\mathcal{ZD}$ for the system (4.93) is the set

$$\mathcal{ZD}^0 = \{p_0 \ : \ \mathbf{u}^*_{\hat{W}}(p_t) = 0 \ \text{ and } \ \mathbf{y}^*_{\hat{W}}(p_t) = 0 \ \text{ for all } \ t \geq 0\},$$

where p_t is the corresponding trajectory of (4.93), and it is assumed that $p_t \in \mathrm{dom}_{\mathrm{pd}}\hat{W}$ for all $t \geq 0$. For the system (4.94) we define

$$\mathcal{ZD} = \{p_0 \ : \ \mathbf{u}^*_{\hat{W}}(p_t) = 0 \ \text{ and } \ \mathbf{y}^*_{\hat{W}}(p_t) = 0 \ \text{ for all } \ t \geq 0 \ \text{ and all } v_1, v_2 \in L_{2,loc}\},$$

where p_t is the corresponding trajectory of (4.94), and it is assumed that $p_t \in \mathrm{dom}_{\mathrm{pd}}\hat{W}$ for all $t \geq 0$.

4.13 List of Properties of the Value Function

In this section of the chapter, we list some of the fundamental properties of the value function W (defined by (4.4)). Chief among these are the structural properties that have been used extensively. The max-plus notation (Appendix C) is used here.

(i) $W(p)$ dominates $\langle p \rangle$:

$$W(p) \geq \langle p \rangle.$$

(ii) Additive homogeneity:

$$W(c \odot p) = c \odot W(p).$$

(iii) Monotonicity:

$$p_1 \geq p_2 \text{ implies } W(p_1) \geq W(p_2).$$

(iv) Max-plus subadditivity:

$$W(p_1 \oplus p_2) \geq W(p_1) \oplus W(p_2).$$

(v) Max-plus sublinearity:

$$W(c_1 \odot p_1 \oplus c_1 \odot p_2) \geq c_1 \odot W(p_1) \oplus c_1 \odot W(p_2).$$

LEMMA 4.13.1 *Monotonicity and max-plus subadditivity are equivalent.*

Proof. Assume monotonicity, and let p_1, p_2 be given. Then $p_1 \oplus p_2 \geq p_1$, and so

$$W(p_1 \oplus p_2) \geq W(p_1).$$

Now also $p_1 \oplus p_2 \geq p_2$, so

$$W(p_1 \oplus p_2) \geq W(p_2).$$

Therefore,

$$W(p_1 \oplus p_2) \geq W(p_1) \oplus W(p_2)$$

proving subadditivity.

Now assume subadditivity and suppose $p_1 \geq p_2$. Then $p_1 \oplus p_2 = p_1$, and by subadditivity

$$W(p_1) = W(p_1 \oplus p_2) \geq W(p_1) \oplus W(p_2) \geq W(p_2)$$

which proves monotonicity. □

4.14 Notes

(i) Information state controllers as used here were defined in [JB95]. The cost to come was defined and used in the context of certainty equivalence for linear systems in [BB91] and for nonlinear systems in [DBB93], [vdS91], [vdS92], [vdS96].

(ii) The method used to prove the dynamic programming equation (4.8) comes from [ES84] and appeared in a similar (although simpler) context in [JB96]. A key point here relates to the fact that $W(p)$ is not finite for all p.

(iii) The ideas and results concerning the dynamic programming PDE (4.16) should be considered to be of a preliminary nature. At present, there is little known about this PDE in the mathematics literature. An attempt at a viscosity definition was made in [JB96], although no uniqueness proof was obtained. This is a crucial issue. What we have done in this chapter is define the concept of p-smoothness, and assuming this property we derived the behavior of the resulting information state controller. The operator $\mathcal{L}^{u,y}$ used in this chapter is defined in Chapter 10 as the generator of a semigroup associated with the minimax problem. As mentioned elsewhere, in discrete time these difficulties of smoothness do not arise [JB95]. In any case, the PDEs are interpreted in an integrated sense (assuming the optimizers exist).

(iv) The coupling condition mentioned here is closely related to the coupling condition for linear systems, which requires that the spectral radius of $X_e Y_e$ is less than γ^2; see Chapter 7. Here, X_e and Y_e are, respectively, the stabilizing solutions to the DGKF Riccati equations. The nonlinear analogue of this condition is $p_e \in q.i.\mathrm{dom}\, W$. The stabilizing property meshes with the quadratic upper limiting property; see Chapter 11.

Chapter 5

State Feedback H^∞ Control

This chapter is concerned with dissipative systems, the bounded real lemmas, and the state feedback H^∞ control problem. Our objective is to provide a summary of some of the results available in the literature concerning the better-understood state feedback problem; this material is provided for use elsewhere in this book as well as to give the reader a familiar reference point to help interpret the output feedback results contained in other chapters of this book.

5.1 Dissipative Systems

The notion of dissipative dynamical systems was introduced by Willems [Wil72], and the associated theory has important applications in the stability analysis of dynamical systems. The theory is a generalization of the behavior of passive electrical circuits and other physical systems that dissipate energy. Central to this theory are the so-called storage functions, which satisfy an inequality known as the dissipation inequality, the infinitesimal version of which is a PDI. If a system possesses a storage function, then this storage function can be used as a (candidate) Lyapunov function in a stability analysis. In general, a dissipative system is stable provided it is observable or detectable. The framework provides a link between input–output and state space stability concepts. See the references [Wil72], [HM76], [HM77], [HM80a], [vdS96], etc., for more information.

The systems we consider are described by models of the form

$$\Sigma : \begin{cases} \dot{x} = A(x) + B(x)w, \quad x(0) = x_0, \\ z = C(x), \end{cases} \tag{5.1}$$

where $x(t) \in \mathbf{R}^n$, $w(t) \in \mathbf{R}^s$, and $z(t) \in \mathbf{R}^r$.

Assumption 5.1.1 We assume that all of the functions appearing in (5.1) are as smooth with bounded first- and second-order partial derivatives, that B is bounded, and that zero is an equilibrium: $A(0) = 0$ and $C(0) = 0$.

Given a control $w : [0, \infty) \to \mathbf{R}^s$, the solution at time $t \geq 0$ with initial condition x_0 is denoted $x(t) = \Phi^w_{t,0}(x_0)$; the corresponding output is $z(t) = C(\Phi^w_{t,0}(x_0))$.

The system (5.1) with *supply rate*

$$\mathrm{s}(w, z) = \tfrac{1}{2}\left(\gamma^2|w|^2 - |z|^2\right) \tag{5.2}$$

is said to be *dissipative* if there exists a nonnegative function $V : \mathbf{R}^n \to \mathbf{R}$, called a *storage function*, with $V(0) = 0$, such that

$$V(x) \geq \sup_{t\geq 0,\, w\in L_{2,T}} \left\{ V(x(t)) - \int_0^t \mathrm{s}(w(r), z(r))\, dr \; : \; x(0) = x \right\}. \tag{5.3}$$

REMARK 5.1.2 Other choices of supply rate are possible, for example, $\mathrm{s}(w, z) = w'z$. The choice (5.2) corresponds to L_2-gain (since $V \geq 0$):

$$\tfrac{1}{2}\int_0^T |z(t)|^2\, dt \leq \gamma^2 \tfrac{1}{2}\int_0^T |w(t)|^2\, dt + V(x_0) \tag{5.4}$$

for all $w \in L_{2,T}$ and all $T \geq 0$ (cf. (2.10)). ▽

The relation (5.3) is known as the *dissipation inequality* and expresses a constraint on the amount of (generalized) energy that can be extracted from the system. $V(x)$ is the amount of energy stored in the system when it is in state x and is a candidate Lyapunov function. In general, a storage function is not uniquely defined, and in fact there is a continuum of storage functions for dissipative systems: $V_a \leq V \leq V_r$, where V_a is the available storage, and V_r is the required supply [Wil72]. The *available storage* is defined by

$$V_a(x) = \sup_{T\geq 0, w\in L_{2,T}} \left\{ -\tfrac{1}{2}\int_0^T [\gamma^2|w(s)|^2 - |z(s)|^2] ds \right\}, \tag{5.5}$$

where the integral is computed corresponding to a system (5.1) trajectory initialized at $x(0) = x$. The *required supply* is defined by

$$V_r(x) = \inf_{0\geq -T, w\in L_2[-T,0]} \left\{ \tfrac{1}{2}\int_{-T}^0 [\gamma^2|w(s)|^2 - |z(s)|^2]\, ds \right\}, \tag{5.6}$$

where the integral is computed corresponding to a system trajectory starting at time $-T$ at $x(-T) = 0$ and terminated at time 0 at $x(0) = x$.

Storage functions can serve as Lyapunov functions in stability analysis. The following basic result is due to Willems [Wil72] (simplified).

THEOREM 5.1.3 *Let V be a continuous storage function for the system* (5.1), *which is quadratically coercive: $V(x) \geq c_0|x|^2$ (for some constant $c_0 > 0$). Then the origin $x = 0$ is a stable equilibrium for the unforced system $\dot{x} = A(x)$.*

Proof. Set $w(\cdot) \equiv 0$ in the dissipation inequality (5.3) to get, integrating along the unforced trajectory,

$$V(x(t)) + \tfrac{1}{2} \int_0^t |z(s)|^2 \, ds \le V(x_0), \quad t \ge 0. \tag{5.7}$$

This implies that $t \mapsto V(x(t))$ is nonincreasing

$$V(x(t)) \le V(x_0)$$

for all $t \ge 0$, and by quadratic coercivity we have

$$c_0 |x(t)|^2 \le V(x_0) \;\text{ for all }\; t \ge 0.$$

This proves that the trajectory is bounded, and hence the unforced system is stable. Explicitly, the bound is

$$|x(t)| \le \sqrt{\frac{V(x_0)}{c_0}} \;\text{ for all }\; t \ge 0.$$

So if $\varepsilon > 0$ is given, select $\eta > 0$ such that $|x_0| \le \eta$ implies $V(x_0) \le c_0 \varepsilon^2$. Then

$$|x(t)| \le \sqrt{\frac{V(x_0)}{c_0}} \le \sqrt{\frac{c_0 \varepsilon^2}{c_0}} = \varepsilon \;\text{ for all }\; t \ge 0.$$

□

REMARK 5.1.4 The quadratic coercivity condition used in Theorem 5.1.3 can be relaxed to *properness*: The level sets $\{x \in \mathbf{R}^n \,:\, 0 \le V(x) \le c\}$ are compact for all $c > 0$. There are many criteria that imply this. ▽

Asymptotic stability requires stronger hypotheses. Recall that the system (5.1) is *strongly zero-state observable* if $w(t) = 0$ for all $t \ge 0$ implies there exists $T_1 > 0$, $c_1 > 0$ such that

$$\tfrac{1}{2} \int_0^T |C(\Phi^0_{t,0}(x_0))|^2 \, dt \ge c_1 |x_0|^2 \;\text{ for all }\; T \ge T_1. \tag{2.27$'$}$$

The following result is due to Hill and Moylan [HM80b] (simplified).

THEOREM 5.1.5 *Let V be a continuous storage function for the system* (5.1), *which is strongly zero-state observable. Then the origin $x = 0$ is an asymptotically stable equilibrium for the unforced system $\dot{x} = A(x)$.*

Proof. Notice first that strong zero-state observability implies the quadratic coerciveness of storage functions (hence properness); thus by Theorem 5.1.3 the origin is stable.

As above, set $w(\cdot) \equiv 0$ in the dissipation inequality (5.3) to get, integrating along the unforced trajectory from $x(t)$ at time t to $x(t+T)$ at time $t+T$,

$$V(x(t+T)) + \tfrac{1}{2} \int_t^{t+T} |z(s)|^2 \, ds \le V(x(t)).$$

Strong zero-state observability implies

$$V(x(t+T)) + c_1|x(t)|^2 \le V(x(t)). \tag{5.8}$$

Now $t \mapsto V(x(t))$ is monotone decreasing and bounded below by zero; therefore the limit $V_\infty = \lim_{t\to\infty} V(x(t))$ exists and is nonnegative. Let $\varepsilon > 0$ be given. There exists $t_\varepsilon > 0$ such that $t \ge t_\varepsilon$ implies

$$V_\infty \le V(x(t)) \le V_\infty + c_1\varepsilon^2. \tag{5.9}$$

Now since $V(x(t+T)) \ge V_\infty$, we combine (5.8) and (5.9) to obtain

$$V_\infty + c_1|x(t)|^2 \le V_\infty + c_1\varepsilon^2,$$

which implies

$$|x(t)| \le \varepsilon \quad \text{for all} \quad t \ge t_\varepsilon. \qquad \square$$

REMARK 5.1.6 In the definition of strong zero-state detectability, the function $r \mapsto c_1 r^2$ can be replaced by any strictly monotone function $\alpha : [0, +\infty) \to [0, +\infty)$ with $\alpha(0) = 0$. ∇

5.2 Bounded Real Lemma

The important bounded real lemma for linear systems provides an algebraic test for the L_2-gain inequality (i.e., dissipativity or $\| \Sigma \|_{H^\infty} < \gamma$) in terms of matrix inequalities, and storage functions can be computed by solving these matrix inequalities; see [AV73]. This result is useful in stability analysis and H^∞ control. The analogous characterization in the nonlinear context is a PDI.

The PDI associated with system (5.1) is

$$\sup_{w\in\mathbf{R}^s} \{\nabla_x V \cdot (A + Bw) + \tfrac{1}{2}(|C|^2 - \gamma^2|w|^2)\} \le 0. \tag{5.10}$$

Evaluating the maximum gives

$$\nabla_x V \cdot A + \gamma^{-2}\tfrac{1}{2}\nabla_x V B B' \nabla_x V' + \tfrac{1}{2}|C|^2 \le 0, \tag{5.10$'$}$$

with optimal disturbance (it depends on the particular storage function V)

$$\mathbf{w}_V^*(x) = \gamma^{-2} B(x)' \nabla_x V(x)'. \tag{5.11}$$

The corresponding PDE is

$$\nabla_x V \cdot A + \gamma^{-2}\tfrac{1}{2}\nabla_x V B B' \nabla_x V' + \tfrac{1}{2}|C|^2 = 0, \tag{5.12}$$

with optimal disturbance $\mathbf{w}_V^*(x) = \gamma^{-2} B(x)' \nabla_x V(x)'$.

In general the PDI is satisfied by storage functions, while the PDE is satisfied by the available storage and required supply. Note that both the PDI and PDE can have many solutions, so when solving a PDI or PDE, one must be careful to select the desired solution. If one is seeking, e.g., the available storage, then the minimal nonnegative solution should be sought. Another issue is that storage functions need not be differentiable, in which case the PDI and PDE can be interpreted in the viscosity sense (see [Jam93], [BH96], Appendix B).

Given an interpretation of solutions to the PDI or PDE, we can state the following theorem.

THEOREM 5.2.1 *System* (5.1) *is dissipative if and only if there exists a solution* $V \geq 0$, $V(0) = 0$, *of the PDI* (5.10) *(viscosity sense).*

Proof. See, e.g., [Wil72], [HM76], [Jam93], [BH96]. □

Using observability conditions, as mentioned above, we can prove that a dissipative system is asymptotically stable. The system (5.1) is *zero-state observable* if $w(t) = 0$, $z(t) = 0$ for all $t \geq 0$ implies $x(t) = 0$ for all $t \geq 0$. The system (5.1) is *zero-state detectable* if $w(t) = 0$, $z(t) = 0$ for all $t \geq 0$ implies $x(t) \to 0$ as $t \to \infty$. A further result using these definitions is as follows (taken from [HM76], [vdS92], [vdS96] with some simplification).

THEOREM 5.2.2 *Let* $V \geq 0$ *be a storage function for system* (5.1).

(i) *If* (5.1) *is zero-state observable, then* $V(x) > 0$ *for all* $x \neq 0$ *(by definition* $V(0) = 0$).

(ii) *Assume* $V \in C^2$ *and* $V(x) > 0$ *for all* $x \neq 0$, $V(0) = 0$, $\nabla_x^2 V(0) \neq 0$, *and* (5.1) *is zero-state detectable. Then* $x = 0$ *is a locally asymptotically stable equilibrium of* $\dot{x} = A(x)$. *If also* V *is proper (i.e., the level sets* $\{x \in \mathbf{R}^n : 0 \leq V(x) \leq c\}$ *are compact for all* $c > 0$), *then* $x = 0$ *is an asymptotically stable equilibrium for* $\dot{x} = A(x)$.

Proof. To establish part (i), suppose $V(x_0) = 0$. Then by (5.4) we have, using $w(\cdot) = 0$,

$$0 \leq \int_0^T |z(t)|^2 \, dt \leq V(x_0) = 0 \quad \text{for all } T \geq 0.$$

This implies $z(\cdot) = 0$, and by zero-state observability it follows that $x(t) = 0$ for all $t \geq 0$. Therefore, $x_0 = x(0) = 0$, and $V(x_0) > 0$ if $x_0 \neq 0$.

Part (ii). Since 0 is a strict minimum for V and since V is smooth, it follows from the mean value theorem that V is locally quadratically coercive:

$$V(x) \geq c_0 |x|^2 \ \text{ if } |x| < r$$

(for some $c_0 > 0$, $r > 0$). The argument in the proof of Theorem 5.1.3 can be applied locally to show that trajectories starting near 0 remain bounded and that 0 is a stable equilibrium (see, e.g., [Vid93, pp. 158–159] for details).

Now it follows from (5.10) (with $w = 0$) that

$$\nabla_x V(x) \cdot A(x) \leq -\tfrac{1}{2} |C(x)|^2 \leq 0,$$

so applying the invariance principle (see, e.g., [Vid93, pp. 177–179]) that for x_0 sufficiently close to 0 we have

$$\Phi^0_{t,0}(x_0) \rightarrow M \text{ as } t \rightarrow \infty,$$

where M is the largest invariant subset of $\{x \in \mathbf{R}^n : \nabla_x V(x) \cdot A(x) = 0\}$. Thus local asymptotic stability will have been proven once it is shown that $M = \{0\}$. To see this, let $x_0 \in M$. Since M is invariant, by definition we get

$$V(\Phi^0_{t,0}(x_0)) = V(x_0) \text{ for all } t \geq 0. \tag{5.13}$$

This implies, via the dissipation inequality (5.7) with $w(t) = 0$ for all $t \geq 0$, that $z(t) = 0$ for all $t \geq 0$. Zero-state detectability implies that $\Phi^0_{t,0}(x_0) \rightarrow 0$ as $t \rightarrow \infty$, and combining this with (5.13), since V is continuous, we get

$$0 = V(0) = \lim_{t\rightarrow\infty} V(\Phi^0_{t,0}(x_0)) = V(x_0).$$

This implies $x_0 = 0$, and hence $M = \{0\}$.

Finally, if V is proper, it follows that 0 is an asymptotically stable equilibrium. □

An important link with the linear theory is given in [vdS91], [vdS92]. Consider the linearization of (5.1) at $x = 0$:

$$\Sigma_{lin} : \begin{cases} \dot{x} = A_{lin}x + B_{lin}w, \quad x(0) = x_0, \\ z = C_{lin}x, \end{cases} \tag{5.14}$$

where the constant matrices A_{lin}, B_{lin}, and C_{lin} are given by

$$A_{lin} = \frac{\partial}{\partial x}A(x)|_{x=0}, \quad B_{lin} = \frac{\partial}{\partial x}B(x)|_{x=0}, \quad C_{lin} = \frac{\partial}{\partial x}C(x)|_{x=0}.$$

THEOREM 5.2.3 (see [vdS91], [vdS92, Theorems 8, 10]). *Assume that A_{lin} is asymptotically stable and $\| \Sigma_{lin} \|_{H^\infty} < \gamma$.* (i) *Then there exists a neighborhood $\mathcal{W}$ of $x = 0$ and a smooth function $V^- \geq 0$ on $\mathcal{W}$ such that* (5.12) *holds in $\mathcal{W}$ and the vector field*

$$A + \gamma^{-2}BB'\nabla_x V^{-\prime}$$

is locally exponentially stable on $\mathcal{W}$. Moreover,

$$\tfrac{1}{2}\int_0^T |z(t)|^2\,dt \leq \gamma^2 \tfrac{1}{2}\int_0^T |w(t)|^2\,dt + V^-(x_0)$$

for all $w \in L_{2,T}$, $T \geq 0$, for which the trajectory $\Phi^w_{t,0}(x_0)$ remains in $\mathcal{W}$ for $0 \leq t \leq T$. (ii) *If (A_{lin}, B_{lin}) is controllable, then there exists a neighborhood $\tilde{\mathcal{W}}$ of $x = 0$ smooth function $V^+ \geq 0$ on $\tilde{\mathcal{W}}$ such that the vector field*

$$-(A + \gamma^{-2}BB'\nabla_x V^{+\prime})$$

is locally exponentially stable on $\tilde{\mathcal{W}}$.

It turns out that

$$V_a = V^-, \quad V_r = V^+ \text{ locally,}$$

or globally under an assumption concerning the existence of global stable and unstable invariant manifolds of the associated Hamiltonian vector field; see [vdS92, Theorem 13]. This means that

$$A + \gamma^{-2} BB' \nabla_x V_a' \text{ is (locally) exponentially stable}$$

and

$$-(A + \gamma^{-2} BB' \nabla_x V_r') \text{ is (locally) exponentially stable.}$$

So the available storage V_a is a *stabilizing* solution of the PDE (5.12), whereas the required supply V_r is an *antistabilizing* solution of the PDE (5.12). The vector fields $A + \gamma^{-2} BB' \nabla_x V_a' = A + B(\gamma^{-2} B' \nabla_x V_a')$ and $A + \gamma^{-2} BB' \nabla_x V_r' = A + B(\gamma^{-2} B' \nabla_x V_r')$ are obtained by using the optimal disturbances $\gamma^{-2} B' \nabla_x V_a'$ and $\gamma^{-2} B' \nabla_x V_r'$, respectively (these are particular instances of (5.11)). Thus the stabilizing/antistabilizing properties provide important characterizations of solutions to the PDE (5.12) and give information on the stability of the optimal-disturbance closed-loop system. Notice that a strict inequality $\| \Sigma_{lin} \|_{H^\infty} < \gamma$ was required in Theorem 5.2.3. These are important themes in H^∞ control and will be discussed in more detail in the next section, which deals with the strict bounded real lemma.

5.3 Strict Bounded Real Lemma

As we have seen, the bounded real lemma provides a characterization of the dissipation property in terms of the solvability of a PDI (or PDE), and by making use of observability conditions, asymptotic stability of the uncontrolled system can be inferred. The strict bounded real lemma (SBRL) [PAJ91] applies to a linear system

$$\begin{cases} \dot{x} = Ax + Bw, \\ z = Cx \end{cases} \tag{5.15}$$

and says that the following three statements are equivalent:

(i) A is *asymptotically stable* and

$$\| C(sI - A)^{-1} B \|_{H^\infty} < 1;$$

(ii) there exists a matrix $\tilde{P} > 0$ such that

$$A'\tilde{P} + \tilde{P}A + \tilde{P}BB'\tilde{P} + C'C < 0; \tag{5.16}$$

(iii) the Riccati equation

$$A'P + PA + PBB'P + C'C = 0 \tag{5.17}$$

has a solution $P \geq 0$ with $A + BB'P$ *asymptotically stable.*

Notice that the SBRL does not make any reachability/observability assumptions but rather cleverly exploits the *strictness* of the H^∞ inequality and asymptotic stability.

The purpose of this section is to generalize this SBRL to the class of nonlinear systems (5.1). Precise statements to this effect are given in the following subsection. These results are used elsewhere in this book and are special cases of results in the paper [YJH98]. Since in general solutions to nonlinear PDI and PDE need not be smooth globally (although they may be locally), one of our aims is to emphasize this point by indicating where smoothness is and is not needed. In the nonlinear case, there are numerous definitions of stability, and sometimes it is necessary to use rather strong forms of stability, and so we emphasize this point also.

5.3.1 Main Results

The system defined by (5.1) is said to have L_2-gain *strictly less than* γ if there exists a finite function $\beta(x) \geq 0$ with $\beta(0) = 0$ and $\varepsilon_0 > 0$ such that for all $0 < \varepsilon \leq \varepsilon_0$,

$$\tfrac{1}{2}\int_0^T |z(t)|^2 dt \leq (1-\varepsilon^2)\gamma^2 \tfrac{1}{2}\int_0^T |w(t)|^2 dt + \beta(x_0) \tag{5.18}$$

for all $w \in L_2[0,T]$ and all $T \geq 0$. G has L_2 gain *less than* γ if (5.18) holds at least for $\varepsilon = 0$.

The generalization of the equivalences (i) $\Leftrightarrow$ (ii) $\Leftrightarrow$ (iii) will be divided into three theorems, as follows. The definition of "viscosity sense" is given in Appendix B.

THEOREM 5.3.1 *Assume* (A, B) *is* L_2 *stable and the system* (5.1) *has* L_2 *gain strictly less than* γ. *Then there exists a lower semicontinuous (l.s.c.) function* $\tilde{V}(x) > 0$ *for* $x \neq 0$ *with* $\tilde{V}(0) = 0$ *such that*

$$\sup_{w\in\mathbf{R}^s} \{\nabla_x \tilde{V}\cdot(A+Bw) - \tfrac{1}{2}[\gamma^2|w|^2 - |C|^2]\} + \eta^2\tfrac{1}{2}|x|^2 \leq 0 \tag{5.19}$$

in the viscosity sense (for some $\eta > 0$*).*

A function $\tilde{V}$ is *locally bounded* if the supremum of $|\tilde{V}(x)|$ is finite as x ranges over any compact subset.

THEOREM 5.3.2 *Assume there exists a (locally bounded) l.s.c. function* $\tilde{V}(x) \geq 0$ *with* $\tilde{V}(0) = 0$ *satisfying the strict PDI* (5.19) *in the viscosity sense. Then there exists a (locally bounded) function* $V(x) \geq 0$ *with* $V(0) = 0$ *satisfying the PDE*

$$\sup_{w\in\mathbf{R}^s} \{\nabla_x V\cdot(A+Bw) - \tfrac{1}{2}[\gamma^2|w|^2 - |C|^2]\} = 0 \tag{5.20}$$

in the viscosity sense. (The available storage is such a function.) Further, if $\tilde{V}$ *is continuous,* V *is smooth, and* $V \in \mathcal{X}_{q,2,b}$, *then the vector field*

$$A^*(x) \triangleq A(x) + \gamma^{-2}B(x)B(x)'\nabla_x V(x)' \tag{5.21}$$

is asymptotically L_2 *stable.*

THEOREM 5.3.3 *Assume there exists a smooth function $V(x) \geq 0$ with $V(0) = 0$ and $V \in \mathcal{X}_{q,2,b}$ satisfying the PDE* (5.20) *in the classical sense, and that the vector field $A^*(x)$ given by* (5.21) *is monotone stable. Then the vector field A is asymptotically L_2 stable and system* (5.1) *has L_2-gain strictly less than γ.*

The circle of ideas embodied in these theorems is not as complete nor as elegant as in the linear case. This is because in the nonlinear case one does not necessarily have smoothness of solutions to PDEs and PDIs, and different types of stability arise naturally (some statements are not true without a sufficiently strong form of stability). However, these theorems contain the overall thrust of the SBRL, and it is likely that they can be sharpened with further effort.

5.3.2 Proofs of Main Results

In the following proofs we set $\gamma = 1$ for simplicity.

Proof of Theorem 5.3.1. The strict gain property (5.18) of (5.1) implies, for $0 < \varepsilon \leq \varepsilon_0$,

$$\tfrac{1}{2}\int_0^T [|z(t)|^2 + \varepsilon^2|w(t)|^2]dt \leq \tfrac{1}{2}\int_0^T |w(t)|^2 dt + \beta(x_0) \tag{5.22}$$

for all $w \in L_2[0,T]$ and all $T \geq 0$. The L_2 stability of A implies there exist $\tilde{\gamma} > 0$ and $\tilde{\beta}(x) \geq 0$, $\tilde{\beta}(0) = 0$ such that

$$\tfrac{1}{2}\int_0^T |x(t)|^2 dt \leq \tilde{\gamma}^2 \tfrac{1}{2}\int_0^T |w(t)|^2 dt + \tilde{\beta}(x_0) \tag{5.23}$$

for all $w \in L_2[0,T]$ and all $T \geq 0$ [FJ96]. Together, (5.22) and (5.23) imply that the system

$$\tilde{\Sigma} : \begin{cases} \dot{x} = A(x) + B(x)w, \\ \tilde{z} = \begin{pmatrix} C(x) \\ \dfrac{\varepsilon}{\tilde{\gamma}} x \end{pmatrix} \end{cases} \tag{5.24}$$

has L_2-gain less than one:

$$\tfrac{1}{2}\int_0^T |\tilde{z}(t)|^2 dt \leq \tfrac{1}{2}\int_0^T |w(t)|^2 dt + \bar{\beta}(x_0), \tag{5.25}$$

where $\bar{\beta}(x) = \beta(x) + \frac{\varepsilon}{\tilde{\gamma}}\tilde{\beta}(x)$. By the bounded real lemma (Theorem 5.2.1) there exists a l.s.c. (storage) function $\tilde{V} \geq 0$ with $\tilde{V}(0) = 0$ solving the PDI

$$\sup_{w \in \mathbf{R}^s} \{\nabla_x \tilde{V} \cdot (A + Bw) - \tfrac{1}{2}[|w|^2 - (|C|^2 + \eta^2|x|^2)]\} \leq 0,$$

where $\eta \triangleq \frac{\varepsilon}{\tilde{\gamma}}$. However, this is just (5.19).

Now the function $\tilde{V}$ satisfies the integral representation (dissipation inequality)

$$\tilde{V}(x) \geq \sup_{T\geq 0, w\in L_2[0,T]} \left\{ \tilde{V}(x(T)) + \tfrac{1}{2}\int_0^T [|z(t)|^2 + \eta|x(t)|^2 - |w(t)|^2]dt : x(0) = x \right\}. \tag{5.26}$$

If $\tilde{V}(x_0) = 0$ for some x_0, then (5.26) implies with $T > 0$ and $w = 0$ that

$$0 = \tilde{V}(x_0) \geq \tfrac{1}{2}\int_0^T \eta|x(t)|^2 dt \geq 0.$$

Hence $x(t) = 0$, $0 \leq t \leq T$. By uniqueness of solutions to ODEs, we must have $x_0 = 0$. □

Proof of Theorem 5.3.2. By Theorem 3.1 of [Jam93], any viscosity solution $\tilde{V} \geq 0$ of the PDI (5.19) is a storage function for the system (5.1), and hence satisfies the integral representation

$$\begin{aligned}\tilde{V}(x) &\geq \sup_{T\geq 0, w\in L_2[0,T]} \left\{ \tilde{V}(x(t)) + \tfrac{1}{2}\int_0^T [|z(t)|^2 + \eta|x(t)|^2 - |w(t)|^2]dt : x(0) = x \right\} \\ &\geq \sup_{T\geq 0, w\in L_2[0,T]} \left\{ \tfrac{1}{2}\int_0^T [|z(t)|^2 + \eta|x(t)|^2 - |w(t)|^2]dt : x(0) = x \right\}.\end{aligned} \tag{5.27}$$

Now define $V(x)$ by the formula

$$V(x) = \sup_{T\geq 0, w\in L_2[0,T]} \left\{ \tfrac{1}{2}\int_0^T [|z(t)|^2 - |w(t)|^2]dt \ : \ x(0) = x \right\}. \tag{5.28}$$

Then from (5.27) and (5.28) we have $0 \leq V(x) \leq \tilde{V}(x)$ and so $V(x)$ exists (is not ∞ for any x). $V(x)$ is called the available storage. The proof of the fact that V is a viscosity solution of (5.20) follows from [Sor96], [McE95b].

Now we suppose that V is a *smooth* solution of the PDE (5.20), and let

$$\mathbf{w}^*(x) = B(x)'\nabla_x V(x)'.$$

Define $S(x) = \tilde{V}(x) - V(x)$ (it is *not* assumed that $\tilde{V}$ is smooth). Then $S(x) \geq 0$ and $S(0) = 0$ (actually $S(x) > 0$ if $x \neq 0$ see below).

Set $w = \mathbf{w}^*$ in (5.27) to get

$$\tilde{V}(x(0)) - \tilde{V}(x(T)) \geq \tfrac{1}{2}\int_0^T [|z(t)|^2 + \eta|x(t)|^2 - |w^*(t)|^2]dt, \tag{5.29}$$

where $w^*(t) = \mathbf{w}^*(x(t))$. Integrate (5.20) to get

$$V(x(0)) - V(x(T)) = \tfrac{1}{2} \int_0^T [|z(t)|^2 - |w^*(t)|^2] dt. \tag{5.30}$$

Subtract the previous two equations to get

$$S(x(T)) \leq -\tfrac{1}{2}\eta^2 \int_0^T |x(s)|^2 ds + S(x(0)), \tag{5.31}$$

so $S(x(T))$ is strictly decreasing. Moreover, since $S(x(T)) \geq 0$ this inequality implies that any initial $x(0)$ produces a trajectory $x(\cdot)$ for which $x(\cdot)$ is in $L_2[0, \infty)$.

Since $V \subset \mathcal{X}_{q,2,b}$, it follows that

$$|\dot{x}| \leq c|x|,$$

for some constant $c \geq 0$, and so $\dot{x}(\cdot) \in L_2[0, \infty)$. Now this bound also shows

$$\begin{aligned} |x(t)|^2 &= |x(s)|^2 + 2\int_s^t \dot{x}(r)' x(r) dr \\ &\leq |x(s)|^2 + c\int_s^t |x(r)|^2 dr \end{aligned} \tag{5.32}$$

for some constant $c \geq 0$. This inequality and (5.31) together with the continuity of S prove stability and asymptotic stability for A^*.

Now we prove strict positivity of S. Suppose $S(x_0) = 0$ for some x_0. Then (5.31) implies

$$\tfrac{1}{2}\eta^2 \int_0^T |x(s)|^2 ds \leq 0;$$

hence $x(s) = 0$, $s \geq 0$. By uniqueness of solutions to ODE, we must have $x_0 = 0$. □

Proof of Theorem 5.3.3. To show that A is asymptotically stable, we write

$$\dot{x}(t) = A(x(t)),\ x(0) = x_0,$$

$$\dot{\xi}(t) = A^*(\xi(t)),\ \xi(0) = \xi_0,$$

and $\zeta(t) = \xi(t) - x(t)$. If $V(x)$ is a solution of (5.20), then with $w = \mathbf{w}^*$ we have

$$\nabla_x V \cdot A + \tfrac{1}{2}|B' \nabla_x V'|^2 + \tfrac{1}{2}|C|^2 = 0,$$

which implies

$$\nabla_x V \cdot A + \tfrac{1}{2}|B' \nabla_x V'|^2 \leq 0.$$

Hence

$$V(x(t)) + \tfrac{1}{2} \int_0^t |B(x(s))' \nabla_x V(x(s))'|^2 ds \leq V(x_0),$$

which implies $B(x(\cdot))' \nabla_x V(x(\cdot))' \in L_2[0, \infty)$.

Next, by monotone stability,

$$\begin{aligned}
\tfrac{d}{dt}\tfrac{1}{2}|\zeta|^2 &= \zeta'\tfrac{d}{dt}\zeta = (A^*(\xi) - A^*(x))\zeta + \zeta' B(x)B(x)'\nabla_x V(x)' \\
&\leq -c|\zeta|^2 + \zeta' B(x)B(x)'\nabla_x V(x)' \\
&\leq -c|\zeta|^2 + |\zeta| \parallel B \parallel_{L^\infty} |B(x)'\nabla_x V(x)'| \\
&\leq -c\tfrac{1}{2}|\zeta|^2 + c_1|B(x)'\nabla_x V(x)'|^2
\end{aligned}$$

for constants $c > 0$ and $c_1 = \parallel B \parallel_{L^\infty} /2c$ (using $ab \leq (\alpha/2)a^2 + (1/2\alpha)b^2$ for any $\alpha > 0$). Integrating, we get

$$\tfrac{1}{2}|\zeta(t)|^2 \leq e^{-ct}\tfrac{1}{2}|\zeta_0|^2 + \tfrac{1}{2}c_1 \int_0^t e^{-c(t-s)}|B(x(s))'\nabla_x V(x(s))'|^2 ds.$$

Therefore, $\zeta(\cdot)$ is in $L_2[0,\infty)$. This implies $x(\cdot) \in L_2$ since $\xi(\cdot) \in L_2$. Also, since $|\dot{x}(\cdot)| \leq c_2|x(\cdot)|$ (some $c_2 \geq 0$), $\dot{x}(\cdot)$ is in L_2, and it follows that A is asymptotically stable.

It follows easily that system (5.1) has L_2-gain less than one, since $V(x)$ serves as a storage function. To see that it is *strictly* less than one needs a bit more work, as follows (cf. [GL95] in the linear case). Consider the auxiliary system

$$L : \begin{cases} \dot{x} = A(x) + B(x)w, \\ v = w - \mathbf{w}^*(x), \end{cases}$$

where $\mathbf{w}^*(x) = B(x)'\nabla_x V(x)'$. The inverse of L is

$$L^{-1} : \begin{cases} \dot{x} = A(x) + B(x)\mathbf{w}^*(x) + B(x)v, \\ w = \mathbf{w}^*(x) + v. \end{cases}$$

Since $A^* = A + B\mathbf{w}^*$, (A^*, B) is incrementally L_2 exponentially stable, there exists $\hat{\gamma} > 0$ and a constant $c_2 \geq 0$ such that

$$\tfrac{1}{2}\int_0^T |w(t)|^2 dt \leq \hat{\gamma}^2 \tfrac{1}{2}\int_0^T |v(t)|^2 dt + c_2|x(0)|^2$$

for all $v \in L_{2,T}$ for all $T \geq 0$.

Fix $T \geq 0$, and $w \in L_{2,T}$, and input this into (5.1), giving trajectory $x(\cdot)$ and $z(\cdot)$. Write $w(t) = \mathbf{w}^*(x(t)) + v(t)$; then $v \in L_{2,T}$ and the previous display holds. That is, on rearranging,

$$-\tfrac{1}{2}\int_0^T |v(t)|^2 dt \leq -\hat{\gamma}^{-2}\tfrac{1}{2}\int_0^T |w(t)|^2 dt + c_2\hat{\gamma}^{-2}|x(0)|^2.$$

By completion of squares, for any $w \in L_{2,loc}$ we have

$$V(x(T)) + \tfrac{1}{2}\int_0^T [|z(s)|^2 - |w(s)|^2]ds = -\tfrac{1}{2}\int_0^T |w(s) - \mathbf{w}^*(x(s))|^2 ds + V(x(0))$$

$$\leq \hat{\gamma}^{-2}\left(-\tfrac{1}{2}\int_0^T |w(s)|^2 ds + c_2|x(0)|^2\right) + V(x(0)).$$

This implies that system (5.1) has finite gain strictly less than one. □

5.4 State Feedback H^∞ Control

In the state feedback case the controller is a function of the state x; i.e., $u = K_{state}(x)$. For simplicity, we assume this to be static state feedback. It is not necessary to use an information state; all that is required is an analogue of the value function $W(p)$ and corresponding dynamic programming equation (4.16), and formulas for the optimal state feedback control and disturbance. In §5.4.5 connections with output feedback control will be discussed.

For an excellent and comprehensive account of the state feedback theory, see [vdS92], [vdS96].

5.4.1 The State Feedback Problem

In the state feedback context the plant G reads

$$G : \begin{cases} \dot{x} = A(x) + B_1(x)w + B_2(x)u, \\ z = C_1(x) + D_{12}(x)u. \end{cases} \tag{5.33}$$

This model is obtained from (2.2) by omitting the output y. The control u is given by static state feedback, $u = K_{state}(x)$. Such a controller is *admissible* if the closed-loop trajectories exist and are unique. This will be the case, e.g., if K_{state} is Lipschitz continuous.

The problem is to find a state feedback controller K^*_{state} such that the closed-loop system (G, K^*_{state}) is γ-dissipative and internally stable (weak internal stability and internal stability are equivalent in the state feedback case).

5.4.2 A State Feedback H^2 Assumption

We consider an H^2 state feedback assumption. This assumption is stronger than the condition (1.6) for linear systems and is very useful for stability. The H_2 assumption concerns the PDE

$$\begin{aligned} &\nabla_x V_{H^2}\left(A - B_2 E_1^{-1} D_{12}' C_1\right) - \tfrac{1}{2}\nabla_x V_{H^2} B_2 E_1^{-1} B_2' \nabla_x V_{H^2}' \\ &\quad + \tfrac{1}{2} C_1'(I - D_{12} E_1^{-1} D_{12}') C_1 = 0 \end{aligned} \tag{5.34}$$

and the vector field

$$A^*_{H^2} \triangleq A - B_2 E_1^{-1} D'_{12} C_1 - B_2 E_1^{-1} B'_2 \nabla_x V'_{H^2}. \tag{5.35}$$

ASSUMPTION 5.4.1 The state feedback H_2 PDE (5.34) has a smooth solution V_{H^2}, which is nonnegative, vanishes at zero, is quadratically coercive

$$V_{H_2}(x) \geq c_{H^2}|x|^2, \tag{5.36}$$

with $c_{H_2} > 0$, and the optimal closed-loop vector field $A^*_{H^2}$ given by (5.35) is incrementally L_2 exponentially stable.

The function V_{H^2} is called the H_2 *state feedback value function* and is given by

$$V_{H^2}(x) = \inf_{K_{state}} \left\{ \frac{1}{2} \int_0^\infty |C_1(\xi(s)) + D_{12}(\xi(s))u(s)|^2 \, ds \ : \ \xi(0) = x, \right. \tag{5.37}$$
$$\left. \lim_{t\to\infty} \xi(t) = 0, \ u(s) = K_{state}(\xi(s)), \ 0 \leq s \leq t \right\},$$

where $\xi(\cdot)$ solves

$$\dot{\xi} = A(\xi) + B_2(\xi)u. \tag{5.38}$$

5.4.3 Necessity

A state feedback value function $V(x)$ can be defined as follows:

$$V(x) = \inf_K \sup_{T\geq 0,\, w\in L_{2,T}} \left\{ \frac{1}{2} \int_0^T [|z(s)|^2 - \gamma^2 |w(s)|^2] \, ds \ : \ x(0) = x \right\}, \tag{5.39}$$

where the infimum is taken over the class of all admissible static state feedback controllers $K_{state}(x)$ achieving dissipation and internal stability. (Under suitable conditions, the value of the infimum is the same if one uses the class of all causal full information controllers $u(t) = K(x_{0,t}, w_{0,t})$.)

We begin with a lemma that relates the H^2 and H^∞ state feedback value functions.

LEMMA 5.4.2 *For all $x \in \mathbf{R}^n$ we have*

$$V(x) \geq V_{H^2}(x), \tag{5.40}$$

and in view of the coercivity of V_{H^2} the value function V is coercive:

$$V(x) \geq c_{H^2}|x|^2 \quad (c_{H^2} > 0). \tag{5.41}$$

Indeed, any bias will be coercive: $\beta_{K_{0,state}}(x) \geq c_{H^2}|x|^2$, where $K_{0,state}$ is any controller solving the state feedback H^∞ control problem.

Proof. For any stabilizing controller K_{state} we have

$$\tfrac{1}{2}\int_0^T |z(s)|^2\,ds \le \sup_{w(\cdot)} \tfrac{1}{2}\int_0^T [|z(s)|^2 - \gamma^2|w(s)|^2]\,ds,$$

where the dynamics are given by (5.33) with $x(0) = x$, and $w(\cdot) = 0$ in the LHS. Since K is stabilizing, this inequality and the definition of V_{H^2} implies

$$V_{H^2}(x) \le \tfrac{1}{2}\int_0^\infty |z(s)|^2\,ds \le \sup_{T\ge 0,\, w(\cdot)} \tfrac{1}{2}\int_0^T [|z(s)|^2 - \gamma^2|w(s)|^2]\,ds.$$

Now minimize the RHS over K to obtain (5.40). □

THEOREM 5.4.3 *Assume that the state feedback γ-dissipative control problem for G is solved by a controller $u = K_{0,state}(x)$. Then the value function defined by formula* (5.39) *enjoys the following properties:*

(i) *$V(x)$ is finite.*

(ii) *$V(x) \ge 0$ for all $x \in \mathbf{R}^n$, and $V(0) = 0$.*

(iii) *If K^ε_{state} is ε-optimal, then for any disturbance input $w \in L_{2,loc}$,*

$$V(x) + \varepsilon \ge V(x(t)), \text{ where } u(t) = K^\varepsilon_{state}(x(t)),\ t \ge 0, \tag{5.42}$$

where $x(\cdot)$ is the solution of (5.33) *with $x(0) = x$. Thus any almost-optimal state feedback controller results in a state trajectory along which $V(x(t))$ is almost decreasing.*

(iv) *The dynamic programming principle holds: for any $t \ge 0$,*

$$V(x) = \inf_{K_{state}} \sup_{w\in L_{2,t}} \left\{ V(x(t)) - \tfrac{1}{2}\int_0^t [\gamma^2|w(s)|^2 - |z(s)|^2]\,ds \right\}, \tag{5.43}$$

where $x(\cdot)$ is the solution of (5.33) *with $x(0) = x$ and $u = K_{state}(x)$.*

(v) *If an admissible optimal state feedback controller K^*_{state} exists, then along any resulting optimal closed-loop state trajectory, for any disturbance input $w \in L_{2,loc}$,*

$$V(x) + \tfrac{1}{2}\gamma^2\int_0^t |w(s)|^2\,ds \ge V(x(t)) + \tfrac{1}{2}\int_0^t |z(s)|^2\,ds, \tag{5.44}$$

$$\text{where } u(t) = K^*_{state}(x(t)),\ t \ge 0,$$

where $x(\cdot)$ is the solution of (5.33) *with $x(0) = x$.*

Therefore, the closed-loop system (G, K^*_{state}) is γ-dissipative.

(a) *If (G, K^*_{state}) is z-detectable, then the closed-loop system is internally stable.*

(b) *If Assumption* 5.4.1 *holds, then the closed-loop system is asymptotically stable.*

Proof. By the definition of V and the hypotheses, we see immediately that

$$0 \leq V(x) \leq \beta_{K_0}(x),$$

where β_{K_0} is the bias for the γ-dissipative controller K_0. This proves items (i) and (ii).

Items (iii) and (iv) follow from dynamic programming considerations; see [McE95b], [Sor96], [vdS92].

The dissipation inequality (5.44) of item (v) follows from (5.43) using the optimal controller stated. Item v(a) follows from Theorem 2.1.3. Item v(b) follows as in the proof of Theorem 5.1.5, since by Lemma 5.4.2 the value function V is coercive. □

REMARK 5.4.4 We see that the state feedback H^2 Assumption 5.4.1 allows us to deduce stability of the closed-loop system without having to invoke the z-detectability assumption. ▽

The PDE satisfied by V is

$$\begin{aligned}\nabla_x V\left(A - B_2 E_1^{-1} D_{12}' C_1\right) + \tfrac{1}{2}\nabla_x V\left(\gamma^{-2} B_1 B_1' - B_2 E_1^{-1} B_2'\right)\nabla_x V' \\ + \tfrac{1}{2} C_1'(I - D_{12} E_1^{-1} D_{12}') C_1 = 0.\end{aligned} \tag{5.45}$$

This equation holds in general in the viscosity sense. If V is sufficiently smooth, the optimal control and disturbance are given by

$$\mathbf{u}^*_{state}(x) = -E_1(x)^{-1}(D_{12}(x)' C_1(x) + B_2(x)' \nabla_x V(x)') \tag{5.46}$$

and

$$\mathbf{w}^*_{state}(x) = \gamma^{-2} B_1(x)' \nabla_x V(x)'. \tag{5.47}$$

In general the value function is not globally smooth. However, Theorem 5.4.5 follows from [Sor96], [McE95b].

THEOREM 5.4.5 *Assume that the state feedback γ-dissipative control problem for G is solved by a controller $u = K_{0,state}(x)$. Then the value function defined by the formula* (5.39) *is a viscosity solution of the PDE* (5.45).

We shall assume that V is smooth in order to define the *state feedback central controller K^*_{state}*:

$$u(t) = K^*_{state}(x(t)) \triangleq \mathbf{u}^*_{state}(x(t)). \tag{5.48}$$

This is illustrated in Figure 5.1.

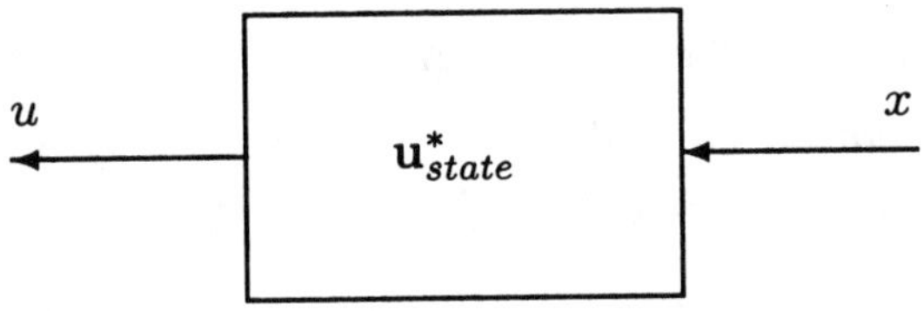

Figure 5.1: State feedback central controller K^*_{state}.

To be precise, the amount of smoothness required of V is C^1 with $\mathbf{u}^*_{state}$ (which depends on $\nabla_x V$) globally Lipschitz, so that solutions to state equation in (5.33) exists and are unique (i.e., K^*_{state} is admissible in the sense that solutions to the closed-loop differential equations exist and are unique).

THEOREM 5.4.6 *Assume that the state feedback γ-dissipative control problem for G is solved by a controller $u = K_{0,state}(x)$. Assume that the value function $V(x)$ defined by* (5.39) *is smooth and the state feedback central controller defined from V via* (5.48) *is admissible. Then* (5.44) *holds, so the closed-loop system (G, K^*_{state}) is γ-dissipative. If (G, K^*_{state}) is z-detectable, then the closed-loop system is internally stable. If Assumption* 5.4.1 *holds, then the closed-loop system is asymptotically stable.*

Proof. By Theorem 5.4.3, we know that V is finite and the conclusions of that theorem hold. If V is smooth, dynamic programming shows that the optimal control is given by (5.48). □

5.4.4 Sufficiency

For dissipation, the dissipation inequality is all that is required. In this subsection we will construct state feedback controllers from smooth solutions of the *state feedback dissipation PDI*:

$$\begin{aligned} &\nabla_x \hat{V}\left(A - B_2 E_1^{-1} D_{12}' C_1\right) + \tfrac{1}{2}\nabla_x \hat{V}\left(\gamma^{-2} B_1 B_1' - B_2 E_1^{-1} B_2'\right)\nabla_x \hat{V}' \\ &\qquad + \tfrac{1}{2} C_1'(I - D_{12} E_1^{-1} D_{12}') C_1 \leq 0. \end{aligned} \tag{5.49}$$

Of course, the state feedback value function V is a particular (and minimal, $V \leq \hat{V}$) solution. The optimal control and disturbance corresponding to $\hat{V}$ are

$$\mathbf{u}^*_{state,\,\hat{V}}(x) = -E_1(x)^{-1}(D_{12}(x)' C_1(x) + B_2(x)' \nabla_x \hat{V}(x)'), \tag{5.50}$$

and

$$\mathbf{w}^*_{state,\,\hat{V}}(x) = \gamma^{-2} B_1(x)' \nabla_x \hat{V}(x)'. \tag{5.51}$$

In this way the function $\hat{V}$ solving the PDI determines a controller $K_{state}^{\mathbf{u}^*_{state,\,\hat{V}}}$.

THEOREM 5.4.7 *Assume that there exists a smooth solution $\hat{V}(x) \geq 0$, $\hat{V}(0) = 0$, of the PDE* (5.49) *and that the controller $K_{state}^{\mathbf{u}^*_{state,\hat{V}}}$ defined by $\mathbf{u}^*_{state,\hat{V}}$ is admissible. Then for any disturbance input $w \in L_{2,loc}$,*

$$\hat{V}(x) + \tfrac{1}{2}\gamma^2 \int_0^t |w(s)|^2\, ds \geq \hat{V}(x(t)) + \tfrac{1}{2}\int_0^t |z(s)|^2\, ds, \tag{5.52}$$

$$\text{where } u(t) = K_{state}^{\mathbf{u}^*_{state,\hat{V}}}(x(t)),\ t \geq 0,$$

where $x(\cdot)$ is the solution of (5.33) *with $x(0) = x$. So the closed-loop system $(G, K_{state}^{\mathbf{u}^*_{state,\hat{V}}})$ is γ-dissipative. If $(G, K_{state}^{\mathbf{u}^*_{state,\hat{V}}})$ is z-detectable, then the closed-loop system is weakly internally stable. If Assumption* 5.4.1 *holds, then the closed-loop system is asymptotically stable.*

Proof. Once the controller is constructed from $\hat{V}$, the proof follows the proofs of Theorems 5.4.3 and 5.4.6. □

THEOREM 5.4.8 *Assume there exists a smooth solution $\hat{V}(x) \geq 0$ of* (5.45) *with $\hat{V}(0) = 0$ and that the vector field*

$$A^*_{state,\hat{V}} \triangleq A - B_2E_1^{-1}D_{12}'C_1 + (\gamma^{-2}B_1B_1' - B_2E_1^{-1}B_2')\nabla_x\hat{V}'$$

is monotone stable. Assume that the state feedback controller $K_{state}^{\mathbf{u}^*_{state,\hat{V}}}(x) = \mathbf{u}^*_{state,\hat{V}}(x)$ is admissible. Then the closed-loop system $(G, K_{state}^{\mathbf{u}^*_{state,\hat{V}}})$ is strictly γ-dissipative and asymptotically L_2 stable.

Proof. Apply Theorem 5.3.3 to the system

$$\dot{x} = A(x) - B_2(x)E_1^{-1}(x)D_{12}(x)'C_1(x) - B_2(x)E_1(x)^{-1}B_2(x)\nabla_x\hat{V}(x)' + B_1(x)w,$$

which has optimal disturbance $\mathbf{w}^* = \gamma^{-2}B_1\nabla_x\hat{V}'$. (See also [YJH98].) □

5.4.5 State Feedback and Its Relation to Output Feedback

Consider the output feedback H^∞ control problem for the plant (2.2), as described in Chapter 2. As described above, the corresponding state feedback problem is obtained by ignoring the measurement equation $y = C_2(x) + D_{21}(x)w$ and providing the controllers with full state access.

THEOREM 5.4.9 *Assume that the state feedback value function V defined by* (5.39) *is a smooth solution of the PDE* (5.45) *with $\mathbf{u}^*_{state}$ stabilizing. Then*

$$W(p) \geq \langle p + V \rangle, \tag{5.53}$$

where W is the output feedback value function for the system (2.2) *defined by* (4.4).

Proof. W is defined to be optimal relative to the class of admissible output feedback controllers achieving dissipation and weak internal stability, while V is optimal over the class of admissible state feedback controllers achieving dissipation and internal stability. Consider the union of these two classes of controllers. Since $\mathbf{u}^*_{state}$ is stabilizing, we see that it attains the infimum over this union. Hence

$$\begin{aligned} W(p) &= \inf_{K_{output}} \sup_{T\geq 0, w(\cdot), x_0} \left\{ p(x_0) + \frac{1}{2}\int_0^T [|z(s)|^2 - \gamma^2 |w(s)|^2]\, ds \right\} \\ &\geq \inf_{K_{output}, K_{state}} \sup_{x_0} \left\{ p(x_0) + \sup_{T\geq 0, w(\cdot)} \left(\frac{1}{2}\int_0^T [|z(s)|^2 - \gamma^2 |w(s)|^2]\, ds \;:\; x_0 \right) \right\} \\ &\geq \sup_{x_0} \{ p(x_0) + V(x_0) \} \end{aligned}$$

as required. □

We next look at the state feedback problem as a special case of the output feedback problem, with $C_2(x) = x$ and $D_{21} = 0$:

$$\begin{aligned} \dot{x} &= A(x) + B_1(x)w + B_2(x)u, \\ z &= C_1(x) + D_{12}(x)u, \\ y &= x. \end{aligned} \tag{5.54}$$

This does not satisfy the standard assumptions, which require D_{21} to be full rank; however, the information state and the value function can still be defined. The key point is that in general, when D_{21} is rank deficient and $C_2(x) = x$, the observation y consists of some state components that are perfectly known, with the remaining ones subject to the influence of measurement disturbances; see [HJM98]. In the case at hand, all state components are perfectly known, and none are subject to measurement disturbances.

THEOREM 5.4.10 *In the state feedback case, we take $C_2(x) = x$ and $D_{21} = 0$, and let the information state p_t and value function W be defined for structure* (5.54). *Let x_0 be a given plant initial condition with disturbance input $w \in L_{2,loc}$ and controller $u = K_{state}(x)$, giving trajectory $x(\cdot)$ of* (2.2) *and output measurement $y(\cdot) = x(\cdot)$. Then the following conditions hold:*

(i) *The information state determined by the signals $u(\cdot)$ and $y(\cdot)$ is given by*

$$p_t(x) = \delta_{x(t)} + p_t(x(t)), \tag{5.55}$$

where $p_0 = \delta_{x_0}$ and

$$p_t(x(t)) = \sup_{\tilde{w}(\cdot)} \left\{ \delta_{x_0}(\xi(0)) + \frac{1}{2}\int_0^t [|C_1(\xi(s)) + D_{12}(\xi(s))u(s)|^2 - \gamma^2 |\tilde{w}(s)|^2]ds \right\}, \tag{5.56}$$

where $\xi(s)$ satisfies

$$\dot{\xi} = A(\xi) + B_2(\xi)u + B_1(\xi)\tilde{w} \tag{2.2$'$}$$

for $0 \leq s \leq t$ *with* $\xi(t) = x(t)$.

(ii) *The value functions are related by*

$$W(\delta_x) = V(x). \tag{5.57}$$

Proof. Consider definition (3.7) of the information state, but with $C_2(x) = x$ and $D_{21} = 0$. The constraints on the trajectories $\xi(\cdot)$ of (2.2) read

$$\xi(t) = x(t), \quad y(s) = \xi(s), \, 0 \leq s \leq t. \tag{5.58}$$

If these constraints are not satisfied, then $p_t(x) = -\infty$.

Let $\tilde{w}(\cdot)$ be given and consider the corresponding trajectory $\xi(\cdot)$ of (2.2)$'$ with $\xi(t) = x$. To satisfy the constraints, we must have $x = x(t)$, and $\tilde{w}$ must produce a trajectory such that $\xi(\cdot) = x(\cdot)$ (there is always one such $\tilde{w}$, viz., w). Then definition (3.7) can be interpreted as (5.55), (5.56).

Next, we have

$$W(\delta_{x_0}) = \inf_K \sup_{T\geq 0,\, \tilde{w},\, \xi_0} \left\{ \delta_{x_0}(\xi_0) + \tfrac{1}{2}\int_0^T [|C_1(\xi(s)) + D_{12}(\xi(s))u(s)|^2 - \gamma^2|\tilde{w}(s)|^2]ds \right\},$$

where $\xi(s)$ satisfies (2.2)$'$ with $\xi(0) = \xi_0$. Then

$$\begin{aligned} W(\delta_{x_0}) &= \inf_K \sup_{T\geq 0,\, \tilde{w}} \left\{ \tfrac{1}{2}\int_0^T [|C_1(\xi(s)) + D_{12}(\xi(s))u(s)|^2 - \gamma^2|\tilde{w}(s)|^2]ds \right\} \\ &= V(x_0), \end{aligned}$$

where $\xi(s)$ satisfies (2.2)$'$ with $\xi(0) = x_0$. □

5.5 Notes

(i) The material on dissipative systems (except for the nonlinear SBRL) is drawn from [Wil72], [HM76], [vdS96].

(ii) For a thorough treatment of the state feedback H^∞ problem, see [vdS96].

Chapter 6

Storage Functions

So far we have discussed dissipation and stability for the output feedback H^∞ problem in terms of the behavior of the information state system, and we have not explicitly used storage functions for the closed-loop system as in the Hill–Moylan–Willems theory of dissipative systems. Indeed the results of Chapter 4 were expressed in terms of functions $\hat{W}$ of the information state, and in particular the value function W. These functions encode dissipation and stability of the closed loop. The purpose of this section is to express our main results in the Hill–Moylan–Willems language and, in particular, we explicitly identify a storage function.

6.1 Storage Functions for the Information State Closed Loop

In Chapter 5 we reviewed some of the main elements of the Hill–Moylan–Willems theory. Let us now apply concepts from this theory to the closed-loop systems $(G, K_{p_0}^{\mathbf{u}^*_{\hat{W}}})$ produced by information state controllers constructed from solutions $\hat{W}$ to the dissipation PDI (4.26). The most important example of such a closed loop is built from the central controller via the value function W and control attractor p_e.

The closed-loop system $(G, K_{p_0}^{\mathbf{u}^*_{\hat{W}}})$ has initial conditions $(x_0, p_0) \in \mathbf{R}^n \times \mathcal{X}_e$. A storage function for this system is a function $V(x, p)$ on the state space $\mathbf{R}^n \times \mathcal{X}_e$ and the *dissipation inequality* reads: for any input $w \in L_{2,t}$, for all $t \geq 0$,

$$V(x(t), p(t)) + \tfrac{1}{2} \int_0^t |z(s)|^2 \, ds \leq V(x_0, p_0) + \gamma^2 \tfrac{1}{2} \int_0^t |w(s)|^2 \, ds \tag{6.1}$$

along the trajectory in the state space $\mathbf{R}^n \times \mathcal{X}_e$ with initial conditions (x_0, p_0) under the action of the controller $K_{p_0}^{\mathbf{u}^*_{\hat{W}}}$. Since (for certain p_0) $K_{p_0}^{\mathbf{u}^*_{\hat{W}}}$ solves the dissipative control problem for G, it is worthwhile investigating and actually constructing storage functions for the closed-loop system. The storage function $V(x, p_0)$ depends on the controller $K_{p_0}^{\mathbf{u}^*_{\hat{W}}}$, which depends in turn both on $\hat{W}$ and p_0, so we write $V^{\hat{W}}(x, p_0)$.

The available storage is defined by

$$V_a^{\hat{W}}(x, p) = \sup_{T \geq 0,\, w \in L_{2,T}} \left\{ -\tfrac{1}{2} \int_0^T [\gamma^2 |w(s)|^2 - |z(s)|^2] ds \right\}, \tag{6.2}$$

where the integral is computed corresponding to $(G, K_{p_0}^{\mathbf{u}^*_{\hat{W}}})$ with system trajectory initialized at $(x(0), p_0) = (x, p)$, and the required supply is defined by

$$V_r^{\hat{W}}(x,p) = - \sup_{T\geq 0,\, w\in L_{2,T}} \left\{ -\tfrac{1}{2}\int_0^T [\gamma^2|w(s)|^2 - |z(s)|^2]ds \right\}, \qquad (6.3)$$

where the integral is computed corresponding to a system trajectory starting at time 0 at $(0, p_0)$ and terminated at time T at $(x(T), p_T) = (x, p)$. The relationship among these storage functions is

$$V_a^{\hat{W}}(x,p) \leq V^{\hat{W}}(x,p) \leq V_r^{\hat{W}}(x,p).$$

A common definition of dissipativity of a system is that it possesses a storage function. Observe that this is *not* precisely the one used in this book, involving β_K; see definition (2.10). However, if K has a state space realization, β_K can be thought of as a restriction of the available storage for (G, K) to the state space of G. This is illustrated by the next theorem, which says this for dissipative information state controllers, as well as giving other information about the available storage.

THEOREM 6.1.1 *Assume that $K_{p_0}^{\mathbf{u}^*_{\hat{W}}}$ solves the γ-dissipative control problem for G, and that the closed-loop system is dissipative in the sense that an available storage function $V_a \geq 0$ exists. Then the following conditions hold:*

(i) *We have*

$$\beta_{K_{p_0}^{\mathbf{u}^*_{\hat{W}}}}(x_0) = V_a^{\hat{W}}(x_0, p_0) \ \ in\ \mathbf{R}^n \times \mathcal{X}_e. \qquad (6.4)$$

(ii) *The available storage satisfies the upper bound*

$$p_0(x_0) + V_a^{\hat{W}}(x_0, p_0) \leq \hat{W}(p_0), \qquad (6.5)$$

and so $V_a^{\hat{W}}(x_0, p_0)$ is finite whenever $p_0 \in \mathrm{dom}\,\hat{W}$ and $p_0(x_0)$ is finite, i.e.,

$$\mathrm{dom}\ V_a^{\hat{W}} \subset \mathrm{support}\ p_0 \times \mathrm{dom}\ \hat{W}.$$

(iii) *The available storage satisfies the lower bound*

$$V_{H_2}(x_0) \leq V_a^{\hat{W}}(x_0, p_0), \qquad (6.6)$$

where V is the state feedback value function (defined by (5.39)*).*

(iv) *The available storage satisfies $V_a^{\hat{W}}(0, p_0) = 0$ whenever $p_0(0) = \hat{W}(p_0)$ (which means that $K_{p_0}^{\mathbf{u}^*_{\hat{W}}}$ is null initialized by Corollary* 4.9.10*).*

(v) *Under the assumptions of Theorem* 4.11.1*, we have*

$$V_a^{\hat{W}}(0, p_e) = 0. \qquad (6.7)$$

Proof. Equation (6.4) follows directly from definitions (2.11) and (6.2). To prove (6.5), we have

$$\begin{aligned} V_a^{\hat{W}}(x_0, p_0) + p_0\ (x_0) &= \sup_{T\geq 0,\, w\in L_{2,T}} \left\{ p_0(x_0) - \tfrac{1}{2}\int_0^T [\gamma^2|w(s)|^2 - |z(s)|^2]ds \right\} \\ &\leq \sup_{T\geq 0,\, w\in L_{2,T},\, x_0\in\mathbf{R}^n} \left\{ p_0(x_0) - \tfrac{1}{2}\int_0^T [\gamma^2|w(s)|^2 - |z(s)|^2]ds \right\} \\ &= \sup_{T\geq 0,\, y\in L_{2,T}} \{\langle p_T\rangle\}. \end{aligned}$$

This last equality follows as in the proof of Theorem 3.1.8. The function $\hat{W}$ is dominating, and satisfies (4.48), so that

$$\langle p_t\rangle \leq \hat{W}(p_T) \leq \hat{W}(p_0).$$

Combining these inequalities we get (6.5). This proves item (ii).

Part (i) and (3.5) imply

$$\begin{aligned} V_a^{\hat{W}}(x_0, p_0) &= \beta_{K_{p_0}^{\mathbf{u}^*_{\hat{W}}}}(x_0) \\ &= J_{\delta_{x_0}}(K_{p_0}^{\mathbf{u}^*_{\hat{W}}}) \\ &\geq W(\delta_{x_0}), \end{aligned}$$

and Theorem 5.4.9 and Lemma 5.4.2 then imply

$$\begin{aligned} W(\delta_{x_0}) &\geq \langle \delta_{x_0} + V\rangle \\ &= V(x_0) \\ &\geq V_{H_2}(x_0). \end{aligned}$$

This proves item (iii).

Part (iv) follows for the available storage thanks to (6.5), and part (v) follows as in Theorem 4.11.1. □

COROLLARY 6.1.2 *If Assumption* 5.4.1 *holds, then* $V_a^{\hat{W}}(x_0, p_0) \geq c_{H_2}|x_0|^2$ *is quadratically coercive, and* $\beta_{K_{p_0}^{\mathbf{u}^*_{\hat{W}}}}(x_0) \geq c_{H_2}|x_0|^2$ *is a coercive bias* (c_{H_2} >0)*. Therefore, we can replace the hypothesis that* $(G, K_{p_0}^{\mathbf{u}^*_{\hat{W}}})$ *is z-detectable in item* (v) *of Theorem* 4.10.3 *to conclude that the plant state in the closed-loop system is asymptotically stable.*

Proof. The proof follows from Theorem 6.1.1 and the proofs of Theorems 5.4.7 and 5.1.5. □

So far we have dealt with integral inequalities for the storage functions. Since in general the bounded real lemma provides a connection between integral and differential forms of the dissipation inequality, we consider the dissipation PDI for storage functions $V^{\hat{W}}(x,p)$; viz.,

$$\sup_w \Big\{ \nabla_x V^{\hat{W}}(x,p) \cdot \big(A(x) + B_1(x)w + B_2(x)\mathbf{u}^*_{\hat{W}}(p)\big)$$

$$+\nabla_p V^{\hat{W}}(x,p)[F(p, \mathbf{u}^*_{\hat{W}}(p), C_2(x) + D_{21}(x)w)] \tag{6.8}$$

$$+\tfrac{1}{2}|C_1(x) + D_{12}(x)\mathbf{u}^*_{\hat{W}}(p)|^2 - \gamma^2 \tfrac{1}{2}|w|^2 \Big\} \leq 0$$

and the corresponding PDE

$$\sup_w \Big\{ \nabla_x V_a^{\hat{W}}(x,p) \cdot \big(A(x) + B_1(x)w + B_2(x)\mathbf{u}^*_{\hat{W}}(p)\big)$$

$$+\nabla_p V_a^{\hat{W}}(x,p)[F(p, \mathbf{u}^*_{\hat{W}}(p), C_2(x) + D_{21}(x)w)] \tag{6.9}$$

$$+\tfrac{1}{2}|C_1(x) + D_{12}(x)\mathbf{u}^*_{\hat{W}}(p)|^2 - \gamma^2 \tfrac{1}{2}|w|^2 \Big\} = 0$$

is satisfied by $V_a^{\hat{W}}$. Note that this inequality and equality are defined on the infinite-dimensional space $\mathbf{R}^n \times \mathcal{X}_e$.

6.2 Explicit Storage Functions

Associated with each function $\hat{W}$ solving the dissipation PDI (4.41) is the *explicit* storage function

$$e^{\hat{W}}(x,p) \triangleq -p(x) + \hat{W}(p). \tag{6.10}$$

THEOREM 6.2.1 *The function $e^{\hat{W}}(x,p)$ is nonnegative and satisfies the dissipation inequality* (6.1) *and so is a storage function for the closed-loop system in $\mathbf{R}^n \times \operatorname{dom} W$ provided W and p are smooth, and*

$$V_a^{\hat{W}}(x,p) \leq e^{\hat{W}}(x,p) \leq V_r^{\hat{W}}(x,p).$$

Further,

$$\operatorname{dom} e^{\hat{W}} \subset \operatorname{support} p \times \operatorname{dom} \hat{W},$$

and $e(0,p) = 0$ whenever $p(0) = \hat{W}(p)$ and, in particular, $e(0,p_e) = 0$ (under the assumptions of Theorem 4.11.1*). The function $e^W(x,p)$ solves the PDE* (6.9)*, and*

$$e^W(x,p) \leq e^{\hat{W}}(x,p) \tag{6.11}$$

for any good solution $\hat{W}$ of the dissipation PDI (4.41).

Proof. We use the fact that $\nabla_x e^{\hat{W}}(x,p) = -\nabla_x p(x)$ and $\nabla_p e^{\hat{W}}(x,p) = -E_x + \nabla_p \hat{W}(p)$ (where E_x is the evaluation operator $E_x[f] = f(x)$):

$$0 \geq \sup_y \{\nabla_p \hat{W}(p)[F(p, \mathbf{u}^*_{\hat{W}}(p), y)]\}$$

$$= \sup_y \Big\{ \sup_w \Big\{ \nabla_x e^{\hat{W}}(x,p) \cdot \big(A(x) + B_1(x)w + B_2(x)\mathbf{u}^*_{\hat{W}}(p)\big)$$

$$+\tfrac{1}{2}|C_1(x) + D_{12}(x)\mathbf{u}^*_{\hat{W}}(p)|^2 - \gamma^2 \tfrac{1}{2}|w|^2 \ : \ y = C_2(x) + D_{21}(x)w \Big\}$$

$$+\nabla_p e^{\hat{W}}(x,p)[F(p, \mathbf{u}^*_{\hat{W}}(p), y)] \Big\}$$

$$= \sup_y \Big\{ \sup_w \Big\{ \nabla_x e^{\hat{W}}(x,p) \cdot \big(A(x) + B_1(x)w + B_2(x)\mathbf{u}^*_{\hat{W}}(p)\big)$$

$$+\tfrac{1}{2}|C_1(x) + D_{12}(x)\mathbf{u}^*_{\hat{W}}(p)|^2 - \gamma^2 \tfrac{1}{2}|w|^2 + \nabla_p e(x,p)[F(p, \mathbf{u}^*_{\hat{W}}(p), y)]$$

$$: \ y = C_2(x) + D_{21}(x)w \Big\}\Big\}$$

$$= \sup_w \Big\{ \nabla_x e^{\hat{W}}(x,p) \cdot \big(A(x) + B_1(x)w + B_2(x)\mathbf{u}^*_{\hat{W}}(p)\big)$$

$$+\tfrac{1}{2}|C_1(x) + D_{12}(x)\mathbf{u}^*_{\hat{W}}(p)|^2 - \gamma^2 \tfrac{1}{2}|w|^2$$

$$+\nabla_p e^{\hat{W}}(x,p)[F(p, \mathbf{u}^*_{\hat{W}}(p), C_2(x) + D_{21}(x)w)] \Big\}. \tag{6.12}$$

(In these expressions, the optimizing y is $\mathbf{y}^*_{\hat{W}}(p)$.) This implies that $e^{\hat{W}}(x,p)$ satisfies the PDI (6.8). Integrating yields the dissipation inequality (6.1).

The inequality $\hat{W}(p) \geq \langle p \rangle$ implies that $e^{\hat{W}}$ is nonnegative. The remaining assertions follow as in Theorem 6.1.1, and from Theorem 4.5.3, $W \leq \hat{W}$. □

REMARK 6.2.2 The above proof used differentials. However, the result could have also been established in integrated form. ▽

ASSUMPTION 6.2.3 We now assume that D_{21} is independent of x.

In the PDE (6.12), the optimizing disturbance $\mathbf{w}^*_{e,\hat{W}}$ is given by

$$\mathbf{w}^*_{e,\hat{W}}(x,p) = -\gamma^{-2}[I - D_{21}'E_2^{-1}D_{21}]B_1(x)'\nabla_x p(x)' + D_{21}'E_2^{-1}(\mathbf{y}^*(p) - C_2(x)). \tag{6.13}$$

We use the subscript in $\mathbf{w}^*_{e,\hat{W}}$ to emphasize that this quantity depends on the function e (defined by (6.10)) and $\hat{W}$. A simple computation shows that if $w = \mathbf{w}^*_{e,\hat{W}}(x,p)$, then $y = C_2(x) + D_{21}\mathbf{w}^*_{e,\hat{W}}(x,p) = \mathbf{y}^*_{\hat{W}}(p)$.

Assumption 6.2.4 We now assume $D_{21} = I$ (1 and 2A block cases).

Under Assumption 6.2.4 we have the simplification

$$\mathbf{w}^*_{e,\hat{W}}(x,p) = \mathbf{y}^*_{\hat{W}}(p) - C_2(x).$$

Recall from §4.12.2 the definition of zero dynamics $\mathcal{ZD}^0$ of the system

$$\dot{p} = F(p, \mathbf{u}^*_{\hat{W}}(p), \mathbf{y}^*_{\hat{W}}(p)), \tag{4.93}$$

and let M_s be the stable manifold for the exponentially hyperbolic vector field $A^\times$.

The next theorem describes a set on which $e^{\hat{W}} = V_a^{\hat{W}}$. This set is rather thin in general.

Theorem 6.2.5 *Assume that* $\hat{W}$ *is a good solution of the dynamic programming PDE* (4.16) *and*

$$p_0 \in \mathcal{ZD}^0 \cap \mathcal{D}^0_{attr}(p_e). \tag{6.14}$$

Assume $\hat{W}(p_e) = 0$. *Then*

$$V_a^{\hat{W}}(x_0, p_0) = e^{\hat{W}}(x_0, p_0) \text{ for all } x_0 \in M_s \cap \operatorname{support} p_0. \tag{6.15}$$

Proof. If $w = \mathbf{w}^*_{e,\hat{W}}$ in the closed-loop system, then $y = \mathbf{y}^*_{\hat{W}}$, and so $\hat{W}(p_0) = \hat{W}(p_t)$ for all $t \geq 0$. Since $p_0 \in \mathcal{ZD}^0$, we have $\mathbf{u}^*_{\hat{W}}(p_t) = 0$ and $\mathbf{y}^*_{\hat{W}}(p_t) = 0$ for all $t \geq 0$. Thus the plant dynamics becomes

$$\dot{x} = A^\times(x),$$

and the controller dynamics is

$$\dot{p} = F(p, 0, 0).$$

If $x_0 \in M_s$, then $x(t) \to 0$ and

$$p_t(x(t)) \to p_0(x_0) + \tfrac{1}{2}\int_0^\infty [|C_1(x(s)) + D_{12}(x(s))u(s)|^2 - \gamma^2|C_2(x(s)) - y(s)|^2]\, ds$$

$$= p_e(0) + c,$$

a finite number, as $t \to \infty$.

Now since $p_0 \in \mathcal{D}^0_{attr}(p_e)$ we have $p_t \Longrightarrow p_e + c$ as $t \to \infty$. By continuity, $\hat{W}(p_0) = \hat{W}(p_t) \to \hat{W}(p_e)$ as $t \to \infty$. Therefore,

$$e^{\hat{W}}(x(t), p_t) = -p_t(x(t)) + \hat{W}(p_t) \to -(p_e(0) + c) + \hat{W}(p_e + c) = 0 \quad (6.16)$$

as $t \to \infty$.

Next, we know from Theorem 6.2.1 that $e^{\hat{W}}(x_0, p_0) \geq V_a^{\hat{W}}(x_0, p_0)$, so we need only verify the opposite inequality. To this end, plug $w = \mathbf{w}^*_{e,\hat{W}}$ into (6.2):

$$V_a^{\hat{W}}(x_0, p_0) \geq -\tfrac{1}{2}\int_0^T [\gamma^2|\mathbf{w}^*_{e,\hat{W}}(s)|^2 - |z(s)|^2]ds.$$

Since $e^{\hat{W}}$ satisfies the PDE (6.9), integration gives

$$-\tfrac{1}{2}\int_0^T [\gamma^2|\mathbf{w}^*_{e,\hat{W}}(s)|^2 - |z(s)|^2]ds = e^{\hat{W}}(x_0, p_0) - e^{\hat{W}}(x(t), p_t),$$

using $w = \mathbf{w}^*_{e,\hat{W}}$. Combining, we get

$$V_a^{\hat{W}}(x_0, p_0) - e^{\hat{W}}(x_0, p_0) \geq -e^{\hat{W}}(x(t), p_t).$$

Now from (6.16) we get

$$V_a^{\hat{W}}(x_0, p_0) - e(x_0, p_0) \geq -\lim_{t\to\infty} e^{\hat{W}}(x(t), p_t) = 0.$$

This implies $V_a^{\hat{W}}(x_0, p_0) \geq e^{\hat{W}}(x_0, p_0)$, completing the proof. □

We next discuss some formulas that connect some of the functions we have been studying. Recall definition (4.4), which can be written as

$$W(p) = \inf_K \sup_{T\geq 0,\, y(\cdot),\, x_0} \left\{ p(x_0) + \tfrac{1}{2}\int_0^T [|C_1(\xi(s)) + u(s)|^2 - \gamma^2|y(s) - C_2(\xi(s))|^2]\, ds \right\},$$

where the trajectory $\xi(\cdot)$ is the solution of

$$\dot{\xi} = A^\times(\xi) + B_1(\xi)y + B_2(\xi)u, \quad (2.6)'$$

with $u(\cdot) = K(y(\cdot))$, and $\xi(0) = x_0$. Now write

$$W(x, p) \triangleq \sup_{T\geq 0,\, y(\cdot)} \left\{ p(x) + \tfrac{1}{2}\int_0^T [|C_1(\xi(s)) + u(s)|^2 - \gamma^2|y(s) - C_2(\xi(s))|^2]\, ds \right\}, \quad (6.17)$$

where $u(\cdot) = K_p^{\mathbf{u}^*_{\hat{W}}}$. Then in fact

$$V_a^{\hat{W}}(x, p) = -p(x) + W(x, p). \quad (6.18)$$

EXAMPLE 6.2.6 (linear systems). The storage function for linear systems is calculated in §7.2.2. ▽

Chapter 7

Special Cases

In this chapter we discuss some important special cases. In particular, the cases considered have special features due to the problem data that simplify the construction of the controller. A class of bilinear systems is discussed in §7.1. These systems are linear systems with an additional term in the state dynamics consisting of a product of the state and control. The information state for this system is finite dimensional, given by a driven Riccati equation, and the dynamic programming PDE, which determines the controller, is correspondingly defined on a finite-dimensional space. Linear systems are special cases of this, and the controller is determined by a Riccati equation. Thus the well-known linear H^∞ controller is obtained from the information state theory as a special case. In section 7.3 systems that satisfy the certainty equivalence principle are discussed. Here, the controller is constructed from a pair of PDEs defined on a finite-dimensional space.

7.1 Bilinear Systems

Consider the following class of bilinear systems:

$$G_{bilinear}: \begin{cases} \dot{x} = A(u)x + B_1 w + B_2 u, \\ z = C_1 x + D_{12} u, \\ y = C_2 x + D_{21} w. \end{cases} \tag{7.1}$$

Here, we take the control input dimension $m = 1$ for simplicity, and

$$A(u) = A + B_3 u.$$

In our general notation (Chapter 2), $B_2(x)$ would read $B_2 + B_3 x$, in terms of the notation of (7.1). The presence of this B_3 term means that the dynamics have a multiplicative term involving both x and u—hence the term *bilinear*. The matrices are of the appropriate sizes (e.g., B_3 is $n \times n$), and the standard assumptions used earlier apply, except that B_2 is no longer bounded.

Let the initial information state be given by

$$p_0(x) = -\gamma^2 \tfrac{1}{2} x' Y_0^{-1} x, \tag{7.2}$$

with $Y_0 > 0$. Then the information state for $t > 0$ is given explicitly by

$$p_t(x) = -\gamma^2 \tfrac{1}{2}(x - \hat{x}(t))' Y(t)^{-1} (x - \hat{x}(t)) + \phi(t), \tag{7.3}$$

where

$$\begin{aligned} \dot{\hat{x}} &= (A(u) - B_1 D_{21}' E_2^{-1} C_2 + Y(\gamma^{-2} C_1' C_1 - C_2' E_2^{-1} C_2))\hat{x} \\ &\quad + (B_1 D_{21}' + Y C_2') E_2^{-1} y + (B_2 + \gamma^{-2} Y C_1' D_{12}) u, \end{aligned} \tag{7.4}$$

$$\begin{aligned} \dot{Y} &= (A(u) - B_1 D_{21}' E_2^{-1} C_2) Y + Y (A(u) - B_1 D_{21}' E_2^{-1} C_2)' \\ &\quad + Y(\gamma^{-2} C_1' C_1 - C_2' E_2^{-1} C_2) Y + B_1 [I - D_{21}' E_2^{-1} D_{21}] B_1' \end{aligned} \tag{7.5}$$

$$\dot{\phi} = \tfrac{1}{2} |C_1 \hat{x} + D_{12} u|^2 - \gamma^2 \tfrac{1}{2} (y - C_2 \hat{x})' E_2^{-1} (y - C_2 \hat{x}). \tag{7.6}$$

The initial conditions are

$$\hat{x}(0) = 0, \quad Y(0) = Y_0, \quad \phi(0) = 0. \tag{7.7}$$

This can be readily verified by differentiation and substitution in (3.9).

The significance of this is that when information state p_0 has the form (7.7) then p_t is given by (7.3) for all $t \geq 0$ and can be *computed by solving the finite set of ODEs* (7.4), (7.5), (7.6). This means that there exists an invariant subspace $\mathcal{F} = \mathbf{R}^n \times \mathbf{R}^{n^2} \times \mathbf{R} \subset \mathcal{X} \subset \mathcal{X}_e$. Note in particular that the Y Riccati equation (7.5) is *driven* by the control signal u.

The minimax cost functional (3.20) can be expressed in terms of these parameters, since

$$\langle p_T \rangle = \phi(T).$$

The problem of computing the optimal control and observation (cf. (4.23)) is also greatly simplified. This can be done by introducing a finite-dimensional "projection" of the value function $W(p)$. Indeed, define

$$\tilde{W}(\hat{x}, Y, \phi) = \inf_{K} \sup_{T \geq 0,\, y \in L_{2,T}} \{\phi(T) \ : \ \hat{x}(0) = \hat{x}, Y(0) = Y, \phi(0) = \phi\}. \tag{7.8}$$

The function $\tilde{W}$ enjoys the following properties: (i) from Theorem 4.3.1, item (ii), we have

$$\nabla_\phi \tilde{W}(\hat{x}, Y, \phi) = 1 \tag{7.9}$$

and

$$\tilde{W}(\hat{x}, Y, \phi) \geq \phi \quad \text{for all } \hat{x}, Y, \phi. \tag{7.10}$$

The PDE (4.16) on the infinite-dimensional space $\mathcal{X}$ converts to a PDE on a finite-dimensional space (i.e., one defined on a subset of $\mathcal{F} = \mathbf{R}^n \times \mathbf{R}^{n^2} \times \mathbf{R}$); viz.,

$$\begin{aligned}\inf_u \sup_y \Big\{ & \nabla_{\hat{x}} \tilde{W}(\hat{x}, Y, \phi) \cdot [(A(u) - B_1 D_{21}' E_2^{-1} C_2 + Y(\gamma^{-2} C_1' C_1 - C_2' E_2^{-1} C_2))\hat{x} \\ & \qquad + (B_1 D_{21}' + Y C_2') E_2^{-1} y + (B_2 + \gamma^{-2} Y C_1' D_{12}) u] \\ & + \nabla_Y \tilde{W}(\hat{x}, Y, \phi) \cdot [(A(u) - B_1 D_{21}' E_2^{-1} C_2) Y + Y (A(u) - B_1 D_{21}' E_2^{-1} C_2)' \\ & \qquad + Y(\gamma^{-2} C_1' C_1 - C_2' E_2^{-1} C_2) Y + B_1 \left(I - D_{21}' E_2^{-1} D_{21}\right) B_1'] \\ & + \nabla_\phi \tilde{W}(\hat{x}, Y, \phi) \cdot [\tfrac{1}{2} |C_1 \hat{x} + D_{12} u|^2 - \gamma^2 (y - C_2 \hat{x})' E_2^{-1} (y - C_2 \hat{x})] \Big\} = 0. \end{aligned} \tag{7.11}$$

In equation (7.11), the gradient $\nabla_{\hat{x},Y,\phi} \tilde{W}$ has components $\nabla_{\hat{x}} \tilde{W}$, $\nabla_Y \tilde{W}$, $\nabla_\phi \tilde{W}$, and the notation $\nabla_Y \tilde{W} \cdot M$ means $\sum_{i,j=1}^n \frac{\partial \tilde{W}}{\partial Y_{ij}} M_{ij}$ (where M is an $n \times n$ matrix). In general it is not possible to solve the PDE explicitly. However, evaluation of the inf and sup in (7.11) gives the following formulas for the optimal control and observation:

$$\begin{aligned} \mathbf{u}^*(\hat{x}, Y, \phi) &= -E_1^{-1} [D_{12}' C_1 \hat{x} + (B_2' + \gamma^{-2} D_{12}' C_1 Y + \hat{x}' B_3') \nabla_{\hat{x}} \tilde{W}(\hat{x}, Y, \phi)' \\ & \qquad + \nabla_Y \tilde{W}(\hat{x}, Y, \phi) \cdot (B_3 Y + Y B_3')] \\ \mathbf{y}^*(\hat{x}, Y, \phi) &= C_2 \hat{x} + (D_{21} B_1' + \gamma^{-2} C_2 Y) \nabla_{\hat{x}} \tilde{W}(\hat{x}, Y, \phi)'. \end{aligned} \tag{7.12}$$

The finite set of ODEs (7.4), (7.5), (7.6) together with the optimal control specified in (7.12) constitutes the optimal information state controller for bilinear systems. This is illustrated in Figure 7.1.

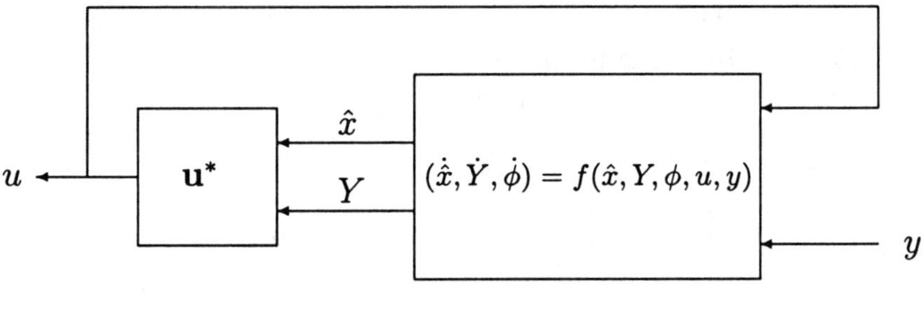

Figure 7.1: Bilinear information state controller. The dynamics f is defined by equations (7.4), (7.5), and (7.6).

7.2 Linear Systems

Consider the following class of linear systems:

$$G_{linear}: \begin{cases} \dot{x} = Ax + B_1 w + B_2 u, \\ z = C_1 x + D_{12} u, \\ y = C_2 x + D_{21} w. \end{cases}$$

This corresponds to a subclass of bilinear systems with $B_3 = 0$, so that multiplicative terms involving both x and u are no longer present. The information state is again given explicitly by (7.3), with the modification $B_3 = 0$. The dynamics are

$$\begin{aligned}\dot{\hat{x}} &= (A - B_1 D_{21}' E_2^{-1} C_2 + Y(\gamma^{-2} C_1' C_1 - C_2' E_2^{-1} C_2))\hat{x} \\ &\quad + (B_1 D_{21}' + Y C_2') E_2^{-1} y + (B_2 + \gamma^{-2} Y C_1' D_{12}) u,\end{aligned} \tag{7.13}$$

$$\begin{aligned}\dot{Y} &= (A - B_1 D_{21}' E_2^{-1} C_2) Y + Y (A - B_1 D_{21}' E_2^{-1} C_2)' \\ &\quad + Y(\gamma^{-2} C_1' C_1 - C_2' E_2^{-1} C_2) Y + B_1 [I - D_{21}' E_2^{-1} D_{21}] B_1'\end{aligned} \tag{7.14}$$

$$\dot{\phi} = \tfrac{1}{2} |C_1 \hat{x} + D_{12} u|^2 - \gamma^2 \tfrac{1}{2} (y - C_2 \hat{x})' E_2^{-1} (y - C_2 \hat{x}). \tag{7.15}$$

REMARK 7.2.1 In the general bilinear case, the Y-Riccati differential equation (7.5) depends on the control u, while in the linear case, (7.14), it does not. ∇

Consider the function $\hat{W}(\hat{x}, Y, \phi)$ defined by

$$\tilde{W}(\hat{x}, Y, \phi) = \tfrac{1}{2} \hat{x}' (I - \gamma^{-2} Y X)^{-1} X \hat{x} + \phi, \tag{7.16}$$

where $X \geq 0$ is a solution of the X-Riccati equation

$$\begin{aligned}&(A - B_2 E_1^{-1} D_{12}' C_1)' X + X (A - B_2 E_1^{-1} D_{12}' C_1) \\ &\quad + X (\tfrac{1}{\gamma^2} B_1 B_1' - B_2 E_1^{-1} B_2') X + C_1' (I - D_{12} E_1^{-1} D_{12}') C_1 = 0.\end{aligned} \tag{7.17}$$

Then $\hat{W}$ solves the PDE (7.11) (with $B_3 = 0$) on a certain domain (see below). This can be checked by differentiation and substitution into (7.11).

The optimal control and observation are given by

$$\begin{aligned}\mathbf{u}^*(\hat{x}, Y, \phi) &= -E_1^{-1} [D_{12}' C_1 \hat{x} + (B_2' + \gamma^{-2} D_{12}' C_1 Y) X (I - \gamma^{-2} Y X)^{-1} \hat{x}] \\ &= -E_1^{-1} [D_{12}' C_1 + B_2' X] (I - \gamma^{-2} Y X)^{-1} \hat{x}, \\ \mathbf{y}^*(\hat{x}, Y, \phi) &= C_2 \hat{x} + \gamma^{-2} (D_{21} B_1' + C_2 Y) X (I - \gamma^{-2} Y X)^{-1} \hat{x} \\ &= [C_2 + \gamma^{-2} D_{21} B_1' X] (I - \gamma^{-2} Y X)^{-1} \hat{x}.\end{aligned} \tag{7.18}$$

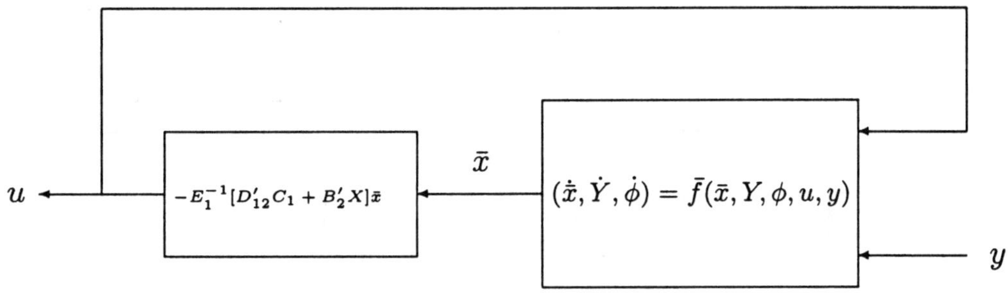

Figure 7.2: Linear optimal information state controller. The dynamics $\bar{f}$ is defined by equations (7.13), (7.14), and (7.15).

Notice that these formulas are independent of ϕ, but depend explicitly on $\hat{x}$, Y.

The finite set of ODEs (7.4), (7.5), (7.6) (with $B_3 = 0$) together with the optimal control specified in (7.18) constitutes the central controller for linear systems. However, the form of this controller can be simplified with the change of variables:

$$\bar{x} = (I - \gamma^{-2}YX)^{-1}\hat{x}. \tag{7.19}$$

Equation (7.4) becomes

$$\begin{aligned}\dot{\bar{x}} = (A &- B_2E_1^{-1}D_{12}'C_1 + (\gamma^{-2}B_1B_1' - B_2E_1^{-1}B_1')X)\bar{x} \\ &+(I - \gamma^{-2}YX)^{-1}(B_1D_{21}' + YC_2')E_2^{-1}(y - [C_2 + D_{21}B_1'X]\bar{x}) \\ &\quad +(I - \gamma^{-2}YX)^{-1}(B_2 + \gamma^{-2}YC_1'D_{12})(u + E_1^{-1}[D_{12}'C_1 + B_2'X]\bar{x}),\end{aligned} \tag{7.20}$$

and the optimal control is simply

$$u = \mathbf{u}^*(\bar{x}, Y, \phi) = \mathbf{u}^*_{state}(\bar{x}) = -E_1^{-1}[D_{12}'C_1 + B_2'X]\bar{x},$$

in complete agreement with the formula (4) for the central controller in [PAJ91] (when one takes $X = X_e$ and $Y = Y_e$ to be the stabilizing solutions). The resulting controller is illustrated in Figure 7.2.

The steady state version of the Y equation (7.14) is

$$\begin{aligned}(A - B_1D_{21}'E_2^{-1}C_2)Y_e + Y_e(A - B_1D_{21}'E_2^{-1}C_2)' \\ +Y_e(\gamma^{-2}C_1'C_1 - C_2'E_2^{-1}C_2)Y_e + B_1[I - D_{21}'E_2^{-1}D_{21}]B_1' = 0,\end{aligned} \tag{7.21}$$

where we seek the stabilizing solution Y_e, meaning

$$A - B_1D_{12}'C_2 + Y_e(\gamma^{-2}C_1C_1' - C_2E_2^{-1}C_2') \text{ asymptotically stable.} \tag{7.22}$$

The central controller $K^*_{p_e}$ uses $X = X_e$, the stabilizing solution of (7.17), meaning

$$A - B_2E_1^{-1}D_{12}'C_1 + (\gamma^{-2}B_1B_1' - B_2E_1^{-1}B_2')X_e \text{ is asymptotically stable} \tag{7.23}$$

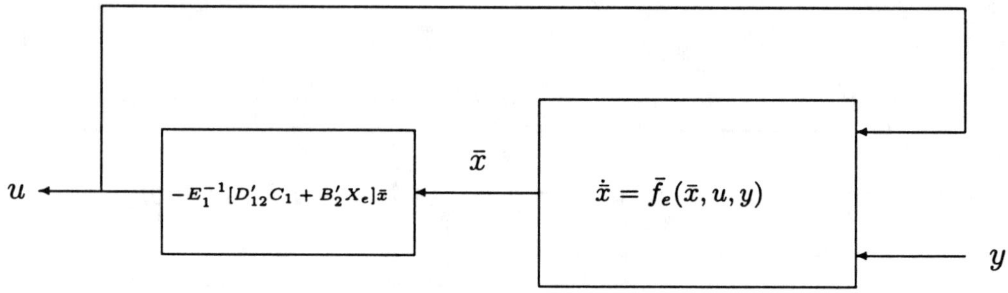

Figure 7.3: Linear central controller. The dynamics $\bar{f}_e$ is defined by equation (7.25).

and is initialized at the equilibrium

$$p_e(x) = \tfrac{1}{2}\gamma^2 x' Y_e^{-1} x, \tag{7.24}$$

with state space description

$$\begin{aligned}\dot{\bar{x}} &= (A - B_2E_1^{-1}D_{21}'C_1 + (\gamma^{-2}B_1B_1' - B_2E_1^{-1}B_1')X_e)\bar{x} \\ &\quad +(I-\gamma^{-2}Y_eX_e)^{-1}(B_1D_{21}' + Y_eC_2')E_2^{-1}(y - [C_2 + D_{21}B_1'X_e]\bar{x}) \\ &\quad +(I-\gamma^{-2}Y_eX_e)^{-1}(B_2 + \gamma^{-2}Y_eC_1'D_{12})(u + E_1^{-1}[D_{12}'C_1 + B_2'X_e]\bar{x}), \\ u &= -E_1^{-1}[D_{12}'C_1 + B_2'X_e]\bar{x}.\end{aligned} \tag{7.25}$$

This controller is illustrated in Figure 7.3.

For detailed results, see [DGKF89], [PAJ91], [GL95], etc. The standard assumptions were given in the introduction, conditions (1.6), (1.7).

7.2.1 Coupling

The coupling condition of [DGKF89] requires that

$$X_eY_e \text{ has spectral radius } < \gamma^2. \tag{7.26}$$

Suppose that $\hat{W}(\hat{x}, T, \phi)$ is defined using the stabilizing solution $X = X_e$, so that $\hat{W} = \tilde{W}$, defined by (7.8). Then setting $Y = Y_e$ we get

$$\tilde{W}(\hat{x}, Y_e, \phi) = \tfrac{1}{2}\hat{x}'(1 - \gamma^{-2}Y_eX_e)^{-1}X_e\hat{x} + \phi \quad \text{is finite.}$$

This corresponds to

$$p_e \in q.i.\text{dom}\, W;$$

recall Theorem 4.11.1. Also note $\hat{W}(0, Y_e, 0) = 0$, corresponding to $W(p_e) = 0$.

If we initialize the controller with $Y_0 \neq Y_e$, we require

$$\phi \leq \hat{W}(\hat{x}, Y_0, \phi) < +\infty,$$

and indeed

$$\phi \leq \hat{W}(\hat{x}(t), Y(t), \phi(t)) < +\infty \quad \text{for all } t \geq 0;$$

cf. §4.9.1.

7.2.2 Storage Function

Take $\gamma = 1$. Then the explicit storage function $e^{\hat{W}}(x,p) = -p(x) + \hat{W}(p)$ of Chapter 6 takes the form

$$e^{\hat{W}}(x,p) = \tfrac{1}{2}(x-\hat{x})'Y^{-1}(x-\hat{x}) + \sup_{\bar{x}}[-\tfrac{1}{2}(\bar{x}-\hat{x})'Y^{-1}(\bar{x}-\hat{x}) + \tfrac{1}{2}\bar{x}'X_e\bar{x}].$$

The supremum occurs at

$$Y^{-1}(\bar{x}-\hat{x}) = X_e\bar{x}$$

and we get

$$e^{\hat{W}}(x,p) = \tfrac{1}{2}(x-\hat{x})'Y^{-1}(x-\hat{x}) + \tfrac{1}{2}\hat{x}X_e[I - YX_e]^{-1}\hat{x}. \tag{7.27}$$

7.3 Certainty Equivalence Principle

In this section we write down the key formulas for the central controller in the special case when the certainty equivalence principle of Basar–Bernhard [BB91] holds.

> ASSUMPTION 7.3.1 In this section we assume that D_{12} and D_{21} are independent of x.

This assumption simplifies the formulas and does not sacrifice generality.

The *minimum stress estimate* is defined by

$$\bar{x}(t) = \operatorname*{argmax}_{x\in\mathbf{R}^n}\{p_t(x) + V(x)\}, \tag{7.28}$$

where $V \geq 0$ is the state feedback value function of (5.45). If this exists and is *unique*, the certainty equivalence controller is given by

$$u^{ce}(t) = \mathbf{u}^*_{state}(\bar{x}(t)), \tag{7.29}$$

where $\mathbf{u}^*_{state}(x)$ is the optimal state feedback controller given by (5.46).

Before presenting explicit formulas, we pause to indicate how this principle fits into the general framework used in this book.

Consider the function

$$\hat{W}(p) = \langle p + V\rangle, \tag{7.30}$$

where $V(x)$ is the state feedback value function (which solves the PDE (5.45)). We define a generalization of the *minimum stress estimate*; viz.,

$$\bar{x}(p) = \operatorname*{argmax}_{x\in\mathbf{R}^n}\{p(x) + V(x)\}, \tag{7.31}$$

which we *assume* to be unique. Then (see [JB96]) $\hat{W}(p)$ is (Gateaux) differentiable, and

$$\nabla_p\hat{W}(p) = E_{\bar{x}(p)}, \tag{7.32}$$

where E_x is the evaluation map $E_x[h] = h(x)$ for $h \in \mathcal{X}$, so that

$$\nabla_p \hat{W}(p)[h] = h(\bar{x}(p)).$$

Most importantly, the function $\hat{W}(p)$ solves the PDE (4.16), and the optimal control and observation (given by (4.23)) simplify to

$$\begin{aligned} \mathbf{u}^*(p) &= -E_1^{-1}(D_{12}'C_1(\bar{x}(p)) - B_2(\bar{x}(p))'\nabla_x p(\bar{x}(p))'), \\ \mathbf{y}^*(p) &= C_2(\bar{x}(p)) - \gamma^{-2} D_{21} B_1(\bar{x}(p))'\nabla_x p(\bar{x}(p))'. \end{aligned} \tag{7.33}$$

Since, if p is nonsingular and smooth,

$$\nabla_x p(\bar{x}(p)) = -\nabla_x V(\bar{x}(p))$$

(first-order necessary condition for a maximum) the certainty equivalence control is, *provided the certainty equivalence assumptions are valid,*

$$u^{ce}(t) = \mathbf{u}^*(p_t) = \mathbf{u}^*_{state}(\bar{x}(p_t)). \tag{7.34}$$

This simplifies the general formula (4.16), (4.23) defining the controller, because under the certainty equivalence assumptions we can use the rather simple solution $\hat{W}(p)$ of equation (4.16), which relates directly to the state feedback controller via the minimum stress estimate. This represents a substantial reduction in complexity.

In order to derive a differential equation for $\bar{x}(t)$, we use the notation

$$\begin{aligned} r_t(x) &= p_t(x) + V(x), \\ \mathbf{y}^*_{state}(x) &= C_2(x) + \gamma^{-2} D_{21} B_1(x)'\nabla_x V(x)', \end{aligned} \tag{7.35}$$

and *assume* that the Hessian matrix satisfies

$$\nabla_x^2 r_t(\bar{x}(t)) > 0, \tag{7.36}$$

meaning that it is positive definite for each $t \geq 0$. Now

$$\nabla_x r_t(\bar{x}(t)) = 0 \ \text{ for all } t \geq 0,$$

and so differentiating we get

$$\nabla_x \frac{\partial r_t}{\partial t}(\bar{x}(t)) + \dot{\bar{x}}(t)'\nabla_x^2 r_t(\bar{x}(t)) = 0.$$

After differentiating the PDEs (3.9) for p and (5.45) for V, we obtain the following differential equation:

$$\begin{aligned} \dot{\bar{x}} &= A - B_2 E_1^{-1} D_{21}' C_1 + (\gamma^{-2} B_1 B_1' - B_2 E_1^{-1} B_2')\nabla_x V' \\ &+ (B_1 D_{21}' - \gamma^2 (\nabla_x^2 r_t)^{-1} \nabla_x y_s^{*\prime}) E_2^{-1} (y - \mathbf{y}^*_{state}) \\ &+ (B_2 + (\nabla_x^2 r_t)^{-1} \nabla_x \mathbf{u}^*_{state}{}' E_1)(u - \mathbf{u}^*_{state}), \end{aligned} \tag{7.37}$$

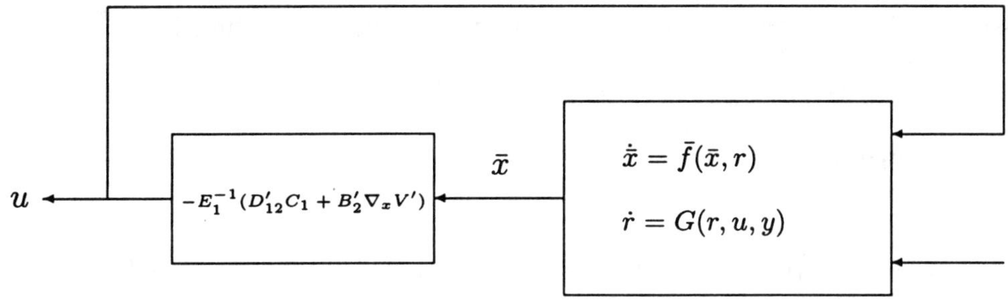

Figure 7.4: Certainty equivalence controller. The dynamics $\bar{f}$ and G are defined by equations (7.37) and (7.38).

where the RHS is evaluated at $x = \bar{x}(t)$, and $u \in L_{2,loc}$, $y \in L_{2,loc}$.

It is important to note that the ODE (7.37) depends on the Hessian of $r = p + V$, and so the certain equivalence controller dynamics is still infinite dimensional. It is possible to write down a PDE for r:

$$\dot{r} = G(r, u, y), \tag{7.38}$$

where

$$\begin{aligned} G(r,u,y) &= -\nabla_x r \cdot [A - B_2 E_1^{-1} D_{12}' C_1 + (\gamma^{-2} B_1 B_1' - B_2 E_1^{-1} B_2')\nabla_x V' \\ &\qquad + B_1 D_{21}' E_2^{-1}(y - \mathbf{y}^*_{state}) + B_2(u - \mathbf{u}^*_{state})] \\ &\quad + \gamma^{-2}\tfrac{1}{2}\nabla_x r B_1 [I - D_{21}' E_2^{-1} D_{21}] B_1' \nabla_x r' \\ &\quad + \tfrac{1}{2}(u - \mathbf{u}^*_{state})' E_1 (u - \mathbf{u}^*_{state}) \\ &\quad - \gamma^2 \tfrac{1}{2}(y - \mathbf{y}^*_{state})' E_2^{-1}(y - \mathbf{y}^*_{state}). \end{aligned} \tag{7.39}$$

This is obtained from the definition $r_t = p_t + V$ and the PDEs (3.9), (5.45), which p_t and V satisfy.

REMARK 7.3.2 Except for u and y, the quantities appearing in (7.39) are functions of x and are *not* evaluated at $\bar{x}$; this includes $\mathbf{u}^*_{state}$ and $\mathbf{y}^*_{state}$. ▽

The structure of the certainty equivalence controller is illustrated in Figure 7.4.

EXAMPLE 7.3.3 (bilinear systems). Connections between certainty equivalence and bilinear systems were studied in [Teo94], [TYJB94]. The state feedback PDE is (5.45) given in Chapter 5. Taking into account the bilinear term, the optimal control and disturbance are given by

$$\begin{aligned} \mathbf{u}^*_{state}(x) &= -E_1^{-1}(D_{12}' C_2 x + (B_2' + x' B_3')\nabla_x V(x)'), \\ \mathbf{w}^*_{state}(x) &= \gamma^{-2} B_1' \nabla_x V(x)'. \end{aligned} \tag{7.40}$$

Assuming the certainty equivalence conditions are satisfied, the function

$$\tilde{\hat{W}}(\hat{x}, Y, \phi) = -\gamma^2 \tfrac{1}{2}(\bar{x} - \hat{x})' Y^{-1} (\bar{x} - \hat{x}) + \phi + V(\bar{x})$$

solves the PDE (7.11), where

$$\bar{x} = \bar{x}(\hat{x}, Y, \phi) = \operatorname{argmax}_x \{-\gamma^2 \tfrac{1}{2}(x - \hat{x})' Y^{-1}(x - \hat{x}) + \phi + V(x)\}$$

satisfies

$$-\gamma^2(\bar{x} - \hat{x})' Y^{-1} + \nabla_x V(\bar{x}) = 0.$$

The certainty equivalence controller is therefore given by

$$u^{ce} = -E_1^{-1}(D_{12}' C_2 \bar{x} + \gamma^2 (B_2' + \bar{x}' B_3') Y^{-1} (\bar{x} - \hat{x})).$$

∇

7.3.1 Breakdown of Certainty Equivalence

Certainty equivalence can break down in two ways. One is the failure of the maximum of $p_t + V$ to be unique. The most dramatic is where the maximum of $p_t + V$ fails to exist. It is this *maximization catastrophe* we discuss in this subsection. Since the functions p_t and V under our assumptions in this book $[p_t + V](x)$ have at most quadratic upper growth, say,

$$[p_t + V](x) \leq c_t |x|^2,$$

for some $c_t > 0$, we can rule out a maximization catastrophe if $c_t \leq 0$ for all $t \geq 0$. But this can fail, so we easily see that the maximum of $p_t + V$ might exist for $t \leq t_0$ and then suddenly it will fail to exist (see §11.2 for examples of situations where this can occur).

At a more abstract level this phenomenon corresponds to our condition

$$p_t \in \operatorname{dom} W$$

holding for $t \leq t_0$ and then suddenly failing. This was discussed in §4.9.1. Key to this is correct initialization of the information state, as specified by quadratic upper limiting property: p_0 satisfying (3.37) implies that p_t satisfies (3.38).

Consider the implications of the results in §4.9.1 in the present context of certainty equivalence. Suppose that $p_0 \leq p_e$, so that (3.38) implies that p_t is bounded above by

$$p_t(x) \leq p_e(x) + k\eta |x|^2 + \frac{c}{\eta} \tag{7.41}$$

for any $\eta > 0$, and any $u, y \in L_2$ (for some constants $k \geq 0$, $c \geq 0$). Assume also that the equilibrium coupling condition holds:

$$p_e(x) + V(x) \leq -\varepsilon_e |x|^2 \tag{7.42}$$

for some $\varepsilon_e > 0$. (This is analogous to the DGKF coupling condition §1.2.) Then if we choose $\eta = \varepsilon_e/(2k)$, by combining (7.41) and (7.42), we have

$$\begin{aligned} p_t(x) + V(x) &\leq p_e(x) + V(x) + k\eta|x|^2 + \frac{c}{\eta} \\ &\leq -(\varepsilon_e - k\eta)|x|^2 + \frac{c}{\eta} \\ &\leq -\tfrac{1}{2}\varepsilon_2|x|^2 + \frac{2kc}{\varepsilon_e} \end{aligned} \tag{7.43}$$

for all $u, y \in L_2$. This implies that *under these circumstances maximization catastrophe cannot occur.*

Similarly, results paralleling the closed-loop theorems of §4.9.1 can be obtained.

7.4 Notes

(i) From an obvious practical point of view, reducing the computational complexity is of vital importance. In the early 1980s, there was considerable interest in the problem of finite-dimensional filters (meaning that the filtering computations are reduced from a PDE to a finite set of (stochastic) differential equations), since the discovery by V. Benes of a class of systems admitting finite-dimensional filters [Ben81]. Bensoussan–Elliott [BE96] obtained finite-dimensional information states for risk-sensitive control, and James–Yuliar [JY95] obtained similar results for differential games; see also [Teo94], [TYJB94], [Yul96]. Other work on bilinear systems includes [SU96], [HJ96a].

(ii) The certainty equivalence principle for risk-sensitive control and dynamic games is due to Whittle [Whi81] and Basar–Bernhard [BB91] (see also [Ber94], [DBB93], [JB96], [Jam94b]).

(iii) The formulas (7.33), (7.34) for the certainty equivalence controller involve $\nabla_x p$, making it difficult to interpret in singular cases. This problem is overcome in Chapter 10.

(iv) Breakdown of certainty equivalence is discussed in [HJ98], [Jam94b].

Chapter 8

Factorization

8.1 Introduction

There are strong connections between H^∞ control and J-inner–outer factorizations of systems; see [BHV91], [Kim97]. In the linear case the H^∞ problem for G is solvable if and only if the reverse-arrow system $\widetilde{G}$ admits a J-inner–outer factorization. This leads to a parameterization of all H^∞ controllers.

In this chapter we show how to extend this to nonlinear systems. Remarkably the information state constructions adapt perfectly to the factorization problem. They fit so naturally that one again in the nonlinear case might suspect that factoring is the most natural way to deal with H^∞ control. Once we obtain appropriate factors we use our formulas to write down large numbers of solutions to the nonlinear H^∞ control problem. Whether or not this generates all solutions remains open.

Our approach to factoring extends the theory for factoring stable systems developed in discrete time by Ball–Helton [BH89], [BH92a] and for continuous time systems by Ball–van der Schaft [BvdS96]. Also there is elegant work due to Baramov and Kimura [BK96]. The key RECIPE presented here was announced and some losslessness properties proved in [HJ94].

8.2 The Problem

8.2.1 Factoring

Typically one says that an operator $\Sigma : \mathbf{S} \to \mathbf{Z}$ factors as

$$\Sigma^I \circ (\Sigma^O)^{-1}$$

if it is the composition

$$\Sigma = \Sigma^I \circ (\Sigma^O)^{-1},$$

where $\Sigma^I : \mathbf{V} \to \mathbf{Z}$ and $\Sigma^O : \mathbf{V} \to \mathbf{S}$. We shall say that Σ has *decomposition* Σ^O, Σ^I provided Σ^I is the composition

$$\Sigma \circ \Sigma^O = \Sigma^I.$$

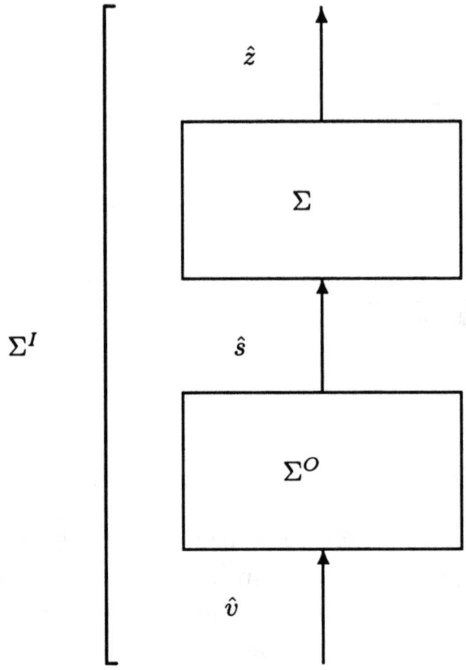

Figure 8.1: The system Σ^I.

This is more directly useful for engineering purposes. The system Σ having a decomposition with Σ^O invertible is equivalent to Σ having a factorization

$$\Sigma = \Sigma^I \circ (\Sigma^O)^{-1}.$$

This is important since as we shall see the control problem calls for a decomposition plus a type of invertible factorization. But we emphasize that its importance to control derives from its relationship to decompositon, as will be seen later. Decomposition is illustrated in Figure 8.1.

There are two ways of thinking about factorizations, depending on whether or not one uses state space realizations for Σ and the two factors Σ^I, Σ^O:

- *Input–output factorization.* In the absence of state space realizations, the operators Σ, Σ^I, and Σ^O are regarded simply as mappings defined on the appropriate spaces, and the above decomposition is defined as the usual composition of mappings.

- *State space factorization.* If a state space realization is given, then it is appropriate to find the state space realizations for the two factors. The state space of the composition is defined on the product of the state spaces of the factors.

Of course, if a state space factorization is given, then input–output factorization can be obtained by suitably initializing the state space systems.

Further, an often smaller state space description of the input–output composition can be obtained from the state space representation of the system composition by cutting it down to the set reachable from a fixed initialization, say, an equilibrium $(0, 0)$. Thus the state space of the input–output composition often has dimension less than or equal to the dimension of the original system.

We will be concerned with state space factorization in this chapter.

8.2.2 The Setup

We wish to factor nonlinear systems Σ given in state space form:

$$\Sigma : \begin{cases} \dot{\hat{x}} = \hat{A}(\hat{x}) + \hat{B}(\hat{x})\hat{s}, \quad \hat{x}(0) = \hat{x}_0, \\ \hat{z} = \hat{C}(\hat{x}) + \hat{D}(\hat{x})\hat{s}. \end{cases} \tag{8.1}$$

In (8.1) $\mathbf{X}$, $\mathbf{S}$, $\mathbf{Z}$ the state, input, and output spaces are vector spaces of finite dimension. These equations define a family of input–output maps $\Sigma_{\hat{x}_0}$ parameterized by the initial condition $\hat{x}_0$. We assume that $\hat{x}_0 = 0$ is an equilibrium $\hat{A}(0) = 0$, $\hat{C}(0) = 0$, with assumptions analogous to Assumption 2.1.1 applying, and write Σ_0 for the input–output map $\hat{s} \mapsto \hat{z}$ with equilibrium initialization.

> ASSUMPTION 8.2.1 Assume $\hat{D}$ is independent of $\hat{x}$ and is invertible, in the sense that
> $$\hat{E} \stackrel{\triangle}{=} \hat{D}' J_{\mathbf{Z}} \hat{D} \text{ invertible.}$$

This assumption will apply throughout this chapter. We will also take $\gamma = 1$.

8.2.3 Dissipation, Losslessness, and Being Outer

Now we list definitions of various sensible notions of dissipation and losslessness. In what follows, J will denote a matrix of the form

$$J \stackrel{\triangle}{=} \begin{bmatrix} I_{n_1} & 0 \\ 0 & -I_{n_2} \end{bmatrix}$$

of appropriate size. These are called *signature matrices*. On $\mathbf{S}$ we select a signature matrix $J_{\mathbf{S}}$, as well as one $J_{\mathbf{Z}}$ on $\mathbf{Z}$. If $\hat{s} \in \mathbf{S}$, it is natural to write it as a sum $\hat{s} = \hat{s}_+ + \hat{s}_-$, where $J_{\mathbf{S}}\hat{s}_+ = \hat{s}_+$ and $J_{\mathbf{S}}\hat{s}_- = -\hat{s}_-$; in other words, with a slight abuse of notation,

$$\hat{s} = \begin{pmatrix} \hat{s}_+ \\ 0 \end{pmatrix} + \begin{pmatrix} 0 \\ \hat{s}_- \end{pmatrix}.$$

It is common to use the misleading notation

$$|\hat{s}|^2_{J_{\mathbf{S}}} \stackrel{\triangle}{=} |\hat{s}_+|^2 - |\hat{s}_-|^2 = (J_{\mathbf{S}}\hat{s}, \hat{s}).$$

Sometimes people call $|\cdot|_{J_{\mathbf{S}}}$ the J-norm even though it does not have the main properties of a norm.

(i) Σ is *J-SS-dissipative* if there exists a finite continuous storage function $e(\cdot) \geq 0$, $e(0) = 0$ on $\mathbf{X}$, such that for all $t_1 \leq t_2$, and inputs $\hat{s} \in L_{2,loc}$,

$$e(\hat{x}(t_2)) + \int_{t_1}^{t_2} |\hat{z}(t)|^2_{J_\mathbf{Z}} dt \leq e(\hat{x}(t_1)) + \int_{t_1}^{t_2} |\hat{s}(t)|^2_{J_\mathbf{S}} dt. \tag{8.2}$$

(ii) Σ is *J-SS-lossless* if there exists a finite continuous storage function $e(\cdot) \geq 0$, $e(0) = 0$ on $\mathbf{X}$, such that for all $t_1 \leq t_2$ and inputs $\hat{s} \in L_{2,loc}$,

$$e(\hat{x}(t_2)) + \int_{t_1}^{t_2} |\hat{z}(t)|^2_{J_\mathbf{Z}} dt = e(\hat{x}(t_1)) + \int_{t_1}^{t_2} |\hat{s}(t)|^2_{J_\mathbf{S}} dt. \tag{8.3}$$

(iii) Σ_0 is *J-IO-dissipative* if for $\hat{x}_0 = 0$ and all $T \geq 0$ and inputs $\hat{s}$,

$$\int_0^T |\hat{z}(t)|^2_{J_\mathbf{Z}} dt \leq \int_0^T |\hat{s}(t)|^2_{J_\mathbf{S}} dt. \tag{8.4}$$

(iv) Σ_0 is *J-IO-lossless* if for initial state $\hat{x}_0 = 0$, all inputs $\hat{s}$ in L_2 that produce L_2 outputs $\hat{z}$ satisfy

$$\int_0^\infty |\hat{z}(t)|^2_{J_\mathbf{Z}} dt = \int_0^\infty |\hat{s}(t)|^2_{J_\mathbf{S}} dt. \tag{8.5}$$

(v) Σ_0 is L_2 *stable* if Σ_0 maps L_2 into L_2.

(vi) Σ_0 is *outer* if both Σ_0 and Σ_0^{-1} are L_2 stable.

The terms *dissipative* and *lossless* (without the J prefix) will have the above meanings with J replaced by the matrix $\mathrm{diag}(I, I)$ as appropriate. All of the conditions (i) through (v) imply that $\dim \mathbf{Z} \leq \dim \mathbf{S}$.

REMARK 8.2.2

(i) J-SS-lossless implies J-SS-dissipative implies J-IO-dissipative.

Proof. This follows immediately from $e \geq 0$ and $e(0) = 0$.

(ii) J-SS-lossless does not necessarily imply J-IO-lossless.

The point is if $x(t) \to 0$, then $e(x(t)) \to 0$, which converts the J-SS-lossless equality as $t \to \infty$ to the J-IO-lossless equality. However, J-SS-lossless systems may not be stable.

(iii) A sad nonproperty of J-dissipation is that it *does not imply stability*. Even if one assumes detectability it does not.

The problem is that $\hat{z}$ can have $\int_0^T |\hat{z}(t)|^2_{J_\mathbf{Z}} dt$ uniformly bounded in T but $\hat{z}$ cannot be in L_2.

∇

8.3 The Information State and Critical Feedback

For fixed $\hat{s} \in L_{2,loc}$, the *information state* $p_t \in \mathcal{X}_e$ *of the system* Σ is defined by

$$p_t(\hat{x}) = p_0(\hat{\xi}(0)) + \tfrac{1}{2}\int_0^t |\hat{C}(\hat{\xi}(s)) + \hat{D}(\hat{\xi}(s))\hat{s}|^2_{J_Z}\, ds, \tag{8.6}$$

where $\hat{\xi}(\cdot)$ is the solution of

$$\dot{\hat{\xi}} = \hat{A}(\hat{\xi}) + \hat{B}(\hat{\xi})\hat{s}, \tag{8.1$'$}$$

with *terminal* condition $\hat{\xi}(t) = \hat{x}$.

The dynamics for p_t, when p_t is a smooth function, is a PDE: for fixed $\hat{s} \in L_{2,loc}$ we have

$$\dot{p}_t = \hat{F}(p_t, \hat{s}(t)), \tag{8.7}$$

where $\hat{F}(p, \hat{s})$ is the differential operator

$$\hat{F}(p, \hat{s}) \stackrel{\Delta}{=} -\nabla_x p \cdot (\hat{A} + \hat{B}\hat{s}) \; + \; \tfrac{1}{2}\, |\hat{C} + \hat{D}\hat{s}|^2_{J_Z}. \tag{8.8}$$

Recall (§3.3 and Chapter 11) that there is an equilibrium p_e with a basin of attraction $\mathcal{D}_{attr}$. The function p_e is usually singular.

Key to constructing our factors is a function $\mathbf{\hat{s}}^*$ on the information states. Later we write down formulas in various cases for $\mathbf{\hat{s}}^*(p)$, but for now we present our results in terms of an unspecified function $\mathbf{\hat{s}}^*(p)$ that is continuous in some sense with respect to p and that has the properties $\mathbf{\hat{s}}^*(p + c) = \mathbf{\hat{s}}^*(p)$ for any constant $c \in \mathbf{R}$. We shall always assume $\mathbf{\hat{s}}^*(p_e) = 0$.

8.4 RECIPE for the Factors

In this section we present formulas for the factors. We present them as differential equations so they require p_t to be smooth functions in order to make sense (classically). This is just to make the equations seem more conventional and intuitive. However, one can integrate the formulas and present them in integral form or appeal to weak sense interpretations. This is superior in that this gives inner–outer factorization formulas when the functions p_t are singular. Indeed then one can work with an initial state p_0, which is singular as is often the case.

Note that the construction makes sense for any function $\hat{F}$ and $\mathbf{\hat{s}}^*$, not just the information state. Pick $\hat{d}$ so that

$$\hat{d}'\hat{E}\hat{d} = J_{\mathbf{V}}, \tag{8.9}$$

and let $\hat{d}^{-L}$ denote a left inverse of it; that is, $\hat{d}^{-L}\hat{d} = I_{\mathbf{V}}$.

(i) Construct Σ^O as follows:

$$\Sigma^O_{p_0} : \begin{cases} \dot{p} \;=\; \hat{F}(p, \mathbf{\hat{s}}^*(p) + \hat{d}\hat{v}), \quad p(0) = p_0, \\ \hat{s} \;=\; \mathbf{\hat{s}}^*(p) + \hat{d}\hat{v}, \\ \mathcal{R}(\Sigma^O_{p_0}) \stackrel{\Delta}{=} \{p \text{ reachable from } p_0 \text{ with inputs } \hat{v} \in L_{2,loc}\}. \end{cases} \tag{8.10}$$

(ii) The J-inner factor Σ^I is given by combining (8.1) and (8.10):

$$\Sigma^I_{\hat{\xi}_0, p_0} : \begin{cases} \dot{p} = \hat{F}(p, \hat{\mathbf{s}}^*(p) + \hat{d}\hat{v}), \quad p(0) = p_0, \\ \dot{\hat{\xi}} = \hat{A}(\hat{\xi}) + \hat{B}(\hat{\xi})(\hat{\mathbf{s}}^*(p) + \hat{d}\hat{v}), \quad \hat{\xi}(0) = \hat{\xi}_0, \\ \hat{z} = \hat{C}(\hat{\xi}) + \hat{D}(\hat{\xi})(\hat{\mathbf{s}}^*(p) + \hat{d}\hat{v}), \\ \mathcal{R}(\Sigma^I_{\hat{\xi}_0, p_0}) \triangleq (\mathbf{X}, \mathcal{R}(\Sigma^O_{p_0})). \end{cases} \tag{8.11}$$

(iii) Take the inverse of Σ^O to give

$$(\Sigma^O_{p_0})^{-1} : \begin{cases} \dot{p} = \hat{F}(p, \hat{s}), \quad p(0) = p_0, \\ \hat{v} = \hat{d}^{-L}(\hat{s} - \hat{\mathbf{s}}^*(p)), \\ \mathcal{R}((\Sigma^O_{p_0})^{-1}) \triangleq \{p \text{ reachable from } p_0 \text{ with inputs } \hat{s} \in L_{2,loc}\}. \end{cases} \tag{8.12}$$

Also interesting is the cutdown of Σ^I:

$$\Sigma^I_{\hat{\xi}_0, p_0|} \triangleq \Sigma^I_{\hat{\xi}_0, p_0} \text{ restricted to } \mathcal{R}(\Sigma^I_{\hat{\xi}_0, p_0|}),$$

where

$$\mathcal{R}(\Sigma^I_{0, p_0|}) \triangleq \{(\hat{\xi}, p) \text{ reachable from } \hat{\xi}_0 = 0, p_0 \text{ and } p_t(\hat{\xi}) \text{ is finite}\}.$$

We often drop the subscripts on the systems if the initial state of the system is obvious from the context.

8.5 Properties of the Factors

In this section we present properties of the factors defined by the formulas given in §8.4. Before doing this we describe necessary properties of the function $\hat{\mathbf{s}}^*$, as well as other assumptions. The function $\hat{\mathbf{s}}^*$ is defined in terms of a PDE (see (8.13) below), which is closely related to the dynamic programming PDE (4.16).

8.5.1 The Factoring PDE

We need a solution $\hat{W}$, $\hat{\mathbf{s}}^*$ to the *factoring PDE*

$$\text{crit}_{\hat{s} \in \mathbf{S}} \nabla_p \hat{W}(p)[\hat{F}(p, \hat{s})] = 0 \tag{8.13}$$

defined on a subset of dom $\hat{W}$ in $\mathcal{X}_e$. Here $\text{crit}_{\hat{s} \in \mathbf{S}}$ stands for a critical point in $\mathbf{S}$ and we assume $\hat{\mathbf{s}}^*(p)$ is such a critical point. It implies that $\hat{W}$ is a function that satisfies

$$\hat{W}(p_t) = \hat{W}(p_0) \tag{8.14}$$

for all $t \geq 0$, where p_t is driven by $\hat{s} = \hat{\mathbf{s}}^*(p_t)$ with an appropriate initialization. The function $\hat{W}$ should satisfy the structural properties (§2.2.2.2). While ultimately understanding (8.13) in weak (e.g., viscosity) senses is important, we now work with smooth $\hat{W}$. Indeed, we will use interpretations as in §4.4.1 to clarify ideas; more general interpretations are left to the reader.

Since $\hat{F}$ is quadratic in $\hat{s}$, there is exactly one critical point that indeed can be computed to be

$$\hat{\mathbf{s}}^*(p) = -\nabla_p \hat{W}(p)[\hat{E}^{-1}(\hat{D}' J_{\mathbf{Z}} \hat{C} - \hat{B}' \nabla_x p')] \tag{8.15}$$

by taking derivatives in (8.13) and setting to 0, where $\hat{E} = \hat{D}' J_{\mathbf{Z}} \hat{D}$. (This formula assumes $\hat{D}$ does not depend on $\hat{x}$.) When we plug $\hat{\mathbf{s}}^*(p)$ into (8.13) we obtain

$$\nabla_p \hat{W}(p)[-\nabla_x p \cdot \hat{A} + \tfrac{1}{2}\hat{C}' J_{\mathbf{Z}} \hat{C}] - \tfrac{1}{2}\hat{\mathbf{s}}^*(p)' \hat{E} \hat{\mathbf{s}}^*(p) = 0. \tag{8.16}$$

See §8.10 for further details.

This differential equation for $\hat{W}$ typically has multiple solutions, and to each one there is an associated $\hat{\mathbf{s}}^*(p)$.

8.5.2 Factoring Assumptions

We make the following assumptions.

ASSUMPTION 8.5.1 We always assume the vector field $\hat{A}$ is incrementally hyperbolic.

ASSUMPTION 8.5.2 $\hat{W}$ is smooth on a domain dom $\hat{W}$, and $\hat{W}$ together with critical point $\hat{\mathbf{s}}^*$ solve (8.15). Also $\hat{W}$ should satisfy the structural conditions (recall §2.2.2.2).

ASSUMPTION 8.5.3 The function p_0 has the property that for all $\hat{s} \in L_{2,loc}$ and all $t \geq 0$ the function p_t solving the information state equation (8.6) exists and belongs to dom $\hat{W}$.

ASSUMPTION 8.5.4 The augmented information state system (8.17)

$$\dot{p} = \hat{F}(p, \hat{\mathbf{s}}^*(p) + \hat{d}\hat{v}) \tag{8.17}$$

when driven by $L_{2,loc}$ input $\hat{v}$ has a solution $p_t \in$ dom $\hat{W}$ for all $t \geq 0$ for suitable initial states p_0.

ASSUMPTION 8.5.5 Assumptions 8.5.3 and 8.5.4 ensure that the dynamics of the systems Σ, Σ_0, Σ^I have solutions for $L_{2,loc}$ inputs. In addition we assume the outputs from these systems are in $L_{2,loc}$.

As we now see these assumptions are sufficient for the RECIPE to produce a J-lossless factor and some of these assumption are necessary. As we shall see Σ^O stable is not necessary for Σ^I to be J-SS-lossless nor is it required by our construction.

8.5.3 The Outer Factor Σ^O

Stability of Σ^O amounts to the *stabilizing* property of $\hat{\mathfrak{s}}^*, p_0$ (recall §4.12.1), which in this context means that the following set is nonempty:

$$\begin{aligned}\mathrm{dom}\,_{cs}\hat{\mathfrak{s}}^* = \{p_0 \in \mathrm{dom}\,\hat{W} \;:\; &\text{the solution } p_t \text{ of (8.10)}\\ &\text{is well defined with } p_t \in \mathrm{dom}\,\hat{W} \text{ and}\\ &t \mapsto \hat{\mathfrak{s}}^*(p_t) \in L_2[0,\infty) \text{ for all } \hat{v} \in L_2[0,\infty)\}.\end{aligned} \tag{8.18}$$

Exactly what assumptions yield this property is not yet understood.

THEOREM 8.5.6 *Under the factoring assumptions, the outer factor $\Sigma^O_{p_0}$ is stable provided $p_0 \in \mathrm{dom}\,_{cs}\hat{\mathfrak{s}}^* \cap \mathcal{D}_{attr}(p_e)$, where p_e is a control attractor, in the sense that $\hat{v} \in L_2$ implies $\Sigma^O_{p_0}\hat{v} \in L_2$ and $p_t \Longrightarrow p_e + c(\hat{v})$.*

Proof. If $p_0 \in \mathrm{dom}\,_{cs}\hat{\mathfrak{s}}^*$, and $\hat{v} \in L_2$, then by definition $t \mapsto \hat{\mathfrak{s}}^*(p_t) \in L_2$. Thus $t \mapsto \hat{\mathfrak{s}}^*(p_t) + \hat{d}\hat{v}(t) = \hat{s}(t) \in L_2$. If also $p_0 \in \mathcal{D}_{attr}(p_e)$, then $p_t \Longrightarrow p_e + c$. □

8.5.4 The Inner Factor Σ^I

The main desirable properties of the inner factor are determined by the factoring PDE. These results are presented in the next theorem.

THEOREM 8.5.7 *Let $p_0 \in \mathrm{dom}\,\hat{W}$ and suppose Assumptions* 8.5.1, 8.5.2, 8.5.3 *and* 8.2.1 *hold. Consider the inner factor $\Sigma^I_{\hat{x}_0, p_0|}$ on the state space $\mathcal{R}(\Sigma^I_{\hat{\xi}_0, p_0|})$. Then*

(i) $\Sigma^I_{\hat{\xi}_0, p_0|}$ *is J-SS-lossless,*

(ii) $\Sigma^I_{\hat{\xi}_0, p_0|}$ *is J-SS-dissipative.*

Now assume in addition that $\hat{W}(p_0) = p_0(0)$; then

(iii) $\Sigma^I_{0, p_0|}$ *is J-IO-dissipative, and*

(iv) *if $p_0 \in \mathrm{dom}\,_{cs}\hat{\mathfrak{s}}^* \cap \mathcal{D}_{attr}(p_e)$, where p_e is a control attractor with $\hat{W}(p_e) = p_e(0)$, and $\hat{e}$ is continuous on its domain in $\mathbf{X} \times \mathcal{X}_e$, the $\Sigma^I_{0, p_0|}$ is J-IO-lossless in the sense that for any input $\hat{v} \in L_2$ that produces $\hat{z} \in L_2$ and $p_t \Longrightarrow p_e + c(\hat{v})$, $\hat{\xi}(t) \to 0$,* (8.5) *holds.*

The proof of this theorem, while not difficult, requires two preliminary lemmas.

LEMMA 8.5.8 *Let $p_0 \in \mathrm{dom}\,\hat{W}$. Then for any $\hat{v}$ in $L_{2,loc}$ such that the information state p_t given by* (8.6) *belongs to* dom $\hat{W}$ *for all $t \geq 0$, we have*

$$\hat{W}(p_t) = \hat{W}(p_0) + \tfrac{1}{2}\int_0^t |\hat{v}(s)|^2_{J_V}\, ds \text{ for all } t \geq 0. \tag{8.19}$$

Proof. By definition of $\hat{\mathbb{s}}^*(p)$ as the critical point in equation (8.13), we have, using also the form of $\hat{F}$,

$$\nabla_p W(p)[\hat{F}(p, \hat{\mathbb{s}}^*(p) + \hat{d}\hat{v})] = \tfrac{1}{2}|\hat{D}\hat{d}\hat{v}|^2_{J_{\mathbf{Z}}} = \tfrac{1}{2}|\hat{v}|^2_{J_{\mathbf{V}}}. \tag{8.20}$$

Integrate this to obtain (8.19). □

The next lemma defines a suitable energy (storage) function; cf. §6.2.

LEMMA 8.5.9 *Suppose that $\hat{W}$ is smooth, satisfies the structural properties, and that $\hat{\mathbb{s}}^*$ is a critical point for* (8.13). *Let* $p_0 \in \mathrm{dom}\,\hat{W}$. *Assume that for any $\hat{v}$ in $L_{2,loc}$ and the augmented information state p_t given by the augmented information state dynamics (first line of* (8.11)) *belongs to* $\mathrm{dom}\,\hat{W}$ *for all* $t \geq 0$. *Then the storage function*

$$\hat{e}(\hat{x}, p) = -p(\hat{x}) + \hat{W}(p)$$

satisfies the energy balance equation for $\Sigma^I_{p_0}$*:*

$$\hat{e}(\hat{\xi}(t), p_t) + \tfrac{1}{2}\int_0^t |\hat{z}(s)|^2_{J_{\mathbf{Z}}}\, ds \;=\; \hat{e}(\hat{\xi}_0, p_0) + \tfrac{1}{2}\int_0^t |\hat{v}|^2_{J_{\mathbf{V}}}\, ds \tag{8.21}$$

for all $\hat{\xi}_0$ *such that* $p_0(\hat{\xi}_0)$ *is finite.*

Proof. First observe, from (8.6), that

$$p_t(\hat{\xi}(t)) = p_0(\hat{\xi}(0)) + \tfrac{1}{2}\int_0^t |\hat{C}(\hat{\xi}(s)) + \hat{D}(\hat{\xi}(s))\hat{s}|^2_{J_{\mathbf{Z}}}\, ds,$$

and so $p_t(\hat{\xi}(t))$ is finite provided $p_0(\hat{\xi}(0))$ is finite, and hence $\hat{e}(\hat{\xi}(t), p_t)$ is finite. Combining the previous display and (8.19), we get the identity

$$\hat{W}(p_t) - p_t(\hat{\xi}(t)) = \hat{W}(p_0) - p_0(\hat{\xi}(0)) + \tfrac{1}{2}\int_0^t |\hat{v}(s)|^2_{J_V}\, ds - \tfrac{1}{2}\int_0^t |\hat{z}(s)|^2_{J_Z}\, ds.$$

But this is just (8.21). □

Proof of Theorem 8.5.7. We know that $\hat{e}(\hat{x}, p) \geq 0$ by Assumption 8.5.2 and by Assumptions 8.5.3, 8.5.4, 8.5.5 that putting $\hat{v} \in L_{2,loc}$ into $\Sigma^I_{p_0}$ gives output in $L_{2,loc}$, so the J-SS-lossless and therefore J-SS-dissipative properties for $\Sigma^I_{p_0}$ and $\Sigma^I_{p_0|}$ follow directly from (8.21) provided $\hat{e}(\hat{x}, p)$ is finite. When we assume p_0 is everywhere finite we get by the definition of p_t that it is everywhere finite. When p_0 is singular $\hat{e}$ no longer is *everywhere* finite, so it no longer serves as a storage function for $\Sigma^I_{p_0}$, but at least the J-SS-dissipative properties for $\Sigma^I_{p_0}$ and $\Sigma^I_{p_0|}$ still follow directly from (8.21) provided $\hat{e}(\hat{x}_0, p_0)$ is finite or, equivalently, $p_0(\hat{x}_0)$ is finite and $p_0 \in \mathrm{dom}\,\hat{W}$. With $\Sigma^I_{p_0|}$ we are restricted to the part of the state space $\mathbf{X}$ on which p_0 is finite, so J-SS-losslessness, etc., are valid as stated. This proves items (i) and (ii).

If $\hat{W}(p_0) = p_0(0)$, then $\hat{e}(0, p_0) = 0$, and item (iii) follows.

Under the hypotheses of item (iv), we have for such an input $\hat{v}$ that $\hat{e}(\hat{\xi}(t), p_t) \to 0$ as $t \to \infty$, and so J-IO-losslessness follows. □

8.5.5 The Inverse Outer Factor $(\Sigma^O)^{-1}$

The stability of the operator $(\Sigma^O_{p_0})^{-1}$ follows provided the initial condition p_0 is suitably chosen and the function $\hat{\mathbf{s}}^*$ is sufficiently well behaved.

THEOREM 8.5.10 *Under the factoring assumptions, the inverse outer factor* $(\Sigma^O_{p_0})^{-1}$ *is stable at least to the extent that if* $p_0 \in \mathcal{D}_{attr}(p_e)$, *where* p_e *is a control attractor, then* $p_t \Longrightarrow p_e + c(\hat{s})$ *for any input* $\hat{s} \in L_2$. *If in addition* $t \mapsto \hat{\mathbf{s}}^*(p_t) \in L_2$, *then* $(\Sigma^O_{p_0})^{-1}\hat{s} \in L_2$.

Proof. This result follows simply from the definitions. □

8.5.6 Necessity of the RECIPE Formulas

Such results will appear later after describing the application of this theory to control. Ironically the control application greatly facilitates proof and presentation of the necessary results.

8.5.7 Singular Cases

The results presented above require $p_0(\hat{\xi}_0)$ finite. More general results for singular cases can be derived along the lines of Chapter 4 using a hyperbolicity assumption; however, we do not do so here.

8.6 Examples

8.6.1 Certainty Equivalence

There are conditions (Chapter 7) under which $\hat{W}$ given by the formula

$$\hat{W}(p) \triangleq \langle p + \hat{V} \rangle \tag{8.22}$$

solves the factoring PDE (8.13). Here $\hat{V}$ denotes the stabilizing solution to *the state feedback factoring PDE*

$$\begin{aligned} &\nabla_{\hat{x}}\hat{V} \cdot (\hat{A} - \hat{B}\hat{E}^{-1}\hat{D}'J_{\mathbf{Z}}\hat{C}) - \tfrac{1}{2}\nabla_{\hat{x}}\hat{V}\hat{B}\hat{E}^{-1}\hat{B}'\nabla_{\hat{x}}\hat{V}' \\ &+\tfrac{1}{2}\hat{C}'J_{\mathbf{Z}}(J_{\mathbf{Z}} - \hat{D}\hat{E}^{-1}\hat{D}')J_{\mathbf{Z}}C = 0, \end{aligned} \tag{8.23}$$

which is nonnegative, $\hat{V}(0) = 0$, and stabilizing, to wit the vector field

$$\hat{A} - \hat{B}\hat{E}^{-1}\hat{D}'J_{\mathbf{Z}}\hat{C} - \hat{B}\hat{E}^{-1}\hat{B}'\nabla_{\hat{x}}\hat{V}'$$

is incrementally L_2 exponentially stable.

The assumptions that make (8.22) true follow:

(i) $\hat{V}$ is finite, smooth, $\hat{V}(0) = 0$, and $\hat{V} \geq 0$.

(ii) The supremum in (8.22) exists and occurs at a unique $\hat{x}$ denoted $\bar{\hat{x}}(p)$.

Then

$$\hat{\mathbf{s}}^*(p) = \hat{\mathbf{s}}^*_{state}(\bar{\hat{x}}(p)),$$

where $\hat{\mathbf{s}}^*_{state}$ is given by

$$\hat{\mathbf{s}}^*_{state}(\hat{x}) \triangleq -E^{-1}[\hat{D}' J_{\mathbf{Z}} \hat{C} + \hat{B}' \nabla_x \hat{V}']_{|\hat{x}}, \tag{8.24}$$

which incidentally solves the "state feedback" factorization problem treated in §8.9. Also note that $\hat{\mathbf{s}}^*_{state}(\hat{x}) = \hat{\mathbf{s}}^*(\delta_{\hat{x}})$ where $\hat{\mathbf{s}}^*$ is given by (8.15). Here we use the fact $\nabla_p \hat{W}(\delta_{\hat{x}})[h] = h(\hat{x})$ for all suitable functions h (see Chapter 10).

Observe that the "stabilizing" property of $\hat{V}$ means that

$$\begin{aligned} \dot{\hat{x}} &= [\hat{A} - \hat{B}\hat{E}^{-1}\hat{D}'\hat{C}](\hat{x}) - \hat{B}E^{-1}\hat{B}'\nabla_x \hat{V}'(\hat{x}) \\ &= \hat{A}(\hat{x}) + \hat{B}\hat{\mathbf{s}}^*_{state}(\hat{x}) \end{aligned}$$

is asymptotically stable.

Assumption (ii) above is what one would call the *certainty equivalence assumption* in this setting. Note that $\hat{V} \geq 0$ implies $\hat{W}(p) = \langle p + \hat{V} \rangle \geq \langle p \rangle$.

The RECIPE in the certainty equivalence case follows:

(i) Outer factor:

$$\Sigma^O_{p_0} : \begin{cases} \dot{p} = \hat{F}(p, \hat{\mathbf{s}}^*_{state}(\bar{\hat{x}}(p)) + \hat{d}\hat{v}), \quad p(0) = p_0, \\ \hat{s} = \hat{\mathbf{s}}^*_{state}(\bar{\hat{x}}(p)) + \hat{d}\hat{v}, \\ \mathcal{R}(\Sigma^O_{p_0}) \triangleq \{p \text{ reachable from } p_0 \text{ with inputs } \hat{v} \in L_{2,loc}\}. \end{cases} \tag{8.25}$$

(ii) The J-inner factor:

$$\Sigma^I_{\hat{\xi}_0, p_0} : \begin{cases} \dot{p} = \hat{F}(p, \hat{\mathbf{s}}^*_{state}(\bar{\hat{x}}(p)) + \hat{d}\hat{v}), \quad p(0) = p_0, \\ \dot{\hat{\xi}} = \hat{A}(\hat{\xi}) + \hat{B}(\hat{\xi})(\hat{\mathbf{s}}^*_{state}(\bar{\hat{x}}(p)) + \hat{d}\hat{v}), \quad \hat{\xi}(0) = \hat{\xi}_0, \\ \hat{z} = \hat{C}(\hat{\xi}) + \hat{D}(\hat{\xi})(\hat{\mathbf{s}}^*(\bar{\hat{x}}(p)) + \hat{d}\hat{v}), \\ \mathcal{R}\left(\Sigma^I_{\hat{\xi}_0, p_0}\right) \triangleq (\mathbf{X}, \mathcal{R}(\Sigma^O_{p_0})). \end{cases} \tag{8.26}$$

(iii) Inverse outer factor:

$$(\Sigma^O_{p_0})^{-1} : \begin{cases} \dot{p} = \hat{F}(p, \hat{s}), \quad p(0) = p_0, \\ \hat{v} = \hat{d}^{-L}(\hat{s} - \hat{\mathbf{s}}^*_{state}(\bar{\hat{x}}(p))), \\ \mathcal{R}((\Sigma^O_{p_0})^{-1}) \triangleq \{p \text{ reachable from } p_0 \text{ with inputs } \hat{s} \in L_{2,loc}\}. \end{cases} \tag{8.27}$$

8.6.2 $\hat{A}$ Stable

The stable equilibrium p_e for the information state equation (8.6) is the singular information state δ_0. The natural initialization in this case is $p_0 = \delta_0$. Then Σ^O has state $p_t = \delta_{\bar{\hat{x}}(t)}$, where $\bar{\hat{x}}(\cdot)$ satisfies

$$\Sigma^O_{\delta_{\bar{\hat{x}}_0}} : \begin{cases} \dot{\bar{\hat{x}}} = \hat{A}(\bar{\hat{x}}) + \hat{B}(\bar{\hat{x}})[\hat{\mathbf{s}}^*_{state}(\bar{\hat{x}}) + \hat{d}\hat{v}], & \bar{\hat{x}}(0) = \bar{\hat{x}}_0, \\ \hat{s} = \hat{\mathbf{s}}^*_{state}(\bar{\hat{x}}) + \hat{d}\hat{v}, \end{cases} \tag{8.28}$$

where $\hat{\mathbf{s}}^*_{state}$ is given by (8.24), which is determined by a solution $\hat{V}$ of the state feedback factoring PDE (8.13). If $\hat{V}$ is stabilizing, then Σ^O is certainly L_2 stable. Note that certainty equivalence holds in this case, since with $p = \delta_{\hat{\xi}}$ we have

$$\bar{\hat{x}}(\delta_{\hat{\xi}}) = \hat{\xi}.$$

Of course $(\Sigma^O)^{-1}$ and Σ^I are derived easily from (8.28). Indeed,

$$(\Sigma^O_{\delta_{\bar{\hat{x}}_0}})^{-1} : \begin{cases} \dot{\hat{x}} = \hat{A}(\hat{x}) + \hat{B}(\hat{x})\hat{v}, & \hat{x}(0) = \hat{x}_0, \\ \hat{v} = \hat{d}^{-L}(\hat{s} - \hat{\mathbf{s}}^*_{state}(\hat{x})). \end{cases} \tag{8.29}$$

If $\hat{A}$ is incrementally L_2 exponentially stable, then $(\Sigma^O_{\delta_{\bar{\hat{x}}_0}})^{-1}$ is L_2 stable. So Σ^O is outer.

The J-inner factor is given by

$$\Sigma^I_{\delta_{\bar{\hat{x}}_0}, \hat{\xi}_0} : \begin{cases} \dot{\bar{\hat{x}}} = \hat{A}(\bar{\hat{x}}) + \hat{B}(\bar{\hat{x}})[\hat{\mathbf{s}}^*_{state}(\bar{\hat{x}}) + \hat{d}\hat{v}], & \bar{\hat{x}}(0) = \bar{\hat{x}}_0, \\ \dot{\hat{\xi}} = \hat{A}(\hat{\xi}) + \hat{B}(\hat{\xi})(\hat{\mathbf{s}}^*_{state}(\bar{\hat{x}}) + \hat{d}\hat{v}), & \hat{\xi}(0) = \hat{\xi}_0, \\ \hat{z} = \hat{C}(\hat{\xi}) + \hat{D}(\hat{\xi})(\hat{\mathbf{s}}^*_{state}(\bar{\hat{x}}) + \hat{d}\hat{v}). \end{cases} \tag{8.30}$$

In the case $\bar{\hat{x}}_0 = \hat{\xi}_0$, this reduces to

$$\Sigma^I_{\delta_{\bar{\hat{x}}_0}, \bar{\hat{x}}_0} : \begin{cases} \dot{\bar{\hat{x}}} = \hat{A}(\bar{\hat{x}}) + \hat{B}(\bar{\hat{x}})[\hat{\mathbf{s}}^*_{state}(\bar{\hat{x}}) + \hat{d}\hat{v}], & \bar{\hat{x}}(0) = \bar{\hat{x}}_0, \\ \hat{z} = \hat{C}(\bar{\hat{x}}) + \hat{D}(\bar{\hat{x}})(\hat{\mathbf{s}}^*_{state}(\bar{\hat{x}}) + \hat{d}\hat{v}). \end{cases} \tag{8.31}$$

8.6.3 $\hat{A}$ Strictly Antistable

The results in the preceding subsection §8.6.2 show how existing results in the literature can be obtained from the information state perspective. The stability of $A^{\times}$ assumed there results in an information state that projects to a finite-dimensional ODE (when correctly initialized). In contrast to this, we mention that the information state framework provides a complete theory when $A^{\times}$ is not stable, for instance if $A^{\times}$ is antistable. When $A^{\times}$ is antistable, the information state is nonsingular and does not in general project to a finite-dimensional system (except in linear and bilinear cases). This case is genuinely infinite dimensional. The energy function $e(x,p)$ is finite for all p's occurring, since they are nonsingular. Further, hyperbolic cases can also be treated using the information state framework, as our results show.

8.6.4 Bilinear Systems

Consider the bilinear system

$$\Sigma : \begin{cases} \dot{\hat{x}} = \hat{A}(\hat{s})\hat{x} + \hat{B}\hat{s}, \quad \hat{x}(0) = \hat{x}_0, \\ \hat{z} = \hat{C}\hat{x} + \hat{D}\hat{s}, \end{cases} \tag{8.32}$$

where $\hat{B}$, $\hat{C}$, and $\hat{D}$ are matrices of appropriate sizes, and $\hat{A}(\hat{s})$ is an affine matrix function of $\hat{s}$. As in §7.1, the information state is given by

$$p_t(\hat{x}) = -\tfrac{1}{2}(\hat{x} - \hat{\bar{x}}(t))'\hat{Y}(t)^{-1}(\hat{x} - \hat{\bar{x}}(t)) + \phi(t), \tag{8.33}$$

where

$$\begin{aligned} \dot{\hat{\bar{x}}} &= (\hat{A}(\hat{s}) + \hat{Y}\hat{C}'J_{\mathbf{Z}}\hat{C})\hat{\bar{x}} + (\hat{B} + \hat{Y}\hat{C}'J_{\mathbf{Z}}\hat{D})\hat{s}, \\ \dot{\hat{Y}} &= \hat{A}(\hat{s})\hat{Y} + \hat{Y}\hat{A}(\hat{s})' + \hat{Y}\hat{C}'J_{\mathbf{Z}}\hat{Y}, \\ \dot{\phi} &= \tfrac{1}{2}|\hat{C}\hat{\bar{x}} + \hat{D}\hat{s}|^2, \end{aligned} \tag{8.34}$$

provided p_0 is such a quadratic form.

The critical feedback $\mathbf{s}^*(\hat{\bar{x}}, \hat{Y})$ is obtained from the bilinear factoring PDE

$$\begin{aligned} &\text{crit}_{\hat{s}}\{\nabla_{\hat{\bar{x}}}\hat{W}(\hat{\bar{x}}, \hat{Y}, \phi) \cdot [(\hat{A}(\hat{s}) + \hat{Y}\hat{C}'J_{\mathbf{Z}}\hat{C})\hat{\bar{x}} + (\hat{B} + \hat{Y}\hat{C}'J_{\mathbf{Z}}\hat{D})\hat{s}] \\ &\quad + \nabla_{\hat{Y}}\hat{W} \cdot [\hat{A}(\hat{s})\hat{Y} + \hat{Y}\hat{A}(\hat{s})' + \hat{Y}\hat{C}'J_{\mathbf{Z}}\hat{Y}] + \nabla_{\phi}\hat{W} \cdot [\tfrac{1}{2}|\hat{C}\hat{\bar{x}} + \hat{D}\hat{s}|^2]\} = 0. \end{aligned} \tag{8.35}$$

This gives the RECIPE for the bilinear case.

(i) Outer factor:

$$\Sigma^O : \begin{cases} \dot{\hat{\bar{x}}} = (\hat{A}(\hat{s}) + \hat{Y}\hat{C}'J_{\mathbf{Z}}\hat{C})\hat{\bar{x}} + (\hat{B} + \hat{Y}\hat{C}'J_{\mathbf{Z}}\hat{D})(\mathbf{s}^*(\hat{\bar{x}}, \hat{Y}) + \hat{d\hat{v}}), \\ \dot{\hat{Y}} = \hat{A}(\mathbf{s}^*(\hat{\bar{x}}, \hat{Y}) + \hat{d\hat{v}})\hat{Y} + \hat{Y}\hat{A}(\mathbf{s}^*(\hat{\bar{x}}, \hat{Y}) + \hat{d\hat{v}})' + \hat{Y}\hat{C}'J_{\mathbf{Z}}\hat{Y}, \\ \hat{s} = \mathbf{s}^*(\hat{\bar{x}}, \hat{Y}) + \hat{d\hat{v}}. \end{cases} \tag{8.36}$$

(ii) Inverse outer:

$$(\Sigma^O)^{-1}: \begin{cases} \dot{\hat{x}} = (\hat{A}(\hat{s}) + \hat{Y}\hat{C}'J_Z\hat{C})\hat{x} + (\hat{B} + \hat{Y}\hat{C}'J_Z\hat{D})\hat{s}, \\ \dot{\hat{Y}} = \hat{A}(\hat{s})\hat{Y} + \hat{Y}\hat{A}(\hat{s})' + \hat{Y}\hat{C}'J_Z\hat{Y}, \\ \hat{v} = \hat{d}^{-L}(\hat{s} - \hat{\mathbf{s}}^*(\hat{x}, \hat{Y})). \end{cases} \tag{8.37}$$

(iii) J-inner:

$$\Sigma^O: \begin{cases} \dot{\bar{\hat{x}}} = (\hat{A}(\hat{s}) + \hat{Y}\hat{C}'J_Z\hat{C})\bar{\hat{x}} + (\hat{B} + \hat{Y}\hat{C}'J_Z\hat{D})(\hat{\mathbf{s}}^*(\bar{\hat{x}}, \hat{Y}) + \hat{d}\hat{v}), \\ \dot{\hat{Y}} = \hat{A}(\hat{\mathbf{s}}^*(\hat{x}, \hat{Y}) + \hat{d}\hat{v})\hat{Y} + \hat{Y}\hat{A}(\hat{\mathbf{s}}^*(\hat{x}, \hat{Y}) + \hat{d}\hat{v})' + \hat{Y}\hat{C}'J_Z\hat{Y}, \\ \dot{\hat{\xi}} = \hat{A}(\hat{\mathbf{s}}^*(\bar{\hat{x}}, \hat{Y}) + \hat{d}\hat{v}) + \hat{B}(\hat{\mathbf{s}}^*(\bar{\hat{x}}, \hat{Y}) + \hat{d}\hat{v}), \\ \hat{z} = \hat{C}\hat{\xi} + \hat{D}(\hat{\mathbf{s}}^*(\bar{\hat{x}}, \hat{Y}) + \hat{d}\hat{v}). \end{cases} \tag{8.38}$$

8.6.5 Linear Systems

Typically one initializes p_0 at p_e, the control attractor. If not p_t will probably converge to p_e anyway. For linear systems, this corresponds to using a solution $\hat{Y}_e \geq 0$ of the Riccati equation

$$\hat{A}\hat{Y}_e + \hat{Y}_e\hat{A}' + \hat{Y}_e\hat{C}'J_Z\hat{Y}_e = 0, \tag{8.39}$$

which satisfies the stabilizing condition

$$A + \hat{Y}_e\hat{C}'J_Z\hat{C} \quad \text{asymptotically stable.} \tag{8.40}$$

In this case $(\Sigma^O)^{-1}$ is stable and is given by

$$(\Sigma^O)^{-1}: \begin{cases} \dot{\hat{x}} = (\hat{A} + \hat{Y}_e\hat{C}'J_Z\hat{C})\hat{x} + (\hat{B} + \hat{Y}_e\hat{C}'J_Z\hat{D})\hat{s}, \\ \hat{v} = \hat{d}^{-L}(\hat{s} + \hat{E}^{-1}(\hat{D}'J_Z\hat{C} + \hat{B}'\hat{X}_e)\hat{x}). \end{cases} \tag{8.41}$$

On the other hand Σ^O is determined from a solution $\hat{X}_e \geq 0$ of the Riccati equation

$$\begin{aligned} &(\hat{A} - \hat{B}\hat{E}^{-1}\hat{D}'J_Z\hat{C})'\hat{X}_e + \hat{X}_e(\hat{A} - \hat{B}\hat{E}^{-1}\hat{D}'J_Z\hat{C}) \\ &-\hat{X}_e\hat{B}\hat{B}'\hat{X}_e + \hat{C}'J_Z(J_Z - \hat{D}\hat{E}^{-1}\hat{D}')J_Z\hat{C} = 0 \end{aligned} \tag{8.42}$$

and is given by

$$\Sigma^O_{p_e}: \begin{cases} \dot{\bar{\hat{x}}} = (\hat{A} - \hat{B}\hat{E}^{-1}\hat{D}'J_Z\hat{C} - \hat{B}\hat{E}^{-1}\hat{B}'\hat{X}_e)\bar{\hat{x}} \\ \qquad +(I - \hat{Y}_e\hat{X}_e)^{-1}(\hat{B} + \hat{Y}_e\hat{C}'J_Z\hat{D})(s + \hat{E}^{-1}(\hat{D}'J_Z\hat{C} + \hat{B}'\hat{X}_e)\bar{\hat{x}}), \\ \hat{s} = -\hat{E}^{-1}(\hat{D}'J_Z\hat{C} + \hat{B}'\hat{X}_e)\bar{\hat{x}} + \hat{d}\hat{v}. \end{cases} \tag{8.43}$$

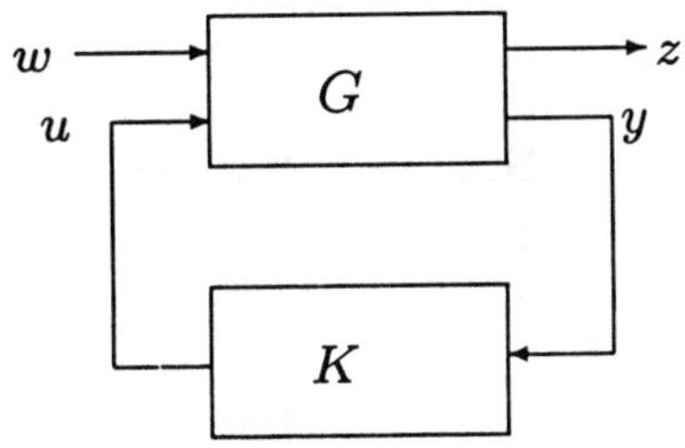

Figure 8.2: Closed-loop system (G, K) for H^∞ control.

This system is stable if and only if $\hat{X}_e$ satisfies the stabilizing condition

$$\hat{A} - \hat{B}\hat{E}^{-1}\hat{D}' J_{\mathbf{Z}} \hat{C} - \hat{B}\hat{E}^{-1}\hat{B}'\hat{X}_e \quad \text{asymptotically stable.} \tag{8.44}$$

However, our constructions require $\hat{X}_e$ to be nonnegative or positive only, but not necessarily stabilizing. Thus the RECIPE is more flexible than traditional J-lossless, outer factorization, which would require that both $\hat{Y}_e$ and $\hat{X}_e$ be stabilizing. Of course, we require the spectral radius of $\hat{Y}_e\hat{X}_e$ to be strictly less than one.

The J-inner factor is given by

$$\Sigma^I_{\hat{\xi}_0,\bar{\hat{x}}_0} : \begin{cases} \dot{\bar{\hat{x}}} = (\hat{A} - \hat{B}\hat{E}^{-1}\hat{D}'J_{\mathbf{Z}}\hat{C} - \hat{B}E^{-1}\hat{B}'\hat{X}_e)\bar{\hat{x}} \\ \qquad +(I - \hat{Y}_e\hat{X}_e)^{-1}(\hat{B} + \hat{Y}_e\hat{C}'J_{\mathbf{Z}}\hat{D})(s + \hat{E}^{-1}(\hat{D}'J_{\mathbf{Z}}\hat{C} + \hat{B}'\hat{X}_e)\bar{\hat{x}}), \\ \qquad \bar{\hat{x}}(0) = \bar{\hat{x}}_0, \\ \dot{\hat{\xi}} = \hat{A}\hat{\xi} + \hat{B}(-\hat{E}^{-1}(\hat{D}'J_{\mathbf{Z}}\hat{C} + \hat{B}'\hat{X}_e)\bar{\hat{x}} + \hat{d}\hat{v}), \quad \hat{\xi}(0) = \hat{\xi}_0, \\ \hat{z} = \hat{C}(\hat{\xi}) + \hat{D}(\hat{\xi})(-\hat{E}^{-1}(\hat{D}'J_{\mathbf{Z}}\hat{C} + \hat{B}'\hat{X}_e)\bar{\hat{x}} + \hat{d}\hat{v}). \end{cases} \tag{8.45}$$

8.7 Factoring and Control

Now we apply factoring to the standard problem of H^∞ control. For simplicity we set $\gamma = 1$. This is the following standard generalized regulator arrangement with nonlinear plant G and nonlinear controller K shown in Figure 8.2. We shall soon see how one solves this by taking Σ to equal the reverse-arrow system $\tilde{G}$ (2.6) of G (2.2) and by factoring Σ. Before pursuing the control solution heavily we first study factoring for its own sake in new coordinates that turn out to be natural to the control problem.

To any signature matrix $J_{\mathbf{V}}$ on a vector space $\mathbf{V}$ there is a natural decomposition of $\mathbf{V}$ into $\mathbf{V}_+$ and $\mathbf{V}_-$, subspaces on which J is I and $-I$, respectively. Thus it is natural to conformably partition the input and output spaces of the system Σ we have been studying:

$$\hat{s} = \begin{pmatrix} u \\ y \end{pmatrix}, \quad \hat{z} = \begin{pmatrix} z \\ w \end{pmatrix}, \quad \hat{v} = \begin{pmatrix} v_2 \\ v_1 \end{pmatrix},$$

so for definiteness

$$u \in \mathbf{S}_+, \quad y \in \mathbf{S}_-, \quad v_2 \in \mathbf{V}_+, \quad \text{and} \quad v_1 \in \mathbf{V}_-,$$

in which case one gets system equations:

$$\Sigma : \begin{cases} \dot{\xi} = A^{\times}(\xi) + B_1(\xi)y + B_2(\xi)u, \\ z = C_1(\xi) + D_{12}u, \\ w = -C_2(\xi) + y. \end{cases} \tag{8.46}$$

Here the notation $A^{\times}$ and $\hat{A}$ are both consistent with previous usage, $A^{\times}$ if we think in terms of $\tilde{G}$ and $\hat{A}$ if we think in terms of Σ. Note that

$$\hat{A} = A^{\times}, \quad \hat{B} = (B_2, B_1), \quad \text{and} \quad \hat{C} = \begin{pmatrix} C_1 \\ -C_2 \end{pmatrix}.$$

This structure assumed of G is the nonlinear analogue of the "2A block problem" (recall Chapter 2) in that we assume the $\hat{D}$ term (as in (8.47)) for Σ is

$$\hat{D} = \begin{pmatrix} D_{12} & 0 \\ 0 & I \end{pmatrix}. \tag{8.47}$$

We shall also assume that $\hat{d}$ in these coordinates is block diagonal, to wit

$$\hat{d} = \begin{pmatrix} \hat{d}_{11} & 0 \\ 0 & I \end{pmatrix}, \tag{8.48}$$

with $\hat{d}_{11} : \mathbf{V}_+ \to \mathbf{S}_+$; e.g., we could take $\hat{d}_{11} = E_1^{-1/2}$.

Then in pictures the J-inner–outer factoring in Figure 8.1 becomes Figure 8.3. Here, $z \in \mathbf{Z}_+$, $w \in \mathbf{Z}_-$, and $v_2 \in \mathbf{V}_+$ and $v_1 \in \mathbf{V}_-$. Thus

$$\left| \begin{pmatrix} z \\ w \end{pmatrix} \right|^2_{J_{\mathbf{Z}}} = \begin{pmatrix} z' & w' \end{pmatrix} J_{\mathbf{Z}} \begin{pmatrix} z \\ w \end{pmatrix} = |z|^2 - |w|^2,$$

$$\left| \begin{pmatrix} v_2 \\ v_1 \end{pmatrix} \right|^2_{J_{\mathbf{V}}} = \begin{pmatrix} v_2' & v_1' \end{pmatrix} J_{\mathbf{V}} \begin{pmatrix} v_2 \\ v_1 \end{pmatrix} = |v_2|^2 - |v_1|^2.$$

The formulas for the factorization of $\Sigma = \tilde{G}$ can be read from the formulas given by the RECIPE in §8.4 with the substitutions given above.

There are strong statements possible about the $+$ and $-$ components of signals centering on the fact that a bound on $|v|_{J_{\mathbf{V}}} = |\hat{v}_+|^2 - |\hat{v}_-|^2$ has little strength because of cancellation, while a bound on $|\hat{v}_2|^2 + |\hat{v}_1|^2$ is a highly substantive statement. The schematic way of looking at this, which is most helpful in exploiting its strengths, is what we think of as *reversing arrows*.

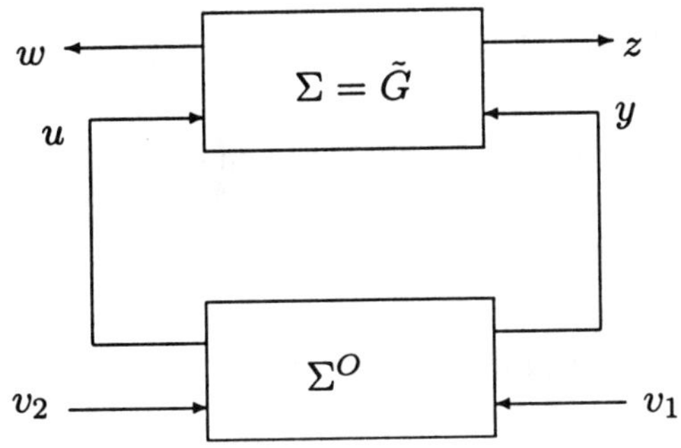

Figure 8.3: Inner factor $\Sigma^I = \Sigma \circ \Sigma^O$.

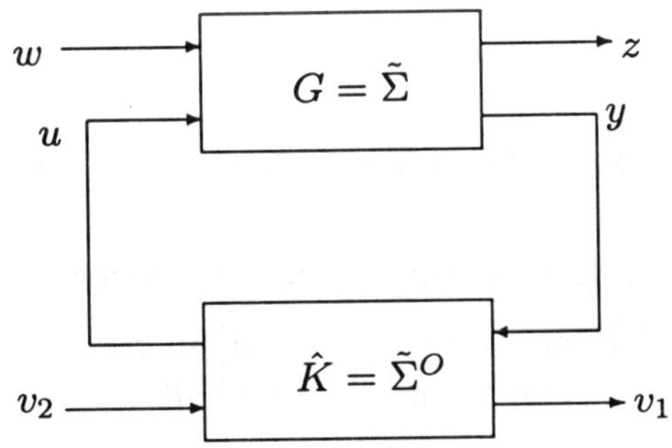

Figure 8.4: Reverse-arrow system $(\tilde{\Sigma}, \tilde{\Sigma}^O) = (G, \hat{K})$.

One can simply reverse the w and y arrows in Figure 8.3. The systems Σ, Σ^O, and system $(\tilde{\Sigma}, \tilde{\Sigma}^I)$ obtained by reversing arrows on $G = \tilde{\Sigma}$ and $\hat{K} = \tilde{\Sigma}^O$ in the figures are uniquely determined one from the other; see Figures 8.3 and 8.4. Note that a controller $K : y \mapsto u$ is obtained by setting $v_2 = 0$ and ignoring v_1, so that $\hat{K} = \tilde{\Sigma}^O$ is in fact an *augmented controller*. Also note

$$F(p, u, y) = \hat{F}(p, \hat{s}) = \hat{F}(p, (u, y)),$$

so the system $\hat{K} = \tilde{\Sigma}^O : (v_2, y) \mapsto (u, v_1)$ is given by

$$\hat{K}_{p_0} = \tilde{\Sigma}^O_{p_0} : \begin{cases} \dot{p} &= F(p, \mathbf{u}^*(p) + \hat{d}_{11}v_2, y), \quad p(0) = p_0, \\ u &= \mathbf{u}^*(p) + \hat{d}_{11}v_2, \\ v_1 &= -\mathbf{y}^*(p) + y, \\ \mathcal{R}(\hat{K}_{p_0}) &= \{p \text{ reachable from } p_0 \text{ with } v_2, y \in L_{2,loc}\}. \end{cases} \tag{8.49}$$

The composition $(G, \hat{K}) = (\tilde{\Sigma}, \tilde{\Sigma}^O)$ is defined in the obvious way on the state space

$$\mathcal{R}(G_{x_0}, \hat{K}_{p_0}) \triangleq \mathbf{R}^n \times \mathcal{R}(\hat{K}_{p_0}).$$

It naturally cuts down to the state space

$$\mathcal{R}(G_{x_0}, \hat{K}_{p_0})_| \triangleq \{(x,p) : \text{ reachable from } (x_0, p_0) \text{ with } p_0(x_0) \text{ finite}\}.$$

The closed-loop system $(G, \hat{K}) = (\tilde{\Sigma}, \tilde{\Sigma}^O)$ is given by

$$(G_{x_0}, \hat{K}_{p_0}) : \begin{cases} \dot{x} = A(x) + B_2(x)\mathbf{u}^*(p) + B_1(x)w, \quad x(0) = x_0, \\ \dot{p} = F(p, \mathbf{u}^*(p) + \hat{d}_{11}v_2, C_2(x) + w), \quad p(0) = p_0, \\ z = C_1(x) + D_{12}\mathbf{u}^*(p), \\ v_1 = -\mathbf{y}^*(p) + C_2(x) + w. \end{cases} \tag{8.50}$$

We shall require the following well-posedness type of assumption.

Assumption 8.7.1 (reversibility). When the system $\Sigma^I_{\hat{\xi}_0, p_0}$ initialized at any state $(\hat{\xi}_0, p_0)$ is driven by any v_1, v_2 in $L_{2,loc}$, then the w and z components of its output are in $L_{2,loc}$ and all $w \in L_2$ occur (as v_1, v_2 sweep all of $L_{2,loc}$).

When the system $(G_{x_0}, \hat{K}_{p_0})$ initialized at any state (z_0, p_0) is driven by any w, v_2 in $L_{2,loc}$, then the v_1 and z components of its output are in $L_{2,loc}$ and all $v_1 \in L_2$ occur (as w, v_2 sweep all of $L_{2,loc}$).

Dissipation properties behave as follows under the reversing-arrows transformation.

Theorem 8.7.2 *Suppose Σ^I is reversible.*[5] *The following statements are true.*

(i) *Σ^I is J-SS-dissipative if and only if $(G, \hat{K})$ is I-SS-dissipative, and likewise for $\Sigma^I_|$ and $(G, \hat{K})_|$.*

(ii) *Σ^I is J-SS-lossless if and only if $(G, \hat{K})$ is I-SS-lossless, and likewise for $\Sigma^I_|$ and $(G, \hat{K})_|$.*

(iii) *If $(G, \hat{K})$ (constructed above) is I-SS-dissipative with energy function e, then it is L_2 stable, and if it is also (z, v_1)-detectable, then it is weakly internally stable. Now assume $p_0 \in \mathcal{D}_{attr}(p_e)$, $e(0, p_0) = 0$, $e(0, p_e) = 0$, where p_e is a control attractor. Then if in addition $(G_{x_0}, \hat{K}_{p_0})$ is I-SS-lossless with continuous storage function e (satisfying $e(x, p + c) = e(x, p)$), then it is I-IO-lossless.*

[5]Here no specific construction of Σ^I or of $e(x,p)$ is needed; this argument works for any J-SS-lossless versus I-SS-lossless system.

Proof. The energy balance equation (8.21) for the $\Sigma^I = (\Sigma, \Sigma^O)$ system (Figure 8.3) in our current notation is

$$e(\hat{\xi}(t), p_t) + \tfrac{1}{2}\int_0^t [|z(s)|^2 - |w(s)|^2]\, ds = e(\hat{\xi}_0, p_0) + \tfrac{1}{2}\int_0^t [|v_2(s)|^2 - |v_1(s)|^2]\, ds. \tag{8.51}$$

This holds for all v_1, v_2 in $L_{2,loc}$. Reversing arrows does not alter this equality, and we write

$$e(\hat{\xi}(t), p_t) + \tfrac{1}{2}\int_0^t [|z(s)|^2 + |v_1(s)|^2]\, ds = e(\hat{\xi}_0, p_0) + \tfrac{1}{2}\int_0^t [|w(s)|^2 + |v_2(s)|^2]\, ds; \tag{8.52}$$

however, for $(G, \hat{K})$, the inputs are w and v_2 and we need to know that the inequality is true whenever w and v_2 sweep all of $L_{2,loc}$. This is not implied by Σ^I being J-SS-lossless alone but is precisely what is implied when we add the reversibility assumption. The I-SS-lossless and J-SS-lossless equivalences for $(G, \hat{K})$ and Σ^I follow. I-SS-dissipation follows from (8.51), (8.52) with inequality:

$$e(\hat{\xi}(t), p_t) + \tfrac{1}{2}\int_0^t [|z(s)|^2 - |w(s)|^2]\, ds \le e(\hat{\xi}_0, p_0) + \tfrac{1}{2}\int_0^t [|v_2(s)|^2 - |v_1(s)|^2]\, ds \tag{8.53}$$

for all v_1, v_2 in $L_{2,loc}$, and reversing arrows

$$e(\hat{\xi}(t), p_t) + \tfrac{1}{2}\int_0^t [|z(s)|^2 + |v_1(s)|^2]\, ds \le e(\hat{\xi}_0, p_0) + \tfrac{1}{2}\int_0^t [|w(s)|^2 + |v_2(s)|^2]\, ds \tag{8.54}$$

for all $w, v_2 \in L_{2,loc}$. This proves conclusions (i) and (ii).

Now suppose that $(G, \hat{K})$ is I-SS-dissipative, so that (8.54) holds for all $w, v_2 \in L_{2,loc}$. If $w, v_2 \in L_2$, we see that $z, v_1 \in L_2$, and so $(G, \hat{K})$ is L_2 stable, and, moreover, if $(G, \hat{K})$ is (z, v_1)-detectable, then it is weakly internally stable by Theorem 2.1.3. Using the remaining hypotheses, we have

$$e(x(t), p_t) + \tfrac{1}{2}\int_0^t [|z(s)|^2 + |v_1(s)|^2]\, ds = e(\hat{\xi}_0, p_0) + \tfrac{1}{2}\int_0^t [|w(s)|^2 + |v_2(s)|^2]\, ds$$

for all $w, v_2 \in L_2$ and all $t \ge 0$. Sending $t \to \infty$ gives I-IO-losslessness. This proves part (iii). □

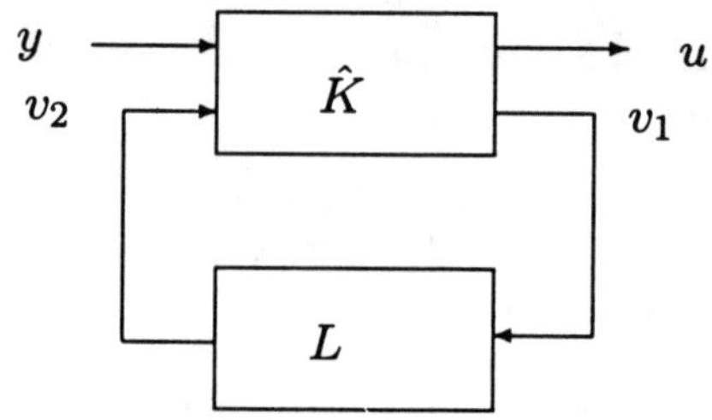

Figure 8.5: The controller $K = (\hat{K}, L)$.

8.7.1 RECIPE for Solving the Control Problem

Recall the H^∞ control problem that opened this book; see Chapter 2. Now we show how factoring produces solutions to it. Recall that we take $\gamma = 1$.

Given the plant G, the recipe goes as follows:

(i) Reverse arrows on G to get $\tilde{G}$. Denote $\tilde{G}$ by Σ.

(ii) Find if possible a decomposition of Σ:

$$\Sigma \circ \Sigma^O = \Sigma^I$$

with Σ^I a J-SS-dissipative reversible system and Σ^O outer. This decomposition could, for example, be obtained using the RECIPE but need not be.

(iii) Rereverse arrows to get $\hat{K}$ from Σ^O and $(G, \hat{K})$ from Σ^I.

(iv) Connect any strict dissipative load L to $\hat{K}$ as in Figure 8.5 to produce a closed-loop system; call it

$$K \triangleq (\hat{K}, L).$$

We assume that this system is well defined in that $L_{2,loc}$ inputs y produce $L_{2,loc}$ signals u, v_1, v_2. This is the desired controller for the H^∞ control problem (illustrated in Figure 8.5).

THEOREM 8.7.3 *Consider any factorization $\Sigma^I = \Sigma \circ \Sigma^O$. Suppose Σ^I is J-SS-dissipative. Make the reversibility Assumption* 8.7.1, *and assume that $K = (\hat{K}, L)$ is well defined, where L is an IO-dissipative system. Then $K = (\hat{K}, L)$ is a controller that solves the H^∞ control problem, provided the closed-loop system (G, K) is z-detectable. That is, the closed-loop system is dissipative and weakly internally stable (meaning that $w \in L_2$ implies $u, y, z, x \in L_2$ and $x(t) \to 0$).*

Proof. $(G, \hat{K})$ is I-SS-dissipative by Theorem 8.7.2, so that

$$e(\hat{\xi}(t), p_t) + \tfrac{1}{2}\int_0^t [|z(s)|^2 + |v_1(s)|^2]\, ds$$

$$\leq e(\hat{\xi}_0, p_0) + \tfrac{1}{2}\int_0^t [|w(s)|^2 + |v_2(s)|^2]\, ds$$

for all $w, v_2 \in L_{2,loc}$. Now L is I-IO-dissipative by hypothesis, so

$$\frac{1}{2}\int_0^t |v_2|^2 ds \le \frac{1}{2}\int_0^t |v_1|^2 ds$$

for all $v_1 \in L_{2,loc}$. Combining these two inequalities we see that (G, K) is dissipative. Now z-detectibility provides the weak internal stability, by Theorem 2.1.3. This means in particular that the signals u, y, z are in L_2 whenever $w \in L_2$, but we cannot conclude that v_1, v_2 are in L_2 without further hypothesis (they are in $L_{2,loc}$). □

Note that the stabilizing property of $\hat{\mathbf{s}}^*$, p_0 is not used here, nor is J-SS-losslessness. Still we are able to produce many solutions to the H^∞ control problem. This, however, will not (at least for linear systems) be all solutions unless we invoke these constraints.

8.7.2 Parameterizing All Solutions

In linear H^∞ control Theorem 8.7.3 parameterizes all solutions to the H^∞ control problem with norm less than one provided that both Σ^O and $(\Sigma^O)^{-1}$ are L_2 stable. For nonlinear systems clearly Theorem 8.7.3 parameterizes many solutions, but whether or not it parameterizes all is an open question.

8.8 Necessity of the RECIPE

The results given in §8.5 we have obtained so far for the general factorization problem specified in §8.2 have been of the sufficiency type, in that the factors were constructed from a solution $\hat{W}$ of the factoring PDE (8.13) and properties of the factors were obtained. In this section, as foreshadowed in §8.5.6, we consider the main problem of necessity, i.e., finding a solution $\hat{W}$ with the required critical feedback function $\hat{\mathbf{s}}^*$. To do this, we make use of the connections between control and factorization explored in §8.7. This will require a slight restriction in generality but with the benefit of efficiency.

THEOREM 8.8.1 *Suppose $\hat{D}$ has the block diagonal form:*

$$\hat{D} = \begin{pmatrix} D_{12} & 0 \\ 0 & I \end{pmatrix}.$$

Suppose that Σ has a decomposition into $\Sigma^I = \Sigma \circ \Sigma^O$, which is J-SS-dissipative, and the reversibility Assumption 8.7.1 holds. Then by Theorem 8.7.3 a solution K to the H^∞ control problem exists; thus the value function W given by equation (4.4) exists. Suppose this W is smooth. Then the critical feedback function is given by

$$\hat{\mathbf{s}}^*(p) = \begin{pmatrix} \mathbf{u}^*(p) \\ \mathbf{y}^*(p) \end{pmatrix},$$

where the functions in the RHS are given by (4.23), and the RECIPE given in §8.4 produces a decomposition Σ^I and Σ^O with Σ^I which is J-SS-lossless, and indeed Assumption 8.5.2 holds.

Proof. Since we assume W is smooth our main necessity Theorem 4.3.1 tells us that W satisfies the dynamic programming principle. Under appropriate smoothness assumptions discussed in Chapter 4 the function W solves the dynamic programming PDE

$$\inf_u \sup_y \nabla_p W(p)[F(p,u,y)] = 0 \tag{4.16$'$}$$

and gives formulas (4.23) for the optimizers $\mathbf{u}^*, \mathbf{y}^*$. The special form of $\hat{D}$ implies the quadratic part of F has decoupled u, y appearing in a way that makes

$$\inf_u \sup_y \nabla_p W(p)[F(p,u,y)] \;=\; \text{crit}_s \nabla_p W(p)[\hat{F}(p,s)]$$

with $s = (u, y)$. Indeed $\hat{\mathbf{s}}^*(p) = \text{crit}_s \nabla_p W(p)[\hat{F}(p,s)]$ is unique and equals

$$\operatorname*{argmin}_u \operatorname*{argmax}_y \nabla_p W(p)[F(p,u,y)].$$

Thus we produce $\hat{\mathbf{s}}^*$ and W satisfying the formulas in §8.5.1, and Assumption 8.5.2 holds. □

The remaining Assumptions listed in §8.5.2 should be checked. This can be approached using the more complicated framework and results given in Chapter 4. A significant point is that our necessity theory (as presented in Chapter 4) does not imply Assumptions 8.5.4 and 8.5.5 which say that solutions p_t exist for the augmented system (8.17) and that the reversibility Assumption 8.7.1 holds. These are significant open questions.

8.9 State Reading Factors

A natural notion is that of factors being able to read the state rather than just a measurement. In this section we give formulas for a state reading J-lossless outer factorization. We associate with this the picture shown in Figure 8.6.

8.9.1 RECIPE for State Reading Factors

The following RECIPE for the state reading factors uses the critical state feedback function $\hat{\mathbf{s}}^*_{state}(x)$ defined by (8.24), which is defined in terms of a stabilizing solution $\hat{V}(\hat{x})$ to the state feedback factoring PDE (8.23).

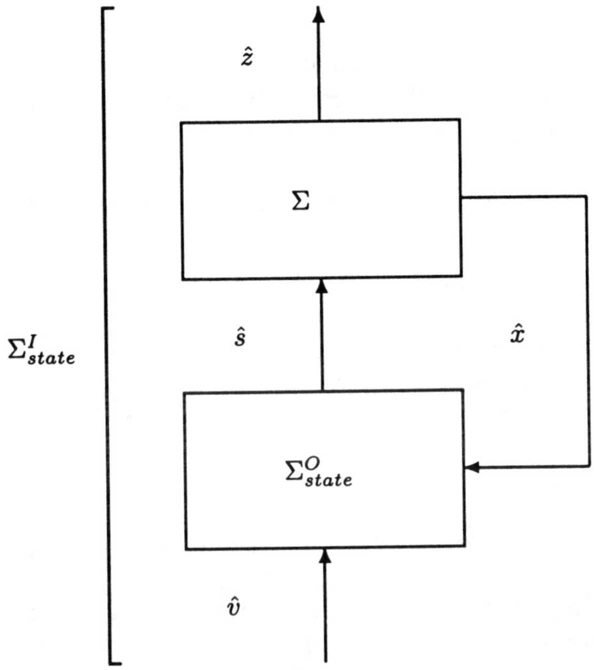

Figure 8.6: State reading inner factor Σ^I.

(i) Construct the factor Σ^O_{state} by

$$\Sigma^O_{state}: \begin{cases} \dot{\bar{\hat{x}}} &= \hat{A}(\bar{\hat{x}}) + \hat{B}(\bar{\hat{x}}), (\hat{\mathbf{s}}^*_{state}(\bar{\hat{x}}) + \hat{d}\hat{v}), \\ \hat{s} &= \hat{\mathbf{s}}^*_{state}(\bar{\hat{x}}) + \hat{d}\hat{v}. \end{cases} \tag{8.55}$$

(ii) Take the inverse of Σ^O_{state} to give

$$(\Sigma^O_{state})^{-1}: \begin{cases} \dot{\hat{x}} &= \hat{A}(\hat{x}) + \hat{B}(\hat{x})\hat{s}, \\ \hat{v} &= \hat{d}^{-1}(\hat{s} - \hat{\mathbf{s}}^*_{state}(\hat{x})). \end{cases} \tag{8.56}$$

(iii) The J-inner factor Σ^I_{state} is given by composing (8.1) and (8.55):

$$\Sigma^I_{state,\,\bar{\hat{x}}_0,\hat{\xi}_0}: \begin{cases} \dot{\bar{\hat{x}}} &= \hat{A}(\bar{\hat{x}}) + \hat{B}(\bar{\hat{x}})(\hat{\mathbf{s}}^*_{state}(\bar{\hat{x}}) + \hat{d}\hat{v}), \quad \bar{\hat{x}}(0) = \bar{\hat{x}}_0, \\ \dot{\hat{\xi}} &= \hat{A}(\hat{\xi}) + \hat{B}(\hat{\xi})(\hat{\mathbf{s}}^*_{state}(\bar{\hat{x}}) + \hat{d}\hat{v}), \quad \hat{\xi}(0) = \hat{\xi}_0, \\ \hat{z} &= \hat{C}(\hat{\xi}) + \hat{D}(\hat{\xi})(\hat{\mathbf{s}}^*_{state}(\bar{\hat{x}}) + \hat{d}\hat{v}). \end{cases} \tag{8.57}$$

As mentioned above, the critical state feedback $\hat{\mathbf{s}}^*_{state}$ is given by the formula (8.24), which is exactly the same as

$$\hat{\mathbf{s}}^*_{state}(\hat{x}) = \hat{\mathbf{s}}^*(\delta_{\hat{x}})$$

with $\hat{\mathbf{s}}^*$ given by (8.15).

In the context of the state feedback control problem (recall Chapter 5), the state reading version $(G, \hat{K}_{state})$ of the corresponding rereversed-arrow system described in (8.49) is still of interest. Here the the system $\hat{K}_{state} : (v_2, y) \mapsto (u, v_1)$ is given by

$$\hat{K}_{state} : \begin{cases} \dot{x} = A^{\times}(x) + B_2(x)(\mathbf{u}^*_{state}(x) + \hat{d}_{11}v_2) + B_1(x)y, \\ u = \mathbf{u}^*_{state}(x) + \hat{d}_{11}v_2, \\ v_1 = -\mathbf{y}^*(x) + y. \end{cases} \tag{8.58}$$

Of course, this can be related to the state feedback control problem, as in §8.7

8.9.2 Properties of State Reading Factors

We now briefly present some basic results about the state reading factors.

THEOREM 8.9.1 *The factor Σ^O_{state} is L_2 stable provided $\hat{V}$ is taken to be the stabilizing solution of the state feedback factoring PDE* (8.23).

Proof. If $\hat{V}$ is stabilizing, then the system $(\hat{A} + \hat{B}\hat{\mathbf{s}}^*_{state}, \hat{B})$ is incrementally exponentially L_2 stable; hence $\Sigma^O_{state}\hat{v} \in L_2$ whenever $\hat{v} \in L_2$. □

REMARK 8.9.2 The inverse $(\Sigma^O)^{-1}_{state}$ is often not stable, since $\hat{A}$ is often not a stable vector field.

THEOREM 8.9.3 *Σ^I_{state} is J-SS-lossless.*

Proof. Since $\hat{V}$ is a solution of the state feedback factoring PDE (8.23), we have

$$\nabla_{\hat{x}}\hat{V}(\hat{x}) \cdot (\hat{A}(\hat{x}) + \hat{B}(\hat{x})(\hat{\mathbf{s}}^*_{state}(\hat{x}) + \hat{d}\hat{v})) = -\tfrac{1}{2}|\hat{C} + \hat{D}(\hat{\mathbf{s}}^*_{state} + \hat{d}\hat{v})|^2_{J_{\mathbf{z}}} + \tfrac{1}{2}|\hat{v}|^2_{J_{\hat{v}}}.$$

Integrating this gives the energy balance equation (8.21). □

8.9.3 Separation Principle

Now we observe a most elegant fact. When certainty equivalence holds, the factoring formulas in the RECIPE of §8.6.1 are actually the composition of a state reading factor and an information state estimator.

Note when the initial conditions for Σ^I_{state} satisfy $x_0 = \xi_0$, the ξ and x dynamics for Σ^I_S are the same. Thus the cutdown $\Sigma^I_{state|}$ given by

$$\Sigma^I_{state|} := \begin{cases} \dot{x} = \hat{A}(x) + \hat{B}(x)(\hat{\mathbf{s}}^*_{state}(x) + \hat{d}\hat{v}), \qquad x(0) = x_0, \\ \hat{z} = \hat{C}(x) + \hat{D}(x)(\hat{\mathbf{s}}^*_{state}(x) + \hat{d}\hat{v}) \end{cases} \tag{8.59}$$

has the same input–output behavior as Σ^I_{state}. Our separation theorem uses $\Sigma_{state|}$ and gives the following theorem.

THEOREM 8.9.4 *Under the hypotheses of* §8.6.1, *the general J-SS-lossless system* Σ^I *given by the RECIPE of* §8.6.1 *can be expressed as the composition*

$$\Sigma^I_{x0,p0} = \Sigma^I_{state|} \circ \Sigma^O_{p0},$$

where Σ^O_{p0} *is the information state type of estimator given by the RECIPE in* §8.4.

Proof. Read the formulas (in the RECIPE of §8.6.1). □

8.10 Nonsquare Factors and the Factoring PDE

In linear control the J-inner–outer factorizations are used to give a formula for all solutions to the H^∞ control problem (recall §8.7.2). Thus they go beyond what is needed merely to find one solution. In nonlinear control one often pays a great price in terms of implementability for achieving more than is absolutely required. As we shall see obtaining one solution corresponds to factoring with those Σ_0 that are nonsquare. So we discuss nonsquare factoring in this section.

8.10.1 Nonsquare Factoring PDE

First notice the construction in the RECIPE has some freedom in it, for example, the choice of $\hat{d}$, and the RECIPE makes perfect sense if dimension $\mathbf{V}$ is less than or equal to that of $\mathbf{Z}$. This freedom is thus reflected in the choice of the space $\mathbf{V}$, and hence the signal $\hat{v}$. In fact, the main assumption we used in the RECIPE was

$$\hat{D}'\hat{d}' J_{\mathbf{Z}} \hat{d}\hat{D} = J_{\mathbf{V}}, \tag{8.9}$$

and in connection with the control problem we needed reversibility.

The main points we wish to make are that the reversibility (recall §8.7) does not use that $\hat{d}_{11}$ is invertible, and therefore Theorem 8.7.2 holds as is. Second, Theorem 8.5.7 on losslessness holds with modifications in the formula (8.15) for $\hat{\mathfrak{s}}^*$ and the factoring PDE (8.16) for $\hat{W}$ in §8.5.1. Third, we would like a derivation of these relations that applies to the directional derivative $\nabla^0_p \hat{W}$, which need not be a linear functional (but is convex).

THEOREM 8.10.1 *Consider the factor* Σ^I *determined by the RECIPE, and define*

$$\Lambda \triangleq \text{range } \hat{d} \subset \mathbf{S},$$

with Λ *not in general equal to* $\mathbf{S}$, *but with* $\hat{d}$ *satisfying* (8.9). *Assume the factoring PDE (equations* (8.16) *and* (8.63) *below) can be interpreted in the sense of directional derivative, with* $\nabla^0_p \hat{W}(p)[\cdot]$ *continuous. Then* Σ^I *is J-SS-lossless if and only if*

$$\nabla^0_p \hat{W}(p)[\hat{F}(p, \hat{\mathfrak{s}}^*(p) + \hat{d}\hat{v})] = \tfrac{1}{2}|\hat{v}|^2_{J_{\mathbf{V}}} \quad \textit{for all } \hat{v} \in \mathbf{V}. \tag{8.60}$$

Assume $\nabla^0_p \hat{W}(p)$ *is linear on the space of functions*

$$\mathbf{S}_{\hat{F}P_\Lambda} = \text{span}\left\{ \frac{d\hat{F}}{d\hat{s}}(p, \hat{\mathfrak{s}}^*(p)) P_\Lambda \hat{s} \;:\; \hat{s} \in \mathbf{S} \right\}. \tag{8.61}$$

Then identity (8.60) *is equivalent to the relations*

$$P_\Lambda \hat{E}\hat{\mathbf{s}}^* = -\nabla_p^0 \hat{W}(p)[P_\Lambda(\hat{D}' J_{\mathbf{Z}} \hat{C} - \hat{B}' \nabla_x p')] \tag{8.62}$$

and

$$\begin{aligned} &\nabla_p^0 \hat{W}(p)[-\nabla_x p\hat{A} + \tfrac{1}{2}\hat{C}' J_{\mathbf{Z}} \hat{C} + (\hat{C}' J_{\mathbf{Z}} \hat{D} - \nabla_x p\hat{B})(I - P_\Lambda)\hat{\mathbf{s}}^*(p)] \\ &\qquad + \hat{\mathbf{s}}^{*\prime}(p)\hat{E}(\tfrac{1}{2}I - P_\Lambda)\hat{\mathbf{s}}^*(p) = 0. \end{aligned} \tag{8.63}$$

Here, $P_\Lambda : \mathbf{S} \to \Lambda \subset \mathbf{S}$ *denotes projection.*

Of course, when we take $\hat{d}$ invertible, with $\mathbf{V} = \Lambda = \mathbf{S}$, we have $P_\Lambda = I$, and so (8.62) and (8.63) reduce to (8.15) and (8.16), respectively.

Linearity of $\nabla_p^0 \hat{W}(p)[\cdot]$ on the subspace Λ amounts to true differentiability of $\hat{W}$ in the directions along the subspace. The following corollary treats this linearity issue.

COROLLARY 8.10.2 *Under the assumptions of Theorem* 8.10.1, *the identity* (8.60) *is equivalent to*

$$\nabla_p^0 \hat{W}(p) \left[\hat{F}(p, \hat{\mathbf{s}}^*(p)) + \frac{d\hat{F}}{d\hat{s}}(p, \hat{\mathbf{s}}^*(p))\tilde{\hat{s}} \right] = 0 \quad \textit{for all} \quad \tilde{\hat{s}} \in \Lambda. \tag{8.64}$$

This identity is also equivalent to

$$\begin{aligned} &\nabla_p^0 \hat{W}(p)[\hat{F}(p, \hat{\mathbf{s}}^*(p))] = 0 \quad \textit{and} \\ &\nabla_p^0 \hat{W}(p) \left[\frac{d\hat{F}}{d\hat{s}}(p, \hat{\mathbf{s}}^*(p)) P_\Lambda \right] = 0. \end{aligned} \tag{8.65}$$

The proofs of these results are deferred to §8.10.3.

8.10.2 Reversing Arrows on One Port

The diagrams in Figure 8.7 concern descriptions of the H^∞ controller, in particular parameterizations using the signal $\hat{v}$. The standard problem of H^∞ control as in Figure 8.7 diagram (a) converts to diagrams (b) and (c) where signals flow in one direction. The operator $\tilde{K}$ in diagram (c) is a nonsquare outer factor.

This is just an instance of the basic principle that we now describe. Consider a system K with a single input y and output u. If we were studying this system in an experiment, we would observe input–output pairs, and the set of all input–output pairs we would observe is the set

$$\{(y, u)\} = \{(y, Ky)\}.$$

Parameterizing this set in terms of y is just one possible construction. Quite possibly we could also parameterize this set as $\{(A(\hat{v}), B(\hat{v}))\}$, where A and B are suitably

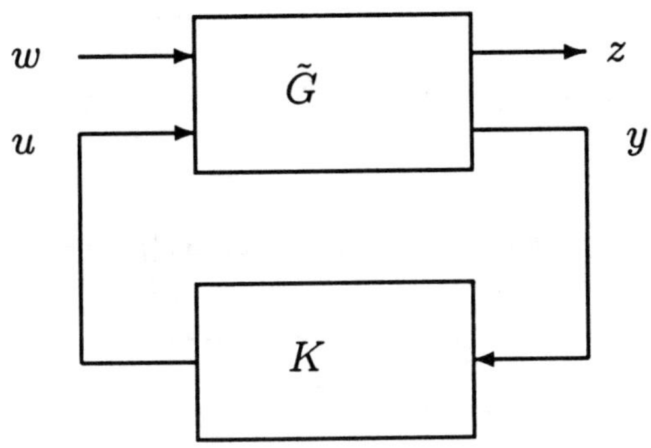

(a) Closed-loop system (G, K) for H^∞ control.

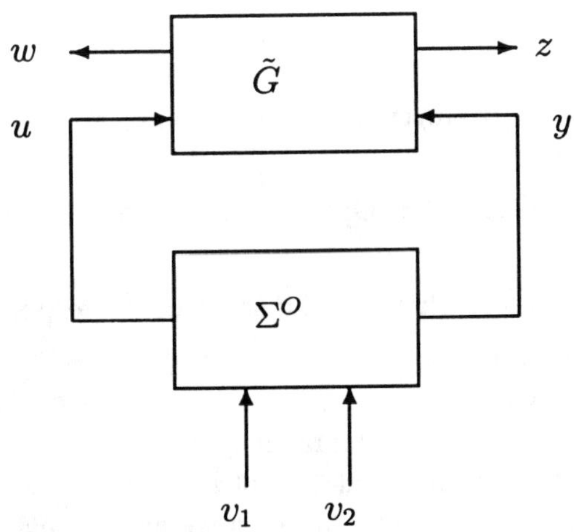

(b) Square inner factor Σ^I showing parameter $\hat{v} = (v_2, v_1)$ entering square Σ^0 (Figure 8.3 repeated).

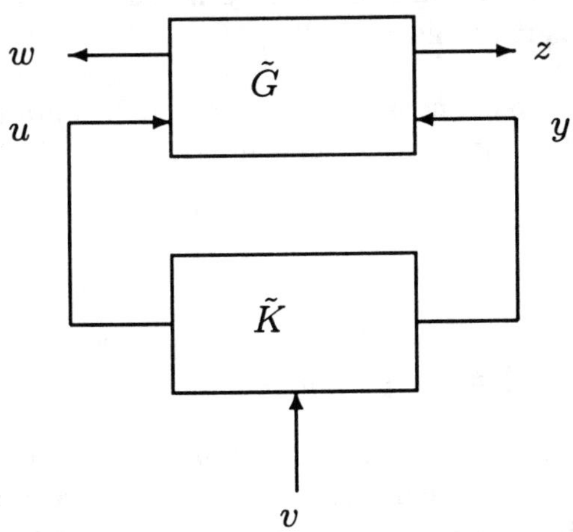

(c) Nonsquare inner factor Σ^I showing parameter $\hat{v} = v$ entering nonsquare Σ^0.

Figure 8.7: Square and nonsquare H^∞ controller parameterizations.

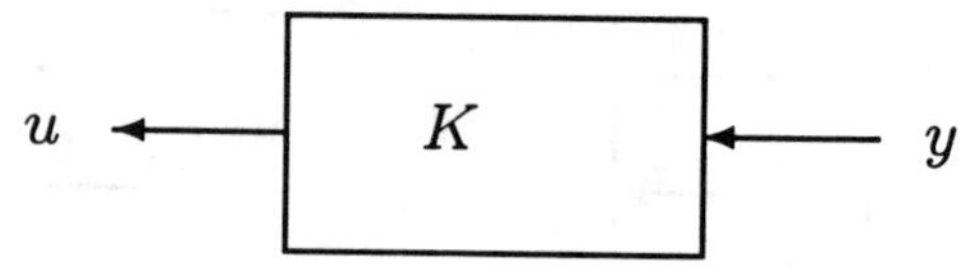

(a) Operator K and input–output pairs $(y, u) = (y, Ky)$.

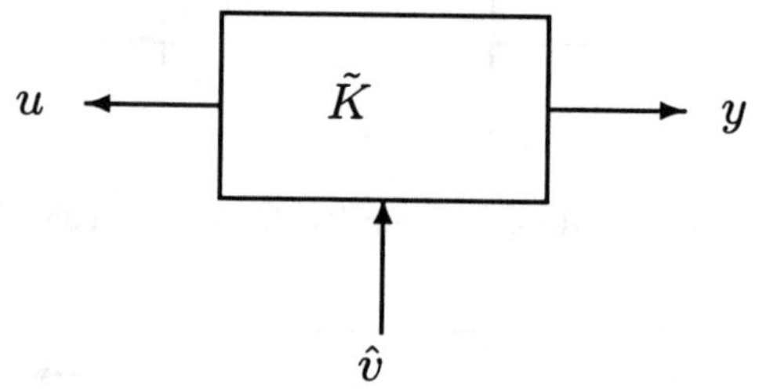

(b) Operator $\tilde{K}$ and parameter $\hat{v}$.

Figure 8.8: Operator parameterizations.

chosen nonlinear operators. This verbal discussion corresponds to Figure 8.8, where $\tilde{K}$ is the map $\tilde{K}(\hat{v}) \triangleq (A(\hat{v}), B(\hat{v}))$.

Clearly, for given K many pairs A, B exist. The converse is true only under some assumptions on A to the effect that its range is onto.

This discussion tells us immediately that if we have the solution to the standard problem of H^∞ control in Figure 8.7(a), then we also have a nonsquare factorization of $\tilde{G}$ as indicated in Figure 8.7(c). This can be seen also in terms of Figure 8.7(b) by setting v_2 to 0; we obtain $\widetilde{K}v_1 = \widetilde{\Sigma}^O(0, v_1)$. Conversely, if we have a factorization whose range meets certain conditions, then we have a solution to the control problem. The next example illustrates this.

EXAMPLE 8.10.3 The famous central controller of H^∞ control (defined in Chapter 4) is an example. Recall that it is given by the top two lines of

$$\tilde{\tilde{K}} : \begin{cases} \dot{p} = F(p, \mathbf{u}^*(p), y), \qquad p(0) = p_0, \\ u = \mathbf{u}^*(p), \\ v_1 = -\mathbf{y}^*(p) + y. \end{cases} \tag{8.66}$$

This is (8.49) with v_2 equal to 0 and $\hat{d}_{11} = 0$. The system is reversible in that $v_1 = -\mathbf{y}^*(p) + y$ can be solved for y in terms of v_1. The appropriate $J_{\mathbf{V}}$ is $J_{\mathbf{V}}v_1 = -v_1$, which is negative definite. Now we write out formulas (8.62) and (8.63).

The ingredients of these formulas are

$$P_\Lambda = \begin{pmatrix} 0 & 0 \\ 0 & I \end{pmatrix}, \quad P_\Lambda \hat{D} = \begin{pmatrix} 0 & 0 \\ 0 & I \end{pmatrix}, \quad P_\Lambda \hat{E} = -\begin{pmatrix} 0 & 0 \\ 0 & I \end{pmatrix},$$

and

$$P_\Lambda(\hat{D}'J_{\mathbf{Z}}\hat{C} - \hat{B}'\nabla_x p') = \begin{pmatrix} 0 \\ C_2 - B_1'\nabla_x p' \end{pmatrix}.$$

Plug this into (8.62) to get

$$\mathbf{y}^*(p) = \nabla_p^o \hat{W}(p)[C_2 - B_1'\nabla_x p'].$$

Turn to the factoring PDE (8.63). We calculate

$$\hat{C}'J_{\mathbf{Z}}\hat{C} = C_1'C_1 - C_2'C_2,$$

$$(\hat{C}'J_{\mathbf{Z}}\hat{D} - \nabla_x p\hat{B})(I - P_\Lambda)\hat{\mathbf{s}}^*(p) = (C_1'D_{12} - \nabla_x pB_2)\mathbf{u}^*(p)$$

and

$$\hat{\mathbf{s}}^{*\prime}(p)\hat{E}(\tfrac{1}{2}I - P_\Lambda)\hat{\mathbf{s}}^*(p) = \tfrac{1}{2}\mathbf{u}^{*\prime}(p)E_1\mathbf{u}^*(p) - \tfrac{1}{2}\mathbf{y}^{*\prime}(p)\mathbf{y}^*(p).$$

Plug these into (8.63) and get

$$\begin{aligned} &\nabla_p^0\hat{W}(p)[-\nabla_x pA^\times + \tfrac{1}{2}(C_1'C_1 - C_2'C_2) + (C_1'D_{12} - \nabla_x pB_2)\mathbf{u}^*(p)] \\ &+\tfrac{1}{2}(\mathbf{u}^{*\prime}(p)E_1\mathbf{u}^*(p) - \mathbf{y}^{*\prime}(p)\mathbf{y}^*(p)) = 0. \end{aligned} \tag{8.67}$$

Note that while $\mathbf{y}^*$ is uniquely determined, $\mathbf{u}^*$ is relatively undetermined. Note that any reversible controller of this form satisfying the previous equation with inequality ≤ 0 is a solution to the dissipative control problem. The standard choice (4.23) for $\mathbf{u}^*$ is obtained by doing the most natural thing, namely, minimizing the left side (Hamiltonian) of (8.67) in $\mathbf{u}^*$. Assuming that $\nabla_p^0\hat{W}(p)[\cdot]$ is linear we get that the minimizer is $\mathbf{u}^* = \nabla_p^0\hat{W}(p)[E_1^{-1}(-D_{12}'C_1 + B_2'\nabla_x p')]$ which is indeed (4.23). Thus once $\hat{d}_{11} = 0$ has been specified our RECIPE determines $\mathbf{y}^*$ uniquely from $\mathbf{V}$, but there may be some freedom in the choice of $\mathbf{u}^*$. ∇

8.10.3 Proof of Factoring PDE

The core of the proof of Theorem 8.10.1 is the next lemma.

LEMMA 8.10.4 *Let ν be a function $\nu : \mathcal{X}_e \to \mathbf{R}$ satisfying the following conditions:*

(i) *ν is convex,*

(ii) *$\nu[p + c] = \nu[p] + c$ if $c \in \mathbf{R}$ is constant, $p \in \mathcal{X}_e$,*

(iii) *$\nu[\lambda p] = \lambda\nu[p]$ if $\lambda \in \mathbf{R}$, $\lambda \geq 0$ (positive homogeneous),*

(iv) *ν is continuous at zero (relative to the topology of $\mathcal{X}_e$).*

Let

$$f(\hat{s}) \triangleq a + b\hat{s} + \tfrac{1}{2}\hat{s}'c\hat{s}$$

be an $\mathcal{X}_e$-valued quadratic function of $\hat{s} \in \mathbf{S}$, where c is a constant self-adjoint invertible matrix. Let $\Lambda \subset \mathbf{S}$ be a subspace. Then a function $\hat{\mathbf{s}}^*(p)$ of $p \in \mathcal{X}_e$ has the property

$$\nu[f(\hat{\mathbf{s}}^* + \tilde{\hat{s}})] = \tfrac{1}{2}\tilde{\hat{s}}'c\tilde{\hat{s}} \text{ for all } \tilde{\hat{s}} \in \Lambda \tag{8.68}$$

if and only if

$$\nu\left[f(\hat{\mathbf{s}}^*) + \frac{df}{d\hat{s}}(\hat{\mathbf{s}}^*)\tilde{\hat{s}}\right] = 0 \quad \text{for all } \tilde{\hat{s}} \in \Lambda. \tag{8.69}$$

Then identity holds if and only if

$$\begin{aligned} &\nu[f(\hat{\mathbf{s}}^*)] = 0 \text{ and} \\ &\nu[\tfrac{df}{d\hat{s}}(\hat{s}^*)P_\Lambda \hat{s}] = 0 \text{ for all } \hat{s} \in \mathbf{S}. \end{aligned} \tag{8.70}$$

Now assume that ν is linear on the subspace of functions

$$\mathbf{S}_{f,P_\Lambda} \triangleq \text{span}\left\{\frac{df}{d\hat{s}}(\hat{\mathbf{s}}^*)P_\Lambda \hat{s} \; : \; \hat{s} \in \mathbf{S}\right\}.$$

Then (8.70) *holds if and only if*

$$\begin{aligned} &\nu[f(\hat{\mathbf{s}}^*)] = 0 \text{ and} \\ &\nu[\tfrac{df}{d\hat{s}}(\hat{s}^*)P_\Lambda] = 0, \end{aligned} \tag{8.71}$$

which holds if and only if $\hat{\mathbf{s}}^*$ and ν satisfy

$$P_\Lambda c\hat{\mathbf{s}}^* = -\nu[P_\Lambda b'] \tag{8.72}$$

and

$$\nu[a + b(1 - P_\Lambda)\hat{\mathbf{s}}^*] + \hat{\mathbf{s}}^{*\prime}c(\tfrac{1}{2}I - P_\Lambda)\hat{\mathbf{s}}^* = 0. \tag{8.73}$$

Since c is invertible we may select $\hat{\mathbf{s}}^*$ in Λ satisfying (8.70), *in which case* (8.72) *and* (8.73) *become*

$$\begin{aligned} &\hat{\mathbf{s}}^* = -(P_\Lambda c P_\Lambda)^{-1}\nu[P_\Lambda b'], \\ &\nu[a] - \tfrac{1}{2}\hat{\mathbf{s}}^{*\prime}c\hat{\mathbf{s}}^* = 0. \end{aligned} \tag{8.74}$$

Proof. We begin the proof with some basic facts. First,

$$f(\hat{\mathbf{s}}^*) = a + b\hat{\mathbf{s}}^* + \tfrac{1}{2}\hat{\mathbf{s}}^{*\prime}c\hat{\mathbf{s}}^*, \quad \text{and} \quad \frac{df}{d\hat{s}}(\hat{\mathbf{s}}^*) = b + \hat{\mathbf{s}}^{*\prime}c, \tag{8.75}$$

and second, by convexity and positive homogeneity,

$$\nu[\alpha + \beta] = 2\nu[\tfrac{1}{2}\alpha + \tfrac{1}{2}\beta] \le 2(\tfrac{1}{2}\nu[\alpha] + \tfrac{1}{2}\nu[\beta]) = \nu[\alpha] + \nu[\beta]. \tag{8.76}$$

This means that ν is sublinear.

Continuing with the proof, since $\tilde{\hat{s}}$ and c are constant, (8.68) is equivalent to

$$\nu\left[f(\hat{\mathbf{s}}^*) + \frac{df}{d\hat{s}}(\hat{\mathbf{s}}^*)\tilde{\hat{s}}\right] + \tfrac{1}{2}\tilde{\hat{s}}'c\tilde{\hat{s}} = \tfrac{1}{2}\tilde{\hat{s}}'c\tilde{\hat{s}} \quad \text{for all } \tilde{\hat{s}} \in \Lambda,$$

and (8.68) is true if and only if (8.69) holds.

Now assume that (8.69) holds. Set $\tilde{\hat{s}} = 0$ in (8.69) to get $\nu[f(\hat{\mathbf{s}}^*)] = 0$. Fix $\tilde{\hat{s}} \in \Lambda$, and let $r > 0$ be a positive constant. Then $r\hat{\mathbf{s}} \in \Lambda$, so by (8.69) we have

$$\nu\left[f(\hat{\mathbf{s}}^*) + r\frac{df}{d\hat{s}}(\hat{\mathbf{s}}^*)\tilde{\hat{s}}\right] = 0,$$

which holds if and only if

$$r\nu\left[\frac{1}{r}f(\hat{\mathbf{s}}^*) + \frac{df}{d\hat{s}}(\hat{\mathbf{s}}^*)\tilde{\hat{s}}\right] = 0,$$

which in turn is equivalent to

$$\nu\left[\frac{1}{r}f(\hat{\mathbf{s}}^*) + \frac{df}{d\hat{s}}(\hat{\mathbf{s}}^*)\tilde{\hat{s}}\right] = 0$$

since $r > 0$. Sending $r \to +\infty$, and making use of the assumed continuity, we get

$$\nu\left[\frac{df}{d\hat{s}}(\hat{\mathbf{s}}^*)\tilde{\hat{s}}\right] = 0$$

for all $\tilde{\hat{s}} \in \Lambda$. Since the projection $P_\Lambda : \mathbf{S} \to \Lambda$ is onto, this is equivalent to

$$\nu\left[\frac{df}{d\hat{s}}(\hat{s}^*)P_\Lambda\hat{s}\right] = 0 \text{ for all } \hat{s} \in \mathbf{S}.$$

This proves that (8.69) implies (8.70).

For the converse, assume (8.70) holds. Let $\tilde{\hat{s}} \in \Lambda$. Then by (8.76), we have

$$\nu\left[f(\hat{\mathbf{s}}^*) + \frac{df}{d\hat{s}}(\hat{\mathbf{s}}^*)\tilde{\hat{s}}\right] \leq \nu\left[f(\hat{\mathbf{s}}^*)\right] + \nu\left[\frac{df}{d\hat{s}}(\hat{\mathbf{s}}^*)\tilde{\hat{s}}\right] = 0.$$

Also, using (8.70),

$$\begin{aligned} 0 &= \nu[f(\hat{\mathbf{s}}^*(p))] \\ &= \nu[(f(\hat{\mathbf{s}}^*(p)) + \tfrac{df}{d\hat{s}}(\hat{\mathbf{s}}^*)\tilde{\hat{s}}) + \tfrac{df}{d\hat{s}}(\hat{\mathbf{s}}^*)(-\tilde{\hat{s}})] \\ &\leq \nu[f(\hat{\mathbf{s}}^*(p)) + \tfrac{df}{d\hat{s}}(\hat{\mathbf{s}}^*)\tilde{\hat{s}}] + 0 \end{aligned}$$

since $-\hat{\mathbf{s}} \in \Lambda$. These inequalities prove that (8.70) implies (8.69).

Assume (8.70). By letting $\hat{s}$ equal basis vectors in $\mathbf{S}$, we see that all components of $\frac{df}{d\hat{s}}(\hat{\mathbf{s}}^*)P_\Lambda$ must satisfy

$$\nu\left[\left(\frac{df}{d\hat{s}}(\hat{\mathbf{s}}^*)P_\Lambda\right)_i\right] = 0.$$

This means that all components of the row vector $\nu[\frac{df}{d\hat{s}}(\hat{\mathbf{s}}^*)P_\Lambda]$ must be zero; but this is just the second line of (8.71).

Now assume that ν is linear on the stated subspace. Then the second line of (8.71) implies

$$\nu\left[\pm\left(\frac{df}{d\hat{s}}(\hat{\mathbf{s}}^*)P_\Lambda\right)_i\right] = 0$$

for each component $g_i = (\frac{df}{d\hat{s}}(\hat{\mathbf{s}}^*)P_\Lambda)_i$ of $\frac{df}{d\hat{s}}(\hat{\mathbf{s}}^*)$. Then as shown at the end of the proof

$$\nu[g\hat{s}] = 0 \tag{8.77}$$

for all $\hat{s} \in \mathbf{S}$. This proves that (8.71) implies (8.70), under the linearity assumption.

We continue making the linearity assumption. Next, we wish to show that (8.71) implies (8.72) and (8.73). By the proven equivalences, set $\tilde{\hat{s}} = -P_\Lambda\hat{\mathbf{s}}^*$ in (8.69) to obtain

$$\nu\left[f(\hat{\mathbf{s}}^*) + \frac{df}{d\hat{s}}(\hat{\mathbf{s}}^*)(-P_\Lambda\hat{\mathbf{s}}^*)\right] = 0, \tag{8.78}$$

and hence (8.73). Using (8.75) and the second line of (8.70) we get (8.72).

Next we assume (8.72), (8.73) and verify (8.71). Equation (8.72) is clearly equivalent to the second line of (8.71). Now

$$\begin{aligned}
\nu[f(\hat{\mathbf{s}}^*)] &= \nu[(f(\hat{\mathbf{s}}^*) + \tfrac{df}{d\hat{s}}(\hat{\mathbf{s}}^*)(-P_\Lambda\hat{\mathbf{s}}^*)) - \tfrac{df}{d\hat{s}}(\hat{\mathbf{s}}^*)(-P_\Lambda\hat{\mathbf{s}}^*)] \\
&\leq \nu[f(\hat{\mathbf{s}}^*) + \tfrac{df}{d\hat{s}}(\hat{\mathbf{s}}^*)(-P_\Lambda\hat{\mathbf{s}}^*)] + \nu[\tfrac{df}{d\hat{s}}(\hat{\mathbf{s}}^*)(P_\Lambda\hat{\mathbf{s}}^*)] \\
&= 0.
\end{aligned}$$

Now (8.73) is equivalent to (8.78), and so

$$\begin{aligned}
0 &= \nu[f(\hat{\mathbf{s}}^*(p)) + \tfrac{df}{d\hat{s}}(\hat{\mathbf{s}}^*(p))(-P_\Lambda\hat{\mathbf{s}}^*(p))] \\
&\leq \nu[f(\hat{\mathbf{s}}^*(p))] + \nu[\tfrac{df}{d\hat{s}}(\hat{\mathbf{s}}^*(p))(-P_\Lambda\hat{\mathbf{s}}^*(p))] \\
&= \nu[f(\hat{\mathbf{s}}^*(p))].
\end{aligned}$$

Thus $\nu[f(\hat{\mathbf{s}}^*(p))] = 0$. This completes the verification of (8.71), assuming (8.72), (8.73).

The identities (8.74) now follow by direct substitution.

Proof of (8.77). The hypothesis is

$$\nu[\pm g_i] = 0$$

for all components of the vector g. We do not use the linearity assumption here. Let $\hat{s} \in \mathbf{S}$. Then

$$\begin{aligned}
\nu[g\hat{s}] &= \nu[\textstyle\sum_i g_i\hat{s}_i] \\
&= \nu[\textstyle\sum_i g_i \,\mathrm{sign}\,(\hat{s}_i)|\hat{s}_i|] \\
&\leq \textstyle\sum_i \nu[g_i \,\mathrm{sign}\,(\hat{s}_i)]|\hat{s}_i| \\
&= 0,
\end{aligned}$$

so $\nu[g\hat{s}] \leq 0$. Also,

$$0 = \nu[0] \leq \nu[a] + \nu[-a]$$

for $a = g\hat{s} = \sum_i g_i \,\mathrm{sign}\,(\hat{s}_i)|\hat{s}_i|$, and

$$\nu[-a] \leq \sum_i \nu[g_i\,(-1)\mathrm{sign}\,(\hat{s}_i)]|\hat{s}_i| = 0,$$

so $\nu[g\hat{s}] \geq 0$. □

Proof of Theorem 8.10.1 *and Corollary* 8.10.2. To prove this note that energy conservation (8.21) is equivalent to (8.19), which is just the integrated form of (8.20), or (8.60).

The remainder of the theorem and the corollary are proven by applying Lemma 8.10.4, with

$$f(\hat{s}) = \hat{F}(p, \hat{s}) \quad \text{and} \quad \nu[\cdot] = \nabla_p^0 \hat{W}(p)[\cdot].$$

Here,

$$a = -\nabla_x p\hat{A} + \tfrac{1}{2}\hat{C}' J_{\mathbf{Z}}\hat{C}, \quad b = -\nabla_x p\hat{B} + \hat{C}' J_{\mathbf{Z}}\hat{D}, \quad \text{and} \quad c = \hat{E} = \hat{D}' J_{\mathbf{Z}}\hat{D}.$$

□

8.11 Notes

(i) J-inner–outer factoring was seen to be related to classical linear H^∞ control at an early stage. The connection came through Nevanlinna–Pick interpolation in that the original solution [ZF81], [ZF83] to the H^∞ control problem relied on such interpolation or generalizations of it. The paper [BH83] showed (in what would now be called a behavioral setting) how Nevannlina–Pick and commutant lifting problems were solved by J-inner–outer factorization. The first (see also Chang–Pearson [CJ84]) paper [FHZ84] to solve the MIMO H^∞ problem actually used such factoring as did the famous proof of [BR87]. Kimura and associates took factoring to great depth in an excellent series of publications; cf. [Kim92], [Kim95], [BK96], [Kim97].

(ii) The first work on extending H^∞ control to nonlinear systems took the factoring approach we have described here. These laid out solutions of varying effectiveness to the discrete time factoring problem when $\hat{A}$ is stable, i.e., when the plant is stable and weights are carefully chosen: the papers [BFHT87b], [BFHT87a] used Volterra expansions, and [BH88c] did it behaviorally. Then came state space. The formulas of §8.6.2 were first described in discrete time in [BH88a], [BH89] with a thorough treatment in [BH92a]. See the book [vdS96] for an excellent treatment of the continuous time case by Ball and van der Schaft. Also there is the elegant work of Baramov and Kimura [BK96]. All of these approaches pertained to the $\hat{A}$ stable situation and used only the HJBI equation. See also [Bar98a], [Bar98b].

(iii) As we saw in this chapter to solve the factoring problem when $\hat{A}$ is not stable requires the information state equation. The approach taken here, the RECIPE and some of its J-lossless properties, was announced in [HJ94], and the bilinear case was studied in [HJ96a].

Chapter 9

The Mixed Sensitivity Problem

9.1 Introduction

In the mixed sensitivity problem one is given a plant P and wishes to find a controller K achieving a given H^∞ performance; see Figure 9.1.

If the weights are chosen correctly for a mixed sensitivity problem, then one gets the standard 2A block problem of H^∞ control. We shall show Result 9.1.1 under strong hypotheses on our setup.

> RESULT 9.1.1 *If the plant P is strongly hyperbolic and has a $k = k_{as}$-dimensional antistable manifold and if the weights are well behaved (defined below), then the key dynamics of the two block problem are strongly hyperbolic and have a k-dimensional antistable manifold. This allows a construction for an information state controller requiring online computation having complexity of the same order as numerical solution of a PDE driven (by inputs) on $k < n$-dimensional space.*

Now it often happens in practice that plants P are stable or at worst have a low dimensional instability. Thus the numerics of constructing the central controller solving a measurement feedback problem might not be much more complex than that of constructing a state feedback controller. We caution the reader that the offline computation in designing an *uncompromised* H^∞ controller would still be huge (by this we mean a full implementation without effort to reduce complexity).

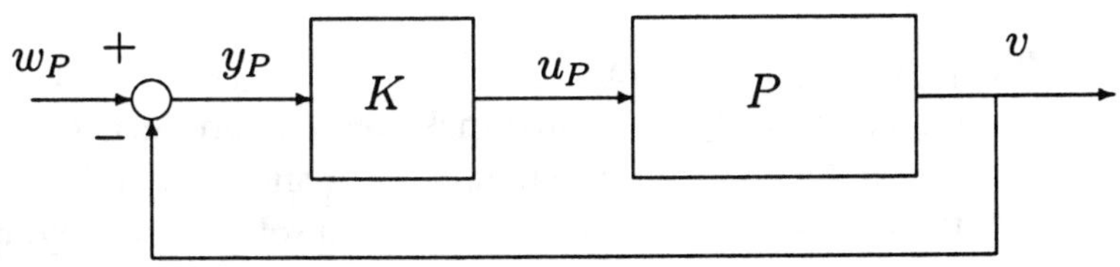

Figure 9.1: Mixed sensitivity setup.

9.2 Notation and Other Details

Now we apply the solution given in the previous sections to the mixed sensitivity problem. The G in standard problem of H^∞ control that we treat in this paper is typically not the plant in classical control problems, such as mixed sensitivity (see Figure 9.1). Rather, G is built from the plant P together with weights.

We treat the mixed sensitivity problem for a nonlinear plant P having state space representation

$$P : \begin{cases} \dot{x}_P &= A_P(x_P) + B_P(x_P)u_P, \\ \\ v &= C_P(x_P). \end{cases} \tag{9.1}$$

The mixed sensitivity problem is to design the compensator K so that (i) the following dissipation inequality holds:

$$\begin{aligned} &\tfrac{1}{2}\int_0^T [|\mathcal{W}_1(y_P)(s)|^2 + |\mathcal{W}_2(v)(s)|^2]\, ds \\ &\le \gamma^2 \tfrac{1}{2}\int_0^T |w_P(s)|^2\, ds + \beta_P(x_P(0)), \end{aligned} \tag{9.2}$$

for all $w_P \in L_{2,T}$, $T \geq 0$, for some finite $\beta_P(x_P(0)) \geq 0$ and, (ii) the closed-loop system is internally stable (see Figure 9.1). Here $\mathcal{W}_1$ and $\mathcal{W}_2$ are systems that weight the signals in the closed-loop system. The relevant system of equations is $u_P = y_P + P(u_P)$, $K(y_P) = u_P$.

To set the problem in the framework of the standard H^∞-problem for G as in (2.2), we should make the following identifications (Figure 9.1) $w = w_P$, $u = u_P$, $y = y_P$, $z = (z_1', z_2')'$, with $z_1 = \mathcal{W}_1(y_P)$, $z_2 = \mathcal{W}_2(v)$. To write down the state space equations for G, we need to specify the form of the state space realizations of the weights. A key issue is the proper choice of weights.

9.3 Choosing the Weights

In the linear case, physically correct choices of the frequency domain weights dictate that for complex numbers s we have

$$\mathcal{W}_1(s) \to 0 \text{ and } \mathcal{W}_2(s)P(s)K(s) \to \text{ constant } \neq 0$$

as $s \to \infty$ [HM98a]. Some perspective on this is given in §§1.8, 1.9. Here we give another. To pick weights correctly, one must make an a priori decision on the rate at which the compensator $K(s)$ must go to 0 at ∞. For simplicity of exposition we make the choice that $K(s) \to c \neq 0$ as $s \to \infty$. One can select other asymptotics by extracting them from the K and placing them in P. Reasonable nonlinear analogues of this with a "proper" plant are given by the following.

We treat the nonlinear analogue of the case where $P(s)$ has a first-order zero at $s = \infty$ and we desire a compensator satisfying $K(s) \to c \neq 0$. Then $\mathcal{W}_2(s)$ must be asymptotic to s as $s \to \infty$. This motivates the state space realizations

$$\mathcal{W}_1 : \begin{cases} \dot{x}_1 = A_{\mathcal{W}_1}(x_1) + B_{\mathcal{W}_1}(x_1)y, \\ z_1 = C_{\mathcal{W}_1}(x_1), \end{cases} \tag{9.3}$$

$$\mathcal{W}_2 : \begin{cases} \dot{x}_2 = A_{\mathcal{W}_2}(x_2) + B_{\mathcal{W}_2}(x_2)v, \\ z_2 = C_{\mathcal{W}_2}(x_2) + D_{\mathcal{W}_2}(x_2)\frac{dv}{dt}. \end{cases} \tag{9.4}$$

Write $n_P = \dim x_P$, $n_1 = \dim x_1$, $n_2 = \dim x_2$, $n = n_P + n_1 + n_2$, and $x = (x_P, x_1, x_2)$. We say that the weights are *well behaved* provided that $\mathcal{W}_1$ and

$$\dot{x}_2 = A^{\mathcal{W}_2}(x_2) + B^{\mathcal{W}_2}(x_2)v, \; z_2 = C^{\mathcal{W}_2}(x_2) + D^{\mathcal{W}_2}(x_2)v$$

are asymptotically stable in a very strong sense. (This is not the same as saying $\mathcal{W}_1$ and $\mathcal{W}_2$ are stable since the $\frac{d}{dt}v$ destabilizes $\mathcal{W}_2$.)

9.4 Standard Form

Using such weights the extended system G (2.2) (of Chapter 2) with state variables $x = (x_P, x_1, x_2)'$ corresponding to (9.1), (9.3), (9.4) is derived from Figure 9.2.

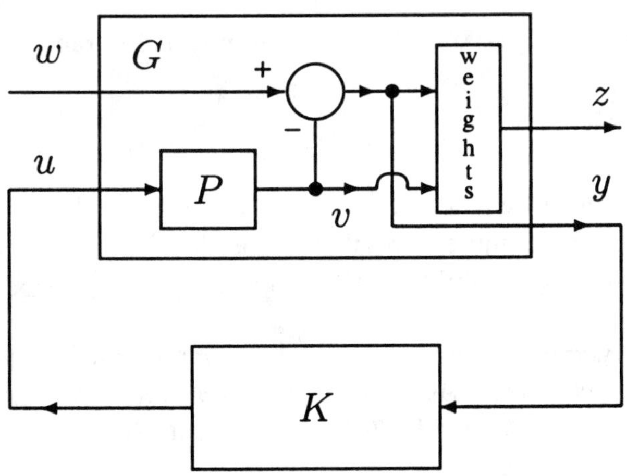

Figure 9.2: Mixed sensitivity embedded in the standard H^∞ problem arrangement (recall Figure 1.1).

The equations for G are

$$\begin{pmatrix} \dot{x}_P \\ \\ \dot{x}_2 \\ \\ \dot{x}_2 \end{pmatrix} = \begin{pmatrix} A_P(x_P) + B_P(x_P)u \\ \\ (A_{\mathcal{W}_1}(x_1) + B_{\mathcal{W}_1}(x_1)(-C_P(x_P) + w)) \\ \\ A_{\mathcal{W}_2}(x_2) + B_{\mathcal{W}_2}(x_2)C_P(x_P) \end{pmatrix},$$

$$\begin{pmatrix} z_1 \\ z_2 \end{pmatrix} = \begin{pmatrix} C_{\mathcal{W}_1}(x_1) \\ \\ C_{\mathcal{W}_2}(x_2) + D_{\mathcal{W}_2}(x_2)\nabla_{x_P}[C_P(x_P)(A_P(x_P) + B_P(x_P)u)] \end{pmatrix},$$

$$y = -C_P(x_P) + w. \tag{9.5}$$

Of course, $u = K(y)$. From (9.5) it is evident that

$$A^{\times}(x_P, x_1, x_2) = \begin{pmatrix} A_P(x_P) \\ \\ A_{\mathcal{W}_1}(x_1) \\ \\ A_{\mathcal{W}_2}(x_2) + B_{\mathcal{W}_2}(x_2)C_P(x_P) \end{pmatrix}, \tag{9.6}$$

and the "D" terms of G satisfy

$$D_{11}(x) = 0, \;\; D_{21}(x) = I, \;\; D_{22}(x) = 0,$$

$$D_{12}(x) = (0, [D_{\mathcal{W}_2}(x_2)\nabla_{x_P}[C_P(x_P)]B_P(x_P)]')'.$$

Thus $D_{21}(x) = I_p$, and if

$$D_{\mathcal{W}_2}(x_2)\nabla_{x_P}[C_P(x_P)]B_P(x_P) \text{ is full rank,}$$

then

$$E_1(x) = D_{12}(x)'D_{12}(x) > 0$$

is invertible, and indeed we have produced the 2A *block problem.*

The stability of the weights ensures that the instabilities of this system are in perfect correspondence with the instabilities of P.

THEOREM 9.4.1 *Assume that the vector field A_P is incrementally hyperbolic,* $\dim M_{as}^P = k_{as}$, *and that the vector fields $A_{\mathcal{W}_1}$, $A_{\mathcal{W}_2}$ are incrementally L_2 exponentially stable. Then the vector field $A^{\times}$ defined by* (9.6) *is incrementally hyperbolic with antistable manifold M_{as} of dimension k_{as}.*

Proof. By the definition of (global) hyperbolicity, we assume that after a coordinate transformation that $\mathbf{R}^{n_P} = M_a^P \times M_{as}^P$, and $x_P = (x_{Ps}, x_{Pas})$. Define

$$M_s = M_s^P \times \mathbf{R}^{n_1} \times \mathbf{R}^{n_2}$$

and

$$M_{as} = M_{as}^P \times \{0_{n_1}\} \times \{0_{n_2}\}.$$

The hypotheses imply that if $x(0) \in M_s$, then $x(t) \to 0$ as $t \to \infty$, where $x(\cdot)$ is the solution to the $A^\times$ dynamics. Similarly, if $x(0) \in M_{as}$, then $x(t) \to 0$ as $t \to -\infty$. By construction, $\dim M_{as} = k_{as} = \dim M_{as}^P$, as required. □

9.5 Formula for the Controller

To get an explicit formula for the controller one merely combines Corollary 10.2.2 and formulas (10.38) describing the controller for the two block problem with formulas (9.6) and (9.5) to give the two block problem in terms of the mixed sensitivity problem. We reiterate that the dynamics of the controller has the numerical complexity of a PDE on a $k = \dim M_{as}^P << n$-dimensional space. Writing them out in $A_\mathcal{W}$ notation is not particularly informative, so we shall not. However, we will be somewhat more explicit in the special case that $A^\times$ is strongly stable. Indeed (see [BH92b]) we have the following corollary.

COROLLARY 9.5.1 *Assume*

(i) $B_1(x_P, x_1, x_2) = (0, B_{\mathcal{W}_1}(x_1)', 0)'$ *and*
$B_2(x_P, x_1, x_2) = (B_P(x_P)', 0, 0)'$ *are independent of* x_P, x_1.

(ii) $A_P(x_P)$ *is incrementally* L_2 *exponentially stable.*

(iii) *There exists a (smooth) solution* $V(x_P, x_1, x_2)$ *solving the PDE* (5.45) *with*

$$A^*(x_P, x_1, x_2) = A - B_2C_1 + (\gamma^{-2}B_1B_1' - B_2B_2')\nabla_x V'$$

incrementally L_2 *exponentially stable.*

Then the controller $K_{\delta_0}^*$ *defined by* (10.39) *solves the mixed sensitivity control problem for* P *with internal stability. This controller is finite dimensional or, from the information state PDE viewpoint, it is given by a PDE on zero-dimensional space.*

9.6 Notes

(i) Result 9.1.1 was given for stable discrete time systems in [BH89], [BH92b]. Coupled with the discrete time version of singular controller [BH92b], this converts solution of the mixed sensitivity problem for stable plants to solving the usual HJBI difference analogue of (5.45). Corollary 9.5.1 in this chapter gives these formulas in continuous time for the special class of systems we study.

Part II

Singular Information States and Stability

Chapter 10

Singular Information States

10.1 Introduction

In the theory of linear H^∞ control, one of the (three) necessary and sufficient conditions for the solvability of the H^∞ control problem is the existence of a (stabilizing) solution $Y_e \geq 0$ of the steady state version of (7.14) (equation (7.21)). In the nonlinear case (and one could argue in the linear case) Y^{-1} rather than Y is what occurs naturally. In the linear case it is easy to change variables to get everything in terms of Y, but in seriously nonlinear cases such a change of variables appears impossible. Thus one seeks an (antistabilizing) solution Y_{Ie} to an appropriate equation. A crucial issue is whether Y_e is invertible and how to make sense of Y_{Ie} when Y_e is not. Problems for which Y_e is not invertible provide a large class of examples with *singular* (equilibrium) information states (such information states are not everywhere finite). Singular information states also occur in the nonlinear case, and the purpose of this chapter is to investigate them and explore their properties and associated controllers.

In §10.2 singular information state dynamics are described, while in §10.3 the dynamic programming PDE for W is interpreted. (This provides a useful reinterpretation of the formula (4.23) for the optimal control $\mathbf{u}^*(p)$ and observation $\mathbf{y}^*(p)$. These formulas involve $\nabla_p W(p)$ and $\nabla_x p(x)$, terms that cannot be understood classically for singular p such as $p = \delta_0$.) Formulas for the central controllers in various cases such as certainty equivalence are given in §10.4.

ASSUMPTION 10.1.1 Throughout this chapter we specialize to the 1 and 2A block cases, and assume D_{12} is independent of x:

$$\begin{aligned} D_{21}(x) &= I \text{ for all } x \in \mathbf{R}^n \text{ and} \\ D_{12}(x) &= D_{12} \text{ for all } x \in \mathbf{R}^n. \end{aligned} \tag{10.1}$$

10.2 Singular Information State Dynamics

10.2.1 Geometrical Description of Information State Dynamics

The information state p_t is well defined (by (3.7) or (3.12)), even if it is not smooth or even if it is singular:

$$p_t(x) = p_0(\xi(0)) + \tfrac{1}{2}\int_0^t [|C_1(\xi(s)) + D_{12}u(s)|^2 - \gamma^2|y(s) - C_2(\xi(s))|^2]ds, \tag{3.12$'$}$$

where

$$\dot{\xi}(s) = A^\times(\xi(s)) + B_1(\xi(s))y + B_2(\xi(s))u \tag{2.6$'$}$$

for $0 \le s \le t$, with terminal condition $\xi(t) = x$. In terms of the transition operator for (2.6), $\xi(s) = \Phi_{s,t}^{u,y}(x)$. Further, as discussed in Chapter 3 (e.g., Lemma 3.1.10), the finiteness of p_t is strongly influenced by its initial value p_0. In the construction of the central controller it is important to use initial information states of the form

$$p_0 = \delta_{M_0} + \breve{p}_0, \tag{10.2}$$

where M_0 is a submanifold of $\mathbf{R}^n$ and $\breve{p}_0 \in C(M_0)$. The function p_0 equals $\breve{p}_0$ on M_0 and equals $-\infty$ everywhere else (recall definition (2.14)). The correct initialization when $A^\times$ is hyperbolic is $p_0 = p_e$, the equilibrium information state, with $M_0 = M_{as}$, the antistable manifold for $A^\times$; see Chapters 4, 11.

Now we give a geometric picture of the information state dynamics with initialization (10.2) (which, for those who find comfort in technical language, evolves on a *fiber bundle*). This describes the dynamics of the information state controller. Let $\mathcal{M}$ denote all (maximal) $k = k_0$-dimensional smooth manifolds embedded in $\mathbf{R}^n$ (i.e., there exists coordinates such that $M_0 = (x_1, \dots, x_k, 0 \dots, 0)$).

Lemma 10.2.1 *If p_0 is of the form* (10.2), *then for any $u, y \in L_{2,loc}$, p_t is of the form* (10.2) *for all $t \ge 0$; i.e.,*

$$p_t(x) = \delta_{M_t}(x) + \breve{p}_t(x), \tag{10.3}$$

where

$$\breve{p}_t(x) = \breve{p}_0(\xi(0)) + \tfrac{1}{2}\int_0^t [|C_1(\xi(s)) + D_{12}u(s)|^2 - \gamma^2|y(s) - C_2(\xi(s))|^2]ds,$$

and $\xi(\cdot)$ is the solution of (2.6), *i.e., with terminal condition $\xi(t) = x \in M_t$, and M_t is the image of M_0 under the flow of* (2.6), *i.e.,*

$$M_t = \{\xi(t) : \xi(0) \in M_0\}.$$

Proof. By definition (3.12),

$$p_t(x) = p_0(\xi(0)) + \breve{p}_t(x),$$

where $\xi(\cdot)$ and $\breve{p}_t(x)$ are as defined in the statement of the lemma. Now if $x \in M_t$, then $\xi(0) \in M_0$ and $\delta_{M_0}(\xi(0)) = 0$; otherwise, if $x \notin M_t$, then $\xi(0) \notin M_0$ and $\delta_{M_0}(\xi(0)) = -\infty$. This proves (10.3). □

The following corollary gives a more explicit geometrical picture.

COROLLARY 10.2.2 *The state space for the information state with initialization* (10.2) *is contained in*

$$\mathcal{F}(\mathcal{M}) \triangleq \{(M, \breve{p}) : M \in \mathcal{M}, \breve{p} \in C(M)\}.$$

(This set is a "fiber bundle" over $\mathcal{M}$.*) The dynamics of the information state for the information state controller with initialization* (10.2) *is given by*

$$M_t = \Phi_{t,0}^{u,y}(M_0), \tag{10.4}$$

where $\Phi_{t,0}^{u,y}$ *is the transition operator associated with* (2.6), *and*

$$\breve{p}_t = \psi_t \circ \Phi_{0,t}^{u,y}, \tag{10.5}$$

where

$$\begin{cases} \dot{\psi}_s &= \frac{1}{2}[|C_1(\cdot) + D_{12}u(s)|^2 - \gamma^2|y(s) - C_2(\cdot)|^2] \circ \Phi_{s,0}^{u,y}, \\ \psi_0 &= \breve{p}_0, \end{cases} \tag{10.6}$$

or in transition operator notation,

$$\breve{p}_t = \hat{\Phi}_{t,0}^{u,y}(\breve{p}_0).$$

REMARK 10.2.3 Note that $\breve{p}_t \in C(M_t)$ is a function of the state $x \in M_t$ at time t, whereas $\psi_t \in C(M_0)$ is a function of the initial state $x_0 \in M_0$. ∇

This geometric picture while strikingly graphic has lots of things moving at once: the information state evolves in a fiber bundle $\mathcal{F}(\mathcal{M})$ with base point $M_t \in \mathcal{M}$ and $\breve{p}_t$ belonging to the fiber $C(M_t)$. Equation (10.5) gives a less dizzying interpretation by "pulling back" onto the initial manifold M_0. Indeed, the formula (10.5) shows that $\breve{p}_t$ is computable via characteristics starting in the k_0-dimensional manifold M_0. The remaining $n - k_0$ dimensions are irrelevant. In this way a significant reduction in complexity is obtained, as discussed in the next subsection.

10.2.2 Computational Complexity

This evolution of functions $\breve{p}_t$ on the $M_0 \equiv \mathbf{R}^{k_0}$ together with the evolution (2.6) constitute the "reduced-dimensional" picture of the controller dynamics (discussed in Chapter 4). Suppose one wishes to approximately compute the compensator state $(M_t, \breve{p}_t)$ via a flow-based method for solving a hyperbolic PDE. This entails using a finite grid on $M_0 \equiv \mathbf{R}^{k_0}$. If N is the number of grid points one wishes to use in each dimension, then the memory requirements scale like

$$nO(N^{k_0}),$$

which is much less than $O(N^n)$ if one had to solve the PDE (3.9) in $\mathbf{R}^n$.

10.3 Interpreting the Dynamic Programming PDE

While we have appealing formulas for the controller dynamics, interpreting the high level formula

$$\mathbf{u}^*(p) = \nabla_p W(p)[E_1^{-1}(-D_{12}'C_1 + B_2'\nabla_x p')] \tag{4.23$'$}$$

for controller output in this singular situation requires technical interpretation, since it involves singular p of the form (10.2). This causes little trouble with $W(p)$, which is a well-defined function on $\mathcal{X}_e$. However, interpretation of $\nabla_p W(p)$ and $u^*(p)$ is difficult (recall the formula (4.23), which involves $\nabla_x p$, a quantity difficult to interpret in singular cases).

The purpose of this section is to give an interpretation of the dynamic programming PDE (4.16), which is suitable for singular information states, viz., (4.40). This is done by defining the operator $\mathcal{L}^{u,y}$ used in Chapter 4.

10.3.1 Transition Operators and Generators

In order to define the operator $\mathcal{L}^{u,y}$, we consider a hierarchy of transition operators and their generators. At the level of the original state space of the plant G we have used the transition operator notation $\xi(s) = \Phi_{s,t}^{u,y}(x)$ to denote the solution of

$$\dot{\xi} = A^\times(\xi) + B_1(\xi)y + B_2(\xi)u \tag{2.6$'$}$$

on the interval $0 \leq s \leq t$ with terminal condition $\xi(t) = x$. The transition operator (or flow) $\Phi_{s,t}^{u,y}$ underlies the definition of the information state p_t and so also the value function W; in fact, the transition operator $\Phi_{s,t}^{u,y}$ induces transition operators at these two levels.

The max-plus framework described in Appendix C will be used. We shall use the notation

$$[[p + f]] \stackrel{\Delta}{=} \operatorname*{argmax}_{x \in \mathbf{R}^n} \{p(x) + f(x)\},$$

$$p(f) = p[f] = \langle p + f \rangle = \sup_{x \in \mathbf{R}^n} \{p(x) + f(x)\}.$$

For $p \in \mathcal{X}_e$ define the transition operator

$$S_t^{u,y} p = p_t, \tag{3.15$'$}$$

where p_t is the information state at time $t \geq 0$ corresponding to the initial state $p_0 = p$ and $u, y \in L_{2,loc}$. So $S_t^{u,y}$ is the operator that maps an initial information state $p_0 = p$ to the information state p_t at time t (recall definition (3.7)). Max-plus linearity was investigated in Lemma 3.1.6, and the following lemma looks at this and other properties.

LEMMA 10.3.1 (i) *If p is quadratically tight, there exists $t_0 > 0$ such that the set $\{S_t^{u,y} p\}_{0 \leq t \leq t_0}$ is uniformly quadratically tight, and $S_t^{u,y} p \Longrightarrow p$ as $t \to 0$.*

(ii) *Let* $p \in \mathcal{X}_e$ *and assume the upper bound* $S_t^{u,y}p(x) \leq C$ *for all* $t \geq 0$, $x \in \mathbf{R}^n$. *Then for each* $t \geq 0$ *we have* $S_t^{u,y}p \in \mathcal{X}_e$. *If in addition* $p_n \Longrightarrow p$ *with* p *quadratically tight, and* $\{S_t^{u,y}p_n\}_{n=0}^{\infty}$ *uniformly quadratically tight, then* $S_t^{u,y}p_n \Longrightarrow S_t^{u,y}p$.

(iii) *The operator* $S_t^{u,y}$ *is max-plus linear:*

$$S_t^{u,y}(p_1 \oplus p_2) = (S_t^{u,y}p_1) \oplus (S_t^{u,y}p_2),$$

$$S_t^{u,y}(c \odot p) = (S_t^{u,y}p) \odot c.$$

(iv) *For each constant pair* $(u, y) \in \mathbf{R}^m \times \mathbf{R}^p$, *one has the semigroup property*

$$S_{s+t}^{u,y}p = S_s^{u,y}S_t^{u,y}p. \tag{10.7}$$

(An analogous relation holds in the general time-inhomogeneous case.)

Proof. Part (i). Let p be quadratically tight:

$$p(x) \leq -c_1|x|^2 + c_2$$

with $c_1 > 0$. By the definition of the information state (3.12), we have

$$S_t^{u,y}p(x) = p(\Phi_{0,t}^{u,y}(x)) + I_t(\Phi_{0,t}^{u,y}(x)), \tag{10.8}$$

where

$$I_t(x_0) = \tfrac{1}{2}\int_0^t [|C_1(\xi(s)) + D_{12}u(s)|^2 - \gamma^2|y(s) - C_2(\xi(s))|^2]\, ds$$

with $\xi(s)$, $0 \leq s \leq t$ the solution of (2.6) with $\xi(0) = x_0$. By standard ODE estimates (A.5), for $0 \leq s \leq t$, we have

$$|\xi(s)| \leq c|\xi(0)| + c \tag{10.9}$$

for some constant $c > 0$ (which in general depends on the time interval $[0, t]$). Using this we can bound the integral as follows:

$$\begin{aligned}|I_t(x_0)| &\leq \int_0^t [k_1|\xi(s)|^2 + k_2(|u(s)|^2 + |y(s)|^2)]\, ds \\ &\leq tk_3|x_0|^2 + k_4(t)\end{aligned}$$

for appropriate constants, where $k_4(\cdot)$ is continuous and $c_4(0) = 0$. Choose t_0 such that $-c_1 + t_0k_3 = -c_1/2$. Then for $0 \leq t \leq t_0$ we have, using these estimates and (10.8),

$$S_t^{u,y}p(x) \leq -\tfrac{1}{2}c_1|\Phi_{0,t}^{u,y}(x)|^2 + c_2 + k_4(t).$$

Also, (10.9) implies

$$-|\Phi_{0,t}^{u,y}(x)|^2 \leq -c_3|x|^2 + c_4$$

for appropriate constants. Therefore,

$$S_t^{u,y}p(x) \leq -\tfrac{1}{2}c_5k_5|x|^2 + c_6$$

for $0 \leq t \leq t_0$, for appropriate constants, proving the required uniform quadratic tightness.

We next prove that $S_t^{u,y}p \Longrightarrow p$ as $t \to 0$. To do this, we will show that $\langle S_t^{u,y}p + f\rangle \to \langle p + f\rangle$ as $t \to 0$ for all $f \in C_b(\mathbf{R}^n)$, as per Theorem C.5.1. Now using the definition of the information state transition operator in the form (10.8) we have for $f \in C_b(\mathbf{R}^n)$,

$$\begin{aligned}\langle S_t^{u,y}p + f\rangle &= \sup_{x\in\mathbf{R}^n}\{(S_t^{u,y})p(x) + f(x)\} \\ &= \sup_{x\in\mathbf{R}^n}\{p(\Phi_{0,t}^{u,y}(x)) + I_t(\Phi_{0,t}^{u,y}(x)) + f(x)\} \qquad (10.10) \\ &= \sup_{\xi\in\mathbf{R}^n}\{p(\xi) + I_t(\xi) + f(\Phi_{t,0}^{u,y}(\xi))\}.\end{aligned}$$

Here, we have interchanged maximization over the final state x with maximization over the initial state ξ. Select $\bar{x} \in [[p + f]]$, so that $\langle p + f\rangle = p(\bar{x}) + f(\bar{x})$. (Note that such point $\bar{x}$ exists since the fact that p is tight and u.s.c. and f is continuous and bounded implies $p + f$ is tight and u.s.c. and therefore maxima exist.) Then setting $\xi = \bar{x}$ gives

$$\langle S_t^{u,y}p + f\rangle \geq p(\bar{x}) + I_t(\bar{x}) + f(\Phi_{t,0}^{u,y}(\bar{x}))$$

and sending $t \to 0$ gives, by continuity,

$$\begin{aligned}\liminf_{t\to 0}\langle S_t^{u,y}p + f\rangle &\geq p(\bar{x}) + I_0(\bar{x}) + f(\Phi_{0,0}^{u,y}(\bar{x})) \\ &= p(\bar{x}) + f(\bar{x}) = \langle p + f\rangle.\end{aligned}$$

Next, let $0 < \varepsilon \leq 1$ be arbitrary, and $0 \leq t \leq t_0$. Then

$$\begin{aligned}&S_t^{u,y}p(x) + f(x) = p(\Phi_{0,t}^{u,y}(x)) + I_t(\Phi_{0,t}^{u,y}(x)) + f(x) \\ &\leq [\varepsilon p(\Phi_{0,t}^{u,y}(x)) + I_t(\Phi_{0,t}^{u,y}(x)) + \varepsilon f(x)] + (1-\varepsilon)[p(\Phi_{0,t}^{u,y}(x)) + f(x)] \\ &\leq [\varepsilon(-c_1|\Phi_{0,t}^{u,y}(x)|^2 + c_2) + (tK_3|\Phi_{0,t}^{u,y}(x)|^2 + K_4(t)) + \varepsilon \parallel f \parallel^{\infty}] \\ &\quad +(1-\varepsilon)[p(\Phi_{0,t}^{u,y}(x)) + f(x)] \\ &\leq -\tfrac{1}{2}c_1|\Phi_{0,t}^{u,y}(x)|^2 + \varepsilon(c_2 + 1 + \parallel f \parallel_{L^\infty}) + (1-\varepsilon)[p(\Phi_{0,t}^{u,y}(x)) + f(x)],\end{aligned}$$

where t_1 is chosen so that $-(\varepsilon c_1 - tK_3) \leq -\frac{1}{2}c_1$ and $K_4(t) \leq \varepsilon$ for $0 \leq t \leq t_1$. Then

$$\begin{aligned}\langle S_t^{u,y}p + f\rangle &\leq \varepsilon(c_2 + 1 + \parallel f \parallel^{\infty}) + (1-\varepsilon)\sup_x\{p(\Phi_{0,t}^{u,y}(x)) + f(x)\}. \\ &= \varepsilon(c_2 + 1 + \parallel f \parallel^{\infty}) + (1-\varepsilon)\sup_\xi\{p(\xi) + f(\Phi_{t,0}^{u,y}(\xi))\}.\end{aligned}$$

Next, the fact that p is tight and f is bounded implies

$$\sup_{\xi \in \mathbf{R}^n} \{p(\xi) + f(\Phi_{t,0}^{u,y}(\xi))\} = \sup_{|\xi| \leq k_1} \{p(\xi) + f(\Phi_{t,0}^{u,y}(\xi))\}$$

for some constant $k_1 > 0$ independent of $0 \leq t \leq t_1$. Also, $f(\Phi_{t,0}^{u,y}(\xi)) \to f(\xi)$ uniformly in $|\xi| \leq k_1$. So select $\bar{\xi}_t \in \mathrm{argmax}_{|\xi| \leq k_1} \{p(\xi) + f(\Phi_{t,0}^{u,y}(\xi))\}$. Then since $|\bar{\xi}_t| \leq k_1$ we have

$$\limsup_{t \to 0} \sup_{|\xi| \leq k_1} \{p(\xi) + f(\Phi_{t,0}^{u,y}(\xi))\} \leq \langle p + f \rangle.$$

Therefore,

$$\limsup_{t \to 0} \langle S_t^{u,y} p + f \rangle \leq \varepsilon(c_2 + 1 + \| f \|^{\infty}) + (1 - \varepsilon)\langle p + f \rangle.$$

Since $\varepsilon > 0$ was arbitrary, we conclude $\limsup_{t \to 0} \langle S_t^{u,y} p + f \rangle \leq \langle p + f \rangle$. Therefore, $S_t^{u,y} p \Longrightarrow p$ as $t \to 0$.

Part (ii). From the assumed upper bound and since $p \in \mathcal{X}_e$ is u.s.c. and $I_t(x)$ is continuous, it follows from (10.8) that $S_t^{u,y} p(x)$ is an u.s.c. function of x and is bounded from above; i.e., $S_t^{u,y} p \in \mathcal{X}_e$.

Let $f \in C_b(\mathbf{R}^n)$. Then, as in (10.10),

$$\langle S_t^{u,y} p_n + f \rangle = \sup_{\xi \in \mathbf{R}^n} \{p_n(\xi) + I_t(\xi) + f(\Phi_{t,0}^{u,y}(\xi))\},$$

and by the assumed tightness,

$$\begin{aligned} p_n(\xi) + I_t(\xi) + f(\Phi_{t,0}^{u,y}(\xi)) &= (S_t^{u,y} p_n)(\Phi_{t,0}^{u,y}(\xi)) + f(\Phi_{t,0}^{u,y}(\xi)) \\ &\leq -c_1 |\Phi_{t,0}^{u,y}(\xi)|^2 + c \end{aligned}$$

for suitable $c \geq 0$ (recall $c_1 > 0$). Now $p_n \Longrightarrow p$ and p is tight, so $p_n \xrightarrow{h} p$ (hypo-convergence, Appendix C); hence, by Theorem C.5.11 we have

$$\sup_{\xi}\{p_n(\xi) + I_t(\xi) + f(\Phi_{t,0}^{u,y}(\xi))\} \to \sup_{\xi}\{p(\xi) + I_t(\xi) + f(\Phi_{t,0}^{u,y}(\xi))\}$$

and so $\langle S_t^{u,y} p_n + f \rangle \to \langle S_t^{u,y} p + f \rangle$; i.e., $S_t^{u,y} p_n \Longrightarrow S_t^{u,y} p$.

Part (iii). Let $p_1, p_2 \in \mathcal{X}_e$. Then

$$\begin{aligned} S_t^{u,y}(p_1 \oplus p_2) &= (p_1 \oplus p_2)(\Phi_{0,t}^{u,y}(x)) + I_t(\Phi_{0,t}^{u,y}(x)) \\ &= (p_1(\Phi_{0,t}^{u,y}(x))) \oplus (p_2(\Phi_{0,t}^{u,y}(x))) + I_t(\Phi_{0,t}^{u,y}(x)) \\ &= (p_1(\Phi_{0,t}^{u,y}(x)) + I_t(\Phi_{0,t}^{u,y}(x))) \oplus (p_2(\Phi_{0,t}^{u,y}(x)) + I_t(\Phi_{0,t}^{u,y}(x))) \\ &= (S_t^{u,y} p_1) \oplus (S_t^{u,y} p_2). \end{aligned}$$

Next, let $p \in \mathcal{X}_e$ and $c \in \mathbf{R}$. Then

$$\begin{aligned} S_t^{u,y}(c \odot p) &= (c \odot p)(\Phi_{0,t}^{u,y}(x)) + I_t(\Phi_{0,t}^{u,y}(x)) \\ &= p(\Phi_{0,t}^{u,y}(x)) + c + I_t(\Phi_{0,t}^{u,y}(x)) \\ &= c \odot (S_t^{u,y} p). \end{aligned}$$

Part (iv). For constant $(u, y) \in \mathbf{R}^m \times \mathbf{R}^p$ the transition operator $\Phi_{s,t}^{u,y}$ is time homogeneous:

$$\Phi_{0,s+t}^{u,y} = \Phi_{0,t}^{u,y} \Phi_{0,s}^{u,y}$$

and $\Phi_{t,s+t}^{u,y} = \Phi_{0,s}^{u,y}$. Then

$$\begin{aligned} S_s(S_t p)(x) &= S_t p(\Phi_{0,s}^{u,y}(x)) + I_s(\Phi_{0,s}^{u,y}(x)) \\ &= p(\Phi_{0,t}^{u,y}\Phi_{0,s}^{u,y}(x)) + I_t(\Phi_{0,t}^{u,y}\Phi_{0,s}^{u,y}(x)) + I_s(\Phi_{0,s}^{u,y}(x)) \\ &= p(\Phi_{0,s+t}^{u,y}(x)) + I_t(\Phi_{0,s+t}^{u,y}(x)) + I_s(\Phi_{t,s+t}^{u,y}(x)) \\ &= p(\Phi_{0,s+t}^{u,y}(x)) + I_{s+t}(\Phi_{0,s+t}^{u,y}(x)) \\ &= (S_{s+t} p)(x). \end{aligned}$$

□

In the nonsingular case, the operator $F(p, u, y)$, defined by (3.10), is the "generator" of the transition operator $S_t^{u,y}$. In the singular case, moving toward an interpretation of the dynamic programming PDE (4.40), we use the following framework.

Let

$$\mathcal{G} = C(\mathcal{X}_e)$$

denote the space of continuous functions $\psi : \mathcal{X}_e \to \mathbf{R}$, where continuity is defined in terms of max-plus convergence in $\mathcal{X}_e$, and

$$\mathcal{G}_b = C_b(\mathcal{X}_e) \subset \mathcal{G} \cap \mathbf{B}_e(\mathcal{X}_e)$$

denotes the subspace of bounded continuous functions (here, $\mathbf{B}_e(\mathcal{X}_e)$ denotes the space of bounded-above functions). We also make use of the space

$$\mathbf{F}(\mathcal{X}_e) = \{\psi \mid \psi : \mathcal{X}_e \to \mathbf{R}\}$$

of real-valued functions ψ defined on $\mathcal{X}_e$. For $\psi \in \mathbf{F}(\mathcal{X}_e)$, write

$$S_t^{u,y}\psi(p) = \psi(S_t^{u,y} p) \tag{10.11}$$

whenever the RHS is defined. In general there will be a domain of functions ψ and points p for which the RHS is defined. Indeed, if $\psi(p)$ is defined for all $p \in \mathcal{X}_e$, then $S_t^{u,y}\psi(p)$ is defined for all $p \in P_t^{u,y}$, where

$$P_t^{u,y} = \{p \in \mathcal{X}_e \;:\; S_t^{u,y} p \in \mathcal{X}_e\}.$$

The next lemma looks at some properties of $S_t^{u,y}$.

LEMMA 10.3.2 (i) *Assume* $S_t^{u,y}p(x) \leq C$ *for all* $t \geq 0$, $x \in \mathbf{R}^n$. *Then for each* $t \geq 0$ *the operator* $\mathcal{S}_t^{u,y}$ *maps* $\mathbf{F}(\mathcal{X}_e) \to \mathbf{F}(\mathcal{X}_e)$ *and* $\mathbf{B}_e(\mathcal{X}_e) \to \mathbf{B}_e(\mathcal{X}_e)$.

(ii) *If* $p_n, p \in \mathcal{X}_e$ *with* $p_n \Longrightarrow p$ *and* p *quadratically tight, and* $\{S_t^{u,y}p_n\}_{n=0}^{\infty} \subset \mathcal{X}_e$ *uniformly quadratically tight, then* $\mathcal{S}_t^{u,y}\psi(p_n) \longrightarrow \mathcal{S}_t^{u,y}\psi(p)$ *for* $\psi \in \mathcal{G}$.

(iii) *If* $\psi_n \to \psi$ *uniformly on* $\mathcal{X}_e$, *then* $\mathcal{S}_t^{u,y}\psi_n \to \mathcal{S}_t^{u,y}\psi$ *uniformly on* $P_t^{u,y}$.

(iv) *For each constant pair* $(u, y) \in \mathbf{R}^m \times \mathbf{R}^p$, $\psi \in \mathcal{F}(\mathcal{X}_e)$, *one has the semigroup property*

$$\mathcal{S}_{s+t}^{u,y}\psi = \mathcal{S}_s^{u,y}\mathcal{S}_t^{u,y}\psi \tag{10.12}$$

whenever the quantities are defined. (An analogous relation holds in the general time-inhomogeneous case.)

Proof. *Part* (i). By Lemma 10.3.1, $S_t^{u,y}p \in \mathcal{X}_e$ when $p \in \mathcal{X}_e$ and so, by definition, if $\psi \in \mathbf{B}_e(\mathcal{X}_e)$,

$$\mathcal{S}_t^{u,y}\psi(p) = \psi(S_t^{u,y}p) \leq \max_{q \in \mathcal{X}_e}\{\psi(q)\} < +\infty;$$

hence $\mathcal{S}_t^{u,y}\psi \in \mathbf{B}_e(\mathcal{X}_e)$.

Part (ii). By Lemma 10.3.1(ii), we have $S_t^{u,y}p_n \Longrightarrow S_t^{u,y}p$. Therefore,

$$\begin{aligned}\mathcal{S}_t^{u,y}\psi(p_n) &= \psi(S_t^{u,y}p_n)\\ &\to \psi(S_t^{u,y}p)\\ &= \mathcal{S}_t^{u,y}\psi(p).\end{aligned}$$

Part (iii). We have

$$\begin{aligned}\mathcal{S}_t^{u,y}\psi_n(p) &= \psi_n(S_t^{u,y}p)\\ &\to \psi(S_t^{u,y}p)\\ &= \mathcal{S}_t^{u,y}\psi(p)\end{aligned}$$

uniformly on P_t.

Part (iv). From Lemma 10.3.1(iv) we have

$$\begin{aligned}\mathcal{S}_s^{u,y}\mathcal{S}_t^{u,y}\psi(p) &= \mathcal{S}_t^{u,y}\psi(S_s^{u,y}p)\\ &= \psi(S_t^{u,y}S_s^{u,y}p)\\ &= \psi(S_{t+s}^{u,y}p)\\ &= \mathcal{S}_{s+t}^{u,y}\psi(p).\end{aligned}$$

□

The transition operator $\mathcal{S}_t^{u,y}$ was defined for fixed signals $u, y \in L_{2,loc}$. By maximizing over y and minimizing over controllers K we obtain an operator with which the value function W (recall (4.4)) can be described. Indeed, two transition operators can be defined:

$$\mathcal{S}_t^u \psi(p) \triangleq \sup_{y \in L_2[0,t]} \mathcal{S}_t^{u,y} \psi(p) \tag{10.13}$$

and

$$\mathcal{S}_t \psi(p) \triangleq \inf_{u \in \mathcal{U}} \sup_{y \in L_2[0,t]} \mathcal{S}_t^{u,y} \psi(p). \tag{10.14}$$

The value function $W(p)$ defined by (4.4) can be expressed in terms of the transition operator $\mathcal{S}_t$ by choosing $\psi(p) = \langle p \rangle$:

$$W(p) = \sup_{t \geq 0} (\mathcal{S}_t \langle \cdot \rangle)(p).$$

Further, the dynamic programming relation (4.8) reads

$$W(p) = (\mathcal{S}_t W)(p)$$

for each $t \geq 0$; i.e., $W = \mathcal{S}_t W$, which means that W is a fixed point of the operator $\mathcal{S}_t$.

We shall say that an operator $\mathcal{L}^{u,y}$ is a *"generator"* for the transition operator $\mathcal{S}^{u,y}$ if there exists a nonempty set $\operatorname{dom} \mathcal{L}^{u,y} \subset \mathcal{G}$ such that for $\psi \in \operatorname{dom} \mathcal{L}^{u,y}$, and each constant pair $(u, y) \in \mathbf{R}^m \times \mathbf{R}^p$ the limit

$$\mathcal{L}^{u,y} \psi(p) = \lim_{t \downarrow 0} \frac{\mathcal{S}_t^{u,y} \psi(p) - \psi(p)}{t} \tag{10.15}$$

exists for p belonging to a nonempty set $\operatorname{dom} \mathcal{L}^{u,y} \psi \subset \mathcal{X}_e$.

In general it will be difficult or impossible to evaluate $\mathcal{L}^{u,y}\psi$ for arbitrary ψ. However, it is possible to evaluate $\mathcal{L}^{u,y}$ for certain types of functions ψ. Let us define the following class of test functions:

$$\hat{\mathcal{G}}_b = \{\psi \in \mathcal{G}_b \ : \ \psi(p) = g(\langle p + f_1 \rangle, \ldots, \langle p + f_k \rangle)$$
$$\text{for some } k \geq 1, \ g \in C_b^1(\mathbf{R}^k), \ f_1, \ldots, f_k \in C_b^1(\mathbf{R}^n)\}.$$

REMARK 10.3.3 W. Fleming pointed out that in the stochastic case, the class of test functions enjoys a density property; see [Fle82]. An analogous property should hold here also, to the effect that continuous functions could be approximated by members of $\hat{\mathcal{G}}_b$ uniformly on uniformly tight subsets of $\mathcal{X}_e$. ▽

The next theorem provides explicit evaluation of $\mathcal{L}^{u,y}\psi$ for $\psi \in \hat{\mathcal{G}}_b$.

THEOREM 10.3.4 *Let $\psi \in \hat{\mathcal{G}}_b$ and let $p \in \mathcal{X}_e$ be quadratically tight. Then for each constant pair $(u, y) \in \mathbf{R}^m \times \mathbf{R}^p$,*

$$\begin{aligned}\mathcal{L}^{u,y}\psi(p) &= \textstyle\sum_{i=1}^{k} \partial_i g(\langle p+f_1\rangle, \ldots, \langle p+f_k\rangle) \\ &\quad \cdot \sup_{\bar{x}\in[[p+f_i]]} \{\nabla_x f_i(\bar{x}) \cdot (A^\times(\bar{x}) + B_1(\bar{x})y + B_2(\bar{x})u) \\ &\qquad + \tfrac{1}{2}|C_1(\bar{x}) + D_{12}u|^2 - \gamma^2 \tfrac{1}{2}|y - C_2(\bar{x})|^2\} \\ &= \textstyle\sum_{i=1}^{k} \partial_i g(\langle p+f_1\rangle, \ldots, \langle p+f_k\rangle) \sup_{\bar{x}\in[[p+f_i]]} \{F(-f_i, u, y)(\bar{x})\}.\end{aligned} \tag{10.16}$$

($\partial_i g$ denotes the ith partial derivative of g.) Thus $\mathcal{L}^{u,y}$ is a "generator" for the transition operator $S^{u,y}$ in the above sense.

REMARK 10.3.5 A remarkable feature of the expression (10.16) is that it does not involve any derivatives of p; the derivatives have been "transferred" to the test functions f_i. Thus (10.16) provides a "weak-sense" interpretation of the formal expression $\nabla_p\psi(p)[F(p,u,y)]$ (recall from the definition (3.10) that $F(p,u,y)$ involves the gradient $\nabla_x p$, which may not exist, whereas $F(-f_i,u,y)$ involves $\nabla_x f_i$, which does exist). ▽

Proof. Let us assume $k = 1$ for simplicity, so that $\psi(p) = g(\langle p+f\rangle)$. We wish to prove that

$$\mathcal{L}^{u,y}\psi(p) = \partial g(\langle p+f\rangle) \sup_{\bar{x}\in[[p+f]]} \{F(-f,u,y)(\bar{x})\}.$$

Now our assumption on the smoothness of g implies that

$$g(a) = g(b) + g'(b)(a-b) + O(|a-b|^2),$$

where g' denotes the derivative and the error term satisfies

$$O(|a-b|^2) \le C|a-b|^2$$

for all $a, b \in \mathbf{R}$. Then, using the notation

$$p_t(f) = \langle p_t + f\rangle = \langle S_t^{u,y}p + f\rangle,$$

we have

$$\begin{aligned}\frac{S_t^{u,y}\psi(p) - \psi(p)}{t} &= \frac{g(p_t(f)) - g(p(f))}{t} \\ &= \partial g(p(f))\frac{p_t(f) - p(f)}{t} + \frac{O(|p_t(f) - p(f)|^2)}{t}\end{aligned}$$

and so (10.16) will be proved (since $p_t(f) = \langle S_t^{u,y} p + f \rangle \to \langle p + f \rangle = p(f)$ as $t \downarrow 0$ by Lemma 10.3.1 (i)) if we can show that

$$\lim_{t\downarrow 0} \frac{p_t(f) - p(f)}{t} = \sup_{\bar{x}\in[[p+f]]} \Big\{ \nabla_x f(\bar{x}) \cdot (A^{\times}(\bar{x}) + B_1(\bar{x})y + B_2(\bar{x})u) + \tfrac{1}{2}|C_1(\bar{x}) + D_{12}u|^2 - \gamma^2 \tfrac{1}{2}|y - C_2(\bar{x})|^2 \Big\}. \tag{10.17}$$

We now prove (10.17). Fix $(u, y) \in \mathbf{R}^m \times \mathbf{R}^p$ and write $\xi(s) = \Phi_{s,t}(x) = \Phi_{s,t}^{u,y}(x)$ for the transition operator of (2.6) with endpoint $\xi(t) = x$. By definition (3.7) we have (cf. the first line of (10.10))

$$p_t(f) = \sup_x \Big\{ p(\Phi_{0,t}(x)) + \int_0^t [\tfrac{1}{2}|C_1(\xi(s)) + D_{12}u|^2 - \gamma^2 \tfrac{1}{2}|y - C_2(\xi(s))|^2]\, ds + f(x) \Big\},$$

where $\xi(t) = x$ for any $t > 0$. Next, subtract $p(f) = \langle p + f \rangle$, divide by t, and add subtract $f(\Phi_{0,t}(x))/t$ to get

$$\frac{p_t(f) - p(f)}{t} = \sup_x \Big\{ \frac{f(x) - f(\Phi_{0,t}(x))}{t} + \frac{1}{t}\int_0^t [\tfrac{1}{2}|C_1(\xi(s)) + D_{12}u|^2 - \gamma^2 \tfrac{1}{2}|y - C_2(\xi(s))|^2]\, ds + \frac{-p(f) + p(\Phi_{0,t}(x)) + f(\Phi_{0,t}(x))}{t} \Big\}. \tag{10.18}$$

Let

$$L = \limsup_{t\to 0} \frac{p_t(f) - p(f)}{t},$$

and let t_i be a subsequence such that

$$\lim_{i\to\infty} t_i = 0, \text{ and } \lim_{i\to\infty} \frac{p_{t_i}(f) - p(f)}{t_i} = L.$$

Select $\bar{x}_i \in \operatorname{argmax}\{p_{t_i}(\cdot) + f(\cdot)\}$. By the tightness of p, and $\lim_{t\to 0} p_t = p$ weakly, we can assume (by selecting a further subsequence if necessary) $\lim_{i\to\infty} \bar{x}_i = \bar{x} \in [[p + f]] = \operatorname{argmax}_x\{p(x) + f(x)\}$ (Theorem C.5.10).

Then since

$$-p(f) + p(\Phi_{0,t}(x)) + f(\Phi_{0,t}(x)) \leq 0$$

for any x, we have, setting $x = \bar{x}_i$ in (10.18),

$$\frac{p_{t_i}(f) - p(f)}{t_i} \leq \frac{f(\bar{x}_i) - f(\Phi_{0,t_i}(\bar{x}_i))}{t_i} + \frac{1}{t_i}\int_0^{t_i} [\tfrac{1}{2}|C_1(\xi(s)) + D_{12}u|^2 - \gamma^2 \tfrac{1}{2}|y - C_2(\xi(s))|^2]\, ds,$$

where $\xi(t_i) = \bar{x}_i$. Sending $i \to \infty$ gives

$$\limsup_{i\to\infty} \frac{p_{t_i}(f) - p(f)}{t_i}$$

$$\leq \nabla_x f(\bar{x}) \cdot (A^\times(\bar{x}) + B_1(\bar{x})y + B_2(\bar{x})u)$$

$$+\tfrac{1}{2}|C_1(\bar{x}) + D_{12}u|^2 - \gamma^2 \tfrac{1}{2}|y - C_2(\bar{x})|^2.$$

This uses the assumed smoothness of f:

$$f(x') = f(x) + \nabla_x f(x)(x' - x) + O(|x - x'|^2)$$

uniformly in $x, x' \in \mathbf{R}^n$. Therefore,

$$\limsup_{t\to 0} \frac{p_t(f) - p(f)}{t} \leq \sup_{\bar{x}' \in [[p+f]]} \Big\{ \nabla_x f(\bar{x}') \cdot (A^\times(\bar{x}') + B_1(\bar{x}')y + B_2(\bar{x}')u)$$

$$+\tfrac{1}{2}|C_1(\bar{x}') + D_{12}u|^2 - \gamma^2 \tfrac{1}{2}|y - C_2(\bar{x}')|^2 \Big\}. \tag{10.19}$$

This proves the upper half of (10.17).

Next, in order to prove the second half of (10.17), let $\bar{\xi} \in [[p + f]]$ be arbitrary. Then

$$-p(f) + p(\bar{\xi}) + f(\bar{\xi}) = 0.$$

Now

$$\frac{p_t(f) - p(f)}{t} = \sup_{\xi} \Big\{ \frac{f(\Phi_{t,0}(\xi)) - f(\xi)}{t}$$

$$+\frac{1}{t}\int_0^t [\tfrac{1}{2}|C_1(\xi(s)) + D_{12}u|^2 - \gamma^2 \tfrac{1}{2}|y - C_2(\xi(s))|^2]\, ds$$

$$+\frac{-p(f) + p(\xi) + f(\xi)}{t} \Big\}, \tag{10.20}$$

where $\xi(0) = \xi$ is the initial state (cf. the second part of (10.10)), $\xi(s) = \Phi_{s,0}(\xi)$, and so, setting $\xi = \bar{\xi}$ in (10.20), we have

$$\frac{p_t(f) - p(f)}{t} \geq \frac{f(\Phi_{t,0}(\bar{\xi})) - f(\bar{\xi})}{t}$$

$$+\frac{1}{t}\int_0^t [\tfrac{1}{2}|C_1(\xi(s)) + D_{12}u|^2 - \gamma^2 \tfrac{1}{2}|y - C_2(\xi(s))|^2]\, ds,$$

where $\xi(0)=\bar{\xi}$. Sending $t\to 0$ we get

$$\begin{aligned}\liminf_{t\to 0}\frac{p_t(f)-p(f)}{t} & \\ &\geq \nabla_x f(\bar{\xi})\cdot(A^{\times}(\bar{\xi})+B_1(\bar{\xi})y+B_2(\bar{\xi})u)\\ &+\tfrac{1}{2}|C_1(\bar{\xi})+D_{12}u|^2-\gamma^2\tfrac{1}{2}|y-C_2(\bar{\xi})|^2\end{aligned}$$

for any $\bar{\xi}\in[[p+f]]$. Therefore,

$$\begin{aligned}\liminf_{t\to 0}\frac{p_t(f)-p(f)}{t} \geq \sup_{\bar{\xi}\in[[p+f]]}\Big\{&\nabla_x f(\bar{\xi})\cdot(A^{\times}(\bar{\xi})+B_1(\bar{\xi})y+B_2(\bar{\xi})u)\\ &+\tfrac{1}{2}|C_1(\bar{\xi})+D_{12}u|^2-\gamma^2\tfrac{1}{2}|y-C_2(\bar{\xi})|^2\Big\}.\end{aligned} \tag{10.21}$$

Combining (10.19) and (10.21) we obtain (10.17). This completes the proof. □

Corresponding to the transition operators $\mathcal{S}_t^u$, $\mathcal{S}_t$, the following two "generators" can be defined:

$$\mathcal{L}^u\psi(p)\triangleq\sup_y\mathcal{L}^{u,y}\psi(p), \tag{10.22}$$

$$\mathcal{L}\psi(p)\triangleq\inf_u\mathcal{L}^u\psi(p). \tag{10.23}$$

With $\psi(p)=g(\langle p+f\rangle)$ a test function ($k=1$) with *unique* maximizer $\bar{x}$ for $\langle p+f\rangle$, for some fixed p, $\partial g(\langle p+f\rangle)\neq 0$, we have

$$\begin{aligned}\mathcal{L}^u\psi(p) &= \partial g(\langle p+f\rangle)[\nabla_x f(\bar{x})\cdot(A^{\times}(\bar{x})+B_2(\bar{x})u)+\tfrac{1}{2}|C_1(\bar{x})+D_{12}u|^2\\ &\quad+\gamma^{-2}\tfrac{1}{2}\nabla_x f(\bar{x})B_1(\bar{x})B_1(\bar{x})'\nabla_x f(\bar{x})']\end{aligned} \tag{10.24}$$

with maximizer

$$y^*(p)=C_2(\bar{x})+\gamma^{-2}B_1(\bar{x})'\nabla_x f(\bar{x})' \tag{10.25}$$

and

$$\begin{aligned}\mathcal{L}\psi(p) &= \partial g(\langle p+f\rangle)[\nabla_x f(\bar{x})\cdot A^{\times}(\bar{x})\\ &\quad+\tfrac{1}{2}\nabla_x f(\bar{x})(\gamma^{-2}B_1(\bar{x})B_1(\bar{x})'-B_2(\bar{x})E_1^{-1}B_2(\bar{x})')\nabla_x f(\bar{x})']\end{aligned} \tag{10.26}$$

with minimizer

$$u^*(p)=-E_1^{-1}(D_{12}'C_1(\bar{x})+B_2(\bar{x})'\nabla_x f(\bar{x})'). \tag{10.27}$$

We will not prove any detailed results concerning these generators, but rather explore in the remainder of this chapter their significance for interpreting the dynamic programming PDE (4.16) for the value function W, and the associated optimal $\mathbf{u}^*$ and $\mathbf{y}^*$. In terms of the operator $\mathcal{L}^{u,y}$ the dynamic programming PDE takes the form

$$\inf_{u\in\mathbf{R}^m} \sup_{y\in\mathbf{R}^p} \{\mathcal{L}^{u,y}W(p)\} = 0 \text{ in } \mathrm{dom}_{\mathrm{pd}} W, \tag{4.40}$$

where $\mathrm{dom}_{\mathrm{pd}} \subset \mathrm{dom}_{\mathrm{smooth}} W \subset \mathrm{dom}\, W$ was defined in §4.5.3. This is a "weak-sense" view of the dynamic programming PDE of nonlinear H^∞ control. The optimal control and observation are given by

$$\begin{aligned} \mathbf{u}^*(p) &\in \operatorname*{argmin}_{u\in\mathbf{R}^m}\{\sup_y \mathcal{L}^{u,y}W(p)\}, \\ \mathbf{y}^*(p) &\in \operatorname*{argmax}_{y\in\mathbf{R}^p}\{\mathcal{L}^{\mathbf{u}^*(p),y}W(p)\}. \end{aligned} \tag{4.43}$$

PROPOSITION 10.3.6 *Assume $p \in \mathrm{dom}_{\mathrm{smooth}} W$. Then for any constant $c \in \mathbf{R}$,*

$$\mathcal{L}^{u,y}W(p \pm c) = \mathcal{L}^{u,y}W(p) \tag{10.28}$$

and

$$\begin{aligned} \mathbf{u}^*(p \pm c) &= \mathbf{u}^*(p), \\ \mathbf{y}^*(p \pm c) &= \mathbf{y}^*(p). \end{aligned} \tag{10.29}$$

Proof. Equality (10.28) follows from definition (10.15) and the additive homogeneity property (recall §2.2.2) enjoyed by W (Theorem 4.3.1(ii)). In view of this, (10.29) follows. □

10.3.2 Certainty Equivalence Case

The certainty equivalence case (see §7.3) corresponds to the function

$$\hat{W}(p) = \langle p + V \rangle$$

solving the infinite-dimensional PDE (4.16); see [JB96]. Here, $V(x)$ is the state feedback value function (Chapter 5). If

$$\bar{x} = \bar{x}(p) = \operatorname*{argmax}_{x\in\mathbf{R}^n}\{p(x) + V(x)\} \tag{7.31$'$}$$

is unique for each $p \in \mathrm{dom}\, W$ (cf. (7.28)), we have with $\psi(p) = \langle p + V \rangle$,

$$\begin{aligned} \inf_u \sup_y \{\mathcal{L}^{u,y}\hat{W}(p)\} &= \mathcal{L}\langle p + V \rangle \\ &= \nabla_x V(\bar{x}) \cdot A^\times(\bar{x}) \\ &\quad + \tfrac{1}{2}\nabla_x V(\bar{x})(\gamma^{-2}B_1(\bar{x})B_1(\bar{x})' - B_2(\bar{x})E_1^{-1}B_2(\bar{x})')\nabla_x V(\bar{x})' \\ &= 0 \end{aligned}$$

by (10.26) and the PDE (5.45) satisfied by V. Furthermore, by (10.25) and (10.27),

$$\begin{aligned} \mathbf{u}^*(p) &= -E_1^{-1}(D_{12}'C_1(\bar{x}(p)) + B_2(\bar{x}(p))'\nabla_x V(\bar{x}(p))'), \\ \mathbf{y}^*(p) &= C_2(\bar{x}(p)) + \gamma^{-2}B_1(\bar{x}(p))'\nabla_x V(\bar{x}(p))'. \end{aligned} \tag{10.30}$$

These formulas are valid for singular p and therefore extend formulas (7.33).

10.3.3 Pure Singular Case

Purely singular information states are of the form

$$p = \delta_\xi + c.$$

This case can be considered in several ways. It is a special case of certainty equivalence since the function $\hat{W}(\delta_\xi) = \langle \delta_\xi + V \rangle = V(\xi)$ always has a unique maximum at ξ (from (7.31)):

$$\bar{x}(\delta_\xi) = \operatorname*{argmax}_x \{\delta_\xi(x) + V(x)\} = \xi.$$

Thus

$$\begin{aligned} \inf_u \sup_y \{\mathcal{L}^{u,y}\hat{W}(\delta_\xi)\} &= \mathcal{L}\langle \delta_\xi + V \rangle \\ &= \nabla_x V(\xi) \cdot A^\times(\xi) \\ &\quad + \tfrac{1}{2}\nabla_x V(\xi)(\gamma^{-2}B_1(\xi)B_1(\xi)' - B_2(\xi)E_1^{-1}B_2(\xi)')\nabla_x V(\xi)' \\ &= 0, \end{aligned}$$

by using the PDE (5.45) satisfied by V, and

$$\begin{aligned} \mathbf{u}^*(\delta_\xi) &= -E_1^{-1}(D_{12}'C_1(\xi) + B_2(\xi)'\nabla_x V(\xi)'), \\ \mathbf{y}^*(\delta_\xi) &= C_2(\xi) + \gamma^{-2}B_1(\xi)'\nabla_x V(\xi)'. \end{aligned} \tag{10.31}$$

Further, the next theorem shows that the PDE for $W(p)$ for p of the form $\delta_\xi + \phi$ projects to a PDE on $\mathbf{R}^n \times \mathbf{R}$, greatly reducing computational complexity.

THEOREM 10.3.7 (i) *Define a function $\hat{W}(\xi, \phi) = W(\delta_\xi + \phi)$, where $\xi \in \mathbf{R}^n$ and $\phi \in \mathbf{R}$. Then $\hat{W}(\xi, \phi)$ satisfies the PDE*

$$\begin{aligned} \inf_u \sup_y \Big\{ &\nabla_\xi \hat{W}(\xi, \phi) \cdot (A(\xi) + B_1(\xi)(y - C_2(\xi)) + B_2(\xi)u) \\ &+ \nabla_\phi \hat{W}(\xi, \phi) \cdot (\tfrac{1}{2}|C_1(\xi) + D_{12}u|^2 - \gamma^2\tfrac{1}{2}|y - C_2(\xi)|^2) \Big\} = 0 \end{aligned} \tag{10.32}$$

in $\mathbf{R}^n \times \mathbf{R}$, *and* $\hat{W}(\xi, \phi) \geq \phi$. *Indeed,*

$$\hat{W}(\xi, \phi) = V(\xi) + \phi, \tag{10.33}$$

where $V(x) \geq 0$ *is a solution of* (5.45) *(for state feedback). Note that* $\delta_\xi \in$ dom W.

(ii) *If* $V(x)$ *is smooth, the controller* $K^*_{\delta_{x_0}}$ *has states* $p_t = \delta_{\xi(t)} + \phi(t)$ *whose dynamics are given by*

$$\begin{aligned} \dot{\xi} &= A(\xi) + B_1(\xi)(y - C_2(\xi)) + B_2(\xi)\mathbf{u}^*_{state}(\xi), \quad \xi(0) = x_0, \\ \dot{\phi} &= \tfrac{1}{2}|C_1(\xi) + D_{12}\mathbf{u}^*_{state}(\xi)|^2 - \gamma^2 \tfrac{1}{2}|y - C_2(\xi)|^2 \end{aligned} \tag{10.34}$$

with controller output

$$\mathbf{u}^*_{state}(\xi) = -E_1^{-1}(D'_{12}C_1(\xi) + B_2(\xi)'\nabla_x V(\xi)') \tag{10.35}$$

(which is independent of ϕ*). (The optimal disturbance is*

$$\mathbf{w}^*_{state}(x) = \tfrac{1}{\gamma^2} B_1(x)'\nabla_x V(x)'.)$$

(iii) *If the initial condition of the plant* G *is* x_0, *then under the action of the controller* $K^*_{\delta_{x_0}}$, *the* γ*-dissipation inequality* (2.10) *holds with* $\beta(x_0) = V(x_0)$*:*

$$\tfrac{1}{2}\int_0^t |z(s)|^2 ds \leq \gamma^2 \tfrac{1}{2}\int_0^t |w(s)|^2\, ds + V(x_0).$$

(iv) *The controller* $K^*_{\delta_0}$ *is null initialized.*

Proof. Using (4.4), (3.20), and (3.7), if $p = \delta_{x_0}$, (10.3) implies

$$\begin{aligned} W(\delta_{x_0}) &= \inf_K \sup_{T\geq 0} \sup_{y \in L_{2,loc}} \Big\{ \tfrac{1}{2}\int_0^T [|C_1(\xi(s)) + D_{12}u(s)|^2 \\ &\qquad -\gamma^2|y(s) - C_2(\xi(s))|^2]\, ds \;:\; \xi(0) = x_0 \Big\} \\ &= \inf_K \sup_{T\geq 0} \sup_{w \in L_{2,loc}} \Big\{ \tfrac{1}{2}\int_0^T [|z(s)|^2 - \gamma^2|w(s)|^2]\, ds \;:\; x(0) = x_0 \Big\}. \end{aligned}$$

This is because knowing x_0, $u(\cdot)$ and $y(\cdot)$, one can determine $w(\cdot)$ and $x(\cdot)$ by integration. However, this last expression is simply the formula for the optimal storage function for a state feedback problem; see, e.g., [vdS92] (optimal version of the bounded real lemma). Call this function $V(x_0)$. Now from (4.11),

$$0 \leq V(x_0) \leq \beta_{K_0}(x_0),$$

and hence also $V(0) = 0$; thus $V \in \mathcal{B}$ (in particular, is finite). It is now routine to show that (see, e.g., [BH96], [Jam93], [McE95b], [Sor96]) V is a viscosity solution of (5.45).

Now consider the function $\hat{W}(\xi, \phi)$ defined in part (ii). Given $u(\cdot)$, $y(\cdot)$, let $\xi(t)$, $\varphi(t)$ be solutions of (2.6) and

$$\dot{\varphi} = \tfrac{1}{2}|C_1(\xi) + D_{12}\mathbf{u}^*_{state}(\xi)|^2 - \gamma^2\tfrac{1}{2}|y - C_2(\xi)|^2$$

with initial conditions $\xi(0) = \xi$, $\varphi(0) = \phi$, respectively. By Lemma 10.2.1, if $p_0 = \delta_\xi + \phi$, then $p_t = \delta_{\xi(t)} + \varphi(t)$ for all $t \geq 0$. Then if $\hat{W}$ is smooth,

$$\frac{d}{dt}\hat{W}(\xi(t), \phi(t)) = \nabla_\xi \hat{W}(\xi(t), \phi(t))\dot{\xi}(t) + \nabla_\phi \hat{W}(\xi(t), \phi(t))\dot{\phi}(t).$$

This together with standard methods from dynamic programming implies that $\hat{W}(\xi, \phi)$ satisfies the PDE (10.32). The remaining assertions of (i) follow directly from the definition and the properties of $V(x)$ (see Chapter 5).

The optimal control $\mathbf{u}^*(\xi, \phi)$ is obtained by minimizing the LHS of (10.32). Noting that

$$\nabla_\xi \hat{W}(\xi, \phi) = \nabla_x V(\xi), \quad \nabla_\phi \hat{W}(\xi, \phi) = 1,$$

one obtains

$$\mathbf{u}^*(\xi, \phi) = -E_1^{-1}(D_{12}'C_1(\xi) + B_2(\xi)'\nabla_x V(\xi)').$$

The formulas for the controller in (iii) now follow directly. Note that $\mathbf{u}^*(\xi, \phi)$ is independent of ϕ, which makes the dynamics of ϕ irrelevant (as far as $\mathbf{u}^*$ is concerned) since it is unobservable (regarding u^* as an output).

To prove (iii), we note that if the initial state of G is x_0, then $\xi(0) = x_0$ implies $\xi(t) = x(t)$ for all $t \geq 0$. Thus for this initial state, the controller $K^*_{\delta_{x_0}}$ yields the dissipation inequality (2.10).

Finally, to prove (iv), we note that if $x_0 = 0$ and $y(\cdot) = 0$, then from (10.34) $\xi(t) = 0$ (since $\nabla_x V(0) = 0$ implies $\mathbf{u}^*_{state}(0) = 0$), and hence from (10.35) $u^*(t) = 0$ for all $t \geq 0$. □

10.4 Formulas for the Central Controller

The *central controller* is the information state controller $K^*_{p_e}$ initialized at the equilibrium p_e (see §3.2 for the definition of p_e and Chapter 11 for results concerning p_e). This section presents formulas for the central controller for 1 and 2A block systems with singular equilibrium p_e, based on the above results. Properties of the central controller were discussed in Chapter 4.

10.4.1 General Singular Case

Let $p_e \in \mathcal{X}_e$ be a possibly singular equilibrium information state, and $K^*_{p_e}$ the associated central controller. The central controller is given by

$$K^*_{p_e} : \begin{cases} \dot{p} = F(p, u, y), \\ u = \mathbf{u}^*(p), \end{cases} \tag{10.36}$$

where $\mathbf{u}^*(p)$ is given by (4.43).

10.4.2 Hyperbolic 1 and 2A Block Systems

The stable equilibrium is (see §11.5.1)

$$p_e = \delta_{M_{as}} + \breve{p}_e, \tag{10.37}$$

where M_{as} is the antistable manifold of $A^\times$ and $\breve{p}_e \in C(M_{as})$.

Using Corollary 10.2.2, we can write down the central controller for $A^\times$ strongly hyperbolic:

$$K^*_{p_e} : \begin{cases} M_t = \Phi^{u,y}_{t,0}(M_{as}), \ \ \breve{p}_t = \hat{\Phi}^{u,y}_{t,0}(\breve{p}), \\ u = \mathbf{u}^*(\delta_{M_t} + \breve{p}_t), \end{cases} \tag{10.38}$$

where $\mathbf{u}^*(\delta_{M_t} + \breve{p}_t)$ is given by formula (4.43). Note that the PDE underlying (10.38) sits on a space of dimension $k_{as} = \dim M_{as} << n$, in general.

10.4.3 Purely Singular 1 and 2A Block Systems

The stable equilibrium is (see §11.5.2)

$$p_e = \delta_0.$$

The central controller for $A^\times$ strongly stable is

$$K^*_{\delta_0} : \begin{cases} \dot{\xi} = A^\times(\xi) + B_1(\xi)y + B_2(\xi)u, \\ u = \mathbf{u}^*_{state}(\xi), \end{cases} \tag{10.39}$$

where $u^*_s(\xi)$ is given by (10.35) (i.e., (4.43)).

10.4.4 Nonsingular 1 and 2A Block Systems

The stable equilibrium is a nonsingular (finite for all $x \in \mathbf{R}^n$, smooth) function p_e (see §11.4).

The central controller for $A^\times$ strongly antistable is

$$K^*_{p_e} : \begin{cases} \dot{p} = F(p, u, y), \\ u = \mathbf{u}^*(p), \end{cases} \tag{10.40}$$

where $\mathbf{u}^*(p)$ is given by (4.23) (a special case of (4.43)).

10.4.5 Certainty Equivalence Controller for Hyperbolic 1 and 2A Block Systems

Consider the stable equilibrium $p_e = \delta_{M_{as}} + \breve{p}_e$, where M_{as} is the antistable manifold of $A^{\times}$ and $\breve{p}_e \in C(M_{as})$ discussed above, and suppose also the certainty equivalence assumptions are valid (see §7.3 and §10.3.2).

The central controller is given by

$$K_{p_e}^{ce} : \begin{cases} M_t = \Phi_{t,0}^{u,y}(M_{as}), \;\; \breve{p}_t = \hat{\Phi}_{t,0}^{u,y}(\breve{p}), \\ u = u^{ce}(\delta_{M_t} + \breve{p}_t), \end{cases} \tag{10.41}$$

where

$$u^{ce}(\delta_M + \breve{p}) = \mathbf{u}^*_{state}(\bar{x}(M, \breve{p}))$$

($\mathbf{u}^*_{state}(x)$ is the optimal state feedback given by (10.35)) and

$$\bar{x}(M, \breve{p}) = \bar{x}(\delta_M + \breve{p}) = \operatorname*{argmax}_{x \in M}\{\breve{p}(x) + V(x)\}$$

(unique by the certainty equivalence assumption).

The significance of this formula is that the controller u^{ce} is determined by an n-dimensional PDE (5.45), and the controller dynamics is less than or equal to n dimensions in general, since it requires the solution of the PDE (3.9). In the light of our results (§10.2), the controller dynamics dimension can be reduced to $k = k_{as} < n$ when $A^{\times}$ is strongly hyperbolic.

Note that when $A^{\times}$ is stable, the certainty equivalence controller (10.41) agrees with the finite-dimensional controller (10.39) (since $M = \{\xi\}$, ϕ a constant, $\bar{x}(\xi, \phi) = \xi$).

10.5 Notes

(i) Singular 2B and 4 block problems can also be considered; see [Jam98].

(ii) In stochastic control theory the information state is a measure-valued stochastic process (in fact a Markov process for constant control), and it satisfies a stochastic partial differential equation (SPDE), called the Duncan–Mortensen–Zakai equation (see, e.g., [FP82]). To define the information state and to describe the SPDE without assuming existence of densities, the information state measures are interpreted in terms of their values on functions (specified by integration). The SPDE is thus expressed in a weak sense. Further, the Nisio transition operator associated with this problem is studied in [Fle82], together with its "generator" (see also [Hij90] and related references). By employing the max-plus framework described in Appendix C, we have begun in this chapter a description of the H^{∞} information state p_t, which is analogous to the probabilistic framework of stochastic control theory. Such a description provides a useful mathematical framework for singular information states.

(iii) The transition operator $\mathcal{S}_t$ is analogous to the transition operator $\mathcal{T}_t$ defined in the stochastic control paper [Fle82].

Chapter 11

Stability of the Information State Equation

11.1 Introduction

In this chapter we present some basic stability results for the information state system

$$\dot{p}_t = F(p_t, u(t), y(t)). \tag{3.9}$$

Critical factors include the following: (i) the initial information state p_0, (ii) the driving signals u and y, (iii) the stabilizability/antistabilizability properties of the reversed-arrow system (2.6), and (iv) issues concerning the solvability of the H^∞ control problem. What we shall prove has different strengths in two different cases. As we shall see in the 1 and 2A block problem p_e is a highly singular function, unless $A^\times$ is antistable (a rare event in practical control). On the other hand for 2B and 4 block problems p_e is often a smooth or at least finite function. While there is a unifying picture in this exposition we shall describe results in terms of these two separate cases.

11.1.1 Nonsingular Cases

We analyze this class of problems under assumptions that allow us to put upper and lower bounds on the information state. A key idea is to compare the H^∞ information state p_t with the more classical H^2 information state $m_t = -\gamma^2 r_t$, which has sign-definite integrand and so is well behaved. Strictness of the H^∞ norm inequality is also used in an essential way. Assuming the assumptions are satisfied, including B_1 and B_2 bounded, the conclusions are of the following nature:

> *There exists an equilibrium solution $p_e \in \mathcal{X}$ of the equilibrium information state equation, which satisfies certain estimates. The resulting p_e may not be smooth; however, if it is, then it will be antistabilizing (see Theorem* 11.4.4).
>
> *If p_e above is smooth and the vector field*

$$A^{p_e} = A^{\times} - \gamma^{-2}\tilde{B}_1\tilde{B}_1'\nabla_x p_e' \tag{3.32}$$

is incrementally L_2 exponentially antistable, then

$$\{p_0 \in \mathcal{X} \ : \ p_0 \leq p_e\} \subset \mathcal{D}_{attr}(p_e)$$

(*see Theorem* 11.4.14).

11.1.2 Singular Cases

In §11.5.1 we utilize the notion of *strict infosubequilibrium* p_{se}, which means roughly that

$$F(p_{se}, 0, 0) < 0$$

in a strict, uniform sense (see §11.5.1).

CONJECTURE 11.1.1 (see Theorem 11.5.2). *Suppose the plant has a* 1 *or* 2B *block structure, B_1 and B_2 are bounded, $A^{\times}$ is exponentially hyperbolic, and p_{se} is a strict infosubequilibrium. Initialize the information state equation at $p_0 \in C(\mathbf{R}^n)$ with $p_0 \leq p_{se}$. Then for all functions u, y in L_2 (or at least decaying reasonably)*

$$\lim_{t\to\infty} p_t = p_e + c$$

uniformly on compact sets. Here p_e has the form by

$$p_e = \delta_{M_{as}} + \breve{p}_e,$$

where $\breve{p}_e \in C(M_{as})$ and $\breve{p}_e(0) = 0$ and c is a real number depending on u, y (and p_0). The limit p_e does not depend on the functions u, y, which are input to the information state equation.

What we actually prove is that the conjecture is true for all $u, y \in L_2$ of compact support (see Theorem 11.5.2). In some special cases, we prove that the conjecture is true for some p_0 and all $u, y \in L_2$, provided we view convergence in the sense of max-plus weak convergence (see Appendix C).

Since it helps exposition to see extreme cases and since what is proven is strong, we call attention to two special cases:

(i) $A^{\times}$ is L_2 exponentially stable: we show that $p_e = \delta_0$ (with $\breve{p}_e = 0$) and $\mathcal{D}_{attr}(\delta_0)$ is nonempty (see Theorems 11.5.5, 11.5.9).

(ii) $A^{\times}$ is L_2 exponentially antistable stable: it follows from Theorem 11.4.14 that $p_e = \breve{p}_e$ is nonsingular ($M_{as} = \mathbf{R}^n$) and that $\mathcal{D}_{attr}(p_e)$ is nonempty.

Convergence from a nonsingular p_0 to a singular limit as in case (ii) above is illustrated in Figure 11.1.

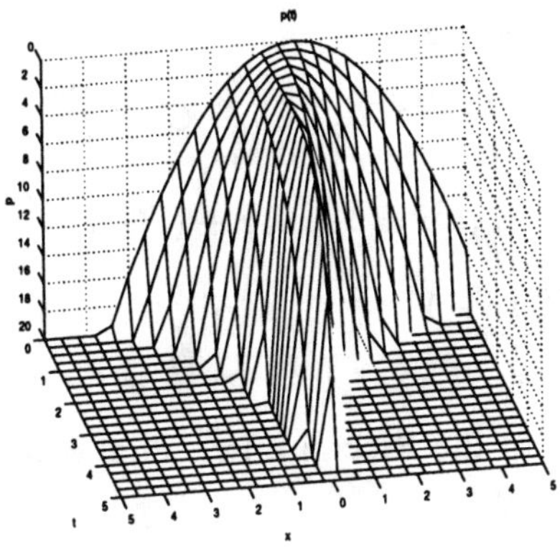

Figure 11.1: Convergence to a singular equilibrium $p_e = \delta_0 + c$

11.1.3 Reader's Guide

The stability of nonlinear (finite- and infinite-dimensional) systems is a complex subject. The stability analysis of the information state system begins in §11.4. This section deals with systems for which the equilibrium state p_e is nonsingular. Then §11.5 considers various singular cases. Before embarking on the stability analysis, the following section presents linear and bilinear examples to help illustrate the variety and nature of the information state dynamics and stability, at least in the context of the relatively simple hyperbolic dynamics (we do not consider multiple equilibria, limit cycles, chaotic attractors, etc.).

11.2 Examples

The purpose of this section is to foreshadow the stability results to follow in this chapter by studying several illustrative examples.

11.2.1 One Block Linear Systems

We consider the case of linear systems:

$$A(x) = Ax, \;\; C_1(x) = C_1 x, \;\; C_2(x) = C_2 x, \;\; D_{12} = I, \;\; D_{21} = I.$$

The vector field $A^\times$ is now a linear vector field

$$A^\times(x) = A^\times x,$$

and we write $A^\times \triangleq A - B_1 C_2$ for this matrix. We shall assume that $A^\times$ is *hyperbolic*, meaning that all eigenvalues have strictly negative or strictly positive real parts (no eigenvalues have zero real part).

The information state is given explicitly by

$$p_t(x) = -\gamma^2 \tfrac{1}{2}(x - \hat{x}(t))' Y(t)^{-1}(x - \hat{x}(t)) + \phi(t) \tag{11.1}$$

(assuming Y is invertible), where

$$\dot{\hat{x}} = (A^\times + Y(\gamma^{-2}C_1'C_1 - C_2'C_2))\hat{x} + (B_1 + YC_2')y + (B_2 + \gamma^{-2}YC_1')u, \tag{11.2}$$

$$\dot{Y} = A^\times Y + YA^{\times\prime} + Y(\gamma^{-2}C_1'C_1 - C_2'C_2)Y, \tag{11.3}$$

$$\dot{\phi} = \tfrac{1}{2}|C_1\hat{x} + u|^2 - \gamma^2\tfrac{1}{2}|y - C_2\hat{x}|^2. \tag{11.4}$$

The stationary version of the Riccati differential equation (11.3) is the algebraic Riccati equation

$$0 = A^\times Y_e + Y_e A^{\times\prime} + Y_e(\gamma^{-2}C_1'C_1 - C_2'C_2)Y_e. \tag{11.5}$$

One of the (three) necessary and sufficient conditions for the solvability of the H^∞ control problem is the existence of a *stabilizing* solution $Y_e \geq 0$ of (11.5), meaning that the matrix

$$A^\times + Y_e(\gamma^{-2}C_1'C_1 - C_2'C_2)$$

has all eigenvalues with strictly negative real parts.

Write $Y_I(t) = Y(t)^{-1}$, which if $Y(t) > 0$, satisfies the Riccati differential equation

$$\dot{Y}_I = -Y_I A^\times - A^{\times\prime} Y_I - (\gamma^{-2}C_1'C_1 - C_2'C_2) \tag{11.6}$$

with corresponding steady state Riccati equation

$$0 = -Y_{Ie}A^\times - A^{\times\prime}Y_{Ie} - (\gamma^{-2}C_1'C_1 - C_2'C_2), \tag{11.7}$$

to which one seeks an *antistabilizing solution* Y_{Ie}, which if Y_{Ie} exists and $Y_{Ie} > 0$ requires that the matrix $A^\times$ be antistable (all eigenvalues have positive real parts). Equation (11.7) corresponds to (3.28). Cases where Y_{Ie} does not exist are of particular interest.

There are three cases to consider: (i) $A^\times$ asymptotically stable; (ii) $A^\times$ asymptotically antistable (i.e., $-A^\times$ asymptotically stable); and (iii) $A^\times$ hyperbolic. In cases (i) and (iii), Y_e is not invertible, while in case (ii) Y_e is invertible. This follows from [PAJ91, Theorem 3.4]. Of course, case (iii) includes (i) and (ii) as special cases.

In case (ii), $Y_e > 0$ and (11.5) implies that the matrix $A^\times + Y_e(\gamma^{-2}C_1'C_1 - C_2'C_2)$ is similar to the matrix $-(A^\times)'$.

If $A^\times$ is hyperbolic, there exist coordinates such that $A^\times$ is equivalent to the block diagonal matrix

$$A^\times \asymp \begin{pmatrix} A_s & 0 \\ 0 & A_{as} \end{pmatrix},$$

corresponding to a stable invariant subspace $M_s = \{(x_1, \ldots, x_{n_s}, 0, \ldots, 0)\}$ and an antistable invariant subspace $M_{as} = \{(0, \ldots, 0, x_{n_s+1}, \ldots, x_n)\}$. In these coordinates, Y_e takes the form

$$Y_e \asymp \begin{pmatrix} 0 & 0 \\ 0 & Y_{as-e} \end{pmatrix}, \tag{11.8}$$

with $Y_{as-e} > 0$.

Our interest is in the asymptotics

$$Y(t)^{-1} \to Y_e^{-1} = \text{“}\, Y_{Ie} \asymp \begin{pmatrix} -\infty & -\infty \\ & \\ -\infty & Y_{as-e} \end{pmatrix} \,\text{”} \tag{11.9}$$

as $t \to \infty$ for various initializations $Y(0) = Y_0 \geq 0$, in the interpretation of this formal expression, and in the subsequent nonlinear generalization to (11.9). Indeed what we find is that the appropriate definition of $(Y_e^{-1}x, x)$ is the singular function

$$p_e(x) = \delta_{M_{as}}(x) + \breve{p}_e(x), \tag{11.10}$$

and where $\breve{p}_e : M_{as} \to \mathbf{R}$; viz., the quadratic form

$$\breve{p}_e(x_s, x_{as}) = -\gamma^2 \tfrac{1}{2} x'_{as} Y_{as-e}^{-1} x_{as}, \tag{11.11}$$

where $x = (x_s, x_{as})$ give coordinates relative to $\mathbf{R}^n = M_s \oplus M_{as}$. Note $\breve{p}_e$ is a finite, smooth function on M_{as}.

This has implications for initializing the information state equation; namely, if $Y_0 = Y_e$ and $u \in L_2$ and $y \in L_2$, then $Y(t) = Y_e$ for all $t \geq 0$ and

$$p_t(x) = \delta_{M_{as}}(x - \hat{x}(t)) - \gamma^2 \tfrac{1}{2}(x_{as} - \hat{x}_{as}(t))' Y_{as-e}^{-1}(x_{as} - \hat{x}_{as}(t)) + \phi(t)$$

$$\to \delta_{M_{as}}(x) - \gamma^2 \tfrac{1}{2} x'_{as} Y_{as-e}^{-1} x_{as} + c$$

as $t \to \infty$ for some constant c.

The equilibrium solution p_e given by (11.10) has both singular and nonsingular components:

$$p_e(x_s, x_{as}) = \delta_{M_{as}}(x_s, x_{as}) - \gamma^2 \tfrac{1}{2} x'_{as} Y_{as-e}^{-1} x_{as}.$$

Observe that in case (i), $Y_e = 0$, and so $x = x_s$ and

$$p_e(x) = \delta_0.$$

This case is *purely singular*. In case (ii), $Y_e > 0$ and so $Y_{Ie} > 0$ exists, $x = x_{as}$ and

$$p_e(x) = -\gamma^2 \tfrac{1}{2} x' Y_{as-e}^{-1} x.$$

This case is *nonsingular*.

As mentioned, an important issue regarding the convergence (11.9) is initialization, i.e., a choice of Y_0. The following two subsections illustrate this by example.

EXAMPLE 11.2.1 Assume that all dimensions are one: $n = m = 1$, etc.
Case (i). $A^{\times} < 0$.

The Y-equation (11.3) has two equilibria: $Y_e = 0$ is stable and

$$Y_{ue} = -2A^{\times}/(\gamma^{-2}C_1^2 - C_2^2)$$

is unstable (these are the two solutions of (11.5)). Now Y_{ue}^{-1} is the unstable equilibrium of (11.6).

Case i(a). $\gamma^{-2}C_1^2 - C_2^2 > 0$, so that $Y_{ue} > 0$.

If $Y_I(0) > Y_{ue}^{-1}$, then $Y_I(t) \to +\infty$ as $t \to \infty$. Hence if $0 < Y_0 < Y_{ue}$, then $Y(t) \to 0$ as $t \to \infty$, and $Y_e = 0$ is the stabilizing solution of (11.3). Of course, if $Y_0 = 0 = Y_e$, then $Y(t) = 0 = Y_e$ for all $t \geq 0$. We can think of the interval $0 \leq Y < Y_{ue}$ as the (positive) domain of attraction $\mathcal{D}_{attr}(Y_e) = \mathcal{D}^0_{attr}(Y_e)$ of the stable equilibrium $Y_e = 0$ of (11.3).

Case i(b). $\gamma^{-2}C_1^2 - C_2^2 < 0$, so that $Y_{ue} < 0$.

Then we can take $\mathcal{D}_{attr}(Y_e) = \mathcal{D}^0_{attr}(Y_e) = \{Y > 0\}$. Figures 11.2 and 11.3 illustrate the trajectories and equilibria of (11.3) and (11.6). Notice the finite escape times for Y, corresponding to the zero crossings for Y_I.

We interpret $Y_e = 0$ as corresponding to the information state

$$p_e(x) = \delta_0(x).$$

The stability of (11.2) implies that the integral defining $\phi(t)$ (see (11.4)) converges for any

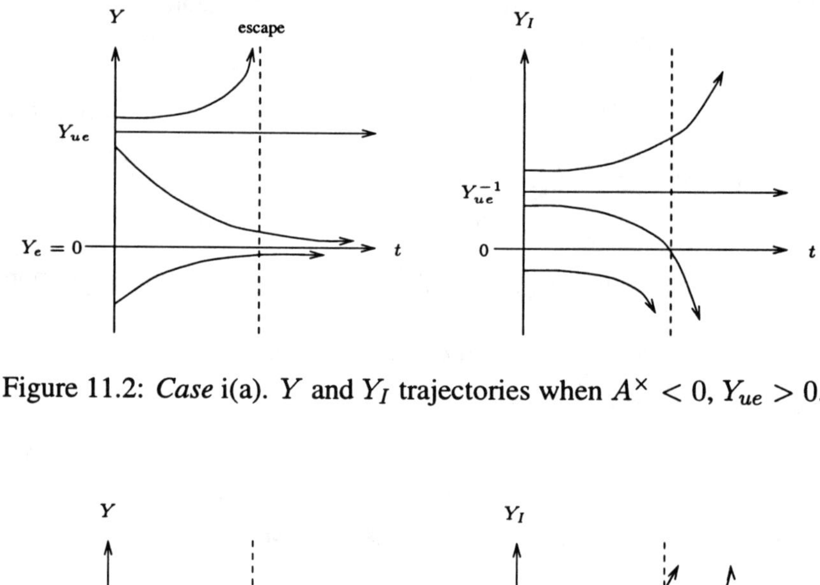

Figure 11.2: *Case* i(a). Y and Y_I trajectories when $A^\times < 0$, $Y_{ue} > 0$.

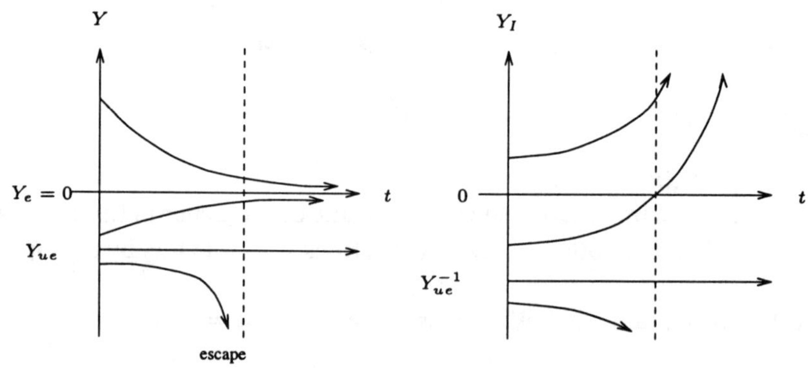

Figure 11.3: *Case* i(b). Y and Y_I trajectories when $A^\times < 0$, $Y_{ue} < 0$.

$u, y \in L_2$; call this number $\psi \in \mathbf{R}$. Thus

$$\lim_{t \to \infty} p_t = \delta_0 + \psi \tag{11.12}$$

provided $p_0(x) = -\gamma^2 \frac{1}{2} Y_0^{-1} x^2$, with $Y_0 \in \mathcal{D}_{attr}(Y_e) = \mathcal{D}^0_{attr}(Y_e)$.

The transition operator for (2.6) can be computed explicitly in this case; indeed if $A^\times = -1$, $B_1 = B_2 = 1$, we have

$$\Phi^{u,y}_{s,t}(x) = e^{t-s}x - \int_s^t e^{r-s}[u(r) + y(r)]dr.$$

From this, the information state can be computed (for $u = y = 0$) to be

$$p_t(x) = -\gamma^2 \tfrac{1}{2} e^{2t} Y_I(0) x^2 + \gamma^2 \tfrac{1}{4}[e^{2t} - 1](\gamma^{-2} C_1^2 - C_2^2) x^2.$$

The coefficient of $\gamma^2 \frac{1}{2} e^{2t} x^2$ is $-Y_I(0) + \frac{1}{2}(\gamma^{-2} C_1^2 - C_2^2)$, which we require to be negative for $p_t(x) \to -\infty$ as $t \to \infty$ ($x \neq 0$). Compare with expression (11.1) for $p_t(x)$. This provides an alternative way of seeing the domain of attraction $\mathcal{D}_{attr}(Y_e)$, since this coefficient is just

$$(A^\times Y_I(0) + (\gamma^{-2} C_1' C_1 - C_2' C_2)) = F(p_0, 0, 0).$$

This foreshadows Theorem 11.5.2. Note that here $\mathcal{D}_{attr}(Y_e) = \mathcal{D}^0_{attr}(Y_e)$ since the Y Riccati equation does not depend on u, y.

Case (ii). $A^\times > 0$.

The equilibrium $Y_{ue} = 0$ is unstable, whereas $Y_e = -2A^\times / (\gamma^{-2} C_1^2 - C_2^2)$ is stable. Now Y_e^{-1} is a stable equilibrium for (11.6), and we see that $Y_I(t) \to Y_e^{-1}$, as $t \to \infty$ for any $Y_I(0)$. Thus if $Y_e > 0$ and $Y_0 > 0$, then $Y(t) \to Y_e$ as $t \to \infty$. If $Y_0 = Y_e$, then $Y(t) = Y_e$ for all $t \geq 0$. The interval $Y > 0$ is the (positive) domain of attraction $\mathcal{D}_{attr}(Y_e) = \mathcal{D}^0_{attr}(Y_e)$ of the stable equilibrium Y_e of (11.3). Figures 11.4 and 11.5 illustrate the trajectories and equilibria of (11.3) and (11.6).

We interpret $Y_e > 0$ as corresponding to the information state

$$p_e(x) = -\gamma^2 \tfrac{1}{2} Y_e^{-1} x^2.$$

The stability of (11.2) implies that the integral defining $\phi(t)$ (see (11.4)) converges for any $u, y \in L_2$; call this number $\psi \in \mathbf{R}$. Thus

$$\lim_{t \to \infty} p_t(x) = -\gamma^2 \tfrac{1}{2} Y_e^{-1} x^2 + \psi \tag{11.13}$$

provided $p_0(x) = -\gamma^2 \frac{1}{2} Y_0^{-1} x^2$, with $Y_0 \in \mathcal{D}_{attr}(Y_e) = \mathcal{D}^0_{attr}(Y_e)$.

If $A^\times = 1$, $B_1 = B_2 = 1$ we have

$$\Phi^{u,y}_{s,t}(x) = e^{s-t}x - \int_s^t e^{s-r}[u(r) + y(r)]dr.$$

From this, the information state can be computed (for $u = y = 0$):

$$p_t(x) = -\gamma^2 \tfrac{1}{2} e^{-2t} Y_I(0) x^2 + \gamma^2 \tfrac{1}{4}[1 - e^{-2t}](\gamma^{-2} C_1^2 - C_2^2) x^2.$$

Then clearly we have $p_t(x) \to \gamma^2 \frac{1}{4}(\gamma^{-2} C_1^2 - C_2^2) x^2 = \gamma^2 \frac{1}{2} Y_e^{-1} x^2$ as $t \to \infty$ for any $Y_I(0)$. ∇

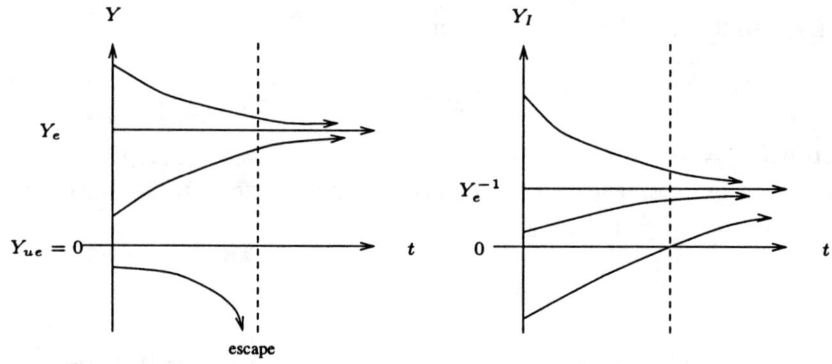

Figure 11.4: *Case* (ii). Y and Y_I trajectories when $A^\times > 0$, $Y_e > 0$.

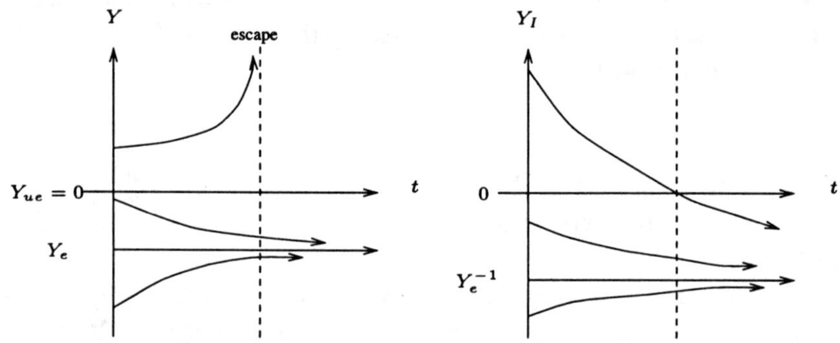

Figure 11.5: *Case* (ii). Y and Y_I trajectories when $A^\times > 0$, $Y_e < 0$.

EXAMPLE 11.2.2 Assume that the state dimension $n = 2$. Consider case (iii), $A^\times$ hyperbolic. To simplify the discussion, we suppose

$$A = \begin{pmatrix} a & 0 \\ 0 & \alpha \end{pmatrix}, \quad B_i = \begin{pmatrix} b_i & 0 \\ 0 & \beta_i \end{pmatrix}, \text{ and } C_i = \begin{pmatrix} c_i & 0 \\ 0 & \varepsilon_i \end{pmatrix}$$

for $i = 1, 2$. Assume that $a - b_1 c_2 < 0$, and $\alpha - \beta_1 \varepsilon_2 > 0$, corresponding to the decomposition mentioned above. Write

$$Y = \begin{pmatrix} Y_{11} & Y_{12} \\ Y_{12} & Y_{22} \end{pmatrix},$$

so that the differential equation (11.3) takes the form

$$\dot{Y}_{11} = 2Y_{11}(a - b_1c_2) + Y_{11}^2(\gamma^{-2}c_1^2 - c_2^2) + Y_{12}^2(\gamma^{-2}\varepsilon_1^2 - \varepsilon_2^2),$$

$$\dot{Y}_{22} = 2Y_{22}(\alpha - \beta_1\varepsilon_2) + Y_{22}^2(\gamma^{-2}\varepsilon_1^2 - \varepsilon_2^2) + Y_{12}^2(\gamma^{-2}c_1^2 - c_2^2),$$

$$\dot{Y}_{12} = Y_{12}(a - b_1c_2 + Y_{11}(\gamma^{-2}c_1^2 - c_2^2)) + Y_{12}(\alpha - \beta_1\varepsilon_2 + Y_{22}(\gamma^{-2}\varepsilon_1^2 - \varepsilon_2^2)). \tag{11.14}$$

Assume that the initial condition $Y_{12}(0) = 0$. Then if $Y_{11}(0) \in \mathcal{D}^0_{attr}(Y_{11-e})$ and $Y_{22}(0) \in \mathcal{D}^0_{attr}(Y_{22-e})$, we have $Y_{11}(t) \to Y_{11-e} = 0$ and $Y_{22}(t) \to Y_{22-e}$ as $t \to \infty$. Thus Y_e takes the form (11.8), when $Y_{22-e} > 0$:

$$Y_e \asymp \begin{pmatrix} 0 & 0 \\ 0 & Y_{22-e} \end{pmatrix}.$$

Of course, the inverse of Y_e is not defined in the usual sense.

We interpret this $Y_e \geq 0$ in terms of the information state

$$p_e(x_1, x_2) = \delta_{M_{as}}(x_1, x_2) - \gamma^2 \tfrac{1}{2} Y_{22-e}^{-1} x_2^2.$$

Here, $M_{as} = \{(x_1, x_2) : x_1 = 0\}$ is the antistable subspace.

If $Y_{12}(0) \neq 0$, then the same limits occur, since $Y_{12}(t) \to 0$ as $t \to \infty$ (this can be seen by studying equation (11.6)).

The stability of (11.2) implies that the integral defining $\phi(t)$ (see (11.4)) converges for any $u, y \in L_2$; call this number $\psi \in \mathbf{R}$. Thus

$$\lim_{t\to\infty} p_t(x_1, x_2) = \delta_{M_{as}}(x_1, x_2) - \gamma^2 \tfrac{1}{2} Y_{22-e}^{-1} x_2^2 + \psi \tag{11.15}$$

provided $p_0(x) = -\gamma^2 \frac{1}{2}(x_1, x_2)' Y_0^{-1}(x_1, x_2)$, with $Y_0 \in \mathcal{D}_{attr}(Y_e) = \mathcal{D}^0_{attr}(Y_e)$. Here,

$$\mathcal{D}_{attr}(Y_e) = \mathcal{D}^0_{attr}(Y_e)$$

$$= \left\{ Y_0 = \begin{pmatrix} Y_{11-0} & Y_{12-0} \\ Y_{12-0} & Y_{22-0} \end{pmatrix} \geq 0 \; : \; Y_{11-0} \in \mathcal{D}^0_{attr}(Y_{11-e}), Y_{22-0} \in \mathcal{D}^0_{attr}(Y_{22-e}) \right\}.$$

∇

11.2.2 One Block Bilinear Systems

The presence of the B_2xu term in bilinear systems (see §7.1) introduces a u-term into the Y-Riccati equation (7.5), and this makes the analysis more complicated than in the linear case, although closely tied to it. Indeed, they share the same equilibria. The Y-equation reads

$$\dot{Y} = A^\times(u)Y + YA^\times(u)' + Y(\gamma^{-2}C_1'C_1 - C_2'C_2)Y, \tag{11.16}$$

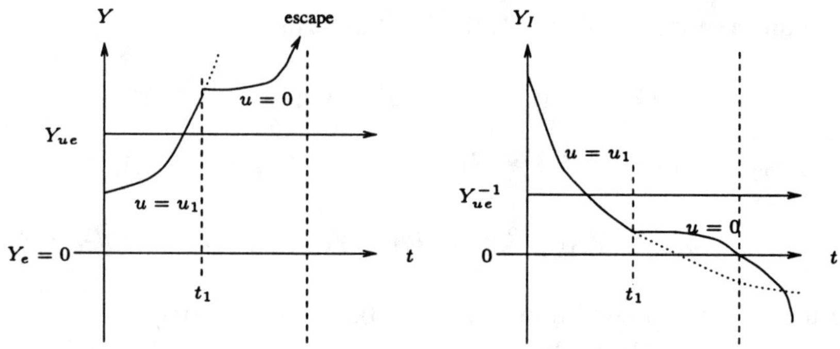

Figure 11.6: Case (i)a. Y and Y_I trajectories for bilinear blowup example.

and the corresponding Y_I-equation is

$$\dot{Y}_I = -Y_I A^{\times}(u) - A^{\times}(u)'Y_I - (\gamma^{-2}C_1'C_1 - C_2'C_2). \tag{11.17}$$

In this subsection we consider a one-dimensional system analogous to the one-dimensional linear system studied in the preceding subsection.

EXAMPLE 11.2.3 Let the state dimension $n = 1$.
Case (i). $A^{\times}(0) = A(0) - B_1C_2 < 0$.

The Y-equation (11.16) has two equilibria: $Y_e = 0$ is stable, and

$$Y_{ue} = -2A^{\times}/(\gamma^{-2}C_1^2 - C_2^2)$$

is unstable. From the linear analysis we know that for $u = 0$, $Y_0 \in \mathcal{D}^0_{attr}(Y_e), Y(t) \to Y_e = 0$ as $t \to \infty$ and (11.12) holds. We now show that *stability definition* (11.12) *can fail in the bilinear case for some inputs u, and some initial values* $Y_0 \in \mathcal{D}^0_{attr}(Y_e)$. In terms of our notation this says that $\mathcal{D}_{attr}(Y_e)$ is a proper subset of $\mathcal{D}^0_{attr}(Y_e)$.

Case i(a). $\gamma^{-2}C_1^2 - C_2^2 > 0$ so that $Y_{ue} > 0$.

Now in this case $\mathcal{D}^0_{attr}(Y_e) = \{0 \le Y < Y_{ue}\}$. Let $0 < Y_0 < Y_{ue}$. Select u_1 such that $A^{\times} + B_3u_1 > 0$. Set

$$u(t) = \begin{cases} u_1 & \text{for } 0 \le t \le t_1, \\ 0 & \text{for } t \ge t_1, \end{cases}$$

where $t_1 > 0$ is selected so that $Y(t_1) > Y_{ue}$ (such a time t_1 exists in view of the above linear analysis). Then clearly $Y_I(t) \to -\infty$ as $t \to \infty$, and $Y(t)$ will suffer escape to $+\infty$ in finite time. Hence if $0 < Y_0 < Y_{ue}$, then $Y_0 \notin \mathcal{D}_{attr}(Y_e)$. This is illustrated in Figure 11.6.

Now

$$p_t(x) = -\gamma^2\tfrac{1}{2}e^{2(t-t_1)}Y_I(t_1)x^2 + \gamma^2\tfrac{1}{4}[e^{2(t-t_1)} - 1](\gamma^{-2}C_1^2 - C_2^2)x^2 + \cdots \quad \text{for } t \ge t_1,$$

where the omitted terms do not depend on t. The coefficient of $\gamma^2 \frac{1}{2} e^{2(t-t_1)} x^2$ is $-Y_I(t_1) + \frac{1}{2}(\gamma^{-2}C_1^2 - C_2^2)$, which is positive, and so $p_t(x) \to +\infty$ as $t \to \infty$ ($x \neq 0$). Therefore (11.12) does not hold in this example.

It is important to note, however, that if $Y_0 = Y_e = 0 \in \mathcal{D}^0_{attr}(Y_e)$, then $Y(t) = Y_e$ for all $t \geq 0$, and $p_t(x) = \delta_{\xi(t)} + \phi_t$ as in Lemma 10.2.1, and (11.12) will hold for any $u, y \in L_2$ (see Theorem 11.5.5 below). Therefore, $\mathcal{D}_{attr} = \{Y_e = 0\}$.

Case i(b). $\gamma^{-2}C_1^2 - C_2^2 < 0$ so that $Y_{ue} < 0$.

Then the (nonnegative) domain of attraction is given by $\mathcal{D}^0_{attr}(Y_e) = \{Y \geq 0\}$. Further, (11.12) will hold for all $u, y \in L_2$, $Y_0 \in \mathcal{D}_{attr}(Y_e)\mathcal{D}^0_{attr}(Y_e) = \{Y \geq 0\}$. □

REMARK 11.2.4 We conclude from this analysis that the initial information state p_0 and the signals u and y can have a dramatic effect on the stability of the information state system (3.9). ▽

11.3 Support of the Attracting Equilibrium p_e

Recall from Chapter 3 that the equilibrium p_e if smooth is a solution of the equilibrium information state equation

$$-\nabla_x p_\infty \cdot (A - B_1 D_{21}' E_2^{-1} C_2) + \frac{1}{2\gamma^2} \nabla_x p_\infty B_1 [I - D_{21}' E_2^{-1} D_{21}] B_1' \nabla_x p_\infty'$$

$$+\frac{1}{2}|C_1|^2 - \gamma^2 \frac{1}{2} C_2' E_2^{-1} C_2 = 0. \qquad (3.28)'$$

Also p_e has the integral characterization, which is its definition:

$$p_e(x) = \sup_{v(\cdot)} \left\{ \lim_{T\to\infty} \frac{1}{2} \int_{-T}^{0} [|C_1(\xi(s))|^2 - \gamma^2(|v(s)|^2 \right.$$

$$+C_2(\xi(s))' E_2^{-1}(\xi(s)) C_2(\xi(s)))] \, ds \qquad (3.30)$$

$$\left. : \lim_{T\to\infty} \xi(-T) = 0, \; \xi(0) = x \right\},$$

where ξ solves the state equation of (2.6) with $u = 0$, $y = 0$:

$$\dot{\xi} = A^\times \xi + \tilde{B}_1 v. \qquad (2.6)'$$

This definition does not require smoothness, and the results we present in this section do not require that p_e be smooth, while most results in the next section do.

The following theorem is a consequence of Lemma 3.2.1 and Theorem B.2.2.

THEOREM 11.3.1 *The L_2 exponentially antistabilizable states $\mathcal{S}_{cas}$ for the system $(A^\times, \tilde{B}_1)$ are contained in the set of states x for which $p_e(x) > -\infty$. Consequently, p_e is*

- *nonsingular if $(A^\times, \tilde{B}_1)$ is L_2 exponentially antistabilizable, and*
- *singular if there exists a state x such that under the action of all controls the trajectory is not square integrable.*

11.4 Information State Attractors: Antistabilizable Case

We begin our detailed analysis of the stability of the information state equation in the case where the equilibrium information state p_e is nonsingular (recall that this means that it is a finite-valued function). Singular information states are left for a subsequent section (§11.5).

Our proof of the convergence of the information state in the $A^\times, \tilde{B}_1$ antistablizable case depends on the existence of a nonsingular function r_e solving an H^2-type estimation equation, and the existence of such a function will be assumed. Indeed, this assumption is merely a strengthening of the stabilizability assumption in the previous section and so is quite plausible.

Under this type of assumption it is proven that the domain of control attraction $\mathcal{D}_{attr}(p_e)$ is nonempty, which here is defined to be the largest subset of $\mathcal{X}_q$ such that $p_t \to p_e + c(u, y, p_0)$ as $t \to \infty$ uniformly on compact subsets. If p_e is quadratically tight, then by Appendix C this implies weak convergence $p_t \Longrightarrow p_e + c(u, y, p_0)$.

11.4.1 Assumptions

Recall that $\mathcal{X}_{q,2,b}$ denotes those twice differentiable functions p that satisfy the bounds $|p(x)| \leq C(1+|x|^2)$, $|\nabla_x p(x)| \leq C(1+|x|)$, and $|\nabla_x^2 p(x)| \leq C$.

The H_2 assumptions involve the PDE

$$\nabla_x r_e \cdot (A - B_1 D_{21}' E_2^{-1} C_2) + \tfrac{1}{2} \nabla_x r_e \tilde{B}_1 \tilde{B}_1' \nabla_x r_e' - \tfrac{1}{2} C_2' E_2^{-1} C_2 = 0 \qquad (11.18)$$

and the vector field

$$A^{r_e} \triangleq A - B_1 D_{21}' E_2^{-1} C_2 + \tilde{B}_1 \tilde{B}_1' \nabla_x r_e'. \qquad (11.19)$$

We present the following assumptions.

ASSUMPTION 11.4.1 B_1 and B_2 are bounded, and D_{12} and D_{21} are constant.

ASSUMPTION 11.4.2 There exists a finite smooth solution $r_e \in \mathcal{X}_{q,2,b}$ of the H_2 estimation equation (11.18) satisfying $r_e(0) = 0$, $r_e(x) \geq 0$, and the system $(A^{r_e}, [B_1 D_{12}' E_2^{-1} : B_2 : \tilde{B}_1])$ is incrementally L_2 exponentially antistable, and the additional coercivity condition $r_e(x) \geq c_{re}|x|^2$ for some $c_{re} > 0$ and all $x \in \mathbf{R}^n$.

Let us call the function r_e a *tame antistabilizing solution to the H^2 estimation equation* and we refer to Assumptions 11.4.1 and 11.4.2 as the H^2 *estimation assumptions.* The last part of Assumption 11.4.2 means that r_e is *coercive*, or *proper*. These assumptions will be in force throughout this section.

The integral version of (11.18) is

$$r_e(x) = \inf_{v(\cdot)}\left\{ r_e(\xi(0)) + \frac{1}{2}\int_0^T [|v(s)|^2 + C_2(\xi(s))' E_2(\xi(s))^{-1} C_2(\xi(s))'] \, ds \right. \\ \left. : \ \xi(T) = x \right\}, \tag{11.20}$$

where ξ solves

$$\dot{\xi} = A^{\times}(\xi) + \tilde{B}_1(\xi) v.$$

The explicit solution satisfying (11.19) and $r_e(0) = 0$, $r_e(x) \geq 0$ is

$$r_e(x) \triangleq \sup_{T \geq 0} \inf_{v(\cdot)} \left\{ \int_0^T [|v(s)|^2 + C_2(\xi(s))' E_2(\xi(s))^{-1} C_2(\xi(s))'] \, ds \right\}. \tag{11.21}$$

To see this, let us write $r_e(x) = \sup_{T \geq 0} r_T(x)$, where $r_T(x)$ is given by the appropriate term in (11.21). Now $r_T(x) \geq 0$, and $r_T(0) = 0$ for all $T \geq 0$, and so $r_e(x) \geq 0$, $r_e(0) = 0$. If every state is $(A^{\times}, \tilde{B}_1)$ antistabilizable, then one can prove the bound $r_T(x) \leq C|x|^2$ for some $C \geq 0$ (simply set $v = 0$ and use the antistability to prove that the integral converges). Further, $r_T(x)$ is monotone increasing in T, hence the limit as $T \to \infty$ exists and equals the supremum indicated. Thus it is fair to say that Assumption 11.4.2 is a strengthened version of the assumption that $(A^{\times}, \tilde{B}_1)$ is antistabilizable.

The coercivity part of Assumption 11.4.2 is a strengthening of the observability or detectability of $(A^{\times}, C_2)$. Indeed, suppose $(A^{\times}, C_2)$ is zero-state observable, and $r_e(x) = 0$ for some $x \in \mathbf{R}^n$. Then for each $T \geq 0$ with the optimal v^*,

$$0 = r_e(x) \geq \int_0^T [|v^*(s)|^2 + C_2(\xi(s))' E_2(\xi(s))^{-1} C_2(\xi(s))'] ds \geq 0,$$

and so $v^*(s) = 0$, $C_2(\xi(s)) = 0$ for all $0 \leq s \leq T$. Zero-state observability then implies $x = 0$, and so is the unique minimizer of r_e.

REMARK 11.4.3 Note that $v = \tilde{B}_1' \nabla_x r_e'$ is an *antistabilizing* input for the system

$$(A^{\times}, \, \tilde{B}_1, \, E_2^{-1/2} C_2).$$

The function r_e may serve as a Lyapunov function. ∇

11.4.2 Main Results

In this subsection we present the main results concerning the existence of and convergence to the equilibrium information state p_e in the antistabilizable case. Theorem 11.4.4 says roughly that if the H^{∞} control problem is strictly solvable and the H^2 filtering assumptions are satisfied, then p_e exists, is quadratically upper limiting, and is the limit (modulo additive constants) of the information state p_t.

THEOREM 11.4.4 *Assume the following conditions:*

(i) *There is a controller K_0 that renders the closed-loop system strictly* γ-dissipative with coercive bias.

(ii) *The H^2 estimation equation* (11.18) *has a tame antistabilizing solution r_e.*

(iii) *Let $p_0 \in \mathcal{X}_q$ satisfy*

$$\langle p_0 + (1+\varepsilon_0^2)\beta_{K_0}\rangle < \infty \tag{11.22}$$

for some $\varepsilon_0 > 0$.

Then the following conditions hold:

(i) *There exists an equilibrium information state p_e satisfying the PDE* (3.28), *$p_e(0) = 0$, and $p_e \geq -\gamma^2 r_e$.*

(ii) *If in addition the function p_e is smooth, then p_e is L_2 asymptotically stable.*

(iii) *If in addition p_e is smooth and $p_e \in \mathcal{X}_{q,2,b}$ is incrementally L_2 exponentially antistabilizing (meaning that the system $[A^{p_e}, [B_1 D_{21}' E_2^{-1} : B_2 : \tilde{B}_1])$ is incrementally L_2 exponentially stable), then*

(a) *for w in L_2 input into the closed-loop system (G, K_0) with initial plant state x_0,*

$$\lim_{t\to+\infty} p_t = p_e + c(w, K_0, x_0, p_0) \tag{11.23}$$

uniformly on compact subsets of $\mathbf{R}^n$, where $c(w, K_0, x_0, p_0)$ is a constant;

(b)

$$-c_1(1+|x|^2) \leq p_e(x) \leq -c_{pe}|x|^2 + c_2, \tag{11.24}$$

and $|\nabla_x p_e(x)| \leq C|x|$, where $c_1, c_2, C \geq 0$, $c_{pe} > 0$;

(c) p_e is given by the formula

$$p_e(x) = \sup_{v(\cdot)} \Bigg\{ \lim_{T\to\infty} \tfrac{1}{2} \int_{-T}^{0} [|C_1(\xi(s))|^2 - \gamma^2(|v(s)|^2$$

$$+C_2(\xi(s))' E_2^{-1}(\xi(s)) C_2(\xi(s)))] \, ds \tag{11.25}$$

$$: \xi(0) = x, \ \lim_{T\to\infty} \xi(-T) = 0 \Bigg\},$$

where ξ solves the state dynamics of (2.6) *with $u = 0$, $y = 0$:*

$$\dot{\xi} = A^\times(\xi) + \tilde{B}_1(\xi) v \tag{2.6$'$}$$

with auxiliary input v;

(d) p_e *is quadratically upper limiting; i.e., there exists a constant* $\bar{\alpha}_0 > 0$ *such that if* $u, y \in L_{2,loc}$, $c_0 \geq 0$, *and*

$$p_0 \leq p_e(x) + \alpha_0|x|^2 + c_0 \tag{3.37}$$

for $0 \leq \alpha_0 \leq \bar{\alpha}_0$, *we have for all* $0 < \eta < 1$,

$$p_t(x) \leq p_e(x) + \frac{c_1(t)}{\eta} + (\eta + c_1(t)\alpha_0)k|x|^2 \text{ for all } t \geq 0, \tag{3.38}$$

where $k \geq 0$ *and* $c_1(t) = c_1(\| u \|_{L_{2,t}}, \| y \|_{L_{2,t}}) \geq 0$ *with* $c_1(\infty)$ *finite whenever* $u, y \in L_2$.

The next theorem is purely about the information state equation and has a life fully independent of the H^∞ control problem. It says roughly that if p_e is an antistabilizing solution to the information state equilibrium equation, then it is a control attractor and the domain of attraction $\mathcal{D}_{attr}(p_e)$ is reasonably large.

THEOREM 11.4.5 *Assume there exists a smooth solution* $p_e \in \mathcal{X}_{q,2,b}$ *of the steady state equation* (3.28) *such that* p_e *is incrementally* L_2 *exponentially stabilizing. Let* $p_0 \in \mathcal{X}_q$ *satisfy* $p_0(0) = 0$, *and let* $u, y \in L_2$. *Then*

(i) p_e *is quadratically upper limiting; i.e., there exists a constant* $\bar{\alpha}_0 > 0$ *such that if* $u, y \in L_{2,loc}$, $c_0 \geq 0$, *and*

$$p_0 \leq p_e(x) + \alpha_0|x|^2 + c_0 \tag{3.37}$$

for $0 \leq \alpha_0 \leq \bar{\alpha}_0$, *we have for all* $0 < \eta < 1$,

$$p_t(x) \leq p_e(x) + \frac{c_1(t)}{\eta} + (\eta + c_1(t)\alpha_0)k|x|^2 \text{ for all } t \geq 0, \tag{3.38}$$

where $k \geq 0$ *and* $c_1(t) = c_1(\| u \|_{L_{2,t}}, \| y \|_{L_{2,t}}) \geq 0$ *with* $c_1(\infty)$ *finite whenever* $u, y \in L_2$, *and*

(ii) *there exists a constant* $0 < \alpha_0 \leq 1$, *such that*

$$\{p_0 \in \mathcal{X}_q \;:\; p_0(x) \leq p_e(x) + \alpha_0|x|^2 + c_0, \text{ for some } c_0 \geq 0\}$$
$$\subset \mathcal{D}_{attr}(p_e) \subset \mathcal{D}^0_{attr}(p_e).$$

The proofs commence in the next section with a careful examination of the convergence of a finite time H^2 filtering function $r_t(x)$ to a limit $r_e(x) + c(u, y, r_0)$ as $t \to \infty$. In particular, detailed estimates concerning quadratic upper and lower bounds are provided. These results are then applied to prove results concerning the existence of an antistabilizing equilibrium p_e and convergence of the information state p_t to $p_e(x) + c(u, y, p_0)$ as $t \to \infty$. Interestingly, when $p_0 = p_e$ the information state p_t never leaves a small "quadratic neighborhood" of p_e.

11.4.3 The H^2 Estimation Problem

The information state of H^∞ control is tailored for the specific game nature of the problem, and as such it has a mixed-sign integrand. Related to this is the information state of H^2 estimation, which is of considerable independent interest. This form of deterministic estimation was developed by Mortensen [Mor68]; see also [Hij80], [BBJ88].

The H^2 *information state* is defined by

$$r_t(x) = \inf_{\tilde{w}(\cdot)} \Big\{ r_0(\xi(0)) + \tfrac{1}{2}\int_0^t |\tilde{w}(s)|^2\, ds$$
$$: \ \xi(t) = x,\ y(s) = C_2(\xi(s)) + D_{21}(\xi(s))\tilde{w}(s),\ 0 \le s \le t \Big\}, \tag{11.26}$$

where $\xi(\cdot)$ satisfies

$$\dot{\xi} = A(\xi) + B_1(\xi)\tilde{w} + B_2(\xi)u. \tag{3.8}$$

The H^2 *information state equation* is

$$\begin{aligned} 0 = {} & \frac{\partial r}{\partial t} + \nabla_x r \cdot (A + B_1 D_{21}' E_2^{-1}(y - C_2) + B_2 u) \\ & + \tfrac{1}{2}\nabla_x r B_1[I - D_{21}' E_2^{-1} D_{21}] B_1' \nabla_x r' - \tfrac{1}{2}(y - C_2)' E_2^{-1}(y - C_2). \end{aligned} \tag{11.27}$$

An alternative integral form of the H_2 information state is

$$r_t(x) = \inf_{v(\cdot)} \Big\{ r_0(\xi(0)) + \tfrac{1}{2}\int_0^t [(|v(s)|^2$$
$$+ (y(s) - C_2(\xi(s)))' E_2^{-1}(\xi(s))(y(s) - C_2(\xi(s))))]\, ds : \ \xi(t) = x \Big\}, \tag{11.28}$$

where $\xi(s)$ is given by

$$\dot{\xi} = A^\times(\xi) + B_1(\xi) D_{21}'(\xi) E_2^{-1}(\xi) y + B_2(\xi) u + \tilde{B}_1(\xi) v. \tag{2.6$'$}$$

The information state and H^2 information state are related by

$$r_t(x) = \lim_{\gamma \to \infty} -\gamma^{-2} p_t(x),$$

where p_t is the information state (given by (3.7)). This limit can be proven in a number of ways (e.g., using viscosity solution methods), although we do not do so here.

THEOREM 11.4.6 *Assume the existence of a tame antistabilizing solution r_e to the H^2 estimation equation. Let $r_0 \in \mathcal{X}_q$ satisfy $r(0) = 0$, let $u, y \in L_2$, and assume that r_t*

is smooth. Then there exists constants $0 < \alpha_0 \leq 1$, $0 < \alpha_1 \leq 1$, $C = C(u, y)$ *such that for all* $0 < \eta \leq \alpha_1$,

$$-\frac{C}{\eta} - C(\eta + \alpha_0)|x|^2 \leq r_t(x) - r_e(x) \leq \frac{C}{\eta} + C(\eta + \| r_0 - r_e \|_q)|x|^2 \quad (11.29)$$

for all $x \in \mathbf{R}^n$, $t \geq 0$, *provided*

$$r_0(x) - r_e(x) \geq -\alpha_0 |x|^2 - c_0, \quad (11.30)$$

for some $c_0 > 0$. *Moreover,*

$$\lim_{t \to \infty} r_t(x) = r_e(x) + c(u, y, r_0) \quad (11.31)$$

uniformly on compact subsets of $\mathbf{R}^n$, *where* $c(u, y, r_0)$ *is a constant depending on* $u, y \in L_2$ *with* $c(0, 0, r_e) = 0$. *If* $r_0 = r_e$, *we can take* $\alpha_0 = 0$, *in which case*

$$-C\left(\frac{1}{\eta} + \eta|x|^2\right) \leq r_t(x) - r_e(x) \leq C\left(\frac{1}{\eta} + \eta|x|^2\right). \quad (11.32)$$

Proof. We will use the following notation: $\Phi_{s,t}^{r_e,v,u,y}(x)$ is the transition operator for the ODE

$$\dot{\xi} = A^{r_e}(\xi) + B_1(\xi) D_{21}' E_2^{-1} y + B_2(\xi) u + \tilde{B}_1(\xi) v, \quad \xi(t) = x,$$

where $(A^{r_e}, [B_1 D_{21}' E_2^{-1} \; : \; B_2 \, \tilde{B}_1])$ (A^{r_e} is given by (11.19)) is incrementally L_2 exponentially antistable and

$$\tilde{B}_1 = B_1 [I - D_{21}' E_2^{-1} D_{21}]. \quad (2.7)$$

Write

$$r_t = r_e + \phi_t.$$

We will analyze the "perturbation" ϕ_t and show that it tends to a constant as $t \to \infty$. Using (11.18) and (11.27), we have the following PDE for ϕ_t:

$$0 = \frac{\partial \phi}{\partial t} + \nabla_x \phi \cdot (A + B_1 D_{21}' E_2^{-1}(y - C_2) + B_2 u + \tilde{B}_1 \tilde{B}_1' \nabla_x r_e')$$
$$+ \tfrac{1}{2} \nabla_x \phi \tilde{B}_1 \tilde{B}_1' \nabla_x \phi' - \tfrac{1}{2} y' E_2^{-1} y + (C_2' + \nabla_x r_e B_1 D_{21}') E_2^{-1} y + \nabla_x r_e B_2 u$$

or

$$0 = \frac{\partial \phi}{\partial t} + \sup_v \{ \nabla_x \phi \cdot (A^{r_e} + B_1 D_{21}' E_2^{-1} y + B_2 u + \tilde{B}_1 v) - \tfrac{1}{2} |v|^2 \}$$
$$- \tfrac{1}{2} y' E_2^{-1} y + (C_2' + \nabla_x r_e B_1 D_{21}') E_2^{-1} y + \nabla_x r_e B_2 u. \quad (11.33)$$

In integral form,

$$\begin{aligned}\phi_t(x) &= \inf_{v(\cdot)}\Big\{\phi_0(\xi(0)) + \int_0^t \Big[\tfrac{1}{2}|v(s)|^2 + \tfrac{1}{2}y(s)'E_2^{-1}y(s) \\ &\quad -\nabla_x r_e(\xi(s))B_2(\xi(s))u(s) \\ &\quad -(C_2(\xi(s))' + \nabla_x r_e(\xi(s))B_1(\xi(s))D_{21}')E_2^{-1}y(s)\Big]\, ds\Big\},\end{aligned} \tag{11.34}$$

where $\xi(s) = \Phi_{s,t}^{r_e,v,u,y}(x)$. Notice that (11.34) is an explicit control-theoretic integral representation of the perturbation ϕ_t.

Since $(A^{r_e}, [B_1D_{21}'E_2^{-1} : B_2 : \tilde{B}_1])$ is L_2 exponentially antistable, we have (by applying the definition to the ODE above)

$$|\Phi_{s,t}^{r_e,v,u,y}(x)| \le Ce^{-c(t-s)/2}|x| + C\left(\int_s^t e^{-c(\tau-s)}\left[|v(\tau)|^2 + |u(\tau)|^2 + |y(\tau)|^2\right]d\tau\right)^{1/2}, \tag{11.35}$$

and since it is also L_2 exponentially stable,

$$\int_T^t |\Phi_{s,t}^{r_e,v,u,y}(x)|^2 ds \le C|x|^2 + C\int_T^t \left[|v(\tau)|^2 + |u(\tau)|^2 + |y(\tau)|^2\right]d\tau, \tag{11.36}$$

where the constants do not depend on x, v, u, y.

Upper bound. Set $v = 0$, a suboptimal control for (11.34). Then

$$\begin{aligned}\phi_t(x) &\le \| \phi_0 \|_q (1 + |\xi(0)|^2) \\ &\quad + \int_0^t \Big[C|y(s)|^2 + C|\xi(s)|(|u(s)| + |y(s)|)\Big]\, ds \\ &\le \| \phi_0 \|_q (1 + |\xi(0)|^2) \\ &\quad + \int_0^t \Big[C|y(s)|^2 + \eta C|\xi(s)|^2 + \frac{C}{\eta}(|u(s)|^2 + |y(s)|^2)\Big]\, ds\end{aligned}$$

for any $\eta > 0$. Now making use of (11.35) and (11.36) we obtain

$$\begin{aligned}
\phi_t(x) \le & \| \phi_0 \|_q \left(1 + Ce^{-ct}|x|^2 + C\int_0^t e^{-c\tau}\left[|u(\tau)|^2 + |y(\tau)|^2\right] d\tau\right) \\
& + \int_0^t \left[C|y(\tau)|^2 + \frac{C}{\eta}(|u(\tau)|^2 + |y(\tau)|^2)\right] d\tau \\
& + \eta\left(C|x|^2 + C\int_0^t \left[|u(\tau)|^2 + |y(\tau)|^2\right] d\tau\right) \\
\le & \; C|x|^2(\eta + e^{-ct} \| \phi_0 \|_q) + \frac{C}{\eta} \\
\le & \; \frac{C}{\eta} + C(\eta + \| \phi_0 \|_q)|x|^2
\end{aligned}$$

for all $0 < \eta \le 1$. This proves the upper half of (11.29).

Lower bound. This is a little more complicated than the upper bound because we need to take into account the effect of $v \neq 0$ on the indefinite terms in the integrand in (11.34). Fortunately, these terms are of first order in x.

Fix any u, y. Let v be any ε-optimal control for the optimal control problem (11.34) for $\phi_t(x)$ (t and x are fixed); i.e., from (11.34)

$$\begin{aligned}
\phi_t(x) + \varepsilon \ge \phi_0(\xi(0)) + \int_0^t \Big[& \tfrac{1}{2}|v(s)|^2 + \tfrac{1}{2}y(s)'E_2^{-1}y(s) \\
& - \nabla_x r_e(\xi(s))B_2(\xi(s))u(s) \\
& - (C_2(\xi(s))' + \nabla_x r_e(\xi(s))B_1(\xi(s))D_{21}')E_2^{-1}y(s)\Big] ds,
\end{aligned}$$

and hence

$$\begin{aligned}
& \varepsilon + \phi_t(x) \\
& \ge - \| \phi_0 \|_q (1 + |\xi(0)|^2) + \int_0^t \left[\tfrac{1}{2}|v(s)|^2 + C|y(s)|^2 - C|\xi(s)|(|u(s)| + |y(s)|)\right] ds \\
& \ge - \| \phi_0 \|_q (1 + |\xi(0)|^2) + \int_0^t \left[\tfrac{1}{2}|v(s)|^2 - \eta C|\xi(s)|^2 - \frac{C}{\eta}(|u(s)|^2 + |y(s)|^2)\right] ds
\end{aligned}$$

for any $\eta > 0$. Again making use of (11.35) and (11.36) we have

$$\varepsilon + \phi_t(x) \geq - \parallel \phi_0 \parallel_q \left(1 + Ce^{-ct}|x|^2 + C\int_0^t e^{-c\tau}[|v(\tau)|^2 + |u(\tau)|^2 + |y(\tau)|^2]d\tau\right)$$

$$+\int_0^t \left[\tfrac{1}{2}|v(s)|^2 - \frac{C}{\eta}(|u(s)|^2 + |y(s)|^2)\right] ds$$

$$-\eta\left(k_0|x|^2 + k_0\int_0^t \left[|v(s)|^2 + |u(s)|^2 + |y(s)|^2\right] ds\right)$$

$$\geq -\tfrac{C}{\eta} - C(\parallel \phi_0 \parallel_q + \eta)|x|^2 + \tfrac{1}{2}(1 - k_0\eta - k_1 \parallel \phi_0 \parallel_q)\int_0^t |v(s)|^2\, ds$$

for all $0 < \eta \leq 1$. Set

$$\alpha_0 = \frac{1}{4k_1}, \alpha_1 = \frac{1}{2k_0}.$$

Then if $\parallel \phi_0 \parallel_q < \alpha_0$ and $0 < \eta < \alpha_1$ we have

$$\begin{aligned} \phi_t(x) + \varepsilon &\geq -\frac{C}{\eta} - C(\alpha_0 + \eta)|x|^2 + \tfrac{1}{8}\int_0^t |v(s)|^2\, ds \\ &\geq -\frac{C}{\eta} - C(\alpha_0 + \eta)|x|^2. \end{aligned} \tag{11.37}$$

This proves the lower bound of (11.29) for $\parallel \phi_0 \parallel_q < \alpha_0$. However, note that if $\phi_0 \geq 0$, then the term $\parallel \phi_0 \parallel_q$ need not appear in these calculations, and so the lower bound is valid for all $\phi_0 > -\alpha_0$, i.e., (11.30). If $r_0 = r_e$, we can take $\alpha_0 = 0$.

Having proved the fundamental estimates (11.29), we now move on to the proof of the limit (11.31). This will entail some additional calculations.

The optimizing v in the PDE (11.33) is denoted

$$\mathbf{v}_s^*(x) = \tilde{B}_1(x)'\nabla_x\phi_s(x)'.$$

So if ϕ_t is smooth, the optimal control on the time interval $[0, t]$ is

$$v_t^*(s; x) = \mathbf{v}_s^*(\xi(s)),$$

where $\xi(s) = \Phi_{s,t}^{r_e,v^*,u,y}(x)$. (Note that (11.34) defines an optimal control problem for each $t \geq 0$, with terminal state x at time t and optimal control and trajectory defined for $0 \leq s \leq t$.) Now the estimate (11.37) applies to v_t^*, and so we have the bound

$$\int_0^t |v_t^*(s; x)|^2\, ds \leq C(x, u, y, \eta) \quad \text{for all } t \geq 0 \tag{11.38}$$

for some constant $C(x, u, y, \eta) = C(u, y)(\frac{1}{\eta} + (\eta + \parallel \phi_0 \parallel_q + \alpha_0)|x|^2)$ $(0 < \eta \leq \alpha_1)$. Applying this estimate to (11.35) and (11.36) we have

$$|\Phi_{s,t}^{r_e,v^*,u,y}(x)|^2 \leq Ce^{-c(t-s)}|x|^2 + C\int_s^t e^{-c(r-s)}\left[|u(r)|^2 + |y(r)|^2\right] dr + C(x, u, y, \eta) \tag{11.39}$$

and

$$\int_0^t |\Phi_{s,t}^{r_e,v^*,u,y}(x)|^2 ds \le C|x|^2 + C\int_0^t [|u(r)|^2 + |y(r)|^2] ds + C(x,u,y,\eta). \quad (11.40)$$

Also, by the definition (2.21) of incremental L_2 exponential stability, we have

$$|\Phi_{s,t}^{r_e,v^*,u,y}(x_1) - \Phi_{s,t}^{r_e,v^*,u,y}(x_2)| \le C(x,u,y,\eta)|x_1 - x_2|e^{-c(t-s)/2}. \quad (11.41)$$

Let $x_1, x_2 \in \mathbf{R}^n$, and $v = v_t^*(\cdot, x_2)$ be optimal for $\phi_t(x_2)$. Denote by $\xi^1(s), \xi^2(s)$ the resulting trajectories with input v and terminal states x_1, x_2. Then, using (11.39), (11.40), and (11.41),

$$\begin{aligned}\phi_t(x_1) - \phi_t(x_2) \le &\left|\phi_0(\xi^1(0)) - \phi_0(\xi^2(0))\right| \\ &+ \left| \int_0^t [(-\nabla_x r_e(\xi^1(s))B_2(\xi^1(s)) + \nabla_x r_e(\xi^2(s))B_2(\xi^2(s)))u(s) \right. \\ &\quad + (-C_2(\xi^1(s))' - \nabla_x r_e(\xi^1(s))B_1(\xi^1(s))D_{21}' \\ &\quad \left. + C_2(\xi^2(s))' + \nabla_x r_e(\xi^2(s))B_1(\xi^2(s))D_{21}')E_2^{-1}y(s)]\, ds \right| \\ &\le Ce^{-ct}(1 + |\xi^1(0)| + |\xi^2(0)|)|x^1 - x^2| \\ &\quad + \int_0^t C(1 + |\xi^1(s)| + |\xi^2(s)|)|\xi^1(s) - \xi^2(s)|(|u(s)| + |y(s)|)\, ds \\ &\le C\left(e^{-ct} + \int_0^t e^{-c(t-s)}(|u(s)| + |y(s)|\right)\, ds)|x^1 - x^2|,\end{aligned}$$

where $C = C(x^1, x^2, u, y, \eta)$. Similarly we get the opposite inequality. Together, the estimates imply the Lipschitz bound

$$|\phi_t(x_1) - \phi_t(x_2)| \le C(x_1, x_2, u, y, \eta)\left(e^{-ct} + \int_0^t (|u(s)| + |y(s)|)e^{-c(t-s)}\, ds\right)|x_1 - x_2|.$$

Since the function ϕ_t is assumed smooth, these estimates imply the gradient bound

$$|\nabla_x \phi_t(x)| \le C(x,u,y,\eta)\left(e^{-ct} + \int_0^t (|u(s)| + |y(s)|)e^{-c(t-s)}\, ds\right),$$

and so $\lim_{t\to+\infty} \nabla_x \phi_t(x) = 0$ uniformly on compact sets and $s \mapsto \nabla_x \phi_s(x)$ is in $L_2[0,\infty)$.

Set $x = 0$ in the PDE for ϕ_t to get

$$\begin{aligned}\dot{\phi}_t(0) = &-\nabla_x \phi_t(0) \cdot (B_1(0)D_{21}'E_2^{-1}y(t) + B_2(0)u(t)) \\ &-\tfrac{1}{2}\nabla_x \phi_t(0)\tilde{B}_1(0)\tilde{B}_1(0)'\nabla_x \phi_t(0)' + \tfrac{1}{2}y(t)'E_2^{-1}y(t).\end{aligned}$$

Integrating,

$$\phi_t(0) = \phi_0(0) + \int_0^t [-\nabla_x\phi_s(0)\cdot(B_1(0)D_{21}'E_2^{-1}y(s) + B_2(0)u(s))$$
$$-\tfrac{1}{2}\nabla_x\phi_s(0)\tilde{B}_1(0)\tilde{B}_1(0)'\nabla_x\phi_s(0)' + \tfrac{1}{2}y(s)'E_2^{-1}y(s)]\,ds.$$

The estimates above imply that this integral converges to a finite number, call it $c(u, y, r_0)$.

Because of the gradient bounds, we now know that $\{\phi_t\}_{t\geq 0}$ is precompact (in the topology of local uniform convergence), and any limit point is a bounded, continuous function. In fact, there is only one limit point and it is the constant $c(u, y, r_0)$. Indeed, by the fundamental theorem of calculus, we have

$$\phi_t(x) = \phi_t(0) + \int_0^1 \nabla_x\phi_t(\lambda x)x\,d\lambda,$$

and so

$$\lim_{t\to\infty}\phi_t(x) = \lim_{t\to\infty}\phi_t(0) + 0 = c(u, y, r_0),$$

uniformly on compact sets, which is independent of x. This completes the proof. □

REMARK 11.4.7 By using directly the representation (11.28) for $r_t(x)$, it is possible to prove the convergence to $r_e(x) + c$ for any initial $r_0(x) \geq 0$. The statement and proof of Theorem 11.4.6 were designed to be adapted to the information state p_t and to provide detailed estimates on the deviation from the equilibrium along trajectories (Theorem 11.4.14 below). ∇

In addition to proving convergence, the estimate (11.29) implies that the quadratic growth rate is almost preserved, and as a consequence positive definiteness is preserved. This is an important compactness result and is stated in the next corollary.

COROLLARY 11.4.8 *Assume the existence of a tame antistabilizing solution r_e of the H_2 filtering equation. Let $u, y \in L_2$. Let $r_0 = r_e$, and recall that r_e is coercive: $r_e(x) \geq c_{re}|x|^2$ ($c_{re} > 0$). Then for all $0 < \varepsilon \leq 1$ we have (for some constant $C = C(u, y)$)*

$$r_t(x) \geq (c_{re} - \varepsilon)|x|^2 - \frac{C}{\varepsilon} \tag{11.42}$$

for all $x \in \mathbf{R}^n$, $t \geq 0$ (i.e., r_t is uniformly coercive*).*

Proof. Select $0 < \varepsilon \leq 1$. Choose $0 < \eta < \varepsilon/C$, where C is the constant in (11.29). Then, using (11.29),

$$\begin{aligned} r_t(x) &= r_e(x) + (r_t(x) - r_e(x)) \\ &\geq c_{re}|x|^2 - C\eta|x|^2 - \tfrac{C}{\eta}. \end{aligned}$$

This proves (11.42). □

11.4.4 Existence of H^∞ Controllers Implies p_e Is Control Attractor

Our goal in this section is to prove the first main theorem, Theorem 11.4.4, of §11.4.3. We state a number of preliminary facts in the following lemmas.

Lemma 11.4.9 uses the H_2 estimation assumptions to obtain a *uniform* lower bound for p_t, assuming that p_0 is everywhere finite and of at most quadratic growth.

LEMMA 11.4.9 *Assume there exists a tame solution r_e of the H^2 filter equation* (11.18). *Let $u, y \in L_2[0,\infty)$, and let $p_0 \in \mathcal{X}_q$ satisfy $p_0(x) \leq 0$. Then there exists a finite constant $C \geq 0$ (depending on γ, u, y, and p_0) such that*

$$-C(1+|x|^2) \leq p_t(x) \text{ for all } t \geq 0. \tag{11.43}$$

Proof. Using the alternative integral representation (3.12) for $p_t(x)$, we have

$$\begin{aligned}
p_t(x) &\geq \sup_{v(\cdot)}\Big\{ p_0(\xi(0)) + \tfrac{1}{2}\int_0^t [-\gamma^2(|v(s)|^2 \\
&\quad -\gamma^2\tfrac{1}{2}(y(s)-C_2(\xi(s)))'E_2^{-1}(\xi(s))(y(s)-C_2(\xi(s))))]\,ds : \ \xi(t)=x\Big\} \\
&\geq -\gamma^2 \inf_{v(\cdot)}\Big\{ r_0(\xi(0)) + \tfrac{1}{2}\int_0^t [(|v(s)|^2 \\
&\quad +(y(s)-C_2(\xi(s)))'E_2^{-1}(\xi(s))(y(s)-C_2(\xi(s))))]\,ds : \ \xi(t)=x\Big\} \\
&= -\gamma^2 r_t(x),
\end{aligned}$$

where $r_0 \geq \max(-p_0/\gamma^2, r_e)$, $r_t(x)$ is defined by (11.28), and the underlying dynamics is (2.6). Thus

$$p_t(x) \geq -\gamma^2 r_t(x). \tag{11.44}$$

Then by Theorem 11.4.6 (note that $r_0 \geq r_e$) we have

$$\begin{aligned}
p_t(x) &\geq -\gamma^2[r_e(x) + \tfrac{C}{\eta} + C(\eta + \| r_0 - r_e \|_q)|x|^2] \\
&\geq -C(1+|x|^2)
\end{aligned}$$

as desired. □

A basic upper bound for p_t was given in Lemma 3.1.10. The next lemma elaborates on this further, making use of strict dissipativity and the H^2 estimation assumptions.

LEMMA 11.4.10 *Let K_0 be a controller that renders the closed-loop system strictly γ-dissipative with coercive bias, and let $p_0 \in \mathcal{X}_q$ satisfy* (11.22) *as in the hypotheses of Theorem* 11.4.4. *Assume there exists a tame antistabilizing solution r_e of the H^2 filtering equation* (11.18).

Let $w \in L_2$, $x_0 \in \mathbf{R}^n$ be given, and let $u = K_0(y)$, y be resulting closed-loop signals. Let p_t be the resulting information state trajectory (driven by the signals $u = K_0(y)$, y) initialized at p_0.

Then p_t is uniformly quadratically tight

$$p_t(x) \leq -c|x|^2 + C \text{ for all } t \geq 0 \tag{11.45}$$

for some constants $c > 0$, $C \geq 0$ (the second depending on γ, u, y, p_0).

Proof. Abbreviate β_{K_0} to β. We express the strict dissipativity achieved by K_0 as follows (recall (2.12)) for some $0 < \varepsilon \leq \varepsilon_{\gamma,K_0}$:

$$\tfrac{1}{2}\int_0^t [|\tilde{z}(s)|^2 - (\gamma^2 - \varepsilon^2)|\tilde{w}(s)|^2]\,ds \leq \beta(\tilde{x}_0) \tag{11.46}$$

for all $\tilde{x}_0 \in \mathbf{R}^n$, $\tilde{w} \in L_{2,T}$, $t \geq 0$. Recall the underlying dynamics is given by (2.2).

Now pick $0 < \varepsilon \leq \min(\varepsilon_{\gamma,K_0}, \varepsilon_0)$, where $\varepsilon_0 > 0$ is as in (11.22). Then add $\varepsilon^2\beta$ to both sides of (11.46) to get

$$\varepsilon^2\left[\beta(\tilde{x}_0) + \tfrac{1}{2}\int_0^t |\tilde{w}(s)|^2\,ds\right] + \tfrac{1}{2}\int_0^t [|\tilde{z}(s)|^2 - \gamma^2|\tilde{w}(s)|^2]\,ds \leq (1+\varepsilon^2)\beta(\tilde{x}_0),$$

where $\tilde{x}(\cdot)$ is given by (2.2), and $\tilde{x}(0) \in \mathbf{R}^n$, $\tilde{w} \in L_{2,e}$.

There exists $0 < \nu \leq 1$ such that

$$\tfrac{1}{\nu}\beta(x) \geq r_e(x).$$

To see this, from the assumptions on r_e and Assumption 11.4.2 we have

$$\beta(x) \geq c|x|^2 \geq \frac{c}{C} r_e(x) \geq \nu r_e(x),$$

where $0 < \nu \leq \min\{\frac{c}{C}, 1\}$. Then

$$\begin{aligned}
\varepsilon^2\left[\beta(\tilde{x}_0) + \tfrac{1}{2}\int_0^t |\tilde{w}(s)|^2\,ds\right] &\geq \varepsilon^2\left[\beta(\tilde{x}_0) + \nu\tfrac{1}{2}\int_0^t |\tilde{w}(s)|^2\,ds\right] \\
&\geq \varepsilon^2\nu\left[\tfrac{1}{\nu}\beta(\tilde{x}_0) + \tfrac{1}{2}\int_0^t |\tilde{w}(s)|^2\,ds\right] \\
&\geq \varepsilon^2\nu\left[r_e(\tilde{x}_0) + \tfrac{1}{2}\int_0^t |\tilde{w}(s)|^2\,ds\right].
\end{aligned}$$

Now in the statement of the lemma $w \in L_2$ and x_0 are fixed, and $u = K_0(y)$, y are thus determined and fixed. The internal stability implies $u, y \in L_2$. Fix (x,t), and observe that

$$\begin{aligned}
r_t(x) = \inf_{\tilde{w}(\cdot)}\Big\{ & r_e(\tilde{x}_0) + \tfrac{1}{2}\int_0^t |\tilde{w}(s)|^2\,ds \\
& : \tilde{x}(t) = x,\ y(s) = C_2(\tilde{x}(s)) + D_{21}\tilde{w}(s),\ 0 \leq s \leq t\Big\},
\end{aligned}$$

where $r_t(x)$ is given by (11.28) with $r_0 = r_e$ (note that the dynamics underlying (11.28) is (2.2), driven by the $u = K_0(y)$, y determined and fixed). Therefore,

$$\varepsilon^2 \nu r_t(x) + \tfrac{1}{2}\int_0^t [|\tilde{z}(s)|^2 - \gamma^2 |\tilde{w}(s)|^2]\, ds \leq (1+\varepsilon^2)\beta(\tilde{x}_0),$$

where $\tilde{x}(t) = x$ and

$$y(s) = C_2(\tilde{x}(s)) + D_{21}\tilde{w}(s),\, s \geq 0. \tag{11.47}$$

Add $p_0(\tilde{x}(0))$ to both sides and take the supremum over $(\tilde{x}_0, \tilde{w})$ satisfying (11.47) to get

$$\varepsilon^2 \nu r_t(x) + p_t(x) \leq (p_0 + (1+\varepsilon^2)\beta) \leq (p_0 + (1+\varepsilon_0^2)\beta) \tag{11.48}$$

(recall definition (3.7) of p_t).

Then, using Theorem 11.4.6, for all $0 < \eta < \alpha_1$,

$$\begin{aligned} p_t(x) &\leq -\varepsilon^2 \nu r_t(x) + C \\ &\leq -\varepsilon^2 \nu r_e(x) + \varepsilon^2 \nu \tfrac{C}{\eta} + \varepsilon^2 \nu C \eta |x|^2 + C \\ &\leq -\varepsilon^2 \nu c_{re} |x|^2 + \varepsilon^2 \nu \tfrac{C}{\eta} + \varepsilon^2 \nu C \eta |x|^2 + C. \end{aligned} \tag{11.49}$$

Now select $\eta > 0$ sufficiently small to obtain (11.45). □

The next lemma contains some important technical results that arise in the course of proving Theorem 11.4.4. These results concern uniform bounds on the information state p_t and its gradient, and these are key to the asymptotic behavior of p_t in particular, $\{p_t\}_{t\geq 0}$ is precompact (relative to the topology of local uniform convergence).

LEMMA 11.4.11 *Make the assumptions of Theorem* 11.4.4. *Then there exist nonnegative constants (with $c > 0$) (depending on K_0, w) such that*

$$-k_1(1+|x|^2) \leq p_t(x) \leq -c|x|^2 + k_1 \;\; \textit{for all } t \geq 0, \tag{11.50}$$

and

$$|\nabla_x p_t(x)| \leq k_2(1+|x|) \;\; \textit{for all } t \geq 0. \tag{11.51}$$

Consequently, $\{p_t\}_{t\geq 0}$ is precompact (local uniform topology).

Proof of Lemma 11.4.11 *and Theorem* 11.4.4.

Outline of proof. The proof is rather more complicated than the proof of our results concerning the H_2 information state (Theorem 11.4.6). This is due to the mixed-sign integrand in the definition of p_t (see equations (3.7), (3.12)). The basic idea is to decompose p_t as $p_t = -\gamma^2 r_e + \phi_t$ and to make use of the antistabilizing property of the H^2 information state r_e and the assumed dissipation arising from the controller K_0 to show that an auxiliary system (running in reverse time) is strictly γ-dissipative (the strict bounded real lemma; Theorems 5.3.1, 5.3.2, and 5.3.3 are applied) with available storage ϕ_e and $p_e = -\gamma^2 r_e + p_e$. To achieve this, some detailed estimates and comparisons are required, as outlined in the following steps.

Step 1. The bound (11.50) on p_t of Lemma 11.4.11 follows immediately from Lemmas 11.4.9 and 11.4.10.

The bound (11.51) on ∇p_t requires the next three steps.

Step 2. Express p_t in terms of the H_2 solution r_e. Its stabilizing property will be used below. Write

$$p_t = -\gamma^2 r_e + \phi_t.$$

Using the assumed growth properties of r_e and (11.50) it follows that

$$|\phi_t(x)| \leq c_1(1+|x|^2) \text{ for all } t \geq 0.$$

This uniform bound will be used in calculations to follow.

Step 3. The function ϕ satisfies the PDE

$$\begin{aligned}\dot{\phi} &= -\nabla_x\phi \cdot (A + B_1 D_{21}' E_2^{-1}(y - C_2) + B_2 u + \tilde{B}_1 \tilde{B}_1' \nabla_x r_e') \\ &\quad + \gamma^{-2}\tfrac{1}{2}\nabla_x\phi \tilde{B}_1 \tilde{B}_1 \nabla_x\phi' + \tfrac{1}{2}|C_1|^2 + \tfrac{1}{2}|D_{12}u|^2 - \gamma^2\tfrac{1}{2}y' E_2^{-1} y \\ &\quad + (C_1' D_{12} + \gamma^2 \nabla_x r_e B_2)u + \gamma^2(C_2' + \nabla_x r_e B_1 D_{21}') E_2^{-1} y\end{aligned} \tag{11.52}$$

and corresponding to this is an integral representation (equation (11.56) below).

Step 4. A gradient estimate for $\nabla_x\phi_t$ is proven using the integral representation for ϕ_t, (see (11.56)). The quadratic term (corresponding to the mixed-sign term in p_t) in the state is handled by using the *strict* dissipation property assumed.

At this point the gradient estimate (11.51) on ∇p_t is proven. This completes the proof of Lemma 11.4.11. We turn now to items (i) and (ii) of Theorem 11.4.4.

Step 5. The PDE (11.52) is almost the dissipation PDE for the following auxiliary system:

$$\begin{cases} \dot{\bar{\xi}} = -[A(\bar{\xi}) - B_1(\bar{\xi})D_{21}' E_2^{-1} C_2(\bar{\xi}) + \tilde{B}_1(\bar{\xi})\tilde{B}_1(\bar{\xi})' \nabla_x r_e(\bar{\xi})' + \tilde{B}_1(\bar{\xi})\bar{v}], \\ \bar{z} = C_1(\bar{\xi}). \end{cases} \tag{11.53}$$

(This observation is a crucial one in the proof.) It is shown that this auxiliary system is strictly γ-dissipative. The minus sign indicates a time reversal.

Step 6. Using the SBRL (Theorem 5.3.2), it is shown that there exists a solution $\phi_e \geq 0$, $\phi_e(0) = 0$ of the steady state version

$$-\nabla_x\phi_e \cdot (A - B_1 D_{21}' E_2^{-1} C_2 + \tilde{B}_1 \tilde{B}_1' \nabla_x r_e') + \gamma^{-2}\tfrac{1}{2}\nabla_x\phi_e \tilde{B}_1 \tilde{B}_1 \nabla_x \phi_e' + \tfrac{1}{2}|C_1|^2 = 0 \tag{11.54}$$

of the PDE (11.52), and the vector field

$$A - B_1 D_{21}' E_2^{-1} C_2 + \tilde{B}_1 \tilde{B}_1{}' \nabla_x r_e' - \gamma^{-2}\tilde{B}_1 \tilde{B}_1 \nabla_x \phi_e' \tag{11.55}$$

is asymptotically L_2 antistable (provided ϕ_e is smooth).

Step 7. Set

$$p_e \triangleq -\gamma^2 r_e + \phi_e.$$

This function is the desired antistabilizing solution of (3.28).

This proves items (i) and (ii). We turn now to item (iii).

Step 8. Item (iii) includes stronger hypotheses concerning the smoothness and antistability of the equilibrium p_e. If the condition (3.37) relating p_0 and p_e holds, then Theorem 11.4.14 implies the quadratic upper limiting and convergence assertions, and the bounds (11.24) and $|\nabla_x p_e(x)| \leq C|x|$ now follow from Lemma 11.4.10.

Since we do not know if condition (3.37) holds, we must use other means to prove these assertions. Step 8 provides the necessary details.

This establishes items iii(a), iii(b), and iii(c).

Step 9. Item iii(d) follows from Theorem B.2.2.

Details of proof.

Step 3. The decomposition of $p_t = -\gamma^2 r_e + \phi_t$ and (3.9) gives the following PDE for ϕ_t:

$$\begin{aligned}
\dot{\phi} &= F(-\gamma^2 r_e + \phi, u, y) \\
&= -\nabla_x \phi \cdot (A + B_1 D_{21}' E_2^{-1}(y - C_2) + B_2 u + \tilde{B}_1 \tilde{B}_1' \nabla_x r_e') \\
&\quad + \gamma^{-2} \tfrac{1}{2} \nabla_x \phi \tilde{B}_1 \tilde{B}_1 \nabla_x \phi' + \tfrac{1}{2}|C_1|^2 + \tfrac{1}{2}|D_{12} u|^2 - \gamma^2 \tfrac{1}{2} y' E_2^{-1} y \\
&\quad + (C_1' D_{12} + \gamma^2 \nabla_x r_e B_2) u + \gamma^2 (C_2' + \nabla_x r_e B_1 D_{21}') E_2^{-1} y \\
&= \sup_v \{ -\nabla_x \phi \cdot (A + B_1 D_{21}' E_2^{-1}(y - C_2) + B_2 u + \tilde{B}_1 \tilde{B}_1' \nabla_x r_e' + \tilde{B}_1 v) \\
&\quad - \gamma^2 \tfrac{1}{2} |v|^2 \} + \tfrac{1}{2}|C_1|^2 + \tfrac{1}{2}|D_{12} u|^2 - \gamma^2 \tfrac{1}{2} y' E_2^{-1} y \\
&\quad + (C_1' D_{12} + \gamma^2 \nabla_x r_e B_2) u + \gamma^2 (C_2' + \nabla_x r_e B_1 D_{21}') E_2^{-1} y.
\end{aligned} \tag{11.52$'$}$$

Integrating, we obtain the following representation for ϕ_t:

$$\begin{aligned}
\phi_t(x) = \sup_{v(\cdot)} \Big\{ & \phi_0(\xi(0)) + \int_0^t \tfrac{1}{2}|C_1(\xi(s))|^2 - \gamma^2 \tfrac{1}{2}|v(s)|^2 + \tfrac{1}{2}|D_{12} u(s)|^2 \\
& - \gamma^2 \tfrac{1}{2} y(s)' E_2^{-1} y(s) + (C_1(\xi(s))' D_{12} + \gamma^2 \nabla_x r_e(\xi(s)) B_2(\xi(s))) u(s) \\
& + \gamma^2 (C_2(\xi(s))' + \nabla_x r_e(\xi(s)) B_1(\xi(s)) D_{21}') E_2^{-1} y(s)] \, ds \Big\},
\end{aligned} \tag{11.56}$$

where $\xi(s) = \Phi_{s,t}^{u,y,v}(x)$ is the solution of the ODE

$$\begin{aligned}
\dot{\xi}(s) &= A(\xi(s)) + B_1(\xi(s)) D_{21}' E_2^{-1}(y(s) - C_2(\xi(s))) + B_2(\xi(s)) u(s) \\
&\quad + \tilde{B}_1(\xi(s)) \tilde{B}_1(\xi(s))' \nabla_x r_e(\xi(s))' + \tilde{B}_1(\xi(s)) v(s) \\
&= A^{r_e}(\xi(s)) + B_1(\xi(s)) D_{21}' E_2^{-1} y(s) + B_2(\xi(s)) u(s) + \tilde{B}_1(\xi(s)) v(s)
\end{aligned} \tag{11.57}$$

with terminal data $\xi(t) = x$, where A^{r_e} is the H_2 vector field (11.19).

Step 4. The optimizing v in the PDE (11.52) is denoted

$$\mathbf{v}_s^*(x) = -\gamma^{-2}\tilde{B}_1(x)'\nabla_x\phi_s(x)'.$$

So if ϕ_t is smooth, the optimal control on the time interval $[0,t]$ is

$$v_t^*(s) = v_t^*(s;x) = \mathbf{v}_s^*(\xi(s)),$$

where $\xi(s) = \Phi_{s,t}^{v^*,u,y}(x)$. We now use the hypothesis that (G, K_0) is strictly γ-dissipative and the incremental L_2 exponential antistability of A^{r_e} to obtain a bound for $\| v_t^* \|_{L_2[0,t]}$. This will furnish the desired gradient bounds.

Since (G, K_0) is strictly γ-dissipative, there exists $0 < \varepsilon \leq \min(\varepsilon_{\gamma,K_0}, \varepsilon_0)$ for which (G, K_0) is $\bar{\gamma}$-dissipative, where $\bar{\gamma}^2 = \gamma^2 - \varepsilon^2$. That is, (2.12) holds. We use overbars "$^-$" to denote quantities corresponding to $\bar{\gamma}$ in what follows.

The results obtained so far also apply to overbar quantities. Then $\bar{p}_t$ also enjoys bounds of the form (11.50), and as a consequence

$$|\bar{\phi}_t| \leq C(1 + |x|^2)$$

for some constant $C \geq 0$. The integral representation for $\bar{\phi}_t$ is

$$\begin{aligned}\bar{\phi}_t(x) = \sup_{v(\cdot)}\Big\{ \bar{\phi}_0(\xi(0)) + \int_0^t [\tfrac{1}{2}|C_1(\xi(s))|^2 - (\gamma^2 - \varepsilon^2)\tfrac{1}{2}|v(s)|^2 + \tfrac{1}{2}|D_{12}u(s)|^2 \\ -(\gamma^2 - \varepsilon^2)\tfrac{1}{2}y(s)'E_2^{-1}y(s) \\ +(C_1(\xi(s))'D_{12} + (\gamma^2 - \varepsilon^2)\nabla_x r_e(\xi(s))B_2(\xi(s)))u(s) \\ +(\gamma^2 - \varepsilon^2)(C_2(\xi(s))' + \nabla_x r_e(\xi(s))B_1(\xi(s))D_{21}')E_2^{-1}y(s)]\, ds\Big\},\end{aligned}$$

where $\xi(s) = \Phi_{s,t}^{u,y,v}(x)$ is the solution of the ODE (11.57). Then plugging $v = v_t^*$ into the representation for $\bar{\phi}_t$ (suboptimal for $\bar{\phi}$) gives

$$\begin{aligned}\bar{\phi}_t(x) \geq \phi_t(x) + \varepsilon^2 \int_0^t [|v_t^*(s)|^2 ds + \varepsilon^2 \int_0^t [|u(s)|^2 + y(s)'E_2^{-1}y(s)]ds \\ +\varepsilon^2 \int_0^t \nabla_x r_e(\xi(s))B_2(\xi(s))u(s)ds \\ -\varepsilon^2 \int_0^t (C_2(\xi(s))' + \nabla_x r_e(\xi(s))B_1(\xi(s))D_{21}')E_2^{-1}y(s)ds\end{aligned}$$

since $\bar{\phi}_0 = \phi_0 + \varepsilon^2 r_e \geq \phi_0$. Next, we lower bound this as follows:

$$\begin{aligned}\bar{\phi}_t(x) &\geq \phi_t(x) + \varepsilon^2 \int_0^t |v_t^*(s)|^2 ds - \varepsilon^2 \int_0^t C|\xi(s)|(|y(s)| + |u(s)|)ds \\ &\geq \phi_t(x) + \varepsilon^2 \int_0^t |v_t^*(s)|^2 ds - \varepsilon^2 C \int_0^t [\nu|\xi(s)|^2 + \tfrac{1}{\nu}(|y(s)|^2 + |u(s)|^2)]ds \\ &\geq \phi_t(x) + \varepsilon^2(1-\nu C)\int_0^t |v_t^*(s)|^2 ds - \varepsilon^2 C\nu(|x|^2 + C(u,y)[1+\tfrac{1}{\nu}])\end{aligned}$$

for all $\nu > 0$. The incremental L_2 exponential antistability of the A^{r_e} system has been used, and the constant $C(u, y)$ depends on the L_2-norm of u and y. This gives

$$\int_0^t |v_t^*(s)|^2 ds \leq C(1 + |x|^2) \text{ for all } t \geq 0 \tag{11.58}$$

provided $\nu > 0$ is chosen sufficiently small (the constant depends on u, y, and ν).

The gradient estimate

$$|\nabla_x \phi_t(x)| \leq C(u, y, \nu)(1 + |x|)\left(1 + e^{-ct} + \int_0^t (|u(s)| + |y(s)|)e^{-c(t-s)}\, ds\right)$$

now follows using the methods employed in the proof of Theorem 11.4.6. Note that the RHS of this inequality does not decay to zero as $t \to \infty$, unlike the analogous quantity in the H_2 case. This involves estimating the difference $\phi_t(x_1) - \phi_t(x_2)$, and here the term absent in the H_2 case is the quadratic term

$$\begin{aligned}&\int_0^t [|C_1(\xi^1(s))|^2 - |C_1(\xi^2(s))|^2]ds \\ &\quad \leq \int_0^t C(|\xi^1(s)| + |\xi^2(s)|)|\xi^1(s) - \xi^2(s)|ds \\ &\quad \leq C(1 + |x_1| + |x_2|)|x_1 - x_2|.\end{aligned}$$

This proves the gradient estimate (11.51).

Step 5. The integral representation (11.56) for ϕ_t is closely related to the representation for the available storage for the auxiliary system (11.53), except for the time direction and the L_2 inputs u and y. It is a straightforward matter to reverse time, and since u and y are in L_2, one would expect their effects to diminish for large time. In this way we use the properties of ϕ_t to prove that the auxiliary system (11.53) is strictly γ-dissipative, with available storage ϕ_e (Step 6). To show *strict* dissipation, we use the function $\bar{\phi}_t$ introduced in Step 4 (defined for $\bar{\gamma}^2 = \gamma^2 - \varepsilon^2$) instead of ϕ_t (defined for γ^2).

The proof will make use of the representation formula for $\bar{\phi}_t$ used in Step 4. In view of Lemma 11.4.11, it will be assumed that $\bar{\phi}_t$ is converging (locally uniformly, through a subsequence) to a limit function $\bar{\phi}^\infty$.

We claim that for some $\alpha > 0$ the available storage (for (11.53))

$$\bar{S}(x) = \sup_{t\geq 0,\bar{v}} \tfrac{1}{2}\int_0^t [-(\gamma^2-\alpha^2)|\bar{v}(r)|^2 + |\bar{z}(r)|^2]\,dr \tag{11.59}$$

is finite, where $\bar{\xi}(r) = \bar{\Phi}^{\bar{v}}_{r,0}(x)$ is the solution of (11.53) with $\bar{\xi}(0) = x$, and $\bar{S}(x) \geq 0$, $\bar{S}(0) = 0$. This will establish the desired strict dissipation.

Using the representation for $\bar{\phi}_t$ and dynamic programming, we have

$$\begin{aligned}\bar{\phi}_{T+t}(x) = \sup_{v(\cdot)}\Big\{ \bar{\phi}_T(\xi(0)) + \int_T^{T+t} & \tfrac{1}{2}|C_1(\xi(s))|^2 - (\gamma^2-\varepsilon^2)\tfrac{1}{2}|v(s)|^2 + \tfrac{1}{2}|D_{12}u(s)|^2 \\ & -(\gamma^2-\varepsilon^2)\tfrac{1}{2}y(s)'E_2^{-1}y(s) \\ & +(C_1(\xi(s))'D_{12} + (\gamma^2-\varepsilon^2)\nabla_x r_e(\xi(s))B_2(\xi(s)))u(s) \\ & +(\gamma^2-\varepsilon^2)(C_2(\xi(s))' + \nabla_x r_e(\xi(s))B_1(\xi(s))D_{21}')E_2^{-1}y(s)]\,ds\Big\},\end{aligned}$$

where $\xi(s) = \Phi^{u,y,v}_{s,T+t}(x)$ is the solution of the ODE (11.57). Below we let $T \to \infty$ and observe the effects of u and y diminish. Then for any $v(\cdot)$ we have

$$\begin{aligned}\tfrac{1}{2}\int_T^{T+t} |C_1(\xi(s))|^2 ds &\leq \tfrac{1}{2}\bar{\gamma}^2\int_T^{T+t}|v(s)|^2 ds - \bar{\phi}_T(\xi(T)) + \bar{\phi}_{T+t}(\xi(T+t)) \\ &\quad +\tfrac{1}{2}\varepsilon_1^2\int_T^{T+t}|v(s)|^2 ds \\ &\quad +C\left(\varepsilon_1^2|x|^2 + (1+\tfrac{1}{\varepsilon_1^2})\int_T^{T+t}[|u(s)|^2 + |y(s)|^2]ds\right) \\ &\leq \tfrac{1}{2}(\gamma^2 - [\varepsilon^2-\varepsilon_1^2])\int_T^{T+t}|v(s)|^2 ds - \bar{\phi}_T(\xi(T)) \\ &\quad +\bar{\phi}_{T+t}(\xi(T+t)) + C\varepsilon_1^2|x|^2 \\ &\quad +C\tfrac{1}{\varepsilon_1^2}\int_T^{T+t}[|u(s)|^2 + |y(s)|^2]ds,\end{aligned} \tag{11.60}$$

where $\varepsilon_1 > 0$ is chosen so that $\varepsilon_2^2 = \varepsilon^2 - \varepsilon_1^2 > 0$ and $\gamma^2 - \varepsilon_2^2 > 0$. Note again that $\xi(s) = \Phi^{v,u,y}_{s,T+t}(x)$ for the solution of the ODE (11.57) on $[T, T+t]$.

We now wish to compare the systems (11.53) and (11.57). To do this, we reverse time on (11.53), so we map the time interval $[0,t]$ for (11.53) to $[T, T+t]$ via $s = T+t-r$.

Let $\bar{v}(r)$ be defined on $[0,t]$ and define

$$v(s) = \bar{v}(T+t-s),\quad \zeta(s) = \bar{\xi}(T+t-s),\quad z(s) = \bar{z}(T+t-s),$$

and so on. Note that

$$\int_0^t |\bar{v}(r)|^2 dr = \int_T^{T+t} |v(s)|^2 ds.$$

By incremental L_2 exponential antistability we have

$$\int_T^{T+t} |\zeta(s) - \xi(s)|^2 ds \le C \int_T^{T+t} [|u(s)|^2 + |y(s)|^2] ds.$$

This yields the estimate

$$\begin{aligned} \tfrac{1}{2}\int_0^t |\bar{z}(r)|^2 dr &\le \tfrac{1}{2}\int_T^{T+t} |C_1(\xi(s))|^2 ds + \eta \int_T^{T+t} |v(s)|^2 ds \\ &+ C\left(\eta|x|^2 + \tfrac{1}{\eta}\int_T^{T+t} [|u(s)|^2 + |y(s)|^2] ds\right) \end{aligned} \tag{11.61}$$

for any $\eta > 0$.

Select η so that $\gamma^2 - [\varepsilon_2^2 - \eta] > 0$ and set $\alpha^2 = [\varepsilon_2^2 - \eta] > 0$. Then combining (11.60) and (11.61) we get

$$\begin{aligned} \tfrac{1}{2}\int_0^t |\bar{z}(r)|^2 dr &\le \tfrac{1}{2}(\gamma^2 - \alpha^2)\int_T^{T+t} |v(s)|^2 ds - \bar{\phi}_T(\xi(T)) + \bar{\phi}_{T+t}(\xi(T+t)) \\ &+ C\left((\varepsilon_1^2 + \eta)|x|^2 + \left(\tfrac{1}{\varepsilon_1^2} + \tfrac{1}{\eta}\right)\int_T^{T+t} [|u(s)|^2 + |y(s)|^2] ds\right). \end{aligned} \tag{11.62}$$

Now $\xi(T) = \Phi^{v,u,y}_{T,T+t}(x)$ and $\xi(T+t) = x$. Applying Lemma A.1.2, we have

$$\lim_{T\to\infty} \Phi^{v,u,y}_{T,T+t}(x) = \Phi^{v,0,0}_{0,t}(x) \tag{11.63}$$

uniformly (recall that v translates with T, but u, y do not).

Now owing to the bounds on $\bar{\phi}_t$ and its gradient, the sequences $\{\bar{\phi}_T\}_{T\ge 0}$ and $\{\bar{\phi}_{T+t}\}_{T\ge 0}$ are precompact with respect to local uniform convergence. So $\bar{\phi}_T$ and $\bar{\phi}_{T+t}$ converge uniformly on compact sets (via subsequences through T) to limit functions $\bar{\phi}^\infty$ and $\bar{\phi}_{\infty,t}$, respectively. Hence we obtain from (11.62) by sending $T \to \infty$ (through the subsequence being used)

$$\tfrac{1}{2}\int_0^t |\bar{z}(r)|^2 dr \le \tfrac{1}{2}(\gamma^2 - \alpha^2)\int_0^t |v(s)|^2 ds - \bar{\phi}^\infty(\Phi^{v,0,0}_{0,t}(x)) + \bar{\phi}_{\infty,t}(x) + C(\varepsilon_1^2 + \eta)|x|^2. \tag{11.64}$$

Referring back fully to the system (11.53) we have

$$\bar{\phi}^\infty(\bar{\xi}(t)) + \tfrac{1}{2}\int_0^t |\bar{z}(r)|^2 dr \le \tfrac{1}{2}(\gamma^2 - \alpha^2)\int_0^t |\bar{v}(r)|^2 dr + \bar{\phi}_{\infty,t}(\bar{\xi}_0) + C(\varepsilon_1^2 + \eta)|\bar{\xi}_0|^2. \tag{11.65}$$

Inequality (11.65) is (almost) the strict γ-dissipation inequality for the auxiliary system (11.53). At this point we cannot conclude that (11.53) is strictly γ-dissipative because we have not produced a storage function with the required properties (bounded below and having $x = 0$ as a global minimum).

Using the definition of $\bar{\phi}_t$ and inequalities (11.44) and (11.49) in the proofs of Lemmas 11.4.9 and 11.4.10 we have (using $\bar{\gamma}$)

$$\bar{\gamma}^2(r_e(x) - r_t^1(x)) \leq \bar{\phi}_t(x) \leq \bar{\gamma}^2 r_e(x) - \varepsilon^2 \nu r_t^2(x) + C,$$

where here r_t^1 has initial condition $r_0^1 \geq \max\{-p_0/\bar{\gamma}^2, r_e\} \geq r_e$ and r_t^2 has initial condition $r_0^2 = r_e$. Send $t \to \infty$ through the subsequence to obtain

$$-\bar{\gamma}^2 c(u, y, r_0^1) \leq \bar{\phi}_\infty(x) \leq (\bar{\gamma}^2 - \varepsilon^2 \nu) r_e(x) - \varepsilon^2 \nu c(u, y, r_0^2) + C. \qquad (11.66)$$

This proves that $\bar{\phi}_\infty$ is bounded below.

Now define the function $\bar{S}(x)$ by (11.59). Then it follows from (11.65) that

$$\begin{aligned} &\bar{\gamma}^2 c(u, y, r_0^1) + \bar{\phi}_{\infty,t}(\bar{\xi}_0) + C(\varepsilon_1^2 + \eta)|\bar{\xi}_0|^2 \\ &\geq \sup_{t \geq 0, \bar{v}} \tfrac{1}{2} \int_0^t [-(\gamma^2 - \alpha^2)|\bar{v}(r)|^2 + |\bar{z}(r)|^2] \, dr \\ &= \bar{S}(x). \end{aligned}$$

Therefore, $\bar{S}(x)$ is finite, and by definition, bounded below (by 0).

By dynamic programming, for any $t \geq 0$,

$$\bar{S}(x) = \sup_{\bar{v}} \left\{ \bar{S}(\bar{\xi}(t)) + \tfrac{1}{2} \int_0^t [-(\gamma^2 - \alpha^2)|\bar{v}(r)|^2 + |\bar{z}(r)|^2] \, dr \right\}.$$

Set $\bar{v} = 0$, $\bar{\xi}_0 = x$, and send $t \to \infty$ to obtain, using the stability of (11.53),

$$\bar{S}(x) \geq \bar{S}(0).$$

This implies that $\bar{S}$ attains a global minimum at $x = 0$. This completes the proof that the auxiliary system (11.53) is strictly γ-dissipative.

Step 6. We now apply the SBRL (Chapter 5). Since $-A^{r_e}$ is L_2 exponentially stable, and the auxiliary system is strictly γ-dissipative, Theorems 5.3.1, 5.3.2 imply the existence of a solution ϕ_e of the PDE (11.52) satisfying $\phi_e(x) \geq 0$, $\phi_e(0) = 0$. If ϕ_e is smooth, then the vector field (11.55) is asymptotically L_2 stable.

Step 7. We define p_e by

$$p_e \stackrel{\Delta}{=} -\gamma^2 r_e + \phi_e.$$

From this it follows immediately that p_e solves the PDE (3.28) (using the PDEs for r_e (11.18) and ϕ_e (11.54), $p_e(0) = 0$ and $p_e \geq -\gamma^2 r_e$ (since $\phi_e \geq 0$)). If p_e is smooth, then it is asymptotically L_2 antistabilizing.

Step 8. In the proof of Theorem 11.4.14, based on the proof of Theorem 11.4.6, condition (3.37) (i.e., (11.67)) plays a crucial role in bounding the optimal auxiliary control energy, which is required for the gradient bounds used to prove the quadratic bounds and convergence. Given that we do not know whether or not (11.67) holds at this stage, we must obtain the bounds using other information.

In Step 4 above we used strict γ-dissipativity and other bounds to obtain the energy bound (11.58). Let us write

$$p_t = -\gamma^2 r_e + \phi_t = -\gamma^2 r_e + \phi_e + \tilde{\phi}_t = p_e + \tilde{\phi}_t$$

so that the optimal controls for ϕ_t and $\tilde{\phi}_t$ (see the proof of Theorem 11.4.14) are related by

$$v_t^*(s;x) = -\gamma^{-2}\tilde{B}_1(\xi(s))'\nabla_x\phi_e(\xi(s))' + \tilde{v}_t^*(s;x)$$

with dynamics

$$\begin{aligned}\dot{\xi}(s) &= A^{r_e}(\xi(s)) + \tilde{B}_1(\xi(s))v_t^*(s;x) \\ &= A^{p_e}(\xi(s)) + \tilde{B}_1(\xi(s))\tilde{v}_t^*(s;x)\end{aligned}$$

and $\xi(t) = x, 0 \leq s \leq t$. Stability implies

$$\int_0^t |\xi(s)|^2 ds \leq C(x,u,y).$$

Therefore, this bound and (11.58) imply the required bound (11.72) needed in the proof of Theorem 11.4.14. The proof can now be completed as in the proofs of Theorems 11.4.6 and 11.4.14. □

Remark 11.4.12 Note that the decomposition $p_e = -\gamma^2 r_e + \phi_e$ expresses p_e as a sum of negative and positive pieces (cf. [Wil71]). ▽

Remark 11.4.13 Since $r_e(x) \geq c_{r_e}|x|^2$, the inequality $p_e \geq -\gamma^2 r_e$ implies the nonlinear analogue of the nonsingularity of Y_e, since in the case of linear systems, we have $Y_e \geq R_e^{-1} > 0$ provided the inverse of Y_{Ie} exists and equals Y_e. ▽

11.4.5 Convergence to the Equilibrium p_e

We return to the convergence of the information state to an equilibrium and assert it under the hypothesis that p_e is antistabilizing. Our goal is to prove Theorem 11.4.5, the second main theorem of this section. The next theorem is close to being the same thing, but provides a slightly different perspective.

Theorem 11.4.14 *Assume there exists a smooth solution p_e of the steady state equation (3.28) such that the system $(A^{p_e}, [B_1 D_{12}' E_2^{-1} : B_2 : \tilde{B}_1])$ is incrementally L_2 exponentially antistable, and $p_e \in \mathcal{X}_{q,2,b}$. Let $p_0 \in \mathcal{X}_{q,2,b}$ satisfy $p_0(0) = 0$, and $u, y \in L_2$. Then there exists constants $0 < \alpha_0 \leq 1$, $C = C(u,y,p_0)$ such that if*

$$p_0(x) - p_e(x) \leq \alpha_0|x|^2 + c_0, \tag{11.67}$$

for some $c_0 \geq 0$, then for all $0 < \eta \leq 1$,

$$\frac{C}{\eta} + C(\eta + \alpha_0)|x|^2 \geq p_t(x) - p_e(x) \geq -\frac{C}{\eta} - C(\eta + \| p_0 - p_e \|_q)|x|^2 \tag{11.68}$$

for all $x \in \mathbf{R}^n$, $t \geq 0$. Moreover,

$$\lim_{t\to+\infty} p_t = p_e + c(u, y, p_0) \tag{11.23$'$}$$

uniformly on compact subsets of $\mathbf{R}^n$, *where* $c(u, y, p_0)$ *is a constant depending on* $u, y \in L_2$ *with* $c(0, 0, p_e) = 0$.

Proof. The proof is similar to that of Theorem 11.4.6, and we sketch only the main ideas. If we write

$$p_t = p_e + \tilde{\phi}_t,$$

we have

$$\begin{aligned}
\dot{\tilde{\phi}} &= F(p_e + \tilde{\phi}, u, y) \\
&= \sup_v \{-\nabla_x \tilde{\phi} \cdot (A^{p_e} + B_1 D_{21}' E_2^{-1} y + B_2 u + \tilde{B}_1 \tilde{v}) - \gamma^2 \tfrac{1}{2} |\tilde{v}|^2\} \\
&+ \tfrac{1}{2} |D_{12} u|^2 - \gamma^2 \tfrac{1}{2} y' E_2^{-1} y + (C_1' D_{12} - \nabla_x p_e B_2) u \\
&\quad + (\gamma^2 C_2' - \nabla_x p_e B_1 D_{21}') E_2^{-1} y
\end{aligned} \tag{11.69}$$

and the corresponding integral representation

$$\begin{aligned}
\tilde{\phi}_t(x) = \sup_{\tilde{v}(\cdot)} \Big\{ &\tilde{\phi}_0(\xi(0)) + \int_0^t [-\gamma^2 \tfrac{1}{2} |\tilde{v}(s)|^2 + \tfrac{1}{2} |D_{12} u(s)|^2 \\
&- \gamma^2 \tfrac{1}{2} y(s)' E_2^{-1} y(s) + (C_1(\xi(s))' D_{12} - \nabla_x p_e(\xi(s)) B_2(\xi(s))) u(s) \\
&+ (\gamma^2 C_2(\xi(s))' - \nabla_x p_e(\xi(s)) B_1(\xi(s)) D_{21}') E_2^{-1} y(s)] \, ds \Big\},
\end{aligned} \tag{11.70}$$

where $\xi(s) = \Phi_{s,t}^{\tilde{v},u,y}(x)$ is the solution of the ODE

$$\dot{\xi}(s) = A^{p_e}(\xi(s)) + B_1(\xi(s)) D_{21}' E_2^{-1} y(s) + B_2(\xi(s)) u(s) + \tilde{B}_1(\xi(s)) \tilde{v}(s). \tag{11.71}$$

The most critical step in the argument proving Theorem 11.4.6 is establishing the bound

$$\int_0^t |\tilde{v}_t^*(s; x)|^2 ds \leq C(x, u, y) \tag{11.72}$$

for the optimizing $\tilde{v}$ (here relative to (11.69), (11.70)). As in the proof of Theorem 11.4.6 this makes use of the condition imposed on the initial information state, viz., (11.67). □

Remark 11.4.15 The detailed proof shows that $\lim_{t\to\infty} p_t = p_e + c(u, y, p_0)$ uniformly on compact sets, where $c(u, y, p_0)$ is the real number

$$\begin{aligned}
c(u, y, p_0) = p_0(0) - p_e(0) + \int_0^\infty &[-\nabla_x \phi_s(0) \cdot (B_1 D_{21}' E_2^{-1} y(s) + B_2 u(s)) \\
&+ \gamma^{-2} \tfrac{1}{2} \nabla_x \phi_s(0) \tilde{B}_1 \tilde{B}_1' \nabla_x \phi_s(0)' + \tfrac{1}{2} |D_{12} u(s)|^2 \\
&- \gamma^2 \tfrac{1}{2} y(s)' E_2^{-1} y(s)] \, ds.
\end{aligned}$$

∇

Theorem 11.4.5 in the introduction to this section is an immediate corollary of Theorem 11.4.14.

The trajectory p_t is uniformly quadratically tight, as the next corollary shows.

COROLLARY 11.4.16 *Assume the existence of a stabilizing solution $p_e \in \mathcal{X}_{q,2,b}$ of the steady state filtering equation* (3.28). *Let $u, y \in L_2$, $p_0 = p_e$, and assume that p_e is* quadratically tight*: for some $c_{pe} > 0$, $p_e(x) \leq -c_{pe}|x|^2$. Then for all $0 < \varepsilon \leq 1$ we have (for some constant $C = C(u, y)$)*

$$p_t(x) \leq -(c_{re} - \varepsilon)|x|^2 + \frac{C}{\varepsilon} \tag{11.73}$$

for all $x \in \mathbf{R}^n$, $t \geq 0$ (i.e., p_t is uniformly quadratically tight*).*

Proof. The proof is similar to that of Corollary 11.4.8 and is omitted. □

11.5 Information State Attractors: Nonantistabilizable Cases

The previous section treated the case where one antistabilizes the pair $(A^\times, \tilde{B}_1)$. Now we look at the purest situation where this is impossible. This case is both more general and more complicated than the nonsingular case treated in §11.4. For example, the equilibrium information state p_e is singular. Thus there are many technical difficulties.

We assume to make the exposition simple that $\tilde{B}_1 = 0$; thus antistabilizing the pair is impossible unless $A^\times$ is already antistable. That is, we specialize to the 1 and 2A block problems. The steady state information state reads

$$-\nabla_x p_\infty \cdot (A - B_1 C_2) + \tfrac{1}{2}|C_1|^2 - \gamma^2 \tfrac{1}{2} C_2' C_2 = 0. \tag{11.74}$$

The equilibrium p_e will solve (11.74) in the sense of (3.29).

Our treatment is in two parts. The next subsection presents the general case where we assume $A^\times$ is hyperbolic. The results we present here offer a preliminary understanding of the stability of the information state in such cases. We believe the reader will not find the analysis complicated, but clearly it is far from complete and the area is fertile ground for future research. The subsection after this treats the informative special case where $A^\times$ is stable. It is opposite the antistabilizable case we already treated and it is distinctive in that our favorite equilibrium p_e is purely singular. This case is also of basic physical importance because in a mixed-sensitivity problem with stable plant one gets $A^\times$ stable.

As we shall see, it is natural to employ the max-plus framework, which interprets the information state in a generalized sense analogous to a measure. The discussion to follow will gradually increase this framework, beginning at a more concrete level. Indeed, we shall see that in some cases one cannot use pointwise convergence (and hence not local uniform convergence) to interpret convergence of the information state, and so we are forced to weaken the notion of convergence to max-plus weak convergence.

The domain of control attraction $\mathcal{D}_{attr}(p_e)$ is defined to be the largest subset of $\mathcal{X}_e$ such that $p_t \Longrightarrow p_e + c(u, y, p_0)$ as $t \to \infty$. The convergence is in the max-plus weak sense.

We assume $A^\times$ to be exponentially hyperbolic (recall the definition in Chapter 2). We will assume in this section that Assumption 11.4.1 is in force.

11.5.1 The Hyperbolic Case

In this section we assume $A^\times$ to be exponentially hyperbolic. We derive existence and some properties of p_e.

LEMMA 11.5.1 *Assume $A^\times$ is exponentially hyperbolic. Then there exists an equilibrium solution p_e of* (11.74) *of the form*

$$p_e = \delta_{M_{as}} + \breve{p}_e \tag{11.75}$$

such that $\breve{p}_e \in C(M_{as})$ and $\breve{p}_e(0) = 0$. Moreover, any other solution of the form $p'_e = \delta_{M_{as}} + \breve{p}'_e$, where $\breve{p}'_e \in C(M_{as})$ satisfies $\breve{p}'_e = \breve{p}_e + c$, for some real number c.

Proof. Let $M_0 \subset \mathbf{R}^n$ be a manifold and $\phi_0 \in C(M_0)$. Consider the initial information state $p_0 = \delta_{M_0} + \phi_0$. Then by Lemma 10.2.1 the information state at time $t \geq 0$ is given by

$$p_t = \delta_{M_t} + \phi_t,$$

where $M_t = \Phi^{u,y}_{t,0}(M_0)$ and $\phi_t \in C(M_t)$ is given by

$$\phi_t(x) = \phi_0(\Phi^{u,y}_{0,t}(x)) + \tfrac{1}{2}\int_0^t [|C_1(\Phi^{u,y}_{s,t}(x))|^2 - \gamma^2|C_2(\Phi^{u,y}_{s,t}(x))|^2]ds \text{ for } x \in M_t.$$

Now set $M_0 = M_{as}$, $\phi_0 = 0$, $u = 0$, and $y = 0$. Then by invariance, $M_t = M_{as}$ for all $t \geq 0$. Further, we see that $\lim_{t\to\infty} \phi_t(x)$ exists for $x \in M_{as}$; call this limit $\breve{p}_e(x)$. The limit is uniform on compact subsets of M_{as}. Actually, $\breve{p}_e \in C^1(M_{as})$. Write $p_e = \delta_{M_{as}} + \breve{p}_e$.

To see that p_e is an equilibrium solution of (11.74), note that

$$\begin{aligned} p_{t+T}(x) &= \delta_{M_{as}}(x) + \phi_{t+T}(x) \\ &= \delta_{M_{as}}(x) + \phi_t(\Phi^{0,0}_{t,t+T}(x)) \\ &\quad + \tfrac{1}{2}\int_t^{t+T} [|C_1(\Phi^{0,0}_{s,t+T}(x))|^2 - \gamma^2|C_2(\Phi^{0,0}_{s,t+T}(x))|^2]ds. \end{aligned}$$

Fixing $x \in M_{as}$, sending $t \to \infty$, and using time invariance, we see that

$$\breve{p}_e(x) = \breve{p}_e(\Phi^{0,0}_{0,T}(x)) + \tfrac{1}{2}\int_0^T [|C_1(\Phi^{0,0}_{s,T}(x))|^2 - \gamma^2|C_2(\Phi^{0,0}_{s,T}(x))|^2]ds,$$

showing the $\breve{p}_e$ is an equilibrium solution of (11.74) on M_{as}.

Finally, the uniqueness assertion follows as in Lemma B.2.1 since $A^{\times}$ is exponentially antistable on M_{as}. □

We say that the smooth function p_{se} on $\mathbf{R}^n$ is a *strict infosubequilibrium* for system (2.6) provided there is an $\varepsilon > 0$ so that p_{se} satisfies the inequality

$$F(p_{se}, 0, 0)(x) + \varepsilon |x|^2 \leq 0 \tag{11.76}$$

for all $x \in \mathbf{R}^n$ and a constant $C > 0$ such that $|\nabla_x p_{se}(x)| \leq C|x|$ for all $x \in \mathbf{R}^n$ and $p_{se}(0) = 0$.

THEOREM 11.5.2 *Assume $A^{\times}$ is exponentially hyperbolic, Assumption* 11.4.1, *and p_{se} is a strict infosubequilibrium. Let $u, y \in L_2[0, \infty)$ have* compact support *and $p_0 \in C(\mathbf{R}^n)$, and assume*

$$p_0 \leq p_{se}. \tag{11.77}$$

Then

$$\lim_{t \to \infty} p_t = p_e + c(u, y, p_0) \tag{11.78}$$

uniformly on compact sets, where p_e is given as in Lemma 11.5.1 *and $c(u, y, p_0)$ is a real number depending on u, y, p_0.*

Proof. Suppose $[0, T]$ contains the support of u and y. Then for $0 \leq s \leq T$,

$$\Phi^{u,y}_{s,t}(x) = \begin{cases} \Phi^{u,y}_{s,T} \Phi^{0,0}_{T,t}(x) & \text{if } 0 \leq s \leq T, \\ \\ \Phi^{0,0}_{s,t}(x) & \text{if } T \leq s \leq t. \end{cases}$$

If $x \in M_{as}$, then $\Phi^{0,0}_{T,t}(x) \to 0$ exponentially as $t \to +\infty$, while if $x \notin M_{as}$, then $|\Phi^{0,0}_{T,t}(x)| \to \infty$ as $t \to +\infty$.

For any continuous p_0, using (3.7), we have

$$\begin{aligned} p_t(x) = p_0(\Phi^{u,y}_{0,T} \Phi^{0,0}_{T,t}(x)) + \tfrac{1}{2} \int_0^T [& |C_1(\Phi^{u,y}_{s,T} \Phi^{0,0}_{T,t}(x)) + u(s)|^2 \\ & - \gamma^2 |y(s) - C_2(\Phi^{u,y}_{s,T} \Phi^{0,0}_{T,t}(x))|^2]\, ds \qquad (11.79) \\ + \tfrac{1}{2} \int_T^t [|C_1(\Phi^{0,0}_{s,t}(x))|^2 & - \gamma^2 |C_2(\Phi^{0,0}_{s,t}(x))|^2]\, ds. \end{aligned}$$

Let $x \in M_{as}$. Then the first and second terms in (11.79) tend (uniformly on compact sets) to a real number $c(u, y, p_0)$ given by

$$c(u, y, p_0) = p_0(\Phi^{u,y}_{0,T}(0)) + \tfrac{1}{2} \int_0^T [|C_1(\Phi^{u,y}_{s,T}(0)) + u(s)|^2 - \gamma^2 |y(s) - C_2(\Phi^{u,y}_{s,T}(0))|^2]\, ds, \tag{11.80}$$

and the third member of (11.79) converges to the function

$$\breve{p}(x) = \lim_{t \to +\infty} \tfrac{1}{2} \int_T^t [|C_1(\Phi^{0,0}_{s,t}(x))|^2 - \gamma^2 |C_2(\Phi^{0,0}_{s,t}(x))|^2]\, ds. \tag{11.81}$$

Note that this uses time invariance. Set $\breve{p}^{u,y}(x) = \breve{p}_e(x) + c(u, y, p_0)$ for $x \in M_{as}$. Therefore,

$$\lim_{t\to\infty} p_t(x) = \breve{p}^{u,y}(x) \text{ for } x \in M_{as}. \tag{11.82}$$

Now suppose $x \notin M_{as}$. Since p_{se} is a strict infosubequilibrium, if $p_0 \le p_{se}$ we have

$$\begin{aligned} F(p_{se}, u, y) &= F(p_{se}, 0, 0) + \tfrac{1}{2}|u|^2 - \gamma^2 \tfrac{1}{2}|y|^2 \\ &\quad + (C_1' - \nabla_x p_{se} B_1)u + (\gamma^2 C_2' - \nabla_x p_{se} B_2)y \\ &\le -(\varepsilon - \eta)|x|^2 + \tfrac{C}{\eta}(|u|^2 + |y|^2) \end{aligned}$$

for any $\eta > 0$, where here (and again) we make use of estimates of the type

$$|\nabla_x p_0(\xi) B_1 y| \le \eta |\xi|^2 + \frac{C}{\eta}|y|^2,$$

valid for any $\eta > 0$. By integration along the trajectory $\Phi_{s,t}^{u,y}(x)$, we therefore have

$$\begin{aligned} p_{se}(x) \ge p_{se}(\Phi_{0,t}^{u,y}(x)) &+ \tfrac{1}{2}\int_0^t [|C_1(\phi_{s,t}^{u,y}(x))|^2 - \gamma^2 |C_2(\phi_{s,t}^{u,y}(x))|^2] ds \\ &+ \int_0^t \left[(\varepsilon - \eta)|\phi_{s,t}^{u,y}(x)|^2 - \frac{C}{\eta}(|u(s)|^2 + |y(s)|^2)\right] ds. \end{aligned}$$

Now by definition and inequality (11.77),

$$\begin{aligned} p_t(x) &= p_0(\phi_{0,t}^{u,y}(x)) + \tfrac{1}{2}\int_0^t [|C_1(\phi_{s,t}^{u,y}(x)) + u(s)|^2 - \gamma^2 |y(s) - C_2(\phi_{s,t}^{u,y}(x))|^2]\, ds \\ &\le p_{se}(\phi_{0,t}^{u,y}(x)) + \tfrac{1}{2}\int_0^t [|C_1(\phi_{s,t}^{u,y}(x))|^2 - \gamma^2 |C_2(\phi_{s,t}^{u,y}(x))|^2]\, ds \\ &\quad + \int_0^t \left[\alpha |\phi_{s,t}^{u,y}(x)|^2 + \frac{C}{\alpha}(|u(s)|^2 + |y(s)|^2)\right] ds \end{aligned}$$

for any $\alpha > 0$. Combining these inequalities we get

$$p_t(x) \le -(\varepsilon - \eta - \alpha) \int_0^t |\phi_{s,t}^{u,y}(x)|^2\, ds + C(u, y) + p_{se}(x), \tag{11.83}$$

where $C(u, y) \ge 0$ is a constant depending on the inputs u, y and on η, α. Select η, α so that $\varepsilon - \eta - \alpha > 0$. Then if $x \notin M_{as}$ we have $\phi_{s,t}^{u,y}(x) \to \infty$ as $t \to \infty$, and hence from (11.83) we get $\limsup_{t\to\infty} p_t(x) \le -\infty$.

We can now conclude that $\delta_{M_{as}} + c(u, y, p_0)$ is the limit of p_t as $t \to \infty$. □

CONJECTURE 11.5.3 *The previous Theorem* 11.5.2 *holds for all functions* u, y *in* L^2 *(or at least decaying reasonably) and gives (fairly strong) convergence of* p_t *as in* (11.78).

11.5.2 The Pure Singular Case

We now assume that $A^\times$ is L_2 exponentially stable. This leads to an equilibrium information state $p_e = \delta_0$, which equals $-\infty$ everywhere except at one point $x = 0$; this case is the opposite extreme to that of §11.4, where p_e was finite everywhere.

LEMMA 11.5.4 *The singular function*

$$p_e = \delta_0 \tag{11.84}$$

is an equilibrium solution of the steady state information state equation (11.74).

Proof. Let $x_0 = 0$, $u = 0$, and $y = 0$. Then the solution of (2.6) is $\xi(t) = 0$, and hence by Lemma 10.2.1 $p_t = \delta_0$ for all $t \geq 0$. This proves that δ_0 is a steady state solution of (11.74). □

The next result gives a stability result for the information state when it is initialized at the singular equilibrium $p_e = \delta_0$.

THEOREM 11.5.5 *Assume $A^\times$ is L_2 exponentially stable. Let $u, y \in L_2[0,\infty)$ and $p_0 = p_e$ given by* (11.84). *Then*

$$p_t \Longrightarrow p_e + c(u, y, p_0) \text{ as } t \to \infty, \tag{11.85}$$

where $c(u, y, p_0)$ is a real number depending on u, y.

Proof. For $t > 0$, p_t is given by (10.3):

$$p_t = \delta_{\xi(t)} + \tfrac{1}{2}\int_0^t [|C_1(\xi(s)) + D_{12}(\xi(s))u(s)|^2 - \gamma^2|y(s) - C_2(\xi(s))|^2]\, ds,$$

and $\xi(\cdot)$ is the solution of

$$\dot{\xi} = A^\times(\xi) + B_1(\xi)y + B_2(\xi)u, \tag{2.6$'$}$$

with initial condition $\xi(0) = 0$. Now the L_2 exponential stability of $A^\times$ implies

$$|\xi(t)|^2 \leq C\int_0^t e^{-c(t-s)}(|u(s)|^2 + |y(s)|^2)ds,$$

where $\xi(t) = \Phi_{t,0}^{u,y}(0)$ is the solution of (2.6)$'$. The RHS is integrable and tends to 0 as $t \to \infty$. Hence the integral in the expression for p_t converges to a limit; call it $c(u, y, p_0)$.

It remains to show that

$$\delta_{\xi(t)} \Longrightarrow \delta_0 \text{ as } t \to \infty. \tag{11.86}$$

This can be done in several ways. We use two methods for illustration.

First, we use directly the definition of hypoconvergence given in Appendix C, which is equivalent to the desired weak convergence. Let $x = 0$. Let t_i be any sequence converging to ∞; then for any sequence x_i converging to 0,

$$\limsup_{i\to\infty} \delta_{\xi(t_i)}(x_i) \leq 0 = \delta_0(0).$$

Now also

$$\liminf_{t\to\infty} \delta_{\xi(t)}(\xi(t)) = 0 \geq 0 = \delta_0(0).$$

This shows that $\delta_{\xi(t)}$ hypoconverges to δ_0 at $x = 0$.

Next, let $x \neq 0$. Let t_i be any sequence converging to ∞; then for any sequence x_i converging to 0,

$$\limsup_{i\to\infty} \delta_{\xi(t_i)}(x_i) \leq -\infty = \delta_0(x)$$

since for large i $|x_i| \geq \eta$ for some $\eta > 0$. Now

$$\liminf_{t\to\infty} \delta_{\xi(t)}(x) \geq -\infty = \delta_0(x).$$

This shows that $\delta_{\xi(t)}$ hypoconverges to δ_0 at $x \neq 0$. This proves (11.86).

Our second proof of (11.86) uses the characterization of weak convergence given in Theorem C.5.1. So select a test function $f \in C_b(\mathbf{R}^n)$. Then

$$\delta_x(f) = \langle \delta_x + f \rangle = f(x),$$

and so

$$\delta_{\xi(t)}(f) = f(\xi(t)) \to f(0) = \delta_0(f) \text{ as } t \to \infty.$$

This also proves (11.86). □

REMARK 11.5.6 Note that weak convergence is needed in Theorem 11.5.5; indeed, pointwise convergence does not hold in general (this corrects an error in the statement of [HJ96b, Theorem 4.4]). Consider $x = 0$ and a generic trajectory $\xi(\cdot)$ (using the notation from the proof). Then $\xi(0) = 0$, but $\xi(t) \neq 0$ for $t > 0$. Hence

$$\delta_{\xi(t)}(0) = -\infty < \delta_0(0) = 0 \text{ for all } t > 0$$

implies that

$$\lim_{t\to\infty} \delta_{\xi(t)}(0) = -\infty \neq \delta_0(0) = 0.$$

The weak convergence relaxes the pointwise convergence criteria in an appropriate way (while strengthening in ways related to uniform convergence). ▽

Theorem 11.5.5 says that $p_e \in \mathcal{D}_{attr}(p_e)$, the domain of control attraction that is important for the central controller. However, it gives no information on other possible points in $\mathcal{D}_{attr}(p_e)$. The next theorem characterizes $\mathcal{D}_{attr}(p_e)$ by making use of the adjoint information state system (3.44).

In the 1 and 2A block cases, the adjoint information state is given by

$$q_{tT}(x) = f(\xi(T)) + \tfrac{1}{2}\int_t^T [|C_1(\xi(s)) + D_{12}u(s)|^2 - \gamma^2|y(s) - C_2(\xi(s))|^2]\, ds, \tag{11.87}$$

where $\xi(s)$ satisfies (2.6) for $t \leq s \leq T$ with $\xi(t) = x$. The steady state adjoint PDE (3.47) in the 2A block case reads

$$-\nabla_x q_\infty \cdot A^\times - \tfrac{1}{2}|C_1|^2 + \gamma^2 \tfrac{1}{2}|C_2|^2 = 0. \tag{11.88}$$

Let q_e denote the unique smooth (by Lemma B.2.1) solution of (11.88) satisfying $q_e(0) = 0$. (See (3.48) for an interpretation.)

REMARK 11.5.7 If q_e is a solution of the steady state adjoint information state equation (11.88), then $-q_e$ is a solution of the steady state information state equation (11.74); i.e., $-q_e$ is an equilibrium information state. ▽

LEMMA 11.5.8 *Assume $A^\times$ is L_2 exponentially stable, and let $f \in C_b(\mathbf{R}^n)$. Let $\Phi^{u,y}_{T,t}(x)$ denote the flow of* (2.6). *Then the following conditions hold:*

(i) *There exists a smooth equilibrium solution q_e of the steady state adjoint information state equation* (11.88) *satisfying $q_e(0) = 0$, $|\nabla_x q_e(x)| \le C|x|$.*

(ii) *For $u = 0$, $y = 0$ we have*

$$q_{tT}(x) = q_e(x) - q_e(\Phi^{0,0}_{T,t}(x)) + f(\Phi^{0,0}_{T,t}(x)). \tag{11.89}$$

Further, since under Assumption 11.4.1 *B_1 and B_2 are bounded, then the following conditions hold:*

(iii) *For $u, y \in L_2[0,\infty)$,*

$$|q_{tT}(x) - q_e(x) + q_e(\Phi^{u,y}_{T,t}(x)) - f(\Phi^{u,y}_{T,t}(x))| \le C(u,y)(1+\tfrac{1}{\varepsilon}) + \varepsilon|x|^2 \tag{11.90}$$

for any $0 < \varepsilon \le 1$.

(iv) *For $u, y \in L_2[0,\infty)$,*

$$\lim_{T\to\infty} q_{tT}(x) = q_e(x) + c(u,y)_t + f(0) \tag{11.91}$$

uniformly on compact sets, where $c(u,y)_t$ is a constant depending on u, y, and t.

Proof. Let $u = 0$, $y = 0$, $f = 0$. Then from (11.87) we have

$$q_{0T}(x) = \int_0^T [|C_1(\xi(s))|^2 - \gamma^2 |C_2(\xi(s))|^2]\, ds.$$

By the L_2 exponential stability of $A^\times$, this integral converges uniformly on compact sets to a limit; call it $q_e(x)$. Now $q_{0T}(0) = 0$, and so $q_e(0) = 0$. It is now routine to check that q_e solves (11.88) with the stated properties.

Let $f \in C_b(\mathbf{R}^n)$. Write

$$q_{tT} = q_e + \phi_{tT}.$$

Then

$$\begin{aligned}\tfrac{\partial \phi_{tT}}{\partial t} &= -\nabla_x \phi_{tT} \cdot (A^\times + B_1 u + B_2 y) \\ &\quad -(C_1' + \nabla_x q_e B_1)u - (C_2' + \nabla_x q_e B_2)y - \tfrac{1}{2}|u| + \gamma^2 \tfrac{1}{2}|y|^2\end{aligned}$$

so that

$$\phi_{tT}(x) = f(\xi(T)) - q_e(\xi(T)) + \int_t^T [-\tfrac{1}{2}|u| + \gamma^2 \tfrac{1}{2}|y|^2$$

$$-(C_1(\xi(s))' + \nabla_x q_e(\xi(s)) B_1(\xi(s))) u(s)$$

$$-(C_2(\xi(s))' + \nabla_x q_e(\xi(s)) B_2(\xi(s))) y(s)]\, ds,$$

where $\xi(s) = \Phi^{u,y}_{s,t}(x)$. Setting $u = 0$, $y = 0$ yields (11.89). If $u, y \in L_2$, the integral is bounded by $C(u,y)(1 + \frac{1}{\varepsilon}) + \varepsilon|x|^2$, using estimates similar to those used in the proof of Theorem 11.4.6 provided B_1 and B_2 are bounded. The convergence also follows as in Theorem 11.4.6 (dual arguments). □

THEOREM 11.5.9 *Assume $A^\times$ is L_2 exponentially stable, and let q_e denote the equilibrium adjoint information state defined above. Assume $q_e \geq 0$. Then*

(i)

$$\{p_0 \in \mathcal{X}_e : p_0(x) + q_e(x) \leq -\eta|x|^2 + C \text{ for some } \eta > 0, C \geq 0\} \subset \mathcal{D}^0_{attr}(p_e). \tag{11.92}$$

(ii) *If $p_0 + q_e$ is quadratically tight, $u = 0$, $y = 0$, then p_t is uniformly quadratically tight, and*

$$p_t \Longrightarrow p_e + c(p_0) \text{ as } t \to \infty, \tag{11.93}$$

where $c(p_0)$ is a constant depending on p_0, with $c(p_e) = 0$.

Assume further in Assumption 11.4.1 that B_1 and B_2 are bounded; then

(iii)

$$\{p_0 \in \mathcal{X}_e : p_0(x) + q_e(x) \leq -\eta|x|^2 + C \text{ for some } \eta > 0, C \geq 0\} \subset \mathcal{D}_{attr}(p_e). \tag{11.94}$$

(iv) *If $p_0 + q_e$ is quadratically tight, $u, y \in L_2$, then p_t is uniformly quadratically tight, and*

$$p_t \Longrightarrow p_e + c(u, y, p_0) \text{ as } t \to \infty, \tag{11.95}$$

where $c(u, y, p_0)$ is a constant depending on u, y, and p_0, with $c(0, 0, p_e) = 0$.

Proof. Let $u, y \in L_2$, $f \in C_b(\mathbf{R}^n)$, and choose $p_0 \in \mathcal{X}_e$ with $p_0(x) + q_e(x) \leq -\eta|x|^2 + C$ (note that $p_0(x) \leq -\eta|x|^2 + C$ since $q_e \geq 0$). Assume B_1 and B_2 are bounded. Using the adjoint equilibrium, we can write

$$p_t(x) = p_0(\Phi^{u,y}_{0,t}(x)) + q_e(\Phi^{u,y}_{0,t}(x)) - q_e(x) + O([1 + \tfrac{1}{\varepsilon}] + \varepsilon|x|^2)$$

$$\leq -(\eta - \varepsilon)|\Phi^{u,y}_{0,t}(x)|^2 - q_e(x)$$

for any $0 < \varepsilon \leq 1$, by Lemma 11.5.8. Select $\varepsilon < \eta/2$. Now by Lemma A.2.2 we have $|x|^2 \leq C|\Phi_{0,t}^{u,y}|^2 + C$, and so

$$p_t(x) \leq -(\eta/2)c|x|^2 + c - q_e(x)$$
$$\leq -(\eta/2)c|x|^2 + c$$

since $q_e \geq 0$. This upper bound is uniform in $t \geq 0$, and so p_t is uniformly quadratically tight.

Let $f \in C_b(\mathbf{R}^n)$ be a "test function," and define the functional

$$\Lambda(f) = \lim_{t\to\infty} p_t(f).$$

We claim that $\Lambda(f)$ exists and is finite.

From (3.46) we have

$$p_t(f) = \langle p_t + f\rangle = \langle p_0 + q_{0t}\rangle,$$

so let us examine the function $p_0 + q_{0t}$:

$$\begin{aligned} p_0(x) + q_{0t}(x) &= p_0(x) + q_e(x) - q_e(\Phi_{t,0}^{u,y}(x)) + f(\Phi_{t,0}^{u,y}(x)) + q_{0T}(x) \\ &\quad -(q_e(x) - q_e(\Phi_{t,0}^{u,y}(x)) + f(\Phi_{t,0}^{u,y}(x))) \\ &\leq p_0(x) + q_e(x) - q_e(\Phi_{t,0}^{u,y}(x)) + f(\Phi_{t,0}^{u,y}(x)) \\ &\quad +C(u,y)(1 + \tfrac{1}{\varepsilon}) + \varepsilon|x|^2 \\ &\leq -(\eta - \varepsilon)|x|^2 + C \end{aligned}$$

for any $0 < \varepsilon \leq 1$, by Lemma 11.5.8, the choice of p_0, and using $q_e \geq 0$. Selecting $\varepsilon < \eta/2$, we see that $p_0 + q_{0t}$ is uniformly quadratically tight. Therefore by Theorem C.5.7, the limit

$$\Lambda(f) = \lim_{t\to\infty} \langle p_0 + q_{0t}\rangle$$

exists and is finite. Indeed, by Lemma 11.5.8, this limit is given by

$$\Lambda(f) = \langle p_0 + q_e\rangle + c(u,y)_0 + f(0).$$

Write $c(u,y,p_0) = \langle p_0 + q_e\rangle + c(u,y)_0$; then clearly

$$\Lambda(f) = \delta_0(f) + \langle p_0 + q_e\rangle + c(u,y)_0 = p_e(f) + c(u,y,p_0).$$

By Theorem C.5.3, this completes the proof of parts (iii) and (iv) of the theorem.

To prove parts (i) and (ii), set $u = y = 0$ and relax the requirement that B_1 and B_2 are bounded. Then a simplification of the above argument completes the proof. □

Remark 11.5.10 Theorem 11.5.9 says that the domain of control attraction $\mathcal{D}_{attr}(p_e)$ contains the *quadratic interior* $q.i.D_{q_e}$ of the set

$$D_{q_e} = \{p_0 \in \mathcal{X}_e \ : \ \langle p_0 + q_e \rangle \text{ is finite}\}.$$

Recall that this means that

$$p_0 \in q.i.D_{q_e} \text{ if and only if } p_0 + \varepsilon|\cdot|^2 \in D_{q_e} \text{ for some } \varepsilon > 0.$$

▽

Example 11.5.11 Consider Case i(a) of Examples 11.2.1 and 11.2.3, concerning one-dimensional one block linear and bilinear systems, respectively. In this case,

$$q_e(x) = \tfrac{1}{2}\gamma^2 Y_{ue}^{-1} x^2 = \tfrac{1}{2}\gamma^2(\gamma^{-2}C_1^2 - C_2^2)x^2 \geq 0.$$

In the bilinear case, $B_1(x) = B_1 + B_3 x$ clearly fails to be bounded in x, while in the linear case $B_1(x) = B_1$ is a constant; clearly bounded. Thus Theorem 11.5.9 explains why in the bilinear case $\mathcal{D}_{attr}(p_e) = \{p_e\} \subset, \neq \mathcal{D}^0_{attr}(p_e) = D^{o,q}_{q_e}$, while in the linear case $\mathcal{D}_{attr}(p_e) = \{p_e\} = \mathcal{D}^0_{attr}(p_e) = D^{o,q}_{q_e}$. When $B_1(x)$ has linear growth in x, the quadratic upper bounds can break down when the information state system is driven by nonzero inputs u, y, and uniform tightness is lost as a result (this is indicated in Figure 11.6 where the $Y_I(t)$ trajectory crosses the $Y_I = 0$-axis, which is where the $Y(t)$ trajectory suffers finite escape). When $u = y = 0$, the B_1 term is not excited and plays no role. ▽

11.6 Note

(i) Results of the type presented in this chapter first appeared in [HJ95], [HJ96b]. The key points are the identification of the singular equilibrium information states, use of max-plus convergence, and the detailed quadratic upper limiting property.

Chapter 12

Time Varying Systems

In this book we treated time invariant systems, although most of the theory can be applied to time varying systems. This was primarily to keep the notation from becoming one notch more involved. In this chapter we state results for a time varying plant G.

12.1 Dissipation and the Control Problem

We use notation that is so consistent with that for the time invariant case that there is no point in repeating all the definitions. For example, the plant coefficients are

$$A(t, x), B_1(x, t), B_2(x, t), C_1(x, t), C_2(x, t), D_{ij}(x, t).$$

The closed-loop system is *γ-dissipative* provided there exist a gain $\gamma > 0$ and a bias $\beta(x, t)$ such that

$$\begin{cases} \frac{1}{2}\int_{t_1}^{t_2} |z(s)|^2 \, ds \;\leq\; \gamma^2 \frac{1}{2}\int_{t_1}^{t_2} |w(s)|^2 \, ds + \beta(x(t_1), t_1) \\ \text{for all } \; w \in L_2[t_1, t_2] \text{ and all } t_2 \geq t_1, \end{cases} \tag{12.1}$$

where the trajectory of the time varying plant (2.2) is initialized at time t_1 at $x(t_1)$.

12.2 The Two Equations

Again there are two equations, the dynamic programming PDE or PDI and the information state equations.

12.2.1 The Information State

The information state equation remains the same except the coefficients now have t's in them; p_t is defined by

$$p_t(x) = \sup_{\tilde{w}(\cdot)} \left\{ p_0(\xi(0)) + \frac{1}{2}\int_0^t [|\tilde{z}(s)|^2 - \gamma^2 |\tilde{w}(s)|^2]\, ds \right.$$

$$\left. : \ \xi(t) = x, y(s) = C_2(\xi(s), s) + D_{21}(\xi(s), s)\tilde{w}(s), \ 0 \le s \le t \right\}, \tag{12.2}$$

where $\xi(\cdot)$ satisfies

$$\dot{\xi}(s) \ = \ A(\xi(s), s) + B_1(\xi(s), s)\tilde{w}(s) + B_2(\xi(s), s)u, \tag{12.3}$$

which was the state equation in (2.2), and $\tilde{z}(s) = C_1(\xi(s), s) + D_{12}(\xi(s), s)u(s)$.

The dynamics are

$$\dot{p}_t = F(p_t, u(t), y(t), t), \tag{12.4}$$

where $F(p, u, y, t)$ is the nonlinear first-order partial differential operator

$$\begin{aligned} F(p, u, y, t) &= -\nabla_x p \cdot (A(\cdot, t) \\ &+ B_1(\cdot, t) D_{21}(\cdot, t)' E_2(\cdot, t)^{-1}(y - C_2(\cdot, t)) + B_2(\cdot, t)u) \\ &+ \frac{1}{2\gamma^2} \nabla_x p B_1(\cdot, t)[I - D_{21}(\cdot, t)' E_2(\cdot, t)^{-1} D_{21}(\cdot, t)] B_1(\cdot, t)' \nabla_x p' \\ &+ \frac{1}{2}|C_1(\cdot, t) + D_{12}(\cdot, t)u|^2 - \gamma^2 \frac{1}{2}(y - C_2(\cdot, t))' E_2(\cdot, t)^{-1}(y - C_2(\cdot, t)). \end{aligned} \tag{12.5}$$

12.2.2 Dynamic Programming Equation

The time varying dynamic programming PDE is no longer a steady state equation, but is time varying. The value function is defined by

$$W(p, t_1) = \inf_K \sup_{t_2 \ge t_1,\, y \in L_2[t_1, t_2]} \{\langle p_{t_2} \rangle : p_{t_1} = p\}. \tag{12.6}$$

The corresponding dynamic programming PDE is

$$\frac{\partial}{\partial t} W(p, t) + \inf_{u \in \mathbf{R}^m} \sup_{y \in \mathbf{R}^p} \nabla_p W(p, t)[F(p, u, y, t)] = 0, \tag{12.7}$$

and PDI

$$\frac{\partial}{\partial t} W(p, t) + \inf_{u \in \mathbf{R}^m} \sup_{y \in \mathbf{R}^p} \nabla_p W(p, t)[F(p, u, y, t)] \le 0. \tag{12.8}$$

Let $\mathbf{u}^*(p, t), \mathbf{y}^*(p, t)$ denote optimizers here:

$$\mathbf{u}^*(p, t) = \nabla_p W(p, t)[E_1^{-1}(\cdot, t)(-D_{12}(\cdot, t)' C_1(\cdot, t) + B_2(\cdot, t)' \nabla_x p')],$$

$$\mathbf{y}^*(p, t) = \nabla_p W(p, t)[C_2(\cdot, t) - \gamma^{-2} D_{21}(\cdot, t) B_1(\cdot, t)' \nabla_x p'].$$

The state feedback PDI reads

$$\begin{aligned}&\frac{\partial}{\partial t}V(x,t)+\nabla_x V(x,t)\left(A(x,t)-B_2(x,t)E_1(x,t)^{-1}D_{12}'(x,t)C_1(x,t)\right)\\&\quad+\tfrac{1}{2}\nabla_x V(x,t)\left(\gamma^{-2}B_1(x,t)B_1(x,t)'-B_2(x,t)E_1(x,t)^{-1}B_2(x,t)'\right)\nabla_x V(x,t)'\\&\qquad+\tfrac{1}{2}C_1(x,t)'(I-D_{12}(x,t)E_1(x,t)^{-1}D_{12}(x,t)')C_1(x,t)\le 0.\end{aligned}\tag{12.9}$$

The optimal state feedback control is

$$\mathbf{u}^*_{state}(x,t)=-E_1(x,t)^{-1}(D_{12}(x,t)'C_1(x,t)+B_2(x,t)'\nabla_x V(x,t)'). \tag{12.10}$$

12.3 The Controller Construction

The information state controller K^*_{p,t_1} initialized at time t_1 at state p is defined by

$$u(t)=K^*_{p,t_1}[y](t)=\mathbf{u}^*(p_t[y],t) \tag{12.11}$$

for $t \geq t_1$, given a solution $W(p,t)$ of (12.8) and having found the corresponding optimizer $\mathbf{u}^*(p,t)$.

12.3.1 Properties of the Controller

The structural properties are again important:

(i) *Domination.* $W(p,t) \geq \langle p \rangle$ for all $p \in \mathcal{X}_e$, and in particular, for all $p \in \operatorname{dom} W(\cdot,t)$.

(ii) *Monotonicity.* If p_1 is in $\operatorname{dom} W(\cdot,t)$, and if $p_1 \geq p_2$, then p_2 is in $\operatorname{dom} W(\cdot,t)$ and

$$W(p_1,t) \geq W(p_2,t).$$

(iii) *Additive homogeneity.* For any constant $c \in \mathbf{R}$, we have

$$W(p \pm c, t) = W(p,t) \pm c. \tag{12.12}$$

Conditions guaranteeing that the construction solves the H^∞ control problem are given in the next theorem.

THEOREM 12.3.1 *Assume that a p-smooth function W satisfies the structural conditions (domination, monotonicity, and additive homogeneity) and solves the PDI* (12.8). *If the information state controller K^*_{p,t_1} constructed from W exists for all $t \geq t_1$, then the closed-loop system (G, K^*_{p,t_1}) is dissipative; i.e.,* (12.15) *holds (with $\beta(x,t) = -p(x) + W(p,t)$).*

Proof. This follows very closely the lines of proofs in Chapter 4. □

The necessity of the conditions in Theorem 12.3.1 is a theory similar to those of Theorem 4.3.1, with similar proofs.

THEOREM 12.3.2 *Assume that the γ-dissipative control problem for a time varying plant G is solved by an admissible controller K_0. Then the value function $W(p,t)$ defined on* $\mathrm{dom}W(\cdot,t) \subset \mathcal{X}_e$ *by* (12.6) *enjoys the following properties:*

(i) $\mathrm{dom}\, W(\cdot,t)$ *is nonempty, and so* $W(\cdot,t)$ *is finite on a nonempty set* $\mathrm{dom}\, W(\cdot,t)$.

(ii) W *satisfies the structural properties.*

(iii) $W(-\beta_{K_0}(\cdot,t),t) = 0$.

(iv) *The dynamic programming principle holds: for any* $t_2 \geq t_1$, $p \in \mathrm{dom}\, W(\cdot,t_1)$,

$$W(p,t_1) = \inf_K \sup_{y \in L_2[t_1,t_2]} \{W(p_{t_2},t_2) : p_{t_1} = p\}. \tag{12.13}$$

(Identity (12.13) *is called the* dynamic programming equation.)

(v) *If for* $p \in \mathrm{dom}\, W(\cdot,t_1)$ *the controller* $\tilde{K}^*_{p,t_1}$ *exists, then* $W(p_t,t)$ *decreases along any optimal closed-loop trajectory determined by* $w \in L_{2,loc}$, $x_{t_1} \in \mathbf{R}^n$*:*

$$W(p_{t_1},t_1) \geq W(p_{t_2},t_2), \tag{12.14}$$

where $u(t) = \tilde{K}^*_p[y](t)$, $p_{t_1} = p$, and $p_{t_2} \in \mathrm{dom}\, W(\cdot,t_2)$ for all $t_2 \geq t_1$.

Moreover, the closed-loop system $(G, \tilde{K}^*_{p,t_1})$ is dissipative for all $x_0 \in \mathbf{R}^n$; i.e.,

$$\begin{cases} \frac{1}{2}\int_{t_1}^{t_2} |z(s)|^2\, ds \;\leq\; \gamma^2 \frac{1}{2}\int_{t_1}^{t_2} |w(s)|^2\, ds - p(x_{t_1},t_1) + W(p,t_1) \\ \textit{for all } \; w \in L_2[t_1,t_2] \textit{ and all } t_2 \geq t_1. \end{cases} \tag{12.15}$$

For perspective observe that these theorems can be proved with a modest portion of Chapter 4. There are two reasons for this simplicity. The most profound is that we have not needed to analyze the asymptotics of the information state equation or of the dynamic programming PDI. To complete the time invariant theory one must do this, so the time invariant theory in some respects is harder than the time varying theory. Also domains in which p_t lies are swept into our assumptions here, while in Chapter 4 they added lots of complication. The second reason for complication is that we have assumed p_0 is everywhere finite and not treated the technically difficult case of singular p's.

Note also that stability can be inferred from dissipation provided a suitable detectability condition is satisfied.

12.3.2 Certainty Equivalence

The certainty equivalence control is given by

$$u_{ce}(t) = \mathbf{u}^*_{state}(\bar{x}(t),t), \tag{12.16}$$

where $\mathbf{u}^*_{state}(x,t)$ is the optimal state feedback given by (12.10), and $\bar{x}(t)$ is the unique minimizer of $p_t(x) + V(x,t)$. The certainty equivalence assumptions and theorems of §7.3 generalize in the obvious way.

12.4 Equilibrium Information States

Equilibrium information states do not make sense for time varying systems without some strong assumptions about the asymptotics of the plant coefficients A, B_1, etc., themselves. We have not pursued this, though it is clearly a topic that could provide someone with endless fun.

Appendix A

Differential Equations and Stability

A.1 Differential Equations

Consider the finite-dimensional nonlinear system

$$\dot{x}(s) = f(x(s)) + g(x(s))v(s), \quad x(t) = x, \tag{A.1}$$

with inputs $v \in L_2[0, \infty)$. We shall assume that f and g are globally Lipschitz continuous:

$$\begin{aligned} |f(x_1) - f(x_2)| &\leq C_f|x_1 - x_2|, \\ |g(x_1) - g(x_2)| &\leq C_g|x_1 - x_2|. \end{aligned} \tag{A.2}$$

The constants $C_f, C_g \geq 0$ do not depend on x_1, $x_2 \in \mathbf{R}^n$. We also assume that $f(0) = 0$ so that $x = 0$ is an equilibrium. In view of these assumptions, f and g grow at most linearly in x:

$$\begin{aligned} |f(x)| &\leq C_f|x|, \\ |g(x)| &\leq C_{gg}(1 + |x|). \end{aligned} \tag{A.3}$$

Let $\Phi^v_{s,t}(x)$ denote the *transition operator* associated with (A.1). Our convention is that $\Phi^v_{s,t}(x)$ denotes the solution at time s from x at time t, so that $\Phi^v_{t,t}(x) = x$ for any t. In particular, if $t \geq 0$, then $\Phi^v_{t,0}(x)$ denotes the solution at time t starting from x at time 0 (Figure 2.1), and $\Phi^v_{0,t}(x)$ denotes the solution at time t starting from x at time 0 (Figure 2.2). From the theory of ODEs, $\Phi^v_{s,t}(x)$ is uniquely defined, is smooth in x (at least C^1) and s, t (at least absolutely continuous), and is a diffeomorphism in x.

The following estimate measures the influence of the input v on the state trajectory for finite time durations.

LEMMA A.1.1 *If f and g are globally Lipschitz continuous with g bounded, then the solution of* (A.1) *satisfies*

$$|x(s)| \leq Ce^{ct}(1 + |x(0)|), \quad 0 \leq s \leq t \tag{A.4}$$

$(c \geq 0,\ C \geq 0)$ *and*

$$|x(t)|^2 \leq 2\left(|x(s)|^2 + C_{gg}^2(t-s)\int_s^t |v(r)|^2 dr\right) e^{2C_f(t-s)} \tag{A.5}$$

$(C_{gg} \geq 0)$.

Proof. These estimates follow from *Gronwall's inequality* (e.g., [MM82, p. 75]): If r, k, l are real and continuous functions that satisfy $r(t) \geq 0$, $k(t) \geq 0$, and

$$r(t) \leq l(t) + \int_a^t k(s)r(s)ds, \quad a \leq t \leq b,$$

then

$$r(t) \leq l(t) + \int_a^t l(s)k(s)\exp\left[\int_s^t k(\tau)d\tau\right] ds, \quad a \leq t \leq b.$$

□

In several proofs in this book we have needed to compare trajectories of two systems of the form (A.1), one of which is driven by inputs that decay and both of which may be driven by inputs that persist. The following lemma presents some estimates needed for such comparisons.

LEMMA A.1.2 *Fix $v \in L_{2,e}$ and $w \in L_2$. For $T \geq 0$ define*

$$v_T(s) = v(s-T),$$

the time delay of v by T. Let $\Phi_{s,t}^{w,v}(x)$ denote the semigroups associated with the system

$$\dot{x}(s) = f(x(s)) + g(x(s))v(s) + g_0(x(s))w(s), \quad x(t) = x, \tag{A.6}$$

where g and g_0 are globally Lipschitz continuous and bounded. Then

$$\lim_{T\to\infty} \Phi_{T+s,T+t}^{w,v_T}(x) = \Phi_{s,t}^{0,v}(x) \tag{A.7}$$

uniformly in x and on compact subsets of $t \geq 0$ for $0 \leq s \leq t$.

Proof. By time shifting, we need to prove that

$$\lim_{T\to\infty} \{\Phi_{s,T+t}^{w,v_T}(x) - \Phi_{s,T+t}^{0,v_T}(x)\} = 0 \tag{A.8}$$

for $T \leq s \leq T+t$.

Write $x(s) = \Phi_{s,T+t}^{0,v_T}(x)$, $\xi(s) = \Phi_{s,T+t}^{w,v_T}(x)$, and $e(s) = x(s) - \xi(s)$. Then for $s \in [T, T+t]$ we have

$$\begin{aligned} \tfrac{d}{ds}|e(s)|^2 &= e(s)'(f(x(s)) - f(\xi(s)) + g(x(s))v_T(s) \\ &\qquad -g(\xi(s))v_T(s) - g_0(\xi(s))w(s)) \\ &\geq -K|e(s)|^2(1 + |v_T(s)|) - K|w(s)|. \end{aligned}$$

This implies, using an integrating factor,

$$|e(s)| \leq \int_s^{T+t} K|w(r)|e^{-K\int_r^{T+t}(1+|v_T(\tau)|)d\tau}dr$$

and hence

$$|e(s)| \leq K(t,v)\left(\int_T^{T+t}|w(r)|^2 dr\right)^{1/2} \to 0 \text{ as } T \to \infty.$$

□

A.2 Stability

The following lemmas use the monotone stability definition (2.18) to obtain various stability and antistability results.

Lemma A.2.1 *Let $\Phi^v_{s,t}(x)$ be the semigroup corresponding to* (A.1) *with inputs $v \in L_2[0,\infty)$. Assume that f is monotone antistable ($-f$ satisfies* (2.18)), *and that f, g satisfy the above assumptions (global Lipschitz continuity* (A.2), *linear growth* (A.3), $f(0) = 0$).

(i) *Then there exist constants $C > 0$, $c > 0$ such that*

$$|\Phi^v_{s,t}(x)| \leq Ce^{-c(t-s)/2}|x| + C\left(\int_s^t e^{-c(r-s)}|v(r)|^2 dr\right)^{1/2}, \quad \text{(A.9)}$$

(ii)

$$\int_T^t |\Phi^v_{s,t}(x)|^2 ds \leq C|x|^2 + C\int_T^t |v(s)|^2 ds. \quad \text{(A.10)}$$

The constants $c > 0$, $C = C(v) \geq 0$ in (A.9), (A.10) *do not depend on x, but $C = C(v)$ does depend continuously on $\| v \|_{L_2}$, and c is independent of v. If g is bounded in x, then C is independent of v.*

(iii)

$$|\Phi^{v_1}_{s,t}(x_1) - \Phi^{v_2}_{s,t}(x_2)|$$

$$\leq C|x_1 - x_2|e^{-c(t-s)/2} + C(\textstyle\int_s^t e^{-c(r-s)}|v_1(r) - v_2(r)|^2 dr)^{1/2}. \quad \text{(A.11)}$$

The constant c in (A.11) *is independent of x_1, x_2, v_1, v_2. The constant $C \geq 0$* (A.11) *may depend continuously on $|x_1|$, $|x_2|$, $\| v_1 \|_{L_2}$, $\| v_2 \|_{L_2}$. If g is bounded in x, then C does not depend on x_1, x_2, and if g is independent of x, then C does not depend on v_1, v_2.*

(iv)

$$|\nabla_x \Phi^v_{s,t}(x)| \leq Ce^{-c(t-s)/2}. \quad \text{(A.12)}$$

The constant $c > 0$ in (A.12) is independent of v, x. The constant $C \geq 0$ in (A.12) may depend continuously on $\| v \|_{L_2}$, $|x|$. If g is independent of x, then C does not depend on x, and if g does not depend on x, then C is independent of v.

Proof. Part (i). Fix $x \in \mathbf{R}^n$, $t > s$, and write $\xi(s) = \Phi^v_{s,t}(x)$. Then ($C > 0$, $c > 0$, etc., will denote generic constants)

$$\begin{aligned}\tfrac{d}{ds}|\xi(s)|^2 &= \xi(s)'(f(\xi(s)) + g(\xi(s))v(s)) \\ &\geq c|\xi(s)|^2 - C|\xi(s)|(1 + |\xi(s)|)|v(s)| \\ &\geq c|\xi(s)|^2 - C|\xi(s)|^2|v(s)| - C\varepsilon|\xi(s)|^2 - \tfrac{C}{\varepsilon}|v(s)|^2 \\ &\geq \tfrac{c}{2}(1 - C_0|v(s)|)|\xi(s)|^2 - C|v(s)|^2\end{aligned}$$

for $\varepsilon > 0$ sufficiently small. Using the integrating factor

$$e^{\int_s^t c(1-C_0|v(\tau)|)d\tau}$$

and integrating we get

$$e^{-ct}\tfrac{1}{2}|x|^2 - e^{\int_s^t c(1-C_0|v(\tau)|)d\tau}\tfrac{1}{2}|\xi(s)|^2 \geq -\int_s^t Ce^{\int_r^t c(1-C_0|v(\tau)|)d\tau}|v(r)|^2 dr$$

and hence

$$|\xi(s)|^2 \leq e^{-\int_s^t c(1-C_0|v(\tau)|)d\tau}|x|^2 + \int_s^t Ce^{-\int_s^r c(1-C_0|v(\tau)|)d\tau}|v(r)|^2 dr.$$

Now

$$\int_s^r |v(\tau)|d\tau \leq (r - s)^{1/2} \| v \|_{L_2},$$

and so (writing $C_1(v) = \| v \|_{L_2}$)

$$\begin{aligned}-\textstyle\int_s^r c(1 - C_0|v(\tau)|)d\tau &\leq -c(r - s) + C_1(v)(r - s)^{1/2} \\ &\leq -c(1 - C_1(v)\eta)(r - s) + C_1(v)/\eta.\end{aligned}$$

If one selects $\eta = \frac{c}{2C_1(v)} > 0$, then

$$-\int_s^r c(1 - C_0|v(\tau)|)d\tau \leq -\frac{c}{2}(r - s) + \frac{2C_1(v)^2}{c}.$$

Therefore,

$$|\xi(s)|^2 \leq e^{-\frac{c}{2}(t-s)}C_2(v)|x|^2 + C_2(v)\int_s^t Ce^{-\frac{c}{2}(r-s)}|v(r)|^2 dr.$$

This proves (A.9) (on redefining the various constants, as appropriate). Notice that the constant $C_2(v) = Ce^{C\|v\|^2_{L_2}}$ is a continuous function of $\| v \|_{L_2}$ and is independent of v when g is bounded.

Part (ii). The second inequality (A.10) follows by integration, where it is noted that

$$\int_T^t \int_s^t Ce^{-c(r-s)}|v(r)|^2 dr ds \leq C \int_T^t |v(r)|^2 dr$$

(this can be seen by integrating from changing the order of integration).

Part (iii). Let $\zeta(s) = \Phi^{v_1}_{s,t}(x_1) - \Phi^{v_2}_{s,t}(x_2)$, $\xi_1(s) = \Phi^{v_1}_{s,t}(x_1)$, and $\xi_1(s) = \Phi^{v_2}_{s,t}(x_2)$. Then we have

$$\begin{aligned}
\tfrac{d}{ds}\tfrac{1}{2}|\zeta(s)|^2 &= \zeta(s)'(f(\xi_1(s)) - f(\xi_2(s)) + g(\xi_1(s))v_1(s) - g(\xi_2(s))v_2(s)) \\
&\geq c|\zeta(s)|^2 - C|\zeta(s)|(|g(\xi_1(s)) - g(\xi_2(s))||v_1(s)| \\
&\qquad + |g(\xi_2(s))||v_1(s) - v_2(s)|) \\
&\geq c|\zeta(s)|^2 - C|\zeta(s)|^2|v_1| - C_1|\zeta(s)||v_1(s) - v_2(s)| \\
&\geq \tfrac{c}{2}(1 - C_0|v_1(s)|)|\zeta(s)|^2 - C|v_1(s) - v_2(s)|^2.
\end{aligned}$$

Then continuing as in part (i), one proves (A.11). (Note that $C_1 = C_1(x_2, v_2)$.)

Part (iv). Let $\zeta(s) = \nabla_x \Phi^v_{s,t}(x)$. Then

$$\frac{d}{ds}\tfrac{1}{2}|\zeta(s)|^2 = \zeta(s)'(\nabla_x f(\Phi^v_{s,t}(x)) + \nabla_x g(\Phi^v_{s,t}(x))v)\zeta(s) \geq \tfrac{1}{2}c|\zeta(s)|^2 - C|\zeta(s)|^2|v|,$$

which is enough to prove (A.12). □

LEMMA A.2.2 *Let $\Phi^v_{s,t}(x)$ be the semigroup corresponding to* (A.1) *with inputs $v \in L_2[0,\infty)$. Assume that f is monotone stable (f satisfies* (2.18)*), and that f, g satisfy the above assumptions (global Lipschitz continuity* (A.2)*, linear growth* (A.3)*, $f(0) = 0$).*

(i) *Then there exist constants $C > 0$, $c > 0$ such that*

$$|\Phi^v_{t,s}(x)| \leq Ce^{-c(t-s)/2}|x| + C\left(\int_s^t e^{-c(t-r)}|v(r)|^2 dr\right)^{1/2}, \qquad \text{(A.13)}$$

(ii)

$$\int_s^T |\Phi^v_{t,s}(x)|^2 dt \leq C|x|^2 + C\int_s^T |v(t)|^2 dt. \qquad \text{(A.14)}$$

The constants $c = c(v) > 0$, $C = C(v) \geq 0$ in (A.13), (A.14) *do not depend on x, but $C = C(v)$ does depend continuously on $\| v \|_{L_2}$, and c is independent of v. If g is bounded in x, then C is independent of v.*

(iii)

$$|\Phi^{v_1}_{t,s}(x_1) - \Phi^{v_2}_{t,s}(x_2)|$$

$$\leq C|x_1 - x_2|e^{-c(t-s)/2} + C(\textstyle\int_s^t e^{-c(t-r)}|v_1(r) - v_2(r)|^2 dr)^{1/2}. \quad \text{(A.15)}$$

The constant c in (A.15) *is independent of x_1, x_2, v_1, v_2. The constant $C \geq 0$ in* (A.15) *may depend continuously on $|x_1|$, $|x_2|$, $\| v_1 \|_{L_2}$, $\| v_2 \|_{L_2}$. If g is bounded in x, then C does not depend on x_1, x_2, and if g is independent of x, then C does not depend on v_1, v_2.*

(iv)

$$|\nabla_x \Phi^v_{s,t}(x)| \leq Ce^{-c(t-s)/2}. \quad \text{(A.16)}$$

The constant $c > 0$ in (A.16) *is independent of v, x. The constant $C \geq 0$ in* (A.16) *may depend continuously on $\| v \|_{L_2}$, $|x|$. If g is independent of x, then C does not depend on x, and if g does not depend on x, then C is independent of v.*

Proof. The proof of this lemma is similar to that of Lemma A.2.1 and is omitted. □

Appendix B

Nonlinear PDE and Riccati Equations

In this book we have made extensive use of nonlinear PDEs of the Hamilton–Jacobi type. For example, the fundamental PDE governing the information state, viz.,

$$\frac{dp}{dt} = -\nabla_x p \cdot (A + B_1 D_{21}' E_2^{-1}(y - C_2) + B_2 u)$$
$$+\tfrac{1}{2\gamma^2}\nabla_x p B_1 [I - D_{21}' E_2^{-1} D_{21}] B_1' \nabla_x p' \qquad (3.9)'$$
$$+\tfrac{1}{2}|C_1 + D_{12}u|^2 - \gamma^2 \tfrac{1}{2}(y - C_2)' E_2^{-1}(y - C_2),$$

is nonlinear and time varying in general, with stationary version

$$-\nabla_x p_\infty \cdot (A - B_1 D_{21}' E_2^{-1} C_2) + \tfrac{1}{2\gamma^2}\nabla_x p_\infty B_1 [I - D_{21}' E_2^{-1} D_{21}] B_1' \nabla_x p_\infty$$
$$+\tfrac{1}{2}|C_1|^2 - \gamma^2 \tfrac{1}{2} C_2' E_2^{-1} C_2 = 0. \qquad (3.28)'$$

Another important class of examples relates to dissipation inequalities and the state feedback PDE

$$\nabla_x V\left(A - B_2 E_1^{-1} D_{12} C_1\right) + \tfrac{1}{2}\nabla_x V\left(\gamma^{-2} B_1 B_1' - B_2 E_1^{-1} B_2'\right)\nabla_x V'$$
$$+\tfrac{1}{2} C_1' (I - D_{12} E_1^{-1} D_{12}') C_1 = 0. \qquad (5.45)$$

Other nonlinear PDE have also been used, e.g., (11.52), etc. The purpose of this appendix is to provide some background information concerning PDEs that have Hamilton–Jacobi structure, such as these examples; in particular, we discuss connections with optimal control and games, viscosity solutions, uniqueness, representations, and so-called nonlinear Riccati PDEs.

B.1 Background

The material in this section is drawn from [FR75], [FS93], [Jam94a]. The notation used is not necessarily related to the notation used in the rest of this book. See also [BCD98].

B.1.1 Optimal Control

We discuss a simple example from optimal control.

Suppose one wants to minimize the cost functional

$$J(t,x;u(\cdot)) = \int_t^{t_1} L(x(s),u(s))\,ds + \psi(x(t_1)), \tag{B.1}$$

where $x(\cdot)$ is the solution of the initial value problem

$$\begin{cases} \dot{x}(s) = f(x(s),u(s)), \quad t \le s \le t_1, \\ x(t) = x. \end{cases} \tag{B.2}$$

Here, $u(\cdot)$ is a control defined on $[t, t_1]$ taking values in, say, $\mathbf{R}^m$, and $x(\cdot)$ is the state trajectory in $\mathbf{R}^n$.

The value function is defined by

$$V(t,x) = \inf_{u(\cdot)} J(t,x;u(\cdot)) \tag{B.3}$$

for $(t,x) \in [t_0, t_1] \times \mathbf{R}^n$, and the *dynamic programming principle* states that for every $r \in [t, t_1]$,

$$V(t,x) = \inf_{u(\cdot)} \left[\int_t^r L(x(s),u(s))\,ds + V(r,x(r)) \right]. \tag{B.4}$$

From this, one can derive formally the equation

$$\frac{\partial}{\partial t}V(t,x) + H(x, \nabla_x V(t,x)) = 0 \text{ in } (t_0,t_1) \times \mathbf{R}^n, \tag{B.5}$$

with terminal data

$$V(t_1,x) = \psi(x) \text{ in } \mathbf{R}^n. \tag{B.6}$$

Here, the *Hamiltonian* is given by

$$H(x,\lambda) = \inf_{v \in \mathbf{R}^m} \{\lambda \cdot f(x,v) + L(x,v)\}. \tag{B.7}$$

The nonlinear first-order PDE (B.5) is the *dynamic programming PDE* or HJB equation. Notice that the Hamiltonian is *concave* in the variable λ (since it is the infimum of linear functions).

Let us see how (B.5) is obtained. Set $r = t + h$, $h > 0$, and rearrange (B.4) to yield

$$\inf_{u(\cdot)} \left[\frac{1}{h}(V(t+h, x(t+h)) - V(t,x)) + \frac{1}{h}\int_t^{t+h} L(x(s),u(s))\,ds \right] = 0.$$

If V and $u(\cdot)$ are sufficiently smooth, then

$$\frac{1}{h}(V(t+h, x(t+h)) - V(t,x)) \to \frac{\partial}{\partial t}V(x,t) + \nabla_x V(x,t) \cdot f(x, u(t)) \text{ as } h \to 0$$

and

$$\frac{1}{h}\int_t^{t+h} L(x(s), u(s))\, ds \to L(x, u(t)) \text{ as } h \to 0.$$

Combining these displays, one is formally led to (B.5). A proof of (B.5) when V is sufficiently smooth requires a careful derivation of two inequalities that combine to give (B.5).

The utility of the value function and the HJB equation is as follows. If there exists a $C^1([t_0, t_1] \times \mathbf{R}^n)$ solution $\tilde{V}$ to (B.5), (B.6) (i.e., the DPE has a classical solution), and if $u^*(\cdot) \in L^\infty$ is such that

$$u^*(s) \in \operatorname*{argmin}_{v \in \mathbf{R}^m} \left\{ \nabla_x \tilde{V}(s, x^*(s)) \cdot f(x^*(s), v) + L(x^*(s), v) \right\},$$

then $u^*(\cdot)$ is optimal, and $x^*(\cdot)$ is the corresponding optimal state trajectory. In addition, $\tilde{V}(t,x) = V(t,x)$. This is the content of the well-known verification theorem of optimal control. The RHS of the preceding display gives a formula for the optimal feedback control.

As remarked earlier, in general $V \notin C^1([t_0, t_1] \times \mathbf{R}^n)$; indeed V may not be everywhere differentiable, in which case (B.5) does not have a classical solution, and the verification theorem does not hold (at least in this form; see [Cla83]). In spite of this, the DPE is still of fundamental importance, and in fact V turns out to be the unique viscosity solution of (B.5), (B.6), in a wide range of applications.

There is an important special case where the control is affine in the drift and quadratic in the cost:

$$f(x,u) = f(x) + g(x)u, \quad L(x,u) = \ell(x) + \tfrac{1}{2}|u|^2.$$

In this case the DPE becomes

$$\frac{\partial}{\partial t}V + \nabla_x V \cdot f - \tfrac{1}{2}\nabla_x V g g' \nabla_x V' + \ell = 0 \tag{B.8}$$

and the optimal feedback control can be explicitly evaluated:

$$u^*(x,t) = -g(x)'\nabla_x V(x,t)'. \tag{B.9}$$

The PDEs appearing in this book are analogous to this type.

In the theory of two-player zero-sum differential games, one has two opposing players, say, u and w. The player u seeks to minimize a cost functional, whereas w aims to maximize it. In this context, the dynamic programming equation has a Hamiltonian that is an $\inf_u$, $\sup_w$ of a linear function and is not concave or convex (see (5.45)). Such a dynamic programming equation is called a *Hamilton–Jacobi–Bellman–Isaacs* (HJBI) equation.

We remark that HJBI-type PDEs arise in many ways and may appear different than (B.5). For instance, signs may be different, or infimums may be replaced by supremums, or time may be reversed.

B.1.2 Nonlinear PDE

The method of characteristics is a classical method for solving equations of the type (B.5), as well as other first-order equations. This method constructs smooth functions that satisfy the PDE locally, in certain regions by integrating along the characteristic curves (obtained by solving Hamilton's equations). In general, a globally defined smooth classical solution cannot be obtained. Typically, the derivatives may suffer discontinuities across certain lower-dimensional sets. This phenomenon is well known in the calculus of variations.

There have been a number of approaches aiming to weaken the notion of solution to nonlinear first-order PDEs. Instead of seeking a solution $\tilde{V}$ in C^1, one may generalize to $C^{0,1}_{loc}$ (locally Lipschitz continuous), since in many cases it turns out that the value function V satisfies this weaker regularity property and hence is differentiable almost everywhere (by Rademacher's theorem). Thus if V is locally Lipschitz, then V satisfies (B.5) at almost every (t, x). Unfortunately, there may be more than one generalized solution in $C^{0,1}_{loc}$, as the following example illustrates.

Consider the equation

$$\begin{cases} -\frac{\partial}{\partial t}V + |\nabla_x V|^2 = 0 \text{ in } (0,1)\times \mathbf{R}, \\ V(x,1) = 0 \text{ in } \mathbf{R}. \end{cases} \tag{B.10}$$

The function

$$V^1(t,x) \equiv 0$$

is a classical solution, while

$$V^2(t,x) = \begin{cases} 0 & \text{if } |x| > 1-t, \\ x-1+t & \text{if } 0 < x < 1-t, \\ -x-1+t & \text{if } -1+t < x < 0, \end{cases}$$

is a locally Lipschitz generalized solution.

A satisfactory notion of solution should enjoy a uniqueness property. One pre-viscosity uniqueness result concerns equations with convex Hamiltonian H, as in the example. Equation (B.10) has a unique semiconcave generalized solution, namely, V^1. In fact, this solution is given by Hopf's formula:

$$V^1(t,x) = \min_{y\in\mathbf{R}} \left[(1-t)\frac{1}{4}\left|\frac{x-y}{1-t}\right|^2 \right] = 0.$$

This variational formula is actually a special case of an optimal control or calculus of variations representation. Equation (B.10) can be interpreted as a DPE for an optimal control problem with $f(x,u) = u$, $L(x,u) = u^2/4$, $\psi = 0$, with value function V^1:

$$V^1(t,x) = \inf_{u(\cdot)} \left\{ \int_t^1 \frac{1}{4}|u(s)|^2\, ds \right\} = 0.$$

Thus variational formulas "pick out the right solution." This observation is relevant to the theory of viscosity solutions, as value functions turn out to be viscosity solutions. The viscosity solution definition also singles out the "right solution," and it is very important to note that convexity or semiconcavity assumptions are not required, and hence the concept applies more generally (in particular, it applies to HJI equations arising in differential games).

B.1.3 Viscosity Solutions

Let us rewrite the DPE as follows:

$$-\frac{\partial}{\partial t}V(t,x) + H(x, \nabla_x V(t,x)) \;=\; 0 \text{ in } (t_0,t_1)\times \mathbf{R}^n, \tag{B.5$'$}$$

with a new definition of the Hamiltonian

$$H(x,\lambda) \;=\; \sup_{v\in\mathbf{R}^m} \{-\lambda\cdot f(x,v) - L(x,v)\}\,. \tag{B.7$'$}$$

The sign convention used in (B.5)$'$ relates to the maximum principle in PDE, and note that the Hamiltonian is now *convex* in λ.

A function $\tilde{V} \in C([t_0,t_1]\times\mathbf{R}^n)$ is a viscosity subsolution (respectively, supersolution) of (B.5)$'$ if for all $w \in C^\infty((t_0,t_1)\times\mathbf{R}^n)$,

$$-\frac{\partial}{\partial t}w(t',x') + H(x', \nabla_x w(t',x')) \;\le\; 0 \;\;(\text{resp.}, \;\ge 0) \tag{B.11}$$

at every point (t',x') where $\tilde{V} - w$ attains a local maximum (respectively, minimum). $\tilde{V}$ is a viscosity solution if it is both a subsolution and a supersolution.

There are a number of equivalent definitions, and the one just stated first appeared in [CEL84] and offers advantages such as ease of use (despite apparent awkwardness). The functions $w \in C^\infty$ play the role of "test functions," and note that the derivatives of w replace those of $\tilde{V}$, which need not exist. Thus differentiation has been transferred to smooth functions, at the expense of a pair of inequalities. (Recall that in the theory of linear PDEs, this is achieved using integration by parts and generalized derivatives or distributions.)

What is happening in (B.11) is that the test functions are characterizing the super- and subdifferentials of $\tilde{V}$: $D^+\tilde{V}$ and $D^-\tilde{V}$. The superdifferential is defined by

$$D^+\tilde{V}(t,x) = \left\{(q,p)\;:\; \limsup_{(s,y)\to(t,x)} \frac{\tilde{V}(s,y)-\tilde{V}(t,x)-q(s-t)-p(y-x)}{|s-t|+|y-x|} \le 0\right\},$$

and analogously for the subdifferential $D^-\tilde{V}$. The super- and subdifferentials can be expressed equivalently in terms of the test functions:

$$D^+\tilde{V}(t,x) \;=\; \Big\{(q,p)\;:\;(q,p) = (\tfrac{\partial}{\partial t}w(t,x), \nabla_x w(t,x)) \text{ where } (t,x) \text{ is a local maximum for } W - w \text{ for some } w \in C^\infty\Big\}.$$

Inequalities (B.11) mean that

$$-q + H(x,p) \leq 0 \text{ (resp., } \geq 0) \text{ for all } (q,p) \in D^+\tilde{V}(t,x) \text{ (resp., } D^-\tilde{V}(t,x)).$$

If $\tilde{V}$ is differentiable at (t',x'), then (B.5) holds at this point, and a classical solution is also a viscosity solution.

Referring to the example in §B.1.2, the function $V^1 \equiv 0$ is a classical and hence a viscosity solution of (B.10). However, V^2 is not a viscosity solution because the supersolution property fails. Indeed, consider the point $(t',x') = (\frac{1}{2},0)$. Then $D^-V(\frac{1}{2},0) = \{(1,p) : p \in [-1,1]\}$, and with $(q,p) = (1,0) \in D^-W(\frac{1}{2},0)$ we see that

$$-1 + 0 \not\geq 0.$$

The subsolution property does hold (in fact, $D^+\tilde{V}(\frac{1}{2},0) = \emptyset$, and there is nothing to check at the point $(\frac{1}{2},0)$).

Existence results are available, using a variety of methods, for example, optimal control and games, vanishing viscosity (see below), and Perron's method. In particular, under very general conditions, value functions for optimal control or game problems are viscosity solutions. The dynamic programming principle is the key to this, and is discussed at length in the book, in both deterministic and stochastic contexts. To see how this works, suppose that $V - w$ attains a maximum (respectively, minimum) at (t',x'). Set $r = t' + h$, $h > 0$, and rearrange (B.4) to yield

$$\sup_{u(\cdot)} \left[-\tfrac{1}{h}(V(t'+h, x(t'+h)) - V(t',x')) - \tfrac{1}{h} \int_t^{t'+h} L(x(s),u(s))\, ds \right] = 0.$$

But also

$$-(w(t'+h,x(t'+h)) - w(t',x')) \leq \text{ (resp., } \geq) - (V(t'+h,x(t'+h)) - V(t',x')).$$

Combining these two displays and sending $h \to 0$ yields (B.11).

Uniqueness theorems are often presented in the form of a comparison theorem as follows. Let $\tilde{V}$ and V be (bounded) viscosity subsolutions and supersolutions, respectively, of (B.5)$'$. Then, assuming some technical conditions are met,

$$\sup_{(t,x)\in[t_0,t_1]\times\mathbf{R}^n} \left(\tilde{V}(t,x) - V(t,x)\right) = \sup_{x\in\mathbf{R}^n} \left(\tilde{V}(t_1,x) - V(t_1,x)\right). \qquad \text{(B.12)}$$

In particular, if $\tilde{V}(t_1,\cdot) \leq V(t_1,\cdot)$, then $\tilde{V} \leq V$. Uniqueness is deduced from (B.12). If $\tilde{V}$ and V are both viscosity solutions of (B.5)$'$ satisfying (B.6), then (B.12) implies $\tilde{V} \leq V$ and $V \leq \tilde{V}$; hence $\tilde{V} = V$. In our example, the function $V^1 \equiv 0$ is the unique viscosity solution of (B.10). In general, the value function V defined by (B.3) is the unique viscosity solution of (B.5)$'$.

B.1.4 Representations

An important issue is the following. Suppose we are given a nonlinear PDE. Can we express the solution (if it is unique) as a value function for an optimal control problem?

This is an issue of *representation*. Such representations can be used as powerful tools for analyzing properties of the solution.

If the PDE has a concave (or convex) Hamiltonian, and if the PDE has a unique viscosity solution, then such optimal control representations can be obtained in a wide range of cases. If the Hamiltonian is neither concave nor convex, then it is often possible to represent the solution as the value function for a dynamic game.

This technique of representation was employed numerous times in this book; e.g., (11.52) has representation (11.56). In the book we have not been concerned to a great extent with viscosity solutions and lack of smoothness. Indeed, we have made numerous assumptions concerning the existence and uniqueness of smooth solutions to PDEs and have made liberal use of optimal control representations. Refer to the comments in §1.12.

Let us explain this representation technique in more detail. Suppose we start with the PDE (B.8) (concave in $\nabla_x V$) with boundary condition (B.6). Assume it has a unique smooth solution. The first step of the representation process is to express the Hamiltonian as a minimum (Legendre transformation):

$$\begin{aligned} H(x,\lambda) &= \lambda \cdot f - \tfrac{1}{2}\lambda g g' \lambda' + \ell \\ &= \inf_{v\in\mathbf{R}^m} \left\{ \lambda\cdot(f+gv) + \ell + \tfrac{1}{2}|v|^2 \right\}. \end{aligned}$$

Now using smoothness, integration yields

$$V(t,x) \leq \left[\int_t^r [\ell(x(s)) + \tfrac{1}{2}|u(s)|^2]\, ds + V(r, x(r)) \right]$$

for any $u(\cdot)$, $r \geq t$, with equality if and only if

$$\begin{aligned} u(s) &= u^*(s) = \operatorname*{argmin}_{v\in\mathbf{R}^m}\{\nabla_x V(s,x(s))\cdot(f(x(s)) + g(x(s))v) \\ &\qquad + \ell(x(s)) + \tfrac{1}{2}|v|^2\} \\ &= u^*(x(s),s) = -g(x(s))'\nabla_x V(x(s),s)'. \end{aligned}$$

Therefore, V satisfies the integrated form of the dynamic programming equation (B.4). Now since V satisfies (B.6), it follows that V has the explicit optimal control representation (B.3).

In the stationary case, the PDE takes the form

$$\nabla_x V \cdot f - \tfrac{1}{2}\nabla_x V g g' \nabla_x V' + \ell = 0.$$

The function V is a function of x only, and the boundary condition (B.6) no longer applies. The representation procedure yields

$$V(x) \leq \left[\int_0^t [\ell(x(s)) + \tfrac{1}{2}|u(s)|^2]\, ds + V(x(r)) \right]$$

for any $u(\cdot)$ with equality if and only if

$$\begin{aligned} u(s) &= u^*(s) = \operatorname*{argmin}_{v \in \mathbf{R}^m} \{\nabla_x V(x(s)) \cdot (f(x(s)) + g(x(s))v) \\ &\qquad\qquad + \ell(x(s)) + \tfrac{1}{2}|v|^2\} \\ &= u^*(x(s)) = -g(x(s))' \nabla_x V(x(s))'. \end{aligned}$$

Therefore, $V = V(x)$ satisfies the following stationary version of the integrated dynamic programming equation: for any $t \geq 0$,

$$V(x) = \inf_{u(\cdot)} \left[\int_0^t [\ell(x(s)) + \tfrac{1}{2}|u(s)|^2] \, ds + V(x(r)) \right]. \tag{B.13}$$

This is an explicit optimal control representation.

B.2 Nonlinear Riccati Equations and Inequalities

The PDEs used in this book arise from control problems that have quadratic performance integrands, and the control terms appear linearly in the state dynamics. The general form of these PDEs is

$$\nabla_x V \cdot f + \nabla_x V R \nabla_x V' + Q = 0 \tag{B.14}$$

in the stationary case. Here, $f \in \mathcal{X}_l$ is a vector field, and $Q, R \in \mathcal{X}_{q,2,b}$. Such equations are sometimes called *nonlinear Riccati equations*. The variety of specific PDEs used in the book arises because of the varying nature of the control problems considered and utilized, and the specific form of the PDE will depend on whether the control problem is defined in forward or reverse time, is a minimium or maximum or minimax problem, etc. In time-varying cases the sign of the time derivative term will also depend on the particular case (see §B.1 for a detailed discussion of a particular case).

Associated with the PDE (B.14) is a pair of PDIs:

$$\nabla_x V \cdot f + \nabla_x V R \nabla_x V' + Q \geq 0 \tag{B.15}$$

and

$$\nabla_x V \cdot f + \nabla_x V R \nabla_x V' + Q \leq 0. \tag{B.16}$$

B.2.1 Classification

Table B.1 classifies the stationary PDEs used in this book in terms of the signs of the parameters R and Q and the direction of time in the underlying optimization problem.

A mixed sign for Q indicates that the cost integrand is sign indefinite, whereas a mixed sign for R means that the optimization problem is a minimax game.

Case	Symbol	Sign of R	Sign of Q	Time	Chap.
H^∞ info. state	p_e	+	mixed	B	3
H^2 info. state	r_e	+	−	B	11
BRL avail.	V	+	+	F	5
H^∞ state f.b.†	V	mixed	+	F	5
H^2 state f.b.	V_{H_2}	−	+	F	5

† State feedback.

Table B.1: Classification of Riccati PDE.

B.2.2 Uniqueness and Representation

A number of important questions arise concerning the PDE and associated PDIs, among which are those concerning uniqueness, representation and comparison. The purpose of this subsection is to present some answers to the questions as they relate to the equations used in this book. In particular, we treat the following special case of (B.14):

$$-\nabla_x V \cdot A + \nabla_x V BB' \nabla_x V' + Q = 0 \tag{B.17}$$

corresponding to the choice $f = -A$, $R = BB'$, with Q of possibly mixed sign, but satisfying

$$A(0) = 0, \quad Q(0) = 0. \tag{B.18}$$

Equation (3.28) for the equilibrium information state p_e is of this type, as is the equation (11.18) for the equilibrium H^2 information state r_e, after appropriate sign changes (equations of the type arising in the bounded real lemma are discussed in Chapter 5). We also consider the PDI

$$-\nabla_x V \cdot A + \nabla_x V BB' \nabla_x V' + Q \leq 0. \tag{B.19}$$

Note that equation (B.17) can be written as

$$\sup_v \{-\nabla_x V \cdot (A + Bv) - \tfrac{1}{2}|v|^2 + Q\} = 0,$$

using the representation method from §B.1.4. The optimal control problems underlying the PDE (B.17) have dynamics

$$\dot{x} = A(x) + B(x)v \tag{B.20}$$

with cost functions defined in terms of the integrand

$$J(v; t_1, t_0) = \int_{t_0}^{t_1} [Q(x(s)) - \tfrac{1}{2}|v(s)|^2] ds. \tag{B.21}$$

Using this, the integral form of the PDE (B.17) is

$$V(x(t_1)) = \sup_{v(\cdot)}\{V(x(t_0)) + J(v; t_1, t_0)\} \tag{B.22}$$

for all $v \in L_2[t_0, t_1]$ and all $t_1 \geq t_0$.

A smooth solution V of (B.17) is called L_2 *stabilizing* if the vector field

$$A^* = A - BB'\nabla_x V \tag{B.23}$$

is asymptotically L_2 stable, or L_2 *antistabilizing* if $-A^*$ is asymptotically L_2 stable.

The following lemma presents important information regarding uniqueness of stationary smooth solutions.

LEMMA B.2.1 *Consider the PDE* (B.17). *If V_1 and V_2 are two smooth L_2 stabilizing solutions then there exists a constant $c \in \mathbf{R}$ such that*

$$V_1 = V_2 + c. \tag{B.24}$$

Similarly, (B.24) *holds if V_1 and V_2 are L_2 antistabilizing smooth solutions.*

Proof. We give the proof for the antistable case; the stable case being similar. Let $\psi = V_1 - V_2$. Then ψ satisfies the equation

$$\begin{aligned} 0 &= -\nabla_x\psi \cdot A + \tfrac{1}{2}(\nabla_x V_1 + \nabla_x V_2)BB'\nabla_x\psi \\ &= -\nabla_x\psi \cdot (A - BB'\nabla_x V_1) - \tfrac{1}{2}\nabla_x\psi BB'\nabla_x\psi \\ &= \inf_v\{-\nabla_x\psi \cdot (A - BB'\nabla_x V_1 + Bv) + \tfrac{1}{2}|v|^2\}. \end{aligned}$$

Integrating, we get

$$\psi(x) = \inf_{v(\cdot)}\left\{\psi(\Phi^v_{s,t}(x)) + \int_0^t \gamma^2\tfrac{1}{2}|v(s)|^2 ds\right\},$$

where $\Phi^v_{s,t}(x)$ is the semigroup for the ODE

$$\dot{\xi} = A(\xi) - B(\xi)B(\xi)'\nabla_x V_1(\xi) + B(\xi)v, \quad \xi(t) = x.$$

Let $v = 0$, so that

$$\psi(x) \leq \psi(\Phi^0_{s,t}(x)) \to \psi(0) \ \text{ as } t \to +\infty$$

using the assumed antistability. This implies

$$V_1(x) - V_2(x) \leq V_1(0) - V_2(0) \triangleq c.$$

However, the argument is symmetric, proving

$$V_1(x) - V_2(x) = V_1(0) - V_2(0) = c.$$

□

This lemma provides a uniqueness result for stabilizing (antistabilizing) smooth solutions, up to an additive constant. We often normalize so that $V(0) = 0$.

Next we define two specific solutions to (B.17) and relate them to other solutions of (B.17) and (B.19). This follows [Wil71] and parallels results for dissipative systems ([Wil72] and Chapter 5). The fact that both Q and $Q - \frac{1}{2}|v|^2$ are sign indefinite makes the analysis somewhat delicate. The two definitions are

$$\overline{V}(x) = \inf_{v \in L_{2,e}[0,\infty)} \left\{ -\lim_{T\to\infty} J(v;T,0) \ : x(0) = x, \ \lim_{T\to\infty} x(T) = 0 \right\} \tag{B.25}$$

and

$$\underline{V}(x) = \sup_{v \in L_{2,e}(-\infty,0]} \left\{ \lim_{T\to\infty} J(v;0,-T) \ : x(0) = x, \ \lim_{T\to\infty} x(-T) = 0 \right\}. \tag{B.26}$$

Any solution V of the PDI (B.19) satisfies the following dissipation inequality (which is just an integral version of (B.19)):

$$V(x(t_1)) \geq V(x(t_0)) + J(v; t_1, t_0) \tag{B.27}$$

for all $v \in L_2[t_0, t_1]$, where $t_1 \geq t_0$. Of course, any solution of the PDE (B.17) is also a solution of the PDI (B.19), and the reader can check that both the functions $\underline{V}$ and $\overline{V}$ are solutions of the PDI. Also, in view of assumption (B.18), $\underline{V}(0) \geq 0$ and $\overline{V}(0) \leq 0$.

THEOREM B.2.2 *Consider the PDE* (B.17) *and the PDI* (B.19).
Then the following conditions hold:

(i) *Any continuous solution V of the PDI* (B.19) *with $V(0) = 0$ satisfies*

$$-\infty < \underline{V}(x) \leq V(x) \tag{B.28}$$

for any L_2 antistabilizable point $x \in \mathcal{S}'_{cas}$.

(ii) *Any continuous solution V of the PDI* (B.19) *with $V(0) = 0$ satisfies*

$$V(x) \leq \overline{V}(x) < +\infty \tag{B.29}$$

for any L_2 stabilizable point $x \in \mathcal{S}'_{cs}$.

(iii) *$\underline{V}(0) \geq 0$ and $\overline{V}(0) \leq 0$, and if there exists a continuous subsolution V with $V(0) = 0$, then $\underline{V}(0) = V(0) = \overline{V}(0) = 0$.*

(iv) *Let $V^+ \in \mathcal{X}_{q,2,b}$ be a stabilizing smooth solution of the PDE* (B.17), *assumed everywhere finite. Then*

$$V^+(x) = \overline{V}(x), \tag{B.30}$$

and $V^+ = \overline{V}$ is the unique L_2 stabilizing solution of equation (B.17) *satisfying $V^+(0) = 0$.*

(v) *Let* $V^- \in \mathcal{X}_{q,2,b}$ *be an* L_2 *antistabilizing smooth solution of the PDE* (B.17), *assumed finite. Then*

$$V^-(x) = \underline{V}(x) \tag{B.31}$$

and $V^- = \underline{V}$ *is the unique* L_2 *antistabilizing solution of* (B.17) *satisfying* $V^-(0) = 0$.

Proof. Let V be any continuous solution of the PDI, so by (B.27) we get, on setting $t_0 = -T$ and $t_1 = 0$,

$$V(x) \geq V(x(-T)) + J(v; 0, -T)$$

for all $v \in L_2[-T, 0]$, $T \geq 0$, where $x(0) = x$. Now let $v(\cdot)$ be any control that antistabilizes the point $x = x(0)$, so that $x(-T) \to 0$ as $T \to \infty$. Then

$$\begin{aligned} V(x) &\geq \lim_{T\to\infty} \{V(x(-T)) + J(v; 0, -T)\} \\ &= \lim_{T\to\infty} J(v; 0, -T). \end{aligned}$$

Taking the supremum over all such antistabilizing controls gives $V(x) \geq \underline{V}(x) > -\infty$. This proves item (i).

Next, again using (B.27) but with $t_0 = 0$, $t_1 = T$, we get

$$V(x) \leq V(x(T)) - J(v; T, 0)$$

for any $v \in L_2[0, T]$, $T \geq 0$, where $x(0) = x$. Let $v(\cdot)$ be any stabilizing control for x, so that $x(T) \to 0$ as $T \to \infty$. Then sending $T \to \infty$ we get

$$V(x) \leq -\lim_{T\to\infty} J(v; T, 0)$$

and hence, by taking the infimum over all such stabilizing controls, $V(x) \leq \overline{V}(x) < +\infty$. This proves item (ii).

It follows now that $\underline{V}(0) = V(0) = \overline{V}(0) = 0$, since $v \equiv 0$ both stabilizes and antistabilizes the point $x = 0$, verifying item (iii).

The proof of item (iv) is similar to that of item (v), so we give only the proof of item (v).

The fact that V^- is an antistabilizing solution of the PDE (B.17) means that for all $-T \leq 0$,

$$V^-(x) = \sup_{v(\cdot)} \{V^-(x(-T)) + J(v; 0, -T)\}$$

with the optimum attained by the antistabilizing control

$$v^*(s) = -B(x(s))'\nabla_x V^{-\prime}(x(s)),$$

where $s \leq 0$ and $x(0) = x$. Since $V^- \in \mathcal{X}_{q,2,b}$, we have

$$|v^*(s)| \leq C|x(s)|,$$

and so since $x(\cdot) \in L_2$, we have $\dot{x}(\cdot) \in L_2$, under the action of this control. Hence $x(-T) \to 0$ as $T \to \infty$. Now

$$V^-(x) = V^-(x(-T)) + J(v^*; 0, -T)$$

for all $T \geq 0$, so sending $T \to \infty$ gives

$$V^-(x) = \lim_{T \to \infty} J(v^*; 0, -T).$$

If we plug $v = v^*$ into the formula for $\underline{V}(x)$ we obtain the inequality

$$\underline{V}(x) \geq V^-(x).$$

To obtain the reverse inequality, we have

$$J(v; 0, -T) \leq V^-(x) - V^-(x(-T))$$

for any antistabilizing control v, and so

$$\lim_{T \to \infty} J(v; 0, -T) \leq V^-(x).$$

This implies

$$\underline{V}(x) \leq V^-(x),$$

completing the proof. □

In the remainder of this subsection we consider singular cases with

$$B = 0 \tag{B.32}$$

(these cases correspond to 1 and 2A block problems). The following lemma deals with uniqueness for singular cases and is analogous to Lemma B.2.1.

LEMMA B.2.3 *Consider the PDE* (B.17) *with* $B = 0$. *Assume that* A *is exponentially hyperbolic. If* V_1 *and* V_2 *are two solutions of the form*

$$V_1 = \delta_{M_{as}} + \breve{V}_1, \quad V_2 = \delta_{M_{as}} + \breve{V}_2,$$

with $\breve{V}_1, \breve{V}_2$ *finite and smooth on* $C(M_{as})$, *then there exists a constant* $c \in \mathbf{R}$ *such that*

$$\breve{V}_1 = \breve{V}_2 + c. \tag{B.33}$$

Proof. Write $\psi = V_1 - V_2$. Then using the convention $-\infty - (-\infty) = 0$ we have $\psi = \breve{V}_1 - \breve{V}_2$. Now for $x \in M_{as}$ we have

$$\psi(x) = \psi(\Phi_{s,t}(x)),$$

where $\Phi_{s,t}(x)$ is the flow on M_{as}. Sending $s \to -\infty$ we get $\psi(x) = \psi(0)$, completing the proof. □

Appendix C

Max-Plus Convergence

C.1 Introduction

This appendix presents basic facts concerning the space $\mathcal{X}_e$ in which the information state takes values such as algebraic structure (in which $\max$ is linear), interpretation in terms of max-plus measures, weak convergence, compactness criteria, limit identification, and metrization of the weak topology. The concepts and results discussed are max-plus analogues of probabilistic concepts and results; so there is an optimization–probability duality. Indeed, from the point of view of the theory of large deviations, this is very natural, since in that theory probability is linked with optimization.

In preparing this appendix, material was drawn from several sources, given in the bibliography. We have attempted to specialize the results from the literature so that they are directly relevant for our purpose of studying the dynamical properties of possibly singular information states for use in developing the nonlinear H^∞ theory, as presented in this book. The principal references used were [AQV94], [Aki95], [Aki96], [AW83], [DE97], [Jia95], [LM95], [Qua90], [Qua92], [MS92]; see also the recent papers [FM], [HM98b].

C.2 The Max-Plus Structure on $\mathbf{R}$

The *max-plus algebra* is defined on the set

$$\mathbf{R}_e = \mathbf{R} \cup \{-\infty\}$$

by the operations

$$a \oplus b = \max(a, b) \qquad [addition]$$

and

$$a \odot b = a + b \qquad [multiplication].$$

There is an additive identity

$$\mathbf{0} = -\infty$$

and a multiplicative identity

$$\mathbf{1} = 0.$$

A key property is

$$a \oplus a = a \qquad [idempotence].$$

The set $\mathbf{R}_e$ has an ordering

$$a \leq_e b \text{ iff } a \oplus b = b$$

and a metric

$$d_e(a, b) = |e^a - e^b|.$$

So $(\mathbf{R}_e, d_e)$ is a complete, separable metric space with an idempotent semiring structure with no additive inverse (but with a multiplicative inverse).

C.3 Max-Plus Functions

We define some basic spaces. The space of *bounded functions* is

$$\mathbf{B}_e(\mathbf{R}^n) = \{p : \mathbf{R}^n \to \mathbf{R}_e \mid \langle p \rangle < +\infty\},$$

where

$$\langle p \rangle = \sup_{x \in \mathbf{R}^n} \{p(x)\} = \| p \|_e$$

is a norm. The space of *bounded u.s.c. functions* is denoted

$$\mathbf{USC}_e(\mathbf{R}^n) = \{p \in \mathbf{B}_e(R^n) \mid p \text{ u.s.c.}\}$$

and the space of *bounded functions with compact level sets* is

$$\mathbf{CLS}_e(\mathbf{R}^n) = \{p \in \mathbf{B}_e(\mathbf{R}^n) \mid \{x : p(x) \geq a\} \text{ compact for all } a \in \mathbf{R}\}.$$

Next we present a simple lemma.

LEMMA C.3.1 *If p has compact level sets, then p is u.s.c.; i.e.,* $\mathbf{CLS}_e(\mathbf{R}^n) \subset \mathbf{USC}_e(\mathbf{R}^n)$.

The proofs of this and all other results are given in §C.6.

We will take as our basic spaces for the information state the following:

$$\mathcal{X}_e = \mathbf{USC}_e(\mathbf{R}^n)$$

and

$$\mathcal{X}_{e,1} = \{p \in \mathcal{X}_e \mid \langle p \rangle = \mathbf{1}(= 0)\}.$$

We shall see that $\mathcal{X}_e$ can be interpreted a space of max-plus measures and $\mathcal{X}_{e,1}$ is a space of max-plus measures with total mass $\mathbf{1}$ (normalized). Further, there is a concept of weak (i.e., narrow) convergence in $\mathcal{X}_e$ that is equivalent to convergence in a metric (see below), and $\mathcal{X}_{e,1}$ is a complete, separable metric space. There are also nice criteria for compactness expressed in terms of uniform tightness.

The max-plus structure is defined on $\mathcal{X}_e$ by

$$(p_1 \oplus p_2)(x) = p_1(x) \oplus p_2(x)$$

and

$$(p \odot c)(x) = p(x) \odot c.$$

Information states are examples of functions in $\mathcal{X}_e$. In Chapter 3 we discussed the max-plus linearity of the information state transition operator $S_t^{u,y}$, which means

$$S_t^{u,y}(c \odot p_0) = c \odot S_t^{u,y}(p_0) \text{ for any constant } c \tag{3.16$'$}$$

and

$$S_t^{u,y}(p_0^1 \oplus p_0^2) = S_t^{u,y}(p_0^1) \oplus S_t^{u,y}(p_0^2). \tag{3.17$'$}$$

C.4 Max-Plus Measures

Let us define max-plus measures and relate them to the above spaces.

Let $\mathcal{B}$ denote the Borel subsets of $\mathbf{R}^n$. Let $\mathcal{G}$ and $\mathcal{F}$, respectively, denote the open and closed subsets of $\mathbf{R}^n$. A function

$$\mu : \mathcal{B} \to \mathbf{R}_e$$

is called a *max-plus measure* if and only if it satisfies

(i) $\mu(\emptyset) = \mathbf{0}(= -\infty)$;

(ii) $\mu(\cup_\alpha A_\alpha) = \sup_\alpha \mu(A_\alpha)$ for all $A_i \in \mathcal{B}$;

(iii) $\mu(A) = \inf\{\mu(G) : G \in \mathcal{G},\ A \subset G\}$;

(iv) for all $G \in \mathcal{G}$, $\mu(G) = \lim_{\varepsilon \to 0} \mu(G^{-\varepsilon})$, where $A^{-\varepsilon} = \{x : d(x, A^c) \geq \varepsilon\}$, and $d(x, A) = \inf_{y \in A} |x - y|$.

Since $\mu(\mathbf{R}^n) \in \mathbf{R}_e$, $\mu(\mathbf{R}^n) < +\infty$, each max-plus measure is *finite*.

A max-plus measure μ has a *density* if there exists $p \in \mathcal{X}_e$ such that

$$\mu(A) = \langle p + \delta_A \rangle = \sup_{x \in A}\{p(x)\},$$

where

$$\delta_A(x) = \begin{cases} \mathbf{1}(= 0) & \text{if } x \in A, \\ \mathbf{0}(= -\infty) & \text{if } x \notin A. \end{cases}$$

Lemma C.4.1

(i) *Every $p \in \mathcal{X}_e$ defines a max-plus measure.*

(ii) *Every max-plus measure has a density in $\mathcal{X}_e$.*

Thus we can identify $\mathcal{X}_e$ with the space of max-plus measures, and we will often use the notation

$$p(A) = \langle p + \delta_A \rangle.$$

$\mathcal{X}_{e,1}$ is identified with the space of max-plus measures with unit mass: $p(\mathbf{R}^n) = \mathbf{1}(=0)$.

Let $\mathcal{K}$ denote the compact subsets of $\mathbf{R}^n$. A max-plus measure μ is *tight* if

$$\inf_{K \in \mathcal{K}} \mu(K^c) = \mathbf{0}(= -\infty).$$

THEOREM C.4.2 (i) *Every* $p \in \mathbf{CLS}_e(\mathbf{R}^n)$ *defines a tight max-plus measure.*

(ii) *Every tight max-plus measure has a density in* $\mathbf{CLS}_e(\mathbf{R}^n)$.

Thus $\mathbf{CLS}_e(\mathbf{R}^n)$ is identified with the space of tight max-plus measures. We will use the notation

$$\mathcal{X}_{et} = \mathbf{CLS}_e(\mathbf{R}^n)$$

and

$$\mathcal{X}_{et,1} = \{p \in \mathcal{X}_{et} \mid \langle p \rangle = \mathbf{1}(=0)\}.$$

Also, we say that $p \in \mathcal{X}_e$ is *quadratically tight* if there exists constants $c_1 > 0$, $c_2 \geq 0$ such that

$$p(x) \leq -c_1|x|^2 + c_2 \quad \text{for all } x \in \mathbf{R}^n.$$

Max-plus integration is defined as follows. The *integral* of $f : \mathbf{R}^n \to \mathbf{R}_e$ with respect to a max-plus measure $p \in \mathcal{X}_e$ is defined by

$$p(f) = \langle p + f \rangle.$$

The function f is *p-integrable* if $p(f) \in \mathbf{R}_e$. Of course, any $f \in \mathbf{B}_e(\mathbf{R}^n)$ is p-integrable for any $p \in \mathcal{X}_e$.

Table C.1 summarizes the various uses of the symbol p for the information state.

Use of symbol p	Meaning
$p(x)$	function of x
$p(A)$	max-plus measure of set A
$p(f)$	max-plus integral of f

Table C.1: Interpretation of the symbol p.

C.5 Max-Plus Convergence

We remark first that there are important situations where familiar modes of convergence do not work. Consider a sequence $\{\delta_{x_n}\} \subset \mathcal{X}_e$, where $x_n \to 0$ as $n \to \infty$. We would like δ_{x_n} to converge to δ_0; however this is not true pointwise. Indeed, if $x_n \neq 0$ for all n, then $\delta_{x_n}(0) = -\infty$ for all n, but $\delta_0(0) = 0$; hence $\delta_{x_n}(0) \not\to \delta_0(0)$.

A sequence $\{p_n\} \subset \mathcal{X}_e$ is said to *converge weakly* to $p_\infty \in \mathcal{X}_e$ if

(i) $\liminf_{n\to\infty} p_n(G) \geq p_\infty(G)$ for all $G \in \mathcal{G}$, and

(ii) $\limsup_{n\to\infty} p_n(F) \leq p_\infty(F)$ for all $F \in \mathcal{F}$.

Weak convergence is denoted $p_n \Longrightarrow p_\infty$.

THEOREM C.5.1 *Let $p_n, p^\infty \in \mathcal{X}_e$. Then $p_n \Longrightarrow p_\infty$ if and only if*

$$p_n(f) \to p_\infty(f)$$

in $\mathbf{R}_e$ for all $f \in C_b(\mathbf{R}^n)$ (continuous, bounded).

REMARK C.5.2 The class of functions $C_b(\mathbf{R}^n)$ in Theorem C.5.1 can be replaced by either (i) $\mathbf{R}$-valued bounded Lipschitz continuous functions, or (ii) $\mathbf{R}_e$-valued functions f such that $e^f \in C_b(\mathbf{R}^n)$. See [DE97] for details. ▽

With the aid of Theorem C.5.1 it is easy to see that $\delta_{x_n} \Longrightarrow \delta_0$. Indeed, for any $f \in C_b(\mathbf{R}^n)$, we have $\delta_{x_n}(f) = f(x_n) \to f(0) = \delta_0(f)$.

A family of max-plus measures $\{p_\alpha\}$ in $\mathcal{X}_e$ is *uniformly tight* if

$$\inf_{K\in\mathcal{K}} \sup_\alpha p_\alpha(K^c) = \mathbf{0}(= -\infty).$$

We say that $\{p_\alpha\}$ is *uniformly quadratically tight* if there exists constants $c_1 > 0$, $c_2 \geq 0$ such that for all p_α,

$$p_\alpha(x) \leq -c_1|x|^2 + c_2 \text{ for all } x \in \mathbf{R}^n.$$

The following result is based on a theorem due to Bryc and is useful for determining weak limits (see [DE97]).

THEOREM C.5.3 *Let $\Lambda(f)$ be a functional defined for every $f \in C_b(\mathbf{R}^n)$ by*

$$\Lambda(f) = \lim_{n\to\infty} p_n(f),$$

where $p_n \in \mathcal{X}_e$ is uniformly tight. Then there exists $p_\infty \in \mathcal{X}_{et}$ such that $p_n \Longrightarrow p_\infty$; indeed,

$$p_\infty(x) = \inf\{\Lambda(f) - f(x) \;:\; f \in C_b(\mathbf{R}^n)\} \tag{C.1}$$

and

$$\Lambda(f) = p_\infty(f) \tag{C.2}$$

for all $f \in C_b(\mathbf{R}^n)$.

The main compactness result is the following [Jia95, Lemma 2.2], [DE97, Theorem 1.3.7].

THEOREM C.5.4 *Let* $p_n \in \mathcal{X}_e$ *be uniformly tight and* $-\infty < c = \limsup_{n\to\infty} p_n(\mathbf{R}^n) < +\infty$. *Then there exists* $p_\infty \in \mathcal{X}_{et,1}$ *and a subsequence* n_k *such that* $p_{n_k} \Longrightarrow p_\infty + c$.

Define a metric ρ_e [Jia95] on $\mathcal{X}_e$ by

$$\rho_e(p,q) = \inf\{\ e^{p(F)} \le e^{q(F^\varepsilon)} + \varepsilon \text{ for all } F \in \mathcal{F}, \text{ and}$$

$$e^{p(G)} \ge e^{q(G^{-\varepsilon})} - \varepsilon \text{ for all } G \in \mathcal{G}\},$$

where $A^\varepsilon = \{x \,:\, d(x, A) < \varepsilon\}$.

THEOREM C.5.5 ρ_e *is a metric on* $\mathcal{X}_e$. *Let* $p_n \in \mathcal{X}_e$, $p^\infty \in \mathcal{X}_{et}$; *then* $\rho_e(p_n, p_\infty) \to 0$ *if and only if* $p_n \Longrightarrow p_\infty$.

THEOREM C.5.6 $(\mathcal{X}_{et,1}, \rho_e)$ *is a complete, separable metric space.*

We say that a sequence $\{p_n\}$ from $\mathcal{X}_e$ *hypoconverges to* p_∞ *at* x if

(i) for all subsequences $\{p_{n_i}\}$ and sequences $x_i \to x$, we have

$$\limsup_{i\to\infty} p_{n_i}(x_i) \le p_\infty(x),$$

and

(ii) there exists a sequence $x_n \to x$ such that

$$\liminf_{n\to\infty} p_n(x_n) \ge p_\infty(x).$$

If (i) and (ii) hold for all x, we say $\{p_n\}$ *hypoconverges to* p_∞, written $h\text{-}\lim_{n\to\infty} p_n = p_\infty$ or $p_n \stackrel{h}{\to} p_\infty$. (*Epiconverge* is similarly defined, with the lim sup and lim inf and the corresponding inequalities interchanged and is written $e\text{-}\lim_{n\to\infty} p_n = p_\infty$ or $p_n \stackrel{e}{\to} p_\infty$.)

THEOREM C.5.7 *The following are basic properties of hypoconvergence (from* [AW83]):

(i) *If* $p_n \stackrel{h}{\to} p_\infty$ *and* $x_n \to x$ *with* $x_n \in \operatorname{argmax} p_n$, *then* (a) $\langle p_n\rangle \to \langle p_\infty\rangle$, *and* (b) $x \in \operatorname{argmax} p_\infty$.

(ii) *Any sequence* $\{p_n\}$ *has a hypoconvergent subsequence (the limit may be trivial; i.e., the limit may equal* $-\infty$ *identically).*

(iii) *Any hypolimit is u.s.c. The hypolimit of a constant sequence* $p_n = p \in \mathbf{B}_e(\mathbf{R}^n)$ *is the u.s.c. envelope* $p^*(x) = \limsup_{x'\to x} p(x')$.

THEOREM C.5.8 *Let* $p_n, p_\infty \in \mathcal{X}_e$, *with* $p_\infty \in \mathcal{X}_{et}$. *Then* $p_n \Longrightarrow p_\infty$ *if and only if* $p_n \stackrel{h}{\to} p_\infty$.

THEOREM C.5.9 *Let* $p_n, p_\infty \in \mathcal{X}_e$, *with* $p_\infty \in \mathcal{X}_{et}$. *If* $p_n \to p_\infty$ *uniformly on compact sets, then* $p_n \Longrightarrow p_\infty$.

THEOREM C.5.10 *Let* $p_n, p_\infty \in \mathcal{X}_e$, *with* $p_\infty \in \mathcal{X}_{et}$ *and* $p_n \Longrightarrow p_\infty$. *Then* $\operatorname{argmax} p_n \to \operatorname{argmax} p_\infty$ *in the sense that given any sequence* $x_n \in \operatorname{argmax} p_n$, *there exists a subsequence* x_{n_k} *and a point* $x_\infty \in \operatorname{argmax} p_\infty$ *such that* $x_{n_k} \to x_\infty$.

THEOREM C.5.11 *Let* $p_n, p_\infty \in \mathcal{X}_e$, *and* $p_n \xrightarrow{h} p_\infty$. *Let* $f \in C(\mathbf{R}^n)$ *and assume* $p_n + f$ *is uniformly tight. Then* $\langle p_n + f\rangle \to \langle p_\infty + f\rangle$.

C.6 Proofs

Proof of Lemma C.3.1. Let $L = \limsup_{y\to x} p(y)$. We want to show $L \le p(x)$. Take $y_i \to x$ with $p(y_i) \to L$. Let $\varepsilon > 0$. Then for all large i, $p(y_i) \ge L - \varepsilon$. Thus $y_i \in \{y : p(y) \ge L-\varepsilon\}$, which is compact. So $y_i \to x$ and $x \in \{y : p(y) \ge L-\varepsilon\}$. Thus $L \le p(x) + \varepsilon$. □

Proof of Lemma C.4.1. Part (i). Let $p \in \mathbf{USC}_e(\mathbf{R}^n)$, and define $\mu(A) = \langle p+\delta_A\rangle$. We claim that μ is a max-plus measure.

Now $\mu(\mathbf{R}^n) = \langle p+\delta_{\mathbf{R}^n}\rangle = \langle p\rangle < +\infty$, so μ is finite. Also, $\mu(\emptyset) = \sup_{x\in\emptyset} p(x) = -\infty$; hence property (i) holds.

Let $A_\alpha \in \mathcal{B}$. Then

$$\mu(\cup_\alpha A_\alpha) = \sup_{x\in\cup_\alpha A_\alpha} p(x) = \sup_\alpha \sup_{x\in A_\alpha} p(x) = \sup_\alpha \mu(A_\alpha).$$

This proves property (ii).

Let $A \in \mathcal{B}$, $\nu(A) = \inf\{\mu(G) : G \in \mathcal{G},\ A \subset G\}$. We want to show $\mu(A) = \nu(A)$. Since μ is monotone, $\mu(A) \le \mu(G)$ for any $G \in \mathcal{G}$, $A \subset G$. Therefore, $\mu(A) \le \nu(A)$. By the u.s.c. property of p, for all $x \in A$ and all $\varepsilon > 0$ there exists an open neighborhood $U_{x,\varepsilon}$ of x such that $y \in U_{x,\varepsilon}$ implies $p(y) \le p(x)+\varepsilon \le p(A)+\varepsilon$. Now $G = \cup_{x\in A} U_{x,\varepsilon}$ is an open set containing A. Thus

$$\nu(A) \le \mu(G) = \sup_{y\in G} p(y) \le \mu(A) + \varepsilon,$$

proving property (iii).

Let $G \in \mathcal{G}$. The sequence $\mu(G^{-\varepsilon})$ is monotone increasing and bounded above by $\mu(G)$; therefore, $\lim_{\varepsilon\to 0} \mu(G^{-\varepsilon}) \le \mu(G)$. Let $\eta > 0$. Choose $x \in G$ such that $p(x) \ge \mu(G) - \eta$. Since G is open, $x \in G^{-\varepsilon}$ for some $\varepsilon > 0$. Then

$$\mu(G^{-\varepsilon}) \ge \mu(G) - \eta,$$

which implies $\limsup_{\varepsilon\to 0} \mu(G^{-\varepsilon}) \ge \mu(G)$, and hence property (iv), completing the proof of Part (i).

Part (ii). Let μ be a finite max-plus measure. Define

$$p(x) = \mu(\{x\}).$$

We claim that $p \in \mathbf{USC}_e(\mathbf{R}^n)$ and $\mu(A) = \langle p + \delta_A \rangle$.

Now $p(x) \leq \mu(\mathbf{R}^n) < +\infty$, so $p \in \mathbf{B}_e(\mathbf{R}^n)$.

To see that p is u.s.c., let $\varepsilon > 0$. There exists $G \in \mathcal{G}$ with $x \in G$ such that

$$\mu(G) \leq \mu(\{x\}) + \varepsilon,$$

by property (iii). Let $y \in G$. Then

$$p(y) = \mu(\{y\}) \leq \mu(G) \leq p(x) + \varepsilon.$$

Therefore, $p \in \mathbf{USC}_e(\mathbf{R}^n)$.

Finally, by property (ii),

$$\mu(A) = \mu(\cup_{x \in A}\{x\}) = \sup_{x \in A} \mu(\{x\}) = \sup_{x \in A} p(x) = \langle p + \delta_A \rangle.$$

□

Proof of Theorem C.4.2. Part (i). Let $p \in \mathcal{X}_e$. By Lemma C.4.1, we know that p defines a finite max-plus measure $p(A) = \langle p + \delta_A \rangle$. We claim that p is tight.

Let $N > 0$, and set $K_N = \{x : p(x) \geq -N\}$. Now K_N is compact, since $p \in \mathbf{CLS}_e(\mathbf{R}^n)$, and so

$$p(K_N^c) < -N.$$

Therefore, $\inf_{K \in \mathcal{K}} \mu(K^c) = -\infty$, as required.

Part (ii). Let μ be a finite, tight max-plus measure. By Lemma C.4.1, we know that μ has a density $p \in \mathbf{USC}_e(\mathbf{R}^n)$. We claim that p has compact level sets.

Let $a \in \mathbf{R}$. Choose $N > 0$ such that $a \geq -N$. There exists a compact set K_N such that

$$\sup_{x \notin K_N} p(x) = \mu(K^c) < -N.$$

Let $p(x) \geq a$. Then $p(x) \geq -N$ and so $x \in K_N$. So $\{x : p(x) \geq a\}$ is a closed subset of the compact set K_N, and hence it is compact. □

Proof of Theorem C.5.1. We follow [Jia95, Lemma 3.3] and [DE97, Theorem 1.2.3].

Assume $p_n \Longrightarrow p_\infty$, and fix $f \in C_b(\mathbf{R}^n)$, $\varepsilon > 0$. There exists closed sets F_i, and points $x_i \in F_i$, $i = 1, \ldots, M$, such that $\mathbf{R}^n = \cup_{i=1}^M F_i$ and $\sup_{x \in F_i} |f(x) - f(x_i)| < \varepsilon$. Then

$$\begin{aligned} p_n(f) &= \max_{i=1}^M \sup_{x \in F_i}\{p_n(x) + f(x)\} \\ &\leq \max_{i=1}^M f(x_i) p_n(F_i) + \varepsilon. \end{aligned}$$

Therefore,

$$\begin{aligned} \limsup_{n \to \infty} p_n(f) &\leq \max_{i=1}^M f(x_i) p_\infty(F_i) + \varepsilon \\ &\leq p_\infty(f) + 2\varepsilon. \end{aligned}$$

Next, fix $x \in \mathbf{R}^n$, $\varepsilon > 0$. There exists $\eta > 0$ such that $\sup_{y \in \{x\}^\eta} |f(y) - f(x)| < \varepsilon$. Then

$$\begin{aligned} p_n(f) &\geq \sup_{y \in \{x\}^\eta}\{p_n(y) + f(y)\} \\ &\geq f(x) p_n(\{x\}^\eta) - \varepsilon. \end{aligned}$$

This implies

$$\liminf_{n\to\infty} p_n(f) \geq f(x)p_\infty(\{x\}^\eta) - \varepsilon$$

$$\geq f(x)p_\infty(x) - \varepsilon.$$

This holds for all x, $\varepsilon > 0$, and so

$$\liminf_{n\to\infty} p_n(f) \geq p_\infty(f).$$

Consequently, we have now shown that $\lim_{n\to\infty} p_n(f) = p_\infty(f)$.

Conversely, assume $\lim_{n\to\infty} p_n(f) = p_\infty(f)$ for all $f \in C_b(\mathbf{R}^n)$. Let F be a closed set and consider $\delta_F \in \mathbf{USC}_e(\mathbf{R}^n)$. Define $\phi_j \in C_b(\mathbf{R}^n)$ by

$$\phi_j(x) = -j\min(d(x, F), 1).$$

From [DE97, Lemma 1.2.4], we have $\phi_j \downarrow \delta_F$ as $j \to \infty$ and

$$\lim_{j\to\infty} p_\infty(\phi_j) = p_\infty(F). \tag{C.3}$$

Now

$$p_n(F) \leq p_n(\phi_j),$$

and so

$$\limsup_{n\to\infty} p_n(F) \leq p_\infty(\phi_j).$$

Sending $j \to \infty$ we obtain the upper bound (ii) in the definition of weak convergence.

For the lower bound (i), let G be an open set, and $x \in G$. There exists $\eta > 0$ such that $\{x\}^\eta \subset G$. Note that $\lim_{n\to\infty} p_n(\mathbf{R}^n) = p_\infty(\mathbf{R}^n)$. Choose $M > 0$ such that

$$p_\infty(\mathbf{R}^n) - M < p_\infty(x).$$

Now define $\phi \in C_b(\mathbf{R}^n)$ by

$$\phi(y) = -M\min\left(\frac{|x-y|}{\eta}, 1\right);$$

then $-M \leq \phi(y) \leq 0$ for all $y \in \mathbf{R}^n$, $\phi(x) = 0$, and $\phi(y) = -M$ if $y \notin \{x\}^\eta$. Then

$$p_n(\phi) = \max(\sup_{y\in\{x\}^\eta}\{p_n(y) + \phi(y)\}, \sup_{y\notin\{x\}^\eta}\{p_n(y) + \phi(y)\})$$

$$\leq \max(\sup_{y\in\{x\}^\eta}\{p_n(y)\}, p_n([\{x\}^\eta]^c) - M)$$

$$\leq \max(p_n(G), p_n(\mathbf{R}^n) - M).$$

Therefore,

$$p_\infty(\phi) = \lim_{n\to\infty} p_n(\phi) \leq \max(\liminf_{n\to\infty} p_n(G), p_\infty(\mathbf{R}^n) - M).$$

Now

$$p_\infty(\phi) \geq p_\infty(x) + \phi(x) = p_\infty(x),$$

and so

$$\liminf_{n\to\infty} p_n(G) \geq p_\infty(x).$$

This inequality holds for all $x \in G$; hence

$$\liminf_{n\to\infty} p_n(G) \geq p_\infty(G).$$

This completes the proof. □

Proof of Theorem C.5.3. We follow the proof of [DE97, Theorems 1.3.7 and 1.3.8].

We can assume without loss of generality that $\Lambda(0) = 0$. Define p_∞ by (C.1). Now for all $f \in C_b(\mathbf{R}^n)$, we have from (C.1)

$$p_\infty(x) \leq \Lambda(f) - f(x),$$

and hence

$$\Lambda(f) \geq p_\infty(f). \tag{C.4}$$

We claim that $p_\infty \in \mathcal{X}_{et}$. Setting $f = 0$ in (C.4) we get

$$p_\infty(x) \leq \Lambda(0) = 0.$$

Thus $p_\infty \in \mathbf{B}_e(\mathbf{R}^n)$.

Fix x and assume $p^\infty(x) > -\infty$ (the other case is treated similarly). Given $\varepsilon > 0$, there exists $f \in C_b(\mathbf{R}^n)$ such that

$$p_\infty(x) + \varepsilon \geq \Lambda(f) - f(x).$$

Since f is continuous, there exists $\eta > 0$ such that $|f(x) - f(y)| < \varepsilon$ for all $y \in \{x\}^\eta$. Then for such y,

$$\begin{aligned} p_\infty(y) &\leq \Lambda(f) - f(y) \\ &\leq \Lambda(f) - f(x) + \varepsilon \\ &\leq p_\infty(x) + 2\varepsilon. \end{aligned}$$

This shows that $p_\infty \in \mathbf{USC}_e(\mathbf{R}^n)$.

Next, we show that p_∞ is tight. Fix $M > 0$. By the uniform tightness of p_n, there exists a compact set K such that

$$\limsup_{n\to\infty} p_n(K^c) \leq -2M.$$

Fix $x \in K^c$. Define $\psi(t) = 1-t$ on $[0,1]$ and $\psi(t) = 0$ on $[0,\infty)$. Let $\eta = d(x,K)/2$ and define

$$f(y) = -2M\psi(d(y,K)/\eta).$$

Then f is uniformly continuous, $f(x) = 0$, $-2M \leq f(y) \leq 0$ for all $y \in \mathbf{R}^n$, and $f(y) = -2M$ for $y \in K$. Then

$$\begin{aligned} \Lambda(f) &= \lim_{n\to\infty} \max(\sup_{y\in K}\{p_n(y) + f(y)\}, \sup_{y\in K^c}\{p_n(y) + f(y)\}) \\ &\leq \limsup_{n\to\infty} \max\{p_n(K) - 2M, p_n(K^c)\} \\ &\leq -2M. \end{aligned}$$

Therefore, $p_\infty(x) \le -2M$, and so $\{y : p_\infty(y) \ge -M\}$ is a closed subset of K. Thus $p_\infty \in \mathcal{X}_{et}$ (tight, u.s.c., bounded above).

We next claim that

$$\Lambda(f) \le p_\infty(f) \text{ for all } f \in C_b(\mathbf{R}^n), \tag{C.5}$$

which together with (C.4) implies (C.2) and hence $p_n \Longrightarrow p_\infty$.

Fix $f \in C_b(\mathbf{R}^n)$ and assume without loss of generality that $f \le 0$. Fix $M > 0$ and select a compact set K such that

$$\limsup_{n\to\infty} p_n(K^c) \le -M.$$

Let $\varepsilon > 0$. Then if $p_\infty(x) > -\infty$, there exists $\phi_x \in C_b(\mathbf{R}^n)$ with $\phi_x(x) = 0$ and

$$p_\infty(x) \ge \Lambda(\phi_x) - \varepsilon,$$

while if $p_\infty(x) = -\infty$, there exists $\phi_x \in C_b(\mathbf{R}^n)$ with $\phi_x(x) = 0$ and

$$\Lambda(\phi_x) \le -M - \| f \|_\infty - 2\varepsilon.$$

For all $x \in K$ there exists $\eta_x > 0$ such that if $y \in \{x\}^{\eta_x}$, then $-\varepsilon \le \phi_x(y)$ and $f(y) \le f(x) + \varepsilon$. Let $x_i \in K, i = 1, \ldots, N$, be a finite set of points such that $\{x_i\}^{\eta_{x_i}}$ is a cover of K (by compactness). Then for $y \in K$ we have

$$f(y) \le \max_{i=1}^{N} (\phi_{x_i}(y) + f(x_i) + 2\varepsilon).$$

Then

$$\begin{aligned}
\Lambda(f) &= \lim_{n\to\infty} \max(\sup_{y\in K}\{p_n(y) + f(y)\}, \sup_{y\in K^c}\{p_n(y) + f(y)\}) \\
&\le \limsup_{n\to\infty} \max(\max_{i=1}^N\{\sup_{y\in K}(p_n(y) + \phi_{x_i}(y) + f(x_i) + 2\varepsilon), p_n(K^c)\}) \\
&\le \limsup_{n\to\infty} \max(\max_{i=1}^N(\{p_n(\phi_{x_i}) + f(x_i) + 2\varepsilon, -M\}) \\
&\le \max(\max_{i=1}^N\{\Lambda(\phi_{x_i}) + f(x_i) + 2\varepsilon, -M\}) \\
&\le \max(\max_{i=1}^N\{p_\infty(x_i) + f(x_i) + 3\varepsilon, -M\}) \\
&\le \max(p_\infty(f) + 3\varepsilon, -M).
\end{aligned}$$

This implies (C.5) on sending $M \to \infty$ and then $\varepsilon \to 0$. □

Proof of Theorem C.5.4. This proof is based on [DE97, Theorem 1.3.7].

There exists a metric m on $\mathbf{R}^n$ such that the space $U_b(\mathbf{R}^n, m)$ of bounded, uniformly continuous functions is separable with respect to the uniform metric $\| \cdot \|^\infty$ defined by m. Let Ξ be a countable, dense subset of $U_b(\mathbf{R}^n, m)$. Assume, without loss of generality, that $c = \lim_{n\to\infty} p_n(\mathbf{R}^n) = 0$. For each $f \in \Xi$, the sequence

$\{p_n(f)\}_{n=0}^{\infty}$ is bounded. Hence by diagonalization, there exists a subsequence n_i such that the limit

$$\Lambda(f) = \lim_{i\to\infty} p_{n_i}(f)$$

exists for all $f \in \Xi$. Also, $\Lambda(0) = 0$. By approximation, this limit exists for all $f \in U_b(\mathbf{R}^n, m)$. Indeed, given any $f \in U_b(\mathbf{R}^n, m)$ there exists $f_k \in \Xi$ with $\| f - f_k \|_\infty \leq 1/k$. Then

$$\Lambda(f_k) - \tfrac{1}{k} \leq \liminf_{i\to\infty} p_{n_i}(f)$$

$$\leq \limsup_{i\to\infty} p_{n_i}(f) \leq \Lambda(f_k) + \tfrac{1}{k}.$$

Sending $k \to \infty$ we infer that the desired limit exists. Further, $\Lambda(f)$ is well defined for all $f \in U_b(\mathbf{R}^n, m)$.

Next, define

$$p_\infty(x) = \inf\{\Lambda(f) - f(x) \ : \ f \in U_b(\mathbf{R}^n, m)\}. \tag{C.6}$$

We claim that $p_n \Longrightarrow p_\infty$. If we knew that $\Lambda(f)$'s were defined for all $f \in C_b(\mathbf{R}^n)$, we could apply Theorem C.5.3 and conclude.

However, the test functions constructed in the proof of Theorem C.5.3 were uniformly continuous and bounded, and the proof actually implies that the function p_∞ defined by (C.6) belongs to $\mathcal{X}_{et}$,

$$\Lambda(f) = p_\infty(f) \ \text{ for all } \ f \in U_b(\mathbf{R}^n, m). \tag{C.7}$$

Setting $f = 0$ we see that $0 = \Lambda(0) = p_\infty(\mathbf{R}^n)$; hence p_∞ is normalized: $p_\infty \in \mathcal{X}_{et,1}$.

To see that $p_n \Longrightarrow p_\infty$, consider the proof of Theorem C.5.1. The test functions ϕ_j and ϕ are bounded and uniformly continuous, and therefore (C.7) is enough to ensure that conditions (i) and (ii) in the definition of weak convergence are satisfied. □

Proof of Theorem C.5.5. Refer to [Jia95, Theorems 2.1 and 2.2]. □

Proof of Theorem C.5.6. Refer to [Jia95, Corollary 2.2]. □

Proof of Theorem C.5.7. Property (i) is given in [AW83, Theorem 2.5].

Property (ii) is given in [AW83, Theorem 2.6].

Property (iii) is from [AW83, §2]. □

Proof of Theorem C.5.8. Assume $p_n \xrightarrow{h} p_\infty$. Let $f \in C_b(\mathbf{R}^n)$. Define $q_n(x) = p_n(x) + f(x)$ and $q_\infty(x) = p_\infty(x) + f(x)$. Then $q_n \xrightarrow{h} q_\infty$. Part i(a) of Theorem C.5.7 implies $p_n(f) = \langle q_n \rangle \to \langle q_\infty \rangle = p_\infty(f)$. Therefore, $p_n \Longrightarrow p_\infty$.

Conversely, assume $p_n \Longrightarrow p_\infty$. Since p_∞ is u.s.c., given $\varepsilon > 0$, there exists $\eta > 0$ such that $p_\infty(y) \leq p_\infty(x) + \varepsilon$ if $y \in \{x\}^\eta$. Let F denote the closure of $\{x\}^{\eta/2}$. Then

$$p_\infty(F) \leq p_\infty(x) + \varepsilon.$$

Let n_i be a subsequence and $x_i \to x$ as $i \to \infty$. There exists I such that $i \geq I$ implies $x_i \in \{x\}^{\eta/4} \subset F$. Now for $i \geq I$,

$$p_{n_i}(x_i) \leq p_{n_i}(F),$$

and sending $i \to \infty$, we have

$$\limsup_{i\to\infty} p_{n_i}(x_i) \le \limsup_{i\to\infty} p_{n_i}(F)$$
$$\le p_\infty(F)$$
$$\le p_\infty(x) + \varepsilon.$$

Since $\varepsilon > 0$ was arbitrary, this proves part (i) of the definition of hypoconvergence.

For part (i) of the definition of hypoconvergence, we use the fact that since p_∞ is tight, $\rho_e(p_n, p_\infty) \to 0$. This implies that there exists $\varepsilon_n \to 0$ such that

$$e^{p_\infty(F)} \le e^{p_n(F^{\varepsilon_n})} + \varepsilon_n \quad \text{for all } F \in \mathcal{F}.$$

Set $F = \{x\}$, so that $F^{\varepsilon_n} = \{x\}^{\varepsilon_n} = \{x' : |x' - x| < \varepsilon_n\}$. For each n, there exists $x_n \in \{x\}^{\varepsilon_n}$ such that

$$e^{p_n(x_n)} \ge e^{p_n(\{x\}^{\varepsilon_n})} - \varepsilon_n.$$

Therefore,

$$e^{p_n(x_n)} \ge e^{p_\infty(\{x\})} - 2\varepsilon_n,$$

and sending $n \to \infty$ gives

$$\liminf_{n\to\infty} e^{p_n(x_n)} \ge e^{p_\infty(x)}$$

as required (on taking logs). □

Proof of Theorem C.5.9. If $p_n \to p_\infty$ uniformly on compact sets, then items (i) and (ii) of the definition of hypoconvergence follow readily for each x; so $p_n \xrightarrow{h} p_\infty$. Since p_∞ is tight, the weak convergence $p_n \Longrightarrow p_\infty$ follows from Theorem C.5.8. □

Proof of Theorem C.5.10. Given any $x \in \mathbf{R}^n$, since $p_n \Longrightarrow p_\infty$, by item (ii) of the definition of weak convergence with $F = \{x\}$ we have, given $\varepsilon > 0$, there exists N such that $n \ge N$ implies

$$p_n(x) \le p_\infty(x) + \varepsilon.$$

Therefore, if $p_n(x) \ge k$, then x is contained in the compact set $K_k = \{x' : p_\infty(x') \ge k - \varepsilon\}$ in view of the tightness of p_∞. This implies that the sets $\operatorname{argmax} p_n$ are nonempty for $n \ge N$ and that all these sets are contained in the compact set K_k for some k.

Thus if $x_n \in \operatorname{argmax}_x\{p_n(x)\}$, then the sequence $\{x_n\}_{n\ge N}$ must have a convergent subsequence, $x_{n_i} \to x_\infty$, for some x_∞, since this sequence is contained in the compact set K_k. Since p_∞ is u.s.c., given $\varepsilon > 0$, there exists $\eta > 0$ such that $p_\infty(y) \le p_\infty(x_\infty) + \varepsilon$ if $y \in \{x_\infty\}^\eta$. Let F denote the closure of $\{x_\infty\}^{\eta/2}$. Then

$$p_\infty(F) \le p_\infty(x_\infty) + \varepsilon.$$

Now there exists I such that $i \ge I$ implies $x_i \in \{x_\infty\}^{\eta/4} \subset F$. Now for $i \ge I$,

$$\langle p_{n_i}\rangle = p_{n_i}(x_i) \le p_{n_i}(F).$$

Also, we have, for i sufficiently large, in view of the weak convergence,

$$\langle p_\infty \rangle - \varepsilon \le \langle p_{n_i} \rangle,$$

and sending $i \to \infty$, we have

$$\begin{aligned}\langle p_\infty \rangle - \varepsilon \le \limsup_{i\to\infty} \langle p_{n_i} \rangle &\le \limsup_{i\to\infty} p_{n_i}(F)\\ &\le p_\infty(F)\\ &\le p_\infty(x_\infty) + \varepsilon.\end{aligned}$$

Since $\varepsilon > 0$ was arbitrary, this proves $x_\infty \in \operatorname{argmax}_x\{p_\infty(x)\}$. □

Proof of Theorem C.5.11. Let $x_\infty \in \operatorname{argmax}_x\{p_n + f\}$. By item (ii) of the definition of hypoconvergence, there exists a sequence such that $x_n \to x_\infty$ and $\liminf_n p_n(x_n) \ge p_\infty(x_\infty)$. Therefore,

$$\langle p_n + f \rangle \ge p_n(x_n) + f(x_n)$$

and sending $n \to \infty$ we get, using also the continuity of f,

$$\liminf_{n\to\infty} \langle p_n + f \rangle \ge p_n(x_\infty) + f(x_\infty) = \langle p_\infty + f \rangle.$$

Conversely, let $x_n \in \operatorname{argmax}_x\{p_n + f\}$. By tightness, there exists a convergent subsequence $x_{n_i} \to x_\infty$ for some limit point x_∞. Then

$$\langle p_{n_i} + f \rangle \ge p_{n_i}(x_{n_i}) + f(x_{n_i})$$

and sending $i \to \infty$ gives

$$\limsup_{i\to\infty} \langle p_{n_i} + f \rangle \le p_\infty(x_\infty) + f(x_\infty) \le \langle p_\infty + f \rangle$$

using item (i) of the definition of hypoconvergence and the continuity of f. This will hold for all subsequences, and so

$$\limsup_{n\to\infty} \langle p_n + f \rangle \le \langle p_\infty + f \rangle.$$

□

Bibliography

[AAK68] V.M. Adamjam, D.Z. Arov, and M.G. Krein. Infinite Hankel matrices and generalized problems of Caratheodory-Fejer and F. Riesz. *Funct. Anal. Appl.*, 2:1–18, 1968.

[AAK72] V.M. Adamjam, D.Z. Arov, and M.G. Krein. Analytic properties of Schmidt pairs for Hankel operator and the generalized Schur-Takagi problem. *Math. USSR-Sb.*, 15:15–78, 1972.

[AAK78] V.M. Adamjam, D.Z. Arov, and M.G. Krein. Infinite block Hankel matrices and related extension problems. *Amer. Math. Soc. Transl.*, 111:133–156, 1978.

[Aki95] M. Akian. Densities of idempotent measures and large deviations. INRIA Report No. 2534, 1995.

[Aki96] M. Akian. A probabilistic viewpoint on the convergence of optimization problems. Preprint, 1996.

[AQV94] M. Akian, J.P. Quadrat, and M. Viot. Bellman processes. In *Lecture Notes in Control and Inform. Sci.*, 199, Springer-Verlag, New York, 1994.

[AV73] B.D.O. Anderson and S. Vongpanitlerd. *Network Analysis and Synthesis.* Prentice-Hall, Englewood Cliffs, NJ, 1973.

[And63] T. Ando. On a pair of commuting contractions. *Acta Sci. Math. (Szeged)*, 24:88–90, 1963.

[AW83] H. Attouch and R.J.B. Wets. A convergence theory for saddle functions. *Trans. Amer. Math. Soc.*, 280(1):1–41, 1983.

[BFHT87a] J.A. Ball, C. Foias, J.W. Helton, and A. Tannenbaum. Nonlinear interpolation theory in H^∞. In Modeling, Robustness and Sensitivity Reduction in Control Systems. NATO Adv. Sci. Inst. Ser. F: Comput. Systems Sci. 34, pp. 31–46. Springer-Verlag, Berlin, New York, 1987.

[BFHT87b] J.A. Ball, C. Foias, J.W. Helton, and A. Tannenbaum. On a local nonlinear commutant lifting theorem. *Indiana Univ. Math. J.*, 36:693–709, 1987.

[BH83] J.A. Ball and J.W. Helton. A Beurling-Lax Theorem for the Lie group $U(m, n)$ which contains most classical interpolation theory. *J. Operator Theory*, 1:107–142, 1983.

[BH88a] J. Ball and J.W. Helton. Factorization of nonlinear systems: Toward a theory for nonlinear H^∞ control. In *Proc. 27th IEEE CDC*, pp. 2376–2381, 1988.

[BH88b] J. Ball and J.W. Helton. Interpolation problems for null and pole structure of nonlinear systems. In *Proc. 27th IEEE CDC*, pp. 14–19, 1988.

[BH88c] J.A. Ball and J.W. Helton. Shift invariant manifolds and nonlinear analytic function theory. *Integral Equations Operator Theory*, 11(5):615–725, 1988.

[BH89] J.A. Ball and J.W. Helton. H^∞ control for nonlinear plants: Connections with differential games. In *Proc. 28th IEEE CDC*, pp. 956–962, 1989.

[BH92a] J.A. Ball and J.W. Helton. Inner-outer factorization of nonlinear operators. *J. Funct. Anal.*, 104:363–413, 1992.

[BH92b] J.A. Ball and J.W. Helton. Nonlinear H_∞ control theory for stable plants. *J. Math. Control, Signals, and Systems*, 5:233–261, 1992.

[BH96] J.A. Ball and J.W. Helton. Viscosity solutions of Hamilton-Jacobi Equations arising in nonlinear H_∞ control. *J. Math. Systems Estim. Control*, 6:1–22, 1996.

[BHV91] J.A. Ball, J.W. Helton, and M. Verma. A factorization principle for stabilization of linear control systems. *Internat. J. Robust Nonlinear Control*, 1:229–294, 1991.

[BHW93] J.A. Ball, J.W. Helton, and M.L. Walker. H^∞ control for nonlinear systems with output feedback. *IEEE Trans. Automat. Control*, 38:546–559, 1993.

[BR87] J.A. Ball and A.C.M. Ran. Optimal Hankel norm model reductions and Wiener–Hopf factorization I: The canonical case. *SIAM J. Control Optim.*, 25(2):362–382, 1987.

[BvdS96] J.A. Ball and A.J. van der Schaft. J-inner outer factorization, J-spectral factorization and robust control for nonlinear systems. *IEEE Trans. Automat. Control*, 41(3), pp. 379–392, 1996.

[Bar98a] L. Baramov. On H_∞ control of nonstandard systems based on factorization. In *Proc. MTNS*, pp. 57–60, 1998.

[Bar98b] L. Baramov. Solutions to a class of nonstandard nonlinear H^∞ control problems. *Internat. J. Control*, to appear.

[BK96] L. Baramov and H. Kimura. Nonlinear local J-lossless conjugation and factorization. *Internat. J. Robust Nonlinear Control*, 6(8):869–893, 1996.

[BBJ88] J.S. Baras, A. Bensoussan, and M.R. James. Dynamic observers as asymptotic limits of recursive filters: Special cases. *SIAM J. Appl. Math.*, 48(5):1147–1158, 1988.

[BJ97] J.S. Baras and M.R. James. Robust and risk-sensitive control for finite state machines and hidden markov models. *J. Math. Systems Estim. Control*, 7:371–374, 1997.

[BK82] J.S. Baras and P.S. Krishnaprasad. Dynamic observers as asymptotic limits of recursive filters. In *Proc. 21st IEEE CDC*, pp. 1126–1127, 1982.

[BCD98] M. Bardi and I. Capuzzo-Dolcetta. *Optimal Control and Viscosity Solutions of Hamilton-Jacobi-Bellman Equations*. Birkhauser, Boston, 1998.

[BO95] T. Basar and G.J. Olsder. *Dynamic Noncooperative Game Theory*. Academic Press, New York, second edition, 1995.

[BB91] T. Basar and P. Bernhard. *H^∞-Optimal Control and Related Minimax Design Problems: A Dynamic Game Approach*. Birkhauser, Boston, 1991 (second ed. 1995).

[Bel57] R. Bellman. *Dynamic Programming*. Princeton University Press, Princeton, NJ, 1957.

[Ben81] V.E. Benes. Exact finite dimensional filters for certain diffusions with nonlinear drift. *Stochastics*, 5:65–92, 1981.

[BE96] A. Bensoussan and R.J. Elliott. A finite-dimensional risk-sensitive control problem. *SIAM J. Control Optim.*, 33(6):1834–1846.

[BvS85] A. Bensoussan and J.H. van Schuppen. Optimal control of partially observable stochastic systems with an exponential-of-integral performance index. *SIAM J. Control Optim.*, 23(4):599–613, 1985.

[Ber91] P. Bernhard. Application of the minimax certainty equivalence principle to sampled data output feedback H_∞ control problem. *Systems Control Lett.*, 16:229–234, 1991.

[Ber94] P. Bernhard. Discrete and continuous time partial information minimax control. Preprint, 1994.

[Ber96] P. Bernhard. A separation theorem for expected value and feared value discrete time games. *ESAIM Control Optim. Calc. Var.*, 1:191–206, 1996. `http://www.emath.fr/cocv/`.

[BR95] P. Bernhard and A. Rapaport. Min-max certainty equivalence principle and differential games. *Internat. J. Robust Nonlinear Control*, 6:825–842, 1996.

[BSS80] M. Bettayeb, L.M. Silverman, and M.G. Safanov. Optimal approximation of continuous time systems. *Proc. 19th IEEE CCDC*, pp. 195–198, 1980.

[CE95] C. Charalambous and R.J. Elliott. Remarks on the explicit solutions for nonlinear partially observable stochastic control problems and relations to H_∞ or robust control. In *Proc. 34th IEEE CDC*, pp. 2858–2863, New Orleans, 1995.

[CS94] D.G. Chichka and J.L. Speyer. An adaptive controller based on disturbance attenuation. In *Proc. 33rd IEEE CDC*, Orlando, FL, pp. 3719–3726, 1994.

[CEL84] M.G. Crandall, L.C. Evans, and P.L. Lions. Some properties of viscosity solutions of Hamilton–Jacobi equations. *Trans. Amer. Math. Soc.*, 282:487–502, 1984.

[CH95] C. Charalambous and J.L. Hibey. Applications of minimum principle for continuous-time partially observable risk-sensitive control problems. In *Proc. 34th IEEE CDC*, New Orleans, pp. 3420–3422, 1995.

[CJ84] B.C. Chang and B. Pearson Jr. Optimal disturbance reduction in linear multivariable systems. *IEEE Trans. Automat. Control.*, AC-29:880–887, 1984.

[Cla83] F.H. Clarke. *Optimization and Non-Smooth Analysis*. Wiley-Interscience, New York, 1983.

[DBB93] G. Didinsky, T. Basar, and P. Bernhard. Structural properties of minimax policies for a class of differential games arising in nonlinear H_∞ control and filtering. In *Proc. 32nd IEEE CDC*, pp. 184–189, 1993.

[DE97] P. Dupuis and R.S. Ellis. *A Weak Convergence Approach to the Theory of Large Deviations*. Wiley, New York, 1997.

[Dei85] K. Deimling. *Nonlinear Functional Analysis*. Springer-Verlag, New York, 1985.

[DFT92] J. Doyle, B. A. Francis, and A. Tannenbaum. *Feedback Control Theory*. Macmillan, New York, 1992.

[DGKF89] J.C. Doyle, K. Glover, P.P. Khargonekar, and B. Francis. State-space solutions to the standard H_2 and H_∞ control problems. *IEEE Trans. Automat. Control*, 34(8):831–847, 1989.

[Doy83] J.C. Doyle. Synthesis of robust controllers and filters. In *Proc. 22nd IEEE CDC*, San Antonio, TX, pp. 109–114, 1983.

[Doy84] J.C. Doyle. Lecture notes in advances in multivariable control. *Honeywell/ONR Workshop*, 1984.

[Dun67] T.E. Duncan. *Probability Densities for Diffusion Processes with Application to Nonlinear Filtering Theory*. Ph.D. thesis, Stanford University, Stanford, CA, 1967.

[DD81a] P. Dewilde and H. Dym. Schur recursions , error formulas and convergence of rational estimators for stationary stochastic processes, *IEEE Trans. Inform. Theory*, 27: 446–461, (1981).

[DD81b] P. Dewilde and H. Dym. Lossless chain scattering matrices and optimum linear prediction: The vector case, *Circuit Theory Appl.*, 9: 135–175, (1981).

[DVK78] P. DeWilde, A. Vieira, and T. Kailath. On a generalized Szegö-Levinson realization algorithm for optimal linear predictors based on a network synthesis approach. *IEEE Trans. Circuit Theory*, 25:663–675, 1978. Special Issue on Math. Foundations of Systems Theory.

[EAM95] R.J. Elliott, L. Aggoun, and J.B. Moore. *Hidden Markov Models: Estimation and Control.* Springer-Verlag, New York, 1995.

[Ell82] R.J. Elliott. *Stochastic Calculus and Applications.* Springer-Verlag, New York, 1982.

[ES84] L.C. Evans and P.E. Souganidis. Differential games and representation formulas for solutions of Hamilton-Jacobi-Isaacs equations. *Indiana Univ. Math. J.*, 33:773–797, 1984.

[FGT95] C. Foias, C. Gu, and A. Tannenbaum. Nonlinear H^∞ optimization: A causal power series approach. *SIAM J. Control Optim.*, 33(1):185–207, 1995.

[FGT96] C. Foias, C. Gu, and A. Tannenbaum. Nonlinearity in H^∞ control theory, causality in the commutant lifting theorem, and extension of intertwining operators. *Integral Equations Operator Theory*, 25:481–489, 1996.

[FGT98] C. Foias, C. Gu, and A. Tannenbaum. On the nonlinear standard H^∞ problem. *Oper. Theory Adv. Appl.*, 1998.

[FHH97] W.H. Fleming and D. Hernández-Hernández. Risk-sensitive control of finite state machines on an infinite horizon I. *SIAM J. Control Optim.*, 35(5):1790–1810, 1997.

[FHH98] W.H. Fleming and D. Hernández-Hernández. Risk-sensitive control of finite state machines on an infinite horizon II. *SIAM J. Control Optim.*, 37(4):1048–1069, 1998.

[FHZ84] B.A. Francis, J.W. Helton, and G. Zames. H^∞-optimal feedback controller for linear multivariable systems. *IEEE Trans. Automat. Control*, AC-29:888–900, 1984.

[FJ96] W.H. Fleming and M.R. James. The risk-sensitive index and the H_2 and H_∞ norms for nonlinear systems. *J. Math. Control, Signals, and Systems*, 8:199–221, 1996.

[Fle82] W.H. Fleming. Nonlinear semigroup for controlled partially observed diffusions. *SIAM J. Control Optim.*, 20(2):286–301, 1982.

[FM] W.H. Fleming and W.M. McEneaney. A max-plus based algorithm for an HJB equation of nonlinear filtering. Preprint. Mathematical Department, North Carolina State University, Raleigh, NC.

[FM92] W.H. Fleming and W.M. McEneaney. Risk-sensitive optimal control and differential games. In *Stochastic Theory and Adaptive Control*, Lecture Notes in Control Inform. Sci. 184, pp. 185–197. Springer-Verlag, Berlin, 1992.

[FM95] W.H. Fleming and W.M. McEneaney. Risk-sensitive control on an infinite time horizon. *SIAM J. Control Optim.*, 33(6):1881–1915, 1995.

[FP82] W.H. Fleming and E. Pardoux. Optimal control for partially observed diffusions. *SIAM J. Control Optim.*, 20(2):261–285, 1982.

[FR75] W.H. Fleming and R.W. Rishel. *Deterministic and Stochastic Optimal Control.* Springer-Verlag, New York, 1975.

[Fra84] B.A. Francis. *A Course in H_∞ Control Theory.* Lecture Notes in Control and Inform. Sci., 88, Springer-Verlag, New York, 1984.

[FS93] W.H. Fleming and H.M. Soner. *Controlled Markov Processes and Viscosity Solutions.* Springer-Verlag, New York, 1993.

[GD88] K. Glover and J.C. Doyle. State-space formulae for all stabilizing controllers that satisfy an H_∞-norm bound and relations to risk-sensitivity. *Systems Control Lett.*, 11(3):167–172, 1988.

[GGLD90] M. Green, K. Glover, D. Limebeer, and J. Doyle. A J-spectral factorization approach to H_∞ control. *SIAM J. Control Optim.*, 28:1350–1371, 1990.

[GL95] M. Green and D.J.N. Limebeer. *Linear Robust Control.* Prentice–Hall, Englewood Cliffs, NJ, 1995.

[Glo84] K. Glover. All optimal Hankel-norm approximations of linear multivariable systems and their L_∞ error bounds. *Internat. J. Control*, 39:1115–1193, 1984.

[Gre92] M. Green. H_∞ controller synthesis by J-lossless-coprime factorization. *SIAM J. Control Optim.*, 30(3):522–547, 1992.

[HBJP87] J.W. Helton, J. Ball, C. Johnson, and C. Palmer. *Operator Theory, Analytic Functions, Matrices and Electrical Engineering*, volume 68 of CBMS *Regional Conference Series in Mathematics*. AMS, Providence, RI, 1987.

[Hel76] J.W. Helton. Operator theory and broadband matching. In *Proc. Eleventh Annual Allerton Conference on Circuits and Systems Theory*, 1976.

[Hel78] J.W. Helton. A mathematical view of broadband impedance matching. In *IEEE Internat. Symp. Circuits and Systems*, pp. 978–980, 1978.

[Hel81] J.W. Helton. Broadbanding gain equalization directly from data. *IEEE Trans. Circuits Systems, CAS*-28, 12:1125–1137, 1981.

[Hel83] J.W. Helton. An H^∞ approach to control. In *IEEE Conference on Decision Control*, San Antonio, TX, pp. 607–612, 1983.

[Hij80] O. Hijab. *Minimum Energy Estimation*. Ph.D. thesis, University of California, Berkeley, 1980.

[Hij90] O. Hijab. Partially observed control of Markov processes III. *Ann. Probab.*, 18:1099–1125, 1990.

[HJ94] J.W. Helton and M.R. James. An information state approach to nonlinear J-inner/outer factorization. In *Proc. 33rd IEEE CDC, Orlando, FL*, pp. 2565–2571, 1994.

[HJ95] J.W. Helton and M.R. James. Reduction of controller complexity in nonlinear H_∞ control. In *Proc. 34th IEEE CDC, New Orleans*, pp. 2233–2238, 1995.

[HJ96a] J.W. Helton and M.R. James. J-inner/outer factorization for bilinear systems. In *Proc 35th IEEE CDC*, Kobe, Japan, pp. 3788–3793, 1996.

[HJ96b] J.W. Helton and M.R. James. On the stability of the information state system. *Systems Control Lett.*, 29:61–72, 1996.

[HJ98] J.W. Helton and M.R. James. On verifying the certainty equivalence assumptions in nonlinear H^∞ control. In *37th IEEE CDC*, pp. 4069–4074, 1998.

[HJM98] J.W. Helton, M.R. James, and W.M. McEneaney. Nonlinear control: The joys of having an extra sensor. In *37th IEEE CDC*, pp. 3518–3524, 1998.

[HK77] R. Hermann and A.J. Krener. Nonlinear controllability and observability. *IEEE Trans. Automat. Control*, 22(5):728–740, 1977.

[HM76] D. Hill and P. Moylan. The stability of nonlinear dissipative systems. *IEEE Trans. Automat. Control*, AC-21(5):708–711, 1976.

[HM77] D. Hill and P. Moylan. Stability results for nonlinear feedback systems. *Automatica*, 13:377–382, 1977.

[HM80a] D. Hill and P. Moylan. Dissipative dynamical systems: Basic input-output and state properties. *J. Franklin Inst.*, 309:327–357, 1980.

[HM80b] D.J. Hill and P.J. Moylan. Connections between finite-gain and asymptotic stability. *IEEE Trans. Automat. Control*, AC-25:931–936, 1980.

[HM98a] J.W. Helton and O. Merino. *Classical Control Using H^∞ Methods: Theory, Optimization, and Design*. SIAM, Philadelphia, 1998.

[HM98b] M. Horton and W.M. McEneaney. Max-plus eigenvector representations for nonlinear H^∞ value functions. Preprint, Dept. Mathematics, North Carolina State University, Raleigh, NC, 1998.

[HV95] J.W. Helton and A.E. Vityaev. A new type of HJBI inequality governing situations where certainty equivalence fails. In *Proc. IFAC NOLCOS*, pp. 715–720, 1995.

[IA92a] A. Isidori and A. Astolfi. Disturbance attenuation and H_∞-control via measurement feedback. *IEEE Trans. Automat. Control*, 37:1283–1293, 1992.

[IA92b] A. Isidori and A. Astolfi. Nonlinear H_∞-control via measurement feedback. *J. Mathematical Systems Estim. Control*, 2(1):31–44, 1992.

[IK95] A. Isidori and W. Kang. H_∞ control via measurement feedback for general nonlinear systems. *IEEE Trans. Automat. Control*, 40(3):466–472, 1995.

[Isa65] R. Isaacs. *Differential Games*. Wiley, New York, 1965 (Kruger, 1975).

[Isi94] A. Isidori. Nonlinear H_∞ control via measurement feedback for affine nonlinear systems. *Internat. J. Robust Nonlinear Control*, 4:553–574, 1994.

[Jac73] D.H. Jacobson. Optimal stochastic linear systems with exponential performance criteria and their relation to deterministic differential games. *IEEE Trans. Automat. Control*, 18:124–131, 1973.

[Jam91] M.R. James. Finite time observer design by probabilistic-variational methods. *SIAM J. Control Optim.*, 29(4):954–967, 1991.

[Jam92] M.R. James. Asymptotic analysis of nonlinear stochastic risk-sensitive control and differential games. *Math. Control Signals Systems*, 5(4):401–417, 1992.

[Jam93] M.R. James. A partial differential inequality for dissipative nonlinear systems. *Systems Control Lett.*, 21(4):315–320, 1993.

[Jam94a] M.R. James. Book review. *Stochastics*, 49:129–137, 1994. (Review of W.H. Fleming and H.M. Soner, *Controlled Markov Processes and Viscosity Solutions*, Springer, New York, 1993.)

[Jam94b] M.R. James. On the certainty equivalence principle and the optimal control of partially observed dynamic games. *IEEE Trans. Automat. Control*, 39(11):2321–2324, 1994.

[Jam98] M.R. James. Nonlinear semigroups for partially observed risk-sensitive control and minimax games. In Stochastic Analysis, Control, Optimization and Applications. W.M. McEneaney et al eds., *Volume in Honor of W.H. Fleming on the Occasion of his 70th Birthday*, Birkhauser, Boston, pp. 57–73, 1999.

[JB88] M.R. James and J.S. Baras. Nonlinear filtering and large deviations: A PDE-control theoretic approach. *Stochastics*, 23(3):391–412, 1988.

[JB95] M.R. James and J.S. Baras. Robust H^∞ output feedback control for nonlinear systems. *IEEE Trans. Automat. Control*, 40:1007–1017, 1995.

[JB96] M.R. James and J.S. Baras. Partially observed differential games, infinite-dimensional Hamilton–Jacobi–Isaacs equations, and nonlinear H^∞ control. *SIAM J. Control Optim.*, 34(4):1342–1364, 1996.

[JBE94] M.R. James, J.S. Baras, and R.J. Elliott. Risk-sensitive control and dynamic games for partially observed discrete-time nonlinear systems. *IEEE Trans. Automat. Control*, 39:780–792, 1994.

[Jia95] T. Jiang and G.L. O'Brien. The metric of large deviations convergence. *J. Theoret. Probab.*, to appear.

[JY95] M.R. James and S. Yuliar. A nonlinear partially observed differential game with a finite dimensional information state. *Systems Control Lett.*, 26:137–145, 1995.

[Kal60] R.E. Kalman. A new approach to linear filtering and prediction theory. *ASME Trans., Series D, J. Basic Engineering*, 82:35–45, 1960.

[KB60] R.E. Kalman and R.S. Bucy. New results in linear filtering and prediction theory. *ASME Trans., Series D, J. Basic Engineering*, 83:95–108, 1961.

[KET75] S.R. Kou, D.L. Elliott, and T.J. Tarn. Exponential observers for nonlinear dynamic systems. *Inform. Cont.*, 29:204–216, 1975.

[Kim84] H. Kimura. Robust stabilizability for a class of transfer functions. *IEEE Trans Automat. Control*, AC-29:788–793, 1984.

[Kim92] H. Kimura. (J, J')-lossless factorization based on conjugation. *Systems Control Lett.*, 19:95–109, 1992.

[Kim95] H. Kimura. Chain scattering representation, J-lossless factorization and H_∞ control. *Math. Systems Estim. Control*, 5:203–255, 1995.

[Kim97] H. Kimura. *Chain Scattering Approach to H^∞-Control.* Birkhauser, Boston, 1997.

[KKK95] M. Krstic, I. Kanellakopoulos, and P. Kokotovic. *Nonlinear and Adaptive Control Design.* Wiley, New York, 1995.

[KR85] A.J. Krener and W. Respondek. Nonlinear observers with linearizable error dynamics. *SIAM J. Control Optim.*, 23(2):197–216, 1985.

[Kre94] A.J. Krener. Necessary and sufficient conditions for nonlinear worst case (H^∞) control and estimation. *J. Math. Systems Estim. Control*, 7(1):81–105, 1997.

[KS89] N.N. Krasouskii and A.I. Subbotin. *Game Theoretical Control Problems.* Springer-Verlag, Berlin, 1989.

[Kus64] H.J. Kushner. On the differential equations satisfied by conditional probability densities of Markov processes, with applications. *SIAM J. Control*, 2(1):106–119, 1964.

[KV86] P.R. Kumar and P. Varaiya. *Stochastic Systems: Estimation, Identification and Adaptive Control.* Prentice-Hall, Englewood Cliffs, NJ, 1986.

[Kwa83] H. Kwakernaak. Robustness optimization of linear feedback systems. In *Proc. IEEE CDC*, pp. 618–624, San Antonio, TX, 1983.

[Kwa86] H. Kwakernaak. A polynomial approach to minimax frequency domain optimization of multivariable systems. *Internat. J. Control*, 44:117–156, 1986.

[LAKG92] D.J.N. Limebeer, B.D.O. Anderson, P.P. Khargonekar, and M. Green. A game theoretic approach to H^∞ control for time-varying systems. *SIAM J. Control Optim.*, 30(2):262–283, 1992.

[LB95] W. Lin and C.I. Byrnes. Discrete-time nonlinear H_∞ control with measurement feedback. *Automatica*, 31(3):419–434, 1995.

[LD94] W.M. Lu and J.C. Doyle. H^∞ control of nonlinear systems via output feedback. *IEEE Trans. Automat. Control*, 39:2517–2521, 1994.

[LH88] D.J.N. Limebeer and G.D. Halikias. A controller degree bound for H^∞-optimal control problems of the second kind. *SIAM J. Control Optim.*, 26(3):646–677, 1988.

[LM95] G.L. Litvinov and V.P. Maslov. Correspondence principle for idempotent calculus and some computer applications. Preprint, 1995.

[LS91] Y. Lin and E.D. Sontag. A universal formula for stabilization with bounded controls. *Systems Control Lett.*, 16:393–397, 1991.

[LS95] Y. Lin and E.D. Sontag. Control-Lyapunov universal formulae for restricted inputs. *Control: Theory Adv. Tech.*, 10:1981–2004, 1995.

[Lue66] D.G. Luenberger. Observers for multivariable systems. *IEEE Trans. Automat. Control*, 11:190–199, 1966.

[Maa96] W.C.A. Maas. *Nonlinear H^∞ Control: The Singular Case*. Ph.D. thesis, University of Twente, the Netherlands, 1996.

[McE95a] W.M. McEneaney. Robust control and differential games on a finite time horizon. *Math. Control Signals Systems*, 8:138–166, 1995.

[McE95b] W.M. McEneaney. Uniqueness for viscosity solutions of nonstationary Hamilton–Jacobi–Bellman equations under some a priori conditions (with applications). *SIAM J. Control Optim.*, 33(5):1560–1576, 1995.

[MM82] R.K. Miller and A.N. Michel. *Ordinary Differential Equations*. Academic Press, New York, 1982.

[Mor66] R.E. Mortensen. *Optimal Control of Continuous-Time Stochastic Systems*. PhD thesis, University of California, Berkeley, 1966.

[Mor68] R.E. Mortensen. Maximum-likelihood recursive nonlinear filtering. *J. Optim. Theory Appl.*, 2:386–394, 1968.

[MS92] V.P. Maslov and S.N. Samborski. *Idempotent Analysis*. AMS Series: Adv. Sov. Math., volume 13, Providence, RI, 1992.

[Nag96] H. Nagai. Bellman equations of risk-sensitive control. *SIAM J. Control Optim.*, 34(1):74–101, 1996.

[Ngu96] S.K. Nguang. Robust nonlinear H_∞ output feedback control. *IEEE Trans. Automat. Control*, 41(7):1003–1007, 1996.

[Nis76] M. Nisio. On a nonlinear semigroup attached to optimal stochastic control. *Publ. Res. Inst. Math. Sci.*, 513–537, 1976.

[PAJ91] I.R. Petersen, B.D.O. Anderson, and E.A. Jonckheere. A first principles solution to the non-singular H^∞ control problem. *Internat. J. Robust Nonlinear Control*, 1(3):171–185, 1991.

[Pet87] I.R. Petersen. Disturbance attenuation and H^∞ optimization: A design method based on the algebraic Riccati equation. *IEEE Trans. Automat. Control*, 32(5):427–429, 1987.

[PMR96] P. Dai Pra, L. Meneghini, and W.J. Runggaldier. Connections between stochastic control and dynamic games. *Math. Control Signals Systems*, 9(4):303–326, 1996.

[Qua90] J.P. Quadrat. Theoremes asymptotiques en programmation dynamique. *C. R. Acad. Sci. Paris*, 311:745–748, 1990.

[Qua92] J.P. Quadrat. Brownian and diffusion decision processes. In *Synchronization and Linearity,* F. Baccelli, G. Cohey, G.J. Olsder, and J.P. Quadrat, Eds., Wiley, New York, 1992. Lecture Notes in Control and Info. Sciences, volume 77.

[RS91] I. Rhee and J.L. Speyer. A game theoretic approach to a finite-time disturbance attenuation problem. *IEEE Trans. Automat. Control*, 36:1021–1032, 1991.

[Run91] T. Runolfsson. The equivalence between infinite horizon optimal control of stochastic systems with exponential of integral performance index and stochastic differential games. Tech. report TR JHU-ECE 91-07, Johns Hopkins University, Baltimore, MD, 1991.

[Sar67] D. Sarason. Generalized interpolation in H^∞. *Trans. Amer. Math. Soc.*, 127:179–203, 1967.

[SNF70] B. Sz.-Nagy and C. Foias. *Harmonic Analysis of Operators on Hilbert Space*. North–Holland, Amsterdam, 1970.

[Sor96] P. Soravia. H_∞ control of nonlinear systems: Differential games and viscosity solutions. *SIAM J. Control Optim.*, 34(3):1071–1097, 1996.

[Sto92] A. A. Stoorvogel. *The H_∞ Control Problem: A State Space Approach.* Prentice–Hall, Englewood Cliffs, NJ, 1992.

[Str65] C. Striebel. Sufficient statistics in the optimal control of stochastic systems. *J. Math. Anal. Appl.*, 12:576–592, 1965.

[Str68] R.L. Stratonovich. *Conditional Markov Processes and Their Application to the Theory of Optimal Control.* Elsevier, New York, 1968.

[SU96] S. Sasaki and K. Uchida. Syntheses of H^∞ output feedback control for bilinear systems. In *Proc. 35th IEEE CDC*, pp. 3282–3287, 1996.

[Tad90] G. Tadmor. Worst-case design in the time domain: The maximum principle and the standard H_∞ problem. *Math. Control Signals Systems*, 3:301–324, 1990.

[Tan80] A. Tannenbaum. Feedback stabilization of plants with uncertainty in the gain factor. *Internat. J. Control*, 32:1–16, 1980.

[Teo94] C. Teolis. *Robust H_∞ Control for Nonlinear Systems*. Ph.D. thesis, ISR, University of Maryland, 1994.

[TYJB94] C. Teolis, S. Yuliar, M.R. James, and J.S. Baras. Robust H^∞ output feedback control of bilinear systems. In *Proc. 33rd IEEE CDC*, pp. 1421–1426, 1994.

[vdS91] A. van der Schaft. On a state space approach to nonlinear H^∞ control. *Systems Control Lett.*, 16(1):1–8, 1991.

[vdS92] A. J. van der Schaft. L_2-gain analysis of nonlinear systems and nonlinear state feedback control. *IEEE Trans. Automat. Control*, 37(6):770–784, 1992.

[vdS96] A.J. van der Schaft. *L_2-Gain and Passivity Techniques in Nonlinear Control*. Springer-Verlag, New York, 1996.

[Vid93] M. Vidyasagar. *Nonlinear Systems Analysis*. Prentice–Hall, Englewood Cliffs, NJ, second edition, 1993.

[Whi81] P. Whittle. Risk-sensitive linear/quadratic/Gaussian control. *Adv. Appl. Probab.*, 13:764–777, 1981.

[Whi90a] P. Whittle. A risk-sensitive maximum principle. *Systems Control Lett.*, 15:183–192, 1990.

[Whi90b] P. Whittle. *Risk-Sensitive Optimal Control*. Wiley, New York, 1990.

[Whi91] P. Whittle. A risk-sensitive maximum principle: The case of imperfect state observation. *IEEE Trans. Automat. Control*, 36:793–801, 1991.

[Wil71] J.C. Willems. Least squares stationary optimal control and the algebraic Riccati equation. *IEEE Trans. Automat. Control*, 16(6):621–634, 1971.

[Wil72] J.C. Willems. Dissipative dynamical systems—Part I: General theory. *Arch. Rational Mech. Anal.*, 45:321–351, 1972.

[Won68] W.M. Wonham. On the separation theorem of stochastic control. *SIAM J. Control*, 6(2):312–326, 1968.

[YJH98] S. Yuliar, M.R. James, and J.W. Helton. Dissipative control systems synthesis with full state feedback. *Math. Control Signals Systems*, 11(4):335–356, 1998.

[You69] L.C. Young. *Lectures on Calculus of Variations and Optimal Control Theory*. Saunders, Philadelphia, 1969.

[YS67] D.C. Youla and M. Saito. Interpolation with positive real functions. *J. Franklin Inst.*, 284:77–108, 1967.

[Yul96] S. Yuliar. *Nonlinear Dissipative Control*. Ph.D. thesis, Australian National University, Canberra, Australia, 1996.

[Zak69] M. Zakai. On the optimal filtering of diffusion processes. *Z. Ver. Gebiete*, 11:230–243, 1969.

[Zam79] G. Zames. Optimal sensitivity and feedback: Weighted seminorms, approximate inverses, and plant invariant schemes. In *Proc. Allerton Conf.*, 1979.

[ZDG96] K. Zhou, J. Doyle, and K. Glover. *Robust and Optimal Control*. Prentice–Hall, Upper Saddle River, NJ, 1996.

[ZF81] G. Zames and B. Francis. Feedback and minimax sensitivity. In *NATO Lecture Notes* 117, 1981. Multivariable Analysis and Design Techniques.

[ZF83] G. Zames and B. Francis. Feedback, minimax sensitivity, and optimal robustness. *IEEE Trans. Automat. Control*, AC-28:585–601, 1983.

Index